MATHEMATICAL
OBJECTS in C++

Computational Tools in a Unified
Object-Oriented Approach

CHAPMAN & HALL/CRC
Numerical Analysis and Scientific Computing

Aims and scope:

Scientific computing and numerical analysis provide invaluable tools for the sciences and engineering. This series aims to capture new developments and summarize state-of-the-art methods over the whole spectrum of these fields. It will include a broad range of textbooks, monographs, and handbooks. Volumes in theory, including discretisation techniques, numerical algorithms, multiscale techniques, parallel and distributed algorithms, as well as applications of these methods in multi-disciplinary fields, are welcome. The inclusion of concrete real-world examples is highly encouraged. This series is meant to appeal to students and researchers in mathematics, engineering, and computational science.

Proposals for the series should be submitted to one of the series editors above or directly to:
CRC Press, Taylor & Francis Group
4th, Floor, Albert House
1-4 Singer Street
London EC2A 4BQ
UK

Published Titles

A Concise Introduction to Image Processing using C++
Meiqing Wang and Choi-Hong Lai

Decomposition Methods for Differential Equations:
   Theory and Applications
Juergen Geiser

Grid Resource Management: Toward Virtual and Services Compliant Grid Computing
Frédéric Magoulès, Thi-Mai-Huong Nguyen, and Lei Yu

Introduction to Grid Computing
Frédéric Magoulès, Jie Pan, Kiat-An Tan, and Abhinit Kumar

Mathematical Objects in C++: Computational Tools in a Unified Object-Oriented Approach
Yair Shapira

Numerical Linear Approximation in C
Nabih N. Abdelmalek and William A. Malek

Numerical Techniques for Direct and Large-Eddy Simulations
Xi Jiang and Choi-Hong Lai

Parallel Algorithms
Henri Casanova, Arnaud Legrand, and Yves Robert

Parallel Iterative Algorithms: From Sequential to Grid Computing
Jacques M. Bahi, Sylvain Contassot-Vivier, and Raphael Couturier

MATHEMATICAL
OBJECTS in C++

Computational Tools in a Unified Object-Oriented Approach

Yair Shapira

CRC Press
Taylor & Francis Group
Boca Raton London New York

CRC Press is an imprint of the
Taylor & Francis Group, an **informa** business

A CHAPMAN & HALL BOOK

CRC Press
Taylor & Francis Group
6000 Broken Sound Parkway NW, Suite 300
Boca Raton, FL 33487-2742

First issued in paperback 2017

ISBN 13: 978-1-138-11376-3 (pbk)
ISBN 13: 978-1-4398-1147-4 (hbk)

Library of Congress Cataloging-in-Publication Data

Shapira, Yair, 1960-
  Mathematical objects in C++ : computational tools in a unified object-oriented approach / Yair Shapira.
      p. cm. -- (CHAPMAN & HALL/CRC numerical analysis and scientific computing)
  Includes bibliographical references and index.
  ISBN-13: 978-1-4398-1147-4 (alk. paper)
  ISBN-10: 1-4398-1147-4 (alk. paper)
  1. Numerical analysis--Data processing. 2. C++ (Computer program language) I. Title.
II. Series.

QA297.S464 2010
518.0285'5133--dc22   2009007343

Visit the Taylor & Francis Web site at
http://www.taylorandfrancis.com

and the CRC Press Web site at
http://www.crcpress.com

Contents

Part IV Introduction to C   247

13 Basics of Programming   251

List of Figures

XVIII

XXIV

Preface

Mathematics can be viewed as the philosophy of abstract objects. Indeed, mathematics studies all sorts of useful objects, from numbers to multilevel hierarchies and more general sets, along with the relations and functions associated with them.

The approach used in this book is to focus on the objects, rather than on the functions that use them. After all, the objects are the main building bricks of the language of mathematics. The C++ implementation of the objects makes them far more understandable and easy to comprehend.

This book shows the strong connection between the theoretical nature of mathematical objects and their practical C++ implementation. For example, the theoretical principle of mathematical induction is used extensively to define useful recursive C++ objects. Furthermore, algebraic and geometrical objects are implemented in several different ways. Moreover, highly unstructured computational objects such as oriented and nonoriented graphs, two- and three-dimensional meshes, and sparse stiffness and mass matrices are implemented in short and well-debugged code segments.

The book is intended for undergraduate and graduate students in math, applied math, computer science, and engineering who want to combine the theoretical and practical aspects. Because the book assumes no background in mathematics, computer science, or any other field, it can serve as a text book in courses in discrete mathematics, computational physics, numerical methods for PDEs, introduction to C for mathematicians and engineers, introduction to C++ for mathematicians and engineers, and data structures.

Parts I–II introduce elementary mathematical objects, such as numbers and geometrical objects. These parts are aimed at beginners, and can be skipped by more experienced readers. Part III provides the required theoretical background, including preliminary definitions, algorithms, and simple results. Part IV teaches C from a mathematical point of view, with an emphasis on recursion. Part V teaches C++ from a mathematical point of view, using templates to implement vectors and linked lists. Part VI implements more complex objects such as trees, graphs, and triangulations. Finally, Part VII implements yet more advanced objects, such as 3-D meshes, polynomials of two and three variables, sparse stiffness and mass matrices to linearize 3-D problems, and 3-D splines.

XXVIII   *Preface*

Each chapter ends with relevant exercises to help comprehend the material. Fully explained solutions are available in the appendix. The original code is also available at www.crcpress.com.

Yair Shapira
August 2008

Part I

Numbers*

Numbers

The most elementary mathematical objects are, no doubt, the numbers, which are accompanied by arithmetic operations such as addition, subtraction, multiplication, and division. In this book, however, we focus on the objects rather than on their functions. Thus, the arithmetic operations between numbers, as well as their other functions, are only presented to characterize the numbers, and shed light about their nature as mathematical objects.

In this part, we discuss five kinds of numbers. We start with the natural numbers, which are defined recursively by mathematical induction. Then, we also introduce the negative counterparts of the natural numbers to produce the set of the integer numbers. Then, we proceed to rational numbers, which are fractions of integer numbers. Then, we proceed to irrational numbers, which can be viewed as limits of sequences of rational numbers. Finally, we discuss complex numbers, which are characterized by an extended interpretation of the arithmetic operations that act upon them.

The convention used throughout the book is that mathematical symbols that are quoted from formulas and explained in the text can be placed in single quotation marks (as in '+'), whereas longer mathematical terms that contain more than one character are paced in double quotation marks (as in "$x + y$").

*This part is for beginners, and can be skipped by more experienced readers.

Chapter 1

Natural Numbers

The most elementary mathematical objects are, no doubt, the natural numbers. In this chapter, we use mathematical induction to construct the natural numbers in the first place. Thanks to this inductive (or recursive) nature, elementary arithmetic operations such as addition and multiplication can also be defined recursively. The conclusion is, thus, that the sum of two natural numbers is a natural number as well, and that the product of two natural numbers is a natural number as well. In other words, the set of natural numbers is closed under addition and multiplication.

Furthermore, we provide recursive algorithms to obtain the decimal and binary representations of natural numbers. Finally, we present recursive algorithms to have the factorization of a natural number as a product of its prime factors and to calculate the greatest common divisor and the least common multiple of two natural numbers.

1.1 The Need for Natural Numbers

Since the dawn of civilization, people had the need to count. In agricultural societies, they had to count fruits, vegetables, and bags of wheat. In shepherd societies, they had to count sheep and cattle. When weight units have been introduced, they also started to count pounds of meat and litters of milk. Furthermore, when money has been introduced, they had to count coins of silver and gold. Thus, counting has served an essential role in trade, and thereby in the development of human civilization.

1.2 Mathematical Induction

The natural numbers start from 1, and then increase by 1 again and again. In other words, the set of natural numbers is

$$1, 2, 3, 4, \ldots$$

The notation "...", however, is not very clear. Until where should this list of natural numbers go on? Does it have an end at all?

A more precise formulation of the natural numbers uses mathematical induction. In mathematical induction, the first mathematical object is first constructed manually:

1 is a natural number.

Then, the induction rule is declared to produce the next natural number from an existing one:

if n is an existing natural number, then $n + 1$ is a natural number as well.

(It is assumed that we know how to add 1 to an existing natural number to obtain the next item in the list of natural numbers.) This means that $2, 3, 4, \ldots$ are only short names for the natural numbers defined recursively as follows:

$$2 \equiv 1 + 1$$
$$3 \equiv 2 + 1$$
$$4 \equiv 3 + 1$$

and so on.

1.3 Unboundedness

The mathematical induction allows to produce arbitrarily large natural numbers by starting from 1 and adding 1 sufficiently many times. Thus, the set of natural numbers is unbounded: there is no "greatest" natural number. Indeed, if N were the greatest natural number, then we could always use mathematical induction to construct the yet greater natural number $N + 1$. As a conclusion, no such N exists, and the set of natural numbers is unbounded.

In the above proof, we have used the method of proof known as proof by contradiction: we have assumed that our assertion was false and showed that this would necessarily lead to a contradiction. The conclusion is, therefore, that our original assertion must be indeed true, and that there is indeed no "greatest" natural number. This method of proof is used often in Euclidean geometry below.

1.4 Infinity

We saw that the set of natural numbers is unbounded. Does this necessarily mean that it is also infinite? The answer is, of course, yes. To prove this, we could again use the method of proof by contradiction. Indeed, if the set of natural numbers were finite, then we could use the maximal number in it as the "greatest" natural number, in violation of the unboundedness of the set of the natural numbers established above. The conclusion must therefore be that the set of natural numbers must be infinite as well.

The concept of infinity is difficult to comprehend. In fact, it is easier to understand positive statements such as "there is an end" rather than negative statements such as "there is no end." This is why, later on in this book, we need to introduce a special axiom about the existence of an infinite set.

1.5 Adding Natural Numbers

Mathematical induction helps us not only to construct the natural numbers in the first place, but also to define arithmetic operations between them. This is the theory of Peano [1].

Given a natural number n, we assumed above that we know how to add 1 to it. Indeed, this is done by the mathematical induction, which produces $n + 1$ as a legitimate natural number as well.

Still, can we add another natural number, m, to n? This is easy when m is small, say $m = 4$:

$$n + m = n + (1 + 1 + 1 + 1) = (((n + 1) + 1) + 1) + 1,$$

where the numbers in the parentheses in the right-hand side are easily produced by the original mathematical induction. But what happens when m is very large? Who could guarantee that $n + m$ can always be calculated and produce a legitimate natural number?

Fortunately, mathematical induction can help not only in the original definition of mathematical objects but also in arithmetic operations between them. Indeed, assume that we already know to add the smaller number $m - 1$ to n. Then, we could use this knowledge to add m to n as well:

$$n + m = n + ((m - 1) + 1) = (n + (m - 1)) + 1.$$

In this right-hand side, $n + (m-1)$ is first calculated using our assumption (the induction hypothesis), and then 1 is added to it using the original induction to produce the next natural number.

1.6 Recursion

In practice, however, we don't know how to add $m - 1$ to n. We must do this by assuming that we know how to add a yet smaller number, $m - 2$, to n:

$$n + (m - 1) = n + ((m - 2) + 1) = (n + (m - 2)) + 1.$$

This is called recursion: m is added to n using a simpler operation: adding $m - 1$ to n. In turn, $m - 1$ is added to n using a yet simpler operation: adding $m - 2$ to n. Similarly, $m - 2$ is added to n using the addition of $m - 3$ to n, and so on, until 2 is added to n using the addition of 1 to n, which is well known from the original induction that produces the natural numbers. This recursion is illustrated schematically in the addition table in Figure 1.1.

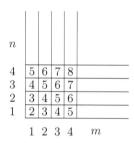

FIGURE 1.1:   The addition table. Each subsquare contains the result $n + m$ where n is the row number and m is the column number of the subsquare.

1.7 The Addition Function

The addition operation may be viewed as a function: a "black-box" machine, which uses one or more inputs to produce a single, uniquely-defined output. In our case, the '+' function uses the inputs n and m to produce the unique result $n + m$ using the above (recursive) list of instructions, or algorithm.

The addition function is illustrated schematically in Figure 1.2. As a matter of fact, the output $n + m$ is written in the appropriate subsquare in Figure 1.1, in the nth row and mth column.

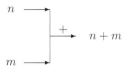

FIGURE 1.2: The addition function uses the inputs n and m to produce the output $n + m$.

1.8 Stack of Calls

The sum $n + m$ cannot be calculated directly by a single application (or call) of the addition function '+'. Before, one must use a recursive application (or call) of the addition function with the smaller inputs (or arguments) n and $m - 1$. Thus, the original call of the addition function to calculate $n + m$ must be placed in an abstract stack until $n + (m - 1)$ is calculated. Once this is done, the original call is taken back out of the stack, and is calculated as

$$n + m = (n + (m - 1)) + 1.$$

Unfortunately, $n + (m - 1)$ also cannot be calculated directly. Therefore, it must also be placed in the stack, on top of the original call. Only once another recursive call to calculate $n + (m - 2)$ is carried out, it can be taken out of the stack and calculated as

$$n + (m - 1) = (n + (m - 2)) + 1.$$

Thus, the recursive calls to the addition function '+' are "pushed" one on top of the previous one in the stack (Figure 1.3). Once the stack is full of recursive calls and the top (mth) call needs only to calculate $n + 1$, the calls "pop" back one-by-one from the stack, and each in turn is calculated using the previous calculation.

1.9 Multiplying Natural Numbers

Multiplying a natural number n by another natural number m is also done by mathematical induction. Indeed, when $m = 1$, the result is clearly $n \cdot 1 = n$.

| $n+1$ |
| $n+2$ |
| $n+3$ |
| $\cdots$ |
| $n+(m-3)$ |
| $n+(m-2)$ |
| $n+(m-1)$ |
| $n+m$ |

FIGURE 1.3:   The stack used for adding n and m. The original call $n+m$ is pushed first, and the recursive calls are pushed one by one on top of it.

Furthermore, assume that we know how to calculate the product $n(m-1)$. Then, we could use this knowledge to calculate the required result

$$nm = n(m-1) + n.$$

In practice, the calculation of $n(m-1)$ is done recursively by

$$n(m-1) = n(m-2) + n.$$

The right-hand side is calculated by yet another recursive call to the multiplication function '$\cdot$', and so on, until the final call to calculate

$$n \cdot 2 = n \cdot 1 + n.$$

The calls are placed one on top of the previous one (Figure 1.4), and then "pop" back one by one from the stack, each calculated and used to calculate the next one. This produces the nth row in the multiplication table in Figure 1.5.

| n |
| $2n$ |
| $3n$ |
| $\cdots$ |
| $n(m-3)$ |
| $n(m-2)$ |
| $n(m-1)$ |
| nm |

FIGURE 1.4:   The stack used for multiplying n by m. The original call nm is pushed first, and the recursive calls are pushed one by one on top of it.

The multiplication function '$\cdot$' accepts two inputs (or arguments), to produce the output (or result) nm. This is illustrated schematically in Figure 1.6.

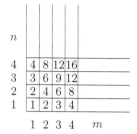

FIGURE 1.5: The multiplication table. Each subsquare contains the result nm where n is the row number and m is the column number of the subsquare.

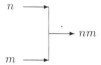

FIGURE 1.6: The multiplication function uses the inputs n and m to produce the output nm.

1.10 One and Zero

The smallest natural number, one (1), is considered as the unit number for the multiplication function in the sense that it satisfies

$$n \cdot 1 = n$$

for every natural number n. In some contexts, however, the yet smaller number zero (0) is also considered as a natural number, and serves as the unit number for the addition function in the sense that it satisfies

$$n + 0 = n$$

for every natural number n. In fact, zero has a special role in the representation of a natural number in the decimal and binary forms below.

1.11 Decimal Representation

In the decimal representation, a natural number n is represented using the ten digits

$$0, 1, 2, 3, 4, 5, 6, 7, 8, 9,$$

which stand for the ten smallest natural numbers (from zero to nine). More precisely, n is represented as a sequence of, say, $k + 1$ digits

$$n = a_k a_{k-1} \ldots a_2 a_1 a_0,$$

where $a_0, a_1, \ldots, a_k$ are the digits (numbers between 0 and 9) used in the expansion of n in powers of 10:

$$n = a_0 + a_1 \cdot 10 + a_2 10^2 + a_3 10^3 + \cdots + a_{k-1} 10^{k-1} + a_k 10^k = \sum_{i=0}^{k} a_i 10^i.$$

In other words, both the value and the position of the digit a_i in the sequence of digits determines its contribution to n: $a_i 10^i$. The least significant digit, a_0, contributes a_0 only, whereas the most significant digit, a_k, contributes $a_k 10^k$. Below we'll see that this representation is indeed unique.

In order to obtain the decimal representation, we need two more arithmetic operations on natural numbers, involving division with residual (or division with remainder). More precisely, if n and m are two natural numbers satisfying $n > m > 0$, then n/m is the result of dividing n by m with residual, and $n \% m$ is that residual (or remainder). For example, $12/10 = 1$, and $12 \% 10 = 2$. With these operations, the decimal representation of a natural number n, denoted by "$decimal(n)$", is obtained recursively as follows.

Algorithm 1.1 *1. If $n \leq 9$, then*

$$decimal(n) = n.$$

2. If, on the other hand, $n > 9$, then

$$decimal(n) = decimal(n/10) \; n \% 10.$$

In other words, if n is a digit between 0 and 9, then its decimal representation is just that digit. Otherwise, the last digit in the decimal representation of n, a_0 (the less significant digit on the far right), is $n \% 10$, and the digits preceding it on the left are obtained from a recursive application of the above algorithm to obtain $decimal(n/10)$ rather than $decimal(n)$. This recursive call is pushed into the stack of calls on top of the original call (Figure 1.7). The calculation of $decimal(n/10)$ must use yet another recursive call to $decimal(n/100)$, which is also pushed in the stack on top of the previous two calls. The process continues until the final call to calculate the most significant digit in n, a_k, is reached; this call uses no recursion, because it is made by Step 1 in the above algorithm. Therefore, it can be used to calculate the top call currently in the stack, and the rest of the calls also "pop" one by one from the stack, each being used in the next one, until the bottom call (the original call $decimal(n)$) also pops out and is obtained as required.

$\cdots$
$decimal(n/100000)$
$decimal(n/10000)$
$decimal(n/1000)$
$decimal(n/100)$
$decimal(n/10)$
$decimal(n)$

FIGURE 1.7: The stack used for writing the decimal representation of n, denoted by $decimal(n)$. The original call $decimal(n)$ is pushed first, and the recursive calls are pushed one by one on top of it.

1.12 Binary Representation

In the decimal representation above, the natural number n is expanded as a sum of powers of 10, e.g., 10^i, with coefficients of the form a_i, which are just digits, that is, numbers between zero and nine. This means that 10 is the base in the decimal representation.

The reason for using 10 as a base is probably that humans have ten fingers. In the early days of the human civilization, people used to count with their fingers. They could use their fingers to count to ten, but then they had to remember that they had already counted one package of ten, and restart using their fingers to count from eleven to twenty, to have two packages of ten, and so on. This naturally leads to the decimal representation of natural numbers, in which the number of packages of ten is placed in the second digit from the right, and the number of extra units is presented by the last digit on the far right.

Not everyone, however, has ten fingers. The computer, for example, has only two "fingers": zero, represented by a bit that is switched off, and one, represented by a bit that is switched on. Indeed, the binary representation uses base 2 rather than base 10 above. In particular, n is represented as a sum of powers of 2 rather than 10, with different coefficients a_i that can now be only zero or one. These new coefficients (the digits in the binary representation of n) are obtained uniquely by modifying the above algorithm to use 2 instead of 10 and 1 instead of 9.

1.13 Prime Numbers

Particularly important natural numbers are the prime numbers. A natural number p is prime if it is divisible only by itself and by one. In other words, for every natural number k between 2 and $p-1$,

$$p \% k > 0.$$

The set of prime numbers is unbounded. This could be proved by contradiction. Indeed, assume that the set of prime numbers were bounded, say, by P. Then, we could construct a yet greater prime number $K > P$, in violation of the assumption that all the prime numbers are bounded by P. Indeed, K could be the product of all the natural numbers up to and including P plus one:

$$K \equiv 1 \cdot 2 \cdot 3 \cdot \ldots \cdot (P-1) \cdot P + 1 = \left(\Pi_{i=1}^{P} i\right) + 1.$$

Clearly, for every natural number n between 2 and P,

$$K \% n = 1.$$

In particular, this is true for every prime number n. This implies that K is indeed prime, in violation of our assumption that P bounds all the prime numbers. This implies that our assumption has been false, and no such P could ever exist.

The unboundedness of the set of prime numbers implies that it is also infinite. This can also be proved by contradiction: indeed, if it were finite, then the maximal prime number in it could also serve as its bound, in violation of the unboundedness of the set of prime numbers proved above.

1.14 Prime Factors

The prime numbers can be viewed as the bricks from which the natural numbers are built. Indeed, every natural number can be written uniquely as a product of prime numbers. This is called factorization by prime factors. For example, 24 is factored as

$$24 = 2 \cdot 2 \cdot 2 \cdot 3 = 2^3 \cdot 3.$$

Below we'll not only give the proof that such a factorization exists, but actually construct it. Indeed, this is done in the following algorithm to factorize the natural number n:

Algorithm 1.2 *1. Let $i > 1$ be the smallest number that divides n in the sense that $n\%i = 0$.*

2. Print: "i is a prime factor."

3. If $i < n$, then apply this algorithm recursively to n/i rather than n.

This algorithm prints out the list of prime factors of n. For example, when $n = 24$, it prints "2 is a prime factor" before being called recursively for $n = 24/2 = 12$. Then, it again prints "2 is a prime factor" before being called recursively for $n = 12/2 = 6$. Then, it prints again "2 is a prime factor" before being called again for $n = 6/2 = 3$. Finally, it prints "3 is a prime factor." Thus, the result is that 2 appears three times in the factorization and 3 appears only once, as required.

1.15 Mathematical Induction in Proofs

Mathematical induction is also useful to prove properties associated with natural numbers. In fact, in order to prove that a particular property holds for every natural number, one should first prove that it holds for the smallest relevant natural number (usually 1). Then, one should also prove that, given a natural number $n > 1$, and assuming that the induction hypothesis holds, that is, that the property holds for every natural number smaller than n, then the property holds for n as well. This is the induction step, which allows us to extend the property from the numbers smaller than n to n itself. Once it is established that the induction step is indeed legitimate, the validity of the property for every natural number n is established as well. Indeed, since it is valid for 1, and since the induction step is legitimate, it is valid for 2 as well. Furthermore, since it is valid for 1 and 2, and since the induction step is legitimate, it is valid for 3 as well. The process continues this way, until the argument that, since the property holds for $1, 2, \ldots, n-1$, and since the induction step is legitimate, the property holds for n as well.

Let us now use mathematical induction to prove that the above algorithm indeed works in general. Indeed, if $n = 2$, then i in Step 1 of the algorithm is also equal to 2, and the algorithm is complete in Step 1. Assume now that $n > 2$, and assume also that the induction hypothesis holds, that is, that the algorithm works for every input smaller than n. Let i be the smallest number that divides n (in the sense that $n\%i = 0$), as in Step 1. If $i = n$, then the algorithm is complete in Step 1. If, on the other hand, $i < n$, then Step 2 in the algorithm is invoked. Since $n/i < n$, and thanks to the induction hypothesis, the recursive call in Step 2 (with n replaced by n/i), prints out the prime factors of n/i, which are also the remaining prime factors of n.

1.16 The Greatest Common Divisor

The greatest common divisor of the natural numbers n and m, denoted by $GCD(n, m)$, is the greatest number that divides both n and m:

$$n \% GCD(n, m) = m \% GCD(n, m) = 0$$

([16]–[17]), where '%' stands for the "mod" operation:

$$n \% m = n \bmod m = n - (n/m)m$$

(where $n \geq m$ are any two natural numbers and n/m stads for integer division with residual).

Here is the recursive algorithm to calculate $GCD(n, m)$ (Euclid's algorithm). The algorithm assumes, without loss of generality, that $n > m$. (Otherwise, just interchange the roles of n and m.)

Algorithm 1.3 *1. If m divides n ($n \% m = 0$), then the output is*

$$GCD(n, m) = m.$$

2. If, on the other hand, $n \% m > 0$, then the output is

$$GCD(n, m) = GCD(m, n \% m).$$

(The right-hand side is calculated by a recursive call to the same algorithm, with smaller arguments.)

For example, this is how the above algorithm is used to calculate the greatest common divisor of 100 and 64:

$$
\begin{aligned}
GCD(100, 64) &= GCD(64, 36) \\
&= GCD(36, 28) \\
&= GCD(28, 8) \\
&= GCD(8, 4) = 4.
\end{aligned}
$$

Let us now show that the above algorithm works in general. This is done by induction on m (for every n). Indeed, when $m = 1$, the algorithm is complete in Step 1, with the output

$$GCD(n, 1) = 1.$$

When $m > 1$, assume that the induction hypothesis holds, that is, that the algorithm works for every second argument smaller than m. Now, if $n\%m = 0$, then the algorithm is complete in Step 1. If, on the other hand, $n\%m > 0$, then $n\%m < m$, so thanks to the induction hypothesis the algorithm can be called recursively to calculate $GCD(m, n\%m)$. All that is now left to show is that

$$GCD(n, m) = GCD(m, n\%m).$$

To show this, let us first show that

$$GCD(n, m) \geq GCD(m, n\%m).$$

This follows from the fact that

$$n = (n/m)m + (n\%m).$$

(Note that n/m is division with residual, so $n > (n/m)m$.) This implies that $GCD(m, n\%m)$ divides both m and n, and therefore cannot be larger than the greatest common divisor of m and n. Furthermore, let us also show that

$$GCD(n, m) \leq GCD(m, n\%m).$$

This follows from the fact that $GCD(n, m)$ divides both m and $n\%m$, hence cannot be larger than the greatest common divisor of m and $n\%m$. The conclusion is, thus, that

$$GCD(n, m) = GCD(m, n\%m),$$

as required.

1.17 Least Common Multiple

The least common multiple of the natural numbers n and m, denoted by $LCM(n, m)$, is the smallest natural number divided by both n and m in the sense that

$$LCM(n, m)\%n = LCM(n, m)\%m = 0.$$

In order to calculate $LCM(n, m)$, let us first write both n and m as products of their prime factors.

Let M be the set of prime factors of m, and N the set of prime factors of n. Then, one can write

$$n = \Pi_{p \in N} p^{l_p} \text{ and } m = \Pi_{p \in M} p^{k_p},$$

where "$p \in M$" means "p belongs to the set M," "$\Pi_{p \in M} p$" means "the product of all the members of the set M," and l_p and k_p denote the powers of the corresponding prime factor p in the factorization of n and m, respectively. In fact, one can also write

$$n = \Pi_{p \in M \cup N} p^{l_p} \text{ and } m = \Pi_{p \in M \cup N} p^{k_p},$$

where $M \cup N$ is the union of the sets M and N, $l_p = 0$ if $p \notin N$, and $k_p = 0$ if $p \notin M$. With these factorizations, we clearly have

$$GCD(n, m) = \Pi_{p \in N \cup M} p^{\min(l_p, k_p)}.$$

Furthermore, $LCM(n, m)$ takes the form

$$LCM(n, m) = \Pi_{p \in N \cup M} p^{\max(l_p, k_p)} = \Pi_{p \in N \cup M} p^{l_p + k_p - \min(l_p, k_p)}.$$

As a result, we have the formula

$$LCM(n, m) = \frac{nm}{GCD(n, m)}.$$

The greatest common divisor and the least common multiple are particularly useful in the rational numbers studied below.

1.18 The Factorial Function

The addition and multiplication functions defined above take each two arguments to produce the required output (Figures 1.2 and 1.6). Similarly, the GCD and LCM functions also take two arguments each to produce their outputs. Other functions, however, may take only one input argument to produce the output. One such function is the factorial function.

The factorial function takes an input argument n to produce the output, denoted by $n!$. This output is defined recursively as follows:

$$n! \equiv \begin{cases} 1 & \text{if } n = 0 \\ (n-1)! \cdot n & \text{if } n > 0. \end{cases}$$

The factorial function is useful in the sequel, particularly in the definition of some irrational numbers below.

1.19 Exercises

1. What is a natural number?

2. Use mathematical induction to define a natural number.
3. Use mathematical induction to show that the set of the natural numbers is closed under addition in the sense that the sum of any two natural numbers is a natural number as well.
4. Use mathematical induction to show that the set of the natural numbers is closed under multiplication in the sense that the product of any two natural numbers is a natural number as well.
5. Write the recursive algorithm that produces the sum of two arbitrarily large natural numbers with arbitrarily long decimal representations.
6. Use mathematical induction on the length of the above decimal representations to show that the above algorithm indeed works and produces the correct sum.
7. Write the recursive algorithm that produces the product of two arbitrarily large natural numbers with arbitrarily long decimal representations.
8. Use mathematical induction on the length of the above decimal representations to show that the above algorithm indeed works and produces the correct product.
9. Repeat the above exercises, only this time use the binary representation instead of the decimal representation.
10. Write the algorithm that checks whether a given natural number is prime or not.
11. Write the algorithm that finds all the prime numbers between 1 and n, where n is an arbitrarily large natural number.
12. Write the recursive algorithm that finds the prime factors of a given natural number. Use mathematical induction to show that it indeed finds the unique representation of the natural number as a product of prime numbers.
13. Use mathematical induction to show that the representation of an arbitrarily large natural number as the product of its prime factors is indeed unique.
14. Write the recursive algorithm that computes the GCD of two given natural numbers. Use mathematical induction to show that it indeed works.
15. Write the recursive algorithm that computes the factorial of a given natural number. Use mathematical induction to show that it indeed produces the correct answer.

Chapter 2

Integer Numbers

Although they have several useful functions, the natural numbers are insufficient to describe the complex structures required in mathematics. In this chapter, we extend the set of the natural numbers to a wider set, which contains also negative numbers: the set of the integer numbers.

2.1 Negative Numbers

So far, we have only discussed nonnegative numbers, that is, numbers that are greater than or at least equal to zero. Here we consider also the negative numbers obtained by adding the minus sign '$-$' before the natural number. In fact, $-n$ is characterized as the number that solves the equation

$$x + n = 0.$$

Clearly, x cannot be a natural number. The only solution to the above equation is $x = -n$.

As a matter of fact, it is sufficient to define the largest negative number -1, because then every negative number of the form $-n$ is obtained as the product

$$-n = (-1) \cdot n.$$

The number -1 is characterized as the number that satisfies

$$(-1) + 1 = 0$$
$$(-1) \cdot 1 = -1$$
$$1 \cdot (-1) = -1$$
$$(-1) \cdot (-1) = 1.$$

Thus, multiplying by -1 can be interpreted as changing the sign of a number. If it was positive, then it becomes negative, and if it was negative, then it becomes positive. In any case, its absolute value remains unchanged:

$$|-n| = |n|,$$

where the absolute value of a (positive or negative) number m is defined by

$$|m| \equiv \begin{cases} m & \text{if } m \geq 0 \\ -m & \text{if } m < 0. \end{cases}$$

In the following, we give an example to show the importance of the negative numbers in general and the largest negative number, -1, in particular.

2.2 The Prisoner Problem

The prisoner problem is as follows. A prisoner escapes from prison, and the guards are after him. He arrives at a T junction, in which only one turn leads to freedom, and the other leads back to prison (Figure 2.1). At the junction, there are two people: one of them is trustworthy, and the other is a pathological liar. Unfortunately, the prisoner doesn't know who is who, and he has no time to ask more than one question. Whom should he approach, and what question should he ask in order to choose the correct turn that would lead him to freedom?

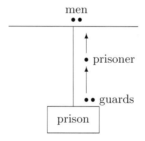

FIGURE 2.1:   The prisoner problem: he can ask only one question in order to know which way leads to freedom. One of the men is a liar.

At first glance, the problem seems unsolvable. Indeed, if the prisoner approached one of the two people at the junction and ask him which turn to take, then he might tell him: "turn right." Still, should the prisoner accept this advice? After all, the man who answered could be the liar, so the right turn might actually lead back to prison. This is indeed a dilemma!

Fortunately, the problem can be solved using a mathematical model, in which a true answer is symboled by 1, and a lie is symboled by -1, because it reverses the truth. Thus, the trustworthy man in the junction can be characterized by 1, whereas the liar can be characterized by -1. Although the prisoner doesn't know who is who, he still knows that the product of these

symbols is always
$$1 \cdot (-1) = (-1) \cdot 1 = -1,$$
regardless of the order in which the two inputs are multiplied. Thus, the prisoner should ask a question that combines the minds of both men in the junction. He should ask one of them: "if I have approached the other man and asked him what turn to take, what would he say?" Now, the answer to this question contains necessarily exactly one lie. Indeed, if the question is directed to the liar, then he would reverse the answer that would be given by the trustworthy man, resulting in a lie $((-1) \cdot 1 = -1)$. If, on the other hand, the question happens to be directed to the trustworthy man, then he would honestly tell the answer of the liar, which also reverses the truth $(1 \cdot (-1) = -1)$. In any case, the prisoner must not follow the answer, but take the other direction to get safely to freedom.

2.3 The Integer Numbers

The set of integer numbers is the set of both natural numbers (including zero) and their negative counterparts. Like the set of the natural numbers, the set of the integer numbers is closed under addition and multiplication. Furthermore, it has the extra advantage that it is also closed under subtraction. In other words, the addition operation is reversible: adding an integer m can be reversed by adding $-m$. Such a set is called a mathematical ring.

2.4 The Number Axis

To have some geometrical insight about the integer numbers, one can place them on a horizontal axis (Figure 2.2). The zero lies in the middle of the axis. To its right, the positive numbers $1, 2, 3, \ldots$ form the positive part of the number axis. To its left, the negative numbers $-1, -2, -3, \ldots$ form the negative part of the axis. The result is an infinite axis, with the integer numbers ordered on it from $-\infty$ (minus infinity) on the far left to ∞ (infinity) on the far right.

· · · −3−2−1 0 1 2 3 · · ·

FIGURE 2.2:   The number axis that contains both the natural numbers to the right of the zero and the negative integer numbers to the left of the zero.

2.5 Angles of Numbers

Here we give a geometric interpretation of integer numbers, which uses the fact that a minus sign before an integer number reverses the direction from which the number faces zero on the number axis. This interpretation is particularly useful in the introduction of complex numbers later on in the book.

-3 3 $+$

FIGURE 2.3:   The arrow from zero to 3 produces a zero angle with the positive part of the axis, and the arrow from zero to -3 produces an angle of 180 degrees with the positive part of the axis.

Each nonzero integer number n can be described by an arrow on the number axis, leading from zero to n (Figure 2.3). If n is positive, then the arrow points to the right, so it forms an angle of 0 degrees (a trivial angle) with the positive part of the number axis. If, on the other hand, n is negative, then the arrow from zero to it forms an angle of 180 degrees with the positive part of the number axis.

Thus, an integer number n can be characterized by two functions: its absolute value $|n|$, and the angle its arrow forms with the positive part of the number axis (0 degrees if $n > 0$ or 180 degrees if $n < 0$).

These functions can help one to interpret the multiplication of integer numbers from a geometrical point of view. For instance, let n and m be two given integer numbers. Now, their product nm can also be interpreted in terms of its two functions: its absolute value and the angle its arrow forms with the positive part of the number axis. Clearly, its absolute value is

$$|nm| = |n| \cdot |m|.$$

Furthermore, the angle associated with nm is obtained by adding the angles associated with n and m. Indeed, if both n and m are positive, then they are associated with the zero angle. Since the sum of the angles is, in this case, zero, the angle associated with nm is zero too, which implies that nm is positive, as required. Furthermore, if one of the numbers n and m is positive and the other is negative, then the angle associated with the positive number is zero, whereas the angle associated with the negative number is 180 degrees. The sum of the angles is therefore 180 degrees, which implies that nm is negative, as required. Finally, if n and m are both negative, then an angle of 180 degrees is associated with both of them. The sum of angles is therefore 360 degrees or zero, which implies that nm is positive, as required.

2.6 Exercises

1. What is an integer number?
2. Show that the set of the integer numbers is closed under addition in the sense that the sum of any two integer numbers is an integer number as well.
3. Show that the set of the integer numbers is closed under multiplication in the sense that the product of any two integer numbers is an integer number as well.
4. Show that, for any integer number n, there is a unique integer number m satisfying $n + m = 0$. (m is called the negative of n, and denoted by $-n$.)
5. Show that 0 is the only integer number that is equal to its negative.

Chapter 3

Rational Numbers

The set of integer numbers is a mathematical ring in the sense that the addition operation is reversible: adding m can be reversed by adding $-m$. In this chapter, we extend it into a yet more complete set: the set of rational numbers. This set is not only a mathematical ring but also a mathematical field, in which not only the addition operation but also the multiplication operation is reversible: multiplying by $m \neq 0$ can be reversed by multiplying by $1/m$.

3.1 Rational Numbers

So far, we have interpreted the symbol '/' as division with residual. This means that, for every two integers n and $m \neq 0$, n/m is an integer. From now on, however, we interpret the symbol '/' as division without residual, so n/m may well be a fraction or a rational number. In particular, the rational number $1/m$ is the unique solution of the equation

$$m \cdot x = 1$$

(see [6]).

As we have seen above, the set of the integer numbers extends the set of the natural numbers into a mathematical ring. Indeed, the set of the integer numbers is the smallest set that contains all the natural numbers and is closed not only under addition and multiplication, but also under subtraction: every two integer numbers can be subtracted from each other to produce a result that is an integer number as well. Similarly, the set of the rational numbers extends the set of the integer numbers into a mathematical field. In fact, by including also the fractions of the form n/m, the set of the rational numbers is closed not only under addition, subtraction, and multiplication, but also under division: each two rational numbers can be divided by each other, and the result is a rational number as well.

3.2 The Unique Form

The rational number n/m has infinitely many equal forms:

$$\frac{n}{m} = \frac{2n}{2m} = \frac{3n}{3m} = \frac{4n}{4m} = \cdots .$$

Therefore, it is important to agree on one particular form to present n/m

$$\frac{n/GCD(n,m)}{m/GCD(n,m)}.$$

This form uses minimal numerator and denominator. In fact, in this form, the greatest common divisor of the numerator and denominator is 1. Thus, this is the unique form of the fraction n/m. In the sequel, it is assumed that the rational numbers are in their unique form.

3.3 Adding Rational Numbers

Here we show that the set of rational numbers is indeed closed under addition and subtraction. Indeed, let n/m and l/k be two rational numbers. Then we have

$$\frac{n}{m} + \frac{l}{k} = \frac{n(k/GCD(m,k))}{m(k/GCD(m,k))}$$
$$+ \frac{l(m/GCD(m,k))}{k(m/GCD(m,k))}$$
$$= \frac{n(k/GCD(m,k)) + l(m/GCD(m,k))}{LCM(m,k)}.$$

This calculation uses minimal numbers in both the numerator and the denominator, to make the calculation as easy as possible. Below we also define the product and ratio of two rational numbers.

3.4 Multiplying Rational Numbers

Furthermore, the set of rational numbers is also closed with respect to the multiplication and division operations. Indeed, for every two rational numbers

n/m and l/k, their product (using minimal numbers in both the numerator and the denominator) is

$$\frac{n}{m} \cdot \frac{l}{k} = \frac{(n/GCD(n,k))(l/GCD(m,l))}{(m/GCD(m,l))(k/GCD(n,k))},$$

and their ratio is

$$\frac{n/m}{l/k} = \frac{n}{m} \cdot \frac{k}{l}.$$

3.5 Periodic Decimal Representation

Some rational numbers have also a finite decimal representation. For example,

$$1/4 = 0.25.$$

This representation can also be interpreted as a periodic infinite decimal representation:

$$1/4 = 0.2500000\ldots = 0.2499999\ldots$$

Here the length of the period is 1, because there is only one digit that is repeated periodically infinitely many times: 0 or 9.

Other rational numbers may have nontrivial periods. Consider, for example, the fraction $1/7$. Using the standard division algorithm, this fraction can be presented as a periodic decimal fraction as follows:

```
0.142857
--------
1        | 7
10
 7
 -
30
28
--
 20
 14
 --
  60
  56
  --
   40
   35
   --
```

$$50$$
$$49$$
$$--$$
$$1$$

and the process repeats again and again. Thus, $1/7$ can be written as the periodic infinite decimal fraction

$$1/7 = 0.142857142857142857\ldots.$$

The six digits in the above period, 142857, are obtained one by one in the division algorithm in a minimal residual approach. Indeed, the rational number $1/7$ is the solution of the equation

$$7x = 1.$$

In other words, $1/7$ minimizes the residual

$$1 - 7x.$$

This is why the first digit right after the decimal point is chosen to be 1; it gives the minimal residual

$$1 - 7 \cdot 0.1 = 0.3.$$

Furthermore, the second digit after the decimal point is chosen to be 4, to give the yet smaller residual

$$1 - 7 \cdot 0.14 = 0.02.$$

(Of course, one must avoid a digit larger than 4, to avoid a negative residual.) The process continues, producing uniquely the six digits in the period, and restarts all over again to produce the same period over and over again in the decimal representation of $1/7$. Thus, the above division algorithm not only proves the existence of the periodic decimal representation of $1/7$, but actually produces it uniquely.

Just like $1/7$, every rational number of the form n/m can be written as a periodic infinite decimal fraction. Indeed, because there are only m natural numbers smaller than m that can serve as residuals under the short horizontal lines in the division algorithm illustrated above for $1/7$, the process used in the division algorithm must repeat itself at some point, leading to a periodic infinite decimal fraction. (In the calculation of $1/7$, for example, the process repeats itself when the residual is again 1 at the bottom.) Thus, the length of the period is at most m digits.

Below we show that the reverse is also true: every periodic infinite decimal fraction is a rational number: it can be written in the form n/m for some integer numbers n and $m \neq 0$. For this, however, we must first introduce the concept of a series that converges to its limit.

3.6 Diverging Series

A series is an object of the form

$$a_1 + a_2 + a_3 + a_4 + \cdots,$$

also denoted by

$$\sum_{n=1}^{\infty} a_n,$$

where the a_n's are some numbers. The mth partial sum of the series (the sum of the m first elements in the series) is denoted by

$$s_m = a_1 + a_2 + a_3 + \cdots + a_{m-1} + a_m = \sum_{n=1}^{m} a_n.$$

We say that the series diverges if the partial sum s_m grows indefinitely when m grows. In other words, given an arbitrarily large number N, one can choose a sufficiently large natural number M such that

$$s_M \geq N$$
$$s_{M+1} \geq N$$
$$s_{M+2} \geq N$$
$$s_{M+3} \geq N$$

and, in fact, $s_m \geq N$ for every $m \geq M$. We denote this by

$$s_m \to_{m \to \infty} \infty$$

or

$$\sum_{n=1}^{\infty} a_n = \infty.$$

Consider, for example, the constant series, in which

$$a_n = a$$

for every $n \geq 1$, for some constant number $a > 0$. In this case, given an arbitrarily large number N, one could choose M as large as $M > N/a$ to guarantee that, for every $m \geq M$,

$$s_m = ma \geq Ma > (N/a)a = N.$$

This implies that the constant series indeed diverges. In the following, we will see a more interesting example of a diverging series.

3.7 The Harmonic Series

The harmonic series is the series

$$\frac{1}{2} + \frac{1}{3} + \frac{1}{4} + \frac{1}{5} + \cdots = \sum_{n=2}^{\infty} \frac{1}{n}.$$

To show that this series indeed diverges, we group the elements in it and bound them from below by $1/2$:

$$\frac{1}{2} = \frac{1}{2}$$

$$\frac{1}{3} + \frac{1}{4} \geq \frac{1}{4} + \frac{1}{4} = \frac{1}{2}$$

$$\frac{1}{5} + \frac{1}{6} + \frac{1}{7} + \frac{1}{8} \geq \frac{1}{8} + \frac{1}{8} + \frac{1}{8} + \frac{1}{8} = \frac{1}{2},$$

and so on. As a result, we have

$$\sum_{n=2}^{\infty} \frac{1}{n} = \sum_{k=1}^{\infty} \sum_{n=2^{k-1}+1}^{2^k} \frac{1}{n}$$

$$\geq \sum_{k=1}^{\infty} \sum_{n=2^{k-1}+1}^{2^k} \frac{1}{2^k}$$

$$= \sum_{k=1}^{\infty} 2^{k-1} \cdot \frac{1}{2^k}$$

$$= \sum_{k=1}^{\infty} \frac{1}{2}$$

$$= \infty.$$

3.8 Converging Series

We say that the series $\sum a_n$ converges to a number s if the partial sums s_m get arbitrarily close to s, or converge to s, in the sense that, for sufficiently large m, $|s_m - s|$ is arbitrarily small. More precisely, given an arbitrarily small number $\varepsilon > 0$, one can choose a natural number M so large that

$$|s_M - s| \leq \varepsilon$$
$$|s_{M+1} - s| \leq \varepsilon$$
$$|s_{M+2} - s| \leq \varepsilon$$

and, in general,

$$|s_m - s| \leq \varepsilon$$

for every $m \geq M$. This is denoted by

$$s_m \rightarrow_{m \to \infty} s,$$

or

$$\sum_{n=1}^{\infty} a_n = s.$$

3.9 Finite Power Series

Consider the finite power series

$$1 + q + q^2 + q^3 + \cdots + q^{m-1} = \sum_{n=0}^{m-1} q^n$$

where $m \geq 1$ is a given natural number, and $q \neq 1$ is a given parameter. Let us use mathematical induction to prove that this sum is equal to

$$\frac{q^m - 1}{q - 1}.$$

Indeed, for $m = 1$, the above sum contains one term only, the first term 1. Thus, we have

$$\sum_{n=0}^{m-1} q^n = 1 = \frac{q^1 - 1}{q - 1} = \frac{q^m - 1}{q - 1},$$

as required.

Assume now that the induction hypothesis holds, that is, that the slightly shorter power series that contains only $m - 1$ terms can be summed as

$$\sum_{n=0}^{m-2} q^n = \frac{q^{m-1} - 1}{q - 1}$$

for some fixed natural number $m \geq 2$. Then, the original series that contains m terms can be split into two parts: the first $m - 2$ terms, and the final term:

$$\sum_{n=0}^{m-1} q^n = \left(\sum_{n=0}^{m-2} q^n\right) + q^{m-1}$$

$$= \frac{q^{m-1} - 1}{q - 1} + q^{m-1}$$

$$= \frac{q^{m-1} - 1 + q^{m-1}(q - 1)}{q - 1}$$

$$= \frac{q^m - 1}{q - 1}.$$

This completes the proof of the induction step. Thus, we have proved by mathematical induction that

$$\sum_{n=0}^{m-1} q^n = \frac{q^m - 1}{q - 1}$$

for every natural number $m \geq 1$.

3.10 Infinite Power Series

The infinite power series is the series

$$1 + q + q^2 + q^3 + \cdots = \sum_{n=0}^{\infty} q^n.$$

Clearly, when $q = 1$, this is just the diverging constant series.

Let us now turn to the more interesting case $q \neq 1$. In this case, we have from the previous section that the partial sum s_{m-1} is equal to

$$s_{m-1} = 1 + q + q^2 + \cdots + q^{m-1} = \sum_{n=0}^{m-1} q^n = \frac{q^m - 1}{q - 1} = \frac{1 - q^m}{1 - q}.$$

By replacing m by $m + 1$, this can also be written as

$$s_m = \frac{1 - q^{m+1}}{1 - q}.$$

For $q = -1$, s_m is either 0 or 1, so s_m neither converges nor diverges, and, hence, the power series $\sum q^n$ neither converges nor diverges. Similarly, when $q < -1$ s_m and $\sum q^n$ neither converge nor diverge. When $q > 1$, on the other hand, the power series diverges:

$$s_m \to_{m \to \infty} \infty,$$

or

$$\sum_{n=0}^{\infty} q^n = \infty.$$

Let us now turn to the more interesting case $|q| < 1$. In this case,

$$q^{m+1} \to_{m\to\infty} 0,$$

so

$$s_m \to_{m\to\infty} \frac{1}{1-q},$$

or, in other words,

$$\sum_{n=0}^{\infty} q^n = \frac{1}{1-q}.$$

In the following, we use this result in periodic infinite decimal fractions.

3.11 Periodic Decimal Fractions

In the above, we have shown that every rational number has a representation as a periodic infinite decimal fraction. Here we show that the reverse is also true: every periodic infinite decimal fraction is actually a rational number. Thus, the set of rational numbers is actually the set of periodic infinite decimal fractions.

Consider a periodic infinite decimal fraction of the form

$$0.a_1 a_2 \ldots a_k a_1 a_2 \ldots a_k \ldots$$

Here the period $a_1 a_2 \ldots a_k$ consists of the k digits $a_1, a_2, \ldots, a_k$, hence is equal to

$$a_1 10^{k-1} + a_2 10^{k-2} + \cdots + a_{k-1} \cdot 10 + a_k.$$

Thus, the periodic infinite decimal fraction is equal to

$$\left(a_1 10^{k-1} + a_2 10^{k-2} + \cdots + a_{k-1} \cdot 10 + a_k.\right) 10^{-k} \sum_{n=0}^{\infty} (10^{-k})^n$$

$$= \left(a_1 10^{k-1} + a_2 10^{k-2} + \cdots + a_{k-1} \cdot 10 + a_k.\right) \frac{10^{-k}}{1 - 10^{-k}}$$

$$= \frac{a_1 10^{k-1} + a_2 10^{k-2} + \cdots + a_{k-1} \cdot 10 + a_k.}{10^k - 1},$$

which is indeed a rational number, as asserted.

3.12 Exercises

1. What is a rational number?
2. Show that the set of the rational numbers is closed under addition in the sense that the sum of any two rational numbers is a rational number as well.
3. Show that the set of the rational numbers is closed under multiplication in the sense that the product of any two rational numbers is a rational number as well.
4. Show that, for any rational number q, there exists a unique rational number w satisfying

$$q + w = 0.$$

   (w is called the negative of q, and is denoted by $-q$.)
5. Show that, for any rational number $q \neq 0$, there exists a unique rational number $w \neq 0$ satisfying

$$qw = 1.$$

   (w is called the reciprocal of q, and is denoted by $1/q$ or q^{-1}.)
6. Show that there is no rational number q satisfying

$$q^2 = 2.$$

7. Show that there is no rational number q satisfying

$$q^2 = 3.$$

8. Show that there is no rational number q satisfying

$$q^2 = p,$$

   where p is a given prime number.
9. Show that there is no rational number q satisfying

$$q^k = p,$$

   where p is a given prime number and k is a given natural number.

Chapter 4

Real Numbers

We have seen above that the set of the integer numbers is the smallest extension of the set of the natural numbers that is closed not only under addition and multiplication but also under subtraction. Furthermore, the set of the rational numbers is the smallest extension of the set of the integer numbers that is closed not only under addition, multiplication, and subtraction, but also under division (with no residual). In this chapter, we present the set of the real numbers, which is the smallest extension of the set of the rational numbers that is also closed under the limit process in the sense that the limit of a converging sequence of real numbers is by itself a real number as well (see [7]). Actually, we prove that the set of the real numbers consists of all the infinite decimal fractions, periodic ones and nonperiodic ones alike.

4.1 The Square Root of 2

The ancient Greeks, although they had no idea about rational numbers, did study ratios between edges of triangles. In particular, they knew that, in a rectangle whose edges have the ratio 4 : 3 between them, the diagonal has the ratios 5 : 4 and 5 : 3 with the edges. Indeed, from Pythagoras' theorem,

$$5^2 = 4^2 + 3^2,$$

which implies that, if the lengths of the edges are four units and three units, then the length of the diagonal must be five units (Figure 4.1). Thus, both the edges and the diagonal can be measured in terms of a common length unit.

Surprisingly, this is no longer the case in a square (Figure 4.2). More explicitly, there is no common length unit with which one can measure both the edge and the diagonal of a square. In other words, the ratio between the diagonal and the edge of a square is not a rational number.

Assume that the length of the edges of the square is 1. From Pythagoras' theorem, we then have that the square of the length of the diagonal is

$$1^2 + 1^2 = 2.$$

In other words, the length of the diagonal is the solution of the equation

FIGURE 4.1:   A right-angled triangle in which all the edges can be measured by a common length unit.

FIGURE 4.2:   A right-angled triangle whose edges have no common length unit. In other words, the lengths of the three edges cannot all be written as integer multiples of any common length unit.

$$x^2 = 2.$$

This solution is denoted by $\sqrt{2}$, or $2^{1/2}$.

Let us prove by contradiction that $\sqrt{2}$ is not a rational number. Indeed, if it were a rational number, then one could write

$$\sqrt{2} = n/m$$

for some nonzero integer numbers n and m. By taking the square of both sides in the above equation, we would then have

$$2m^2 = n^2.$$

Consider now the prime factorization of both sides of this equation. In the right-hand side, the prime factor 2 must appear an even number of times. In the left-hand side, on the other hand, the prime factor 2 must appear an odd number of times. This contradiction implies that our assumption is indeed false, and that $\sqrt{2}$ cannot be written as a ratio n/m of two integer numbers, so it is not a rational number: it is rather an irrational number.

4.2 The Least-Upper-Bound Axiom

Below we'll show that $\sqrt{2}$ indeed exists as a real number, namely, as a limit of a sequence of rational numbers. For this, however, we need first to present the least-upper-bound axiom. This axiom says that numbers that are all bounded by a common bound also have a least upper bound.

The least-upper-bound axiom implies that a monotonically increasing sequence must either diverge or converge to its least upper bound. Indeed, let

$$s_1 \leq s_2 \leq s_3 \leq \cdots \leq s_{m-1} \leq s_m \leq \cdots$$

be a monotonically increasing sequence. If it is unbounded, then for every given (arbitrarily large) number N there is a sufficiently large number M such that

$$s_M > N.$$

Because the sequence is monotonically increasing, this also implies that

$$s_m > N$$

for every $m \geq M$, which means that the sequence diverges.

If, on the other hand, the sequence is bounded, then, by the least-upper-bound axiom, it has a least upper bound s. This means that, for a given (arbitrarily small) number ε, there exists a sufficiently large number M for which

$$0 \leq s - s_M \leq \varepsilon.$$

Thanks to the fact that the sequence $\{s_m\}$ is monotonically increasing, this implies that

$$|s - s_m| \leq \varepsilon$$

for every $m \geq M$. This implies that

$$s_m \to_{m \to \infty} s,$$

as asserted.

By multiplying all the elements in the set by -1, a bounded set becomes bounded from below. Thus, the least-upper-bound axiom also has a reversed form: numbers that are all bounded from below also have a greatest lower bound. In particular, this implies that a monotonically decreasing sequence of numbers that are all bounded from below converges to their greatest lower bound.

4.3 The Real Numbers

The set of the real numbers is the smallest extension of the set of the rational numbers that is closed not only under addition, subtraction, multiplication, and division, but also under the limit process: every limit of a sequence of real numbers is by itself a real number as well.

Let us show that every decimal fraction of the form

$$0.a_1a_2a_3\ldots = \sum_{n=1}^{\infty} a_n 10^{-n}$$

is indeed a real number. For this, it is sufficient to show that it is indeed the limit of a sequence of rational numbers. This is indeed true: it is the limit of the partial sums s_m of the above series. In fact, these sums are all bounded (e.g., by 1); furthermore, they form a monotonically increasing sequence. Therefore, they converge to their least upper bound, s:

$$s_m \to_{m\to\infty} s,$$

also denoted as the sum of the infinite series:

$$s = \sum_{n=1}^{\infty} a_n 10^{-n}.$$

So far, we have shown that every decimal fraction is indeed a real number in the sense that it is indeed the limit of a sequence of rational numbers. Furthermore, the reverse is also true: every real number, i.e., the limit s of a sequence of the rational numbers s_m, can be presented as a decimal fraction. Indeed, if s is a finite decimal fraction, then it clearly has a finite decimal representation. If, on the other hand, s is not a finite decimal number, then for every (arbitrarily large) k, there is a yet larger number $l > k$ such that

$$|s - d| \geq 10^{-l}$$

for every finite decimal fraction d with at most k digits behind the decimal point. Now, let M be so large that, for every $m \geq M$,

$$|s_m - s| \leq 10^{-l}/2.$$

Clearly, these s_m's (with $m \geq M$), as well as s itself, have the same k digits behind the decimal point. By doing this for arbitrarily large k, one obtains the infinite decimal representation of s uniquely.

We've therefore established not only that every decimal fraction is a real number, but also that every real number can be represented as a decimal fraction. Thus, the set of real numbers is equivalent to the set of (periodic and nonperiodic) decimal fractions.

4.4 Decimal Representation of $\sqrt{2}$

One can now ask: since $\sqrt{2}$ cannot be presented as a rational number of the form n/m, where n and m are some integer numbers, does it exist at all? In other words, does the equation

$$x^2 = 2$$

have a solution at all?

In the following, we not only prove that the required solution exists, but also provide an algorithm to construct its (infinite) decimal representation. Of course, this representation must be nonperiodic, or the solution would be rational, which it is not. Thus, the solution exists as a nonperiodic infinite decimal fraction, or an irrational real number.

The algorithm to construct the decimal representation of $\sqrt{2}$ produces the digits in it one by one in such a way that the residual

$$2 - x^2$$

is minimized. For example, the first digit in the decimal representation of $\sqrt{2}$ (the unit digit) must be 1, because it gives the minimal residual

$$2 - 1^2 = 1.$$

Furthermore, the first digit right after the decimal point must be 4, because it gives the minimal residual

$$2 - (1.4)^2 = 0.04.$$

Moreover, the next digit must be 1, because it gives the yet smaller residual

$$2 - (1.41)^2 = 0.0119.$$

Note that one must avoid a negative residual, which would mean that the chosen digit is too large. Thus, the algorithm determines uniquely the digits in the infinite decimal fraction.

Let us denote the digits produced in the above algorithm by

$$a_0 = 1$$
$$a_1 = 4$$
$$a_2 = 1$$

and so on. Thus, the above algorithm actually produces the infinite series

$$a_0 + a_1 10^{-1} + a_2 10^{-2} + \cdots = \sum_{n=0}^{\infty} a_n 10^{-n}.$$

Define the partial sums

$$s_m \equiv \sum_{n=0}^{m} a_n 10^{-n}.$$

As we have seen above, this sequence converges to the real number s:

$$s_m \to_{m \to \infty} s,$$

or, in other words,

$$s = \sum_{n=0}^{\infty} a_n 10^{-n}.$$

Does s satisfy the equation

$$s^2 = 2?$$

Yes, it does. Indeed, the residuals

$$2 - s_m^2$$

are all greater than or equal to zero, so, by the least-upper-bound axiom, they must also have a greatest lower bound $d \geq 0$. Furthermore, because these residuals are monotonically decreasing, d is also their limit:

$$2 - s_m^2 \to_{m \to \infty} d.$$

Let us now prove by contradiction that $d = 0$. Indeed, if d were positive, then, there would exist a sufficiently large n, for which one could increase a_n by 1 to obtain a residual smaller than the residual obtained in our algorithm, in violation of the definition of the algorithm. Thus, we have established that $d = 0$, and, hence,

$$2 - s^2 = \lim_{m \to \infty} (2 - s_m^2) = 0.$$

The conclusion is that

$$\sqrt{2} = s$$

indeed exists as a real number, namely, as the limit of a sequence of the rational numbers s_m.

4.5 Irrational Numbers

In the above, we have shown that $\sqrt{2}$ cannot be written as the ratio n/m of two integer numbers n and m, or, in other words, that $\sqrt{2}$ is irrational. In a similar way, one can show that $\sqrt{p}$ is irrational for every prime number

p. Furthermore, one could also show that $p^{1/k}$ is irrational for every prime number p and natural number $k > 1$. Moreover, for every two natural numbers $l > 1$ and $k > 1$, $l^{1/k}$ is either a natural number or an irrational number. Indeed, assume that $l^{1/k}$ was a rational number:

$$l^{1/k} = n/m.$$

Then, we could take the kth power of both sides of the above equation to get

$$lm^k = n^k.$$

Now, in the right-hand side, every prime factor must appear a multiple of k times. Therefore, the same must also hold in the left-hand side. As a consequence, each prime factor of l must appear a multiple of k times in the prime factorization of l. This means that $l^{1/k}$ must be a natural number.

Alternatively, the original assumption that $l^{1/k}$ can be written as a fraction of the form n/m is false. This implies that $l^{1/k}$ is irrational. This is indeed the case, e.g., when l is prime. Indeed, in this case the only prime factor of l is l itself, and the prime factorization of l is just $l = l^1$. Because 1 can never be a multiple of k, our original assumption that $l^{1/k}$ is rational must be false.

4.6 Transcendental Numbers

The above irrational numbers are the solutions of equations of the form

$$x^k = p,$$

where $k > 1$ is a natural number and p is a prime number. The solution $p^{1/k}$ of such an algebraic equation is called an algebraic number.

There are, however, other irrational numbers, which do not solve any such equation. These numbers are called transcendental numbers. Such a number is π, the ratio between a circle and its diameter. Another such number is the natural exponent e defined below.

4.7 The Natural Exponent

Suppose that your bank offers you to put your money in a certified deposit (CD) for ten years under the following conditions. In each year, you'll receive an interest of 10%. Thus, in ten years, you'll receive a total interest of 100%, so you'll double your money.

This, however, is not a very good offer. After all, at the end of each year you should already have the extra 10%, so in the next year you should have received interest not only for your original money but also for the interest accrued in the previous year. Thus, after ten years, the original sum that you put in the CD should have multiplied not by 2 but actually by

$$\left(1 + \frac{1}{10}\right)^{10} > 2.$$

Actually, even this offer is not sufficiently good. The interest should actually be accrued at the end of each month, so that in the next month you'll receive interest not only for your original money but also for the interest accrued in the previous month. This way, at the end of the ten years, your original money should actually be multiplied by

$$\left(1 + \frac{1}{120}\right)^{120}.$$

Some (more decent) banks give interest that is accrued daily. This way, each day you receive interest not only for your original money but also for the interest accrued in the previous day. In this case, at the end of the ten years your money is multiplied by

$$\left(1 + \frac{1}{43800}\right)^{43800}.$$

Ideally, the bank should give you your interest in every (infinitely small) moment, to accrue more interest in the next moment. Thus, at the end of the ten years, your money should ideally be multiplied by

$$\lim_{n \to \infty} \left(1 + \frac{1}{n}\right)^n = 2.718\ldots = e.$$

Here the natural exponent e is the limit of the sequence $(1 + 1/n)^n$ in the sense that, for every given (arbitrarily small) number $\varepsilon > 0$, one can choose a sufficiently large number N such that

$$\left|\left(1 + \frac{1}{n}\right)^n - e\right| \le \varepsilon$$

for every $n \ge N$. From the discussion in the previous sections, e is indeed a real number.

It is now clear that the original offer of your bank to double your money after ten years is not that good. The bank should actually offer to multiply your original money by $e = 2.718\ldots$ rather than just by 2. You should thus consider other investment options, in which the interest accrues momentarily rather than yearly or ten-yearly.

The natural exponent e can also be present as an infinite sum:

$$e = 2.5 + \frac{1}{6} + \frac{1}{24} + \frac{1}{120} + \cdots = \frac{1}{0!} + \frac{1}{1!} + \frac{1}{2!} + \frac{1}{3!} + \frac{1}{4!} + \cdots = \sum_{n=0}^{\infty} \frac{1}{n!}.$$

4.8 Exercises

1. What is a limit of a given sequence of numbers?
2. Show that the limit of a given sequence of numbers is unique.
3. What is a real number?
4. Interpret a real number as the unique limit of a sequence of rational numbers.
5. Show that the set of the real numbers is closed under addition in the sense that the sum of any two real numbers is a real number as well. (Hint: show that this sum is the limit of a sequence of rational numbers.)
6. Show that the set of the real numbers is closed under multiplication in the sense that the product of any two real numbers is a real number as well. (Hint: show that this product is the limit of a sequence of rational numbers.)
7. Show that the set of the real numbers is closed under the limit process in the sense that the limit of a sequence of real numbers is a real number as well. (Hint: show that this limit is also the limit of a sequence of rational numbers.)
8. Show that, for any real number r, there exists a unique real number w satisfying

$$r + w = 0.$$

(w is called the negative of r, and is denoted by $-r$.)
9. Show that, for any real number $r \neq 0$, there exists a unique real number $w \neq 0$ satisfying

$$rw = 1.$$

(w is called the reciprocal of r, and is denoted by $1/r$ or r^{-1}.)
10. Write the algorithm that finds a real number r satisfying

$$r^2 = 2.$$

Show that this solution is unique.
11. Write the algorithm that finds a real number r satisfying

$$r^2 = 3.$$

Show that this solution is unique.

12. Write the algorithm that finds a real number r satisfying

$$r^2 = p,$$

where p is a given prime number. Show that this solution is unique.

13. Write the algorithm that finds a real number r satisfying

$$r^k = p,$$

where p is a given prime number and k is a given natural number. Show that this solution is unique.

Chapter 5

Complex Numbers

The set of the natural numbers, although closed under both addition and multiplication, is incomplete because it contains no solution for the equation

$$x + n = 0,$$

where n is a positive natural number. This is why the set of the integer numbers is introduced: it is the smallest extension of the set of the natural numbers in which every equation of the above form has a solution.

Unfortunately, the set of the integer numbers is still incomplete in the sense that it contains no solution for the equation

$$mx = 1,$$

where $m > 1$ is a natural number. This is why the set of the rational numbers is introduced: it is the smallest extension of the set of the integer numbers in which every equation of the above form has a solution.

complex numbers	$x^2 = -1$
∪	
real numbers	$x^k = p$
∪	
rational numbers	$mx = 1$
∪	
integer numbers	$x + n = 0$
∪	
natural numbers	

FIGURE 5.1: Hierarchy of sets of numbers. Each set contains the solution to the equation to its right. The symbol '⊂' means inclusion of a set in a yet bigger set. Thus, each set is a subset of the set in the next higher level.

Still, the set of the rational numbers is also incomplete in the sense that it has no solution for the nonlinear equation

$$x^k = p,$$

where $k > 1$ is a natural number and p is a prime number. This is why the set of the real numbers is introduced: it is the smallest extension of the set of the rational numbers that contains the solution of each equation of the above form and is still closed under the arithmetic operations.

Still, the set of the real numbers contains no solution to the equation

$$x^2 = -1.$$

This problem is fixed in the set of the complex numbers, which is the smallest extension of the set of the real numbers that contains a solution for the above equation and is still closed under the arithmetic operations (Figure 5.1). In fact, the set of the complex numbers is the smallest mathematical field in which every algebraic equation can be solved (see [16]–[17]).

5.1 The Imaginary Number

The number axis illustrated in Figure 2.2 above originally contains only the integer numbers. Now, however, we can place in it also the rational numbers and even the real numbers. In fact, every real number x can be characterized by the arrow (or vector) leading from zero in the middle of the number axis to x. This is why the number axis is also called the real axis (Figure 5.2).

x 0 the real axis

FIGURE 5.2:   The real axis. The arrow leading from zero to the negative number x produces an angle of 180 degrees (or π) with the positive part of the real axis.

As illustrated in Figures 2.3 and 5.2, every positive number x is characterized by the zero angle between the arrow leading from zero to it and the positive part of the real axis, whereas every negative number $x < 0$ is characterized by the 180-degree angle between the arrow leading from zero to it and the positive part of the real axis.

The angle associated with the product of two numbers is just the sum of the angles associated with the numbers. This is why the angle associated with $x^2 = x \cdot x$ is always zero. Indeed, if $x > 0$, then the angle associated with it is zero, so the angle associated with x^2 is $0 + 0 = 0$ as well. If, on the other hand, $x < 0$, then the angle associated with it is of 180 degrees, so the angle associated with x^2 is of 360 degrees, which is again just the zero angle. Thus, $x^2 \geq 0$ for every real number x.

The conclusion is, thus, that there is no real number x for which

$$x^2 = -1.$$

Still, this equation does have a solution outside of the real axis!

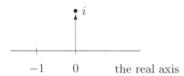

$$-1 \qquad 0 \qquad \text{the real axis}$$

FIGURE 5.3: The imaginary number i. The arrow leading from the origin to i produces a right angle with the positive part of the real axis. This angle is doubled in i^2 to produce the required result -1.

Indeed, let us extend the real axis to the number plain (Figure 5.3). This plane is based on two axes: the horizontal real axis and the vertical imaginary axis. Let us mark the point 1 on the imaginary axis at distance 1 above the origin. This point marks the number i – the imaginary number.

The imaginary number i is characterized by the absolute value 1 (its distance from the origin) and the right angle of 90 degrees (the angle between the arrow leading from the origin to it and the positive part of the real axis). Therefore, the square of i, i^2, is characterized by its absolute value

$$|i^2| = |i| \cdot |i| = 1 \cdot 1 = 1$$

and the angle between the arrow leading from the origin to it and the positive part of the real axis:

$$90 + 90 = 180.$$

This means that

$$i^2 = -1,$$

or i is the solution to the original equation

$$x^2 = -1.$$

5.2 The Number Plane

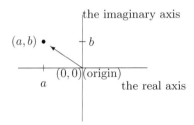

FIGURE 5.4:   The complex plane. The complex number $a + bi$ is represented by the point (a, b).

So far, we have introduced only one number outside of the real axis: the imaginary number i. However, in order to have a set that is closed under multiplication, we must also introduce the numbers of the form bi for every real number b. This means that the entire imaginary axis (the vertical axis in Figure 5.4) must be included as well.

Furthermore, in order to have a set that is also closed under addition, numbers of the form $a + bi$ must also be included for every two real numbers a and b. These numbers are represented by the points (a, b) in the plane in Figure 5.4, where a is the horizontal coordinate (the distance from the imaginary axis) and b is the vertical coordinate (the distance from the real axis). This produces the entire number plane, in which each number is complex: it is the sum of a real number of the form a and an imaginary number of the form bi. This is why this number plane is also called the complex plane.

5.3 Sine and Cosine

The complex number $a + bi$, represented by the point (a, b) in the complex plane in Figure 5.4, is characterized by two parameters: a, its real component, and b, its imaginary component. Furthermore, it can also be represented by two other parameters: $r = \sqrt{a^2 + b^2}$, its absolute value [the length of the arrow leading from the origin $(0,0)$ to (a,b)], and θ, the angle between that arrow and the positive part of the real axis. The representation that uses these two parameters is called the polar representation.

In fact, θ itself can be characterized by two possible functions:

$$\cos(\theta) = \frac{a}{\sqrt{a^2 + b^2}},$$

or

$$\sin(\theta) = \frac{b}{\sqrt{a^2 + b^2}}.$$

Using these definitions, we have

$$a + bi = r(\cos(\theta) + \sin(\theta)i).$$

It is common to denote an angle of 180 degrees by π, because it is associated with one half of the unit circle (the circle of radius 1), whose length is indeed π. One half of this angle, the right angle, is denoted by $\pi/2$. With these notations, we have

$$\sin(0) = \sin(\pi) = 0, \ \cos(\pi/2) = \cos(3\pi/2) = 0,$$

$$\cos(0) = \sin(\pi/2) = 1, \ \text{and} \ \cos(\pi) = \sin(3\pi/2) = -1.$$

As a matter of fact, because angles are defined only up to a multiple of 2π, the angle $3\pi/2$ is the same as the angle $-\pi/2$. In the following, we use the arithmetic operations between complex numbers to get some more information about the nature of the sine and cosine functions.

5.4 Adding Complex Numbers

The complex plane is closed under addition in the sense that the sum of two complex numbers is a complex number as well. In fact, the addition of two complex numbers is done coordinate by coordinate:

$$(a + bi) + (c + di) = (a + c) + (b + d)i.$$

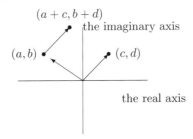

FIGURE 5.5:   Adding the complex numbers $a + bi$ and $c + di$ by the parallelogram rule produces the result $a + c + (b + d)i$.

Geometrically, this means that the addition uses the parallelogram rule (Figure 5.5).

In particular, if we choose $c = -a$ and $d = -b$, then we have

$$(a + bi) + (c + di) = 0.$$

This means that $c + di$ is the negative of $a + bi$:

$$c + di = -(a + bi).$$

5.5 Multiplying Complex Numbers

So far, we only know how to multiply two real numbers and how to multiply i by itself:

$$i^2 = -1.$$

This means that the angle associated with i, the right angle, is doubled to obtain the angle of 180 degrees (or the angle π) associated with -1. Furthermore, the absolute value of -1 is the square of the absolute value of i: $1 \cdot 1 = 1$.

This multiplication is now extended linearly to the entire complex plane. For example,

$$i(a + bi) = ai + bi^2 = -b + ai.$$

Geometrically, this means that the point (a, b) has been rotated by a right angle, the angle associated with i. This produces the new point $(-b, a)$, which has the same absolute value as the original point (a, b), only its angle is $\theta + \pi/2$ rather than θ.

More generally, we have that the product of two general complex numbers $a + bi$ and $c + di$ is

$$(a + bi)(c + di) = ac + bdi^2 + adi + bci = (ac - bd) + (ad + bc)i.$$

Geometrically, this means that the original point (c, d) has been rotated by the angle θ, and the arrow leading to it has been stretched by factor $r = \sqrt{a^2 + b^2}$. In fact, if the original complex numbers $a + bi$ and $c + di$ have the polar representations (r, θ) and (q, ϕ), then their product has the polar representation $(rq, \theta + \phi)$. In particular, if we choose $q = 1/r$ and $\phi = -\theta$, then we have the reciprocal:

$$c + di = \frac{1}{a + bi}.$$

This reciprocal, whose polar representation is $(1/r, -\theta)$, is obtained by choosing

$$c = \frac{a}{a^2 + b^2} \text{ and } d = \frac{-b}{a^2 + b^2}.$$

In summary, the product of two complex numbers is a complex number as well. This means that the complex plane is closed not only under addition and subtraction but also under multiplication and division. This means that the complex plane is actually a mathematical field. In fact, it is the smallest mathematical field in which every algebraic equation, including

$$x^2 = -1,$$

has a solution.

5.6 The Sine and Cosine Theorems

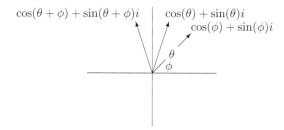

FIGURE 5.6: Multiplying the two complex numbers $\cos(\theta) + \sin(\theta)i$ and $\cos(\phi) + \sin(\phi)i$ results in the complex number $\cos(\theta + \phi) + \sin(\theta + \phi)i$.

The unit circle is the set of complex numbers with absolute value 1, namely, with distance 1 from the origin (Figure 5.6). Consider two complex numbers on this circle: $\cos(\theta) + \sin(\theta)i$ and $\cos(\phi) + \sin(\phi)i$. The first of these numbers produces the angle θ between the arrow leading from the origin to it and the positive part of the real axis, whereas the second one produces the angle ϕ between the arrow leading from the origin to it and the positive part of the real axis. Thus, their product is the complex number on the unit circle that produces the angle $\theta + \phi$ between the arrow leading from the origin to it and the positive part of the real axis (Figure 5.6).

This is a geometric observation; from an algebraic point of view, on the other hand, this product can also be obtained by an arithmetic calculation, using the linear nature of the multiplication operation:

$$\cos(\theta + \phi) + \sin(\theta + \phi)i$$
$$= (\cos(\theta) + \sin(\theta)i)(\cos(\phi) + \sin(\phi)i)$$
$$= (\cos(\theta)\cos(\phi) - \sin(\theta)\sin(\phi)) + (\sin(\theta)\cos(\phi) + \cos(\theta)\sin(\phi))i.$$

By comparing the corresponding real and imaginary parts in the above equation, we obtain the following two formulas:

$$\cos(\theta + \phi) = \cos(\theta)\cos(\phi) - \sin(\theta)\sin(\phi),$$

known as the cosine theorem, and

$$\sin(\theta + \phi) = \sin(\theta)\cos(\phi) + \cos(\theta)\sin(\phi),$$

known as the sine theorem.

5.7 Exercises

1. What is a complex number?
2. Interpret the arithmetic operations between complex numbers in geometrical terms.
3. Let $a \neq 0$, b, and c be some given real parameters. Show that the equation

$$ax^2 + bx + c = 0$$

   has at least one and at most two complex solutions. Hint: rewrite

$$ax^2 + bx + c = a(x + b/(2a))^2 + (c - b^2/(4a)).$$

4. Find a necessary and sufficient condition to guarantee that the above solutions are also real.
5. Let $z = a + ib$ be some complex number. Show that

$$|z|^2 = z\bar{z}.$$

6. Assume that z has the polar representation
$$z = r(\cos(\theta) + i \cdot \sin(\theta)).$$
Show that its complex conjugate $\bar{z}$ has the polar representation
$$\bar{z} = r(\cos(-\theta) + i \cdot \sin(-\theta)).$$
Use the rules for multiplying complex numbers by their polar representations to obtain
$$z\bar{z} = r^2.$$

7. Let the 2 by 2 matrix A be the table of four given numbers $a_{1,1}$, $a_{1,2}$, $a_{2,1}$, and $a_{2,2}$:
$$A = \begin{pmatrix} a_{1,1} & a_{1,2} \\ a_{2,1} & a_{2,2} \end{pmatrix}.$$
Similarly, let B be the 2×2 matrix
$$B = \begin{pmatrix} b_{1,1} & b_{1,2} \\ b_{2,1} & b_{2,2} \end{pmatrix},$$
where $b_{1,1}$, $b_{1,2}$, $b_{2,1}$, and $b_{2,2}$ are four given numbers. Define the product of matrices AB to be the 2×2 matrix
$$AB = \begin{pmatrix} a_{1,1}b_{1,1} + a_{1,1}b_{2,1} & a_{1,1}b_{1,2} + a_{1,2}b_{2,2} \\ a_{2,1}b_{1,1} + a_{2,2}b_{2,1} & a_{2,1}b_{1,2} + a_{2,2}b_{2,2} \end{pmatrix}.$$
Furthermore, let the complex number $a + ib$ be associated with the matrix
$$a + ib \sim \begin{pmatrix} a & -b \\ b & a \end{pmatrix}.$$
Similarly, let the complex number $c + id$ be associated with the matrix
$$c + id \sim \begin{pmatrix} c & -d \\ d & c \end{pmatrix}.$$
Show that the product of complex numbers is associated with the product of matrices:
$$(a + ib)(c + id) \sim \begin{pmatrix} a & -b \\ b & a \end{pmatrix} \begin{pmatrix} c & -d \\ d & c \end{pmatrix}.$$

8. Furthermore, for the above complex number $a + ib$ define r and θ that satisfy
$$r = \sqrt{a^2 + b^2} \quad \text{and} \quad \tan(\theta) = a/b.$$
Show that the matrix associated with $a + ib$ can also be written as
$$\begin{pmatrix} a & -b \\ b & a \end{pmatrix} = r \begin{pmatrix} \cos(\theta) & -\sin(\theta) \\ \sin(\theta) & \cos(\theta) \end{pmatrix}.$$
Conclude that
$$r(\cos(\theta) + i\sin(\theta)) = a + ib$$
is the polar representation of the original complex number $a + ib$.

9. Let $a+ib$ above be a fixed complex number, and $c+id$ above be a complex variable in the complex plane. Show that the multiplication by $a + ib$ is a linear operation on the complex plane.

10. Consider the particular case $c = 1$ and $d = 0$, in which $c + id$ is represented in the complex plane by the horizontal standard unit vector $(1, 0)$. Show that the multiplication of this complex number by $a + ib$ on the left amounts to rotating this vector by angle θ and then stretching it by factor r.

11. Consider the particular case $c = 0$ and $d = 1$, in which $c+id$ is represented in the complex plane by the vertical standard unit vector $(0, 1)$. Show that the multiplication of this complex number by $a + ib$ on the left amounts to rotating this vector by angle θ and then stretching it by factor r.

12. Use the linearity property to show that the multiplication of any complex number in the complex plane by $a + ib$ on the left amounts to rotating it by angle θ and then stretching it by factor r.

13. Conclude that a complex number can be viewed as a transformation of the entire 2-dimensional plane, which is composed of a rotation followed by stretching by a constant factor.

Part II

Geometrical Objects*

Geometrical Objects

So far, we have dealt with numbers, along with the functions associated with them: arithmetic operators that take two numbers to return the output (their sum, product, etc.), and relations like '$<$', '$>$', and '$=$', which again take two arguments to either "true" (if the relations indeed holds) or "false" (otherwise). Here we turn to another field in mathematics, which uses different kind of objects: the field of geometry.

The mathematical objects used in geometry are no longer numbers but rather pure geometrical objects: point, line, line segment, angle, circle, etc. The relations between these objects are actually functions of two variables. For example, the relations "the point lies on the line" takes two arguments, some particular point and line, and return 1 (or true) if the point indeed lies on the line or 0 (false) otherwise.

In Euclidean geometry, no numbers are used. The elementary objects, along with some axioms associated with them that are accepted because they make sense from a geometrical point of view, are used to prove theorems using elementary logics. In analytic geometry, on the other hand, real numbers are used to help define the geometric objects as sets of points, and help prove the theorems associated with them.

*This part is for beginners, and can be skipped by more experienced readers.

Chapter 6

Euclidean Geometry

The theory of Euclidean geometry is completely independent of the numbers defined and discussed above. Furthermore, it is also independent of any geometric intuition: it is based on pure logics only [6] [12].

The theory assumes the existence of some abstract objects such as points, lines, line segments, angles, and circles. It also assumes some relations between these objects, such as "the point lies on the line" or "the line passes through the point." Still, these objects and relations don't have to be interpreted geometrically; they can be viewed as purely abstract concepts.

The axioms used in Euclidean geometry make sense from a geometrical point of view. However, once they are accepted, they take a purely logical form. This way, they can be used best to prove theorems.

The human geometrical intuition is often biased and inaccurate, hence can lead to errors. Euclidean geometry avoids such errors by using abstract objects and pure logics only.

6.1 Points and Lines

We have seen above that each real number can be represented as a point on the real axis. In fact, the real axis is just the set of all the real numbers, ordered in increasing order.

In Euclidean geometry, on the other hand, numbers are never used. In fact, a point has no numerical or geometrical interpretation: it is merely an abstract object. Moreover, a line doesn't consist of points: it is merely an abstract object as well. Still, points and lines may relate to each other: a point may lie on a line, and a line may pass through a point. Two distinct lines may cross each other at their unique joint point, or be parallel to each other if they have no joint point.

There are two elementary axioms about the relations between lines and points. The first one says that, if two distinct points are given, then there is exactly one line that passes through both of them. The second one says that, if a line and a point that doesn't lie on it are given, then there is another line that passes through the point and is also parallel to the original line. These axioms indeed make sense from a geometrical point of view. However, since they are accepted, their geometric meaning is no longer material: they are used as purely logical statements in the proofs of theorems.

6.2 Rays and Intervals

Let A and B be two distinct points on a particular line. Then, **AB** is the ray that starts at A and goes towards B, passes it, and continues beyond it. Furthermore, AB is the line segment leading from A to B on the original line.

6.3 Comparing Intervals

Although no numbers are used in Euclidean geometry to measure the size of AB (or the distance from A to B), it can still be compared to the size of other line segments. For example, if $A'B'$ is some other line segment, then either $AB > A'B'$ or $AB < A'B'$ or $AB = A'B'$. For this, however, we first need to know how line segments can be mapped.

FIGURE 6.1:   Mapping the line segment AB onto its image $A'B''$. This mapping implies that $A'B'' < A'B'$.

A line segment of the form AB can be mapped onto the ray **A'B'** using a compass as follows (Figure 6.1). First, one should place one leg of the compass at A and the other at B. This way, the length of AB is stored in the compass. Then, one should place the main leg of the compass at A', and draw an arc that crosses **A'B'** at a new point, say B''. This way, we have

$$AB = A'B''$$

in the sense that the length of these line segments is the same.

The above mapping allows one to compare the lengths of the two original line segments AB and $A'B'$. In fact, if the new point B'' lies in between A' and B', then

$$AB = A'B'' < A'B'.$$

If, on the other hand, B' lies in between A' and B'', then

$$AB = A'B'' > A'B'.$$

Finally, if B'' coincides with B', then

$$AB = A'B'' = A'B'.$$

Thus, although in Euclidean geometry line segments don't have any numerical length, they can still be compared to each other to tell which one is longer. Furthermore, the above mapping can also be used to compute the sum of two line segments. In fact, the sum of two line segments is a line segment into which both of the original line segments can be mapped disjointly with no remainder. For example, the sum of line segments of the form AB and DE can be computed by mapping DE onto the new line segment BC, which lies on the ray **AB** and does not overlap with AB. This way, we have

$$AB + DE = AC.$$

6.4 Ratios between Intervals

As discussed above, in Euclidean geometry there is no length function that assigns to each line segment a number to measure its length. Nevertheless, as we'll see below, it is still possible to compare not only lengths of line segments but also the ratios (or proportions) of pairs of line segments.

Let AB and $A'B'$ be some given line segments. Since their length is unavailable, how can we define their ratio $A'B'/AB$? Fortunately, this can be done by an algorithm in the spirit of Euclid's algorithm for computing the greatest common divisor of two natural numbers.

Let us illustrate how a ratio can be characterized by a sequence of integer numbers. Suppose that we are interested in the ratio between a line segment whose length is 7.5 units and a line segment whose length is one unit only. Because the second line segment can be embedded seven times in the first one, the first member in the sequence should be 7. Now, although the line segment whose length is one unit cannot be embedded in the line segment remainder (whose length is half a unit only), their roles may be interchanged: the line segment remainder can be embedded twice in the line segment of one unit.

To pay for this interchange, one should also assign a minus sign to the next member in the sequence, -2.

Suppose now that the first line segment is of length 7.4 units rather than 7.5 units. In this case, one extra member must be added to the above sequence, to count the number of times that the final line segment remainder, whose length is 0.2 units, can be embedded in the previous line segment remainder, whose length is 0.4 units. But the new ratio 7.4/1 must be smaller than the old ratio 7.5/1, which implies that ∞ should be added as the final member in the sequence corresponding to 7.5/1, to have

$$\frac{7.5}{1} \sim \{7, -2, \infty\} > \{7, -2, 2, -\infty\} \sim \frac{7.4}{1},$$

where the inequality sign '>' between the sequences above is in terms of the usual lexicographical order. In this order, comparing sequences is the same as comparing their first members that are different from each other. Because the third member in the first sequence above, ∞, is greater than the third member in the second sequence, 2, the first sequence above is also greater than the second sequence, implying that the first ratio, 7.5/1, is also greater than the second one, 7.4/1, as indeed required.

The alternating signs used in the above sequences is particularly suitable for comparing sequences using the lexicographical order. For example,

$$\frac{7.5}{1} \sim \{7, -2, \infty\} > \{7, -4, \infty\} \sim \frac{7.25}{1},$$

as indeed required.

Let us give the above ideas a more precise formulation. Let ∞ be a symbol that is greater than every integer number. Similarly, let $-\infty$ be a symbol that is less than every integer number. The ratio between $A'B'$, the line segment numerator, and AB, the line segment denominator, is a sequence of symbols that are either integer numbers or $\pm\infty$. More precisely, the sequence must be either infinite or finite with its last member being ∞ or $-\infty$. In the following, we also use the term "sequence to the power -1" to denote the sequence obtained by reversing the signs of the members in the original sequence.

In the following algorithm, we assume that $A'B' \geq AB$. [Otherwise, we use the definition

$$\frac{A'B'}{AB} = \left(\frac{AB}{A'B'}\right)^{-1}.]$$

Here is the recursive algorithm to obtain the required sequence.

Algorithm 6.1 *1. If AB is the trivial line segment, that is, if A coincides with B, then the sequence is just ∞ (the sequence with only one member, ∞).*

2. Otherwise, map AB into $A'B'$ $k+1$ times, so that

$$AB = A'A_1 = A_1A_2 = A_2A_3 = \cdots = A_{k-1}A_k = A_kA_{k+1},$$

where $A_1, A_2, \ldots, A_k, A_{k+1}$ are the endpoints of the mapped line segment, and $k+1$ is the smallest natural number for which A_{k+1} lies outside $A'B'$.

3. The required sequence is

$$k, \left(\frac{AB}{A_k B'}\right)^{-1},$$

that is, the sequence in which the first member is k, and the remaining members are the members of the sequence $AB/A_k B'$ with their signs changed.

Note that, if the sequence produced by this algorithm is finite, then the final line segment remainder (denoted by $A_k B'$) used in the final recursive call to the algorithm is trivial in the sense that A_k coincides with B'. Therefore, the final recursive call uses the first rather than second step in the algorithm, leading to its termination and to the member $\pm\infty$ at the end of the sequence. Furthermore, the line segment remainder of the form $A_k B'$ used in the recursive call that precedes the final recursive call, on the other hand, can be used as a common length unit for both AB and $A'B'$. (For example, a unit of length 1 can serve as a common length unit for the edges of the triangle in Figure 4.1.) Indeed, both AB and $A'B'$ can be viewed as multiples of that line segment.

If, on the other hand, the above sequence is infinite, then there is no common length unit, as in Figure 4.2. In this case, the above algorithm never terminates. Still, each particular member in the sequence can be computed in a finite number of steps. Indeed, for every natural number n, the nth member can be computed in $n-1$ recursive calls to the algorithm.

Now, the comparison between two ratios, or two such sequences, is done in the usual lexicographical order: a sequence is greater than another sequence if, for some natural number n, the nth member in it is greater than the nth member in the other sequence, whereas all the previous $n-1$ members are the same as in the other sequence. In particular, two ratios are the same if all the members in their sequences are the same.

In the lexicographical order, there are no sequences between the sequence $\{-1, \infty\}$ and the sequence $\{1, -\infty\}$, so they can both be viewed as the same sequence, denoted by the number 1:

$$\{-1, \infty\} = \{1, -\infty\} = 1.$$

Clearly, these sequences are obtained only when the line segments have the same length:

$$AB = A'B'.$$

Thus, the equation

$$\frac{A'B'}{AB} = \frac{AB}{A'B'} = 1$$

can be interpreted not only in the above sense but also in the usual sense.

The above definition of a ratio allows one to find out which of two different ratios is the greater one. Indeed, since they are different from each other, for some natural number n the nth member in one sequence must be greater than the nth member in the other sequence. The above algorithm can find these members in $n - 1$ recursive calls and compare them to tell which ratio is the greater one. If, on the other hand, the ratios are equal to each other, then we can know this for sure only if they have finite sequences, because we can then compare the corresponding members one by one and verify that each member in one sequence is indeed the same as the corresponding member in the other sequence. If, on the other hand, the sequences are infinite, then there is no practical algorithm that can tell us for sure that the ratios are indeed equal to each other. Indeed, since the above algorithm never terminates, one never gets a final answer about whether or not the entire sequences are exactly the same.

Still, even when one can never decide in finite time whether or not two ratios are equal to each other, one can assume that this information is given from some other source. Thus, it makes sense to assume *a priori* that the information that two ratios are equal to each other is available. Such assumptions are often made in the study of similar triangles below.

6.5 Angles

An angle is the region confined in between two rays of the form **BC** and **BA** that share the same starting point B. This angle is vertexed at B, and is denoted by $\angle ABC$. In this notation, the middle letter, B, denotes the vertex, whereas the left and right letters, A and C, denote some points on the rays.

Note that the term "region" is used loosely above only to give some geometric intuition. In fact, it is never really used in the formal definition of the angle $\angle ABC$. Indeed, $\angle ABC$ can be obtained as a formal function of the three points A, B (the vertex), and C and an extra binary variable to tell which side of $\angle ABC$ is relevant, because there are actually two angles that can be denoted by $\angle ABC$, the sum of which is 360 degrees or 2π.

6.6 Comparing Angles

An angle can be mapped to another place in the plane using two compasses. This is done as follows. Suppose that we want to map $\angle ABC$ onto a new angle that uses the new vertex B' and the new ray **B'C'**, where B' and C' are some given points. To do this, we use a compass to draw an arc around B that crosses **BC** at C'' and **BA** at A''.

Then, we open a second compass at the distance between C'' and A''. Now, we've got all the information about the angle stored in the compasses, and we are therefore ready to draw the mapped angle. To this end, we use the first compass to draw an arc around B', which crosses **B'C'** at the new point C'''. Then, we use the second compass to draw an arc around C''' so it crosses the previous arc at the new point A'. The required mapped angle is, therefore,

$$\angle A'B'C' = \angle ABC.$$

Note that in the above equation, the order of letters is important. The middle letters, B and B', mark the vertices of the original angle and the mapped angle. The rest of the letters, on the other hand, are just any points on the rays (Figure 6.2).

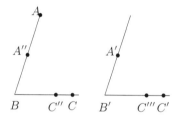

FIGURE 6.2:   The angle vertexed at B is mapped onto the angle vertexed at B' using one compass to store the distance $BC'' = BA''$ and another compass to store the distance $C''A''$.

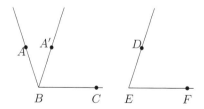

FIGURE 6.3:   The angle DEF is smaller than the angle ABC, because it can be mapped onto the angle $A'BC$, which lies inside the angle ABC.

The above mapping allows one to compare two given angles $\angle ABC$ and

$\angle DEF$ to realize which angle is the greater one (Figure 6.3). For example, one could map $\angle DEF$ onto the vertex B and the ray $\mathbf{BC}$, so that the $\mathbf{EF}$ is mapped onto $\mathbf{BC}$ and $\mathbf{ED}$ is mapped onto a new ray $\mathbf{BA'}$, so that there are three possibilities: if $\mathbf{BA'}$ lies within $\angle ABC$, then we have

$$\angle DEF = \angle A'BC < \angle ABC.$$

If on the other hand, $\mathbf{BA}$ lies within $\angle A'BC$, then we have

$$\angle ABC < \angle A'BC = \angle DEF.$$

Finally, if $\mathbf{BA'}$ coincides with $\mathbf{BA}$, then we have

$$\angle ABC = \angle A'BC = \angle DEF.$$

Thus, although in Euclidean geometry angles don't have any function to specify their size numerically, they can still be compared to each other to decide which angle is greater. This relation between angles will be used further below.

The above mapping is also useful in computing the sum of two angles. In fact, the sum of two angles of the form $\angle ABC$ and $\angle DEF$ is the new angle $\angle A'BC$ obtained by mapping $\angle DEF$ onto $\angle A'BA$ in such a way that $\mathbf{EF}$ coincides with $\mathbf{BA}$ and $\mathbf{ED}$ coincides with $\mathbf{BA'}$ (Figure 6.4).

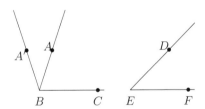

FIGURE 6.4:   The sum of the angle ABC and the angle DEF is obtained by mapping the latter onto the new angle $A'BA$ to produce the joint angle $A'BC$.

An angle of the form $\angle ABC$ is called a straight angle if all the three points A, B, and C lie on the same line. If the sum of two angles is a straight angle, then we say that their sum is also equal to 180 degrees or π. We then say that the two angles are supplementary to each other. If the two angles are also equal to each other, then we also say that they are right angles, or equal to 90 degrees or $\pi/2$.

6.7 Corresponding and Alternate Angles

Let a and b be two distinct lines. Let O be a point on a, and Q a point on b. From our original axiom, we know that there is exactly one line, say c, that passes through both O and Q.

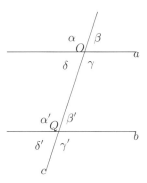

FIGURE 6.5: Pairs of corresponding angles between the parallel lines a and b: $\alpha = \alpha'$, $\beta = \beta'$, $\gamma = \gamma'$, and $\delta = \delta'$.

The line c produces with a four angles vertexed at O: α, β, γ, and δ (Figure 6.5). Similarly, c produces with b four corresponding angles vertexed at Q: α', β', γ', and δ'.

The vertical-angle theorem says that vertical angles are equal to each other, because they are both supplementary to the same angle. For example,

$$\alpha = \pi - \beta = \gamma \text{ and } \beta = \pi - \alpha = \delta.$$

The corresponding-angle axiom says that, if a and b are parallel to each other $(a\|b)$, then corresponding angles are equal to each other. In other words, $a\|b$ implies that

$$\alpha = \alpha', \ \beta = \beta', \ \gamma = \gamma', \text{ and } \delta = \delta'.$$

The alternate-angle theorem says that, if $a\|b$, then alternate angles are equal to each other:

$$\alpha = \gamma', \ \beta = \delta', \ \gamma = \alpha', \text{ and } \delta = \beta'.$$

Indeed, this theorem follows from the corresponding-angle axiom and the vertical-angle theorem. For example,

$$\alpha = \alpha' = \gamma' \text{ and } \delta = \delta' = \beta'.$$

In the above, we have accepted the corresponding-angle axiom, and used it to prove the alternate-angle theorem. However, we could also interchange their roles: we could accept the alternate-angle theorem as an axiom, and use it to prove the equality of corresponding angles:

$$\alpha = \gamma = \alpha', \ \ \beta = \delta = \beta',$$

and so on. This means that not only the corresponding-angle axiom implies the alternate-angle theorem, but also that an alternate-angle axiom would imply a corresponding-angle theorem. In other words, the corresponding-angle axiom is equivalent to the alternate-angle theorem.

It is common, though, to accept the corresponding-angle axiom, and use it to prove the alternate-angle theorem and other theorems as well. In fact, in the following we use it to prove also the reversed corresponding-angle theorem.

6.8 The Reversed Corresponding-Angle Theorem

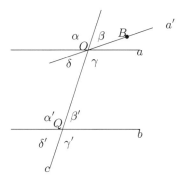

FIGURE 6.6:   Proving the reversed corresponding-angle theorem by contradiction: if a were not parallel to b, then one could draw another line a' that would be parallel to b.

The reversed corresponding-angle theorem says that, if corresponding angles are equal to each other, then the lines a and b are necessarily parallel to each other. In other words, $\gamma = \gamma'$ implies that $a\|b$. This theorem can be proved by contradiction. Indeed, assume that a and b were not parallel to each other. Then, using our original axiom, there is another line, say a', that passes through O and is parallel to b (Figure 6.6). Let B be a point on the right part of a'. From the corresponding-angle axiom, we have that

$$\angle BOQ = \gamma'.$$

Using our original assumption that $\gamma = \gamma'$, we therefore have that

$$\angle BOQ = \gamma' = \gamma.$$

But this contradicts our method of comparing angles, from which it is obvious that

$$\angle BOQ \neq \gamma.$$

Thus, our original assumption must have been false: a must indeed be parallel to b, as asserted.

6.9 Parallel Lines – The Uniqueness Theorem

Our original axiom says that, given a line b and a point O outside it, there exists a line a that passes through O and is also parallel to b. Here we show that the uniqueness of this line also follows from the corresponding-angle axiom.

The uniqueness of a is actually a negative assertion: it means that there is no other line, say a', that also passes through O and is also parallel to b. A negative statement is usually hard to prove in a direct way; could we possibly check every hypothetic line a' to verify that it is not parallel to b? Thus, the best way to prove the uniqueness of a is by contradiction, because this method introduces a new positive assumption, which can be easily denied: the existence of a hypothetic second line, a', with the same properties as a.

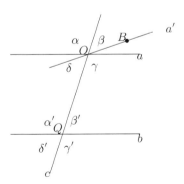

FIGURE 6.7: Proving the uniqueness of the parallel line a by contradiction: if it were not, then one could draw another line a' that is parallel to b as well.

In a proof by contradiction, we assume momentarily that our assertion was false, and seek a contradiction. Once a contradiction is established, we can safely conclude that our momentary assumption must have been false, so our original assertion must indeed be true. In the present case, the momentary assumption is that there is another line, say a', that also passes through O and is also parallel to b. The existence of such an a' is a positive hypothesis, which can be easily denied: indeed, if there were such an a', then we could draw the line c as in Figure 6.7, and use the corresponding-angle axiom to have

$$\gamma = \gamma' = \angle BOQ,$$

in violation of our method to compare angles, which clearly implies that

$$\gamma \neq \angle BOQ.$$

This contradiction implies that no such a' could ever exist, as asserted.

In some texts, the uniqueness of a is also accepted as an axiom. In this case, one could also accept the reversed corresponding-angle theorem as an axiom, and use both of these axioms to prove the corresponding-angle axiom, so it takes the status of a theorem rather than an axiom.

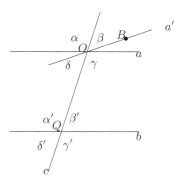

FIGURE 6.8:   Proving the corresponding-angle axiom by contradiction: if γ were smaller than γ', then one could draw another line a' so that the new angle BOQ is equal to γ'.

To prove the corresponding-angle axiom, let us show that, if $a\|b$, then $\gamma = \gamma'$ in Figure 6.8. This is again proved by contradiction: assume momentarily that this was false, that is, that $\gamma \neq \gamma'$. (Without loss of generality, assume that $\gamma < \gamma'$.) Then, we could draw a line a' in such a way that

$$\angle BOQ = \gamma'$$

(Figure 6.8). But the reversed corresponding-angle theorem would then imply that a' is also parallel to b, in violation of the uniqueness axiom assumed above. This necessarily leads to the conclusion that our momentary assumption was false, so $\gamma = \gamma'$, as asserted.

In summary, we have shown not only that the corresponding-angle axiom implies both the uniqueness theorem and the reversed corresponding-angle theorem, but also that it is implied by them. Thus, it is actually equivalent to them. Still it is common to accept it as an axiom, whereas they take the status of theorems.

6.10 Triangles

A triangle is based on three points (vertices), say A, B, and C, which do not lie on the same line. These points are then connected to form the edges of the triangle: AB, BC, and CA. The triangle is then denoted by $\triangle ABC$.

Note that the list of vertices in the triangle, A, B, and C, can be extended periodically, so C is again followed by A. This is why the third edge of the triangle is denoted by CA (rather than AC) to preserve this periodic order. Furthermore, this periodic order induces the notations of the angles in the triangle: $\angle CAB$, $\angle ABC$, and $\angle BCA$ (Figure 6.9).

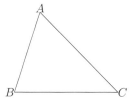

FIGURE 6.9:   The triangle $\triangle ABC$ with the interior angles vertexed at A, B, and C.

Let us prove that the sum of these angles is 180 degrees (or π, or a straight angle). For this, let us draw the unique line that passes through A and is also parallel to BC (Figure 6.10). Using the alternate-angle theorem, we have

$$\angle CAB + \angle ABC + \angle BCA = \angle CAB + \theta + \phi = \pi,$$

which completes the proof.

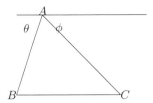

FIGURE 6.10: Proving that the sum of the angles in a triangle is always π, using the alternate-angle theorem.

6.11 Similar and Identical Triangles

Two triangles, say $\triangle ABC$ and $\triangle A'B'C'$, are said to be similar to each other if their corresponding angles are equal to each other, that is,

$$\angle CAB = \angle C'A'B', \ \angle ABC = \angle A'B'C', \text{ and } \angle BCA = \angle B'C'A',$$

and the ratios between their corresponding edges are the same:

$$\frac{AB}{A'B'} = \frac{BC}{B'C'} = \frac{CA}{C'A'}.$$

When these five conditions hold, the triangles are similar:

$$\triangle ABC \sim \triangle A'B'C'.$$

Loosely speaking, $\triangle A'B'C'$ is obtained from putting the original triangle, $\triangle ABC$, in a photocopier that may increase or decrease its size but not change the proportions in it.

If, in addition, the above ratios are also equal to 1, that is,

$$\frac{AB}{A'B'} = \frac{BC}{B'C'} = \frac{CA}{C'A'} = 1,$$

then the above "photocopier" neither increases nor decreases the size of the triangle, so the triangles are actually identical:

$$\triangle ABC \simeq \triangle A'B'C'.$$

Thus, identical triangles satisfy six conditions: three equal corresponding angles and three equal corresponding edges.

Although similarity of triangles involves five conditions and identity of triangles involves six conditions, in some cases it is sufficient to verify two conditions only to establish similarity, and three conditions only to establish identity. This is summarized in the following four axioms, known as the similarity axioms and the identity axioms.

1. If two edges in the triangles have the same ratio, that is,

$$\frac{AB}{A'B'} = \frac{BC}{B'C'},$$

and the corresponding angles in between them are the same, that is,

$$\angle ABC = \angle A'B'C',$$

then the triangles are similar to each other:

$$\triangle ABC \sim \triangle A'B'C'.$$

If, in addition, the above ratios are also equal to 1, that is,

$$AB = A'B' \text{ and } BC = B'C',$$

then the triangles are also identical:

$$\triangle ABC \simeq \triangle A'B'C'.$$

2. If there are two angles in one triangle that are equal to their counterparts in the other triangle, e.g.,

$$\angle CAB = \angle C'A'B' \text{ and } \angle ABC = \angle A'B'C',$$

then the triangles are similar to each other. If, in addition, the ratio of two corresponding edges is equal to 1, e.g.,

$$\frac{AB}{A'B'} = 1 \text{ or } AB = A'B',$$

then the triangles are also identical to each other.

3. If the three ratios of corresponding edges are the same, that is,

$$\frac{AB}{A'B'} = \frac{BC}{B'C'} = \frac{CA}{C'A'},$$

then the triangles are similar to each other. If, in addition, these ratios are also equal to 1, then the triangles are also identical to each other.

4. If the ratios in two pairs of corresponding edges are the same, that is,

$$\frac{AB}{A'B'} = \frac{CA}{C'A'},$$

and the angle that lies in one triangle across from the longer of the two edges in it that take part in the above ratios is the same as its counterpart in the other triangle, that is,

$$\angle ABC = \angle A'B'C' \text{ and } CA \geq AB,$$

then the triangles are similar to each other. If, in addition, the above ratios are also equal to 1, then the triangles are also identical to each other.

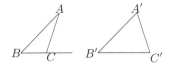

FIGURE 6.11:   Two triangles that satisfy the equations in the fourth axiom, but not the inequality in it, hence are neither identical nor similar to each other.

In each of the first three axioms, two equations are sufficient to imply similarity, and one extra equation is needed to imply identity. In the fourth axiom, on the other hand, there is also an extra inequality that must hold: the angles that are known to be equal to each other must lie across from the longer of the edges (in each triangle) that are used in the equations about the ratios. In other words, we must have $CA \geq AB$.

Let us illustrate why this inequality is so important. Indeed, if it was violated, then we could have a case as in Figure 6.11, in which the three identity conditions in the fourth axiom are satisfied, yet the triangles are neither identical nor similar to each other. The reason for this is that, unlike in the fourth axiom, here $CA < AB$. As a result, no similarity axiom applies, and the triangles may well be dissimilar.

6.12 Isosceles Triangles

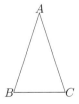

FIGURE 6.12:   An isosceles triangle, in which $CA = AB$.

An isosceles triangle is a triangle with two equal edges. For example, in Figure 6.12,

$$CA = AB.$$

The third edge, BC, which is not necessarily equal to the two other edges, is called the base. The angles near it, $\angle ABC$ and $\angle BCA$, are called the base angles. The third angle, $\angle CAB$, is called the head angle.

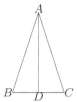

FIGURE 6.13: Dividing the head angle in the isosceles triangle into two equal parts to prove that the base angles are equal to each other.

Let us use the above identity axioms to show that, in an isosceles triangle, the base angles are equal to each other. For this, let us divide the head angle into two equal angles by drawing the line segment AD (Figure 6.13):

$$\angle DAB = \angle DAC.$$

(The line segment AD is then called the bisector of the angle $\angle CAB$.) Thanks to the fact that the triangle is isosceles, we have that

$$AB = AC.$$

Furthermore, we also have trivially that

$$DA = DA.$$

Thus, the first identity axiom implies that

$$\triangle ABD \simeq \triangle ACD.$$

Note that the order of letters in both triangles in this formula, when extended periodically, agrees with the order of letters in the above equations about angles and edges. Because identical triangles have equal edges and angles (respectively), we can conclude that the base angles are indeed equal to each other, as asserted:

$$\angle ABD = \angle ACD.$$

Note that the order of letters in both of these angles again agrees with the order of letters used to denote the triangles to whom they belong. Note also that, in the above proof, we only translated the language in which the assumption is written into the language in which the assertion is written. In fact, the assumption is given in terms of equal edges in the original isosceles triangle. The assertion, on the other hand, is given in terms of equal angles: the base angles. Thanks to the second identity axiom, we were able to transfer the known equality of edges to the asserted equality of angles.

Let us now prove the reversed theorem, which assumes that the base angles in some triangle are equal to each other, and claims that the triangle must then be an isosceles triangle. In other words, assuming that

$$\angle ABC = \angle BCA$$

in Figure 6.12, we have to show that

$$AB = CA.$$

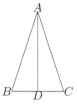

FIGURE 6.14:   Dividing the head angle into two equal parts to prove that the triangle is isosceles.

To do this, let us use Figure 6.14. Note that, in the present case, the assumption is given in terms of equal angles, whereas the assertion is given in terms of equal edges. In order to transfer the information in the assumption into the required language used in the assertion, we use this time the second identity axiom. Indeed, since

$$\angle ABD = \angle ACD, \; \angle DAB = \angle DAC, \; \text{and} \; DA = DA,$$

the second identity axiom implies that

$$\triangle ABD \simeq \triangle ACD.$$

Since corresponding edges in identical triangles are equal to each other, we have that

$$AB = AC,$$

as indeed asserted.

6.13 Pythagoras' Axiom

FIGURE 6.15:   In a right-angled triangle, the hypotenuse CA is the largest edge.

Pythagoras' axiom says that, in a right-angled triangle as in Figure 6.15, in which $\angle ABC$ is a right angle, the hypotenuse (the edge that lies across from the right angle) is greater than each other edge:

$$CA > AB \text{ and } CA > BC.$$

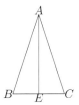

FIGURE 6.16:   Using the fourth identity axiom to prove that the height in an isosceles triangle divides the head angle into two equal parts.

Let us use Pythagoras' axiom to prove that, in an isosceles triangle $\triangle ABC$ as in Figure 6.16, the height AE to the base BC (which makes a right angle with BC) is also a bisector of the head angle, that is, it divides the head angle into two equal parts:

$$\angle EAB = \angle EAC.$$

Indeed, the height makes a right angle with the base BC:

$$\angle BEA = \angle CEA = \pi/2.$$

Furthermore, from Pythagoras' axiom we also have that

$$AB > EA \text{ and } AC > EA.$$

Using also the original assumption that $AB = AC$ and the trivial equation $EA = EA$, we have from the fourth identity axiom that

$$\triangle ABE \simeq \triangle ACE.$$

In particular, we have that

$$\angle EAB = \angle EAC,$$

as indeed asserted.

6.14 Sum of Edges

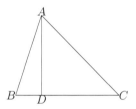

FIGURE 6.17:   Using the height AD and Pythagoras' axiom to prove that the sum of the edges CA and AB is greater than the third edge, BC.

Let us also use Pythagoras' axiom to prove that, in a triangle $\triangle ABC$ as in Figure 6.9, the sum of two edges is greater than the third edge:

$$CA + AB > BC.$$

To this end, let us draw the height AD that meets BC in a right angle (Figure 6.17):

$$\angle ADB = \angle ADC = \pi/2.$$

(Without loss of generality, one can assume that D, the endpoint of the height, is indeed on the line segment BC; otherwise, the proof is even easier, because it follows immediately from Pythagoras' axiom.) From Pythagoras' axiom for the right-angled triangle $\triangle ADB$, we have that

$$AB > BD.$$

Similarly, from Pythagoras' axiom for $\triangle ADC$, we have that

$$CA > DC.$$

By adding these equations, we have

$$AB + CA > BD + DC = BC,$$

as indeed asserted.

6.15 The Longer Edge

Let us use the above theorem to prove that, in a triangle $\triangle ABC$ as in Figure 6.9, the longer edge lies across from the larger angle. In other words, if

$$\angle ABC > \angle BCA,$$

then

$$CA > AB.$$

To do this, note that the first inequality, the assumption, is given in terms of angles, whereas the second inequality, the assertion, is written in terms of edges. Thus, to prove the theorem we must translate the known information about angles into terminology of edges. To do this, we must first understand better the meaning of the given information, so that we can use it properly.

The assumption is given as an inequality between angles. This means that we can map the smaller angle, $\angle BCA$, and embed it in the larger one, $\angle ABC$. This is done in Figure 6.18, yielding the equation

$$\angle BCD = \angle DBC,$$

where D is the point at which the ray onto which **CA** is mapped crosses CA.

Fortunately, this equation can be easily translated into an equation about edges. Indeed, it implies that $\triangle DBC$ is an isosceles triangle, or

$$DB = CD.$$

Thus, the edge CA in the original triangle $\triangle ABC$ can be written as the sum

$$CA = CD + DA = BD + DA.$$

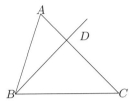

FIGURE 6.18:   The smaller angle, vertexed at C, is mapped onto the new angle DBC which lies inside the larger angle, ABC.

Fortunately, the latter sum uses two edges in the new triangle $\triangle ABD$, hence is necessarily greater than the third edge in this triangle:

$$BD + DA > AB.$$

This implies that

$$CA > AB,$$

as indeed asserted.

6.16 Tales' Theorem

The fourth similarity axiom requires an extra condition: the angle must be across from the larger edge. Indeed, if this condition is violated, one could construct dissimilar triangles, as in Figure 6.11. These dissimilar triangles have an important role in the proof of Tales' theorem.

Tales' theorem says that, in a triangle $\triangle ABC$ as in Figure 6.9, if the line segment BD is a bisector of $\angle ABC$, that is, it divides it into the two equal parts

$$\angle DBA = \angle DBC$$

as in Figure 6.19, then the following ratios are equal to each other:

$$\frac{AB}{CB} = \frac{DA}{DB}.$$

In this theorem, the information is given in terms of angles, whereas the assertion uses terms of ratios. How can we translate the language used in the assumption into language of ratios? The obvious way is to use a similarity axiom.

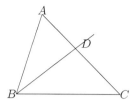

FIGURE 6.19: Tales' theorem: if BD divides the angle ABC into two equal parts, then $AB/BC = AD/DC$.

Unfortunately, the naive candidates for similarity, the triangles $\triangle DBA$ and $\triangle DBC$, are not similar to each other. Indeed, BD produces two supplementary angles, $\angle ADB$ and $\angle CDB$, which are not necessarily equal to each other. In Figure 6.19, for example,

$$\angle ADB < \pi/2 < \angle CDB.$$

Thus, in order to prove Tales' theorem, we must turn from $\triangle CDB$ to a triangle that is related to it in the same way as the triangles in Figure 6.11 are related to each other. For this purpose, we use a compass to draw an arc of radius CD around C, so it crosses the ray **BD** at the new point E, producing the isosceles triangle $\triangle CDE$ (Figure 6.20).

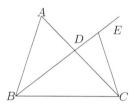

FIGURE 6.20: Tales' theorem follows from the similarity $\triangle ECB \sim \triangle DAB$, which follows from the second similarity axiom and the fact that $CD = CE$.

Note that $\triangle DCB$ and $\triangle ECB$ are related to each other in the same way as the triangles in Figure 6.11 are related to each other. Thus, $\triangle ECB$ may be a better candidate to be similar to $\triangle DAB$. To show this, recall that, since $\triangle CDE$ is an isosceles triangle, its base angles are equal to each other:

$$\angle CDE = \angle DEC.$$

Furthermore, since $\angle BDA$ and $\angle CDE$ are vertical angles, we have

$$\angle BDA = \angle CDE = \angle DEC.$$

Using also the assumption and the second similarity axiom, we have that

$$\triangle DAB \sim \triangle ECB.$$

Since in similar triangles all the ratios between corresponding edges are the same, we have that

$$\frac{AB}{CB} = \frac{DA}{EC} = \frac{DA}{DC},$$

as indeed asserted. This completes the proof of Tales' theorem.

6.17 The Reversed Tales' Theorem

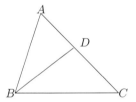

FIGURE 6.21:   The reversed Tales' theorem: if $AB/BC = AD/DC$, then BD divides the angle ABC into two equal parts.

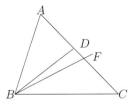

FIGURE 6.22:   Proving the reversed Tales' theorem by contradiction: if the angle DBA were smaller than the angle DBC, then there would be a point F on CA such that BF divides the angle ABC into two equal parts.

Let us now use Tales' theorem to prove the reversed theorem, in which the roles of the assumption and the assertion in Tales' theorem are interchanged. More explicitly, this theorem assumes that

$$\frac{AB}{CB} = \frac{DA}{DC}$$

in Figure 6.21, and asserts that

$$\angle DBA = \angle DBC.$$

Let us prove this theorem by contradiction. For this purpose, assume momentarily that BD is not a bisector of the angle $\angle ABC$, that is,

$$\angle DBA \neq \angle DBC.$$

Then, one could divide $\angle ABC$ into two equal parts by the bisector BF:

$$\angle ABF = \angle CBF,$$

where F is a point on CA that is different from D (Figure 6.22). (This can be done by using a compass to draw an arc of radius BA around B, so it crosses $\mathbf{BC}$ at the new point G, and then using a ruler to draw the height in the isosceles triangle $\triangle BAG$ from B to the base AG.) From Tales' theorem, we now have that

$$\frac{BA}{BC} = \frac{FA}{FC} \neq \frac{DA}{DC},$$

in contradiction to our original assumption. Thus, our momentary assumption as if

$$\angle ABD \neq \angle CBD$$

must have been false, which completes the proof of the theorem.

6.18 Circles

A circle is an object based on two more elementary objects: a point O to denote the center of the circle, and a line segment r to denote the radius of the circle. A point P lies on the circle if its distance from its center is equal to the radius:

$$OP = r.$$

A chord is a line segment that connects two distinct points that lie on the circle. For example, if A and C lie on the circle, then AC is a chord in this circle.

A central angle is an angle that is vertexed at the center of the circle. For example, if AC is a chord, then $\angle AOC$ is a central angle. In this case, we say that the chord AC subtends the central angle $\angle AOC$.

An inscribed angle is an angle that is vertexed at some point that lies on the circle. For example, if AC is a chord and B is a point on the circle, then $\angle ABC$ is an inscribed angle. In this case, we say that the chord AC subtends the inscribed angle $\angle ABC$.

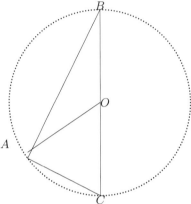

FIGURE 6.23:   The central angle AOC is twice the inscribed angle ABC subtended by the same chord AC. The first case, in which the center O lies on the leg BC of the inscribed angle ABC.

Let us show that a central angle is twice the inscribed angle subtended by the same chord:
$$\angle AOC = 2\angle ABC.$$

To prove this, let us consider three possible cases. In the first case, the center O lies on one of the legs of $\angle ABC$, say BC (Figure 6.23). To prove the assertion, we must first understand that the main assumption is that the angles are embedded in a circle, which is an object that is characterized by the property that every point that lies on it is of the same distance from its center. In particular, this is true for the points A and B, implying that

$$AO = BO,$$

or that $\triangle OAB$ is an isosceles triangle.

So far, we have only interpreted the assumption in terms of line segments and the relation between them. The assertion, on the other hand, is written as a relation between angles. Therefore, we must translate the assumption into a property of angles rather than edges. Indeed, thanks to the fact that $\triangle OAB$ is an isosceles triangle, we know that its base angles are equal to each other:

$$\angle OAB = \angle ABO.$$

Thus, the central angle is related to the inscribed angle by

$$\angle AOC = \pi - \angle AOB = \angle OAB + \angle ABO = 2\angle ABO = 2\angle ABC,$$

as indeed asserted.

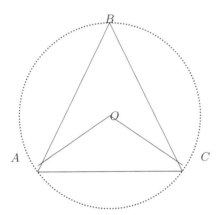

FIGURE 6.24: The central angle AOC is twice the inscribed angle ABC subtended by the same chord AC. The second case, in which the center O lies inside the inscribed angle ABC.

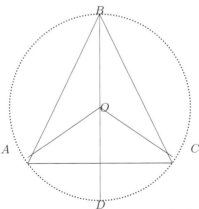

FIGURE 6.25: Proving that the central angle AOC is twice the inscribed angle ABC (subtended by the same chord AC) by drawing the diameter BD that splits it into two angles.

Let us now consider the case in which the center O lies within the inscribed angle $\angle ABC$, as in Figure 6.24. To prove the assertion in this case, let us draw the diameter BD that passes through the center O (Figure 6.25). From the previous case, we have that

$$\angle AOD = 2\angle ABD$$

and that

$$\angle COD = 2\angle CBD.$$

By adding these equations, we have that

$$\angle AOC = 2\angle ABC,$$

as required.

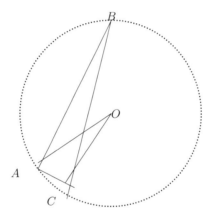

FIGURE 6.26:   The central angle AOC is twice the inscribed angle ABC subtended by the same chord AC. The third case, in which the center O lies outside the inscribed angle ABC.

Finally, let us consider the case in which the center O lies outside the inscribed angle $\angle ABC$, as in Figure 6.26. Again, let us draw the diameter BD, which passes through O (Figure 6.27). From the first case above, we have that

$$\angle AOD = 2\angle ABD$$

and that

$$\angle COD = 2\angle CBD.$$

By subtracting this equation from the previous one, we have that

$$\angle AOC = 2\angle ABC,$$

as asserted.

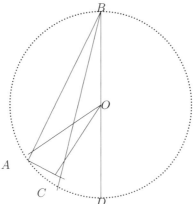

FIGURE 6.27: Proving that the central angle AOC is twice the inscribed angle ABC (subtended by the same chord AC) by drawing the diameter BD and using the first case.

6.19 Tangents

A tangent is a line that shares only one point with a circle. More precisely, if P is the only point that lies on both the circle and the line, then we say that the line is tangent to the circle at P.

Let us show that the tangent makes a right angle with the radius OP. In other words, if Q is some point on the tangent, then

$$\angle OPQ = \pi/2.$$

Let us prove this by contradiction. Indeed, assume momentarily that the tangent makes with OP an acute angle:

$$\angle OPQ = \alpha < \pi/2.$$

Then, we could draw the line segment OU that crosses the tangent at U and makes an angle of $\pi - 2\alpha$ with OP:

$$\angle UOP = \pi - 2\alpha$$

(Figure 6.28). But then the new triangle $\triangle OPU$ has the equal base angles

$$\angle OPU = \angle PUO = \alpha,$$

so it must be also an isosceles triangle in the sense that

$$OP = OU.$$

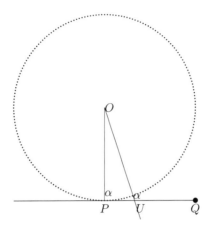

FIGURE 6.28:   Proving by contradiction that the tangent makes a right angle
with the radius OP. Indeed, if α were smaller than $\pi/2$, then the triangle OPU
would be an isosceles triangle, so both P and U would lie on both the circle and
the tangent, in violation of the definition of a tangent.

But this implies that U lies not only on the tangent but also on the circle, so
the tangent shares with the circle not one but rather two points, P and U, in
contradiction to the definition of a tangent. The conclusion must therefore be
that our momentary assumption is false, and that our assertion that

$$\angle OPQ = \pi/2$$

is true.

Conversely, let us now show that a line that passes through a point P
on the circle and makes a right angle with the radius OP must also be a
tangent. (This is the reversed theorem, in which the roles of the assumption
and the assertion in the above theorem are interchanged.) Let us prove this
by contradiction. Indeed, let us assume momentarily that the line were not a
tangent. Then, there would be another point, say U, shared by the circle and
the tangent (Figure 6.29). This means that the new triangle $\triangle OPU$ would
have been an isosceles triangle in the sense that

$$OP = OU.$$

But, from our original assumption, this triangle is also a right-angled triangle
in the sense that

$$\angle OPU = \pi/2,$$

which implies, in conjunction with Pythagoras' axiom, that

$$OP < OU.$$

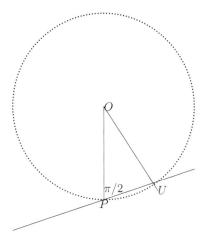

FIGURE 6.29: Proving by contradiction that a line that passes through P and makes a right angle with the radius OP must be a tangent. Indeed, if it were not (as in the figure), then the triangle OPU would be an isosceles triangle, in violation of Pythagoras' axiom.

This contradiction implies that our momentary assumption must have been false, and the line is indeed a tangent, as asserted.

In summary, a line is tangent to the circle at P if, and only if, it makes a right angle with the radius OP. In other words, making a right angle with the radius is not only a mere property of the tangent, but actually a characterization that could be used as an alternative definition: a tangent to the circle at P is the line that makes a right angle with the radius OP. Below we'll use this characterization to study further the tangent and its properties.

6.20 Properties of the Tangent

The above characterization implies immediately that the tangent to the circle at P is unique. Indeed, from the definition of an angle, there can be only one line that passes through P and makes a right angle with the radius OP. Below, we'll use this characterization to prove more theorems.

Let us show that the angle produced by the tangent and a chord AP is the same as the inscribed angle subtended by this chord. In other words,

$$\angle EPA = \angle ABP$$

in Figure 6.30, where BP is chosen to be a diameter that passes through O. (Because every inscribed angle subtended by the chord AP is equal to half

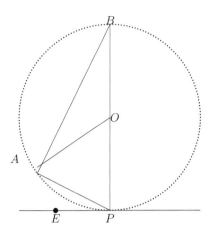

FIGURE 6.30:   The angle produced by the tangent and the chord AP is the same as the inscribed angle subtended by the chord AP.

the central angle subtended by this chord, they are all equal to each other, so one may choose the location of B on the circle in such a way that BP is indeed a diameter.)

Indeed, the straight angle $\angle BOP$ can be viewed as a central angle subtended by the chord BP (the diameter). Because the inscribed angle $\angle BAP$ is also subtended by the chord BP, we have that

$$\angle BAP = \frac{1}{2}\angle BOP = \pi/2.$$

This means that the triangle $\triangle BAP$ is actually a right-angled triangle. Furthermore, from our characterization of a tangent, it follows that $\angle BPE$ is a right angle as well. As a result, we have that

$$\angle APE = \angle BPE - \angle BPA = \pi/2 - \angle BPA = \angle ABP,$$

as indeed asserted.

Finally, let us use the above characterization of a tangent to show that the two tangents to the circle at the two distinct points P and Q produce an isosceles triangle $\triangle UPQ$ in the sense that

$$UP = UQ,$$

where U is their crossing point (Figure 6.31). Indeed, from the characterization of a tangent it follows that

$$\angle OPU = \angle OQU = \pi/2.$$

From Pythagoras' axiom, it therefore follows that

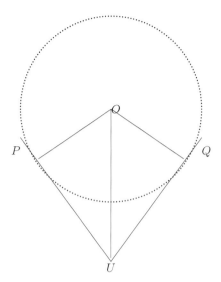

FIGURE 6.31: Two tangents that cross each other at the point U: proving that $UP = UQ$ by using the fourth identity axiom to show that the triangle OPU is identical to the triangle OQU.

$$OU > OP = OQ.$$

From the fourth identity axiom, it therefore follows that

$$\triangle OPU \simeq \triangle OQU.$$

As a result, we have that

$$PU = QU,$$

as indeed asserted.

6.21 Exercises

1. What is a point in Euclidean geometry?
2. What is the possible relation between points in Euclidean geometry?
3. What is a line in Euclidean geometry?
4. What are the possible relations between lines in Euclidean geometry?
5. What are the possible relations between points and lines in Euclidean geometry?
6. Interpret points and lines in Euclidean geometry as mere abstract objects that have some possible relations and satisfy some axioms.

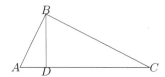

FIGURE 6.32:   Assume that angle ADB is a right angle. Show that angle ABC is a right angle if and only if $AD/DB = BD/DC$.

7. What is a line segment in Euclidean geometry?
8. What are the possible relations between line segments? How are they checked?
9. Show that the $<=$ relation between line segments is complete in the sense any two line segments have this relation between them.
10. What is an angle in Euclidean geometry?
11. What are the possible relations between angles? How are they checked?
12. Show that the $<=$ relation between angles is complete in the sense any two angles have this relation between them.
13. What is a triangle in Euclidean geometry?
14. What are the possible relations between triangles in Euclidean geometry? How are they checked?
15. Say the four identity and similarity axioms by heart.
16. Assume that, in Figure 6.32, the angles $\angle ABC$ and $\angle ADB$ are right angles. Show that

$$\frac{AD}{DB} = \frac{BD}{DC}.$$

(Note that the assertion is written in terms of ratios, so the assumptions should also be transformed to the language of ratios, using a similarity axiom.)
17. Conversely, assume that $\angle ADB$ is a right angle and that

$$\frac{AD}{DB} = \frac{BD}{DC}.$$

Show that $\angle ABC$ is a right angle as well.
18. What is a circle in Euclidean geometry?
19. What are the possible relations between lines and circles?

Chapter 7

Analytic Geometry

In Euclidean geometry, elementary objects such as points and lines are never defined explicitly; only the relations between them are specified. These objects are then used to obtain more complex objects, such as rays, line segments, angles, and triangles. Finally, circles are obtained from points and line segments.

In analytic geometry, on the other hand, the only elementary object is the point, which is defined in terms of its two coordinates x and y to specify its location in the plane [30]. All other objects, including lines and circles, are just sets of points that satisfy some prescribed algebraic equation. Thus, analytic geometry depends strongly on real numbers and the arithmetic operations between them.

7.1 God and the Origin

The ancient Greeks based their mathematical theory of geometry on a few elementary objects such as point, line, angle, etc., which may have relations between them, such as a point lying on a line, a line passing through a point, etc. Euclidean geometry also uses a few natural axioms, such as that only one line can pass through two given points and that only one line passes through a given point in such a way that it is also parallel to a given line. These axioms are then used to prove further theorems and establish the entire theory to give a complete picture about the nature and properties of shapes in the two-dimensional plane.

It was Rene Descartes in the 17th century who observed how helpful it could be to use an axes system in the plane, with the x- and y-axes that are perpendicular to each other. This way, each point can be described by its coordinates (x, y), and a line is just a set of points whose coordinates are related by some linear equation. This approach, known as analytic geometry, provides straightforward proofs to many theorems in geometry.

The interesting question is: why didn't the ancient Greeks think about the Cartesian geometry in the first place? After all, they were well aware of numbers and their properties. Why then didn't they use them to describe sets of points analytically?

One possible answer is that the ancient Greeks couldn't accept relativism. Indeed, in order to use Cartesian geometry, one must first introduce an axes system. But this can be done in infinitely many ways. Who is authorized to determine where the axes, and in particular the origin, should be?

The ancient Greeks probably expected God to determine the absolute axes system. Unfortunately, God never did this. They were therefore forced to stick to their original definition of a line as a mere object rather than a set of points. In fact, in Euclidean geometry, the line object can relate to a point by passing through it, but is never viewed as a set of points.

Only Descartes, with his more secular views, could introduce a more relative axes system. Indeed, in his/her theory, everyone can define his/her own axes system as he/she sees fit. Then, they could go ahead and prove theorems according to their private axes system. Fortunately, the proofs are independent of the particular axes system in use, thus are relevant not only to the one who writes them originally but also to everyone else, who may well use them in their own axes systems as well.

Thus, Descartes' secular views led to the introduction of an axes system that is no longer absolute but rather relative to the person who defines it in the first place. The geometrical objects and the proofs of theorems are then also relative to this axes system. Still, the theory is not limited to this particular axes system, because equivalent proofs can be written in any other axes system as well.

In Descartes' view, God may thus take a rather nontraditional interpretation: no longer as a transcendental force that creates and affects our materialistic world from the outside, but rather as an abstract, personal, psychological drive that represents one's mind, spirit, or conscience, and lies in one's soul or consciousness. This God is also universal, because it lies in each and every human being, regardless of their nationality, race, gender, or religion. This personal God may well introduce a private axis system for personal use; the analytic geometry developed from this axis system is independent of it, and may be easily transferred to any other axis system introduced by any other person.

7.2 Numbers and Points

In Euclidean geometry, points and lines are the most elementary objects: they are defined only implicitly, using their relations and axioms. In analytic geometry, on the other hand, they are no longer the most elementary objects. In fact, they are defined explicitly in terms of yet more elementary objects: numbers.

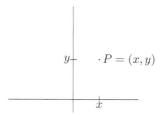

FIGURE 7.1: The point $P = (x, y)$ whose x-coordinate is x and y-coordinate is y.

Indeed, in analytic geometry points and lines are no longer mere abstract objects imposed on us by God or any other external force or mind. On the contrary: they are defined explicitly, using real numbers in some axis system. In particular, a point P is just a pair of real numbers of the form

$$P = (x, y),$$

where x is the real number denoting the x-coordinate of the point P and y is the real number denoting the y-coordinate of P (Figure 7.1).

7.3 Lines – Sets of Points

Furthermore, a line l is now defined as a set of all the points whose x- and y-coordinates are related by some linear equation:

$$l = \{(x, y) \mid a_l x + b_l y = c_l\},$$

where a_l, b_l, and c_l are some given real parameters that specify the line l, and a_l and b_l do not both vanish.

Similarly, one could define another line m by

$$m = \{(x, y) \mid a_m x + b_m y = c_m\}.$$

The two lines l and m are parallel to each other if there is a nonzero real number q that relates both a_l and b_l to a_m and b_m:

$$a_l = q a_m \text{ and } b_l = q b_m.$$

If, in addition, c_l and c_m are also related to each other in the same way, that is, if

$$c_l - q c_m$$

as well, then l and m are actually identical to each other:

$$l = m.$$

If, on the other hand, l and m are neither parallel nor identical to each other, that is, if no such number q exists, then it can be shown algebraically that l and m cross each other at a unique point, that is, there is exactly one point that lies on both l and m. In other words, there is exactly one pair of real numbers (x_{lm}, y_{lm}) that solves both equations

$$a_l x_{lm} + b_l y_{lm} = c_l \text{ and } a_m x_{lm} + b_m y_{lm} = c_m.$$

7.4 Hierarchy of Objects

Thus, analytical geometry is based on a hierarchy of objects. The most elementary objects, the real numbers, lie at the bottom of the hierarchy. These objects are then used to define more complex objects, points, at the next higher level in the hierarchy. Furthermore, points are then used to define yet more complex objects, lines, at the next higher level of the hierarchy, and so on.

This hierarchy may be extended further to define yet more complex objects such as angles, triangles, circles, etc. The definitions of these objects as sets of points may then be used to prove theorems from Euclidean geometry in an easier way.

7.5 Half-Planes

Let us use the above hierarchy of objects to define angles. For this, however, we need first to define half-planes.

The half-plane H_l defined from the line l defined above is the set of all the points whose coordinates are related by a linear inequality of the form:

$$H_l = \{(x, y) \mid a_l x + b_l y \geq c_l\}$$

(Figure 7.2).

Note that the above definition is somewhat ambiguous. Indeed, if the line l takes the equivalent form

$$l = \{(x, y) \mid -a_l x - b_l y = -c_l\},$$

FIGURE 7.2:   The half-plane H_l that lies in the north-western side of the line l.

then the half-plane H_l associated with it is not the above half-plane but rather its complementary half-plane:

$$H_l = \{(x,y) \mid -a_l x - b_l y \geq -c_l\}.$$

Thus, the precise definition of the correct half-plane H_l depends not only on the line l but also on its parameterization, that is, on the way in which it is defined. In the sequel, one should determine which H_l is used (the original one or its complementary) according to the requirements in the object constructed from it.

7.6 Angles

An angle is now defined as the intersection of the two half-planes associated with two nonparallel lines l and m:

$$H_l \cap H_m$$

(Figures 7.3–7.4).

Clearly, if l and m are the same line, then the above definition gives a straight angle

$$H_l \cap H_l = H_l.$$

Thus, by choosing the lines l and m properly, one could construct every angle that is smaller than or equal to π. If an angle larger than π is required, then one should also consider unions of half planes of the form

$$H_l \cup H_m.$$

$$H_m$$

FIGURE 7.3:   The second half-plane H_m that lies in the north-eastern side of
the line m.

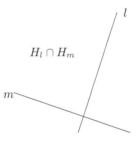

FIGURE 7.4:   The angle $H_l \cap H_m$ created by the lines l and m is the
intersection of the half-planes H_l and H_m.

7.7 Triangles

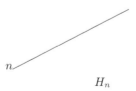

FIGURE 7.5:   The third half-plane H_n that lies in the south-eastern side of
the line n.

A triangle can now be defined as the intersection of three half-planes:

$$H_l \cap H_m \cap H_n,$$

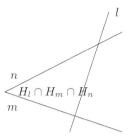

FIGURE 7.6:   The triangle $H_l \cap H_m \cap H_n$ created by the lines l, m, and n is the intersection of the half-planes H_l, H_m, and H_n.

where l m, and n are distinct nonparallel lines (Figures 7.5–7.6). Of course, the right sides of these lines should be used to avoid a trivial triangle.

7.8 Circles

In analytic geometry, a circle is defined in terms of two given parameters: a point $O = (x_o, y_o)$ to mark its center, and a positive real number r to denote its radius. Using Pythagoras' theorem, the circle c is then defined as the set of all the points whose distance from O is equal to r:

$$c = \{(x, y) \mid (x - x_o)^2 + (y - y_o)^2 = r^2\}.$$

Thus, unlike in Euclidean geometry, in which the radius is a line segment, here the radius is a positive real number. Furthermore, the circle is no longer a mere elementary object, but is rather the set of all the points that satisfy the quadratic equation that stems from Pythagoras' theorem. This algebraic interpretation gives the opportunity to prove theorems more easily than in Euclidean geometry.

7.9 Exercises

1. What is a point in analytic geometry?
2. What is the possible relation between points? How is it checked?
3. What is a line in analytic geometry?
4. What are the possible relations between lines? How are they checked?
5. What are the possible relations between points and lines? How are they checked?

6. What is a line segment in analytic geometry?
7. What are the possible relations between line segments? How are they checked?
8. What is an angle in analytic geometry?
9. What are the possible relations between angles? How are they checked?
10. What is a circle in analytic geometry?
11. What are the possible relations between circles? How are they checked?
12. What are the possible relations between lines and circles? How are they checked?
13. Let the 2 by 2 matrix A be the table of four given numbers $a_{1,1}$, $a_{1,2}$, $a_{2,1}$, and $a_{2,2}$:

$$A = \begin{pmatrix} a_{1,1} & a_{1,2} \\ a_{2,1} & a_{2,2} \end{pmatrix}.$$

The matrix A can be viewed as a transformation of the Cartesian plane. Indeed, if a point (x, y) in the Cartesian plane is viewed as a 2-dimensional column vector with the 2 components x and y, then A transforms it into the 2-dimensional column vector

$$A \begin{pmatrix} x \\ y \end{pmatrix} = \begin{pmatrix} a_{1,1}x + a_{1,2}y \\ a_{2,1}x + a_{2,2}y \end{pmatrix}.$$

Show that this transformation is linear in the sense that, for any point p in the Cartesian plane and any scalar α,

$$A(\alpha p) = \alpha A p,$$

and, for any two points p and q in the Cartesian plane,

$$A(p + q) = Ap + Aq.$$

14. Consider the special case, in which

$$A = \begin{pmatrix} \cos(\theta) & -\sin(\theta) \\ \sin(\theta) & \cos(\theta) \end{pmatrix}$$

for some $0 \le \theta < 2\pi$. Show that A transforms the point $(1, 0)$ in the Cartesian plane by rotating it by angle θ. Furthermore, show that A also transforms the point $(0, 1)$ by rotating it by angle θ. Moreover, use the representation

$$(x, y) = x(1, 0) + y(0, 1)$$

and the linearity of the transformation to show that A transforms any point (x, y) in the Cartesian plane by rotating it by angle θ.
15. Conclude that A transforms the unit circle onto itself.

Part III

Introduction to Composite Mathematical Objects

Introduction to
Composite Mathematical Objects

The mathematical objects discussed so far can be divided into two classes: elementary objects and composite objects. Elementary objects can be viewed as atoms that cannot be divided and are defined only in terms of the axioms associated with them. Composite objects, on the other hand, can be defined in terms of simpler objects. Thus, in the hierarchy of objects, the elementary objects are placed in the lowest level, whereas the composite objects are placed in the higher levels, so their definitions and functions may use more elementary objects from the lower levels.

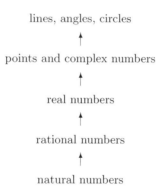

FIGURE 7.7: Hierarchy of mathematical objects, from the most elementary ones at the lowest level to the most complicated ones at the highest level.

For example, natural numbers can be viewed as elementary objects, because their definition and arithmetic operations are based on the induction axiom only. This is why natural numbers should be placed in the lowest level in the hierarchy of mathematical objects (Figure 7.7). Rational numbers, on the other hand, can be viewed as composite objects. Indeed, a rational number is actually a pair of two natural numbers: its numerator and its denominator.

A pair is actually an ordered set of two elements: the first element and the second element. Thus, a rational number can be viewed as an ordered set of two natural numbers. In other words, a rational number can be viewed as a composite object, consisting of two natural numbers. This is why rational numbers should be placed in the next higher level in the hierarchy of mathematical objects.

Furthermore, a real number can be viewed as an infinite set containing the rational numbers that approach it. Thus, the real number is a yet more complicated object, and should be placed in the next higher level in the hierarchy of mathematical objects.

Furthermore, both a complex number and a point in the Cartesian plane are actually pairs of real numbers, and should thus be placed in the next higher level in the hierarchy of mathematical objects. Furthermore, geometrical objects such as lines, angles, triangles, and circles in analytic geometry are actually infinite sets of points, and should thus be placed in the next higher level in the hierarchy of mathematical objects.

It is thus obvious that the concept of set is most important in constructing mathematical objects. The first chapter in this part is devoted to this concept. The next chapters in this part use further the concept of set to construct several kinds of useful composite objects, with their special functions that give them their special properties and nature.

Chapter 8

Sets

As we have seen above, the notion of set is necessary to construct composite mathematical objects from more elementary ones [15]. For example, a rational number is actually an ordered set of two natural numbers, a real number is actually an infinite set of rational numbers, a point in the Cartesian plane is an ordered set of two real numbers, a line in analytic geometry is an infinite set of points, and so on.

Thus, the notion of set is most important in constructing more and more complex objects in the hierarchy of mathematical objects. This is why we open this part with a chapter about sets and their nature.

8.1 Alice in Wonderland

In one of the scenes in Lewis Carroll's *Alice in Wonderland*, Alice is invited to a tea party. The host of this party, the Mad Hatter, declares that he is going to pour tea to guests who don't pour for themselves, but not for guests who do pour for themselves. But then he is puzzled about whether or not he should pour for himself: after all, if he pours for himself, then he too is a guest who pours for himself, so according to his own declaration he shouldn't pour for himself. If, on the other hand, he doesn't pour for himself, then he is a guest who doesn't pour for himself, so according to his declaration he should pour for himself. This paradox is the basis for set theory, as is shown below.

8.2 Sets and Containers

The above paradox is equivalent to Russell's paradox in set theory, which can be viewed as the basis for the concept of the set. In the naive approach, a set is a container that can contain elements of any kind. Below, however, we show that this is not entirely true: not every container is also a set.

Let us introduce a few useful notations about sets. First, if the set S contains the element E, then this is denoted by

$$E \in S$$

(E belongs to S). Furthermore, it is assumed that every subset of the set S is by itself a set as well. In other words, if T contains some of the elements in S, then T is a set as well. This is then denoted by

$$T \subset S$$

(T is a subset of S).

So far, we haven't specified the type of the elements in the set S. Actually, they can be of any type; in fact, an element $E \in S$ may well be a set in its own right. Still, this doesn't mean that $E \subset S$, because we have no information about the elements in E, so we cannot check whether they belong to S as well. What we do know is that $\{E\}$ (the set whose only element is E) is a subset of S:

$$\{E\} \subset S.$$

Indeed, the only element in $\{E\}$, E, belongs to S as well.

8.3 Russell's Paradox

However, a set is not just a container of elements. In other words, not every container of elements is a set. Consider for example the container S that contains everything. Let us show by contradiction that S is not a set. Indeed, if it were a set, then one could extract from it the following subset:

$$T = \{A \in S \mid A \notin A\}.$$

In other words, T contains only the containers that do not contain themselves as an element.

Now, does T contain itself as an element? If it does, then it is not among the A's in the right-hand side in the above definition, so it cannot belong to T. If, on the other hand, it doesn't, then it is among the A's in the right-hand side in the above definition, so it must belong to T as an element. In any case, we have a contradiction, so we can have neither $T \in T$ nor $T \notin T$. This is Russell's paradox [9]; it leads to the conclusion that T is not a set, which implies that S too is not a set.

8.4 The Empty Set

We have seen above that the container that contains everything is too big to serve as a set. Thus, not every container can be accepted as a set. We must start from scratch: define small sets first, and use them to construct bigger and bigger sets gradually and patiently.

The smallest set is the empty set: the set with no elements, denoted by $\emptyset$. Although it may seem trivial, this set has a most important role in set theory. In fact, just as the zero number is the additive unit number in the sense that

$$r + 0 = r$$

for every number r, $\emptyset$ is the additive unit set in the sense that

$$S \cup \emptyset = S$$

for every set S, where '$\cup$' denotes the union operation:

$$A \cup B = \{E \mid E \in A \text{ or } E \in B\}$$

for every two sets A and B.

8.5 Sets and Natural Numbers

As we have seen above, the empty set $\emptyset$ is equivalent to the number zero. In fact, it can also be given the name '0'. Similarly, the set $\{\emptyset\}$ (the set whose only element is $\emptyset$) can be given the name '1'. Furthermore, the set $\{\{\emptyset\}\}$ can be given the name '2', and so on. In fact, this leads to an equivalent inductive definition of the natural numbers:

$$0 \equiv \emptyset$$

and

$$n + 1 \equiv \{n\}$$

for $n = 0, 1, 2, 3, \dots$.

8.6 The Order of the Natural Numbers

Using the above alternative definition of natural numbers, one can easily define the order relation '$<$' between two natural numbers m and n: $m < n$ if

$m \in n$ or $m < n-1$ (where $n-1$ is the element in the set n). More concisely, the recursive definition of the '<' order says that $m < n$ if $m \leq n-1$, precisely as in the original definition of natural numbers in the beginning of this book.

8.7 Mappings and Cardinality

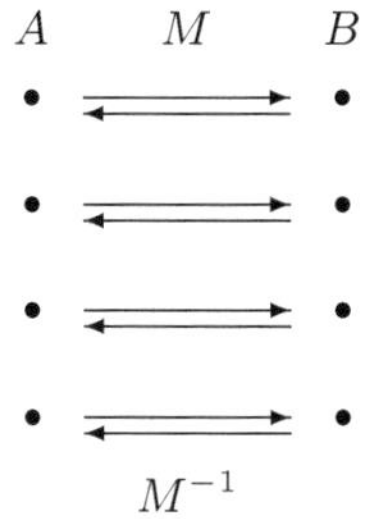

FIGURE 8.1:   Two sets A and B are equivalent to each other if there exists a one-to-one mapping M from A onto B. Because each element $b \in B$ has a unique element $a \in A$ that is mapped to it, one can also define the inverse mapping M^{-1} from B onto A by $M^{-1}(b) = a$.

Consider the set that contains the first n natural numbers:

$$T = \{1, 2, 3, \ldots, n\}.$$

This is indeed a set, because it can be written as the union of a finite number of sets of one element each:

$$T = \{1, 2, 3, \ldots, n\} = \{1\} \cup \{2\} \cup \{3\} \cup \cdots \cup \{n\}.$$

Furthermore, this set is equivalent to any other set of the form

$$S = \{a_1, a_2, a_3, \ldots, a_n\}$$

in the sense that there exists a one-to-one mapping M from S onto T (Figure 8.1). Here, by "one-to-one" we mean that every two distinct elements in S are mapped to two distinct elements in T, and by "onto" we mean that every element in T has an element in S that is mapped to it. More explicitly, the mapping can be defined by

$$M(a_i) \equiv i, \quad 1 \leq i \leq n.$$

Thanks to the "one-to-one" and "onto" properties of the mapping M, one can also define the inverse mapping M^{-1} from T back to S:

$$M^{-1}(i) = a_i, \quad 1 \leq i \leq n.$$

The cardinality of a set is a measurement of its size. For finite sets such as S and T above, the cardinality is just the number of elements in each of them:

$$|S| = |T| = n.$$

The equivalence of S and T, or the existence of the one-to-one mapping M from S onto T, implies that their cardinality must be the same, as is indeed apparent from the above equation.

Later on, we'll define cardinality also for infinite sets. Their cardinality, however, can no longer be a number, because they contain infinitely many elements. Instead, it is a symbol that is the same for all sets that are equivalent to each other in the sense that there exists a one-to-one mapping from one onto the other.

8.8 Ordered Sets and Sequences

The set T that contains the first n natural numbers is an ordered set. Indeed, for every two elements i and j in it, either $i < j$ or $i > j$ or $i = j$. Furthermore, this order also induces an order in the set S defined above as follows: if a and b are two distinct elements in S, then a is before b if $M(a) < M(b)$ and b is before a if $M(b) < M(a)$ (Figure 8.2). Thus, the order in S is

$$a_1, a_2, a_3, \ldots, a_n.$$

In fact, with this order, S is not only a set but actually a sequence of n numbers, denoted also by

$$\{a_i\}_{i=1}^{n}.$$

In particular, when $n = 2$, S is a sequence of two numbers, or a pair. This pair is then denoted by

$$(a_1, a_2).$$

Pairs are particularly useful in the definitions of points in the Cartesian plane below.

8.9 Infinite Sets

Consider the container that contains all the natural numbers:

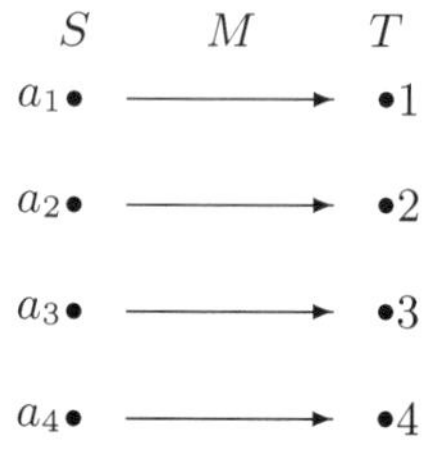

FIGURE 8.2:   The order in S is defined by the one-to-one mapping $M : S \to T$. For every two distinct elements $a, b \in S$, a is before b if $M(a) < M(b)$.

$$N = \{1, 2, 3, \ldots\}.$$

More precisely, $\mathbb{N}$ is defined inductively: first, it is assumed that 1 is in $\mathbb{N}$; then, it is also assumed that, for $n = 1, 2, 3, 4, \ldots$, if n is in $\mathbb{N}$, then $n + 1$ is in $\mathbb{N}$ as well.

Is $\mathbb{N}$ a set? This is not a trivial question: after all, we have already met above a container that is not a set. To show that $\mathbb{N}$ is indeed a set, we have to use an axiom from set theory.

This axiom says that there exists an infinite set, say S. This implies that every subset of S is a set as well. Now, to show that $\mathbb{N}$ is indeed a set, it is sufficient to show that it is equivalent to a subset of S.

For this purpose, define the following one-to-one mapping M from the subset $S_0 \subset S$ onto $\mathbb{N}$ as follows. Pick an element $s_1 \in S$ and map it to $1 \in \mathbb{N}$. Then, pick another element $s_2 \in S$ and map it to $2 \in \mathbb{N}$, and so on. As a matter of fact, this is an inductive definition: assuming that $s_1, s_2, s_3, \ldots, s_n \in S$ have been mapped to $1, 2, 3, \ldots, n \in \mathbb{N}$, pick another element $s_{n+1} \in S$ and map it to $n+1 \in \mathbb{N}$. (Because S is infinite, this is always possible.) As a result, $\mathbb{N}$ is indeed equivalent to

$$S_0 \equiv \{s_1, s_2, s_3, \ldots\} \subset S,$$

so it is indeed a set, as asserted.

8.10 Enumerable Sets

The cardinality of $\mathbb{N}$, the set of natural numbers, is called $\aleph_0$. This cardinality is not a number but merely a symbol to characterize sets that are equivalent to $\mathbb{N}$ in the sense that they can be mapped onto $\mathbb{N}$ by a one-to-one mapping. Such a set is called an enumerable set, because the elements in it can be given natural indices and ordered in an infinite sequence (Figure 8.3).

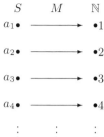

FIGURE 8.3:   The infinite set S is called enumerable if it can be mapped by a one-to-one mapping M onto $\mathbb{N}$, the set of the natural numbers.

The cardinality $\aleph_0$ is greater than every finite number n. Indeed, every finite set of n elements can be mapped by a one-to-one mapping into $\mathbb{N}$, but not onto $\mathbb{N}$. Furthermore, $\aleph_0$ is the minimal cardinality that is greater than every finite number, because (as we've seen above) every infinite set contains a subset of cardinality $\aleph_0$.

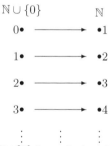

FIGURE 8.4:   The set $\mathbb{N} \cup \{0\}$ is equivalent to $\mathbb{N}$ by the one-to-one mapping $i \to i + 1$ that maps it onto $\mathbb{N}$.

It is common to assume that 0 is not a natural number, so the set of natural numbers takes the form
$$\mathbb{N} = \{1, 2, 3, \ldots\}.$$
In the following, we mostly use this convention. Actually, it makes little difference whether 0 is considered as a natural number or not, because the above induction to define the natural numbers could equally well start from 1 rather than from 0. Furthermore, the one-to-one mapping

$$i \to i + 1$$

implies that

$$|\mathbb{N} \cup \{0\}| = |\mathbb{N}| = \aleph_0$$

(Figure 8.4).

In the following, we consider some important sets of numbers, and check whether they are enumerable or not.

8.11 The Set of Integer Numbers

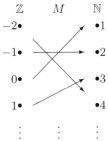

FIGURE 8.5:   The one-to-one mapping M that maps $\mathbb{Z}$ (the set of the integer numbers) onto $\mathbb{N}$: the negative numbers are mapped onto the even numbers, and the nonnegative numbers are mapped onto the odd numbers.

Let $\mathbb{Z}$ denote the set of the integer numbers. Let us show that $\mathbb{Z}$ is enumerable. For this purpose, we need to define a one-to-one mapping M from $\mathbb{Z}$ onto $\mathbb{N}$. This mapping is defined as follows:

$$M(i) = \begin{cases} 2i + 1 & \text{if } i \geq 0 \\ 2|i| & \text{if } i < 0. \end{cases}$$

Clearly, M is a one-to-one mapping from $\mathbb{Z}$ onto $\mathbb{N}$, as required (Figure 8.5).

8.12 Product of Sets

Let A and B be some given sets. Their product is defined as the set of pairs with a first component from A and a second component from B:

$$AB \equiv \{(a, b) \mid a \in A, \ b \in B\}.$$

This definition leads to the definition of the product of cardinalities:

$$|A| \cdot |B| \equiv |AB|.$$

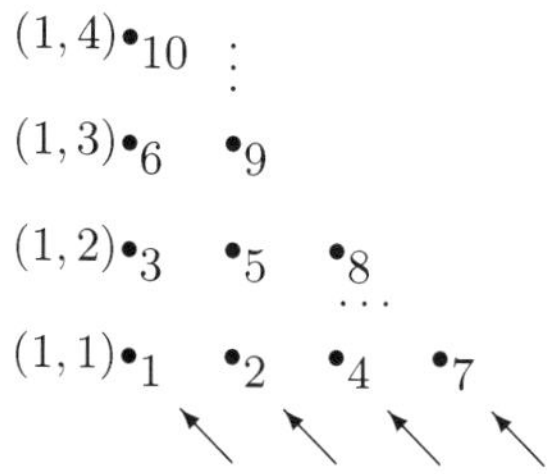

$(1,4)\bullet_{10}$ $\vdots$

$(1,3)\bullet_6$ $\bullet_9$

$(1,2)\bullet_3$ $\bullet_5$ $\bullet_8$

$(1,1)\bullet_1$ $\bullet_2$ $\bullet_4$ $\bullet_7$

FIGURE 8.6: The infinite grid $\mathbb{N}^2$ is enumerable (equivalent to $\mathbb{N}$) because it can be ordered diagonal by diagonal in an infinite sequence. The index in this sequence is denoted in the above figure by the number to the right of each point in the grid.

Let us use this definition to obtain the square of $\aleph_0$:

$$\aleph_0^2 = \aleph_0 \cdot \aleph_0 = |\mathbb{N}| \cdot |\mathbb{N}| = |\mathbb{N}^2|.$$

Recall that $\mathbb{N}^2$ is the set of pairs of natural numbers:

$$\mathbb{N}^2 = \{(m,n) \mid m,n \in \mathbb{N}\}.$$

In fact, this set can be viewed as an infinite grid of points in the upper-right quarter of the Cartesian plane (Figure 8.6). The points in this grid can be ordered diagonal by diagonal, where the first diagonal contains the single point $(1,1)$, the second diagonal contains the two points $(2,1)$ and $(1,2)$, the third diagonal contains the three points $(3,1)$, $(2,2)$, and $(1,3)$, and so on. As a result, the points in the grid are ordered in a sequence:

$$(1,1), \ (2,1), \ (1,2), \ (3,1), \ (2,2), \ (1,3), \ \ldots.$$

Thus, $\mathbb{N}^2$ is an enumerable set, with cardinality $\aleph_0$. Therefore, we have

$$\aleph_0^2 = |\mathbb{N}^2| = \aleph_0.$$

8.13 Equivalence of Sets

The equivalence relation between sets is indeed a mathematical equivalence relation in the sense that it has three important properties: it is reflective, symmetric, and transitive. Indeed, it is reflective in the sense that every set A is equivalent to itself through the identity mapping $I(a) = a$ that maps each element to itself. Furthermore, it is symmetric in the sense that if A is equivalent to B by the one-to-one mapping M that maps A onto B, then B

is also equivalent to A by the inverse mapping M^{-1} that maps B back onto A. Finally, it is transitive in the sense that if A is equivalent to B by the one-to-one mapping M that maps A onto B and B is equivalent to C by the one-to-one mapping M' that maps B onto C, then A is also equivalent to C by the composite mapping $M'M$ that maps A onto C:

$$M'M(a) \equiv M'(M(a)), \quad a \in A.$$

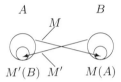

FIGURE 8.7:   The sets A and B are equivalent if A is equivalent to a subset of B (by the one-to-one mapping M from A onto the range $M(A) \subset B$) and B is equivalent to a subset of A (by the one-to-one mapping M' from B onto the range $M'(B) \subset A$).

To establish that two sets A and B are equivalent to each other, it is actually sufficient to show that A is equivalent to a subset of B and B is equivalent to a subset of A (Figure 8.7). Indeed, assume that A is equivalent to $M(A) \subset B$ (the range of M) by the one-to-one mapping M, and that B is equivalent to the subset $M'(B) \subset A$ (the range of M') by the one-to-one mapping M'. Let us define the mapping K

$$K : A \rightarrow M'(B)$$

from A to $M'(B)$ as follows:

$$K(a) \equiv \begin{cases} M'M(a) & \text{if } a \in \cup_{i=0}^{infty}(M'M)^i(A \setminus M'(B)) \\ a & \text{otherwise,} \end{cases}$$

where $A \setminus M'(B)$ contains all the elements of A that are not in the range of M', and $\cup_{i=0}^{\infty}$ means the infinite union over every $i \geq 0$:

$$\cup_{i=0}^{infty}(M'M)^i(A \setminus M'(B))$$
$$= (A \setminus M'(B)) \cup M'M(A \setminus M'(B)) \cup (M'M)(M'M)(A \setminus M'(B)) \cdots .$$

Clearly, this infinite union is invariant under $M'M$ in the sense that every element in it is mapped by $M'M$ to an element in it. This implies that K is indeed a one-to-one mapping. Furthermore, K is a mapping from A onto $M'(B)$. Indeed, every element $a \in M'(B)$ is either in the above union or not. If it is, then it must lie in $(M'M)^i(A \setminus M'(B))$ for some $i \geq 1$; if, on the other hand, it is not, then it must satisfy $K(a) = a$. In either case, it is in the range of K, as required. Thus, K is indeed a one-to-one mapping from A onto $M'(B)$.

The conclusion is, thus, that A is equivalent to $M'(B)$ by the one-to-one mapping K. As a consequence, A is also equivalent to B by the composite mapping $(M')^{-1}K$. This completes the proof of the equivalence of A and B.

The assumption that A is equivalent to a subset of B can be written symbolically as

$$|A| \leq |B|.$$

With this notation, one may say that we have actually proved that the inequalities

$$|A| \leq |B| \text{ and } |B| \leq |A|$$

imply that

$$|A| = |B|.$$

In the following, we'll use this theorem to obtain the cardinality of the set of rational numbers.

8.14 The Set of Rational Numbers

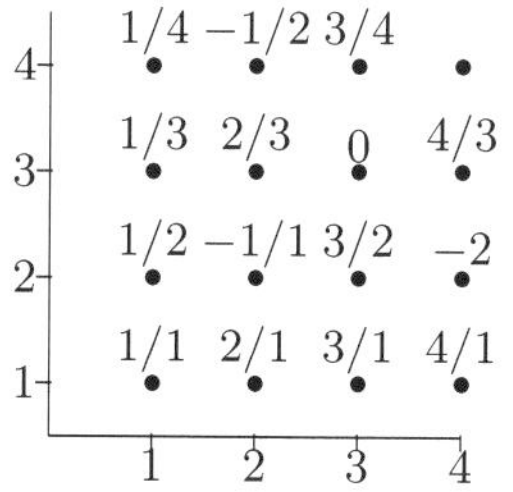

FIGURE 8.8: The set of the rational numbers, $\mathbb{Q}$, is embedded in the infinite grid $\mathbb{N}^2$, to imply that $|\mathbb{Q}| \leq |\mathbb{N}^2| = \aleph_0$. In particular, $m/n \in \mathbb{Q}$ is embedded in $(m,n) \in \mathbb{N}^2$, $-m/n \in \mathbb{Q}$ is embedded in $(2m, 2n) \in \mathbb{N}^2$, and 0 is embedded in $(3,3)$, as denoted by the fractions just above the points in the above figure.

The set of the rational numbers, denoted by $\mathbb{Q}$, can be embedded by a one-to-one mapping in $\mathbb{N}^2$ (Figure 8.8). Indeed, every positive rational number can be written in the form m/n for some natural numbers m and n satisfying $GCD(m, n) = 1$. Thus, m/n is mapped to the point $(m, n) \in \mathbb{N}^2$. For example, $1 = 1/1$ is mapped to the point $(1, 1)$. Furthermore, the negative counterpart of m/n, $-m/n$ can then be safely embedded in $(2m, 2n)$. For example, $-1 = -1/1$ is mapped to $(2, 2)$. Finally, 0 can be mapped to $(3, 3)$. Using the results in the previous sections, we then have

$$|\mathbb{Q}| \leq |\mathbb{N}^2| = \aleph_0.$$

On the other hand, $\mathbb{N}$ is a subset of $\mathbb{Q}$, so we also have

$$|\mathbb{Q}| \geq |\mathbb{N}| = \aleph_0.$$

Using the result in the previous section, we therefore have

$$|\mathbb{Q}| = \aleph_0.$$

In other words, the set of the rational numbers, although seeming larger than the set of the natural numbers, is actually of the same size: it is enumerable as well.

8.15 Arbitrarily Long Finite Sequences

The enumerability of $\mathbb{Q}$ implies that every infinite subset of it is enumerable as well. For example, consider the set S of all arbitrarily long finite sequences of 0's and 1's. Because S is infinite, we clearly have

$$|S| \geq |\mathbb{N}| = \aleph_0.$$

Furthermore, each finite sequence in S, once placed after the decimal point, represents uniquely a finite decimal fraction, which is actually a rational number. Thus, we also have

$$|S| \leq |\mathbb{Q}| = \aleph_0.$$

By combining these results, we conclude that

$$|S| = \aleph_0.$$

In the following, we'll meet sets that are genuinely greater than $\mathbb{N}$ in the sense that they have cardinality greater than $\aleph_0$. Such sets are called nonenumerable. For example, we'll see below that the set of infinite sequences of 0's and 1's is nonenumerable.

8.16 Function Sets

A function can be viewed as a mapping that maps each element in one set to an element in another set, or a machine that takes an input to produce an output. Commonly, if x denotes the input element and f denotes the function, then $f(x)$ denotes the output produced by f from x, or the target element to which x is mapped by f.

The most elementary functions are Boolean functions, which map each element either to 0 or to 1. More explicitly, a Boolean function f from a set S to the set that contains the two elements 0 and 1,

$$f : S \to \{0,1\},$$

is defined for each element $s \in S$ to produce the output $f(s)$, which can be either 0 or 1, according to the particular definition of the function.

The set of all such functions can be imagined as an infinite "list" of duplicate copies of $\{0,1\}$, each associated with a particular element from S:

$$\{0,1\}, \ \{0,1\}, \ \{0,1\}, \ \ldots \quad (|S| \text{ times}).$$

A particular function f picks a particular number, either 0 or 1, from each duplicate copy of $\{0,1\}$ associated with each $s \in S$ to define $f(s)$. This is why the set of all such functions is denoted by

$$\{0,1\}^S \equiv \{f \mid f : S \to \{0,1\}\}.$$

For example,

$$\{0,1\}^{\mathbb{N}} = \{0,1\}, \ \{0,1\}, \ \{0,1\}, \ \ldots \quad (\aleph_0 \text{ times})$$

can be viewed as the set of all the infinite sequences of 0's and 1's. Indeed, each sequence of the form

$$a_1, a_2, a_3, \ldots$$

in which a_i is either 0 or 1 for all $i \geq 1$ can be interpreted as a function $f : \mathbb{N} \to \{0,1\}$ for which $f(i) = a_i$ for all $i \geq 1$. In the following, we'll show that the function set $\{0,1\}^{\mathbb{N}}$ is nonenumerable.

8.17 Cardinality of Function Sets

Clearly, S is equivalent to a subset of $\{0,1\}^S$. Indeed, one can define a one-to-one mapping M from S onto a subset of $\{0,1\}^S$ by letting $M(s)$ (for each element $s \in S$) be the function that produces the value 1 only for the input s:

$$M(s)(t) \equiv \begin{cases} 1 & \text{if } t = s \\ 0 & \text{if } t \in S, \ t \neq s. \end{cases}$$

In other words,

$$|\{0,1\}^S| \geq |S|.$$

Let us show that this inequality is actually strict, namely, that

$$|\{0,1\}^S| > |S|.$$

This is proved by contradiction. Assume momentarily that there were a one-to-one mapping M from S onto $\{0,1\}^S$. For each element $s \in S$, $M(s)$ would then be a function from S to $\{0,1\}$. In particular, when it takes the input s, this function produces the output $M(s)(s)$, which is either 0 or 1. Let us define a new function $f : S \to \{0,1\}$ that disagrees with this output:

$$f(s) \equiv 1 - M(s)(s) \neq M(s)(s), \quad s \in S.$$

As a consequence, f cannot be in the range of M, in contradiction to our momentary assumption that M is onto $\{0,1\}^S$. Thus, our momentary assumption must have been false, which implies that

$$|\{0,1\}^S| > |S|,$$

as asserted.

Using the notation

$$|A|^{|B|} \equiv |A^B|$$

for any two sets A and B, we therefore have

$$2^{|S|} = |\{0,1\}^S| > |S|.$$

Below we use this result to obtain nonenumerable sets.

8.18 Nonenumerable Sets

Consider the set S of infinite sequences of 0's and 1's. As we have seen above,

$$S = \{0,1\}^{\mathbb{N}}.$$

Therefore, we have

$$|S| = 2^{\aleph_0} > \aleph_0.$$

This cardinality is also denoted by

$$\aleph \equiv 2^{\aleph_0}.$$

In the following, we'll show that this is also the cardinality of the set of the real numbers.

8.19 Cardinality of the Real Axis

To obtain the cardinality of the real axis $\mathbb{R}$, we first need to have the cardinality of the unit interval that lies in it. Let $[0, 1]$ denote the closed unit interval of real numbers (with the endpoints):

$$[0, 1] \equiv \{x \in \mathbb{R} \mid 0 \leq x \leq 1\}.$$

Furthermore, let $(0, 1)$ denote the open unit interval (without the endpoints):

$$(0, 1) \equiv \{x \in \mathbb{R} \mid 0 < x < 1\}.$$

Note that round parentheses are used to denote an open interval, whereas square parentheses are used to denote a closed interval. Both of these kinds of parentheses are different from the ones used in $\{0, 1\}$, the set that contains the two elements 0 and 1 only.

Now, each infinite sequence of 0's and 1's, once placed after the decimal point, represents uniquely an infinite decimal fraction in $[0, 1]$. Thus, we have

$$|[0, 1]| \geq |\{0, 1\}^{\mathbb{N}}| = 2^{\aleph_0} = \aleph.$$

On the other hand, using basis 2, each real number in $[0, 1]$ can be represented as a binary fraction, with an infinite sequence of 0's and 1's behind the point. As a matter of fact, infinite binary fractions are represented uniquely by such a sequence, whereas finite binary fractions can be represented by two possible sequences: one ends with infinitely many 0's, and the other ends with infinitely many 1's. Thus, there exists a one-to-one mapping from $[0, 1]$ onto a subset of $\{0, 1\}^{\mathbb{N}}$, or

$$|[0, 1]| \leq |\{0, 1\}^{\mathbb{N}}| = 2^{\aleph_0} = \aleph.$$

Combining the above two inequalities, we have

$$|[0, 1]| = |\{0, 1\}^{\mathbb{N}}| = 2^{\aleph_0} = \aleph.$$

Furthermore, $[0, 1]$ is also equivalent to the entire real axis $\mathbb{R}$. Indeed, $[0, 1]$ is a subset of $\mathbb{R}$, which implies that

$$|\mathbb{R}| \geq |[0, 1]| = \aleph.$$

Furthermore, the function

$$\tan(\pi(x - 1/2)) \ : \ (0, 1) \to \mathbb{R}$$

is a one-to-one mapping from the open unit interval $(0, 1)$ onto $\mathbb{R}$. Thus, $\mathbb{R}$ is equivalent to a subset of $[0, 1]$, or

$$|\mathbb{R}| \leq |[0, 1]| = \aleph.$$

By combining these two inequalities, we have

$$|\mathbb{R}| = |[0, 1]| = \aleph.$$

8.20 Cardinality of the Plane

At first glance, it would seem as if the Cartesian plane $\mathbb{R}^2$ has a larger cardinality than the real axis $\mathbb{R}$. Below we will see that this is not so.

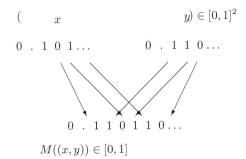

$$M((x,y)) \in [0,1]$$

FIGURE 8.9:   The point (x,y) in the unit square is mapped to the number $M((x,y))$ in the unit interval whose binary representation is combined from the binary representations of x and y.

To this end, let us first show that the closed unit square

$$[0,1]^2 = \{(x,y) \in \mathbb{R}^2 \mid 0 \le x, y \le 1\}$$

has the same cardinality as the closed unit interval $[0,1]$. Indeed, since $[0,1]$ is equivalent to the lower edge of $[0,1]^2$, we have that

$$|[0,1]^2| \ge |[0,1]| = \aleph.$$

Furthermore, each of the coordinates x and y in each point $(x,y) \in [0,1]^2$ can be represented uniquely as an infinite binary fraction with an infinite sequence of 0's and 1's behind the point that doesn't end with infinitely many 1's. Now, these infinite sequences can be combined to form a new infinite sequence of 0's and 1's whose odd-numbered digits are the same as the digits in the binary representation of x, and its even-numbered digits are the same as the digits in the binary representation of y. Once placed behind the point, this new sequence represents a unique number in $[0,1]$. This forms a one-to-one mapping from $[0,1]^2$ onto a subset of $[0,1]$, or

$$|[0,1]^2| \le |[0,1]| = \aleph.$$

By combining these two inequalities, we have

$$|[0,1]^2| = |[0,1]| = \aleph.$$

The above proof uses the binary representation of real numbers between 0 and 1. In particular, it combines two such representations to produce a new representation for a new real number between 0 and 1. This method of proof can also be used for natural numbers. In fact, the binary representations of two natural numbers can be combined to form a new binary representation for a new natural number. This may lead to a one-to-one mapping from $\mathbb{N}^2$ into $\mathbb{N}$, or to an alternative proof for the inequality

$$|\mathbb{N}^2| \le |\mathbb{N}|,$$

which has already been proved in Figure 8.6 above.

Let us now show that the entire Cartesian plane is also of cardinality $\aleph$. Indeed, since the closed unit square is a subset of the Cartesian plane, we have

$$|\mathbb{R}^2| \ge |[0,1]^2| = \aleph.$$

Furthermore, the open unit square is equivalent to the entire Cartesian plane by the one-to-one mapping

$$\tan(\pi(x - 1/2))\tan(\pi(y - 1/2)) \; : \; (0,1)^2 \to \mathbb{R}^2,$$

which means that

$$|\mathbb{R}^2| \le |[0,1]^2| = \aleph.$$

By combining these two inequalities, we have

$$|\mathbb{R}^2| = |[0,1]^2| = \aleph.$$

8.21 Cardinality of the Multidimensional Space

The above method of proof can be used to show that the finite-dimensional space $\mathbb{R}^n$ is also of cardinality $\aleph$ for every fixed natural number n. However, this method of proof is not good enough to obtain also the cardinality of the infinite-dimensional space $\mathbb{R}^\mathbb{N}$, which can be imagined as an infinite list of duplicate copies of $\mathbb{R}$:

$$\mathbb{R}, \, \mathbb{R}, \, \mathbb{R}, \, ldots \quad (\aleph_0 \text{ times})$$

or, in other words, the set of all the functions from $\mathbb{N}$ to $\mathbb{R}$:

$$\mathbb{R}^\mathbb{N} = \{f \mid f : \mathbb{N} \to \mathbb{R}\}.$$

In order to have the cardinality of both finite-dimensional and infinite-dimensional spaces, we need to develop more general tools.

Let A, B, and C be some given sets. Let us show that

$$|(A^B)^C| = |A^{BC}|.$$

To this end, let us define a one-to-one mapping M from

$$(A^B)^C = \{f \mid f : C \to A^B\}$$

onto

$$A^{BC} = \{f \mid f : BC \to A\}.$$

To do this, consider a particular function $f : C \to A^B$ in $(A^B)^C$. For each element $c \in C$, $f(c)$ is by itself a function from B to A, i.e.,

$$f(c)(b) \in A, \quad b \in B.$$

Then, the mapped function $M(f) : BC \to A$ is defined naturally by

$$M(f)((b, c)) = f(c)(b).$$

Clearly, M is indeed a one-to-one mapping onto A^{BC}, as required. This completes the proof of the above equality of cardinalities:

$$|(A^B)^C| = |A^{BC}|.$$

Let us use this result to obtain the cardinality of multidimensional spaces. First, the cardinality of the finite-dimensional space $\mathbb{R}^n$ is

$$|\mathbb{R}^n| = |(\{0,1\}^{\mathbb{N}})^n| = |\{0,1\}^{n\mathbb{N}}| = |\{0,1\}^{\mathbb{N}}| = \aleph.$$

Furthermore, the cardinality of the infinite-dimensional space is

$$|\mathbb{R}^{\mathbb{N}}| = |(\{0,1\}^{\mathbb{N}})^{\mathbb{N}}| = |\{0,1\}^{\mathbb{N}^2}| = |\{0,1\}^{\mathbb{N}}| = \aleph.$$

More concisely, we have

$$\aleph^{\aleph_0} = \aleph^n = \aleph.$$

Thus, the cardinality of both finite-dimensional and infinite-dimensional spaces is not larger than that of the original real axis. To have cardinalities larger than $\aleph$, we must therefore turn to sets of functions defined on $\mathbb{R}$.

8.22 Larger Cardinalities

The next cardinality, which is larger than $\aleph$, is the cardinality of the set of binary functions defined on the real axis:

$$2^{\aleph} = |\{0,1\}^{\mathbb{R}}| > |\mathbb{R}| = \aleph.$$

Is there a yet larger cardinality? Well, $\aleph^{\aleph}$, the cardinality of the set of functions

$$\{f \mid f : \mathbb{R} \to \mathbb{R}\},$$

seems to be a good candidate. However, it is not really larger than $2^{\aleph}$:

$$\aleph^{\aleph} = |\mathbb{R}^{\mathbb{R}}| = |(\{0,1\}^{\mathbb{N}})^{\mathbb{R}}| = |\{0,1\}^{\mathbb{N}\mathbb{R}}| \leq |\{0,1\}^{\mathbb{R}^2}| = |\{0,1\}^{\mathbb{R}}| = 2^{\aleph}.$$

Thus, to have a cardinality larger than $2^{\aleph}$ we must turn to its exponent, namely, the cardinality of the set of functions defined on it:

$$2^{2^{\aleph}} > 2^{\aleph}.$$

Is there a yet greater cardinality? Let's try $(2^{\aleph})^{2^{\aleph}}$:

$$(2^{\aleph})^{2^{\aleph}} = 2^{\aleph \cdot 2^{\aleph}} \leq 2^{2^{\aleph} \cdot 2^{\aleph}} = 2^{(2^{\aleph})^2} = 2^{2^{2\aleph}} = 2^{2^{\aleph}}.$$

Thus, $(2^{\aleph})^{2^{\aleph}}$ is not really a greater cardinality. To have a cardinality greater than $2^{2^{\aleph}}$, we must therefore turn to its exponent:

$$2^{2^{2^{\aleph}}} > 2^{2^{\aleph}},$$

and so on.

8.23 Sets of Zero Measure

Enumerable sets have zero measure in the sense that, given an arbitrarily small number $\varepsilon > 0$, they can be covered completely by open intervals (or open squares in 2-D) whose total size is no more than ε. For example, the natural numbers in $\mathbb{N}$ can be covered as follows (Figure 8.10): The first number, 1, is covered by the open interval

$$(1 - \varepsilon/4, 1 + \varepsilon/4).$$

The second number, 2, is covered by the open interval

$$(2 - \varepsilon/8, 2 + \varepsilon/8).$$

The third number, 3, is covered by the open interval

$$(3 - \varepsilon/16, 3 + \varepsilon/16),$$

and so on. The total size (or length) of the intervals is

$$\frac{\varepsilon}{2} + \frac{\varepsilon}{4} + \frac{\varepsilon}{8} + \cdots = \frac{\varepsilon}{2} \sum_{i=0}^{\infty} (1/2)^i = \frac{\varepsilon}{2} \frac{1}{1-1/2} = \varepsilon.$$

Thus, $\mathbb{N}$ has been covered completely by open intervals with total length ε. Because ε can be chosen to be arbitrarily small, $\mathbb{N}$ has the same size as that of one point on the real axis. This means that $\mathbb{N}$ has indeed a zero measure in the real axis.

FIGURE 8.10:   The set of the natural numbers is of zero measure in the real axis because, for an arbitrarily small $\varepsilon > 0$, it can be covered by open intervals with total length as small as ε.

Similarly, $\mathbb{N}^2$ can be covered completely by open squares whose total area is no more than ε (Figure 8.11). Indeed, each point $(i,j) \in \mathbb{N}^2$ can be covered by a small open square of dimensions

$$\sqrt{\varepsilon}/2^i \text{ by } \sqrt{\varepsilon}/2^j.$$

Clearly, the total area of these squares is

$$\varepsilon \sum_{(i,j)\in\mathbb{N}^2} (1/2)^{i+j} = \varepsilon \sum_{i=1}^{\infty} (1/2)^i \sum_{j=1}^{\infty} (1/2)^j = \varepsilon.$$

Below we'll see that there are not only enumerable sets but also nonenumerable sets of zero measure.

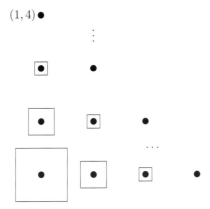

FIGURE 8.11: The infinite grid $\mathbb{N}^2$ is of zero measure in the Cartesian plane because, for an arbitrarily small $\varepsilon > 0$, it can be covered by open squares with total area as small as ε.

8.24 Cantor's Set

The smallest nonenumerable set that we've encountered so far is the unit interval. The measure of this interval in the real axis (or its length) is equal to 1. Is there a yet smaller nonenumerable set whose measure in the real axis is as small as zero?

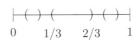

FIGURE 8.12: Cantor's set is obtained from the closed unit interval by dropping from it the open subinterval $(1/3, 2/3)$, then dropping the open subintervals $(1/9, 2/9)$ and $(7/9, 8/9)$ from the remaining closed subintervals $[0, 1/3]$ and $[2/3, 1]$, and so on.

Yes, there exists such a set: Cantor's set. This set is constructed as follows. Consider the closed unit interval $[0, 1]$. Drop from it its middle third, namely, the open subinterval $(1/3, 2/3)$ (Figure 8.12). In other words, if each point in the original interval is represented in base 3 by an infinite sequence of 0's, 1's, and 2's behind the point, then dropping this subinterval would mean eliminating the numbers with the digit 1 right after the point. For example, the number $1/3$, which has not been dropped, can be represented in base 3 by the infinite fraction

$$1/3 = 0.022222\ldots.$$

Furthermore, the number $2/3$, which has not been dropped, can be represented in base 3 as

$$2/3 = 0.200000\ldots.$$

The remaining subintervals are $[0, 1/3]$ and $[2/3, 1]$. In the next stage, the middle third is dropped from each of them. More explicitly, the open intervals $(1/9, 2/9)$ and $(7/9, 8/9)$ are dropped. In base 3, this means that the numbers whose second digit after the point is 1 are eliminated too.

The remaining intervals are the closed intervals $[0, 1/9]$, $[2/9, 3/9]$, $[2/3, 7/9]$, and $[8/9, 1]$. In the next stage, the open subintervals that are the middle third of each of these intervals are dropped as well. In terms of the base-3 representation, this means eliminating all the fractions whose third digit after the point is 1.

The process goes on and on in a similar way. In each stage, twice as many open subintervals are dropped. However, the length of each dropped subinterval is one third of the length of an interval that has been dropped in the previous stage. Thus, the total lengths of all the dropped subintervals is $2/3$ times the total lengths of the subintervals dropped in the previous stage.

In terms of the base-3 representation, the fractions dropped in the ith stage are those fractions whose ith digit after the point is 1. Thus, the fractions that remain once this infinite process is complete are those that are represented in base 3 by a sequence of 0's and 2's after the point.

These remaining numbers form the set known as Cantor's set. Clearly, it is equivalent to $\{0, 1\}^{\mathbb{N}}$ by the mapping that replaces each digit 2 by digit 1. Furthermore, its measure in the real axis is as small as zero. Indeed, let us sum up the lengths of the subintervals that have been dropped throughout the entire process:

$$\frac{1}{3}\sum_{i=0}^{\infty}(2/3)^i = \frac{1}{3}\cdot\frac{1}{1-2/3} = 1.$$

Since the process has started from the unit interval, the set that remains after all these subintervals have been dropped must be of size zero. Thus, Cantor's set is indeed a nonenumerable set of zero measure, as required.

8.25 Exercises

1. Interpret elements and sets as some abstract objects with the relation $\in$ between them: $e \in A$ means that the element e belongs to the set A.
2. Interpret sets as some abstract objects with the relation $\subset$ between them: $A \subset B$ means that every element in A lies in B as well.
3. Show that $A = B$ if and only if $A \subset B$ and $B \subset A$.
4. Let A, B, and C be some sets. Show that

$$A \cap (B \cup C) = (A \cap B) \cup (A \cap C).$$

5. Furthermore, show that

$$A \cup (B \cap C) = (A \cup B) \cap (A \cup C).$$

6. Show that the cardinality of the even numbers is the same as the cardinality of all the natural numbers.
7. Show that the cardinality of the odd numbers is the same as the cardinality of all the natural numbers.
8. Use the mapping $x \to (b - a)x + a$ to show that the cardinality of any closed interval of the form $[a, b]$ is the same as the cardinality of the closed unit interval $[0, 1]$.
9. Use the mapping $x \to (b - a)x + a$ to show that the cardinality of any open interval of the form (a, b) is the same as the cardinality of the open unit interval $(0, 1)$.
10. Use the mapping $x \to \tan(x)$ to show that the open interval $(-\pi/2, \pi/2)$ is equivalent to the entire real axis.
11. Show in two different ways that the set of the rational numbers is enumerable.
12. Show that the unit interval is nonenumerable.
13. Show in two different ways that the unit square is equivalent to the unit interval.
14. What is a function?
15. Show that the set of the binary functions defined on the unit interval has a larger cardinality than the unit interval.
16. Show that the set of the rational numbers has a zero measure in the real axis.

Chapter 9

Vectors and Matrices

As we have seen above, sequences are ordered enumerable sets. This is the basis for the definition of vectors and matrices [28].

A vector is a finite sequence of numbers (usually real numbers). For example, in the Cartesian plane, a vector can be viewed as an arrow from the origin to some point $(x, y) \in \mathbb{R}^2$. Thus, the vector can be denoted simply by (x, y). Thus, in the plane, a vector is a very short sequence of two components only: x and y. In the three-dimensional space, on the other hand, a vector is denoted by a longer sequence of three components: (x, y, z).

Vectors, however, are more than mere sequences: they also have linear algebraic operations defined on them, such as addition and multiplication. Furthermore, below we also define another special kind of finite sequences: matrices, along with some useful linear algebraic operations between matrices and matrices and between matrices and vectors.

9.1 Two-Dimensional Vectors

Vectors are basically finite sets of numbers. More precisely, they are ordered finite sets, or finite sequences.

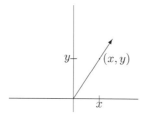

FIGURE 9.1: The vector (x, y) starts at the origin $(0, 0)$ and points to the point (x, y) in the Cartesian plane.

For example, a vector in the Cartesian plane is an arrow leading from the origin $(0,0)$ to some point $(x, y) \in \mathbb{R}^2$ (Figure 9.1). Thus, the vector can be denoted simply by the sequence (x, y) containing the two components x (the horizontal coordinate) and then y (the vertical coordinate).

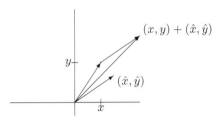

FIGURE 9.2:   Adding the vectors (x, y) and $(\hat{x}, \hat{y})$ using the parallelogram rule.

9.2 Adding Vectors

Two vectors (x, y) and $(\hat{x}, \hat{y})$ can be added to each other according to the parallelogram rule. More specifically, the original vectors (x, y) and $(\hat{x}, \hat{y})$ are completed in to a parallelogram, whose diagonal starts at the origin and points to the new point $(x, y) + (\hat{x}, \hat{y})$ (Figure 9.2). This diagonal is the required vector, the sum of the two original vectors.

From an algebraic point of view, the parallelogram rule means that the vectors are added component by component:

$$(x, y) + (\hat{x}, \hat{y}) \equiv (x + \hat{x}, y + \hat{y}).$$

In other words, each coordinate in the sum vector is just the sum of the corresponding coordinates in the original vectors. This algebraic definition is most useful in calculations, and can be easily extended to spaces of higher dimension below.

9.3 Multiplying a Vector by a Scalar

A vector can also be multiplied by a numerical factor (scalar). In this operation, the vector is stretched by this factor, while keeping its original direction unchanged. In other words, the length of the vector is multiplied by the factor, but the proportion between its coordinates remains the same (Figure 9.3).

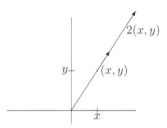

FIGURE 9.3:   Multiplying the vector (x, y) by the scalar 2, or stretching it by factor 2, to obtain the new vector $2(x, y)$, which is twice as long.

In algebraic terms, this leads to the formula

$$a(x, y) \equiv (ax, ay),$$

where a is a real number. This algebraic formula is particularly useful in calculations, and can be easily extended to the multidimensional case below.

9.4 Three-Dimensional Vectors

In the three-dimensional Cartesian space, a vector is an arrow leading from the origin $(0, 0, 0)$ to some point $(x, y, z) \in \mathbb{R}^3$ (Figure 9.4). Thus, the vector can be denoted by the sequence (x, y, z), which contains three components: first x (the horizontal coordinate), then y (the vertical coordinate), and finally z (the height coordinate).

As in the plane, two vectors in the space are added coordinate by coordinate:

$$(x, y, z) + (\hat{x}, \hat{y}, \hat{z}) \equiv (x + \hat{x}, y + \hat{y}, z + \hat{z}).$$

Furthermore, a vector is multiplied by a scalar coordinate by coordinate:

$$a(x, y, z) \equiv (ax, ay, az).$$

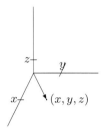

FIGURE 9.4:   The vector (x, y, z) starts at the origin $(0, 0, 0)$ and points to the point (x, y, z) in the three-dimensional Cartesian space.

Below we extend these definitions also to n-dimensional spaces for any natural number n.

9.5 Multidimensional Vectors

In the general case, an n-dimensional vector is a finite sequence of n real numbers:

$$v \equiv (v_1, v_2, v_3, \ldots, v_n),$$

where n is some fixed natural number (the dimension) and $v_1, v_2, \ldots, v_n$ are some real numbers. We then say that v is a vector in the n-dimensional space $\mathbb{R}^n$. This way, the above 2-dimensional Cartesian plane and 3-dimensional Cartesian space are just special cases of $\mathbb{R}^n$, with $n = 2$ and $n = 3$, respectively.

As in the previous sections, the addition of another vector

$$u \equiv (u_1, u_2, \ldots, u_n) \in \mathbb{R}^n$$

to v is done component by component:

$$v + u \equiv (v_1 + u_1, v_2 + u_2, \ldots, v_n + u_n).$$

Furthermore, this operation is linear in the sense that the commutative law applies:

$$
\begin{aligned}
v + (u + w) &= (v_1 + (u_1 + w_1), v_2 + (u_2 + w_2), \ldots, v_n + (u_n + w_n)) \\
&= ((v_1 + u_1) + w_1, (v_2 + u_2) + w_2, \ldots, (v_n + u_n) + w_n) \\
&= (v + u) + w,
\end{aligned}
$$

where $w \equiv (w_1, w_2, \ldots, w_n)$ is a vector in $\mathbb{R}^n$ as well.

The zero vector (the analogue of the origin in the 2-d Cartesian plane and in the 3-d Cartesian space) is the vector whose all components vanish:

$$\mathbf{0} \equiv (0,0,0,\ldots,0).$$

This vector has the special property that it can be added to every vector without changing it whatsoever:

$$\mathbf{0} + v = v + \mathbf{0} = v$$

for every vector $v \in \mathbb{R}^n$.

Moreover, the multiplication of a vector by a scalar $a \in \mathbb{R}$ is also done component by component:

$$av \equiv (av_1, av_2, \ldots, av_n).$$

This operation is commutative in the sense that

$$b(av) = b(av_1, av_2, \ldots, av_n) = (bav_1, bav_2, \ldots, bav_n) = (ba)v.$$

Furthermore, it is distributive in terms of the scalars that multiply the vector:

$$\begin{aligned}
(a+b)v &= ((a+b)v_1, (a+b)v_2, \ldots, (a+b)v_n) \\
&= (av_1 + bv_1, av_2 + bv_2, \ldots, av_n + bv_n) \\
&= av + bv,
\end{aligned}$$

as well as in terms of the vectors that are multiplied by the scalar:

$$\begin{aligned}
a(v+u) &= (a(v_1+u_1), a(v_2+u_2), \ldots, a(v_n+u_n)) \\
&= (av_1 + au_1, av_2 + au_2, \ldots, av_n + au_n) \\
&= av + au.
\end{aligned}$$

This completes the definition of the vector space $\mathbb{R}^n$ and the linear algebraic operations in it. Similarly, one could define the vector space $\mathbb{C}^n$: the only difference is that in $\mathbb{C}^n$ the components in the vectors, as well as the scalars that multiply them, can be not only real numbers but also complex numbers. It can be easily checked that the above commutative and distributive laws apply to $\mathbb{C}^n$ as well, so the algebraic operations in it remain linear. In fact, the algebraic operations in $\mathbb{C}^n$ can be viewed as extensions of the corresponding operations in $\mathbb{R}^n$. Thus, the larger vector space $\mathbb{C}^n$ can be viewed as an extension of the smaller vector space $\mathbb{R}^n \subset \mathbb{C}^n$.

9.6 Matrices

An m by n (or $m \times n$) matrix A is a finite sequence of n m-dimensional vectors:

$$A \equiv \left(v^{(1)} \mid v^{(2)} \mid v^{(3)} \mid \cdots \mid v^{(n)} \right)$$

where $v^{(1)}, v^{(2)}, \ldots, v^{(n)}$ are vectors in $\mathbb{R}^m$ for some fixed natural numbers m and n.

The jth vector, $v^{(j)}$ $(1 \leq j \leq n)$, is called the jth column of the matrix A, and takes the column form:

$$v^{(j)} \equiv \begin{pmatrix} v_1^{(j)} \\ v_2^{(j)} \\ v_3^{(j)} \\ \vdots \\ v_m^{(j)} \end{pmatrix}.$$

The ith component in $v^{(j)}$ is called the (i, j)th element in the matrix A, and is denoted by

$$a_{i,j} \equiv v_i^{(j)}.$$

For example, if $m = 3$ and $n = 4$, then A takes the form

$$A = \begin{pmatrix} a_{1,1} \ a_{1,2} \ a_{1,3} \ a_{1,4} \\ a_{2,1} \ a_{2,2} \ a_{2,3} \ a_{2,4} \\ a_{3,1} \ a_{3,2} \ a_{3,3} \ a_{3,4} \end{pmatrix}.$$

In this form, A may also be viewed as a sequence of three rows, each containing four numbers.

9.7 Adding Matrices

An $m \times n$ matrix

$$B \equiv \left(u^{(1)} \mid u^{(2)} \mid u^{(3)} \mid \cdots \mid u^{(n)} \right)$$

is added to the above matrix A column by column:

$$A + B \equiv \left(v^{(1)} + u^{(1)} \mid v^{(2)} + u^{(2)} \mid v^{(3)} + u^{(3)} \mid \cdots \mid v^{(n)} + u^{(n)} \right).$$

In other words, if B is denoted in its elementwise form

$$B = (b_{i,j})_{1 \leq i \leq m, 1 \leq j \leq n},$$

then it is added to A element by element:

$$A + B = (a_{i,j} + b_{i,j})_{1 \leq i \leq m, 1 \leq j \leq n}.$$

It is easy to check that this operation is commutative in the sense that

$$(A + B) + C = A + (B + C),$$

where C is another $m \times n$ matrix.

9.8 Multiplying a Matrix by a Scalar

The multiplication of a matrix A by a real number $r \in \mathbb{R}$ is done element by element:

$$rA \equiv (ra_{i,j})_{1 \leq i \leq m, 1 \leq j \leq n}.$$

Clearly, this operation is commutative:

$$q(rA) \equiv q(ra_{i,j})_{1 \leq i \leq m, 1 \leq j \leq n} = (qra_{i,j})_{1 \leq i \leq m, 1 \leq j \leq n} = (qr)A,$$

where $q \in \mathbb{R}$ is another scalar. Furthermore, it is distributive both in terms of the scalars that multiply the matrix:

$$(q + r)A = qA + rA,$$

and in terms of the matrix multiplied by the scalar:

$$r(A + B) = rA + rB.$$

9.9 Matrix times Vector

Let us define the operation in which the matrix

$$A = \left(v^{(1)} \mid v^{(2)} \mid \cdots \mid v^{(n)} \right)$$

multiplies the column vector

$$w = \begin{pmatrix} w_1 \\ w_2 \\ \vdots \\ w_n \end{pmatrix}.$$

The result of this operation is the m-dimensional column vector that is obtained from summing the columns of the matrix A after they have been multiplied by the scalars that are the components of w:

$$Aw \equiv w_1 v^{(1)} + w_2 v^{(2)} + \cdots + w_n v^{(n)} = \sum_{j=1}^{n} w_j v^{(j)}.$$

Thanks to the fact that the dimension of w is the same as the number of columns in A, this sum is indeed well defined. Note that the dimension of Aw

is not necessarily the same as the dimension of the original vector w. Indeed, $w \in \mathbb{R}^n$, whereas $Aw \in \mathbb{R}^m$.

Let us now look at the particular components in Aw. In fact, the ith component in Aw $(1 \le i \le m)$ is

$$(Aw)_i = \sum_{j=1}^{n} w_j v_i^{(j)} = \sum_{j=1}^{n} a_{i,j} w_j.$$

Clearly, the matrix-times-vector operation is commutative in the sense that, for a scalar $r \in \mathbb{R}$,

$$A(rw) = r(Aw) = (rA)w.$$

Furthermore, it is distributive both in terms of the matrix that multiplies the vector:

$$(A + B)w = Aw + Bw,$$

and in terms of the vector that is multiplied by the matrix:

$$A(w + u) = Aw + Au,$$

where u is another n-dimensional vector.

9.10 Matrix times Matrix

Let B be an $l \times m$ matrix, where l is a natural number. The product B times A is obtained by multiplying the columns of A one by one:

$$BA \equiv \left(Bv^{(1)} \mid Bv^{(2)} \mid \cdots \mid Bv^{(n)} \right),$$

where the $v^{(j)}$'s are the columns of A. Thanks to the fact that the number of rows in A is the same as the number of columns in B, these products are indeed well defined, and the result BA is an $l \times n$ matrix.

Let i and k be natural numbers satisfying $1 \le i \le l$ and $1 \le k \le n$. From the above, the (i, k)th element in BA is

$$(BA)_{i,k} = (Bv^{(k)})_i = \sum_{j=1}^{m} b_{i,j} v_j^{(k)} = \sum_{j=1}^{m} b_{i,j} a_{j,k}.$$

From this formula, it follows that the product of matrices is distributive both in terms of the matrix on the left:

$$(B + \hat{B})A = BA + \hat{B}A$$

(where $\hat{B}$ is another $l \times m$ matrix), and in terms of the matrix on the right:

$$B(A + \hat{A}) = BA + B\hat{A}$$

(where $\hat{A}$ is another $m \times n$ matrix).

Let us show that the product of matrices is also commutative. To this end, let

$$C = (c_{i,j})$$

be a $k \times l$ matrix, where k is a fixed natural number. Because the number of columns in C is the same as the number of rows in B and in BA, the products $C(BA)$ and $(CB)A$ are well defined. In particular, let us calculate the (s,t)th element in $C(BA)$, where s and t are natural numbers satisfying $1 \leq s \leq k$ and $1 \leq t \leq n$:

$$
\begin{aligned}
(C(BA))_{s,t} &= \sum_{i=1}^{l} c_{s,i} \sum_{j=1}^{m} b_{i,j} a_{j,t} \\
&= \sum_{i=1}^{l} \sum_{j=1}^{m} c_{s,i} b_{i,j} a_{j,t} \\
&= \sum_{j=1}^{m} \sum_{i=1}^{l} c_{s,i} b_{i,j} a_{j,t} \\
&= \sum_{j=1}^{m} \left(\sum_{i=1}^{l} c_{s,i} b_{i,j} \right) a_{j,t} \\
&= \sum_{j=1}^{m} (CB)_{s,j} a_{j,t} \\
&= ((CB)A)_{s,t}.
\end{aligned}
$$

Since this is true for every element (s,t) in the triple product, we have

$$C(BA) = (CB)A.$$

In other words, the multiplication of matrices is not only distributive but also commutative.

9.11 The Transpose of a Matrix

The transpose of A, denoted by A^t, is the $n \times m$ matrix whose (j,i)th element $(1 \leq i \leq m, 1 \leq j \leq n)$ is the same as the (i,j)th element in A:

$$A^t_{j,i} = a_{i,j}.$$

For example, if A is the 3×4 matrix

$$A = \begin{pmatrix} a_{1,1} \; a_{1,2} \; a_{1,3} \; a_{1,4} \\ a_{2,1} \; a_{2,2} \; a_{2,3} \; a_{2,4} \\ a_{3,1} \; a_{3,2} \; a_{3,3} \; a_{3,4} \end{pmatrix},$$

then A^t is the 4×3 matrix

$$A^t = \begin{pmatrix} a_{1,1} \; a_{2,1} \; a_{3,1} \\ a_{1,2} \; a_{2,2} \; a_{3,2} \\ a_{1,3} \; a_{2,3} \; a_{3,3} \\ a_{1,4} \; a_{2,4} \; a_{3,4} \end{pmatrix}.$$

From the above definition, it clearly follows that

$$(A^t)^t = A.$$

Note that, if the number of columns in B is the same as the number of rows in A, then the number of columns in A^t is the same as the number of rows in B^t, so the product $A^t B^t$ is well defined. Let us show that

$$(BA)^t = A^t B^t.$$

To this end, consider the (k, i)th element in $(BA)^t$ for some $1 \le i \le l$ and $1 \le k \le n$:

$$(BA)^t_{k,i} = (BA)_{i,k} = \sum_{j=1}^{m} b_{i,j} a_{j,k} = \sum_{j=1}^{m} A^t_{k,j} B^t_{j,i} = (A^t B^t)_{k,i}.$$

9.12 Symmetric Matrices

So far, we have considered rectangular matrices, whose number of rows m is not necessarily the same as the number of columns n. In this section, however, we focus on square matrices, whose number of rows is the same as the number of columns: $m = n$. This number is then called the order of the matrix.

A square matrix A is symmetric if it is equal to its transpose:

$$A = A^t.$$

In other words, for $1 \le i, j \le n$, the (i, j)th element in A is the same as the (i, j)th element in A^t:

$$a_{i,j} = a_{j,i}.$$

The main diagonal in the square matrix A contains the elements $a_{i,i}$, whose column index is the same as their row index. The identity matrix of order n, denoted by I, is the square matrix whose main-diagonal elements are all equal to 1, whereas its other elements (the off-diagonal elements) are all equal to 0:

$$I \equiv \begin{pmatrix} 1 & & & 0 \\ & 1 & & \\ & & \ddots & \\ 0 & & & 1 \end{pmatrix},$$

where the blank spaces in the above matrix contain zero elements as well.

Clearly, the identity matrix I is symmetric. In fact, I is particularly important, because it is a unit matrix in the sense that it can be applied to any n-dimensional vector v without changing it whatsoever:

$$Iv = v.$$

Moreover, I is also a unit matrix in the sense that it can multiply any matrix A of order n without changing it whatsoever:

$$IA = AI = A.$$

9.13 Hermitian Matrices

So far, we have considered real matrices, whose elements are real numbers in $\mathbb{R}$. This concept can be extended to complex matrices, whose elements may well be complex numbers in $\mathbb{C}$. For such matrices, all the properties discussed so far in this chapter remain valid. The concept of the transpose, however, should be replaced by the more general notion of the adjoint or Hermitian conjugate.

Recall that the complex conjugate of a complex number

$$c = a + ib$$

(where a and b are some real numbers and $i = \sqrt{-1}$) is defined by

$$\bar{c} \equiv a - ib.$$

This way, we have

$$c\bar{c} = (a + ib)(a - ib) = a^2 - i^2 b^2 = a^2 + b^2 = |c|^2,$$

where the absolute value of the complex number c is defined by

$$|c| \equiv \sqrt{a^2 + b^2}.$$

Note that if c happens to be a real number ($b = 0$), then it remains unchanged under the complex-conjugate operation:

$$\bar{c} = a = c.$$

Thus, the complex-conjugate operator is reduced to the identity operator on the real axis $\mathbb{R} \subset \mathbb{C}$.

Recall also that the complex-conjugate operation is linear in the sense that the complex conjugate of the sum of two complex numbers c and d is the sum of the complex conjugate of c and the complex conjugate of d:

$$\overline{c + d} = \bar{c} + \bar{d},$$

and the complex conjugate of their product is equal to the product of their complex conjugates:

$$\overline{cd}\bar{c} \cdot \bar{d}.$$

Indeed, the latter property can be proved easily using the polar representation of complex numbers.

The Hermitian conjugate of the $m \times n$ matrix A, denoted by A^h, is the $n \times m$ matrix whose (j, i)th element ($1 \le i \le m$, $1 \le j \le n$) is the complex conjugate of the (i, j)th element in A:

$$A^h_{j,i} = \bar{a}_{i,j}.$$

Because the complex-conjugate operation has no effect on real numbers, this definition agrees with the original definition of the transpose operator for real matrices, and can thus be considered as a natural extension of it to the case of complex matrices.

As in the case of the transpose matrix, it is easy to see that

$$(A^h)^h = A$$

and that

$$(BA)^h = A^h B^h.$$

When complex square matrices of order $m = n$ are considered, it makes sense to introduce the notion of an Hermitian matrix, which is equal to its Hermitian conjugate:

$$A = A^h.$$

In other words, the complex square matrix A is Hermitian if its (i, j)th element (for every $1 \le i, j \le n$) is equal to the (i, j)th element in its Hermitian conjugate:

$$a_{i,j} = \bar{a}_{j,i}.$$

In particular, the main-diagonal elements in a Hermitian matrix must be real:

$$a_{i,i} = \bar{a}_{i,i}.$$

9.14 Inner Product

A column vector of dimension n can actually be viewed as an $n \times 1$ matrix. For example, if u and v are two column vectors in $\mathbb{C}^n$, then

$$u^h = (\bar{u}_1, \bar{u}_2, \ldots, \bar{u}_n).$$

In other words, u^h is an $1 \times n$ matrix. Since the number of columns in this matrix is the same as the number of rows in the $n \times 1$ matrix v, the product $u^h v$ is well defined as a product of two matrices. The result of this product is a scalar (complex number), known as the inner product of u and v, denoted by

$$(u, v) \equiv u^h v = \sum_{j=1}^{n} \bar{u}_j v_j.$$

(Note that, when u is a real vector with real components, the above inner product is equal to the so-called real inner product, defined by

$$u^t v = \sum_{j=1}^{n} u_j v_j.)$$

From the above definition, it follows that, for every complex scalar $c \in \mathbb{C}$,

$$(cu, v) = \bar{c}(u, v)$$

and

$$(u, cv) = c(u, v).$$

Note that the inner product is a skew-symmetric operation in the sense that changing the order of the vectors in the inner product yields the complex conjugate of the original inner product:

$$(v, u) = \sum_{j=1}^{n} \bar{v}_j u_j = \overline{\sum_{j=1}^{n} \bar{u}_j v_j} = \overline{(u, v)}.$$

Furthermore, if u and v happen to be real vectors ($u, v \in \mathbb{R}^n$), then their inner product is a real number:

$$(u, v) = \sum_{j=1}^{n} \bar{u}_j v_j = \sum_{j=1}^{n} u_j v_j \in \mathbb{R}.$$

Note that the inner product of v with itself is

$$(v, v) = \sum_{j=1}^{n} \bar{v}_j v_j = \sum_{j=1}^{n} |v_j|^2 \geq 0.$$

In fact, (v, v) vanishes if and only if all the components v_j vanish:

$$(v, v) = 0 \quad \Leftrightarrow \quad v = \mathbf{0}.$$

Therefore, it makes sense to define the norm of v, denoted by $\|v\|$, as the square root of its inner product with itself:

$$\|v\| \equiv \sqrt{(v, v)}.$$

This way, we have

$$\|v\| \geq 0$$

and

$$\|v\| = 0 \quad \Leftrightarrow \quad v = \mathbf{0}.$$

This definition of norm has the desirable property that, for any complex scalar $c \in \mathbb{C}$, stretching v by factor c leads to enlarging the norm by factor $|c|$:

$$\|cv\| = \sqrt{(cv, cv)} = \sqrt{\bar{c}c(v, v)} = \sqrt{|c|^2(v, v)} = |c|\sqrt{(v, v)} = |c| \cdot \|v\|.$$

In particular, if v is a nonzero vector, then $\|v\| > 0$, so one may choose $c = 1/\|v\|$ to obtain the (normalized) unit vector $v/\|v\|$, namely, the vector of norm 1 that is proportional to the original vector v.

9.15 Norms of Vectors

The norm $\|v\|$ defined above is also called the l_2-norm of v and denoted by $\|v\|_2$, to distinguish it from other useful norms: the l_1-norm, defined by

$$\|v\|_1 \equiv \sum_{i=1}^{n} |v_i|,$$

and the l_∞- or maximum norm, defined by

$$\|v\|_\infty \equiv \max_{1 \leq i \leq n} |v_i|.$$

Here, however, we use mostly the l_2-norm defined in the previous section. This is why we denote it simply by $\|v\|$ rather than $\|v\|_2$.

9.16 Inner Product and the Hermitian Conjugate

Let A be an $m \times n$ matrix, and assume that u is an m-dimensional vector and that v is an n-dimensional vector. Then, the product Av is well defined.

As a matter of fact, Av is an m-dimensional vector, so the inner product (u, Av) is a well defined scalar. Furthermore, A^h is an $n \times m$ matrix, so the product $A^h u$ is well defined. In fact, $A^h u$ is an n-dimensional vector, so the inner product $(A^h u, v)$ is well defined. Using the commutativity of the triple product of matrices, we therefore have

$$(u, Av) = u^h(Av) = (u^h A)v = (A^h u)^h v = (A^h u, v).$$

In particular, if $m = n$ and A is Hermitian, then

$$(u, Av) = (Au, v)$$

for any two n-dimensional vectors u and v.

9.17 Orthogonal Matrices

Two n-dimensional vectors u and v are orthogonal to each other if their inner product vanishes:

$$(u, v) = 0.$$

Furthermore, u and v are also orthonormal if they are not only orthogonal to each other but also unit vectors in the sense that their norm is equal to 1:

$$\|u\| = \|v\| = 1.$$

A square matrix A of order n is called orthogonal if its columns are orthonormal in the sense that, for $1 \le i, j \le n$,

$$(v^{(i)}, v^{(j)}) = 0$$

and

$$\|v^{(j)}\| = 1,$$

where the $v^{(j)}$'s are the columns of A. In other words, the (i, k)th element in the product $A^h A$ is equal to

$$
\begin{aligned}
(A^h A)_{i,k} &= \sum_{j=1}^{n} (A^h)_{i,j} a_{j,k} \\
&= \sum_{j=1}^{n} \bar{a}_{j,i} a_{j,k} \\
&= \sum_{j=1}^{n} \bar{v}_j^{(i)} v_j^{(k)} \\
&= (v^{(i)}, v^{(k)}) = \begin{cases} 1 & \text{if } i = k \\ 0 & \text{if } i \ne k. \end{cases}
\end{aligned}
$$

In other words, if A is orthogonal, then $A^h A$ is the identity matrix:

$$A^h A = I.$$

This can also be written as

$$A^h = A^{-1}$$

or

$$A = (A^h)^{-1}.$$

Therefore, we also have

$$AA^h = I.$$

9.18 Eigenvectors and Eigenvalues

Let A be a square matrix of order n. A nonzero vector $v \in \mathbb{C}^n$ is called an eigenvector of A if there exists a scalar $\lambda \in \mathbb{C}$ such that

$$Av = \lambda v.$$

The scalar λ is then called an eigenvalue of A, or, more specifically, the eigenvalue associated with the eigenvector v.

Note that, for every nonzero complex scalar $c \in \mathbb{C}$, cv is an eigenvector as well:

$$A(cv) = cAv = c\lambda v = \lambda(cv).$$

In particular, thanks to the assumption that v is a nonzero vector, we have $\|v\| > 0$ so we can choose $c = 1/\|v\|$:

$$A(v/\|v\|) = \lambda(v/\|v\|).$$

This way, we obtain the (normalized) unit eigenvector $v/\|v\|$, namely, the eigenvector of norm 1 that is proportional to the original eigenvector v.

9.19 Eigenvalues of a Hermitian Matrix

Assume also that A is Hermitian. Then, we have

$$\lambda(v, v) = (v, \lambda v) = (v, Av) = (Av, v) = (\lambda v, v) = \bar{\lambda}(v, v).$$

Because v is a nonzero vector, we must also have

$$(v, v) > 0.$$

Therefore, we must also have

$$\lambda = \bar{\lambda},$$

or

$$\lambda \in \mathbb{R}.$$

The conclusion is, thus, that the eigenvalues of a Hermitian matrix must be real.

9.20 Eigenvectors of a Hermitian Matrix

Let u and v be two eigenvectors of the Hermitian matrix A:

$$Au = \mu u \quad \text{and} \quad Av = \lambda v,$$

where μ and λ are two distinct eigenvalues of A. We then have

$$\mu(u, v) = \bar{\mu}(u, v) = (\mu u, v) = (Au, v) = (u, Av) = (u, \lambda v) = \lambda(u, v).$$

Because we have assumed that $\mu \neq \lambda$, we can conclude that

$$(u, v) = 0,$$

or that u and v are orthogonal to each other. Furthermore, we can normalize u and v to obtain the two orthonormal eigenvectors $u/\|u\|$ and $v/\|v\|$.

9.21 The Sine Transform

A diagonal matrix is a square matrix of order n whose all off-diagonal elements vanish:

$$\Lambda \equiv \begin{pmatrix} \lambda_1 & & & \\ & \lambda_2 & & \\ & & \ddots & \\ & & & \lambda_n \end{pmatrix},$$

where the blank spaces in the matrix stand for zero elements. This matrix is also denoted by

$$\Lambda = diag(\lambda_1, \lambda 2, \dots, \lambda_n) = diag(\lambda_i)_{i=1}^n.$$

A tridiagonal matrix is a square matrix of order n that has nonzero elements only in its main diagonal or in the two diagonals that lie immediately above and below it. For example,

$$T \equiv \begin{pmatrix} 2 & -1 & & & \\ -1 & 2 & -1 & & \\ & \ddots & \ddots & \ddots & \\ & & -1 & 2 & -1 \\ & & & -1 & 2 \end{pmatrix}$$

is a tridiagonal matrix with 2's on its main diagonal, -1's on the diagonals immediately below and above it, and 0's elsewhere. This matrix is also denoted by

$$T = tridiag(-1, 2, -1).$$

Let us find the eigenvectors and eigenvalues of T. In fact, for $1 \le j \le n$, the eigenvectors are the column vectors $v^{(j)}$ whose components are

$$v_i^{(j)} = \sqrt{\frac{2}{n}} \sin(ij\pi/(n+1)),$$

for $1 \le i \le n$. From this definition, it follows that $v^{(j)}$ is indeed an eigenvector of T:

$$Tv^{(j)} = \lambda_j v^{(j)},$$

where

$$\lambda_j = 2 - 2\cos(j\pi/(n+1)) = 4\sin^2(j\pi/(2(n+1))).$$

Thanks to the fact that T is a symmetric matrix with n distinct eigenvalues, we have that its eigenvectors are orthogonal to each other. Furthermore, the eigenvectors $v^{(j)}$ are also unit vectors whose norm is equal to 1. Thus, the matrix A formed by these column vectors

$$A \equiv \left(v^{(1)} \mid v^{(2)} \mid \cdots \mid v^{(n)} \right)$$

is an orthogonal matrix:

$$A^{-1} = A^t.$$

Furthermore, A is symmetric, so

$$A^{-1} = A^t = A.$$

The matrix A is called the sine transform. Thanks to the above properties of A and its definition, one can write compactly

$$TA = A\Lambda,$$

or

$$T = A\Lambda A^{-1} = A\Lambda A^t = A\Lambda A.$$

This is called the diagonal form, or the diagonalization, of T in terms of its eigenvectors.

9.22 The Cosine Transform

Assume now that the corner elements in the tridiagonal matrix T defined above are changed to read

$$T_{1,1} = T_{n,n} = 1$$

instead of 2. Although this T is still symmetric, its eigenvectors are different from before. In fact, the ith component in the column eigenvector $v^{(j)}$ is now

$$v_i^{(j)} = \cos((i - 1/2)(j - 1)\pi/n),$$

for $1 \leq i, j \leq n$. Furthermore, the eigenvalue of T associated with this $v^{(j)}$ is now

$$\lambda_j = 2 - 2\cos((j - 1)\pi/n) = 4\sin^2((j - 1)\pi/(2n)).$$

As before, thanks to the fact that T is a symmetric matrix with n distinct eigenvalues, the matrix A composed from the normalized column vectors $v^{(j)}/\|v^{(j)}\|$ is orthogonal:

$$A^{-1} = A^t.$$

Thus, the matrix A, known as the cosine transform, can be used to obtain the diagonal form of T:

$$T = A\Lambda A^{-1} = A\Lambda A^t.$$

9.23 Determinant of a Square Matrix

Let A be a square matrix of order $n > 1$. For each $1 \leq i, j \leq n$, the (i, j)th minor of A is the $(n - 1) \times (n - 1)$ matrix obtained from A by dropping its ith row and jth column. In the sequel, we denote the (i, j)th minor by $A^{(i,j)}$.

The above definition is useful in defining the determinant of a square matrix A. In fact, the determinant of a square matrix is a function $det : \mathbb{R}^{n^2} \rightarrow \mathbb{R}$ defined by induction on $n \geq 1$ as follows:

$$det(A) \equiv \begin{cases} a_{1,1} & \text{if } n = 1 \\ \sum_{j=1}^n (-1)^{j+1} A_{1,j} \, det(A^{(1,j)}) & \text{if } n > 1. \end{cases}$$

This definition is indeed inductive: for $n = 1$, the determinant of A, denoted by $det(A)$, is the same as the only element in A, $a_{1,1}$. For $n > 1$, on the other hand, the determinant of A is defined in terms of the determinant of its minors, which are matrices of the smaller order $n - 1$.

The determinant of the square matrix A is useful in calculating its inverse matrix, A^{-1}.

9.24 Inverse of a Square Matrix

The inverse of a square matrix A of order n is a square matrix, denoted by A^{-1}, satisfying

$$A^{-1}A = AA^{-1} = I,$$

where I is the identity matrix of order n. If the inverse matrix indeed exists, then we say that A is nonsingular. In this case, $\det(A) \neq 0$, and A^{-1} is unique. If, on the other hand, no such matrix A^{-1} exists, then we say that A is singular. In this case, $\det(A) = 0$.

Kremer's formula for calculating A^{-1} for a nonsingular matrix A is as follows:

$$(A^{-1})_{i,j} = (-1)^{i+j} \frac{\det(A^{(j,i)})}{\det(A)}.$$

9.25 Vector Product

The determinant function defined above is also useful in defining the vector product of two vectors in the three-dimensional Cartesian space.

Let us define the three standard unit vectors in the 3-d Cartesian space:

$$\mathbf{i} = (1,0,0)$$
$$\mathbf{j} = (0,1,0)$$
$$\mathbf{k} = (0,0,1).$$

These standard unit vectors are now used to define the vector product, which is actually a function from $\mathbb{R}^3 \times \mathbb{R}^3 \to \mathbb{R}^3$. Let

$$u = (u_1, u_2, u_3)$$
$$v = (v_1, v_2, v_3)$$

be two vectors in the 3-d Cartesian space. The vector product of u and v is the vector defined as follows:

$$u \times v \equiv \det\left(\begin{pmatrix} \mathbf{i} & \mathbf{j} & \mathbf{k} \\ u_1 & u_2 & u_3 \\ v_1 & v_2 & v_3 \end{pmatrix}\right)$$
$$= \mathbf{i}(u_2 v_3 - u_3 v_2) - \mathbf{j}(u_1 v_3 - u_3 v_1) + \mathbf{k}(u_1 v_2 - u_2 v_1).$$

9.26 Exercises

1. Let A be the 2×2 matrix

$$A = \begin{pmatrix} \cos(\theta) & -\sin(\theta) \\ \sin(\theta) & \cos(\theta) \end{pmatrix}$$

   for some $0 \leq \theta < 2\pi$. Show that the columns of A are orthonormal.

2. Conclude that A is orthogonal.

3. Verify that A indeed satisfies

$$A^t A = A A^t = I$$

   (where I is the 2×2 identity matrix).

4. Let A be the $n \times n$ matrix of powers

$$A = tridiag(-1, 2, -1) + B,$$

   where B is the $n \times n$ matrix with the only nonzero elements

$$b_{0,n-1} = b_{n-1,0} = -1$$

   (the other elements of B vanish). In other words, the elements in A are

$$a_{i,j} = \begin{cases} 2 & \text{if } i = j \\ -1 & \text{if } i - j = \pm 1 \bmod n \\ 0 & \text{otherwise.} \end{cases}$$

   Let W be the $n \times n$ matrix of powers

$$W = (n^{-1/2} w^{ij})_{0 \leq i, j < n},$$

   where

$$w = \cos(2\pi/n) + i\sin(2\pi/n).$$

   Show that the jth column of W $(0 \leq j < n)$ is an eigenvector of A, with the corresponding eigenvalue

$$\lambda_j = 2 - (w^j + w^{-j}) = 2 - 2\cos(j \cdot 2\pi/n) = 4\sin^2(j\pi/n).$$

5. Use the symmetry of A to conclude that the columns of W are orthogonal to each other.

6. Show that the columns of W are orthonormal.

7. Conclude that W is orthogonal. (W is known as the discrete Fourier transform.)

8. Verify that W indeed satisfies

$$W\bar{W}^t = W\bar{W} = I$$

and

$$\bar{W}^t W = \bar{W}W = I,$$

where I is the $n \times n$ identity matrix.

9. Conclude that

$$W A \bar{W} = \Lambda, A = W \Lambda \bar{W},$$

where Λ is the $n \times n$ diagonal matrix

$$\Lambda = diag(\lambda_0, \lambda_1, \ldots, \lambda_{n-1}).$$

10. Let $K \equiv (k_{i,j})_{0 \le i,j < n}$ be the $n \times n$ matrix with 1's on the secondary diagonal and 0 elsewhere:

$$k_{i,j} = \begin{cases} 1 & \text{if } i+j = n-1 \\ 0 & \text{otherwise.} \end{cases}$$

Show that K is both symmetric and orthogonal. Conclude that

$$K^2 = K^t K = I.$$

11. Show that

$$\bar{W} = WK.$$

Conclude that

$$A = W \Lambda W K.$$

Chapter 10

Multilevel Objects

The concept of multilevel is useful not only in constructing composite mathematical objects but also in developing a systematic way of thinking towards solving problems and forming scientific theories. In fact, multilevel is a fundamental philosophical tool not only in mathematics but also in human culture in general.

Consider, for example, the problem of having suitable units to measure quantities such as weight, distance, time, etc. You could ask your grocer to have 1700 grams of apples, and he/she would no doubt reply that the cost is 400 cents; it would, however, make more sense to ask 1.7 kilogram and pay 4 dollars for it. By grouping 1000 grams into one kilogram and 100 cents into one dollar, we turn from the too fine level of grams and cents into the coarser, and more suitable, level of kilograms and dollars.

Returning to mathematics, we introduce here the concept of multilevel, and use it to construct multilevel objects. These objects not only stem from the philosophy of multilevel but also contribute back to it to enrich it and enlighten new ways of thinking.

10.1 Induction and Deduction

The fundamentals of logical and analytical thinking, introduced by the ancient Greeks, are based on induction and deduction. Assume that you are given a concrete engineering problem: say, to build a road between two particular cities. For this purpose, you are given the precise information about your resources, the topography of the area, etc.

You could think hard and find an efficient way to solve your particular problem. This way, however, you may be misled by the specifics in your problem, and driven away from the optimal solution.

A better approach is based on induction: generalize your particular problem into a more general problem by introducing the required concepts and giving them appropriate names. Writing the problem in general terms may clarify the subject and lead to general theory that may provide the optimal solution. Furthermore, this approach may develop a new useful and general terminology, which by itself may contribute to a better understanding of the fundamentals and concepts behind the problem.

Now, deduce from the general solution the required solution for your particular problem by replacing the general characteristics by the particular characteristics of the original application.

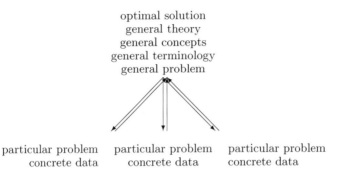

FIGURE 10.1:   The tree of problems: the concrete problems in the fine (low) level are solved by climbing up the tree to form the general problem (induction), solving it optimally by introducing general concepts, terminology, and theory, and then going back to the original problem (deduction).

This way of thinking may be viewed as a two-level approach. The particular applications are placed in the fine level (Figure 10.1). The first stage, the induction, forms the general problem in the higher level. Once an optimal solution is found in this level, one may return to the original fine level in the deduction stage, and adopt the optimal solution in the original application.

This two-level structure may also be interpreted in mathematical terms. Indeed, it can be viewed as a tree, with its head at the high level, in which the general problem is placed, and its leaves at the fine (low) level, in which the particular applications are placed (Figure 10.1).

Furthermore, the induction-deduction process of climbing up and down the tree may be viewed as a V-cycle (Figure 10.2). This cycle is called the V-cycle because it contains two legs, as in the Latin letter 'V'. In the first (left) leg of the V-cycle, the induction leg, one goes down from the particular problem to the general problem, which is written in general terms, hence is clearer and easier to solve optimally by introducing the required mathematical concepts, terminology, and theory. Once this is done, one may climb up the second (right) leg in the V-cycle, the deduction leg, to return to the original application and use the optimal solution in it (Figure 10.2).

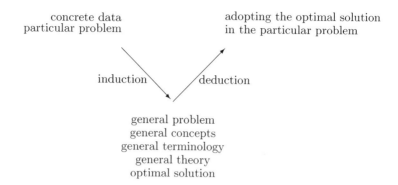

concrete data
particular problem

adopting the optimal solution
in the particular problem

induction   deduction

general problem
general concepts
general terminology
general theory
optimal solution

FIGURE 10.2:   The V-cycle: the concrete problem is solved by forming the general problem (induction) in the left leg of the 'V', solving it optimally by introducing general concepts, terminology, and theory, and then going back to the original problem (deduction) in the right leg of the 'V'.

10.2 Mathematical Induction

In the induction process described above, the original concrete problem, which is often obscured by too many details, is rewritten in a general form using general terminology, so its fundamentals become clearer. This gives one the opportunity to develop the required theory and solve the general problem optimally. The general solution is then used also in the original problem: this is the deduction stage.

This induction-deduction process is not limited to mathematics: it is relevant to other fields as well. Mathematical induction, on the other hand, is a special kind of induction, which is relevant for enumerable sets only. In fact, in mathematical induction, a property that is well known for a finite number of elements in the enumerable set is generalized to the entire set. Then, in the deduction stage, one may use this property in each particular element in the set.

Furthermore, mathematical induction can be used not only to establish that each and every element in the enumerable set enjoys some property, but also to create the infinitely many elements in the set in the first place. This is how the natural numbers are created in the beginning of the book.

To start the mathematical induction, one must know that the first object exists (e.g., the first number 1), or that the property holds for it. Then, one assumes that the induction hypothesis holds, that is, that the $(n-1)$st element in the set exists (e.g., the natural number $n-1$), or that the property holds for it. If one can use the induction hypothesis to prove the induction step, that is, that the nth element exists (e.g., $n = (n - 1) + 1$), or that the property holds for it, then the mathematical-induction axiom implies that the entire enumerable set of elements exists (e.g., $\mathbb{N}$, the set of the natural numbers), or that the property applies to each and every element in it.

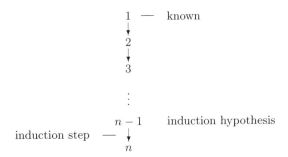

FIGURE 10.3:   Mathematical induction as a multilevel process. In the first level at the top, it is known that the property holds for 1. For $n = 2, 3, 4, \ldots$, the induction hypothesis assumes that the property holds for $n - 1$. In the induction step, the induction hypothesis is used to prove that the property holds for n as well. Thus, one may "climb" level by level from 1 down to any arbitrarily large number, and show that the property holds for it as well.

The nth element in the set can be viewed as the nth level in the multilevel hierarchy formed by the induction steps (Figure 10.3). The induction hypothesis may then be interpreted to say that the $(n - 1)$st level exists, or that the property holds for it. The induction step may be interpreted to say that it is possible to "climb" in the multilevel hierarchy, that is, to construct the nth level from the $(n-1)$st level, or to prove that the property holds for it as well. Since it has been assumed that the first level at the top exists (or that the property holds for it) the induction step actually means that infinitely many levels exist (or that the property holds for them as well); indeed, one can start at the first level and climb downwards an arbitrarily large number of steps in the multilevel hierarchy.

The original purpose of mathematical induction is to create the natural numbers. However, it is also useful to construct many other mathematical objects and discover their properties, as is illustrated below.

10.3 Trees

An important object defined by mathematical induction is the tree. In Figure 10.1 above, we have already seen a two-level tree; here we extend it into a multilevel tree.

The definition of the multilevel tree is done by mathematical induction. Let a one-level tree be the trivial tree that contains one node only: its head. Assume that the induction hypothesis holds, that is, that, for $n = 2, 3, 4, \ldots$, we already know how to define a k-level tree for every k between 1 and $n - 1$. Then, an n-level tree is obtained by letting a node serve as the head of the tree, issuing some edges (branches) from this head, and placing at the end of each branch a k-level tree $(1 \leq k \leq n - 1)$. For example, the two-level tree in Figure 10.1 is a special case deduced from this inductive definition: indeed, it is obtained by using $n = 2$, issuing three branches from the head, and placing a one-level tree (or just a node) at the end of each branch.

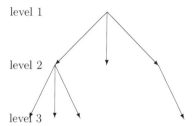

FIGURE 10.4: A three-level tree: three branches are issued from the node at the top (the head). The middle branch ends with a trivial one-level tree or a leaf. The right and left branches, on the other hand, end with two-level trees with one to three branches.

In Figure 10.4, we illustrate a three-level tree that can also be deduced from the above inductive definition by using $n = 3$, issuing three branches from it, and placing a one-level tree (or a node, or a leaf) at the end of the middle branch, and two-level trees at the end of the left and right branches.

10.4 Binary Trees

In the above definition of a tree, there is no bound on the number of branches issued from the head: it may be arbitrarily large. This produces a general tree; here, however, we modify the above definition to obtain a binary tree.

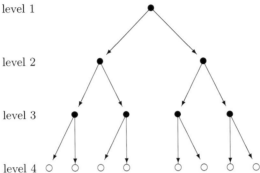

FIGURE 10.5:   A four-level binary tree: the arrows represent branches, the circles at the lowest level stand for leaves, and the bullets stand for nodes that are not leaves.

The definition of a binary tree is similar to the above definition of the general tree, except for the following additional constraint imposed in the induction step: the number of branches issued from the head must be either zero or two. It is easy to prove (by mathematical induction) that this implies that this constraint applies not only to the head but also to every node in the tree: the number of branches issued from it (not including the branch leading to it) is either zero (for the leaves at the bottom of the tree) or two (for nodes that are not leaves). Figure 10.5 illustrates this property in a four-level tree.

10.5 Arithmetic Expressions

The trees defined above are particularly useful to model arithmetic expressions. The symbol of the arithmetic operation of the least priority (usually the last + or − in the arithmetic expression) is placed at the head of the tree. Then, the arithmetic expression to the left of this symbol is placed in the subtree at the end of the left branch issued from the head, and the arithmetic expression to the right of this symbol is placed in the subtree at the end of the right branch issued from the head.

The value of the arithmetic expression can be calculated bottom to top inductively: assuming that the subexpressions in the subtrees have already been calculated recursively, the value of the original arithmetic expression is calculated by applying the symbol in its head to the values of these subexpressions.

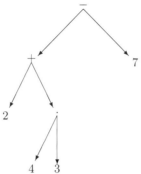

FIGURE 10.6: Modeling the arithmetic expression $2 + 4 \cdot 3 - 7$ in a four-level binary tree. The calculation is carried out bottom to top: the top-priority arithmetic operation, $4 \cdot 3$, is carried out in the third level. The next operation, $2 + 4 \cdot 3$, is carried out in the second level. Finally, the least-priority operation, $2 + 4 \cdot 3 - 7$, is carried out at the top of the tree.

For example, the arithmetic expression

$$2 + 4 \cdot 3 - 7$$

is modeled by the four-level binary tree in Figure 10.6: the subtraction symbol '$-$', which is of least priority, is placed at the top of the tree, to be applied last; the addition symbol '$+$', which is of intermediate priority, is placed in the second level in the tree; and the multiplication symbol '$\cdot$', which is of top priority, is placed in the third level, so it is performed first on its arguments in the leaves at the bottom of the tree.

10.6 Boolean Expressions

Boolean expressions are obtained from Boolean variables (variables that may have only two possible values: 1 for true or 0 for false) by applying to them the "and" operation (denoted by the symbol '$\wedge$') or the "or" operation (denoted by the symbol '$\vee$'). Like arithmetic expressions, these expressions can be modeled in trees. The symbol of the least priority (the last '$\vee$' in the

expression) is placed at the top of the tree, to be applied last; the arguments
of this symbol, the left subexpression and the right subexpression, are placed
in the left and right subtrees, respectively.

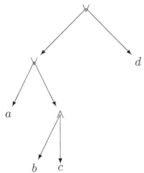

FIGURE 10.7:   Modeling the Boolean expression $a \vee b \wedge c \vee d$ in a four-level
binary tree. The calculation is carried out bottom to top: the top-priority Boolean
operation, $b \wedge c$, is carried out in the third level. The next operation, $a \vee b \wedge c$, is
carried out in the second level. Finally, the least-priority operation, $a \vee b \wedge c \vee d$, is
carried out at the top of the tree.

As in arithmetic expressions, calculating the value of the expression (0 or 1)
is done bottom to top by mathematical induction: assuming that the values
of the left and right subexpressions have already been calculated recursively,
the symbol at the head of the tree is applied to them to yield the value of the
entire expression. For example, the value of the Boolean expression

$$a \vee b \wedge c \vee d$$

(which means "either a is true or both b and c are true or d is true," where a,
b, c, and d are some Boolean variables), is calculated as follows (Figure 10.7):
first, the top priority symbol '$\wedge$' at the third level is applied to its argument
in the leaves to calculate $b \wedge c$; then, the '$\vee$' in the second level is applied to
calculate $a \vee b \wedge c$; and finally, the '$\vee$' at the top of the tree, the symbol of
least priority, is applied to calculate the original expression $a \vee b \wedge c \vee d$.

We say that a tree is full if it contains leaves in its lowest level only. For
example, the tree in Figure 10.5 is full, because its leaves lie in its fourth level
only. The trees in Figures 10.6–10.7, on the other hand, are not full, because
they contain leaves not only in the fourth level but also in the third level (the
left node) and the second level (the right node).

10.7 The Tower Problem

The binary tree defined above is useful to model not only mathematical expressions but also more abstract notions such as mathematical problems and algorithms. Consider, for example, the following problem, known as the tower problem. Suppose that three columns are given, denoted by column 1, column 2, and column 3. A tower of n rings, one on top of the other, is placed on column 1, where n is some (large) natural number. The radius of the rings decreases from bottom to top: the largest ring lies at the bottom of column 1, a smaller ring lies on top of it, a yet smaller ring lies on top of it, and so on, until the smallest (nth) ring at the top of column 1.

The task is to use the minimal possible number of moves to transfer the entire tower of n rings from column 1 to column 3, while preserving the following rules:

1. In each move, only one ring is moved from one column to another column.
2. A ring that another ring lies on top of it cannot be moved.
3. A ring cannot lie on a smaller ring.

The solution of this problem is found by mathematical induction. Indeed, for $n = 1$, the tower contains only one ring, so it can be moved to column 3, and the problem is solved in one move only. Now, assume that the induction hypothesis holds, that is, that we know how to transfer a slightly smaller tower of $n-1$ rings from one column to another, while preserving the original order from the largest ring at the bottom to the smallest ring at the top. Let us use this hypothesis to transfer also the original tower of n rings from column 1 to column 3, while preserving this order (the induction step). For this, we first use the induction hypothesis to transfer the $n-1$ top rings from column 1 to column 2 (while preserving the original order). The only ring left on column 1 is the largest ring. In the next move, this ring is moved to column 3. Then, we use the induction hypothesis once again to move the remaining $n-1$ rings from column 2 to column 3 (while preserving their original order), and placing them on top of the largest ring. This indeed completes the task and the proof of the induction step.

Let us prove by mathematical induction that the total number of moves used in the above algorithm is $2^n - 1$. Indeed, for $n = 1$, the number of moves is

$$2^1 - 1 = 1.$$

Assume that the induction hypothesis holds, that is, that for $n-1$, the required number of moves is

$$2^{n-1} - 1.$$

Since transferring n rings requires two transfers of $n-1$ rings plus one move of the largest ring, we have that the total number of moves required to transfer n rings is

$$2(2^{n-1} - 1) + 1 = 2^n - 1,$$

as indeed asserted. This completes the proof of the induction step.

10.8 The Tree of the Tower Problem

The above algorithm to solve the tower problem can be modeled as a full n-level binary tree, where n is the number of rings in the tower. This can be proved by mathematical induction of n. Indeed, the move of the largest ring from column 1 to column 3 is placed in the head of the tree. Furthermore, when $n > 1$, the original algorithm to transfer the entire tower of n rings requires two applications (or recursive calls) of the same algorithm itself to transfer two slightly smaller towers of $n-1$ rings from column to column. From the induction hypothesis, these recursive calls can themselves be modeled by full $(n - 1)$-level binary trees, which can be placed at the end of the left and right branches issued from the head to serve as the left and right subtrees. This completes the construction of the full n-level binary tree associated with the original algorithm to transfer the entire tower of n rings.

In each of these recursive calls made in the original algorithm, two other recursive calls are made to transfer two yet smaller towers of $n - 2$ rings from one column to another. The full $(n - 2)$-level binary trees associated with these recursive calls are placed in the next (lower) level in the tree to serve as subtrees, and so on. Figure 10.8 illustrates the case $n = 4$.

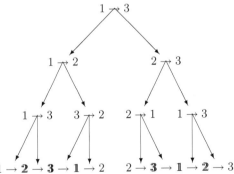

FIGURE 10.8:   The four-level binary tree with the moves required to transfer a tower of four rings from column 1 to column 3. The algorithm is carried out bottom-left to top. Each node contains a particular move. For example, the first move in the lower-left node moves the top ring in column 1 to the top of column 2.

Note that each node in the above binary tree represents one move of one ring. For example, the trivial tower of one ring is represented by a one-level tree, whose only node stands for the move of the ring from column 1 to column 3. Furthermore, for a tower of two rings, a large ring and a small ring, the above algorithm is modeled by a two-level binary tree of three nodes: the lower-left node stands for moving the small ring from column 1 to column 2, the top node stands for moving the large ring from column 1 to column 3, and the lower-right node stands for moving the small ring from column 2 to column 3, which completes the task.

For larger n, the equivalence between the nodes in the tree and the moves in the algorithm can be proved by mathematical induction. Indeed, assume that this equivalence holds for $n - 1$. Then, for n, the corresponding tree contains n levels. In particular, the second level contains two heads of two subtrees of $n - 1$ levels each. From the induction hypothesis, each node in each of these subtrees represents one move of one ring in the transfer of the subtowers of $n - 1$ rings. Furthermore, the top node in the original tree represents moving the largest (nth) ring from column 1 to column 3. Thus, each node in the original tree stands for one move in the original algorithm to transfer the original tower of n rings. This completes the proof of the induction step.

From the representation of the algorithm as a tree, we have another method to calculate the total number of moves used in it. Indeed, this number must be the same as the number of the nodes in the n-level binary tree, which is

$$\sum_{i=1}^{n} 2^{i-1} = \sum_{i=0}^{n-1} 2^i = \frac{2^n - 1}{2 - 1} = 2^n - 1,$$

which indeed agrees with the number of moves as calculated inductively before.

10.9 Pascal's Triangle

In the binary tree studied above, the number of nodes is doubled when turning from a particular level to the next level below it. Here we consider another kind of a multilevel object, in which the number of entries only increases by 1 from some level to the next lower level.

Pascal's triangle (Figure 10.9) consists of lines of oblique subsquares that contain numbers (entries). These lines (or levels) are numbered by the indices $0, 1, 2, 3, \ldots$. For example, the 0th level at the top of the triangle contains one entry only, the next level just below it contains two entries, the next level just below it contains three entries, and so on.

The numbers (or entries) that are placed in the subsquares in Pascal's triangle are defined by mathematical induction level by level. In particular, the number 1 is placed in the 0th level at the top of the triangle. Assume now

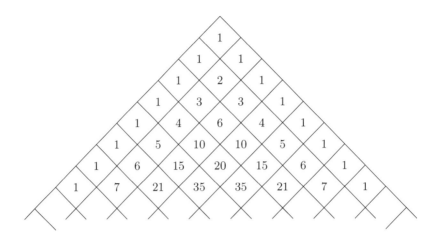

FIGURE 10.9:   Pascal's triangle: each entry is equal to the sum of the two
entries in the upper-left and upper-right subsquares (if exist).

that n entries have already been placed in the n subsquares in the $(n-1)$st
level (the induction hypothesis). Let us use these entries to define also the
entries in the nth level (the induction step).

This is done as follows: in each subsquare in the nth level, one places the
sum of the two entries in the upper-left and upper-right subsquares that lie
just above it. Of course, if one of these subsquares is missing, then this sum
reduces to the value in the existing subsquare only. This way, the entry in the
first subsquare in the nth level is the same as the entry in the first subsquare
in the $(n-1)$st level, and the entry in the last subsquare in the nth level is the
same as the entry in the last subsquare in the $(n-1)$st level. (In other words,
all the entries at the edges of the triangle are equal to 1.) This completes the
induction step to define the entries in the subsquares in Pascal's triangle.

The above mathematical induction gives an algorithm to define the entries
in Pascal's triangle recursively level by level. However, it gives no explicit
formula for these entries. Below we provide such an explicit formula in terms
of the binomial coefficients.

10.10 The Binomial Coefficients

For two nonnegative integer numbers $n \geq k \geq 0$, the binomial coefficient $\binom{n}{k}$ is defined by

$$\binom{n}{k} \equiv \frac{n!}{k!(n-k)!},$$

where, for every nonnegative integer number $n \geq 0$, the factorial function, denoted by '!', is defined recursively by

$$n! \equiv \begin{cases} 1 & \text{if } n = 0 \\ (n-1)! \cdot n & \text{if } n > 0. \end{cases}$$

Like the entries in Pascal's triangle, the binomial coefficient $\binom{n}{k}$ also enjoys the property that it can be written as the sum of two $(n-1)$'level binomial coefficients:

$$\binom{n}{k} = \binom{n-1}{k-1} + \binom{n-1}{k}$$

for every $n > 0$ and every $0 < k < n$. Indeed,

$$\binom{n-1}{k-1} + \binom{n-1}{k} = \frac{(n-1)!}{(k-1)!(n-k)!} + \frac{(n-1)!}{k!(n-k-1)!}$$

$$= \frac{k}{n} \cdot \frac{n!}{k!(n-k)!} + \frac{n-k}{n} \cdot \frac{n!}{k!(n-k)!}$$

$$= \frac{k+n-k}{n} \cdot \binom{n}{k}$$

$$= \binom{n}{k}.$$

We can now use this formula to show that the entries in Pascal's triangle are the same as the binomial coefficients. More precisely, we'll use mathematical induction to show that the kth entry $(0 \leq k \leq n)$ in the nth level in Pascal's triangle $(n \geq 0)$ is equal to the binomial coefficient $\binom{n}{k}$. Indeed, for $n = 0$, the only entry at the top of Pascal's triangle is 1, which is also equal to the binomial coefficient for which $n = k = 0$:

$$\binom{0}{0} = \frac{0!}{0! \cdot (0-0)!} = \frac{1}{1 \cdot 1} = 1.$$

Furthermore, assume that the induction hypothesis holds, that is, that the kth entry in the $(n-1)$st level in Pascal's triangle $(0 \leq k < n)$ is equal to the binomial coefficient $\binom{n-1}{k}$. Then, this hypothesis can be used to prove that the kth entry in the nth level in Pascal's triangle ($0 \le k \le n$) is also equal to the binomial coefficient $\binom{n}{k}$.

Indeed, for $k = 0$,

$$\binom{n}{0} = \frac{n!}{0!(n-0)!} = \frac{n!}{1 \cdot n!} = 1,$$

exactly as in the first (0th) entry in the nth level in Pascal's triangle.

Furthermore, for $k = n$,

$$\binom{n}{n} = \frac{n!}{n!(n-n)!} = \frac{n!}{n! \cdot 1} = 1,$$

exactly as in the last (nth) entry in the nth level in Pascal's triangle.

Finally, for $1 < k < n$,

$$\binom{n}{k} = \binom{n-1}{k-1} + \binom{n-1}{k}$$

is, by the induction hypothesis, the sum of the two entries in the upper-left and upper-right subsquares, which is, by definition, the kth entry in the nth level in Pascal's triangle. This completes the proof of the induction step for every $0 \le k \le n$, establishing the assertion that all the entries in Pascal's triangle are indeed equal to the corresponding binomial coefficients.

10.11 Paths in Pascal's Triangle

The entry that lies in a particular subsquare in Pascal's triangle can be interpreted as the number of distinct paths that lead from the top of the triangle to this particular subsquare, where a path is a sequence of consecutive steps from some subsquare to the adjacent lower-left or lower-right subsquare (Figure 10.10).

More precisely, a path from the top of the triangle to the kth subsquare in the nth level of the triangle may be represented by an n-dimensional vector whose components are either 0 or 1: 0 for a down-left step, and 1 for a down-right step. We assert that the kth entry in the nth level in Pascal's triangle is the number of distinct paths leading to it. Clearly, such a path must contain exactly $n - k$ down-left steps and k down-right steps, so the n-dimensional vector representing it must contain exactly $n - k$ 0's and k 1's, so its squared norm must be exactly k.

Thus, the number of distinct paths leading to the kth entry in the nth level in Pascal's triangle is the cardinality of the set of vectors

$$\left\{ p \in \{0, 1\}^n \ \middle| \ \|p\|^2 = k \right\}.$$

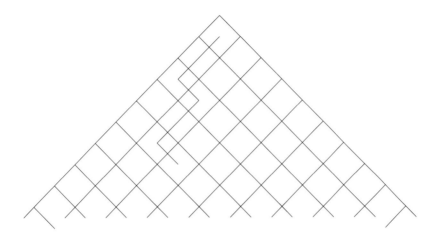

FIGURE 10.10: The path leading from the top subsquare in Pascal's triangle to the subsquare $k = 2$ in level $n = 6$. This path corresponds to the 6-dimensional vector $(0, 0, 1, 0, 0, 1)$, because it contains down-right moves in the third and sixth steps only, and down-left moves elsewhere.

Thus, our assertion also means that the kth entry in the nth level in Pascal's triangle is equal to the cardinality of this set.

Let us prove this assertion by induction on the levels in Pascal's triangle. Indeed, for $n = 0$, there is only one path that leads from the top of the triangle to itself: the trivial path that contains no steps at all (represented by the empty vector or the empty set). Furthermore, let us assume that the induction hypothesis holds, that is, that the entries in the subsquares in the $(n-1)$st level are equal to the number of distinct paths leading to them. Now, each path leading to the kth subsquare in the nth level ($0 \le k \le n$) must pass either through the upper-left subsquare (the $(k-1)$st subsquare in the $(n-1)$st level) or through the upper-right subsquare (the kth subsquare in the $(n-1)$st level). (If one of these subsquares lies outside the triangle, then it of course doesn't count.) Thus, the total number of distinct paths leading to the kth subsquare in the nth level is the sum of the numbers of distinct paths leading to the $(k-1)$st and kth subsquares in the $(n-1)$st level (if exist). By the induction hypothesis, this is just the sum of the $(k-1)$st and kth entries in the $(n-1)$st level, which is, by definition, just the kth entry in the nth level. This completes the proof of the induction step.

10.12 Paths and the Binomial Coefficients

In the above sections, we have proved that the binomial coefficient $\binom{n}{k}$ is equal to the entry in the kth subsquare in the nth level in Pascal's triangle, which is equal to the number of distinct paths leading from the top of the triangle to it, or to the number of distinct n-dimensional vectors with $n-k$ 0 components and k 1 components:

$$\binom{n}{k} = \left|\{p \in \{0,1\}^n \mid \|p\|^2 = k\}\right|.$$

Let us prove this equality more directly, without mathematical induction. This is done by counting the vectors in the above set. To choose a particular vector in this set, we must decide where to place the k components whose value is 1 among the n components in the vector.

How many different ways are there to do this?

We have k components whose value is 1 to place in the vector. Let us start with the first component whose value is 1. Clearly, there are n different ways to place it in the vector: it can lie in the first, second, ..., or nth coordinate in the vector.

Let us now turn to the next component whose value is 1. It can be placed in either of the $n-1$ coordinates that are left in the vector. For example, if the first 1 has been placed in the ith coordinate in the vector ($1 \leq i \leq n$), then the second component of value 1 can be placed in either of the $n-1$ components j that satisfy $1 \leq j \leq n$ and $j \neq i$.

It seems, therefore, that there are $n(n-1)$ possibilities to place the two first components of value 1 in the vector. Still, are all these possibilities genuinely different from each other? After all, placing the first component of value 1 in the ith coordinate and the second component of value 1 in the jth coordinate is the same as placing the first component of value 1 in the jth coordinate and the second component of value 1 in the ith coordinate. Indeed, both possibilities yield the same vector, with the value 1 at the ith and jth coordinates and 0 elsewhere. Thus, the total number of distinct vectors with exactly two components of value 1 is $n(n-1)/2$ rather than $n(n-1)$.

By repeating this process, one can easily see that there are $n-2$ possibilities to place the third component of value 1 in the vector. Indeed, it can be placed in every coordinate l that satisfies $1 \leq l \leq n$, $l \neq i$, and $l \neq j$, where i and j are the coordinates where the first two components of value 1 have been placed.

This yields the vector with the value 1 at the ith, jth, and lth coordinates and 0 elsewhere.

Still, there are three possible choices that lead to this vector: choosing the first two components to be placed at the ith and jth coordinates and the third component of value 1 at the lth coordinate, choosing the first two components to be placed at the ith and lth coordinates and the third component of value 1 at the jth coordinate, and choosing the first two components to be placed at the jth and lth coordinates and the third component of value 1 at the ith coordinate. Thus, in order to count distinct vectors only, one should multiply the number of distinct vectors with two components of value 1 not by $n - 2$ but rather by $(n - 2)/3$, yielding

$$\frac{n}{1} \cdot \frac{n - 1}{2} \cdot \frac{n - 2}{3}$$

as the total number of distinct vectors with exactly 3 components of value 1 and 0 elsewhere.

By repeating this process, one has that the number of distinct vectors with k components of value 1 and $n - k$ components of value 0 is

$$\frac{n}{1} \cdot \frac{n - 1}{2} \cdot \frac{n - 2}{3} \cdot \ldots \cdot \frac{n - k + 1}{k} = \frac{n!}{(n - k)!k!} = \binom{n}{k},$$

as indeed asserted.

Below we show how useful the paths and the binomial coefficients introduced above can be in practical applications.

10.13 Newton's Binomial

The paths studied above are particularly useful in Newton's binomial [19], which is the formula that allows one to open the parentheses in the expression

$$(a + b)^n = (a + b)(a + b)(a + b) \cdots (a + b) \quad n \text{ times}$$

(where n is some given nonnegative integer number and a and b are given parameters), and rewrite it as a sum of products rather than the product of the factors $(a+b)$. In fact, when the parentheses in this expression are opened, one gets the sum of products of the form $a^k b^{n-k}$, where k $(0 \le k \le n)$ is the number of factors of the form $(a + b)$ from which a is picked, and $n - k$ is the number of factors of the form $(a + b)$ from which b is picked.

Now, how many distinct possible ways are there to pick a from k factors of the form $(a + b)$ and b from the remaining $n - k$ factors? Since each such way can be characterized by an n-dimensional vector of 0's and 1's, with 0 standing for factors from which b is picked and 1 standing for factors from which a is picked, the total number of such ways is

$$\left|\{p \in \{0,1\}^n \mid \|p\|^2 = k\}\right|. = \binom{n}{k}.$$

Thus, when the parentheses in the original expression $(a + b)^n$ are opened, the term $a^k b^{n-k}$ appears $\binom{n}{k}$ times, once for each possible way to pick k a's and $n - k$ b's from the n factors of the form $(a + b)$ in $(a + b)^n$. These terms sum up to contribute

$$\binom{n}{k} a^k b^{n-k}$$

to the sum of products obtained from the original expression $(a + b)^n$ when the parentheses are opened. Since this can be done for each k between 0 and n, opening the parentheses yields the formula

$$(a + b)^n = \sum_{k=0}^{n} \binom{n}{k} a^k b^{n-k}.$$

This formula is known as Newton's binomial. The coefficients $\binom{n}{k}$ (the binomial coefficients) can be obtained from the nth level in Pascal's triangle.

10.14 Brownian Motion

Here we describe another application of the binomial coefficients and the paths associated with them. This is the Brownian motion in Stochastics [18].

Consider a particle that lies on the real axis and moves on it step by step either one unit to the right or one unit to the left. In the beginning, the particle lies at the origin 0. In each step, it moves by 1 either to the right (from l to $l + 1$) or to the left (from l to $l - 1$).

The process is nondeterministic: we don't know for sure where the particle goes in each step. Still we know that in each particular step, there is a probability a that the particle goes to the left and a probability b that it goes to the right, where a and b are given positive parameters satisfying $a + b = 1$.

Where will the particle be after n steps? Of course, we cannot tell this for sure. Nevertheless, we can calculate the probability that it would then be at some point on the real axis.

In the first n steps, the particle must go k moves to the left and $n - k$ moves to the right, where k is some integer number between 0 and n. The k moves to the left move the particle by a total amount of $-k$, whereas the $n - k$ moves to the right move it by a total amount of $n - k$. Thus, after these n steps are complete, the particle will lie at the point $n - 2k$ on the real axis.

For a fixed k, $0 \leq k \leq n$, what is the probability that the particle will indeed be at the point $n - 2k$ after n steps? Well, this of course depends on the number of distinct possible ways to make k moves to the left and $n - k$ moves to the right during the first n steps. As we've seen above, this number is the binomial coefficient $\binom{n}{k}$.

Furthermore, since the steps are independent of each other, the probability that the particle makes a particular path of k moves to the left and $n - k$ moves to the right is $a^k b^{n-k}$. Thus, the total probability that the particle takes any path of k moves to the left and $n - k$ moves to the right is

$$\binom{n}{k} a^k b^{n-k}.$$

This is also the probability that the particle would lie at the point $n - 2k$ on the real axis after n steps.

So far, we've calculated the probability that the particle would make k moves to the left and $n - k$ moves to the right during the first n steps, where k is a fixed number between 0 and n. In fact, the particle must make k moves to the left and $n - k$ moves to the right for some $0 \leq k \leq n$. Thus, the sum of the probabilities calculated above must be 1. This indeed follows from Newton's binomial:

$$\sum_{k=0}^{n} \binom{n}{k} a^k b^{n-k} = (a + b)^n = 1^n = 1.$$

When $a = b = 1/2$, that is, when the probability that the particle moves to the right is the same as the probability that it moves to the left, the above process models diffusion along a one-dimensional axis, or Brownian motion. In this case, the probabilities for the location of the particle after $n = 5$, 6, and 7 steps are illustrated in Figures 10.11–10.13.

When, on the other hand, $a < b$, the above process models the diffusion when a slight wind blows to the right. Finally, when $a > b$, the above process models the diffusion when a slight wind blows to the left.

10.15 Counting Integer Vectors

Let us use the binomial coefficients to count the total number of k-dimensional vectors with nonnegative integer components whose sum is at most n. In other words, we are interested in the cardinality of the set

$$\left\{ (v_1, v_2, \ldots, v_k) \in (\mathbb{Z}^+)^k \;\middle|\; \sum_{i=1}^{k} v_i \leq n \right\},$$

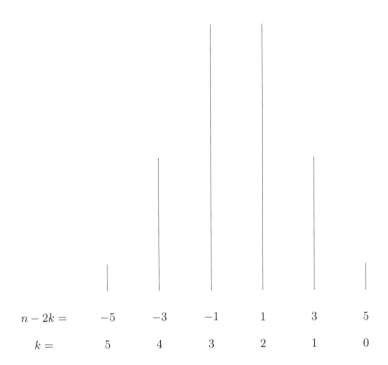

$n - 2k =$	-5	-3	-1	1	3	5
$k =$	5	4	3	2	1	0

FIGURE 10.11: Brownian motion ($a = b = 1/2$), distribution diagram after $n = 5$ steps: the columns in the diagram represent the probability of the particle to reach the point $n - 2k$ ($0 \le k \le n$) after $n = 5$ steps. (This requires $n - k$ moves to the right and k moves to the left.)

where $k \ge 1$ and $n \ge 0$ are given integers, and

$$\mathbb{Z}^+ \equiv \mathbb{N} \cup 0$$

is the set of nonnegative integers. In the following, we'll show that the cardinality of the above set is

$$\left| \left\{ (v_1, v_2, \dots, v_k) \in (\mathbb{Z}^+)^k \ \Big| \ \sum_{i=1}^{k} v_i \le n \right\} \right| = \binom{n+k}{k}.$$

The proof is by induction on $k \ge 1$, in which the induction step is by itself proved by an inner induction on $n \ge 0$. Indeed, for $k = 1$, the above vectors

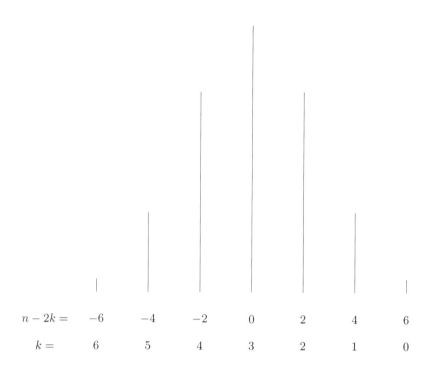

$n - 2k =$	-6	-4	-2	0	2	4	6
$k =$	6	5	4	3	2	1	0

FIGURE 10.12: Brownian motion $(a = b = 1/2)$, distribution diagram after $n = 6$ steps: the columns in the diagram represent the probability of the particle to reach the point $n - 2k$ $(0 \le k \le n)$ after $n = 6$ steps. (This requires $n - k$ moves to the right and k moves to the left.)

are actually scalars. Clearly, for every $n \ge 0$, the total number of integer numbers between 0 and n is

$$n + 1 = \binom{n+1}{1},$$

as required, Furthermore, let us use the induction hypothesis to prove the asserted formula for every $k \ge 2$ as well. According to this hypothesis, the above formula holds for $k - 1$. In other words, for every $n \ge 0$, the total number of vectors in $(\mathbb{Z}^+)^{k-1}$ whose component sum is at most n is

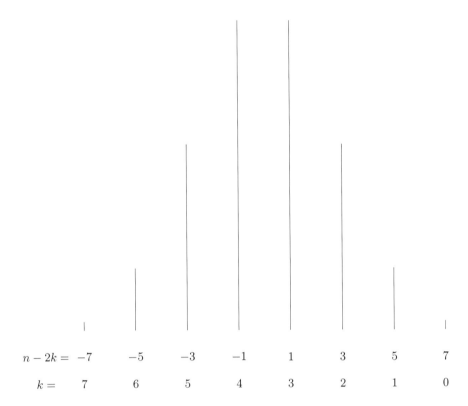

$n - 2k =$	-7	-5	-3	-1	1	3	5	7
$k =$	7	6	5	4	3	2	1	0

FIGURE 10.13:　Brownian motion $(a = b = 1/2)$, distribution diagram after $n = 7$ steps: the columns in the diagram represent the probability of the particle to reach the point $n - 2k$ $(0 \leq k \leq n)$ after $n = 7$ steps. (This requires $n - k$ moves to the right and k moves to the left.)

$$\binom{n + k - 1}{k - 1}.$$

Now, to prove the assertion for k as well, let us use an inner induction on $n \geq 0$. Clearly, for $n = 0$, the number of vectors in $(\mathbb{Z}^+)^k$ whose components sum is no more than 0 is

$$1 = \binom{0 + k}{k},$$

as required. Assume now that the assertion holds for $n - 1$, that is, that the total number of vectors in $(\mathbb{Z}^+)^k$ whose component sum is at most $n - 1$ is

$$\binom{n-1+k}{k}.$$

Let us use this assumption to count the vectors in $(\mathbb{Z}^+)^k$ whose component sum is at most n. These vectors can be of two possible kinds: those whose last (kth) component vanishes, and those whose last component doesn't vanish. The number of vectors in the first subset is the same as the number of vectors in the set considered in the induction hypothesis on k, namely,

$$\binom{n+k-1}{k-1}.$$

Furthermore, the number of vectors in the second subset can be calculated as follows. Each vector in this subset can be obtained by adding 1 to the last (kth) component in a unique corresponding vector from the set considered in the induction hypothesis in the induction step on n. Therefore, the number of vectors in the second subset is the same as the number of vectors in the set considered in the induction hypothesis on n, namely,

$$\binom{n-1+k}{k}.$$

Thus, the required total number is just the sum of these two numbers:

$$\binom{n+k-1}{k-1}+\binom{n-1+k}{k}=\frac{k}{n+k}\binom{n+k}{k}+\frac{n}{n+k}\binom{n+k}{k}=\binom{n+k}{k},$$

as asserted.

Note that the above induction step assumes that the assertion is true for a smaller k (with n being the same) and for a smaller n (with k being the same). In both cases, the sum $n+k$ is smaller. Thus, the above nested induction can actually be viewed as an induction on $n + k = 1, 2, 3, \ldots$. This way, the induction is carried out diagonal by diagonal in the number plane that contains pairs of the form (n, k) with $k \geq 1$ and $n \geq 0$.

With this approach, there is actually a more natural proof for the above result, which also uses a diagonal-by-diagonal induction rather than a nested induction. In this proof, the initial conditions are the same as before: for either $n = 0$ or $k = 1$, the assertion holds trivially as above. To show that the assertion holds in the entire n-k plane as well, we must prove the induction step on $n+k \geq 2$. For this, we may assume that the induction hypothesis holds, that is, that the assertion holds for the pairs $(n - 1, k)$ and $(n, k - 1)$, which belong to the previous (lower) diagonal in the n-k plane. Moreover, let us use the splitting of the original set of k-dimensional vectors whose component sum is at most n as the union of two disjoint subsets: the subset of vectors whose component sum is at most $n - 1$, and the subset of vectors whose component sum is exactly n:

$$\left\{ (v_1, v_2, \ldots, v_k) \in (\mathbb{Z}^+)^k \;\Big|\; \sum_{i=1}^{k} v_i \le n \right\}$$

$$= \left\{ (v_1, v_2, \ldots, v_k) \in (\mathbb{Z}^+)^k \;\Big|\; \sum_{i=1}^{k} v_i \le n - 1 \right\}$$

$$\cup \left\{ (v_1, v_2, \ldots, v_k) \in (\mathbb{Z}^+)^k \;\Big|\; \sum_{i=1}^{k} v_i = n \right\}.$$

Thanks to the induction hypothesis, the number of vectors in the first subset is

$$\binom{n - 1 + k}{k}.$$

Thus, all that is left to do is to count the vectors in the second subset. For this, observe that each vector in it can be transformed uniquely into a $(k-1)$-dimensional vector whose component sum is at most n by just dropping the kth component. In fact, this transformation is reversible, because this kth component can be added back in a unique way. As a result, we have

$$\left| \left\{ (v_1, v_2, \ldots, v_k) \in (\mathbb{Z}^+)^k \;\Big|\; \sum_{i=1}^{k} v_i \le n \right\} \right|$$

$$= \left| \left\{ (v_1, v_2, \ldots, v_k) \in (\mathbb{Z}^+)^k \;\Big|\; \sum_{i=1}^{k} v_i \le n - 1 \right\} \right|$$

$$+ \left| \left\{ (v_1, v_2, \ldots, v_k) \in (\mathbb{Z}^+)^k \;\Big|\; \sum_{i=1}^{k} v_i = n \right\} \right|$$

$$= \left| \left\{ (v_1, v_2, \ldots, v_k) \in (\mathbb{Z}^+)^k \;\Big|\; \sum_{i=1}^{k} v_i \le n - 1 \right\} \right|$$

$$+ \left| \left\{ (v_1, v_2, \ldots, v_{k-1}) \in (\mathbb{Z}^+)^{k-1} \;\Big|\; \sum_{i=1}^{k-1} v_i \le n \right\} \right|$$

$$= \binom{n - 1 + k}{k} + \binom{n + k - 1}{k - 1}$$

$$= \binom{n + k}{k},$$

as required.

From this proof, we also have as a by-product the formula

$$\left| \left\{ (v_1, v_2, \ldots, v_k) \in (\mathbb{Z}^+)^k \;\Big|\; \sum_{i=1}^{k} v_i = n \right\} \right| = \binom{n + k - 1}{k - 1}.$$

As a result, we also have

$$\binom{n+k}{k}$$

$$= \left| \left\{ (v_1, v_2, \ldots, v_k) \in (\mathbb{Z}^+)^k \ \Big| \ \sum_{i=1}^{k} v_i \le n \right\} \right|$$

$$= \left| \cup_{m=0}^{n} \left\{ (v_1, v_2, \ldots, v_k) \in (\mathbb{Z}^+)^k \ \Big| \ \sum_{i=1}^{k} v_i = m \right\} \right|$$

$$= \sum_{m=0}^{n} \left| \left\{ (v_1, v_2, \ldots, v_k) \in (\mathbb{Z}^+)^k \ \Big| \ \sum_{i=1}^{k} v_i = m \right\} \right|$$

$$= \sum_{m=0}^{n} \binom{m+k-1}{k-1}.$$

10.16 Mathematical Induction in Newton's Binomial

Newton's binomial formula

$$(a+b)^n = \sum_{k=0}^{n} \binom{n}{k} a^k b^{n-k}$$

can be proved most compactly by mathematical induction. Indeed, for $n = 0$, we trivially have

$$(a+b)^0 = 1 = \binom{0}{0} a^0 b^0 = \sum_{k=0}^{0} \binom{0}{k} a^k b^{0-k} = \sum_{k=0}^{n} \binom{n}{k} a^k b^{n-k},$$

as required. Furthermore, for $n \ge 1$, let us assume that the induction hypothesis holds, that is,

$$(a+b)^{n-1} = \sum_{k=0}^{n-1} \binom{n-1}{k} a^k b^{n-1-k}.$$

Using this hypothesis, we have

$$(a+b)^n = (a+b)(a+b)^{n-1}$$

$$= (a+b) \sum_{k=0}^{n-1} \binom{n-1}{k} a^k b^{n-1-k}$$

$$= a \sum_{k=0}^{n-1} \binom{n-1}{k} a^k b^{n-1-k} + b \sum_{k=0}^{n-1} \binom{n-1}{k} a^k b^{n-1-k}$$

$$= \sum_{k=0}^{n-1} \binom{n-1}{k} a^{k+1} b^{n-(k+1)} + \sum_{k=0}^{n-1} \binom{n-1}{k} a^k b^{n-k}$$

$$= \sum_{k=1}^{n} \binom{n-1}{k-1} a^k b^{n-k} + \sum_{k=0}^{n-1} \binom{n-1}{k} a^k b^{n-k}$$

$$= a^n + \sum_{k=1}^{n-1} \binom{n-1}{k-1} a^k b^{n-k} + \sum_{k=1}^{n-1} \binom{n-1}{k} a^k b^{n-k} + b^n$$

$$= a^n + \sum_{k=1}^{n-1} \left(\binom{n-1}{k-1} + \binom{n-1}{k} \right) a^k b^{n-k} + b^n$$

$$= a^n + \sum_{k=1}^{n-1} \binom{n}{k} a^k b^{n-k} + b^n$$

$$= \sum_{k=0}^{n} \binom{n}{k} a^k b^{n-k},$$

as required.

10.17 Factorial of a Sum

The above method of proof can also be used to prove another interesting formula. For this, however, we need some more notations.

For any number a and a nonnegative number n, define

$$C_{a,n} \equiv a(a-1)(a-2)\cdots(a-(n-1)) = \begin{cases} 1 & \text{if } n = 0 \\ aC_{a-1,n-1} & \text{if } n \geq 1. \end{cases}$$

In particular, if a is a nonnegative integer number, then

$$C_{a,n} = \begin{cases} \frac{a!}{(a-n)!} & \text{if } a \geq n \\ 0 & \text{if } a < n. \end{cases}$$

With this notation, let us prove the formula

$$C_{a+b,n} = \sum_{k=0}^{n} \binom{n}{k} C_{a,k} C_{b,n-k}.$$

Indeed, the formula is clearly true for $n = 0$. Let us use mathematical induction to show that it is true for every $n \geq 1$ as well. For this, let us assume the induction hypothesis, that is,

$$C_{a+b,n-1} = \sum_{k=0}^{n-1} \binom{n-1}{k} C_{a,k} C_{b,n-1-k}.$$

Indeed, from this formula we have

$$
\begin{aligned}
C_{a+b,n} &= (a+b)C_{a+b-1,n-1} \\
&= aC_{(a-1)+b,n-1} + bC_{a+(b-1),n-1} \\
&= a \sum_{k=0}^{n-1} \binom{n-1}{k} C_{a-1,k} C_{b,n-1-k} + b \sum_{k=0}^{n-1} \binom{n-1}{k} C_{a,k} C_{b-1,n-1-k} \\
&= \sum_{k=0}^{n-1} \binom{n-1}{k} C_{a,k+1} C_{b,n-(k+1)} + \sum_{k=0}^{n-1} \binom{n-1}{k} C_{a,k} C_{b,n-k} \\
&= \sum_{k=1}^{n} \binom{n-1}{k-1} C_{a,k} C_{b,n-k} + \sum_{k=0}^{n-1} \binom{n-1}{k} C_{a,k} C_{b,n-k} \\
&= C_{a,n} + \sum_{k=1}^{n-1} \binom{n-1}{k-1} C_{a,k} C_{b,n-k} + \sum_{k=1}^{n-1} \binom{n-1}{k} C_{a,k} C_{b,n-k} + C_{b,n} \\
&= C_{a,n} + \sum_{k=1}^{n-1} \left(\binom{n-1}{k-1} + \binom{n-1}{k} \right) C_{a,k} C_{b,n-k} + C_{b,n} \\
&= C_{a,n} + \sum_{k=1}^{n-1} \binom{n}{k} C_{a,k} C_{b,n-k} + C_{b,n} \\
&= \sum_{k=0}^{n} \binom{n}{k} C_{a,k} C_{b,n-k},
\end{aligned}
$$

as required.

When both a and b are nonnegative integers satisfying $a + b \geq n$, we have the special case

$$\frac{(a+b)!}{(a+b-n)!} = \sum_{k=\max(0,n-b)}^{\min(a,n)} \binom{n}{k} \frac{a!}{(a-k)!} \cdot \frac{b!}{(b-(n-k))!}.$$

10.18 The Trinomial Formula

By applying the binomial formula twice, we obtain the following trinomial formula:

$$
\begin{aligned}
(a+b+c)^n &= ((a+b)+c)^n \\
&= \sum_{k=0}^{n} \binom{n}{k} (a+b)^k c^{n-k} \\
&= \sum_{k=0}^{n} \binom{n}{k} \left(\sum_{l=0}^{k} \binom{k}{l} a^l b^{k-l} \right) c^{n-k} \\
&= \sum_{k=0}^{n} \sum_{l=0}^{k} \binom{n}{k} \binom{k}{l} a^l b^{k-l} c^{n-k} \\
&= \sum_{k=0}^{n} \sum_{l=0}^{k} \frac{n!}{k!(n-k)!} \cdot \frac{k!}{l!(k-l)!} a^l b^{k-l} c^{n-k} \\
&= \sum_{k=0}^{n} \sum_{l=0}^{k} \frac{n!}{l!(k-l)!(n-k)!} a^l b^{k-l} c^{n-k} \\
&= \sum_{0 \le l,j,m \le n,\ l+j+m=n} \frac{n!}{l!j!m!} a^l b^j c^m.
\end{aligned}
$$

(In the last equality, we have substituted j for $k-l$ and m for $n-k$.)

Similarly, by applying the formula in Section 10.17 twice, we have the following formula:

$$C_{a+b+c,n} = C_{(a+b)+c,n}$$

$$= \sum_{k=0}^{n} \binom{n}{k} C_{a+b,k} C_{c,n-k}$$

$$= \sum_{k=0}^{n} \binom{n}{k} \left(\sum_{l=0}^{k} \binom{k}{l} C_{a,l} C_{b,k-l} \right) C_{c,n-k}$$

$$= \sum_{k=0}^{n} \sum_{l=0}^{k} \binom{n}{k} \binom{k}{l} C_{a,l} C_{b,k-l} C_{c,n-k}$$

$$= \sum_{k=0}^{n} \sum_{l=0}^{k} \frac{n!}{k!(n-k)!} \cdot \frac{k!}{l!(k-l)!} C_{a,l} C_{b,k-l} C_{c,n-k}$$

$$= \sum_{k=0}^{n} \sum_{l=0}^{k} \frac{n!}{l!(k-l)!(n-k)!} C_{a,l} C_{b,k-l} C_{c,n-k}$$

$$= \sum_{0 \le l,j,m \le n,\ l+j+m=n} \frac{n!}{l!j!m!} C_{a,l} C_{b,j} C_{c,m}.$$

These formulas will be useful later on in the book.

10.19 Multiscale

The multilevel objects considered so far are composite objects defined inductively level by level, with levels that contain more and more entries (or scalars, or numbers). Furthermore, the principle of multilevel can be used not only in composite objects but also in more elementary objects such as the scalars themselves. In this context, however, a more suitable name for the multilevel structure is multiscale.

The notion of multiscale is used in every kind of measuring. Indeed, small units such as centimeters, grams, and seconds are suitable to measure small quantities of distance, weight, and time, whereas large units such as meter, kilogram, and hour are suitable to measure large quantities. An accurate measure must often combine (or sum up) the large units that measure the core of the quantity (coarse-scale measuring) with the small units that measure the remainder of the quantity (fine-scale measuring) to yield an accurate measurement of the entire quantity.

For example, one may say that the distance between two points is 17.63 meters (17 meters plus 63 centimeters), the weight of some object is 8.130 kilograms (8 kilograms plus 130 grams), and the time span is 5 : 31 : 20 hours (5 hours, 31 minutes, and 20 seconds). In the first two examples, large-scale units (meters and kilograms, respectively) are combined with small-scale units (centimeters and grams, respectively). In the third example, on the other hand, large-scale units (hours), intermediate-scale units (minutes), and small-scale units (second) are all combined to come up with an accurate multiscale measurement of time.

10.20 The Decimal Representation

The decimal representation of natural numbers can also be viewed as a multiscale representation. For example,

$$178 = 100 + 70 + 8 = 1 \cdot 100 + 7 \cdot 10 + 8 \cdot 1$$

may actually be viewed as the combination (or sum) of large-scale, intermediate-scale, and small-scale units: the largest (coarsest) unit is the hundreds (multiples of 100), the intermediate unit is the tens (multiples of 10), and the smallest (finest) unit is the unit digit (multiples of 1).

The decimal representation of fractions uses yet finer and finer scales: multiples of 10^{-1}, 10^{-2}, 10^{-3}, and so on, yielding better and better accuracy.

10.21 The Binary Representation

Similarly, the binary representation of a number can also be viewed as a multiscale representation. Here, however, the powers of 10 used in the decimal representation above are replaced with powers of 2. For example, the number 1101.01 in base 2 is interpreted as

$$1 \cdot 2^3 + 1 \cdot 2^2 + 0 \cdot 2^1 + 1 \cdot 2^0 + 0 \cdot 2^{-1} + 1 \cdot 2^{-2}.$$

This representation combines (sums) six different scales, or powers of 2: from 2^3 (the largest or coarsest scale) to 2^{-2} (the smallest or finest scale). The coefficients of these powers of 2 are the digits in the base-2 representation 1101.01.

10.22 The Sine Transform

In Chapter 9 (Section 9.21) above, we've introduced the Sine transform, which is the $n \times n$ matrix

$$A \equiv \left(v^{(1)} \mid v^{(2)} \mid v^{(3)} \mid \cdots \mid v^{(n)} \right),$$

where, for $1 \leq j \leq n$, the column vector $v^{(j)}$ is the n-dimensional vector

$$v^{(j)} \equiv \sqrt{\frac{2}{n}} \begin{pmatrix} \sin(j\pi/(n+1)) \\ \sin(2j\pi/(n+1)) \\ \sin(3j\pi/(n+1)) \\ \vdots \\ \sin(nj\pi/(n+1)) \end{pmatrix}.$$

The vector $v^{(j)}$ can be viewed as a sample of the function $\sqrt{2/n}\sin(j\pi x)$ in the uniform grid consisting of the points

$$x = 1/(n+1),\ 2/(n+1),\ 3/(n+1),\ \ldots,\ n/(n+1)$$

in the unit interval $0 < x < 1$. The function $\sqrt{2/n}\sin(j\pi x)$ is rather smooth for $j = 1$, and oscillates more and more frequently as j increases. This is why the number j is called the wave number or the wave frequency.

We have also proved in Chapter 9, Section 9.21, that A is both symmetric and orthogonal, so

$$AA = A^t A = I$$

(the identity matrix of order n). Thus, every n-dimensional vector u can be decomposed as a combination (sum) of the $v^{(j)}$'s, with coefficients that are the components of Au:

$$u = Iu = (AA)u = A(Au) = \sum_{j=1}^{n} (Au)_j v^{(j)}.$$

In other words, u is decomposed as the combination of uniform samples of more and more oscillatory functions of the form $\sin(j\pi x)$, multiplied by the corresponding coefficient

$$(Au)_j \sqrt{2/n}$$

(the amplitude of the jth wave in the decomposition of u). In particular, the smoothest part of u is the multiple of the first discrete wave

$$(Au)_1 v^{(1)},$$

the next more oscillatory part is

$$(Au)_2 v^{(2)},$$

and so on, until the most oscillatory part

$$(Au)_n v^{(n)}.$$

This decomposition of the original n-dimensional vector u can actually be interpreted as a multiscale decomposition. In fact, the first discrete wave $v^{(1)}$ can be viewed as the coarsest scale, in which the original vector u is approximated roughly. The second discrete wave, $v^{(2)}$, approximates the remainder

$$u - (Au)_1 v^{(1)}$$

roughly, contributing a finer term to a better approximation of the original vector u. The next discrete wave, $v^{(3)}$, approximates the remainder

$$u - (Au)_1 v^{(1)} - (Au)_2 v^{(2)}$$

roughly, providing a yet finer term to the yet better approximation for the original vector u. This process continues until the nth discrete wave, $v^{(n)}$, contributes the finest term to yield the exact decomposition of u in terms of the finer and finer scales obtained from the more and more frequent waves $v^{(1)}$, $v^{(2)}$, $v^{(3)}$, ..., $v^{(n)}$.

10.23 Exercises

1. Use mathematical induction on $n \geq 0$ to show that the number of nodes in the nth level in a binary tree ($n \geq 0$) is 2^n. (Assume that the head of the tree lies in the 0th level.)
2. Use mathematical induction on $n \geq 0$ to show that the total number of nodes in an n-level binary tree is $2^{n+1} - 1$.
3. Define the set of n-dimensional binary vectors by

$$V \equiv V(n) \equiv \{0, 1\}^{\{1,2,...,n\}}$$
$$= \{(v_1, v_2, \ldots, v_n) \mid v_i = 0 \text{ or } v_i = 1, \ 1 \leq i \leq n\}.$$

   Use mathematical induction on $n \geq 1$ to show that the number of distinct vectors in V is

$$|V| = 2^n.$$

4. Let $V_k \equiv V_k(n)$ be the subset of V that contains vectors with exactly k nonzero components:

$$V_k \equiv \{v \in V \mid \|v\|_2^2 = k\}.$$

   Use mathematical induction on $k = 0, 1, 2, \ldots, n$ to show that the number of distinct vectors in V_k is

$$|V_k| = \binom{n}{k}.$$

5. Show that, for two different numbers $k \neq l$, V_k and V_l are disjoint sets of vectors:

$$V_k \cap V_l = \Phi.$$

6. Show that V can be written as the union of disjoint sets

$$V = \cup_{k=0}^{n} V_k.$$

7. Conclude (once again) that the number of distinct vectors in V is

$$|V| = \sum_{k=0}^{n} |V_k| = \sum_{k=0}^{n} \binom{n}{k} = 2^n.$$

8. Show that

$$\binom{n}{0} = \binom{n}{n} = 1.$$

9. Show that for any natural numbers n and k satisfying $0 < k < n$,

$$\binom{n}{k} = \binom{n-1}{k} + \binom{n-1}{k-1}.$$

10. Conclude that the number of paths leading from the head of Pascal's triangle (in its 0th level) to the kth entry in its nth level ($0 \le k \le n$) is the Newton binomial

$$\binom{n}{k}.$$

11. Why is the total number of distinct paths in the previous exercise the same as the total number of vectors in V_k above?

12. Define an invertible mapping that identifies each path leading from the head of Pascal's triangle to its nth level with a particular vector in V_k.

13. Show in a yet different way, this time using mathematical induction on n rather than on k, that the number of distinct vectors in V_k is indeed

$$|V_k| = \binom{n}{k}.$$

14. Why is the total number of vectors in V_k the same as the total number of possible choices to pick k a's and $n - k$ b's to form a product of the form $a^k b^{n-k}$ in Newton's binomial formula to open the parentheses in $(a+b)^n$?

15. Define an invertible mapping that identifies each particular choice of k a's and $n - k$ b's in the previous exercise with a particular vector in V_k.

16. Write the algorithm that computes an arbitrarily long arithmetic expression. The solution can be found in Chapter 14, Section 14.8.

17. Write the algorithm that transforms the decimal representation of a natural number to its binary representation. The solution can be found in Chapter 14, Section 14.6.

Chapter 11

Graphs

The tree object introduced above is defined recursively: a node is placed at the top of the tree, and branches (edges) are issued from it. Then, a subtree is placed at the end of each branch. These subtrees are defined recursively by the same definition.

Here we introduce a more general object: the graph [13] [14] [34]. Unlike the tree, the graph not necessarily has a recursive structure. In fact, the graph consists of two sets: N, the set of the nodes used in the graph, and E, the set of edges that connect nodes in the graph.

To show how graphs are handled, we consider the node-coloring problem. In this problem, the nodes in the graph should be colored in such a way that every two nodes that are connected by an edge have distinct colors. The challenge is to complete this task using only a small number of colors.

Furthermore, we also consider the edge-coloring problem, in which the edges in E should be colored in such a way that every two edges that share a node have distinct colors.

Finally, we consider a special kind of graphs: triangulations or meshes of triangles. In this case, we also consider the triangle-coloring problem, in which the triangles should be colored in such a way that every two adjacent triangles have distinct colors.

11.1 Oriented Graphs

A graph consists of two sets: N, the set of nodes used in the graph, and E, the set of edges that may connect two nodes to each other. In the following, we describe the nature of the nodes and the edges.

The nodes in N may have some geometrical interpretation. For example, they may be 2-D points in the Cartesian plane (vectors in $\mathbb{R}^2$) or 3-D points in the Cartesian space (vectors in $\mathbb{R}^3$). This interpretation, however, is completely immaterial in the present discussion. Indeed, here the nodes are considered as mere abstract elements in N that may be connected to each other by edges in E. This abstract formulation allows one to deal with general graphs, independent of any geometrical interpretation they may have.

An edge in E may be viewed as an ordered pair (2-D vector) of nodes in N. For example, if i and j are two nodes in N, then the edge that leads from i to j may be denoted by the ordered pair (i, j).

Recall that the set of all such pairs is denoted by

$$N^2 = N \times N = \{(i, j) \mid i, j \in N\}.$$

Thus, the set of edges is actually a subset of N^2:

$$E \subset N^2.$$

The order of the components i and j in the edge $(i, j) \in E$ reflects the fact that the edge leads from i to j rather than from j to i. This is why general graphs are also called oriented graphs: each edge in E not only connects two nodes in N to each other, but also does it in a specific order: it leads from the node in its first component to the node in its second component. An oriented graph is illustrated in Figure 11.1.

Below we also consider nonoriented graphs, in which the edges have no direction. In other words, the edges can be viewed as unordered pairs rather than ordered pairs of nodes.

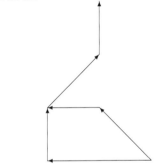

FIGURE 11.1:   An oriented graph.

11.2 Nonoriented Graphs

In the oriented graphs introduced above, the edges in E are ordered pairs of the form (i, j), where i and j are some nodes in N. The fact that i comes before j in (i, j) means that the edge (i, j) leads from i to j rather than from j to i.

A nonoriented graph, on the other hand, consists of the following two sets: N, the set of nodes, and E, the set of unordered pairs of the form $\{i, j\}$, where i and j are some nodes in N. In an edge of this form, there is no order:

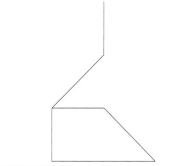

FIGURE 11.2: A nonoriented graph.

both of the nodes i and j have equal status in it. Thus, the edge neither leads from i to j nor leads from j to i; it merely connects i and j to each other. A nonoriented graph is illustrated in Figure 11.2.

The problems and algorithms described below are written in uniform terms, so they apply to nonoriented as well as oriented graphs.

11.3 The Node-Coloring Problem

In the node-coloring problem, one has to color the nodes in N in such a way that every two nodes that are connected by an edge in E are colored by distinct colors. More precisely, a color can be viewed as a natural number assigned to certain nodes in N. For example, if the graph is colored by C colors (for some natural number C), then the number 1 is assigned to the nodes that are colored by the first color, the number 2 is assigned to the nodes that are colored by the second color, and so on, until the number C, which is assigned to the remaining nodes in N that are still uncolored.

This coloring may be viewed as a function $c : N \to \mathbb{N}$, in which $c(i)$ denotes the index of the color by which the node i is colored, and

$$C \equiv \max(\{c(i) \mid i \in N\})$$

is the number of colors used in the coloring.

The challenge is to find a coloring with C as small as possible. Unfortunately, finding the optimal coloring in this sense is too difficult. (In fact, the coloring problem belongs to the class of the so-called NP-complete-problems, which can probably be solved only in exponentially long time.)

Here we present an algorithm to obtain a rather good coloring, for which the number of colors C is not too large.

11.4 The Node-Coloring Algorithm

The node-coloring algorithm is defined by mathematical induction on $|N|$, the number of nodes in N. Clearly, when $|N| = 1$, the graph contains only one node, so the color assigned to it must be the first color, denoted by 1, so the total number of colors is $C = 1$.

Let us now define the induction step. Let $n \in N$ be some node in the graph. Define the set of edges that use the node n by

$$edges(n) \equiv \{(i,j) \in E \mid i = n \text{ or } j = n\}.$$

Furthermore, define the set of nodes that are connected to n by an edge by

$$neighbors(n) \equiv \{i \in N \mid (i,n) \in E \text{ or } (n,i) \in E\}.$$

Moreover, for any set of nodes $N' \subset N$, define the set of colors used in it by

$$c(N') \equiv \{c(i) \mid i \in N'\}.$$

In particular, $c(neighbors(n))$ contains the colors used in the nodes that are connected to n by an edge, or the natural numbers assigned to them.

Now, to make the induction step, we consider the slightly smaller graph, from which n and the edges issued from or directed to it are excluded. In other words, this graph has the slightly smaller set of nodes $N \setminus \{n\}$ and the slightly smaller set of edges $E \setminus edges(n)$. By the induction hypothesis, we may assume that this smaller graph has already been colored. Thus, all that is left to do is to color n as well, or to define $c(n)$ properly.

This is done as follows. The remaining node n is colored by a color that has not been used in any of the nodes that are connected to n by an edge:

$$c(n) \equiv \min \left(\mathbb{N} \setminus c(neighbors(n)) \right).$$

This completes the induction step. This completes the definition of the node-coloring algorithm.

The above algorithm is illustrated in Figure 11.3. In this example, two colors are sufficient to color all the nodes in the graph.

It is easy to prove (by mathematical induction) that the coloring produced by the above algorithm indeed satisfies the requirement that every two nodes that are connected by an edge are colored by distinct colors, or

$$(i,j) \in E \implies c(i) \neq c(j).$$

Nevertheless, the above coloring depends on the choice of the particular node $n \in N$ picked to perform the induction step. In other words, it depends on the particular order in which the nodes in N are colored. To have the optimal

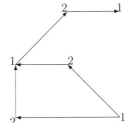

FIGURE 11.3:   The node-coloring algorithm uses only two colors to color a graph with six nodes.

coloring (the coloring with the minimal number of colors C) one needs to repeat the above algorithm for every possible order of nodes in N and choose the particular order that produces the optimal coloring. Because the total number of possible orders is $|N|!$, this task is prohibitively time consuming.

Still, one may be rather happy with the coloring produced by the above node-coloring algorithm, because it uses a moderate number of colors C. Below we use a similar algorithm to color the edges in the graph.

11.5 The Edge-Coloring Problem

In the previous sections, we have discussed the problem of coloring the nodes in N. Here we turn to the problem of coloring the edges in E, or assigning a natural number $c(e)$ to each and every edge $e \in E$ in such a way that two adjacent (node sharing) edges are colored by distinct colors:

$$c((i,j)) \neq c((k,l)) \text{ if } i = k \text{ or } j = l \text{ or } i = l \text{ or } j = k.$$

As before, the challenge is to find a coloring for which the total number of colors, defined by

$$C \equiv \max \{c(e) \mid e \in E\},$$

is as small as possible. Again, it is too difficult to find the optimal coloring in this sense; still, the algorithm presented below produces an edge-coloring with a moderate number of colors C.

11.6 The Edge-Coloring Algorithm

Like the node-coloring algorithm, the edge-coloring algorithm is also defined by mathematical induction on $|N|$, the number of nodes in N. Clearly, when N contains only one node, E can contain at most one edge: the edge that leads from this node to itself. Thus, the edge coloring requires one color only, so $C = 1$.

Furthermore, to define the induction step, let us apply the induction hypothesis to a slightly smaller graph. For this purpose, let us pick some node $n \in N$. Now, consider the graph obtained from excluding the node n and the edges that are issued from it or lead to it, namely, the graph whose set of nodes is $N \setminus \{n\}$ and whose set of edges is $E \setminus edges(n)$. By the induction hypothesis, one may assume that the edges in this smaller graph (namely, the edges in $E \setminus edges(n)$) have already been colored properly by the edge-coloring algorithm. Thus, all that is left to do is to color properly also the edges in $edges(n)$.

This task is completed as follows.

1. Initially, assign to every edge $e \in edges(n)$ the zero color:

$$c(e) \equiv 0$$

   for every $e \in edges(n)$.
2. Then, scan the edges in $edges(n)$ one by one. For each edge $e \in edges(n)$ encountered, e must be either of the form $e = (i, n)$ or of the form $e = (n, i)$ for some node $i \in neighbors(n)$. Therefore, e is assigned its final color to be a color that has not been used in any edge issued from or directed to either n or i:

$$c(e) \leftarrow \min\left(\mathbb{N} \setminus c(edges(n)) \setminus c(edges(i))\right),$$

   where '$\leftarrow$' stands for substitution,

$$c(edges(n)) \equiv \{c(e) \mid e \in edges(n)\},$$

   and

$$c(edges(i)) \equiv \{c(e) \mid e \in edges(i)\}.$$

This way, the final color assigned to e is different from the colors assigned previously to the edges that are issued from or lead to either of its endpoints n or i. This completes the induction step. This completes the definition of the edge-coloring algorithm.

It is easy to prove by mathematical induction on the number of nodes in N [and, within it, by an inner mathematical induction on the number of edges in $edges(n)$] that the coloring produced by the above edge-coloring algorithm indeed satisfies the requirement that every two adjacent (node sharing) edges are indeed colored by different colors.

Like the node-coloring algorithm, the edge-coloring algorithm also depends on the particular node $n \in N$ picked in the induction step. In other words, it depends on the order of nodes in N. Furthermore, it also depends on the order in which the edges in $edges(n)$ are scanned. Thus, there is no guarantee that the algorithm would produce the optimal coloring that uses the minimal number of colors C. Still, the above edge-coloring algorithm produces a fairly good coloring with a rather small number of colors C. This is illustrated in Figure 11.4, where three colors are used to color a graph with six edges.

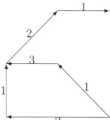

FIGURE 11.4:   The edge-coloring algorithm uses only three colors to color a graph with six edges. It is assumed that the nodes are ordered counter-clockwise in the algorithm.

Below we show that the edge-coloring algorithm may actually be viewed as a special case of the node-coloring algorithm applied to a new graph.

11.7 Graph of Edges

We say that two edges in E are adjacent to each other if they share the same node as their joint endpoint. Let E' be the set of unordered pairs of adjacent edges:

$$E' \equiv \left\{ \{(i,j),(k,l)\} \ \middle| \ \begin{array}{l} (i,j) \in E, \ (k,l) \in E, \ \text{and} \\ \text{either } i = k \text{ or } j = l \text{ or } i = l \text{ or } j = k \end{array} \right\}.$$

This definition helps to form a new (nonoriented) graph, whose nodes are the elements in E, and whose edges are the unordered pairs in E'. In other words, each edge in the original graph serves as a node in the new graph, and each two adjacent edges in the original graph are connected by an edge in the new graph.

Now, the edge-coloring problem in the original graph is equivalent to the node-coloring problem in the new graph. Therefore, the node-coloring algorithm could actually be applied to the new graph to provide an edge-coloring algorithm for the original graph.

As a matter of fact, the edge-coloring algorithm described in the previous section can also be viewed as a node-coloring algorithm in the new graph (the graph of edges). To see this, note that the edges in the edge-coloring algorithm are colored one by one in a particular order. This order can be defined by mathematical induction on $|N|$: assuming that the induction hypothesis holds, that is, that the order has already been defined on the edges in $E \setminus edges(n)$, it is also defined somehow on the edges in $edges(n)$ (the induction step). With this order, applying the node-coloring algorithm to the new graph is the same as applying the edge-coloring algorithm to the original graph.

This point of view is used below to color the triangles in the special graph known as triangulation.

11.8 Triangulation

triangulation = conformal mesh of triangles

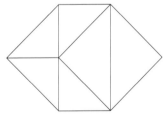

FIGURE 11.5:   A triangulation, or a conformal mesh of triangles.

A triangulation is a special kind of graph, which looks like a mesh of triangles (Figure 11.5). Below we present the axioms that produce this kind of mathematical object. For this, however, we need some preliminary definitions.

A circle in a graph is a list of k edges in E of the form

$$(i_1, i_2), \ (i_2, i_3), \ (i_3, i_4), \ \ldots, \ (i_{k-1}, i_k), \ (i_k, i_1),$$

where k is some natural number, and i_1, i_2, ..., i_k are some k nodes in N. In particular, if $k = 3$, then this circle is called a triangle.

A subcircle of the above circle is a circle based on a subset of the set of k nodes $\{i_1, i_2, \ldots, i_k\}$. For example, if (i_2, i_4), (i_4, i_7), and (i_7, i_2) are edges in E, then they form a subcircle of the above circle. Because it contains three edges, this particular subcircle is also a triangle.

Let us now use these definitions to define a triangulation. A triangulation is a graph with the following properties:

1. The graph is nonoriented.
2. The graph contains no edges of the form (i, i).
3. Each node in N is shared by at least two different edges in E as their joint endpoint. (This guarantees that there are no dangling nodes or edges.)
4. The graph is embedded in the Cartesian plane in the sense that each node $i \in N$ is interpreted geometrically as a point $(x(i), y(i)) \in \mathbb{R}^2$, and each edge of the form $(i, j) \in E$ is interpreted as the line segment leading from $(x(i), y(i))$ to $(x(j), y(j))$.
5. With the above geometric interpretation, two edges in E cannot cross each other in $\mathbb{R}^2$. (This means that the graph is a planar graph.)
6. Each circle of $k > 3$ edges must contain a triangle as a subcircle. (This guarantees that the graph is indeed a mesh of triangles.)
7. In the above geometrical interpretation in $\mathbb{R}^2$, each node that lies on an edge must also serve as one of its endpoints. (This guarantees that the triangulation is conformal in the sense that two triangles that share an edge must share it in its entirety and must also share its endpoints as their joint vertices.)

This formulation illustrates clearly how a mathematical object is created in a purely abstract way. Indeed, with the above mathematical axioms, the abstract notion of a graph as a set of nodes and a set of edges takes the much better visualized form of a conformal mesh of triangles. Below we use this form to color the triangle in the triangulation.

11.9 The Triangle-Coloring Problem

In the triangle-coloring problem, one has to assign a color (or a natural number) to each triangle in the triangulation in such a way that two adjacent triangles (triangles that share an edge as their joint side and also share its endpoints as their joint vertices) have distinct colors. As before, a good coloring uses a moderate number of colors.

Below we introduce an algorithm to have a good triangle coloring. This is done by introducing a new graph, in which the original triangle-coloring problem takes the form of a node-coloring problem.

Indeed, the triangles in the triangulation may also serve as nodes in a new nonoriented graph. In this new graph, two triangles are connected by an edge if they are adjacent in the original triangulation, that is, if they share an edge as their joint side and also share its endpoints as their joint vertices.

Clearly, the node-coloring algorithm applied to this new graph yields a triangle-coloring algorithm for the original triangulation. Because each triangle has at most three neighbors, the resulting triangle coloring uses at most four colors, which is nearly optimal. For example, in Figure 11.6 three colors are used to color a triangulation with six triangles.

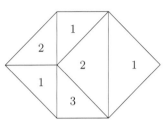

FIGURE 11.6:   The triangle-coloring algorithm uses three colors to color a triangulation with six triangles. It is assumed that the triangles are ordered counter-clockwise in the algorithm.

11.10 Weighted Graphs

In a weighted graph, a positive number (or weight) is assigned to each edge in E. The weight assigned to an edge of the form $(j, i) \in E$ is denoted by $a_{i,j}$. For completeness, we also define $a_{i,j} \equiv 0$ whenever $(j, i) \notin E$.

The sum of the weights associated with edges that are issued from a particular node $j \in N$ is equal to 1:

$$\sum_{i=1}^{|N|} a_{i,j} = 1.$$

Thanks to this formula, $a_{i,j}$ can be interpreted as the probability that a particle that initially lies at node j would use the edge leading from j to i to move to node i. Below we give an example in which this probability is uniform.

In this example,

$$a_{i,j} \equiv \begin{cases} \frac{1}{|outgoing(j)|} & \text{if } (j, i) \in E \\ 0 & \text{if } (j, i) \notin E, \end{cases}$$

where $outgoing(j)$ is the set of edges issued from j:

$$outgoing(j) \equiv \{(j, i) \in E \mid i \in N\},$$

and $|outgoing(j)|$ is its cardinality (number of edges in it). Indeed, these weights sum to 1:

$$\sum_{i=1}^{|N|} a_{i,j} = \sum_{\{i \in N \;|\; (j,i) \in E\}} \frac{1}{|outgoing(j)|} = |outgoing(j)| \frac{1}{|outgoing(j)|} = 1.$$

The above weights allow one to define a flow in the graph, performed step by step. To this end, assume that the nodes in N are also assigned nonnegative numbers, called masses. However, unlike the weights assigned to the edges, these masses may change step by step.

For example, assume that the node $j \in N$ is assigned initially the mass $u_j \geq 0$. Then, in the first step, this mass is transferred through edges of the form (j, i) to their endpoints i. The node i at the end of such an edge receives from j the mass $a_{i,j} u_j$. This way, the mass u_j that was originally concentrated at the node j has been divided among the nodes i that serve as endpoints of edges of the form (j, i).

This flow preserves the original mass u_j. Indeed, the total amount of mass transferred to the nodes of the form i is equal to the original mass concentrated at j:

$$\sum_{i=1}^{|N|} a_{i,j} u_j = u_j \sum_{i=1}^{|N|} a_{i,j} = u_j.$$

Clearly, after the above step, the node j contains no mass any more, unless there exists a reflexive edge of the form (j, j) in E, in which case j also receives the mass $a_{j,j} u_j$ after the above step.

The above procedure may repeat: in the next step, each mass received by node i flows further to the endpoints of the edges issued from i. Thanks to the property of mass preservation discussed above, the total mass in the graph (the sum of the masses at all the nodes in N) remains u_j after any number of steps.

The mass distribution among the nodes in N is called a state. For example, in the initial state above, the mass is concentrated at node j (in the amount u_j), and vanishes at all the other nodes in N. In the state obtained from this state after one step, on the other hand, the mass is no longer concentrated at one node only: it spreads to all the nodes i that lie at the end of an edge of the form (j, i) (with the amount $a_{i,j} u_j$).

In a more general state, each and every node $j \in N$ may contain mass in the amount u_j. After one step, each such node contributes the amount $a_{i,j} u_j$ to every node i that lies at the end of an edge of the form (j, i). Therefore, in the next state, the node i will have mass in the total amount contributed to it from all the nodes j, namely,

$$\sum_{j=1}^{|N|} a_{i,j} u_j.$$

The total amount of mass in this state is still the same as in the previous state. Indeed, the total mass in this state is

$$\sum_{i=1}^{|N|}\sum_{j=1}^{|N|} a_{i,j}u_j = \sum_{j=1}^{|N|}\sum_{i=1}^{|N|} a_{i,j}u_j = \sum_{j=1}^{|N|} u_j \sum_{i=1}^{|N|} a_{i,j} = \sum_{j=1}^{|N|} u_j,$$

which is the total mass in the previous state. Thus, preservation of mass holds for a general state as well.

Below we describe the flow in the graph in algebraic terms, using matrices and vectors.

11.11 Algebraic Formulation

Assume that the nodes in N are numbered by the natural numbers

$$1, 2, 3, \ldots, |N|.$$

With this numbering, the weights $a_{i,j}$ in the weighted graph form the $|N| \times |N|$ matrix

$$A \equiv (a_{i,j})_{1 \le i,j \le |N|}.$$

Furthermore, the masses u_j in the general state discussed above form the $|N|$-dimensional vector u, whose jth component, u_j, is the mass located at the node $j \in N$. With these notations, the state obtained from u after one step of the flow is represented by the $|N|$-dimensional vector Au. Indeed, the ith component in this vector is

$$(Au)_i = \sum_{j=1}^{|N|} a_{i,j}u_j,$$

which is indeed the mass at node i after one step.

Note that the columns of A sum to 1:

$$\sum_{i=1}^{|N|} a_{i,j} = 1, \ 1 \le j \le |N|.$$

This implies that the rows of the transpose matrix, A^t, also sum to 1. In other words, the $|N|$-dimensional vector w whose all components are equal to 1 ($w_j = 1$, $1 \le j \le |N|$) is an eigenvector of A^t corresponding to the eigenvalue 1:

$$A^t w = w.$$

As a consequence, 1 is also an eigenvalue of the original matrix A.

Below we also discuss the eigenvector of A corresponding to this eigenvalue. This eigenvector represents the steady state of the flow.

11.12 The Steady State

Here we assume that the weighted graph has no invariant subset of nodes (subset of N out of which no mass can flow away). Under this assumption, we discuss the steady state of the flow.

The steady state of the flow is a state that is represented by an eigenvector v of the matrix A that corresponds to the eigenvalue 1:

$$Av = v.$$

Clearly, the vector v is defined only up to multiplication by a scalar. Indeed, for any nonzero number α, αv represents a steady state as well:

$$A(\alpha v) = \alpha A v = \alpha v.$$

From Peron-Frobenius theory [31], it follows that all the components of v are positive:

$$v_j > 0, \quad 1 \leq j \leq |N|,$$

and that 1 is the largest eigenvalue of A: every other eigenvalue λ satisfies

$$|\lambda| < 1.$$

Let w be an $|N|$-dimensional vector that is the sum of the (pseudo-) eigenvectors of A corresponding to eigenvalues smaller than 1 in magnitude. Because the eigenvector v does not participate in this sum, w satisfies

$$\|A^k w\| \rightarrow_{k \to \infty} 0.$$

Consider now an arbitrary initial state represented by the nonzero $|N|$-dimensional vector u. Let us describe how, after sufficiently many steps, the process converges to a steady state. Indeed, u can be written as

$$u = \alpha v + (u - \alpha v),$$

where $u - \alpha v$ can be written as the sum of the (pseudo-) eigenvectors of A corresponding to eigenvectors smaller than 1 in magnitude. Now, the next state is represented by the $|N|$-dimensional vector

$$Au = A(\alpha v + (u - \alpha v)) = A(\alpha v) + A(u - \alpha v), = \alpha v + A(u - \alpha v).$$

Similarly, by applying A k times, we have that the kth state [the state after the $(k-1)$st step] is

$$A^k u = A^k(\alpha v + (u - \alpha v)) = A^k(\alpha v) + A^k(u - \alpha v) = \alpha v + A^k(u - \alpha v).$$

Because

$$\|A^k(u - \alpha v)\| \to_{k\to\infty} 0,$$

we have that the flow converges to the steady state αv:

$$A^k u \to_{k\to\infty} \alpha v.$$

Thus, the conclusion is that, as the number of steps grows, the mass distribution among the nodes in N approaches a steady state represented by a scalar times v, regardless of the initial state u. More precisely, thanks to the fact that the total mass is preserved, the coefficient α satisfies

$$\alpha = \frac{\sum_{j=1}^{|N|} |u_j|}{\sum_{j=1}^{|N|} v_j}.$$

11.13 Exercises

1. Show that the maximal number of edges in an oriented graph with $|N|$ nodes is $|N|^2$.

2. Show that the maximal number of edges that connect a node in N to itself in an oriented graph is $|N|$.

3. Show in two different ways that the maximal number of edges that connect two distinct nodes in N in an oriented graph is $|N|(|N| - 1)$.

4. Show that the maximal number of edges that connect two distinct nodes in N in a nonoriented graph is

$$\binom{|N|}{2} = \frac{|N|(|N| - 1)}{2}.$$

5. Show that the maximal number of edges that connect a node in N to itself in a nonoriented graph is $|N|$.

6. Conclude that the maximal number of edges in a nonoriented graph with $|N|$ nodes is $|N|(|N| + 1)/2$.

7. Solve the previous exercise in yet another way: add a dummy $(|N| + 1)$st node to the graph, and replace each edge connecting a node to itself in the original graph by an edge connecting the node to the dummy node in the new graph. Since the new graph is of $|N| + 1$ nodes, what is the maximal number of edges that connect two distinct nodes in it?

8. Show that the edge-coloring problem can be viewed as a special case of the node-coloring problem for a nonoriented graph.

9. Show that the triangle-coloring problem can be viewed as a special case of the node-coloring problem for a nonoriented graph.

10. Show that the edge-coloring algorithm can be viewed as a special case of the node-coloring algorithm for a nonoriented graph.

11. Show that the triangle-coloring algorithm can be viewed as a special case of the node-coloring algorithm for a nonoriented graph.

12. Write the node-coloring algorithm for an oriented graph in matrix formulation. The solution can be found in Chapter 19, Section 19.2.

13. Write the edge-coloring algorithm for an oriented graph in matrix formulation. The solution can be found in Chapter 19, Section 19.3.

14. Write the node-coloring algorithm for a nonoriented graph in matrix formulation. The solution can be found in Chapter 19, Section 19.8.

15. Write the edge-coloring algorithm for a nonoriented graph in matrix formulation. The solution can be found in Chapter 19, Section 19.8.

16. Show that the triangle-coloring algorithm uses at most four colors to color any triangulation.

17. Let m be the maximal number of edges issued from or directed to a node in a given graph. Show that the node-coloring algorithm uses at most $m + 1$ colors to color the nodes in the graph.

18. A color may be viewed as a maximal subset of nodes that are decoupled from each other (are not connected by an edge). (By "maximal" we mean here that no node can be added to it.) Write an algorithm that produces a color as follows: initially, all the nodes are colored. Then, the nodes are scanned one by one; for each colored node encountered, the color is erased from all of its neighbors. (In other words, the neighbors are uncolored, and dropped from the set of colored nodes.)

19. Repeat the above algorithm also for the set of the nodes that have remained uncolored, to produce a new color that is disjoint from the first color.

20. Repeat the above algorithm until all the nodes are colored by some color. This yields an alternative node-coloring algorithm.

21. Highlight the disadvantage of the above alternative node-coloring algorithm: since it is not based on mathematical induction, it cannot be used to add an extra node to an existing colored graph and color it properly too.

22. Use the interpretation of the edge-coloring problem as a node-coloring problem in the graph of edges to extend the above algorithm into an alternative edge-coloring algorithm.

23. Highlight the disadvantage of the above alternative edge-coloring algorithm: since it is not based on mathematical induction, it cannot be used to add an extra node to an existing colored graph and color the edges that are issued from or directed to it properly as well.

24. Generalize the above definition of a triangulation into the definition of a mesh of cells of k vertices (where k is a given natural number). Make sure that your definition of a mesh of cells of 3 vertices agrees with the original definition of a triangulation.

Chapter 12

Polynomials

A real polynomial is a function $p : \mathbb{R} \to \mathbb{R}$ defined by

$$p(x) \equiv a_0 + a_1 x + a_2 x^2 + \cdots + a_n x^n = \sum_{i=0}^{n} a_i x^i,$$

where n is a nonnegative integer called the degree of the polynomial, and $a_0, a_1, a_2 \ldots, a_n$ are given real numbers called the coefficients.

Thus, to define a concrete polynomial it is sufficient to specify its coefficients $a_0, a_1, a_2 \ldots, a_n$. Thus, the polynomial is equivalent to the $(n+1)$-dimensional vector

$$(a_0, a_1, a_2 \ldots, a_n).$$

A complex polynomial is different from a real polynomial in that the coefficients $a_0, a_1, a_2 \ldots, a_n$, as well as the variable x, can be not only real but also complex. This makes the polynomial a complex function $p : \mathbb{C} \to \mathbb{C}$.

12.1 Adding Polynomials

The interpretation of polynomials as $(n+1)$-dimensional vectors is useful in some arithmetic operations. For example, if another polynomial of degree $m \leq n$

$$q(x) \equiv \sum_{i=0}^{m} b_i x^i$$

is also given, then the sum of p and q is defined by

$$(p + q)(x) \equiv p(x) + q(x) = \sum_{i=0}^{n} (a_i + b_i) x^i,$$

where, if $m < n$, one also needs to define the fictitious zero coefficients

$$b_{m+1} = b_{m+2} = \cdots = b_n \equiv 0.$$

(Without loss of generality, one may assume that $m \leq n$; otherwise, the roles of p and q may interchange.)

In other words, the vector of coefficients associated with the sum $p+q$ is the sum of the individual vectors associated with p and q (extended by leading zero components if necessary, so that they are both $(n+1)$-dimensional). Thus, the interpretation of polynomials as vectors of coefficients helps us to add them by just adding their vectors.

12.2 Multiplying a Polynomial by a Scalar

The representation of the polynomial as a vector also helps to multiply it by a given scalar a:

$$(ap)(x) \equiv a \cdot p(x) = a \sum_{i=0}^{n} a_i x^i = \sum_{i=0}^{n} (aa_i) x^i.$$

Thus, the vector of coefficients associated with the resulting polynomial ap is just a times the original vector associated with the original polynomial p.

12.3 Multiplying Polynomials

Here we consider the task of multiplying the two polynomials p and q to obtain the new polynomial pq. Note that we are not only interested in the value $p(x)q(x)$ for a given x. Indeed, this value can be easily obtained by calculating $p(x)$ and $q(x)$ separately and multiplying the results. We are actually interested in much more than that: we want to have the entire vector of coefficients of the new polynomial pq. This vector is useful not only for the efficient calculation of $p(x)q(x)$, but also for many other purposes as well.

Unfortunately, two vectors cannot be multiplied to produce a new vector. Thus, algebraic operations between vectors are insufficient to produce the required vector of coefficients of the product pq: a more sophisticated approach is required.

The product of the two polynomials p (of degree n) and q (of degree m) is defined by

$$(pq)(x) \equiv p(x)q(x) = \sum_{i=0}^{n} a_i x^i \sum_{j=0}^{m} b_j x^j = \sum_{i=0}^{n} \sum_{j=0}^{m} a_i b_j x^{i+j}.$$

Note that this sum scans the $n+1$ by $m+1$ grid

$$\{(i,j) \mid 0 \leq i \leq n, \ 0 \leq j \leq m\}.$$

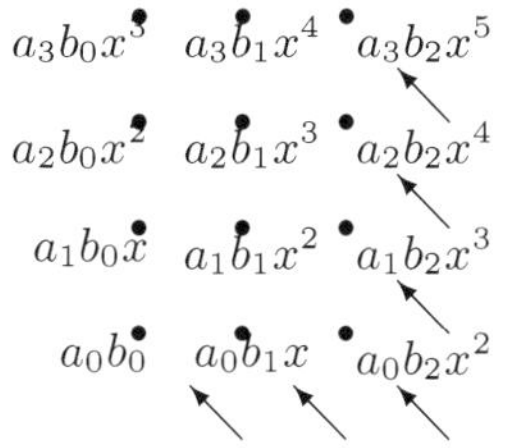

$$a_3b_0x^3 \quad a_3b_1x^4 \quad a_3b_2x^5$$

$$a_2b_0x^2 \quad a_2b_1x^3 \quad a_2b_2x^4$$

$$a_1b_0x \quad a_1b_1x^2 \quad a_1b_2x^3$$

$$a_0b_0 \quad a_0b_1x \quad a_0b_2x^2$$

FIGURE 12.1: Multiplying the polynomial $p(x) = a_0 + a_1x + a_2x^2 + a_3x^3$ by the polynomial $q(x) = b_0 + b_1x + b_2x^2$ by summing the terms diagonal by diagonal, where the kth diagonal ($0 \le k \le 5$) contains terms with x^k only.

However, it makes more sense to calculate this sum diagonal by diagonal, scanning diagonals with constant powers of x (Figure 12.1). These diagonals are indexed by the new index $k = i + j = 0, 1, 2, \ldots, n + m$ in the following sum:

$$(pq)(x) = \sum_{i=0}^{n}\sum_{j=0}^{m} a_ib_jx^{i+j} = \sum_{k=0}^{n+m} \sum_{i=\max(0,k-m)}^{\min(k,n)} a_ib_{k-i}x^k.$$

Thus, the product polynomial pq is associated with the vector of coefficients

$$(c_1, c_2, c_3, \ldots, c_{n+m}) = (c_k)_{k=0}^{n+m},$$

where the coefficients c_k are defined by

$$c_k \equiv \sum_{i=\max(0,k-m)}^{\min(k,n)} a_ib_{k-i}.$$

Note that, when q is a trivial polynomial of degree $m = 0$, that is, when q is the constant function

$$q(x) \equiv b_0,$$

the above definition agrees with the original definition of a scalar times a polynomial. Indeed, in this case, the coefficients c_k reduce to

$$c_k = \sum_{i=k}^{k} a_ib_{k-i} = a_kb_0$$

($0 \le k \le n$), so

$$(pq)(x) = \sum_{k=0}^{n} c_kx^k = \sum_{k=0}^{n} b_0a_kx^k = b_0\sum_{i=0}^{n} a_ix^i = b_0 \cdot p(x),$$

as in the original definition of the scalar b_0 times the polynomial $p(x)$.

12.4 Computing a Polynomial

A common problem is to compute the value of $p(x)$ for a given x. The naive approach to do this requires three stages. First, calculate the powers of x

$$x^2, x^3, x^4, \ldots, x^n.$$

Then, multiply these powers by the corresponding coefficients to obtain

$$a_1 x, a_2 x^2, a_3 x^3, \ldots, a_n x^n.$$

Finally, sum up these terms to obtain

$$a_0 + a_1 x + a_2 x^2 + \cdots + a_n x^n = p(x).$$

The first stage above can be done recursively:

$$x^i = x \cdot x^{i-1}$$

for $i = 2, 3, 4, \ldots, n$. This requires $n - 1$ multiplications. Furthermore, the second stage above requires another n multiplications, and the third stage requires n additions. Thus, the total cost of computing $p(x)$ for a given x is $2n - 1$ multiplications and n additions.

Can this calculation be done more efficiently? Yes, it can. For this, one needs to introduce parentheses and take a common factor out of them.

Consider the problem of computing

$$ab + ac,$$

where a, b, and c are given numbers. At first glance, it would seem that this calculation requires two multiplications to calculate ab and ac, and then one addition to calculate the required sum $ab+ac$. However, this can be done more efficiently by using the distributive law to introduce parentheses and take the common factor a out of them:

$$ab + ac = a(b + c).$$

Indeed, the right-hand side in this equation can be calculated more efficiently than the left-hand side: it requires only one addition to calculate $b + c$, and then one multiplication to calculate $a(b + c)$.

The same idea also works in the efficient calculation of

$$p(x) = \sum_{i=0}^{n} a_i x^i$$

for a given x. Here, the task is to sum up not only two terms as in $ab + ac$, but rather $n + 1$ terms. Although these terms contain no common factor, the n final terms

$$a_1 x, a_2 x^2, a_3 x^3, \ldots, a_n x^n$$

share the common factor x. This common factor can indeed be taken out of parentheses to yield a more efficient computation method. In fact, this leads to the representation of the original polynomial $p(x)$ as

$$p(x) = a_0 + x p_1(x),$$

where $p_1(x)$ is defined by

$$p_1(x) = a_1 + a_2 x + a_3 x^2 + \cdots + a_n x^{n-1} = \sum_{i=0}^{n-1} a_{i+1} x^i.$$

Now, thanks to the fact that $p_1(x)$ is a polynomial of degree $n-1$ only, the value of $p_1(x)$ can be calculated recursively by the same method itself. This is Horner's algorithm for computing the value of $p(x)$ efficiently [17].

Let us show that the calculation of $p(x)$ by the above algorithm requires n multiplications and n additions only. This is done by mathematical induction on the degree n. Indeed, for $n = 0$, $p(x)$ is the just the constant function $p(x) \equiv a_0$, so its calculation requires 0 multiplications and 0 additions. Assume now that the induction hypothesis holds, that is, that we already know that the calculation of a polynomial of degree $n-1$ requires $n-1$ multiplications and $n-1$ additions. In particular, the calculation of the polynomial $p_1(x)$ defined above requires $n-1$ multiplications and $n-1$ additions. In order to calculate

$$p(x) = a_0 + x p_1(x),$$

one needs one extra multiplication to calculate $x p_1(x)$, and one extra addition to calculate $a_0 + x p_1(x)$. Thus, in summary, $p(x)$ has been calculated in a total number of n multiplications and n additions, as asserted.

12.5 Composition of Polynomials

Horner's algorithm is useful not only in computing the value of a polynomial, but also in the composition of two polynomials. The composition of the two polynomials p and q is defined by

$$(p \circ q)(x) \equiv p(q(x)).$$

Note that we are not only interested here in the calculation of the value of $(p \circ q)(x)$ for a given x. Indeed, this value can be easily calculated by calculating $q(x)$ first, and then using the result as the argument in the polynomial p to calculate $p(q(x))$. Here, however, we are interested in much more than that:

we want to have the entire vector of coefficients of the new polynomial $p \circ q$. This vector is most useful in many applications.

The algorithm to obtain the entire vector of coefficients of $p \circ q$ is defined by mathematical induction on the degree of p, n. Indeed, for $n = 0$, p is just the constant function $p(x) \equiv a_0$, so

$$(p \circ q)(x) = p(q(x)) = a_0$$

as well. Assume now that we know how to obtain the entire vector of coefficients of $p_1 \circ q$ for any polynomial p_1 of degree at most $n - 1$. In particular, this applies to the polynomial p_1 defined in the previous section. Furthermore, we also know how to multiply the two polynomials q and $p_1 \circ q$ to obtain the entire vector of coefficients of the product

$$q \cdot (p_1 \circ q).$$

Finally, we only have to add a_0 to the first coefficient in this vector to obtain the required vector of coefficients of the polynomial

$$(p \circ q) = p(q(x)) = a_0 + q(x)p_1(q(x)) = a_0 + q(x)(p_1 \circ q)(x).$$

12.6 Natural Numbers as Polynomials

The notion of the polynomial is also useful in the representation of natural numbers. Indeed, consider the natural number k, $10^n \leq k < 10^{n+1}$ for some natural number n. The common decimal representation of k as a sequence of digits

$$a_n a_{n-1} a_{n-2} \cdots a_1 a_0$$

can also be viewed as the polynomial

$$k = a_0 + a_1 \cdot 10 + a_2 \cdot 10^2 + \cdots + a_n \cdots 10^n = \sum_{i=0}^{n} a_i \cdot 10^i = p(10).$$

In other words, the decimal representation of k is nothing but a polynomial in the argument 10 (the decimal base), with vector of coefficients consisting of the digits used to form the decimal representation of k.

Similarly, consider the natural number l, $2^n \leq l < 2^{n+1}$ for some natural number n. The binary representation of l can also be viewed as the value of a polynomial p in the argument 2 (the binary base), with vector of coefficients

$$a_0, a_1, a_2, \ldots, a_n$$

that are either 0 or 1:

$$l = a_0 + a_1 \cdot 2 + a_2 \cdot 2^2 + \cdots + a_n \cdot 2^n = p(2).$$

12.7 Computing a Monomial

As we have seen above, each natural number l can be written as $l = p(2)$, where p is a polynomial with coefficients that are either 0 or 1. Let us use this binary representation in the efficient calculation of the value of the monomial x^l for a given x.

The recursive algorithm to complete this task is rather expensive. Indeed, it requires $l - 1$ multiplications:

$$x^i = x \cdot x^{i-1}, \quad i = 2, 3, 4, \ldots, l.$$

The advantage of this approach is that it computes not only x^l but also all the powers of the form x^i, $1 < i \leq l$. Still, what if we don't need all these powers? Is there an algorithm that computes x^l alone more efficiently?

Yes, there is: Horner's algorithm can be used to compute the value of x^l in at most $2n$ multiplications. Since $2^n \leq l < 2^{n+1}$, $2n$ is usually far smaller than l, which leads to a considerable reduction in the total cost.

From Horner's algorithm, we have

$$p(2) = a_0 + 2p_1(2).$$

Thus,

$$x^l = x^{p(2)} = x^{a_0 + 2p_1(2)} = x^{a_0} x^{2p_1(2)} = x^{a_0} (x^2)^{p_1(2)} = \begin{cases} x \cdot (x^2)^{p_1(2)} & \text{if } a_0 = 1 \\ (x^2)^{p_1(2)} & \text{if } a_0 = 0. \end{cases}$$

The calculation of the above right-hand side requires one multiplication to calculate x^2 (to be used in the recursive application of the algorithm to calculate $(x^2)^{p_1(2)}$), and then at most one other multiplication to multiply the result by x if $a_0 = 1$. It is easy to prove by induction that the entire recursive algorithm indeed requires at most $2n$ multiplications, where n is the degree of p, or the number of digits in the binary representation of l.

Note that, in the above representation of l as the binary polynomial

$$l = p(2) = a_0 + 2p_1(2),$$

a_0 is just the binary unit digit. Thus, if l is even, then

$$a_0 = 0 \quad \text{and} \quad p_1(2) = l/2.$$

If, on the other hand, l is odd, then

$$a_0 = 1 \quad \text{and} \quad p_1(2) = (l-1)/2.$$

Thus, the above recursive algorithm to calculate x^l can be written more informatively as

$$x^l = \begin{cases} x \cdot (x^2)^{(l-1)/2} & \text{if } l \text{ is odd} \\ (x^2)^{l/2} & \text{if } l \text{ is even.} \end{cases}$$

This compact form is useful below.

12.8 Derivative

The derivative of the polynomial

$$p(x) = \sum_{i=0}^{n} a_i x^i$$

of degree n is the polynomial of degree $n - 1$

$$p'(x) \equiv \begin{cases} 0 & |mboxifn = 0 \\ \sum_{i=1}^{n} a_i i x^{i-1} & \text{if } n > 0 \end{cases}$$

([7] [19]).

Furthermore, because the derivative $p'(x)$ is by itself a polynomial in x, it can be derived as well, to yield the so-called second derivative of p:

$$(p'(x))' = p''(x).$$

12.9 Indefinite Integral

The indefinite integral of the polynomial

$$p(x) = \sum_{i=0}^{n} a_i x^i$$

is defined to be the polynomial of degree $n + 1$

$$P(x) = \sum_{i=0}^{n} \frac{a_i}{i+1} x^{i+1}.$$

Note that the indefinite integral $P(x)$ is characterized by the property that its derivative is the original polynomial $p(x)$:

$$P'(x) = p(x).$$

12.10 Integral over an Interval

The indefinite integral is most useful to calculate areas. In particular, the area of the 2-d region bounded by the x-axis, the graph of $p(x)$, and the verticals to the x-axis at the points $x = a$ and $x = b$ is given by

$$\int_a^b p(x)dx = P(b) - P(a).$$

In particular, when $a = 0$ and $b = 1$, this is the integral of the function $p(x)$ over the unit interval $[0, 1]$:

$$\int_0^1 p(x)dx = P(1) - P(0) = P(1).$$

12.11 Sparse Polynomials

Horner's algorithm for computing the value of $p(x)$ for a given x is efficient for a dense polynomial, with many nonzero coefficients, but not necessarily for a sparse polynomial, with only a few nonzero coefficients. Consider, for example, the polynomial

$$p(x) = x^l = x \cdot x \cdot x \cdot \dots \cdot x \quad (l \text{ times}).$$

For computing the value of this polynomial, Horner's algorithm reduces to the naive algorithm, which multiplies x by itself $l - 1$ times. As we have seen above, this algorithm is far less efficient than the recursive algorithm

$$x^l = \begin{cases} x & \text{if } l = 1 \\ x \cdot (x^2)^{(l-1)/2} & \text{if } l > 1 \text{ and } l \text{ is odd} \\ (x^2)^{l/2} & \text{if } l > 1 \text{ and } l \text{ is even.} \end{cases}$$

Let us study Horner's algorithm, and why it is not always efficient. Well, as discussed above, the idea behind Horner's algorithm is to introduce parentheses in the original polynomial and take the common factor x out of them. This makes sense for dense polynomials, but not for sparse polynomials: indeed, in sparse polynomials, it makes much more sense to take the much larger common factor x^l out of the parentheses, yielding a much more efficient recursive algorithm.

Indeed, the sparse polynomial $p(x)$ of degree n can be written as the sum of a few monomials, each multiplied by its corresponding coefficient:

$$p(x) = a_0 + a_l x^l + a_k x^k + \dots + a_n x^n,$$

where a_0 is either zero or nonzero, $l > 0$ is the index of the next nonzero coefficient $a_l \neq 0$, $k > l$ is the index of the next nonzero coefficient $a_k \neq 0$, and $n > k$ is the index of the final nonzero coefficient $a_n \neq 0$.

Here we introduce a modified version of Horner's algorithm for computing the value of this polynomial for a given x. In this version, the factor x^l is taken out of parentheses:

$$p(x) = a_0 + x^l p_1(x),$$

where

$$p_1(x) \equiv a_l + a_k x^{k-l} + \cdots + a_n x^{n-l}.$$

Thanks to the fact that the degree of p_1 is smaller than n, the value of $p_1(x)$ can be calculated recursively by the same algorithm itself. Thus, the modified Horner's algorithm for computing the value of the sparse polynomial $p(x)$ for a given x reads as follows:

$$p(x) = \begin{cases} a_0 + x^l p_1(x) & \text{if } a_0 \neq 0 \\ x^l p_1(x) & \text{if } a_0 = 0, \end{cases}$$

where x^l is computed efficiently by the recursive algorithm:

$$x^l = \begin{cases} x & \text{if } l = 1 \\ x \cdot (x^2)^{(l-1)/2} & \text{if } l > 1 \text{ and } l \text{ is odd} \\ (x^2)^{l/2} & \text{if } l > 1 \text{ and } l \text{ is even}. \end{cases}$$

This completes the definition of the modified Horner's algorithm for computing the value of the sparse polynomial $p(x)$ for a given x. Below, we use a version of this algorithm also to obtain the entire vector of coefficients of the composition $p \circ q$, where p is a sparse polynomial as above.

12.12 Composition of Sparse Polynomials

Similarly, when p is a sparse polynomial, the composition of the two polynomials p and q may not necessarily benefit from the traditional Horner's algorithm. Indeed, consider again the extreme case, in which

$$p(x) = x^l.$$

In this case,

$$(p \circ q)(x) = p(q(x)) = q^l(x).$$

Now, the traditional Horner's algorithm for obtaining the entire vector of coefficients of this composition of polynomials reduces to the naive algorithm, which multiplies the polynomial q by itself $l - 1$ times. A far more efficient approach is to use the recursive algorithm

$$q^l = \begin{cases} q & \text{if } l = 1 \\ q \cdot (q^2)^{(l-1)/2} & \text{if } l > 1 \text{ and } l \text{ is odd} \\ (q^2)^{l/2} & \text{if } l > 1 \text{ and } l \text{ is even}. \end{cases}$$

Now, consider the sparse polynomial p of degree n discussed in the previous section. Recall that this polynomial can be represented as

$$p(x) = a_0 + x^l p_1(x),$$

where p_1 is a polynomial of degree $n - l$. This representation can be used to compute the entire vector of coefficients of the composition of p and q. Indeed, for any argument x, we have

$$(p \circ q)(x) = p(q(x)) = a_0 + q^l(x)p_1(q(x)) = a_0 + q^l(x)(p_1 \circ q)(x).$$

More compactly, we have

$$p \circ q = a_0 + q^l \cdot (p_1 \circ q).$$

Thanks to the fact that the degree of p_1 is $n - l$ only, its vector of coefficients can be obtained recursively by the same algorithm itself. Furthermore, the vector of coefficients of q^l can be obtained as in the beginning of this section. Then, q^l and $p_1 \circ q$ are multiplied to yield the vector of coefficients of the product $q^l \cdot (p_1 \circ q)$. Finally, if $a_0 \neq 0$, then a_0 should be added in the beginning of this vector of coefficients, to yield the required vector of coefficients of $p \circ q$. This completes the definition of the modified Horner's algorithm for producing the entire vector of coefficients of the composition of the sparse polynomial p with q.

12.13 Polynomials of Two Variables

A real polynomial of two variables is a function $p : \mathbb{R}^2 \to \mathbb{R}$ that can be written in the form

$$p(x, y) = \sum_{i=0}^{n} a_i(x)y^i,$$

where x and y are real numbers, and $a_i(x)$ ($0 \leq i \leq n$) is a real polynomial in the variable x. We also refer to $p(x, y)$ as a 2-d polynomial.

Similarly, a complex polynomial in two variables is a function $p : \mathbb{C}^2 \to \mathbb{C}$ with the same structure, except that x and y may be complex numbers, and the polynomials $a_i(x)$ may also be complex polynomials.

The arithmetic operation between polynomials of two variables are similar to those defined above for polynomials of one variable only. The only difference is that here the coefficients a_i are no longer scalars but rather polynomials in the variable x, so the arithmetic operations between the two original polynomials of two variables involve not just sums and products of scalars but rather sums and products of polynomials of one variable. For example, if

$$p(x, y) = \sum_{i=0}^{n} a_i(x)y^i \quad \text{and} \quad q(x, y) = \sum_{j=0}^{m} b_j(x)y^j$$

are two polynomials of two variables for some natural numbers $m \leq n$, then

$$(p+q)(x,y) = p(x,y) + q(x,y) = \sum_{i=0}^{n}(a_i + b_i)(x)y^i,$$

where

$$(a_i + b_i)(x) = a_i(x) + b_i(x)$$

is the sum of the two polynomials a_i and b_i, and, if $m < n$, then

$$b_{m+1} = b_{m+2} = \cdots = b_n \equiv 0$$

are some dummy zero polynomials.

Furthermore, the product of p and q is

$$(pq)(x,y) = p(x,y)q(x,y)$$

$$= \left(\sum_{i=0}^{n} a_i(x)y^i\right)\left(\sum_{j=0}^{m} b_j(x)y^j\right)$$

$$= \sum_{i=0}^{n}\left(\sum_{j=0}^{m}(a_i b_j)(x)y^{i+j}\right),$$

which is just the sum of n polynomials of two variables.

Later on in the book, we implement this formula to multiply two polynomial objects of two variables by each other.

12.14 Partial Derivatives

The partial derivative of the polynomial of two variables

$$p(x,y) = \sum_{i=0}^{n} a_i(x)y^i$$

with respect to the first variable x is the polynomial of two variables $p_x(x,y)$ obtained by viewing the second variable y as if it were a fixed parameter and deriving $p(x,y)$ as if it were a polynomial of the only variable x:

$$p_x(x,y) \equiv \sum_{i=0}^{n} a_i'(x)y^i,$$

where $a_i'(x)$ is the derivative of $a_i(x)$.

Similarly, the partial derivative of $p(x, y)$ with respect to the second variable y is the polynomial of two variables $p_y(x, y)$ obtained by viewing the first variable x as if it were a fixed parameter and deriving $p(x, y)$ as if it were a polynomial of the only variable y:

$$p_y(x, y) \equiv \begin{cases} 0 & \text{if } n = 0 \\ \sum_{i=1}^{n} a_i(x) i y^{i-1} & \text{if } n > 0. \end{cases}$$

12.15 The Gradient

The gradient of $p(x, y)$, denoted by $\nabla p(x, y)$, is the 2-d vector whose first coordinate is $p_x(x, y)$ and second coordinate is $p_y(x, y)$:

$$\nabla p(x, y) \equiv \begin{pmatrix} p_x(x, y) \\ p_y(x, y) \end{pmatrix}.$$

Thus, the gradient of p is actually a vector function that not only takes but also returns a 2-d vector:

$$\nabla p : \mathbb{R}^2 \to \mathbb{R}^2,$$

or

$$\nabla p \in (\mathbb{R}^2)^{\mathbb{R}^2}.$$

12.16 Integral over the Unit Triangle

FIGURE 12.2: The unit triangle.

Consider the unit (right-angled) triangle

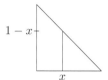

FIGURE 12.3: Integration on the unit triangle: for each fixed x, the integration is done over the vertical line $0 \leq y \leq 1 - x$.

$$t \equiv \{(x, y) \mid 0 \leq x, y, x + y \leq 1\}$$

(see Figure 12.2). The integral over this triangle of the polynomial of two variables

$$p(x, y) = \sum_{i=0}^{n} a_i(x) y^i$$

is the volume of the 3-d region bounded by the triangle, the graph (or manifold) of $p(x, y)$, and the planes that are perpendicular to the x-y plane at the edges of the triangle. This volume is denoted by

$$\int \int_t p(x, y) dx dy.$$

To calculate this integral, let $0 \leq x < 1$ be a fixed parameter, as in Figure 12.3. Let $P(x, y)$ be the indefinite integral of $p(x, y)$ with respect to y:

$$P(x, y) = \sum_{i=0}^{n} \frac{a_i(x)}{i + 1} y^{i+1}.$$

Note that $P(x, y)$ is characterized by the property that its partial derivative with respect to y is the original polynomial $p(x, y)$:

$$P_y(x, y) = p(x, y).$$

Consider now the line segment that is vertical to the x-axis and connects the points $(x, 0)$ to $(x, 1 - x)$ (see Figure 12.3). Let us construct the plane that is perpendicular to the x-y plane at this line segment. This plane cuts a slice from the above 3-d region that lies on top of this line segment and is also bounded by the graph of $p(x, y)$ and the verticals to the x-y plane at the points $(x, 0)$ and $(x, 1 - x)$. The area of this slice is

$$\int_0^{1-x} p(x, y) dy = P(x, 1 - x) - P(x, 0) = P(x, 1 - x).$$

The desired volume of the above 3-d region is now obtained by repeating the above for each fixed $0 \leq x < 1$, multiplying each slice by an infinitesimal width dx, and summing up the volumes of all the individual slices:

$$\int\int_t p(x,y)\,dx\,dy = \int_0^1 P(x, 1-x)\,dx.$$

12.17 Second Partial Derivatives

Because the partial derivative is by itself a polynomial of two variables, it can be derived as well, to yield the so-called second partial derivatives of the original polynomial $p(x,y)$. For example, the partial derivative of p_x with respect to y is

$$p_{xy}(x,y) = (p_x(x,y))_y.$$

Clearly, the order in which the partial derivation is carried out doesn't matter:

$$p_{xy}(x,y) = p_{yx}(x,y).$$

This second partial derivative is also referred to as the $(1,1)$th partial derivative of the original polynomial p, because it can be written as

$$p_{x^1 y^1}(x,y).$$

With this terminology, the $(0,0)$th partial derivative of p is nothing but p itself:

$$p_{x^0 y^0}(x,y) = p(x,y).$$

Moreover, one can derive a second partial derivative once again to produce a derivative of order three. For example, the $(2,1)$th partial derivative of p is

$$p_{x^2 y^1}(x,y) = p_{xxy}(x,y).$$

In general, the order of the (i,j)th partial derivative of p is the sum $i+j$. From Chapter 10, Section 10.15, it follows that the total number of different partial derivatives of order up to n is

$$\binom{n+2}{2} = \frac{(n+2)!}{2(n!)}.$$

Furthermore, from the formulas at the end of Section 10.15 it follows that the total number of different partial derivatives of order n exactly is

$$\binom{n+2-1}{2-1} = \binom{n+1}{1} = n+1.$$

12.18 Degree

Let us write each polynomial $a_i(x)$ above explicitly as

$$a_i(x) = \sum_j a_{i,j} x^j,$$

where the $a_{i,j}$'s are some scalars. The polynomial of two variables can now be written as

$$p(x, y) = \sum_i a_i(x) y^i = \sum_i \sum_j a_{i,j} x^j y^i.$$

The degree of p is the maximal sum $i + j$ for which a monomial of the form $a_{i,j} x^j y^i$ (with $a_{i,j} \neq 0$) appears in p. From Chapter 10, Section 10.15, it follows that the total number of monomials of the form $a_{i,j} x^j y^i$ (with either $a_{i,j} \neq 0$ or $a_{i,j} = 0$) that can appear in a polynomial of degree n is

$$\binom{n+2}{2} = \frac{(n+2)!}{n! \cdot 2!}.$$

12.19 Polynomials of Three Variables

Similarly, a polynomial of three variables is obtained by introducing the new independent variable z:

$$p(x, y, z) = \sum_{i=0}^{n} a_i(x, y) z^i,$$

where the coefficients a_i are now polynomials of the two independent variables x and y. We also refer to $p(x, y, z)$ as a 3-d polynomial.

The arithmetic operations between such polynomials is defined in the same way as before, with the only change that the sums and products of the a_i's are now interpreted as sums and products of polynomials of two variables. The full implementation is given later in the book.

12.20 Partial Derivatives

The partial derivative of the polynomial of three variables

$$p(x, y, z) = \sum_{i=0}^{n} a_i(x, y) z^i$$

with respect to the first variable x is the polynomial of three variables $p_x(x, y, z)$ obtained by viewing the second and third variables, y and z, as if they were fixed parameters, and deriving $p(x, y, z)$ as if it were a polynomial of the only variable x:

$$p_x(x, y, z) \equiv \sum_{i=0}^{n} (a_i)_x(x, y) z^i.$$

Similarly, the partial derivative of $p(x, y, z)$ with respect to the second variable y is the polynomial of three variables $p_y(x, y, z)$ obtained by viewing the first and third variables, x and z, as if they were fixed parameters, and deriving $p(x, y, z)$ as if it were a polynomial of the only variable y:

$$p_y(x, y, z) \equiv \sum_{i=0}^{n} (a_i)_y(x, y) z^i.$$

Finally, the partial derivative of $p(x, y, z)$ with respect to the third variable z is the polynomial of three variables $p_z(x, y, z)$ obtained by viewing the first and second variables x and y as if they were fixed parameters, and deriving $p(x, y, z)$ as if it were a polynomial of the only variable z:

$$p_z(x, y, z) \equiv \begin{cases} 0 & \text{if } n = 0 \\ \sum_{i=1}^{n} a_i(x, y) i z^{i-1} & \text{if } n > 0. \end{cases}$$

12.21 The Gradient

The gradient of $p(x, y, z)$, denoted by $\nabla p(x, y, z)$, is the 3-d vector whose first coordinate is $p_x(x, y, z)$, its second coordinate is $p_y(x, y, z)$, and its third coordinate is $p_z(x, y, z)$:

$$\nabla p(x, y, z) \equiv \begin{pmatrix} p_x(x, y, z) \\ p_y(x, y, z) \\ p_z(x, y, z) \end{pmatrix}.$$

Thus, the gradient of p is actually a vector function that not only takes but also returns a 3-d vector:

$$\nabla p : \mathbb{R}^3 \to \mathbb{R}^3,$$

or

$$\nabla p \in (\mathbb{R}^3)^{\mathbb{R}^3}.$$

12.22 Integral over the Unit Tetrahedron

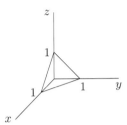

FIGURE 12.4:   The unit tetrahedron.

The unit (right-angled) tetrahedron in the 3-d Cartesian space is defined by

$$T \equiv \{(x, y, z) \mid 0 \le x, y, z, x + y + z \le 1\}$$

(see Figure 12.4). The integral of the 3-d polynomial

$$p(x, y, z) \equiv \sum_{i=0}^{n} a_i(x, y) z^i$$

over this tetrahedron is calculated as follows:

$$\int \int \int_T p(x, y, z) dx dy dz = \int \int_t P(x, y, 1 - x - y) dx dy,$$

where t is the unit triangle in Figure 12.2, and $P(x, y, z)$ is the indefinite integral of $p(x, y, z)$ with respect to the z spatial direction:

$$P(x, y, z) = \sum_{i=0}^{n} \frac{a_i(x, y)}{i + 1} z^{i+1}.$$

The computer code that calculates this integral is provided later on in the book.

12.23 Directional Derivatives

Let **n** be a fixed vector in $\mathbb{R}^3$:

$$\mathbf{n} \equiv \begin{pmatrix} n_1 \\ n_2 \\ n_3 \end{pmatrix}.$$

Assume also that $\mathbf{n}$ is a unit vector:

$$\|\mathbf{n}\|_2 \equiv \sqrt{n_1^2 + n_2^2 + n_3^2} = 1.$$

The directional derivative of $p(x, y, z)$ in the direction specified by $\mathbf{n}$ is the inner product of the gradient of p at (x, y, z) with $\mathbf{n}$:

$$(\mathbf{n}, \nabla p(x, y, z)) = \mathbf{n}^t \nabla p(x, y, z) = n_1 p_x(x, y, z) + n_2 p_y(x, y, z) + n_3 p_z(x, y, z).$$

12.24 Normal Derivatives

Assume now that $\mathbf{n}$ is normal (or orthogonal, or perpendicular) to a particular line or plane in $\mathbb{R}^3$, that is, $\mathbf{n}$ produces a zero inner product with the difference of any two distinct points on the line or on the plane. For example, consider the line

$$\{(x, y, 0) \mid x + y = 1\}.$$

(This line contains one of the edges in the unit tetrahedron in Figure 12.4.) In this case, $\mathbf{n}$ could be either

$$\mathbf{n} = \begin{pmatrix} 0 \\ 0 \\ 1 \end{pmatrix}$$

or

$$\mathbf{n} = \frac{1}{\sqrt{2}} \begin{pmatrix} 1 \\ 1 \\ 0 \end{pmatrix}$$

(or any linear combination of these two vectors, normalized to have l_2-norm that is equal to 1). In fact, if $\partial/\partial x$ denotes the operator of partial derivation with respect to x and $\partial/\partial y$ denotes the operator of partial derivation with respect to y, then the above normal derivation can be denoted by

$$\frac{1}{\sqrt{2}} \left(\frac{\partial}{\partial x} + \frac{\partial}{\partial y} \right).$$

For yet another example, consider the plane

$$\{(x, y, z) \mid x + y + z = 1\}.$$

(This plane contains the largest side in the unit tetrahedron.) In this case, the normal vector $\mathbf{n}$ is

$$\mathbf{n} = \frac{1}{\sqrt{3}} \begin{pmatrix} 1 \\ 1 \\ 1 \end{pmatrix}.$$

In fact, if $\partial/\partial z$ denotes the operator of partial derivation with respect to z, then the above normal derivation can be denoted by

$$\frac{1}{\sqrt{3}} \left(\frac{\partial}{\partial x} + \frac{\partial}{\partial y} + \frac{\partial}{\partial z} \right).$$

Because the normal derivative is by itself a polynomial in three variables, it has normal derivatives as well. The normal derivatives of a normal derivative are called second normal derivatives. Similarly, for $n = 1, 2, 3, \ldots$, the normal derivative of the nth normal derivative is called the $(n+1)$st normal derivative.

For example, the nth normal derivative of a monomial of the form $x^a y^b$ in the direction

$$\frac{1}{\sqrt{2}} \begin{pmatrix} 1 \\ 1 \\ 0 \end{pmatrix}$$

is

$$\left(\frac{1}{\sqrt{2}} \left(\frac{\partial}{\partial x} + \frac{\partial}{\partial y} \right) \right)^n (x^a y^b)$$

$$= \frac{1}{2^{n/2}} \sum_{k=0}^{n} \binom{n}{k} \left(\frac{\partial}{\partial x} \right)^k \left(\frac{\partial}{\partial y} \right)^{n-k} (x^a y^b)$$

$$= \frac{1}{2^{n/2}} \sum_{k=0}^{n} \binom{n}{k} C_{a,k} C_{b,n-k} x^{a-k} y^{b-(n-k)},$$

where $C_{a,k}$ is as in Chapter 10, Section 10.17. Furthermore, from the formula proved there, it follows that, at the point $(x, y, z) = (1/2, 1/2, 0)$, this nth normal derivative is equal to

$$\frac{C_{a+b,n}}{2^{n/2+a+b-n}}.$$

Similarly, from Chapter 10, Section 10.18, it follows that the nth normal derivative of a monomial of the form $x^a y^b z^c$ in the direction

$$\frac{1}{\sqrt{3}} \begin{pmatrix} 1 \\ 1 \\ 1 \end{pmatrix}$$

is

$$\left(\frac{1}{\sqrt{3}} \left(\frac{\partial}{\partial x} + \frac{\partial}{\partial y} + \frac{\partial}{\partial z} \right) \right)^n (x^a y^b z^c)$$

$$= \frac{1}{3^{n/2}} \sum_{0 \leq l,j,m \leq n,\ l+j+m=n} \frac{n!}{l!j!m!} \left(\frac{\partial}{\partial x} \right)^l \left(\frac{\partial}{\partial y} \right)^j \left(\frac{\partial}{\partial z} \right)^m (x^a y^b z^c)$$

$$= \frac{1}{3^{n/2}} \sum_{0 \leq l,j,m \leq n,\ l+j+m=n} \frac{n!}{l!j!m!} C_{a,l} C_{b,j} C_{c,m} x^{a-l} y^{b-j} z^{c-m}.$$

Furthermore, from the formula at the end of Section 10.18, at the point $(x, y, z) = (1/3, 1/3, 1/3)$ this nth normal derivative is equal to

$$\frac{C_{a+b+c,n}}{3^{n/2+a+b+c-n}}.$$

12.25 Tangential Derivatives

Assume now that $\mathbf{n}$ is parallel to a line or a plane in the 3-d Cartesian space, that is, it is orthogonal (perpendicular) to any vector that is normal to the line or the plane. Then the directional derivative in direction $\mathbf{n}$ is also called the tangential derivative to the line or the plane in direction $\mathbf{n}$.

12.26 High-Order Partial Derivatives

Because a partial derivative is by itself a polynomial of three variables, it can be derived as well. For example, the derivative of p_x with respect to z is

$$p_{xz}(x, y, z) = (p_x(x, y, z))_z.$$

Clearly, the order in which the derivation takes place is immaterial:

$$p_{xz}(x, y, z) = p_{zx}(x, y, z).$$

Furthermore, the (i, j, k)th partial derivative of p is

$$p_{x^i y^j z^k}(x, y, z) = \left(\left(\frac{\partial}{\partial x} \right)^i \left(\frac{\partial}{\partial y} \right)^j \left(\frac{\partial}{\partial z} \right)^k p \right)(x, y, z).$$

For example, the $(2, 1, 0)$th partial derivative of p is

$$p_{x^2 y^1 z^0}(x, y, z) = p_{xxy}(x, y, z).$$

Furthermore, the $(0,0,0)$th partial derivative of p is p itself:

$$p_{x^0 y^0 z^0}(x, y, z) = p(x, y, z).$$

The order of the (i, j, k)th derivative is the sum $i + j + k$. From Chapter 10, Section 10.15, it follows that the total number of different partial derivatives of order up to n (where n is a given natural number) is

$$\binom{n+3}{3} = \frac{(n+3)!}{6(n!)}.$$

Furthermore, from the formulas at the end of Section 10.15, it follows that the total number of different partial derivatives of order n exactly is

$$\binom{n+3-1}{3-1} = \binom{n+2}{2} = \frac{(n+2)!}{2(n!)}.$$

12.27 The Hessian

Let $p(x, y, z)$ be a polynomial of three variables. Let us denote the row vector that is the transpose of the gradient of p by

$$\nabla^t p(x, y, z) \equiv (\nabla p(x, y, z))^t = (p_x(x, y, z), p_y(x, y, z), p_z(x, y, z)).$$

More compactly, this can be written as

$$\nabla^t p \equiv (\nabla p)^t = (p_x, p_y, p_z).$$

The Hessian is the 3×3 matrix that contains the second partial derivatives of p:

$$(\nabla\nabla^t p)(x, y, z) \equiv \nabla(\nabla^t p)(x, y, z) = \nabla(p_x(x, y, z), p_y(x, y, z), p_z(x, y, z)).$$

More compactly, this can be written as

$$\nabla\nabla^t p = (\nabla p_x \mid \nabla p_y \mid \nabla p_z).$$

Thanks to the property that partial derivation is independent of the order in which it is carried out, that is,

$$p_{xy} = p_{yx}$$
$$p_{xz} = p_{zx}$$
$$p_{yz} = p_{zy},$$

we have that the Hessian is a symmetric matrix.

12.28 Degree

As in Section 12.18 above, the polynomials of two variables $a_i(x, y)$ can be written as

$$a_i(x, y) = \sum_j \sum_k a_{i,j,k} x^k y^j,$$

where $a_{i,j,k}$ are some scalars. Using this formulation, the original polynomial of three variables can be written as

$$p(x, y, z) = \sum_i \sum_j \sum_k a_{i,j,k} x^k y^j z^i.$$

The degree of p is the maximal sum $i+j+k$ for which a monomial of the form $a_{i,j,k} x^k y^j z^i$ (with $a_{i,j,k} \neq 0$) appears in p. From Chapter 10, Section 10.15, it follows that the total number of monomials of the form $a_{i,j,k} x^k y^j z^i$ (with either $a_{i,j,k} \neq 0$ or $a_{i,j,k} = 0$) that can appear in a polynomial of degree n is

$$\binom{n+3}{3} = \frac{(n+3)!}{n! \cdot 3!}.$$

12.29 Degrees of Freedom

In order to specify a polynomial $p(x, y, z)$ of degree n, one can specify its coefficients $a_{i,j,k}$ $(i+j+k \leq n)$. For $n = 5$, for example, this requires specifying

$$\binom{5+3}{3} = 56$$

coefficients of the form $a_{i,j,k}$ $(i + j + k \leq 5)$.

This explicit approach, however, is not the only way to specify the polynomial $p(x, y, z)$ of degree 5. One can equally well characterize p more implicitly by giving any 56 independent pieces of information about it, such as its values at 56 independent points in the 3-d Cartesian space. Better yet, one can specify p by specifying not only its values but also the values of its partial derivatives at a smaller number of points. A sensible approach, for example, would be to specify the partial derivatives of p up to (and including) order 2 at the four corners of the unit tetrahedron. Since the total number of such partial derivatives is

$$\binom{2+3}{3} = 10,$$

this makes a total of 40 independent pieces of information that have been specified about p. More precisely, the values of these 10 partial derivatives of p should be specified to be the parameters

$$p^{(i)} \qquad (0 \leq i < 10)$$

at the origin $(0,0,0)$, the parameters

$$p^{(i)} \qquad (10 \leq i < 20)$$

at the corner $(1,0,0)$, the parameters

$$p^{(i)} \qquad (20 \leq i < 30)$$

at the corner $(0,1,0)$, and the parameters

$$p^{(i)} \qquad (30 \leq i < 40)$$

at the corner $(0,0,1)$.

So far, we have specified a total number of 40 independent pieces of information (or degrees of freedom) about p. In order to specify p uniquely, however, we must specify 16 more degrees of freedom. For this purpose, it makes sense to specify two normal derivatives at the midpoint of each of the six edges of the unit tetrahedron (which makes 12 more degrees of freedom). These degrees of freedom are given by the parameters $p^{(40)}, p^{(41)}, \ldots, p^{(51)}$.

This makes a total of 52 degrees of freedom that have been specified so far. The four final degrees of freedom are the normal derivative at the midpoint of each of the four sides of the tetrahedron. These degrees of freedom are given in the parameters $p^{(52)}, p^{(53)}, p^{(54)}$, and $p^{(55)}$. This makes a total of 56 degrees of freedom, as required.

Thus, in order to specify p uniquely, it is sufficient to specify the above parameters $p^{(0)}, p^{(1)}, \ldots, p^{(55)}$. Below we use this method to specify some special and important polynomials, called the basis functions.

12.30 Basis Functions in the Unit Tetrahedron

A basis function in the unit tetrahedron T in Figure 12.4 is a polynomial $p_i(x, y, z)$ $(0 \leq i < 56)$ that is obtained by specifying the ith parameter above to be 1, whereas all the other parameters vanish. More precisely, the 56 parameters (or degrees of freedom) $p_i^{(0)}, p_i^{(1)}, \ldots, p_i^{(55)}$ required to specify the polynomial $p_i(x, y, z)$ uniquely are given by

$$p_i^{(j)} \equiv \begin{cases} 1 & \text{if } j = i \\ 0 & \text{if } j \neq i \end{cases}$$

($0 \leq j < 56$). In fact, every polynomial $p(x, y, z)$ of degree at most 5 can be written uniquely as the sum

$$p(x, y, z) = \sum_{i=0}^{55} p^{(i)} p_i(x, y, z).$$

Indeed, both the left-hand side and the right-hand side of this equation are polynomials of degree at most 5 with the same 56 degrees of freedom $p^{(0)}, p^{(1)}, \ldots, p^{(55)}$.

12.31 Computing the Basis Functions

The triplets of the form (i, j, k) used to index the coefficients $a_{i,j,k}$ of a polynomial of degree 5 form a discrete tetrahedron:

$$T^{(5)} \equiv \{(i, j, k) \mid 0 \leq i, j, k, i + j + k \leq 5\}.$$

Let us order these triplets in the lexicographic order, that is (i, j, k) is prior to (l, m, n) if either (a) $i < l$ or (b) $i = l$ and $j < m$ or (c) $i = l$ and $j = m$ and $k < n$. Let us use the index $0 \leq \hat{j}_{i,j,k} < 56$ to index these triplets in this order. Using this index, we can form the 56-dimensional vector of coefficients **x**:

$$\mathbf{x}_{\hat{j}_{i,j,k}} \equiv a_{i,j,k}, \qquad (i, j, k) \in T^{(5)}.$$

The same can be done for polynomials of degree 2. In this case, the index $0 \leq \hat{i}_{l,m,n} < 10$ can be used to index the corresponding triplets of the form (l, m, n) in the discrete tetrahedron

$$T^{(2)} = \{(l, m, n) \mid 0 \leq l, m, n, l + m + n \leq 2\}.$$

To compute a basis function p_q for some given integer $0 \leq q < 56$, we need to specify its coefficients $a_{i,j,k}$, or the components of **x**. Unfortunately, these coefficients are rarely available explicitly. In order to find them, we must use every data we may have about p_q, including its available degrees of freedom. In a more formal language, we must solve a linear system of equations, in which **x** is the vector of unknowns:

$$B\mathbf{x} = I^{(q)},$$

where $I^{(q)}$ is the qth column of the identity matrix of order 56, and B is the 56×56 matrix defined by the relations

$$p_q^{(i)} = \sum_{j=0}^{55} B_{i,j} \mathbf{x}_j = \begin{cases} 1 & \text{if } i = q \\ 0 & \text{if } i \neq q, \end{cases}$$

$0 \leq i < 56$.

The above formula tells us how the elements $B_{i,j}$ should be defined. Indeed, the ith parameter $p_q^{(i)}$ is obtained from a partial derivative of p_q at one of the corners (or the edge midpoints, or the side midpoints) of the unit tetrahedron. Because the partial derivatives are linear operations, they are actually applied to each monomial in p_q, evaluated at the relevant point (either a corner or an edge midpoint or a side midpoint of the unit tetrahedron), multiplied by the corresponding (unknown) coefficient (stored in $\mathbf{x}_j$), and summed up. Thus, an element in B must be obtained by a particular partial derivative applied to a particular monomial and then evaluated at a particular point in the unit tetrahedron.

More precisely, for $0 \leq j < 56$, let us substitute $j \leftarrow \hat{j}_{i,j,k}$, where $(i,j,k) \in T^{(5)}$. Furthermore, for the first 10 equations in the above linear system ($0 \leq i < 10$), let us substitute $i \leftarrow \hat{i}_{l,m,n}$, where $(l,m,n) \in T^{(2)}$. Then the parameter $p_q^{(\hat{i}_{l,m,n})}$ is the value of the (n,m,l)th partial derivative of p_q at the origin $(0,0,0)$. Thus, $B_{\hat{i}_{l,m,n},\hat{j}_{i,j,k}}$ must be the value of the (n,m,l)th partial derivative of the monomial $x^k y^j z^i$ at the origin:

$$B_{\hat{i}_{l,m,n},\hat{j}_{i,j,k}} \equiv \begin{cases} i!j!k! & \text{if } i=l \text{ and } j=m \text{ and } k=n \\ 0 & \text{otherwise.} \end{cases}$$

Furthermore, in the next 10 equations in the above linear system, the partial derivatives are evaluated at the corner $(1,0,0)$ rather than at the origin. Therefore, the parameter $p_q^{(10+\hat{i}_{l,m,n})}$ is the (n,m,l)th partial derivative of p_q at $(1,0,0)$, so $B_{10+\hat{i}_{l,m,n},\hat{j}_{i,j,k}}$ must be defined as the (n,m,l)th partial derivative of the monomial $x^k y^j z^i$ at $(1,0,0)$:

$$B_{10+\hat{i}_{l,m,n},\hat{j}_{i,j,k}} \equiv \begin{cases} i!j!\frac{k!}{(k-n)!} & \text{if } i=l \text{ and } j=m \text{ and } k \geq n \\ 0 & \text{otherwise.} \end{cases}$$

Similarly, in the next 10 equations, the partial derivatives are evaluated at $(0,1,0)$, yielding the definitions

$$B_{20+\hat{i}_{l,m,n},\hat{j}_{i,j,k}} \equiv \begin{cases} i!k!\frac{j!}{(j-m)!} & \text{if } i=l \text{ and } j \geq m \text{ and } k=n \\ 0 & \text{otherwise.} \end{cases}$$

Similarly, in the next 10 equations, the partial derivatives are evaluated at $(0,0,1)$, yielding the definitions

$$B_{30+\hat{i}_{l,m,n},\hat{j}_{i,j,k}} \equiv \begin{cases} j!k!\frac{i!}{(i-l)!} & \text{if } i \geq l \text{ and } j=m \text{ and } k=n \\ 0 & \text{otherwise.} \end{cases}$$

This completes the definition of the first 40 rows in B, indexed from 0 to 39.

The next two equations are obtained from the two normal derivatives (the y- and z-partial derivatives) at the edge midpoint $(1/2,0,0)$ (see Section 12.24 above):

$$B_{40,\hat{j}_{i,j,k}} \equiv \begin{cases} 2^{-k} & \text{if } j = 1 \text{ and } i = 0 \\ 0 & \text{otherwise} \end{cases}$$

$$B_{41,\hat{j}_{i,j,k}} \equiv \begin{cases} 2^{-k} & \text{if } j = 0 \text{ and } i = 1 \\ 0 & \text{otherwise.} \end{cases}$$

Similarly, the next two equations are obtained from the two normal derivatives (the x- and z-partial derivatives) at the edge midpoint $(0, 1/2, 0)$:

$$B_{42,\hat{j}_{i,j,k}} \equiv \begin{cases} 2^{-j} & \text{if } k = 1 \text{ and } i = 0 \\ 0 & \text{otherwise} \end{cases}$$

$$B_{43,\hat{j}_{i,j,k}} \equiv \begin{cases} 2^{-j} & \text{if } k = 0 \text{ and } i = 1 \\ 0 & \text{otherwise.} \end{cases}$$

Similarly, the next two equations are obtained from the two normal derivatives (the x- and y-partial derivatives) at the edge midpoint $(0, 0, 1/2)$:

$$B_{44,\hat{j}_{i,j,k}} \equiv \begin{cases} 2^{-i} & \text{if } k = 1 \text{ and } j = 0 \\ 0 & \text{otherwise} \end{cases}$$

$$B_{45,\hat{j}_{i,j,k}} \equiv \begin{cases} 2^{-i} & \text{if } k = 0 \text{ and } j = 1 \\ 0 & \text{otherwise.} \end{cases}$$

Furthermore, the next two equations are obtained from the two normal derivatives (the z-partial derivative and the sum of the x- and y-partial derivatives, divided by $\sqrt{2}$) at the edge midpoint $(1/2, 1/2, 0)$ (see Section 12.24):

$$B_{46,\hat{j}_{i,j,k}} \equiv \begin{cases} 2^{-j-k} & \text{if } i = 1 \\ 0 & \text{otherwise} \end{cases}$$

$$B_{47,\hat{j}_{i,j,k}} \equiv \begin{cases} (j+k)2^{-j-k+1}/\sqrt{2} & \text{if } i = 0 \\ 0 & \text{otherwise.} \end{cases}$$

Similarly, the next two equations are obtained from the two normal derivatives (the y-partial derivative and the sum of the x- and z-partial derivatives, divided by $\sqrt{2}$) at the edge midpoint $(1/2, 0, 1/2)$:

$$B_{48,\hat{j}_{i,j,k}} \equiv \begin{cases} 2^{-i-k} & \text{if } j = 1 \\ 0 & \text{otherwise} \end{cases}$$

$$B_{49,\hat{j}_{i,j,k}} \equiv \begin{cases} (i+k)2^{-i-k+1}/\sqrt{2} & \text{if } j = 0 \\ 0 & \text{otherwise.} \end{cases}$$

Similarly, the next two equations are obtained from the two normal derivatives (the x-partial derivative and the sum of the y- and z-partial derivatives, divided by $\sqrt{2}$) at the edge midpoint $(0, 1/2, 1/2)$:

$$B_{50,\hat{j}_{i,j,k}} \equiv \begin{cases} 2^{-i-j} & \text{if } k = 1 \\ 0 & \text{otherwise} \end{cases}$$

$$B_{51,\hat{j}_{i,j,k}} \equiv \begin{cases} (i+j)2^{-i-j+1}/\sqrt{2} & \text{if } k = 0 \\ 0 & \text{otherwise.} \end{cases}$$

The final four equations are obtained from evaluating the normal derivatives at the side midpoints. The next equation is obtained from the normal derivative (the z-partial derivative) at the side midpoint $(1/3, 1/3, 0)$:

$$B_{52, \hat{j}_{i,j,k}} \equiv \begin{cases} 3^{-j-k} & \text{if } i = 1 \\ 0 & \text{otherwise.} \end{cases}$$

Similarly, the next equation is obtained from the normal derivative (the y-partial derivative) at the side midpoint $(1/3, 0, 1/3)$:

$$B_{53, \hat{j}_{i,j,k}} \equiv \begin{cases} 3^{-i-k} & \text{if } j = 1 \\ 0 & \text{otherwise.} \end{cases}$$

Similarly, the next equation is obtained from the normal derivative (the x-partial derivative) at the side midpoint $(0, 1/3, 1/3)$:

$$B_{54, \hat{j}_{i,j,k}} \equiv \begin{cases} 3^{-i-j} & \text{if } k = 1 \\ 0 & \text{otherwise.} \end{cases}$$

The final equation is obtained from the normal derivative (the sum of the x-, y-, and z-partial derivatives, divided by $\sqrt{3}$) at the side midpoint $(1/3, 1/3, 1/3)$ (see Section 12.24):

$$B_{55, \hat{j}_{i,j,k}} \equiv (i + j + k)3^{-i-j-k+1}/\sqrt{3}.$$

This completes the definition of the matrix B.

In order to solve the above linear system for $\mathbf{x}$, one can use three possible approaches. The direct approach is to compute the inverse matrix B^{-1} explicitly, and obtain $\mathbf{x}$ as the qth column in it. A more indirect approach is to solve the above linear system iteratively by a Krylov-subspace method, such as GMRES [23]. Finally, one may also multiply the above linear system by the transpose matrix B^t to obtain the so-called normal equations

$$B^t B\mathbf{x} = B^t \cdot I^{(q)},$$

and apply to them the preconditioned-conjugate-gradient iterative method [22].

12.32 Composite Functions in a General Tetrahedron

Consider now a general tetrahedron t in the 3-d Cartesian space, with vertices (corners) denoted by $\mathbf{k}, \mathbf{l}, \mathbf{m}, \mathbf{n} \in \mathbb{R}^3$. In this case, we write

$$t = (\mathbf{k}, \mathbf{l}, \mathbf{m}, \mathbf{n}),$$

where the order of vertices in this notation is determined arbitrarily in advance.

Let

$$|\mathbf{k}| = 0$$
$$|\mathbf{l}| = 10$$
$$|\mathbf{m}| = 20$$
$$|\mathbf{n}| = 30$$

denote the indices of the corners of t in the list of the 56 degrees of freedom in t, to be specified below. In fact, each basis function in t is going to be a polynomial of three variables with 55 degrees of freedom (partial derivatives at the corners, edge midpoints, or side midpoints of t) that are equal to 0, and only one degree of freedom that is equal to 1.

Let S_t be the 3×3 matrix whose columns are the 3-d vectors leading from $\mathbf{k}$ to the three other corners of t:

$$S_t \equiv (\mathbf{l} - \mathbf{k} \mid \mathbf{m} - \mathbf{k} \mid \mathbf{n} - \mathbf{k}).$$

Let us use the matrix S_t to define a mapping from T onto t:

$$E_t \left(\begin{pmatrix} x \\ y \\ z \end{pmatrix} \right) \equiv \mathbf{k} + S_t \begin{pmatrix} x \\ y \\ z \end{pmatrix}.$$

Indeed, the corners of the unit tetrahedron T are clearly mapped by E_t onto the corresponding corners of t:

$$E_t \left(\begin{pmatrix} 0 \\ 0 \\ 0 \end{pmatrix} \right) = \mathbf{k}$$

$$E_t \left(\begin{pmatrix} 1 \\ 0 \\ 0 \end{pmatrix} \right) = \mathbf{l}$$

$$E_t \left(\begin{pmatrix} 0 \\ 1 \\ 0 \end{pmatrix} \right) = \mathbf{m}$$

$$E_t \left(\begin{pmatrix} 0 \\ 0 \\ 1 \end{pmatrix} \right) = \mathbf{n}.$$

Clearly, the inverse mapping maps t back onto T:

$$E_t^{-1} \left(\begin{pmatrix} x \\ y \\ z \end{pmatrix} \right) = S_t^{-1} \left(\begin{pmatrix} x \\ y \\ z \end{pmatrix} - \mathbf{k} \right).$$

The basis functions in Section 12.31 can now be composed with E_t^{-1} to form the corresponding functions

$$r_{i,t} \equiv p_i \circ E_t^{-1}$$

$(0 \leq i < 56)$, which are defined in t rather than in T.

Unfortunately, these functions are not basis functions, as they may have more than one nonzero partial derivative in t. Still, they may be used to form the desired basis functions in t as follows.

12.33 The Chain Rule

The so-called "chain rule" tells us how to compute partial derivatives of the composition of two functions. In our case, it gives the gradient of the composed function $r_{i,t}$ in terms of the gradient of its first component, p_i:

$$\nabla r_{i,t} = \nabla(p_i \circ E_t^{-1}) = S_t^{-t}((\nabla p_i) \circ E_t^{-1})$$

(where S_t^{-t} is the transpose of the inverse of S_t). Furthermore, by taking the transpose of both sides of the above equation, we have

$$\nabla^t r_{i,t} == ((\nabla^t p_i) \circ E_t^{-1}) S_t^{-1}.$$

As a result, if the gradient of p_i vanishes at some point $(x, y, z) \in T$, then the gradient of $r_{i,t}$ vanishes at $E_t(x, y, z) \in t$. In particular, if the gradient of p_i vanishes at some corner of T, then the gradient of $r_{i,t}$ vanishes at the corresponding corner of t.

12.34 Directional Derivative of a Composite Function

The above formulas can be used to write the directional derivative of $r_{i,t}$ in terms of that of p_i. To see this, let $\mathbf{n} \in \mathbb{R}^3$ be a unit vector. Define also the unit vector

$$\mathbf{w} \equiv \frac{S_t \mathbf{n}}{\|S_t \mathbf{n}\|_2}.$$

With this notation, the directional derivative of $r_{i,t}$ in direction $\mathbf{w}$ is

$$\mathbf{w}^t \nabla r_{i,t} = \frac{1}{\|S_t\mathbf{n}\|_2}(S_t\mathbf{n})^t \nabla r_{i,t}$$

$$= \frac{1}{\|S_t\mathbf{n}\|_2}\mathbf{n}^t S_t^t S_t^{-t}((\nabla p_i)\circ E_t^{-1})$$

$$= \frac{1}{\|S_t\mathbf{n}\|_2}\mathbf{n}^t((\nabla p_i)\circ E_t^{-1}).$$

In other words, the directional derivative of $r_{i,t}$ in direction $\mathbf{w}$ at some point $(x,y,z)\in t$ is proportional to the directional derivative of p_i in direction $\mathbf{n}$ at the corresponding point $E_t^{-1}(x,y,z)\in T$.

12.35 The Hessian of a Composite Function

The formula at the end of Section 12.33 that gives $\nabla^t r_{i,t}$ in terms of $\nabla^t p_i$ is also useful to have the Hessian of $r_{i,t}$. in terms of that of p_i. Indeed, by applying ∇ to both sides of the this formula, we have

$$\nabla\nabla^t r_{i,t} = \nabla((\nabla^t p_i)\circ E_t^{-1})S_t^{-1} = S_t^{-t}((\nabla\nabla^t p_i)\circ E_t^{-1})S_t^{-1}.$$

12.36 Basis Functions in a General Tetrahedron

As a result of the above formula, if the Hessian of p_i vanishes at some point $(x,y,z)\in T$, then the Hessian of $r_{i,t}$ vanishes at $E_t(x,y,z)\in t$. In particular, if the Hessian of p_i vanishes at some corner of T, then the Hessian of $r_{i,t}$ vanishes at the corresponding corner of t. As a consequence, we can define four basis functions in t:

$$R_{i,t}\equiv r_{i,t}, \qquad i=0,10,20,3.$$

Indeed, from the above, each of these functions has the value 1 at one of the corners of t, whereas its partial derivatives vanish at the corners, edge midpoints, and side midpoints of t. For example, since

$$|\mathbf{k}| = 0$$
$$|\mathbf{l}| = 10$$
$$|\mathbf{m}| = 20$$
$$|\mathbf{n}| = 30$$

are the indices of the corners of t in the list of 56 degrees of freedom in it, the partial derivatives of $R_{0,t}$ (of order 0, 1, or 2) vanish at $\mathbf{k}$, $\mathbf{l}$, $\mathbf{m}$, and $\mathbf{n}$ (as well as at the edge and side midpoints), except for the 0th partial derivative at $\mathbf{k}$, for which

$$R_{0,t}(\mathbf{k}) = r_{0,t}(\mathbf{k}) = p_0(E_t^{-1}(\mathbf{k})) = p_0(0,0,0) = 1.$$

Furthermore, let us define basis functions in t with only one first partial derivative that does not vanish at only one corner of t. For this purpose, let $\mathbf{i}$ be some corner of t, and define the three basis function in t by

$$\begin{pmatrix} R_{|\mathbf{i}|+\hat{\imath}_{0,0,1},t} \\ R_{|\mathbf{i}|+\hat{\imath}_{0,1,0},t} \\ R_{|\mathbf{i}|+\hat{\imath}_{1,0,0},t} \end{pmatrix} \equiv S_t \begin{pmatrix} r_{|\mathbf{i}|+\hat{\imath}_{0,0,1},t} \\ r_{|\mathbf{i}|+\hat{\imath}_{0,1,0},t} \\ r_{|\mathbf{i}|+\hat{\imath}_{1,0,0},t} \end{pmatrix}.$$

Indeed, by applying ∇^t to both sides of the above equation, we obtain the 3×3 matrix equation

$$\nabla^t \begin{pmatrix} R_{|\mathbf{i}|+\hat{\imath}_{0,0,1},t} \\ R_{|\mathbf{i}|+\hat{\imath}_{0,1,0},t} \\ R_{|\mathbf{i}|+\hat{\imath}_{1,0,0},t} \end{pmatrix} = S_t \nabla^t \begin{pmatrix} r_{|\mathbf{i}|+\hat{\imath}_{0,0,1},t} \\ r_{|\mathbf{i}|+\hat{\imath}_{0,1,0},t} \\ r_{|\mathbf{i}|+\hat{\imath}_{1,0,0},t} \end{pmatrix} = S_t \left(\begin{pmatrix} \nabla^t p_{|\mathbf{i}|+\hat{\imath}_{0,0,1}} \\ \nabla^t p_{|\mathbf{i}|+\hat{\imath}_{0,1,0}} \\ \nabla^t p_{|\mathbf{i}|+\hat{\imath}_{1,0,0}} \end{pmatrix} \circ E_t^{-1} \right) S_t^{-1}.$$

Clearly, at every corner of t that is different from $\mathbf{i}$ the middle term in the above triple product is just the 3×3 zero matrix, which implies that the three basis functions have zero partial derivatives there, as required. At the corner $\mathbf{i}$, on the other hand, the middle term in the above triple product is the 3×3 identity matrix I, which implies that

$$\nabla^t \begin{pmatrix} R_{|\mathbf{i}|+\hat{\imath}_{0,0,1},t} \\ R_{|\mathbf{i}|+\hat{\imath}_{0,1,0},t} \\ R_{|\mathbf{i}|+\hat{\imath}_{1,0,0},t} \end{pmatrix} = S_t S_t^{-1} = I,$$

as required.

Moreover, let us now define the six basis functions in t corresponding to the second partial derivatives at the corners of t. For this, however, we need some new notations.

12.37 Tensors

Let A be a 3×3 symmetric matrix. Thanks to the symmetry property, A is well defined in terms of six elements only, and can therefore be represented uniquely as the six-dimensional column vector

$$A \equiv \begin{pmatrix} a_{1,1} \\ a_{2,1} \\ a_{2,2} \\ a_{3,1} \\ a_{3,2} \\ a_{3,3} \end{pmatrix}.$$

Let S be some 3×3 matrix. Consider the linear mapping

$$A \rightarrow SAS^t,$$

which maps A to the 3×3 symmetric matrix SAS^t. More specifically, for every two indices $1 \leq l, m \leq 3$, we have

$$\left(SAS^t\right)_{l,m} = \sum_{j=1}^{3} = \sum_{k=1}^{3} S_{l,j} a_{j,k} (S^t)_{k,m} = \sum_{j,k=1}^{3} S_{l,j} S_{m,k} a_{j,k}.$$

In this formulation, A can be viewed as a nine-dimensional vector, with the 2-d vector index (j, k) rather than the usual scalar index. Furthermore, $S_{l,j} S_{m,k}$ can be viewed as a matrix of order 9, with the 2-d vector indices (l, m) (row index) and (j, k) (column index) rather than the usual scalar indices.

A matrix with this kind of indexing is also called a tensor. Here the tensor $S \otimes S$ is defined by

$$(S \otimes S)_{(l,m),(j,k)} \equiv S_{l,j} S_{m,k}.$$

Clearly, the transpose tensor is obtained by interchanging the roles of the row index (l, m) and the column index (j, k). Thus, the $((l, m), (j, k))$th element in the transpose tensor is

$$S_{j,l} S_{k,m} = (S^t)_{l,j} (S^t)_{m,k}.$$

As a result, the transpose tensor is associated with the transpose mapping

$$A \rightarrow S^t A S.$$

Thanks to the linearity of the original mapping that maps the symmetric matrix A to the symmetric matrix SAS^t, it can also be represented more economically as a mapping of six-dimensional column vectors:

$$\begin{pmatrix} a_{1,1} \\ a_{2,1} \\ a_{2,2} \\ a_{3,1} \\ a_{3,2} \\ a_{3,3} \end{pmatrix} \rightarrow Z \begin{pmatrix} a_{1,1} \\ a_{2,1} \\ a_{2,2} \\ a_{3,1} \\ a_{3,2} \\ a_{3,3} \end{pmatrix},$$

where $Z \equiv Z(S)$ is a suitable 6×6 matrix. In fact, Z can be constructed as follows: the six components in its first column are the same as the six elements in the lower triangular part of

$$S \begin{pmatrix} 1\,0\,0 \\ 0\,0\,0 \\ 0\,0\,0 \end{pmatrix} S^t,$$

the six components in its second column are the same as the six elements in the lower triangular part of

$$S \begin{pmatrix} 0\,1\,0 \\ 1\,0\,0 \\ 0\,0\,0 \end{pmatrix} S^t,$$

and so on.

As a matter of fact, the mapping defined by Z can also be viewed as a mapping of six-dimensional row vectors:

$$(a_{1,1}, a_{2,1}, a_{2,2}, a_{3,1}, a_{3,2}, a_{3,3}) \rightarrow (a_{1,1}, a_{2,1}, a_{2,2}, a_{3,1}, a_{3,2}, a_{3,3}) \, Z^t.$$

12.38 Hessian-Related Basis Functions

In the present study of the tetrahedron t, the above matrix S is set to be the transpose of the inverse of S_t:

$$S \leftarrow S_t^{-t},$$

so the 6×6 matrix Z takes the concrete value

$$Z = Z(S_t^{-t}).$$

This special case is particularly relevant to the second partial derivatives of a composite function defined in t. Indeed, for any function $v(x, y, z)$ with continuous second partial derivatives, $\nabla \nabla^t v$ is a 3×3 symmetric matrix. Therefore, $\nabla \nabla^t v$ can also be represented uniquely as the six-dimensional row vector

$$(v_{xx}, v_{xy}, v_{yy}, v_{xz}, v_{yz}, v_{zz}).$$

We are now ready to define the basis functions in t whose partial derivatives vanish at the corners and edge and side midpoints of t, except for only one second partial derivative at one corner $\mathbf{i}$ that takes the value 1. In fact, the six basis functions are defined compactly by

$$\begin{pmatrix} R_{|\mathbf{i}|+\hat{\imath}_{0,0,2},t} \\ R_{|\mathbf{i}|+\hat{\imath}_{0,1,1},t} \\ R_{|\mathbf{i}|+\hat{\imath}_{0,2,0},t} \\ R_{|\mathbf{i}|+\hat{\imath}_{1,0,1},t} \\ R_{|\mathbf{i}|+\hat{\imath}_{1,1,0},t} \\ R_{|\mathbf{i}|+\hat{\imath}_{2,0,0},t} \end{pmatrix} \equiv Z^{-t} \begin{pmatrix} r_{|\mathbf{i}|+\hat{\imath}_{0,0,2},t} \\ r_{|\mathbf{i}|+\hat{\imath}_{0,1,1},t} \\ r_{|\mathbf{i}|+\hat{\imath}_{0,2,0},t} \\ r_{|\mathbf{i}|+\hat{\imath}_{1,0,1},t} \\ r_{|\mathbf{i}|+\hat{\imath}_{1,1,0},t} \\ r_{|\mathbf{i}|+\hat{\imath}_{2,0,0},t} \end{pmatrix}.$$

Indeed, the Hessian operator can be applied separately to each of these six functions, to map it to the six-dimensional row vector of its second partial derivatives and form the following 6×6 matrix:

$$\nabla\nabla^t \begin{pmatrix} R_{|\mathbf{i}|+\hat{i}_{0,0,2},t} \\ R_{|\mathbf{i}|+\hat{i}_{0,1,1},t} \\ R_{|\mathbf{i}|+\hat{i}_{0,2,0},t} \\ R_{|\mathbf{i}|+\hat{i}_{1,0,1},t} \\ R_{|\mathbf{i}|+\hat{i}_{1,1,0},t} \\ R_{|\mathbf{i}|+\hat{i}_{2,0,0},t} \end{pmatrix} = Z^{-t} \begin{pmatrix} \nabla\nabla^t r_{|\mathbf{i}|+\hat{i}_{0,0,2},t} \\ \nabla\nabla^t r_{|\mathbf{i}|+\hat{i}_{0,1,1},t} \\ \nabla\nabla^t r_{|\mathbf{i}|+\hat{i}_{0,2,0},t} \\ \nabla\nabla^t r_{|\mathbf{i}|+\hat{i}_{1,0,1},t} \\ \nabla\nabla^t r_{|\mathbf{i}|+\hat{i}_{1,1,0},t} \\ \nabla\nabla^t r_{|\mathbf{i}|+\hat{i}_{2,0,0},t} \end{pmatrix}$$

$$= Z^{-t} \begin{pmatrix} (\nabla\nabla^t p_{|\mathbf{i}|+\hat{i}_{0,0,2}}) \circ E_t^{-1} \\ (\nabla\nabla^t p_{|\mathbf{i}|+\hat{i}_{0,1,1}}) \circ E_t^{-1} \\ (\nabla\nabla^t p_{|\mathbf{i}|+\hat{i}_{0,2,0}}) \circ E_t^{-1} \\ (\nabla\nabla^t p_{|\mathbf{i}|+\hat{i}_{1,0,1}}) \circ E_t^{-1} \\ (\nabla\nabla^t p_{|\mathbf{i}|+\hat{i}_{1,1,0}}) \circ E_t^{-1} \\ (\nabla\nabla^t p_{|\mathbf{i}|+\hat{i}_{2,0,0}}) \circ E_t^{-1} \end{pmatrix} Z^t.$$

Clearly, at the corner $\mathbf{i}$ of t, the middle term in the above triple product is just the 6×6 identity matrix I, leading to

$$\nabla\nabla^t \begin{pmatrix} R_{|\mathbf{i}|+\hat{i}_{0,0,2},t}(\mathbf{i}) \\ R_{|\mathbf{i}|+\hat{i}_{0,1,1},t}(\mathbf{i}) \\ R_{|\mathbf{i}|+\hat{i}_{0,2,0},t}(\mathbf{i}) \\ R_{|\mathbf{i}|+\hat{i}_{1,0,1},t}(\mathbf{i}) \\ R_{|\mathbf{i}|+\hat{i}_{1,1,0},t}(\mathbf{i}) \\ R_{|\mathbf{i}|+\hat{i}_{2,0,0},t}(\mathbf{i}) \end{pmatrix} = Z^{-t} I Z^t = I,$$

as required.

It is easy to see that the second partial derivatives of these functions vanish at every corner other than $\mathbf{i}$. Furthermore, their first partial derivatives vanish at every corner and edge and side midpoint of t, as required. This guarantees that they are indeed basis functions in t.

12.39 Basis Functions at Edge Midpoints

Finally, let us define the remaining basis functions $R_{i,t}$ ($40 \leq i < 56$) in t. Consider, for example, the two degrees of freedom $40 \leq i, i+1 < 56$ corresponding to the normal derivatives at the edge midpoint $w \in e$, where e is some edge in the unit tetrahedron T. Let us denote these normal directions by the 3-d unit vectors $\mathbf{n}^{(1)}$ and $\mathbf{n}^{(2)}$. Let $\mathbf{n}^{(3)}$ be the unit vector that is tangent to e. As follows from Section 12.42 below, both basis functions p_i and p_{i+1} must vanish in e, so their tangential derivative along e (and, in particular, at its midpoint) must vanish as well:

$$\nabla^t p_i(w)\mathbf{n}^{(3)} = \nabla^t p_{i+1}(w)\mathbf{n}^{(3)} = 0.$$

Define the 3×3 matrix formed by these three unit column vectors:

$$\mathbf{N}_e \equiv \left(\mathbf{n}^{(1)} \mid \mathbf{n}^{(2)} \mid \mathbf{n}^{(3)} \right).$$

Although $\mathbf{n}^{(1)}$ and $\mathbf{n}^{(2)}$ are the directions specified in T, they are not necessarily the directions we want to be used in the derivatives in t. In fact, in what follows we choose more proper direction vectors in t.

Suppose that, at the edge midpoint $E_t(w) \in E_t(e) \subset t$, we want to use the directions specified by the columns of the 3×2 matrix $S_t \mathbf{N}_e Y_{t,e}$, where $Y_{t,e}$ is a 3×2 matrix chosen arbitrarily in advance. The only condition that must be satisfied by $S_t \mathbf{N}_e Y_{t,e}$ is that its columns span a plane that is not parallel to the edge $E_t(e)$, that is, it is crossed by the line that contains $E_t(e)$. For example, if $Y_{t,e}$ contains the two first columns in $(S_t \mathbf{N}_e)^{-1}$, then

$$S_t \mathbf{N}_e Y_{t,e} = \begin{pmatrix} 1 & 0 \\ 0 & 1 \\ 0 & 0 \end{pmatrix},$$

so the relevant basis functions defined below in t correspond to the x- and y-partial derivatives at $E_t(w) \in t$. This is our default choice.

The above default choice is possible so long as $E_t(e)$ is not parallel to the x-y plane. If it is, then one may still choose a matrix $Y_{t,e}$ that contains the first and third (or the second and third) columns in $(S_t \mathbf{N}_e)^{-1}$. In this case,

$$S_t \mathbf{N}_e Y_{t,e} = \begin{pmatrix} 1 & 0 \\ 0 & 0 \\ 0 & 1 \end{pmatrix},$$

so the relevant basis functions defined below in t correspond to the x- and z-partial derivatives at $E_t(w) \in t$.

Fortunately, the condition that $E_t(e)$ is not parallel to the x-y plane is a geometric condition. Therefore, it can be checked not only from t but also from any other tetrahedron that uses $E_t(e)$ as an edge. Thus, the desirable basis function whose x- (or y-) partial derivative is 1 at $E_t(w)$ can be defined not only in t but also in every other tetrahedron that shares $E_t(e)$ as an edge. Furthermore, the individual basis functions defined in this way in the individual tetrahedra that share $E_t(e)$ as an edge can be combined to form a global piecewise-polynomial basis function in the entire mesh, whose x- (or y-) partial derivative at $E_t(w)$ is 1, whereas all its other degrees of freedom in each tetrahedron vanish. As we'll see below, this defines a continuous basis function in the entire mesh, with a continuous gradient across the edges.

Returning to our individual tetrahedron t, let $\hat{Y}_{t,e}$ be the 2×2 matrix that contains the first two rows in $Y_{t,e}$. In other words,

$$Y_{t,e} = \begin{pmatrix} \hat{Y}_{t,e} \\ \beta \quad \gamma \end{pmatrix},$$

where β and γ are some scalars.

Let us show that $\hat{Y}_{t,e}$ is nonsingular (invertible). Indeed, let $e^{(1)}$ and $e^{(2)}$ denote the endpoints of e, so

$$w = \frac{e^{(1)} + e^{(2)}}{2}.$$

Then we have

$$S_t^{-1}\left(E_t(e^{(2)}) - E_t(e^{(1)})\right) = S_t^{-1}\left(S_t(e^{(2)}) - S_t(e^{(1)})\right) = e^{(2)} - e^{(1)} = \alpha n^{(3)}$$

for some nonzero scalar α. Thanks to the fact that $n^{(1)}$, $n^{(2)}$, and $n^{(3)}$ are orthonormal, we have

$$\mathbf{N}_e^t S_t^{-1}\left(E_t(e^{(2)}) - E_t(e^{(1)})\right) = \alpha \mathbf{N}_e^t n^{(3)} = \begin{pmatrix} 0 \\ 0 \\ \alpha \end{pmatrix}.$$

Furthermore, thanks to the fact that $\mathbf{N}_e$ is orthogonal, we also have

$$(S_t \mathbf{N}_e)^{-1} = \mathbf{N}_e^{-1} S_t^{-1} = \mathbf{N}_e^t S_t^{-1}.$$

Using also our default choice, we have

$$(S_t \mathbf{N}_e)^{-1}\left(E_t(e^{(2)}) - E_t(e^{(1)}) \mid \begin{matrix} 1 & 0 \\ 0 & 1 \\ 0 & 0 \end{matrix} \right) = \begin{pmatrix} 0 \\ 0 \mid Y_{t,e} \\ \alpha \end{pmatrix} = \begin{pmatrix} 0 & \hat{Y}_{t,e} \\ 0 \mid \beta \quad \gamma \\ \alpha \end{pmatrix}.$$

Using also our default assumption that $E_t(e)$ is not parallel to the x-y plane, or that

$$\left(E_t(e^{(2)}) - E_t(e^{(1)})\right)_3 \neq 0,$$

we have that the above left-hand side is a triple product of three nonsingular matrices. As a consequence, the matrix in the right-hand side is nonsingular as well, which implies that $\hat{Y}_{t,e}$ is nonsingular as well, as asserted.

Using the above result, we can now define the basis functions in t by

$$\begin{pmatrix} R_{i,t} \\ R_{i+1,t} \end{pmatrix} \equiv \hat{Y}_{t,e}^{-1} \begin{pmatrix} r_{i,t} \\ r_{i+1,t} \end{pmatrix}.$$

Let us verify that these are indeed basis functions in t in the sense that that they take the value 1 only for one of the directional derivatives at $E_t(w)$

(in the direction specified by one of the columns in $S_t N_e Y_{t,e}$) and zero for the other one:

$$\begin{pmatrix} \nabla^t R_{i,t} \\ \nabla^t R_{i+1,t} \end{pmatrix} S_t N_e Y_{t,e} = \hat{Y}_{t,e}^{-1} \begin{pmatrix} \nabla^t r_{i,t} \\ \nabla^t r_{i+1,t} \end{pmatrix} S_t N_e Y_{t,e}$$

$$= \hat{Y}_{t,e}^{-1} \begin{pmatrix} (\nabla^t p_i) \circ E_t^{-1} \\ (\nabla^t p_{i+1}) \circ E_t^{-1} \end{pmatrix} S_t^{-1} S_t N_e Y_{t,e}$$

$$= \hat{Y}_{t,e}^{-1} \begin{pmatrix} (\nabla^t p_i) \circ E_t^{-1} \\ (\nabla^t p_{i+1}) \circ E_t^{-1} \end{pmatrix} N_e Y_{t,e}.$$

At the edge midpoint $E_t(w) \in E_t(e) \subset t$, the middle term in this triple product takes the value

$$\begin{pmatrix} (\nabla^t p_i) \circ E_t^{-1}(E_t w) \\ (\nabla^t p_{i+1}) \circ E_t^{-1}(E_t w) \end{pmatrix} N_e = \begin{pmatrix} \nabla^t p_i(w) \\ \nabla^t p_{i+1}(w) \end{pmatrix} N_e = \begin{pmatrix} 1 & 0 & 0 \\ 0 & 1 & 0 \end{pmatrix},$$

so the entire triple product is just the 2×2 identity matrix, as required.

12.40 Basis Functions at Side Midpoints

The basis function associated with the midpoint of the side $E_t(s) \subset t$ (where s is a side of T) is defined in a similar way, except that here $\mathbf{n}^{(1)}$ is normal to s, $\mathbf{n}^{(2)}$ and $\mathbf{n}^{(3)}$ are orthonormal vectors that are tangent (parallel) to s, $\mathbf{N}_s$ is the orthogonal matrix whose columns are $n^{(1)}$, $n^{(2)}$, and $n^{(3)}$, $Y_{t,s}$ is a 3×1 matrix (a 3-d column vector) rather than a 3×2 matrix, and $\hat{Y}_{t,s}$ is a 1×1 matrix (a mere scalar) rather than a 2×2 matrix. [Here $Y_{t,s}$ must be chosen in such a way that the column vector $S_t N_s Y_{t,s}$ is not parallel to $E_t(s)$; for example, if $E_t(s)$ is not parallel to the x-axis, then $Y_{t,s}$ could be the first column in $(S_t N_s)^{-1}$.] This is our default choice, provided that $E_t(s)$ is not parallel to the x-axis, or that

$$\left(S_t n^{(2)} \times S_t n^{(3)} \right)_1 = \det \left(\begin{pmatrix} \left(S_t n^{(2)} \right)_2 & \left(S_t n^{(3)} \right)_2 \\ \left(S_t n^{(2)} \right)_3 & \left(S_t n^{(3)} \right)_3 \end{pmatrix} \right) \neq 0.$$

(As above, this condition guarantees that $\hat{Y}_{t,s} \neq 0$.) This completes the definition of the basis functions in t.

12.41 Continuity

Let t_1 and t_2 be two neighbor tetrahedra that share the joint side s. (Without loss of generality, assume that s is not parallel to the x-axis, so its edges are not parallel to the x-axis as well.) Furthermore, let u_1 be a basis function in t_1 and u_2 a basis function in t_2. Assume that u_1 and u_2 share the same degrees of freedom in s, that is, they have zero degrees of freedom in s, except at most one degree of freedom, say

$$(u_1)_x(\mathbf{j}) = (u_2)_x(\mathbf{j}) = 1,$$

where $\mathbf{j}$ is either a corner or a midpoint or an edge midpoint of s.

Thus, both u_1 and u_2 agree on every partial derivative (of order at most 2) in every corner of s. Furthermore, they also agree on the normal derivatives at the edge midpoints of s. Below we'll use these properties to show that they must agree on the entire side s.

Let us view u_1 and u_2 as functions of two variables defined in s only. As such, these functions agree on six tangential derivatives in s (of orders 0, 1, and 2) at each corner of s. Furthermore, they also agree on some tangential derivative in s at each edge midpoint of s (in a direction that is not parallel to this edge). Thus, the total number of degrees of freedom for which both of these functions agree in s is

$$3 \cdot 6 + 3 = 21.$$

Fortunately, this is also the number of monomials in a polynomial of two variables of degree five:

$$\binom{5+2}{2} = 21.$$

Therefore, a polynomial of two variables of degree five is determined uniquely by 21 degrees of freedom. Since both u_1 and u_2 are polynomials of degree five in s that share the same 21 degrees of freedom in s, they must be identical in the entire side s.

The above result can be used to define the function

$$u(x, y, z) \equiv \begin{cases} u_1(x, y, z) & \text{if } (x, y, z) \in t_1 \\ u_2(x, y, z) & \text{if } (x, y, z) \in t_2. \end{cases}$$

This function is continuous in the union of tetrahedra

$$t_1 \cup t_2.$$

Furthermore, it is a polynomial of degree five in t_1 and also a (different) polynomial of degree five in t_2. Such functions will be used later in the book to define continuous basis functions in the entire mesh of tetrahedra.

12.42 Continuity of Gradient

Assume now that t_1 and t_2 share an edge e rather than a side. In this case, u_1 and u_2 are basis functions in t_1 and t_2 (respectively) that share the same degrees of freedom in e: they have the same tangential derivatives along e (up to and including order 2) at the endpoints of e, and also have the same directional derivatives at the midpoint of e in some directions that are not parallel to e. More precisely, they have zero degrees of freedom at these points, except at most one degree of freedom, say

$$(u_1)_x(\mathbf{j}) = (u_2)_x(\mathbf{j}) = 1,$$

where $\mathbf{j}$ is either an endpoint or a midpoint of e (assuming that e is not parallel to x-axis).

Let us view u_1 and u_2 as polynomials of one variable in e. As such, they share six degrees of freedom in e: the tangential derivatives along e (of order 0, 1, and 2) at the endpoints of e. Thanks to the fact that a polynomial of one variable of degree five has six monomials in it, it is determined uniquely by six degrees of freedom. As a result, both u_1 and u_2 must be identical in the entire edge e.

Clearly, because u_1 and u_2 are identical in e, they also have the same tangential derivative along it. Below we'll see that they also have the same normal derivatives in e, so that in summary they have the same gradient in e.

Indeed, consider a unit vector $\mathbf{n}$ that is not parallel to e. The directional derivatives $\nabla^t u_1 \cdot \mathbf{n}$ and $\nabla^t u_2 \cdot \mathbf{n}$ can be viewed as polynomials of one variable of degree four in e. As such, they must be identical in the entire edge e, since they agree on five degrees of freedom in it: the tangential derivatives along e (of order 0 and 1) at the endpoints, and also the value of function itself at the midpoint.

The above results can be used to define the function

$$u(x, y, z) \equiv \begin{cases} u_1(x, y, z) & \text{if } (x, y, z) \in t_1 \\ u_2(x, y, z) & \text{if } (x, y, z) \in t_2. \end{cases}$$

This function is not only continuous in the union of tetrahedra

$$t_1 \cup t_2,$$

but also has a continuous gradient across the joint edge e.

12.43 Integral over a General Tetrahedron

The linear mapping $E_t : T \to t$ can also be used in a formula that helps computing the integral of a given function $F(x, y, z)$ over the general tetrahedron t:

$$\int \int \int_t F(x, y, z)\,dxdydz = |\det(S_t)| \int \int \int_T F \circ E_t(x, y, z)\,dxdydz.$$

Here the original function F is defined in t, so the composite function $F \circ E_t$ is well defined (and therefore can indeed be integrated) in the unit tetrahedron T. For example, if F is a product of two functions

$$F(x, y, z) = f(x, y, z) g(x, y, z),$$

then the above formula takes the form

$$\int \int \int_t f g \, dx dy dz = |\det(S_t)| \int \int \int_T (f \circ E_t)(g \circ E_t) dx dy dz.$$

Furthermore, the inner product of the gradients of f and g can be integrated in t by

$$\int \int \int_t \nabla^t f \nabla g \, dx dy dz$$

$$= |\det(S_t)| \int \int \int_T ((\nabla^t f) \circ E_t)((\nabla g) \circ E_t) dx dy dz$$

$$= |\det(S_t)| \int \int \int_T ((\nabla^t (f \circ E_t \circ E_t^{-1})) \circ E_t)((\nabla(g \circ E_t \circ E_t^{-1})) \circ E_t) dx dy dz$$

$$= |\det(S_t)| \int \int \int_T$$

$$((\nabla^t (f \circ E_t) \circ E_t^{-1}) \circ E_t) S_t^{-1} S_t^{-t} ((\nabla(g \circ E_t) \circ E_t^{-1}) \circ E_t) dx dy dz$$

$$= |\det(S_t)| \int \int \int_T \nabla^t (f \circ E_t) S_t^{-1} S_t^{-t} \nabla(g \circ E_t) dx dy dz.$$

This formula will be most helpful in the applications later on in the book.

12.44 Exercises

1. Let $u = (u_i)_{0 \le i \le n}$ be an $(n+1)$-dimensional vector, and $v = (v_i)_{0 \le i \le m}$ be an $(m+1)$-dimensional vector. Complete both u and v into $(n+m+1)$-dimensional vectors by adding dummy zero components:

$$u_{n+1} = u_{n+2} = \cdots = u_{n+m} = 0$$

and

$$v_{m+1} = v_{m+2} = \cdots = v_{n+m} = 0.$$

Define the convolution of u and v, denoted by $u * v$, to be the $(n+m+1)$-dimensional vector with the components

$$(u * v)_k = \sum_{i=0}^{k} u_i v_{k-i}, \qquad 0 \le k \le n+m.$$

Show that

$$u * v = v * u.$$

2. Let p be the polynomial of degree n whose vector of coefficients is u:

$$p(x) = \sum_{i=0}^{n} u_i x^i.$$

Similarly, let q be the polynomial of degree m whose vector of coefficients is v:

$$q(x) = \sum_{i=0}^{m} v_i x^i.$$

Show that the convolution vector $u * v$ is the vector of coefficients of the product polynomial pq.

3. The infinite Taylor series of the exponent function $\exp(x) = e^x$ is

$$\exp(x) = 1 + x + \frac{x^2}{2!} + \frac{x^3}{3!} + \frac{x^4}{4!} + \cdots = \sum_{n=0}^{\infty} \frac{x^n}{n!}$$

(see [7] [19]).

For moderate $|x|$, this series can be approximated by the Taylor polynomial of degree k, obtained by truncating the above series after $k + 1$ terms:

$$\exp(x) \doteq \sum_{n=0}^{k} \frac{x^n}{n!}.$$

Write a version of Horner's algorithm to compute this polynomial efficiently for a given x. The solution can be found in Chapter 14, Section 14.10.

4. The infinite Taylor series of the sine function $\sin(x)$ is

$$\sin(x) = x - \frac{x^3}{3!} + \frac{x^5}{5!} - \frac{x^7}{7!} + \cdots = \sum_{n=0}^{\infty} (-1)^n \frac{x^{2n+1}}{(2n+1)!}.$$

For moderate $|x|$, this series can be approximated by the Taylor polynomial of degree $2k + 1$, obtained by truncating the above series after $k + 1$ terms:

$$\sin(x) \doteq \sum_{n=0}^{k} (-1)^n \frac{x^{2n+1}}{(2n+1)!}.$$

Write a version of Horner's algorithm to compute this polynomial efficiently for a given x.

5. The infinite Taylor series of the cosine function $\cos(x)$ is

$$\cos(x) = 1 - \frac{x^2}{2!} + \frac{x^4}{4!} - \frac{x^6}{6!} + \cdots = \sum_{n=0}^{\infty} (-1)^n \frac{x^{2n}}{(2n)!}.$$

For moderate $|x|$, this series can be approximated by the Taylor polynomial of degree $2k$, obtained by truncating the above series after $k + 1$ terms:

$$\cos(x) \doteq \sum_{n=0}^{k} (-1)^n \frac{x^{2n}}{(2n)!}.$$

Write a version of Horner's algorithm to compute this polynomial efficiently for a given x.

6. Use the above Taylor series to show that, for a given imaginary number of the form ix (where $i = \sqrt{-1}$ and x is some real number),

$$\exp(ix) = \cos(x) + i \cdot \sin(x).$$

7. Use the above Taylor series to show that the derivative of $\exp(x)$ is $\exp(x)$ as well:

$$\exp'(x) = \exp(x).$$

8. Use the above Taylor series to show that the derivative of $\sin(x)$ is

$$\sin'(x) = \cos(x).$$

9. Use the above Taylor series to show that the derivative of $\cos(x)$ is

$$\cos'(x) = -\sin(x).$$

10. Conclude that

$$\cos''(x) = -\cos(x).$$

11. Conclude also that

$$\sin''(x) = -\sin(x).$$

12. Show that the Hessian is a 3×3 symmetric matrix.
13. Show that the polynomials computed in Section 12.31 are indeed basis functions in the unit tetrahedron T.
14. Show that, with the default choice at the end of Section 12.40,

$$\hat{Y}_{t,s} \neq 0.$$

15. In what case the default choice for $Y_{t,s}$ at the end of Section 12.40 cannot be used? What choice should be used instead? Why is the above inequality still valid in this case as well?
16. Use the above inequality to define the four basis functions associated with the side midpoints in a general tetrahedron t.
17. Show that the $R_{i,t}$'s ($0 \le i < 56$) in Sections 12.36–12.40 are indeed basis functions in the general tetrahedron t.

Part IV

Introduction to C

Introduction to C

So far, we have studied the theoretical background of some useful mathematical objects, including numbers, geometrical objects, and composite objects such as sets and graphs. Still, in order to comprehend these objects and their features fully, it is advisable to implement them on a computer.

In other words, the implementation of the mathematical objects on a computer is useful not only to develop algorithms to solve practical problems but also to gain a better understanding of the objects and their functions.

For this purpose, object-oriented programming languages such as C++ are most suitable: they focus not only on the algorithms to solve specific problems, but mainly on the objects themselves and their nature. This way, the process of programming is better related to the original mathematical background. In fact, the programming helps to get a better idea about the objects and why and how they are designed. The objects can thus help not only to develop algorithms and solve problems but also to comprehend mathematical ideas and theories.

Before we can use C++ to define new mathematical objects, we must first learn a more elementary programming language: C. Strictly speaking, C is not really an object-oriented programming language. However, it serves as a necessary framework, on top of which C++ is built.

The main advantage of C is in the opportunity to define new functions that take some input numbers (or arguments) to produce the output number (the returned value). The opportunity to use functions makes the computer program (or code) much more modular: each function is responsible only for a particular task, for which only a few instructions (or commands) are needed. These commands may by themselves invoke (or call) other functions.

This modularity of the computer program is achieved by multilevel programming: The original task to solve the original problem is accomplished by calling the main function. (This can be viewed as the highest level of the code.) For this purpose, the main function uses (or calls) some other functions to accomplish the required subtasks and solve the required subproblems. This can be viewed as the next lower level in the code, and so on, until the most elementary functions are called in the lowest level to make simple calculations without calling any further functions.

This multilevel programming is also helpful in debugging: finding errors in the code. For this purpose, one only needs to find the source of the error in the main function. This source may well be a command that calls some other function. Then, one needs to study the call to this function (with the specific input arguments used in it) to find the source of the error. The debugging process may then continue recursively, until the exact source of the error is found in the lowest level of the code.

Functions are also important to make the code elegant and easy to read. Since each function is responsible only to a very specific task and contains only the few commands required to accomplish it, it is easy to read and understand.

Thus, other programmers may modify it if necessary and use it in their own applications. Furthermore, even the writer of the original code would find it much easier to remember what its purpose was when reading it again after a time.

In other words, a code should be efficient not only in terms of minimum computing resources, but also in terms of minimum human resources. Indeed, even a code that requires little computer memory and runs fairly quickly on it to solve the original problem may be useless if it is so complicated and hard to read that nobody can modify it to solve other problems or even other instances of the original problem. A well-written code, on the other hand, may provide not only the main function required to solve the original problem but also many other well-written functions that can be used in many other applications.

Chapter 13

Basics of Programming

In the first half of the book, we have studied some elementary mathematical objects such as numbers, geometrical objects, sets, trees, and graphs. In this half of the book, we are going to implement them on the computer. This is done not only for solving practical problems, but also (and mainly) to get a better idea about these abstract objects, and get a better insight about their nature and features. For this purpose, it is particularly important that the mathematical objects used in the code enjoy the same features, functions, operations, and notations as in their original mathematical formulation. This way, the program can use the same point of view used in the very definition of the mathematical objects. Furthermore, the programming may even contribute to a better understanding and feeling of the mathematical objects, and help develop both theory and algorithms.

The most natural treatment of mathematical objects is possible in object-oriented programming languages such as C++. First, however, we study the C programming language, which is the basis for C++. Strictly speaking, C is not an object-oriented programming language; still, integer and real numbers are well implemented in it. Furthermore, the variables that can be defined in it are easily referred to by their addresses. Moreover, C supports recursion quite well, which enables a natural implementation of recursive functions.

13.1 The Computer and its Memory

The computer is a tool to solve computational problems that are too difficult or big to be solved by humans. Thanks to its large memory and strong computational power, the computer has quite a good chance to solve problems that are prohibitively large for the human mind. Still, even the most powerful computer won't be able to solve a computational problem in real (acceptable) time, unless it has an efficient algorithm (method, or list of instructions) to do this.

Furthermore, the algorithm must be implemented (written) in a language understandable by the computer, namely, as a computer program or code. The program must not only be efficient in the sense that it avoids redundant computations and data fetching from the memory, but also modular and elegant to make it easy to debug from errors and also easy to improve and extend to other applications whenever needed.

The computer consists of three basic elements: memory to store data, processor to fetch data from the memory and use them to perform calculations, and input/output (I/O) devices, such as keyboard, mouse, screen, and printer to obtain data and return answers to the user.

The memory of the computer is based on semiconductors. Each semiconductor can be set to two possible states, denoted by 0 and 1. Using the binary representation, several semiconductors (or bits) can form together a natural number. Furthermore, by adding an extra bit to contain the sign of the number, integer numbers can be implemented as well.

Moreover, for some fixed number n, say $n = 4$ or $n = 8$, two lists of bits $a_1 a_2 \cdots a_n$ and $b_1 b_2 \cdots b_n$ can form a fairly good approximation to any real number in the form

$$\pm 0.a_1 a_2 \cdots a_n \cdot \exp(\pm b_1 b_2 \cdots b_n).$$

This is called the "float" implementation of the real number. A slightly more accurate approximation to the real number can be obtained by the "double" implementation, in which the above lists of bits are of length $2n$ rather than n.

The memory of the computer is divided into two main parts: the primary memory, a small device near the processor, which can be accessed easily and quickly to perform immediate calculations, and the secondary memory, a big and slow device, which lies farther away from the processor, and contains big files, such as data files and programs. To perform calculations, the processor must first fetch the required data from the secondary memory and place it in the primary memory for further use. It is therefore advisable to exploit data that already lie in the primary memory as much as possible, before they are being returned to the secondary memory. This way, the processor may avoid expensive (time consuming) data fetching from the secondary memory whenever possible.

13.2 The Program or Code

In order to solve a computational problem, one must have an algorithm (or method): a list of instructions (or commands) that should be executed to produce the desired solution. Most often, the algorithm contains too many instructions for a human being to perform; it must be therefore fed into the computer which has a sufficiently large computational power.

Unfortunately, the computer doesn't understand English or any other natural (human) language. The algorithm must therefore be written in a formal, unambiguous context-free language: a programming language.

Actually, the computer understands only a very explicit programming language, which tells it very specifically what datum to fetch from the memory, what arithmetic operation to perform on it, and where exactly to store the result. This is called the low-level (or machine) language. Writing in such a tedious manner would be quite impractical even for the best of programmers. Fortunately, the programmer doesn't have to write his/her program in the machine language, but rather in a high-level programming language, such as C.

The high-level programming language is much easier to use. It uses certain words from English, called keywords or reserved words, to refer to common programming tools such as logical conditions, loops, etc. Furthermore, it uses certain characters to denote arithmetic and logical Boolean operations.

Once the high-level code is complete, it is translated to machine language by the compiler. The compiler is a software that can be applied to a code that has been written properly in a high-level programming language,   to produce the required machine-language code executable by the computer. For example, the C compiler can be applied to a well-written C code to produce the final machine code ready to be executed (run) on the computer.

The stage in which the compiler translates the high-level code written by the programmer into the machine language executable by the computer is called compilation time. The next stage, in which this translated program is actually executed by the computer, is called run time. Variables (or memory cells) that are unspecified in compilation time and are assigned meaningful values only in run time are called dynamic variables.

13.3 The Code Segments in this Book

The code segments in this book are fully debugged and tested. They are compiled with the standard GNU compiler. To use this compiler, the program must be placed in a file called "program.cxx". On the UNIX operating system, this compiler is then invoked by the commands

```
>> g++ program.cxx
>> a.out
```

The output of the program "program.cxx" is then printed onto the screen. When the program produces a lot of output, it can also be printed into a file named "Output" by the commands

```
>> g++ program.cxx
```

```
>> a.out > Output
```

The output can then be read from the file by using, e.g., the "vi" editor:

```
>> vi Output
```

One can also use the Windows operating system to compile and run C++ programs, but this requires some extra linking commands.

The GNU compiler used here is one of the most widely used C++ compilers. Other compilers may require slight modifications due to some other restrictions, requirements, or properties. In principle, the code is suitable for other compilers as well.

Our convention is that words quoted from code are placed in quotation marks. Double quotation marks are used for strings (e.g., "const"), and single quotation marks are used for single characters (e.g., 'c'). When the word quoted from the code is a function name, it is often followed by "()", e.g., "main()".

Each command in the code ends with the symbol ';'. Commands that are too long are broken into several consecutive code lines. These code lines are interpreted as one continuous code line that lasts until the symbol ';' at the end of the command.

The code segments are presented in nested-block style; that is, an inner block is shifted farther to the right than the outer block that contains it. A code line that belongs to a particular inner block is also shifted to the right in the same way even when it is on its own to indicate that it is not just an isolated code line but actually belongs to this block.

Let us now introduce the main elements and tools required to write a proper C program.

13.4 Variables and Types

A variable is a small space in the computer memory to store a fixed amount of data, which is then interpreted as the current value of the variable. As discussed below, there may be several ways to interpret this data, depending on the particular type of the variable.

Thus, the type of a variable is actually the way the data stored in it is interpreted. For example, in a variable of type "int" (integer), the first bit determines the sign of the integer number (plus or minus), whereas the rest of the bits are interpreted as the binary digits (0 or 1) that produce the binary representation of the number.

In a variable of type "float", on the other hand, the first 4 (or 8) bits are interpreted as the binary digits that form a binary fraction (the coefficient), whereas the next 4 (or 8) bits form the binary representation of the exponent.

The product of this coefficient and this exponent provide a good approximation to the required real number.

A yet better approximation to the real number under consideration can be stored in a variable of type "double". Indeed, in such a variable, the 8 (or even 16) first bits are used to form the binary fraction (the coefficient), whereas the next 8 (or 16) bits are used to form the exponent. These larger numbers of bits provide a better precision in approximating real numbers on the computer.

The user of the variables, however, is completely unaware of the different interpretations that may be given to the data stored in the bits. After all, the user uses the variables only through the functions available in the programming language. These functions form the interface by which the user can access and use the variables. For example, the user can call the function '/' to divide two variables named i and j by writing i/j. As we'll see below, the function invoked by the compiler depends on the type of i and j: if they are variables of type "int", then division with residual is invoked, so the returned value (the result i/j) is an integer variable as well. If, on the other hand, they are variables of type "float" or "double", then division without residual is invoked, returning the "float" or "double" variable i/j. In both cases, the value i/j is returned by the division function in a variable with no name, which exists only in the code line in which the call to the function is made, and disappears right after the ';' symbol that marks the end of this code line.

Thus, the type of the variable becomes relevant to the user only when interface functions, such as arithmetic and Boolean operators, are applied to it. The types that are used often in C are "int", "float", "double", and "char" (character). The "char" variable is stored as an unsigned integer number. More precisely, it contains 8 bits to form a binary number between 0 and 255. Each number is interpreted as one of the keys on the keyboard, including low-case letters, upper-case letters, digits, and special symbols. As we'll see below, variables of type "char" have some extra functions to read them from the screen (or from a file) and to print them onto the screen (or onto a file).

Variables of the above types can be viewed as objects, which can be manipulated by interface functions, such as arithmetic and logical operators and read/write functions. Below we'll see that in C++ the user can define his/her own objects, with their special interface functions to manipulate and use them. This is why C++ can be viewed as an object-oriented extension of C. Here, however, we stick to the above four types available in C and to their interface functions that are built in the C compiler.

13.5 Defining Variables

A variable in C is defined (allocated memory) by writing the type of the variable (say "int") followed by some name to refer to it. For example, the code line "int i;" allocates sufficient space in the computer memory for a variable of type integer. The data placed in this space, that is, the value assigned to the variable 'i', can then be accessed through the name of the variable, namely, 'i'.

The command line (or instruction to the computer) ends with the symbol ';'. A short command line such as "int i;" can be written in one code line in the program. Longer instructions, on the other hand, may occupy several code lines. Still, the instruction ends only upon reaching the ';' symbol. For example, one could equally well write "int" in one code line and "i;" in the next one, with precisely the same meaning as before.

Similarly, one can define variables of types "float", "double", and "char":

```
int i;
float a;
double x;
char c;
```

When this code is executed, sufficient memory is allocated to store an integer number, a real number, a double-precision real number, and a character.

The integer variable may take every integer value, may it be negative, positive, or zero. Both "float" and "double" variables may take every real value. The character variable may take only nonnegative integer values between 0 and 255. Each of these potential values represents a character on the keyboard, such as a letter in English (a lowercase or a capital letter), a digit, an arithmetic symbol, or any other special symbol on the keyboard.

13.6 Assignment

As discussed above, variables are referred to by their names ('i', 'a', 'x', and 'c' in the above example). Here we show how these names can be used to assign values to variables.

Upon definition, variables are assigned meaningless random values. More meaningful values can be then assigned in assignment commands:

```
i = 0;
a = 0.;
x = 0.;
```

```
c = 'A';
```

Note that '0' stands for the integer number zero, whereas "0." stands for the real number zero.

A command in C is also a function that not only carries out some operation but also returns a value. In particular, the assignment operator '=' used above not only assigns a value to a particular variable but also returns this value as an output for future use. This feature can be used to write

```
x = a = i = 0;
```

This command is executed from right to left. First, the integer value 0 is assigned to 'i'. This assignment also returns the assigned integer number 0, which in turn is converted implicitly to the real number 0. and assigned to the "float" variable 'a'. This assignment also returns the (single-precision) real number 0., which in turn is converted implicitly to the (double-precision) real number 0. and assigned to the "double" variable 'x'. Thus, the above command is the same as

```
i = 0;
a = 0.;
x = 0.;
```

13.7 Initialization

The above approach, in which the variables initially contain meaningless values before being assigned meaningful values, is somewhat inefficient. After all, why not assign to them with meaningful values immediately upon definition? Fortunately, one can indeed avoid the above assignment operation by defining and initializing the variables at the same time:

```
int i = 0;
float a = 0.;
double x = 0.;
char c = 'A';
```

Here, the '=' symbol stands not for an assignment to an existing variable as before but rather for an initialization of a new variable that is being defined now.

Unlike assignment, initialization returns no value, so it is impossible to write

```
double x = double y = 0.;   /* error! */
```

Here, the characters "/*"   indicate the start of a comment, which ends with
the characters "*/". Such comments are skipped and ignored by the C compiler; their only purpose is to explain the code to the reader. (C++ has another
form of comment: the characters "//" indicate the start of a comment line
ignored by the C++ compiler.)

Usually, comments are used to explain briefly what the code does. Here,
however, the comment is used to warn the reader that the code is wrong.
Indeed, because the initialization "double y = 0." on the right returns no
value, it cannot be used to initialize 'x' on the left.

Initialization can also be used to define "read-only" types. Such types are
obtained by writing the reserved word "const" before the type name:

```
const int i=1;
```

This way, 'i' is of type constant integer. Therefore, it must be initialized upon
definition, and its value can no longer change throughout the block in which
it is defined.

13.8 Explicit Conversion

We have seen above that the value returned from a function can be converted implicitly to a type that can be assigned to another variable. In this
section, we see that variables can also be converted not only implicitly but
also explicitly.

Conversion is a function that takes a variable as an argument and returns
its value, converted to the required type. In this process, the original variable
never changes: both its type and its value remain the same. Thus, the term
"conversion" is somewhat misleading: no real conversion takes place, and everything remains as before. Still, we keep using this term loosely, assuming
that everybody is aware of its inaccuracy.

Explicit conversion can be used as follows:

```
i = 5;
x = (double)i;  /* or: x = double(i) */
i = (int)3.4;   /* or: i = int(3.4)  */
```

First, the integer variable 'i' is assigned the integer value 5. Then, the prefix
"(double)" before 'i' invokes the explicit-conversion function available in the
C compiler to return the (double-precision) real number 5., which in turn is
assigned to the "double" variable 'x'. Finally, the prefix "(int)" invokes yet
another explicit conversion, which uses the real argument 3.4 to return the
truncated integer number 3, which in turn is assigned back to 'i'.

13.9 Implicit Conversion

As a matter of fact, the explicit conversion used in the above code is completely unnecessary: the same results could be obtained without the prefixes "(double)" and "(int)", because the C compiler would invoke the required conversions implicitly. Indeed, because 'x' is of type "double", the compiler understands that only (double-precision) real numbers can be assigned to it. Similarly, because 'i' is of type "int", the compiler understands that only integer values can be assigned to it. Therefore, in both assignments, implicit conversion is used whenever the relevant explicit conversion is missing.

13.10 Arithmetic Operations

The C compiler also supports standard binary arithmetic operations such as addition (denoted by '+'), subtraction (denoted by '−'), multiplication (denoted by '*'), and division (denoted by '/'), Furthermore, it also supports the unary positive ('+') and negative ('−') operators. These arithmetic operators can be viewed as functions of two (or one in unary operators) variables that return a result of the same type as the type of its arguments. For example, when the compiler encounters "i + j" for some integer variables 'i' and 'j', it invokes the integer-plus-integer version of the '+' binary operator to produce the integer sum of 'i' and 'j'. When, on the other hand, it encounters "x + y" for some "double" variables 'x' and 'y', it invokes the double-plus-double version, to produce the required "double" sum.

If variables of different types are added, then the variable of lower type is converted implicitly to the higher type of the other variable before being added to it. For example, to calculate the sum "i + y", 'i' is converted implicitly to "double" before being added to 'y'.

The arithmetic operations are executed in the standard priority order (see Chapter 10, Section 10.5): multiplication and division are prior to the modulus operator (%) (that returns the residual in integer division), which in turn is prior to both addition and subtraction.

Furthermore, operations of the same priority are executed left to right. For example, $1 + 8/4*2$ is calculated as follows. First, the division operator is invoked to calculate $8/4 = 2$. Then, the multiplication operator is invoked to calculate $2*2 = 4$. Finally, the addition operator is invoked to calculate $1 + 4 = 5$. (Round parentheses can be introduced to change this standard priority order if required.)

Integer variables are divided with residual. This residual can then be obtained by the modulus operator, denoted by '%'. For example, 10/3 is 3, and

10%3 is the residual in this division, namely, 1.

For the sake of readability and clarity, arithmetic symbols are often separated from the arguments by blank spaces. For example, "a + b" is the same
as "a+b", and is also slightly easier to read. When multiplication is used,
however, one must be careful to use the blank spaces symmetrically; that is,
either use them from both sides of the arithmetic symbol, or not use them
at all. For example, both "a * b" and "a*b" mean 'a' times 'b', but "a* b"
and "a *b" mean a completely different thing, which has nothing to do with
multiplication.

The result of an arithmetic operation, as well as the output returned from
any other function, is stored in a temporary variable that has no name. This
temporary variable exists only in the present command line, and is destroyed
at the ';' symbol that ends it. Thus, if the returned value is needed in forthcoming commands as well, then it must be stored in a properly defined variable
in an assignment or an initialization. For example, the command line

```
int i = 3 * 4;
```

initializes the proper variable 'i' with the value of the temporary variable
that contains the output of the binary multiplication operator, applied to the
arguments 3 and 4.

The C compiler also supports some special arithmetic operations:

```
x += 1.;
x -= 1.;
x *= 1.;
x /= 1.;
++i;
--i;
i++;
i--;
```

In fact, these operations are the same as

```
x = x + 1.;
x = x - 1.;
x = x * 1.;
x = x / 1.;
i = i + 1;
i = i - 1;
i = i + 1;
i = i - 1;
```

respectively.

Although both of the above code segments do the same, the first one is more
efficient. Indeed, each command line in it uses one operation only (update

of an existing variable) rather than two (binary arithmetic operation and assignment).

The "+=", "− =", "∗=", and "/ =" operators can also be viewed as functions that not only update their first argument (the variable on the left) but also return the new (up-to-date) value as an output. This property can be used to write commands like

```
a = x += 1.;
j = ++i;
j = --i;
```

Each of these commands is executed right to left: first, the functions on the right-hand side is invoked to update 'x' or 'i'; then, the value returned from this function, which is the up-to-date value of 'x' or 'i', is assigned to the variable on the left-hand side ('a' or 'j').

The unary operators "++" and "−−" can be used in two possible versions: if the symbols are placed before the variable name (as in "++i" and "−−i"), then the returned value is indeed the up-to-date value of the variable, as described above. If, on the other hand, the symbols are placed after the variable (as in "i++" and "i−−"), then the returned value is the old value of the variable.

13.11 Functions

In the above, we have seen that arithmetic operators can be viewed as functions that use their arguments to calculate and return the required output. Furthermore, in C, the assignment operator and the update operators ("+ =", "++", etc.) can also be viewed as functions that not only modify their first argument (the variable on the left) but also return a value (usually the assigned value or the up-to-date value).

The above functions are standard C functions: they are built in the C compiler, and are available to any user. Here we see how programmers can also write their own original functions. Once a function is defined (written) by the programmer, it can be used throughout the program; in fact, it is invoked by the C compiler just like standard C functions.

Functions must be defined in a certain format. In particular, the function header (the first code line in the function definition) must contain the type of the returned value (also referred to as the function type), followed by the function name and a list of arguments in round parentheses. If no function type is specified, then it is assumed that an integer is returned. A function may also be of type "void", to indicate that no value is returned.

The function name that follows the function type is determined arbitrarily by the programmer. This name may then be used later on to invoke the function.

The function name is followed by a list of arguments, separated by commas. Each argument is preceded by its types.

The function header is followed by the function block, which starts with '{' and ends with '}'. The function block (or the body of the function) contains the list of instructions to be carried out when the function is invoked (called) later on.

The convention is to write the symbol '{' that starts the function block at the end of the header, right after the list of arguments, and to write the symbol '}' that concludes the function block in a separate code line, right under the first character in the header. This makes it clear to any reader where the function block starts and where it ends. The body of the function (the instructions to be executed when the function is actually called) is then written in between these symbols. Each instruction is written in a separate command line, shifted two blank spaces to the right. This way, the reader can easily distinguish the function block from the rest of the code.

Here is a simple example:

```
int add(int i, int j){
  return i+j;
}
```

The function "add" [or, more precisely, "add()"] returns the sum of its two integer arguments. Because this sum is an integer as well, the function type (the first word in the header) is the reserved word "int". The integers 'i' and 'j' that are added in the body of the function are referred to as local (or dummy) arguments (or variables), because they exist only within the function block, and disappear as soon as the function terminates. In fact, when the function is actually called, the dummy arguments are initialized with the values of the corresponding concrete arguments that are passed to the function as input.

The "return" command in the function block creates an unnamed variable to store the value returned by the function. The type of this variable is the function type, specified in the first word in the header ("int" in the present example). This new variable is temporary: it exists only in the command line that calls the function, and disappears soon after.

The "return" command also terminates the execution of the function, regardless of whether the end of the function block has been reached or not. In fact, even functions of type "void", which return no value, may use a "return" command to halt the execution whenever necessary. In this case, the "return" command is followed just by ';'.

When the C compiler encounters a definition of a function, it creates a finite state machine or an automaton (a process that takes input to executes commands and produce an output) that implements the function in machine language. This automaton may have input lines to take concrete arguments and an output line to return a value.

When the compiler encounters a call to the function, it uses the concrete arguments as input to invoke this automaton. This approach is particularly economic, as it avoids unnecessary compilation: indeed, the automaton is compiled once and for all in the definition of the function, and is then used again and again in each and every call.

Here is how the above "add()" function is called:

```
int k=3, m=5, n;
n = add(k,m);
```

Here 'k' and 'm' are the concrete arguments that are passed to the function. When the function is called with these arguments, its local arguments 'i' and 'j' are initialized to have the same values as 'k' and 'm', respectively. The "add()" function then calculates and returns the sum of its local arguments 'i' and 'j', which is indeed equal to the required sum of 'k' and 'm'.

Because it is returned in a temporary (unnamed) variable, this sum must be assigned to the well-defined variable 'n' before it disappears with no trace.

The "add()" function can also be called as follows:

```
int i=3, j=5, n;
n = add(i,j);
```

The concrete arguments 'i' and 'j' in this code are not the same as the dummy arguments in the definition of the "add()" function above. In fact, they are well-defined variables that exist before, during, and after the call, whereas the dummy arguments exist only during the call.

Although the dummy arguments have the same names as the concrete arguments, they are stored elsewhere in the computer memory. This storage is indeed released once the function terminates. The concrete 'i' and 'j', on the other hand, which have been defined before the call, continue to occupy their original storage after the call as well.

Thus, there is no ambiguity in the names 'i' and 'j'. In the call "add(i,j)", they refer to the external variables that are passed as concrete arguments. In the definition of the function, on the other hand, they refer to the local (or dummy) variables. It is thus allowed and indeed recommended to use the same names for the external and the local variables to reflect the fact that they play the same roles in the mathematical formula to calculate the sum "i+j".

13.12 The "Main" Function

Every C program must contain a function whose name is "main()". This function doesn't have to be called explicitly; it is called only implicitly upon running the program, and the commands in it are executed one by one. The rest of the functions in the program, on the other hand, are not executed until they are called in one of the commands in "main()".

The "main()" function returns an integer value (say, 0) that is never used in the present program. The main purpose of the "return" command is thus to halt the run whenever necessary.

13.13 Printing Output

Here we show how the "main" function can be used to make a few calculations and print the results onto the screen. For this purpose, the "include" command is placed at the beginning of the program grants access to the standard I/O (Input/Output) library that contains all sorts of useful functions, including functions to read/write data. In particular, it contains the standard "printf()" function to print data onto the screen.

The "printf()" function takes the following arguments. The first argument is the string to be printed onto the screen. The string appears in double quotation marks, and often ends with the character '\n', which stands for "end of line." The string may also contain the symbols "%d" (integer number) and "%f" (real number). These numbers are specified in the rest of the arguments passed to the "printf()" function.

In the following program, the "printf()" function is used to illustrate the difference between integer division and real division. For this purpose, the program first prints the result and the residual in the integer division of 10/3. Then, it prints the result of the real division of 10./3., in which the integer 3 is converted implicitly to the real number 3. before being used to divide the real number 10.:

```
#include<stdio.h>
int main(){
  printf("10/3=%d,10 mod 3=%d,10./3.=%f\n",10/3,10%3,10./3.);
  return 0;
}
```

Furthermore, since the assignment operator not only assigns but also returns the assigned value, one can also write

```
int i;
printf("10/3=%d.\n",i = 10/3);
```

to assign the value $10/3 = 3$ to 'i' and to print it onto the screen at the same time.

Initialization, on the other hand, returns no value, so one can't write

```
printf("10/3=%d.\n",int i = 10/3);
            /*  wrong!!! no returned value  */
```

Here is a useful function that just prints its "double" argument onto the screen:

```
void print(double d){
  printf("%f; ",d);
}  /*  print a double variable  */
```

With this function, the user can print a "double" variable 'x' onto the screen just by writing "print(x)".

13.14 Comparison Operators

The C compiler also supports the binary comparison operators '<', '>', "==", "<=", and ">=". In particular, if 'i' and 'j' are variables of the same type (that is, either both are of type "int" or both are of type "float" or both are of type "double"), then

- "i<j" returns a nonzero integer (say, 1) if and only if 'i' is indeed smaller than 'j'; otherwise, it returns the integer 0.
- "i>j" returns a nonzero integer (to indicate true) if and only if 'i' is indeed greater than 'j'; otherwise, it returns the integer 0 (to indicate false).
- "i==j" returns a nonzero integer if and only if 'i' is indeed equal to 'j'; otherwise, it returns the integer 0.
- "i! =j" returns a nonzero integer if and only if 'i' is indeed different from 'j'; otherwise, it returns the integer 0.
- "i<=j" returns a nonzero integer if and only if 'i' is indeed smaller than or equal to 'j'; otherwise, it returns the integer 0.
- "i>=j" returns a nonzero integer if and only if 'i' is indeed greater than or equal to 'j'; otherwise, it returns the integer 0.

Be careful not to confuse the above "equal to" operator, "==", with the assignment operator '=', which has a completely different meaning.

13.15 Boolean Operators

The C compiler also supports the Boolean (logical) operators used in mathematical logics. For example, if 'i' and 'j' are integer variables, then

- "i&&j" is a nonzero integer (say, 1) if and only if both 'i' and 'j' are nonzero; otherwise, it is the integer 0.
- "i||j" is the integer 0 if and only if both 'i' and 'j' are zero; otherwise, it is a nonzero integer (say, 1).
- "!i" is a nonzero integer (say, 1) if and only if 'i' is zero; otherwise, it is the integer 0.

The priority order of these operators is as in mathematical logics (see Chapter 10, Section 10.6): the unary "not" operator, '!', is prior to the binary "and" operator, "&&", which in turn is prior to the binary "or" operator, "||". Round parentheses can be introduced to change this standard priority order if necessary.

13.16 The "?:" Operator

The "?:" operator takes three arguments, separated by the '?' and ':' symbols. The first argument, an integer, is placed before the '?' symbol. The second argument is placed after the '?' symbol and before the ':' symbol. Finally, the third argument, which is of the same type as the second one, is placed right after the ':' symbol.

Now, the "?:" operator works as follows. If the first argument is nonzero, then the second argument is returned. If, on the other hand, the first argument is zero, then the third argument is returned.

Note that the output returned from the "?:" operator is stored in a temporary variable, which disappears at the end of the present command line. Therefore, it must be stored in a properly defined variable if it is indeed required later in the code:

```
double a = 3., b = 5.;
double max = a > b ? a : b;
```

Here 'a' is smaller than 'b' so "a>b" is false or zero. Therefore, the "?:" operator that uses it as its first argument returns its third argument, 'b'. This output is used to initialize the variable "max", which stores it safely for future use.

13.17 Conditional Instructions

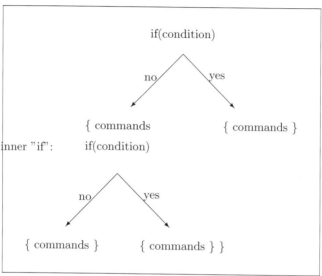

FIGURE 13.1: The if-else scheme. If the condition at the top holds, then the commands on the right are executed. Otherwise, the commands on the left are executed, including the inner if-else question.

The reserved words "if" and "else" allow one to write conditional instructions as follows: if some integer variable is nonzero (or if something is true), then do some instruction(s); else, do some other instruction(s) (see Figure 13.1). For example, the above code can also be implemented by

```
double a = 3., b = 5.;
double max;
if(a > b)
  max = a;
else
  max = b;
```

In this code, if the value returned by the '<' operator is nonzero ('a' is indeed greater than 'b'), then the instruction that follows the "if" question is executed, and "max" is assigned with the value of 'a'. If, on the other hand, 'a' is smaller than or equal to 'b', then the instruction that follows the "else" is executed, and "max" is assigned with the value of 'b'.

The "else" part is optional. If it is missing, then nothing is done if the integer that follows the "if" is zero.

In the above code segment, there is only one instruction to be executed if the condition that follows the "if" holds, and another instruction to be executed if it does not. This, however, is not always the case: one may wish to carry out a complete block of instruction if the condition holds, and another one if it does not. For example,

```
double a = 3., b = 5.;
double nim, max;
if(a > b){
  max = a;
  min = b;
}
else{
  max = b;
  min = a;
}
```

13.18 Scope of Variables

A variable in C exists only throughout the block in which it is defined. This is why the variables "max" and "min" are defined in the beginning of the above code segment, before the "if" and "else" blocks. This way, the variables exist not only in these blocks but also after they terminate, and the values assigned to them can be used later on.

To illustrate this point, consider the following strange code, in which both "max" and "min" are local variables that are defined and exist only within the "if" and "else" blocks:

```
double a = 3., b = 5.;
if(a>b){
  double max = a;/* bad programming!!!  */
  double min = b;/*  local variables     */
}
else{
  double max = b; /* bad programming!!! */
  double min = a; /* local variables    */
}
```

This code is completely useless: indeed, both "max" and "min" are local variables that disappear at the '}' symbol that ends the block in which they are defined, before they could be of any use. This is why it makes more sense to define them before the blocks, as before. This way, they continue to exist even after the blocks terminate, and can be of further use.

When defined properly before the if-else blocks, "min" and "max" contain the minimum and maximum (respectively) of the original variables 'a' and 'b'. Unfortunately, the above code must be rewritten every time one needs to compute the minimum or maximum of two numbers. To avoid this, it makes sense to define functions that return the minimum and the maximum of their two arguments.

These functions can be written in two possible versions: functions that take integer arguments to return an integer output,

```
int max(int a, int b){
  return a>b ? a : b;
}  /* maximum of two integers */

int min(int a, int b){
  return a<b ? a : b;
}  /* minimum of two integers */
```

and functions that take real arguments to return a real output:

```
double max(double a, double b){
  return a>b ? a : b;
}  /* maximum of real numbers */

double min(double a, double b){
  return a<b ? a : b;
}  /* minimum of real numbers */
```

With these functions, the user can write commands like "max(c,d)" to compute the maximum of some variables 'c' and 'd' defined beforehand.

When the compiler encounters such a command, it first looks at the type of 'c' and 'd'. If they are of type "int", then it invokes the integer version of the "max" function to compute their integer maximum. If, on the other hand, they are of type "double", then it invokes the "double" version to return their "double" maximum. In both cases, the local (dummy) arguments 'a' and 'b' are initialized in the function block to have the same values as 'c' and 'd', respectively, and disappear at the '}' symbol that marks the end of the function block, after being used to construct the temporary variable returned by the function.

As a matter of fact, if the user defines two variables 'a' and 'b', he/she can compute their maximum by writing "max(a,b)". In this case, there is no confusion between the external variables 'a' and 'b' that are passed to the function as concrete arguments and exist before, during, and after the call and the local variables 'a' and 'b' that are initialized with their values at the beginning of the function block and disappear soon after having been used to compute the required maximum at the end of the function block.

Another useful function returns the absolute value of a real number:

```
double abs(double d){
   return d > 0. ? d : -d;
}   /* absolute value */
```

This function is actually available in the standard "math.h" library with the slightly different name "fabs()". This library can be included in a program by writing

```
#include<math.h>
```

in the beginning of it. The user can then write either "abs(x)" or "fabs(x)" to have the absolute value of the well-defined variable 'x'.

13.19 Loops

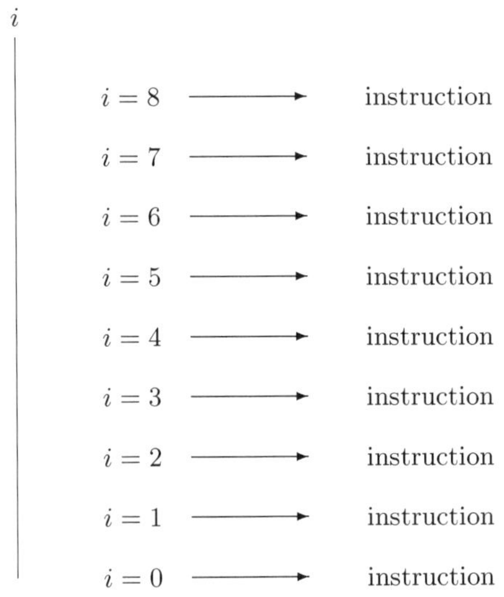

i

$i = 8$	instruction
$i = 7$	instruction
$i = 6$	instruction
$i = 5$	instruction
$i = 4$	instruction
$i = 3$	instruction
$i = 2$	instruction
$i = 1$	instruction
$i = 0$	instruction

FIGURE 13.2:   A loop: the same instruction is repeated for $i = 0, 1, 2, 3, 4, 5, 6, 7, 8$.

The main advantage of the computer over the human mind is its ability to perform many calculations quickly, without forgetting anything or getting tired. The human, however, must first feed it with the suitable instructions.

Thus, it would be counterproductive to write every instruction explicitly. In a high-level programming language such as C, there must be a mechanism to instruct the computer to repeat the same calculation time and again in a few code lines only. This is the loop (Figure 13.2).

The loop uses an index, which is usually an integer variable (say, 'i') that changes throughout the loop to indicate what data should be used. For example, the following loop prints onto the screen the natural numbers from 1 to 100:

```
int i = 1;
while(i<=100){
  printf("%d\n",i);
  i++;
}
```

This loop consists of two parts: the header, which contains the reserved word "while" followed by a temporary integer variable in round parentheses, and a block of instructions. When the loop is entered, the instructions in the block are executed time and again so long as the temporary integer variable in the header is nonzero. When this variable becomes zero, the loop terminates, and the execution proceeds to the next command line that follows the block of instructions, if any.

In the above example, the header invokes the '<' operator to check whether 'i' is still smaller than or equal to 100 or not. If it is, then 'i' is printed onto the screen and incremented by 1, as required. The loop terminates once 'i' reaches the value of 101, which is not printed any more, as required.

Thanks to the fact that 'i' has been defined before the block of instructions, it continues to exist after it as well. This way, the value 101 is kept in it for further use.

Thanks to the fact that the function "i++" not only increments 'i' by 1 but also returns the old value of 'i', the two instructions in the above block could actually be united into a single instruction. This way, the '{' and '}' symbols that mark the start and the end of the block of instructions can be omitted:

```
int i = 1;
while(i<=100)
  printf("%d\n",i++);
```

Furthermore, the above "while" loop can also be written equivalently as a "do-while" loop:

```
int i = 1;
do
  printf("%d\n",i++);
```

```
while(i<=100);
```

Moreover, the loop can be written equivalently as a "for" loop:

```
int i;
for(i=1;i<=100;i++)
  printf("%d\n",i);
```

The header in the "for" loop contains the reserved word "for", followed by round parentheses, which contain three items separated by ';' symbols. The first item contains a command to be executed upon entering the loop. The second item contains a temporary integer variable to terminate the loop when becoming zero. The third item contains a command to be repeated right after each time the instruction (or the block of instructions) that follows the header is repeated in the loop.

In the above example, 'i' is printed time and again onto the screen (as in the instruction) and incremented by 1 (as in the third item in the header), from its initial value 1 (as in the first item in the header) to the value 100. When 'i' becomes 101, it is no longer printed or incremented, and the loop terminates thanks to the second item in the header, as required. Furthermore, thanks to the fact that 'i' has been defined before the loop, it continues to exist after it as well, and the value 101 is kept in it for further use.

Because the command in the first item in the header is carried out only once at the beginning of the loop, it can also be placed before the loop, leaving the first item in the header empty. Furthermore, because the command in the third item in the header is carried out right after the instruction(s), it can also be placed at the end of the instruction block, leaving the third item in the header empty:

```
int i=1;
for(;i<=100;){
  printf("%d\n",i);
  i++;
}
```

Still, it makes more sense to place commands that have something to do with the index in the header rather than before or after it. In particular, if 'i' is no longer needed after the loop, then it makes more sense to place its entire definition in the first item in the header:

```
for(int i=1;i<=100;i++)
  printf("%d\n",i);
```

This way, 'i' is a local variable that exists throughout the loop only, and disappears soon after.

13.20 The Power Function

In the following sections, we give some examples to show the usefulness of loops. The following function calculates the power base^{exp}, where "base" is an integer number and "exp" is a natural number (a variable of type "unsigned"):

```
int power(int base, unsigned exp){
  int result = 1;
  for(int i=0; i<exp; i++)
    result *= base;
  return result;
} /* "base" to the "exp" */
```

Indeed, the local variable "result" is initially 1, and is then multiplied by "base" in a loop that is repeated "exp" times to produce the required result base^{exp}.

13.21 Integer Logarithm

Another nice example of using loops is the following "log()" function, which returns $\lfloor \log_2 n \rfloor$ (the largest integer that is smaller than or equal to $\log_2 n$), where n is a given natural number:

```
int log(int n){
  int log2 = 0;
  while(n>1){
    n /= 2;
    log2++;
  }
  return log2;
} /* compute log(n) */
```

Indeed, the local variable "log2" is initially zero, and is incremented successively by one each time the dummy argument 'n' is divided (an integer division) by two. This is repeated in the loop until 'n' cannot be divided any longer, when "log2" is returned as the required result.

13.22 The Factorial Function

A loop is also useful in the implementation of the factorial function, defined by

$$n! = 1 \cdot 2 \cdot 3 \cdots n,$$

where n is a natural number. (The factorial is also defined for $n = 0$: $0! = 1$.) The implementation is as follows:

```
int factorial(int n){
   int result = 1;
   for(int i=1; i<=n; i++)
      result *= i;
   return result;
} /* compute n! using a loop */
```

Indeed, the local variable "result" is initially 1, and is then multiplied successively by the index 'i', until 'i' reaches the value 'n', when "result" is returned as the required result.

13.23 Nested Loops

The instruction block in the loop can by itself contain a loop. This is called a nested loop or a double loop (Figure 13.3).

Here is a program that uses a nested loop to print a checkerboard:

```
#include<stdio.h>
int main(){
   for(int i=0;i<8;i++){
      for(int j=0;j<8;j++)
         printf("%c ",((i+j)%2)?'*':'o');
      printf("\n");
   }
   return 0;
} /* print a checkerboard */
```

Indeed, the index 'i' (the row number in the checkerboard) is incremented in the main (outer) loop, whereas the index 'j' (the row number in the checkerboard) is incremented in the secondary (inner) loop. The instruction block in the inner loop contains one instruction only: to print a particular symbol if the sum of 'i' and 'j' is even (to indicate a red subsquare in the checkerboard) or another symbol if it is odd (to indicate a black subsquare in the checkerboard).

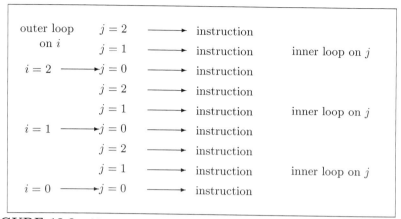

FIGURE 13.3: Nested loops: the outer loop uses the index $i = 0, 1, 2$; for each particular i, the inner loop uses the index $j = 0, 1, 2$.

Finally, the instruction block of the outer loop contains one extra instruction to move on to the next row in the printed image of the checkerboard.

13.24 Reversed Number

Here a loop is used in a function that takes an arbitrarily long natural number and produces another natural number with a reversed order of digits. For example, for the concrete argument 123, the function returns the output 321.

```
int reversed(int number){
   int result=0;
   while (number){
      result *= 10;
      result += number % 10;
      number /= 10;
   }
   return result;
} /*  reversing an integer number  */
```

Indeed, the integer local variable "result" is initially zero. In the loop, the most significant digits at the head of the dummy argument "number" are transferred one by one to the tail of "result" to serve as its least significant digits, until "number" vanishes and "result" is returned as the required reversed number.

The reversed number is produced above in the usual decimal base. A slightly different version of the above function produces the reversed number in an-

other base. This is done just by introducing the extra argument "base" to
specify the required base to represent the reversed number:

```
int reversed(int number, int base){
  int result=0;
  while (number){
    result *= 10;
    result += number % base;
    number /= base;
  }
  return result;
}  /*  reversed number in any base */
```

The only difference is that in this version the digits are truncated from the
dummy argument "number" in its representation in base "base", and are
added one by one to "result" in its representation in base "base" as well.

13.25 Binary Number

The above version of the "reversed()" function is particularly useful to
obtain the binary representation of an arbitrarily long natural number. Indeed,
to have this representation, the "reversed()" function is called twice: once to
obtain the reversed binary representation, and once again to obtain the well-
ordered binary representation.

Note that, when the concrete argument that is passed to the "reversed()"
function ends with zeroes, they get lost in the reversed number. For example,
3400 is reversed into 43 rather than 043. This is why an even number must
be incremented by 1 to become odd before being passed to the "reversed()"
function. Once the binary representation of this incremented number is ready,
it is decremented by 1 to yield the correct binary representation of the original
number, as required.

```
int binary(int n){
  int last = 1;
```

The local variable "last" is supposed to contain the last binary digit in the
original number 'n'. Here we have used the default assumption that 'n' is odd,
so "last" is initialized with the value 1. If, on the other hand, 'n' is even, then
its last binary digit is 0 rather than 1. As discussed above, in this case 'n' must
be incremented by 1, so its last binary digit must be first stored in "last" for
safekeeping:

```
if(!(n%2)){
```

```
    last = 0;
    n += 1;
}
```

By now, the dummy argument 'n' is odd, so its binary representation ends with 1. Therefore, 'n' can be safely reversed, with no fear that leading zeroes in the reversed number would be lost. In fact, the first call to the "reversed()" function uses 2 as a second argument to produces the reversed binary representation of 'n'. The second call, on the other hand, uses 10 as a second argument, so it makes no change of base but merely reverses the number once again. Finally, if "last" is zero, which indicates that the original number is even, then 1 must be subtracted before the final result is returned:

```
    return reversed(reversed(n,2),10) - (1-last);
}   /* binary representation */
```

13.26 Pointers

A pointer is an integer variable that may contain only the address of a variable of a particular type. For example, pointer-to-"double" may contain only the address of a "double" variable, whereas pointer-to-integer may contain only the address of an integer variable.

Most often, the pointer is initialized with a random meaningless value or with the zero value. We then say that the pointer points to nothing. When, on the other hand, the pointer contains a meaningful address of an existing variable, we say that the pointer "points" to the variable.

Here is how pointers are defined:

```
    double *p;
    int *q;
```

Here '*' stands not for multiplication but rather for dereferencing. The dereferencing operator takes a pointer to return its content, or the variable it points to. In the above code, the content of 'p', "*p", is defined to be a "double" variable. This means that 'p' is defined at the same time to be a pointer-to-"double". The content of 'q', on the other hand, is defined to be an integer variable, which also defines 'q' at the same time to be a pointer-to-integer.

The '*' symbol can also be shifted one space to the left with the same meaning:

```
    double* p;
    int* q;
    char* w;
```

In this style, "double*" can be viewed as the pointer-to-"double" type, (used to define 'p'), "int*" can be viewed as the pointer-to-integer type (used to define 'q'), and "char*" can be viewed as the pointer-to-character (or string) type (used to define 'w'). Since the pointers 'p', 'q', and 'w' are not initialized, they contain random meaningless values or the zero value, which means that they point to nothing.

13.27 Pointer to a Constant Variable

One can also define a pointer to a constant variable by writing

```
const int* p;
```

Here, the content of 'p' is a constant integer (or, in other words, 'p' points to a constant integer). Thus, once the content of 'p' is initialized to have some value, it can never change throughout the block in which 'p' is defined. For this reason, 'p' can never be assigned to (or be used to initialize) a pointer-to-nonconstant-integer. Indeed, since 'p' contains the address of a constant variable, if one wrote

```
int* q = p;/* error!!! p points to a read-only integer */
```

then the new pointer 'q' would contain the same address as well, so the constant variable could change through 'q', which is of course in contradiction to its very definition as a constant variable. For this reason, the compiler would never accept such a code line, and would issue a compilation error.

13.28 The Referencing Operator

As we have seen above, the dereferencing operator '*' takes a pointer to returns its content. The referencing operator '&', on the other hand, takes a variable to return its address:

```
double* p;
double v;
p = &v;
```

Here the address of 'v', "&v", is assigned to the pointer 'p'.

Both the referencing and the dereferencing operators are used often later on in the book.

13.29 Arrays

An array in C is a pointer to the first variable in a sequence of variables of the same type and size that are placed continuously in the computer memory. For example, the command

```
double a[10];
```

defines an array of 10 variables of type "double", which are allocated consecutive memory and referred to as the entries of 'a' or "a[0]", "a[1]", "a[2]", ..., "a[9]". Thanks to the fact that the entries are stored consecutively in the computer memory, their addresses are 'a', "a+1", "a+2", ..., "a+9", respectively. (This is referred to as pointer arithmetics.)

13.30 Two-Dimensional Arrays

The entries in the array must all be of the same type and have the same size (occupy the same amount of memory). This implies that one can define an array of arrays, that is, an array whose entries are arrays of the same length. Indeed, since these entries occupy the same amount of memory, they can be placed one by one in an array, to produce a two-dimensional array.

For example, one can define an array of five entries, each of which is an array of ten "double" variables:

```
double a[5][10];
```

The "double" variables in this array are ordered row by row in the computer memory. More precisely,k the first row, "a[0][0]", "a[0][1]", ..., "a[0][9]", is stored first, the second row, "a[1][0]", "a[1][1]", ..., "a[1][9]" is stored next, and so on. This storage pattern is particularly suitable for scanning the two-dimensional array in nested loops: the outer loop "jumps" from row to row by advancing the first index in the two-dimensional array, whereas the inner loop scans each individual row by advancing the second index. This way, the variables in the two-dimensional array are scanned in their physical order in the computer memory to increase efficiency.

The name of the two-dimensional array, 'a', points to its first entry, the "double" variable "a[0][0]". Moreover, thanks to the above storage pattern in which the entries in the two-dimensional array are stored row by row, the "double" variable "a[i][j]" is stored at the address (or pointer) "a+10*i+j" ($0 \leq i < 5$, $0 \leq j < 10$). (This is referred to as pointer arithmetics.)

When the size of the two-dimensional array is not known in advance in compilation time, it can still be defined as a pointer-to-pointer-to-double rather than array-of-array-of-doubles:

```
double** a;
```

13.31 Passing Arguments to Functions

The pointers introduced above are particularly useful in passing arguments to functions. Consider, for example, the following function, which takes an integer argument and returns its value plus one:

```
int addOne(int i){
   return ++i;
} /* return value plus one */
```

The concrete argument 'i' passed to this function, however, remains unchanged. For example, if a user calls the function by writing

```
int k=0;
addOne(k); /* k remains unchanged */
```

then the value of 'k' remains zero. This is because 'k' is passed to the function by value (or by name). In other words, when the function is called with 'k' as a concrete argument, it defines a local variable, named 'i', and initializes it with the value of 'k'. It is 'i', not 'k', that is used and indeed changed throughout the function block. Unfortunately, 'i' is only a local variable that disappears at the end of this block, so all the changes made to it are lost.

Many functions are not supposed to change their arguments but merely to use them to calculate an output. In such functions, passing arguments by name is good enough. In some other functions, however, there may be a need not only to read the value of the concrete argument but also to change it permanently. In such cases, the concrete argument must be passed by address rather than by name:

```
int addOne(int *q){
   return ++(*q);
} /* add one */
```

In this version, the function takes an argument of type pointer-to-integer rather than integer. When it is called, the function creates a local copy of this pointer, named 'q', and initializes it with the same address as in the concrete pointer. The address in 'q' is then used not only to read but also to actually change the content of 'q' (which is also the content of the concrete pointer), as required.

The user can now call the function by writing

```
int k=0;
addOne(&k);   /* k is indeed incremented by 1 */
```

In this call, the address of 'k' is passed as a concrete argument. The function then creates a local variable of type pointer-to-integer, named 'q', and initializes it with the address of 'k'. Then, 'k' is incremented by 1 through its address in 'q'. Although 'q' is only a local pointer that disappears when the function terminates, the effect on 'k' remains valid, as required.

13.32 Input/Output (I/O)

The first step in using the computer is to feed it with input data. Among these data, there is also the program to be executed by the computer. Once the computer completes the execution, it provides the user with output data. The only way for the user to make sure that there is no mistake (bug) in the original program is to examine these data and verify that they indeed make sense.

So far, we have only used a function that prints output: the standard "printf" function, which prints onto the screen a string of characters, including the required output calculated throughout the execution of the program. Below we also use a function that reads input to the computer: this is the standard "scanf" function, which reads a string of characters from the screen.

The first argument in this function is this string to be read. The rest of the arguments are the addresses of the variables in which the input data should be stored for further use. These arguments must indeed be passed by address (rather than by name) to store properly the values that are read into them:

```
#include<stdio.h>
#include<stdlib.h>
int main(){
    int i=0;
    double x=0.;
    scanf("%d %f\n",&i,&x);
    printf("i=%d, x=%f\n",i,x);
```

Indeed, to run this code, the user must type an integer number and a "double" number onto the screen, and then hit the "return" key on the keyboard. Once the "scanf" function is called, it creates local copies of the addresses of the variables 'i' and 'x' defined before, which are then used to read these numbers from the screen into 'i' and 'x'. Although these local copies disappear at the end of the call, the external variables 'i' and 'x' still exist, with the input values stored safely in them. To verify this, these values are then printed back to the screen at the end of the above code.

13.33 Input/Out with Files

So far, we have seen standard functions to read from and write to the screen. However, the screen can contain only a limited amount of data. Much larger storage resources are available in files, which are stored in the computer's secondary memory.

It is thus important to have functions to read from and write onto files rather than the screen. These are the "fscanf" and "fprintf" standard functions.

Clearly, to use a file, one must first have access to it and a way to refer to it. The "fopen" standard function opens a file for reading or writing and returns its address in the computer's secondary memory. This address is stored in a variable of type pointer-to-file ("FILE*"), where "FILE" is a reserved word to indicate a file variable.

The "fopen" function takes two string arguments: the first string is the name of the file in the computer memory, and the second string is either 'r' (to read from the file) or 'w' (to write on it):

```
FILE* fp = fopen("readFile","r");
```

Here the pointer-to-file "fp" is defined and initialized with the address of the file "readFile" in the secondary memory. This way, "fp" can be passed to the "fscanf" function as its first argument. In fact, "fscanf" can be viewed as an advanced version of "scanf", which reads from the file whose address is in its first argument rather than from the screen:

```
fscanf(fp,"%d %f\n",&i,&x);
```

In this example, "fscanf" reads two numbers from the file "readFile": an integer number into 'i', and a double-precision real number into 'x'.

To verify that the integer and real numbers in "readFile" have indeed been read properly, we now print them onto the file "writeFile". To open this file for writing, the "fopen" function must be called this time with the string 'w' as its second argument:

```
fp = fopen("writeFile","w");
```

The pointer-to-file "fp", which contains now the address of "writeFile", is passed to the "fprintf" function as its first argument. This function may be viewed as an advanced version of "printf", which prints onto the file whose address is in its first argument rather than onto the screen:

```
    fprintf(fp,"i=%d, x=%f\n",i,x);
    return 0;
}
```

Thus, the values of 'i' and 'x' are printed onto the file "writeFile", which is stored in the directory in which the program runs.

13.34 Exercises

1. The standard function "sizeof()" takes some type in C and returns the number of bytes required to store a variable of this type in the computer memory. For example, on most computers, "sizeof(float)" returns 4, indicating that four bytes are used to store a "float" number. Since each byte stores two decimal digits, the precision of type "float" is eight digits. On the other hand, "sizeof(double)" is usually 8, indicating that "double" numbers are stored with a precision of sixteen decimal digits. Write a code that prints "sizeof(float)" and "sizeof(double)" to find out what the precision is on your computer.

2. Verify that arithmetic operations with "double" numbers are indeed more precise than those with "float" numbers by printing the difference $x_1 - x_2$, where $x_1 = 10^{10} + 1$ and $x_2 = 10^{10}$. If x_1 and x_2 are defined as "double" variables, then the result is 1, as it should be. If, however, they are defined as "float" numbers, then the result is 0, due to finite machine precision.

3. As we've seen in Chapter 3, Section 3.7, the harmonic series $\sum 1/n$ diverges. Write a function "harmonic(N)" that returns the sum of the first N terms in the harmonic series. (Make sure to use "1./n" in your code rather than "1/n", so that the division is interpreted as division of real numbers rather than division of integers.) Verify that the result of this function grows indefinitely with N.

4. On the other hand, the series $\sum 1/n^2$ converges. Indeed,

$$\sum_{n=1}^{\infty} \frac{1}{n^2} \leq \sum_{n=1}^{\infty} \frac{2}{n(n+1)} = 2 \sum_{n=1}^{\infty} \left(\frac{1}{n} - \frac{1}{n+1} \right) = 2.$$

Write the function "series(N)" that calculates the sum of the first N terms in this series. Verify that the result of this function indeed converges as N increases.

5. Write a function "board(N)" that prints a checkerboard of size $N \times N$, where N is an integer argument. Use '+' to denote red cells and '-' to denote black cells on the board.

6. Run the code segments of the examples in Sections 13.20–13.22 and verify that they indeed produce the required results.

7. Define a two-dimensional array that stores the checkerboard in Section 13.23. Scan it in a nested loop and print it to the screen row by row. Verify that the output is indeed the same as in Section 13.23.

8. Rewrite the function "reversed()" in Section 13.24 more elegantly, so its block contains three lines only. The solution can be found in Section 28.1 in the appendix.

9. Generalize the function "binary()" in Section 13.25 into a more general function "changeBase()" that accepts two integer parameters 'n' and "base" and returns 'n' as represented in base "base". Make sure to write the "changeBase()" function so elegantly that its block contains two lines only. The solution can be found in Section 28.1 in the appendix.

Chapter 14

Recursion

C may be viewed as a "function-oriented" programming language. Indeed, a C command is not only an instruction to the computer but also a function that may return a value for further use. This is also why C is so good at recursion: a function written in C can easily call itself recursively, using input calculated in the original call.

Recursion may be viewed as a practical form of mathematical induction. The only difference is that mathematical induction works in the standard forward direction, starting from the simplest object and advancing gradually to more and more complex objects, whereas recursion works in the backward direction, applying a function to a complex object by calling it recursively to simpler and simpler objects.

Thus, mathematical induction and recursion are the two sides of the same thing. Indeed, the recursive call to a C function uses the induction hypothesis to guarantee that it is indeed valid. Furthermore, the innermost recursive call uses the initial condition in the corresponding mathematical induction to start the process.

Below we show how mathematical functions and algorithms can be implemented easily and transparently using recursion. This can be viewed as a preparation work for the study of recursive mathematical objects in C++, later on in the book.

14.1 Recursive Functions

Recursion is a process in which a function is called in its very definition. The recursive call may use not only new arguments that are different from those used in the original call but also new data structures that must be allocated extra memory dynamically in run time. This is why C, which supports dynamic memory allocation, is so suitable for recursion.

In the sequel, we use recursion to reimplement some functions that have already been implemented with loops in Chapter 13. The present implementation is often more transparent and elegant, as it is more in the spirit of the original mathematical formulation.

14.2 The Power Function

We start with the "power" function, implemented with a loop in Chapter 13, Section 13.20. The present recursive implementation is more natural, because it follows the original mathematical definition of the power function, which uses mathematical induction as follows:

$$\text{base}^{\text{exp}} = \begin{cases} \text{base} \cdot \text{base}^{\text{exp}-1} & \text{if } \exp \geq 1 \\ 1 & \text{if } \exp = 0. \end{cases}$$

This formulation is indeed translated into a computer code in the following implementation:

```
int power(int base, unsigned exp){
  return exp ? base * power(base,exp-1) : 1;
} /* "base" to the "exp" (with recursion) */
```

Indeed, if the integer exponent "exp" is greater than zero, then the induction hypothesis is used to calculate the required result recursively. If, on the other hand, "exp" is zero, then the initial condition in the mathematical induction is used to produce the required result with no recursive call.

The very definition of the "power" function uses a recursive call to the same function itself. When the computer encounters this call in run time, it uses the very definition of the "power" function to execute it. For this purpose, it allocates the required memory for the arguments and the returned value in this recursive call.

The recursive call can by itself use a further recursive call with a yet smaller argument "exp". This nested process continues until the final (innermost) recursive call that uses a zero "exp" argument is reached, in which no further recursive calls are made.

The extra memory allocation for arguments and returned values required throughout the recursive process may make the recursive implementation slightly less efficient than the original one. This overhead, however, is well worth it to have an elegant code that follows the original mathematical formula, particularly when more complicated algorithms and structures are considered later on in the book.

A version of the "power" function can also be written for a real base:

```
double power(double basis, unsigned exp){
  return exp ? basis * power(basis,exp-1) : 1.;
} /* double "basis" to the "exp" */
```

14.3 Integer Logarithm

The "log()" function in Chapter 13, Section 13.21 can also be reimplemented recursively as follows:

```
int log(int n){
   return n>1 ? log(n/2)+1 : 0;
}  /* compute log(n) recursively */
```

The correctness of this code can indeed be proved by mathematical induction on the input 'n'.

14.4 The Factorial Function

The "factorial" function in Chapter 13, Section 13.22 can also be implemented most naturally using recursion. This implementation uses the original definition of the factorial function by mathematical induction:

$$n! = \begin{cases} 1 & \text{if } n = 0, \\ n((n-1)!) & \text{if } n > 0. \end{cases}$$

This definition is indeed translated into a computer code as follows:

```
int factorial(int n){
   return n ? n * factorial(n-1) : 1;
}  /*  compute  n! using recursion  */
```

As discussed before, this implementation may be slightly less efficient than the original one in Chapter 13, Section 13.22 due to the extra memory allocation required for the arguments and return values in each recursive call. Still, it is particularly short and elegant, and follows the spirit of the original mathematical definition.

14.5 Ordered Arrays

Here we give another example to show how useful recursion may be. Consider the following problem: given an array a of length n that contains n integer numbers in increasing order, that is,

$$a[0] < a[1] < a[2] < \cdots < a[n-1],$$

find out whether a given integer number k is contained in a, that is, whether there exists at least one particular index i ($0 \le i < n$) for which

$$a[i] = k.$$

Of course, this task can be completed easily using a loop to scan the cells in a one by one. However, this approach may cost n comparisons to check whether k is indeed equal to an item in a. Furthermore, the above approach never uses the property that the items in the array are ordered in increasing order. Is there a better way to complete the task?

Fortunately, there is an algorithm that uses the order in the array to solve the above problem in $\log_2 n$ comparisons only. This algorithm uses recursion as follows. First, it divides a into two subarrays of length $n/2$ each: the first half, which contains the items indexed by $0, 1, 2, \ldots, n/2 - 1$, and the second half, which contains the items indexed by $n/2, n/2 + 1, n/2 + 2, \ldots, n - 1$. Then, it compares k with the middle item in the original array, $a[n/2]$. Now, if $k < a[n/2]$, then k can lie only in the first subarray, so the algorithm should be applied recursively to it. If, on the other hand, $k \ge a[n/2]$, then k can lie only in the second subarray, so the algorithm should be applied recursively to it.

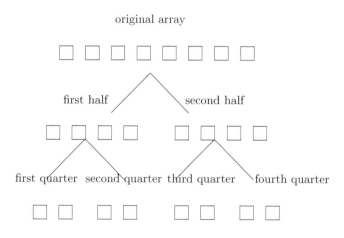

FIGURE 14.1:   Finding whether a given number k lies in a given array of n cells. In each level, the algorithm is applied recursively either to the left subarray or to the right subarray.

The above algorithm forms a virtual binary tree of $\log_2 n$ levels (Chapter 10, Section 10.4). In each level, only one comparison is used to decide whether the algorithm should proceed to the left or right subtree (Figure 14.1). Therefore, the total cost of the algorithm is $\log_2 n$ comparisons only.

Here is the function "findInArray()" that implements the above algorithm. The function uses two nested "?:" questions to return the correct logical answer: 1 if k is indeed contained in the array a (of length n), or 0 if it is not. To make the code easy to understand, it is typed with the same rules as in "if-else" questions: the instructions that should be carried out whether the condition in the "?:" question is satisfied or not are shifted two blank spaces to the right.

```
int findInArray(int n, int* a, int k){
   return n > 1 ?
             k < a[n/2] ?
               findInArray(n/2,a,k)
             :
               findInArray((n+1)/2 ,a+n/2,k)
           :
           k == a[0];
} /* find out whether k lies in a[] */
```

Note that the outer question, "$n > 1$?", is associated with the case $n = 1$ in the mathematical induction in Chapter 10, Section 10.2. Indeed, if $n = 1$, then the array a contains one item only, so k should be simply compared to it, as is indeed done at the end of the function. If, on the other hand, $n > 1$, then the inner question is invoked to form the induction step and decide whether the recursive call should apply to the first subarray,

$$a[0], a[1], a[2], \ldots, a[(n+1)/2 - 1],$$

or to the second subarray,

$$a[n/2], a[n/2 + 1], a[n/2 + 2], \ldots, a[n - 1].$$

In other words, the inner question decides whether the algorithm should proceed to the left or right subtree in Figure 14.1.

Let us use the above code to check whether a given natural number has a rational square root or not. As discussed in Chapter 4, Section 4.5, a natural number may have either a natural square root or an irrational square root. In other words, a natural number k may have a natural square root only if it is of the form

$$k = i^2$$

for some natural number i; otherwise, its square root must be irrational.

Let us list the natural numbers of the form i^2 in the array a:

$$a[0] = 0$$
$$a[1] = 1$$
$$a[2] = 4$$
$$a[3] = 9$$
$$\cdots$$
$$a[i] = i^2$$
$$\cdots \quad .$$

Then, k may have a natural square root only if it lies in the array a. Because the items are ordered in a in an increasing order, the "findInArray()" function can be used to check whether k indeed lies in a, and, hence, has a natural square root, or doesn't lie in a, and, hence, must have an irrational square root:

```c
#include<stdio.h>

int main(){
  int n;
  printf("length=");
  scanf("%d",&n);
  int a[n];
  for(int i=0; i<n; i++)
    a[i] = i*i;
  int k;
  printf("input number=");
  scanf("%d",&k);
  if(findInArray(n,a,k))
    printf("%d is the square of a natural number\n",k);
  else
    printf("%d is the square of an irrational number\n",k);
  return 0;
}
```

14.6 Binary Representation

Here we show how useful recursion can be to produce the binary representation of a natural number n, namely, the unique representation of n as a polynomial in the base 2:

$$n = \sum_{i=0}^{\lfloor \log_2 n \rfloor} a_i 2^i,$$

where the coefficient a_i is either 0 or 1 (see Chapter 12, Section 12.6).

Unfortunately, this formula is not easy to implement using loops; indeed, the sequence of coefficients a_i must be reversed and reversed again in the code in Chapter 13, Section 13.25 before the binary representation is obtained. The recursive implementation, on the other hand, is much more natural, because it is based on the following mathematical induction:

$$n = 2(n/2) + (n\%2),$$

where $n/2$ means integer division, with residual $n\%2$. This formula is now translated into a recursive computer code, to produce the required binary representation of n:

```
int binary(int n){
  return n>1 ? 10*binary(n/2)+n%2 : n%2;
}  /* binary representation */
```

14.7 Pascal's Triangle

As we've seen in Chapter 10, Section 10.9, Pascal's triangle can be viewed as a multilevel object defined by mathematical induction. Although the present implementation avoids recursive functions to save the extra cost they require, it still uses the original mathematical induction implicitly in the nested loop used to produce the binomial coefficients.

More specifically, Pascal's triangle is embedded in the lower-left part of a two-dimensional array (see Figure 14.2). This way, it is rotated in such a way that its head lies in the bottom-left corner of the two-dimensional array.

The following implementation is based on the original mathematical definition in Chapter 10, Section 10.9:

```
#include<stdio.h>
int main(){
  const int n=8;
  int triangle[n][n];
  for(int i=0; i<n; i++)
    triangle[i][0]=triangle[0][i]=1;
  for(int i=1; i<n-1; i++)
    for(int j=1; j<=n-1-i; j++)
      triangle[i][j] = triangle[i-1][j]+triangle[i][j-1];
  return 0;
}  /*  filling Pascal's triangle  */
```

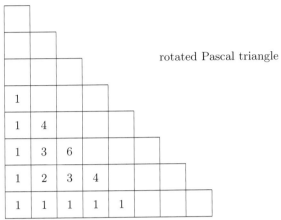

rotated Pascal triangle

FIGURE 14.2: Pascal's triangle rotated in such a way that its head lies in the lower-left corner of the two-dimensional array.

14.8 Arithmetic Expression

As we've seen in Chapter 10, Section 10.5, an arithmetic expression can be viewed as a binary tree, with the least-priority arithmetic symbol in its head, more and more prior arithmetic symbols in its nodes, and numbers in its leaves (see Figure 10.6). As a matter of fact, this is a recursive (or inductive) form: the least-priority arithmetic symbol divides the entire arithmetic expression into two subexpressions that are placed in the left and the right subtrees.

Here we use this observation to implement an algorithm that reads an arbitrarily long arithmetic expression, prints it in the postfix and prefix formats, and calculates its value.

The postfix (respectively, prefix) format of a binary arithmetic expression places the arithmetic symbol before (respectively, after) its two arguments. For example, the arithmetic expression $2 + 3$ has the postfix format $+23$ and the prefix format $23+$. (No parentheses are used.)

These tasks are carried out recursively as follows. The original arithmetic expression is stored in a string or an array of digits and arithmetic symbols like '+', '-', etc. This string is fed as an input into the function "fix()" that carries out the required task, may it be printing in prefix or postfix format or calculating the value of the arithmetic expression.

In the "fix()" function, the string is scanned in the reversed order (from right to left). Once an arithmetic symbol of least priority is found, the original string is split into two substrings, and the function is applied recursively to each of them (see Figure 14.3).

To implement this process, we first need to write a preliminary function

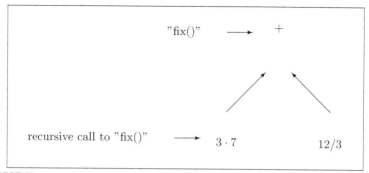

FIGURE 14.3: The "fix()" function calculates $3 \cdot 7 + 12/3$ by scanning this expression backward until the least-priority symbol '+' is found and splitting the original expression into the two subexpressions $3 \cdot 7$ and $12/3$, which are then calculated recursively separately and added to each other.

that copies the first 'n' characters from a string 's' to a string 't':

```
#include<stdio.h>
void copy(char* t, char* s, int n){
  for(int i=0;i<n;i++)
    t[i]=s[i];
  t[n]='\n';
}  /* copy n first characters from s to t  */
```

The function "fix()" defined below carries out one of three possible tasks: printing in postfix format, printing in prefix format, or computing the value of the arithmetic expression. The particular task to be carried out is specified by the third (and last) argument of the function, the integer "task".

The argument "task" may have three possible values to indicate the particular task that is carried out. If "task" is zero, then the task is to print in the postfix format. If, on the other hand, "task" is one, then the task is to print in the prefix format. Finally, if "task" is two, then the task is to calculate the value of the arithmetic expression.

The first two arguments in the "fix()" function are of type string (pointer-to-character) and integer. The first argument contains the arithmetic expression, and the second argument contains its length (number of characters).

When the argument "task" is either zero or one, the characters in the original string are printed (in the postfix or prefix format) onto the screen using the standard "printf" function, with the symbol "%c" to indicate a variable of type character. Here is the complete implementation of the "fix()" function:

```
int fix(char* s, int length, int task){
```

This is the loop that scans the original string backward (from the last character to the first one):

```
for(int i=length-1;i>=0;i--)
```

In this loop, we look for the least-priority arithmetic operation. Clearly, if the symbol '+' or '-' is found, then this is the required least-priority symbol, at which the original expression 's' should be split into the two subexpressions "s1" and "s2":

```
if((s[i]=='+')||(s[i]=='-')){
    char s1[i+1];
    char s2[length-i];
```

Here the first subexpression is copied from 's' to "s1":

```
copy(s1,s,i);
```

Then, pointer arithmetic (as in Chapter 13, Section 13.29) is used to locate the end of the first subexpression in 's', and then the second subexpression is copied from 's' to "s2":

```
copy(s2,s+i+1,length-i-1);
```

Now, the present task (which depends on the particular value of the argument "task") is applied recursively to each subexpression. First, it is assumed that "task" is two, so the task is to calculate the arithmetic expression:

```
if(task==2){
    if(s[i]=='+')
        return fix(s1,i,task) + fix(s2,length-i-1,task);
    else
        return fix(s1,i,task) - fix(s2,length-i-1,task);
}
```

Next, it is assumed that "task" is zero, so the task is to print in the postfix format. This is why the present symbol, may it be '+' or '-', is printed before the recursive calls:

```
if(task==0)printf("%c",s[i]);
fix(s1,i,task);
fix(s2,length-i-1,task);
```

Next, it is assumed that "task" is one, so the task is to print in the prefix format. In this case the present symbol, may it be '+' or '-', is printed after the recursive calls:

```
if(task==1)printf("%c",s[i]);
```

The rest of the work will now be done in the recursive calls, so the original call can safely terminate:

```
        return 0;
    }
```

If, however, there is no '+' or '-' in the original string, then we must use another loop to look for a symbol that stands for an arithmetic operation of the next priority, '%':

```
for(int i=length-1;i>=0;i--)
  if(s[i]=='%'){
    char s1[i+1];
    char s2[length-i];
    copy(s1,s,i);
    copy(s2,s+i+1,length-i-1);
    if(task==2)
      return fix(s1,i,task) % fix(s2,length-i-1,task);
    if(task==0)printf("%c",s[i]);
    fix(s1,i,task);
    fix(s2,length-i-1,task);
    if(task==1)printf("%c",s[i]);
    return 0;
  }
```

If, however, there is also no '%' in the original string, then we must use yet another loop to look for symbols that stand for arithmetic operations of the highest priority, '*' and '/':

```
for(int i=length-1;i>=0;i--)
  if((s[i]=='*')||(s[i]=='/')){
    char s1[i+1];
    char s2[length-i];
    copy(s1,s,i);
    copy(s2,s+i+1,length-i-1);
    if(task==2){
      if(s[i]=='*')
        return fix(s1,i,task) * fix(s2,length-i-1,task);
      else
        return fix(s1,i,task) / fix(s2,length-i-1,task);
    }
    if(task==0)printf("%c",s[i]);
    fix(s1,i,task);
    fix(s2,length-i-1,task);
    if(task==1)printf("%c",s[i]);
    return 0;
  }
```

Finally, if there are no arithmetic symbols in the string at all, then we must have reached an innermost recursive call applied to a trivial subexpression (in the very bottom of the binary tree, that is, in one of the leaves in Figure 10.6), which contains one number only. This is why the string must be scanned once again, and this time we look for digits, which are either combined to form and return a natural number (if "task" is two)

```
if(*s == '\n'){
  printf("error");
  return 0;
}
if(task==2){
  int sum=0;
  int exp=1;
  for(int i=length-1;i>=0;i--){
    if((s[i]>='0')&&(s[i]<='9')){
      sum += (s[i]-'0') * exp;
      exp *= 10;
    }
    else{
      printf("error");
      return 0;
    }
  }
  return sum;
}
```

or printed onto the screen (if "task" is either zero or one):

```
for(int i=0;i<length;i++){
  if((s[i]>='0')&&(s[i]<='9'))
    printf("%c",s[i]);
  else{
    printf("error");
    return 0;
  }
}
return 0;
} /* calculate or print in prefix/postfix format */
```

This completes the "fix()" function that calculates an arbitrarily long arithmetic expression or prints it in the postfix or prefix format.

Here is the "main()" function that reads the arithmetic expression [using the standard "getchar()" function to read each individual character], prints it in its prefix and postfix formats, and calculates it:

```
int main(){
  char s[80];
  int i;
```

```
for(i=0; (s[i]=getchar()) != '\n'; i++);
```

Although this "for" loop is empty (contains no instruction), it still manages to read the entire arithmetic expression from the screen character by character. Indeed, the standard "getchar()" function invoked in the header reads the next input character from the screen, which is then placed in the next available character in the string 's'. Thanks to the fact that the index 'i' is defined before the loop, it continues to exist after it as well, and contains the length of 's' (the number of characters in it) for further use in the calls to the "fix()" function:

```
    fix(s,i,0);
    printf("\n");
    fix(s,i,1);
    printf("\n");
    printf("%d\n",fix(s,i,2));
    return 0;
}
```

To run this program, all the user needs to do is to type the original arithmetic expression onto the screen and hit the "return" key on the keyboard.

14.9 Static Variables

As discussed above, a variable that is defined inside the block of a function (a local variable) disappears when the function terminates with no trace. Still, it may be saved by declaring it as "static". For example, if we wrote in the above "fix()" function

```
    static FILE* fp = fopen("writeFile","w");
    fprintf(fp,"length of subexpression=%d\n",length);
```

then the length of the original arithmetic expression, as well as the lengths of all the subexpressions, would be printed onto the file "writeFile".

Here "fp" is a pointer to a static file variable rather than to a local file variable. This file is created and initialized at the first time the function is called, and exists throughout the entire run. This is why data from all the calls to the function, including subsequent recursive calls, are printed on it continuously.

14.10 The Exponent Function

In this section, we define the exponent function $\exp(x)$ (or e^x), defined by

$$e^x \equiv 1 + x + \frac{x^2}{2!} + \frac{x^3}{3!} + \frac{x^4}{4!} + \cdots = \sum_{n=0}^{\infty} \frac{x^n}{n!}.$$

This infinite series is called the Taylor expansion of the exponent function around $x = 0$.

Although the exponent function is available in the standard "math.h" library, we implement it here explicitly as a good exercise in using loops and recursion. Furthermore, the present implementation can be extended to compute the exponent of a square matrix (Chapter 16, Section 16.18 below).

We approximate the exponent function by the truncated Taylor series (or the Taylor polynomial)

$$T_K(x) = \sum_{n=0}^{K} \frac{x^n}{n!}.$$

Here, K is some predetermined integer, say $K = 10$.

The Taylor polynomial T_K is a good approximation to the exponent function when x is rather small in magnitude. When x is large in magnitude, $\exp(x)$ can still be approximated by picking a sufficiently large integer m in such a way that $x/2^m$ is sufficiently small in magnitude and approximating

$$\exp(x) = \exp(x/2^m)^{2^m}$$

by

$$\exp(x) \doteq (T_K(x/2^m))^{2^m}.$$

This formula is implemented in the function "expTaylor(arg)", which calculates the exponent of its argument "arg":

```
double expTaylor(double arg){
    const int K=10;
    double x=arg;
```

First, we need to find an appropriate integer m. For this, the function uses a local variable 'x', which initially has the same value as the argument "arg", and is then divided successively by 2 m times. This is done in a loop in which x is successively divided by 2 until its magnitude is sufficiently small, say smaller than 0.5. The total number of times x has been divided by 2 is the value assigned to m:

```
    int m=0;
    while(abs(x)>0.5){
```

```
x /= 2.;
m++;
}
```

As a matter of fact, this loop can be viewed as a recursive process to realize the recursive formula

$$x/2^m = \begin{cases} x & \text{if } m = 0 \\ \left(x/2^{m-1}\right)/2 & \text{if } m > 0, \end{cases}$$

as can be proved easily by mathematical induction on $m = 0, 1, 2, 3, \ldots$. This recursion is indeed implemented in the above loop.

Once x has been divided by 2^m, its magnitude is sufficiently small. Therefore, we can return to the evaluation of the Taylor polynomial $T_K(x)$. This can be done most efficiently in the spirit of Horner's algorithm:

$$T_K(x) = \left(\cdots \left(\left(\left(\frac{x}{K} + 1 \right) \frac{x}{K-1} + 1 \right) \frac{x}{K-2} + 1 \right) \cdots \right) x + 1.$$

More precisely, this formula should actually be rewritten as a recursive formula. To do this, define for $0 \le i \le k$ the polynomial of degree $k - i$

$$P_{k,i}(x) \equiv \frac{i!}{x^i} \sum_{j=i}^{k} \frac{x^j}{j!}.$$

Clearly, the first term in this polynomial is the constant 1. Furthermore, when $i = k$, the polynomial reduces to a polynomial of degree 0 (or merely a constant):

$$P_{k,k}(x) \equiv 1.$$

Moreover, when $i = 0$, we obtain the original polynomial T_k:

$$P_{k,0}(x) = T_k(x).$$

Let us now design a recursive formula to lead from the trivial case $i = k$ to the desirable case $i = 0$:

$$P_{k,i-1}(x) = 1 + \frac{x}{i} P_{k,i}(x).$$

Indeed, by mathematical induction on $i = k, k - 1, k - 2, \ldots, 2, 1$, we have

$$P_{k,i-1}(x) = \frac{(i-1)!}{x^{i-1}} \sum_{j=i-1}^{k} \frac{x^j}{j!}$$

$$= 1 + \frac{(i-1)!}{x^{i-1}} \sum_{j=i}^{k} \frac{x^j}{j!}$$

$$= 1 + \frac{x}{i} \cdot \frac{i!}{x^i} \sum_{j=i}^{k} \frac{x^j}{j!}$$

$$= 1 + \frac{x}{i} P_{k,i}(x).$$

The implementation of this formula uses a loop with a decreasing index $i = k, k-1, \ldots, 2, 1$, in which the variable "sum" takes the initial value $P_{k,k} = 1$, before being successively multiplied by x/i and incremented by 1:

```
double sum=1.;
for(int i=K; i>0; i--){
  sum *= x/i;
  sum += 1.;
}
```

Once this loop has been complete, the local variable "sum" has the value $T_K(\text{arg}/2^m)$. The required output, the 2^m-power of $T_K(\text{arg}/2^m)$, is obtained from a third loop of length m, in which "sum" is successively replaced by its square:

```
for(int i=0; i<m; i++)
  sum *= sum;
return sum;
} /* calculate exp(arg) using Taylor series */
```

As a matter of fact, this loop can also be viewed as a recursive process to realize the recursive formula

$$\text{sum}^{2^i} = \begin{cases} \text{sum} & \text{if } i = 0 \\ \left(\text{sum}^{2^{i-1}}\right)^2 & \text{if } 0 < i \le m, \end{cases}$$

as can be proved easily by mathematical induction on $i = 0, 1, 2, 3, \ldots m$. This recursion is indeed implemented in the above loop to make the required substitution

$$\text{sum} \leftarrow \text{sum}^{2^m},$$

or

$$T_k(\text{arg}/2^m) \leftarrow (T_k(\text{arg}/2^m))^{2^m},$$

which is the required approximation to e^x.

14.11 Exercises

1. Use mathematical induction on n to prove the correctness of the code in Section 14.5.

2. Run the code in Section 14.5 to check whether a given natural number k is also the square of some natural number i (for which $k = i^2$). (To run the code, first enter some number $n > k$ to serve as the length of the array, then hit the "return" key on the keyboard, then enter k, and then hit the "return" key once again.)

3. Modify the code in Section 14.6 to produce the representation of an integer number in base "base", where "base" is another integer argument.
4. Run the code in Section 14.7 that constructs Pascal's triangle and verify that the number in cell (k, l) is indeed Newton's binomial coefficient

$$\binom{k + l}{k}.$$

Furthermore, calculate the sum of the entries along the nth diagonal $\{(k, l) \mid k + l = n\}$, and verify that it is indeed equal to

$$\sum_{k=0}^{n} \binom{n}{k} = \sum_{k=0}^{n} \binom{n}{k} 1^{k} 1^{n-k} = (1 + 1)^{n} = 2^{n}.$$

5. Use recursion to write a function that prints the prime factors of an arbitrarily large integer number (Chapter 1, Section 1.14). The solution can be found in Section 28.2 in the appendix.
6. Use recursion to implement Euclid's algorithm to find the greatest common divisor of two natural numbers m and n ($m > n$) (Chapter 1, Section 1.16). The solution can be found in Section 28.3 in the appendix.
7. Use mathematical induction on m to show that the above code indeed works.
8. Use recursion to implement the function

$$C_{a,n} = \frac{a!}{(a - n)!}$$

(where a and n are nonnegative integers satisfying $a \geq n$) defined in Chapter 10, Section 10.17. The solution can be found in Section 28.4 in the appendix.
9. Modify the code in Section 14.8 so that the results are printed to a static file defined in the "fix()" function.
10. Modify the code in Section 14.8 to read arithmetic expressions with parentheses and also print them in the prefix and postfix forms with parentheses.
11. Compare the results of the functions in Section 14.10 to the result of the "exp(x)" function available in the "math.h" library.
12. The sine function $\sin(x)$ has the Taylor expansion

$$\sin(x) = x - \frac{x^3}{3!} + \frac{x^5}{5!} - \frac{x^7}{7!} + \cdots = \sum_{n=0}^{\infty} (-1)^n \frac{x^{2n+1}}{(2n + 1)!}.$$

Modify the above "expTaylor" function to produce the "sinTaylor" function, which computes $\sin(x)$ for a given real number x. Compare the results of this function to those of the "sin(x)" function, available in the "math.h" library.

13. The cosine function $\cos(x)$ has the Taylor expansion

$$\cos(x) = 1 - \frac{x^2}{2!} + \frac{x^4}{4!} - \frac{x^6}{6!} + \cdots = \sum_{n=0}^{\infty} (-1)^n \frac{x^{2n}}{(2n)!}.$$

Modify the above "expTaylor" function to produce the "cosTaylor" function, which computes $\cos(x)$ for a given real number x. Compare the results of this function to those of the "cos(x)" function, available in the "math.h" library.

14. Use the above Taylor expansions to show that, for a given imaginary number of the form ix (where $i = \sqrt{-1}$ and x is some real number),

$$\exp(ix) = \cos(x) + i \cdot \sin(x).$$

Part V

Introduction to C++

Introduction to C++

C++ is built on top of C in the sense that every reserved word (keyword) in C is available in C++ as well. This includes if-else conditions, loops, numerical types such as "int" and "double", logic and arithmetic operations, etc. Thus, C++ enjoys all the good features in C.

Furthermore, C++ gives the programmer the opportunity to define new objects, to be used later as if they were standard types [8] [26]. In this sense, C++ is a dynamic language, which can be enriched by the programmer to support more and more types.

The focus in C++ is on the objects rather than on the algorithms or the applications that use them. The purpose of the program is to introduce new objects and the functions associated with them. These objects can then be used in the present application as well as in other applications solved by other users who have permission to access to them.

Together with the objects, the programmer can also define operators to manipulate them. Once these operators are well defined, the objects can be treated in much the same way as in the original mathematical formulas. This feature has not only a practical value to help implementing algorithms in the same spirit as in their original mathematical formulation, but also a theoretical value to help thinking about the mathematical objects in their original format and continue developing the mathematical theory associated with them.

Thus, an object-oriented programming language such as C++ serves not only as a mere tool to communicate with the computer, but also as a mechanism to enrich the programmer's world of thoughts, give them more insight about what they are doing, and preserve a constant contact between their theoretical knowledge and their practical work. After all, this is the main purpose of any language, may it be formal or human: to communicate not only with others, but also with one's own mind, to express ideas in terms of suitable objects.

Chapter 15

Objects

As discussed above, C may be viewed as a function-oriented programming language: each command in C is also a function that returns a temporary variable, which can be used in the same code line. Furthermore, the programmer may define his/her own functions, which take input as arguments to produce and return an output.

An object-oriented programming language such as C++, on the other hand, focuses on the objects rather than on the functions that use them. Indeed, once mathematical objects are well defined, including the operators and functions that manipulate them, the programmer and other users who have permission to access them can use them as if they were standard numerical types available in C. Furthermore, the objects can then be used in much the same spirit as in their original mathematical formulation, without bothering with technical details such as storage.

Once the mathematical objects are well implemented, they can be placed in a library of objects. Every user who has permission to access this library can then use them to define more complicated composite objects. This may form a hierarchy of libraries of more and more complex objects, to enrich the programming language and give future users proper tools not only to realize their mathematical ideas and algorithms in a practical way but also to think about them and keep developing and improving them.

15.1 Classes

A new object in C++ is defined in a class. The class contains a header with the object name, followed by a block (the class block), in which data fields that belong to the object are declared, and functions associated with the object are declared and defined.

Suppose, for example, that the programmer wants to implement a point in the two-dimensional Cartesian plane to be used by any user as if it were a standard type in C, with no need to bother with any implementation detail such as storage or arithmetic operations. This would indeed make the programming language richer, and would free the mind of the user to concentrate on his/her particular application or algorithm.

In particular, users should be able to write commands like

```
point P;
point Q=P;
```

to define a point 'P' and use it to initialize another point 'Q', exactly as one can do with numerical types such as "int" and "double" in C. As we'll see below, this objective can indeed be achieved by defining a "point" class with its interface functions that can be called by any user.

15.2 Private and Public Members

Here is how a class can be used to define a new object in C++:

```
class point{
  public:
    double x;
    double y;  // not object oriented
};
```

The symbol "//" indicates the start of a comment line, which is skipped and ignored by the C++ compiler. Here, the comment tells us that this is not a good object-oriented programming.

Indeed, the block of the "point" class above contains two data fields of type "double", 'x' and 'y', to store the x and y coordinates of a point object. Unfortunately, the reserved word "public" before these fields implies that they are public in the sense that they can be accessed and even changed by the user. In fact, a user who defines a point object 'P' can access (and indeed change) its coordinates by writing "P.x" and "P.y".

This is not what we want in object-oriented programming. In fact, we want the user to deal with (and indeed think about) the point object as a complete unit, and never deal directly with its private coordinates. This way, the user's mind will be free to really benefit from the point object and its properties in analytic geometry.

It is more in the spirit of object-oriented programming to declare the data fields 'x' and 'y' as private (not accessible to users) by placing the word "public" after their declarations in the class block. This way, point objects will be protected from any inadvertent change by inexperienced users. Furthermore, the programmer who has written the original "point" class will be able to modify it whenever necessary, with no need to alert users about this. In fact, the users will remain completely unaware of any change in the implementation, and won't have to change their code at all.

The default in C++ is that fields are private unless declared otherwise. To make the above code a more object-oriented code, it is sufficient to place the reserved word "public" after rather than before the 'x' and 'y' fields. This way, only the functions that follow the word "public" are accessible to users, but not the 'x' and 'y' fields that appear before it:

```
class point{
    double x;
    double y;  //  object-oriented implementation
public:
```

This way, a user who defines a point object 'P' can no longer access its coordinates simply by writing "P.x" or "P.y". Still, the programmer of the "point" class may give users permission to read the 'x' and 'y' coordinates through public interface functions as follows:

```
double X() const{
  return x;
}  //  read x

double Y() const{
  return y;
}  //  read y
```

This way, the user can write "P.X()" or "P.Y()" to invoke the public interface functions "X()" or "Y()" to read the 'x' or 'y' coordinate of the point object 'P'. Thanks to the reserved word "const" before the blocks of these functions, the point object 'P' with which they are called (referred to as the current object or variable) is protected from any change through these functions: they can only read a datum, but not change it. In fact, any attempt to change any data field in the current object would invoke a compilation error, to alert the programmer about an inadvertent change.

Furthermore, the calls "P.X()" and "P.Y()" are not more expensive than the corresponding calls "P.x" and "P.y" in the original (bad) implementation. Indeed, the functions contain only one command line each, and create no new objects.

Still, the programmer may elect to give users permission even to change data fields in a careful and well-controlled way:

```
void zero(){
    x=y=0.;
}  //  set to zero
};
```

This way, the user can writ "P.zero()" to set the point 'P' to zero. Note that the "zero()" function lacks the word "const" before its block to allow changing the current point object and sets it to zero.

The symbols "};" complete the block of the class. Before these symbols, the programmer may add more public functions for users to call with their own "point" objects. Furthermore, the programmer might want to write the reserved word "private:" and also add private functions before the end of the class block for his/her own use only.

15.3 Interface Functions

The advantage of interface functions like "X()", "Y()", and "zero" is in the opportunity to modify them at any time (with no need to notify the users), provided that they still take the same arguments and return the same output as before. In fact, the users remain completely unaware of the particular implementation or of any change in it. All they need to know is how to call the interface functions. In fact, they can think of a "point" variable like 'P' indeed as a point in the two-dimensional Cartesian plane. The interface functions associated with it can also be thought of as operations on points in the Cartesian plane, in the spirit of their original mathematical interpretation.

As we've seen above, interface functions are often placed inside the class block, right after the definitions of data fields. This style is suitable for short functions that contain a few code lines only. These functions are then recompiled every time the function is called.

A more efficient style, which is suitable for longer functions as well, only declares the function inside the class block, leaving its actual definition until later. The definition is placed outside the class block, with the function name preceded by a prefix containing the class name followed by the symbol "::", to indicate that this is indeed an interface function from this class. This way, the definition is treated as if it were inside the class block. For example, the "point" class could have been written equivalently as follows:

```
class point{
    double x;
    double y;
  public:
    double X() const;
    double Y() const;  //  declarations only
    void zero();
};
```

This completes the class block. The three interface functions declared in it are now defined in detail. In these definitions, the function name is preceded by the prefix "point::" to indicate that this is a definition of an interface function declared in the block of the "point" class:

```
double point::X() const{
   return x;
} //  definition of X()

double point::Y() const{
   return y;
} //  definition of Y()

void point::zero(){
   x=y=0.;
} //  definition of "zero()"
```

This way, each definition is compiled only once to create a finite state machine (automaton). This machine is then invoked every time the function is called, with the concrete arguments that are passed to it as input and the returned value as output.

The prefix "point::" may actually be viewed as an operator that "transfers" the definition back into the class block. This somewhat more complicated style, however, is unnecessary in the present "point" example, which uses short definitions only. The original style, in which the complete definitions are placed inside the class block, is therefore preferable.

15.4 Information and Memory

The dilemma whether to compile a function once and for all and store the resulting state machine (or automaton) in the memory for further use or to recompile it over and over again each and every time it is called is analogous to the dilemma whether to remember mathematical formulas by heart or to reformulate them whenever needed. Consider, for example, the algebraic formula

$$(a + b)^2 = a^2 + 2ab + b^2.$$

Because it is so useful, most of us know this formula by heart in an easily accessed part of our brains. Still, it occupies valuable memory, which could have been used for more vital purposes. Wouldn't it be better to release this valuable memory, and reformulate the formula whenever needed?

$$\begin{aligned}
(a + b)^2 &= (a + b)(a + b) \\
&= a(a + b) + b(a + b) \\
&= a^2 + ab + ba + b^2 \\
&= a^2 + 2ab + b^2.
\end{aligned}$$

After all, this way we'd train our brains in using the logics behind the formula, and also realize that it holds only in mathematical rings that support the distributive and commutative laws.

Still, memorizing the formula by heart also has its own advantage. Indeed, this way the formula itself becomes an individual object in the mathematical world. As such, it can be placed in any trees used to prove any mathematical theorem.

For a trained mathematician there is probably not much difference between these two approaches. In fact, the formula becomes an integral part of his/her vocabulary, to be used in many mathematical proofs. Furthermore, it goes hand in hand with its own proof, using the distributive and commutative laws. This is indeed an object-oriented thinking: the formula is an object in its own right, which contains yet another abstract object in it: its proof.

In the above approach, the formula becomes an object in one's immediate mathematical language, with its own interpretation or proof. There is, however, a yet better and more insightful approach: to use an induction process to place the formula in a more general context. Indeed, the formula corresponds to the second row in Pascal's triangle (Chapter 10, Section 10.9). By studying the general framework of Pascal's triangle and understanding the reason behind it (the induction stage), we not only obtain the above formula as a special case (the deduction stage, Chapter 10, Section 10.1), but also develop a complete theory, with applications in probability and stochastics as well (Chapter 10, Sections 10.9–10.14).

15.5 Constructors

We want the user to be able to define a point object simply by writing

```
point P;
```

Upon encountering such a command, the C++ compiler looks in the class block for a special interface function: the constructor. Thus, the programmer of the "point" class must write a proper constructor in the class block to allocate the required memory for the data fields in every "point" object defined by the user.

If no constructor is written, then the C++ compiler invokes its own default constructor to allocate memory for the data fields and initialize them with random values. Still, the programmer is well advised to write his/her own explicit constructor, not only to define the data fields but also to initialize them with suitable values.

The constructor must be a public interface function available to any user. The name of the constructor function must be the same as the name of the class. For example, the programmer could write in the block of the "point" class a trivial empty constructor, also referred to as the default constructor:

```
point(){
} // default constructor
```

Upon encountering a command like "point P;", the compiler would then invoke this constructor to allocate memory to the "double" data fields "P.x" and "P.y" (in the order in which they appear in the class block) and to initialize them with random values.

The above compiler, however, is too trivial. In fact, it only does what the default constructor available in the C++ compiler would do anyway. It is therefore a better idea to write a more sophisticated constructor, which not only allocates memory for the data fields but also initializes them with more meaningful values:

```
point(double xx,double yy){
  x=xx;
  y=yy;
}
```

Indeed, with this constructor, the user can write commands like

```
point P(3.,5.);
```

to define a new point object 'P' with the x-coordinate 3 and the y-coordinate 5.

15.6 Initialization List

The above constructor first initializes the data fields 'x' and 'y' with meaningless random values, and then assigns to them the more meaningful values passed to it as arguments. This is slightly inefficient; after all, it would make more sense to initialize the data fields with their correct values immediately upon definition. This can indeed be done by using an initialization list as follows:

```
point(double xx, double yy):x(xx),y(yy){
} // constructor with initialization list
```

The initialization list follows the character ':' that follows the list of arguments in the header of the constructor. The initialization list contains the names of data fields from the class block, separated by commas. Each data field in the initialization list is followed by the value with which it is initialized (in round parentheses). This way, when the compiler encounters a command like

```
point P(3.,5.);
```

it initializes the data fields "P.x" and "P.y" immediately with their correct values 3 and 5, respectively. The order in which the data fields are defined and initialized is the order in which they appear in the class block.

15.7 Default Arguments

Better yet, the constructor may also assign default values to its local (dummy) arguments:

```
point(double xx=0.,double yy=0.):x(xx),y(yy){
} //  arguments with default values
```

This way, if no concrete arguments are passed to the constructor, then its local arguments "xx" and "yy" take the zero value. This value is then used to initialize the data fields 'x' and 'y' in the initialization list. Thus, if the compiler encounters a command like

```
point P;
```

then it constructs the point object 'P'= $(0,0)$ (the origin in the Cartesian plane). Thus, there is no longer a need to write a default constructor in the block of the "point" class; the above constructor serves as a default constructor as well.

Furthermore, if the compiler encounters a command like

```
point P(3.);  //  or: point P = 3.;
```

then it assumes that the second local argument, "yy", whose value is missing, takes the default zero value. Thus, it constructs the point object 'P'$(3,0)$.

15.8 Explicit Conversion

The constructor defined above can also serve as an explicit-conversion operator from type "double" to type "point". Indeed, when the compiler encounters code like "point(3.)" or "(point)3.", it invokes the constructor as in the above example to produce the temporary "point" object $(3,0)$.

15.9 Implicit Conversion

When the compiler expects a "point" object but encounters a "double" number instead, it invokes the above constructor implicitly to convert the "double" object into the required "point" object. This feature is particularly useful in functions that take "point" arguments. Indeed, when a "double" concrete argument is passed to such a function, it is converted implicitly into the required "point" object before being assigned to the dummy "point" argument used throughout the function.

Implicit conversion has advantages and disadvantages. On one hand, it may make code more transparent and straightforward; on the other hand, it may be rather expensive.

Indeed, implicit conversion uses an extra call to the constructor, which requires extra time and storage. Although this overhead is rather negligible, it may easily accumulate into a more significant overhead when implicit conversion is repeated in long loops or when objects that are much bigger than the present "point" are converted.

To avoid implicit conversion altogether, one could just decline to specify default values for the dummy arguments, as in the original version in Section 15.6:

```
point(double xx, double yy):x(xx),y(yy){
} //  constructor with no implicit conversion
```

This way, the compiler would never invoke the constructor to convert implicitly a "double" object into a "point" object. The user would have to do this explicitly wherever necessary, and to pass "point" arguments to functions that expect them.

15.10 The Default Copy Constructor

Users of the "point" class may also wish to initialize a new "point" object to be the same as an existing one. For example, they may want to write code like

```
point P(3.,5.);
point Q(P); //  or point Q=P;
```

to initialize the point 'Q' to be the same as 'P'. This is done by the copy constructor, defined in the class block.

The copy constructor constructs (allocates memory for) a new object and initializes it with the value of the object that is passed to it as a concrete argument. The memory allocation and initialization of the particular data fields is done in the order in which they appear in the class block. In the above example, the x-coordinate of 'Q' is allocated memory and initialized to be the same as the x-coordinate of 'P', and then the y-coordinate of 'Q' is allocated memory and initialized to be the same as the y-coordinate of 'P'.

If no copy constructor is defined in the class block, then the compiler invokes the default copy constructor available in it. This constructor just allocates the required memory for each data field in the constructed object and initializes it with the corresponding data field in the argument. Loosely speaking, we say that the data fields are copied from the argument to the constructed object, or that the entire concrete argument is copied.

Still, the programmer is well advised to write his/her own copy constructor in the class block, and not rely on the default copy constructor available in the C++ compiler, which may do the wrong thing. We'll return to this subject in Section 15.20.

The copy constructor is invoked implicitly every time an object is passed to a function by value. Indeed, the local (dummy) object must be constructed and initialized to be the same as the concrete argument. Loosely speaking, the concrete argument is "copied" by the copy constructor to the local object used in the function block only.

Consider, for example, the following ordinary (noninterface) function, written outside of the class block:

```
const point negative(const point p){
  return point(-p.X(),-p.Y());
}
```

When the compiler encounters a call of the form "negative(P)" for some well-defined "point" object 'P', it first invokes the copy constructor to copy the concrete object 'P' into the dummy object 'P' used in the function block, then it invokes the constructor in Section 15.7 to construct $-P$ as in the function block, and finally it invokes the copy constructor once again to copy $-P$ into the constant temporary "point" object returned by the function, as indicated by the words "const point" at the beginning of the header of the function.

This seems to be a rather expensive process. Fortunately, some compilers support a compilation option that avoids the third construction. Furthermore, in Section 15.19, we'll see how the concrete argument can be passed by reference rather than by value to avoid the first call to the copy constructor as well.

The "negative" function returns a temporary (unnamed) "point" variable that exists only in the command line in which the call is made and disappears soon after. Why is this variable declared as constant in the beginning of the header of the function?

Temporary variables have no business to change, because they disappear anyway at the end of the present command line. Declaring them as constants protects them from inadvertent changes.

Indeed, a nonconstant temporary object can serve as a current object associated with an interface function, even when it could change there. However, it cannot be passed by address as a concrete argument to a function that could change it: the C++ compiler would refuse to create a local pointer that points to a nonconstant temporary object, out of fear that it would change in the function, which makes no sense because it is going to disappear anyway at the end of the command line in which the call is made. The compiler would therefore suspect that this isn't the real intention of the programmer, and would therefore issue a compilation error to alert him/her.

For example, the nonconstant temporary object "point(1.)" returned by the constructor of the "point" class cannot be passed by address to any function with a pointer-to-(nonconstant)-point argument, out of fear that it would undergo changes that make no sense.

Declaring the temporary object returned from the "negative" function as a constant (by placing the reserved word "const" at the beginning of the header of the function) solves this problem. Indeed, since this temporary object is constant, there is no fear that it would change inadvertently by any function, may it be an interface or an ordinary function. It may thus serve not only as a current object in an interface function but also as a concrete argument (passed either by value or by address, with either a constant or a nonconstant local pointer) in any function.

Note also that the above "negative()" function can also be called with a "double" argument, e.g., "negative(1.)" or "negative(a)", where 'a' is a "double" variable. Indeed, in such calls, the "double" argument is converted implicitly into a temporary "point" object before being used as a concrete argument in the "negative" function.

15.11 Destructor

At the end of the block of a function, the local variables are destroyed, automatically, and the memory that they occupy is released for future use. This is done by the destructor function, invoked implicitly by the compiler.

If no destructor is defined in the class block, then the default destructor available in the C++ compiler is invoked. This destructor goes over the data fields in the object that should be destroyed, and removes them one by one, in the reversed order to the order in which they appear in the class block. For example, in a "point" object, the 'y' data field is removed before the 'x' data field.

The default destructor, however, does not always do the right thing. It is thus advisable to write an explicit destructor in the class block as follows:

```
~point(){
```

```
}  //  default destructor
```

This destructor has an empty block, because everything is done implicitly. Indeed, at the '}' symbol that marks the end of the block, the data fields in the current object with which the destructor is called are destroyed one by one in the backward order: first the 'y' field, and then the 'x' field. Thus, this destructor works precisely as the default destructor available in the C++ compiler. This is why it is also referred to as the default destructor. In fact, it makes no difference whether it is written or not: in either case the compiler would use the same method to destroy an old object that is no longer needed.

The default destructor is good enough for the simple "point" object, which contains standard data fields of type "double" only, but not for more complicated objects with fields of type pointer-to-some-object (pointer fields). Indeed, if the default destructor encounters a pointer field, then it destroys only the address it contains, not the variable stored in it. As a result, although this variable is no longer accessible because its address is gone, it still occupies valuable computer memory.

Thus, in classes that contain not only standard data fields but also pointer fields, the block of the destructor can no longer remain empty. It must contain commands with the reserved word "delete" followed by the name of the pointer field. This command invokes implicitly the destructor of the class of the variable stored in this address to release the memory it occupies before its address is lost forever.

15.12 Member and Friend Functions

Interface functions may be of two possible kinds: member functions, which are called in association with a current object, and friend functions, which only take standard arguments that are passed to them as input. Some basic functions, such as constructors, destructors, and assignment operators, must be applied to a current object, hence must be member functions. More optional functions such as "X()", "Y()", and "zero" can be either defined in the class block as above to act upon the current object 'P' with which they are called [as in "P.X()", "P.Y()", and "P.zero()"], or as friend functions below (with no current objects) to act upon their concrete argument.

Since member functions are defined inside the class block, they are "unaware" of any code written outside the class block. Therefore, their definitions can use (call) only interface functions, that is, functions that are declared in the class block; they cannot use ordinary (noninterface) functions defined outside the class block, unless these functions are also declared as friends in the class block.

When a member function is called, it is assumed that the data fields mentioned in its block belong to the current object with which it is called. For example, in a call of the form "P.zero()", the "zero()" member function is invoked with 'x' interpreted as "P.x" and 'y' interpreted as "P.y".

Friend functions, on the other hand, have no current object; they can only take arguments in the usual way.

The most important property of member functions is that they have access to every field in the class block, private or public, including data fields of any object of this type (including the current object) and any function declared in the class block.

15.13 The Current Object and its Address

How is access to the current object granted? When a member function is called, there is one extra argument that is passed to it implicitly, even though it is not listed in the list of arguments. This argument, stored in a local pointer referred to by the reserved word "this", is of type constant-pointer-to-object, and points to the current object with which the member function is called.

This way, the current object can be accessed in the block of the member function through its address "this". In member functions of the present "point" class, for example, the 'x' field in the current "point" object can be accessed by writing "this− >x" or "(*this).x" (which means the 'x' field in the "point" object in "this") or 'x' for short, and the 'y' field in the current "point" object can be accessed by writing "this− >y" or "(*this).y" (which means the 'y' field in the "point" object in "this") or 'y' for short.

When a member function is actually called, as in "P.zero()" (for some concrete "point" object 'P'), "this" takes the value "&P" (the address of 'P'). As a result, 'x' in the function block is interpreted as "P.x" (the x-coordinate in 'P'), and 'y' in the function block is interpreted as "P.y" (the y-coordinate in 'P').

In constant member functions like "X()" and "Y()" that have the reserved word "const" written right before the function block, "this" is not only of type constant-pointer-to-a-point but actually of type constant-pointer-to-a-constant-point, which implies that both of the data fields in the current object can never change throughout the function, so they are actually read-only functions.

As a matter of fact, the command line in the block of the "X()" member function could have been written equivalently as

```
return this->x;  // or: return (*this).x;
```

In nonconstant functions like "zero" that lack the word "const" before their blocks, on the other hand, "this" is a constant-pointer-to-nonconstant-point rather than a constant-pointer-to-constant-point. This is why the 'x' and 'y' coordinates of the (nonconstant) current object (or "this— >x" and "this— >y") can indeed be changed through the "this" pointer and set to zero, as required.

When "P.zero()" is called by the user, the local pointer "this" takes the value "&P", so "this— >x" and "this— >y" are interpreted as "P.x" and "P.y". Once these nonconstant coordinates are set to zero, the nonconstant object 'P' is set to the origin $(0, 0)$.

15.14 Returned Pointer

The local "this" pointer can also be used to return the current object by address. For example, the following version of the "zero()" function not only sets the current object to $(0, 0)$ but also returns its address:

```
point* zero(){
    x=y=0.;
    return this;
}  //  returns pointer-to-current-point
```

This way, a temporary pointer-to-point variable is created at the end of the function block, initialized with the address in "this", and returned for further use in the same command line in which the "zero()" function is called.

15.15 Pointer to a Constant Object

Because the pointer returned from this version of the "zero" function exists only temporarily, it makes little sense to change its content. In fact, variables should change only through permanent, well-defined pointers rather than through temporary ones. This is why it is better yet to declare the returned pointer as a pointer-to-constant-point, so it can never be used to change its content:

```
const point* zero(){
    x=y=0.;
    return this;
}  //  returns a pointer-to-constant-point
```

Indeed, the reserved word "const" at the beginning of the header makes sure that the returned pointer points to a constant point that can undergo no change.

The returned pointer can now be used to print the x-coordinate onto the screen:

```
int main(){
  point P;
  printf("P.x=%f\n",P.zero()->X());
  return 0;
} //  print P.x after P has been set to (0,0)
```

Indeed, the call "P.zero()" not only sets 'P' to $(0,0)$ but also returns its address. By adding the suffix ">x", we have the 'x' field of the content of this address, or "P.x", which is then printed onto the screen in the standard "printf" function.

Later on we'll also see how the "zero" function can also be rewritten as a "friend" function rather than a member function. For this, the reserved word "friend" should be placed at the beginning of the declaration of the function in the class block. In this implementation, no current object or "this" pointer is available; objects must be passed explicitly as arguments, as in ordinary functions.

In the sequel, we'll see that arguments should better pass not by name (value) but rather by reference (address).

15.16 References

Instead of returning (or taking) a pointer, a function may also return (or take) a reference to an object. In this style, the compiler does the same thing as before in the sense that the object is returned (or passed) by address; still, the function becomes easier to read, understand, and use, as it avoids dealing with pointers and addresses.

In C++, one can define a reference to an existing variable. This means that the variable can be referred to not only by its original name but also by an alternative name:

```
point p;
point& q = p;
```

The prefix "point&" indicates that 'q' is not an independent "point" object but merely a reference-to-point, initialized to refer to the existing "point" object 'p'. In fact, no copy constructor is used; what the compiler really does is create a new pointer-to-point, named "&q", and initialize it with the address of 'p', "&p". The convenient thing about this is that the user no longer needs to use the symbol '&' every time he/she wants to change the original variable 'p' or pass it by address; he/she can just change 'q', and 'p' would automatically change at the same time as well.

15.17 Passing Arguments by Reference

In the previous chapter, we've seen that an argument must be passed to a function by address if it is supposed to change in it. Here we see that this can equally well be done by passing the argument by reference, as in the following version of the above "zero()" function. First, the function is declared as a "friend" in the class block, to have access to the private data fields of any "point" object:

```
friend const point* zero(point&);
```

It is already clear from the type "point&" in the round parentheses that the argument is passed by reference rather than by address. The declaration, however, specifies no name for this local reference; this is done only in the actual definition, in which the local reference to the argument takes the name 'p' for further use:

```
const point* zero(point&p){
  p.x=p.y=0.;
  return &p;
} //  set the "point" argument to zero
```

Thanks to the symbol '&' in the round parentheses, 'p' is not a copy of the concrete argument but rather a reference to it. Thus, the concrete argument is passed to the function not by value but rather by reference, and every change to the local reference 'p' effects it as well. This is why when a user makes a call of the form "zero(P)" for some well-defined "point" object 'P', 'P' is really changed and set to $(0,0)$, as required.

The advantage of declaring "zero()" as a friend rather than a member of the "point" class is in the opportunity to declare it also as a friend of other classes as well, to be able to access private fields of other objects as well whenever necessary. This is why the actual definition is better placed outside the class block, to be recognized in other classes as well.

15.18 Returning by Reference

Better yet, the "zero()" function could be rewritten to return a reference rather than a pointer:

```
friend const point& zero(point&p){
  p.x=p.y=0.;
  return p;
```

```
}
```

Indeed, thanks to the words "const point&" before the function name and
the final command "return p", it returns a reference to the concrete argument
[that has just been set to $(0,0)$] for further use in the same command line in
which the function is called:

```
printf("P.x=%f\n",zero(P).X());
```

Here, the reference to the existing "point" object 'P' returned by "zero(P)"
is further used to invoke the "X()" member function to return and print the
x-coordinate of 'P', which has just been set to zero.

Friend functions are usually used to read private data fields from one or
more classes. The "zero()" function that actually changes the data fields, on
the other hand, is better implemented as a public member function in the
block of the "point" class as follows:

```
const point& zero(){
    x=y=0.;
    return *this;
}
```

In this style, the "zero()" function still returns a reference to its current object,
which lies in the address "this", and hence is referred to as "*this". Thus,
the user can still set set the existing "point" object 'P' to $(0,0)$ and print its
x-coordinate at the same command line:

```
printf("P.x=%f\n",P.zero().X());
```

Here, the reference to 'P' returned by the call "P.zero()" serves as a current
object for the next call to the member function "X()" to return and print the
x-coordinate of 'P'.

15.19 Efficiency in Passing by Reference

Clearly, passing by reference is much more efficient than passing by value,
as it avoids an (implicit) call to the copy constructor. For example, the
"negative" function in Section 15.10 can be rewritten as

```
const point negative(const point& p){
    return point(-p.X(),-p.Y());
} // passing argument by reference
```

This way, thanks to the '&' symbol in the round parentheses, 'p' is only a
local reference to the concrete argument, so it requires no call to the copy
constructor.

Even in its improved version, the "negative()" function still requires two
calls to constructor: first, the constructor that takes two "double" arguments is
called explicitly in the function block to create −p. Then, the copy constructor
is called implicitly to copy −p into the constant temporary "point" object
returned by the function, as is indeed indicated by the words "const point"
before the function name.

Unfortunately, this call cannot be avoided. Indeed, if one had inserted the
symbol '&' before the function name to return a reference to −p rather than
a copy of it as in

```
const point& negative(const point& p){
  return point(-p.X(),-p.Y());
} // wrong!!! returns reference to nothing
```

then the function would return a reference to the local "point" object −p that
has already disappeared. This is why the local object −p must be copied into
the temporary object returned by the function before it is gone, as is indeed
indicated by the words "const point" (rather than "const point&") before the
function name in the correct versions.

15.20 Copy Constructor

As mentioned in Section 15.10, it is advisable to define an explicit copy
constructor in the class block as follows:

```
point(const point& p):x(p.x),y(p.y){
} // copy constructor
```

Here, the data fields in the copied "point" object 'p' are used in the initialization list to initialize the corresponding data fields in the new object.

Actually, this copy constructor works exactly the same as the default copy
constructor available in the C++ compiler. This is why it is also sometimes
referred to as the default copy constructor. Still, it is a good practice to write
your own copy constructor rather than to rely on a standard one, which may
do unsuitable things.

With the above copy constructor, the user can write

```
point Q = P;  // same as  point Q(P);
```

to construct a new "point" object 'Q' and initialize it to be the same as
the existing "point" object 'P'. Yet more importantly, the copy constructor

is invoked implicitly to return a "point" object by value, as in the above "negative" function (in its correct version).

15.21 Assignment Operators

Users of the "point" class may also want to assign the value of an existing object 'Q' to the existing objects 'W' and 'P' by writing

```
point P,W,Q(1.,2.);
P=W=Q;
```

To allow users to write this, an assignment operator must be written in the class block in such a way that it not only assigns the value of the concrete argument 'Q' into the current object 'W' but also returns a reference to it, which can then be assigned to 'P' as well:

```
const point& operator=(const point& p){
```

This is the header of the assignment operator, named "operator=". (This name is necessary to allow users to make calls like "W = Q" to invoke it to assign 'Q' to 'W'.) The "point" argument is passed to the function by reference, thanks to the symbol '&' in the round parentheses above. Furthermore, the current object (which exists before, during, and after the call to the function) is also returned by reference, thanks to the symbol '&' before the function name. This helps to avoid unnecessary calls to the copy constructor.

Moreover, both the argument and the returned object are declared as constants to protect them from any inadvertent change. This way, the function can take even constant (or temporary) arguments, with no fear that they would undergo inappropriate changes through their local reference.

We are now ready to start the function block. If the current object is the same as the argument (or, more precisely, the address of the current object, "this", is the same as that of the argument, "&p"), then nothing should be assigned. If, on the other hand, they are not the same,

```
if(this != &p){
```

then the data fields should be assigned one by one:

```
    x = p.x;
    y = p.y;
}
```

Finally, the current object (which lies in the address "this") is also returned for further use:

```
        return *this;
   }  //  point-to-point assignment operator
```

We refer to this operator as the point-to-point assignment operator. This operator is invoked whenever the compiler encounters a command like "W = Q" to assign the value of 'Q' to 'W' as well. Furthermore, when the compiler encounters a command like

```
   W=1.;
```

it first invokes implicitly the constructor that takes "double" arguments to create the temporary "point" object $(1, 0)$ before assigning it to 'W'. The "zero()" function in Section 15.2 is, thus, no longer necessary: one can set 'W' to zero simply by writing "W = 0.".

To avoid this extra construction, one may write a special "double"-to-point assignment operator in the class block:

```
        const point& operator=(double xx){
          x = xx;
          y = 0.;
          return *this;
        }  //  double-to-point assignment operator
```

This assignment operator is invoked whenever the compiler encounters a command of the form "W = 1.", because it can take the "double" argument on the right-hand side with no implicit conversion. Furthermore, the current object 'W' returned by reference from this call can then be assigned to yet another object 'P':

```
        P.operator=(W.operator=(0.));  //  same as P=W=0.;
```

This is exactly the same as the original form "P = W = 0.", which is more transparent and elegant and more in the spirit of object-oriented programming, since it is as in standard C types.

15.22 Operators

One may also define more operators, using all sorts of arithmetic or logical symbols. In fact, operators are functions that can be called not only by their names but also by their symbols. For example, one may choose to use the '&' to denote inner product in the Cartesian plane:

```
        double operator&&(const point&p, const point&q){
          return p.X() * q.X() + p.Y() * q.Y();
        }  //  inner product
```

With this operator, the user may write just "P && Q" to have the inner product of the "point" objects 'P' and 'Q'.

Thus, in the context of "point" objects, the "&&" has nothing to do with the logical "and" operator in C. All that it inherits from it is the number of arguments (which must be two) and the priority order with respect to other operators.

Note that the above operator is implemented as an ordinary (nonmember, nonfriend) function, because it needs no access to any private member of the "point" class. In fact, it accesses the 'x' and 'y' fields in 'p' and 'q' through the public member functions "X()" and "Y()".

Note also that, as in the point-to-point assignment operator above, the arguments are passed by reference, rather than by value, to avoid unnecessary calls to the copy constructor. Furthermore, they are also declared as constants, so that the function can be applied even to constant (or temporary) concrete arguments, with no fear that they would undergo inappropriate changes through their local references.

15.23 Inverse Conversion

One may also define in the class block a public member inverse-conversion operator to convert the current object into another object. This operator is special, because its name is not a symbol but rather the type into which the object is converted.

Here is how an inverse-conversion operator can be defined in the block of the "point" class:

```
operator double() const{
  return x;
} //  inverse conversion
```

With this public member function, the user can write "double(P)" or "(double)P" to read the x-coordinate of the "point" object 'P' without changing it at all (as indicated by the reserved word "const" before the function block). Furthermore, the user can pass a "point" object to a function that takes a "double" argument. Indeed, in this case, the compiler would invoke the inverse-conversion operator implicitly to pass to the function the x-coordinate of the "point" object.

This, however, may not at all be what the user wanted to do. In fact, it might be just a human error, and the user would much rather the compiler to issue a compilation error to alert him/her. Therefore, the careful programmer may decline to write an inverse-conversion operator, to prevent the compiler from accepting code that might contain human errors.

15.24 Unary Operators

The "negative" function in Section 15.10 can also take the form of a unary operator that takes a "point" object and returns its negative. For this purpose, it is enough to change the name of the "negative" function to "operator−". With this new name, the user can invoke the function simply by writing "−P", where 'P' is a well-defined "point" object.

The "operator−" can be even more efficient than the original "negative" function. Indeed, with the original function, the code

```
point W = negative(1.);
```

requires an implicit conversion of the "double" number 1 into the "point" object $(1,0)$, which is passed to the "negative" function to produce the point $(-1,0)$. The copy constructor is then called to construct and initialize 'W'. With the new "operator−", on the other hand, this code takes the form

```
point W = -1.;
```

which only uses one call to the constructor to form directly the required point 'W'= $(-1,0)$.

The reserved word "operator" that appears in the function name can be omitted from the actual call, leaving only the symbol that follows it. Thus, instead of writing "operator−(P)", the user can simply write "−P", as in the original mathematical formulation.

15.25 Update Operators

Here we consider update operators that use their argument to update their current object with which they are called. In particular, the "+ =" operator below adds its argument to its current object. In its main version, this operator is defined as a member function inside the class block:

```
const point& operator+=(const point& p){
   x += p.x;
   y += p.y;
   return *this;
} //  adding a point to the current point
```

With this operator, the user can simply write "P + = Q" to invoke the above operator with 'P' on the left being the current object and 'Q' on the right being the argument. In fact, this call adds 'Q' to 'P' and stores the sum back in 'P'. Furthermore, it also returns a constant reference to the current object 'P', for further use in the same command line in which the call is made. This way, the user can write "W = P + = Q" to store the sum of 'P' and 'Q' not only in 'P' but also in 'W'.

If, on the other hand, the user writes a command like "P + = 1.", then the double number 1 is first converted implicitly to the "point" object $(1, 0)$ before being added to 'P'. To avoid this conversion, one could write in the class block a special version of "operator+ =" that takes "double" rather than "point" argument:

```
const point& operator+=(double xx){
   x += xx;
   return *this;
} //  add a real number to the current point
```

This version is then invoked whenever the compiler encounters a call like "P + = 1.", because it can take it with no implicit conversion.

15.26 Friend Update Operators

The natural implementation of "operator+ =" is as a member function, which adds its argument to its current object. However, it could also be implemented as a "friend" function, with an extra (nonconstant) argument instead of the current object:

```
friend const point&
    operator+=(point&P,const point& p){
   P.x += p.x;
   P.y += p.y;
   return P;
}
```

This version can still be called simply by "P + = Q" as before. Indeed, in this call, 'P on the left is passed as the first (nonconstant) argument, whereas 'Q' on the right is passed as the second (constant) argument that is added to it.

The "friend" implementation, although correct, is somewhat unnatural in the context of object-oriented programming. Indeed, it has the format of an ordinary function that changes its argument. In object-oriented programming, however, we prefer to think in terms of objects that have functions to express their features, rather than in terms of functions that act upon objects. This concept is better expressed in the original implementation as a member function.

Furthermore, the "friend" version has yet another drawback: it wouldn't take a temporary object as its first (nonconstant) argument, out of fear that changing it makes no sense and must be a human error. For example, it would issue a compilation error whenever encountering calls like "point(P) + = Q". The original member implementation, on the other hand, would take a temporary object as its current object, thus would accept such calls as well. Below we'll see that such calls are particularly useful, so one should better stick to the original member implementation.

15.27 Binary Operators

One can also define binary operators that take two "point" arguments to calculate and return the required output. In particular, the '+' operator is defined below outside of the class block as an ordinary function that needs no access to any private data field:

```
const point
operator+(const point& p, const point& q){
  return point(p.X()+q.X(),p.Y()+q.Y());
} //  add two points
```

Unlike the "+=" operator, this operator doesn't change its arguments, which are both passed (by reference) as constant "point" objects. Their sum, however, can't be returned by reference, because it would then be a reference to a local "point" object that disappears when the function ends. It is rather returned by value (as is indeed indicated by the words "const point" in the beginning of the header), so that this local "point" object is copied to the temporary objects returned by the function before it ends.

The above '+' can be defined as the programmer wishes, and has nothing to do with the '+' arithmetic operation on integer or real numbers. Still, it makes sense to define the '+' operator as in the common mathematical formulation. Indeed, with the above definition, the user can write simply "P + Q" to have the sum of the points 'P' and 'Q'.

There are only two things that the '+' operator defined above does inherit from the '+' arithmetic operation in C: the number of arguments (it must be a binary operator that takes two arguments) and the priority order with respect to other operators. For example, if the programmer also defines a '*' operator to somehow multiply two point objects, then it is prior to the above '+' operator.

As mentioned at the end of Section 15.23, we assume that no inverse conversion is available, because the "operator double()" that converts "point" to "double" is dropped. Therefore, since both arguments in "operator+" are of type reference-to-constant-point, "operator+" can be called not only with two "point" arguments (as in "P + Q") but also with one "point" argument and one "double" argument (as in "P + 1." or "1. + P"). Indeed, thanks to the implicit double-to-point conversion in Section 15.9, the "double" number 1 is converted to the point $(1, 0)$ before being added to 'P'. Furthermore, thanks to the lack of inverse conversion, there is no ambiguity, because it is impossible to convert 'P' to "double" and add it to 1 as "double" numbers.

To avoid the above implicit conversion, though, one can write explicit versions of "operator+" to add "double" and "point" objects:

```
const point operator+(const point& p, double xx){
   return point(p.X()+xx,p.Y());
} //  point plus real number

const point operator+(double xx, const point& p){
   return point(p.X()+xx,p.Y());
} //  real number plus point
```

15.28 Friend Binary Operators

Since the '+' operator is defined outside of the class block, it cannot be called from functions inside the class block, unless declared in the class block as a friend:

```
friend const point operator+(const point&, const point&);
```

With this declaration, the '+' operator can also be called from inside the class block as well. Furthermore, it has access to the private data fields 'x' and 'y' of its "point" arguments. In fact, it could equally well be defined inside the class block as follows:

```
friend const point operator+(
        const point& p, const point& q){
   return point(p.x+q.x,p.y+q.y);
} //  defined as "friend" in the class block
```

15.29 Member Binary Operators

The '+' operator could also be implemented as a member function inside the class block as follows:

```
const point operator+(const point& p) const{
    return point(x+p.x,y+p.y);
}  //  defined as "member" in the class block
```

With this implementation, the user can still call the operator simply by writing
"P + Q". In this call, 'P' is the (constant) current object, and 'Q' is the
(constant) argument used to calculate and return the sum of 'P' and 'Q'.

This, however, is a rather nonsymmetric implementation. Indeed, implicit
conversion can take place only for the second argument, but not for the first
one, the current object. Therefore, "P + 1." is legal, but "1. + P" is not. This
nonsymmetry makes no apparent sense.

15.30 Ordinary Binary Operators

The original implementation of "operator+" as an ordinary function outside
of the class block is also more in the spirit of object-oriented programming.
Indeed, it avoids direct access to the private data fields 'x' and 'y' in "point"
objects, and uses only indirect access through the public member functions
"X()" and "Y()" to read them. This way, the '+' operator is independent of
the internal implementation of "point" objects.

The original implementation of the '+' operator as an ordinary function
can also be written in a more elegant way, using the "operator+=" member
function defined in Section 15.25:

```
const point
operator+(const point& p, const point& q){
    return point(p) += q;
}  //  point plus point
```

Indeed, thanks to the fact that the "+=" operator is defined in Section 15.25
as a member (rather than a mere friend) of the "point" class, it takes even
temporary "point" objects [like the temporary object "point(p)" returned
from the copy constructor] as its current objects.

15.31 Complex Numbers

The "complex" numbers introduced in Chapter 5 are available as a standard type (along with the arithmetic operations between them) in a standard
library that can be included in the program. Here, however, we prefer not to

rely on this library and define our own "complex" class, as a good exercise in implementing and using mathematical objects.

Like the "point" class, the "complex" class should contain two private data fields to store the real and imaginary parts of a complex number. Furthermore, it contains some member operators to implement arithmetic operations between complex numbers. Moreover, thanks to the constructor and the copy constructor, the user of the "complex" class can define complex variables simply by writing commands like "complex c", "complex c(0.,0.)", "complex d(c)", or "complex d = c".

The "complex" class is implemented as follows:

```
class complex{
    double real;
    double image;
```

These are the private "double" fields that store the real and imaginary parts of the "complex" object. In the following constructor, these data fields are initialized in the initialization list to have their required values immediately upon definition:

```
public:
    complex(double r=0.,double i=0.):real(r), image(i){
    } //  constructor
```

Furthermore, thanks to the default values given to the dummy arguments, the user can define a complex variable not only by writing "complex c(0.,0.)" but also by writing "complex c", "complex c(0.)", or "complex c = 0.". Moreover, thanks to the default values, the above constructor also supports explicit conversion as in "complex(0.)" or "(complex)0.", and even implicit "double"-to-complex conversion whenever a "double" argument is passed to a function that expects a "complex" argument.

The copy constructor is implemented in a similar way:

```
    complex(const complex&c):real(c.real),image(c.image){
    } //  copy constructor
```

With this constructor, the user can define a new "complex" variable 'd' and initialize it immediately upon definition to have the same value as the existing "complex" variable 'c' simply by writing "complex d(c)" or "complex d = c".

The (default) destructor defined below has an empty block, because the "image" and "real" fields are destroyed implicitly (in this order) at the '}' symbol that makes the end of the following block:

```
    ~complex(){
    } //  destructor
```

Because the data fields "real" and "image" are declared before the reserved word "public:", they are by default private members of the class. Therefore, only members and friends of the class can access them. Users and ordinary functions, on the other hand, can only read them indirectly through the following public member functions:

```
double re() const{
  return real;
} //  read real part

double im() const{
  return image;
} //  read imaginary part
```

The assignment operator is defined as follows:

```
const complex&operator=(const complex&c){
  real = c.real;
  image = c.image;
  return *this;
} //  assignment operator
```

Here the current object is not only assigned a value from the argument (which is passed by a constant reference) but also returned by a constant reference. This allows users to write commands like "e = d = c" (where 'c', 'd', and 'e' are well-defined "complex" objects) to assign 'c' not only to 'd' but also to 'e'.

15.32 Member Arithmetic Operators with Complex Numbers

Next, we define some member arithmetic operators that update the current "complex" object.

```
const complex&operator+=(const complex&c){
  real += c.real;
  image += c.image;
  return *this;
} //  add complex to the current complex
```

With this operator, the user can write "e = d += c" to add 'c' to 'd' and place the sum in both 'd' and 'e'.

The "-=" operator is defined in a similar way:

```
const complex&operator-=(const complex&c){
  real -= c.real;
  image -= c.image;
  return *this;
} //  subtract complex from the current complex
```

With this operator, the user can write "e = d -= c" to subtract 'c' from 'd' and place the sum in both 'd' and 'e'.

The following operator multiplies the current object by the argument:

```
const complex&operator*=(const complex&c){
  double keepreal = real;
  real = real*c.real-image*c.image;
  image = keepreal*c.image+image*c.real;
  return *this;
} //  multiply the current complex by a complex
```

With this operator, the user can write "e = d *= c" to multiply 'd' by 'c' and place the sum in both 'd' and 'e'.

The following operator divides the current "complex" object by the real number 'd':

```
const complex&operator/=(double d){
  real /= d;
  image /= d;
  return *this;
} //  divide the current complex by a real number
```

Later on, we'll also define yet another version of "operator/ =" that divides the current "complex" object by a "complex" (rather than a "double") argument. As a member function, this version will recognize no ordinary function defined outside of the class block unless declared as a friend in the class block. This is why the following two functions, which are going to be used in it, are indeed declared as friends in the class block:

The following "operator!" returns the complex conjugate of a complex number. Although it uses the '!' symbol, it has nothing to do with the logical "not" operator in C. The only common property in these two operators is that both are unary, that is, take one argument only.

The function requires no current object; this is why it is defined as a friend rather than a member of the "complex" class. The second word in the header, "complex", indicates that the output is returned by value rather than by reference, to avoid referring to a local variable that no longer exists when the function terminates.

```
friend complex operator!(const complex&c){
  return complex(c.re(),-c.im());
} //  conjugate of a complex
```

With this operator, the user can just write "!c" to have the complex conjugate of 'c'.

The following "abs2()" function returns the square of the absolute value of the complex number:

```
friend double abs2(const complex&c){
   return c.re() * c.re() + c.im() * c.im();
} // square of the absolute value of a complex
```

These two friend functions are now used in the following member function that divides the current "complex" object by the "complex" argument:

```
const complex&operator/=(const complex&c){
   return *this *= (!c) /= abs2(c);
} // divide the current complex by a complex
};
```

Indeed, the code line in the function block is executed from right to left: first, the original version of "operator/ =" is invoked to divide the complex conjugate of 'c' by the real number "abs2(c)". (As a member function, it can take the temporary nonconstant object "!c" as its current object.) Then, the output of this division is used to multiply the current "complex" object in "this", as required. Finally, the current object in "this" is also returned by reference for further use. This way, the user can write "e = d /= c" to divide 'd' by 'c' and place the result in both 'd' and 'e'. This completes the block of the "complex" class.

15.33 Ordinary Arithmetic Operators with Complex Numbers

Here we implement some ordinary (nonmember, nonfriend) arithmetic operators on complex numbers. The following unary "operator−" returns the minus of a complex number:

```
const complex
operator-(const complex&c){
   return complex(-c.re(),-c.im());
} // negative of a complex number
```

With this operator, the user can write "-c" to have the minus of the complex number 'c'.

The following binary "operator−" returns the difference between two complex numbers:

```
const complex
operator-(const complex&c,const complex&d){
  return complex(c.re()-d.re(),c.im()-d.im());
} //  subtraction of two complex numbers
```

With this operator, the user can write "c - d" to have the difference between 'c' and 'd'. There is no ambiguity between the two "operator−" versions: the C++ compiler uses the binary version when the '−' symbol is placed in between two "complex" objects and the unary version when it is placed before a single "complex" object.

The following "operator+" returns the sum of two complex numbers:

```
const complex
operator+(const complex&c,const complex&d){
  return complex(c.re()+d.re(),c.im()+d.im());
} //  addition of two complex numbers
```

Note that, as is indicated by the words "const complex" at the beginning of the header, the output is returned by value rather than by reference, to avoid referring to a local variable that no longer exists at the end of the function.

The following operator returns the product of two complex numbers:

```
const complex
operator*(const complex&c,const complex&d){
  return complex(c) *= d;
} //  multiplication of two complex numbers
```

Indeed, as a member function, the "*=" operator can take the temporary non-constant "complex" object "complex(c)" returned from the copy constructor, use it as its current object, multiply it by 'd', and return it by value, as required.

Similarly, the following operator returns the ratio between two complex numbers:

```
const complex
operator/(const complex&c,const complex&d){
  return complex(c) /= d;
} //  division of two complex numbers
```

Finally, we also define the unary "operator+" to return the complex conjugate of a complex number. This way, the user can write "+t" for any variable 't' of a numerical type, may it be either real or complex. Indeed, if 't' is real, then "+t" is the same as 't', as required; if, on the other hand, 't' is complex, then "+t" is the complex conjugate of 't', as required.

```
complex operator+(const complex&c){
  return complex(c.re(),-c.im());
} //  conjugate complex
```

This concludes the arithmetic operations with complex numbers. Finally, we define a function that prints a complex number to the screen:

```
void print(const complex&c){
  printf("(%f,%f)\n",c.re(),c.im());
} //  printing a complex number
```

15.34 Exercises

1. Write the class "point3" that implements a point in the three-dimensional Cartesian space. Write the constructors, destructor, assignment operators, and arithmetic operators for this class.

2. Write a function that uses the polar representation of a complex number

$$c \equiv r(\cos(\theta) + i \cdot \sin(\theta))$$

   to calculate its square root $\sqrt{c}$. (Here $0 \leq \theta < 2\pi$ is the angle with the positive x-axis, see Chapter 5.) The solution can be found in Section 28.5 in the appendix.

3. In the previous exercise, the angle in the polar representation of c, θ, lies between 0 and 2π, so the angle in the polar representation of $\sqrt{c}$, $\theta/2$, lies between 0 and π. Unfortunately, this way the square root function is discontinuous at the positive part of the real axis. Indeed, when θ approaches 0^+ ($\theta > 0$) $\theta/2$ approaches 0, as required; but, when θ approaches $2\pi^-$ ($\theta < 2\pi$), $\theta/2$ approaches π, yielding the negative of the required square root. To fix this, modify your code so that the original angle, θ, lies between $-\pi$ and π, so $\theta/2$ lies between $-\pi/2$ and $\pi/2$. This way, $\theta/2$ approaches 0 whenever θ approaches 0, regardless of whether $\theta > 0$ or $\theta < 0$. The solution can be found at the end of Section 28.5 in the appendix.

4. Implement complex numbers in polar coordinates: a "complex" object contains two fields, 'r' and "theta", to store the parameters $r \geq 0$ and $0 \leq \theta < 2\pi$ used in the polar representation

$$r \exp(i\theta) = r \left(\cos(\theta) + i \cdot \sin(\theta) \right).$$

   Redefine the above constructors, destructor, and assignment operators in this implementation. Furthermore, redefine and test the required arithmetic operations.

5. Do users of the "complex" class have to be informed about the modification made above? Why?

Chapter 16

Vectors and Matrices

The most important objects in linear algebra are vectors and matrices (Chapter 9). Here we implement these objects, along with the operators that use them. Users who include this code in their programs can define vectors and matrices just by writing commands like "vector v;" or "matrix m;".

Furthermore, the various versions of the '*' operator defined below allow the user to write commands like "v*v", "m*v", and "m*m" to multiply vector times vector (inner product), matrix times vector, and matrix times matrix (respectively), exactly as in the corresponding mathematical formulation.

Thus, with the present implementation, users can use vectors and matrices as if they were standard objects available in C. This way, users can think and write about vectors and matrices in their original mathematical spirit, without bothering with any storage details. This helps not only to implement complicated algorithms in short, elegant, and well-debugged codes, but also to understand better the nature of the objects and continue to develop and improve the algorithms and applications that use them.

16.1 Induction and Deduction in Object-Oriented Programming

In Chapter 10, Section 10.1, we have discussed the principle of induction and deduction in solving problems. This principle says that, in order to solve a particular problem, it is sometimes useful to generalize it into a more general problem, written in more general terms. In these general terminology and framework, it is sometimes much easier to solve the problem, because the technical details that characterize the original problem and may obscure the essential components in it are moved out of the way. Furthermore, the general problem obtained from the original particular problem may have a general theory, based on fundamental objects and ideas.

The generalization of the original particular problem into a more general problem is called induction. Once a general framework has been defined and a suitable theory has been developed, the general problem can be solved. Now, the solution of the original problem is immediately obtained as a special case. This is called deduction. The result is, thus, that not only the original problem has been solved as required, but also that a general theory has been developed to solve the general problem.

The principle of induction and deduction can also be used in object-oriented programming. In this case, however, no explicit problem is given. Instead, the task is to define and implement a particular object, along with its useful functions. Here the principle of induction indicates that one should better define a more general object, from which the required object can be obtained as a special case.

This approach may help to define the general object more correctly. Indeed, in the general form of the object, the immaterial details are moved out of the way, leaving only its essential characters and features. In the general implementation, these characters and features are written in terms of (member) functions that make the object look as it indeed should look. Once the implementation of the general object is complete, the required object can be obtained as a special case.

Consider, for example, the task to implement a point in the Cartesian plane. Such a point can be viewed as a vector with two coordinates. In this case, one is well advised to use induction to generalize the original task into the more general task of implementing an n-dimensional vector (as defined in Chapter 9, Section 9.5), where n is any natural number. Once this general object is implemented, the original task is also complete by deduction: that is, by using the special case $n = 2$.

16.2 Templates

A powerful programming tool to use induction and deduction is the template. Indeed, one can write a template class to define a general object that depends on some yet unspecified parameter, say n. Then, one can write all sorts of interface functions using the parameter n. This way, the user of this template class can define a concrete object by specifying the parameter n. For example, if the user wishes to define a two-dimensional vector, then they can define an n-dimensional vector with n being specified to be 2.

Arithmetic operations with vectors are better implemented on general n-dimensional vectors (as in the induction stage above) than on specific two-dimensional or three-dimensional vectors. Once they have been implemented in their general form, it is particularly easy to obtain the desirable concrete form by setting $n = 2$ for points in the Cartesian plane or $n = 3$ for points in the Cartesian space (the deduction stage).

Thus, the template class can define general n-dimensional vectors, along with their arithmetic operations as well as other useful functions. It is only later that the user specifies n to suit his/her concrete application. The arithmetic operators still apply, because they are written in a most general form suitable for every n.

16.3 The Vector Object

Here we implement n-dimensional vectors in a template class that uses the integer 'N' to stand for the dimension n. Furthermore, the template class also uses the parameter 'T' to stand for the particular type of the components in the vector. For example, if the user specifies 'T' to be "float" or "double", then real-valued vectors (vectors with real components) are implemented. If, on the other hand, the user specifies 'T' to be the "complex" class in Chapter 15, Section 15.31, then complex-valued vectors (vectors with complex components) are implemented.

With the present "vector" class, there is no longer any need to implement points in the Cartesian plane as in the "point" class in Chapter 15. Indeed, such points can be obtained as a special case (deduction) by setting 'N' to be 2 and 'T' to be "double". This approach avoids a lot of unnecessary programming work.

In the following implementation, the words "template<class T, int N>" before the definitions of the class and the functions indicate that they indeed take two template parameters: 'T' to specify the type of the components in the vector, and 'N' to specify the number of components in a vector. In the definitions, these parameters are not yet specified. In fact, it is easier to define the class and the functions with 'T' being an unspecified type and 'N' being an unspecified natural number (the induction stage). As a matter of fact, 'T' and 'N' are specified only when the user defines a concrete "vector" object later on (the deduction stage).

```
template<class T, int N> class vector{
    T component[N];
```

The private data field "component" is an array of 'N' entries of type 'T'. Here are the declarations of the public member functions: a constructor that takes a 'T' argument,

```
public:
    vector(const T&);
```

a constructor that takes a 'T', an integer, and a string:

```
vector(const T&,int,char*);
```

a copy constructor,

```
vector(const vector&);
```

an assignment operator,

```
const vector& operator=(const vector&);
```

and an assignment operator that takes a 'T' argument:

```
const vector& operator=(const T&);
```

These functions will be defined in detail later on.

Next, we define the destructor. The function block is empty, because the data field "component" is destroyed automatically at the '}' symbol that marks its end.

```
~vector(){
} // destructor
```

Because it does the same thing as the default destructor available in the C++ compiler, this destructor is also referred to as the default destructor.

Furthermore, because the components in the vector are private members, the user can access them only through public member functions:

```
const T& operator[](int i) const{
  return component[i];
} //read ith component
```

Thanks to the reserved word "const" before the function block, the current "vector" object is protected from any inadvertent change. After all, the purpose of this operator is only to read the 'i'th entry, not to change it. In fact, the 'i'th entry is returned by reference rather than by value, to avoid an unnecessary call to the copy constructor of the 'T' class, whatever it may be. Furthermore, the returned entry is declared as constant in the beginning of the header, so that it cannot be passed (neither as an argument nor as a current object) to any function that may change its value, a change that makes no sense.

With the above operator, the user can just write "v[i]" to read the 'i'th component in the "vector" object 'v'. This is indeed in the spirit of the original mathematical formulation of vectors.

Moreover, here is the member function that sets the 'i'th entry to have the same value as the 'T' argument:

```
void set(int i,const T& a){
  component[i] = a;
} // change ith component
```

In fact, the only way the user can change the value of the 'i'th component in 'v' into the scalar 'a' is to apply this function to the current object 'v' by writing "v.set(i,a)".

Moreover, here are the declarations of some member update operators: adding a vector to the current vector,

```
    const vector& operator+=(const vector&);
```

subtracting a vector from the current vector,

```
    const vector& operator-=(const vector&);
```

multiplying the current vector by a scalar,

```
    const vector& operator*=(const T&);
```

and dividing the current vector by a nonzero scalar:

```
    const vector& operator/=(const T&);
};
```

This concludes the block of the "vector" template class.

16.4 Constructors

Next, we define explicitly the member functions that are only declared in the class block above. Each header starts with the words "template<class T,int N>" to indicate that this is indeed a template function that, although defined in terms of the general parameter 'T' and 'N' (the induction stage), does require their explicit specification when actually called by the user (the deduction stage). Furthermore, the prefix "vector<T,N>::" before the function name in the header indicates that this is indeed the definition of a member function (with a current "vector" object) that actually belongs in the class block.

Here is the constructor that takes a scalar argument 'a' to set the values of the components:

```
template<class T, int N>
vector<T,N>::vector(const T& a = 0){
    for(int i = 0; i < N; i++)
        component[i] = a;
} // constructor
```

Here is a yet more sophisticated constructor that takes not only the scalar 'a' but also the integer 'n' to construct the 'n'th unit vector, that is, the vector whose components vanish, except of the 'n'th component, whose value is 'a'.

```
template<class T, int N>
vector<T,N>::vector(const T& a = 0,int n,char*){
    for(int i = 0; i < N; i++)
        component[i] = 0;
    component[n] = a;
} // constructor of a standard unit vector
```

To construct the unit vector $(1, 0)$, for example, the user can just write "vector<double,2> x(1.,0,"x")". The third argument, the string, is present only to let the compiler know that this constructor is indeed the one that is called.

Similarly, the copy constructor is defined as follows:

```
template<class T, int N>
vector<T,N>::vector(const vector<T,N>& v){
  for(int i = 0; i < N; i++)
    component[i] = v.component[i];
}  //  copy constructor
```

16.5 Assignment Operators

The assignment operator is defined as follows:

```
template<class T, int N>
const vector<T,N>& vector<T,N>::operator=(
        const vector<T,N>& v){
```

If the assignment is indeed nontrivial, that is, the argument 'v' is not the same object as the current object,

```
if(this != &v)
```

then assign the components of 'v' one by one to the corresponding components of the current object:

```
for(int i = 0; i < N; i++)
    component[i] = v.component[i];
```

Finally, return a constant reference to the current object, as is indeed indicated by the words "const vector<T,N>&" in the header:

```
return *this;
}  //  assignment operator
```

This way, the user can write commands like "w =u =v" to assign the vector 'v' to both vectors 'u' and 'w', provided that they are all of the same dimension.

Here is yet another version of the assignment operator, which takes a scalar (rather than a vector) argument 'a', and assigns it to all the components of the current vector:

```
template<class T, int N>
const vector<T,N>& vector<T,N>::operator=(const T& a){
   for(int i = 0; i < N; i++)
      component[i] = a;
   return *this;
} // assignment operator with a scalar argument
```

This version is invoked whenever the user writes commands like "u = 0.".

16.6 Arithmetic Operators

Here we define the member update operators that are only declared in the class block above. The "+=" operator, which adds the vector argument 'v' to the current vector, is defined as follows:

```
template<class T, int N>
const vector<T,N>&
vector<T,N>::operator+=(const vector<T,N>&v){
    for(int i = 0; i < N; i++)
       component[i] += v[i];
    return *this;
} // adding a vector to the current vector
```

In this code, it is assumed that the current vector has the same dimension as the vector 'v' that is passed as an argument. It is advisable to verify this in an "if" question in the beginning of the function, and issue an error message if this is not the case. These details are left to the reader.

The above "+=" operator is now used to define the binary '+' operator to calculate the sum of the vectors 'u' and 'v' of the same dimension:

```
template<class T, int N>
const vector<T,N>
operator+(const vector<T,N>&u, const vector<T,N>&v){
   return vector<T,N>(u) += v;
} // vector plus vector
```

Indeed, as a member function, the "+=" operator can indeed take the temporary vector "vector<T,N>(u)" (returned from the copy constructor) as its current vector, add 'v' to it, and return the sum by value, as is indeed indicated by the words "const vector<T,N>" in the header, just before the function name. The '+' operator is an ordinary function that can be invoked by the user simply by writing "u + v" to calculate the sum of the two existing vectors 'u' and 'v' (which are of the same dimension).

The implementation of the rest of the arithmetic operators with vectors (such as multiplying a vector by a scalar, etc.) is left as an exercise. In the sequel, we assume that these operators are already available.

The following ordinary function is the unary operator that returns the negative of the argument:

```
template<class T, int N>
const vector<T,N>
operator-(const vector<T,N>&u){
   return vector<T,N>(u) *= -1;
} // negative of a vector
```

With this unary operator, the user can just write "−v" to have the negative of the well-defined vector 'v', as in standard mathematical formulas.

The following ordinary function returns the inner product of two vectors (defined in Chapter 9, Section 9.14):

```
template<class T, int N>
const T
operator*(const vector<T,N>&u, const vector<T,N>&v){
    T sum = 0;
    for(int i = 0; i < N; i++)
       sum += (+u[i]) * v[i];
    return sum;
} // vector times vector (inner product)
```

Indeed, if 'T' is "float" or "double", then "+u[i]" is just the 'i'th component in the vector 'u'. If, on the other hand, 'T' is the "complex" class, then "+u[i]" is the complex conjugate of the 'i'th component of 'u', as required.

The above function is also used in the following ordinary function to return the squared l_2-norm of a vector:

```
template<class T, int N>
T squaredNorm(const vector<T,N>&u){
    return u*u;
} // squared l2-norm
```

Similarly, the following ordinary function returns the l_2-norm of a vector:

```
template<class T, int N>
const T l2norm(const vector<T,N>&u){
    return sqrt(u*u);
} // l2 norm
```

Finally, here is the ordinary function that prints a vector component by component onto the screen:

```
template<class T, int N>
void print(const vector<T,N>&v){
  printf("(");
  for(int i = 0;i < N; i++){
    printf("v[%d]=",i);
    print(v[i]);
  }
  printf(")\n");
} //  printing a vector
```

16.7 Points in the Cartesian Plane and Space

The induction step, in which general 'N'-dimensional vectors are implemented, is now complete. The "point" class implemented in Chapter 15 is no longer necessary; indeed, it can be obtained from the 'N'-dimensional vector as a special case, that is, in the deduction that sets 'N' to be 2:

```
typedef vector<double,2> point;
```

Indeed, the "typedef" command tells the C compiler that two terms are the same. As a result, "point" can be used as short for "vector<double,22", namely, a point in the two-dimensional Cartesian plane.

Similarly, the command

```
typedef vector<double,3> point3;
```

allows the user to use the term "point3" as short for "vector<double,3>", for a point in the three-dimensional Cartesian space.

16.8 Inheritance

An object in C++ may "have" yet another object contained in it. For example, the "vector" class indeed has components of type 'T' in it. This is the "has a" approach. These components are private members that can be accessed from member and friend functions only.

On the other hand, an object in C++ may actually be viewed as another object as well. For example, the "complex" object may actually be viewed as a "point" object. Indeed, as discussed in Chapter 5, the complex number may also be interpreted geometrically as a point in the Cartesian plane. This is the "is a" approach.

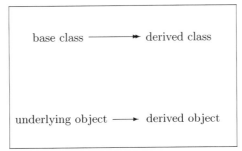

FIGURE 16.1:   The principle of inheritance.

The principle of inheritance in C++ is based on the "is a" approach (Figure 16.1). With this tool, new objects can be derived from existing (base) objects, inheriting their features, including data fields and functions. In the derived class, one may also define new functions or new versions of functions to override their original versions in the base class. For example, the "complex" class could actually be derived from the "point" class, inheriting some functions from it and rewriting some other functions whenever necessary to override their original version in the base "point" class. This way, the "complex" object is nothing but a "point" object with some extra algebraic features. (See an exercise at the end of this chapter.)

The "is a" approach is more in the spirit of object-oriented programming. Indeed, it focuses on the object and its nature, including its interpretation in terms of other mathematical fields, such as geometry. This way, the description of the object becomes more modular and closer to its original mathematical definition: its elementary features are described in the base class, whereas its more advanced features are described in the derived class. In fact, this may lead to a multilevel hierarchy of more and more sophisticated objects built on top of each other, from the most elementary object at the lowest level, which can be interpreted in elementary mathematical terms only, to the most advanced object at the highest level, which enjoys advanced mathematical features as well.

16.9 Public Derivation

In the definition of the derived class (derivation), the class name in the header is followed by the reserved word ":public", followed by the name of the base class. In the block that follows, the extra data fields in the derived class are defined, and its extra member and friend functions are declared or even defined explicitly, if appropriate.

There are more restrictive derivation patterns, in which the word ":public" before the base-class name is replaced by ":private" or ":protected". However, because these patterns are not used in the applications in this book, we prefer to omit them from the present discussion.

16.10 Protected Members of the Base Class

The derived class has no access to the private (field or function) members of the base class. However, it does have access to "half private" members, namely, members that are declared as "protected" in the base-class block by writing the reserved word "protected:" before their names. These members are accessible from derived classes only, but not from ordinary functions or classes that are not declared as friends of the base class. In fact, if the "component" field had been declared as "protected" in the base "vector" class, then it could have been accessed from any derived class, including the "matrix" class below.

Unfortunately, with the present implementation of the "vector" class in Section 16.3, the "component" field is private by default, as its definition is preceded by no "public:" or "protected:" statement. This is why it can be accessed from derived classes such as the "matrix" class below indirectly only, by calling the public "set" function.

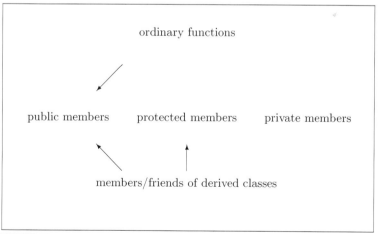

FIGURE 16.2: The three possible kinds of members of a class (public, protected, and private) and their access patterns.

In summary, the members of a class can be of three possible kinds: (a) public members that can be accessed by everyone; (b) private members that are accessible only to members and friends; and (c) protected members that are accessible to members and friends of derived classes, but not to ordinary functions (see Figure 16.2).

16.11 Constructing a Derived Object

When the derived object is constructed, the data fields inherited from the base class are constructed first by the default constructor of the base class. This is why the derived-class constructor cannot use an initialization list to initialize these fields; after all, they have already been initialized to their default values. All that the derived-class constructor can do is, therefore, to modify these data fields in its block.

If these data fields are only protected base-class members, then this modification can be done directly. If, on the other hand, they are private base-class members, then they can be modified only indirectly, using public member functions such as the "set" member function in the "vector" class.

16.12 Functions of Derived Objects

When a derived object is used as a current object in a call to some function, the compiler first looks for this function among the derived-class member functions. If, however, there is no such derived-class member function, then the compiler interprets the object as a mere base-class object, and looks for the function among the base-class member functions.

Similarly, when a derived object is passed as a concrete argument to a function, the compiler first looks for this function among the functions that indeed take a derived-class argument. Only if no such function is found, the compiler interprets the passed object as a mere base-class object, and looks for the function among the functions that take a base-class argument. The concrete argument may then undergo a default inverse conversion, in which it is converted implicitly from a derived object into its base-class counterpart.

16.13 Destroying a Derived Object

When the derived-class destructor is called, the data members inherited from the base class are destroyed last (in a reversed order to the order in which they are defined in the base-class block) by an implicit call to the base-class default destructor. Thus, if there are no more data fields in the derived class, the block of its destructor can remain empty.

16.14 Inherited Member Functions

The members inherited from the base class are also considered as members of the derived class in the sense that they can be called freely in the derived-class block as well. Furthermore, since the derivation is public, the public members of the base class remain public also in the derived class in the sense that they can be called from any ordinary function as well.

16.15 Overridden Member Functions

The derived class doesn't have to inherit all member functions from the base class. In fact, it can rewrite them in new versions. In this case, the new version in the derived class overrides the original version in the base class, and is invoked whenever the function is called in conjunction with derived objects.

One can still call the original version in the base class (even with derived objects) by adding to the function name the prefix that contains the base-class name followed by "::" to let the compiler know that this name refers to the original version in the base-class block rather than to the new version in the derived-class block.

16.16 The Matrix Object

The matrix defined in Chapter 9, Section 9.6 can be viewed as a finite sequence of column vectors. This is why a possible way to implement it is as a vector whose components are no longer scalars but rather vectors in their own right. This is where both templates and inheritance prove to be most useful.

In fact, the matrix can be derived from a vector of vectors. Several arithmetic operators are then inherited from the base "vector" class. Some operators, however, must be rewritten in the derived "matrix" class.

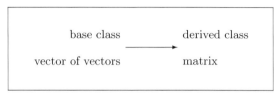

FIGURE 16.3:   Inheritance from the base class "vector<vector>" to the derived class "matrix".

More precisely, the "matrix" class is derived below from the "vector<T>" class, with 'T' being interpreted no longer as a scalar but rather as a column vector or a "vector" object in its own right (Figure 16.3).

Some elementary arithmetic operations, such as addition and subtraction, can be done column by column, hence can be inherited from the base "vector" class, with the columns being treated like standard components in the base "vector" class. More specific arithmetic operations like vector-matrix, matrix-vector, and matrix-matrix multiplication, on the other hand, must be implemented exactly in the "matrix" class, following their original mathematical formulation in Chapter 9, Sections 9.9–9.10.

The derived "matrix" class is also a template class that uses three parameters: 'T' to specify the type of the matrix elements, 'N' to specify the number of rows, and 'M' to specify the number of columns. Thus, the 'N'×'M' matrix is implemented as a finite sequence of 'M' 'N'-dimensional column vectors:

```
template<class T, int N, int M>
class matrix : public vector<vector<T,N>,M>{
```

The reserved word ":public" that follows the name of the derived class indicates that this is indeed a public derivation. The name of the base class that follows indicates that this is indeed an 'M'-dimensional vector, with components that are 'N'-dimensional vectors, as required.

```
public:
    matrix(){
    } // default constructor
```

With this default constructor, the user can define a "matrix" object with no arguments whatsoever. At the '{' symbol that marks the start of the above empty block, the compiler invokes implicitly the constructor of the base "vector" class in Section 16.4. Since no argument is specified, the zero default value is assigned there to the individual components. Since these components are vectors in their own right, the assignment operator at the end of Section 16.5 is invoked to assign the zero value to every component of each column vector. As a result, the above default constructor constructs the 'N'×'N' zero matrix, as required.

Furthermore, the following constructor converts implicitly a base "vector" object into the derived "matrix" object:

```
matrix(const vector<vector<T,N>,M>&){
} // implicit converter
```

With this converter, functions of type "matrix" (that are supposed to return a "matrix" object) may return a mere "vector" object in their definitions, because it will be converted implicitly to the required "matrix" object upon returning.

Next, we turn to more meaningful constructors that also take vector arguments and use them as columns in the constructed matrix. The following constructor assumes that the number of columns, 'M', is equal to 2:

```
matrix(const vector<T, N>&u, const vector<T,N>&v){
```

As before, the 'N'×2 zero matrix is constructed implicitly at the '{' symbol at the end of this header. Then, the 'N'-dimensional vector arguments 'u' and 'v' are assigned to the columns of this matrix, using the public "set" function inherited from the base "vector" class:

```
set(0,u);
set(1,v);
} // constructor with 2 columns
```

Similarly, the following constructor takes three vector arguments to construct an 'N'×3 matrix:

```
matrix(const vector<T, N>&u,
    const vector<T,N>&v,
    const vector<T,N>&w){
set(0,u);
set(1,v);
set(2,w);
} // constructor with 3 columns
```

Next, we define an operator to read an element from the matrix:

```
const T& operator()(int i,int j) const{
  return (*this)[j][i];
} // read the (i,j)th matrix element
```

Indeed, in the command line in the above block, the "operator[]" inherited from the base "vector" class is applied twice to the current "matrix" object "*this": the first time with the argument 'j' to have the 'j'th column, and the second time with the argument 'i' to read the 'i'th component in this column, or the (i,j)th matrix element, as required. Thanks to the reserved word "const" at the end of the header (just before the above block), the current "matrix" object can never change by this operator: it is a read-only operator, as required.

Finally, we declare some member update operators that will be defined later on. Although these operators do the same thing as their original counterparts in the base "vector" class, they must be redefined in the derived "matrix" class as well only to let the compiler know that the original version in the "base" class should be applied to the underlying "vector" object:

```
const matrix& operator+=(const matrix&);
const matrix& operator-=(const matrix&);
const matrix& operator*=(const T&);
const matrix& operator/=(const T&);
};
```

This completes the block of the derived "matrix" class. No copy constructor, assignment operator, or destructor needs to be defined here, because the original versions in the base "vector" class work just fine.

The arithmetic update operators to add or subtract a matrix and multiply or divide by a scalar, on the other hand, must be rewritten, because the original versions in the base "vector" class return "vector" objects rather than the required "matrix" objects. The actual definitions of these operators are left as exercises, with detailed solutions in Section 28.7 in the appendix.

The implementation of the vector-times-matrix, matrix-times-vector, and matrix-times-matrix multiplication operators is also left as an exercise, with a detailed solution in Section 28.7 in the appendix.

In the applications studied later in the book, we often use 2×2 and 3×3 matrices. Therefore, we introduce the following new types:

```
typedef matrix<double,2,2> matrix2;
typedef matrix<double,3,3> matrix3;
```

This way, "matrix2" stands for a 2×2 matrix, and "matrix3" stands for a 3×3 matrix.

16.17 Power of a Square Matrix

Here we rewrite the function "power()" of Chapter 14 as a template function, so it can be used to compute not only the power x^n of a given scalar x but also the power A^n of a given square matrix A. Furthermore, we improve on the algorithm used in Chapter 14 by using the efficient algorithm introduced in Chapter 12, Section 12.7, which requires at most $2 \log_2(n)$ multiplications. The required template function is implemented as follows.

```
template<class T>
const T
power(const T&x, int n){
  return n>1 ?
           (n%2 ?
             x * power(x * x,n/2)
           :
              power(x * x,n/2))
         :
           x;
} //  compute a power recursively
```

Note that the function returns the output of a nested "?:" question. The outer question checks whether $n > 1$ or not. If it is, then the inner question (which is shifted two blank spaces to the right to make the code easy to read) checks whether n is odd or even, to proceed as in the efficient algorithm in Chapter 12, Section 12.7. If, on the other hand, $n = 1$, then the returned value is simply $x^1 = x$, as indeed returned in the last code line in the function.

With the above template function, the user can just write "power(A,n)" to have the power A^n of a square matrix A. This shows very clearly how templates can be used to write elegant and easy-to-read code.

16.18 Exponent of a Square Matrix

The exponent of a square matrix A of order N is defined by the converging infinite series

$$\exp(A) = I + A + \frac{A^2}{2!} + \frac{A^3}{3!} + \cdots = \sum_{n=0}^{\infty} \frac{A^n}{n!},$$

where I is the identity matrix of order N.

This function can be approximated by the Taylor approximation in Chapter 14, Section 14.10 above, provided that the scalar x used there is replaced by by the matrix A. For this purpose, the present "matrix" class is most helpful.

In the sequel, we use the term "l_2-norm of the matrix A" (denoted by $\|A\|_2$) to refer to the square root of the maximal eigenvalue of $A^h A$ (where A^h is the Hermitian adjoint of A, see Chapter 9, Section 9.13).

As in [33] and in Chapter 14, Section 14.10, in order to approximate well the above infinite Taylor series by a Taylor polynomial, one must first find a sufficiently large integer m such that the l_2-norm of $A/2^m$ is sufficiently small (say, smaller than $1/2$). Since the l_2-norm is not available, we estimate it in terms of the l_1- and l_∞-norms:

$$\|A\|_2 \leq \sqrt{\|A\|_1 \|A\|_\infty},$$

where the l_1- and l_∞-norms are given by

$$\|A\|_1 = \max_{0 \leq j < N} \sum_{i=0}^{N-1} |A_{i,j}|,$$

$$\|A\|_\infty = \max_{0 \leq i < N} \sum_{j=0}^{N-1} |A_{i,j}|.$$

Thus, by finding an integer m so large that

$$2\sqrt{\|A\|_1 \|A\|_\infty} < 2^m,$$

we guarantee that the l_2-norm of $A/2^m$ is smaller than $1/2$, as required.

The algorithm to approximate $\exp(A)$ proceeds as in the algorithm in Chapter 14, Section 14.10, which uses the Taylor polynomial to approximate the original infinite Taylor series. The scalar x used there is replaced by the square matrix A. The code used there can be easily adapted to apply also to square matrices, provided that the required arithmetic operations between matrices are well defined. This is indeed done in the exercises below by rewriting the original function used there as a template function.

16.19 Exercises

1. Implement complex numbers as a template class "complex<T>", where 'T' is the type of the real and imaginary parts. Define the required arithmetic operations and test them on objects of type "complex<float>" and "complex<double>".

2. Complete the missing arithmetic operators in Section 16.6, such as subtraction of vectors and multiplication and division by a scalar. The solutions are given in Section 28.6 in the appendix.

3. Use the observation that a complex number can also be interpreted geometrically as a point in the Cartesian plane to reimplement the "complex" class using inheritance: derive it from the "point" class in Section 16.7, and rewrite the required arithmetic operations to override their original "point" versions.

4. Implement the operators that add and subtract two matrices. The solutions are given in Section 28.7 in the appendix.

5. Implement the vector-matrix, matrix-vector, matrix-matrix, and scalar-matrix products that are missing in the code in Section 16.16. The solutions are given in Section 28.7 in the appendix.

6. Write functions that return the transpose, determinant, and inverse of 2×2 matrices (Chapter 9, Sections 9.23–9.24). The solution can be found in Section 28.8 in the appendix.

7. Write functions that return the transpose, determinant, and inverse of 3×3 matrices (Chapter 9, Sections 9.23–9.24). The solution can be found in Section 28.9 in the appendix.

8. Redefine the "matrix" class to have a two-dimensional array to store its elements, using the "has a" rather than the "is a" approach. What are the advantages and disadvantages of each implementation?

9. Write an "operator&" function that takes two 3-d vectors and returns their vector product (Chapter 9, Section 9.25). The solution can be found in Section 28.10 in the appendix.

10. Rewrite the "expTaylor" function in Chapter 14, Section 14.10, as a template function that takes an argument of type 'T'. The solution can be found in Section 28.11 in the appendix.

11. Use your code with 'T' being the "complex" type. Verify that, for an imaginary argument of the form ix (where $i = \sqrt{-1}$ and x is some real number), you indeed get

$$\exp(ix) = \cos(x) + i \cdot \sin(x).$$

12. Apply your code to an argument A of type "matrix" to compute $\exp(A)$. Make sure that all the required arithmetic operations between matrices are available in your code.

13. Apply your code also to objects of type "matrix<complex,4,4>", and verify that, for a complex parameter λ,

$$\exp\left(\begin{pmatrix} \lambda & & & \\ 1 & \lambda & & \\ & 1 & \lambda & \\ & & 1 & \lambda \end{pmatrix}\right) = \exp(\lambda) \begin{pmatrix} 1 & & & \\ 1/1! & 1 & & \\ 1/2! & 1/1! & 1 & \\ 1/3! & 1/2! & 1/1! & 1 \end{pmatrix}$$

(the blank spaces in the above matrices indicate zero elements).

Chapter 17

Dynamic Vectors and Lists

In object-oriented languages such as C++, the programmer has the opportunity to implement new objects. These objects can then be used as if they were part of the standard language. In fact, every user who has permission can use them in his/her own application. Furthermore, users can use these objects to implement more and more complicated objects to form a complete hierarchy of useful objects. This is called multilevel programming.

Well-implemented objects must be ready to be used by any user easily and efficiently. In particular, they must be sufficiently flexible to give the user the freedom to use them as he/she may wish. For example, in many cases the user might want to specify the object size in run time rather than in compilation time. This way, the user can use information available in run-time to specify the object size more economically. This kind of dynamic objects is discussed in this chapter.

The dynamic vector introduced below improves on the standard vector implemented above in the opportunity to determine its dimension dynamically in run time rather than in compilation time. Unfortunately, both the standard vector and the dynamic vector must contain entries that are all of the same size. To have the freedom to use entries of different sizes, one must introduce a yet more flexible object: the list.

Below we introduce two kinds of lists: the standard list, whose number of entries must be determined once and for all, and the more flexible linked list, which can grow by taking more items and then shrink again by dropping items dynamically in run time. Furthermore, the recursive definition of the linked list is the key to many important features and functions, which will prove most useful later on in the book.

17.1 Dynamic Vectors

The implementation of the "vector" object in Chapter 16, Section 16.3 requires the *a priori* knowledge of its dimension 'N' in compilation time. In many cases, however, the dimension is available in run time only. Dynamic vectors whose dimension is specified only in run time are clearly necessary.

In this section, we implement dynamic vectors, whose dimension is no longer a template parameter but is rather stored in a private data field. This way, the dynamic-vector object contains not only the components of the vector but also an extra integer to store the dimension of the vector. This integer field can be set in run time, as required.

Because the dimension of the dynamic vector is not yet available in compilation time, the memory required to implement it cannot be allocated as yet. This is why the memory must be allocated in run time, using the reserved word "new". The "new" function, available in the C++ compiler, allocates sufficient memory to store a specified object, and returns its address. For example, the command line

```
double* p = new double;
```

allocates memory for a "double" variable and stores its address n the pointer-to-double 'p'. To access this variable, one then needs to write "*p".

In dynamic vectors, templates are used only to specify the type of the components. (This type is denoted by 'T' in the "dynamicVector" class below.)) The number of components, on the other hand, is determined dynamically during run time, hence requires no template parameter.

The data fields in the "dynamicVector" class below are declared as "protected" (rather than strictly private) to make them accessible from derived classes to be defined later. Two data fields are used: the integer "dimension" that indicates the dimension of the vector and the pointer "component" that points to the components of the vector.

Because the dimension is not yet available, the "component" field must be declared as a pointer-to-T rather than the array-of-T's used in the "vector" class in Chapter 16, Section 16.3. Only upon constructing a concrete dynamic-vector object the number of components is specified and sufficient memory to store them is allocated.

```
template<class T> class dynamicVector{
  protected:
    int dimension;
    T* component;
  public:
```

First, we declare the constructors and assignment operators. (The detailed definition is deferred until later.)

```
dynamicVector(int, const T&);
dynamicVector(const dynamicVector&);
const dynamicVector& operator=(const dynamicVector&);
const dynamicVector& operator=(const T&);
```

Because the destructor is very short, it is defined here in the class block. In fact, it contains only one command line, to delete the pointer "component" and free the memory it occupies:

```
~dynamicVector(){
  delete [] component;
}  //  destructor
```

In fact, the "delete[]" command available in the C++ compiler deletes the entire "component" array and frees the memory occupied by it for future use. Note that the "dimension" field doesn't have to be removed explicitly. Indeed, because it is not a pointer, it is removed implicitly by the default destructor of the C++ compiler, which is invoked automatically at the end of every call to the above destructor.

Because the "dimension" field is declared as "protected", we need a public function to access it even from ordinary classes that are not derived from the "dynamicVector" class:

```
int dim() const{
  return dimension;
}  //  return the dimension
```

This public function can be called even from ordinary functions to read the dimension of dynamic-vector objects.

Furthermore, the "operator()" defined below returns a nonconstant reference to the 'i'th component in the current dynamic vector, which can be used not only to read but also to change this component. The "operator[]" defined next, on the other hand, returns a constant (rather than nonconstant) reference to the 'i'th component, so it can be used only to read it:

```
T& operator()(int i){
  return component[i];
}  //  read/write ith component

const T& operator[](int i) const{
  return component[i];
}  //  read only ith component
```

Finally, we declare some member arithmetic operators on the current dynamic vector. The detailed definition is deferred until later.

```
      const dynamicVector& operator+=(const dynamicVector&);
      const dynamicVector& operator-=(const dynamicVector&);
      const dynamicVector& operator*=(const T&);
      const dynamicVector& operator/=(const T&);
};
```

This concludes the block of the "dynamicVector" class.

Next, we define the functions that are only declared in the class block above. In the constructor defined below, in particular, the memory required to store the array of components is allocated dynamically in the initialization list, using the "new" command:

```
template<class T>
dynamicVector<T>::dynamicVector(
        int dim = 0, const T& a = 0)
        : dimension(dim), component(dim ? new T[dim] : 0){
   for(int i = 0; i < dim; i++)
      component[i] = a;
} // constructor
```

A similar approach is used in the copy constructor:

```
template<class T>
dynamicVector<T>::dynamicVector(const dynamicVector<T>& v)
   : dimension(v.dimension),
     component(v.dimension ? new T[v.dimension] : 0){
   for(int i = 0; i < v.dimension; i++)
     component[i] = v.component[i];
} // copy constructor
```

The assignment operator is defined as follows:

```
template<class T>
const dynamicVector<T>&
dynamicVector<T>::operator=(const dynamicVector<T>& v){
```

There is one case, though, in which the assignment operator needs to do nothing. This is the case in which the user writes a trivial command of the form "u = u". To exclude this case, we use the following "if" question:

```
if(this != &v){
```

This "if" block is entered only if the current dynamic vector is different from the dynamic vector that is passed to the function as an argument, which is indeed the nontrivial case, invoked by a user who writes a meaningful assignment command of the form "u = v". Once we have made sure that the assignment operator is not called by a trivial call of the form "u = u", we must also modify the dimension of the current vector to be the same as that of the argument vector 'v':

```
    if(dimension != v.dimension){
        delete [] component;
        component = new T[v.dimension];
    }
```

This way, the array "component[]" contains the same number of entries as the array "v.component[]", and is ready to be filled with the corresponding values in a standard loop:

```
   for(int i = 0; i < v.dimension; i++)
     component[i] = v.component[i];
   dimension = v.dimension;
 }
 return *this;
} // assignment operator
```

This completes the definition of the assignment operator that takes a dynamic-vector argument. Next, we also define another assignment operator that takes a scalar argument, and assigns the same value to all the components in the current dynamic vector:

```
template<class T>
const dynamicVector<T>&
dynamicVector<T>::operator=(const T& a){
  for(int i = 0; i < dimension; i++)
    component[i] = a;
  return *this;
} // assignment operator with a scalar argument
```

Furthermore, we implement some useful member arithmetic operators on the current dynamic vector:

```
template<class T>
const dynamicVector<T>&
dynamicVector<T>::operator+=( const dynamicVector<T>&v){
    for(int i = 0; i < dimension; i++)
      component[i] += v[i];
    return *this;
}// adding a dynamicVector to the current one
```

With this operator, the user can write commands like "u + = v", where 'u' and 'v' are dynamic vectors of the same dimension.

We don't bother here to verify that the dimension of the argument vector is the same as that of the current vector before they are added to each other. The careful programmer is advised to make sure in the beginning of the function block that this is indeed the case, to avoid all sorts of bugs.

The above member operator is now used in an ordinary operator to add two dynamic vectors:

```
template<class T>
const dynamicVector<T>
operator+(const dynamicVector<T>&u,
          const dynamicVector<T>&v){
  return dynamicVector<T>(u) += v;
} // dynamicVector plus dynamicVector
```

With this operator, the user can write "u + v", where 'u' and 'v' are dynamic vectors of the same dimension.

Next, we implement the unary negative operator:

```
template<class T>
const dynamicVector<T>
operator-(const dynamicVector<T>&u){
  return dynamicVector<T>(u) *= -1.;
} // negative of a dynamicVector
```

With this operator, the user can write "−u" to have the negative of the dynamic vector 'u'.

Finally, we implement a function that prints a dynamic vector to the screen:

```
template<class T>
void print(const dynamicVector<T>&v){
  print("(");
  for(int i = 0;i < v.dim(); i++){
    printf("v[%d]=",i);
    print(v[i]);
  }
  print(")\n");
} // printing a dynamicVector
```

Other useful arithmetic operations, such as subtraction, multiplication and division by scalar, and inner product are implemented in the exercises. Assuming that they are already available, one can write all sorts of vector operations as follows:

```
int main(){
  dynamicVector<double> v(3,1.);
  dynamicVector<double> u;
  u=2.*v;
  printf("v:\n");
  print(v);
  printf("u:\n");
  print(u);
  printf("u+v:\n");
  print(u+v);
  printf("u-v:\n");
  print(u-v);
  printf("u*v=%f\n",u*v);
  return 0;
}
```

17.2 Ordinary Lists

The vector and dynamic-vector above are implemented as arrays of components of type 'T'. By definition, an array in C (as well as other programming languages) must contain entries of the same type and size. No array can contain entries that occupy different amounts of memory.

In many applications, however, one needs to use sequences of objects of different sizes. For example, one may need to implement a sequence of vectors of different dimensions. Such sequences cannot be implemented in (even dynamic) vectors; a more flexible data structure is necessary.

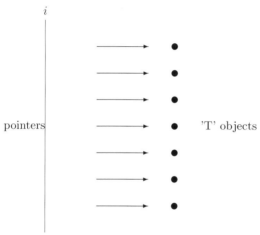

FIGURE 17.1: A list: the arrows stand for pointers that point to the bullets, which stand for objects of type 'T' (to be specified later in compilation time).

The required data structure is implemented in the "list" class below. This class contains an array of entries of type pointer-to-'T' rather than 'T' (see Figure 17.1). Although objects of type 'T' may have different sizes (e.g., when 'T' is a dynamic vector), their addresses are all of the same type and size: in fact, they are just integer numbers associated with certain places in the computer memory.

Although the template parameter 'T' must be specified during compilation time, concrete objects of type 'T' are placed in the addresses in the array in the list during run time only. For example, if 'T' is specified in compilation time as "dynamicVector", then the dimensions of the "dynamicVector" objects in the list are specified only during run time, using the constructor of the "dynamicVector" class. Then, their addresses are placed in the array in the list.

The length of the list (the number of pointers-to-'T' in the array in it) can also be determined dynamically in run time. Indeed, as in the "dynamicVector" class, an extra integer field stores the number of items in the list. Thus, the "list" class contains two protected data fields: "number" to indicate the number of items in the list, and "item" to store their addresses. These fields are declared as "protected" (rather than private) to allow accessing them from derived classes later on.

```
template<class T> class list{
  protected:
    int number;
    T** item;
  public:
```

The first constructor takes only one integer parameter, 'n', to be used in the initialization list to determine the number of items to be contained in the future list. If 'n' is zero, then nothing is constructed. (This is also the default case.) If, on the other hand, 'n' is nonzero, then the "new" command is used to allocate sufficient memory for 'n' pointers-to-'T'. These pointers point to nothing as yet, that is, they contain no meaningful address of any concrete item.

```
list(int n=0):number(n), item(n ? new T*[n]:0){
} // constructor
```

Next, we move on to a more meaningful constructor, which takes two arguments: 'n', to specify the number of items in the constructed list, and 't', to initialize these items with a meaningful value. First, 'n' is used in the initialization list as above to allocate sufficient memory for "item", the array of addresses. Then, 't' is used in the block of the function to initialize the items in the list, using the "new" command and the copy constructor of the 'T' class:

```
list(int n, const T&t)
  : number(n), item(n ? new T*[n] : 0){
  for(int i=0; i<number; i++)
    item[i] = new T(t);
} // constructor with a T argument
```

Next, we declare the copy constructor and the assignment operator. The detailed definitions are deferred until later.

```
list(const list<T>&);
const list<T>& operator=(const list<T>&);
```

Next, we define the destructor. First, the "delete" command available in the C++ compiler is called in a standard loop to delete the pointers in the array in the list. Once such a pointer is deleted, the destructor of the 'T' class is invoked automatically implicitly to destroy the item pointed at by it. Once deleted, these pointers point to nothing, or contain meaningless (or zero) addresses. Still, they occupy valuable memory, which should be released for future use. For this, the "delete" command is called once again to delete the entire array of pointers:

```
~list(){
  for(int i=0; i<number; i++)
    delete item[i];
  delete [] item;
} //   destructor
```

Because the "number" field is protected, it cannot be read from ordinary (nonmember, nonfriend) functions. The only way to read it is through the following public "size()" function:

```
int size() const{
  return number;
} //  size of list
```

Similarly, the items in the list, whose addresses are stored in the protected "item" field, can be accessed from ordinary functions only through the public "operator()" (to read/write) or "operator[]" (to read only) below:

```
T& operator()(int i){
  if(item[i])return *(item[i]);
} //  read/write ith item

const T& operator[](int i)const{
  if(item[i])return *(item[i]);
} //  read only ith item
};
```

The user should be careful to call "l(i)" or "l[i]" only for a list 'l' that contains at least 'i' items; otherwise, a bug can be encountered. To avoid such bugs, the careful programmer is advised to verify in the beginning of these operators that 'i' is indeed smaller than the number of items in the current list, and issue an error message if it is not. These details are left as an exercise.

This concludes the block of the "list" class. The definition of the copy constructor declared in it is similar to that of the above constructor:

```
template<class T>
list<T>::list(const list<T>&l):number(l.number),
     item(l.number ? new T*[l.number] : 0){
  for(int i=0; i<l.number; i++)
    if(l.item[i])
      item[i] = new T(*l.item[i]);
```

```
       else
           item[i] = 0;
   }   //  copy constructor
```

Here is also the definition of the assignment operator declared above:

```
   template<class T>
   const list<T>&
   list<T>::operator=(const list<T>& l){
```

If the user has inadvertently made a trivial call such as "l = l", then nothing should be assigned. If, on the other hand, a nontrivial call is made, then the following "if" block is entered:

```
       if(this != &l){
```

First, we modify the current list object to contain the same number of items as the list 'l' that is passed to the function as an argument.

```
         if(number != l.number){
            delete [] item;
            item = new T*[l.number];
         }
```

We are now ready to copy the items in 'l' also to the current "list" object:

```
         for(int i = 0; i < l.number; i++)
            if(l.item[i])
      item[i] = new T(*l.item[i]);
   else
      item[i] = 0;
         number = l.number;
       }
```

Finally, we also return a constant reference to the current "list" object:

```
       return *this;
   }  //  assignment operator
```

With this implementation, the user can write a code line of the form "l3 = l2 = l1" to assign the list "l1" to both the lists "l2" and "l3".

The following ordinary function prints the items in the list to the screen:

```
   template<class T>
   void print(const list<T>&l){
     for(int i=0; i<l.size(); i++){
       printf("i=%d:\n",i);
       print(l[i]);
     }
   }  //  printing a list
```

17.3 Linked Lists

The "list" object in Section 17.2 is implemented as an array of pointers to (or addresses of) objects of type 'T' (to be specified later in compilation time). Unfortunately, in many cases an array is not flexible enough for this purpose. Indeed, new items cannot be easily inserted to the list, and old items are not easily removed from it. Furthermore, the number of items in the list is determined once and for all upon construction, and cannot be easily changed later on. These drawbacks make the "list" class unsuitable for many important applications. A more flexible kind of list is clearly necessary. This is the linked list.

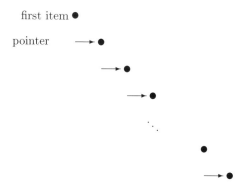

null (zero) pointer   ⟶ ○   nothing

FIGURE 17.2: A linked list: each item (denoted by a bullet) contains a pointer (denoted by an arrow) to point to the next item [except of the last item, which contains the null (or zero) pointer].

Unlike the ordinary lists implemented above, the linked list doesn't use an array at all. Instead, each item in the linked list contains a pointer that points to the next item (see Figure 17.2). As we'll see below, this structure allows inserting any number of new, as well as dropping old ones.

Accessing items, on the other hand, is more complicated in a linked list than in an ordinary list. Indeed, in an ordinary list, the ith item can be accessed through its address, stored in the ith entry in the array of addresses. In a linked list, on the other hand, one must start from the first item and "jump" from item to item until the required item is reached. Clearly, this access pattern is more complicated and expensive. Fortunately, as we'll see below, most functions operate on the linked list as a complete object, avoiding accessing individual items.

The access patterns used in both ordinary and linked lists are called "indirect indexing," because the individual items are accessed indirectly through their addresses. Clearly, this approach is less efficient than the direct indexing used in arrays, in which the items are stored consecutively in the computer memory, hence can be scanned one by one in efficient loops. Nevertheless, the advantages of indirect indexing far exceed this disadvantage. Indeed, the freedom to insert and remove items, which is a prominent feature in linked lists, is essential in implementing many useful (recursive) objects later on in the book.

The linked list is implemented in the template class "linkedList" defined below. The items in the linked list are of type 'T', to be defined later in compilation time.

The "linkedList" class contains two data fields: "item", to contain the first item in the linked list, and "next", to contain the address of the rest of the linked list. Both data fields are declared as "protected" (rather than strictly private) to be accessible from derived classes as well.

The linked list is implemented as a recursive object. Indeed, its definition uses mathematical induction: assuming that its "tail" (the shorter linked list that contains the second item, the third item, the fourth item, etc.) is well defined in the induction hypothesis, the original linked list is defined to contain the first item (in the field "item") and a pointer to this tail (in the field "next"):

```
template<class T> class linkedList{
  protected:
    T item;
    linkedList* next;
  public:
```

First, we define a default constructor:

```
linkedList():next(0){
} // default constructor
```

With this constructor, the user can write commands like "linkedList<double> l" to define a trivial linked list 'l' with no items in it.

Next, we define a more meaningful constructor that takes two arguments: an argument of type 'T' to specify the first item in the constructed list, and a pointer-to-linked-list to define the rest of the constructed list:

```
linkedList(T&t, linkedList* N=0)
  : item(t),next(N){
} // constructor
```

In particular, when the second argument in this constructor is missing, it is assumed to be the zero (meaningless) pointer. This way, the user can write commands like "linkedList<double> l(1.)" to construct a linked list 'l' with the only item 1. in it.

The first item in the linked list can be read (although not changed) and used in the definitions of ordinary (nonmember, nonfriend) functions through the public member function "operator()":

```
const T& operator()() const{
  return item;
} // read item field
```

Indeed, with this operator, the user can write "l()" to obtain a constant reference to the first item in the linked list 'l'.

Similarly, the rest of the linked list can be read (although not changed) through the following public function:

```
const linkedList* readNext() const{
  return next;
} // read next
```

Indeed, with this function, the user can write "l.readNext()" to have the address of the linked list that contains the tail of 'l'.

Next, we declare the assignment operator:

```
const linkedList& operator=(const linkedList&);
```

to be defined later on.

17.4 The Copy Constructor

The recursive pattern of the linked list is particularly useful in the copy constructor defined below. Indeed, thanks to this pattern, one only needs to copy the first item in the linked list (using the copy constructor of the 'T' class) and to use recursion to copy the tail:

```
linkedList(const linkedList&l):item(l()),
    next(l.next ? new linkedList(*l.next):0){
} // copy constructor
```

In fact, everything is done in the initialization list: first, the first item in 'l', "l()", is copied into "item", the first item in the constructed list. Then, the tail of 'l', pointed at by "l.next", is copied recursively into a new tail, whose address is stored in "next". This completes the copying of the entire linked list 'l' into a new linked list, as required.

17.5 The Destructor

The recursive structure of the linked list is also useful in the destructor. Indeed, once the pointer "next" is deleted, the destructor is invoked automatically to destroy recursively the entire tail. Then, the destructor of the 'T' class is invoked implicitly to destroy the first item, "item", as well:

```
~linkedList(){
   delete next;
   next = 0;
} // destructor
```

17.6 Recursive Member Functions

The recursive pattern of the linked list is also useful to define a function that returns the last item (embedded in a trivial linked list of one item only):

```
linkedList& last(){
   return next ? next->last() : *this;
} // last item
```

Indeed, it can be shown by mathematical induction on the total number of items that this function indeed returns the last item: if the total number of items is one, then "next" must be the zero pointer, so the current linked list, which contains the only item "item", is returned, as required. If, on the other hand, the total number of items is greater than one, then the induction hypothesis implies that the last item in the tail is returned. This item is also the last item in the original linked list, as required.

A similar proof shows that the following function counts the total number of items in the linked list:

```
int length() const{
   return next ? next->length() + 1 : 1;
} // number of items
```

The above two functions are now used to form a function that appends a new item at the end of the linked list:

```
void append(T&t){
   last().next = new linkedList(t);
} // append a new item
```

17.7 Inserting New Items

The recursive structure of the linked list is also useful in inserting new items into it. The following function places the new item right after the first item in the linked list:

```
void insertNextItem(T&t){
  next = new linkedList(t,next);
} // insert item in the second place
```

The above command line uses the second constructor in Section 17.3 to replace the current tail by a slightly longer one, with the new item 't' in its beginning, as required.

The following function places the new item at the beginning of the current linked list:

```
void insertFirstItem(T&t){
```

First, the second constructor in Section 17.3 is used to duplicate the first item, "item":

```
next = new linkedList(item,next);
```

The first copy of "item" is now replaced by the new item, 't', using the assignment operator of the 'T' class:

```
  item = t;
} // insert a new item at the beginning
```

The block of the "linkedList" class ends with the declaration of some more member functions, to be defined later:

```
  void dropNextItem();
  void dropFirstItem();
  const linkedList& operator+=(linkedList&);
  linkedList& order(int);
};
```

This concludes the block of the "linkedList" class.

One may ask here: why isn't the 'T' argument in the above functions (and in the above constructor) declared as a constant? After all, it is never changed when inserted into the current linked list! Declaring it as a constant could protect it from inadvertent changes, couldn't it?

The answer is that later on we'll derive from the "linkedList" class a nonstandard class with a constructor that does change its argument. Furthermore, in this class, an inserted object may change as well. This is why this class can't inherit the above functions unless they use a nonconstant argument.

17.8 The Assignment Operator

The assignment operator defined below also uses the recursive structure of the linked list: first, the first item in the linked list is assigned. Then, the tail of the linked list is assigned recursively:

```
template<class T>
const linkedList<T>&
linkedList<T>::operator=(const linkedList<T>&L){
```

If the user writes inadvertently a trivial assignment such as "l = l", then nothing should be done. If, on the other hand, the assignment is nontrivial, then the following "if" block is entered:

```
if(this != &L){
```

First, the first item in the current linked list is assigned the same value as in the first item in the argument 'L':

```
item = L();
```

Next, the tail of 'L' is assigned to the tail of the current linked list as well. To do this, there are two cases: if the current linked list does have an old tail, then the assignment operator is called recursively (that is, provided that 'L' indeed has a nontrivial tail):

```
if(next){
  if(L.next)
    *next = *L.next;
  else{
    delete next;
    next = 0;
  }
}
```

If, on the other hand, the current linked list has no tail, then the tail of 'L' is appended to it using the copy constructor and the "new" command:

```
else
  if(L.next)next = new linkedList(*L.next);
}
```

Finally, a constant reference to the current linked list is also returned:

```
return *this;
} //  assignment operator
```

This way, the user can write "l3 = l2 = l1" to assign the linked list "l1" to both "l2" and "l3".

17.9 Dropping Items

Unlike in ordinary lists, in linked lists one can not only insert new items but also drop old ones if required easily and efficiently. For example, here is the function that drops the second item in the linked list:

```
template<class T>
void linkedList<T>::dropNextItem(){
```

First, we must ask if there is indeed a second item to drop:

```
if(next){
```

Now, there are two possible cases: if there is also a third item in the linked list,

```
if(next->next){
```

then we must also make sure that it is not lost when the second item (which points to it) is removed. For this purpose, we keep the address of the tail of the linked list also in a local pointer, named "keep":

```
linkedList<T>* keep = next;
```

This way, we can replace the original tail by a yet shorter tail that doesn't contain the second item in the original list:

```
next = next->next;
```

The removed item, however, still occupies valuable memory. Fortunately, we have been careful to keep track of it in the local pointer "keep". This pointer can now be used to destroy it completely, using the destructor of the 'T' class:

```
keep->item.~T();
}
```

This way, "next" points to the third item in the original list rather than to the second one, as required.

If, on the other hand, there is no third item in the original list, then the entire tail (which contains the second item only) is removed:

```
else{
    delete next;
    next = 0;
  }
}
```

If, on the other hand, there is no second item to drop, then an error message is printed to the screen:

```
    else
      printf("error: cannot drop nonexisting next item\n");
  }  //  drop the second item from the linked list
```

This completes the function that drops the second item in the linked list. In fact, this function can also be used to drop the third item by applying it to the tail of the current linked list, pointed at by "next", and so on.

Furthermore, the above function is also useful in dropping the first item in the linked list:

```
    template<class T>
    void linkedList<T>::dropFirstItem(){
```

Indeed, if there is a second item in the current linked list,

```
      if(next){
```

then a duplicate copy of it is stored in the first item as well:

```
        item = next->item;
```

Then, the "dropNextItem" function is invoked to drop the second duplicate copy, as required:

```
        dropNextItem();
      }
```

If, on the other hand, there is no second item in the original list, then the first item cannot be dropped, or the list would remain completely empty. Therefore, an error message should be printed:

```
      else
        printf("error: cannot drop the only item.\n");
  }  //  drop the first item in the linked list
```

So, we see that the linked-list object must contain at least one item; any attempt to remove this item would end in a bug. Thus, the user must be careful not to do it. Although not ideal, this implementation is good enough for our applications in this book. Below, however, we also introduce the stack object, which improves on the linked list, and can be completely emptied if necessary.

The recursive structure of the linked list is also useful to print it onto the screen. Indeed, after the first item has been printed, the "print" function is applied recursively to the tail, to print the rest of the items in the linked list as well:

```
    template<class T>
    void print(const linkedList<T>&l){
      printf("item:\n");
```

```
    print(l());
    if(l.readNext())print(*l.readNext());
}  //  print a linked list recursively
```

Here is an example that shows how linked lists are used in practice:

```
int main(){
    linkedList<double> c(3.);
    c.append(5.);
    c.append(6.);
    c.dropFirstItem();
    print(c);
    return 0;
}
```

This code first produces a linked list of items of type "double", with the items 3, 5, and 6 in it. (Because $3 < 5 < 6$, such a list is referred to as an ordered list.) Then, the first item from the list. Finally, the remaining items 5 and 6 are printed onto the screen.

17.10 The Merging Problem

Assume that the class 'T' used in the linked list has a well-defined "operator<" in it to impose a complete order on objects of type 'T'. For example, if 'T' is interpreted as the integer type, then '<' could have the usual meaning: $m < n$ means that m is smaller than n. An ordered list is a list that preserves this order in the sense that a smaller item (in terms of the '<' operator in class 'T') must appear before a larger item in the list. For example, $3, 5, 6$ is an ordered list, whereas $5, 3, 6$ is not.

In the merging problem, two ordered lists should be merged efficiently into one ordered list. As we'll see below, the linked-list object is quite suitable for this purpose, thanks to the opportunity to introduce new items and drop old ones easily.

In the following, it is assumed that the type 'T' supports a complete priority order; that is, every two 'T' objects can be compared by the '<', '>', or "==" binary operators. It is also assumed that the current "linkedList" object and the "linkedList" argument that is merged into it are well ordered in the sense that each item is smaller than the next item.

It is also assumed that the 'T' class supports the "+ =" operator. This way, if two equal items are encountered during the merging process, then they can be simply added to each other to form a new item that contains their sum. This operator is used in the "+ =" operator defined below in the "linkedList" class, which merges the argument linked list into the current linked list.

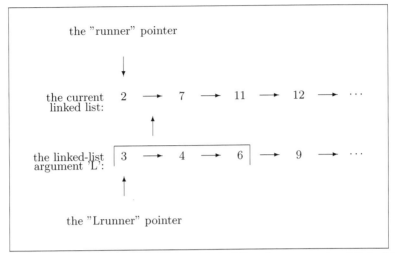

FIGURE 17.3:   Merging two ordered linked lists into one ordered linked list.
The items in the top linked list (the current object) are scanned by the pointer
"runner" in the outer loop. The items in the bottom linked list 'L' (the argument)
are scanned by the pointer "Lrunner" in the inner loop, and inserted into the right
places.

   The code that implements the merging uses two pointers to scan the items
in the linked lists (see Figure 17.3). The main pointer, named "runner", scans
the items in the current linked list. The room between the item pointed at
by "runner" and the next item should be filled by items from the second
linked list that is passed to the function as an argument. For this purpose,
a secondary pointer, named "Lrunner", is used to scan the second linked list
and find those items that indeed belong there in terms of order. These items
are then inserted one by one into this room.
   In case an item in the second linked list has the same priority order '<' as
an existing item in the current linked list, that is, it is equal to it in terms
of the "==" operator of the 'T' class, it is just added to it, using the "+="
operator of the 'T' class.

```
template<class T>
const linkedList<T>&
linkedList<T>::operator+=(linkedList<T>&L){
  linkedList<T>* runner = this;
  linkedList<T>* Lrunner = &L;
```

Here, the local pointers are defined before the loops in which they are used,
because they'll be needed also after these loops terminate. In particular, the
local pointer "runner" points to the current linked list. Furthermore, the local
pointer "Lrunner" points initially to the linked list 'L' that is passed to the
function as an argument. However, in order to start the merging process, we

must first make sure that the first item in the current linked list "item", is prior (in terms of the priority order '<') to the first item in 'L', "L.item". If this is not the case, then the merging process must start by placing "L.item" at the beginning of the current linked list and advancing "Lrunner" to the tail of 'L':

```
if(L.item < item){
  insertFirstItem(L.item);
  Lrunner = L.next;
}
```

We are now ready to start the merging process. For this, we use an outer loop to scan the items in the current linked list. The local pointer used in this loop, "runner", already points to the first relevant item in the current linked list. It is then advanced from item to item in it, until it points to the shortest possible tail that contains only the last item in the original list:

```
for(; runner->next; runner=runner->next){
```

We are now ready to fill the room between the item pointed at by "runner" and the next item in the current linked list with items from 'L' that belong there in terms of the '<' priority order. The first item considered for this is the item pointed at by "Lrunner". Indeed, if this item is equal to the item pointed at by "runner" (in terms of the "==" operator in the 'T' class), then it is just added to it (using the "+ =" operator of the 'T' class), and "Lrunner" is advanced to the next item in 'L':

```
if(Lrunner&&(Lrunner->item == runner->item)){
  runner->item += Lrunner->item;
  Lrunner = Lrunner->next;
}
```

Furthermore, an inner loop is used to copy from 'L' the items that belong in between the item pointed at by "runner" and the next item in the current linked list. Fortunately, the pointer used in this loop, "Lrunner", already points to the first item that should be copied:

```
for(; Lrunner&&(Lrunner->item < runner->next->item);
        Lrunner = Lrunner->next){
  runner->insertNextItem(Lrunner->item);
```

Furthermore, in this inner loop, once an item from 'L' has been copied to the current linked list, the "runner" pointer must be advanced to skip it:

```
  runner = runner->next;
}
```

This concludes the inner loop. By the end of it, "Lrunner" points either to nothing or to an item in 'L' that is too large to belong in between the item pointed at by "runner" and the next item in the current linked list. In fact, such an item must be copied the next time the inner loop is entered, that is, in the next step in the outer loop:

```
}
```

This concludes the outer loop as well.

Still, 'L' may contain items larger than or equal to the largest item in the current linked list. The short tail that contains these items only (if they exist) is now pointed at by "Lrunner". Furthermore, "runner" points to the last item in the current linked list. Thus, the remaining items in 'L' can be either added to the last item in the current linked list by

```
if(Lrunner&&(Lrunner->item == runner->item)){
  runner->item += Lrunner->item;
  Lrunner = Lrunner->next;
}
```

or appended to the end of the current linked list by

```
if(Lrunner)
  runner->next = new linkedList<T>(*Lrunner);
```

Finally, the function also returns a constant reference to the current linked list in its up-to-date state, with 'L' merged into it:

```
return *this;
} //  merge two linked lists while preserving order
```

17.11 The Ordering Problem

Assume that a disordered list of items is given. The ordering problem is to reorder the items according to the priority order '<', so that a given item is smaller than the next item. Clearly, this must be done as efficiently as possible.

Still, having an efficient algorithm to order the list properly is not enough. Indeed, the algorithm must also be implemented efficiently on the computer. For this purpose, the linked-list object proves to be most suitable, thanks to the opportunity to insert new items and drop old ones efficiently.

The recursive ordering algorithm is as follows (see Figure 17.4). First, the original list is split into two sublists. Then, the ordering algorithm is used recursively to put these sublists in the correct order. Finally, the sublists are

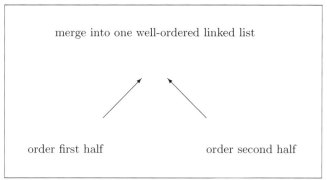

merge into one well-ordered linked list

order first half   order second half

FIGURE 17.4:   The ordering algorithm: the original list is split into two sublists, which are first ordered properly by a recursive call and then merged into one well-ordered list.

merged into one well-ordered list. For this, the "+ =" operator defined above is most useful.

This recursion is implemented in the "order()" member function below. The integer argument "length" passed to the function stands for the number of items in the current linked list. This is why when the "order()" function is applied recursively to the sublists it should take the smaller argument "length"/2 or so.

```
template<class T>
linkedList<T>&
linkedList<T>::order(int length){
```

If the list contains one item only, then it is already well ordered, so nothing should be done. If, on the other hand, it contains more than one item, then it should be split into two sublists. Clearly, the first sublist is pointed at by "this", the address of the current linked list. Finding the address of the second sublist, however, is more tricky. To do this, we define the local pointer "runner", which is initialized as "this", and is then advanced gradually from item to item until it reaches the middle of the original list:

```
if(length>1){
  linkedList<T>* runner = this;
  for(int i=0; i<length/2-1; i++)
    runner = runner->next;
```

By the end of this loop, "runner" points to the last item in the first subloop. Thus, it can be used to define the local pointer "second", which points to the second sublist:

```
linkedList<T>* second = runner->next;
```

Furthermore, "runner" can also be used to remove the second sublist from the original list, so that "this" points to the first sublist only, rather than to the entire list:

```
runner->next = 0;
```

We are now ready for the recursion. Indeed, the present "order()" function is applied recursively to the first sublist, pointed at by "this":

```
order(length/2);
```

Furthermore, thanks to the fact that (as we'll see below) the "order()" function also returns a reference to the well-ordered list, the second sublist can be ordered recursively and merged into the first sublist (pointed at by "this"), to form one well-ordered list, as required:

```
    *this += second->order(length-length/2);
}
```

Finally, the function also returns a reference to the current linked list in its final correct order:

```
    return *this;
}  //  order a disordered linked list
```

17.12 Stacks

The stack object discussed in Chapter 1, Section 1.8, can now be derived from the "linkedList" class. This would indeed provide the stack object with its desired properties: having an unlimited capacity in terms of number of items, and being able to push new items one by one at the top of it and pop items one by one out of it.

Clearly, in order to pop an item out of the stack, one must first check whether the stack is empty or not. For this purpose, the "stack" class should contain one extra field that is not included in the base "linkedList" class. This integer field, named "empty", takes the value 1 if the stack is empty and 0 otherwise.

Thanks to the fact that the fields "item" and "next" are declared as protected (rather than private) in the base "linkedList" class, they can also be accessed from the derived "stack" class. Here is how this derivation is done:

```
template<class T>
class stack : public linkedList<T>{
        int empty;
```

```
public:
    stack():empty(1){
    } // default constructor
```

This default constructor uses an initialization list to set the field "empty" to 1, to indicate that an empty stack is constructed. Furthermore, it invokes implicitly the default constructor of the base "linkedList" class, which initializes the field "next" with the zero value, to indicate that this is a trivial stack that points to no more items.

Next, we implement another public member function that can be used to verify whether the stack is empty or not:

```
    int isEmpty() const{
        return empty;
    } // check whether the stack is empty or not
```

Next, we implement the member function that pushes an item at the top of the stack. If the stack is empty, then this is done simply by setting the field "item" inherited from the base "linkedList" class to the required value:

```
    void push(const T&t){
        if(empty){
            item = t;
            empty = 0;
        }
```

If, on the other hand, the stack is not empty, then one would naturally like to use the "insertFirstItem" function inherited from the base "linkedList" class. This function, however, must take a nonconstant argument; this is why a local nonconstant variable must be defined and passed to it as an argument:

```
        else{
            T insert = t;
            insertFirstItem(insert);
        }
    } // push an item at the top of the stack
```

Next, we implement the member function that pops the top item out of the stack. Of course, if the stack is empty, then this is impossible, so an error message must be printed onto the screen:

```
    const T pop(){
        if(empty)
            printf("error: no item to pop\n");
        else
```

If, on the other hand, the stack is not empty, then there are still two possibilities: if the stack contains more than one item, then one would naturally like to use the "dropFirstItem" function, inherited from the base "linkedList" class:

```
if(next){
    const T out = item;
    dropFirstItem();
    return out;
}
```

If, on the other hand, the stack contains one item only, then the "drop-FirstItem" function inherited from the base "linkedList" class cannot be used. Instead, one simply sets the "empty" field to 1 to indicate that the stack will soon empty, and returns the "item" field that contains the only item in the stack, which has just popped out from it:

```
    else{
        empty = 1;
        return item;
    }
} //  pop the top item out of the stack

};
```

This completes the block of the "stack" class. Since the entire mathematical nature of the stack object is expressed fully in the public member functions "isEmpty", "push", and "pop" implemented above every user can now use it in his/her own application, with no worry about its internal data structure. This is illustrated in the exercises below.

17.13 Exercises

1. Complete the missing arithmetic operators in the implementation of the dynamic vector in Section 17.1, such as subtraction and multiplication and division by scalar. The solution is given in Section 28.12 in the appendix.
2. Implement Pascal's triangle in Chapter 14, Section 14.7, as a list of diagonals. The diagonals are implemented as dynamic vectors of increasing dimension. (The first diagonal is of length 1, the second is of length 2, and so on.) The components in these vectors are integer numbers. Verify that the sum of the entries along the nth diagonal is indeed 2^n and that the sum of the entries in the first, second, ..., nth diagonals is indeed $2^{n+1} - 1$.
3. Define the template class "triangle<T>" that is derived from a list of dynamic vectors of increasing dimension, as above. The components in

these vectors are of the unspecified type 'T'. Implement Pascal's triangle as a "triangle<int>" object. Verify that the sum of the entries in the nth diagonal is indeed 2^n.

4. Apply the "order()" function in Section 17.11 to a linked list of integer numbers and order it with respect to absolute value. For example, verify that the list

$$(-5, 2, -3, 0, \ldots)$$

is reordered as

$$(0, 2, -3, -5, \ldots).$$

5. Why is it important for the "push" function in the "stack" class to take a constant (rather than nonconstant) argument?

6. In light of your answer to the above exercise, can the user of the "stack" class define a class 'S' of integer numbers and push the number 5 at the top of it simply by writing "S.push(5)"? If yes, would this also be possible had the "push" function taken a nonconstant argument?

7. Use the above "stack" class to define a stack of integer numbers. Push the numbers 6, 4, and 2 into it (in this order), print it, and pop them back out one by one from it. Print each number that pops out from the stack, as well as the remaining items in the stack. The solution can be found in Section 28.13 in the appendix.

8. Implement the stack in Figure 1.3 above, and use it to add two natural numbers by Peano's theory.

9. Implement the stack in Figure 1.4 above, and use it to multiply two natural numbers by Peano's theory.

Part VI

Implementation of Computational Objects

Implementation of Computational Objects

The vectors and lists implemented above are now used to implement composite mathematical objects such as graphs, matrices, meshes, and polynomials. The implementation of these objects uses efficient algorithms in the spirit of their original mathematical formulation in the first half of this book. This way, further users of these objects can deal with them as in the original mathematical theory, with no need to bother with technical details of storage, etc. Such a natural implementation is helpful not only to implement advanced algorithms in elegant and easily-debugged codes but also to understand better the nature of the original mathematical concepts, improve the algorithms, develop more advanced theory, and introduce more advanced mathematical objects for future use.

The mathematical objects implemented in this part are used here in the context of graph theory only. Still, they can be used in scientific computing as well (see [26] and Chapters 26–27 in the next part).

Chapter 18

Trees

As we have seen above, the tree is a multilevel object, whose definition is based on mathematical induction. Indeed, assuming that the node at the head of the tree is given, then the entire tree is constructed by issuing edges (branches) from the head, and placing a smaller tree (a subtree) at the other end of each branch. Thanks to the induction hypothesis, one may assume that the definition of the subtrees is available. As a result, the entire tree is also well defined.

This mathematical induction is useful also in the recursive implementation of the tree object on the computer. Indeed, to construct a tree object, it is sufficient to construct its head and the branches that are issued from it. All the rest, that is, the construction of the subtrees, can be done recursively in the same way. This completes the construction of the entire tree object.

This procedure is best implemented in an object-oriented programming language such as C++. Indeed, the tree object needs to contain only one field to store the node in the head, and a few other fields to store the pointers to (or the addresses of) its subtrees, which are tree objects in their own right, although smaller than the original tree.

The recursive nature of the tree object is also useful in many (member, friend, and ordinary) functions that are applied to it. This is particularly useful to solve the tower problem below.

18.1 Binary Trees

The linked list implemented above may be viewed as a unary tree, in which each node has exactly one son, or one branch that is issued from it, except of the leaf at the lowest level, which has no sons at all. This is a rather trivial tree, with one node only at each level. Here we consider the more interesting case of a binary tree, in which each node has two sons (except of the leaves at the lowest level, which have no sons), so each level contains twice as many nodes as the previous level above it.

18.2 Recursive Definition

In C++, the definition of the binary-tree object is particularly straightforward and elegant. In fact, it has to contain only three fields: one to store the node at its head, and two other fields to store the pointers to (or the addresses of) its left and right subtrees.

Here one may ask: since the definition of the binary tree is not yet complete, how will the subtrees be defined? After all, they are trees in their own right, so they must be defined by the very definition of the tree object, which is not yet available?!

The answer is that, in the definition of the tree object, the subtrees are never defined. Only their addresses are defined, which are just integer numbers ready to be used as addresses for trees that will be defined later. Therefore, when the compiler encounters a definition of a binary tree, it assigns room for three variables in the memory of the computer: one variable to store the head, and two integer variables to store the addresses of the subtrees, whenever they are actually created.

Initially, these two addresses are assigned zero or just meaningless random numbers, to indicate that they point to nothing. Only upon construction of the concrete subtrees will these addresses take meaningful values and point to real subtrees.

18.3 Implementation of Binary Tree

The binary-tree object is implemented as a template class, with a yet unspecified type 'T', used to denote the type of the nodes in the tree. This parameter remains unspecified throughout the block of the class, including the member, friend, and ordinary functions associated with it. It is specified only by a user who wants to use the definition to construct a concrete binary-tree object. For example, if the user writes "binaryTree<int>", then the compiler invokes the constructor defined in the block of the "binaryTree" template class, with 'T' interpreted as the integer type. The compiler would then allocate memory for an integer variable for every variable of type 'T' in the definition, or for each node in the tree, resulting in a binary tree of integer numbers.

Here is the explicit definition of the "binaryTree" class:

```
template<class T> class binaryTree{
  T head;
  binaryTree *left;
```

```
binaryTree *right;
```

The binary-tree object contains thus three fields: "head", a variable of type 'T' (to be specified later upon construction) to store the value of the node at the top of the tree, and "left" and "right", variables of type pointer-to-binary-tree, to store the addresses of the left and right subtrees, if exist (Figure 18.1).

Since these fields are private, we need public member functions to read them. The first function returns the value of the node at the top of the tree. Before the function name, the type of the returned value is specified: "const T&". The character '&' means that the returned value is only a reference to the head of the tree. This avoids constructing a copy of the node at the top of the tree, which might be rather expensive when 'T' stands for a big object. Furthermore, thanks to the word "const" before the function type, the value at the head can only be read, but not changed, which avoids inadvertent changes.

```
public:
  const T& operator()() const{
    return head;
  } //  read head
```

Note that the second "const" word just before the block of the above function guarantees that the entire current object, with which the function is called by the user, also remains unchanged. This protects the object from any inadvertent change, and saves a lot of worrying from the user. Once the user constructs a concrete tree 't', he/she can simply write "t()" to invoke the above operator and read the head of the tree.

Similar functions are defined to read the addresses of the left and right subtrees. If, however, no subtree exists in the current binary-tree object with which the functions are called, then they return the zero address, which contains nothing:

```
const binaryTree* readLeft() const{
  return left;
} //  read left subtree
const binaryTree* readRight() const{
  return right;
} //  read right subtree
```

Next, we define the default constructor of the binary-tree object. This constructor takes only one argument of type "const T&". This way, when the constructor is actually called by the user, its argument is passed by reference, avoiding any need to copy it. Furthermore, the argument is also protected from any inadvertent change by the word "const". If, on the other hand, no argument is specified by the user, then the constructor gives it the default value zero, as indicated by the characters "=0" after the argument's name.

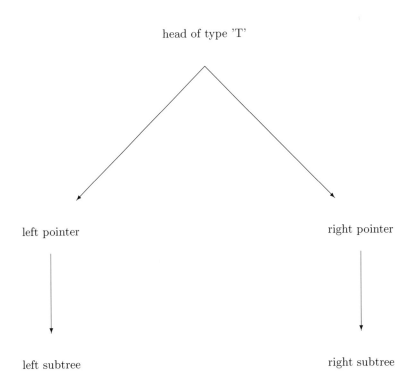

FIGURE 18.1:   The recursive structure of the binary-tree object.

This value is assigned to the argument by the constructor of the 'T' class if no explicit value is specified.

The three fields in the constructed binary-tree object are initialized in the initialization list: "head", the node at the top of the tree, takes the same value as the argument, and "left" and "right", the addresses of the subtrees, are initialized with the zero value, because no subtrees exist as yet.

```
binaryTree(const T&t=0):head(t),left(0),right(0){
} //  default constructor
```

Furthermore, we also define a copy constructor. This constructor takes as an argument a constant reference to a binary-tree object. Using a reference avoids copying the argument: only its address is actually passed to the constructor function. Furthermore, the word "const" before the argument's name protects it from any inadvertent change.

The three fields in the new tree object are initialized in the initialization list. The node at the head is initialized with the same value as that of the head of the copied tree that is passed as an argument. Furthermore, the left and right subtrees are copied from the left and right subtrees in the argument tree, using the copy constructor recursively (Figure 18.2).

If, however, any of these subtrees is missing in the argument tree, then the corresponding field in the constructed tree object takes the value zero, to indicate that it points to nothing:

```
binaryTree(const binaryTree&b):head(b.head),
    left(b.left?new binaryTree(*b.left):0),
    right(b.right?new binaryTree(*b.right):0){
} //  copy constructor
```

18.4 The Destructor

In the destructor of the binary-tree object, the command "delete" is used to erase the subtrees and release the memory occupied by them. This is done implicitly recursively: indeed, to apply the command "delete" to a pointer or an address, the compiler first applies the destructor recursively to the content of the address (Figure 18.3). This erases the subtrees in the "left" and "right" addresses. Then, the compiler releases the memory that had been occupied by these subtrees for future use. Furthermore, each deleted pointer is assigned the value zero, to indicate that it points to nothing. Finally, the compiler applies the destructor of class 'T' implicitly to erase the field "head" and release the memory occupied by it for future use.

```
~binaryTree(){
```

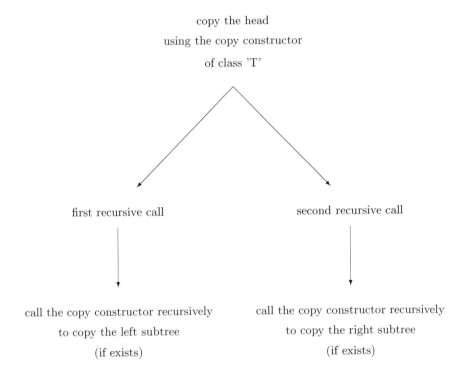

FIGURE 18.2:   The copy constructor: first, the head is copied using the copy constructor of the 'T' class, whatever it may be. Then, the left and right subtrees are copied (if exist), using recursive calls to the copy constructor.

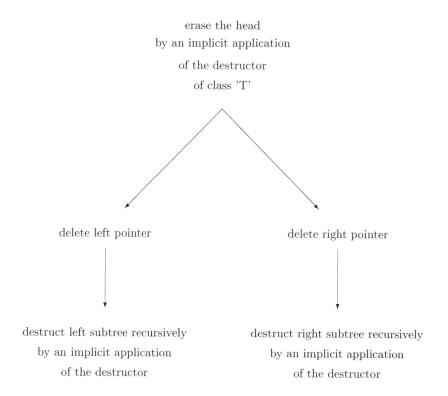

erase the head
by an implicit application
of the destructor
of class 'T'

delete left pointer

delete right pointer

destruct left subtree recursively
by an implicit application
of the destructor

destruct right subtree recursively
by an implicit application
of the destructor

FIGURE 18.3:   The destructor: first, the left and right pointers are deleted. This invokes implicit recursive applications of the destructor to the left and right subtrees. Finally, the "head" field is erased by an implicit application of the destructor of the 'T' class, whatever it may be.

```
      delete left;
      delete right;
      left = right = 0;
   }   //  destructor
```

18.5 The Tower Constructor

Let us use the binary-tree object to solve the tower problem in Chapter 10, Section 10.7. This binary tree contains "move" objects in its nodes, where a "move" object is an ordered pair of integer numbers. For example, the move $(1, 2)$ means that the top ring in column 1 is moved to the top of column 2. Thus, the "move" object can be implemented as a 2-D vector, with two integer components:

```
      typedef vector<int,2> move;
```

Once this "typedef" command is placed before the "binaryTree" class, the term "move" can be used to denote a 2-D vector of integers. The first component of this vector stands for the number of the column from which the top ring is moved, whereas the second component of the vector stands for the number of the column on top of which the ring is placed.

Let us now complete the block of the "binaryTree" class. The last function in this block is yet another constructor, specifically designed to solve the tower problem. Indeed, it takes two arguments: an integer argument n to indicate the number of rings in the tower problem, and a move argument to indicate the original column on which the tower is placed initially and the destination column to which it should be transferred. If no second argument is specified, then it is assumed that the tower should be transferred from column 1 to column 3, as in the original tower problem.

Now, the constructor does the following. First, it finds the number of the column that is used neither in the first nor in the second component of the "move" object that is passed to it as the second argument, that is, the column that is neither the original column on which the tower is placed initially, nor the destination column to which it should be transferred. The number of this column is placed in the integer variable named "empty", to indicate the intermediate column, which is only used as a temporary stop to help transfer the rings. For example, if the second argument in the call to the constructor is the move $(1, 2)$, then "empty" is assigned the value 3.

Then, the constructor is called recursively twice to construct the subtrees. For this, the variable "empty" is most useful. Indeed, in the left subtree, the $n-1$ top rings are transferred to column "empty", using a recursive call to the constructor with the first argument being $n-1$ and the second argument being

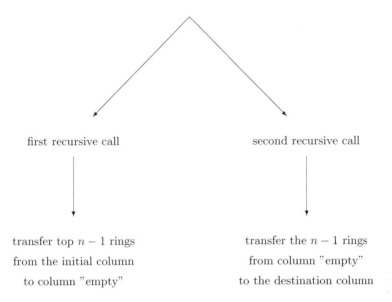

FIGURE 18.4: The tower constructor: the left subtree contains the moves required to transfer the top $n - 1$ rings from the initial column to column "empty", the head contains the move required to move the bottom ring from the initial column to the destination column, and the right subtree contains the moves required to transfer the above $n - 1$ rings from column "empty" to the destination column.

the move from the original column to column "empty". [In the above example, the second argument in this recursive call is $(1,3)$.] Then, the move at the head is used to move the bottom ring from the initial column to the destination column. Finally, the right subtree is constructed by another recursive call to the constructor, this time to transfer the above subtower of $n - 1$ rings from column "empty" to its final destination (Figure 18.4). This is why this second recursive call uses $n-1$ as a first argument and the move from column "empty" to the destination column as a second argument. [In the above example, the second argument in this second recursive call is $(3,2)$.]

```
binaryTree(int n, const move&m=move(1,3)):
    head(m),left(0),right(0){
  if(n>1){
    int empty=1;
    while((empty==head[0])||(empty==head[1]))empty++;
    left = new binaryTree(n-1,move(head[0],empty));
    right = new binaryTree(n-1,move(empty,head[1]));
  }
} // tower constructor
};
```

[Clearly, when $n = 1$, no subtrees are needed, since a tower that contains one ring only can be transferred in one move (the move at the head) to its final destination.] This recursive constructor produces the tree of moves required to solve the tower problem, as discussed in Chapter 10, Section 10.8.

To invoke the above tower constructor, the user just needs to write, e.g., "binaryTree<move>(4)". This way, the unspecified type 'T' in the "binaryTree" template class is interpreted as the "move" class. Furthermore, the above tower constructor is called with the first argument being $n = 4$ and the second argument being its default value $(1,3)$. As a result, this call produces the four-level tree in Figure 10.8, which contains the moves required to transfer a tower of four rings.

This completes the block of the "binaryTree" class. Below we use it to print the entire list of moves required to solve the tower problem.

18.6 Solving the Tower Problem

The tower constructor defined above uses mathematical induction on n, the number of rings in the tower, Indeed, it assumes that the left subtree contains the moves required to transfer the top $n - 1$ rings from the original column to column "empty", and that the right subtree contains the moves required to transfer these $n - 1$ rings from column "empty" to the destination column. Using these hypotheses, all that is left to do is to place at the head of the tree the move that is passed as the second argument to the constructor, to move the bottom ring from the original column to the destination column. In particular, when the second argument takes its default value $(1, 3)$, the entire tree produced by the tower constructor contains the moves required to transfer the entire tree from column 1 to column 3, as in Figure 10.8.

This mathematical induction can be used thanks to the recursive structure of the binary-tree object. Furthermore, this recursive nature enables one to define many useful functions.

Here we use this recursive nature to print the nodes in the binary tree. This can be done by mathematical induction on the number of levels in the tree. Indeed, since the subtrees have fewer levels than the original tree, the induction hypothesis can be used to assume that their nodes can be printed by recursive calls to the same function itself. Thus, all that is left to do is to print the head as well.

Furthermore, mathematical induction can also be used to show that, in a binary-tree object that has been constructed by the above tower constructor, the nodes (namely, the moves) are printed in the order in which they should be carried out to solve the tower problem (Figure 18.5). Indeed, by the induction hypothesis, it can be assumed that the first recursive call indeed prints the moves in the left subtree in the order in which they should be carried out to transfer the top $n - 1$ rings from the original column to column "empty". Then, the function prints the move required to move the bottom ring from the original column to the destination column. Finally, again by the induction hypothesis, the second recursive call prints the moves required to move the above $n-1$ rings from column "empty" to the destination column. This indeed guarantees that the list of moves is indeed printed in the proper order in which they should be carried out.

The implementation of the function that prints the nodes in the binary-tree object in the above order is, thus, as follows:

```
template<class T>
void print(const binaryTree<T>&b){
  if(b.readLeft())print(*b.readLeft());
  print(b());
  if(b.readRight())print(*b.readRight());
} // print binary tree
```

Here is how the user can use this function to print the list of moves required to transfer a tower of 16 rings from column 1 to column 3:

```
int main(){
  binaryTree<move> tower(16);
  print(tower);
  return 0;
}
```

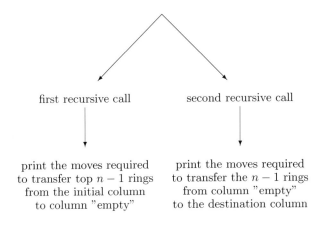

FIGURE 18.5:   Printing the tree constructed by the tower constructor: first, the function is applied recursively to the left subtree to print the moves required to transfer the top $n-1$ rings from the initial column to column "empty". (By the induction hypothesis, this is done in the correct order.) Then, the move in the head is printed to move the bottom ring from the initial column to the destination column. Finally, the moves in the right subtree are printed in the correct order required to transfer the $n-1$ rings from column "empty" to the destination column.

18.7 General Trees

So far, we have considered binary trees, in which each node has exactly two sons or two branches that are issued from it, except of the leaves at the lowest level, from which no branches are issued. This structure allows one to use the recursive implementation in Figure 18.1 above.

General trees, on the other hand, may have any number of branches issued from any node. Therefore, the tree object must have a linked list of tree objects to stand for its sons or subtrees. More precisely, the tree object should have a pointer to (or the address of) the linked list containing its sons. Thanks to the fact that a linked list may contain any number of items, the head of the tree may have any number of branches issued from it to its subtrees. Furthermore, a tree may also have no sons at all, in which case its list of sons is empty, so the pointer that points to it takes the value zero.

Here is how the tree object should be implemented in the "tree" template class:

```
template<class T> class tree{
  protected:
    T item;
    linkedList<tree<T> >* branch;
  public:

    . . .

};
```

The details are left to the reader.

Below we also discuss the implementation of general graphs, which lack the recursive nature of a tree. In a general graph, the notion of "son" is irrelevant: a node may well be both the son and the parent of another node. This is why a graph cannot be implemented like a tree. Indeed, if a graph had been implemented using the above "tree" class, then the linked list in it may well contain nodes that have already been defined previously and don't have to be defined again but rather to be referred to. To avoid infinite recursion, the linked list in the graph object mustn't contain new graphs, but merely references to existing graphs or, better yet, references to the nodes that are connected to the current node in the graph. This implementation is discussed below.

18.8 Exercises

1. Use the above binary-tree object to store and compute arbitrarily long arithmetic expressions.

2. Implement the required arithmetic operations between such binary-tree objects. For example, add two binary trees by using them as subtrees in a new tree, with '+' in its head.

3. Write a function that prints the arithmetic expression properly (with parentheses if necessary).

4. Write a recursive function that computes the value of the arithmetic expression stored in the binary tree.

5. Use the binary-tree object also to store and compute arbitrarily long Boolean expressions.

6. Implement the required logic operations between such binary-tree objects. For example, define the conjunction of two binary trees by using them as subtrees in a new binary tree, with '∧' in its head.

7. Write a function that prints the Boolean expression properly (with parentheses if necessary).

8. Write a recursive function that computes the true value (0 or 1) of the Boolean expression stored in the binary tree.

9. Introduce a static integer variable in the above tower constructor to count the total number of moves used throughout the construction. Verify that the number of moves is indeed $2^n - 1$, where n is the number of levels in the binary tree.

10. Explain the implicit recursion used in the destructor in the "binaryTree" class. How does it release the entire memory occupied by a binary-tree object?

Chapter 19

Graphs

The implementation of the tree object above is inherently recursive. The binary tree, for example, is defined in terms of its left and right subtrees, which are trees in their own right.

General graphs, on the other hand, have no recursive structure that can be used in their implementation. Therefore, a graph must be implemented as a mere set of nodes and a set of edges issued from or directed to each node.

The lack of any recursive nature makes the graph particularly difficult to implement on the computer. Indeed, as discussed at the end of the previous chapter, its naive (direct) implementation would require to attach to each node a linked list of references to the nodes that are connected to it by an edge. To avoid this extra complication, it makes much more sense to use the algebraic formulation of the graph as discussed below.

19.1 The Matrix Formulation

To have a transparent and useful implementation, the graph could be formulated as a matrix. In this form, the edges issued from node j are represented by the nonzero elements in the jth column in the matrix. With this formulation, the coloring algorithms take a particularly straightforward and easily implemented form.

In Chapter 10, Sections 11.10–11.11, we have seen that a weighted graph is associated with the square matrix of order $|N|$:

$$A \equiv (a_{i,j})_{1 \leq i,j \leq |N|},$$

where N is the set of nodes in the graph, $|N|$ is its cardinality (the total number of nodes), and the nodes in N are numbered by the natural numbers

$$1, 2, 3, \ldots, |N|.$$

The element $a_{i,j}$ in the matrix A is the weight assigned to the edge leading from node j to node i.

An unweighted graph can easily be made weighted by assigning the uniform weight 1 to each edge in it. The matrix A associated with it takes then the form

$$a_{i,j} \equiv \begin{cases} 1 & \text{if } (j,i) \in E \\ 0 & \text{if } (j,i) \notin E, \end{cases}$$

where E is the set of the edges in the graph. In this formulation, the coloring algorithms in Chapter 10, Sections 11.4 and 11.6, above take a particularly straightforward form.

19.2 The Node-Coloring Algorithm

In the above matrix formulation, the node-coloring algorithm takes a particularly transparent form [25]: for $j = 1, 2, 3, \ldots, |N|$, define

$$c(j) \equiv \min(\mathbb{N} \setminus \{c(i) \mid i < j, \ a_{i,j} \neq 0\} \setminus \{c(i) \mid i < j, \ a_{j,i} \neq 0\}.$$

This way, the color $c(j)$ assigned to node j is different from any color assigned previously to any node i that is connected to it by an edge of either the form (j, i) or the form (i, j).

Later on, we'll see how this algorithm can be easily implemented on a computer.

19.3 The Edge-Coloring Algorithm

Let us use the above matrix formulation also to have a straightforward representation of the edge-coloring algorithm. For this, we first need the following definitions.

For $1 \leq n \leq |N|$, let A_n denote the $n \times n$ submatrix in the upper-left corner of A:

$$(A_n)_{i,j} \equiv a_{i,j}, \quad 1 \leq i, j \leq n.$$

Furthermore, let

$$c_{i,j} \equiv c(a_{i,j})$$

be the color assigned to the element $a_{i,j}$ in A, or to the edge (j, i) in E. Moreover, let $c_i(A_n)$ be the set of colors assigned to the elements in the ith row in the matrix A_n:

$$c_i(A_n) \equiv \{c_{i,j} \mid 1 \leq j \leq n\}.$$

Finally, recall that

$$A_n^t \equiv (a_{i,j}^t)_{1 \leq i,j \leq n}$$

is the transpose of A_n, defined by

$$a_{i,j}^t \equiv a_{j,i}.$$

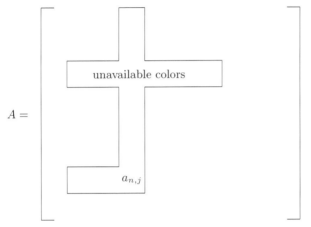

$A =$

unavailable colors

$a_{n,j}$

FIGURE 19.1: Step 1 in the edge-coloring algorithm, in which edges of the form $a_{n,j}$ (for $j = 1, 2, 3, \ldots, n$) are colored in a color that has not been used previously in the area marked "unavailable colors."

The edge-coloring algorithm can now be formulated as follows. For $n = 1, 2, 3, \ldots, |N|$, do the following:

1. For $j = 1, 2, 3, \ldots, n$, define

$$c_{n,j} \equiv \min(\mathbb{N} \setminus \{c_{n,k} \mid k < j\} \setminus c_j(A_{n-1}) \setminus c_j(A_{n-1}^t)).$$

This way, the elements in the nth row of A_n are colored properly; indeed, $c_{n,j}$, the color used to color the edge leading from j to n, has never been used before to color any edge issued from or directed to either j or n (see Figure 19.1).

2. For $j = 1, 2, 3, \ldots, n - 1$, define

$$c_{j,n} \equiv \min(\mathbb{N} \setminus c_n(A_n) \setminus \{c_{k,n} \mid k < j\} \setminus c_j(A_{n-1}) \setminus c_j(A_{n-1}^t)).$$

This way, the elements in the nth column of A_n are colored properly as well (see Figure 19.2).

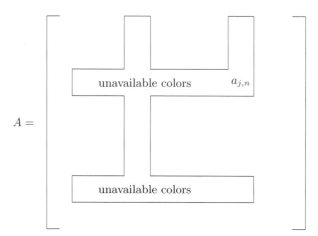

FIGURE 19.2:   Step 2 in the edge-coloring algorithm, in which edges of the form $a_{j,n}$ (for $j = 1, 2, 3, \ldots, n-1$) are colored in a color that has not been used previously in the area marked "unavailable colors."

The above definition indicates that the transpose matrix A^t should be stored as well as A. Indeed, the transpose submatrix A^t_{n-1} is often used throughout the above algorithm. Below we discuss the efficient implementation of A, A^t, and the above algorithms.

19.4 Sparse Matrix

In most graphs, each node has only a few nodes to which it is connected. In other words, most of the elements $a_{i,j}$ in A vanish. This means that A is sparse.

Thus, it makes no sense to store all the elements in A, including the zero ones. Indeed, this would require prohibitively large time and storage resources, to complete the above coloring algorithms. Indeed, these algorithms would then have to check in each step what colors have been used previously to color not only real edges $(j, i) \in E$, for which $a_{i,j} \neq 0$, but also nonexisting dummy edges $(j, i) \notin E$, for which $a_{i,j} = 0$. This redundant work would make the algorithms prohibitively expensive.

A much better implementation uses selective storage: to store only the nonzero elements in A, which correspond to real edges in E. This approach not only saves valuable storage, but also reduces significantly the time required to complete the above coloring algorithms: indeed, in each step, one only needs to scan existing edges in E (or nonzero elements in A) and check what colors have been used in them.

A sparse matrix can be implemented efficiently by storing only its nonzero elements. Indeed, it can be stored in a list of $|N|$ items, each of which corresponds to a node in N, or a row in A. Because the number of nonzero elements in each row in A is unknown in advance, the ith item in the list should store the ith row in A as a linked list of nonzero elements of the form $a_{i,j}$. This way, there is no *a priori* limit on the number of nonzero elements stored in each row, and the storage requirement is kept to a minimum.

Here one could ask: why not store the sparse matrix in a mere $|N|$-dimensional vector, whose components are the rows in A? The answer is that, since the rows are implemented as linked lists of variable length, they have different sizes, thus cannot be kept in an array. Only their addresses, which are actually mere integers, can be kept in an array; this means that the rows themselves must be stored in a list rather than a vector.

19.5 Data Access

Let us first consider a naive implementation of a graph in terms of its set of nodes N and its set of edges E. Later on, we'll see how this implementation improves considerably when formulated in terms of the sparse matrix A associated with the graph, rather than in terms of the original graph.

In order to apply the coloring algorithms, each node in N must "know" to what nodes it is connected. By "know" we mean here "have access to" or "know the address of" or "have a reference to." In other words, the node object that represents a node $j \in N$ must contain at least two linked lists of addresses or references: one list to know from which nodes an edge leads to it (the nonzero elements in the jth row in A), and another list to know the nodes to which edges are issued from it (the nonzero elements in the jth column in A). Furthermore, the node object must also contain an integer field to store the index of the color assigned to it.

The references-to-nodes in the above linked lists actually stand for edges in E, or nonzero elements in A. Because these edges should be also colored in the edge-coloring algorithm, they should actually be replaced by objects with two fields: one to contain the required reference-to-node, and another one to contain the index of the color in which the corresponding edge is colored. This is the edge object.

The edge object that represents an edge of the form $(j, i) \in E$ (or an element

$a_{i,j} \neq 0$) is duplicated in the above implementation. Indeed, it appears not only in the first linked list in the node object i to stand for an edge leading to it, but also in the second linked list of the node object j to stand for an edge issued from it. In order to prevent a conflict between the colors assigned to these two edge objects, they must be one and the same object. Therefore, one of the linked lists in the node object, say the second one, must be not a list of concrete edges but rather a list of references-to-edges.

In the above implementation, the edge object that stands for the edge (j, i) contains information about the node j from which it is issued, but not about the node i to which it leads. This approach has been considered because we have assumed that this information is already known to the user, because the entire linked list of edges is stored in the ith item in the list of nodes. However, this assumption is not always valid; indeed, we have just seen that a reference to this edge is placed also in the jth item in the list of nodes, to let node j know that it is connected to node i. Unfortunately, this information is denied from node j, since there is no field to store the number i in the edge object.

In order to fix this, the implementation of the edge object must be yet more complicated. In fact, it must contain three fields: the first field, of type reference-to-node, to store a reference to the node j from which it is issued; the second one, of type reference-to-node too, to store a reference to the node i to which it leads; and, finally, an integer field to store its color.

Below we see how this implementation simplifies considerably when translated to matrix language.

19.6 Virtual Addresses

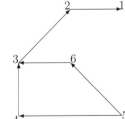

FIGURE 19.3:   A node indexing in an oriented graph. The six nodes are numbered $1, 2, 3, 4, 5, 6$.

In the matrix formulation, each node in N is assigned a number between 1 and $|N|$ to serve as its index. This indexing scheme induces an order on the nodes in N: they are now ordered one by one from the first node, indexed by 1, to the last node, indexed by $|N|$.

For example, the nodes in the graph in Figure 19.3 are ordered counter-clockwise, and indexed by the numbers $1, 2, 3, 4, 5, 6$. With this indexing, the 6×6 matrix A associated with the graph takes the form

$$A = \begin{pmatrix} 0 & 1 & 0 & 0 & 0 & 0 \\ 0 & 0 & 1 & 0 & 0 & 0 \\ 0 & 0 & 0 & 1 & 0 & 1 \\ 0 & 0 & 0 & 0 & 1 & 0 \\ 0 & 0 & 0 & 0 & 0 & 0 \\ 0 & 0 & 0 & 0 & 1 & 0 \end{pmatrix}.$$

The indexing scheme is particularly useful to store the colors assigned to the nodes in the node-coloring algorithm. Indeed, these colors can now be stored in an $|N|$-dimensional vector v, whose ith component v_i stores the color assigned to node i:

$$v_i = c(i).$$

The index assigned to the node can be thought of as its virtual address, through which it can be referred. Unlike the physical address in the memory of the computer, the virtual address is not only straightforward and continuous, but also reflects well the mathematical properties of the node. Indeed, its index i can be used to refer directly to the ith row in A, which stores all the edges (j, i) leading to it (or the nonzero elements $a_{i,j} \neq 0$). The node j from which such an edge is issued is stored in an integer field in the edge object, in which the index j (the column of A in which $a_{i,j}$ is placed) is stored.

Thus, in the matrix formulation, an edge (or a nonzero element $a_{i,j} \neq 0$) is an object that contains two integer fields only: one to store the column j, and another one to store the color assigned to the edge in the edge-coloring algorithm. The row number i doesn't have to be stored in this object, because the entire linked list of these objects (or the entire ith row in A) is placed in the ith item in the list of rows that form the entire sparse matrix A, so it is already known to the user.

In summary, the sparse matrix A is implemented as a list of $|N|$ row objects (see Figure 19.4). The ith item in the list contains the ith row in A, which is actually a linked list of row-element objects. Each row-element object contains two integer fields: one to store the column number j (the virtual address of the node j), and another one to store the color $c_{i,j}$ assigned to $a_{i,j}$ in the edge-coloring algorithm.

Still, both coloring algorithms require scanning not only the rows in A but also the columns in it. For this purpose, one must store not only the original matrix A but also its transpose A^t. Indeed, scanning the jth linked list in the implementation of A^t amounts to scanning the jth column in A as required.

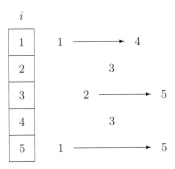

FIGURE 19.4:   A 5×5 sparse matrix, implemented as a list of five row objects. Each row object is a linked list of row-element objects. Each row-element object contains an integer field to indicate the column in which it is placed in the matrix.

The rows in A^t (or the columns in A) correspond to the second linked list in the above (naive) implementation of a node in the original graph, which stores the edges that are issued from it. As discussed above, these edges have already been implemented, so the second linked list mustn't contain new edge objects but rather references to existing edges. This should prevent conflicts in coloring the edges later.

Similarly, in the present implementation of the graph in terms of the sparse matrix A, the transpose matrix A^t should be implemented not as a list of linked lists of concrete edges but rather as a list of linked lists of references-to-edges. This way, an element $a^t_{j,i}$ in A^t is implemented not as a new row-element object but rather as a reference to the existing row-element object $a_{i,j}$. This way, the color $c_{i,j}$ assigned to $a_{i,j}$ automatically serves also as the color assigned to $a^t_{j,i}$, as required.

Unfortunately, since the row-element $a_{i,j}$ is now referred to not only from the ith row in A but also from the jth row in A^t, it must contain an extra integer field to store the integer number i. This somewhat complicates the implementation of the row-element object.

To avoid the above extra integer field in the implementation of the row-element object , one may stick to the original implementation of A^t as a list of row objects. In this approach, however, one must make sure that any change to $a_{i,j}$ influences $a^t_{j,i}$ as well. In particular, once a color $c_{i,j}$ is assigned to $a_{i,j}$

in the edge-coloring algorithm, it must be assigned immediately to $a_{j,i}^t$ as well, to avoid a conflict between these two duplicate representations of the same edge. This is indeed the approach used later in the book.

19.7 Advantages and Disadvantages

The indices (or virtual addresses) $i = 1, 2, 3, \ldots, |N|$ assigned to the nodes in N are continuous in the sense that they proceed continuously from 1 to $|N|$, skipping no number in between. This way, the colors assigned to the nodes can be stored in an $|N|$-dimensional vector v, whose components v_i store the color assigned to the node i:

$$v_i = c(i).$$

Thus, the continuous index $i = 1, 2, 3, \ldots, |N|$ is used to refer to the ith node in N, and assign to it the color stored in the ith component in v. This helps to avoid an extra integer field in each node object to store its color.

The node object should thus contain only the information about the edges that lead to it. The explicit implementation of the node object can thus be avoided: in fact, the ith node in N is implemented implicitly as the row object that stores the ith row in A. Indeed, this row contains the elements $a_{i,j}$ that represent the edges that lead to node i, as required.

Furthermore, the above indexing scheme is also useful to refer to all the nodes from which an edge leads to node i. Indeed, these nodes are referred to by their index j, or the column in A in which the corresponding element $a_{i,j}$ is placed. These row elements are placed in the linked list that forms the ith row in A. These row objects are now placed in a list indexed $i = 1, 2, 3, \ldots, |N|$ to form the entire sparse matrix A associated with the graph.

The row-element objects used to implement the $a_{i,j}$'s should preserve increasing column order. Indeed, if $j < k$, then $a_{i,j}$ should appear before $a_{i,k}$ in the linked list that implements the row object that stores the ith row in A.

The drawback in this structure is that it is somewhat stiff: it would be rather difficult to add an extra node to an existing graph, because this may require changing the entire indexing scheme. Indeed, the list of row objects that are used to store the rows in A is fixed and contains $|N|$ items only. There is no guarantee that sufficient room in the memory of the computer is available to enlarge this list to contain more than $|N|$ items. Therefore, even adding one new node at the end of the list of nodes is not always possible, unless the list of row objects to implement A is replaced by a linked list of row objects.

To avoid this extra complication, we assume here that the graph under consideration is complete, and no extra node is expected to be added to it. Only at this final point are the indices $i = 1, 2, 3, \ldots, |N|$ assigned to the nodes in the graph. Under this assumption, one can safely use a list of $|N|$ row objects to form the entire $|N| \times |N|$ sparse matrix A associated with the graph.

19.8 Nonoriented Graphs

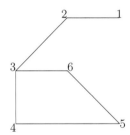

FIGURE 19.5:   Node indexing in an nonoriented graph. The six nodes are numbered $1, 2, 3, 4, 5, 6$.

So far, we have discussed oriented graphs, in which an edge is an ordered pair of the form (i, j), where i and j are some (indices of) nodes. In nonoriented graphs, on the other hand, an edge $\{i, j\}$ is an unordered pair, or a set of two nodes, in which both nodes i and j take an equal status: no one is prior to the other. In terms of the matrix formulation, $a_{i,j} \neq 0$ if and only if $a_{j,i} \neq 0$. In other words, A is symmetric:

$$A^t = A.$$

For example, the symmetric 6×6 matrix A associated with the nonoriented graph in Figure 19.5 takes the form

$$A = \begin{pmatrix} 0\,1\,0\,0\,0\,0 \\ 1\,0\,1\,0\,0\,0 \\ 0\,1\,0\,1\,0\,1 \\ 0\,0\,1\,0\,1\,0 \\ 0\,0\,0\,1\,0\,1 \\ 0\,0\,1\,0\,1\,0 \end{pmatrix}.$$

Thanks to the symmetry of A, the node-coloring algorithm in the nonoriented graph takes the simpler form: for $j = 1, 2, 3, \ldots, |N|$, define

$$c(j) \equiv \min(\mathbb{N} \setminus \{c(i) \mid i < j, \ a_{j,i} \neq 0\}).$$

In terms of matrix elements, one should bear in mind that the only relevant elements in A are those that lie in its lower triangular part, that is, the nonzero elements $a_{i,j}$ for which $i \geq j$. Indeed, for $i \geq j$, the element $a_{j,i}$ refers to the same edge as $a_{i,j}$, so there is no need to consider it as an independent matrix element. Instead, it may be considered as a mere reference to $a_{i,j}$.

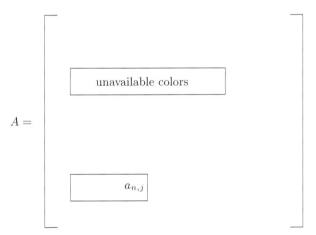

FIGURE 19.6: Step 1 in the edge-coloring algorithm for a nonoriented graph, in which edges of the form $a_{n,j}$ (for $j = 1, 2, 3, \ldots, n$) are colored in a color that has not been used previously in the area marked "unavailable colors."

Thus, for nonoriented graphs, the edge-coloring algorithm takes a far simpler form than in Section 19.3 above: For $n = 1, 2, 3, \ldots, |N|$, do the following:

1. For $j = 1, 2, 3, \ldots, n$, define

$$c_{n,j} \equiv \min(\mathbb{N} \setminus \{c_{n,k} \mid k < j\} \setminus c_j(A_{n-1})$$

(see Figure 19.6).

2. For $j = 1, 2, 3, \ldots, n - 1$, define

$$c_{j,n} \equiv c_{n,j}.$$

In order to implement this algorithm, there is no longer a need to store the transpose matrix A^t. Still, as can be seen from the final step in the above edge-coloring algorithm, one should be careful to update the color assigned to an element of the form $a_{i,j}$ ($i \geq j$) not only in the row-element object that stores $a_{i,j}$ but also in the row-element object that stores $a_{j,i}$, to avoid a conflict between these two objects, which actually refer to the same edge $\{i, j\}$.

In the next chapter, we give the full code to implement the sparse-matrix object and the coloring algorithms.

19.9 Exercises

1. Implement the node-coloring algorithm for an oriented graph in its matrix formulation, using the "matrix" template class in Chapter 16, Section 16.16.

2. Introduce an integer variable "count" to count the number of checks required in the entire coloring algorithm. Test your code for several graphs, and observe how large "count" is in each application.

3. Implement the edge-coloring algorithm for an oriented graph in its matrix formulation, using the "matrix" template class in Chapter 16, Section 16.16.

4. Introduce an integer variable "count" to count the number of checks required in the entire coloring algorithm. Test your code for several graphs, and observe how large "count" is in each application.

5. Implement the node-coloring algorithm for a nonoriented graph in its matrix formulation, using the "matrix" template class in Chapter 16, Section 16.16.

6. Introduce an integer variable "count" to count the number of checks required in the entire coloring algorithm. Test your code for several graphs, and observe how large "count" is in each application.

7. Implement the edge-coloring algorithm for a nonoriented graph in its matrix formulation, using the "matrix" template class in Chapter 16, Section 16.16.

8. Introduce an integer variable "count" to count the number of checks required in the entire coloring algorithm. Test your code for several graphs, and observe how large "count" is in each application.

9. What are the advantages of using sparse matrices rather than the "matrix" object above?

10. By how much would "count" decrease if sparse matrices had been used in the above applications?

11. For oriented graphs, why is it necessary to store also the transpose of the sparse matrix to carry out the coloring algorithms?

12. Write the alternative coloring algorithms in the last exercises at the end of Chapter 11 in terms of the matrix formulation.

Chapter 20

Sparse Matrices

Here we introduce a multilevel hierarchy of mathematical objects: from the most elementary object, the matrix-element object in the lowest level, through the more complicated object, the row object implemented as a linked list of matrix elements, to the matrix object in the highest level of the hierarchy, implemented as a list of row objects. This is a good example for how object-oriented programming can be used to form more and more complicated mathematical objects.

The sparse-matrix object implemented in the highest level of the above hierarchy is particularly flexible in the sense that arbitrarily many matrix elements can be used in each row in it. Furthermore, it is particularly easy to add new matrix elements and drop unnecessary ones, without any need to waste storage requirements on dummy elements.

As discussed in Chapter 19 above, the sparse-matrix object is most suitable to implement both oriented and nonoriented graphs. Furthermore, the node-coloring and edge-coloring algorithms are implemented in an elegant and transparent code, along the guidelines in Chapter 19 above.

20.1 The Matrix-Element Object

The most elementary object in the hierarchy, the matrix element, is implemented in the "rowElement" class below, which contains two data fields: the first (of type 'T', to be specified upon construction later on in compilation time) to store the value of the element, and the second (of type "int") to store the index of the column in which the element is placed in the matrix. These fields are declared as "protected" (rather than the default "private" status) to make them accessible not only from definitions of members and friends of the class but also from definitions of members and friends of any class that is derived from it.

```
template<class T> class rowElement{
  protected:
    T value;
    int column;
```

The following constructor uses an initialization list to initialize these fields with proper values immediately upon construction, in the same order in which they are listed in the class block:

```
public:
  rowElement(const T& val=0, int col=-1)
      : value(val),column(col){
  } // constructor
```

With this constructor, the user can write "rowElement<double> e" to construct a new matrix element 'e' with the default value 0 and the meaningless default column index −1.

The copy constructor is defined in a similar way:

```
rowElement(const rowElement&e)
    : value(e.value),column(e.column){
} // copy constructor
```

With this copy constructor, the user can write "rowElement<double> d(e)" or "rowElement<double> d=e" to construct yet another matrix element 'd' with the same data fields as 'e'.

Unlike the constructor, which construct a new matrix element, the following assignment operator assigns the value of an existing matrix element to an existing (the current) matrix element. This is why no initialization list can be used: the data fields must be assigned their proper values in the function block:

```
const rowElement& operator=(const rowElement&e){
```

As is indicated by the words "const rowElement&" in these round parentheses, the argument is passed by (constant) reference, to avoid invoking the copy constructor. Now, if the argument is indeed not the same object as the current object, then the data fields are assigned one by one:

```
if(this != &e){
  value = e.value;
  column = e.column;
}
```

Finally, as indicated by the words "const rowElement&" in the beginning of the above header, the current object is also returned by (constant) reference:

```
return *this;
} // assignment operator
```

This way, the user can write commands like "c = d= e" to assign the value of the matrix element 'e' to both matrix elements 'd' and 'c'.

The destructor below has an empty block. Indeed, because there are no pointer fields in the class, no "delete" commands are needed; the data fields are destroyed implicitly at the '}' symbol that marks the end of the function block, in an order reversed to the order in which they are listed in the class.

```
~rowElement(){
} // destructor
```

As discussed above, the data fields are declared as protected rather than private. This way, they can be also accessed from classes derived from the "rowElement" class later on in the book. Furthermore, they can be read (although not changed) even from ordinary functions by the public "getValue" and "getIndex" member functions:

```
const T& getValue() const{
  return value;
} // read the value
```

Here, thanks to the words "const T&" at the beginning of its header, the function returns the value of the current matrix element by (constant) reference. Furthermore, the reserved word "const" before the function block guarantees that the current object never changes in the function. The user can thus safely write "e.getValue()" to read the value of the matrix element 'e'.

Similarly, the column index is read as follows:

```
int getColumn() const{
  return column;
} // return the column
```

Here the output is returned by value (rather than by address or by reference) because it is only an integer, so copying it is no more expensive than copying its address.

20.2 Member Arithmetic Operators

Next, we define some member arithmetic operations. In particular, the "+=" operator below allows one to form a linked list of matrix elements later on in this chapter and apply to it the merging and ordering functions in Chapter 17, Section 17.10–17.11.

The "+=" operator has two versions. The first version adds the 'T' argument to the value of the current "rowElement" object:

```
const rowElement&
operator+=(const T&t){
  value += t;
  return *this;
} // adding a T
```

With this version, the user can write "d = e += 1." to increment the value of the matrix element 'e' by 1 and assign the up-to-date 'e' in the matrix element 'd' as well.

The following version, on the other hand, takes a "rowElement" (rather than a 'T') argument:

```
const rowElement&
operator+=(const rowElement<T>&e){
  value += e.value;
  return *this;
}  //  adding a rowElement
```

With this version, the user can write "c = d+= e" to increment the value of 'd' by the value of 'e' and assign the updated matrix element 'd' in 'c' as well.

Similarly, the "-=" operator also has two versions:

```
const rowElement& operator-=(const T&t){
  value -= t;
  return *this;
}  //  subtracting a T

const rowElement&
operator-=(const rowElement<T>&e){
  value -= e.value;
  return *this;
}  //  subtracting a rowElement
```

The following operator multiplies the value of the current matrix element by the argument:

```
const rowElement&
operator*=(const T&t){
  value *= t;
  return *this;
}  //  multiplying by a T
```

Similarly, the following operator divides the value of the current matrix element by the (nonzero) argument:

```
const rowElement& operator/=(const T&t){
  value /= t;
  return *this;
}  //  dividing by a T
};
```

This completes the block of the "rowElement" class.

20.3 Comparison in Terms of Column Index

Next, we define the binary '<', '>', and "==" ordinary (nonmember, non-friend) operators to compare two matrix elements in terms of their column indices. For example, with these operators, "e < f" returns 1 if the column index of 'e' is indeed smaller than that of 'f', and 0 otherwise. These operators allow one to form linked lists of matrix elements later on in this chapter and order and merge them (while preserving increasing column-index order) as in Chapter 17, Section 17.10–17.11,

```
template<class T>
int
operator<(const rowElement<T>&e, const rowElement<T>&f){
  return e.getColumn() < f.getColumn();
} //   smaller column index
```

Similarly, the '>' operator is defined by

```
template<class T>
int
operator>(const rowElement<T>&e, const rowElement<T>&f){
  return e.getColumn() > f.getColumn();
} //   greater column index
```

With this operator, "e > f" is 1 if the column index of 'e' is indeed greater than that of 'f', and 0 otherwise.

Similarly, the "==" operator is defined by

```
template<class T>
int
operator==(const rowElement<T>&e, const rowElement<T>&f){
  return e.getColumn() == f.getColumn();
} //   same column
```

With this operator, "e == f" is 1 if both 'e' and 'f' have the same column index, and 0 otherwise.

20.4 Ordinary Arithmetic Operators

Here we define some ordinary binary arithmetic operators with a "rowElement" object and a 'T' object:

```
template<class T>
const rowElement<T>
operator+(const rowElement<T>&e, const T&t){
  return rowElement<T>(e) += t;
} //   rowElement plus a T
```

Note that the arguments are passed to the operator by (constant) reference, to avoid unnecessary calls to the copy constructors of the "rowElement" and 'T' classes. Because the required sum is placed in a (temporary) local "rowElement" variable that disappears at the end of the function, it cannot be returned by reference; it must be returned by value, as is indeed indicated by the word "rowElement<T>" (rather than "rowElement<T>&") in the header, just before the function name. This way, just before it disappears, it is copied by the copy constructor of the "rowElement" class into the temporary object returned by the function.

With this operator, "e + t" returns a matrix element whose column index is as in 'e' and whose value is the sum of the value in 'e' and the scalar 't'.

The following version of the '+' operator takes the arguments in the reversed order:

```
template<class T>
const rowElement<T>
operator+(const T&t, const rowElement<T>&e){
  return rowElement<T>(e) += t;
} //   T plus rowElement
```

With this operator, "t + e" is the same as "e + t" above.

The following operator returns the "rowElement" object whose column index is as in the first argument, and its value is the difference between the value in the first argument and the second argument:

```
template<class T>
const rowElement<T>
operator-(const rowElement<T>&e, const T&t){
  return rowElement<T>(e) -= t;
} //   rowElement minus T
```

With this operator, "e - t" returns the matrix element whose column index is as in 'e' and whose value is the difference between the value in 'e' and the scalar 't'.

Similarly, the following '*' operator calculates the product of a "rowElement" object and a 'T' object:

```
template<class T>
const rowElement<T>
operator*(const rowElement<T>&e, const T&t){
  return rowElement<T>(e) *= t;
} //   rowElement times a T
```

The following version of the '*' operator is the same as the original version, except that it takes the arguments in the reversed order:

```
template<class T>
const rowElement<T>
operator*(const T&t, const rowElement<T>&e){
  return rowElement<T>(e) *= t;
} //   T times rowElement
```

With these versions, both "e * t" and "t * e" return the "rowElement" object whose column index is as in 'e' and whose value is the product of the value in 'e' and the scalar 't'.

Similarly, the next operator calculates the ratio between a matrix element and a scalar:

```
template<class T>
const rowElement<T>
operator/(const rowElement<T>&e, const T&t){
  return rowElement<T>(e) /= t;
} //   rowElement divided by a T
```

Indeed, with this operator, "e / t" returns the "rowElement" object whose column index is as in 'e' and whose value is the ratio between the value in 'e' and the scalar 't'.

Finally, the following ordinary function prints the data fields in a "rowElement" object onto the screen:

```
template<class T>
void print(const rowElement<T>&e){
  print(e.getValue());
  printf("column=%d\n",e.getColumn());
} //   print a rowElement object
```

20.5 The Row Object

We are now ready to implement a row in a sparse matrix as a linked list of matrix elements. More precisely, the "row" class below is derived from the "linkedList" class in Chapter 17, Section 17.3, with its template 'T' specified to be the "rowElement" class (Figure 20.1).

The "row" class is a template class as well: it uses the symbol 'T' to stand for the type of the value of the matrix element. This type is specified only upon constructing a concrete "row" object later on in compilation time.

Since we focus here on a single row in a sparse matrix, we refer in the sequel to its elements as row elements (or just elements) rather than matrix

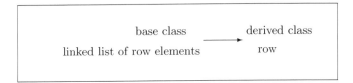

FIGURE 20.1:   Inheritance from the base class "linkedList<rowElement>" to the derived class "row".

elements. Furthermore, it is assumed hereafter that the elements in the "row" object are ordered in increasing column order. To guarantee this, the user is well advised to be careful to construct concrete "row" objects according to this order.

```
template<class T>
class row : public linkedList<rowElement<T> >{
  public:
      row(const T&val=0,int col=-1){
        item = rowElement<T>(val,col);
      } //  constructor
```

This constructor takes two arguments, a 'T' argument and an integer argument, and uses them to call the constructor of the "rowElement" class to form the first element in the row. Thanks to the fact that the "item" field in the base "linkedList" class is declared as protected (rather than private) in Chapter 17, Section 17.3, it can be accessed from the derived "row" class to set the first element in the row.

Fortunately, the copy constructor, assignment operator, and destructor are inherited properly from the base "linkedList" class, so they don't have to be rewritten here.

20.6 Reading the First Element

The following public member function can only read (but not change) the first element in the current "row" object:

```
const rowElement<T>& operator()() const{
  return item;
} //  read first element
```

With this function, the user can write "r()" to read the first element in the row 'r'.

Furthermore, the next member function reads the value of the first element in the current "row" object:

```
const T& getValue() const{
  return item.getValue();
} //  read value of first element
```

Moreover, the next function reads the column index of the first element in the current "row" object:

```
int getColumn() const{
  return item.getColumn();
} //  read column-index of first element
```

20.7 Inserting a New Element

As discussed above, since the "row" class is derived from the "linkedList" class, it can use its public and protected members. In particular, it can use the functions that insert or drop items. Still, the "row" class has its own versions of the "insertNextItem", "insertFirstItem", and "append" functions. These versions are different from the original versions in Chapter 17 only in that they take two arguments to specify the value and the column index of the inserted element. In fact, the definitions of each new version calls the corresponding original version, using the prefix "linkedList::" before the function name to indicate that the original version in the base "linkedList" class is called:

```
void insertNextItem(const T&val, int col){
  rowElement<T> e(val,col);
```

First, the arguments "val" and "col" are used to invoke the row-element constructor to construct the new row element 'e'. This row element is then inserted as a second element in the current "row" object by using the "::" operator to invoke the original "insertNextItem" function of the base "linkedList" class:

```
linkedList<rowElement<T> >::insertNextItem(e);
} //  insert a rowElement as second item
```

The same approach is used to insert a new row element at the beginning of the current "row" object:

```
void insertFirstItem(const T&val, int col){
  rowElement<T> e(val,col);
  linkedList<rowElement<T> >::insertFirstItem(e);
} // insert a rowElement at the beginning
```

This approach is used to append a new row element at the end of the current "row" object:

```
void append(const T&val, int col){
  rowElement<T> e(val,col);
  linkedList<rowElement<T> >::append(e);
} // append a rowElement at the end of row
```

20.8 Recursive Functions

The recursive structure of the base "linkedList" class is particularly useful in the definitions of the member functions below, which are applied recursively to the "next" field that points to the tail of the row (the row elements that follow the first row element). However, the "next" field inherited from the base "linkedList" class is of type pointer-to-linkedList rather than pointer-to-row. Therefore, it must be converted explicitly to pointer-to-row before the recursive call can take place. This is done by inserting the prefix "(row*)" before it.

Usually, this conversion is considered risky because in theory "next" could point to a "linkedList" object or to any object derived from it, with a completely different version of the recursively called function. Fortunately, here "next" always points to a "row" object, so the conversion is safe.

Below we use recursion to implement the "operator[]" member function, which takes an integer argument 'i' to return a copy of the value of the element in column 'i', if exists in the current row, or 0, if it doesn't. First, the "column" field in the first element is examined. If it is equal to 'i', then the required element has been found, so its value is returned, as required. If, on the other hand, it is greater than 'i', then the current (well-ordered) row has no element with column index 'i' in it, so 0 is returned. Finally, if it is smaller than 'i', then the "operator[]" is applied recursively to the tail of the row.

As before, the "next" field must be converted explicitly from mere pointer-to-linkedList to pointer-to-row before recursion can be applied to it. This is done by inserting the prefix "(row*)" before it.

The "operator[]" function returns its output by value rather than by reference [as indeed indicated by the words "const T" (rather than "const T&") before its name], so that it can also return the local constant 0 whenever appropriate.

Furthermore, unlike "operator()" in Section 20.6, "operator[]" must always take exactly one argument, as is indeed the case in the implementation below:

```
const T operator[](int i) const{
  return (getColumn() == i) ?
           getValue()
         :
         next&&(getColumn() < i) ?
```

At this stage, it is assumed that the column index of the first element in the current row is different from 'i', so the required element has not yet been found. Now, if it is smaller than 'i', then the required element may still lie ahead in the row, so "operator[]" should be applied recursively to the tail, which lies in the "next" field inherited from the base "linkedList" class. However, since "operator[]" applies to "row" objects only, "next" must first be converted explicitly from mere pointer-to-linked-list to pointer-to-row:

```
(*(row*)next)[i]
```

If, on the other hand, the column index of the first element is greater than 'i', then the current (well-ordered) row contains no element with column index 'i', so 0 should be returned:

```
         :
         0.;
} // read the value at column i
```

20.9 Update Operators

Recursion is also used in the member "*=" operator that multiplies the current row by a scalar:

```
const row& operator*=(const T&t){
  item *= t;
  if(next) *(row*)next *= t;
  return *this;
} // multiply by a T
```

Indeed, the first element in the current row, "item", is first multiplied by the scalar argument 't' by the "*=" operator in Section 20.2. Then, recursion is used to multiply the tail of the current row (which lies in the field "next" inherited from the base "linkedList" class) by 't' as well. However, the present function applies to "row" objects only; this is why "next" must be converted

explicitly from a mere pointer-to-linked-list to pointer-to-row before the recursive call can be made.

The same approach is also used to divide the elements in the current row
by the nonzero scalar 't':

```
const row& operator/=(const T&t){
  item /= t;
  if(next) *(row*)next /= t;
  return *this;
} //  divide by a T
```

There is no need to implement here the "+=" operator to add a row to
the current row, because this operator is inherited properly from the base
"linkedList" class.

20.10 Member Binary Operators

Recursion is also useful in binary arithmetic operators, such as the '*' operator below that calculates the (real) inner product of the current (constant)
row and the dynamic-vector argument 'v' (Chapter 9, Section 9.14):

```
const T
operator*(const dynamicVector<T>&v) const{
```

Indeed, if the current row has a nontrivial tail, then recursion can be used to
calculate its inner product with 'v'. To this, one only needs to add the value
of the first element in the current row times the corresponding component in
'v':

```
    return
      next ?
  getValue() * v[getColumn()]
        + *(row*)next * v
```

If, on the other hand, the current row has a trivial tail, then no recursion is
needed, and the required result is just the value of the only element in the
current row times the corresponding component in 'v':

```
      :
  getValue() * v[getColumn()];
  } //  row times vector (real inner product)
};
```

This completes the block of the "row" class.

Note that functions that use recursion may call themselves many times. Therefore, one should be careful to avoid expensive operations in them, such as the construction of big objects like dynamic vectors.

20.11 Ordinary Binary Operators

Here we use the above update operators to define some ordinary binary arithmetic operators between rows and scalars:

```
template<class T>
const row<T>
operator*(const row<T>&r, const T&t){
  return row<T>(r) *= t;
} //  row times T
```

Indeed, as a member function, the "*=" operator can take the temporary "row" object "row<T>(r)" returned by the copy constructor inherited from the base "linkedList" class as its current object, multiply it by 't', and return the result by value, as required.

Furthermore, the following version of the binary '*' operator works in the same way, except it takes the arguments in the reversed order: first the scalar, and then the row.

```
template<class T>
const row<T>
operator*(const T&t, const row<T>&r){
  return row<T>(r) *= t;
} //  T times row
```

With these versions, the user can write either "r * t" or "t * r" to calculate the product of the well-defined row 'r' with the scalar 't'.

The same approach is now used to divide the row argument 'r' by the nonzero scalar 't':

```
template<class T>
const row<T>
operator/(const row<T>&r, const T&t){
  return row<T>(r) /= t;
} //  row divided by a T
```

20.12 The Sparse-Matrix Object

As discussed in Chapter 19 above, the most efficient way to implement a
sparse matrix is as a list of linked lists of elements, or as a list of "row" objects.
Indeed, this way, only the nonzero matrix elements are stored and take part
in the calculations, whereas the zero elements are ignored.

Although linked lists use indirect indexing, in which data are not stored
consecutively in the computer memory, which may lead to a longer computation time due to a slower data access, this overhead is far exceeded by the
saving gained by avoiding the unnecessary zero matrix elements. Furthermore,
in some cases, it is possible to map the linked list onto a more continuous data
structure and make the required calculations there.

The multilevel hierarchy of objects used to implement the sparse matrix
is displayed in Figure 20.2. The "sparseMatrix" object at the top level is
implemented as a list of "row" objects in the next lower level, which are in
turn implemented as linked lists of "rowElement" objects, which are in turn
implemented in a template class, using the unspecified class 'T' at the lowest
level to store the value of the matrix element.

The "sparseMatrix" class below is derived from a list of "row" objects (see
Figure 20.3). Thanks to this, it has access not only to public but also to
protected members of the base "list" class in Chapter 17.

```
template<class T>
class sparseMatrix : public list<row<T> >{
  public:
    sparseMatrix(int n=0){
      number = n;
      item = n ? new row<T>*[n] : 0;
      for(int i=0; i<n; i++)
        item[i] = 0;
    } // constructor
```

At the '{' symbol that marks the start of the block of this constructor, the
default constructor of the base "list" class is called implicitly. Unfortunately,
this default constructor assumes the default value "n= 0" in Chapter 17,
Section 17.2 to construct a trivial list with no items in it. This is why the
present constructor must assign its own integer argument 'n' (which may well
be nonzero) to "number" (the first field inherited from the base "list" class),
and also allocate memory for the array "item" (the second field inherited from
the base "list" class), to contain 'n' zero pointers that point to no row as yet.

The next constructor, on the other hand, also takes yet another scalar
argument, 'a', and uses it (in a loop) to call the constructor of the "row" class
and form a diagonal matrix with the value 'a' on its main diagonal:

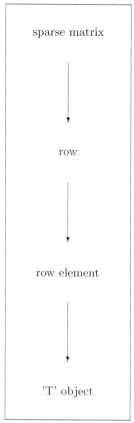

FIGURE 20.2: The multilevel hierarchy of objects used to implement a sparse matrix: the "sparseMatrix" object is a list of "row" objects, which are linked lists of "rowElement" objects, which use the template 'T' to store the value of the matrix elements.

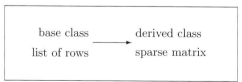

FIGURE 20.3: Inheritance from the base class "list<row>" to the derived class "sparseMatrix".

```
sparseMatrix(int n, const T&a){
  number = n;
  item = n ? new row<T>*[n] : 0;
  for(int i=0; i<n; i++)
    item[i] = new row<T>(a,i);
} // constructor of a diagonal matrix
```

The following constructor, on the other hand, takes yet another integer argument to serve as a column index at all the rows in the matrix. This way, a column matrix (a matrix that contains one nonzero column only) is formed:

```
sparseMatrix(int n, const T&a, int col){
  number = n;
  item = n ? new row<T>*[n] : 0;
  for(int i=0; i<n; i++)
    item[i] = new row<T>(a,col);
} // constructor of a column matrix
```

Fortunately, no copy constructor or assignment operator needs to be defined, because the corresponding functions inherited from the base "list" class work just fine.

The following destructor needs to do nothing, because the underlying "list" object is destroyed properly by the destructor of the base "list" class invoked implicitly at the '}' symbol that marks the end of its block:

```
~sparseMatrix(){
} // destructor
```

Finally, we also declare a constructor with a "mesh" argument, to be defined in detail later on in the book:

```
sparseMatrix(mesh<triangle>&);
```

In the three-dimensional applications at the end of the book, some more versions of constructors that take a "mesh" argument are defined. These versions should be declared here as well.

20.13 Reading a Matrix Element

The following function reads the "(i,j)"th matrix element in the current sparse matrix:

```
const T operator()(int i,int j) const{
  return (*item[i])[j];
} // (i,j)th element (read only)
```

Indeed, thanks to the fact that the array "item" inherited from the base "list" class is declared as "protected" (rather than the default "private" status) in Chapter 17, it can be accessed from the present function. Furthermore, "item[i]", the 'i'th entry in the array "item", points to the 'i'th row, "*item[i]". Once the "operator[]" in Section 20.8 is applied to this row, the required matrix element is read. Still, since this operator may well return the zero value if the 'j'th element in this row vanishes, the output must be returned by value rather than by reference, as indeed indicated by the words "const T" (rather than "const T&") at the beginning of the header.

20.14 Some More Member Functions

The following function returns the number of rows in the current sparse matrix:

```
int rowNumber() const{
  return number;
} // number of rows
```

Furthermore, we also declare a function that returns the number of columns [assuming that there are no zero columns at the end (far right) of the matrix]. The detailed definition of this function is left as an exercise, with a solution in the appendix.

```
int columnNumber() const;
```

Finally, assuming that the above function has already been defined properly, we define a function that returns the order of a square matrix:

```
int order() const{
  return max(rowNumber(), columnNumber());
} // order of square matrix
```

Finally, we also declare some more member and friend functions:

```
const sparseMatrix& operator+=(const sparseMatrix<T>&);
const sparseMatrix& operator-=(const sparseMatrix<T>&);
const sparseMatrix<T>& operator*=(const T&);
friend const sparseMatrix<T>
    operator*<T>(const sparseMatrix<T>&,
        const sparseMatrix<T>&);
friend const sparseMatrix<T>
    transpose<T>(const sparseMatrix<T>&);
const dynamicVector<int> colorNodes() const;
```

```
            const sparseMatrix<T> colorEdges() const;
            void colorEdges(sparseMatrix<T>&,sparseMatrix<T>&) const;
    };
```

These functions are only declared here; their detailed definitions are provided later on in this chapter and in the appendix. This completes the block of the "sparseMatrix" class.

In particular, the function "transpose()" that returns the transpose of a sparse matrix is defined in detail in the appendix. In what follows, however, we assume that it has already been defined and can be called from other functions.

Furthermore, the matrix-times-vector multiplication implemented in the appendix uses the '*' operator in Section 20.10 to calculate the real inner product (defined in Chapter 9, Section 9.14) between each row and the vector.

Finally, to implement matrix-times-matrix multiplication, the algorithm described in Chapter 9, Section 9.10, is not very useful, because it uses mainly column operations. A more suitable algorithm is the following one, which uses row operations only.

Let A be a matrix with N rows, and let B be a matrix with N columns. Let $B^{(i)}$ be the ith row in B. Then the ith row in BA can be written as

$$(BA)^{(i)} = B^{(i)}A.$$

Thus, each row in BA is the linear combination of rows in A with coefficients from $B^{(i)}$. This linear combination can be calculated using row operations only: the '*' operator in Section 20.11 to multiply a row by a scalar, and the "+=" operator inherited from the base "linkedList" class to add rows: The detailed implementation is left as an exercise, with a solution in the appendix.

20.15 The Node-Coloring Code

The best way to color the nodes in a graph is by implementing the algorithm in Chapter 19, Section 19.2, using the formulation of the graph as a (sparse) matrix of the form

$$A \equiv (a_{i,j})_{0 \le i,j < |N|}$$

(where N is the set of the nodes in the graph, and $|N|$ is its cardinality, or the total number of nodes). This is indeed done below, using the sparse-matrix object defined above.

In order to implement this algorithm, we first need to define an ordinary function that takes a vector as an input argument and returns the index of its first zero component. In other words, if the input vector is denoted by v, then the function returns the nonnegative integer number i for which

$$v_0 \neq 0, \; v_1 \neq 0, \; v_2 \neq 0, \; \ldots, \; v_{i-1} \neq 0, \; v_i = 0.$$

```
int firstZero(const dynamicVector<int>&v){
   int i=0;
   while((i<v.dim())&&v[i])i++;
   return i;
} //  first zero component
```

This function is useful in the node-coloring code below. Indeed, suppose that we have a vector named "colors" whose dimension is the same as the number of nodes in the graph, or the number of rows in the matrix associated with the graph. Our aim is to store the color in which the ith node is colored in the ith component in "colors". For this, we use an outer loop on the rows in the matrix, using the index $n = 0, 1, 2, \ldots, |N|$. For each particular n, we have to color the nth node by assigning the color number to the corresponding component in "colors", that is, to "colors[n]". To find a suitable color, we must scan the nonzero matrix elements of the form $a_{n,j}$ and $a_{j,n}$ $(0 \leq j < n)$ to exclude the colors assigned to node j from the set of colors that can be used to color node n:

```
template<class T>
const dynamicVector<int>
sparseMatrix<T>::colorNodes() const{
   int colorsNumber = 0;
   dynamicVector<int> colors(rowNumber(),-1);
   sparseMatrix<T> At = transpose(*this);
   for(int n=0; n<rowNumber(); n++){
      dynamicVector<int> usedColors(colorsNumber,0);
```

The vector "usedColors" contains the numbers of the colors that are candidates to color the node n. Initially, all the components in this vector are set to zero, which means that all the colors are good candidates to color the node n. Then, an inner loop is used to eliminate the colors that cannot be used because they have already been used to color neighbor nodes, that is, nodes j for which either $a_{n,j} \neq 0$ or $a_{j,n} \neq 0$:

```
for(const row<T>* j = item[n];
         j&&(j->getColumn()<n);
         j = (const row<T>*)j->readNext())
      usedColors(colors[j->getColumn()]) = 1;
```

The nature of the pointer 'j' used to scan the nth row will be explained later. Anyway, the above loop excludes the colors of nodes j in this row from the set of colors that can be used to color node n. Next, a similar loop is used to exclude colors that have been used to color nodes in the nth column of the current matrix, or the nth row in its transpose, 'At':

```
for(const row<T>* j = At.item[n];
        j&&(j->getColumn()<n);
        j = (const row<T>*)j->readNext())
    usedColors(colors[j->getColumn()]) = 1;
```

In both of the above loops, 'j' is a pointer-to-row that points to the nonzero elements in the nth row in A (or A^t). Furthermore, "j− >getColumn()" is the index of the column in which this element is placed in the matrix. The color that has been used to color the node corresponding to this index cannot be used to color the node n, hence must be eliminated from the list of available colors, which means that the component of "usedColors" associated with it must be set to 1.

Once these inner loops are complete, all that is left to do is to find an available color, say the first component in "usedColors" whose value is still zero, which means that it has never been used to color any neighbor of the node n:

```
int availableColor = firstZero(usedColors);
```

This color can now be used to color the node n. Only when "usedColors" contains no zero component, which means that no color is available, must a new color be introduced, and the total number of colors increases by 1:

```
        colors(n) = availableColor<usedColors.dim() ?
            availableColor : colorsNumber++;
    }
    return colors;
}  //  color nodes
```

Finally, the "colorNodes" function returns the $|N|$-dimensional vector whose components are the colors of the nodes in N.

20.16 Edge Coloring in a Nonoriented Graph

A somewhat more difficult task is to color the edges in a graph. We start with the easier case, in which the graph is nonoriented, so it can be formulated as a symmetric matrix. This formulation proves to be most helpful in implementing the edge-coloring algorithm in Chapter 19, Section 19.8.

```
template<class T>
const sparseMatrix<T>
sparseMatrix<T>::colorEdges() const{
    int colorsNumber = 0;
    sparseMatrix<T> c(rowNumber());
```

The sparse matrix 'c' produced and returned by the "colorEdges" function contains the numbers $c_{i,j}$, which are the colors assigned to the edges $a_{i,j}$. For this purpose, we use a triple nested loop: the outer loop scans the nodes in the graph, or the rows in the original matrix A. For this purpose, we use the index n, starting at the minimal n for which the $n \times n$ upper-left submatrix A_n doesn't vanish.

```
int n=0;
while((n<rowNumber())&&(n<item[n]->getColumn()))n++;
for(; n<rowNumber(); n++){
  dynamicVector<int> usedColorsN(colorsNumber,0);
```

For each such n, we use an inner loop to color the nonzero elements of the form $a_{n,j}$ $(0 \le j \le n)$.

```
for(const row<T>* j = item[n];
        j&&(j->getColumn()<=n);
        j = (const row<T>*)j->readNext()){
    dynamicVector<int> usedColorsJ(usedColorsN.dim(),0);
```

For each particular j, we use a yet inner loop to eliminate the colors that have been used to color any element in the jth row in A_{n-1}, and not use it to color $a_{n,j}$. These colors are eliminated from the set of available colors by setting the corresponding component in the vector "usedColorsJ" to 1.

```
if(j->getColumn()<n)
  for(const row<int>* k = c.item[j->getColumn()];
      k&&(k->getColumn()<n);
      k = (const row<int>*)k->readNext())
    usedColorsJ(k->getValue()) = 1;
```

Furthermore, the colors that have been used previously to color any of the elements $a_{n,0}, a_{n,1}, \ldots, a_{n,j-1}$ must also be excluded from the candidates to color $a_{n,j}$; this will be done below by setting the corresponding components of the vector "usedColorsN" to 1.

Thus, the best candidate to color $a_{n,j}$ is the one whose number is the index of the first component that vanishes in both "usedColorsJ" and "usedColorsN". Indeed, this color has never been used to color any element in either the jth or the nth rows of A_n, hence is a good candidate to color $a_{n,j}$:

```
int availableColor=firstZero(usedColorsN+usedColorsJ);
```

This number is denoted by "availableColor" in the above code. If, however, no such color is available, a new color must be introduced, and the total number of colors must increase by 1:

```
int colorNJ =
    availableColor<usedColorsN.dim() ?
```

```
availableColor : colorsNumber++;
if(colorNJ<usedColorsN.dim())usedColorsN(colorNJ) = 1;
```

The number of the color by which $a_{n,j}$ is colored, may it be new or old, is now placed in the variable "colorNJ". This number is also placed now in the (n, j)th element in the matrix c produced in the function. For this, however, we must first check whether the nth row in the matrix c has already been constructed. If it has, then the new element $c_{n,j}$ is added at the end of it by the "append" function; otherwise, the nth row in the matrix c must be constructed using the constructor of the "row" class and the "new" command:

```
if(c.item[n])
  c.item[n]->append(colorNJ,j->getColumn());
else
  c.item[n] = new row<T>(colorNJ,j->getColumn());
```

Finally, if $j < n$, then $c_{j,n}$ is also defined symmetrically by

$$c_{j,n} \equiv c_{n,j} :$$

```
if(j->getColumn()<n){
  if(c.item[j->getColumn()])
    c.item[j->getColumn()]->append(colorNJ,n);
  else
    c.item[j->getColumn()] = new row<T>(colorNJ,n);
  }
 }
}
return c;
} // color edges in a nonoriented graph
```

This way, the symmetric sparse matrix c returned by the function contains the colors $c_{i,j}$ used to color the edges $a_{i,j}$ of a nonoriented graph, as in the algorithm in Chapter 19, Section 19.8. Below we also implement the algorithm to color the edges in a general oriented graph.

20.17 Edge Coloring in an Oriented Graph

In the above, we have implemented the algorithm to color the edges in a nonoriented graph, using its formulation as a symmetric matrix and the edge-coloring algorithm in Chapter 19, Section 19.8 above. Here we turn to the slightly more complicated task of coloring the edges in a general oriented graph, using its formulation as a sparse matrix and the algorithm in Chapter 19, Section 19.3.

Because the matrix of an oriented graph is in general nonsymmetric, the next version of the "colorEdge" function is different from the previous version in that it produces not only the matrix c that contains the colors $c_{i,j}$ used to color the edges $a_{i,j}$ in the original graph but also its transpose c^t. This transpose matrix is used throughout the function to scan the rows in the sparse matrix c^t, which are actually the columns in the matrix c.

The present version of the function "colorEdges" requires two arguments: the sparse matrices c and c^t. It is assumed that these matrices have the same number of rows as A, the original matrix of the graph (which is assumed to be the current sparse-matrix object), and that these rows have not yet been constructed, so they have initially zero addresses. Thanks to the fact that both c and c^t are passed to the function as nonconstant references to sparse matrices, they can be changed throughout the function to take their final values as the required matrix of colors $c = (c_{i,j})$ and its transpose $c^t = (c_{j,i})$.

```
template<class T>
void
sparseMatrix<T>::
colorEdges(sparseMatrix<T>&c, sparseMatrix<T>&ct) const{
   int colorsNumber = 0;
   const sparseMatrix<T> At = transpose(*this);
```

The transpose of A, A^t, is constructed using the function "transpose()" to allow one to scan not only the rows of A but also the rows of A^t, or the columns of A, as indeed required in the algorithm in Chapter 19, Section 19.3.

Next, a triple nested loop is carried out in much the same way as in the nonoriented case above:

```
int n=0;
while((n<rowNumber())&&(n<item[n]->getColumn())
       && (n<At.item[n]->getColumn()))n++;
for(; n<rowNumber(); n++){
  dynamicVector<int> usedColorsN(colorsNumber,0);
  for(const row<T>* j = item[n];
          j&&(j->getColumn()<=n);
          j = (const row<T>*)j->readNext()){
    dynamicVector<int> usedColorsJ(usedColorsN.dim(),0);
    if(j->getColumn()<n){
      for(const row<int>* k = c.item[j->getColumn()];
          k&&(k->getColumn()<n);
          k = (const row<int>*)k->readNext())
      usedColorsJ(k->getValue()) = 1;
```

Unlike in the nonoriented case, however, here c is in general nonsymmetric; this is why one must also scan the jth row in c^t (or the jth column in c) to eliminate the colors used previously in it from the set of candidates to color the current edge $a_{n,j}$:

```
for(const row<int>* k = ct.item[j->getColumn()];
    k&&(k->getColumn()<n);
    k = (const row<int>*)k->readNext())
  usedColorsJ(k->getValue()) = 1;
}
int availableColor=firstZero(usedColorsN+usedColorsJ);
int colorNJ =
    availableColor<usedColorsN.dim() ?
    availableColor : colorsNumber++;
if(colorNJ<usedColorsN.dim())usedColorsN(colorNJ) = 1;
if(c.item[n])
  c.item[n]->append(colorNJ,j->getColumn());
else
  c.item[n] = new row<T>(colorNJ,j->getColumn());
```

Furthermore, once the best candidate to color $a_{n,j}$ has been found and placed in the variable "colorNJ", it is placed not only in $c_{n,j}$ but also in $c^t_{j,n}$:

```
if(ct.item[j->getColumn()])
  ct.item[j->getColumn()]->append(colorNJ,n);
else
  ct.item[j->getColumn()] = new row<T>(colorNJ,n);
}
```

Moreover, a similar inner loop is carried out once again to color the elements $a_{j,n}$ ($0 \le j < n$) in the nth column of A (or in the nth row of A^t), as in Step 2 in the algorithm in Chapter 19, Section 19.3:

```
for(const row<T>* j = At.item[n];
    j&&(j->getColumn()<n);
    j = (const row<T>*)j->readNext()){
  dynamicVector<int> usedColorsJ(usedColorsN.dim(),0);
  for(const row<int>* k = ct.item[j->getColumn()];
      k&&(k->getColumn()<n);
      k = (const row<int>*)k->readNext())
    usedColorsJ(k->getValue()) = 1;
```

Again, one must exclude from the list of candidates to color $a_{j,n}$ not only the colors that have been used previously to color the jth row in c^t (or the jth column in c) but also those that have been used to color the jth row in c:

```
for(const row<int>* k = c.item[j->getColumn()];
    k&&(k->getColumn()<n);
    k = (const row<int>*)k->readNext())
  usedColorsJ(k->getValue()) = 1;
```

Clearly, one must also exclude colors that have been used previously to color any element in the nth row or the nth column of A; this will be done further below.

The best candidate to color $a_{j,n}$ is, thus, the one that still has the value 0 in both "usedColorsJ" (the vector that indicates what colors have been used in the jth row and jth column in c) and "usedColorsN" (the vector that indicates what colors have been used in the nth row and nth column in c):

```
int availableColor=firstZero(usedColorsN+usedColorsJ);
```

The number of this color is now placed in the variable "availableColor". Next, it is also placed in the variable "colorJN". If, however, all the colors have been excluded from the set of candidates to color $a_{j,n}$, then a new color must be introduced, and the total number of colors must increase by 1. In this case, "colorJN" takes the number of this new color:

```
int colorJN =
    availableColor<usedColorsN.dim() ?
    availableColor : colorsNumber++;
if(colorJN<usedColorsN.dim())usedColorsN(colorJN) = 1;
```

The desired color number in "colorJN" is now placed in its proper place in $c^t_{n,j}$ and $c_{j,n}$:

```
if(ct.item[n])
  ct.item[n]->append(colorJN,j->getColumn());
else
  ct.item[n] = new row<T>(colorJN,j->getColumn());
if(c.item[j->getColumn()])
  c.item[j->getColumn()]->append(colorJN,n);
else
  c.item[j->getColumn()] = new row<T>(colorJN,n);
}
```

Note, however, that the element $c^t_{n,n}$ has been constructed in the first inner loop, before the elements of the form $c^t_{n,j}$ $(j < n)$ have been constructed in the second inner loop. As a result, the elements in the nth row in c^t may be not in the required increasing-column order. To fix this, we apply the "order()" function inherited from the base "linkedList" class:

```
ct.item[n]->order(ct.item[n]->length());
  }
} // color edges
```

This completes the implementation of the algorithm in Chapter 19, Section 19.3, to color the edges in a general oriented graph.

The user can now call either version of the "colorEdges" functions defined above. If the call is made with no arguments, then the compiler would invoke the first version to color a nonoriented graph. If, on the other hand, two arguments of type "sparseMatrix" are used, then the compiler would invoke the second version to color a general oriented graph. These arguments would then change throughout the call, and would eventually contain the required matrix of colors c and its transpose c^t.

Here is how the above functions can be used to color the nodes and the edges in a graph represented by a 5×5 matrix 'B', with nonzero elements on its diagonal and on its second and fifth columns only:

```
int main(){
    sparseMatrix<int> B(5,1);
    sparseMatrix<int> Col1(5,1,1);
    sparseMatrix<int> Col4(5,1,4);
    sparseMatrix<int> B += Col1 + Col4;
    print(B.colorNodes());
    sparseMatrix<int> C(B.rowNumber());
    sparseMatrix<int> Ct(B.rowNumber());
    B.colorEdges(C,Ct);
    print(C);
    print(Ct);
    return 0;
}
```

20.18 Exercises

1. Implement arithmetic operations with sparse matrices, such as addition, subtraction, multiplication by scalar, matrix times dynamic vector, and matrix times matrix. The solution can be found in Section 28.14 in the appendix.

2. Implement the "columnNumber" member function that returns the number of columns (the maximum column index in the elements in the rows). The solution can be found in Section 28.14 in the appendix.

3. Write the "transpose" function that takes a sparse matrix A as a constant argument and returns its transpose A^t. The solution can be found in Section 28.14 in the appendix.

4. Does this function have to be a friend of the "sparseMatrix" class? Why?

5. Use the sparse-matrix object to implement the alternative coloring algorithms in the last exercises at the end of Chapters 11 and 19.

6. Test which edge-coloring algorithm is more efficient: the original one or the alternative one.

Chapter 21

Meshes

In the implementation of a graph as a sparse matrix, the node is the key object. In fact, the nodes in the graph are indexed by the index $i = 1, 2, 3, \ldots, |N|$, where N is the set of nodes and $|N|$ is its cardinality (the total number of nodes). The index i of the ith node serves then as its virtual address in the list of nodes, which allows one to access the information about its role in the graph. Indeed, this information can be found in the ith row in the matrix of the graph, in which the matrix elements indicate from which nodes edges are issued towards the ith node. Furthermore, the ith row in the transpose of the matrix of the graph indicates towards which nodes edges are issued from the ith node.

Thus, in the above implementation, a node is implemented only virtually as a row index. All the information about the ith node, namely, the edges issued from or directed to it, is stored in the ith row as a linked list of indices (or virtual addresses) of nodes that are connected to it in the graph. This is only natural: after all, a graph is a purely abstract mathematical concept, so no wonder it is implemented in terms of virtual indices rather than any physical or geometrical terms.

This is no longer the case with more concrete graphs such as meshes and triangulations, which have a more concrete geometrical interpretation. Indeed, in a mesh or a triangulation, a node has a concrete geometrical interpretation as a point in the Cartesian plane. Furthermore, the axioms by which the mesh or the triangulation is defined (Chapter 11, Section 11.8 and the last exercise) imply that not the node but rather the cell (or the triangle) is the basic object with which the mesh (or the triangulation) is built. Indeed, these axioms imply that nodes serve not as basic objects but rather as mere vertices in cells.

Thus, the cells, whose existence is guaranteed by the axioms, are the basic brick with which the mesh is built. Indeed, the cells are the objects that are colored in the coloring problem and algorithm. Therefore, the cells are the ones that should be indexed; the nodes don't have to be indexed any more, unless the node-coloring problem should be solved as well. For the node-coloring problem, however, the mesh object is useless; the matrix of the graph must be constructed, and the node-coloring algorithm must be applied to it as in Chapter 20, Section 20.15.

In summary, the mesh is produced by a multilevel hierarchy of mathematical objects. The node object at the lowest level carries the geometrical information about its location in the Cartesian plane. This node can be used as a vertex in

many cells, so it must be pointed at from each such cell. Thus, the cell object
is implemented as a list of pointers-to-nodes. In particular, the triangle in a
triangulation is implemented as a list of three pointers-to-nodes. Finally, the
mesh object is implemented as a linked list of cells, to allow a high degree of
flexibility in introducing new cells and dropping old ones. The implementation
of the mesh as a linked list of cells is also most suitable to implement the cell-
coloring algorithm in a short and transparent code.

21.1 The Node Object

Here we introduce the node object, the most elementary object in the multilevel hierarchy used in our object-oriented framework. Later on, we'll use the
"node" object to define more complicated objects, such as cells and meshes.

In a mesh, a node may be shared by two or more cells. This is why the node
object must contain information not only about its geometric location but
also about the number of cells that share it. For this, the node object must
contain three data fields: the first to specify its geometric location, the second
to specify its index in the list of nodes in the entire mesh, and the third to
specify the number of cells that share it as their joint vertex.

```
template<class T> class node{
   T location;
   int index;
   int sharingCells;
```

The type of the first data field, named "location", is 'T', to be specified later
(upon the construction of a concrete node in compilation time) as a point in
the Cartesian plane or the three-dimensional Cartesian space.

The two remaining data fields, the index of the node and the number of
cells that share it, can be set only when the entire mesh is ready. When an
individual node is originally constructed, these fields are initialized with the
trivial values -1 and 0, to indicate that this is indeed an isolated new node
that hasn't yet been placed in a mesh or even in a cell:

```
public:
   node(const T&loc=0., int ind=-1, int sharing=0)
   : location(loc),index(ind),sharingCells(sharing){
   } // constructor
```

With this constructor, the user can write commands like "node<point> n" to
construct the isolated node 'n' at the origin $(0,0)$. Indeed, upon encountering
such a command, the compiler invokes the first constructor in Chapter 16,

Section 16.4 to initialize the dummy "point" argument "loc" with the value $(0, 0)$.

An initialization list is also used in the copy constructor:

```
node(const node&n):location(n.location),index(n.index),
  sharingCells(n.sharingCells){
} //  copy constructor
```

As before, the data fields in the constructed "node" object are initialized in the initialization list in the order in which they are listed in the class block. In particular, the field "index" in the constructed node is initialized with the value "n.index", and the field "sharingCells" in the constructed node is initialized with the value "n.sharingCells". This seems unnecessary, since the constructed node belongs to no cell as yet. Therefore, it may make more sense to initialize these fields with the trivial values -1 and 0 (respectively), as in the previous constructor. This way, a node argument that is passed by value to a function would be copied into a dangling local (dummy) node that belongs to no cell, which seems to make more sense. This version, however, is left to the reader to check. Fortunately, in the present applications node arguments are passed by reference only, so it makes little difference what version of the copy constructor is actually used.

The assignment operator is only declared here; the detailed definition will be given later on.

```
const node& operator=(const node&);
```

Since there are no pointer fields, the block of the destructor remains empty. Indeed, the data fields are destroyed implicitly automatically one by one (in an order reversed to the order in which they are listed in the class block) at the '}' symbol that marks the end of the following block:

```
~node(){
} //  destructor
```

21.2 Reading and Accessing Data Fields

The following member function reads the location of the current "node" object:

```
const T& operator()() const{
  return location;
} //  read the location
```

With this operator, the user can write "n()" to read the location of the well-defined "node" object 'n'.

The following member function reads the "index" field in the current "node" object:

```
int getIndex() const{
  return index;
} //  read index
```

The following member function sets the "index" field of the current "node" object to have the value 'i':

```
void setIndex(int i){
  index = i;
} //  set index
```

So far, we have dealt with the first data field in the "node" object, "location", and with the second field, "index". Next, we consider the third data field, "sharingCells", which will be most useful when the individual node is placed in a mesh. Indeed, the value of "sharingCells" increases by 1 whenever a new cell that shares the node is created, and decreases by 1 whenever such a cell is destroyed. Therefore, we have to define public member functions to increase, decrease, and read the value of the "sharingCells" data field:

```
int getSharingCells() const{
  return sharingCells;
} //  read number of cells that share this node
```

This function reads the private "sharingCells" fields in the current "node" object. The next member function increases the value of this field by one:

```
void moreSharingCells(){
  sharingCells++;
} //  increase number of cells that share this node
```

The next member function, on the other hand, decreases the "sharingCells" field by one. Furthermore, it also returns 0 so long as the node still serves as a vertex in at least one cell, and 1 once it becomes an isolated node shared by no cell:

```
int lessSharingCells(){
  return
    sharingCells ?
      !(--sharingCells)
    :
      1;
} //  decrease number of cells that share this node
```

Finally, the following member function checks whether or not the current node is indeed an isolated node shared by no cell. Indeed, like the previous function, it returns a nonzero value if and only if "sharingCells" vanishes:

```
int noSharingCell() const{
  return !sharingCells;
} //  an isolated node
};
```

This completes the block of the "node" class.

To complete the implementation, here is also the detailed definition of the assignment operator declared in the block of the "node" class:

```
template<class T>
const node<T>&
node<T>::operator=(const node<T>&n){
  if(this != &n){
    location = n.location;
    index = n.index;
    sharingCells = n.sharingCells;
  }
  return *this;
} //  assignment operator
```

Finally, the following ordinary function prints the data fields in its "node" argument onto the screen:

```
template<class T>
void print(const node<T>&n){
  print(n());
  printf("index=%d; %d sharing cells\n",
      n.getIndex(),n.getSharingCells());
} //  print a node
```

21.3 The Cell – a Highly Abstract Object

As we have seen in Chapter 11 above, a graph is defined in terms of nodes and edges only. These objects are absolutely abstract: they have no geometrical interpretation whatsoever.

The first time that a geometrical sense is introduced is in Section 11.8 in Chapter 11, in which the triangulation is defined. Indeed, the nodes in the triangulation can be viewed as points in the Cartesian plane, and the edges can be viewed as line segments that connect them to each other. Still, the triangles in the triangulation are never defined explicitly; their existence only follows implicitly from the axioms in Chapter 11, Section 11.8.

The concept of triangulation is extended in the last exercise in Chapter 11 into a more general mesh of cells. When the cells are triangles, we have a triangulation. When the cells are squares, we have a mesh of squares, and so on. Again, the cells are never defined explicitly or used to define the graph: they only exist in our imagination.

Thus, the cells are even more abstract than the nodes: they are never used or needed to construct the graph. One may even say that they don't really exist.

Why then are the cells still important and useful? Because they can serve as abstract containers to contain the more concrete objects, the nodes. Indeed, the vertices of a particular cell are listed one by one in it, to indicate that they indeed belong to it and indeed connected by edges in the original graph. More precisely, the cell contains only references to its vertices. This way, a node can be referred to from every cell that uses it as a vertex.

This approach indeed agrees with the high level of abstraction of the cell object. Indeed, the cell object is highly abstract not only in the mathematical sense, which means that it is only used in the axioms required in a triangulation (or a mesh of cells) rather than in the original definition of the concrete graph, but also in the practical sense, which means that it contains no concrete physical nodes but merely references to existing nodes. We can thus see clearly how the implementation is indeed in the spirit of the original mathematical formulation.

Because the cell object is so abstract, it must also be implemented in a rather nonstandard way. Indeed, as discussed above, objects that contain pointer fields must use the "new" command in their constructors to allocate sufficient memory for the objects pointed at by them, and the "delete" command in their destructors to release this memory. This is why one is advised to write his/her own constructors and destructors, and not rely on the default constructor and destructor available in the C++ compiler, which never allocate memory for, initialize, or release the memory occupied by the variables pointed at by pointer fields.

In the constructors of the cell object, on the other hand, no new memory needs to be allocated for any new node. In fact, only a new reference to an existing node is defined, rather than a new physical node. This nonstandard nature of the cell object is discussed further below.

21.4 The Cell Object

As discussed above, the "cell" object contains no concrete "node" objects but rather pointers to nodes only. This way, two cells can share a node as their joint vertex by containing a pointer to it. This way, each cell can access the node through its own pointer field.

More precisely, the "cell" object contains 'N' pointers to nodes, and each node has a "location" field of type 'T', where both 'N' and 'T' are template parameters, to be specified upon construction later on in compilation time. In a triangulation, for example, 'T' is specified to be the "point" class, and 'N' is specified to be 3, the number of vertices in each triangle.

```
template<class T, int N> class cell{
    node<T>* vertex[N];
    int index;
```

The first data field, "vertex", is an array of 'N' pointers to nodes to point to the 'N' vertices of the cell. The second field, "index", will be used to store the index of the cell in the list of cells in the entire mesh.

In the following default constructor, "index" is set to -1, to indicate that the constructed cell has not been placed in any mesh as yet. Furthermore, the vertices are all set to lie at the origin, using the "new" command and the constructor of the "node" class in Section 21.1:

```
public:
  cell():index(-1){
    for(int i=0; i<N; i++)
      vertex[i] = new node<T>(0.,-1,1);
  } // default constructor
```

Clearly, this is a trivial meaningless cell. Next, we declare a more meaningful constructor that takes three "node" arguments to serve as its vertices. This constructor will be defined in detail later on.

```
cell(node<T>&,node<T>&,node<T>&);
```

Next, we also declare a constructor that takes four "node" arguments to serve as its vertices. This constructor will be particularly useful in the three-dimensional applications at the end of the book to construct a tetrahedron with four vertices.

```
cell(node<T>&,node<T>&,node<T>&,node<T>&);
```

This constructor will also be defined later on. Furthermore, the copy constructor, assignment operator, and destructor declared next will all be defined explicitly later on.

```
cell(cell<T,N>&);

const cell<T,N>&
    operator=(cell<T,N>&);

~cell();
```

21.5 Reading and Accessing Vertices

The following member operator takes the integer argument 'i' and returns a nonconstant reference to the 'i'th vertex in the cell:

```
node<T>& operator()(int i){
   return *(vertex[i]);
} //  read/write ith vertex
```

Indeed, "vertex[i]" (the 'i'th entry in the data field "vertex") is a pointer to the 'i'th vertex in the cell. Its content, "*vertex[i]", is therefore the 'i'th vertex itself. Thanks to the word "node<T>&" at the beginning of the header, this vertex is indeed returned by nonconstant reference, so it can be accessed (and indeed changed if necessary) in the same code line in which the operator is called. For example, with the above operator, the user can write "c(i) = n" to assign the node 'n' to the 'i'th vertex in the cell 'c' as well.

The next, operator, on the other hand, returns a constant reference to the 'i'th vertex in the current cell (as is indeed indicated by the words "const node<T>&" at the beginning of the header), which can be used for reading but not for changing. Indeed, the reserved word "const" just before the function block guarantees that the current "cell" object can never change in it:

```
const node<T>&
operator[](int i)const{
   return *(vertex[i]);
} //  read only ith vertex
```

The following member function sets the "index" field in each individual vertex in the cell to its default value −1: Since this field is private in the "node" class, this must be done by calling the public "setIndex" function of this class:

```
void resetIndices(){
   for(int i=0; i<N; i++)
     vertex[i]->setIndex(-1);
} //  reset indices to -1
```

The negative indices -1 assigned to the individual vertices in the cell indicates that these vertices have not yet been indexed properly in the list of nodes in the entire mesh. The following function, on the other hand, gives meaningful indices to every vertex that has not yet been indexed:

```
void indexing(int&count){
```

Indeed, its integer argument "count" stands for the number of nodes that have been indexed so far in the entire mesh. The following loop scans the vertices in the current cell one by one, indexes (by a continuously increasing index) each vertex that has not yet been indexed, and increments "count" by one:

```
for(int i=0; i<N; i++)
  if(vertex[i]->getIndex()<0)
    vertex[i]->setIndex(count++);
} //  indexing the unindexed vertices
```

Fortunately, "count" has been passed to the function by (nonconstant) reference, so it indeed changes throughout the function to store the up-to-date number of nodes that have been indexed so far in the entire mesh. This new value can then be used to reapply the function to the next cell in the mesh to index the vertices in it that have not been indexed as yet.

So far, we have considered the indices of the individual vertices in the cell. Next, we consider the index of the current cell itself (in the list of cells in the entire mesh), stored in the integer data field "index" in the "cell" class. The following member function sets the value of this field to the integer argument 'i':

```
void setIndex(int i){
  index = i;
} //  set the index of the cell to i
```

Furthermore, the following function reads this field:

```
int getIndex() const{
  return index;
} //  read the index of the cell
};
```

This completes the block of the "cell" class. The member functions that are only declared in it are defined explicitly next.

21.6 Constructors

Here is the constructor that takes three "node" arguments to construct a cell with three vertices, or a triangle:

```
template<class T, int N>
cell<T,N>::cell(
    node<T>&a, node<T>&b, node<T>&c):index(-1){
```

The "node" arguments are now considered one by one. If the first "node" argument, 'a', is not used as a vertex in any other cell, then its "sharing-Cells" field must vanish, so its member "noSharingCell" function must return a nonzero value. In this case, the "new" command and the copy constructor of the "node" class are invoked to allocate memory for the first vertex in the constructed cell and copy 'a' into it. If, on the other hand, 'a' already serves as a vertex in some other cell, then it doesn't have to be reconstructed; instead, its address is placed in the first vertex in the constructed cell to indicate that it serves as a vertex in this cell as well:

```
vertex[0] = a.noSharingCell() ? new node<T>(a) : &a;
```

The same approach is now used for the remaining "node" arguments 'b' and 'c':

```
vertex[1] = b.noSharingCell() ? new node<T>(b) : &b;
vertex[2] = c.noSharingCell() ? new node<T>(c) : &c;
```

Finally, the "sharingCells" field in each vertex in the newly constructed cell is incremented by 1 to indicate that it serves as a vertex in one more cell:

```
    for(int i=0; i<N; i++)
      vertex[i]->moreSharingCells();
} // constructor with 3 node arguments
```

This is also why the "node" arguments passed to this function must be non-constant. After all, they may change in the above command line.

The same approach is also used in the next constructor, which takes four "node" arguments to construct a cell with four vertices, such as a tetrahedron in a three-dimensional mesh:

```
template<class T, int N>
cell<T,N>::cell(node<T>&a, node<T>&b,
        node<T>&c, node<T>&d){
  vertex[0] = a.noSharingElement() ? new node<T>(a) : &a;
  vertex[1] = b.noSharingElement() ? new node<T>(b) : &b;
  vertex[2] = c.noSharingElement() ? new node<T>(c) : &c;
  vertex[3] = d.noSharingElement() ? new node<T>(d) : &d;
  for(int i=0; i<N; i++)
    vertex[i]->moreSharingElements();
} // constructor with 4 node arguments
```

The copy constructor below takes a well-defined, nontrivial "cell" argument 'e'. First, the address of each vertex of 'e' is also placed in the corresponding pointer-to-node in the constructed cell, to indicate that this node serves as its vertex as well:

```
template<class T, int N>
cell<T,N>::cell(cell<T,N>&e):index(e.index){
  for(int i=0; i<N; i++){
    vertex[i] = e.vertex[i];
```

Then, the "sharingCells" field in each vertex is incremented by 1 to indicate that this node is now shared by one more cell:

```
    vertex[i]->moreSharingCells();
  }
} // copy constructor
```

This is also why the "cell" argument passed to this function must be non-constant. After all, it changes in the above command line.

21.7 The Assignment Operator

The assignment operator is defined as follows:

```
template<class T, int N>
const cell<T,N>&
cell<T,N>::operator=(cell<T,N>&e){
  if(this != &e){
    index = e.index;
```

First, the current cell is "removed" by scanning its vertices and "removing" them one by one by invoking the "lessSharingCells" to decrease by one the number of cells that share them. Only if the "lessSharingCells" returns a nonzero value (which indicates that the node under consideration is shared by no cell) is it removed physically by the destructor of the "node" class, invoked implicitly when the "delete" command is applied to its address:

```
    for(int i=0; i<N; i++)
      if(vertex[i]->lessSharingCells())
        delete vertex[i];
```

Once the current cell has been removed, it is reconstructed again as in the copy constructor above:

```
      for(int i=0; i<N; i++){
        vertex[i] = e.vertex[i];
        vertex[i]->moreSharingCells();
```

```
      }
    }
    return *this;
} // assignment operator
```

The destructor also uses the same loop as in the first part of the above
assignment operator:

```
template<class T, int N>
cell<T,N>::~cell(){
  for(int i=0; i<N; i++)
    if(vertex[i]->lessSharingCells())
      delete vertex[i];
} //   destructor
```

21.8 Nodes in a Cell

The ordinary "operator<" function defined below takes a "node" argument
'n' and a "cell" argument 'e', and checks whether 'n' indeed serves as a vertex
in 'e'. If it does, then the function returns the index on 'n' in the array "vertex"
in 'e' plus 1. If, on the other hand, it doesn't, then it returns 0.

```
template<class T, int N>
int
operator<(const node<T>&n, const cell<T,N>&e){
```

In the following loop, the vertices in 'e' are scanned:

```
for(int i=0; i<N; i++)
```

By using the "operator[]" in Section 21.5, we have that "e[i]" returns the 'i'th
vertex in 'e' by reference. If the address of this vertex (or the entry "vertex[i]"
in the array-of-pointers "vertex" in 'e') is indeed the same as the address of
'n', then 'n' is indeed the 'i'th vertex in 'e', so 'i'+1 is returned:

```
if(&n == &(e[i]))
  return i+1;
```

If, on the other hand, no vertex has been found with the same address as 'n',
then the conclusion is that 'n' is not a vertex in 'e', so 0 is returned:

```
    return 0;
} //   check whether a node n is in a cell e
```

With this operator, the user can write "n < e" to check whether the well-defined node 'n' is indeed a vertex in the well-defined cell 'e'. In fact, the '<' symbol is chosen here because it is similar to the standard '$\in$' symbol used often in set theory.

Finally, we define an ordinary function that prints the vertices in the cell onto the screen:

```
template<class T, int N>
void print(const cell<T,N>&e){
  for(int i=0; i<N; i++)
    print(e[i]);
}  //  printing a cell
```

The "typedef" command is now used to define short and convenient types: "triangle" for a cell with three vertices in the Cartesian plane,

```
typedef cell<point,3> triangle;
```

and "tetrahedron" for a cell with four vertices in the three-dimensional Cartesian space.

```
typedef cell<point3,4> tetrahedron;
```

21.9 Edge-Sharing Cells

The following operator checks whether two cells share an edge or not. It takes two arguments (of type constant references to cells), denoted by 'e' and 'f'. Then, it uses a nested double loop (with distinct indices denoted by 'i' and 'j') to scan the vertices in 'e'. In the inner loop, the '<' operator in Section 21.8 is invoked twice to find two distinct vertices "e[i]" and "e[j]" that belong not only to 'e' but also to 'f'. If such vertices are indeed found, then this means that the cells 'e' and 'f' do indeed share an edge, so the function returns the output 1 (true). If, on the other hand, no such vertices have been found, then the function returns the output 0 (false):

```
template<class T, int N>
int operator&(const cell<T,N>&e, const cell<T,N>&f){
  for(int i=1; i<N; i++)
    for(int j=0; j<i; j++)
      if((e[i] < f)&&(e[j] < f)) return 1;
  return 0;
}  //  edge-sharing cells
```

With this operator, the user can write just "e & f" to check whether the well-defined cells 'e' and 'f' indeed share an edge. This operator is used later on to color the cells in the mesh.

21.10 The Mesh Object

The "mesh" template class is derived below from a linked list of objects of type 'T', to be specified later (upon construction of a concrete "mesh" object) as a triangle (in a triangulation) or a tetrahedron (in a three-dimensional mesh). This derivation (Figure 21.1) allows one to insert new cells or remove unnecessary cells easily and efficiently by calling the suitable functions inherited from the base "linkedList" class.

In the multilevel hierarchy of objects used to implement the mesh (Figure 21.2), the "mesh" object at the top level is a linked list of "cell" objects in the next lower level, each of which is a list of (pointers to) "node" objects, each of which contains a "point" object at the lowest level to store its location in the Cartesian plane.

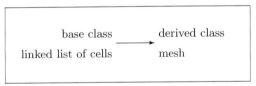

FIGURE 21.1:   Inheritance from the base class "linkedList" to the derived class "mesh".

```
template<class T>
class mesh : public linkedList<T>{
```

The "mesh" class contains no data field but the data fields "item" and "next" inherited from the base "linkedList" class in Chapter 17, Section 17.3.

The default constructor below has an empty block. At the '{' symbol that marks the start of this block, the default constructor of the base "linkedList" class in Chapter 17, Section 17.3 is called implicitly automatically. However, this constructor has an empty block as well. All that it does is to call implicitly the default constructor of class 'T' to construct its "item" field, and then to initialize the "next" pointer (as in its initialization list) with the zero value to indicate that it points to no item.

```
public:
```

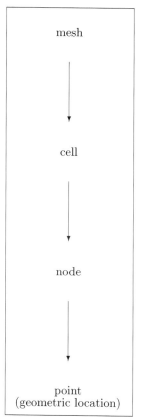

FIGURE 21.2: The multilevel hierarchy of objects used to implement the mesh as a linked list of "cell" objects, each of which is a list of (pointers to) "node" objects, each of which contains a "point" object to indicate its location in the Cartesian plane.

```
mesh(){
} // default constructor
```

For example, if 'T' is the "cell" class, then the default constructor in Section 21.4 produces a cell "item" whose vertices lie at the origin and have "sharingCells" fields that are equal to 1 and "index" fields that are equal to −1.

The next constructor, on the other hand, does a little more. At the beginning of its block, it calls the default constructor of the base "linkedList" class as above. For example, if 'T' is the "cell" class, then "item" is a trivial cell whose vertices lie at the origin and have "sharingCells" fields that are equal to 1 and "index" fields that are equal to −1. Fortunately, "item" is declared as mere "protected" (rather than private) in the base "linkedList" class. Therefore, it is accessible from the derived "mesh" class as well, and can be set here by the assignment operator of the 'T' class to have the same value as the argument 'e'.

For example, if 'T' is the "cell" class, then the assignment operator in Section 21.7 is invoked to decrement the "sharingCells" fields in the vertices of "item" from 1 to 0 and remove the trivial nodes in these vertices. Then, the vertices of "item" are assigned with the addresses in the corresponding vertices in 'e', as required. This is also why 'e' must be nonconstant: the "sharingCells" fields in its vertices increase by 1 in this process.

```
mesh(T&e){
  item = e;
} // constructor
```

Fortunately, the copy constructor, assignment operator, and destructor inherited from the base "linkedList" class work just fine, so they don't have to be rewritten here.

The member functions declared below will be defined explicitly later on.

```
int indexing();

int indexingCells();

void refineNeighbor(node<point>&,
    node<point>&,node<point>&);

void refineNeighbors(node<point3>&,
    node<point3>&, node<point3>&);

void refine();
};
```

This completes the block of the "mesh" class. Next, we define some of the member functions that were only declared in the class block above.

21.11 Indexing the Nodes

Assume that the template parameter 'T' in the "mesh" is specified to be the "cell" class. The following member function of the "mesh" class assigns indices to the nodes in the mesh:

```
template<class T>
int mesh<T>::indexing(){
```

The following loop scans the cells in the mesh (or the items inherited from the base "linkedList" class) and sets the indices of their vertices to the dummy value -1 by invoking the public "resetIndices()" member function of the "cell" class:

```
for(mesh<T>* runner = this;
     runner; runner=(mesh<T>*)runner->next)
   runner->item.resetIndices();
```

In this loop, the cells in the mesh are scanned by the pointer-to-mesh "runner", which is advanced time and again to the address of the next item in the underlying linked list. Indeed, this address is stored in the pointer "next", which is declared as mere "protected" (rather than private) in the base "linkedList" class, which makes it accessible from the derived "mesh" class as well.

Unfortunately, "next" is a mere pointer-to-linkedList rather than a pointer-to-mesh. This is why it must be converted explicitly to type pointer-to-mesh before it can be assigned to "runner", as is indeed indicated by the prefix "(mesh<T>*)".

Once the indices of all the nodes in the mesh have been set to -1, yet another loop is used to index the nodes properly. Indeed, to each cell encountered in this loop, the "indexing()" member function of the "cell" class (Section 21.5) is invoked to index its (yet unindexed) vertices:

```
int count=0;
for(mesh<T>* runner = this;
     runner; runner=(mesh<T>*)runner->next)
   runner->item.indexing(count);
```

Furthermore, the "indexing()" member function of the "cell" class also increases its argument, "count", by the number of vertices indexed by it. Thus, at the end of the above loop, all the nodes in the mesh have been indexed, so "count" is the total number of nodes in the entire mesh:

```
   return count;
} //  indexing the nodes in the mesh
```

21.12 An Example

In this example, we construct a mesh of three triangles. First, we use the constructor of the "node" class (Section 21.1) to construct the nodes 'a', 'b', 'c', 'd', and 'e':

```
int main(){
  node<point> a(point(1,1));
```

```
node<point> b(point(2,2));
node<point> c(point(2,0));
node<point> d(point(3,1));
node<point> e(point(3,3));
```

Then, the nodes 'a', 'b', and 'c' are used in the first constructor in Section 21.6 to construct the triangle "t1":

```
triangle t1(a,b,c);
```

Next, the nodes 'b', 'c', and 'd' are used to form yet another triangle, "t2". However, 'b' has already been copied to the second vertex of "t1", as in the first constructor in Section 21.6. It is this copy (rather than the original node 'b', which is only a dangling node that belongs to no cell) that should be used as the first vertex in "t2".

There are two possible ways to refer to the second vertex in "t1": "t1(1)" invokes the "operator()" member function of the "cell" class (Section 21.5) to return a nonconstant reference to the second vertex in "t1", whereas "t1[1]" invokes "operator[]" to return a constant reference. Here, since the constructor of the "cell" class in Section 21.6 increases the "sharingCells" field in its "node" arguments, "t1(1)" must be used rather than "t1[1]". This way, when "t2" is constructed below, the "sharingCells" field in the "node" object pointed at by "t1(1)" can indeed increase from 1 to 2, as required.

Similarly, "t1(2)" (the third vertex in "t1") is used instead of the original node 'c' to form the new triangle "t2":

```
triangle t2(t1(1),t1(2),d);
```

This way, the vertices of "t2" are indeed copies of 'b', 'c', and 'd', as required.

In fact, the copy of 'b' in the mesh can now also be referred to as the first vertex of "t2", or "t2(0)", and the copy of 'd' in the mesh can now also be referred to as the third vertex in "t2", or "t2(2)". These nonconstant references are now used to form yet another triangle, "t3", vertexed at 'b', 'd', and 'e':

```
triangle t3(t2(0),t2(2),e);
```

Next, the second constructor in Section 21.10 is used to construct a mesh 'm' with one triangle only (namely, "t1") in it:

```
mesh<triangle> m(t1);
```

Furthermore, the "append()" function inherited from the base "linkedList" class (Chapter 17, Section 17.6) is now used to append "t2" to 'm' as well. In fact, this call invokes the constructor of the base "linkedList" class (Chapter 17, Section 17.3) to form the tail of 'm' with the only item "t2" in it. This constructor invokes in turn the copy constructor of the "cell" class in Section 21.6, which creates no physical node, but only increments by 1 the "sharingCells" fields in the vertices of "t2":

```
m.append(t2);
```

Similarly, "t3" is appended to 'm' as well:

```
m.append(t3);
```

Once "t1" has been assigned to the first cell in 'm', it has completed its job and can be removed by the destructor of the "cell" class at the end of Section 21.7. In fact, this destructor removes no physical node, but only decrements by 1 the "sharingCells" fields in the vertices of "t1":

```
t1.~triangle();
```

Similarly, once "t2" and "t3" have been appended to 'm', they can be removed as well:

```
t2.~triangle();
t3.~triangle();
```

Furthermore, the nodes in 'm' are indexed:

```
m.indexing();
```

Finally, 'm' is printed onto the screen, using the "print()" function (Chapter 17, end of Section 17.9) applied to the underlying linked list:

```
print(m);
return 0;
}
```

21.13 Indexing the Cells

Here is the definition of the "indexingCells" member function that assigns indices to the individual cells in the mesh, and returns the total number of cells in the entire mesh:

```
template<class T>
int mesh<T>::indexingCells(){
    int count=0;
```

The indexing is done by scanning the cells in the mesh by the pointer-to-mesh named "runner". Indeed, "runner" jumps from item to item in the linked list that contains the cells in the mesh.

```
for(mesh<T>* runner = this; runner;
        runner = (mesh<T>*)runner->next)
```

To make "runner" jump to the next item, it is assigned the address in the "next" field in the loop above header. However, since "next" is inherited from the base "linkedList" class, it is of type pointer-to-linked-list rather than pointer-to-mesh; therefore, "next" must be converted explicitly into pointer-to-mesh before being assigned to "runner", as is indeed indicated by the prefix "mesh<T>*" in the loop header.

In the body of the loop, the cell pointed at by "runner" is assigned a successively increasing index, using the "setIndex" member function of the "cell" class:

```
runner->item.setIndex(count++);
```

Finally, the function also returns the total number of cells in the mesh:

```
    return count;
} // indexing the cells
```

Once the user calls this function, the cells in the mesh take their indices in the same order in which they are ordered in the underlying linked list. This indexing scheme is used later on to assign colors to the cells.

21.14 Exercises

1. In what sense is the cell object nonstandard?
2. In what way are its constructors and destructor nonstandard?
3. What is the risk in this practice?
4. How can one protect oneself from these risks?
5. What field in the node object may change when it is passed as an argument to the constructor in the "cell" class?
6. Why isn't the node argument that is passed to the constructor in the "cell" class declared as constant?
7. Can a temporary unnamed node object be passed as an argument to the constructor of the cell object? Why?
8. Why isn't the cell argument that is passed to the constructor in the "mesh" class declared as constant?
9. Can a temporary unnamed cell object be passed as an argument to the constructor of a mesh object? Why?
10. Why isn't the 'T' argument that is passed to the constructor in the base "linkedList" template class in Chapter 17, Section 17.3, declared as constant?
11. Why isn't the 'T' argument that is passed to the "append" member function in the base "linkedList" class in Chapter 17, Section 17.3, declared as constant?

Chapter 22

Triangulation

The triangulation introduced in Chapter 11, Section 11.8, is a special kind of graph. As discussed above, a graph is defined in terms of nodes and edges. In a general graph, the nodes and edges are purely abstract objects, with no geometrical meaning whatsoever. In a triangulation, on the other hand, they also have a concrete geometrical interpretation as points and line segments in the Cartesian plane.

The triangles in the triangulation are even more abstract and less concrete than the nodes and edges. Indeed, they are never used in the original definition of the graph. In fact, they are only used indirectly in the axioms required in a triangulation.

The abstract nature of the triangles is apparent not only from the mathematical formulation but also from the object-oriented implementation. Indeed, once the nodes have been defined and stored in the memory of the computer as points in the Cartesian plane $\mathbb{R}^2$, the triangles don't have to be stored any more. Indeed, a triangle only has to "know" (or has access to) its vertices. This is why a triangle is implemented as a list of three pointers-to-nodes rather than three nodes. This way, a node can be shared (or pointed at) by more than one triangle, as is often the case in a triangulation.

The present framework can thus be summarized as follows: in the implementation of the triangulation, the triangles are more abstract than the nodes. Indeed, the definition of the triangle object uses pointer fields only, whereas the definition of the node object uses concrete data fields to store its x and y coordinates in the Cartesian plane.

This kind of implementation is no surprise: after all, the triangles are only implicitly defined by the axioms listed in Chapter 11, Section 11.8. It is thus only natural to implement them as virtual objects that only refer to existing nodes but contain no concrete nodes.

The present approach is a nice example to show how the actual implementation agrees with the original mathematical background to decide on the proper level of abstraction suitable for a particular object. Indeed, because the triangle is defined only implicitly in terms of the axioms in Chapter 11, Section 11.8, it is also implemented only virtually with pointers-to-nodes rather than physical nodes.

Although we focus here on a two-dimensional triangulation, the discussion extends most naturally to a more general mesh of cells (including the three-dimensional mesh of tetrahedra implemented at the end of this book), defined in Chapter 11, last exercise. Indeed, although this coloring code below is applied here to a two-dimensional triangulation only, it is written in terms of a general mesh of cells, so that it is as general as possible.

Since the triangulation is just a special kind of a graph, it makes sense to form the sparse matrix that represents this graph. This matrix formulation can then be used to color not only the cells but also the nodes and the edges in the triangulation.

22.1 Triangulation of a Domain

The triangulation is a particularly useful tool to approximate a two-dimensional domain with a curved boundary, such as the unit circle. Indeed, one may use rather big triangles in the middle of the domain, and smaller and smaller triangles next to the curved boundary. Still, one must be careful to preserve conformity by following the axioms in Chapter 11, Section 11.8. This is done best by the iterative refinement method discussed below.

22.2 Multilevel Iterative Mesh Refinement

In multilevel iterative refinement [20] [21], one starts with a coarse mesh that approximates the domain poorly. In the present application, for example, the unit circle is initially approximated by a coarse mesh with two triangles only. In the next iteration (refinement step), the mesh is refined by dividing each triangle into two subtriangles. This is done by connecting the midpoint of one of the edges to the vertex that lies across from it. If there exists a neighbor triangle that also shares this edge, then it is also divided in the same way by connecting this midpoint to the vertex that lies across from it in this triangle as well. This way, conformity is preserved, as required in the axioms in Chapter 11, Section 11.8.

If, on the other hand, there is no neighbor triangle on the other side of this edge, then it must be a boundary edge (an edge that lies next to the boundary). In this case, two small triangles are introduced between the divided edge and the boundary, to yield a yet better approximation to the curved boundary. Once all the triangles in the original coarse mesh have been divided, the refinement step is complete, and the refined mesh is ready for the next iteration (or the next refinement step).

Clearly, the mesh produced from the above refinement step approximates the original domain better than the original coarse mesh. Furthermore, it can now serve as a coarse mesh in yet another refinement step, to produce a yet better approximation to the original domain. By repeating this process iteratively, one can produce finer and finer triangulations that approximate the original domain better and better.

Let us now turn to the actual implementation of a refinement step.

22.3 Dividing a Triangle and its Neighbor

The "refineNeighbor" member function of the "mesh" class takes three "node" arguments to represent nI, nJ, and nIJ in Figure 22.1, and uses them to search, find, and divide the adjacent (neighbor) triangle as well.

Thanks to the fact that the concrete "node" arguments (including the midpoint nIJ) already serve as vertices in well-defined triangles in the mesh, they can also be used to divide the neighbor triangle into two subtriangles. This is also why they must be passed by nonconstant reference (or reference-to-nonconstant-node), so that their "sharingCells" fields could change when the neighbor triangle is divided.

To find the neighbor triangle, we use the '<' operator in Chapter 21, Section 21.8, which checks whether a given node is indeed a vertex in a given triangle. If it is, then this operator returns the index of this node in the list of vertices of this triangle plus one (so the output is either one or two or three). Otherwise, it returns zero.

The '<' operator is called twice for each triangle to check whether both nI and nJ serve as vertices in it. If they do, then it must indeed be the required neighbor triangle. The third vertex in it is then located by a straightforward elimination process: after all, it is the only vertex in the neighbor triangle that is different from both nI and nJ. The neighbor triangle is then replaced by two subtriangles, denoted by "t1" and "t2".

To find the neighbor triangle, the recursive structure of the mesh object (which is actually a linked list of triangles) is particularly helpful: if the first triangle in the mesh (or the first item in the underlying linked list of triangles) isn't the required neighbor triangle, then the "refineNeighbor" function is called recursively to look for the neighbor triangle among the rest of the triangles in the mesh. For this purpose, the "refineNeighbor" function is applied recursively to the "tail" of the original (current) mesh, pointed at by the "next" pointer.

Fortunately, both "item" (the first cell in the mesh) and "next" (the pointer to the tail of the mesh) are declared as mere "protected" (rather than private) members of the base "linkedList" class in Chapter 17, Section 17.3. Therefore, both can be accessed from the derived "mesh" class as well.

the coarse mesh:

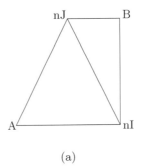

(a)

dividing the triangle:

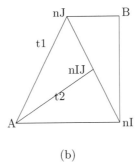

(b)

dividing the neighbor:

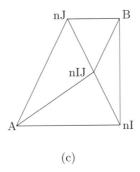

(c)

FIGURE 22.1:　The coarse triangle vertexed at A, nI, and nJ [see (a)] is divided into two smaller triangles by the new line leading from A to nIJ [see (b)]. Furthermore, its neighbor triangle on the upper right is also divided by a new line leading from nIJ to B [see (c)].

Unfortunately, the "next" field is inherited from the base "linkedList" class as a mere pointer-to-linkedList rather than a pointer-to-mesh. Therefore, it must first be converted explicitly into a pointer-to-mesh before the "refineNeighbor" function can be applied recursively to it.

Usually, this would be considered as a risky practice, because in theory "next" could point to a "linkedList" object or to any other object derived from the "linkedList" class, which might have a completely different version of a "refineNeighbor" function that might do completely different things. Fortunately, here "next" must point to a "mesh" object as well, so the recursive call is safe.

```
void mesh<triangle>::refineNeighbor(node<point>&nI,
        node<point>&nJ, node<point>&nIJ){
   int ni = nI < item;
   int nj = nJ < item;
```

Here, the arguments "nI" and "nJ" are the node objects corresponding to the endpoints nI and nJ in Figure 22.1, respectively. If both nI and nJ are indeed vertices in the first triangle in the mesh, "item", then the integers "ni" and "nj" are the corresponding indices of nI and nJ in the list of vertices in "item" plus 1. These integers are now used to identify the third vertex in "item":

```
   if(ni&&nj){
```

If "item" is indeed the required neighbor triangle that shares both nI and nJ as its own vertices, then all that is left to do is to identify the third vertex in it and divide it into two subtriangles. This is done as follows. First, we identify the integer "nk", the index of the third vertex in the list of vertices in "item":

```
      ni--;
      nj--;
      int nk = 0;
      while((nk==ni)||(nk==nj))
        nk++;
```

Next, "nk" is used to form two subtriangles "t1" and "t2" to replace the original neighbor triangle "item":

```
      triangle t1(nI,nIJ,item(nk));
      triangle t2(nJ,nIJ,item(nk));
      insertNextItem(t2);
      insertNextItem(t1);
      dropFirstItem();
   }
   else{
```

If, on the other hand, "item" is not the required neighbor triangle, then the "refineNeighbor" function is applied recursively to the remaining triangles in the mesh to search for the required neighbor triangle. For this, however, the "next" field must first be converted explicitly from a mere pointer-to-linkedList into a pointer-to-mesh:

```
if(next)
  ((mesh<triangle>*)next)->refineNeighbor(nI,nJ,nIJ);
```

If, however, we have reached the innermost recursive call that is applied to the trivial linked list that contains only the last triangle in the mesh, then it is clear that there is no neighbor triangle, so the edge that leads from nI to nJ must be a boundary edge. In this case, two extra triangles, which are also vertexed at nIJ and at the boundary point nIJ/‖nIJ‖, are added to the mesh:

```
else{
  node<point>
    newNode((1./sqrt(squaredNorm(nIJ())))) * nIJ());
  triangle t1(nI,nIJ,newNode);
  triangle t2(nJ,nIJ,t1(2));
  insertNextItem(t2);
  insertNextItem(t1);
  }
}
} //  refine also the neighbor of a refined triangle
```

This final "else" block deals with the case in which no neighbor triangle has been found in the entire mesh. In this case, the edge leading from nI to nJ must be a boundary edge, that is, an edge that lies next to the boundary of the domain approximated by the triangulation. Thus, in order to improve the approximation, one needs to add two more triangles to the mesh, between this boundary edge and the boundary of the domain: one triangle vertexed at nI, nIJ, and the point "newNode" that lies on the boundary, and the other triangle vertexed at nJ, nIJ, and "newNode". In the above code, it is assumed that the domain approximated by the triangulation is the unit circle, so a natural choice for "newNode" is

$$\text{newNode} = \frac{\text{nIJ}}{\|\text{nIJ}\|_2}.$$

22.4 Refining the Mesh

Here we define the member function "refine()" of the "mesh" class that applies one refinement step to the mesh by dividing the triangles in it, along with their neighbors. Because the mesh object is actually a linked list of triangles, it is only natural to use recursion for this purpose: first, the first triangle in the mesh (the first item in the underlying linked list), "item", is divided into two subtriangles: "t1", with vertices A, nIJ, and nI, and "t2", with vertices A, nIJ, and nJ, as in Figure 22.1. The "refineNeighbor" function is then called to divide the neighbor triangle as well, if exists. The first triangle, "item", is then replaced by the two subtriangles "t1" and "t2".

The "refine()" function is then called recursively to divide the rest of the triangles in the tail of the original (current) mesh as well. In particular, this recursive call may also divide the edge leading from A to nI in "t1" or the edge leading from A to nJ in "t2".

The new edges that emerge from nIJ, on the other hand, are not divided in "refine()" any more. This is guaranteed by the "index" field in the new node nIJ, which is assigned the dummy value −1, and is therefore excluded from any further dividing in the present refinement step.

The vertices in each triangle 't' in the mesh can be accessed by "t(0)", "t(1)", and "t(2)". This access method uses the "operator()" (rather than the "operator[]") of the "cell" class that returns a nonconstant (rather than constant) reference-to-node, because the "sharingCells" fields in the nodes may change in the refinement step.

```
void mesh<triangle>::refine(){
    for(int i=0; i<3; i++)
        for(int j=2; j>i; j--)
            if((item[i].getIndex() >= 0)
            &&(item[j].getIndex() >= 0)){
```

We are now in the beginning of a nested loop over the vertices in the first triangle in the mesh, "item". By now, we have found a pair of distinct vertices of "item", indexed by 'i' and 'j', which are connected by an edge that has not yet been divided in the present refinement step. Indeed, these vertices have nonnegative "index" fields, which indicates that they are old vertices that have existed even before the beginning of the present refinement step. Thus, we proceed to define the midpoint of this edge:

```
node<point> itemij = (item[i]()+item[j]())/2.;
```

Here we apply the "operator()" of the "node" class to the nodes "item[i]" and "item[j]" to have their locations in the Cartesian plane, namely, the 2-d points nI and nJ. The point that lies in between nI and nJ, nIJ, is then converted implicitly into the required node object "itemij".

Furthermore, we can now find the third vertex in the triangle, indexed by 'k':

```
int k=0;
while((k==i)||(k==j))
    k++;
```

These points are now used to construct the two halves of the triangle "item":

```
triangle t1(item(i),itemij,item(k));
triangle t2(item(j),t1(1),item(k));
```

Note that, once the new node "itemij" has been copied to the new triangle "t1", it must be referred to as "t1(1)" (namely, the second node in the new triangle "t1") rather than its original name "itemij" which is just a dangling node that belongs to no triangle. This way, when "t1(1)" is used to construct the second new triangle, "t2", its "sharingCells" field increases to 2, as required.

The small new triangles "t1" and "t2" are now used to find the triangle adjacent to "item" and divide it as well:

```
if(next)
  ((mesh<triangle>*)next)->
              refineNeighbor(item(i),item(j),t1(1));
```

Then, "t1" and "t2" are placed in the mesh instead of the original triangle "item":

```
insertNextItem(t2);
insertNextItem(t1);
dropFirstItem();
```

By now, we have divided the first triangle in the mesh and its neighbor. The mesh has therefore changed, and new triangles have been introduced, which need to be considered for refinement as well. Therefore, we have to call the "refine()" function recursively here. This call can only divide edges that do not use the new node "itemij", whose "index" field is −1:

```
refine();
return;
}
```

Finally, if no edge that should be divided has been found in the first triangle in the mesh, then the "refine()" function is applied recursively to the rest of the triangles in the mesh (contained in the "next" variable), not before this variable is converted explicitly from pointer-to-linkedList to pointer-to-mesh:

```
if(next)((mesh<triangle>*)next)->refine();
}  //  refinement step
```

22.5 Approximating a Circle

In this example, the above function is used to form a triangulation to approximate the unit circle. First, a coarse triangulation that provides a poor approximation is constructed as in Figure 22.2, using two triangles only: the upper triangle "t1" [vertexed at $(1,0)$, $(0,1)$, and $(-1,0)$], and the lower triangle "t2" [vertexed at $(-1,0)$, $(0,-1)$, and $(1,0)$]. (Note that the vertices in each triangle are ordered counterclockwise, as in the positive mathematical direction.)

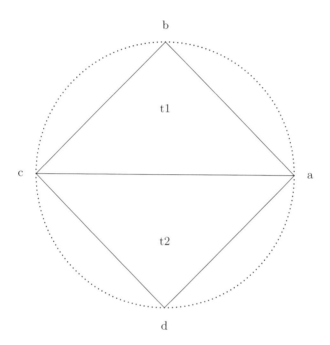

FIGURE 22.2: The coarse triangulation that approximates the unit circle poorly.

These triangles are constructed by the first constructor in Chapter 21, Section 21.6, which takes three "node" arguments. However, since this constructor may change the "sharingCells" fields of its arguments, it cannot take temporary objects, because it makes little sense to change them, so the compiler assumes that this must be a human error and refuses to accept this. This is why the user must define the nodes properly as permanent variables ('a', 'b', 'c', and 'd' below) before they are passed to the "cell" constructor to construct "t1" and "t2":

```
int main(){
  node<point> a(point(1,0));
```

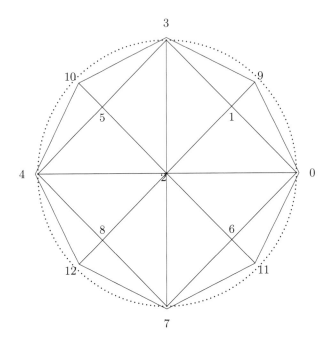

FIGURE 22.3: The finer triangulation obtained from one refinement step applied to the original coarse triangulation above. The nodes are indexed from 0 to 12 by the "indexing()" function.

```
node<point> b(point(0,1));
node<point> c(point(-1,0));
node<point> d(point(0,-1));
triangle t1(a,b,c);
triangle t2(t1(2),d,t1(0));
```

(See Chapter 21, Section 21.12, for more explanations.)

The coarse triangulation 'm' that contains "t1" and "t2" is now formed by the second constructor in Chapter 21, Section 21.10, and the "append()" function inherited from the base "linkedList" class:

```
mesh<triangle> m(t1);
m.append(t2);
```

Once the original (coarse) triangles "t1" and "t2" have been copied into the mesh 'm', they have completed their job and can be removed:

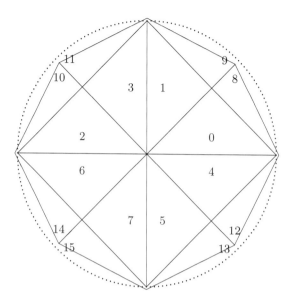

FIGURE 22.4:   The triangles are indexed from 0 to 15, by the "indexingCells" function, in the order in which they appear in the underlying linked list.

```
t1.~triangle();
t2.~triangle();
```

One refinement step is now applied to 'm':

```
m.indexing();
m.refine();
```

The improved mesh (with nodes indexed as in Figure 22.3) can now be printed onto the screen:

```
m.indexing();
print(m);
return 0;
}
```

The order in which the fine triangles are created in the mesh is displayed in Figure 22.4. This order has some effect on the coloring of the triangles below.

22.6 The Cell-Coloring Code

Here we implement the triangle-coloring algorithm in Chapter 11, Section 11.9 above. Actually, the algorithm is implemented in a more general way, so it can be used to color not only triangles but also any kind of cell in a mesh.

For this purpose, the function "colorCells" below is defined as a template function, with the yet unspecified type 'T' standing for the type of cell in the mesh. This way, the function can be called by any user not only for a mesh of triangles, in which case 'T' takes the "triangle" type, but also for any kind of mesh in the Cartesian plane, such as mesh of squares or even more complicated geometrical shapes.

The function returns a vector (named "colors") whose dimension (number of components) is the same as the number of cells in the mesh. This number is returned by the function "indexingCells" applied to the mesh:

```
template<class T>
const dynamicVector<int>
colorCells(mesh<T>&m){
   int colorsNumber = 0;
   dynamicVector<int> colors(m.indexingCells(),-1);
```

The coloring of the cells is done by a double nested loop on the cells in the mesh. In the outer loop, the cells are scanned by the pointer-to-mesh "runner", and in the inner loop, they are scanned again by the pointer-to-mesh "neighbor". Both "runner" and "neighbor" are advanced from cell to cell using the "readNext" member function of the base "linkedList" class. However, since this function returns a pointer to linked list, they must be converted explicitly to type pointer to mesh:

```
for(const mesh<T>* runner = &m; runner;
        runner=(const mesh<T>*)runner->readNext()){
   dynamicVector<int> usedColors(colorsNumber,0);
   for(const mesh<T>* neighbor = &m;
          neighbor&&(neighbor != runner);
          neighbor=(const mesh<T>*)neighbor->readNext())
```

The cells pointed at by "runner" and "neighbor" can be obtained by invoking the "operator()" member function of the base "linkedList" class, which returns the first item in a linked list. In the body of the inner loop, it is checked whether these cells share an edge by invoking the "operator&" function:

```
if((*neighbor)() & (*runner)())
   usedColors(colors[(*neighbor)().getIndex()]) = 1;
```

The colors that have been used to color any cell that shares an edge with the cell pointed at by "runner" are now excluded from the potential colors to color this cell. The color that is chosen to color this cell is the one that has never been used to color its neighbors. If, however, no such color is available, then a new color must be introduced, and the total number of colors must increase by 1:

```
    int availableColor = firstZero(usedColors);
    colors((*runner)().getIndex()) =
            availableColor<usedColors.dim() ?
            availableColor : colorsNumber++;
  }
  return colors;
} //   color cells
```

This completes the proper coloring of the cells in the mesh. The user can now color a triangulation by applying the above function to a mesh of triangles: For example, adding the command

```
print(colorCells(m));
```

at the end of the "main()" function in the previous section produces the coloring in Figure 22.5, which uses three colors to color sixteen triangles. This is a suboptimal number of colors: indeed two colors would be sufficient if the triangles were ordered counter-clockwise rather than in the order in Figure 22.4. Still, three colors is a moderate number of colors, as can be expected from the triangle-coloring algorithm.

22.7 The Matrix Formulation

Here we implement the constructor that takes a triangulation as an argument, and produces the sparse matrix associated with it, namely, the symmetric sparse matrix A whose element $a_{i,j}$ is nonzero if and only if the nodes indexed by i and j are connected by an edge in the triangulation:

```
template<class T>
sparseMatrix<T>::sparseMatrix(mesh<triangle>&m){
   item = new row<T>*[number = m.indexing()];
```

This code line allocates memory for the field "item" inherited from the base "list" class. Indeed, in the derived "sparseMatrix" class, this field is actually an array of pointers-to-rows. The number of rows (or the number of items in the underlying list) is the same as the number of nodes in the triangulation, returned by the "indexing()" function.

Initially, the pointers-to-rows in the list are initialized with the trivial zero value, that is, they point to nothing:

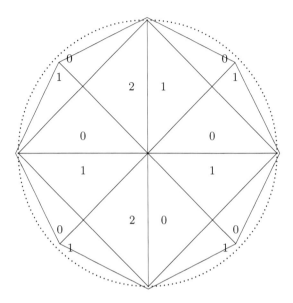

FIGURE 22.5:   The coloring produced by the triangle-coloring code uses three colors to color the fine triangulation. A better coloring, which uses two colors only, would result from the code if the triangles had been ordered counter-clockwise.

```
for(int i=0; i<number; i++)
    item[i] = 0;
```

In order to give the rows meaningful values, the triangles in the mesh 'm' are scanned in the following loop:

```
for(const mesh<triangle>* runner = &m; runner;
        runner=(const mesh<triangle>*)runner->readNext()){
```

The pointer-to-mesh "runner" used in this loop is advanced by using the "readNext" member function of the base "linkedList" class. However, since this function returns a pointer to linked list, it must be converted explicitly into a pointer to mesh. The first cell in the mesh pointed at by "runner" can now be obtained by invoking the "operator()" member function of the base "linkedList" class. Furthermore, the individual nodes in the cell can be referred to by invoking the "operator[]" member function in the "cell" class. Moreover,

the index of each such node in the list of nodes in the triangulation can also
be obtained by the "getIndex" member function of the "node" class. Thanks
to these properties, one can now carry out a double loop on the vertices in
each cell to define a nonzero element $a_{I,J}$ in the matrix A, where 'I' and 'J'
are the indices associated with these vertices:

```
for(int i=0; i<3; i++){
    int I = (*runner)()[i].getIndex();
    for(int j=0; j<3; j++){
        int J = (*runner)()[j].getIndex();
```

This way, the vertices indexed by 'i' and 'j' in the individual cell pointed at
by "runner" are also indexed by the global indices 'I' and 'J' in the global list
of nodes in the entire mesh.

These global indices are now used to give the corresponding matrix element $a_{I,J}$ a nonzero value, as required. For this, however, we first need
to check whether the 'I'th row in the current sparse matrix has already
been constructed. If it has, then the "+=" operator inherited from the base
"linkedList" class should be used to add the nonzero element $a_{I,J}$ to it; otherwise, the "new" command should be used instead:

```
if(item[I]){
    row<T> r(1,J);
    *item[I] += r;
}
else
    item[I] = new row<T>(1,J);
    }
   }
  }
 } //  the matrix of the triangulation
```

This completes the constructor of the matrix associated with the graph of the
triangulation. (As a matter of fact, the above function can be easily extended
to apply not only to a triangulation but also to a more general mesh of cells
of any shape.) Furthermore, once the symmetric matrix has been constructed,
one can apply to it the "colorNodes" and "colorEdges" functions to color the
nodes and the edges in the nonoriented graph of the triangulation.

22.8 The Code-Size Rule

Object-oriented programming is particularly useful to limit the amount of
code required in advanced applications. Roughly speaking, with no object-
oriented programming, the code size may grow exponentially as the problem

becomes more and more complicated, whereas with object-oriented programming it grows only linearly.

Indeed, assume that we already have a code that works well for some particular case (for example, the above code that works well for a triangulation), and we need to generalize it to apply to a yet more complicated case as well (for example, to a 3-D mesh of tetrahedra as in the next chapter). If our original code were not an object-oriented code, then we would have to modify each and every function in it to apply to the more advanced (3-D rather than 2-D) case as well. This would no doubt require more code (as well as more programming work) in each function, increasing the total code size (as well as the total programming time) by a constant factor greater than one.

Fortunately, our code is indeed an object-oriented code. Therefore, all that is required is to implement properly the more general objects, along with their own functions (for example, the 3-D mesh of tetrahedra and its own refinement functions below). With these new objects, the original algorithm is still well-implemented in the original code, with at most some minor adjustments.

Thus, with object-oriented programming, one only needs to add to the original code a fixed amount of new code to implement the new objects required in the more general (and complicated) case. Thus, the total code size (as well as the total amount of programming work) increases only linearly as the problem is written in more and more general terms, to apply to more and more difficult cases.

22.9 Exercises

1. Explain why the node object, defined in terms of its concrete location in the Cartesian plane, is less abstract than the triangle object, defined as a list of three pointers-to-nodes.

2. How does the high level of abstraction of the triangle object in its actual implementation agrees with its original mathematical definition, which only follows from the axioms in Chapter 11, Section 11.8?

3. Why must the pointer "runner" used in the above loops be converted explicitly from type pointer-to-linked-list to type pointer-to-mesh?

4. A user defines the nodes 'a', 'b', and 'c' in a code. Can these names be passed to the constructor in the "cell" class to form the triangle "t1(a,b,c)"? Can any of these nodes, say 'a', be passed once again to construct yet another triangle, say "t2", intended to be placed in the same mesh? Why must the user refer to 'a' as "t1(0)" rather than by its original name 'a'? (Hint: what happens to the value of the "sharingCells" field in the node when it is passed as an argument to the cell constructor?)

5. Apply the "refine" function to the above triangulation of the unit circle

to obtain a yet finer and more accurate triangulation.

6. Color the above triangulation by the cell-coloring code. How many colors are used? Why must the number of colors be at most four?

7. Use the above code to obtain the matrix formulation of the graph of the above triangulation.

8. Check that the above matrix is indeed symmetric.

9. Apply to the above matrix the node-coloring code for nonoriented graphs in Chapter 20, Section 20.15 to obtain the coloring of the nodes in the above triangulation. How many colors are used?

10. Apply to the above matrix the edge-coloring code of nonoriented graphs in Chapter 20, Section 20.16 to obtain the coloring of the edges in the above triangulation. How many colors are used?

Part VII

Three-Dimensional Applications

Three-Dimensional Applications

In this part, we extend the above mesh of triangles (or triangulation) into its three-dimensional counter-part: a mesh of tetrahedra. For this complicated mesh, we implement an iterative refinement algorithm, in which each tetrahedron is refined along with its neighbor (edge-sharing) tetrahedra, to preserve conformity.

Furthermore, we define 1-d, 2-d, and 3-d polynomial objects, along with their arithmetic operations, composition, and integration. Moreover, we also implement efficiently sparse polynomials, which may contain a lot of zero coefficients.

Finally, the three-dimensional mesh of tetrahedra, along with the three-dimensional polynomials, is used to form the stiffness and mass matrices, which are often used in practical applications in computational physics and engineering.

Chapter 23

Mesh of Tetrahedra

The mesh implemented in this chapter may be viewed as a generalization of the mesh implemented in the previous chapter to three spatial dimensions. Indeed, it is implemented as a linked list of tetrahedra, each of which may be viewed as the three-dimensional generalization of a triangle. Furthermore, the iterative refinement process preserves conformity: once a particular tetrahedron is refined by dividing one of its edges and connecting its midpoint to the corner that lies across from it to form two subtetrahedra, all the neighbor tetrahedra that share this edge are also refined in the same way, using the midpoint as a new corner in the new subtetrahedra.

In the following, we'll see the detailed code that realizes this procedure to produce the fine conformal mesh of tetrahedra.

23.1 The Mesh Refinement

The tetrahedron object defined in Chapter 21, Section 21.8, is actually a cell in the three-dimensional Cartesian space, with four corners that form four triangular sides. Once the template symbol 'T' in the "mesh" class is specified to be the tetrahedron object, one obtains the required three-dimensional mesh, implemented as a linked list of tetrahedra.

Usually, the original mesh is too coarse to approximate well the curved three-dimensional domain under consideration. A refinement step is necessary to produce a finer and more accurate mesh.

To make sure that the fine mesh is conformal as well, one must be careful to refine not only each coarse tetrahedron but also its neighbors (edge sharing tetrahedra). This way, the midpoint of the edge divided in the refinement step serves not only as a corner in the two subtetrahedra of the original coarse tetrahedron but also as a corner in all the subtetrahedra of the neighbor tetrahedra, as required to preserve conformity. This is implemented next.

23.2 Refining the Neighbor Tetrahedra

Suppose that we have decided to refine a particular tetrahedron in the mesh by dividing one of its edges and connecting its midpoint to the corner that lies across from it to form two subtetrahedra. Let us denote the two original endpoints of this edge by "nI" and "nJ", and its midpoint by "nIJ". These nodes are passed as arguments to the "refineNeighbors" function below to search for any neighbor tetrahedra that also share this edge as their common edge, and refine them accordingly to preserve conformity:

```
void mesh<tetrahedron>::
    refineNeighbors(node<point3>&nI,
    node<point3>&nJ, node<point3>&nIJ){
```

In the case of triangulation discussed in the previous chapter, there may be at most one neighbor triangle to refine. Here, on the other hand, there may be several neighbor tetrahedra that share the edge leading from "nI" to "nJ". All of these neighbor tetrahedra must be found and refined as well. To do this, it makes more sense to scan the tetrahedra in the reversed order, from the last tetrahedron backward. This way, once a neighbor tetrahedron has been found and divided into two subtetrahedra, the scanning proceeds backward, avoiding the need to scan these two new subtetrahedra that have just been added to the linked list. This is why the recursive call is made in the beginning of the function, to start dividing neighbor tetrahedra from the far end of the underlying linked list:

```
if(next)
    ((mesh<tetrahedron>*)next)->
        refineNeighbors(nI,nJ,nIJ);
```

(Note that the "next" field is a mere pointer to a linked list of tetrahedra; this is why it must be converted explicitly to a pointer to a mesh before the recursive call can be made.)

Thus, the first thing that the "refineNeighbors" function does is the recursive call to search the "tail" of the linked list (which contains all the tetrahedra but the first one) for potential neighbors. The first thing that this recursion does is yet another recursion, and so on. Thus, the first thing that is actually done in the entire recursive process is to examine the last tetrahedron in the linked list and refine it if it is indeed a neighbor. Only then is the previous recursive call actually executed to examine the previous tetrahedron and refine it if appropriate, and so on, until the first tetrahedron is checked and refined if necessary:

```
int ni = nI < item;
int nj = nJ < item;
```

This is how the first tetrahedron, "item", is checked whether or not it is a neighbor that shares the edge leading from "nI" to "nJ": first, the "<" operator of Chapter 21, Section 21.8, is invoked to check whether "nI" and "nJ" are indeed nodes in "item". If they are, then this operator also returns their indices in the list of nodes in "item" plus one. This extra one can now be subtracted to obtain the indices "ni" and 'nj' of "nI" and "nJ" in the list of nodes in "item":

```
if(ni&&nj){
  ni--;
  nj--;
```

Furthermore, the two other nodes in "item" also have some indices 'nk' and 'nl' (that are different from "ni" and "nj") in its list of nodes, which can be found by simple loops:

```
int nk = 0;
while((nk==ni)||(nk==nj))
  nk++;
int nl = 0;
while((nl==ni)||(nl==nj)||(nl==nk))
  nl++;
```

These indices are now used to form the two subtetrahedra of "item" "t1" and "t2":

```
tetrahedron t1(nI,nIJ,item(nk),item(nl));
tetrahedron t2(nJ,nIJ,item(nk),item(nl));
```

These new tetrahedra are now added to the linked list instead of the original neighbor tetrahedron "item":

```
    insertNextItem(t2);
    insertNextItem(t1);
    dropFirstItem();
  }
} //  refine also the neighbor tetrahedra
```

This completes the function that refines also the neighbor tetrahedra that share the edge leading from "nI" to "nJ". In the sequel, we'll see how this function is used in the refinement step.

23.3 The Refinement Step

The following function implements the entire refinement step: it scans the tetrahedra in the mesh one by one, and refines each of them by dividing one of its edges:

```
void mesh<tetrahedron>::refine(){
  for(int i=0; i<4; i++)
    for(int j=3; j>i; j--)
```

The integers 'i' and 'j' represent distinct indices of nodes in the first tetrahedron, "item". The edge that connects these nodes is a good candidate for being divided only if it is indeed an original edge that has never been divided before in the present refinement step, or if these nodes indeed have nonnegative indices in the entire list of nodes in the entire mesh:

```
if((item[i].getIndex() >= 0)&&
   (item[j].getIndex() >= 0)){
```

In this case, their midpoint, denoted by "itemij", is used to form the two required subtetrahedra. For this purpose, however, we first need to find the two other vertices in "item". More precisely, we must find a way to refer to them by finding their indices 'k' and 'l' in the list of vertices in "item":

```
node<point3> itemij =
  (item[i]()+item[j]())/2.;
int k=0;
while((k==i)||(k==j))
  k++;
int l=0;
while((l==i)||(l==j)||(l==k))
  l++;
```

The midpoint "itemij", as well as the indices 'k' and 'l' of the two other vertices in "item", are now used to form the two subtetrahedra "t1" and "t2":

```
tetrahedron t1(item(i),
    itemij,item(k),item(l));
tetrahedron t2(item(j),
    t1(1),item(k),item(l));
```

Note that, once "itemij" has been used as a vertex in "t1", it is referred to as "t1(1)" (a legitimate node in the mesh) rather than by its original name "itemij", which stands for a dangling node that belongs to no cell.

Now, the "refineNeighbors" function is called to refine not only "item" but also its neighbor tetrahedra that share the edge under consideration. Clearly, such neighbors can be found only in the rest of the linked list of tetrahedra, because all the previous ones have already been refined fully:

```
if(next)
  ((mesh<tetrahedron>*)next)->
    refineNeighbors(item(i),item(j),t1(1));
```

Note that the "next" field is a mere pointer to a linked list of tetrahedra; this is why it must be converted explicitly into a pointer to an actual mesh before the "refineNeighbors" function can be called.

The two new tetrahedra "t1" and "t2" can now be added to the linked list instead of the original tetrahedron "item":

```
insertNextItem(t2);
insertNextItem(t1);
dropFirstItem();
```

Once "item" has been divided successfully and replaced by its two subtetrahedra "t1" and "t2", the rest of the tetrahedra in the linked list (new as well as old) are also refined by a recursive call to the "refine" function:

```
refine();
return;
}
```

This recursive call actually replaces the original call; indeed, the "return" commands that follows it closes the original call, leaving only the recursive call active. This recursive call continues to refine the up-to-date linked list of tetrahedra, starting from "t1", "t2", etc.

Although "t1" and "t2" have one edge that cannot be divided any more in this refinement step (because it has just been produced as one half of an original edge), they may still have old (original) edges that may still be divided in the present refinement step. This is why the recursive call must consider them as well.

Indeed, only original edges that belong to the original coarse mesh are considered for being divided, thanks to the "if" question in the beginning of the function that makes sure that their endpoints indeed have nonnegative indices, to indicate that they indeed belong to the original coarse mesh.

If, on the other hand, no more original edges have been found in "item", then it cannot be divided any more. In this case, the recursive call to the "refine" function is made from the next tetrahedron in the linked list:

```
if(next)
  ((mesh<tetrahedron>*)next)->refine();
} // adaptive refinement
```

This completes the refinement step to produce the finer mesh of tetrahedra.

23.4 Exercises

1. Write the unit cube

$$[0, 1] \times [0, 1] \times [0, 1] = \{(x, y, z) \mid 0 \le x, y, z \le 1\}$$

as the union of six disjoint tetrahedra. Combine these tetrahedra to form a complete mesh. The solution can be found in Section 28.15 in the appendix.

2. Apply the "refine" function to the above mesh and print the resulting fine mesh.
3. Apply the cell-coloring code to the above mesh of tetrahedra.
4. Write the constructor that produces the sparse matrix of the graph of the three-dimensional mesh of tetrahedra (it is analogue to the constructor in Chapter 22, Section 22.7).
5. Apply the above constructor to the above mesh of tetrahedra that covers the unit cube.
6. Apply the edge-coloring code to the resulting sparse matrix.

Chapter 24

Polynomials

In this chapter, we implement polynomials of the form

$$\sum_{i=0}^{n} a_i x^i,$$

along with some useful arithmetic operations (Chapter 12). Because the type of the independent variable x is unspecified, it is best denoted by the symbol 'T' in a template class. This way, x can be specified later by the user to be either a real or a complex variable.

Thus, in order to implement the polynomial, it is sufficient to store its coefficients

$$a_0, \; a_1, \; a_2, \; \ldots, \; a_n.$$

The naive way to do this is in an $(n+1)$-dimensional vector. This approach, however, is not sufficiently general and flexible. Indeed, in a vector object, the coefficients would actually be stored in an array, which requires that all of them are of the same size. This is good enough for polynomials of one variable, in which the coefficients are scalars, but not for polynomials of two or more variables, in which the coefficients are themselves polynomials that may have variable sizes.

A better way to implement the polynomial is, thus, as a list of coefficients. Indeed, the major advantage of the list is that its items don't have to be of the same size. This provides the extra flexibility required in the polynomial object implemented below.

24.1 The Polynomial Object

Here we implement the polynomial object as a list of objects of type 'T', to be specified later by the user:

```
template<class T> class polynomial:public list<T>{
public:
  polynomial(int n=0){
```

```
    number = n;
    item = n ? new T*[n] : 0;
    for(int i=0; i<n; i++)
      item[i] = 0;
} // constructor
```

Thanks to the fact that the fields "number" and "item" are declared as protected (rather than private) in the base "list" class, they can be accessed also from the derived "polynomial" class. this property is used in the default constructor above. Indeed, after the "number" and "item" fields have been set to zero by the default constructor of the base "list" class invoked automatically at the beginning of the above function, they are further updated to take more meaningful values: "number" takes the value 'n' passed to the above function as an argument, and, if 'n' is nonzero, then "item" is also constructed as an array of 'n' null pointers.

The next constructor is yet more specific: if 'n' is nonzero, then it also fills the array "item" with pointers to some objects of type 'T':

```
    polynomial(int n, const T&a){
      number = n;
      item = n ? new T*[n] : 0;
      for(int i=0; i<n; i++)
        item[i] = new T(a);
    } // constructor with 'T' argument
```

This constructor uses the "new" command and the copy constructor of the 'T' class to loop on the array "item" and initialize its components as pointers to the same value, 'a'.

The next constructor assumes that n= 2; it fills the two components in "item" with pointers to two prescribed objects, 'a' and 'b'.

```
    polynomial(const T&a, const T&b){
      number = 2;
      item = new T*[2];
      item[0] = new T(a);
      item[1] = new T(b);
    } // constructor with 2 'T' arguments
```

The destructor needs to do nothing, because everything is done by the default destructor of the base "list" class, invoked implicitly at the end of the following empty block:

```
    ~polynomial(){
    } // destructor
```

The following function returns the degree of the polynomial, which is just the number of coefficients minus one:

```
int degree() const{
  return number-1;
} // degree of polynomial
```

The next function allows the user to access the 'i'th coefficient in a polynomial 'p' simply by writing "p(i)":

```
T& operator()(int i){
  return list<T>::operator()(i);
} // read/write ith coefficient
```

Finally, we declare some more member functions, to be defined later in detail. Note that the arguments passed to these functions are of type 'S', which is not necessarily the same as 'T':

```
template<class S>
  const T
  HornerArray(T** const&, int, const S&) const;
template<class S>
  const T operator()(const S&) const;
template<class S>
  const S operator()(const S&, const S&) const;
template<class S>
  const S operator()(const S&,
      const S&, const S&) const;
};
```

This completes the block of the "polynomial" class. Next, we define some useful arithmetic operators with polynomials.

24.2 Adding Polynomials

The "+ =" operator is implemented as an ordinary (nonmember) function, which takes two polynomial arguments, and adds the second one to the first one. This is why only the second one is declared as constant, whereas the first one is not. With this operator, users who have defined two polynomials 'p' and 'q' may just write "p+ =q" to add 'q' to 'p'.

```
template<class T>
const polynomial<T>&
operator+=(polynomial<T>& p, const polynomial<T>&q){
```

If 'p' is of a larger degree than 'q', then the coefficients in 'q' are added one by one to the corresponding coefficients in 'p':

```
        if(p.degree() >= q.degree())
          for(int i=0; i<=q.degree(); i++)
            p(i) += q[i];
```

If, on the other hand, 'q' is of a larger degree than 'p', then it cannot be added to it by a straightforward loop. Instead, the "+ =" operator must be called recursively, with the roles of 'p' and 'q' interchanged. However, since 'q' is a constant polynomial, we must first define a nonconstant polynomial, "keepQ", which is the same as 'q'. The recursive call to the "+ =" operator then adds 'p' to "keepQ", and the result is not only placed in 'p', as required, but also returned as the output of the function:

```
        else{
          polynomial<T> keepQ = q;
          p = keepQ += p;
        }
        return p;
      }  //  add polynomial
```

The above function is now used to define a '+' operator:

```
      template<class T>
      const polynomial<T>
      operator+(const polynomial<T>& p,
          const polynomial<T>&q){
        polynomial<T> keep = p;
        return keep += q;
      }  //  add two polynomials
```

24.3 Multiplication by a Scalar

Here we define ordinary (nonmember) operators to multiply a polynomial by a scalar of type 'S', to be specified later by the user. First, we define the "*=" operator, which multiplies its first argument, the nonconstant polynomial 'p', by its second argument, 'a', an object of type 'S':

```
      template<class T, class S>
      const polynomial<T>&
      operator*=(polynomial<T>& p, const S&a){
        for(int i=0; i<=p.degree(); i++)
          p(i) *= a;
        return p;
      }  //  multiplication by scalar
```

The "*=" operator invoked in the above loop is interpreted to multiply an object of type 'T' by an object of type 'S', whatever these types may be. Thus, all the coefficients in 'p' are multiplied one by one by 'a', as required. With the above function, users can write "p *= d" to multiply a polynomial 'p' by a scalar 'd' of any type.

The above function is now used to define scalar-times-polynomial and polynomial-times-scalar multiplications:

```
template<class T, class S>
const polynomial<T>
operator*(const S&a, const polynomial<T>&p){
  polynomial<T> keep = p;
  return keep *= a;
} //  scalar times polynomial

template<class T, class S>
const polynomial<T>
operator*(const polynomial<T>&p, const S&a){
  polynomial<T> keep = p;
  return keep *= a;
} //  polynomial times scalar
```

With these functions, the user can write "p * d" or "d * p" to multiply a polynomial 'p' by a scalar 'd' of any type.

24.4 Multiplying Polynomials

The above operators are now used to multiply two polynomials, in light of the algorithm in Chapter 12, Section 12.3:

```
template<class T>
polynomial<T>
operator*(const polynomial<T>&p,
    const polynomial<T>&q){
```

First, we define and initialize to zero the polynomial "result", which will contain the required product of the polynomials 'p' and 'q':

```
polynomial<T>
    result(p.degree()+q.degree()+1,0);
for(int i=0; i<=result.degree(); i++)
```

The outer loop uses the index 'i' to form the 'i'th coefficient of "result". The inner loop, on the other hand, uses the index 'j' to obtain the 'i'th coefficient in "result" as a sum of products of coefficients in 'p' and coefficients in 'q':

```
for(int j=max(0,i-q.degree());
        j<=min(i,p.degree()); j++){
```

At the start of the inner loop, the 'i'th coefficient in "result" doesn't exist as yet. This is why the '=' operator is used to initialize it:

```
if(j == max(0,i-q.degree()))
        result(i) = p[j] * q[i-j];
```

In the rest of the inner loop, on the other hand, the 'i'th coefficient in "result" already exists and is already initialized. This is why the "+ =" operator is used to update it:

```
else
        result(i) += p[j] * q[i-j];
    }
  return result;
} //  multiply 2 polynomials
```

With this function, users can just write "p * q" to obtain the product of the polynomials 'p' and 'q'.

The above "operator*" is now used also to form the "*=" operator, which multiplies its first (nonconstant) polynomial argument by its second one:

```
template<class T>
polynomial<T>&
operator*=(polynomial<T>&p,
    const polynomial<T>&q){
  return p = p * q;
} //  multiply by polynomial
```

With this function, users can write "p *= q" to multiply the polynomial 'p' by the polynomial 'q'.

24.5 Calculating a Polynomial

Here we use the algorithms in Chapter 12, Section 12.4, to calculate the value $p(x)$ of a given polynomial p at a given argument x. First, we implement the naive algorithm, which calculates the powers x^i ($2 \leq i \leq n$), multiplies them by the corresponding coefficients a_i, and sums up:

```
template<class T>
const T
calculatePolynomial(const polynomial<T>&p, const T&x){
```

```
    T powerOfX = 1;
    T sum=0;
    for(int i=0; i<=p.degree(); i++){
      sum += p[i] * powerOfX;
      powerOfX *= x;
    }
    return sum;
  } //  calculate a polynomial
```

Next, we implement the more efficient Horner algorithm. The recursion required in this algorithm is implemented in a loop:

```
  template<class T>
  const T
  HornerLoop(const polynomial<T>&p, const T&x){
    T result = p[p.degree()];
    for(int i=p.degree(); i>0; i--){
      result *= x;
      result += p[i-1];
    }
    return result;
  } //  Horner algorithm to calculate a polynomial
```

24.6 Composition of Polynomials

Furthermore, the above code can be slightly modified to yield the composition $p \circ q$ of the given polynomials p and q, in light of the algorithm in Chapter 12, Section 12.5:

```
  template<class T>
  const polynomial<T>
  operator&(const polynomial<T>&p,
      const polynomial<T>&q){
    polynomial<T> result(1,p[p.degree()]);
    for(int i=p.degree(); i>0; i--){
      result *= q;
      result += polynomial<T>(1,p[i-1]);
    }
    return result;
  } //  Horner algorithm to compose p and q
```

24.7 Recursive Horner Code

In the previous sections, Horner's algorithm is implemented in a loop. This way, one avoids the expensive construction of the polynomial p_1 used in the recursion in Chapter 12, Section 12.4. Here, however, we show how one can stick to the original recursive formulation of Horner's algorithm, and yet avoid the explicit construction of p_1. This is done by observing that p_1 is already available in the array of coefficients of p, provided that the first coefficient is disregarded.

The function "HornerArray" that implements the above idea is declared as a member function of the "polynomial" class to allow other member functions to call it, which will prove useful below. Furthermore, the type of the argument 'x' is declared as the template class 'S', to be specified later by the user. This provides extra flexibility, since 'S' doesn't have to be the same as 'T'.

```
template<class T>
template<class S>
const T
polynomial<T>::HornerArray(T** const&p,
          int n, const S&x) const{
   return n = =0 ?
             *p[0]
          :
             *p[0] + x * HornerArray(p+1,n-1,x);
} // Horner algorithm for an array p
```

Here the first argument, 'p', stands for the array of pointers-to-coefficients in the polynomial p of degree 'n'. Therefore, "p+1" is nothing but the array of pointers-to-coefficients of the polynomial p_1 of degree "n−1" used in the recursion in chapter 12, Section 12.4. This is why this array is used in the recursive call to the "HornerArray" function above.

Thanks to the fact that the "item" field is declared as protected (rather than private) in the base "list" class, it can be accessed from the derived "polynomial" class as well. The "operator()" function defined below uses this property to apply the "HornerArray" function to the array "item" to produce the required value of the current polynomial at the argument 'x':

```
template<class T>
template<class S>
const T
polynomial<T>::operator()(const S&x) const{
   return HornerArray(item,degree(),x);
} // Horner algorithm to calculate a polynomial
```

In the functions defined further below, we'll use this operator to write "p(x)" to obtain the value of the polynomial 'p' at a given argument 'x'.

24.8 Polynomials of Two Variables

The template class 'T' that denotes the type of the coefficients in the polynomial is not necessarily a scalar. Here we'll indeed see that it may well be an object of variable size, e.g., a polynomial.

Indeed, in the polynomial of two variables introduced in chapter 12, Section 12.13, the coefficients $a_i(x)$ are polynomials in the independent variable x rather than mere scalars. Here we'll use this implementation to calculate the value of a polynomial of two variables $p(x, y)$ at some given arguments x and y.

```
template<class T>
template<class S>
const S
polynomial<T>::operator()(const S&x,
    const S&y) const{
  return (*this)(y)(x);
} //  compute p(x,y)
```

This function calls the original "operator()" of Section 24.7 above twice. In the first call, the argument y is used to obtain the polynomial $p(\cdot, y)$, in which y is a fixed number. In the second call, the argument x is used to compute the required output $p(x, y)$.

24.9 Polynomials of Three Variables

Similarly, following their definition in Chapter 12, Section 12.19, polynomials of three variables are implemented as polynomials with coefficients of the form $a_i(x, y)$, which are polynomials of two variables rather than mere scalars. Here is how this implementation is used to compute the value $p(x, y, z)$ at some given arguments x, y, and z:

```
template<class T>
template<class S>
const S
polynomial<T>::operator()(const S&x,
    const S&y, const S&z) const{
```

```
      return (*this)(z)(y)(x);
   }  //  compute p(x,y,z)
```

In this function, the original "operator()" in Section 24.7 above is called three times: the first time to calculate the polynomial $p(\cdot, \cdot, z)$ at the fixed argument z, the second time to calculate the polynomial $p(\cdot, y, z)$ at the fixed arguments y and z, and the third time to calculate the required value $p(x, y, z)$ at the given arguments x, y, and z.

24.10 Indefinite Integral

In Chapter 12, Section 12.9, we have introduced the indefinite integral of the polynomial $p(x)$, denoted by $P(x)$. Here is the ordinary (nonmember) function that returns this polynomial:

```
template<class T>
const polynomial<T>
indefiniteIntegral(const polynomial<T>&p){
  polynomial<T> result(p.degree()+2,0);
  for(int i=0; i<=p.degree(); i++)
    result(i+1) = (1./(i+1)) * p[i];
  return result;
} //  indefinite integral
```

24.11 Integral on the Unit Interval

In Chapter 12, Section 12.10, we have described a method to calculate the integral of a given polynomial $p(x)$ over an interval of the form $[a, b]$ and, in particular, over the unit interval $[0, 1]$. Here is the ordinary (nonmember) function that calculates this integral:

```
template<class T>
const T
integral(const polynomial<T>&p){
  return indefiniteIntegral(p)(1.);
} //  integral on the unit interval
```

24.12 Integral on the Unit Triangle

In Chapter 12, Section 12.16, we have described a method to calculate the integral of a given 2-d polynomial $p(x, y)$ over the unit triangle. Here is the ordinary (nonmember) function that calculates this integral:

```
template<class T>
const T
integral(const polynomial<polynomial<T> >&p){
```

First, the polynomial $1 - x$ is constructed using the constructor that takes two "double" coefficients, 1 and -1. This polynomial is then stored in the polynomial object named "oneMinusx":

```
polynomial<T> oneMinusx(1.,-1.);
```

Next, the "indefiniteIntegral" function is applied to the original polynomial 'p' to produce the indefinite integral with respect to y, denoted by $P(x, y)$ in Chapter 12, Section 12.16. Then, the "operator()" of Section 24.7 above is applied to $P(x, y)$ (with the 1-d polynomial argument $1 - x$) to substitute $1 - x$ for y and produce $P(x, 1 - x)$. Finally, the original version of "integral()" is applied to the 1-d polynomial $P(x, 1 - x)$ to calculate its integral over the unit interval:

```
return
    integral(indefiniteIntegral(p)(oneMinusx));
} //  integral on the triangle
```

Note the order in which the functions are called in the above code line. First, "indefiniteIntegral()" is called to produce $P(x, y)$. Then, "operator()" is applied to it with the polynomial argument "oneMinusx" to produce $P(x, 1 - x)$. This is done by the "operator()" of Section 24.7, with both 'S' and 'T' being polynomials of one variable. Finally, the original "integral()" version [rather than the present one] is applied to the polynomial of one variable $P(x, 1 - x)$ to calculate its integral over the unit interval, which is the desired result.

24.13 Integral on the Unit Tetrahedron

In Chapter 12, Section 12.22, we have described a method to integrate a polynomial of three variables $p(x, y, z)$ over the unit tetrahedron T in Figure 12.4. Here is the ordinary (nonmember) function that calculates this integral:

```
template<class T>
const T
integral(const
    polynomial<polynomial<polynomial<T> > >&p){
```

First, we need to produce the polynomial of two variables $1 - x - y$. This polynomial has two coefficients: $a_0(x) = 1 - x$ and $a_1(x) = -1$. We start by constructing $a_1(x)$, using the constructor that takes the integer argument 1 to indicate that this is a polynomial of degree 0 and the "double" argument -1 to indicate that its only coefficient is -1:

```
polynomial<T> minus1(1,-1.);
```

Next, we construct the polynomial $a_0(x) = 1 - x$ using the constructor that takes two "double" arguments to set the two coefficients, 1 and -1:

```
polynomial<T> oneMinusx(1.,-1.);
```

Finally, we construct the required polynomial $1 - x - y$ using the constructor that takes the two polynomial arguments $a_0(x) = 1 - x$ and $a_1(x) = -1$:

```
polynomial<polynomial<T> >
        oneMinusxMinusy(oneMinusx,minus1);
```

The "indefiniteIntegral" function is then applied to the original polynomial $p(x, y, z)$ to produce its indefinite integral with respect to the z spatial direction, denoted by $P(x, y, z)$ in Chapter 12, Section 12.22. The original "operator()" function in Section 24.7 above (with both 'S' and 'T' being polynomials of two variables) is then applied to it, to calculate it at the fixed argument $z = 1 - x - y$ and produce $P(x, y, 1 - x - y)$. Finally, the "integral()" version of the previous section is applied to $P(x, y, 1 - x - y)$ to produce the required output:

```
    return
      integral(indefiniteIntegral(p)(oneMinusxMinusy));
  } //  integral on the tetrahedron
```

24.14 Exercises

1. Use the "integral()" function in Section 24.12 above to compute the area of the unit triangle t in Figure 12.2 by integrating the constant 2-d polynomial $p(x, y) \equiv 1$ over it:

$$\int \int_t dx dy = 1/2.$$

2. The nodal basis functions in the unit triangle are the 2-d polynomials that have the value 1 at one vertex and 0 at the two other vertices:

$$p_0(x, y) = 1 - x - y$$
$$p_1(x, y) = x$$
$$p_2(x, y) = y.$$

Use the above "integral()" function to compute the integral of these functions over the unit triangle, and show that it is the same:

$$\int\int_t p_0 dx dy = \int\int_t p_1 dx dy$$
$$= \int\int_t p_2 dx dy$$
$$= 1/6.$$

3. Furthermore, show that the integral over the unit triangle of the squares of the nodal basis functions is also the same:

$$\int\int_t p_0^2 dx dy = \int\int_t p_1^2 dx dy$$
$$= \int\int_t p_2^2 dx dy$$
$$= 1/12.$$

4. Furthermore, show that the integral over the unit triangle of any product of two different nodal basis functions is the same:

$$\int\int_t p_0 p_1 dx dy = \int\int_t p_1 p_2 dx dy$$
$$= \int\int_t p_2 p_0 dx dy$$
$$= 1/24.$$

5. Use the "integral" function in Section 24.13 above to compute the volume of the unit tetrahedron T by integrating the constant 3-d polynomial $p(x, y, z) \equiv 1$ over it:

$$\int\int\int_T dx dy dz = 1/6.$$

6. The nodal basis functions in the unit tetrahedron are the polynomials that have the value 1 at one corner and 0 at the three other corners:

$$p_0(x, y, z) = 1 - x - y - z$$
$$p_1(x, y, z) = x$$
$$p_2(x, y, z) = y$$
$$p_3(x, y, z) = z.$$

Use the above "integral()" function to compute the integral of these function over the unit tetrahedron, and show that it is the same:

$$\int \int \int_T p_0 dx dy dz = \int \int \int_T p_1 dx dy dz$$

$$= \int \int \int_T p_2 dx dy dz$$

$$= \int \int \int_T p_3 dx dy dz$$

$$= 1/24.$$

7. Use the above "integral()" function to show that the integral of the square of the nodal basis functions is also the same:

$$\int \int \int_T p_0^2 dx dy dz = \int \int \int_T p_1^2 dx dy dz$$

$$= \int \int \int_T p_2^2 dx dy dz$$

$$= \int \int \int_T p_3^2 dx dy dz$$

$$= 1/60.$$

8. Use the above "integral()" function to show that the integral of products of nodal basis functions over the tetrahedron is insensitive to any permutation of the indices $\{0, 1, 2, 3\}$:

$$\int \int \int_T p_0 p_1 dx dy dz = \int \int \int_T p_1 p_2 dx dy dz$$

$$= \int \int \int_T p_2 p_3 dx dy dz$$

$$= \int \int \int_T p_3 p_0 dx dy dz$$

$$= \int \int \int_T p_0 p_2 dx dy dz$$

$$= \int \int \int_T p_1 p_3 dx dy dz$$

$$= 1/120$$

and

$$\int \int \int_T p_0 p_1 p_2 dx dy dz = \int \int \int_T p_1 p_2 p_3 dx dy dz$$
$$= \int \int \int_T p_2 p_3 p_0 dx dy dz$$
$$= \int \int \int_T p_3 p_0 p_1 dx dy dz$$
$$= 1/720.$$

The solution can be found in Section 28.16 in the appendix.

9. Write a function "d()" that takes a polynomial object p and an integer k to produce the kth derivative of p. The solution can be found in Section 28.17 in the appendix.

10. Write a function "d()" that takes a polynomial of two variables p and two integers j and k to produce the (j, k)th partial derivative of p. The solution can be found in Section 28.17 in the appendix.

11. Write a function "d()" that takes a polynomial of three variables p and three integers i, j, and k to produce the (i, j, k)th partial derivative of p. The solution can be found in Section 28.17 in the appendix.

Chapter 25

Sparse Polynomials

The polynomial object is implemented in the previous chapter under the assumption that it is rather dense, that is, that most of its coefficients, $a_0, a_1, \ldots, a_n$, are nonzero, hence have to be stored. In this chapter, however, we are particularly interested in sparse polynomials, in which most of the coefficients vanish, and only a few of them are nonzero. Thus, it makes no sense to store all of the coefficients: it may save a lot of time and storage resources to ignore the zero coefficients and store only the nonzero ones.

Sparse polynomials may contain any number of nonzero coefficients, which is not always known in advance in compilation time. Therefore, the best way to implement the sparse polynomial is in a linked list of nonzero coefficients. The recursive nature of the linked list is particularly helpful in defining arithmetic operations between sparse polynomials, including addition, multiplication, and composition.

25.1 The Monomial Object

In the standard implementation of a polynomial as a vector of coefficients, the ith coefficient a_i is placed in the ith component of the vector. This way, the coefficient a_i can be easily addressed through its virtual address: the index i that indicates its place in the vector of coefficients.

In the present implementation of a sparse polynomial as a linked list of nonzero coefficients, on the other hand, each nonzero coefficient a_i must be also accompanied with another field of type integer to contain the index i. In other words, the sparse polynomial must be implemented as a linked list of monomials of the form

$$a_i x^i.$$

Each such monomial must be implemented as a pair of two fields: a field of the yet unspecified type 'T' to contain the (real or complex) coefficient a_i, and an integer field to contain the index i, the power of x in the monomial.

Fortunately, we already have such an object available: the row-element object in Chapter 20, Section 20.1. The required monomial object can thus be derived from the base row-element object:

```
template<class T>
class monomial : public rowElement<T>{
public:
```

The "monomial" object inherits two fields from the base row-element object: the first field "value" (of type 'T') and the second field "column" (of type integer). Thanks to the fact that these fields are declared as protected (rather than the default private status) in the base "rowElement" class, they can be accessed from the derived "monomial" class. Still, it makes much more sense to access the second field, "column", by a new member function, that reflects its new meaning as the power i in the monomial $a_i x^i$:

```
int getPower() const{
  return column;
} // power in the monomial
```

The monomial object also inherits the default and copy constructors from the base row-element object. Nevertheless, the constructor that takes two arguments of type 'T' and "int" is not inherited properly and must be rewritten:

```
monomial(const T&coefficient=0, int power=0){
  value = coefficient;
  column = power;
} // constructor
```

Upon calling this constructor, the underlying row-element object is constructed by its own default constructor, so the "value" and "column" fields take the default values assigned to them in the default constructor of the "rowElement" class. These fields are then assigned more meaningful values in the body of the above constructor.

Furthermore, we also define a function that "converts" the monomial object to a 'T' object. In other words, this function just returns the field "value" in the base row-element object:

```
operator T() const{
  return value;
} // converter
```

With this converter, the user can define, say, a monomial 'm' (a variable of type "monomial<double>"), and then obtain its coefficient (the "value" field in the underlying "rowElement" object) by just writing "double(m)" or "(double)m".

Below we define some more operators that are special to the monomial object and reflect its nature.

25.2 Multiplying Monomials

In the base "rowElement" class in Chapter 20, Section 20.1, there is no operator to multiply two row-element objects by each other, as such an operation would make no mathematical sense. In the derived "monomial" class, on the other hand, such an operation indeed makes a lot of sense. For example, the product of the two monomials $a_i x^i$ and $b_j x^j$ yields the monomial

$$\left(a_i x^i\right)\left(b_j x^j\right) = a_i b_j x^{i+j}.$$

In other words, the coefficient in the product monomial is the product of the original coefficients, and the power in the product monomial is the sum of the original powers. This is indeed implemented in the following "*=" operator, which multiplies the current monomial $a_i x^i$ by an argument monomial of the form $b_j x^j$:

```
const monomial&operator*=(const monomial&m){
   value *=  m.value;
   column += m.column;
   return *this;
} // multiplying by a monomial
};
```

This completes the block of the "monomial" class. The above member operator can now be used to define an ordinary operator to multiply two monomial objects by each other:

```
template<class T>
const monomial<T>
operator*(const monomial<T>&m1,
        const monomial<T>&m2){
   return monomial<T>(m1) *= m2;
} // multiplying two monomials
```

The monomial object can now be used to define the sparse-polynomial object.

25.3 The Sparse-Polynomial Object

Once the monomial object is well defined, we can define the sparse-polynomial object as a linked list of monomial objects. This way, there is no limit on the number of monomials used in a sparse polynomial. Furthermore, monomials can be added to and dropped from an existing sparse polynomial in run time using member functions inherited from the base linked-list class.

```
template<class T>
class sparsePolynomial :
    public linkedList<monomial<T> >{
public:
```

First, we define a constructor that takes a monomial argument:

```
sparsePolynomial(const monomial<T>&m){
  item = m;
} // constructor with a monomial argument
```

Upon calling this constructor, the underlying linked-list object is constructed by the default constructor of the base "linkedList" class. The first item in this linked list, the field "item" inherited from the base "linkedList" class, is then set to have the same value as the monomial argument in the above constructor, using the assignment operator inherited from the base "rowElement" class.

Similarly, we also define a constructor that takes 'T' and integer arguments:

```
sparsePolynomial(const T&t=0, int n=0){
  item = monomial<T>(t,n);
} // constructor with T and integer arguments
```

Furthermore, we define a constructor that takes a linked-list argument:

```
sparsePolynomial(linkedList<monomial<T> >&){
} // trivial constructor
```

Upon calling this constructor, the copy constructor of the base "linkedList" class is invoked to initialize the underlying linked-list object with the argument passed to the constructor. Since this completes the construction of the sparse-polynomial object, no further action is needed, so the body of the above constructor is empty.

Furthermore, we define functions that read the first monomial in the sparse-polynomial object (the "item" field in the underlying linked list of monomials), the coefficient a_i in it (the "value" field in this monomial object), and its power i (the "column" field in the monomial "item"):

```
const monomial<T>& operator()() const{
  return item;
} // read first monomial

const T& getValue() const{
  return item.getValue();
} // read the coefficient in the first monomial

int getPower() const{
  return item.getColumn();
} // the power in the first monomial
```

Moreover, we define member functions not only to append but also to construct and append a new monomial at the end of the current sparse polynomial:

```
void append(const T&t, int n){
  monomial<T> mon(t,n);
  linkedList<monomial<T> >::append(mon);
} // construct and append a monomial

void append(monomial<T>&m){
  linkedList<monomial<T> >::append(m);
} // append a monomial
```

Finally, we also declare some member functions, to be defined later in detail: Note that the scalar argument passed to these functions is not necessarily of the same type as the coefficients in the monomials. This is why it is denoted by the extra template symbol, 'S', which may be different from 'T':

```
template<class S>
  const sparsePolynomial& operator*=(const S&);
template<class S>
  const T modifiedHorner(const S&, int) const;
template<class S>
  const T operator()(const S&) const;
template<class S>
  const S operator()(const S&, const S&) const;
template<class S>
  const S operator()(const S&, const S&,
      const S&) const;
};
```

This completes the block of the "sparsePolynomial" class.

Below we define the member operator that multiplies the current sparse-polynomial object by a scalar of type 'S'. (The common application of this operator is with 'S' being a monomial.) This operator is then used to multiply two sparse polynomials by each other.

25.4 Multiplying a Sparse Polynomial by a Scalar

The member operator that multiplies the current sparse-polynomial object by an object of type 'S' (which is usually a monomial) uses fully the recursive nature of the underlying linked-list object:

```
template<class T>
template<class S>
const sparsePolynomial<T>&
sparsePolynomial<T>::operator*=(const S&m){
   item *= m;
```

Indeed, first of all the first monomial, "item", is multiplied by the argument 'm' by invoking the relevant "*=" operator of the "monomial" class. Then, the present "*=" operator is called recursively to multiply the rest of the monomials in the current sparse-polynomial object by 'm' as well:

```
   if(next) *(sparsePolynomial<T>*)next *= m;
   return *this;
} //  current sparse polynomial times a scalar
```

Note that the field "next" that points to the rest of the monomials in the current sparse-polynomial object is of type pointer-to-linked-list; this is why it must be converted explicitly to type pointer-to-sparse-polynomial before the recursive application of the "*=" function can be used in the above code.

The above member operator is now used to define ordinary (nonmember) '*' operators to multiply sparse polynomial and scalar:

```
template<class T, class S>
const sparsePolynomial<T>
operator*(const S&m, const sparsePolynomial<T>&p){
   return sparsePolynomial<T>(p) *= m;
} //  scalar times sparse polynomial

template<class T, class S>
const sparsePolynomial<T>
operator*(const sparsePolynomial<T>&p, const S&m){
   return sparsePolynomial<T>(p) *= m;
} //  sparse polynomial times scalar
```

Below we use these operators to define the operator that multiplies two sparse polynomials by each other.

25.5 Multiplying Sparse Polynomials

The recursive nature of the underlying linked-list object is also used to define the operator that multiplies two sparse polynomials (denoted by 'p' and 'q') by each other:

```
template<class T>
const sparsePolynomial<T>
operator*(const sparsePolynomial<T>&p,
         const sparsePolynomial<T>&q){
  sparsePolynomial<T> result = q() * p;
```

Indeed, first of all, the operator defined in the previous section is used to multiply 'p' by the first monomial in 'q', "q()", returned by the "operator()" inherited from the base "linkedList" class. The result of this multiplication is stored in the local sparse polynomial named "result".

Then, the present function is called recursively to multiply 'p' by the rest of the monomials in 'q'. The result of this multiplication is stored in the local sparse polynomial named "rest":

```
if(q.readNext()){
  sparsePolynomial<T> rest =
     *(const sparsePolynomial<T>*)q.readNext() * p;
```

Note that the field "next" in the sparse-polynomial object 'q' that points to the rest of the monomials in 'q' is of type pointer-to-linked-list; therefore, it must be converted explicitly to type pointer-to-sparse-polynomial before the present function can be applied recursively to it.

Finally, the sum of the local sparse polynomials "result" and "rest" is returned as the required output:

```
  result += rest;
}
return result;
} // sparse polynomial times sparse polynomial
```

Below we also define the operator that adds two sparse polynomials to each other.

25.6 Adding Sparse Polynomials

The linked-list class in Chapter 17, Section 17.3, serves as a base class not only for the present sparse-polynomial class but also for the "mesh" class in Chapter 21, Section 21.10. Indeed, the mesh object is defined as a linked list of cells, where a cell object contains pointers to its nodes. Each node object in each cell contains an integer field named "sharingCells" to count how many cells in the mesh share the node as their joint vertex.

The mesh object is thus a very special object. Indeed, when a cell is added to it or used to construct it in the first place, the cell must be nonconstant, so

that the "sharingCells" fields of its nodes may increase when more and more
cells share them.

This is also why the base "linkedList" class in Chapter 17, Section 17.3, uses
nonconstant arguments in its member functions. This way, these functions can
be used in the derived "mesh" class as well.

The above feature also affects the present sparse-polynomial class, derived
from the base linked-list class. Indeed, in the "operator+" function that adds
two sparse polynomials to each other, one would naturally like to use constant arguments, to allow also the addition of temporary unnamed sparse-
polynomial objects by potential users of the function. Unfortunately, the
"operator+" function calls the "+ =" operator inherited from the base linkedlist class, which takes nonconstant argument only. As a result, constant arguments passed to the "operator+" function cannot be passed to the inner call
to the "+ =" operator: indeed, the compiler would refuse to accept them, out
of fear that the constant argument might change throughout the execution
of the "+ =" function. To overcome this problem, local nonconstant well-
named copies of these constant arguments must be used for this purpose. The
sum of these copies can then be converted implicitly from a mere linked list
to the required sparse-polynomial object (using the relevant constructor in
the "sparsePolynomial" class) and be returned as the desired output of the
function.

```
template<class T>
const sparsePolynomial<T>
operator+(const sparsePolynomial<T>&p,
          const sparsePolynomial<T>&q){
  sparsePolynomial<T> p2 = p;
  sparsePolynomial<T> q2 = q;
  p2 += q2;
  return p2;
}  //  adding two sparse polynomials
```

The above operator is used further in the modified Horner code below to
calculate the value of a sparse polynomial and the composition of two sparse
polynomials.

25.7 The Modified Horner Code

Here we implement the modified Horner algorithm in Chapter 12, Section
12.11 to compute the value of a given sparse polynomial p at a given argument
x. For this, we use the template function "power()" in Chapter 16, Section
16.17, which calculates the power x^n for any argument x of any type.

The two template symbols 'S' and 'T' provide extra flexibility to users of the function defined below. Indeed, they can use it not only with both 'S' and 'T' being scalars to compute the value $p(x)$ of a polynomial of one independent variable but also to compute the polynomial $p(x, y_0)$ obtained by fixing $y = y_0$ in a given polynomial $p(x, y)$ of two independent variables. These possibilities will be useful later on.

```
template<class T>
template<class S>
const T
sparsePolynomial<T>::modifiedHorner(const S&x,
          int n) const{
```

The function also takes an extra integer argument n; this parameter is subtracted from all the powers in the monomials in the current sparse polynomial. For example, if the current sparse polynomial is

$$p(x) = a_l x^l + a_k x^k + \cdots$$

for some $n \leq l < k$, then we can actually work with the more general sparse polynomial

$$a_l x^{l-n} + a_k x^{k-n} + \cdots.$$

Clearly, one can always use $n = 0$ to obtain the original polynomial $p(x)$ as a special case. This more general formulation that uses the extra parameter n, though, is particularly helpful in the recursion below.

As in the definition in Chapter 12, Section 12.11, the modified Horner algorithm distinguishes between two possible cases: the case in which the power $l - n$ in the first monomial is positive, in which the common factor x^{l-n} is taken out of parentheses,

```
return
  getPower() > n ?
    power(x, getPower() - n)
    * modifiedHorner(x, getPower())
```

and the case in which the first monomial is just a scalar (that is, $l = n$), in which it is just added to the result of the recursive call:

```
    :
    readNext() ?
      getValue() + ((sparsePolynomial<T>*)
      readNext())->modifiedHorner(x,n)
      :
  getValue();
} // modified Horner algorithm
```

As mentioned above, the original application of the "modifiedHorner" function to the original polynomial p uses the parameter $n = 0$. We therefore define an "operator()" member function that takes an argument x to calculate $p(x)$:

```
template<class T>
template<class S>
const T
sparsePolynomial<T>::operator()(const S&x) const{
  return modifiedHorner(x,0);
} //  compute p(x)
```

This operator makes life particularly easy for potential users of the sparse-polynomial class. Indeed, once they have defined a sparse polynomial 'p', they can just write "p(x)" to have its value for any scalar argument 'x'. Below we'll see how helpful this operator is also in computing the value of a polynomial $p(x, y)$ at given arguments x and y.

25.8 Polynomials of Two Variables

Here we implement sparse polynomials of two independent variables x and y of the form

$$p(x, y) \equiv \sum a_i(x)y^i,$$

where $a_i()$ is by itself a sparse polynomial in x rather than a mere scalar (Chapter 12, Section 12.13). In fact, we already have the required framework: $p(x, y)$ can be implemented as "sparsePolynomial<T>", where the template parameter 'T' is by itself the sparse-polynomial object that stores the polynomial $a_i(x)$. Here is how this framework is used to calculate the value of $p(x, y)$ for some given numbers x and y. This is done below in the member "operator()" function that takes two arguments, x and y. This function uses two template symbols: 'S', which stands for a scalar, and 'T', which stands for a polynomial in a variable of type 'S':

```
template<class T>
template<class S>
const S
sparsePolynomial<T>::operator()(const S&x,
    const S&y) const{
  return (*this)(y)(x);
} //  compute p(x,y)
```

For a current sparse polynomial $p(x, y)$ of two variables, this function uses two calls to the previous version of "operator()" that uses one argument only.

The first call uses the argument y to produce the polynomial $p(\cdot, y)$ (where y is now a fixed number rather than a variable). The second call uses then the argument x to produce the required number $p(x, y)$.

25.9 Polynomials of Three Variables

Similarly, a sparse polynomial of three variables of the form

$$p(x, y, z) \equiv \sum_i a_i(x, y) z^i$$

(where $a_i(x, y)$ is now a sparse polynomial of two variables) can be implemented as a sparse polynomial whose coefficients are no longer scalars but rather sparse polynomials of two variables. With this implementation, the calculation of $p(x, y, z)$ for a current sparse polynomial p and three given arguments x, y, and z is particularly easy:

```
template<class T>
template<class S>
const S
sparsePolynomial<T>::operator()(const S&x,
    const S&y, const S&z) const{
  return (*this)(z)(y)(x);
} //  compute p(x,y,z)
```

This function uses three calls to the original version of "operator()" with one argument only. The first call produces the sparse polynomial of two variables $p(\cdot, \cdot, z)$, where z is the fixed argument. The second call produces in turn the polynomial of one variable $p(\cdot, y, z)$, where both y and z are fixed arguments. Finally, the third call produces the required number $p(x, y, z)$, where x, y, and z are the given arguments. With this operator, the users can now define a sparse polynomial 'p' of three independent variables and obtain its value simply by writing "p(x,y,z)" for any suitable scalars 'x', 'y', and 'z'.

25.10 Exercises

1. Use the sparse-polynomial template class to define the polynomial

$$p(x) \equiv x + 2x^3.$$

2. Use the "operator*" function to calculate the product polynomial

$$p^2 = p \cdot p.$$

3. Use the "operator()" member function to calculate $p(x)$ for $x = 3, 5, 7$, and 9. The solution can be found in Section 28.19 in the appendix.

4. Implement the modified Horner algorithm in Chapter 12, Section 12.12, in the ordinary function "operator&" that takes two sparse-polynomial arguments to produce their composition. (With this function, the users can simply write "p&q" to obtain the composition of the sparse polynomials 'p' and 'q'.) The solution can be found in Section 28.18 in the appendix.

5. Use the above "operator&" function to calculate the composition polynomial $p \circ p$ for the above concrete polynomial $p(x) = x + 2x^3$.

6. Furthermore, use "operator&" twice to calculate $p \circ p \circ p$. (Do this in a single code line.) The solution can be found in Section 28.19 in the appendix.

7. The above exercise can be solved only thanks to the fact that the "modifiedHorner" and "operator()" member functions are declared as constant functions that cannot change the current sparse-polynomial object. Explain why this statement is true.

8. Use your code to verify that

$$p(p(p(x))) = (p \circ p \circ p)(x)$$

for $x = \pm 0.5$ and $x = \pm 1.5$.

9. Implement the polynomial of two variables

$$p_2(x, y) = (x + 2x^3)(y + y^3)$$

as a sparse polynomial with coefficients that are by themselves sparse polynomials in x. Use the "operator()" defined in Section 25.8 above to calculate $p_2(2, 3)$. The solution can be found in Section 28.19 in the appendix.

Chapter 26

Stiffness and Mass Matrices

In this chapter, we use the three-dimensional mesh of tetrahedra, along with the three-dimensional polynomials, to form the so-called stiffness and mass matrices, which are particularly useful in scientific computing [4] [29]. In fact, the mathematical objects developed throughout the book are combined here to form an advanced practical tool in applied science and engineering.

26.1 The Neumann Matrix

Let M be a mesh of tetrahedra, and N the set of nodes in it. Let $|N|$ denote the number of nodes in N. Here we define the $|N| \times |N|$ Neumann matrix A, which is most important in realistic applications in applied science and engineering.

Furthermore, let T be the unit tetrahedron in Figure 12.4. Recall from the exercises at the end of Chapter 24 that the volume of this tetrahedron is $1/6$.

In the following, we use notation like $\mathbf{i}$ for a node in N, and the corresponding integer i $(1 \leq i \leq |N|)$ to denote its index in the list of nodes in the mesh.

For each tetrahedron t in M, denote

$$t = (\mathbf{k}, \mathbf{l}, \mathbf{m}, \mathbf{n}),$$

where $\mathbf{k}$, $\mathbf{l}$, $\mathbf{m}$, and $\mathbf{n}$ are the 3-d vectors that are the corners of t, in some order that is determined arbitrarily in advance. Define the 3-d polynomials

$$
\begin{aligned}
P_{k,t}(x, y, z) &= 1 - x - y - z \\
P_{l,t}(x, y, z) &= x \\
P_{m,t}(x, y, z) &= y \\
P_{n,t}(x, y, z) &= z,
\end{aligned}
$$

where k, l, m, and n are the corresponding indices of $\mathbf{k}$, $\mathbf{l}$, $\mathbf{m}$, and $\mathbf{n}$ in the list of nodes in N. Furthermore, define the constant 3-d vectors (the gradients of these polynomials) by

$$g_{k,t} = \nabla P_{k,t} = \begin{pmatrix} -1 \\ -1 \\ -1 \end{pmatrix}$$

$$g_{l,t} = \nabla P_{l,t} = \begin{pmatrix} 1 \\ 0 \\ 0 \end{pmatrix}$$

$$g_{m,t} = \nabla P_{m,t} = \begin{pmatrix} 0 \\ 1 \\ 0 \end{pmatrix}$$

$$g_{n,t} = \nabla P_{n,t} = \begin{pmatrix} 0 \\ 0 \\ 1 \end{pmatrix}.$$

As in Chapter 12, Section 12.32, define the 3×3 matrix

$$S_t \equiv (\mathbf{l} - \mathbf{k} \mid \mathbf{m} - \mathbf{k} \mid \mathbf{n} - \mathbf{k}).$$

For every $1 \leq i, j \leq |N|$, the above notations are now used to define the element $a_{i,j}$ in A:

$$
\begin{aligned}
a_{i,j} &\equiv \sum_{t \in M,\ \mathbf{i,j} \in t} |\det(S_t)| g^t_{j,t} S_t^{-1} S_t^{-t} g_{i,t} \int \int \int_T dx\,dy\,dz \\
&= \frac{1}{6} \sum_{t \in M,\ \mathbf{i,j} \in t} |\det(S_t)| g^t_{j,t} S_t^{-1} S_t^{-t} g_{i,t},
\end{aligned}
$$

where $g^t_{j,t}$ is the row vector that is the transpose of the column vector $g_{j,t}$, and S_t^{-t} is the transpose of the inverse of the 3×3 matrix S_t. This completes the definition of the $|N| \times |N|$ Neumann matrix A, the major part in the stiffness matrix defined below.

26.2 The Boundary Matrix

Here we define the so-called "boundary" matrix $A^{(b)}$, the second component in the stiffness matrix defined below. Indeed, this $|N| \times |N|$ matrix will be added to the Neumann matrix A defined above to form the stiffness matrix below.

Let T denote the unit triangle in Figure 12.2. Furthermore, let ∂M denote the boundary of the mesh M, that contains only triangles that distinguish between M and the outside 3-D space that surrounds it. In fact, a triangle $t \in \partial M$ also satisfies

$$t \subset \mathbb{R}^3 \setminus \partial M.$$

In other words, a triangle t in ∂M cannot serve as a side in more than one tetrahedron in M:

$$|\{q \in M \mid t \subset q\}| = 1,$$

where q denotes the (only) tetrahedron in M that uses the triangle t as a side.

Let $b \subset \partial M$ be a subset that contains some of these boundary triangles. We denote a triangle in b by

$$t \equiv \triangle(\mathbf{k}, \mathbf{l}, \mathbf{m}),$$

where $\mathbf{k}$, $\mathbf{l}$, and $\mathbf{m}$ are the 3-d vectors that are the nodes in N that lie in the vertices of this triangle, ordered in some arbitrary order that is determined in advance.

Moreover, define the 2-d polynomials

$$q_{k,t}(x, y) = 1 - x - y$$
$$q_{l,t}(x, y) = x$$
$$q_{m,t}(x, y) = y,$$

where k, l, and m are the corresponding indices in the list of nodes in N. For every distinct indices $1 \leq i \neq j \leq |N|$, the (i, j)th element in $A^{(b)}$ is defined by

$$A_{i,j}^{(b)} \equiv \sum_{t=\triangle(\mathbf{i},\mathbf{j},\mathbf{k})\in b} \|(\mathbf{j}-\mathbf{i}) \times (\mathbf{k}-\mathbf{i})\|_2 \int\int_T q_{i,t} q_{j,t} dx dy$$

$$= \frac{1}{24} \sum_{t=\triangle(\mathbf{i},\mathbf{j},\mathbf{k})\in b} \|(\mathbf{j}-\mathbf{i}) \times (\mathbf{k}-\mathbf{i})\|_2,$$

where '$\times$' stands for the vector product defined in Chapter 9, Section 9.25.

This completes the definition of the off-diagonal elements in $A^{(b)}$. The main-diagonal elements are now defined to be the same as the sum of the corresponding off-diagonal elements in the same row:

$$A_{i,i}^{(b)} \equiv \sum_{j\neq i} A_{i,j}^{(b)}.$$

This completes the definition of the boundary matrix $A^{(b)}$.

26.3 The Stiffness Matrix

The stiffness matrix $A^{(s)}$ is now defined as the sum of the two matrices defined above:

$$A^{(s)} \equiv A + \alpha A^{(b)},$$

where α is some given parameter.

The stiffness matrix defined above is particularly important in practical applications in computational physics and engineering. Next, we define another important matrix, called the mass matrix.

26.4 The Mass Matrix

Here we define the mass matrix $A^{(m)}$. In the sequel, we use T to denote the unit tetrahedron in Figure 12.4, and use the 3-d integrals calculated over it in the exercises at the end of Chapter 24.

For $1 \leq i, j \leq |N|$, the (i, j)th element in $A^{(m)}$ is defined by

$$A^{(m)}_{i,j} = \sum_{t \in M, \; \mathbf{i}, \mathbf{j} \in t} |\det(S_t)| \int \int \int_T P_{i,t} P_{j,t} dx dy dz$$

$$= \begin{cases} \frac{1}{120} \sum_{t \in M, \; \mathbf{i},\mathbf{j} \in t} |\det(S_t)| & \text{if } i \neq j \\ \frac{1}{60} \sum_{t \in M, \; \mathbf{i},\mathbf{j} \in t} |\det(S_t)| & \text{if } i = j. \end{cases}$$

This completes the definition of the mass matrix $A^{(m)}$. In the next section, we extend this definition to obtain the so-called Newton's mass matrix.

26.5 Newton's Mass Matrix

Let u be a grid function that returns the real number $u_{\mathbf{i}}$ for every node $\mathbf{i}$ in N. In other words,

$$u : N \to \mathbb{R},$$

or

$$u \in \mathbb{R}^N.$$

We now use the 3-d polynomials defined in Section 26.1 above to define another grid function, denoted by $f(u)$:

$$f(u)_{\mathbf{i}} \equiv \sum_{t \in M, \; \mathbf{i} \in t = (\mathbf{k},\mathbf{l},\mathbf{m},\mathbf{n})} |\det(S_t)| \int \int \int_T$$

$$(u_{\mathbf{k}} P_{k,t} + u_{\mathbf{l}} P_{l,t} + u_{\mathbf{m}} P_{m,t} + u_{\mathbf{n}} P_{n,t})^3 P_{i,t} dx dy dz$$

(for every node $\mathbf{i} \in N$). In fact, this defines a function (or operator)

$$f : \mathbb{R}^N \to \mathbb{R}^N,$$

which takes a grid function u to produce the new grid function $f(u)$.

Clearly, f is a nonlinear function. Still, for every fixed grid function v, one may define a linear function that approximates f best in a small neighborhood around v. This linear function from $\mathbb{R}^N$ to $\mathbb{R}^N$ is called the linearization of f at v. Clearly, it depends on the particular grid function v that has been picked. To define it, we need first to define Newton's mass matrix.

For any fixed grid function v, Newton's mass matrix, $A^{(n)}(v)$ is defined as follows. For every $1 \leq i, j \leq |N|$, the (i, j)th element in $A^{(n)}(v)$ is defined by [partial derivation (under the integral sign) of $f(v)_i$ with respect to v_j]

$$A^{(n)}(v)_{i,j} \equiv \sum_{t \in M, \ \mathbf{i}, \mathbf{j} \in t = (\mathbf{k}, \mathbf{l}, \mathbf{m}, \mathbf{n})} |\det(S_t)|$$
$$\cdot \int \int \int_T 3(v_{\mathbf{k}} P_{k,t} + v_{\mathbf{l}} P_{l,t} + v_{\mathbf{m}} P_{m,t} + v_{\mathbf{n}} P_{n,t})^2 P_{i,t} P_{j,t} dx dy dz.$$

The matrix $A^{(n)}(v)$ is called the Jacobian matrix of f at v. Using it, one can now define the linearization of f at v as the linear mapping

$$u \to A^{(n)}(v)u, \qquad u \in \mathbb{R}^N.$$

26.6 Helmholtz Mass Matrix

Consider now a complex-valued grid function $v : N \to \mathbb{C}$, or $v \in \mathbb{C}^N$. In fact, v can be written in terms of its real and imaginary parts:

$$v_{\mathbf{i}} = \Re v_{\mathbf{i}} + \sqrt{-1} \Im v_{\mathbf{i}}.$$

[Such a separation is used in [3] to apply a finite-difference scheme to the one-dimensional nonlinear Helmholtz equation.] Furthermore, the complex number $v_{\mathbf{i}}$ can be interpreted as a point in the 2-d Cartesian plane:

$$v_{\mathbf{i}} = \begin{pmatrix} \Re v_{\mathbf{i}} \\ \Im v_{\mathbf{i}} \end{pmatrix}$$

(Chapter 5). In this interpretation, v can be viewed as a function from N to $\mathbb{R}^2$, or

$$v \in (\mathbb{R}^2)^N.$$

This is why the function v is also called a vector grid function.

Consider now a function (or operator) $f : \mathbb{C}^N \to \mathbb{C}^N$ that transforms a complex-valued grid function v into another complex-valued grid function

$f(v)$. In fact, $f(v)$ can also be written in terms of its real and imaginary parts.

$$f(v)_{\mathbf{i}} = \Re f(v)_{\mathbf{i}} + \sqrt{-1}\,\Im f(v)_{\mathbf{i}}, \qquad \mathbf{i} \in N.$$

Furthermore, $f(v)_{\mathbf{i}}$ can also be interpreted as a point in the 2-d Cartesian plane:

$$f(v)_{\mathbf{i}} = \begin{pmatrix} \Re f(v)_{\mathbf{i}} \\ \Im f(v)_{\mathbf{i}} \end{pmatrix}.$$

This way, $f(v)$ is interpreted as a vector grid function as well, and f is actually interpreted as an operator on the set of all vector grid functions:

$$f : (\mathbb{R}^2)^N \to (\mathbb{R}^2)^N.$$

Let us now define some helpful notation. For every tetrahedron $t = (\mathbf{k}, \mathbf{l}, \mathbf{m}, \mathbf{n}) \in M$, let us denote for short

$$P_t(v) = v_{\mathbf{k}} P_{k,t} + v_{\mathbf{l}} P_{l,t} + v_{\mathbf{m}} P_{m,t} + v_{\mathbf{n}} P_{n,t}.$$

In fact, this 3-d polynomial can also be written in terms of its real and imaginary parts:

$$
\begin{aligned}
P_t(v) &= \Re P_t(v) + \sqrt{-1}\,\Im P_t(v) \\
&= \Re v_{\mathbf{k}} P_{k,t} + \Re v_{\mathbf{l}} P_{l,t} + \Re v_{\mathbf{m}} P_{m,t} + \Re v_{\mathbf{n}} P_{n,t} \\
&\quad + \sqrt{(-1)}\left(\Im v_{\mathbf{k}} P_{k,t} + \Im v_{\mathbf{l}} P_{l,t} + \Im v_{\mathbf{m}} P_{m,t} + \Im v_{\mathbf{n}} P_{n,t}\right).
\end{aligned}
$$

Let us use this notation to define f more explicitly as a generalization of the function f in Section 26.5 above:

$f(v)_{\mathbf{i}}$

$$
\begin{aligned}
&\equiv \sum_{t \in M,\ \mathbf{i} \in t} |\det(S_t)| \int\!\!\int\!\!\int_T |P_t(v)|^2 P_t(v) P_{i,t}\,dxdydz \\
&= \sum_{t \in M,\ \mathbf{i} \in t} |\det(S_t)| \int\!\!\int\!\!\int_T \left(\Re^2 P_t(v) + \Im^2 P_t(v)\right) \Re P_t(v) P_{i,t}\,dxdydz \\
&\quad + \sqrt{-1} \sum_{t \in M,\ \mathbf{i} \in t} |\det(S_t)| \int\!\!\int\!\!\int_T \left(\Re^2 P_t(v) + \Im^2 P_t(v)\right) \Im P_t(v) P_{i,t}\,dxdydz
\end{aligned}
$$

(for every node $\mathbf{i} \in N$). This way, f is written in terms of its real and imaginary parts, as before.

The so-called Helmholtz mass matrix can be viewed as a generalization of Newton's mass matrix, with elements that are no longer scalars but rather 2×2 block matrices. More specifically, the (i,j)th 2×2 block in the Helmholtz mass matrix $A^{(h)}(v)$ is defined as follows. Its upper left element is defined by [partial derivation (under the integral sign) of $\Re f(v)_{\mathbf{i}}$ with respect to $\Re v_{\mathbf{j}}$]

$$\left(A^{(h)}(v)_{i,j}\right)_{1,1}$$
$$\equiv \sum_{t\in M,\ \mathbf{i}\in t} |\det(S_t)| \int\int\int_T \left(3\Re^2 P_t(v) + \Im^2 P_t(v)\right) P_{i,t} P_{j,t} dxdydz.$$

Furthermore, its upper right element is defined by [partial derivation (under the integral sign) of $\Re f(v)_{\mathbf{i}}$ with respect to $\Im v_{\mathbf{j}}$]

$$\left(A^{(h)}(v)_{i,j}\right)_{1,2} \equiv \sum_{t\in M,\ \mathbf{i}\in t} |\det(S_t)| \int\int\int_T 2\Re P_t(v)\Im P_t(v) P_{i,t} P_{j,t} dxdydz.$$

Similarly, its lower left element is defined by [partial derivation (under the integral sign) of $\Im f(v)_{\mathbf{i}}$ with respect to $\Re v_{\mathbf{j}}$]

$$\left(A^{(h)}(v)_{i,j}\right)_{2,1} \equiv \sum_{t\in M,\ \mathbf{i}\in t} |\det(S_t)| \int\int\int_T 2\Re P_t(v)\Im P_t(v) P_{i,t} P_{j,t} dxdydz$$
$$= \left(A^{(h)}(v)_{i,j}\right)_{1,2}.$$

Finally, its lower right element is defined by [partial derivation (under the integral sign) of $\Im f(v)_{\mathbf{i}}$ with respect to $\Im v_{\mathbf{j}}$]

$$\left(A^{(h)}(v)_{i,j}\right)_{2,2}$$
$$\equiv \sum_{t\in M,\ \mathbf{i}\in t} |\det(S_t)| \int\int\int_T \left(\Re^2 P_t(v) + 3\Im^2 P_t(v)\right) P_{i,t} P_{j,t} dxdydz.$$

This completes the definition of the Helmholtz mass matrix $A^{(h)}(v)$ at the given vector grid function v.

The power of object-oriented programming is apparent in the implementation of this matrix. Indeed, it can be implemented in a sparse-matrix object, with 'T' being a "matrix2" object.

The Helmholtz mass matrix can now be used to linearize f at v. Indeed, when v is interpreted as a function in $(\mathbb{R}^2)^N$ and f is interpreted as a (non-linear) operator

$$f : (\mathbb{R}^2)^N \to (\mathbb{R}^2)^N$$

that maps every vector grid function $u \in (\mathbb{R}^2)^N$ into a vector grid function $f(u) \in (\mathbb{R}^2)^N$, the best linear approximation to f in a neighborhood of v is the linear mapping

$$u \to A^{(h)}(v)u, \qquad u \in (\mathbb{R}^2)^N.$$

26.7 Helmholtz Matrix

Recall from the exercises at the end of Chapter 5 that every complex number $c = a + \sqrt{-1}b$ can be implemented as the 2×2 matrix

$$\begin{pmatrix} a & -b \\ b & a \end{pmatrix}.$$

In particular, when c is actually a real number with no imaginary part ($b = 0$), it can be realized as the diagonal 2×2 matrix

$$\begin{pmatrix} a & 0 \\ 0 & a \end{pmatrix}.$$

Using this representation, the elements in the Neumann matrix A, the boundary matrix $A^{(b)}$, and the mass matrix $A^{(m)}$ can now be replaced by diagonal 2×2 blocks. This way, these elements are interpreted not merely as real numbers in the real axis in Figure 5.2 but rather as real numbers (with no imaginary parts) in the complex plane in Figure 5.4.

This is not merely a semantic difference. In fact, it also makes a practical difference: the matrices A, $A^{(b)}$, and $A^{(m)}$ can now be multiplied by a constant complex coefficient, represented as a 2×2 matrix as well.

The above implementation allows one to define the so-called Helmholtz matrix

$$H(v) \equiv A - \sqrt{-1}\beta A^{(b)} - \beta^2 A^{(m)} - \beta^2 A^{(h)}(v),$$

where β is a complex number with a positive real part. (In most applications, β is a positive integer number.) Here β is represented by the 2×2 matrix

$$\beta = \begin{pmatrix} \Re\beta & -\Im\beta \\ \Im\beta & \Re\beta \end{pmatrix},$$

the elements in A, $A^{(b)}$, and $A^{(m)}$ are implemented as the 2×2 blocks that represent the original real elements embedded in the complex plane, and the imaginary number $\sqrt{-1}$ is implemented as the 2×2 matrix

$$\sqrt{-1} = \begin{pmatrix} 0 & -1 \\ 1 & 0 \end{pmatrix}.$$

The construction of this matrix is left to the exercises, with solutions in the appendix.

26.8 Newton's Iteration

Consider the nonlinear mapping $F : (\mathbb{R}^2)^N \to (\mathbb{R}^2)^N$ defined by

$$F(u) \equiv \left(A - \sqrt{-1} \beta A^{(b)} - \beta^2 A^{(m)} \right) u - \beta^2 f(u),$$

$u \in (\mathbb{R}^2)^N$. Consider the following problem: find a vector grid function $v \in (\mathbb{R}^2)^N$ for which

$$F(v) = \mathbf{0}$$

(where $\mathbf{0}$ is the $2|N|$-dimensional zero vector). To solve this problem, we'll use the property that the nonlinear mapping

$$u \rightarrow F(u)$$

is linearized at v by the linear mapping

$$u \rightarrow H(v)u.$$

Indeed, this property allows one to add suitable correction terms in an iterative method that converges rapidly to the desired solution v.

The Newton iteration that converges to the required solution v is defined as follows:

1. Let $v^{(0)} \in (\mathbb{R}^2)^N$ be a suitable initial guess.
2. For $i = 0, 1, 2, \ldots$,
   a) Solve the linear system

$$H(v^{(i)})x = F(v^{(i)})$$

   for the unknown vector x.
   b) Define

$$v^{(i+1)} \equiv v^{(i)} - x.$$

Below we'll see how the Newton iteration can also be applied to a restrained problem.

26.9 Dirichlet Boundary Conditions

Here we consider a restrained problem, in which the values of the unknown grid function v are prescribed in advance at a subset G of boundary nodes. These restraints are called Dirichlet boundary conditions.

Let

$$G \equiv N \cap (\partial M \setminus b)$$

be the set of boundary nodes that do not lie in the set b (defined in Section 26.2). Let

$$R \equiv N \setminus G$$

be the complementary set of nodes that are not in G. The splitting

$$N = R \cup G$$

in terms of the disjoint subsets R and G induces (after a suitable reordering of components) the representation of any vector grid function $u \in (\mathbb{R}^2)^N$ in terms of $2|R|$-dimensional and $2|G|$-dimensional subvectors:

$$u = \begin{pmatrix} u_R \\ u_G \end{pmatrix}.$$

Furthermore, the nonlinear mapping $F(u)$ defined above can also be written as

$$F(u) = \begin{pmatrix} F_R(u) \\ F_G(u) \end{pmatrix}.$$

Moreover, this splitting also induces (after a suitable reordering of rows and columns) the block representation

$$H(u) = \begin{pmatrix} H_{RR}(u) & H_{RG}(u) \\ H_{GR}(u) & H_{GG}(u) \end{pmatrix}.$$

Consider now the following restrained problem. Let D_G be a given $2|G|$-dimensional vector. Find a vector grid function $v \in (\mathbb{R}^2)^N$ that satisfies both

$$v_G = D_G$$

and

$$F_R(v) = \mathbf{0}$$

(where $\mathbf{0}$ is the zero $2|R|$-dimensional vector).

This formulation uses $2|R|$ algebraic equations. Still, one could add to it other $2|G|$ dummy (trivial) equations to come up with an equivalent unrestrained formulation that uses a total of $2|N|$ algebraic equations: find $v \in (\mathbb{R}^2)^N$ that satisfies

$$\tilde{F}(v) \equiv \begin{pmatrix} F_R(v) \\ v_G - D_G \end{pmatrix} = \mathbf{0}$$

(where $\mathbf{0}$ is the $2|N|$-dimensional zero vector).

Clearly, the nonlinear mapping

$$u \to \tilde{F}(u)$$

is linearized at v by the linear mapping

$$u \to \begin{pmatrix} H_{RR}(v) & H_{RG}(v) \\ 0 & I \end{pmatrix} u,$$

where I is the identity matrix of order $2|G|$. Thus, the Newton iteration to solve this problem takes the following form:

1. Let $v^{(0)} \in (\mathbb{R}^2)^N$ be a suitable initial guess of the form

$$v^{(0)} = \begin{pmatrix} v_R^{(0)} \\ D_G \end{pmatrix}.$$

2. For $i = 0, 1, 2, \ldots,$
   a) Solve the linear system

$$\begin{pmatrix} H_{RR}(v^{(i)}) & H_{RG}(v^{(i)}) \\ 0 & I \end{pmatrix} x = \tilde{F}(v^{(i)}),$$

where I is the $2|G| \times 2|G|$ identity matrix. In fact, since (by induction on i) the second subvector in both $\tilde{F}(v^{(i)})$ and x vanishes, this system is equivalent to

$$\begin{pmatrix} H_{RR}(v^{(i)}) & 0 \\ 0 & I \end{pmatrix} x = \tilde{F}(v^{(i)}).$$

(The matrix on the left-hand side is called the Dirichlet matrix.)
   b) Define

$$v^{(i+1)} \equiv v^{(i)} - x.$$

26.10 Exercises

1. Show that the Neumann matrix A defined in Section 26.1 above is symmetric.
2. Show that the row sums in A are all equal to zero.
3. Conclude that the column sums in A are equal to zero as well.
4. Let $w \in \mathbb{R}^N$ be the constant grid function:

$$w_{\mathbf{i}} = C, \qquad \mathbf{i} \in N,$$

for some constant number C. Show that

$$Aw = 0.$$

5. Conclude that A is singular (A^{-1} doesn't exist).
6. Conclude that

$$\det(A) = 0.$$

7. Show that all the elements in the boundary matrix $A^{(b)}$ are nonnegative.
8. Show that the boundary matrix $A^{(b)}$ is symmetric.
9. Conclude that the stiffness matrix $A^{(s)}$ is symmetric as well.

10. Write the constructor that takes the 3-d mesh M as an argument and produces the stiffness matrix $A^{(s)}$ (with $\alpha = 1$) in a sparse-matrix object. It is assumed that the mesh M covers the unit cube, as in Section 28.15 in the appendix. Furthermore, it is also assumed that b, the subset of the boundary ∂M, contains five sides of the unit cube:

$$b = \{(x, y, z) \mid (1-x)y(1-y)z(1-z) = 0\}.$$

Moreover, it is assumed that at least one refinement step has been applied to M; otherwise, coarse triangles in the bottom side of the cube may contribute to $A^{(b)}$, although they should be excluded from b. The solution can be found in Section 28.20 in the appendix.

11. Show that the mass matrix $A^{(m)}$ is symmetric.

12. Show that all the elements in $A^{(m)}$ are nonnegative.

13. Assume that the definition of the above grid function w uses the constant $C = 1/sqrt3$. Show that the mass matrix can be obtained as a special case of Newton's mass matrix:

$$A^{(m)} = A^{(n)}(w).$$

14. Write the constructor that takes the 3-d mesh M as an argument and produces Newton's mass matrix in a sparse-matrix object. The 3-d polynomials $P_{k,t}$, $P_{l,t}$, $P_{m,t}$, and $P_{n,t}$ of Section 26.1 should be passed to the constructor by reference in the 3-d polynomial objects "P[0]", "P[1]", "P[2]", and "P[3]" defined in Section 28.16 in the appendix. Note that the constructor should also have two "dynamicVector" arguments that are passed to it by reference: one to take the input grid function v, and another one to store the output grid function $f(v)$ that is also produced by the constructor. The solution can be found in Section 28.21 in the appendix.

15. Modify the above constructor to produce Helmholtz mass matrix. The solution can be found in Section 28.22 in the appendix.

16. Modify the constructor of the stiffness matrix $A^{(s)}$ above to produce 2×2 block elements of the form

$$\begin{pmatrix} a_{i,j} & 0 \\ 0 & a_{i,j} \end{pmatrix} - \begin{pmatrix} 0 & -1 \\ 1 & 0 \end{pmatrix} \begin{pmatrix} \Re\beta & -\Im\beta \\ \Im\beta & \Re\beta \end{pmatrix} \begin{pmatrix} A^{(b)}_{i,j} & 0 \\ 0 & A^{(b)}_{i,j} \end{pmatrix}$$

rather than the original scalar elements

$$a_{i,j} + \alpha A^{(b)}_{i,j}$$

in the original constructor.

17. Show that the mass matrix of Section 26.4 above can be represented equivalently with 2×2 block elements as

$$A^{(m)} = A^{(h)}(w) + A^{(h)}(\sqrt{-1}w),$$

where w can be interpreted as the constant vector grid function

$$w_{\mathbf{i}} \equiv \begin{pmatrix} 1/\sqrt{3} \\ 0 \end{pmatrix},$$

and $\sqrt{-1}w$ can be interpreted as the constant vector grid function

$$\sqrt{-1}w_{\mathbf{i}} \equiv \begin{pmatrix} 0 \\ 1/\sqrt{3} \end{pmatrix},$$

for every node $\mathbf{i} \in N$.

18. Combine the above modified constructors to produce the Helmholtz matrix

$$H(v) \equiv A - \sqrt{-1}\beta A^{(b)} - \beta^2 A^{(m)} - \beta^2 A^{(h)}(v),$$

where A, $A^{(b)}$, and $A^{(m)}$ are interpreted to have 2×2 block elements that are the same as the original real elements embedded in the complex plane, the complex number β is implemented as the 2×2 matrix

$$\beta = \begin{pmatrix} \Re\beta & -\Im\beta \\ \Im\beta & \Re\beta \end{pmatrix},$$

and the imaginary number $\sqrt{-1}$ is implemented as the 2×2 matrix

$$\sqrt{-1} = \begin{pmatrix} 0 & -1 \\ 1 & 0 \end{pmatrix}.$$

19. Show that the nonlinear mapping

$$u \to F(u) \equiv \left(A - \sqrt{-1}\beta A^{(b)} - \beta^2 A^{(m)} \right) u - \beta^2 f(u)$$

[where $u \in (\mathbb{R}^2)^N$] is linearized at the fixed vector grid function $v \in (\mathbb{R}^2)^N$ by the linear mapping

$$u \to H(v)u \equiv \left(A - \sqrt{-1}\beta A^{(b)} - \beta^2 A^{(m)} - \beta^2 A^{(h)}(v) \right) u$$

$(u \in (\mathbb{R}^2)^N)$.

20. Show that the nonlinear mapping

$$u \to \tilde{F}(u)$$

$(u \in (\mathbb{R}^2)^N)$ defined in Section 26.9 above is linearized at the fixed vector grid function $v \in (\mathbb{R}^2)^N$ by the linear mapping

$$u \to \begin{pmatrix} H_{RR}(v) & H_{RG}(v) \\ 0 & I \end{pmatrix} u$$

$(u \in (\mathbb{R}^2)^N)$, where I is the identity matrix of order $2|G|$.

21. Use mathematical induction on $i \geq 0$ to show that, in the Newton iteration in Section 26.9 above,

$$v_G^{(i)} = \mathbf{0}$$

(where $\mathbf{0}$ is the $2|G|$-dimensional zero vector).

22. Conclude that, in the Newton iteration in Section 26.9, the system

$$\begin{pmatrix} H_{RR}(v^{(i)}) & H_{RG}(v^{(i)}) \\ 0 & I \end{pmatrix} x = \tilde{F}(v^{(i)})$$

is equivalent to the system

$$\begin{pmatrix} H_{RR}(v^{(i)}) & 0 \\ 0 & I \end{pmatrix} x = \tilde{F}(v^{(i)}).$$

23. Write a function that takes the Helmholtz matrix as an argument and produces the Dirichlet matrix associated with it. For the solution, see a similar exercise at the end of Chapter 27.

24. Assume that there is a computer function "solve()" that takes a sparse matrix C and a right-hand-side vector r as arguments, solves the linear system

$$Cx = r,$$

and returns the vector solution x. Use this function to implement the Newton iteration in Section 26.8 above. The solution will be placed on the website of the book.

25. Use the above function also to implement the Newton iteration in Section 26.9 above. The solution will be placed on the website of the book.

Chapter 27

Splines

In the previous chapter, we have considered polynomials that are linear in the tetrahedron, and used them to form the matrix that linearizes a nonlinear operator defined on $\mathbb{R}^N$ or $\mathbb{C}^N$ (the set of real-valued or complex-valued grid functions). In this chapter, we extend this framework to the more difficult case of functions that, in each tetrahedron in the mesh, can be viewed as a polynomial of degree five. Furthermore, the functions are rather smooth: they are continuous in the entire mesh and have continuous gradients across edges in the entire mesh. They are therefore suitable to help extend a given grid function defined on the individual nodes into a smooth function defined in the entire mesh. This smooth extension is called the spline of the original grid function [24] [32].

The values of the original grid function at the individual nodes in the mesh may be viewed as Dirichlet conditions or data, which must be observed when the required smooth function is sought. In fact, the spline problem discussed below is to find a function with minimum energy (or minimal gradients and variation) in the entire mesh, which agrees with the original values of the original grid function (the Dirichlet data) at the individual nodes. The solution of this problem is given below in terms of the solution to a linear system of algebraic equations.

27.1 The Indexing Scheme

In Chapter 12, Section 12.30 above, we have seen that 10 degrees of freedom (or values of partial derivatives that one needs to specify) are associated with each corner of the unit tetrahedron T in Figure 12.4. This makes a total of 40 degrees of freedom for the four corners. Furthermore, two more degrees of freedom are associated with the midpoint of each of the six edges of the tetrahedron. Moreover, one more degree of freedom is associated with the midpoint of each of the four sides of the tetrahedron. This makes the total amount of 56 degrees of freedom required to specify a polynomial of degree up to five uniquely. In fact, once the polynomial and its partial derivatives are specified accordingly at the corners, edge midpoints, and side midpoints, the polynomial is well defined in the tetrahedron and indeed in the entire 3-d Cartesian space.

Here, however, we are interested in the definition of the polynomial in the tetrahedron only. In fact, in each individual tetrahedron in the mesh we may want to define the function differently, according to the degrees of freedom in this particular tetrahedron. This would make a piecewise polynomial function, namely, a function whose restriction to each tetrahedron in the mesh is a polynomial of degree up to five. Furthermore, thanks to the fact that the degrees of freedom are the same in common corners, edges, and sides of neighbor tetrahedra in the mesh, the function would be fairly smooth.

Using the above, we extend the notion of a basis function in an individual tetrahedron to the more general notion of a basis function defined in the entire mesh of tetrahedra. For this, we assume that each node in the mesh has 10 degrees of freedom associated with it, each edge midpoint in the mesh has two degrees of freedom associated with it, and each side midpoint in the mesh has one degree of freedom associated with it. A basis function in the mesh is obtained by specifying only one degree of freedom to be of value 1, whereas all the others vanish.

To be more precise, we must have an indexing scheme to index the degrees of freedom in the mesh. From the above, we have a total of

$$K \equiv 10|N| + 2|E| + |L|$$

degrees of freedom in the entire mesh, where N is the set of nodes in the mesh (so $|N|$ is the total number of nodes), E is the set of edges in the mesh (so $|E|$ is the total number of edges), and L is the set of sides in the mesh (so $|L|$ is the total number of sides).

The degrees of freedom in the entire mesh can now be indexed from 0 to $K - 1$. Specifying one of them to be of value 1 whereas all the others vanish defines uniquely a particular basis function in the mesh.

27.2 Basis Functions in the Mesh

Here we use the above indexing scheme of the degrees of freedom in the entire mesh and the basis functions defined in each particular tetrahedron t in it to define basis functions in the entire mesh. For this purpose, let us first introduce some notations.

For any node $\mathbf{i} \in N$, let $0 \le |\mathbf{i}| < |N|$ denote its index in the list of nodes in the mesh. Furthermore, let us write

$$t = (\mathbf{k}, \mathbf{l}, \mathbf{m}, \mathbf{n}),$$

where $\mathbf{k}$, $\mathbf{l}$, $\mathbf{m}$, and $\mathbf{n}$ are the corners of t in some arbitrary order that is determined in advance.

Let us now use the basis functions $R_{i,t}$ defined in some tetrahedron t in Chapter 12, Sections 12.36–12.40 to define in t corresponding basis functions of the form $p_{n,t}$, using the above indexing scheme of the degrees of freedom in the entire mesh. In fact, the new polynomial $p_{n,t}$ is identical to some polynomial $R_{i,t}$; the only difference is in the index:

$$p_{10|\mathbf{k}|+q,t} = R_{q,t}$$
$$p_{10|\mathbf{l}|+q,t} = R_{10+q,t}$$
$$p_{10|\mathbf{m}|+q,t} = R_{20+q,t}$$
$$p_{10|\mathbf{n}|+q,t} = R_{30+q,t}$$

(where q is an integer between 0 and 9). This defines 40 basis functions corresponding to the first 40 basis functions in Chapter 12, Sections 12.30–12.31.

Next, we define basis functions corresponding to two normal derivatives at the midpoint of each of the six edges of T. For this, however, we must first index the edges in t in the list of edges in the entire mesh. Indeed, for any edge of the form $(\mathbf{i}, \mathbf{j})$ in t (with endpoints $\mathbf{i}$ and $\mathbf{j}$ that also serve as corners in t), let $|(\mathbf{i}, \mathbf{j})|$ denote its index in the list of edges in E ($0 \le |(\mathbf{i}, \mathbf{j})| < |E|$). With this notation, we can now define the basis functions associated with edge midpoints in t by

$$p_{10|N|+2|(\mathbf{k},\mathbf{l})|+q,t} = R_{40+q,t}$$
$$p_{10|N|+2|(\mathbf{k},\mathbf{m})|+q,t} = R_{42+q,t}$$
$$p_{10|N|+2|(\mathbf{k},\mathbf{n})|+q,t} = R_{44+q,t}$$
$$p_{10|N|+2|(\mathbf{l},\mathbf{m})|+q,t} = R_{46+q,t}$$
$$p_{10|N|+2|(\mathbf{l},\mathbf{n})|+q,t} = R_{48+q,t}$$
$$p_{10|N|+2|(\mathbf{m},\mathbf{n})|+q,t} = R_{50+q,t},$$

where q is either 0 or 1. Finally, let us define the four basis functions that correspond to side midpoints. For this, however, we must first have an index for the sides in the mesh. For every side of the form $(\mathbf{i}, \mathbf{j}, \mathbf{h})$ in t (whose corners $\mathbf{i}$, $\mathbf{j}$, and $\mathbf{h}$ are also corners in t), let $|(\mathbf{i}, \mathbf{j}, \mathbf{h})|$ denote its index in the list of sides in the mesh $0 \le |(\mathbf{i}, \mathbf{j}, \mathbf{h})| < |L||$. Then, the basis functions associated with side midpoints are defined by

$$p_{10|N|+2|E|+|(\mathbf{k},\mathbf{l},\mathbf{m})|,t} = R_{52,t}$$
$$p_{10|N|+2|E|+|(\mathbf{k},\mathbf{l},\mathbf{n})|,t} = R_{53,t}$$
$$p_{10|N|+2|E|+|(\mathbf{k},\mathbf{m},\mathbf{n})|,t} = R_{54,t}$$
$$p_{10|N|+2|E|+|(\mathbf{l},\mathbf{m},\mathbf{n})|,t} = R_{55,t}.$$

We also say that the indices in the left-hand sides above are associated with t, as they index degrees of freedom associated with one of the corners, edge midpoints, or side midpoints in t. This property is denoted by

$$10|\mathbf{k}| + q \sim t$$
$$10|\mathbf{l}| + q \sim t$$
$$10|\mathbf{m}| + q \sim t$$
$$10|\mathbf{n}| + q \sim t$$

$(0 \le q < 10)$,

$$10|N| + 2|(\mathbf{k}, \mathbf{l})| + q \sim t$$
$$10|N| + 2|(\mathbf{k}, \mathbf{m})| + q \sim t$$
$$10|N| + 2|(\mathbf{k}, \mathbf{n})| + q \sim t$$
$$10|N| + 2|(\mathbf{l}, \mathbf{m})| + q \sim t$$
$$10|N| + 2|(\mathbf{l}, \mathbf{n})| + q \sim t$$
$$10|N| + 2|(\mathbf{m}, \mathbf{n})| + q \sim t$$

$(0 \le q < 2)$, and

$$10|N| + 2|E| + |(\mathbf{k}, \mathbf{l}, \mathbf{m})| \sim t$$
$$10|N| + 2|E| + |(\mathbf{k}, \mathbf{l}, \mathbf{n})| \sim t$$
$$10|N| + 2|E| + |(\mathbf{k}, \mathbf{m}, \mathbf{n})| \sim t$$
$$10|N| + 2|E| + |(\mathbf{l}, \mathbf{m}, \mathbf{n})| \sim t.$$

For each fixed $0 \le i < K$, we can use these basis functions defined in t to define basis functions in the entire mesh M:

$$\phi_i \equiv \begin{cases} p_{i,t} & \text{if } i \sim t \\ 0 & \text{if } i \nsim t \end{cases}$$

(for every $t \in M$).

From Chapter 12, Sections 12.41–12.42, it follows that ϕ_i is continuous in M and also has a continuous gradient across the edges in M. Furthermore, ϕ_i has the property that it has the value 1 only for one degree of freedom in M, that is, it takes the value 1 only for one partial derivative at one node in the mesh or for one directional derivative at one edge midpoint or side midpoint in the mesh. Thus, every function f that (a) is continuous in M, (b) is a polynomial of degree five in each and every particular tetrahedron $t \in M$, and (c) has a continuous gradient across every edge in M can be written uniquely as a sum of basis functions:

$$f = \sum_{i=0}^{K-1} f^{(i)} \phi_i,$$

where $f^{(i)}$ is the ith degree of freedom of f (the value of the corresponding partial derivative of f at the corresponding point). Indeed, in each tetrahedron t in the mesh, both the left-hand side and the right-hand side of the above equation have the same 56 degrees of freedom, hence must be identical in t. Since this is true for every $t \in M$, they must be identical in the entire mesh M as well.

Usually, not all the $f^{(i)}$'s are available. Below we consider a problem in which only some of them are given, and the rest should be found in an optimal way.

27.3 The Neumann Matrix

Suppose that the values of f are given at the nodes in N. In other words, suppose that the parameters $f^{(i)}$ are available only for

$$i = 0, 10, 20, \ldots, 10(|N| - 1).$$

How should one specify the other $K - |N|$ parameters $f^{(i)}$ (or degrees of freedom) to have a function f that is as smooth as possible?

To answer this question, we need first to define the $K \times K$ Neumann matrix A. The definition is analogous to the one in Chapter 26, Section 26.1. In fact, for every $0 \le i, j < K$, the formula in Chapter 12, Section 12.43 is used to define the (i, j)th element in A as follows:

$$
\begin{aligned}
a_{i,j} &\equiv \int \int \int_M \nabla^t \phi_j \nabla \phi_i dx dy dz \\
&= \sum_{t \in M} \int \int \int_t \nabla^t \phi_j \nabla \phi_i dx dy dz \\
&= \sum_{t \in M,\ i,j \sim t} \int \int \int_t \nabla^t p_{j,t} \nabla p_{i,t} dx dy dz \\
&= \sum_{t \in M,\ i,j \sim t} |\det(S_t)| \int \int \int_T \nabla^t (p_{j,t} \circ E_t) S_t^{-1} S_t^{-t} \nabla (p_{i,t} \circ E_t) dx dy dz.
\end{aligned}
$$

This completes the definition of the Neumann matrix.

Now, in order to consider the above question, let us reorder the components of any K-dimensional vector v so that the components

$$v_0, v_{10}, v_{20}, \ldots, v_{10(|N|-1)}$$

appear in a second subvector v_N, whereas all the other components appear in a first subvector v_Q:

$$v = \begin{pmatrix} v_Q \\ v_N \end{pmatrix},$$

where v_N is an $|N|$-dimensional subvector, and v_Q is a $(K - |N|)$-dimensional subvector. Similarly, reorder the rows and columns in A so that rows that are indexed by

$$i = 0, 10, 20, \ldots, 10(|N| - 1)$$

appear in the bottom of the matrix, and columns indexed by

$$j = 0, 10, 20, \ldots, 10(|N| - 1)$$

appear in the far right of the matrix. This way, A takes the block form

$$A = \begin{pmatrix} A_{QQ} & A_{QN} \\ A_{NQ} & A_{NN} \end{pmatrix}.$$

Note that, because A is symmetric,

$$A_{NQ}^t = A_{QN}.$$

27.4 The Spline Problem

Using the above Neumann matrix A, the above problem can be reformulated more precisely as follows [27]: find a K-dimensional vector x that minimizes $x^t A x$, subject to the constraint that its second subvector, x_N, must agree with the corresponding components that are available for f:

$$x_N = f_N$$

(Dirichlet conditions). This is the spline problem: to extend f from N to the entire mesh M as smoothly as possible, or with as little energy as possible.

In order to solve this problem, note that, thanks to the above block formulation, A can be written as the following triple product:

$$A = \begin{pmatrix} A_{QQ} & A_{QN} \\ A_{NQ} & A_{NN} \end{pmatrix} = \begin{pmatrix} A_{QQ} & 0 \\ A_{NQ} & I \end{pmatrix} \begin{pmatrix} A_{QQ}^{-1} & 0 \\ 0 & S \end{pmatrix} \begin{pmatrix} A_{QQ} & A_{QN} \\ 0 & I \end{pmatrix},$$

where I is the identity matrix of order $|N|$, and S is the Schur-complement submatrix:

$$S \equiv A_{NN} - A_{NQ} A_{QQ}^{-1} A_{QN}.$$

With this representation, the spline problem takes the form: find $x \in \mathbb{R}^K$ that minimizes

$$x^t \begin{pmatrix} A_{QQ} & 0 \\ A_{NQ} & I \end{pmatrix} \begin{pmatrix} A_{QQ}^{-1} & 0 \\ 0 & S \end{pmatrix} \begin{pmatrix} A_{QQ} & A_{QN} \\ 0 & I \end{pmatrix} x$$

$$= \left(\begin{pmatrix} A_{QQ} & A_{QN} \\ 0 & I \end{pmatrix} x \right)^t \begin{pmatrix} A_{QQ}^{-1} & 0 \\ 0 & S \end{pmatrix} \begin{pmatrix} A_{QQ} & A_{QN} \\ 0 & I \end{pmatrix} x$$

$$= \begin{pmatrix} A_{QQ}x_Q + A_{QN}x_N \\ x_N \end{pmatrix}^t \begin{pmatrix} A_{QQ}^{-1} & 0 \\ 0 & S \end{pmatrix} \begin{pmatrix} A_{QQ}x_Q + A_{QN}x_N \\ x_N \end{pmatrix}$$

$$= (A_{QQ}x_Q + A_{QN}x_N)^t A_{QQ}^{-1} (A_{QQ}x_Q + A_{QN}x_N) + x_N^t S x_N$$

$$= (A_{QQ}x_Q + A_{QN}f_N)^t A_{QQ}^{-1} (A_{QQ}x_Q + A_{QN}f_N) + f_N^t S f_N.$$

27.5 The Dirichlet Matrix

Clearly, we have no control over the second term above. In oder to make the first term as small as zero, x_Q must satisfy

$$A_{QQ}x_Q + A_{QN}f_N = \mathbf{0},$$

where $\mathbf{0}$ is the $(K - |N|)$-dimensional zero vector. In other words, the solution x must satisfy the linear system of equations

$$\begin{pmatrix} A_{QQ} & 0 \\ 0 & I \end{pmatrix} x = \begin{pmatrix} -A_{QN}f_N \\ f_N \end{pmatrix}.$$

The matrix on the left-hand side is called the Dirichlet matrix, because it is obtained from A by using the condition $x_N = f_N$ to eliminate the unknowns in the second subvector and to replace the corresponding equations by trivial ones.

27.6 Exercises

1. Index the edges in the mesh M as follows. Use an outer loop on the tetrahedra in M, and an inner loop on the edges in each tetrahedron. Consider the nonoriented graph $\hat{G}$ whose nodes are these edges. Assume that two nodes in $\hat{G}$ are not connected in $\hat{G}$ if and only if they represent the same edge in M. Apply the node-coloring algorithm in Chapter 19, Section 19.8 to the nonoriented graph $\hat{G}$. The solution can be found in Section 28.23 in the appendix.

2. Show that the above algorithm indeed provides a proper indexing of the edges in M.

3. Index the sides in the mesh M as follows. Use an outer loop on the tetra-hedra in M, and an inner loop on the sides in each tetrahedron. Consider the nonoriented graph $\hat{G}$ whose nodes are these sides. Assume that two nodes in $\hat{G}$ are not connected in $\hat{G}$ if and only if they represent the same side in M. Apply the node-coloring algorithm in Chapter 19, Section 19.8 to the nonoriented graph $\hat{G}$. The solution can be found in Section 28.24 in the appendix.

4. Show that the above algorithm indeed provides a proper indexing of the sides in M.

5. Show that the Neumann matrix A is symmetric.

6. Conclude that the Dirichlet matrix is symmetric as well.

7. Write a constructor that produces the sparse matrix B in Chapter 12, Section 12.31. The solution can be found in Section 28.25 in the appendix.

8. Modify the above constructor so that the first ten rows in B are reordered in such a way that the first row corresponds to the zeroth partial derivative (the function itself), the next three rows correspond to the first partial derivatives (the gradient), and the next six rows correspond to the second partial derivatives (the Hessian). Apply the same reordering also to the next 30 rows in B, so that the ordering of degrees of freedom is the same at the four corners of the tetrahedron. (With this ordering, the basis functions are ready for the transformations in Chapter 12, Sections 12.36–12.40.) The solution can be found in Section 28.25 in the appendix.

9. Write a function that produces the polynomial of three variables p_q from the 56-dimensional vector $\mathbf{x}$ in Chapter 12, Section 12.31. The solution can be found in Section 28.25 in the appendix.

10. Assume that there is a function "solve()" that takes the sparse matrix B and the right-hand side vector $I^{(q)}$ in Chapter 12, Section 12.31 as arguments, solves the linear system

$$Bx = I^{(q)},$$

and returns the vector solution $\mathbf{x}$. Use this function and the code in the previous exercises to write a constructor that produces the Neumann ma-trix A. The solution will be placed on the website of the book.

11. Write a function that takes the Neumann matrix A and the Dirichlet data f_N and produces the Dirichlet matrix and its right-hand side $-A_{QN}f_N$. The solution can be found in Section 28.26 in the appendix.

12. Let v be a K-dimensional complex vector, in which each component v_i corresponds to the ith degree of freedom in the above indexing scheme. For each $t \in M$, redefine $P_t(v)$ in Chapter 26, Section 26.6 by

$$P_t(v) \equiv \sum_{i \sim t} v_i p_{i,t} \circ E_t.$$

Use this new definition to redefine and reconstruct the $2K \times 2K$ Helmholtz mass matrix. The solution will be placed on the website of the book.

Chapter 28

Appendix: Solutions of Exercises

28.1 Representation of Integers in any Base

The following code can be used to represent an integer number in any base. In particular, when the chosen base is 2, the code is equivalent to that in Sections 13.24–13.25 in Chapter 13.

```
int reverse(int n, int base){
   int reversed = n%base;
   while(n /= base)
     reversed = reversed * 10 + n%base;
   return reversed;
}  //  convert to base "base" and reverse
```

This function serves two purposes: to convert the given integer number to base "base", and to write this representation in a reversed order of digits. Unfortunately, leading zeroes get lost in the process. For example, 2300 is reversed to 32. In order to avoid this, one must add 1 to the original number. This extra 1 is subtracted once the original number has been reversed and reversed again by two successive calls to the above function to transform the original number to the required representation in base "base":

```
int changeBase(int n, int base){
   int even = n%base ? 0 : 1;
   return reverse(reverse(n+even,base),10)-even;
}  //  change base from 10 to "base"
```

28.2 Prime Factors

In order to have the list of prime factors of an integer number, we first write a simple function that checks whether a number is prime or not:

```
int prime(int j){
```

```
     for (int i=2; i<j; i++)
       if(j%i==0)return 0;
     return 1;
} // prime or not
```

This function is now used in the recursion that prints the prime factors of an integer number. Indeed, given an arbitrarily large integer number n, a loop is used to find its smallest factor i (the smallest number that divides n). (Clearly, i is prime.) Then, both i and n/i are factored recursively, and the loop is terminated. (Clearly, the first recursive call only prints the prime factor i.) The second recursive call, on the other hand, prints the prime factors of n/i, which are indeed the required prime factors of n:

```
void primeFactors(int n){
  if(prime(n))printf("%d ",n);
  else
    for(int i=2; i<n; i++)
      if(n%i==0){
        primeFactors(i);
        primeFactors(n/i);
        return;
      }
} // print prime factors
```

This is how the above function is actually called in a program:

```
int main(){
  int n;
  scanf("%d",&n);
  printf("%d is the product of  ",n);

  primeFactors(n);
  return 0;
}
```

28.3 Greatest Common Divisor

Here is the code that uses recursion to implement Euclid's algorithm to find the greatest common divisor of the two positive integer numbers m and n. (Without loss of generality, it is assumed that $m > n$.)

```
int GCD(int m, int n){
  return m%n ? GCD(n,m%n) : n;
} // greatest common divisor
```

Here is how this function is used:

```
int main(){
    int m,n;
    scanf("%d %d",&m,&n);
    printf("the GCD of %d and %d is %d\n",m,n,GCD(m,n));
    return 0;
}
```

28.4 Recursive Implementation of $C_{a,n}$

Here we use recursion to implement the function

$$C_{a,n} = \frac{a!}{(a-n)!}$$

(where a and n are nonnegative integers satisfying $a \geq n$) defined in Chapter 10, Section 10.17. The implementation can be done in two equivalent ways: either by

```
int C(int a, int n){
    return n ? a * C(a-1,n-1) : 1;
}  //  a!/(a-n)!
```

or by

```
int C(int a, int n){
    return n ? (a-n+1) * C(a,n-1) : 1;
}  //  a!/(a-n)!
```

28.5 Square Root of a Complex Number

Here is the function that uses the polar representation of a given complex number c to calculate its square root:

$$\sqrt{c} = \sqrt{r(\cos\theta + i \cdot \sin\theta)} = \sqrt{r}(\cos(\theta/2) + i \cdot \sin(\theta/2)).$$

Here it is assumed that $0 \leq \theta < 2\pi$, so $0 \leq \theta/2 < \pi$. Furthermore, it is assumed that the "math.h" library is included in the code, so the sine ("sin"), cosine ("cos"), arc-cosine ("acos"), and square root ("sqrt") functions of a real number are available.

In order to implement the required square root function of a complex number, we first define the "fabs" function, which returns the absolute value of a complex number:

```
double fabs(const complex&c){
    return sqrt(abs2(c));
} //   absolute value of a complex number
```

The naive implementation of the square root function of a complex number calculates θ and $\theta/2$ explicitly:

```
const complex sqrt(const complex&c){
    double r=fabs(c);
    double theta = acos(c.re()/r);
```

Unfortunately, the variable "theta" defined above is not always the same as the original angle θ. Indeed, because the "acos" function used above always returns an angle between 0 and π, "theta" is equal to θ only for $0 \leq \theta \leq \pi$. For $\pi < \theta < 2\pi$, on the other hand, "theta" is equal to $2\pi - \theta$ rather than θ. Thus, the following substitution must be used to correct this and make sure that "theta" is indeed the same as θ:

```
if(c.im() < 0.) theta = 4. * acos(0.) - theta;
```

The required square root of c can now be formed and returned:

```
    r = sqrt(r);
    theta /= 2.;
    return complex(r*cos(theta), r*sin(theta));
} //   square root of a complex number
```

A more sophisticated approach, on the other hand, never calculates θ or $\theta/2$ explicitly, avoiding using the "sin", "cos", and "acos" functions. Instead, it uses the cosine theorem (Chapter 5, Section 5.6) to have

$$\cos(\theta) = \cos^2(\theta/2) - \sin^2(\theta/2),$$

or, using Pythagoras' theorem,

$$\cos(\theta) = 2\cos^2(\theta/2) - 1 = 1 - 2\sin^2(\theta/2).$$

As a result, one can have both $\cos(\theta/2)$ and $\sin(\theta/2)$ in terms of $\cos(\theta)$, the real part of c:

$$\cos(\theta/2) = \pm\sqrt{\frac{1 + \cos(\theta)}{2}}$$

and

$$\sin(\theta/2) = \sqrt{\frac{1 - \cos(\theta)}{2}}.$$

Note that, in the formula for $\cos(\theta/2)$ above, the plus sign should be used for $0 \leq \theta \leq \pi$, whereas the minus sign should be used for $\pi < \theta < 2\pi$. The above formulas give rise to the following code to calculate the required $\sqrt{c}$:

```
const complex sqrt(const complex&c){
    double r=fabs(c);
    double cosTheta = c.re()/r;
    double sign = c.im() < 0. ? -1. : 1.;
    double cosTheta2 = sign * sqrt((1.+cosTheta)/2.);
    double sinTheta2 = sqrt((1.-cosTheta)/2.);
    r = sqrt(r);
    return complex(r*cosTheta2,r*sinTheta2);
} //  square root of a complex number
```

Unfortunately, the above definition of the square root function is discontinuous at the positive part of the real axis. Indeed, when θ approaches 0^+ (that is, θ is positive and approaches 0 from above), $\theta/2$ approaches 0 as well. When, on the other hand, θ approaches $2\pi^-$ ($\theta < 2\pi$), $\theta/2$ approaches π, yielding the negative of the required square root. To avoid this problem, let us assume that the original angle θ lies between $-\pi$ and π rather than between 0 and 2π. This way, $\theta/2$ approaches 0 whenever θ approaches 0, regardless of whether $\theta > 0$ or $\theta < 0$. As a result, the above formulas take the form

$$\cos(\theta/2) = \sqrt{\frac{1 + \cos(\theta)}{2}}$$

and

$$\sin(\theta/2) = \pm\sqrt{\frac{1 - \cos(\theta)}{2}},$$

where the plus sign in the above formula is used when $\theta \geq 0$, and the minus sign is used when $\theta < 0$. As a consequence, the definitions of "cosTheta2" and "sinTheta2" in the above code should be modified to read

```
double cosTheta2 = sqrt((1.+cosTheta)/2.);
double sinTheta2 = sign * sqrt((1.-cosTheta)/2.);
```

The rest of the code remains the same as before.

28.6 Operations with Vectors

Here is the detailed implementation of some arithmetic operators of the "vector" class that have been left as exercises:

```
template<class T, int N>
const vector<T,N>&
vector<T,N>::operator-=(const vector<T,N>&v){
    for(int i = 0; i < N; i++)
      component[i] -= v[i];
    return *this;
} // subtracting a vector from the current vector

template<class T, int N>
const vector<T,N>&
vector<T,N>::operator*=(const T& a){
    for(int i = 0; i < N; i++)
      component[i] *= a;
    return *this;
} // multiplying the current vector by a scalar

template<class T, int N>
const vector<T,N>&
vector<T,N>::operator/=(const T& a){
    for(int i = 0; i < N; i++)
      component[i] /= a;
    return *this;
} // dividing the current vector by a scalar

template<class T, int N>
const vector<T,N>
operator-(const vector<T,N>&u, const vector<T,N>&v){
  return vector<T,N>(u) -= v;
} // vector minus vector

template<class T, int N>
const vector<T,N>
operator*(const vector<T,N>&u, const T& a){
  return vector<T,N>(u) *= a;
} // vector times scalar

template<class T, int N>
const vector<T,N>
operator*(const T& a, const vector<T,N>&u){
  return vector<T,N>(u) *= a;
} // T times vector

template<class T, int N>
const vector<T,N>
operator/(const vector<T,N>&u, const T& a){
```

```
    return vector<T,N>(u) /= a;
}  //  vector divided by scalar
```

28.7 Operations with Matrices

Here is the implementation of the operators that add two matrices, subtract two matrices, and multiply scalar times matrix, vector times matrix, matrix times vector, and matrix times matrix, as declared in Chapter 16 above.

```
template<class T, int N, int M>
const matrix<T,N,M>&
matrix<T,N,M>::operator+=(const matrix<T,N,M>&m){
  vector<vector<T,N>,M>::operator+=(m);
  return *this;
}  //  adding a matrix

template<class T, int N, int M>
const matrix<T,N,M>&
matrix<T,N,M>::operator-=(const matrix<T,N,M>&m){
  vector<vector<T,N>,M>::operator-=(m);
  return *this;
}  //  subtracting a matrix

template<class T, int N, int M>
const matrix<T,N,M>&
matrix<T,N,M>::operator*=(const T&a){
  for(int i=0; i<M; i++)
    set(i,(*this)[i] * a);
  return *this;
}  //  multiplication by scalar

template<class T, int N, int M>
const matrix<T,N,M>&
matrix<T,N,M>::operator/=(const T&a){
  for(int i=0; i<M; i++)
    set(i,(*this)[i] / a);
  return *this;
}  //  division by scalar

template<class T, int N, int M>
const matrix<T,N,M>
operator*(const T&a,const matrix<T,N,M>&m){
```

```
    return matrix<T,N,M>(m) *= a;
} // scalar times matrix

template<class T, int N, int M>
const matrix<T,N,M>
operator*(const matrix<T,N,M>&m, const T&a){
  return matrix<T,N,M>(m) *= a;
} // matrix times scalar

template<class T, int N, int M>
const matrix<T,N,M>
operator/(const matrix<T,N,M>&m, const T&a){
  return matrix<T,N,M>(m) /= a;
} // matrix divided by scalar

template<class T, int N, int M>
const vector<T,M>
operator*(const vector<T,N>&v,const matrix<T,N,M>&m){
  vector<T,M> result;
  for(int i=0; i<M; i++)
    result.set(i, v * m[i]);
  return result;
} // vector times matrix

template<class T, int N, int M>
const vector<T,N>
operator*(const matrix<T,N,M>&m,const vector<T,M>&v){
  vector<T,N> result;
  for(int i=0; i<M; i++)
    result += v[i] * m[i];
  return result;
} // matrix times vector

template<class T, int N, int M, int K>
const matrix<T,N,K>
operator*(const matrix<T,N,M>&m1,const matrix<T,M,K>&m2){
  matrix<T,N,K> result;
  for(int i=0; i<K; i++)
    result.set(i,m1 * m2[i]);
  return result;
} // matrix times matrix

template<class T, int N, int M, int K>
const matrix<T,N,K>&
operator*=(matrix<T,N,M>&m1,
```

```
      const matrix<T,M,K>&m2){
    return m1 = m1 * m2;
} //  multiplying a matrix by a matrix

template<class T, int N, int M>
const matrix<T,N,M>
operator+(const matrix<T,N,M>&m1,
    const matrix<T,N,M>&m2){
  return matrix<T,N,M>(m1) += m2;
} //  matrix plus matrix

template<class T, int N, int M>
const matrix<T,N,M>
operator-(const matrix<T,N,M>&m1,
    const matrix<T,N,M>&m2){
  return matrix<T,N,M>(m1) -= m2;
} //  matrix minus matrix
```

28.8 Determinant, Inverse, and Transpose of 2×2 Matrix

Here are some more functions that compute the determinant, inverse, and transpose of 2×2 matrices of class "matrix2" in Chapter 16 above:

```
typedef matrix<double,2,2> matrix2;
double det(const matrix2&A){
  return A(0,0)*A(1,1) - A(0,1)*A(1,0);
} //  determinant of a 2 by 2 matrix
```

The above "det()" function is now used to compute A^{-1} by Kremer's formula:

$$\begin{pmatrix} A_{0,0} & A_{0,1} \\ A_{1,0} & A_{1,1} \end{pmatrix}^{-1} = \det(A)^{-1} \begin{pmatrix} A_{1,1} & -A_{0,1} \\ -A_{1,0} & A_{0,0} \end{pmatrix}.$$

This formula is implemented as follows:

```
const matrix2 inverse(const matrix2&A){
  point column0(A(1,1),-A(1,0));
  point column1(-A(0,1),A(0,0));
  return matrix2(column0,column1)/det(A);
} //  inverse of 2 by 2 matrix
```

Finally, the transpose of a 2×2 matrix is computed as follows:

```
const matrix2 transpose(const matrix2&A){
  return matrix2(point(A(0,0),A(0,1)),
      point(A(1,0),A(1,1)));
} // transpose of 2 by 2 matrix
```

28.9 Determinant, Inverse, and Transpose of 3×3 Matrix

Here are also the analogue functions that compute and return the determinant, inverse, and transpose of 3×3 matrices, using the definitions in Chapter 9, Sections 9.23–9.24:

```
double det(const matrix3&A){
  return A(0,0) * (A(1,1)*A(2,2)-A(1,2)*A(2,1))
    - A(0,1) * (A(1,0)*A(2,2)-A(1,2)*A(2,0))
    + A(0,2) * (A(1,0)*A(2,1)-A(1,1)*A(2,0));
} // determinant of matrix3

const matrix3 inverse(const matrix3&A){
  point3 column0(A(1,1)*A(2,2)-A(1,2)*A(2,1),
      -(A(1,0)*A(2,2)-A(1,2)*A(2,0)),
       A(1,0)*A(2,1)-A(1,1)*A(2,0));
  point3 column1(-(A(0,1)*A(2,2)-A(0,2)*A(2,1)),
       A(0,0)*A(2,2)-A(0,2)*A(2,0),
      -(A(0,0)*A(2,1)-A(0,1)*A(2,0)));
  point3 column2(A(0,1)*A(1,2)-A(0,2)*A(1,1),
      -(A(0,0)*A(1,2)-A(0,2)*A(1,0)),
       A(0,0)*A(1,1)-A(0,1)*A(1,0));
  return
    matrix3(column0,column1,column2)/det(A);
} // inverse of matrix3

const matrix3 transpose(const matrix3&A){
  return
    matrix3(point3(A(0,0),A(0,1),A(0,2)),
        point3(A(1,0),A(1,1),A(1,2)),
        point3(A(2,0),A(2,1),A(2,2)));
} // transpose of a matrix3
```

28.10 Vector Product

Here is the function that computes and returns the vector product of two
3-d vectors, defined in Chapter 9, Section 9.25:

```
const point3 operator&(const point3&u,
        const point3&v){
  point3 i(1.,0.,0.);
  point3 j(0.,1.,0.);
  point3 k(0.,0.,1.);
  return i * (u[1]*v[2]-u[2]*v[1])
       - j * (u[0]*v[2]-u[2]*v[0])
       + k * (u[0]*v[1]-u[1]*v[0]);
} //  vector product
```

With this operator, users can write "u&v" to have the vector product of the
two 3-d vectors 'u' and 'v'.

28.11 The Matrix Exponent Function

Here we rewrite the "expTaylor" function as a template function, so it is
applicable not only to a scalar argument x to calculate $\exp(x)$ but also to a
matrix argument A to calculate $\exp(A)$. For this, however, we need first to
define a few ordinary functions.

The first function is the function that returns the l_1-norm of a vector, or
the sum of the absolute value of its components:

```
template<class T, int N>
const T L1Norm(const vector<T,N>&v){
    T sum = 0;
    for(int i = 0; i < N; i++)
      sum += abs(v[i]);
    return sum;
} //  L1 norm of a vector
```

The next function calculates the l_1-norm of a matrix as the maximum of the
l_1-norm of its columns:

```
template<class T, int N, int M>
const T L1Norm(const matrix<T,N,M>&m){
  T maxColumn=0.;
  for(int i=0; i<M; i++)
```

```
    maxColumn = max(maxColumn,L1Norm(m[i]));
  return maxColumn;
} // L1 norm of a matrix
```

The next function returns the transpose of a matrix:

```
template<class T, int N, int M>
const matrix<T,M,N>
transpose(const matrix<T,N,M>&m){
  matrix<T,M,N> result;
  for(int i=0; i<N; i++){
    vector<T,M> column;
    for(int j=0; j<M; j++)
      column.set(j,m(i,j));
    result.set(i,column);
  }
  return result;
} // transpose of a matrix
```

The next function gives an upper bound for the l_2-norm of a matrix. This upper bound is the square root of the l_1-norm of the matrix times the l_1-norm of its transpose:

```
template<class T, int N, int M>
const T abs(const matrix<T,N,M>&m){
  return sqrt(L1Norm(m) * L1Norm(transpose(m)));
} // estimate for the L2 norm of a matrix
```

The function "abs()" used in the "expTaylor" function can now refer not only to a scalar to give its absolute value but also to a matrix to give an estimate for its l_2-norm.

Furthermore, we also define the ordinary "*=" operator to multiply the current matrix by another matrix:

```
template<class T, int N, int M, int K> const matrix<T,N,K>&
operator*=(matrix<T,N,M>&m1,const matrix<T,M,K>&m2){
  return m1 = m1 * m2;
} // current matrix times a matrix
```

Once this operator is defined, we are ready to rewrite the "expTaylor" function as a template function, with the "double" type used in the original version replaced by the yet unspecified type 'T':

```
template <class T>
T expTaylor(const T& arg){
  const int K=10;
  T x=arg;
  int m=0;
```

```
while(abs(x)>0.5){
  x /= 2.;
  m++;
}
T sum=T(1.);
```

This way, if 'T' is a matrix, then "sum" takes initially the value of the identity matrix I.

```
for(int i=K; i>0; i--){
  sum *= x/double(i);
```

This way, if 'T' is a matrix, then the matrix "sum" is multiplied by the matrix 'x' divided by the real number i. Then, the identity matrix I is added to the matrix "sum", using the "+ =" operator inherited from the base "vector" class:

```
  sum += T(1.);
}
for(int i=0; i<m; i++)
  sum *= sum;
return sum;
} // exponent of a 'T'
```

Thanks to the "operator*=" defined above, the final loop in this code applies not only to scalars but also to matrices. Thus, the above template function can be used to calculate the exponent of matrices as well as the exponent of scalars.

28.12 Operations with Dynamic Vectors

Here is the detailed implementation of some arithmetic operators of the "dynamicVector" class that have been left as exercises (subtraction, multiplication, and division by scalar, inner product, etc.):

```
template<class T>
const dynamicVector<T>&
dynamicVector<T>::operator-=( const dynamicVector<T>&v){
    for(int i = 0; i < dimension; i++)
      component[i] -= v[i];
    return *this;
} // subtract a dynamicVector from the current one

template<class T>
```

```
const dynamicVector<T>&
dynamicVector<T>::operator*=(const T& a){
    for(int i = 0; i < dimension; i++)
      component[i] *= a;
    return *this;
} // multiply the current dynamicVector by a scalar

template<class T>
const dynamicVector<T>&
dynamicVector<T>::operator/=(const T& a){
    for(int i = 0; i < dimension; i++)
      component[i] /= a;
    return *this;
} // divide the current dynamicVector by a scalar

template<class T>
const dynamicVector<T>
operator-(const dynamicVector<T>&u,
        const dynamicVector<T>&v){
  return dynamicVector<T>(u) -= v;
} // dynamicVector minus dynamicVector

template<class T>
const dynamicVector<T>
operator*(const dynamicVector<T>&u, const T& a){
  return dynamicVector<T>(u) *= a;
} // dynamicVector times scalar

template<class T>
const dynamicVector<T>
operator*(const T& a, const dynamicVector<T>&u){
  return dynamicVector<T>(u) *= a;
} // T times dynamicVector

template<class T>
const dynamicVector<T>
operator/(const dynamicVector<T>&u, const T& a){
  return dynamicVector<T>(u) /= a;
} // dynamicVector divided by a scalar

template<class T>
T operator*(const dynamicVector<T>&u,
        const dynamicVector<T>&v){
    T sum = 0;
    for(int i = 0; i < u.dim(); i++)
```

```
      sum += (+u[i]) * v[i];
   return sum;
}  //  inner product
```

28.13 Using the Stack Object

Here is how the user can use the "stack" class to construct a stack with
the three numbers 2, 4, and 6 in it, print it onto the screen using the "print"
function inherited from the base "linkedList" class, and pop the items one by
one back out of it:

```
int main(){
   stack<int> S;
   S.push(6);
   S.push(4);
   S.push(2);
   print(S);
   print(S.pop());
   printf("\n");
   print(S);
   print(S.pop());
   printf("\n");
   print(S);
   print(S.pop());
   printf("\n");
   return 0;
}
```

28.14 Operations with Sparse Matrices

Here is the detailed implementation of the member arithmetic operators
and functions of the "sparseMatrix" class that have been left as exercises.

```
template<class T>
int sparseMatrix<T>::columnNumber() const{
   int maxColumn = -1;
   for(int i=0; i<rowNumber(); i++)
     if(item[i])maxColumn =
       max(maxColumn, item[i]->last()().getColumn());
```

```
            return maxColumn + 1;
     }  //   number of columns
```

Note that the function "columnNumber()" defined above returns the maximal column index used in any of the rows in the matrix. This number is also the number of columns in the matrix, provided that its last column is nonzero, which is assumed to be the case.

The number of columns in a matrix is particularly useful to construct its transpose. This is done in the function "transpose()" defined later in this section. However, one must be careful not to use this function for a matrix whose last column is zero, or it would be dropped from the transpose matrix.

Here are some more functions defined on sparse matrices:

```
     template<class T>
     const sparseMatrix<T>&
     sparseMatrix<T>::operator+=(const sparseMatrix<T>&M){
       for(int i=0; i<rowNumber(); i++)
         *item[i] += *M.item[i];
       return *this;
     }  //   add a sparse matrix

     template<class T>
     const sparseMatrix<T>&
     sparseMatrix<T>::operator-=(const sparseMatrix<T>&M){
       for(int i=0; i<rowNumber(); i++){
         row<T> minus = -1. * *M.item[i];
         *item[i] += minus;
       }
       return *this;
     }  //   subtract a sparse matrix

     template<class T>
     const sparseMatrix<T>&
     sparseMatrix<T>::operator*=(const T&t){
       for(int i=0; i<rowNumber(); i++)
         *item[i] *= t;
       return *this;
     }  //   multiply by T
```

Here are some nonmember arithmetic operators:

```
     template<class T>
     const sparseMatrix<T>
     operator+(const sparseMatrix<T>&M1,
               const sparseMatrix<T>&M2){
       return sparseMatrix<T>(M1) += M2;
```

```
}  //  matrix plus matrix

template<class T>
const sparseMatrix<T>
operator-(const sparseMatrix<T>&M1,
        const sparseMatrix<T>&M2){
  return sparseMatrix<T>(M1) -= M2;
}  //  matrix minus matrix

template<class T>
const sparseMatrix<T>
operator*(const T&t, const sparseMatrix<T>&M){
  return sparseMatrix<T>(M) *= t;
}  //  scalar times sparse matrix

template<class T>
const sparseMatrix<T>
operator*(const sparseMatrix<T>&M, const T&t){
  return sparseMatrix<T>(M) *= t;
}  //  sparse matrix times scalar

template<class T>
const dynamicVector<T>
operator*(const sparseMatrix<T>&M,
        const dynamicVector<T>&v){
  dynamicVector<T> result(M.rowNumber(),0.);
  for(int i=0; i<M.rowNumber(); i++)
    result(i) = M[i] * v;
  return result;
}  //  matrix times vector
```

Here is the implementation of some friend functions of the "sparseMatrix" class. The "operator*" function returns the product of two sparse matrices. (The calculation uses the algorithm described in Chapter 20.)

```
template<class T>
const sparseMatrix<T>
operator*(const sparseMatrix<T>&M1,
        const sparseMatrix<T>&M2){
    sparseMatrix<T> result(M1.rowNumber());
    for(int i = 0; i < M1.rowNumber(); i++){
      result.item[i] =
          new row<T>(M1.item[i]->getValue() *
          *M2.item[M1.item[i]->getColumn()]);
        for(const row<T>* runner =
          (const row<T>*)M1.item[i]->readNext();
```

```
            runner; runner =
            (const row<T>*)runner->readNext()){
          row<T> r =
              runner->getValue() *
                *M2.item[runner->getColumn()];
          *result.item[i] += r;
        }
      }
      return result;
    }  //  matrix times matrix
```

Furthermore, the "transpose" function returns the transpose of a sparse matrix:

```
    template<class T>
    const sparseMatrix<T>
    transpose(const sparseMatrix<T>&M){
      sparseMatrix<T> Mt(M.columnNumber());
      for(int i=0; i<M.rowNumber(); i++)
        for(const row<T>* runner = M.item[i]; runner;
            runner = (const row<T>*)runner->readNext()){
          if(Mt.item[runner->getColumn()])
            Mt.item[runner->getColumn()]->
                append(runner->getValue(),i);
          else
            Mt.item[runner->getColumn()] =
                new row<T>(runner->getValue(),i);
        }
      return Mt;
    }  //  transpose of sparse matrix
```

Note that the matrix 'M' that is passed (by reference) as an argument to the "transpose()" function above is assumed to have a nonzero last column, so the function "columnNumber()" indeed returns its correct number of columns, to serve as the number of rows in its transpose matrix.

28.15 Three-Dimensional Mesh

Here is how the unit cube can be covered by a mesh of six disjoint tetrahedra. First, the eight corners of the unit cube are defined as node objects:

```
    main(){
      node<point3> a000(point3(0,0,0));
```

```
node<point3> a100(point3(1,0,0));
node<point3> a010(point3(0,1,0));
node<point3> a001(point3(0,0,1));
node<point3> a011(point3(0,1,1));
node<point3> a111(point3(1,1,1));
node<point3> a101(point3(1,0,1));
node<point3> a110(point3(1,1,0));
```

Now, these nodes are used to form the six required tetrahedra. Note that, once a node has been used in a tetrahedron, it is referred to as a vertex in that tetrahedron, rather than by its original name, which refers to a dangling node that belongs to no tetrahedron as yet.

```
tetrahedron t1(a000,a100,a010,a001);
tetrahedron t2(a111,a011,a101,a110);
tetrahedron t3(t2(1),t2(2),t1(1),t2(3));
tetrahedron t4(t2(1),t1(2),t1(1),t2(3));
tetrahedron t5(t2(1),t2(2),t1(1),t1(3));
tetrahedron t6(t2(1),t1(2),t1(1),t1(3));
```

Next, the constructor of the "mesh" class is called to form a mesh 'm' with only one tetrahedron, "t1". Then, the rest of the tetrahedra, "t2", "t3", "t4", "t5", and "t6" are appended to it one by one, to form the required mesh of six tetrahedra that covers the entire unit cube.

```
mesh<tetrahedron> m(t1);
m.append(t2);
m.append(t3);
m.append(t4);
m.append(t5);
m.append(t6);
```

Once the tetrahedra have been placed in the mesh 'm', the original tetrahedra referred to as $t_1, t_2, \ldots, t_6$ can be removed. This way, the "sharingCells" fields in their vertices only count the cells in 'm' that share them, but not the original dangling tetrahedra "t1", ..., "t6".

```
t1.~tetrahedron();
t2.~tetrahedron();
t3.~tetrahedron();
t4.~tetrahedron();
t5.~tetrahedron();
t6.~tetrahedron();
```

The mesh 'm' is now ready for a refinement step. The finer mesh is then printed onto the screen:

```
    m.refine();
    print(m);
    return 0;
}
```

28.16 Integrals over the Tetrahedron

The following code defines the nodal basis functions in the tetrahedron, that is, the polynomials of three variables that have the value 1 at one corner of the tetrahedron and 0 at the three other corners:

$$p_0(x, y, z) = 1 - x - y - z$$
$$p_1(x, y, z) = x$$
$$p_2(x, y, z) = y$$
$$p_3(x, y, z) = z.$$

```
int main(){
  polynomial<double> zero(1,0.);
  polynomial<polynomial<double> > Zero(1,zero);
  polynomial<double> one(1,1.);
  polynomial<polynomial<double> > One(1,one);
  polynomial<double> minus1(1,-1.);
  polynomial<polynomial<double> > Minus1(1,minus1);
  polynomial<double> oneMinusx(1.,-1.);
  polynomial<polynomial<double> >
      oneMinusxMinusy(oneMinusx,minus1);
  polynomial<polynomial<double> > yy(zero,one);
  polynomial<double> x1(0.,1.);
  polynomial<polynomial<double> > xx(1,x1);
```

Note that "x1" and "xx" are different kinds of polynomials: "x1" is the 1-d polynomial $p(x) = x$, whereas "xx" is the 2-d polynomial $p(x, y) = x$, which doesn't depend on y at all. Similarly, "xxx" is the 3-d polynomial $p(x, y, z) = x$, which depends neither on y nor on z:

```
  polynomial<polynomial<polynomial<double> > >
      xxx(1,xx);
  list<polynomial<polynomial<polynomial<double> > > >
      P(4,xxx);
  P(0) = polynomial<polynomial<
    polynomial<double> > >(oneMinusxMinusy,Minus1);
  P(2) = polynomial<polynomial<
```

```
    polynomial<double> > >(1,yy);
  P(3) = polynomial<polynomial<
    polynomial<double> > >(Zero, One);
```

This completes the definition of the nodal basis functions p_0, p_1, p_2, and p_3. These definitions are now used to calculate the integral over the tetrahedron of the triple products of three nodal basis functions:

```
    print(integral(P[0]*P[1]*P[2]));
    printf("\n");
    print(integral(P[1]*P[2]*P[3]));
    printf("\n");
    print(integral(P[2]*P[3]*P[0]));
    printf("\n");
    print(integral(P[3]*P[0]*P[1]));
    printf("\n");
    return 0;
}
```

28.17 Computing Partial Derivatives

Here we define the function that computes and returns the kth derivative of its polynomial argument. For this, we use the function "C()" defined in Section 28.4.

```
    template<class T>
    const polynomial<T>
    d(const polynomial<T>&p, int k){
       if(k>p.degree())
          return polynomial<T>(1,0.);
```

Clearly, if k is larger than the degree of the polynomial, then the zero polynomial is returned, and the function terminates. If, on the other hand, k is smaller than or equal to the degree of the polynomial, then the kth derivative is returned in the polynomial object "dp":

```
    polynomial<T> dp(p.degree()+1-k,0.);
    for(int n=0; n<=dp.degree(); n++)
      dp(n) = C(n+k,k) * p[n+k];
    return dp;
  } //  kth derivative
```

The above function is now used to compute the (j, k)th partial derivative of a polynomial of two variables:

```
template<class T>
const polynomial<polynomial<T> >
d(const polynomial<polynomial<T> >&p, int j, int k){
  polynomial<T> zero(1,0.);
  if(k>p.degree())
    return polynomial<polynomial<T> >(1,zero);
```

Note that here "p.degree()" is not the degree of p in the usual sense but rather the largest y-power in $p = \sum_n a_n(x)y^n$. Clearly, if k is larger than this number, then the zero polynomial is returned and the function terminates. Otherwise, the function proceeds to calculate the (j,k)th partial derivative, using the previous function to calculate the jth derivative of the $a_n(x)$'s:

```
  polynomial<polynomial<T> > dp(p.degree()+1-k,zero);
  for(int n=0; n<=dp.degree(); n++)
    dp(n) = C(n+k,k) * d(p[n+k],j);
  return dp;
} //  (j,k)th partial derivative
```

Finally, the following function uses the previous function to calculate the (i,j,k)th partial derivative of a polynomial of three variables:

```
template<class T>
const polynomial<polynomial<polynomial<T> > >
d(const polynomial<polynomial<polynomial<T> > >&p,
        int i, int j, int k){
  polynomial<T> zero(1,0.);
  polynomial<polynomial<T> > Zero(1,zero);
  if(k>p.degree())
    return polynomial<polynomial<polynomial<T> > >(1,Zero);
```

Again, here "p.degree()" is the largest z-power in $p = \sum_n a_n(x,y)z^n$. Clearly, if k is larger than this number, then the zero polynomial is returned and the function terminates. Otherwise, the function proceeds to calculate the (i,j,k)th partial derivative, using the previous function to calculate the (i,j)th partial derivative of the a_n's:

```
  polynomial<polynomial<polynomial<T> > >
      dp(p.degree()+1-k,Zero);
  for(int n=0; n<=dp.degree(); n++)
    dp(n) = C(n+k,k) * d(p[n+k],i,j);
  return dp;
} //  (i,j,k)th partial derivative
```

28.18 Composing Sparse Polynomials

Here we use the "operator()" function in Chapter 25, Section 25.7 to produce the composition $p \circ q$ of two given sparse polynomials $p(x)$ and $q(x)$. This is done as follows. First, $p = \sum a_i x^i$ is transformed into the dummy polynomial of "two" variables

$$p_2(x, y) = \sum_i a_i(x) y^i,$$

with the trivial coefficients

$$a_i(x) \equiv a_i$$

that actually do not depend on x whatsoever. Then, the original "operator()" function is applied to this dummy polynomial to compute its polynomial value at the fixed argument $y = q$:

$$p_2(x, q) = \sum a_i q^i = p \circ q.$$

```
template<class T>
const sparsePolynomial<T>
operator&(const sparsePolynomial<T>&p,
        const sparsePolynomial<T>&q){
  sparsePolynomial<T> first(p.getValue(),0);
  sparsePolynomial<sparsePolynomial<T> >
      p2(first,p.getPower());
```

By now, the dummy polynomial of two variables "p2" contains one monomial only, the first monomial in the original polynomial 'p'. Next, a loop on the rest of the monomials in 'p' is used to construct the trivial polynomials

$$a_i(x) = a_i x^0$$

and append them one by one to "p2" as well:

```
if(p.readNext())
  for(const sparsePolynomial<T>* runner =
       (const sparsePolynomial<T>*)p.readNext();
      runner;
      runner = (const sparsePolynomial<T>*)
        runner->readNext()){
    sparsePolynomial<T> coef((*runner).getValue(),0);
    p2.append(coef,(*runner).getPower());
  }
```

Finally, the original "operator()" is invoked to return the required polynomial

$$p_2(x, q) = \sum_i a_i q^i = p \circ q$$

(where x is actually immaterial):

```
    return p2(q);
  } //  compose p&q
```

28.19 Calculations with Sparse Polynomials

Here is how sparse polynomials are defined and used:

```
main(){
    sparsePolynomial<double> p(monomial<double>(1.,1));
```

This command line constructs the temporary monomial x, and then uses it to construct the sparse polynomial $p(x) = x$. The cubic term $2x^3$ is then added to this polynomial by the "append()" function:

```
    monomial<double> mon(2.,3);
    p.append(mon);
```

Note that, as defined in the linked-list class, the "append()" function takes a nonconstant argument. (This is done on purpose to enable the derivation of the mesh object, which may change the cell objects appended to it by increasing the "sharingCells" fields in their nodes.) As a consequence, temporary unnamed objects cannot be passed to it as arguments, out of fear that they would be changed throughout it, which makes no sense. This is why the well-named monomial "mon" is defined above and passed as an argument to the "append()" function.

At this stage, the sparse polynomial p takes its final form

$$p(x) = x + 2x^3.$$

The sparse polynomials p and p^2 are then printed to the screen, using the "operator*" function to multiply sparse polynomials and the "print" function inherited from the base linked-list class:

```
    print(p);
    print(p * p);
```

Furthermore, the "operator()" member function of the sparse-polynomial class is invoked with 'S' being interpreted as the "double" type to calculate $p(3)$:

```
print(p(3.));
```

Furthermore, the "operator&" of the previous section is used to compute and print the polynomial $p \circ p$:

```
print(p&p);
```

Finally, we check that

$$p(p(p(x))) = (p \circ p \circ p)(x).$$

The left-hand side is calculated by three applications of the "operator()" function. The right-hand side, on the other hand, is calculated by two applications of the "operator&" function to calculate the composition $p \circ p \circ p$, followed by one application of "operator()" to calculate $(p \circ p \circ p)(x)$.

```
print(p(p(p(1.))));
print((p&p&p)(1.));
```

Note that "p&p&p" above returns a temporary sparse-polynomial object that contains the composition $p \circ p \circ p$. Thanks to the fact that "operator()" is declared as a constant member function that cannot change its current sparse-polynomial object, it can be safely applied to this temporary object, with no fear that it may change throughout the execution of the function. This is why "p&p&p" can be safely passed as the current argument of the "operator()" function to calculate the required value.

The above polynomial 'p' can now be used as a monomial in the polynomial of two variables

$$p_2(x, y) = (x + 2x^3)y + (x + 2x^3)y^3 :$$

```
monomial<sparsePolynomial<double> > mon2(p,1);
monomial<sparsePolynomial<double> > mon3(p,3);
sparsePolynomial<sparsePolynomial<double> >
    p2(mon2);
p2.append(mon3);
```

This implementation can now be used to calculate and print $p_2(2, 3)$ simply by writing

```
print(p2(2.,3.));
return 0;
}
```

28.20 The Stiffness Matrix

Here we implement the constructor that takes a 3-d mesh as an argument and produces the stiffness matrix $A^{(s)}$ defined in Chapter 26, Sections 26.1–26.3. (For simplicity, we assume that $\alpha = 1$ there.) It is assumed that the mesh M covers the unit cube (as in Section 28.15 above), and that b contains five sides of this cube:

$$b = \{(x, y, z) \mid (1 - x)y(1 - y)z(1 - z) = 0\}.$$

Furthermore, it is also assumed that at least one refinement step has been applied to M, so the triangles that cover the bottom side of the cube are not too big. This guarantees that these triangles are indeed excluded from b and do not contribute to $A^{(b)}$, as we'll see in the code below.

```
template<class T>
sparseMatrix<T>::sparseMatrix(
    mesh<tetrahedron>&m){
  item = new row<T>*[number = m.indexing()];
```

Because the "sparseMatrix" class is derived from a list of rows, it inherits the field "item", which is an array of pointers to "row" objects. In the above code line, this array is set to contain $|N|$ pointers to rows, where $|N|$, the number of nodes in the mesh, is returned from the "indexing" function applied to the "mesh" argument 'm'.

Next, we define an $|N|$-dimensional vector of integers, to indicate whether a particular node in the mesh is a boundary node or not:

```
dynamicVector<int> boundary(number,0);
```

This $|N|$-dimensional vector (or grid function) is initialized to zero. Later on, it will take the value 1 at indices i corresponding to boundary nodes $\mathbf{i} \in b$.

```
for(int i=0; i<number; i++)
  item[i] = 0;
point3 gradient[4];
gradient[0] = point3(-1,-1,-1);
gradient[1] = point3(1,0,0);
gradient[2] = point3(0,1,0);
gradient[3] = point3(0,0,1);
for(const mesh<tetrahedron>* runner = &m;
  runner;
  runner =
    (const mesh<tetrahedron>*)runner->readNext()){
```

We have just entered the loop over the tetrahedra of the form $t \in M$. The pointer "runner", which points to the rest of the tetrahedra in the mesh, is converted explicitly from a mere pointer to a linked list of tetrahedra to a pointer to a concrete mesh of tetrahedra. This way, one can refer not only to the first tetrahedron in the "tail" of the linked list by writing "(*runner)()" [invoking the "operator()" of the underlying linked-list object] but also to member functions of the derived "mesh" class as well.

Next, the 3×3 matrix S_t is defined:

```
matrix3
  S((*runner)()[1]() - (*runner)()[0](),
    (*runner)()[2]() - (*runner)()[0](),
    (*runner)()[3]() - (*runner)()[0]());
matrix3 Sinverse = inverse(S);
matrix3 weight =
  fabs(det(S)/6.) *
      Sinverse * transpose(Sinverse);
for(int i=0; i<4; i++){
  int I = (*runner)()[i].getIndex();
```

We have just entered the inner loop, which runs on the corners **i** of the tetrahedron t pointed at by "runner". For each corner **i** encountered in this inner loop, the integer 'I' denotes its index i in the list of nodes N. Then, it is checked whether **i** is a boundary point in b. If it is, then it sets the corresponding component in the dynamic vector "boundary" to 1:

```
if(((*runner)()[i]()[0] >= 1. - 1.e-6)
    ||((*runner)()[i]()[1] <= 1.e-6)
    ||((*runner)()[i]()[1] >= 1. - 1.e-6)
    ||((*runner)()[i]()[2] <= 1.e-6)
    ||((*runner)()[i]()[2] >= 1. - 1.e-6))
  boundary(I) = 1;
}
for(int i=0; i<4; i++){
  int I = (*runner)()[i].getIndex();
  for(int j=0; j<4; j++){
    int J = (*runner)()[j].getIndex();
```

We have just entered the double nested loop on the corners of the tetrahedron t, pointed at by "runner". The indices 'I' and 'J' stand for the indices i and j used in Chapter 26, Section 26.1, to index the corners **i** and **j**. Thus, we actually compute the contribution to the ('I','J')th element in A from the tetrahedron t. For this, we distinguish between two cases: if the 'I'th row in the constructed matrix already exists, then this contribution has just to be added to it:

```
if(item[I]){
```

```
row<T>
  r(gradient[j]*weight*gradient[i],J);
*item[I] += r;
}
```

If, on the other hand, this is the first contribution ever to the 'I'th row, then it must be constructed using the "new" command:

```
else
  item[I] = new
    row<T>(gradient[j]*weight*gradient[i],J);
```

Next, we also add the potential contributions from the tetrahedron t to the boundary matrix $A^{(b)}$, the second term in the stiffness matrix $A^{(s)}$. This is done by a yet inner loop on the corners of the form $\mathbf{k} \in t$, which, together with the corners $\mathbf{i}$ and $\mathbf{j}$ looped upon in the outer loops, make a triangle of the form $\triangle(\mathbf{i}, \mathbf{j}, \mathbf{k}) \in b$:

```
for(int k=0; k<4; k++){
  int K = (*runner)()[k].getIndex();
  if((i!=j)&&(j!=k)&&(k!=i)
     &&boundary[I]
       &&boundary[J]
       &&boundary[K]){
```

This "if" question makes sure that $\triangle(\mathbf{i}, \mathbf{j}, \mathbf{k})$ is indeed a boundary triangle in b. The contribution from it to the ('I','J')th element in $A^{(b)}$ is now calculated, using the vector-product operator in Section 28.10 above:

```
point3 jMinusi =
  (*runner)()[j]() - (*runner)()[i]();
point3 kMinusi =
    (*runner)()[k]() - (*runner)()[i]();
T boundaryTerm =
    l2norm(jMinusi&kMinusi) / 24.;
row<T> boundaryTermJ(boundaryTerm,J);
*item[I] += boundaryTermJ;
```

The same contribution to the ('I','J')th element in $A^{(b)}$ must also go to the corresponding main-diagonal element, that is, to the ('I','I')th element in $A^{(b)}$:

```
row<T> boundaryTermI(boundaryTerm,I);
*item[I] += boundaryTermI;
        }
      }
    }
  }
}
} // constructing the stiffness matrix
```

This completes the construction of the stiffness matrix $A^{(s)}$ with $\alpha = 1$.

28.21 Newton's Mass Matrix

Here we implement the constructor that takes four arguments (by reference): the mesh 'm', the list of four polynomials "P[0]", "P[1]", "P[2]", and "P[3]" of Section 28.16 above, the fixed grid function v, and its image under the nonlinear function f, $f(v)$. Actually, the dynamic vector f passed to the constructor is initially the zero vector; it takes its desired value $f(v)$ during the execution of the constructor. This is why it must be passed to it by reference: this way, the required value $f(v)$ is indeed saved. The main output of the constructor, however, is the sparse matrix that contains Newton's mass matrix, $A^{(n)}(v)$.

```
template<class T>
sparseMatrix<T>::sparseMatrix(mesh<tetrahedron>&m,
    const
    list<polynomial<polynomial<polynomial<T> > > >&P,
    const dynamicVector<T>&v, dynamicVector<T>&f){
  item = new row<T>*[number = m.indexing()];
  f = dynamicVector<T>(number,0.);
  for(int i=0; i<number; i++)
    item[i] = 0;
  polynomial<T> zero(1,0.);
  polynomial<polynomial<T> > Zero(1,zero);
```

The loop below runs over all the tetrahedra of the form $t = (\mathbf{k}, \mathbf{l}, \mathbf{m}, \mathbf{n})$ in the mesh:

```
for(const mesh<tetrahedron>* runner = &m;
    runner;
    runner =
    (const mesh<tetrahedron>*)runner->readNext()){
  matrix3 S((*runner)()[1]() - (*runner)()[0](),
            (*runner)()[2]() - (*runner)()[0](),
            (*runner)()[3]() - (*runner)()[0]());
```

This defines the 3×3 matrix S_t. The 3-d polynomial 'V' defined below is first set to zero, and then is reset to its desired value $v_{\mathbf{k}}P_{k,t}+v_{\mathbf{l}}P_{l,t}+v_{\mathbf{m}}P_{m,t}+v_{\mathbf{n}}P_{n,t}$ in an inner loop over the corners in the tetrahedron t pointed at by "runner". The 3-d polynomials $P_{k,t}$, $P_{l,t}$, $P_{m,t}$, and $P_{n,t}$ used in this sum are stored in the 3-d polynomial objects "P[0]", "P[1]", "P[2]", and "P[3]" that are calculated in Section 28.16 above and passed to the present constructor by reference in the list of polynomials 'P'.

```
polynomial<polynomial<polynomial<T> > >
  V(1,Zero);
for(int i=0; i<4; i++){
  int I = (*runner)()[i].getIndex();
  V += v[I] * P[i];
}
polynomial<polynomial<polynomial<T> > >
    V2 = V * V;
polynomial<polynomial<polynomial<T> > >
        V3 = V2 * V;
T detS = fabs(det(S));
for(int i=0; i<4; i++){
  int I = (*runner)()[i].getIndex();
  f(I) += detS * integral(V3 * P[i]);
  for(int j=0; j<4; j++){
    int J = (*runner)()[j].getIndex();
    T Jacobian =
      detS * 3. * integral(V2 * P[i] * P[j]);
    if(item[I]){
      row<T> JacobianJ(Jacobian,J);
      *item[I] += JacobianJ;
    }
    else
      item[I] = new row<T>(Jacobian,J);
  }
}
}
} // constructing Newton's mass matrix
```

This completes the construction of Newton's mass matrix $A^{(n)}(v)$ at the given grid function v. As a byproduct, $f(v)$ is also computed, and stored in the dynamic vector f that is passed to the constructor by reference.

28.22 Helmholtz Mass Matrix

Here is the constructor that produces Helmholtz mass matrix $A^{(h)}(v)$, where v is a given vector grid function in $(\mathbb{R}^2)^N$. As a byproduct, the constructor also produces the vector grid function $f(v) \in (\mathbb{R}^2)^N$, the image of v under f. This output is saved in the argument 'f'; thanks to the fact that this is a nonconstant dynamic vector [with 2-d vector components to store $f(v)_i \in \mathbb{R}^2$] that is passed by reference, the output $f(v)$ is indeed saved in it for further use.

It is assumed that the user should call this constructor with 'T' being
"matrix2", to store the required 2×2 blocks $A_{i,j}^{(h)}$.

```
template<class T>
sparseMatrix<T>::sparseMatrix(mesh<tetrahedron>&m,
    const
    list<polynomial<polynomial<polynomial<double> > > >&P,
    const dynamicVector<point>&v,
    dynamicVector<point>&f){
  item = new row<T>*[number = m.indexing()];
  f = dynamicVector<point>(number,0.);
  for(int i=0; i<number; i++)
    item[i] = 0;
  polynomial<double> zero(1,0.);
  polynomial<polynomial<double> > Zero(1,zero);
  for(const mesh<tetrahedron>* runner = &m;
      runner;
      runner =
      (const mesh<tetrahedron>*)runner->readNext()){
    polynomial<polynomial<polynomial<double> > >
        ReV(1,Zero);
    polynomial<polynomial<polynomial<double> > >
        ImV(1,Zero);
    matrix3 S((*runner)()[1]() - (*runner)()[0](),
              (*runner)()[2]() - (*runner)()[0](),
              (*runner)()[3]() - (*runner)()[0]());
    for(int i=0; i<4; i++){
      int I = (*runner)()[i].getIndex();
```

In this loop, "ReV" is assigned the real part of the 3-d polynomial

$$P_t(v) = v_{\mathbf{k}} P_{k,t} + v_{l} P_{l,t} + v_{\mathbf{m}} P_{m,t} + v_{\mathbf{n}} P_{n,t}.$$

For example, to add the term $\Re v_{\mathbf{k}} P_{k,t}$ contributed from the first corner in the
tetrahedron t, the loop uses the index 'i'= 0. For this index, 'I' contains the
index k of the node $\mathbf{k}$ in the list of nodes in N. Thus, "v[I]" is the point object
that stores the two numbers

$$(\Re v_{\mathbf{k}}, \Im v_{\mathbf{k}}).$$

The first of these numbers, stored in "v[I][0]", is used to add the required term
to "ReV" to form $\Re P_t(v)$. The second number, stored in "v[I][1]", is then used
to add the required term to "ImV" to form $\Im P_t(v)$:

```
      ReV += v[I][0] * P[i];
      ImV += v[I][1] * P[i];
    }
```

Once the 3-d polynomial objects "ReV" and "ImV" that store the real and imaginary parts of the complex-valued 3-d polynomial $P_t(v)$ have been properly defined, their powers can be defined as well for future use:

```
polynomial<polynomial<polynomial<double> > >
    Re2V = ReV * ReV;
polynomial<polynomial<polynomial<double> > >
    Im2V = ImV * ImV;
polynomial<polynomial<polynomial<double> > >
    Re2VplusIm2V = Re2V + Im2V;
polynomial<polynomial<polynomial<double> > >
    threeRe2VplusIm2V = 3. * Re2V + Im2V;
polynomial<polynomial<polynomial<double> > >
    Re2VplusThreeIm2V = Re2V + 3. * Im2V;
polynomial<polynomial<polynomial<double> > >
    twoReVImV = 2. * ReV * ImV;
double detS = fabs(det(S));
for(int i=0; i<4; i++){
  int I = (*runner)()[i].getIndex();
```

These polynomials are now used to add the contribution from the tetrahedron t to the real and imaginary parts of $f(v)_i$ for any corner **i**:

```
f(I) += detS *
    point(integral(Re2VplusIm2V * ReV * P[i]),
          integral(Re2VplusIm2V * ImV * P[i]));
```

Furthermore, they are also used to compute the contribution from t to the 2×2 block $(A^{(h)}(v))_{i,j}$ for any two corners **i** and **j**:

```
for(int j=0; j<4; j++){
  int J = (*runner)()[j].getIndex();
  point
    J1(integral(threeRe2VplusIm2V * P[i] * P[j]),
       integral(twoReVImV * P[i] * P[j]));
  point J2(integral(twoReVImV * P[i] * P[j]),
     integral(Re2VplusThreeIm2V * P[i] * P[j]));
  T Jacobian(J1,J2);
  if(item[I]){
    row<T> JacobianJ(detS * Jacobian,J);
    *item[I] += JacobianJ;
  }
  else
    item[I] = new row<T>(detS * Jacobian,J);
  }
 }
}
```

```
}  //  constructing Helmholtz mass matrix
```

28.23 Indexing the Edges in the Mesh

Here we give each of the six edges in each tetrahedron in the mesh an index
to indicate its place in the list of edges in the entire mesh. More precisely,
each edge is given two consecutive indices to index the two degrees of freedom
associated with the two normal derivatives at its midpoint.

For this purpose, every tetrahedron must contain an array of 56 integers
to store the indices of its degrees of freedom in the list of degrees of freedom
in the entire mesh. For this purpose, the block of the "cell" class must be
modified as follows:

```
template<class T, int N> class cell{
  node<T>* vertex[N];
  int index[56];
public:
  int readMeshIndex(int i) const{
    return index[i];
  }  //  read-only the indices

  int& meshIndex(int i){
    return index[i];
  }  //  read/write the indices

  ...
```

This way, the private array "index" that contains 56 integers is added to each
"cell" object, along with the public functions "readMeshIndex" (to read the
entries in it) and "meshIndex" (to read/write the entries in it). The array
"index" will contain the indices assigned to the degrees of freedom in the
current tetrahedron to indicate their place in the list of degrees of freedom in
the entire mesh.

It is also a good idea to initialize the entries in the above "index" array to
their default value -1 by adding at the end of the constructors of the "cell"
class the following loop:

```
for(int i=0; i<56; i++)
  index[i] = -1;
```

Similarly, one should better add at the end of the copy constructor and the
assignment operator of the "cell" class the loop

```
for(int i=0; i<56; i++)
  index[i] = e.index[i];
```

More meaningful indices should be assigned to the "cell" object only once it has been embedded in a complete mesh. This is why the "edgeIndexing" function that assigns these indices must be a member of the "mesh" class, so it can be applied to a current "mesh" object. This is also why this function must first be declared in the block of the "mesh" class:

```
int edgeIndexing(int);
```

Here is how this function is actually defined. The function takes the integer argument "nodes" that stores the total number of nodes in the mesh:

```
template<class T>
int mesh<T>::edgeIndexing(int nodes){
  int edges = 0;
```

The integer variable "edges" counts the edges in the mesh. Now, the tetrahedra in the mesh are scanned in an outer loop:

```
for(mesh<T>* runner = this; runner;
    runner=(mesh<T>*)runner->next){
```

First, each vertex in each tetrahedron encountered in this loop must be assigned ten entries in the array "index" to store the indices of the ten degrees of freedom associated with it. Fortunately, these vertices are already indexed in the list of nodes in the entire mesh. In fact, the index of each vertex in the list of nodes in the entire mesh can be read by the public "getIndex" function of the "node" class:

```
for(int i=0; i<10; i++){
  runner->item.meshIndex(i) =
    10 * runner->item[0].getIndex() + i;
  runner->item.meshIndex(10+i) =
    10 * runner->item[1].getIndex() + i;
  runner->item.meshIndex(20+i) =
    10 * runner->item[2].getIndex() + i;
  runner->item.meshIndex(30+i) =
    10 * runner->item[3].getIndex() + i;
}
```

Furthermore, in each tetrahedron encountered in the above outer loop, the six edges are scanned in a nested inner loop as follows. Let the four vertices of the tetrahedron be denoted by the numbers 0, 1, 2, and 3. Then, the edges of the tetrahedron are scanned in the following order:

$$(0,1), \ (0,2), \ (0,3), \ (1,2), \ (1,3), \ \text{and} \ (2,3).$$

In other words, the edge leading from vertex i to vertex j in the tetrahedron is represented as (i, j), where $0 \leq i < j \leq 3$:

```
int I = -1;
for(int i=0; i<3; i++)
  for(int j=i+1; j<=3; j++){
    I++;
```

In this nested loop, the integer variable 'I'$= 0, 1, 2, 3, 4, 5$ counts the edges in the current tetrahedron. Moreover, a yet inner loop is now used to scan all the previous tetrahedra in the mesh to check whether the current edge (i, j) has already been indexed in any of them:

```
for(mesh<T>* previous = this;
    previous&&(previous != runner)
      &&(runner->item.readMeshIndex(40+2*I) < 0);
    previous=(mesh<T>*)previous->next){
```

To check this, we use the "operator<" of the "cell" class that checks whether a given node indeed serves as a vertex in a given cell:

```
int ni = runner->item[i] < previous->item;
int nj = runner->item[j] < previous->item;
if(ni&&nj){
  ni--;
  nj--;
```

If both nodes i and j are indeed shared by a previous tetrahedron in the mesh, then the current edge (i, j) can also be denoted by (ni, nj), where "ni" and "nj" are the indices of these nodes in the list of four vertices in that previous tetrahedron. Furthermore, we can find the index $0 \leq J < 6$ of (ni, nj) in the list of six edges in that previous tetrahedron:

```
int J = min(ni,nj) == 0 ?
        max(ni,nj) - 1
      :
        min(ni,nj) == 1 ?
          max(ni,nj) + 1
      :
        5;
```

This way, 'J' is the (local) index of the edge (ni, nj) in the list of edges in the previous tetrahedron. Furthermore, 'J' is now used to assign the (global) index of (ni, nj) in the list of edges in the entire mesh to (i, j) as well:

```
runner->item.meshIndex(40+2*I)
  = previous->item.readMeshIndex(40+2*J);
runner->item.meshIndex(40+2*I+1)
```

```
            = previous->item.readMeshIndex(40+2*J+1);
        }
    }
```

If, however, no previous tetrahedron that shares the edge (i, j) has been found, then this edge must be assigned a new index in the list of edges in the mesh, and the edge counter "edges" must be incremented by 1:

```
        if(runner->item.readMeshIndex(40+2*I) < 0){
            runner->item.meshIndex(40+2*I) =
                10 * nodes + 2 * edges;
            runner->item.meshIndex(40+2*I+1) =
                10 * nodes + 2 * edges++ + 1;
        }
    }
}
```

Finally, the function also returns the total number of edges in the mesh:

```
    return edges;
}  //  indexing the edges in the mesh
```

This completes the edge indexing in the entire mesh.

28.24 Indexing the Sides in the Mesh

A similar function, named "sideIndexing", is used to index the sides of the tetrahedra in the entire mesh. First, the function is declared in the block of the "mesh" class:

```
    int sideIndexing(int, int);
```

Here is how this function is actually defined. It takes two integer arguments to store the total numbers of nodes and edges in the entire mesh:

```
    template<class T>
    int mesh<T>::sideIndexing(int nodes, int edges){
        int sides = 0;
```

The integer variable "sides" counts the sides in the mesh. As in the previous section, an outer loop is used to scan the tetrahedra in the mesh:

```
        for(mesh<T>* runner = this; runner;
            runner=(mesh<T>*)runner->next){
```

For each tetrahedron encountered in this loop, the four vertices can again be denoted by 0, 1, 2, and 3. With this notation, the four sides in the tetrahedron are then scanned in the following order:

$$(0, 1, 2), \ (0, 1, 3), \ (0, 2, 3), \ (1, 2, 3).$$

This is indeed done in the following triple loop, in which the sides are denoted by (i, j, k) (where $0 \leq i < j < k \leq 3$), and the integer variable 'I' counts them:

```
int I = -1;
for(int i=0; i<=1; i++)
  for(int j=i+1; j<=2; j++)
    for(int k=j+1; k<=3; k++){
      I++;
```

As before, a yet inner loop is used to check whether (i, j, k) also serves as a side of the form (ni, nj, nk) in a yet previous tetrahedron in the mesh:

```
for(mesh<T>* previous = this;
        previous&&(previous != runner)
        &&(runner->item.readMeshIndex(40+2*6+I) < 0);
        previous=(mesh<T>*)previous->next){
    int ni = runner->item[i] < previous->item;
    int nj = runner->item[j] < previous->item;
    int nk = runner->item[k] < previous->item;
```

If it does, then the (local) index of the fourth vertex in the list of vertices in that previous tetrahedron is identified and placed in the variable "nl":

```
if(ni&&nj&&nk){
      ni--;
      nj--;
      nk--;
      int nl = 0;
      while((nl==ni)||(nl==nj)||(nl==nk))
        nl++;
```

This fourth vertex, "nl", is now used to define the (local) index 'J' of (ni, nj, nk) in the list of four sides in that previous tetrahedron:

```
int J = 3 - nl;
```

The (local) index 'J' is now being used to assign the (global) index of (ni, nj, nk) in the list of sides in the entire mesh to (i, j, k) as well:

```
runner->item.meshIndex(40+2*6+I)
        = previous->item.readMeshIndex(40+2*6+J);
    }
  }
```

If, however, no previous tetrahedron that shares the side (i, j, k) has been found, then (i, j, k) must be assigned a new index in the list of sides in the entire mesh, and the side counter "sides" must be incremented by 1:

```
        if(runner->item.readMeshIndex(40+2*6+I) < 0)
            runner->item.meshIndex(40+2*6+I) =
                10 * nodes + 2 * edges + sides++;
        }
    }
```

Finally, the function also returns the total number of sides in the mesh:

```
    return sides;
} //  indexing the sides in the mesh
```

This completes the indexing of the sides in the entire mesh.

28.25 Computing Basis Functions

Here we define the constructor that produces the 56×56 sparse matrix B in Chapter 12, Section 12.31. For this, we use the function "C()" defined in Section 28.4.

Since the constructor must be a member function, it should first be declared in the block of the "sparseMatrix" class as follows:

```
    sparseMatrix(char*);
```

Note that the constructor takes a dummy string argument to distinguish it from the default constructor. Here is the complete definition:

```
    template<class T>
    sparseMatrix<T>::sparseMatrix(char*){
        item = new row<T>*[number = 56];
        for(int i=0; i<number; i++)
            item[i] = 0;
        int J = -1;
```

Once the 56 rows have been initialized to be the zero pointers, the appropriate matrix elements can be set in nested loops. The outer loop is over the 56 columns in the matrix. More precisely, since these columns correspond to the points in the discrete tetrahedron $T^{(5)}$, the outer loop is actually a triple loop over the triplets of the form (i, j, k) in this tetrahedron:

```
for(int i=0; i<=5; i++)
  for(int j=0; j<=5-i; j++)
    for(int k=0; k<=5-i-j; k++){
    J++;
```

The integer 'J' contains the index $\hat{j}_{i,j,k}$ in Chapter 12, Section 12.31. Now, in order to loop over the rows in the matrix, the first 40 rows are divided into four groups of ten rows each, associated with partial derivatives evaluated at the four corners of the unit tetrahedron. Each inner loop on a group of ten rows is actually implemented as a triple loop on the points $(l, m, n) \in T^{(2)}$

```
int I = -1;
for(int l=0; l<=2; l++)
  for(int m=0; m<=2-l; m++)
    for(int n=0; n<=2-l-m; n++){
    I++;
```

The integer 'I' contains the index $\hat{i}_{l,m,n}$ in Chapter 12, Section 12.31.

The first ten rows are associated with partial derivatives evaluated at the origin $(0, 0, 0)$. Clearly, the (n, m, l)th partial derivative vanishes at $x = y = z = 0$ for every monomial, except for the monomial $x^n y^m z^l$, for which its value is the scalar $n!m!l!$:

```
if((i==l)&&(j==m)&&(k==n)){
  T coef = factorial(i) * factorial(j) *
      factorial(k);
  if(item[I])
    item[I]->append(coef,J);
  else
    item[I] = new row<T>(coef,J);
}
```

Furthermore, the next ten rows are associated with partial derivatives evaluated at the corner $(1, 0, 0)$. Clearly, at this corner, the (n, m, l)th partial derivative is nonzero only for monomials that their y-power is exactly m, their z-power is exactly l, and their x-power is at least n:

```
if((i==l)&&(j==m)&&(k>=n)){
  T coef = factorial(i) *
            factorial(j) * C(k,n);
  if(item[10+I])
    item[10+I]->append(coef,J);
  else
    item[10+I] = new row<T>(coef,J);
}
```

Similarly, the next ten rows are associated with partial derivatives evaluated at the corner $(0, 1, 0)$. Clearly, at this corner, the (n, m, l)th partial derivative is nonzero only for monomials that their x-power is exactly n, their z-power is exactly l, and their y-power is at least m:

```
if((i==1)&&(j>=m)&&(k==n)){
  T coef = factorial(i) *
            C(j,m) * factorial(k);
    if(item[20+I])
      item[20+I]->append(coef,J);
    else
      item[20+I] = new row<T>(coef,J);
}
```

Similarly, the next ten rows are associated with partial derivatives evaluated at the corner $(0, 0, 1)$. Clearly, at this corner, the (n, m, l)th partial derivative is nonzero only for monomials that their x-power is exactly n, their y-power is exactly m, and their z-power is at least l:

```
if((i>=1)&&(j==m)&&(k==n)){
  T coef = C(i,l) * factorial(j) *
            factorial(k);
    if(item[30+I])
      item[30+I]->append(coef,J);
    else
      item[30+I] = new row<T>(coef,J);
  }
}
```

This completes the first 40 rows in the matrix.

In the above code, each ten consecutive rows in the constructed sparse matrix are ordered according to the lexicographical ordering of partial derivatives of order at most 2: x-, xx-, y-, yx, yy, z-, zx, zy-, and zz-partial derivatives. However, the above code can be modified to produce a more useful row ordering, in which the first partial derivatives appear before the second ones: x-, y-, z-, xx-, yx-, yy-, zx-, zy-, and zz-partial derivatives. For this, the above inner triple loop should be modified to read

```
int I = -1;
for(int q=0; q<=2; q++)
  for(int l=0; l<=2; l++)
    for(int m=0; m<=2-l; m++)
      for(int n=0; n<=2-l-m; n++)
        if(l+m+n==q){
          I++;

          . . .
```

and the rest is as before.

The next 12 rows are associated with the two normal derivatives evaluated at the six midpoints of the six edges of the unit tetrahedron. In order to define these rows, we use the function "power()" in Chapter 14, Section 14.2, and the results in Chapter 12, Section 12.24.

The first two rows correspond to the y- and z-partial derivatives evaluated at $(1/2, 0, 0)$:

```
if((i==0)&&(j==1)){
  T coef = 1. / power(2,k);
  if(item[40])
      item[40]->append(coef,J);
  else
    item[40] = new row<T>(coef,J);
}
if((i==1)&&(j==0)){
  T coef = 1. / power(2,k);
  if(item[41])
      item[41]->append(coef,J);
  else
    item[41] = new row<T>(coef,J);
}
```

Similarly, the next two rows correspond to the x- and z-partial derivatives evaluated at $(0, 1/2, 0)$:

```
if((i==0)&&(k==1)){
  T coef = 1. / power(2,j);
  if(item[42])
      item[42]->append(coef,J);
  else
    item[42] = new row<T>(coef,J);
}
if((i==1)&&(k==0)){
  T coef = 1. / power(2,j);
  if(item[43])
      item[43]->append(coef,J);
  else
    item[43] = new row<T>(coef,J);
}
```

Similarly, the next two rows correspond to the x- and y-partial derivatives evaluated at $(0, 0, 1/2)$:

```
if((j==0)&&(k==1)){
  T coef = 1. / power(2,i);
  if(item[44])
```

```
            item[44]->append(coef,J);
         else
            item[44] = new row<T>(coef,J);
      }
      if((j==1)&&(k==0)){
        T coef = 1. / power(2,i);
        if(item[45])
            item[45]->append(coef,J);
        else
          item[45] = new row<T>(coef,J);
      }
```

Furthermore, the next two rows correspond to the edge midpoint $(1/2, 1/2, 0)$. The normal derivatives at this point are the z-partial derivative and the sum of the x- and y-partial derivatives, divided by $\sqrt{2}$ (see Chapter 12, Section 12.24):

```
      if(i==1){
        T coef = 1. / power(2,j+k);
        if(item[46])
            item[46]->append(coef,J);
        else
          item[46] = new row<T>(coef,J);
      }
      if(!i&&(j||k)){
        T coef = (j+k) / sqrt(2.) / power(2,j+k-1);
        if(item[47])
            item[47]->append(coef,J);
        else
          item[47] = new row<T>(coef,J);
      }
```

Similarly, the next two rows correspond to the edge midpoint $(1/2, 0, 1/2)$. The normal derivatives at this point are the y-partial derivative and the sum of the x- and z-partial derivatives, divided by $\sqrt{2}$:

```
      if(j==1){
        T coef = 1. / power(2,i+k);
        if(item[48])
            item[48]->append(coef,J);
        else
          item[48] = new row<T>(coef,J);
      }
      if(!j&&(i||k)){
        T coef = (i+k) / sqrt(2.) / power(2,i+k-1);
        if(item[49])
```

```
            item[49]->append(coef,J);
        else
            item[49] = new row<T>(coef,J);
    }
```

Similarly, the next two rows correspond to the edge midpoint $(0, 1/2, 1/2)$. The normal derivatives at this point are the x-partial derivative and the sum of the y- and z-partial derivatives, divided by $\sqrt{2}$:

```
        if(k==1){
            T coef = 1. / power(2,i+j);
            if(item[50])
                item[50]->append(coef,J);
            else
                item[50] = new row<T>(coef,J);
        }
        if(!k&&(i||j)){
            T coef = (i+j) / sqrt(2.) / power(2,i+j-1);
            if(item[51])
                item[51]->append(coef,J);
            else
                item[51] = new row<T>(coef,J);
        }
```

The final four rows correspond to midpoints of sides of the unit tetrahedron. In particular, the next row corresponds to the normal derivative (or the z-partial derivative) at $(1/3, 1/3, 0)$:

```
        if(i==1){
            T coef = 1. / power(3,j+k);
            if(item[52])
                item[52]->append(coef,J);
            else
                item[52] = new row<T>(coef,J);
        }
```

Furthermore, the next row correspond to the normal derivative (or the y-partial derivative) at $(1/3, 0, 1/3)$:

```
        if(j==1){
            T coef = 1. / power(3,i+k);
            if(item[53])
                item[53]->append(coef,J);
            else
                item[53] = new row<T>(coef,J);
        }
```

Similarly, the next row corresponds to the normal derivative (or the x-partial derivative) at $(0, 1/3, 1/3)$:

```
if(k==1){
  T coef = 1. / power(3,i+j);
  if(item[54])
      item[54]->append(coef,J);
  else
      item[54] = new row<T>(coef,J);
}
```

Finally, the last row corresponds to the midpoint of the largest side of the unit tetrahedron, at $(1/3, 1/3, 1/3)$. The normal derivative at this point is the sum of the x-, y-, and z-partial derivatives, divided by $\sqrt{3}$ (see Chapter 12, Section 12.24):

```
if(i||j||k){
  T coef = (i+j+k) / sqrt(3.) / power(3,i+j+k-1);
  if(item[55])
      item[55]->append(coef,J);
  else
      item[55] = new row<T>(coef,J);
  }
 }
} //  construct the 56*56 matrix B
```

This completes the definition of the 56×56 matrix B in Chapter 12, Section 12.31. Once the linear system

$$B\mathbf{x} = I^{(q)}$$

is solved for the vector of unknowns $\mathbf{x}$, the basis function p_q can be produced from $\mathbf{x}$ by the following function:

```
template<class T>
const polynomial<polynomial<polynomial<T> > >
producePolynomial(const dynamicVector<T>&x){
  polynomial<T> zero(1,0.);
  polynomial<polynomial<T> > Zero(1,zero);
  polynomial<polynomial<polynomial<T> > > p(6,Zero);
  int J = -1;
  for(int i=0; i<=5; i++){
    polynomial<polynomial<T> > A(6-i,zero);
    for(int j=0; j<=5-i; j++){
      polynomial<T> a(6-i-j,0.);
      for(int k=0; k<=5-i-j; k++){
        J++;
```

```
          a(k) = x[J];
      }
      A(j) = a;
    }
    p(i) = A;
  }
  return p;
}  //  produce a polynomial of degree 5 from x
```

Indeed, in this function, the coefficients $a_{i,j,k}$ that are listed in **x** in the lexicographical order are placed in their proper places in the polynomial of three variables returned by the function. This can be tested in the following "main()" function, which prints all the 56 degrees of freedom of p_q:

```
int main(){
  sparseMatrix<double> B("56");
  dynamicVector<double> Iq(56,0.);
  Iq(55) = 1.;
  dynamicVector<double> x = solve(B,Iq);
  polynomial<polynomial<polynomial<double> > >
    pq = producePolynomial(x);
  for(int i=0; i<=2; i++)
    for(int j=0; j<=2-i; j++)
      for(int k=0; k<=2-i-j; k++){
        print(d(pq,k,j,i)(0.,0.,0.));
        print(d(pq,k,j,i)(1.,0.,0.));
        print(d(pq,k,j,i)(0.,1.,0.));
        print(d(pq,k,j,i)(0.,0.,1.));
        printf("\n");
      }
  print(d(pq,0,1,0)(.5,0.,0.));
  print(d(pq,0,0,1)(.5,0.,0.));
  print(d(pq,1,0,0)(0.,.5,0.));
  print(d(pq,0,0,1)(0.,.5,0.));
  print(d(pq,1,0,0)(0.,0.,.5));
  print(d(pq,0,1,0)(0.,0.,.5));
  printf("\n");
  print(d(pq,0,0,1)(.5,.5,0.));
  print((d(pq,1,0,0)(.5,.5,0.)+
         d(pq,0,1,0)(.5,.5,0.))/sqrt(2.));
  print(d(pq,0,1,0)(.5,0.,.5));
  print((d(pq,0,0,1)(.5,0.,.5)+
         d(pq,1,0,0)(.5,0.,.5))/sqrt(2.));
  print(d(pq,1,0,0)(0.,.5,.5));
  print((d(pq,0,0,1)(0.,.5,.5)+
         d(pq,0,1,0)(0.,.5,.5))/sqrt(2.));
```

```
    printf("\n");
    print(d(pq,0,0,1)(1./3.,1./3.,0.));
    print(d(pq,0,1,0)(1./3.,0.,1./3.));
    print(d(pq,1,0,0)(0.,1./3.,1./3.));
    print((d(pq,0,0,1)(1./3.,1./3.,1./3.)
        +d(pq,0,1,0)(1./3.,1./3.,1./3.)
        +d(pq,1,0,0)(1./3.,1./3.,1./3.))/sqrt(3.));
    return 0;
}
```

28.26 Setting Dirichlet Conditions

The Dirichlet matrix is obtained from the Neumann matrix A by eliminating the Dirichlet unknowns, that is, the unknowns whose values are already available (Chapter 27, Section 27.5). In order to produce this matrix, we need to add two member functions to the "row" class.

The purpose of these functions is to drop row elements that lie in columns that correspond to the Dirichlet unknowns. These unknowns are specified in a vector of integers (named "mask") passed to the functions by reference. In this vector, nonzero components indicate Dirichlet unknowns that should be eliminated, whereas zero components indicate meaningful unknowns that should be solved for.

Here is how the two new functions should be declared in the block of the "row" class:

```
    template<class S>
    const S maskTail(const dynamicVector<int>&,
        const dynamicVector<S>&);
    template<class S>
    const S maskAll(const dynamicVector<int>&,
        const dynamicVector<S>&);
```

Note that the functions use not only the template 'T' of the "row<T>" class but also the template 'S', which may be different from 'T'. It is assumed, though, that 'T'-times-'S' is a legitimate operation. In most applications, 'T' and 'S' are both scalars. In more advanced applications, however, such as in the Helmholtz matrix, 'T' can be interpreted as "matrix2", whereas 'S' is interpreted as "point".

The function "maskTail" defined below considers the "tail" of the current row object (from the second element onward). More precisely, the function drops every element in this tail that lies in a column for which the vector of integers "mask" that is passed to the function by reference has a nonzero

component. Furthermore, the function also returns the sum of the products
of each dropped element times the corresponding component in the vector 'f'
that is passed to the function by reference.

```
template<class T>
template<class S>
const S row<T>::maskTail(
    const dynamicVector<int>& mask,
    const dynamicVector<S>&f){
  S sum = 0.;
  if(next){
```

In order to drop an element, it is not a good idea to use the "dropFirstItem"
function inherited from the base "linkedList" class, because this function
would never drop the last element in the row. A better idea is to use a "look
ahead" strategy and consider the next row element (the first element in the
tail). This is indeed done in the following "if" question:

```
if(mask[(*next)().getColumn()]){
```

If the next row element (the second element in the row, or the first element
in the tail) should indeed drop, then this "if" block is entered, and the element drops by a call to the "dropNextItem" function inherited from the base
"linkedList" class:

```
sum = (*next)().getValue() *
        f[(*next)().getColumn()];
dropNextItem();
```

This way, the second row element drops, and its place is occupied by the
third row element, which becomes now the second row element. Thus, we
have a new (shorter) tail, which stretches from the (new) second row element
onward. The function is now applied recursively to this new tail to consider
the elements in it for dropping:

```
    sum += maskTail(mask,f);
}
```

If, on the other hand, the first element in the original tail (the second element
in the original row) should not drop (because the corresponding component
in the vector "mask" vanishes), then the above procedure should be applied
to a yet shorter tail that starts from the third row element onward. This is
done by applying the function recursively to the content of the "next" field
inherited from the base "linkedList" class, not before its type is converted
explicitly from mere pointer-to-linked-list to concrete pointer-to-row:

```
    else
        sum = (*(row<T>*)next).maskTail(mask,f);
```

```
    }
    return sum;
  } //  mask tail
```

The "maskTail" function is now used in the "maskAll" function below to consider for dropping not only the tail but also the first row element:

```
    template<class T>
    template<class S>
    const S row<T>::maskAll(
        const dynamicVector<int>& mask,
        const dynamicVector<S>&f){
      S sum = maskTail(mask,f);
```

This call considers for dropping all the elements except of the first one. Now, the first row element is considered for dropping too, unless it is the only element left, which never happens in practice:

```
    if(next&&mask[getColumn()]){
      sum += getValue() * f[getColumn()];
      dropFirstItem();
    }
    return sum;
  } //  mask all row elements
```

The above function is now used in the function "setDirichlet", which produces the Dirichlet matrix. As a member of the "sparseMatrix" class, this function must be declared in the block of this class:

```
    template<class S>
    void setDirichlet(dynamicVector<S>&,
        dynamicVector<int>&);
```

This function not only changes the current "sparseMatrix" object from the Neumann matrix to the Dirichlet matrix, but also sets its first argument, the nonconstant vector 'f', to be the required right-hand side of the spline problem. Indeed, 'f' is changed throughout the function from its initial value

$$f = \begin{pmatrix} 0 \\ f_N \end{pmatrix}$$

to its final value

$$f = \begin{pmatrix} -A_{QN}f_N \\ f_N \end{pmatrix}.$$

The second argument passed to the function by reference, the vector of integers "Dirichlet", plays the same role as the vector of integers "mask" above. In fact, it is assumed that it contains nonzero components for available Dirichlet unknowns, and zero components for meaningful unknowns.

As before, the templates 'T' and 'S' may be different from each other. Still, it is assumed that 'T'-times-'S' is a valid operation.

```
template<class T>
template<class S>
void sparseMatrix<T>::setDirichlet(
        dynamicVector<S>&f,
        dynamicVector<int>&Dirichlet){
  for(int i=0; i<number; i++){
    if(!Dirichlet[i])
```

Here the block A_{QN} in Chapter 27, Section 27.4, drops, and f_Q takes its correct value $f_Q = -A_{QN}f_N$:

```
        f(i) -= item[i]->maskAll(Dirichlet,f);
    else
```

Here, on the other hand, A_{NQ} drops, and A_{NN} is set to the identity matrix of order $|N|$:

```
        *item[i] = row<T>(1.,i);
    }
} //  set Dirichlet matrix
```

References

[1] Adamowicz, Z. and Zbierski, P.: *Logic of Mathematics: A Modern Course of Classical Logic* (third edition). John Wiley and Sons, 1997.

[2] Baruch, G., Fibich, G., and Tsynkov, S.: High-order numerical solution of the nonlinear Helmholtz equation with axial symmetry. *J. Comput. Appl. Math.* 204 (2007), pp. 477–492.

[3] Baruch, G., Fibich, G., and Tsynkov, S.: High-order numerical method for the nonlinear Helmholtz equation with material discontinuities in one space dimension. *J. Comput. Physics* 227 (2007), pp. 820-850.

[4] Brenner, S.C. and Scott, L.R.: *The Mathematical Theory of Finite Element Methods. Texts in Applied Mathematics*, 15, Springer-Verlag, New York, 2002.

[5] Cheney, E.W.: *Introduction to Approximation Theory* (second edition). Chelsea, 1982.

[6] Corry, L.: *Modern Algebra and the Rise of Mathematical Structures* (Science Networks Vol. 17, second edition). Birkhaer Verlag, Basel and Boston, 2004.

[7] Courant, R. and John, F.: *Introduction to Calculus and Analysis* (vol. 1–2). Springer, New York, 1998–1999.

[8] Dench, D. and Prior, B.: *Introduction to C++*. Chapman & Hall, 1994.

[9] Drucker, T.: *Perspectives on the History of Mathematical Logic*. American Mathematical Society, Birkhaer, 2008.

[10] Fibich, G., Ilan, B., and Tsynkov, S.: Computation of nonlinear backscattering using a high-order numerical method. *J. Sci. Comput.* 17 (2002), pp. 351-364.

[11] Fibich, G. and Tsynkov, S.: High-order two-way artificial boundary conditions for nonlinear wave propagation with backscattering. *J. Comput. Phys.* 171 (2001), pp. 632–677.

[12] Gibson, C.C.: *Elementary Euclidean Geometry: An Introduction*. Cambridge University Press, 2003.

[13] Gross, J.L. and Yellen, J.: *Graph Theory and Its Applications* (second edition). CRC Press, 2006.

[14] Hansen, P. and Marcotte, O.: *Graph Colouring and Applications*. AMS Bookstore, 1999.

[15] Hausdorff, F.: *Set Theory* (third edition). AMS Bookstore, 1978.

[16] Henrici, P.: *Applied and Computational Complex Analysis*, Vol. 1–2. Wiley-Interscience, New York, 1977.

[17] Henrici, P. and Kenan, W.R.: *Applied and Computational Complex Analysis: Power Series-Integration-Conformal Mapping-Location of Zeros.* Wiley-IEEE, 1988.

[18] Karatzas, I. and Shreve, S.E.: *Brownian Motion and Stochastic Calculus* (second edition). Springer, New York, 1991.

[19] Kuratowski, K., and Musielak, J.: *Introduction to Calculus.* Pergamon Press, 1961.

[20] Layton, W., Lee, H.K., and Peterson, J.: Numerical solution of the stationary Navier–Stokes equations using a multilevel finite element method. *SIAM J. Sci. Comput.* 20 (1998), pp. 1–12.

[21] Mitchell, W.F.: Optimal multilevel iterative methods for adaptive grids. *SIAM J. Sci. Stat. Comput.* 13 (1992), pp. 146–167.

[22] Ortega J.M.: *Introduction to Parallel and Vector Solution of Linear Systems.* Plenum Press, New York, 1988.

[23] Saad, Y. and Schultz, M.H.: GMRES: a generalized minimal residual algorithm for solving nonsymmetric linear systems. *SIAM J. Sci. Stat. Comput.* 7 (1986), pp. 856–869.

[24] Schumaker, L.L.: *Spline Functions: Basic Theory* (third edition). Cambridge University Press, 2007.

[25] Shahinyan, M.: Algorithms in graph theory with application in computer design. M.Sc. thesis, Faculty of Applied Mathematics, Yerevan State University, Yerevan, Armenia (1986). (Advisor: Dr. S. Markossian.)

[26] Shapira, Y.: *Solving PDEs in C++.* SIAM, Philadelphia, 2006.

[27] Strang, G.: *Introduction to Applied Mathematics.* Wellesley-Cambridge Press, 1986.

[28] Strang, G.: *Introduction to Linear Algebra* (third edition). SIAM, Philadelphia, 2003

[29] Strang, G. and Fix, G.: *An Analysis of the Finite Element Method.* Prentice–Hall, Englewood Cliffs, NJ, 1973.

[30] Vaisman, I.: *Analytical Geometry.* World Scientific, 1997.

[31] Varga, R.: *Matrix Iterative Analysis.* Prentice-Hall, Englewood Cliffs, NJ, 1962.

[32] Wang, R.: *Multivariate Spline Functions and Their Applications.* Springer, New York, 2001.

[33] Ward, R.C.: Numerical computation of the matrix exponential with accuracy estimate. *SIAM J. Numer. Anal.* 14 (1977), pp. 600–610.

[34] West, D.B.: *Introduction to Graph Theory* (second edition). Prentice Hall, Englewood Cliffs, NJ, 2000.

Index

Printed and bound by CPI Group (UK) Ltd, Croydon, CR0 4YY

25/10/2024

01779328-0001

Dedicated to my wife Polymnia Sideris-Korres for her love, support and encouragement. I would also like to dedicate this work to the memory of my parents Emmanuel and Sofia Korres who taught me the value of hard and honest work but above all how to pursue my dreams with dignity.

Nicholas E. Korres

Preface

In light of public concerns about sustainable food production, the necessity of human and environmental protection, along with the evolution of herbicide-resistant weeds, a review of current weed control strategies is needed. Sustainable weed control requires an integrated approach based on knowledge of each crop and the weeds that threaten it.

Important issues of weed science are thoroughly discussed in the first section of the book. Integrated weed control in relation to weed management along with herbicide and weed management effects on soil and freshwater ecosystems and insects are critically discussed. Occupational hazards due to non-judicious use of herbicides along with hygiene practices, herbicide storage and herbicide regulation are carefully discussed.

The second section of the book is divided into seven sub-sections or crops, namely cereals, row, cash crops, plantations, orchards and grape-yards and root crops. Major weeds and weed control of twenty-two crops of these cropping systems are discussed in terms of mechanical or physical, cultural, preventive and chemical weed approaches. Evaluation of weed control sustainability for each crop within cropping system is also discussed. The use of aromatic plants and essential oils for sustainable weed control along with weed control in grassland and organic farming systems are discussed under miscellaneous cropping systems, the last sub-section of the book.

This book will be an invaluable source of information for scholars, growers, consultants, researchers and other stakeholders dealing with agronomic, horticultural, and grassland-based production systems. The uniqueness of this book comes from the balanced coverage of the best weed control practices of the most important cropping systems worldwide that minimize herbicide effects on humans and the environment. Furthermore, it amalgamates and discusses the most appropriate, judicious and suitable weed control strategies for a wide range of crops. It reviews the available information and suggests solutions that are not merely feasible but also optimal. The reader will gain in-depth knowledge of both cropping systems and their related weed control. He/she will also be able to learn the principles of sustainable weed management, which are now more needed than before, and of alternative non-chemical weed control strategies for a wide range of crops around the world.

Despite the great effort that authors, editors, and reviewers have invested in this work, mistakes may have been made. We would like to ask readers to inform us of any mistakes or omissions that they find, as well as suggestions for future improvements by mailing us at the following e-mail addresses quoting **"Weed Control: Sustainability, Hazards and Risks in Cropping Systems Worldwide"**.

Nicholas E. Korres (nkorres@yahoo.co.uk)
Nilda R. Burgos (nburgos@uark.edu)
Stephen O. Duke (Stephen.Duke@ARS.USDA.GOV)

Contents

List of Reviewers

1. **Dr. Renan Aguero**
   Estacion Experimental Agricola Fabio Baudrit Moreno, Facultad de Ciencias Agroalimentarias, Universidad de Costa Rica, San Jose, Costa Rica.
   E-mail: ragueroster@gmail.com

2. **Dr. Mohammad Taghi Alebrahim**
   University of Mohaghegh Ardabili, Islamic Republic of IRAN.
   E-mail: m_ebrahim@uma.ac.ir

3. **Dr. Cezar F. Araujo Jr.**
   Agronomic Institute of Parana – IAPAR, Rodovia Celso Garcia Cid, km 375, Caixa Postal 10.030, CEP 86.047-902, Londrina, Parana, Brazil. E-mail: cezar_araujo@iapar.br

4. **Dr. Bruce Auld**
   Orange Agricultural Institute, Forest Road, Orange, New South Wales 2800, Australia.
   E-mail: bruce.auld@agric.nsw.gov.au

5. **Dr. Duane Bartholomew**
   University of Hawaii at Manoa, Dept. of Tropical Plant & Soil Science, Honolulu, HI 96822, USA. E-mail: duaneb@hawaii.edu

6. **Dr. Mirco Bundschuh**
   Institute for Environmental Sciences, University of Koblenz-Landau, Landau Germany, Department of Aquatic Sciences and Assessment, Swedish University of Agricultural Sciences, Uppsala, Sweden. E-mail: mirco.bundschuh@slu.se

7. **Dr. Nilda R. Burgos**
   University of Arkansas, Dept. of Crop, Soil and Environmental Sciences, 1366 W. Altheimer Lab, Fayetteville, AR, 72704, Arkansas, USA. Email: nburgos@uark.edu

8. **Dr. Gillian Champion**
   Weed Scientist. Suffolk, UK. E-mail: gillianchampion1@gmail.com

9. **Dr. Jayanta Deka**
   Assam Agricultural University, Dept. of Agronomy, Jorhat, 785013, Assam, India.
   E-mail: jayantadeka.2008@rediffmail.com

10. **Dr. Peter J. Dittmar**
    University of Florida Institute of Food & Agricultural Sciences 1233 Fifield Hall, P.O. Box 110690, Gainesville, FL 32611-0690, USA. Email: pdittmar@ufl.edu

11. **Dr. Doug Doohan**
    Ohio State University, College of Food, Agricultural and Environmental Sciences, 205 Gourley Hall, OH, USA. E-mail: doohan.1@osu.edu

12. **Dr. Stephen O. Duke**
Research Leader, Natural Products Utilization Research Unit, USDA, ARS, University, MS 38677, USA. E-mail: Stephen.Duke@ARS.USDA.GOV

13. **Dr. Fredrik Fogelberg**
Research Institutes of Sweden, Enheten for jordbruk & livsmedel. Besöksadress: Ultunaallén 4, 756 51 Uppsala. Postadress: Box 7033, 750 07 Uppsala.
E-mail: fredrik.fogelberg@ri.se

14. **Dr. Robert P. Freckleton**
Director of Research and Innovation, Faculty of Science, Department of Animal & Plant Sciences, University of Sheffield, Sheffield S10 2TN, UK.
E-mail: r.freckleton@sheffield.ac.uk

15. **Dr. Thomas K. Gitsopoulos**
Institute of Plant Breeding and Genetic Resources, Hellenic Agricultural Organization-Demeter, GR-57001, Thermi, Thessaloniki, Greece. E-mail: gitsopoulos@yahoo.gr

16. **Dr. David Granatstein**
Washington State University, Center for Sustaining Agriculture and Natural Resources, 1100 N. Western Ave., Wenatchee, WA 98801, USA. E-mail: granats@wsu.edu

17. **Dr. W. J. Grichar**
Texas A&M AgriLife Research Station, 3507 Highway 59E, Beeville, TX 78102, USA.
E-mail: w-grichar@tamu.edu

18. **Dr. A. D. Nuwan Gunarathne**
University of Sri Jayewardenepura, Nugegoda, Sri Lanka. E-mail: adnuwan@gmail.com

19. **Dr. Geert Haesaert**
Head of Department, Applied Biosciences, Ghent University, Faculty of Bioscience Engineering, Valentin Vaerwyckweg 1, B- 9000 Gent, Belgium.
E-mail: Geert.Haesaert@UGent.be

20. **Dr. K. Neil Harker**
Agriculture and Agri-Food Canada, 6000 C and E Trail, Lacombe, Alberta T4L 1W1, Canada. Email: neil.harker@agr.gc.ca

21. **Dr. Josie A. Hugie**
Ag Division Crop Research Manager, Wilbur-Ellis Company, 345 California street, 27th Floor, San Francisco, California, USA. E-mail: jhugie@wilburellis.com

22. **Dr. Chuck Ingels**
University of California Cooperative Extension, 4145 Branch Center Rd., Sacramento, CA 95827-3823, USA. E-mail: caingels@ucanr.edu

23. **Dr. Nicholas E. Korres**
University of Arkansas, Dept. of Crop, Soil and Environmental Sciences, 1366 W. Altheimer Lab, Fayetteville, AR, 72704, Arkansas, USA. E-mail: nkorres@yahoo.co.uk; korres@uark.edu

24. **Dr. Erik A. Lehnhoff**
New Mexico State University, MSC 3BE, Las Cruces, NM 88003, USA.
E-mail: lehnhoff@nmsu.edu

25. **Dr. María de L. Lugo-Torres**
University of Puerto Rico, College of Agriculture Science, Department of Agro-Environmental Sciences, Mayaguez Campus. E-mail: maria.lugo15@upr.edu

26. **Dr. Gulsham Mahajan**
Sr. Agronomist (Rice), Department of Plant Breeding & Genetics, Punjab Agricultural University, Ludhiana-141004, India. E-mail: mahajangulshan@rediffmail.com

27. **Dr. Mike Marshall**
Clemson University, Edisto Research and Education Center, 64 Research Road, Blackville, SC 29817, USA. E-mail: marsha3@clemson.edu

28. **Dr. Joshua McGinty**
Department of Soil and Crop Sciences, Texas A&M AgriLife Extension Service, 10345 State Hwy 44, Corpus Christi, TX 78406, USA. E-mail: joshua.mcginty@ag.tamu.edu

29. **Dr. Husrev Mennan**
Ondokuz Mayıs University, Faculty of Agriculture, Plant Protection Department, Atakum/Samsun, Turkey. E-mail: hmennan@omu.edu.tr

30. **Dr. Blessing Mhlanga**
University of Zimbabwe, P. O. Box MP 167, Mount Pleasant, Harare, Zimbabwe E-mail: blessing.mhlangah@gmail.com

31. **Dr. Camilla Moonen**
Scuola Superiore Sant'Anna di Pisa, Italy. E-mail: moonen@sssup.it

32. **Dr. J. Michael Moore**
University of Georgia, College of Agricultural and Environmental Sciences, 2360 Rainwater Rd., Tifton, GA 31793-5766, USA. E-mail: jmmoore@uga.edu

33. **Dr. Vijay K. Nandula**
Crop Production Systems Research Unit, USDA-ARS, 141 Experiment Station Road, P. O. Box 350, Stoneville, MS 38776, USA. E-mail: vijay.nandula@ars.usda.gov

34. **Dr. Euro Pannacci**
Dept. of Agricultural, Food and Environmental Sciences, Research Unit of Agronomy and Crop Sciences, Borgo XX Giugno, 74, 06121 Perugia, Italy. E-mail: euro.pannacci@unipg.it

35. **Dr. Ed R. Peachey**
Oregon State University, Horticulture Dept., ALS 4045, Corvallis, OR 97331, USA. E-mail: ed.peachey@oregonstate.edu

36. **Dr. Joao Portugal**
Polytechnic Institute of Beja, Dept. of Biosciences, Beja, Portugal. E-mail: jportugal@ipbeja.pt

37. **Dr. John A. Roncoroni**
UCCE Farm Advisor, 1710 Soscol Ave suite #4, Napa, CA 94559, USA. E-mail: jaroncoroni@ucanr.edu

38. **Dr. Angelina Sanderson Bellamy**
School of Geography and Planning, Cardiff University, 33 Park Place, Cathays, Cardiff, CF10 3BA, UK. E-mail: bellamya1@cardiff.ac.uk

39. **Dr. Ronnie W. Schnell**
351C Heep Center, Soil and Crop Science Department, Texas A&M AgriLife Extension, College Station, TX 77843-2474, USA. E-mail: ronschnell@tamu.edu

40. **Dr. R. F. Smith**
Vegetable Crops and Weed Science Farm Advisor, UC Cooperative Extension, 1432 Abbott Street, Salinas, CA 93901, USA. E-mail: rifsmith@ucdavis.edu

41. **Dr. Gladys Stephenson**
Adjunct Professor and Special Graduate Faculty, School of Environmental Sciences, Ontario Agriculture College, Room 2123 Bovey Building (#81), University of Guelph, 50 Stone Rd West Guelph, ON, N1G 2W1, Canada. Email: stepheng@uoguelph.ca

42. **Dr. Tami Stubbs**
Conservation Ag. Coordinator, Palouse Conservation District, 1615 NE Eastgate Blvd. Pullman, WA, 99163, USA. E-mail: TamiS@palousecd.org

43. **Dr. Curtis Thompson**
Kansas State University, 2017C Throckmorton Ctr., Manhattan, KS 66506, USA. E-mail: cthompso@ksu.edu

44. **Dr. Vijay K. Varanasi**
University of Arkansas, 1366 W. Altheimer Dr., Fayetteville, AR, 72704, USA.
E-mail: Vijaya.varanasi@gmail.com

45. **Dr. Ionnis Vasilakoglou**
Technological Education Institute Larisa, Dept. of Plant production, 41110, Larisa, Greece.
E-mail: vasilakoglou@teilar.gr

46. **Dr. Jose Aires Ventura**
Instituto Capixaba de Pesquisa, Assistencia Tecnica e Extensao Rural (INCAPER), Vitoria, ES, Brazil. E-mail: ventura@incaper.es.gov.br

47. **Dr. Maurizio Vurro**
Director of Research, Institute of Sciences of Food Production, National Research Council, via Amendola 122/O – 70125, Bari, Italy. E-mail: maurizio.vurro@ispa.cnr.it

48. **Dr. Paula Westerman**
Faculty of Agricultural and Environmental Sciences University of Rostock Satower Str. 48. D-18051 Rostock, Germany. E-mail: paula.westerman@uni-rostock.de

49. **Dr. Lisa Woodruff**
USDA-Agricultural Research Service, Crop Production Systems Research Unit, 141 Experiment Station Road, Stoneville, MS 38776, USA. E-mail: lisa.woodruff@ars.usda.gov

50. **Dr. Hanwen Wu**
Department of Primary Industries, Pine Gully Road | Wagga Wagga NSW 2650, Australia. E-mail: hanwen.wu@dpi.nsw.gov.au

51. **Dr. Bernard Zandstra**
Michigan State University, 1066 Bogue St.,East Lansing, MI 48824, USA.
E-mail: zandstra@msu.edu

SECTION I

Weed Science—Sustainability, Hazards and Risks

CHAPTER

1

Sustainable Agriculture and Integrated Weed Management

Fabian D. Menalled

719 Leon Johnson Hall, Department of Land Resources and Environmental Sciences,
Montana State University, Bozeman, MT 59717-3120, USA
E-mail: menalled@montana.edu

Agricultural Sustainability: Finding Consensus in Food Production

Regardless of social background, political affiliation, or personal beliefs, there is a unifying consensus across our society on the need to advance the sustainability of food production systems. Goals such as producing healthy food to sustain local communities, improving food security in impoverished regions of the world, ensuring farmers make a living of sustainably-grown food, advocating for biodiversity and environmental health, and building up local and regional sustainable food systems are shared by multinational agricultural corporations, national grassroots organizations, and local farmers' markets (Farmers Market Coalition 2017, Monsanto 2017, National Sustainable Agriculture Coalition 2017). But, how do we define sustainable food production systems? What bio-physical, social, and economic characteristics secure the sustainability of food systems in the face of increasing population growth, shifts in consumer demands, and unprecedented climate scenarios? What knowledge and technological breakthroughs are needed to apply agroecological principles in the design and management of sustainable food systems? What role should weed science play in the development and adoption of sustainable practices?

While concepts related to sustainability date back to the oldest surviving writings from China, Greece, and Rome, interest in how these ideas relate to food systems can be traced to the environmental movements of the 1950s and 1960s (Pretty 2008). Despite its relatively short life, the concept of sustainable food systems has been evaluated from contrasting and sometimes conflicting perspectives within academic, applied, and legislative frameworks and little agreement has been reached (Zimdahl 2006). Recently, to summarize previous characterizations of sustainable agriculture into a wide and inclusive definition, the National Research Council (2010) identified "four generally agreed-upon goals" to help define sustainable agriculture:

- Satisfy human food, feed, and fiber needs, and contribute to biofuel needs.
- Enhance environmental quality and the resource base.
- Sustain the economic viability of agriculture.
- Enhance the quality of life of farmers, farm workers and the society as a whole.

Embedded in this definition of sustainable agriculture are three overarching concepts. First, the idea that all agricultural systems aim at obtaining usufruct in the form of food, feed, fiber, or energy. However, and in contrast with conventional approaches to farming where yield maximization is perceived as the overarching goal, the design of a sustainable agricultural systems also considers the provision of a diverse array of ecosystem services including aesthetic open space and recreation, conservation of biodiversity, provision of clean water and air, soil health, and carbon sequestration. At the same time, sustainable agricultural systems aim at minimizing side effects and externalities, i.e. environmental, social, and health costs, such as production of greenhouse gases, soil erosion, and water eutrophication that are imposed by agriculture on other sectors (Pretty 2008, Mortensen et al. 2012).

Second, is the concept of intergenerational equity such that current agricultural practices do not compromise the ability of future generations to meet their needs (Robertson 2015). This requirement necessitates a temporal fairness commitment so that future generations are recognized as entitled to a non-deteriorated ecological and economic capacity (Padilla 2002). From an ecological perspective, the intergenerational equity concept requires that sustainable systems incorporate practices that increase the resilience, or ability to buffer shocks and stresses, as well as the long-term persistence of their bio-physical production basis. However, while there is abundant literature on fairness across generations, there is no consensus on empirical approaches to operationalize this idea to the point where it can be incorporated in the decision-making process (Pannell and Schilizzi 1999).

Finally, is the notion that identifying a pathway to sustainability is a complex task that requires a major paradigm shift in our relationship with the many ecosystem services provided by agricultural landscapes. While guidelines for the process of conversion to sustainable agroecosystems exists (Gliessman 1998), the sustainability of any agricultural system should not be evaluated through adoption of a determined technological package or as reaching a particular end state, but as a dynamic and flexible trajectory aimed at achieving each one and all of the aforementioned goals (Gliessman 2016). As such, the pursuit of sustainability requires agricultural professionals to jointly evaluate, within an adaptive framework, a wide range of concepts ranging from the bio-physical processes that determine agricultural productivity to an understanding of the socio-economic barriers and cultural norms that influence the adoption of specific management practices (Swinton et al. 2015).

As an integral component of sustainable agriculture, weed science is not exempt to these ideas.

Weeds, Weed Science, and Agricultural Sustainability

To most growers, yield reduction due to weed competition represents a persistent, undesirable, and almost inevitable problem. This threat is not new and the negative impacts of weedy plants on farmers' livelihood have played a central role in agriculture for thousands of years (Zimdahl 2004). It should not be surprising then that weed scientists have conducted substantial applied research aimed at the eradication of weeds through 'silver bullet' tactics, such as herbicides and cultivation (Lewis et al. 1997), in an approach that has been characterized as a 'ruthless fight to the last weed' (Zimdahl 1994). Three unifying, but many times untested, principles drive this approach to weed control: (1) the precautionary principle is based on the perception that the potential impact that weeds could have on crop yield and quality warrants the use of the most lethal treatment possible, even if in doubt about the actual level of threat weeds convey, (2) rapid and effective response: regardless of the temporal and spatial abundance and dynamics of the weed species in consideration, management must be quick, effective, and deadly to eliminate the chances that weed abundance and spread gets too big to handle, and (3) eradication expectation: the pest must be removed, including the weed seedbank (Menalled et al. 2016).

These principles lead much of the weed science research agenda that occurred since the beginning of the chemical era of agriculture that began in the 1950's, when newly developed synthetic pesticides and fertilizers became widely available (Zimdahl 2006). Born out of advances in chemistry that emerged following the Second World War, weed science was initially perceived as a discipline that should provide tools to allow farmers an undefeatable approach to control weeds. For example, Sagar (1968) commented that, "there is widespread impression that a weed-free environment is almost with us. This is an exaggeration, but the rate of progress towards it is rapid and there is no *a priori* reason why it should not be technically possible within, say, the next 10–20 years". Unfortunately, and despite much research, the goal of weed-free fields is as elusive today as it was 50 years ago.

As a relatively new discipline, modern weed science developed in close association with agronomy and crop science. In this context, for the last 60 years weed scientists have focused their attention on minimizing weed competition within genetically uniform landscapes through heavy reliance on herbicides. Resulting from this narrow and applied focus, weed science has evolved with no theoretical foundation rooted in evolutionary and ecological disciplines (Neve et al. 2009). For example, a recent survey of the membership of the Weed Science Society of America (WSSA), a leading professional organization, indicated that the largest group of respondents (41%) received support principally from private industry sources to conduct research on issues related to herbicide efficacy. In the same survey, only 22% of the respondents received public funding to evaluate a wide variety of non-chemically based research topics including invasion biology, ecological weed management, ecosystem restoration, and assessing the genetics, molecular biology, and physiology basis of weedy traits (Davis et al. 2009). In accordance, Harker and O'Donovan (2013) observed that while information exists on non-herbicidal and integrated approaches to manage weeds, most of the published research centers on herbicide-based tactics. The herbicide-centered research priority did not arise in isolation, but can be regarded as both cause and consequence of the mainstream attitudes towards agricultural production practices.

In 2014, 11 major companies allocated $2.52 billion to conduct research and development on chemical control tactics but only $180 million was used to assess the potential efficacy of biological control products (Young et al. 2017). Worldwide, 2,563 million kg of active ingredients of pesticides are used every year in semi-natural and managed ecosystems, 37% of them herbicides (Pretty 2008). The annual value of the global pesticide market reaches US $25 billion (Pretty 2005) of which US $15.5 billion are used each year to control weeds (Grube et al. 2011). In the USA, 97% of corn (*Zea mays* L.), 98% of soybean (*Glycine max* L.), and 61% of wheat (*Triticum aestivum* L.) crops are regularly treated with herbicides (NASS 2017). In Western Canada, 86% of the area seeded in cereals and annual legumes, as well as 93% of the area seeded in oilseed crops, received at least one herbicide application in 2006 (Shanner and Beckie 2014). The relatively recent development and wide-scale implementation of transgenic herbicide tolerant crops, has further simplified the spatial and temporal complexity of agricultural landscapes and secured these trends.

While successful in terms of securing yield and maximizing farm labour efficiency, the "silver bullet" approach to managing weeds has, over the long-term, consistently failed (Lewis et al. 1997, Mortensen et al. 2000). This failure is the resultant of several ecological, evolutionary, economic, and environmental factors (Menalled et al. 2016). Smith et al. (2006) summarizes the main shortcomings of conventional weed control strategies in agroecosystems. First, the notion that weeds are external and damaging factors to crop production. Although weeds are integral components of the associated biodiversity occurring within crop fields (Marshall et al. 2003), the goal of eradicating weeds within agricultural landscapes drove much of the weed science research agendas (Appleby 2005). As a result, weed scientists have not systematically assessed the biological, evolutionary, and ecological principles driving the spread and abundance of local weed populations within agricultural landscapes (Neve et al. 2009). Second, the fact that most of the weed control strategies are applied at the individual species level. This species-centered approach to weed management has three major limitations: (1) it fails to consider the inherent genetic variability occurring at the population and metapopulation level, many times leading to expensive management practices being applied much more widely than necessary

(Shaw 2005), (2) it does not consider the mechanisms of potential shifts in weed communities that could result from changes in agricultural practices (Booth and Swanton 2002), and (3) it limits the possibility of manipulating synergistic interactions among species that may help reduce the need for intensive weed control practices (White et al. 2007, Sciegienka et al. 2011).

Biodiversity reduction, soil and water contamination, herbicide drift, direct and indirect impacts of weed management tactics on the environment and human health, and increased cost of production are among the critiques herbicide-based weed management has received (Robinson and Sutherland 2002, Robertson and Swinton 2005). More recently, the selection of weed biotypes that survive the exposure to herbicide families targeting the same site of action, a phenomenon known as cross resistance, or different sites of action, referred to as multiple resistance, resulted in major management challenges (Menalled et al. 2016, Heap 2017). These failures sparked a debate among weed scientists: Are we asking the right questions and assessing the appropriated management tools? What are the drivers determining the weed science research agenda? What new collaborations should weed scientists embrace in order to remain relevant in the task of increasing the sustainability of food production systems? What ecological role do weeds play within agroecosystems? Which changes in government policies, market variables, and social infrastructure are needed to advance farmers' adoption of sustainable weed management programs? (Ward et al. 2014, Liebman et al. 2016).

Answering these questions will allow weed science to be an active player in the development of management tactics that aim at augmenting the ecological processes that provide the functions necessary for sustained production. In this context, weed scientists should focus not only at reducing the spread and impact of weeds, but should also ponder the impact that the proposed management tactics have on ecosystem services including nutrient cycling, pollination, pest suppression, regulation of soil temperature and moisture, and detoxification of noxious chemicals (Robertson and Swinton 2005).

Weeds as Intrinsic and Inevitable Components of Sustainable Agroecosystems

Advancing the integration of weed science and sustainable agriculture requires evaluating weeds not just for their detrimental impacts on yield and crop quality, but as integral components of agroecosystems that interact with organisms at the same and other trophic levels. As weed scientists have begun to acknowledge the constraints of a herbicide-based research discipline, they are re-evaluating the dominant paradigm regarding role of weeds and their impact on agroecosystems. The traditional definition of a weed as "a plant growing where it is not wanted" (Roberts et al. 1982) is being replaced by one where they are regarded as "plants that are especially successful at colonizing disturbed, but potentially productive, sites and at maintaining their abundance under conditions of repeated disturbances" (Mohler 2001). This shift reflects the need to replace a prescriptive weed science with one that strives at understanding the ecological and evolutionary drivers of weed abundance, distribution, and role within agroecosystems. Alongside, weed science is replacing the concept of 'weed control' with the idea of 'weed management' (Harker and O'Donovan 2013). While the first concept aims at eradicating weeds, the later focuses at preventing the causes that lead to weed problems (Buhler et al. 2000) and strives at understanding the agroecological functions of weeds (Jordan and Vatovec 2004) while exploring management options. In this process, weed scientists are gaining new insights into the functional role of weeds within agroecosystems (Smith et al. 2006). More recently, advances in eco-evolutionary biology, a discipline that seeks to understand how local population dynamics arise from phenotypic variation, habitat distribution, and propagule dispersal provide a framework to re-think weed management (Gross et al. 2015, Menalled et al. 2016).

Ecological Role of Agricultural Weeds

Although Darwin (1859) cited agricultural cases to support his hypothesis that diversity affects productivity (see McNaughton 1993), much of the experimental research relating diversity and ecosystem function has been conducted in unmanaged systems and experimental grasslands. However, the conditions of these systems are fundamentally different from those commonly found in most agroecosystems, where ecological disturbances in the form of management practices aim at reducing overall weed density and diversity (Hector and Bagchi 2007, Savory 1988, Tilman and Lehman 2001). Despite these limitations, theoretical concepts and empirical evidence developed in non-cropping systems provide an insight into the role that biodiversity may play in agroecosystems (reviewed in Marshall et al. 2003, Jordan and Vatovec 2004, Gross et al. 2015).

The suite species that comprise managed ecosystems, including agroecosystems, are part of either the planned or the associated biodiversity. Planned biodiversity consists of the species that the land manager intentionally includes in the system. Associated biodiversity includes insect pests, weeds, pathogens, and beneficial organisms as well as neutral species, individuals that while not intentionally introduced into the system by the land manager, live in or colonise it from adjacent habitats (Altieri 1999). In agroecosystems, the planned biodiversity and the associated set of practices utilised to manage it, determines the abundance, dynamics, and function of the associated biodiversity (Matson et al. 1997, Menalled et al. 2001, Pollnac et al. 2009).

There is evidence that farmland bird and game species are associated with weedy structures (Hoft and Gerowitt 2006). Weeds may act as cover crops (Gliessman 1998) and affect nematode dynamics (Thomas et al. 2005). Leguminous weeds can fix atmospheric nitrogen, helping farmers reduce the need for fertilizer inputs (Jordan and Vatovec 2004). By providing pollen and nectar resources for parasitoids and pollinators, increased weed diversity could enhance pest biological control and pollination services (Norris and Kogan 2005, Landis et al. 2005). Finally, through changes in soil microbial communities (Jordan et al. 2000, Ishaq et al. 2017), weeds can impact subsequent crop growth and crop competitive ability in predictable patterns (Miller and Menalled 2015, Johnson et al. 2017).

Certainly, not all the functional effects of weed diversity are likely to be beneficial. For example, the role of weeds in increasing incidence of plant pathogens or by harboring insect pests could be detrimental. In a study conducted in continuous winter wheat systems in Montana, USA, Miller et al. (2013) determined that *Bromus tectorum* L. (cheatgrass, downy brome), a winter annual weed that colonizes small grain systems, can increase snow mold disease incidence and damage, reducing crop stands below replanting thresholds. Similarly, *Arctium minus* Bernh (common burdock) and *Taraxacum officinale* F. H. Wigg. (dandelion) may play an important role in the epidemiology of Iris yellow spot virus—a damaging pathogen in onion fields.

Multi-trophic interactions are common within agroecosystems, but seldom studied. Many times, complex ecological relationships mandate that weed management decisions should consider direct and indirect biological interactions occurring within agroecosystems. For example, Keren et al. (2015) observed that the oviposition behavior of *Cephus cinctus* (wheat stem sawfly), a dominant insect pest in wheat fields, varies as a function of *B. tectorum* pressure. Furthermore, this study determined that crop yield was more readily explained by the joint effects of management tactics on both categories of pests and their interactions than just by the direct impact of any management scheme on yield. Similarly, Dosdall et al. (2003) determined that early weed removal favored canola (*Brassica napus* L.) yield but increased root maggot (*Delia* spp.) egg deposition and subsequent damage to canola taproots roots.

Integrated Weed Management: Kindling the "Science" within Weed Science

Shifting from a weed science centered on the eradication of weeds species to one that jointly considers the risks and benefits of the entire weed community and assesses multiple tactics of management requires novel approaches to the study of agroecosystems. The development of integrated weed management (IWM) tactics represent a potential avenue to reach this goal. Initially conceived in the early 80s as a component of pest management strategies that emphasize "an agroecosystem approach for the management and control of weeds. . . (that) includes the use of multiple-pest resistant, high yielding, well-adapted varieties that resist competition". In its initial definition, IWM advocated for the "... precision placement of fertilizer. . . , and timing the fertilizer application for maximum stimulation of the crop and minimum stimulation of the weed population" (Shaw 1982). This definition was further expanded to "the integration of effective, environmentally safe, and sociologically acceptable control tactics that reduce weed interference" (Thill et al. 1991). More recently, Norsworthy et al. (2012) highlighted the importance of incorporating an IWM perspective in the design of best management practices to prevent weed spread and manage herbicide resistance.

Though there has been a disagreement among weed scientists about the core concepts of IWM (Owen et al. 2015) and its usefulness to prevent multiple herbicide resistance (Menalled et al. 2016), there is a consensus that the integration of complementary tactics should include the biological and ecological characteristics of agricultural weeds and the agroecological context where they occur (Swanton and Weise 1991). Instead of targeting one specific weed species in a single year, IWM should focus on developing a preventive holistic approach to weed management that can be maintained through multiple years. To achieve this temporal sustainability, the development of an IWM program needs to consider ecological, environmental, and social issues including propagule movement, soil protection, water and soil quality, as well as impacts on biodiversity, human wellbeing, and economics.

Implicit in the concept of IWM is the idea that additive or synergist interactions occur between management tactics, but this assumption is seldom tested (Boydston and Williams 2004). Additive interactions take place when management tactics have independent effects on weed growth and/or mortality. For example, Sciegienka et al. (2011) observed an additive relationship between two biological control agents and a herbicide to manage *Cirsium arvens* (L.) Scop. (Canada thistle). Synergy refers to the condition when the joint impact of two or more factors causes a relatively greater reduction on the targeted weeds than would be expected from their independent, additive effects. Synergism might arise if a weed biological control agent benefits from a herbicide through increased plant quality (Messersmith and Adkins 1995, Wilson et al. 2004) or if the biological control agent enhances herbicide uptake (Nelson and Lym 2003). Unfortunately, antagonist interactions between management tactics can also occur to the point of rendering an IWM plan ineffective. For example, although insects are usually resilient to herbicide applications (e.g., Trumble and Kok 1980, Lindgren et al. 1998), antagonism between herbicides and biological control agents could occur via toxicity or through a reduction in plant quality (Messersmith and Adkins 1995, Paynter 2003, Boydston and Williams 2004). Also, timing of herbicide applications relative to the phenology of the biological control agent can be crucial in determining antagonism between management tactics (Paynter 2003, Story and Stougaard 2006).

Recently, the concept of IWM has expanded into contrasting views. On one side, the 'many little hammers' concept suggests that weed management should be based on the integration of concatenated control tactics and ecological interactions of whose individual effects may be small, but its cumulative impacts will help reduce weed abundance and competitive ability (Liebman and Gallandt 1997). In this framework, management tactics should aim at minimizing propagule production and distributions, enhancing crop competitiveness, reducing weed survivorship and biomass production, as well as increasing losses to the weed seedbank due to germination, predation, and decay. Combining field data and a matrix population dynamic model Westerman et al. (2005) provided support to the 'many little hammers' concept by demonstrating that exploiting multiple stress and mortality factors can result in effective weed suppression with less reliance on herbicides. Similarly, Anderson (2003) determined that combining multiple management practices with appropriated crop rotations improved crop competitiveness against weeds.

On the other side, there is a call to develop IWM programs that take advantage of recent technological advances in crop breeding, remote sensing, and decision support systems (Young et al. 2017). The viability and utility of the proposed technologies have not been fully explored and several unanswered questions remain: what extension/outreach programs are needed to secure growers learning of the skills required for the implementation of technologically-based IWM programs? What are the economic and social costs associated with the adoption of such technological packages? What advantages could emerge from the integration of the 'many little hammers' concept with technologically-based IWM programs?

The Next Steps

To help advance agricultural sustainability, weed science must address new and complex challenges. Among them is the selection of multiple herbicide resistant biotypes, the need for reducing environmental and social externalities associated with IWM, and adapting to shift in food systems associates with technological advancements, population growth, shifts in market demands, and climate challenges. Shifting from a prescriptive weed science research agenda centered on the eradication of weeds to a preventive approach that jointly considers the risks and benefits of an IWM program represents a fundamental step in such direction (Radosevich et al. 2007). Sustainable IWM considerations should not only focus on changes in weed abundance and concomitant yield patterns, but should consider interactions among the economic, environmental, and social dimensions of agroecosystems. To reach such goal, weed science must move away from a mono-disciplinary perspective and must embrace collaborations with multiple ecological, environmental, and social disciplines (Ward et al. 2014).

Sustainable IWM is critically important to secure agricultural yields and help maintain ecosystem health in both agricultural and non-agricultural systems. As such, weed scientists should undertake research on complex IWM problems within projects that evaluate multidisciplinary aspects of sustainable agriculture (Jordan et al. 2016). To do that, weed science needs to be informed by the broad social and environmental context where weeds exist and should aim at advancing agricultural sustainability. While this idea could be daunting, Liebman et al. (2016) provide empirical examples across the world of how utilizing multiple tactics, enhancing crop competitiveness, and tailoring weed management strategies to better accommodate weed spatial distributions can help in the development of sustainable farming systems.

IWM plays a fundamental role in advancing the sustainability of cropping systems and it should aim at sustaining the bio-physical infrastructure required for production. Doing that requires a new interdisciplinary approach to address critical production issues. Ultimately, the implementation of alternative weed management practices will depend on individual and societal priorities, and education efforts should aim at easing the adoption of innovative IWM programs.

REFERENCES

Altieri, M.A. 1999. The ecological role of biodiversity in agroecosystems. Agriculture, Agric. Ecosys. Environ. 74: 19–31.

Anderson, R.L. 2003. An ecological approach to strengthen weed management in the semiarid Great Plains. Adv. Agron. 80: 33–62.

Appleby, A.P. 2005. A history of weed control in the United States and Canada – a sequel. Weed Sci. 53: 762–768.

Booth, B.D. and C.J. Swanton. 2002. Assembly theory applied to weed communities. Weed Sci. 50: 2–13.

Boydston, R.A. and M.M. Williams. 2004. Combined effects of Aceria malherbae and herbicides on field bindweed (*Convolvulus arvensis*) growth. Weed Sci. 52: 297–301.

Buhler, D.D., M. Liebman and J.J. Obrycki. 2000. Theoretical and practical challenges to an IPM approach to weed management. Weed Sci. 48: 274–280.

Darwin, C. 1859. On the Origins of Species by Natural Selection. The Modern Library, Random House, New York, USA.

Davis, A.S., J.C. Hall, M. Jasieniuk, M.A. Locke, E.C. Luschei, D.A. Mortensen, D.E. Riechers, R.G. Smith, T.M. Sterling and J.H. Westwood. 2009. Weed Science Research and Funding: a call to action. Weed Sci. 57: 442–448.

Dosdall, L.M., G.W. Clayton, K.N. Harker, J.T. O'Donovan and F.C. Stevenson. 2003. Weed control and root maggots: making canola pest management strategies compatible. Weed Sci. 51: 576–585.

Farmers Market Coalition. 2017. Farmers Markets Promote Sustainability. Available at: https://farmersmarketcoalition.org/education/farmers-markets-promote-sustainability/ (Accessed on February 11, 2017).

Gliessman, S.R. 1998. Agroecology: Ecological Processes in Sustainable Agriculture. Sleeping Bear Press, Chelsea, MI, USA.

Gliessman, S.R. 2016. Agroecology: it's not the destination, it's the journey. Agroecol. Sustain. Food Syst. 40: 893–894.

Gross, K.L., S. Emery, A.S. Davis, R.G. Smith and T.M.P. Robinson. 2015. Plant community dynamics in agricultural and successional fields. pp. 158–187. *In:* Hamilton, S.K., Doll, J.E. and Robertson, G.P. (Eds.) The Ecology of Agricultural Landscapes. Long-Term Research on the Path to Sustainability. Oxford University Press, New York, New York, USA.

Grube, A., D. Donaldson, T. Kiely and L. Wu. 2011. Pesticides Industry Sales and Usage – 2006 and 2007 Market Estimates. US Environmental Protection Agency: Washington, DC, USA. Available at: https://nepis.epa.gov/Exe/ZyPDF.cgi/P100A8DN.PDF?Dockey=P100A8DN.PDF#ga=1.47126055.1632740713.1445374245 (Accessed on February 10 , 2017).

Harker, K.N. and J.T. O'Donovan. 2013. Recent weed control, weed management and integrated weed management. Weed Technol. 27: 1–11.

Heap, I. 2017. International Survey of Herbicide Resistant Weeds. Available at: http://www.weedscience. com/summary/home.aspx (Accessed on February 8, 2017).

Hector, A. and R. Bagchi. 2007. Biodiversity and ecosystem multifunctionality. Nature 448: 188–186.

Hoeft, A. and B. Gerowitt. 2006. Rewarding weeds in arable farming – traits, goals and concepts. J. Plant Dis. Prot. XX: 517–526.

Ishaq, S., S. Johnson, Z. Miller, E. Lehnhoff, S. Olivo, C. Yeoman and F. Menalled. 2017. Impact of cropping systems, soil inoculum, and plant species identity on soil bacterial community structure. Microb. Ecol. 73: 417–434.

Johnson, S., Z. Miller, P. Miller, E. Lehnhoff and F. Menalled. 2017. Cropping systems modify soil biota effects on wheat (*Triticum aestivum* L.) growth and competitive ability. Weed Res. 57: 6–15.

Jordan, N., M. Schut, S. Graham, J.N. Barney, D.Z. Childs, S. Christensen, R.D. Cousens, A.S. Davis, H. Eizenberg, D.E. Ervin, C. Fernandez-Quintanilla, L.J. Harrison, M.A. Harsch, S. Heijting, M. Liebman, D. Loddo, S.B. Mirsky, M. Riemens, P. Neve, D.A. Peltzer, M. Renton, M. Williams, J. Recasens and M. Sonderskov. 2016. Transdisciplinary weed research: new leverage on challenging weed problems? Weed Res. 56: 345–358.

Jordan, N.R., J. Zhang and S. Huerd. 2000. Arbuscular-mycorrhizal fungi: potential roles in weed management. Weed Res. 40: 397–410.

Jordan, N. and C. Vatovec. 2004. Agroecological benefits from weeds. pp. 137–158. *In:* Inderjit (Ed.) Weed biology and management. Kluwer Academic Publishers, Netherlands.

Keren, I.N., F.D. Menalled, D.K. Weaver and J.F. Robison-Cox. 2015. Interacting agricultural pests and their effect on crop yield: application of a Bayesian decision theory approach to the joint management of *Bromus tectorum* and *Cephus cinctus*. Plos One 10.

Landis, D.A., F.D. Menalled, A.C. Costamagna and T.K. Wilkinson. 2005. Manipulating plant resources to enhance beneficial arthropods in agricultural landscapes. Weed Sci. 53: 902–908.

Lewis, W.J., J.C. van Lenteren, S.C. Phatak and J.H. Tumlinson. 1997. A total system approach to sustainable pest management. Proc. Nat. Acad. Sci. USA 94: 12243–12248.

Liebman, M., B. Baraibar, Y. Buckley, D. Childs, S. Christensen, R. Cousens, H. Eizenberg, S. Heijting, D. Loddo, A. Merotto, Jr. , M. Renton and M. Riemens. 2016. Ecologically sustainable weed management: How do we get from proof-of-concept to adoption? Ecol. Appl. 26: 1352–1369.

Liebman, M. and E.R. Gallandt. 1997. Many little hammers: Ecological management of crop-weed interactions. pp. 291–343. *In:* L.E. Jackson (Ed.) Ecology in Agriculture. Academic Press, San Diego California, USA.

Lindgren, G.J., T.S. Gabor and H.R. Murkin. 1998. Impact of triclopyr amine on *Galerucella calmariensis* L. (Coleoptera : Chrysomelidae) and a step toward integrated management of purple loosestrife *Lythrum salicaria* L. Biol. Control 12: 14–19.

Marshall, E.J.P., V.K. Brown, N.D. Boatman, P.J.W. Lutman, G.R. Squire and L.K. Ward. 2003. The role of weeds in supporting biological diversity within crop fields. Weed Res. 43: 77–89.

Matson, P.A., W.J. Parton, A.G. Power and M.J. Swift. 1997. Agricultural intensification and ecosystem properties. Science 277: 504–509.

McNaughton, S.J. 1993. Biodiversity and function of grazing ecosystems. pp. 361–383. *In:* D. Schulze and H.A. Mooney (Eds.) Biodiversity and Ecosystem Function. Springer-Verlag, Berlin, Germany.

Menalled, F.D. , K.L. Gross and M. Hammond. 2001. Weed aboveground and seedbank community responses to agricultural management systems. Ecol. Appl. 11: 1586–1601.

Menalled, F., R.K. Peterson, R.G. Smith, W.S. Curran, D.J. Perez and B.D. Maxwell. 2016. The eco-evolutionary imperative: revisiting weed management in the midst of a herbicide resistance crisis. Sustainability 8(12): 1297, doi:10.3390/su8121297

Messersmith, C.G. and S.W. Adkins. 1995. Integrating weed-feeding insects and herbicides for weed control. Weed Technol. 9: 199–208.

Miller, Z.J. and F.D. Menalled 2015. Impact of species identity and phylogenetic relatedness on biologically-mediated plant-soil feedbacks in a low and a high intensity agroecosystem. Plant Soil 389: 171–183.

Miller, Z.J., F.D. Menalled and M. Burrows. 2013. Winter annual grassy weeds increase over-winter mortality in autumn-sown wheat. Weed Res. 53: 102–109.

Mohler, C.L. 2001. Weed life history: identifying vulnerabilities. pp. 40–98. *In:* M. Liebman, C.L. Mohler and C.P. Staver (Eds.), Ecological Management of Agricultural Weeds. Cambridge University Press, Cambridge, Great Britain.

Monsanto. 2017. Available at: http://www. monsanto. com/ (Accessed on February 11, 2017).

Mortensen, D.A., L. Bastiaans and M. Sattin. 2000. The role of ecology in the development of weed management systems: an outlook. Weed Res. 40: 49–62.

Mortensen, D.A., J.F. Egan, B.D. Maxwell, M.R. Ryan and R.G. Smith. 2012. Navigating a critical juncture for sustainable weed management. BioSci. 61: 75–84.

National Research Council. 2010. Towards sustainable agricultural systems in the 21st century. The National Academies Press. Washington, D.C., USA.

National Sustainable Agriculture Coalition. 2017. Available at: http://sustainableagriculture. net/ (Accessed on February 11, 2017).

Nelson, J.A. and R.G. Lym. 2003. Interactive effects of *Aphthona nigriscutis* and picloram plus 2,4-D in leafy spurge (*Euphorbia esula*). Weed Sci. 51: 118–124.

Neve, P., M. Vila-Aiub, and F. Roux. 2009. Evolutionary-thinking in agricultural weed management. New Phytol. 184: 783–793.

Norris, R.F. and M. Kogan. 2005. Ecology of interactions between weeds and arthropods. Annu. Rev. Entomol. 50: 479–503.

North American Statistics Service (NASS). 2017. Agricultural Chemical Use Program. United States Department of Agriculture. Available at: http://www.nass.usda.gov/Surveys/Guide_to_NASS_Surveys/Chemical_Use/index.php (Accessed on February 10, 2017).

Norsworthy, J.K., S.M. Ward, D.R. Shaw, R.S. Llewellyn, R.L. Nichols, T.M. Webster, K.W. Bradley, G. Frisvold, S.B. Powles, N.R. Burgos, W.W. Witt and M. Barrett. 2012. Reducing the risks of herbicide resistance: best management practices and recommendations. Weed Sci. 60: 31–62.

Owen, M.D.K., H.J. Beckie, J.Y. Leeson, J.K. Norsworthy and L.E. Steckel. 2015. Integrated pest management and weed management in the United States and Canada. Pest Managem. Sci. 71: 357–376.

Padilla, E. 2002. Intergenerational equity and sustainability. Ecol. Econ. 41: 69–83.

Pannell, D.J. and S. Schilizzi. 1999. Sustainable agriculture: a matter of ecology, equity, economic efficiency or expedience? J. Sust. Agric. 13: 57–66.

Paynter, Q. 2003. Integrated weed management: effect of herbicide choice and timing of application on the survival of a biological control agent of the tropical wetland weed, *Mimosa pigra*. Biol Control 26: 162–167.

Pollnac, F.W., B.D. Maxwell and F.D. Menalled. 2009. Weed community characteristics and crop performance: a neighbourhood approach. Weed Res. 49: 242–250.

Pretty, J. 2005. The pesticide detox. Earthscan, London. United Kingdom.

Pretty, J. 2008. Agricultural sustainability: concepts, principles and evidence. Philos. Trans. R. Soc. Lond. B Biol. Sci. 363: 447–465.

Radosevich, S.R., J.S. Holt and C.M. Ghersa. 2007. Ecology of Weeds and Invasive Plants: Relationship to Agriculture and Natural Resource Management. John Wiley & Sons, New Jersey. USA.

Roberts, H.A., R.J. Chancellor and T.A. Hill. 1982. The biology of weeds. pp. 1—36. *In:* H.A. Roberts (Ed.) Weed Control Handbook: Principles. Blackwell Scientific, Oxford, Great Britain.

Robertson, G.P. 2015. A Sustainable Agriculture? Daedalus 144: 76–89.

Robertson, G.P. and S.M. Swinton. 2005. Reconciling agricultural productivity and environmental integrity: a grand challenge for agriculture. Frontiers Ecol. Environ. 3: 38–46.

Robinson, R.A. and W.J. Sutherland. 2002. Post-war changes in arable farming and biodiversity in Great Britain. J. Appl. Ecol. 39: 157–176.

Sagar, G.R. 1968. Weed biology a future. Netherlands J. Agri. Sci. 16: 155–164.

Savory, A. 1988. Holistic Resource Management. Island Press. Washington, D.C. 2009. 564 pp.

Sciegienka, J.K., E.N. Keren and F.D. Menalled. 2011. Interactions between two biological control agents and an herbicide for Canada Thistle (*Cirsium arvense*) suppression. Invasive Plant Sci. Manage. 4: 151–158.

Shaner, D.L. and H.J. Beckie. 2014. The future for weed control and technology. Pest Manage. Sci. 70: 1329–1339.

Shaw, D.R. 2005. Remote sensing and site-specific weed management. Frontiers Ecol. Environ. 3: 526–532.

Shaw, W.C. 1982. Integrated weed management-systems technology for pest management. Weed Science 30: 2–12.

Smith, R.G., B.D. Maxwell, F.D. Menalled and L.J. Rew. 2006. Lessons from agriculture may improve the management of invasive plants in wildland systems. Frontiers Ecol. Environ. 4: 428–434.

Story, J.M. and R.N. Stougaard. 2006. Compatibility of two herbicides with *Cyphocleonus achates* (Coleoptera: Curculionidae) and *Agapeta zoegana* (Lepidoptera: Tortricidae), two root insects introduced for biological control of spotted knapweed. Environ. Entomol. 35: 373–378.

Swanton, C.J. and S.F. Weise. 1991. Integrated weed management – The rational and approach. Weed Technol. 5: 657–663.

Swinton, S.M., N. Rector, G.P. Robertson, C.B. Jolejole-Foreman and F. Lupi. 2015. Farmer decisions about adopting environmentally beneficial practices. pp. 340–359. *In:* Hamilton, S.K., J.E. Doll and G.P. Robertson (Eds.). The Ecology of Agricultural Landscapes. Long-Term Research on the Path to Sustainability. Oxford University Press New York, New York, USA.

Thill, D.C., J.M. Lish, R.H. Callihan and E.J. Bechinski. 1991. Integrated weed management: a component of integrated pest management: A critical review. Weed Technol. 5: 648–656.

Tilman, D. and C. Lehman. 2001. Biodiversity, composition and ecosystems processes: theory and concepts. pp. 9–41. *In:* Kinzig, A.P., S.W. Pacala and D. Tilman (Eds.) The Functional Consequences of Biodiversity: Empirical Progress and Theoretical Extensions. Princeton University Press, Princeton, New Jersey, USA.

Thomas, S.H., J. Schroeder and L.W. Murray. 2005. The role of weeds in nematode management. Weed Sci. 53: 923–928.

Trumble, J.T. and L.T. Kok. 1980. Impact of 2,4-D on *Ceuthorhynchidius-Horridus* (Coleoptera: Curculionidae) and their compatibility for integrated control of *Carduus* thistles. Weed Res. 20: 73–75.

Ward, S.M., R.D. Cousens, M.V. Bagavathiannan, J.N. Barney, H.J. Beckie, R. Busi, A.S. Davis, J.S. Dukes, F. Forcella, R.P. Freckleton, E.R. Gallandt, L.M. Hall, M. Jasieniuk, A. Lawton-Rauh, E.A. Lehnhoff, M. Liebman, B.D. Maxwell, M.B. Mesgaran, J.V. Murray, P. Neve, M.A. Nunez, A.

Pauchard, S.A. Queenborough and B.L. Webber. 2014. Agricultural weed research: a critique and two proposals. Weed Sci. 62: 672–678.

Weddle, P.W., S.C. Welter and D. Thomson. 2009. History of IPM in California pears – 50 years of pesticide use and the transition to biologically intensive IPM. Pest Manage. Sci. 65: 1287–1292.

Westerman, P., M. Liebman, F.D. Menalled, A.H. Heggenstaller, R.G. Hartzler and P.M. Dixon. 2005. Are many little hammers effective? —— Velvetleaf (*Abutilon theophrasti*) population dynamics in two- and four-year crop rotation systems. Weed Sci. 53: 382–392.

White, S.S., K.A. Renner, F.D. Menalled and D.A. Landis. 2007. Feeding preferences of weed seed predators and effect on weed emergence. Weed Sci. 55: 606–612.

Wilson, R., K.G. Beck and P. Westra. 2004. Combined effects of herbicides and Sphenoptera jugoslavica on diffuse knapweed (*Centaurea diffusa*) population dynamics. Weed Sci. 52: 418–423.

Young, S.L., S.K. Pitla, F.K. Van Evert, J.K. Schueller and F.J. Pierce. 2017. Moving integrated weed management from low level to a truly integrated and highly specific weed management system using advanced technologies. Weed Res. 57: 1–5.

Zimdahl, R.L. 1994. Who are you and where are you going. Weed Technol. 8: 388–391.

Zimdahl, R.L. 2004. Weed–Crop Competition: A Review (2nd Edition). Blackwell Publishing. Ames, Iowa, USA.

Zimdahl, R.L. 2006. Agriculture's Ethical Horizon. Elsevier. Amsterdam, The Netherlands.

Herbicide Effects on Humans: Exposure, Short and Long-term Effects and Occupational Hygiene

Nicholas E. Korres

Department of Crop, Soil and Environmental Sciences, University of Arkansas,
1366 W Altheimer Drive, Fayetteville, 72704, Arkansas, USA
E-mail: nkorres@yahoo.co.uk; korres@uark.edu

Herbicides and Human Health

Herbicide benefits are well documented and undoubtedly their usage has contributed greatly to food security. US farmers are estimated to spend over $3.5 billion annually on chemical weed control and over $2.5 billion for non-chemical weed management. The loss in food and fiber without the use of herbicides and the likely substitution of alternatives (e.g., non-chemical control methods) is worth of $13.3 billion in 2003 dollars (Cahoon et al. 2016, Korres, 2018). The economic benefits seem to support the use of herbicides; however, these numbers do not evaluate the human health and environmental costs.

According to Pimental et al. (1992) there are an estimated 20,000 unintentional deaths and one million poisonings each year worldwide, whereas in the United States, each year, a minimum of 67,000 poisonings and 27 accidental deaths are caused by pesticides (i.e., herbicides, insecticides and fungicides in this chapter). More recently, Cole (2006) reported an estimated 2 to 5 million pesticide acute poisonings per year worldwide that result in 40,000 deaths. One World Health Organization (WHO) study indicates that 3 per cent of agricultural workers in developing countries suffering a poisoning incidence each year (Jeyaratham 1990). Each year in the United States, an estimated 7.7 kg of pesticides per citizen will be incorporated into the environment for a total of 2.1 billion kg of pesticides (Aspelin 1999). According to Atwood and Paisley-Jones (2017) 44.6% of the 2.4 billion kg used in US agriculture between 2008 and 2012 is herbicides. However, insecticides generally have considerably more acute toxicity than herbicides because many of them target aspects of the insect nervous system that are common to mammals including humans. Of these products an estimated 85–90 per cent will not reach their target organisms (Repetto and Baliga 1996), thus they will most probably enter the air, water and soil (Tyler and Locke 2018-Chapter 3 in this book; Arts and Hanson 2018-Chapter 4 in this book; Lutman 2018-Chapter 17 in this book).

Adverse impacts usually caused by herbicide mis- and non-judicious use or unprotected exposure include effects on agro- and natural ecosystems (Chapters 3-5 in this book) and the evolution of herbicide-resistant weeds (Chapter 6 in this book), along with direct effects on human health. The direct effects on humans can occur from contact with herbicides during mixing, application, or from contact through field operations (Korres 2005). When used improperly, an applicator can receive an exposure orders of magnitude more than a consumer would be exposed to in the food supply.

Karabelas et al. (2009) reported that 84 of the 276 active ingredients approved as plant protection products in Europe in 2008; 25 of the 87 herbicides, 32 of the 76 fungicides and 24 of the 66 insecticides— had at least one deleterious effect on human health following acute and/or chronic exposure. These effects included acute toxicity, carcinogenicity, reproductive and neurodevelopmental disorders and endocrine disruption (Bourguet and Guillemaud 2016). Human health and health determinants amongst others are considered in risk assessments concerning the use of pesticides in agriculture (Nicolopoulou-Stamati et al. 2016).

This chapter describes various types and sources of human exposure to herbicides, briefly summarizes symptomology of acute and chronic effects of a few herbicides and discusses occupational hygiene practices.

Types and Sources of Herbicide Exposure

People can be exposed to herbicides by inhalation, through skin contact (dermal exposure) (Figure 2.1) or by ingestion (oral exposure) (Korres 2005).

Depending on the situation, herbicides could enter the body by any one or all of these routes. Sources of pesticide exposure include contaminated food (on the surface of fresh fruits and vegetables, within fresh crops which have taken up herbicides, and within processed foods) consumption (Kumar et al. 2000, Shin et al. 2011, Nikolopoulou-Stamati et al. 2016), ground or surface water systems that feed drinking water supplies (Aktar et al. 2009, Lutman 2018-Chapter 17 in this book), by air through sprays or air-application of herbicides (Bourguet and Guillemaud 2016) and through handling herbicides at work where farmers, herbicide applicators, or others who work with herbicides can be exposed to them (Alif et al. 2017).

The exposure to four herbicide families, i.e., chlorophenoxy, organophosphate (glyphosate), bipyridylium (paraquat and diquat) and triazines during 2000–14 for each across US, herbicide family and year is shown in Figures 2.2 and 2.3.

California, Texas, Florida, Washington, North Carolina, Ohio, Georgia, and Pennsylvania are the States with the higher number of exposures to glyphosate and chlorophenoxy herbicides followed by Michigan, Illinois, Oregon, Indiana, New York, Arizona, Virginia, and Utah, the States with a relative moderate sum of exposures (>4000, <5000). In Missouri, Tennessee, Minnesota, Alabama and Colorado the sum of exposures was recorded between 3000 and 4000 during the period 2000–14 (Figure 2.2).

The exposure to glyphosate and chlorophenoxy compounds was permanently recorded at higher levels throughout 2000–14 period. There is a noticeable decline in exposure incidences for the period 2010–14 for both herbicides (Figure 2.3).

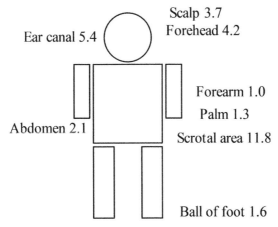

Figure 2.1. Comparative rates (i.e., absorption rates compared to forearm which is 1.0) of dermal absorption of herbicides for different parts of the body. Adapted from Korres, 2005.

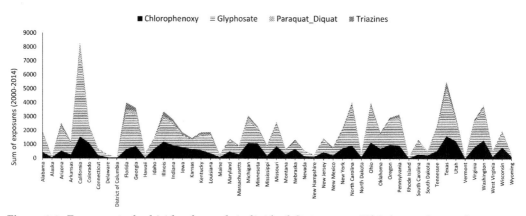

Figure 2.2. Exposure to herbicides for each individual State across USA (sum of annual exposure during 2000-2014) as collected by the Centres for Disease Control and Prevention: Environmental Public Health Tracking Network. www.cdc.gov.ephtracking (accessed February 15, 2018). Data are provided by the American Association of Poison Control Centers (AAPCC).

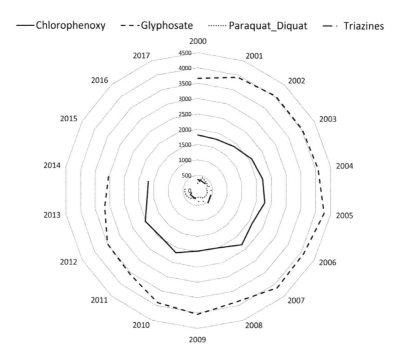

Figure 2.3. Annual exposure to four herbicide families for the period 2000-2014 (sum of US State exposure during 2000–14) as collected by the Centers for Disease Control and Prevention: Environmental Public Health Tracking Network. www.cdc.gov.ephtracking (accessed February 15, 2018). Data are provided by the American Association of Poison Control Centers (AAPCC). Data for 2015–17 are not available.

As stated by the Environmental Public Health Tracking Network, exposures to herbicides/pesticides are self-reported unless very serious and exposure requires hospitalization or results in a major health outcome treated by a general practitioner or emergency department. Herbicide/pesticide poisonings are likely underreported because of difficulty with diagnosis, incomplete reporting to surveillance systems (Calvert et al. 2004) and symptoms not being recognized as a poisoning (Texas Department of State Health Services 2010). Chronic illnesses from long-term low-dose exposures, which may be more typical of pesticides exposures, are not included in poison control center data. State-specific data has various fine distinctions from state to state. Utilization of PCs per unit population varies widely across the U.S., so it is very hard to compare between states. Another limitation is that poison center practices vary by poison center, so their coding of reported exposures may be different.

Acute and Chronic Herbicide, Insecticide and Fungicide Poisoning Symptomology

Research of pesticides on human-health effects serve as the basis for determining the need for preventive actions. Early research was focused primarily on acute effects such as poisoning but more recently, interest in chronic effects has increased (Roberts and Reigart 2013, CDC 2000).

Acute Effects

Acute pesticide poisoning is commonly reported by farmers, and particularly by those making pesticide applications (Kishi et al. 1995, Calvert et al. 2008, Lee et al. 2012). Information on non-fatal poisoning is more limited than that for fatalities, and many symptoms undoubtedly go unreported or are misdiagnosed (Blair et al. 1996). Nevertheless, Jeyaratnam (1990) reported that the global problem of acute poisoning has been confirmed as extensive by a variety of independent estimates, especially in developing countries. A summary of some of the most common acute pesticide, including herbicide, poisoning symptoms is presented below (Thundiyil et al. 2008, Korres 2005). Individuals exposed to mild pesticide poisoning may experience headache, fatigue, skin irritation, loss of appetite, dizziness, weakness, nervousness, nausea, perspiration, diarrhoea, eye irritation, insomnia, thirst, restlessness, irritation of nose and/or changes of mood. Moderate poisoning which may be the beginning of severe symptoms can cause nausea, trembling, muscular incoordination, excessive salivation, blurring of vision, feeling of constriction in the throat and chest, difficulty breathing, flushed or yellow skin, abdominal cramps, vomiting, diarrhoea, mental confusion, twitching of muscles, weeping, excessive perspiration, profound weakness, rapid pulse, and/or coughing. In case of severe poisoning symptoms could be observed include vomiting, loss of reflexes, inability to breathe, uncontrollable muscular twitching, constriction of pupils, convulsions, unconsciousness, severe secretion from respiratory tract, fever, intense thirst, and/or increased rate of breathing (Korres 2005).

Chronic or Delayed Effects

Chronic effects are more difficult to evaluate than acute effects because they do not occur immediately after exposure. Chronic diseases of special concern include cancer and diseases of the nervous, immune, and reproductive systems. In this chapter only chronic effects related to herbicide poisoning will be discussed.

Carcinogenic Effects

Despite the fact that farmers have a low risk of cancer in comparison with other groups, most probably due to a healthier environment, several incidents have initiated research on the relationship between cancer and pesticide exposure (Blair et al. 1996). Suspected as carcinogenic are the products belong to triazines (Kettles et al. 1997) and bipyridylium herbicides (Anderson and Scerri 2003). In addition, a report published by IARC (2015) classifies phosphonate herbicides, such as glyphosate as a probable carcinogenic to humans. The need for long-term health effects of glyphosate on humans has been suggested by De Roos et al. (2005) and Samsel and Seneff (2013) since the association of the herbicide exposure with multiple myeloma (De Roos et al. 2005) and celiac sprue and gluten intolerance incidences (Samsel and Seneff 2013) can lead to various types of cancer. Nevertheless, Gasnier et al. (2009), Williams et al. (2000), and Mink et al. (2011) along with other research reports stated that glyphosate has no carcinogenicity or genotoxic potential, is not associated with non-Hodgkin's lymphoma or multiple myeloma and is unlikely to pose a carcinogenic risk to humans (Williams et al. 2016, Brusick et al. 2016, Acquavella et al. 2016).

Neurologic Disorders

Diseases of the nervous system resulting from pesticide exposure are of special concern (Luft and Bode 2002). Neurological symptoms due to pesticide exposure include tremors, anorexia, muscular weakness, insomnia, convulsions and depression (Barrett 2005). Neurologic disorders have occurred with arsenic pesticides (Roberts and Reigart 2013) and diquat (Olson 1994). Parkinsonism has also been reported due to paraquat and diquat poisoning (Liou et al. 1997, Sechi et al. 1992). However, Kamel and Hoppin (2004) suggested that further research is needed to substantiate the results of pesticides on neurological dysfunctions.

Immune System Diseases

The immune system acts to protect the body against foreign invaders. It is composed of a number of cellular and chemical components (Blair et al. 1996). Lower doses of some xenobiotics have been found to affect the immune system than would affect target organs (Burns et al. 1996). Some immunotoxicological studies have suggested that the action of some pesticides on the immune system is through the alteration of proteins to haptens (Pan et al. 1989). Herbicides reported to cause immune dysfunction are diquat and paraquat (Street and Sharman 1978), although Donald and Marr (2011) reported no evidence of an immunosuppressant effect on the humoral immune system due to diquat administration on mice.

Reproductive System

There is increasing evidence that certain pesticides mimic the action of animal and human hormones which may result in malformations and cancers or upset estrogen levels and cause abnormal sexual development and impaired reproduction (Colborn and vomSaal 1993)

According to Shape and Skakkebaek (1993) exposure to exogenous estrogens during foetal and neonatal life can lead to an increase in reproductive disorders. Nice et al. (2003) discussed the possible effects of herbicides, among other chemicals, on oysters' imposex, a pseudohermaphroditic condition in which females acquire male sex characteristics. On the contrary, Smialowizc et al. (1985) found no effects of a thiocarbamate herbicide Ordram (S-ethyl hexahydro-1H-azepine-1-carbothioate) on mice immune system.

Use and Toxicity Profile of Herbicide Families

Arsenic Compounds

Arsenical herbicides are used post-emergence with limited translocation for the control of grass weeds in turfgrass, cotton, rice and orchards (Meharg and Hartley-Whitaker 2002) although some uses of these compounds as herbicides were discontinued in 2009 (U.S. EPA 2009). EPA reregistered only monosodium mathanearsonate for use on cotton but decided to stop the use of it on golf courses, sod farms, and right-of-ways in 2013 (Hughes et al. 2011). However, EPA agreed to reevaluate new information on the carcinogenicity of inorganic arsenic, the environmental degradation product of the organic arsenical products in 2012 (Hughes et al. 2011).

Toxicity

Arsenical active ingredients are mainly consisting of methylated compounds which are considered to exert less toxic hazard (50–100 times) than the inorganic forms of arsenic (Fishel 2005). Acute arsenic poisoning symptoms appear within one to several hours of exposure and consists of garlic odour of the breath and faeces, a metallic taste in mouth and adverse gastrointestinal symptoms (Roberts and Reigart 2013). The gastrointestinal symptoms of acute sodium arsenite, an aquatic herbicide (Tanner and Clayton 1990) poisoning caused by ingestion were characterised by abdominal pain, nausea, vomiting and diarrhoea (Watterson 1988, Bartolome et al. 1999). The central nervous system is also commonly affected with acute exposure and the symptoms include dizziness, headache, drowsiness and confusion (Roberts and Reigart 2013) and may progress to muscle weakness and spasms, hypothermia, lethargy, delirium, coma and convulsions (Malachowski 1990). Chronic exposures from repeated absorption of toxic amounts are more difficult to determine but neurological symptoms (peripheral neuropathy manifested by paresthesia, anaesthesia, paresis and ataxia) are usually more common than gastrointestinal effects which are more closely associated with acute poisoning (Roberts and Reigart 2013). In epidemiological studies and case reports chronic exposure to arsenic have been associated with the development of several types of skin cancer (Tay 1974) including Bowen's disease, squamous cell carcinoma (IARC 1973, 1980, Li 1986), keratoacanthoma (Wong et al. 1998) and basal cell epithelioma (Alain et al. 1993). Nevertheless, IARC (1987) concluded that there was limited evidence for the carcinogenicity of inorganic arsenic compounds in experimental animals.

Bipyridylium Herbicides

These herbicides are used almost exclusively for post-emergence weed control (Ashton and Crafts 1973). Paraquat and diquat, the two members of this herbicide group with commercial value, are well known as potent, non-selective herbicides that cause a rapid wilting of foliage with a frostbitten appearance (Cobb 1992) soon after their contact with the plant tissue. Their mode of action is diversion of energy from photosystem I to molecular oxygen, resulting in the destruction of plant cell membranes by reactive oxygen species, particularly in sunlight often only a few hours after application. They are quickly absorbed by the plant tissue, and their translocation to other tissues is very limited, most probably due to the quick death of tissues to which they make direct contact (Fishel 2005). Diquat is used as an aquatic herbicide, but primarily is used for desiccation purposes in certain crops (e.g., oilseed rape, clover, potatoes, peas, lupines, field beans, grain sorghum) (Whitehead 1995). It can also be used for direct spraying in trees and fruit crops and other sites where non-selective control is desired. Both diquat and paraquat have no residual activity, as they are quickly adsorbed onto soil surfaces, rendering them biologically unavailable (EC 2003, Zimdahl 1993) although the potential for desorption does exist (Watts 2011). The majority of diquat products are formulated as soluble concentrates. Paraquat has pre-harvest desiccant uses in potatoes, row crops, sugar beet, orchards, blackcurrants, strawberries (Whitehead 1995). Unlike diquat, no formulations for aquatic use are registered for paraquat, which is one of the most widely used herbicide in the world (Kudsk and Streibig 2003).

Toxicity

Both products are potentially toxic to humans (Li et al. 2004, Jones and Vale 2000). The major acute symptoms for diquat and paraquat poisoning consist of mucosal irritation, nausea and vomiting, transient renal/hepatic function impairment, vomiting and diarrhoea, oropharyngeal and gastrointestinal ulcerations, cardiac arrhythmias, coma, death within several days after poisoning due to multiple organ failure [Mahieu et al. (1984), Vale et al. (1987), Watterson (1988), Reigart and Roberts (1999), Saeed et al. (2001)]. As Litchfield et al. (1973) and Reigart and Roberts (1999) reported the main primary target organ of paraquat, which shows the highest mortality rates but lower cases of poisoning amongst other pesticides in the USA (Sittipunt 2005), is the lung, leading to pulmonary fibrosis which is usually fatal (Saeed et al. 2001). Diquat targets the central nervous system resulting in neurological injuries including nervousness, irritability, restlessness, disorientation, inability to recognise family members or friends and diminished reflexes (Reigart and Roberts 1999). Prolonged contact with paraquat will cause erythema, blistering, abrasion and ulceration. Anderson and Scerri (2003) reported a case of a 58-year-old strawberry farmer who, five months after his first visit to hospital, had developed about 100 skin lesions which were suspected to be squamous and basal cell carcinomas. There was no family history or clinical features that would suggest Gorlin's syndrome, but he had a strong occupational history of pesticide exposure to demeton-O-methyl (an insecticide) and paraquat. The authors, based on this case and previous reports of squamous cell carcinoma development due to paraquat exposure (Bowra et al. 1982), concluded that the direct contact exposure to paraquat was most likely the cause of localized multiple skin cancers in the farm worker.

Chlorophenoxy Herbicides

A group of herbicides consisting of phenoxy compound was first introduced in 1942 in the USA and UK (Zimdahl 1993). The most well-known active ingredients 2,4-D, 2,4-DB, MCPA, mecoprop (MCPP) and dichlorprop of this family are used for the control of annual and perennial broadleaf weeds in cereals, soybean, grasses as well as in conifer plantations, forestry, fence rows and waterside areas (Whitehead 1995, Fishel 2005). Various products of 2,4-D, amongst the many which exist in the market, are formulated as emulsifiable acids, amine salts, mineral salts and esters. Amine forms of 2,4-D are soluble in water with less volatility and consequently reduced potential causing harm to desirable vegetation. Esters, in contrast to amine formulations, which are preferred for controlling difficult vegetation (e.g., woody species), are essentially insoluble in water and relatively volatile and they might cause injuries to non-targeted plants. MCPA is less injurious to cereal grain crops and has been labelled for weed control in small-seeded legumes (e.g., clover). 2,4-DB is used as a foliar treatment to peanut and soybean. It is also formulated as amines and low-volatile esters. MCCP (mecoprop) acts relatively slowly compared to the other chlorophenoxy herbicides. In most cases it is packaged with other herbicidal active ingredients (Fryer and Makepeace 1978, Zimdahl 1993, Whitehead 1995, Fishel 2005).

Toxicity

These compounds are regarded as a family with low to moderate toxicity in terms of acute and chronic effects (Korres 2005, Fishel 2005). According to Watterson (1988) and Roberts and Reigart (2013), these chemicals can be absorbed through skin or by inhaling them resulting in nose, eye and throat bronchi irritation, burning sensations and cough, dizziness and unsteady gait, chest or abdominal pains, diarrhoea, fatigue, muscle twitching or tenderness and weakness. In case of ingestion, the above symptoms can be observed as well as hypersalivation, mental confusion, speech difficulties, and general reduction in motor activity, incoordination, gradual loss of reflexes and in severe case coma and death. Sharma and Kaur (1990) concluded that chlorophenoxy herbicides are potent sensitizers since 12% of farmers tested with contact dermatitis were using chlorophenoxy products during their farming practices; allergic reactions to control group were not observed. Blair (1996) reported that 2,4-D could provoke non-Hodgkin lymphoma especially for the farmers that rarely wear personal protective equipment although evidence of carcinogenicity in animals does not confirm carcinogenic potential to humans. Some tests on fish have revealed the potential of these chemicals, particularly 2,4-D salts, to cause delays or termination in the development of embryos in the early stages of lake fish *Alburnus alburnus* (Biro 1979). The same author reported that the effects of 2,4-D salts can be attributed to different emulsifiable agents that exist in the final product. According to Lerda and Rizzi (1991), a significantly higher prevalence of asthenozoospermia, necrozoospermia and teratozoospermia was found in farm workers applying 2,4-D compared with unaffected members of a control group, although Gerber et al. (1988) after the examination of 382 consecutive patients of an infertility clinic in Iowa found no correspondence between occupation, exposure to chemical agents and infertility. Finally, a MCPP can induce cramping syndromes with muscular or nervous mechanism (Meulenbelt et al. 1988-cited in Parisi et al. 2003).

Triazines

As a chemical family, triazines are a group of herbicides with a wide range of uses. Most are used in selective weed control programs (Cobb 1992). Triazines may be used alone or in combination with other herbicide active ingredients to increase the weed control spectrum (Zimdahl 1993). As a family, their chemical structures are heterocyclic, composed of carbon and nitrogen in their rings. Herbicide members of this family include atrazine, hexazinone, metribuzin, prometon, prometryn and simazine (Zimdahl 1993). Atrazine is used in maize and for selective weed control in turfgrass, sugarcane, sweet corn and sorghum. Some of its uses are classified as restricted because of ground and surface water pollution problems. Hexazinone is used primarily in alfalfa, pastures, pine plantations and industrial sites. Prometryn is used in cotton, celery and several other species while metribuzin in soybean and some vegetables for selective weed control. Simazine is used on grapes and certain berries since it is usually classified as restricted (Whitehead 1995).

Toxicity

Systemic toxicity is unlikely unless large amounts have been ingested. Some of the triazines are moderately irritating to the eyes, skin and respiratory tract. Atrazine is considered slightly to moderately toxic to humans (EPA has established a lifetime health advisory level in drinking water of 3 µg/l). Water containing atrazine at or below this level is acceptable for drinking every day over the course of one's lifetime. Long-term consumption of high levels of atrazine has caused adverse health effects in animals, such as tremors, changes in organ weights and damage to the liver and heart. Hexazinone is not considered to be acutely toxic but can cause serious and irreversibly eye damage (Fishel 2005).

Glyphosate

Glyphosate is a broad-spectrum, non-selective systemic herbicide structurally similar to organophosphate pesticides, and is usually formulated as an isopropylamine salt (Szekaca and Darvas 2011). This product has the highest production volumes of all herbicides and is currently used worldwide in agriculture, forestry, urban and home application for the control of broad-leaf and grass weeds (Botelho et al. 2011, Szekaca and Darvas 2011, Campbell 2014, IARC 2017).

In addition, the introduction of genetically engineering crops, particularly maize, cotton and soybean has enhanced the use of glyphosate significantly (Woodburn 2000, Bonny 2008, Duke and Powles 2009).

Toxicity

The World Health Organization's (WHO) International Agency for Research on Cancer (IARC) released a summary of monograph 112 (diazinon, glyphosate, malathion, parathion and tetrachlorvinphos) on March 20, 2015. In this report glyphosate was classified as a category 2A 'probably carcinogenic to humans'. According to IARC (2015) this category is used when there is limited evidence of carcinogenicity in humans and sufficient evidence of carcinogenicity in experimental animals. Limited evidence means that a positive association has been observed between exposure to the agent and cancer but that other explanations for the observations (known as chance, bias, or confounding) could not be excluded. Category 2A is also used when there is limited evidence of carcinogenicity in humans and strong data on how the agent causes cancer. Glyphosate acts by inhibiting aromatic amino acid biosynthesis in plants (Amrhein et al. 1980) by disrupting the shikimate pathway (Herman and Weaver 1999) which is responsible for the synthesis of aromatic amino acids and critical plant metabolites (Szekaca and Darvas 2011). Glyphosate exerts this effect by inhibiting the activity of the 5-enolpyruvyl shikimate 3-phosphate synthase (EPSPS) enzyme which catalyses the transformation of phosphoenol pyruvate (PEP) to shikimate-3-phosphate (S3P) (Amrhein et al. 1980). This metabolic pathway exists in plants, fungi, and bacteria, but not in animals (Kishore and Shah 1988). It has mentioned that glyphosate through its disruption in shikimate pathway might affect the metabolism of human gut bacteria and the supply of essential amino acids to humans as a secondary effect (Samsel and Seneff 2015) although no clear evidence exists to confirm this assumption. Swanson et al. (2014) and Thongprakaisang et al. (2013) reported possible endocrine-disrupting glyphosate activity whereas Kwiatkowska et al. (2014) reported *in vitro* effects of glyphosate on human erythrocytes. There are other cases in the literature where toxic effects of glyphosate, other than those mentioned earlier, have been reported (Slager et al. 2009, Benachour and Seralini 2009, Richard et al. 2005, Anadon et al. 2009, Marc et al. 2002). Commercial formulations of the glyphosate-based herbicides might be more toxic to humans than the glyphosate alone (Thongprakaisang et al. 2013, Kwiatkowska et al. 2014, Pieniazek et al. 2004).

Nevertheless, as mentioned earlier issues regarding the safety of this product are controversial. Williams et al. (2000), based on studies performed for regulatory purposes as well as published research reports, reported that Roundup, one of the commercial formulations of glyphosate, presents no human health threats. In addition, Williams et al. (2016) and Acquavella et al. (2016) based on assessments of epidemiological data reported that there is no causal relationship between glyphosate exposure and non-Hodgkin's lymphoma or multiple myeloma. Brusick et al. (2016) in line with Williams et al. (2016) and Acquavella et al. (2016) concluded that evidence relating to an oxidative stress mechanism of genotoxic carcinogenicity caused by glyphosate or glyphosate products were unconvincing. Solomon (2016) argued about the accuracy of the IARC report since glyphosate exposures were found to be less than the reference dose and the acceptable daily intakes proposed by several regulatory agencies. In addition, European Union (EU) including the European Food Safety Authority (EFSA) and the German Federal Institute for Risk Assessment (BfR) along with Food and Agriculture Organization of the United Nations (FAO), the United States Environmental Protection Agency (US EPA), Health Canada and the Australian Pesticides and Veterinary Medicines Authority (APVMA) questioned the risk or carcinogenic and genotoxic potential of glyphosate to humans (APVMA 2016).

Occupational Hygiene Practices for Using Pesticides (i.e., herbicides, insecticides and fungicides)

Considering the information provided above, the importance of minimising the risk of pesticide exposure, as much as possible, hence to secure human and environmental health is becoming very demanding. Cole (2006) having the chronic effects caused by pesticides mainly carbofuran, metamidofos and mancozeb, in Carchi, Equator, potato growers as an example, highlights the importance of occupational hygiene practices (e.g., the use of personal protective equipment during the preparation and application of pesticides, storage of pesticides and personal hygiene after application) as well as the need for appropriate training of applicators on 'safe use' of these products especially those under "restricted use" regulations. Another case has been reported by Poskitt et al. (1994) regarding the development of chloracne, palmoplantat keratoma and scleroderma after many years of exposure to a variety of pesticides either because the applicator was not following the appropriate guidelines during application, mixing, loading or storage of the pesticides used. Curwin et al. (2003) reported that agricultural workers harvesting tobacco manually have the potential for high dermal exposure to pesticides (e.g., acephate), particularly on the hands. Often gloves are not worn as it hinders the harvesters' ability to harvest the tobacco leaves. To enable harvesters to remove pesticide residue from the hands and decrease absorbed doses, the Environmental Protection Agency (EPA) Worker Protection Standard requires growers to have hand-wash stations available in the field or to follow the appropriate hygiene precautions.

A major pathway for transport of pesticides into groundwater is leaching from soils following pesticide application in accordance to the persistence and mobility of these compounds. Drains, soak ways, sumps and fractures can serve as rapid transport routes to groundwater which bypass some of the natural attenuation potential provided by the soil and shallow unsaturated zone (Lapworth et al. 2006, Arts and Hanson 2018-Chapter 4 in this book, Lutman 2018-Chapter 17 in this book).

EPA (1993) has published a guide that makes an extensive reference on standards of worker's protection of agricultural pesticides. Also the British Crop Protection Council (1996) has issued a very useful guide about safety and effective pesticide spraying. In both publications the obligation to follow, blindly, practices that aim in the significant prevention of unjustified use of pesticides; hence promoting human, environmental and wildlife health is emphasized.

Personal Protective Equipment

As Nater et al. (1979) reports amongst other exposure routes to pesticides (i.e., ingestion and inhalation) skin contact is by far the most important one since over 97% of the pesticides to which the body is subjected are deposited on the skin, especially when liquid pesticides are used. It is quite clear that the use of personal protective equipment (PPE) is of great importance. PPE are items made of material or substances that hinder the passage of pesticides resulting in the protection of the wearer from contamination in the normal circumstances of use. PPE should only be considered as a method of control after all other measures have been employed as far as is reasonably practicable or where engineering or other control measures do not provide adequate control of exposure. Even for pesticides with no product label hazard warning or classification, the use of PPE constitutes recommended occupational hygiene. Coveralls, footwear, head gear, face shields, gloves, goggles and respiratory protective equipment are all considered necessary protective items for pesticide applicators and/or farm-workers (Korres 2005).

Calibration and Maintenance of the Pesticide Application Machinery

Equipment calibration is a process that ensures that the correct amount of pesticide over a target area is applied uniformly. It involves adjustments of the sprayer, as well as calculations of delivery rate (output) (Korres 2005). Improper calibration and damaged or malfunctioned application equipment cause unsatisfactory pest control due to an incorrect pesticide application rate and uneven pesticide distribution. Excessive pesticide residues cause crop injuries, environmental pollution and waste of time and money. Additionally, increased exposure with high probability of severe pesticide poisoning for the applicator or the bystander can occur (Anderson and Scerri 2003, Corazza et al. 2003).

Disposal of Obsolete Pesticides and Empty Containers

Whenever possible, the choice of the product should be made based on the ease of disposal of the containers, for example, refillable containers, those in water-soluble bags or tablet formulations. Generally, empty pesticide containers should never be re-used for any purpose, except possibly if they are well preserved and are to be refilled with an identical pesticide from another deteriorated or leaking container. When containers have been used:

- Should be thoroughly triple or pressure rinsed (particularly if they contained liquid pesticides), according to the instructions provided by the label of the product or the appropriate environmental authority.
- After their proper cleaning the containers should be punctured in several places, or crushed to make them unusable. The perforated or crashed containers should either be stored in a secure compound, preferably not a pesticide store, pending their disposal, or taken immediately to a waste disposal site.

If the product's label indicates that empty containers can be buried, the burial site must be chosen carefully in order to minimise the risk of polluting people, buildings, animals, water courses or susceptible plants. Records must be kept regarding the place of burial, the type and the quantity of waste buried. When waste containers are to be burnt (the appropriate authorities should be contacted to ensure that open burning of pesticide containers is permitted), precautions should be taken to ensure that:

- Burning occurs in an open space away from any public highway and not in a location where the smoke is likely to drift over people or livestock, or move towards any highway, housing, or business premises, particularly in the case of herbicides (orchards or other crops nearby can be harmed).
- The containers are opened, and they must be disposed in a very hot fire, a few at a time.
- The fire is vigorous and under continuous supervision.
- Care is taken to avoid breathing any smoke produced.
- The fire is extinguished before being left and the ashes taken to an appropriate disposal site or buried, if this is allowed.

The problems and potential dangers of the disposal of the waste containers, if specific safety guidelines are not followed, can be:

- Potential contamination of ground and/or surface waters if the used containers are left lying near wells, streams, dugouts, or in areas with high water tales.
- Many residues left in the containers are toxic to fish and other forms of wildlife.
- Potential access by members of the public, especially children, to dangerous chemicals.

Note: The text above is adopted from Korres (2005).

Transportation—Spillage Management

Pesticide containers should be kept in an upright position during transportation for prevention of spillage. Spilled products may release toxic vapors, contaminate other goods and cause poisoning to people or animals exposed to them. Isolated pesticides should be placed in a separate container away from food, clothing or any other materials. Pesticide containers must be secured and separated from people by as great a distance as possible. Some containers can be dangerous when transported inside a car or truck (they should always be transported secured in the boot of a car or in the bed of a truck, placed in a box to ensure that they will not move or break if the vehicle comes to a sudden stop. Chemicals in paper packing should be kept dry and in a separate box, since a wet paper package can be ripped off more easily, allowing the product to spill from its paper bag container. If a major spill occurs, especially on a main road, the appropriate environmental authorities, agencies or a representative of the manufacturer should be informed for proper decontamination (Korres 2005).

Accidental spills can occur in transport, storage, mixing, loading or application activities. The following actions can minimise or prevent the dangers of spillage (British Crop Protection Council 1996, Korres 2005).

- All pesticide wastes must be disposed of carefully by following the instructions provided on the label of the product.
- An area, if it is possible, should be designated for mixing.
- Discharge to sewers must be licensed.
- Spills that are left to dry on any working area may be washed into drains by the next rainfall.

When spillage occurs the following practices can be applied.

- The scene of the accident must be secured.
- Action must be taken to prevent spillage from spreading.
- Small spills can be absorbed by using plenty of dry soil, sand or any other absorbent material available and the material should be kept in a leak-proof container until its proper disposal.
- Major spills must be controlled by banking with plenty of soil or other absorbent until the appropriate authorities arrive in the contaminated area.
- In any case the contaminated area should not be washed.
- Neutralisation of the contaminated area might be necessary.

Pesticide Storage

Proper storage of pesticides can enable them to retain their properties for a considerable time period. Low temperatures, excessive heat and other inappropriate storage conditions can alter the physical and chemical properties of the product. Additionally, certain hygiene rules specify the store place conditions and structure. Hence, according to Korres (2005) regarding the location and construction of the store place:

- The storage shed or buildings should be kept locked, clearly marked and away from people, animals, feed and food places.
- It should be away from any watercourse, drain or areas prone to flooding.
- It should have a cement floor without floor drains and be equipped with a ventilation system.
- It should be equipped with facilities to guard against fire.

  Furthermore, the following points should be respected concerning pesticide storage:

- They should be stored in their original containers, with their labels securely attached and plainly visible.
- They should be stored separately, based on their purpose.
- Products that are not approved, or unlabelled products, should not be stored.
- Pesticides should be stored on shelves (the heavy drums or containers on lower shelves or pallets, the lighter packs above).
- Rotation of the stock (to keep the oldest products at the front) is good management practice.
- A complete inventory of readily accessible products should be kept in a separate place.
- Volatile products should be stored preferably in a separate place to prevent cross-contamination of other chemicals.

Pesticide Drift Avoidance

The physical movement of pesticide out of target area to any non-target site at the time of the spraying operation or soon thereafter. Pesticide drift is a significant and costly problem which may result in damage of susceptible non-target crops in neighboring properties, contamination of the people or environment and application at a lower rate than was intended to the target crops, hence a reduced pesticide efficacy. Two types of drift can be distinguished namely airborne or droplets drift (the movement of droplets of pesticide from the target during the application because of wind, air current or other factors such as droplet size) and vapor drift (the movement of vapor that results from volatilization or evaporation of pesticide). Hence, a simple set of rules, such as control of droplet size through spray pressure and nozzle type, appropriate width of spray fan (wider angles of the fan result in smaller droplet size), right boom/nozzle height (as height increases the possibility of drift increases), correct spraying speed (high speeds may create turbulence and hence greater possibilities for drift) or spraying when weather (wind speed, temperatures, evaporation) conditions are suitable, can minimize the risk of pesticide drift (Korres 2005, ECPA 2006).

Product Label and Re-entry Interval

Product Label

Product label is considered any written, printed or graphic information on, or attached to, the pesticide container or wrapped of the retail package. Labelling is any printed information, supplementary to the label printed information, such as brochures, booklets, and/or flyers about the particular pesticide provided by the manufacturer or supplier of the product. Both label and labelling are legal documents and must be followed accurately. The information provided by the product label can be classified into the following three categories: i) statutory, ii) risk assessments and safety precautions and iii) advisory (Korres 2005). Adequate knowledge and rigid application of safe practices are fundamental for the protection of operating personnel, the public and the environment during and after application since the most accessible and important source of information for proper and efficient application and safety is the product label. Severe effects caused by improper operational procedures and lack of personal hygiene during and after pesticide handling were attributed to mis- or under-understanding of the product label (Cole 2006).

Re-entry Interval

Re-entry intervals is the elapse time after pesticide application before workers can safely re-enter the treated area without the need to wear protective clothing or equipment (Korres 2005). Violating the specified re-entry interval (in most cases product label indicates a re-entry interval) may cause significant exposure to pesticide residues, hence potential danger of pesticide poisoning (Levy and Wegman 2000).

Concluding Remarks

Use of herbicides has increased and secured food production by minimizing yield losses due to weed interference, reducing production costs and facilitating harvesting operations. The constant increase of global population is accompanied with increased demand for food production. Weed management is essential for agricultural production and chemical management will hold an important role in determining whether future food production requirements will be met; herbicides will continue to be used, though perhaps in a more limited fashion (Westwood et al. 2018). Independently the extent of herbicide practise into food production systems their judicious use is necessary. The direct and indirect effects of herbicides on humans and environment can be seized if herbicides used properly. Adherence to label instructions of these products along with their proper application procedures can significantly minimize their acute and long-term effects. Finally, much of the toxicology literature is difficult to interpret for the following reasons: These studies often use formulated herbicides, so that the effects of the herbicide cannot be separated from those of the formulation; the doses of the herbicide used are often much higher than those that a human could have expected to ever encounter; the method of administration to the test subject is often unrelated to how a human would come in contact with the compound; the test subject is often one that does not relate well to humans, i.e., not a mammal.

REFERENCES

Acquavella, J., D. Garabrant, G. Marsh, T. Sorahan and D.L. Weed. 2016. Glyphosate epidemiology expert panel review: a weight of evidence systematic review of the relationship between glyphosate exposure and non-Hodgkin's lymphoma or multiple myeloma. Crit. Rev. Toxicol. 46: 28–43.

Aktar, W., D. Sengupta and A. Chowdhury. 2009. Impact of pesticides use in agriculture: their benefits and hazards. Interdisc Toxicol. 2: 1–12.

Alain, G., J. Tousignant and E. Rozenfarb. 1993. Chronic arsenic toxicity. Int. J. Dermatol. 32: 899–901.

Alif, S.M., S.C. Dharmage, G. Benke, M. Dennekamp, J.A. Burgess, J.L. Perret, C.L. Lodge, S. Morrson, D.P. Johns, G.G. Gilles, L.C. Gurrin and P.S. Thomas et al. 2017.Occupational exposure to pesticides are associated with fixed airflow obstruction in middle-age. Thorax 72: 990–997.

Amrhein, N., B. Deus, P. Gehrke and H.C. Steinrucken. 1980. The site of the inhibition of the shikimate pathway by glyphosate. II. Interference of glyphosate with chorismate formation in vivo and in vitro. Plant Physiol. 66: 830–834.

Anadon, A., M.R. Martinez-Larranaga, M.A. Martinez, V.J. Castellano, M. Martinez, M.T. Martin, M.J. Nozaland and J.L. Bernal. 2009. Toxicokinetics of glyphosate and its metabolite aminomethyl phosphonic acid in rats. Toxicol. Letters 190: 91–95.

Anderson, K.D. and G.V. Scerri. 2003. A case of multiple skin cancers after occupational exposure to pesticides. British J. Dermatol. 149: 1088–1089.

APVMA (Australian Pesticides and Veterinary Medicines Authority). 2016. Review of IARC Monograph 112. Office of Chemical Safety, Department of Health, pp. 35. Available at: www. apvma.gov.au (Accessed on January 15, 2018).

Arts, G. and M. Hanson. 2018. Effects of herbicides on freshwater ecosystems. *In:* Korres, N.E., Burgos, N.R. and Duke, S.O. (Eds.), Weed Control. Sustainability, Hazards and Risks in Cropping Systems Worldwide. CRC Press, Boca Raton, FL.

Ashton, F.M. and A.S. Crafts. 1973. Mode of Action of Herbicides. A Wiley Interscience Publication, John Wiley & Sons. New York.

Atwood, D. and C. Paisley-Jones. 2017. Pesticides industry sales and usage 2008-2012 estimates. Biological and Economic Analysis Division, Office of Pesticide Programs, Office of Chemical Safety and Pollution Prevention, United States Environmental Protection Agency.

Barrett, J.R. 2005. More concerns for farmers: neurologic effects of chronic pesticide exposure. Environ. Health Perspect. 113: A472.

Bartolome, B., S. Cordoba, S. Nieto, J. Fernandez-Herrera and A. Carcia-Diez. 1999. Acute arsenic poisoning: clinical and histopathological features. British J. Dermatol. 141: 1106–1109.

Benachour, N. and G.E. Seralini. 2009. Glyphosate formulations induce apoptosis and necrosis in human umbilical, embryonic, and placental cells. Chem. Res. Toxicol. 22: 97–105.

Biro, P. 1979. Acute effects of the sodium salt of 2,4-D on the early developmental stages of bleak, *Alburnus alburnus*. J. Fish Biol. 14: 101–109.

Blair, A., M. Francis and S. Lynch. 1996. Occupational exposures to pesticides and their effects on human health. Proceedings of the Third National IPM Symposium/Workshop: Broadening Support for 21 Century IPM. Washington, D.C. pp. 59–75.

Bonny, S. 2008. Genetically modified glyphosate-tolerant soybean in the USA: adoption factors, impacts and prospects. A review. Agron. Sustain. Dev. 28: 21–32.

Botelho, R.G., J.P. Cury, V.L. Tornisieloand and J.B. dos Santos. 2011. Herbicides and the aquatic environment. pp. 149–165. *In:* M.N.A. El-Ghany Hasaneen (Ed.) Herbicides-Properties, Synthesis and Control of Weeds. Intech, Rijeka, Croatia.

Bourguet, D. and T. Guillemaud. 2016. The hidden and external costs of pesticide use. Sustain. Agric. Rev. 18: 35–120.

Bowra, G.T., D.P. Duffield, A.J. Osbornand and I.F.H. Purchase. 1982. Premalignant and neoplastic skin lesions associated with occupational exposure to 'tarry' byproducts during manufacture of 4,4'-bipyridyl. Brit. J. Ind. Med. 39: 76-81.

British Crop Protection Council. 1996. Using pesticides: a complete guide to safe, effective spraying. Pub. British Crop Protection Council & ATB-Landbase. pp. 11.3–11.7.

Brusick, D., M. Aardema, L. Kier, D. Kirklandand and G. Williams. 2016. Genotoxicity Expert Panel review: weight of evidence evaluation of the genotoxicity of glyphosate, glyphosate-based formulations, and aminomethylphosphonic acid. Crit. Rev. Toxicol. 46: 56–74.

Burns, L.A., B.J. Meade and A.E. Munson. 1996. Toxic responses of the immune system. pp. 355–402. *In:* Klaassen, C.D. Casarett and Doull's Toxicology: The Basic Science of Poisons, 5th edition. McGraw Hill, NY.

CDC. 2000. Occupational Fatalities associated with 2,4-Dichlorophenol (2,4-DCP) exposure, 1980–1998. Office of Pollution Prevention and Toxics, US EPA, Occupational Safety and Health Administration. Div. of Surveillance, Hazard Evaluations and Field Studies for Occupational Safety and Health, CDC.

Cahoon, C.W., M.L. Flessner, R.L. Ritter, B.A. Majek, W.S. Curran, R. Chandran and M. VanGessel. 2016. Weed control in field crops. Introduction to weed management, pp. 5–3. *In:* Pest management guide, Field crops 2016 and 2017. Virginia Cooperative Extension, Virginia Tech. Available at: https://pubs.ext.vt.edu/456/456-016/456-016.html (Accessed on January 2018).

Calvert, G.M., J. Karnik, L. Mehler, J. Beckman, B. Morrissey, J. Sievert, R. Barrett, M. Lackovic, L. Mabee, A. Schwartz, Y. Mitchelland and S. Moraga-McHaley. 2008. Acute pesticide poisoning among agricultural workers in the United States, 1998–2005. Am. J. Ind. Med. 51: 883–898.

Calvert, G.M., D.K. Plate, R. Das, R. Rosales, O. Shafey, C. Thomsen, D. Male, J. Beckman, E. Arvizuand and M. Lackovic. 2004. Acute occupational pesticide-related illness in the US, 1998–1999: surveillance findings from the SENSOR-Pesticides program. Am. J. Ind. Med. 45: 14–23.

Campbell, A.W. 2014. Glyphosate: its effects on humans. Altern. Ther. Health Med. 20: 9–11.

Cobb, A. 1992. Herbicides and Plant Physiology. Pub. Chapman & Hall, London. pp. 51–53.

Colborn, T. and F.S. vomSaal. 1993. Developmental effects of endocrine-disrupting chemicals in wildlife and humans. Environ. Health Perspec. 101: 378–384.

Cole, D. 2006. Occupational health hazards of agriculture. Focus 13, Brief 8 of 16. *In:* Hawkes, C. and M.T. Ruel. Understanding the links between agriculture and health. International Food Policy Research Institute. 2020 Vision for Food, Agriculture and the Environment.

Corazza, M., G. Zinna and A. Virgili. 2003. Allergic contact dermatitis due to 1,3-dichloropropene soil fumigant. Contact Dermatitis 48: 337–349.

Curwin, B.D., M.J. Hein, W.T. Sanderson, M. Nishiokaand and W. Buhler. 2003. Acephate exposure and decontamination on tobacco harvesters' hands. J. Exposure An. Environ. Epidemiol. 13: 203–210.

De Roos, A.J., A. Blair, J.A. Rusiecki, J.A. Hoppin, M. Svec, M. Dosemeci, D.P. Sandier and M.C. Alavanja. 2005. Cancer incidence among glyphosate-exposed pesticide applicators in the agricultural health study. Environ. Health Perspec. 113: 49–54.

Donald, E. and C. Marr. 2011. Diquat dibromide-A 28 day oral (dietary) immunotoxicity study in mice using sheep red blood cells as the antigen. Charles River, Tranent, Edinburgh, Scotland, United Kingdom. Laboratory Report No. 31974. Unpublished. Syngenta File No. PP901/10815. Submitted to WHO by Syngenta Crop Protection AG, Basel, Switzerland.

Duke, S.O. and S.B. Powles. 2009. Glyphosate-resistant crops and weeds: now and in the future. AgBioForum 12: 346–357.

EC. 2003. Review report for the active substance paraquat. SANCO/10382/2002-final. 3 October 2003. Health & Consumer Protection Directorate-General, European Commission, Brussels. Available at: http://ec.europa.eu/food/plant/protection/evaluation/existactive/list_ araquat.pdf (Accessed on December 28, 2017).

ECPA (European Crop Protection Association) 2006. Protecting crops–protecting wildlife. Perspectives on crop protection & crop science. European Crop Protection Association Newsletters, May 2006.

Fishel, F.M. 2005. Pesticide toxicity profile. Documents PI50-PI91. *In:* Series of the Pesticide Information Office, Florida Cooperative Extension Service, Institute of Food and Agricultural Sciences, University of Florida, USA.

Fryer, J.D. and R.J. Makepeace (Eds.). 1978. Weed Control Handbook. Vol. II. Recommendations including Plant Growth Regulators (8th Edition). Issued by the British Crop Protection Council. Pub. Blackwell Scientific Publications, Oxford, UK.

Gasnier, C., C. Dumont, N. Benachour, E. Clair, M.C. Chagnonand and G.E. Seralini. 2009. Glyphosate-based herbicides are toxic and endocrine disruptors in human cell lines. Toxicology 262: 184–191.

Gerber, W.L., V.E. de la Penaand and W.C. Mobley. 1988. Infertility, chemical exposure, and farming in Iowa: absence of an association. Urology 31: 46–50.

Herman, K.M. and L.M. Weaver. 1999. The shikimate pathway. Ann. Rev. Plant Physiol. Plant Mol. Biol. 50: 473–503.

Hughes, M.F., B.D. Beck, Y. Chen, A.S. Lewis and D.J. Thomas. 2011. Arsenic Exposure and Toxicology: a historical perspective. Toxicological Sciences 123: 305–332.

IARC (International Agency for Research on Cancer). 1973. Some inorganic and organometallic compounds. IARC Monographs, Vol. 2. Lyon, France: International Agency for Research on Cancer, 181 p.

IARC (International Agency for Research on Cancer). 2015. Evaluation of five organophosphate insecticides and herbicides. IARC monographs, Vol. 112. Available at: www.iarc.fr/en/media-centre/iarcnews/pdf/MonographVolume112.pdf (Accessed on January 21, 2018).

IARC (International Agency for Research on Cancer). 2017. Some organophosphate insecticides and herbicides. IARC Monographs on the Evaluation of Carcinogenic Risk of Chemicals to Humans, Vol. 112, Lyon 3–10 March 2015, France. World Health Organization, 452 p.

Jeyaratham, J. 1990. Acute pesticide poisoning: a major problem. World Health Statistics Quarterly 43: 139–144.

Jones, G.M. and J.A. Vale. 2000. Mechanisms of toxicity, clinical features and management of diquat poisoning: a review. Journal of Toxicology: Clinical Toxicology 38: 123–128.

Kamel, F. and J.A. Hoppin. 2004. Association of pesticide exposure with neurologic dysfunction and disease. Environ. Health Persec. 112: 950–958.

Karabelas, A.J., K.V. Plakas, E.S. Solomou, V. Drossouand and D.A. Sarigiannis. 2009. Impact of European legislation on marketed pesticides: a view from the standpoint of health impact assessment studies. Environ. Int. 35: 1096–1107.

Kettles, M.K., S.R. Browning, T.S. Princeand and S.W. Horstman. 1997. Triazine herbicide exposure and breast cancer incidence: an ecologic study of Kentucky counties. Environ. Health Perspect. 105: 1222–1227.

Kishi, M., N. Hirschhorn, M. Djajadisastra, L.N. Satterlee, S. Strowmanand and R. Dilts. 1995. Relationship of pesticide spraying to signs and symptoms in Indonesian farmers. Scand. J. Work Environ. Health 21: 124–133.

Kishore, G.M. and D.M. Shah. 1988. Amino acid biosynthesis inhibitors as herbicides. Ann. Rev. Biochem. 57: 627–663.

Korres, N.E. 2018. Agronomic weed control: a trustworthy approach for sustainable weed management. pp. 97–114. *In:* Jabran, K. and Chauhan, B.S. (Eds.) Non-chemical weed control. Academic Press, Elsevier.

Korres, N.E. 2005. Encyclopaedic Dictionary of Weed Science: theory and digest. Lavoisier SAS and Intercept Ltd, Andover, UK, 724 p.

Kudsk, P. and J.C. Streibig. 2003. Herbicides – a two edged sword. Weed Res. 43: 90–102.

Kumar, R., N. Pant and S.R. Srivastava. 2000. Chlorinated pesticides and heavy metals in human semen. Int. J. Androl. 23: 145–149.

Kwiatkowska, M., B. Hurasand and B. Bukowska. 2014. The effect of metabolites and impurities of glyphosate on human erythrocytes (*in vitro*). Pestic. Biochem. Physiol. 109: 34–43.

Lapworth, D.J., D.C. Goody, P.J. Chilton, G. Cachandt, M. Knappand and S. Bishop. 2006. Pesticides in groundwater: some observations on temporal and spatial trends. Water Environ. J. 20: 55–64.

Lee, W.J., E.S. Cha, J. Park, Y. Ko, H.J. Kimand and J. Kim. 2012. Incidence of acute occupational pesticide poisoning among male farmers in South Korea. Am. J. Ind. Med. 55: 799–807.

Lerda, D. and R. Rizzi. 1991. Study of reproductive function in persons occupationally exposed to 2,4-dichlorophenocyacetic acid (2,4-D). Mutation Res. 262: 47–50.

Levy, B.S. and D.H. Wegman. 2000. Occupational health: recognizing and preventing work-related diseases and injury. Lippincott, 4th Edition. Philadelphia, USA.

Lewis, D. 1998. Applications of a theoretic model of information exposure to health interventions. Human Com. Res. 24: 454–468.

Li, W.M. 1986. The role of pesticides in skin disease. Int. J. Dermatol. 25: 295–297.

Li, S., P.A. Crooks, X. Wei and J. Leon. 2004. Toxicity of dipyridyl compounds and related compounds. Clin. Rev. in Toxicol. 34: 447–460.

Liou, H.H., M.C. Tsai, C.J. Chen, J.S. Jeng, Y.C. Chang, S.Y. Chen and R.C. Chen. 1997. Environmental risk factors and Parkinson's disease: a case-control study in Taiwan. Neurology 48: 1583–1588.

Litchfield, M.H., J.W. Danieland and S. Longshaw. 1973. The tissue distribution of the bipyridylium herbicides diquat and paraquat in rats and mice. Toxicol. 1: 155–165.

Luft, J. and G. Bode. 2002. Integration of safety pharmacology endpoints into toxicology studies. Fundam. Clinical Pharmacol. 16: 91–103.

Lutman, P.J.W. 2018. Sustainable weed control in oilseed rape. *In:* Korres, N.E., Burgos, N.R., Duke, S.O. (Eds.) Weed Control. Sustainability, Hazards and Risks in Cropping Systems Worldwide. CRC Press, Boca Raton, FL.

Mahieu, P., Y. Bonduelleand and A. Bernard. 1984. Acute diquat intoxication interest of its repeated determination in urine and the evaluation of renal proximal tubule integrity. J. Toxicol. Clin. Toxicol. 22: 363–369.

Malachowski, M.E. 1990. An update on arsenic. Clin. Lab. Medic. 10: 459–472.

Meharg, A.A. and J. Hartley-Whitaker. 2002. Arsenic uptake and metabolism in arsenic resistant and nonresistant plant species. New Phytologist 154: 29–43.

Mink, P.J., J.S. Mandel, J.I. Lundinand and B.K. Sceurman. 2011. Epidemiologic studies of glyphosate and non-cancer health outcomes. Regul. Toxicol. Pharmacol. 61: 172–184.

Nater, J.P., H. Terpstraand and B. Bluemink. 1979. Allergic contact sensitization to the fungicide Maneb. Contact Dermatitis 5: 24–26.

Nice, H.E., D. Morritt and M. Thorndyke. 2003. Long-term and transgenerational effects of nonylphenol exposure at a key stage in the development of *Crassostrea gigas*. Possible endocrine disruption? Mar. Ecol. Progr. Ser. 256: 293–300.

Nicolopoulou-Stamati, P., S. Maipas, C. Kotampasi, P. Stamatis and L. Hens. 2016. Chemical pesticides and human health: the urgent need for a new concept in agriculture. Front. Public Health 4:148. doi: 10.3389/fpubh.2016.00148.

Olson, K.R. 1994. Paraquat and Diquat. Poisoning and Drug Overdose (2nd edition). Appelton and Lange, Norwalk. pp. 245–246.

Pan, Y., W.J. Rea, A.R. Johnson and E.J. Fenyves. 1989. Formaldehyde sensitivity. Clinical Ecol. 6: 79–84.

Parisi, L., F. Pierelli, G. Amabile, G. Valente, E. Calandriello, F. Fattapposta, P. Rossi and M. Serrao. 2003. Muscular cramps: proposals for a new classification. Acta Neurologica Scand. 107: 176–186.

Pieniazek, D., B. Bukowskaand and W. Duda. 2004. Comparison of the effect of Roundup Ultra 360 SL pesticide and its active compound glyphosate on human erythrocytes. Pestic. Biochem. Physiol. 79: 58–63.

Pimental, D., H. Acduay, M. Biltonen, P. Rice, M. Silva, J. Nelson, V. Lipnor, S. Giordano, A. Horowitz and M. D'amore. 1992. Environmental and economic impacts of pesticide use. pp. 277–278. *In:* S.A. Briggs and the Staff of Rachel Carson Council. Basic Guide to Pesticides: Their Characteristics and Hazards. Rachel Carson Council, Taylor & Francis, Washington DC, USA.

Poskitt, L.B., M.B. Duffilland and M. Rademaker. 1994. Chloracne, palmoplantar ketatoma and localised scleroderma in a weed sprayer. Clin. Experim. Dermatol. 19: 264–267.

Reigart, R.J. and Roberts J.R. (Ed.). 1999. Recognition and Management of Pesticide Poisonings (5th Edition). US Environmental Protection Agency, Publication EPA-735-R-98-003.

Repetto, R. and S. Baliga. 1996. Pesticides and the Immune System: The Public Health Risk. World Resources Institute.

Richard, S., S. Moslemi, H. Sipahutar, N. Benachourand and G.E. Seralini. 2005. Differential effects of glyphosate and roundup on human placental cells and aromatase. Environ. Health Perspec. 113: 716–720.

Roberts, J.R. and R.J. Reigart (Ed.) . 1999. Recognition and Management of Pesticide Poisonings (6th Edition). US Environmental Protection Agency, Publication EPA-735-K-13001. Available at: http://www2.epa.gov/pesticide-worker-safety)Accessed on May 11, 2017).

Saeed, S.A.S., M.F. Wilksand and M. Coupe. 2001. Acute diquat poisoning with intracerebral bleeding. Postgrad. Medical J. 77: 329–332.

Samsel, A. and S. Seneff. 2013. Glyphosate, pathways to modern diseases II: celiac sprue and gluten intolerance. Interdisc. Toxicol. 6: 159–184.

Samsel, A. and S. Seneff. 2015. Glyphosate, pathways to modern diseases III: manganese, neurological diseases, and associated pathologies. Surg. Neurol. Int. 6: 45.

Sechi, G.P., V. Agnetti, M. Piredda, M. Canu, H.A. Omarand and G. Rosati. 1992. Acute and persistent parkinsonism after use of diquat. Neurology 42: 261–263.

Sharma, V.K. and S. Kaur. 1990. Contact sensitization by pesticides in farmers. Contact Dermatitis 23: 77–80.

Sharpe, R.M. and N.E. Skakkebaek. 1993. Are oestrogens involved in falling sperm counts and disorders of the male reproductive tract? Lancet 341: 1392–1395.

Shin, E.H., J.H. Choi, A.M. Abd El-Aty, S. Khay, S.J. Kim, M. Im, C.H. Kwon and J.H. Shim. 2011. Simultaneous determination of three acidic herbicide residues in food crops using HPLC and confirmation via LC-MS/MS. Biomed. Chromatogr. 25: 124–135.

Sittipunt, C. 2005. Paraquat Poisoning. Respir. Care 50(3): 383–385.

Slager, R.E., J.A. Poole, T.D. LeVan, D.P. Sandler, M.C.R. Alavanjaand and J.A. Hoppin. 2009. Rhinitis associated with pesticide exposure among commercial pesticide applicators in the agricultural health study. Occupat. Environ. Medic. 66: 718–724.

Smialowicz, R.J., R.W. Luebke, R.R. Rogers, M.M. Riddleand and D.G. Rowe. 1985. Evaluation of immune function in mice exposed to ordram. Toxicol. 37: 307–314.

Solomon, K.R. 2016. Glyphosate in the general population and in applicators: a critical review of studies on exposures. Crit. Rev. Toxicol. 46: 21–27.

Street, J.C. and R.P. Sharma. 1975. Alteration of induced cellular and humoral immune responses by pesticides and chemicals of environmental concern: quantitative studies of immunosuppression by DDT, aroclor 1254, carbaryl, carbofuran, methylparathion. Toxicol. Appl. Pharmacol. 32: 587–602.

Swanson, N.L., A. Leu, J. Abrahamsonand and B. Wallet. 2014. Genetically engineered crops, glyphosate and the deterioration of health in the United States of America. J. Org. Syst. 9: 6–37.

Szekacs, A. and B. Darvas. 2011. Forty years with glyphosate. pp. 149–165. *In:* M.N.A. El–Ghany Hasaneen (Ed.) Herbicides – Properties, Synthesis and Control of Weeds, Intech, Rijeka, Croatia.

Tanner, C.C. and J.S. Clayton. 1990. Persistence of arsenic 24 years after sodium arsenite herbicide application to Lake Rotoroa, Hamilton, New Zealand. New Zealand J. Mar. Freshwater Res. 24: 173–179.

Tay, C.H. 1974. Cutaneous manifestations of arsenic poisoning due to certain Chinese herbal medicines. Australasia J. Dermatol. 15: 121–131.

Thongprakaisang, S., A. Thiantanawat, N. Rangkadilok, T. Suriyoand and J. Satayavivad. 2013. Glyphosate induces human breast cancer cells growth via estrogen receptors. Food Chem. Toxicol. 59: 129–136.

Thundiyil, J.G., J. Stoberand and J. Pronczuk. 2008. Acute pesticide poisoning: a proposed classification tool. Bulletin of the World Health Organization 86: 205–209.

Tyler, H.L. and M.A. Locke. 2018. Effects of weed management on soil ecosystems. *In:* Korres, N.E., Burgos, N.R., Duke, S.O. (Eds.), Weed Control. Sustainability, Hazards and Risks in Cropping Systems Worldwide. CRC Press, Boca Raton, FL.

U.S. EPA. 2009. Amendment to Organic Arsenicals RED Available at: http:// www.epa.gov/ oppsrrd1/REDs/organic-arsenicals-amended.pdf)Accessed on January 31, 2018).

Watterson, A. 1988. Pesticide User's Health and Safety Handbook: An international guide. Gower Technical, Aldershot.

Watts, M. 2011. Paraquat. Pesticide Action Network Asia and the Pacific, 2011. Available at: http:// wssroc.agron.ntu.edu.tw/note/Paraquat.pdf)Accessed on January 23, 2018).

Westwood, J.H., R. Charudattan, S.O. Duke, S.A. Fennimore, P. Marrone , D.C. Slaughter, C. Swantonand and R. Zollinger. 2018. Weed management in 2050: Perspectives on the future of Weed Science. Weed Sci. 10.1017/wsc.2017.78.

Whitehead, R. (Ed.) 1995. The UK Pesticide Guide. CAB International and British Crop Protection Council.

Williams, G.M., R. Kroesand and I.C. Munro. 2000. Safety evaluation and risk assessment of the herbicide roundup and its active ingredient, glyphosate, for humans. Regul. Toxicol. Pharmacol. 31: 117–165.

Williams, G.M., M. Aardema, J. Acquavella, S.C. Berry, D. Brusick, M.M. Burns, J.L. Viana de Camargo, D. Garabrant, H.A. Greim, L.D. Kier, D.J. Kirkland, G. Marsh, K.R. Solomon, T. Sorahan, A. Robertsand and D.L. Weed. 2016. A review of the carcinogenic potential of glyphosate by four independent expert panels and comparison to the IARC assessment. Crit. Rev. Toxicol. 46: 3–20.

Woodburn, A.T. 2000. Glyphosate: production, pricing and use worldwide. Pest Manag. Sci. 56: 309–312.

Wong, S.S., K.C. Tan and C.L. Goh (1998). Cutaneous manifestations of chronic arsenicism: review of seventeen cases. J. Am. Acad. Dermatol. 38: 179–185.

Zimdahl, R.L. 1993. Fundamentals of Weed Science. Academic Press, Inc. San Diego, California.

3

Effects of Weed Management on Soil Ecosystems

Heather L. Tyler[*1] and Martin A. Locke[2]

[1] Research Microbiologist, USDA Agricultural Research Service, 141 Experiment Station Road, Stoneville, MS 38776, USA

[2] Director, USDA Agricultural Research Service, National Sedimentation Laboratory, 598 McElroy, Oxford, MS 38655, USA

Introduction

The soil as an ecosystem is a community of living organisms within the soil matrix interacting in conjunction with the nonliving components of soil. Like any ecosystem, soil ecosystems exist at varying scales. A soil pedon is the smallest, three-dimensional unit that is considered to be soil. A similar term is that of 'soil individual' (Johnson 1963). A soil profile is the vertical sequence of distinct horizons. A soil series is a taxonomic or mapping unit that is the most basic category of the United States Department of Agriculture Natural Resource Conservation Service soil classification system. Within the landscape, soil ecosystems could consist of several soil series.

Roles that agricultural soil ecosystems play include (1) structural, e.g., as a medium for plant growth, (2) filter and reservoir, e.g., clean and store water, (3) fertility, e.g., nutrient cycling, (4) biodiversity conservation, e.g., reservoir for biological diversity, and (5) climate regulation, e.g., carbon sequestration and greenhouse gas regulation (Dominati et al. 2010). Soil ecosystems are complex and provide many services to meet, in part, human needs, i.e., ecosystem services. Dominati et al. (2010) proposed a framework for classifying and quantifying soil ecosystem services (cultural, regulating, and provisioning) that helps link ecosystem services with inherent soil properties as well as those properties influenced by management.

There are a variety of techniques used for the control of weed populations in agricultural fields, all of which have the potential to influence soil ecosystem services. Historically, mechanical methods, such as tillage, that physically disrupt weed plants are among the older methods employed by farmers. Meanwhile, herbicide use has been increasing steadily since its introduction in the early 1900s (Timmons 1970). With the more recent emphasis on conservation practices aimed at minimizing the impact of agriculture on the environment, other methods, such as crop rotations with cover crops that inhibit weed growth, are also being employed. Details of these methods are described elsewhere (Timmons 1970, Locke et al. 2002, Teasdale et al. 2007, Miller 2016). In this chapter, we will discuss the impacts that some of the more common weed management practices, including tillage, cover crops, and herbicide application, have on soil ecosystems in agricultural landscapes.

---

*Corresponding author: heather.tyler@ars.usda.gov

Conservation Management

The term 'conservation' in the context of soil is used to convey that losses are minimized or reduced, whether the loss is of the physical soil itself, components of soil, such as nutrients or water, or the loss of a soil characteristic that impedes its ability to function, e.g., loss of soil fertility. Conservation management practices refer to a variety of processes that conserve soil. Some of the more common practices related to weed management include conservation tillage and use of cover crops. There are other conservation practices, but the two mentioned here have a strong relationship with weed management. Previous reviews have been published on similar topics, e.g., conservation management and weed management or ecology (Locke and Bryson 1997, Lal and Kimble 1997, Matson et al. 1997, Kladivko 2001, Locke and Zablotowicz 2004, Doran et al. 2006, Locke et al. 2006, Alletto et al. 2010, Locke et al. 2010).

Tillage

Tillage has been used in agriculture since humans began to cultivate crops for food and fiber. Some form of the moldboard plow has been used in farming for centuries in Asia and Europe to turn under the topsoil. Later versions developed in the last two centuries have greatly enhanced the ability of the plow to mix the soil. Cultivating the soil by tillage provides benefits, such as aeration, redistribution and mixing of plant nutrients, bed preparation for planting, and weed management. However, tillage can have a negative impact on soil as well. In the United States, tilling large tracts of semi-arid land for crop use led to the destructive effects of drought and wind erosion during the Dust Bowl era in the 1930s. In many tropical regions of the world, slash, burn, and till management systems have denuded vast tracts of lush rainforest rendering the soil vulnerable to erosion. These are but a few examples of the devastating impacts tillage has on soil. The problem is worldwide, and effects of tillage on the soil landscape have been evident throughout the history of mankind.

Previous to the introduction and widespread use of synthetic chemicals in the middle of the twentieth century, the primary methods of weed management included tillage, smother crops, or use of inorganic chemicals such as arsenic. Proliferation of organic herbicide use in industrial nations in the decades following World War II, together with increased mechanization of farm equipment, led to large efficiencies of scale in farm management. Soil-applied herbicides supplemented by post-emergent herbicides were the standard prescription for weed management in the decades following the initiation of the Green Revolution. Even with herbicides, tillage was still an integral component of the toolbox for most weed management systems during this period. However, the availability of herbicides gave farmers more options as to the intensity, timing, and frequency of tillage. In many developing countries where herbicides are less available or are too expensive for subsistence farmers, burning and tilling are still the primary means of weed management.

Weed management has undergone another revolution over the last 30 years. Weed resistance began to limit the use of many herbicides. Environmental and health concerns about the use of agrochemicals have led to regulation and reduction in the use of or phasing out of many 'old guard' herbicides. Fewer new herbicides are coming to market to replace those that were lost. The success of glyphosate resistant crops also resulted in plummeting sales for other herbicides, to the point that it was no longer profitable to sustain their registration. The widespread use of transgenic crops in the last 15 years in some regions has resulted in profound changes with respect to both weed and soil management. In other regions, transgenic crops have not been implemented due to health or ecological concerns. In those regions where transgenic crops have been used, adaptations to weed management have ensued. Initially, in systems where genetically modified glyphosate resistant crops were grown, glyphosate with no additional soil-applied herbicide was recommended due to its broad-spectrum activity. With the resulting development and proliferation of glyphosate resistant weeds due to this practice, the recommendation was modified to also include other herbicides, particularly those that were soil applied. Another observed trend with the use of transgenic crops is the facilitation of conservation tillage practices (soil cultivation methods that leave crop residue for the conservation of soil and water). Using herbicide to burn winter vegetation prior to planting has long been a common practice to prepare land for the new crop season. However, using herbicide resistant crops provides flexibility to avoid or minimize tillage (both before and after planting the crop) and apply herbicide over the plant canopy after emergence.

With respect to tillage, two opposite weed management strategies can be employed. The first strategy, tillage to manage existing weeds, has been a common practice reaching back throughout the history of agriculture. This strategy can be used with or without herbicides. The second strategy usually involves the use of post-emergent herbicides or surface applied pre-emergent herbicides with little or no tillage. Reduction in tillage, sometimes called conservation tillage, helps to (1) maintain integrity of a soil crust and preserve plant residue coverage to inhibit weed germination, and (2) minimize new exposure of buried weed seeds that would germinate in freshly tilled soil. This section on tillage will address the following question: What effects do either the strategy of (1) tilling the soil to kill existing weeds or (2) minimizing tillage (conservation tillage) to maintain a soil crust and inhibit weed germination have on soil ecosystems? To facilitate discussion of the effects of tillage on soil ecosystem health, factors will be separated into the categories of physical, chemical, and biological characteristics.

Physical Characteristics

Soil structure is the arrangement of a soil system into separate units called aggregates. Soil structure is a key physical characteristic influencing the soil ecosystem. It affects many processes and soil characteristics, including water infiltration, moisture content, aeration, drainage, and carbon and nutrient sequestration. Soil structure is influenced by many factors, including inherent soil texture and particle size distribution within the soil profile, plant residue decomposition, activity of soil biota, regional climate, inherent soil mineralogy, soil chemistry, and location within the landscape. These factors and processes are highly interrelated. For example, the rate of plant residue decomposition is greatly influenced by climate (i.e., primarily moisture and temperature) and activity of soil biota.

Development of soil structure is a long-term process (Benjamin et al. 2007). However, soil structure is disrupted by any form of tillage. Therefore, one incidence of tillage can compromise many years of soil structural formation. The extent of harm to soil structure depends on the intensity, frequency, and depth of tillage. Several studies have reported that due to stratification with soil depth, most effects of tillage occur in the surface soil (Helgason et al. 2009). Spatial differences in tillage effects are also observed. In a strip tillage system, where tillage was performed in narrow bands, soil characteristics were distributed along a spatial gradient (Overstreet and Hoyt 2008). For example, bulk density was greater in the inter-row area due to increased compaction from wheel traffic of heavy implements moving through the field.

An increasing prevalence in size and number of aggregates in soil is an indicator of soil structural stability (Sainju et al. 2009). Reducing tillage often results in an increase in size and stability of aggregates (macro- versus micro-aggregates), particularly in the soil surface (Blanco-Canqui et al. 2009a, Pikul et al. 2006, Liebig et al. 2004, Shaver et al. 2003, Mitchell et al. 2017). Macro-aggregates form around particles of non-decomposed plant residues, partially decomposed litter, and micro-aggregates, thereby sequestering carbon and nutrients. Significant quantities of carbon and nutrients are often sequestered within macro-aggregates (Mikha and Rice 2004, Muñoz-Romero et al. 2017, Six et al. 2000). In conservation tillage soils, elevated populations of microbes, particularly fungi contribute to the development of aggregates (Helgason et al. 2010, Frey et al. 1999).

Wheel traffic from farming operations is a major source of compaction in row crop soils. If the same rows for traffic are used multiple times each year, compaction may result in poor aeration and lower water infiltration. Compaction from wheel traffic can occur in both conventionally tilled and reduced tillage soils (Locke et al. 2013). Tillage can be used to mitigate effects of compaction. Subsoiling may be used to alleviate compaction, but one study showed it was not necessary to subsoil on a regular basis (Hill and Meza-Montalvo 1990). If a strictly no-tillage system is used, management options include either reducing the number of trips across the field, rotating rows used for driving equipment, or using lighter equipment. If the conservation tillage system is not strictly no-tillage, establishing permanent lanes for traffic may help to minimize the extent of compaction. In a ridge tillage system, Liebig et al. (1993) found that bulk density, penetration resistance, aggregate mean weight diameter, and water content (both field capacity and wilting point) were highest in the traffic interrow and lowest in the row. This resulted in lower saturated water content and gravitational water in the traffic interrow. When the traffic interrow was compared with the non-traffic interrow and row, penetration resistance was higher and saturated hydraulic conductivity was lower.

Effects of management on compaction can be site specific and vary depending on inherent soil properties and organic matter content. In some cases, conservation tillage may result in higher compaction with associated negative effects. Blanco-Canqui et al. (2010) observed that bulk density, cone index, shear strength, and aggregate tensile strength increased in wheel traffic areas of a no-tillage soil. This resulted in a reduction of cumulative infiltration, logarithm of saturated hydraulic conductivity, soil water retention, plant available water, effective porosity, and volume of >50-µm pores. Dao (1996) observed both a lower soil bulk density with decreasing tillage intensity and a reduction of wheel traffic in conservation tillage soils. On coarse textured soils in another study, reduced tillage systems had higher bulk density, lower total porosity, higher microporosity, and higher field capacity (water retention capacity) than tilled systems (Raczkowski et al. 2012). A higher intensity of tillage and subsequent destruction of soil structure may offset the positive benefits of tilling. Intensive tillage breaks down soil aggregates into smaller sizes, resulting in degraded soil structure (loss of soil aggregation) and increased compaction when compared to a no-tillage soil (Govaerts et al. 2007).

Factors, such as higher organic matter, may buffer some of the negative effects due to soil compaction. Benjamin et al. (2007) observed that over a 15-year period, saturated hydraulic conductivity increased and bulk density decreased in several no-tillage rotation cropping systems. Blanco-Canqui et al. (2009b) attributed reduction in bulk density to increases in organic carbon in no-tillage soils as compared to conventional tillage. Kumar et al. (2012) measured lower bulk density and higher SOC and percentage of water stable aggregates in the no-tillage soil as compared to minimum tillage and plow tillage.

Tillage removes, destroys, or enhances the decomposition of surface plant residues, but bare, tilled soil surfaces are vulnerable to erosion. The physical presence of a plant residue cover reduces the energy of rainfall on soil and thus the susceptibility to erosion. Maintaining plant residues on the surface may also improve macroporosity in conservation tillage soils, resulting in enhanced infiltration, and, thus, less runoff (Dao 1996). Soils with a stable soil structure are more resistant to soil erosive forces. The prevalence of macro-aggregates (>2 µm) in the surface soil was found to be negatively correlated with soil erosion (Barthès and Roose 2002). Combinations of no-tillage and cover crop enhanced surface plant residue coverage, reduced loss of nutrients, pesticides, and sediment in runoff from a cotton crop (Krutz et al. 2009, Locke et al. 2015). In an Ohio study, runoff losses from either a chisel tillage or no-tillage treatment were dependent on the timing of rain events and fertilizer application management (Shipitalo et al. 2013). No-tillage reduced sediment losses, but runoff of total P and nitrate were similar between no-tillage and chisel tillage (Shipitalo et al. 2013).

Biological Properties

Several factors contribute to favorable conditions for biological activity and biodiversity in conservation tillage soils. One important factor is improved soil moisture conditions. Reducing tillage increases soil moisture because the soil crust and surface plant residues in untilled soils lower evaporation rates. Also, with less surface water runoff in conservation tillage systems, more water is available for infiltration into the soil profile. Elevated organic carbon in conservation tillage soils provides substrate for organisms. Improved and more stable soil structure provides a platform for root growth, promotes carbon sequestration, and facilitates water movement. These environmental conditions also help to buffer conservation tillage soils from extremes in temperature.

Many studies have documented greater overall abundance of microbial populations in conservation tillage soils, including total microbial biomass, bacteria, and fungi (Helgason et al. 2010, Feng et al. 2003, Helgason et al. 2009, Lupwayi et al. 2017, White and Rice 2009). Helgason et al. (2009) noted greater bacterial and fungal PLFA and higher bacterial and total microbial biomass in soils that were less intensively tilled in most cases, although this was not always the case. White and Rice (2009) found higher total Gram positive and Gram negative bacteria in the surface of no-tillage soils. Pankhurst et al. (2002) observed increases in microbial biomass when conventionally tilled areas were converted to direct-drilled, stubble-retained systems. Feng et al. (2003) measured increased microbial biomass carbon in the soil surface.

Soil conditions under conservation tillage may also alter the composition of microbial communities. Several studies have demonstrated that fungal populations are sometimes favored under conservation tillage management (Helgason et al. 2010, Drijber et al. 2000, Fontana et al. 2015, Frey et al. 1999, Lupwayi et al. 2017, Muruganandam et al. 2009). Undisturbed soils provide a more stable environment conducive to growth of fungal hyphae. In turn, fungal hyphae and associated proteins, such as glomalin contribute to the development of soil macroaggregates. Helgason et al. (2010) observed arbuscular mycorrhizal fungi were 40–60% greater in various sizes of aggregates in no-till soils as compared to tilled soils. When stubble-retained systems were converted to conventional tillage, Pankhurst et al. (2002) measured a decline in ratios of fungal to bacterial fatty acids in the surface soil. Fungal hyphal length was 1.9 to 2.5 times higher in the surface of no-tillage, and the proportion of microbial biomass consisting of fungi was greater (Frey et al. 1999).

Biological populations are sensitive to environmental conditions, and timing of soil sampling may influence results. In a wheat-fallow study comparing no-till and tilled treatments, Feng et al. (2003) noted microbial communities were influenced by different factors depending on the season. Physical and chemical conditions (e.g., increased infiltration, water holding capacity, cooler and more stable temperatures, higher organic matter) were the predominant factors attributed to tillage influencing microbial activity during the fallow season. However, during the growing season, microbial communities were influenced more by the growing crop, e.g., exudates from roots, than from tillage. Similarly, Drijber et al. (2000) found that biomarkers for arbuscular mycorrhizal fungi declined in tilled soils, but the strongest tillage differences were observed during the fallow season rather than when wheat was growing. They concluded that during the wheat growing season, microbes were more influenced by inputs from wheat, but during the fallow season, the physicochemical environment resulting from long-term tillage management had a greater effect.

Soil exoenzyme activities are often used as a measure of the level of microbial activity in soil. Many studies have measured enhanced soil enzyme activities in no-tillage systems (Locke et al. 2013, Lupwayi et al. 2017, Muruganandam et al. 2009, Acosta-Martínez et al. 2007). Enzyme activities (e. g., N-acetyl-β-glucosaminidase, arylamidase, L-Asparaginase L-Glutaminase) were positively correlated with higher fungal populations in conservation tillage (Muruganandam et al. 2009). Activity of β-glucosidase activity was 50% greater under conservation management than conventional management (Lupwayi et al. 2017). Fontana et al. (2015) measured increased dehydrogenase activity under reduced compared to conventional tillage. Several studies have shown that in the surface of reduced tillage soils, exoenzyme activity is increased (Acosta-Martínez et al. 2007, Acosta-Martínez et al. 2003, Babujia et al. 2016, Jia et al. 2016).

Studies have demonstrated that tillage can also influence faunal populations in soil. Categories for soil fauna include macrofauna (e.g., small mammals, such as moles, earthworms, insects), mesofauna (e.g., nematodes, arthropods), and microfauna (e.g., protozoa). Fauna, such as earthworms, can physically modify the soil environment by creating macropores. They also contribute nutrients through fecal depositions and exudates (Blouin et al. 2013). Due to challenges with respect to sampling methodology and spatial variability, results from studies evaluating effects of conservation tillage on earthworm populations are varied (Chan 2001, Joschko et al. 2009). Enhanced earthworm abundance and biomass were observed in conservation tillage systems in some cases (Castellanos-Navarrete et al. 2012, Ernst and Emmerling 2009, Hubbard et al. 1999, Rovira et al. 1987), while in others, there was no change, reduced populations, or variability in results (Umiker et al. 2009, Doube et al. 1994, Crittenden et al. 2014, Kladivko et al. 1997, Locke et al. 2013). Ernst and Emmerling (2009) found species differences due to tillage as densities of anecic earthworms were negatively impacted in tilled areas when compared to reduced tillage, while endogeic earthworm species were seen to increase with tillage. In earlier work, Emmerling (2001) found increased richness of earthworm species with reduced tillage. Errouissi et al. (2011) used quadrat and pitfall trap methods to assess several species of soil invertebrates in wheat under two tillage systems and found that no-tillage favored more diverse communities. They determined that no-till improved species richness (34 species compared to 26 species in conventional tillage) and abundance (319 specimens versus 61, respectively). Predators, detritivores, and herbivores were more abundant in no-till. Studies demonstrated inconsistent effects of tillage on nematode populations. Plant parasitic nematodes were slightly favored in tilled soils in some studies (Liphadzi et al. 2005, Locke et al. 2013, Rahman et al. 2007), while in other studies, there was either no effect due to tillage or there were higher populations in conservation tillage soils (Gavassoni et al. 2001, Govaerts et al. 2007, López-Fando and Bello 1995, Gallaher et al. 1988).

Depending on the crop and locale, reducing tillage may have some negative effects on soil biology, including increased incidence of populations that cause plant diseases. Cooler, more moist conditions in conservation tillage soils may contribute to the enhancement of disease-causing organisms (Peters et al. 2003, Govaerts et al. 2007). Slow germination and early growth of crops in conservation tillage systems may also lead to vulnerability to disease. Increased incidence of *Fusarium* was observed in reduced tillage soils (Ahmed et al. 2012). Almeida et al. (2003) observed either no difference between conventional tillage and no-tillage or a higher incidence of charcoal rot (caused by *Macrophomina phaseolina*) in conventional tillage soybeans. Gossen and Derksen (2003) assessed the impact of tillage and rotation on *Ascochyta* blight (*Ascochyta lentis*) in lentil and determined that although the incidence and severity of the blight was higher in no tillage than conventional tillage in continuous lentil, it could be managed by rotating with non-host crops. Similarly, Govaerts et al. (2007) reported there was a higher incidence of root rot in no-tillage corn but that crop rotation helped to mitigate this effect.

Chemical Characteristics

Reducing the frequency and intensity of tillage promotes the accumulation of plant residues on the soil surface where they are left to weather and decompose. In many cases, as plant residues decompose, soil organic carbon and nutrients accumulate and leach into the surface soil (e.g., Moore et al. 2014, Nawaz et al. 2017). Initially, as residues decompose, component organic carbon and nutrients exist in transient, dynamic pools that can vary widely in composition over short-term periods (e.g., 60% within one year (Kochsiek et al. 2013)). Li et al. (2015) found particulate (>53 µm) organic matter carbon and nitrogen increased more than 145% with conservation tillage management as compared to 45–50% increase in total (all pools) organic carbon and nitrogen, and 20% increases in fine organic matter carbon and nitrogen (<53 µm).

In some cases, the carbon and nutrient fractions in the dynamic soil pools are labile or available for plant uptake or further processing. Halpern et al. (2010) observed that in the surface of Canadian soils, labile fractions of soil organic carbon in microbial biomass and potentially mineralizable carbon and nitrogen pools were greater in conservation tillage soils as compared to that of conventional tillage after 16 years. Values of soil labile nitrogen fractions, potentially mineralizable nitrogen, and both coarse and fine particulate organic matter nitrogen were higher under no-tillage management in Argentina (Martínez et al. 2017). Sometimes, when inherent fertility of a soil is low, immobilization of nutrients in soil organic components may limit nutrient uptake by plants. For example, nutrient mineralization, crop nutrient uptake, and yields on a soil in southern Africa were lower in areas managed with a less intensive tillage treatment (ripper tillage) than with more intensive plough tillage (Masvaya et al. 2017). The physical mixing of soil during tillage redistributes surface soil particles and soluble nutrients, mixes accumulated surface plant residues into the soil surface, and breaks apart soil aggregates. More aerated conditions after tillage and increased surface area contact of plant residues with soil enhance decomposition of the plant material. Aeration and oxidative soil conditions resulting from tillage may increase solubility of some nutrients. The destruction of soil aggregates exposes more surface area and enhances mineralization or degradation of carbon and labile chemicals. This may result in a greater proportion of more recalcitrant fractions of soil organic carbon in tilled soils (Muñoz-Romero et al. 2017). Pankhurst et al. (2002) observed a decline in organic carbon, total nitrogen, and nitrate-nitrogen when converting a direct drill, stubble-retained system to a conventional tillage system in Australia.

Chemical dynamics and availability of nutrients in soils are influenced by interactions among soil physical characteristics, biological activity, and quality and composition of the soil mineral and organic components. As discussed previously, general characteristics of conservation tillage soils with respect to those managed under conventional tillage include enhanced moisture availability and storage capacity, higher biological activities, more stable aggregation, and higher levels of organic carbon and plant residues. Soon and Arshad (2004) attributed increased availability of nitrogen to improved soil moisture conditions. Exoenzyme activities associated with microbial populations in conservation tillage soils are correlated with mineralization of nutrients. Muruganandam et al. (2009) measured elevated activities of enzymes involved in both the aminization and ammonification in no-tillage soils that corresponded with increased total nitrogen concentrations. Similarly, Mina et al. (2008) associated increased levels of nitrogen, phosphorus, and potassium in no-tillage with higher enzyme activities (alkaline phosphatase, protease, dehydrogenase, and cellulase). Another factor influencing chemical dynamics includes the longevity of conservation tillage management. The duration that a soil is managed in conservation tillage can determine the quantity, composition, and distribution of labile organic carbon and chemical fractions (Mishra et al. 2010, Spargo et al. 2012). After 11 years of continuous no-till management, crop needs for nitrogen were increasingly supplied by mineralization of labile organic nitrogen, reducing the need for fertilizer nitrogen (Spargo et al. 2012). Surface soil in conventional tillage from a 20-year study contained humic acid with more recalcitrant functional groups than that from a no-tillage soil (Ding et al. 2002).

A common observation in conservation tillage soils is the stratification of soil organic carbon and chemicals within the soil profile. A frequent trend is that of enhanced organic carbon, nutrients, or pesticides in the surface of these soils as compared to conventionally tilled soils, with a sharp reduction in concentrations below surface depths (Locke et al. 2013, Ding et al. 2002, Martínez et al. 2016, Muñoz-Romero et al. 2017, Sainju et al. 2014, Umiker et al. 2009, Mitchell et al. 2017). In soils that are tilled, more of the organic carbon and chemicals may be distributed near the soil surface, but declines in organic carbon or chemical concentrations with depth are often less pronounced, i.e., more uniform distribution, at, within, or just below the plow layer. Because the organic carbon or chemical concentrations in the surface of conservation tillage soils tend to be higher than that of tilled soils, it might be assumed that overall sequestration of total carbon and chemicals is greater for the soil profile as a whole. If stratification occurs, tillage effects may occur in the distribution of carbon and nutrients within the soil profile; however, the effect of tillage on total carbon or chemical sequestration within the entire soil profile may be diminished (Blanco-Canqui and Lal 2008, Gál et al. 2007, Hou et al. 2012, Lupwayi et al. 2006, Martínez et al. 2016).

Stratification of organic carbon and chemicals in conservation tillage soils result from several factors. First, there is less mixing of soil when tillage is eliminated or reduced. Second, chemicals, such as fertilizer and pesticides, are applied on the surface either as a broadcast application or are incorporated into the soil by knifing. Some downward migration of more soluble nutrients or pesticides might occur, but most chemicals remain in the surface soil horizon where they can be taken up by plants or dissipate *in situ*. For those less soluble chemicals, enhanced soil organic carbon might provide a greater surface area for sorption of chemicals, i.e., increased number of exchange sites, thus inhibiting mobility. In addition to fertilizer application, a source of nutrients is from mineralization of decomposed plant residues. Decomposition of plant residues remaining on the soil surface is slower than when they are mixed with the soil. As nutrients from plant residue decomposition of surface residues are mineralized, they tend to remain in a relatively thin band at the soil surface. Al-Kaisi et al. (2005) measured higher soil organic carbon and total nitrogen in no-till as compared with chisel plow, but concluded that higher levels were due to decreased mineralization rates in the surface and not stratification, since tillage differences were not observed at lower depths. Plant residues on the surface also can intercept some applied chemicals and sequester them within the residue. A higher proportion of organic components in the surface with a high carbon to nitrogen ratio can immobilize chemicals resulting in a slower release through mineralization (Mishra et al. 2010). Soil organic carbon and chemicals sequestered within aggregates receive a measure of protection from degradation or mineralization (Mikha and Rice 2004). Enhanced formation of stable aggregates in conservation tillage soils might buffer plant residues and associated chemicals from decomposition and result in a slower mineralization, i.e., release of nutrients (Six et al. 2000). Muñoz-Romero et al. (2017) measured higher soil organic carbon in the surface of no-tillage soil as compared to conventional tillage. For soils in both tillage treatments, most of the organic carbon resided in a recalcitrant fraction. However, the labile carbon fraction in no-tillage was greater than that of conventional tillage since tilling the soil resulted in greater mineralization of labile carbon.

A nutrient-rich soil with elevated organic carbon provides an environment that could influence the fate of pesticides. Healthy and diverse microbial populations in reduced tillage soils might lead to accelerated degradation of herbicides. As previously stated, conservation tillage soils possess enhanced microbial biomass, and soil with larger microbial biomass displays higher degradation rates of 2,4-D and dicamba (Voos and Groffman 1997). Levanon et al. (1994) found atrazine mineralization was more rapid in no tillage compared to conventional tillage, and attributed this to the greater microbial activities. Alachor mineralization has also been observed to be greater in no tillage compared to conventional tilled soil (Locke et al. 1996). Zablotowicz et al. (2000) found that although fluometuron degraded more rapidly under conventional tillage compared to no tillage plots, a ryegrass cover crop stimulated fluometuron degradation in both tillage treatments. The lower degradation of fluometuron under no-tillage (with no cover crop) was attributed to the increased sorptive capacity in no-tillage soils, rendering the herbicide less bioavailable for degradation.

Cover Crops

In row crop systems, cover crops are typically planted in the fall and grown until early to late spring, depending on the timing for planting the cash crop. Resulting biomass covers the soil surface and promotes conditions unfavorable for weed establishment (Teasdale et al. 1991). However, cover crops can also compete with summer crops for resources, such as soil moisture and nutrients, if not killed early enough prior to planting in the spring. Given the added labor, cost of seed, and herbicide application to kill cover crops prior to planting summer crops, many farmers are hesitant to employ this method of weed management unless the economic benefits (reduced herbicide use during the summer, increased crop yield) outweigh the costs. Despite this, cover crops offer numerous benefits for maintaining soil health including shading the soil surface, decreasing erosion, increasing infiltration, and adding biomass to soil.

Plant biomass from cover crops enhances a variety of soil chemical properties, from soil moisture, soil organic matter, and soil nutrient levels. Both rye (*Secale cereal* L.) and Balansa clover (*Trifolium michelianum* Savi var. *balansae* [Boiss.] Azn.) yielded more soil carbon compared to non-cover cropped soils in a cotton field system (Locke et al. 2013). Rye and hairy vetch (*Vicia villoa* Roth) also resulted in higher soil total organic carbon compared to no winter cover plots in soybean fields (Zablotowicz et al. 2010). In the case of soil nitrogen, soils from leguminous cover crop plots tend to have higher nitrogen levels than both non-cover and non-legume based cover crops. Soil from Balansa clover cover cropped fields had higher total nitrogen (Locke et al. 2013), while hairy vetch, another legume, has been found to increase soil nitrate levels by two fold (Zablotowicz et al. 2010). Soil from crimson clover (*Trifolium incarnatum*) plots also had higher nitrate, as well as sulfate and manganese, when compared to rye cover crop or no cover crop (Reddy et al. 2003). The beneficial effects of cover crops on soil nutrients appear to be seasonal, vary by sampling time, and are likely dependent upon the time frame of biomass degradation in soil. Fernandez et al. (2016) found both vetch and rye increase soil nitrate levels in mid-summer, but not late spring, due to slow mineralization as plant residues decompose.

Cover crops can also influence biological parameters in soil, although the nature and extent of such changes can vary by environments, season, and species of cover crops used. Microbial communities in soils of cover crop plots are generally enhanced. Both cereal rye and Trios (*Triticale* × *Triosecale*) have been shown to increase microbial biomass C and N compared to non-cover soils (Steenwerth and Belina 2008b, a). Ryegrass as a cover crop in cotton farming systems also resulted in higher numbers of viable microbial cells in the upper 2 cm of both tilled and no till plots, while both heterotrophic bacteria and fungi were higher in the 2-10 cm depth in tilled plots (Zablotowicz et al. 2007). In soybean fields, winter cover crops of crimson clover and cereal rye also significantly increased the numbers of total bacterial and fungal cells in surface soils (0–5 cm) (Reddy et al. 2003). Increases in soil microbial communities are not always observed in response to cover crop implementation. Cover crops can also cause shifts in microbial community composition, even if the total numbers or biomass in the soil remains the same. Mbuthia et al. (2015) found that neither hairy vetch nor winter wheat cover crops changed the abundance of bacteria in cotton fields, but they did cause shifts in community composition, i.e., Gram positive bacteria increased in abundance while mycorrhizae fungi decreased. However, effects of cover crops on community composition are often minor compared to the spatial effects, such as between field sites or sampling location. Fernandez et al. (2016) found the variation in community composition associated with cover crops was small compared to the impacts of location and even bulk vs. rhizosphere soil. Similarly, Locke et al. (2013) noted significant effects of cover crops (cereal rye and Balansa clover) on microbial community structure in soil compared to tillage treatments, while Zablotowicz et al. (2010) reported tillage had a greater impact on community composition than cover crops (cereal rye and hairy vetch). Both studies [Locke et al. (2013) and Zablotowicz et al. (2010)] were performed in the Mid-South area of the United States, further highlighting the strong influence that field site locations can have on microbial community composition, even within the same region.

With the potential for shifts in the size and composition of soil microbial communities, cover crops also influence activities involved in soil nutrient cycling. Measures of general hydrolytic activity, such as fluoresce in diacetate (FDA) hydrolysis, demonstrate how cover crops can increase overall microbial activities in soil. Cereal rye and crimson clover also resulted in significantly higher microbial activity, as measured by FDA hydrolysis, in soil compared to no cover crop treatments in both soybean (Reddy et al. 2003) and cotton production systems (Locke et al. 2013). Ryegrass cover also resulted in significantly higher microbial activities (FDA hydrolysis and aryl acylamidase) as well as more rapid degradation of fluometuron in soils (Zablotowicz et al. 2007). Such stimulation of microbial activities in soil can also lead to more rapid degradation of pesticides applied to fields. Locke et al. (2005) found fluometuron degradation was more rapid in plots with ryegrass as a winter cover compared to non-cover treatments.

With substrate specific activities, results can be more variable. Oilseed radish, cereal rye, and buckwheat have been associated with increased β-glucosidase activity, although this response is not consistent across different study sites (Fernandez et al. 2016). Hairy vetch has been documented to either increase (Mbuthia et al. 2015) or have no significant effect on β-glucosidase activities (Fernandez et al. 2016). Rye cover crops have been associated with higher N-acetylglucosaminidase (NAGase) activity, but only at one time point and sampling site, stressing the strong effect that differences in soil physiochemical parameters between field sites have on these functions (Fernandez et al. 2016). In regards to specific nutrient cycling processes, nitrification potential, nitrogen mineralization, and denitrification (Steenwerth and Belina 2008a), as well as carbon dioxide efflux (Steenwerth and Belina 2008b) were all higher in Trios and cereal rye cover crop treatments compared to control plots using traditional cultivation methods, demonstrating how cover crops can influence carbon and nitrogen dynamics in agricultural ecosystems.

Effects of Herbicides

Weed management using herbicides has increased over the last several decades due to their efficiency and lower cost compared to other weed control methods as well as the recent introduction of genetically modified herbicide-resistant crops (Green 2014). Herbicides can end up in soil through direct application, although without incorporation into the soil, impacts may be small and limited to surface residues. Herbicides applied to plant foliage may enter the soil through wash off from leaves (Reddy et al. 1995, Reddy and Locke 1996) or as exudates from roots into the rhizosphere (Neumann et al. 2006, Coupland and Caseley 1979, Laitinen et al. 2007). The impact of herbicide application on soil health can vary greatly depending on the chemical used, and a variety of other factors, including but not limited to soil texture, pH, organic matter, and dominant plant species present. Soil microorganisms are a major contributor to soil functions and nutrient cycling, and how they react to herbicide application can have a major impact on ecosystem function. As such, this section will focus on effects of herbicide application on soil microbial communities, including effects on microbial biomass, microbially-mediated processes, microbial community composition, and implications for plant disease. The impact of transgenic crops and degradation of herbicides in soil will also be discussed.

Impact of Herbicides on Soil Biological Parameters

Microbial Biomass

Not all bacteria respond to herbicides in the same way. Some are inhibited while others are stimulated. However, microbial communities present in agricultural soils are highly diverse, and it is important to consider how the entire soil community responds to herbicide application. Microbial biomass is a measure of the amount of live microorganisms in soil and can be used to indicate soil health. Even with microbial biomass, the impact of herbicide application can be variable depending on study conditions. In the case of glyphosate, some studies have reported increases in microbial biomass upon glyphosate application (Gupta and Joshi 2009, Haney et al. 2002) while several other studies have found it does not tend to impact microbial biomass, even when applied at greater than label rates (Haney et al. 2000, Lancaster et al. 2006, Liphadzi et al. 2005). In contrast to these reports, microbial biomass in soybean rhizospheres decreased after glyphosate application in soils that had no prior exposure to the herbicide (Lane et al. 2012) as well as in bulk soil from fields under continuous soybean production (Gomez et al. 2009).

Possible reasons for variability in soil microbial biomass response to glyphosate are differences in soil conditions, exposure concentration, and exposure time between experiments. In addition, glyphosate can have an inhibitory effect on some soil bacteria and fungi through the same mechanism as in plants, the inhibition of 5-enolpyruvylshikimate-3-phosphate (EPSP) synthase in the shikimate pathway (Fischer et al. 1986, Moorman et al. 1992). Using meta-analysis and modeling, Nguyen et al. (2016) found that, during short exposure times (less than 100 days), low pH (less than 5.5), and high glyphosate concentration (greater than 200 mg per kg), microbial biomass tended to increase, while it tended to decrease under long term exposure in soil with a neutral pH and lower exposure concentrations. However, it should be noted that application rates similar to those used in field crops do not significantly impact microbial biomass (Nguyen et al. 2016).

Given the increased use of glyphosate following the introduction of Roundup™, fewer studies have examined the impacts of other herbicides on microbial biomass in soil in recent years. Results of older studies are discussed elsewhere (Moorman 1989, Wardle 1992), while we will focus on studies conducted in the last 20 years. Lupwayi et al. (2004) found different herbicides vary in their effect on microbial biomass with metribuzin, imazamox/imazethapyr, triasulfuron, and metsulfuron lowering MBC compared to glufosinate and sethoxydim. Gupta and Joshi (2009) found that 2,4-D application decreased levels of MBC and MBP in soils as exposure concentrations increased. García-Orenes et al. (2010) found that MBC was lower in oxyfluorfen treated fields compared to controls, while paraquat had no effect. As has been observed with glyphosate, multiple studies found atrazine had no consistent or substantial impact on MBC in treated soils (Ghani et al. 1996, Mahía et al. 2008). However, Ros et al. (2006) found higher application rates of 100 and 1000 mg kg $^{-1}$ atrazine enhanced bacterial numbers in soil, likely due to their ability to utilize it as a nutrient source. In contrast, both rimsulfuron and imazethapyr had minimal effects when applied at field rates but significantly decreased MBC and MBN when applied at ten times the recommended rate (Perucci et al. 2000). Similarly, Sofo et al. (2012) found higher application rates of the triazinyl-sulfonylurea herbicides, cinosulfuron, prosulfuron, thifensulfuron methyl, and triasulfuron, increased their negative effects on MBC and MBN. Organic soil amendments have been demonstrated to enhance MBC in herbicide treated soils, and such measures could help minimize negative effects and maintain soil health (Singh and Ghoshal 2010).

Microbially Mediated Processes Following Herbicide Application

Soil organisms play a pivotal role in the cycling of nutrients, such as carbon, nitrogen, and phosphorus, in agricultural ecosystems. As such, it is important to understand how herbicide application influences microbial activities and behavior in soil. A number of different approaches have been used to assess microbial activities and nutrient metabolism in soils, from microbial respiration for determining broad impacts on the overall microbial community, to enzyme assays for specific activities and functional diversity assessment, as well as measurements of different processes in the nitrogen cycle. We will discuss some of the more commonly used approaches below.

Microbial Respiration

Several herbicides have been found to alter microbial respiration. The most notable effects are seen in response to high herbicide application rates. Numerous studies have found glyphosate stimulates soil respiration when applied at levels above the recommended field application rate, most likely due to metabolism of glyphosate by soil microbes (Wardle and Parkinson 1990, Zabaloy and Gómez 2008, Araújo et al. 2003, Partoazar et al. 2011). In fact, meta-analyses of data from multiple studies have demonstrated glyphosate concentration has a strong influence on soil microbial respiration (Nguyen et al. 2016). When it comes to lower applications rates as would be seen in the field, any impact of glyphosate on soil respiration appears to be transient (Zabaloy and Gómez 2008, Zabaloy et al. 2008). High rates of 2,4-D application have also been shown to stimulate microbial respiration, while the impact of lower rates is small and transient (Zabaloy and Gómez 2008, Wardle and Parkinson 1990).

When looking at substrate induced respiration, high rates of glyphosate (Wardle and Parkinson 1990) and mesotrione (Crouzet et al. 2010) have also been found to have a stimulatory effect, while no significant impacts were observed when herbicides were applied at recommended rates. In contrast, 2,4-D has an inhibitory effect, but only transiently and at high doses above the recommended rate (Zabaloy et al. 2008, Wardle and Parkinson 1990). When evaluating the environmental impact of a herbicide on ecosystem functions, it is important to consider the relative magnitude of these effects compared to other herbicides and weed control alternatives. Evidence presented thus far has demonstrated that glyphosate can transiently impact microbial respiration at high enough doses. However, Liphadzi et al. (2005) found glyphosate did not alter substrate induced respiration or substrate utilization profiles when compared to the 'conventional' herbicide treatments it has replaced, including mixtures of S-metolachlor and sulfentrazone in soybean and alachlor plus atrazine in corn.

Soil Enzyme Activities

Overall microbial activity has also been assessed through the use of various enzyme assays.

Dehydrogenase is a measure of overall microbial activity, plays a role in the oxidation of soil organic matter (Wolińska and Stępniewska 2012), and is frequently used to assess soil health. The impact of herbicide application on dehydrogenase activity is highly variable with inconsistencies between studies. A number of studies have noted a stimulatory effect on dehydrogenase in response to high rates of herbicide application, including atrazine (Moreno et al. 2007), 2,4-D (Zabaloy et al. 2008), and glyphosate (Partoazar et al. 2011). In contrast, Sebiomo et al. (2011) noted an inhibitory effect by atrazine, glyphosate, and paraquat.

Moreno et al. (2007) found that high application rates of atrazine stimulated dehydrogenase activity, while Sebiomo et al. (2011) noted an inhibitory effect. The disparities between these studies might be due to differences in experimental approach, as Moreno et al. (2007) followed activity after spiking the soil with atrazine, while Sebiomo et al. (2011) performed multiple applications at field rates over the course of several weeks. Using another assay for microbial activity, FDA hydrolysis, Araújo et al. (2003) found both short- and long-term glyphosate exposure resulted in higher activities compared to untreated soils, while Jenkins et al. (2017) found glyphosate had no impact on FDA hydrolysis in soil from either conventional or no-till systems. However, at three times the recommended application rate, glyphosate can have an inhibitory effect on FDA hydrolysis (Weaver et al. 2007).

More substrate specific enzyme assays have been used to assess how herbicide application may alter certain nutrient cycling processes, although as they are performed under laboratory conditions with artificial substrates, they are generally considered more of an indicator of potential rather than a direct measure of nutrient cycling processes in soil. Elucidating impacts of herbicide application on nutrient cycling can be complex, as different herbicides can have different effects on soil activities, and different enzymes activities might respond differently to the same herbicide. In many instances, herbicide application can have an inhibitory impact on many microbial activities associated with nutrient cycling. Acid phosphatase in soil can be inhibited by 2,4-D (Bécaert et al. 2006, Oleszczuk et al. 2014, Gupta and Joshi 2009), and dicamba (Oleszczuk et al. 2014). The impact of glyphosate on acid phosphatase activities is more variable, with some studies reporting an inhibitory effect (Sannino and Gianfreda 2001, Yu et al. 2011), while others found no significant impact (Cherni et al. 2015, Nakatani et al. 2014, Gupta and Joshi 2009, Jenkins et al. 2017). The response of alkaline phosphatase activities to herbicide application is also variable. Dicamba is reported to enhance alkaline phosphatase activity (Oleszczuk et al. 2014), and Gupta and Joshi (2009) found that glyphosate has no significant impact. In the case of 2,4-D, contradictory reports indicate this herbicide can inhibit (Bécaert et al. 2006), enhance (Oleszczuk et al. 2014), or have no effect (Gupta and Joshi 2009) on alkaline phosphatase activities in soil.

Functional Diversity

Functional diversity can be described as the range of functions carried out by organisms in a given ecosystem (Petchey and Gaston 2006). Given the wide array of activities carried out in soil and the complex interactions involved in nutrient cycling, understanding the impacts of herbicides on functional processes in soil can be a challenge. Some studies use high throughput systems to determine a variety of different substrate utilization profiles in order to assess functional diversity, rather than assaying individual enzyme activities in soils. As with soil respiration, a dosage effect is common, where herbicides applied at field rates have little or no effect, while higher application rates do impact functional diversity. Higher rates of atrazine application (100 and 1000 mg kg^{-1}) were found to decrease functional diversity in soils when compared to lower rates, with significant impacts noted on activities of carbohydrate and amine breakdown (Ros et al. 2006). Glyphosate can also have an inhibitory effect on functional diversity, with two applications of glyphosate reducing the functional diversity of the microbial community in bulk soils when applied to canola (Lupwayi et al. 2009). In contrast, glyphosate can stimulate functional diversity 15 days after application to soils with triticale and pea, but there is no consistent effect on microbial activities after 30 days (Mijangos et al. 2009). Zabaloy et al. (2008) noted a transient increase in functional diversity, but at a rate of 100 times the application rate of glyphosate. The caveat of these functional diversity approaches is that many of them (Biolog plates) are culture dependent and involve the separation of bacterial cells from soil particles. Thus, they do not represent conditions seen in soil and also might underrepresent activities of bacterial community members that cannot grow under laboratory conditions.

Nitrogen Cycle

Soil bacteria play a crucial role in nitrogen cycling in the soil ecosystem, from nitrification (i.e., bacteria converting soil nitrogen to forms available for plant uptake) to denitrification (i.e., converting soil nitrates to atmospheric nitrogen). Glyphosate has been found to inhibit nitrification at higher application rates (Carlisle and Trevors 1986). Similarly, Tu (1994) reported that 2,4-D, dicamba, and glyphosate significantly inhibited nitrification in sandy soils, though only for a period of two weeks. Martens and Bremner (1993) found amitrole, chlorpropham, 2,4-D amine, dinoseb, propham, and propanil reduced nitrification rates in coarse soil when applied at a rate of 5 mg kg^{-1} of soil. Olson and Lindwall (1991) also demonstrated that 2,4-D and glyphosate could lower nitrification rates when applied at twice and ten times the recommended rate. However, 2,4-D (Olson and Lindwall 1991), dicamba (Ratnayake and Audus 1978), and glyphosate (Olson and Lindwall 1991) do not appear to inhibit nitrification in soils at normal field concentrations.

Denitrification in soils responds similarly to herbicide application as nitrification, with responses mainly seen at higher concentrations. Yeomans and Bremner (1985) found neither atrazine, dicamba, 2,4-D amine, or 2,4-D ester altered denitrification rates when applied at 10 μg g^{-1}. In the case of glyphosate, Carlisle and Trevors (1986) reported no impact on nitrous oxide reduction in unamended soils, but higher concentrations inhibited denitrification. This study also found that glyphosate did not impact nitrogen fixation at field rates and that stimulation of nitrate reduction in soil was dose-dependent, except at low levels where no response to glyphosate was observed. The conclusion was that glyphosate did not impact nitrogen cycling activities in the soil at realistic field application rates.

Microbial Community Composition in Herbicide Treated Soils

Given the variability in the response of different species and strains of soil microorganisms, herbicide application has the potential to alter community composition, thereby altering key ecosystem functions carried out in the soil. One such function includes phosphate cycling and phosphorus availability to plants. Correlations have been observed between rhizosphere community structure and phosphorus accumulation in plant tissue (George et al. 2009). In regards to manganese availability, glyphosate application decreases the manganese reducer: manganese oxidizer ratio in the soybean rhizosphere (Zobiole et al. 2010, Kremer and Means 2009). Zobiole et al. (2010) suggested this shift could be due to suppression of select bacterial groups by glyphosate (e.g., *Pseudomonas* spp.) or changes in carbon substrates secreted in root exudates of treated plants. As such, it is important to be aware of the potential effects of herbicide use on soil communities in agricultural fields that could potentially impact nutrient cycling or other ecosystem services. As there is an extensive array of herbicides available in the market, we will focus on a selection of commonly used herbicides for pre-emergent (atrazine) and post-emergent (glyphosate, 2,4-D, and dicamba) applications. These herbicides are used across a spectrum of crops and each has a solid history of research to draw upon in relation to other herbicides on the market.

Glyphosate can have toxic effects on some soil bacteria, with varying degrees of susceptibility depending on species and strain (Moorman et al. 1992, Fischer et al. 1986, Zabaloy and Gómez 2005). Glyphosate sensitivity in soil fungi is also variable (Tanney and Hutchison 2010). Given its potential for inhibiting soil bacteria, the impacts of glyphosate on microbial community composition has been the subject of many studies.

Glyphosate's effect on microbial communities in soil varies, with some studies reporting shifts in community composition while others report no impact. In bulk soil, Zabaloy et al. (2016) reported that a high rate of glyphosate, did not alter overall bacterial or ammonia-oxidizing bacterial community structure. In contrast, other studies have observed shifts in the microbial community associated with glyphosate application and cropping systems. Locke et al. (2008) found that five years of growing glyphosate resistant crops resulted in statistically significant shifts in microbial community composition compared to fields with non-glyphosate resistant crops, although differences were small in magnitude. Cherni et al. (2015) found glyphosate increased total bacterial and Rhizobiaceae community diversity, with an increased impact observed at ten times the application rate. Lancaster et al. (2010) found that repeated application of glyphosate increased the levels of Proteobacteria. In contrast, Acidobacteria and Firmicutes decreased in abundance following both one and five applications. The response of Actinobacteria was more variable, with abundance increasing after one application, but decreasing after five applications (Lancaster et al. 2010). This variability within Actinobacteria response is evident between other studies. Araújo et al. (2003) found glyphosate application increased levels of actinomycetes, while Sebiomo et al. (2011) found it lowered the levels of actinomycetes in soils. These two studies also reported conflicting results with regard to fungal abundance in soils. Potential contributors to these differences might include site history and location (i.e., cassava farm in Nigeria with no pesticide application history versus peach orchards in Brazil with or without glyphosate application history).

Glyphosate is secreted through root exudates of treated plants, therefore, microorganisms in the rhizosphere might be more affected by its application than those in bulk soil. Given the important roles that rhizosphere bacteria play in nutrient acquisition, plant growth promotion, and disease development, any potential impacts of herbicide application on rhizosphere bacteria could impact the ecosystem in cropping systems. Studies have found glyphosate has no consistent or significant effect on microbial community composition in the rhizospheres of corn (Hart et al. 2009), cotton (Barriuso and Mellado 2012), pea+triticale (Mijangos et al. 2009), and soybean (Lane et al. 2012, Weaver et al. 2007). The only differences observed in these studies were the result of applications above recommended rates (Mijangos et al. 2009, Weaver et al. 2007) or differences in between sampling sites (Lane et al. 2012). In contrast, Barriuso et al. (2010) reported glyphosate altered community composition in the corn rhizosphere by decreasing the relative abundance of Actinobacteria and Grammaproteobacteria in untreated compared to glyphosate treated soils, while increasing the abundances of Beta- and Delta-proteobacteria. However, for long-term (e.g., three years) studies, with repeated glyphosate applications, no significant impact on community composition in the corn rhizosphere was observed in treated compared to untreated fields (Barriuso et al. 2011).

Even small changes in microbial community composition in the rhizosphere can impact ecosystem function, depending on which members of the microbial community are impacted. Only modest perturbation of rhizosphere community composition occurred with glyphosate treatment of susceptible and resistant biotypes of giant ragweed (*Ambrosia trifida*) (Schafer et al. 2014). Examination of specific taxa indicate that plant growth promoting *Burkholderia* increase following glyphosate application of both cotton (Schafer et al. 2014) and ragweed (Lancaster et al. 2010). Arango et al. (2014) found bacterial diversity in the soybean rhizosphere was also resilient to glyphosate application. However, in contrast to the above studies, glyphosate application resulted in a decrease in *Burkholderia* sp. abundance in the soybean rhizosphere. Such variability in the response of specific bacterial groups among studies might be attributable to species-specific differences.

Other plant growth promoting bacteria can also be influenced by glyphosate. Glyphosate decreased the levels of auxin producing bacteria inhabiting the soybean rhizosphere (Zobiole et al. 2010). As auxins are plant hormones commonly produced in many plant growth promoting rhizobacteria, it has been suggested that inhibition of auxin producing bacteria in the rhizosphere might lower the growth promoting effect of these bacteria, possibly contributing to a subsequent reduction in plant biomass observed in glyphosate treated crops. Decreases in auxin production may be one explanation for the decrease in soybean root mass resulting from increased concentrations of glyphosate application, with smaller root systems potentially decreasing the plant's ability to take up nutrients and making it more susceptible to infection (Zobiole et al. 2010). Glyphosate application can also influence beneficial fungi, since its application reduces root colonization of *Lolium multiflorum* Lam and spore viability of arbuscular mycorrhizal fungi (Druille et al. 2013).

Given the widespread use of glyphosate due to Round-Up Ready™ crops, there is relatively less information on the impacts of other commonly used herbicides. In the case of 2,4-D, beneficial soil bacteria including *Azospirillum brasilense* Cd (Rivarola et al. 1992) and some strains of *Rhizobium* sp. (Fabra et al. 1997) are sensitive to 2,4-D inhibition. Several studies have noted significant impacts of 2,4-D at high application rates, while lower rates have little to no effect on soil community composition. Application of 2,4-D to soil can significantly reduce the diversity of dominant bacteria (Macur et al. 2007), as well as decrease total bacterial abundance (Durga Devi et al. 2008, Zhang et al. 2010). Meanwhile, fungal abundance in soil increases following 2,4-D treatment (Durga Devi et al. 2008, Zhang et al. 2010). At lower concentrations, 2,4-D has been shown to increase total bacteria and actinobacteria in soil (Zhang et al. 2010). In addition, the ratio of Gram positive to Gram negative bacteria decreases with increasing 2,4-D application rates (Zhang et al. 2010). Kraiser et al. (2013) reported 2,4-D application resulted in significant changes in the rhizosphere community and increased 2,4-D degrading bacteria in the rhizosphere. The effects of 2,4-D on microbial community composition appear to be temporary, with populations returning to normal seven (Gonod et al. 2006) and 30 days (Durga Devi et al. 2008) after application. This shift in microbial communities is correlated with increased 2,4-D degradation (Gonod et al. 2006), with recovery to normal corresponding to a point when 2,4-D is no longer detectable in soils (Durga Devi et al. 2008).

Toxicity of bacteria to dicamba appears to be variable. Zabaloy and Gómez (2005) found that dicamba sensitivity of 76 different rhizobial isolates varied significantly. Ratnayake and Audus (1978) reported differing levels of dicamba toxicity against two nitrifying soil bacteria, *Nitrobacter wingradskii* and *Nitrosomonas eurpeae*. Dicamba has also been found to inhibit bacterial growth and nitrogen fixation in the common soil bacterium, *Azotobacter vinelandii* (Gonzalez-Lopez et al. 1984). However, it should be noted that many of these studies on bacterial toxicity were conducted on pure cultures, which might not reflect how bacteria will respond in soil and under natural conditions. While correlations between dicamba inhibition in soil vs. pure culture were seen with *Nitrobacter*, no correlation was observed with *Nitrosomonas* in pure culture and soil (Ratnayake and Audus 1978). When examining microbial community composition in soil, Ławniczak et al. (2016) found both herbicidal ionic liquid and commercial formulations of dicamba increased the abundance of Proteobacteria, Actinitobacteria, and Bacteriodetes, while Firmicutes abundance decreased. Given that applications of dicamba resulted in changes in the plant community present at sampling sites, differences in bacterial community composition between treated and untreated soil cannot solely be attributed to dicamba; the presence or absence of a specific weed species was a potential confounding factor (Ławniczak et al. 2016).

Atrazine's impact on soil community composition is similar to that seen with the other herbicides. Sebiomo et al. (2011) found atrazine decreased the numbers of culturable bacteria, actinomycetes, and fungi in soils. Ros et al. (2006) observed a dosage effect, where higher atrazine concentrations of 10, 100, and 100 mg kg^{-1} altered microbial community composition when compared to untreated controls, while the effects of a 1 mg kg^{-1} treatment were minimal. Much of the research on atrazine's impact on soil communities has focused on the bacteria that can degrade this herbicide. In fact, atrazine is a prime example of how microbial communities can respond to repeated application of herbicides and adapt to rapidly degrade it. Fang et al. (2015) found microbial community shifts indicating such adaptations; the relative abundance of atrazine degrading genera (including *Nocardioides*, *Arthrobacter*, *Methylobacterium*, and *Bradyrhizobium*) with genes for the complete atrazine degradation pathway increased with increasing atrazine application frequency.

Overall, the common theme among all assessments of microbial community composition is the small, transitory nature of a herbicide's effects when applied at recommended field rates.

Influence on Plant Disease Development

Herbicide applications can directly impact plant pathogens that inhabit the soil (Sanyal and Shrestha 2008). Mekwatanakarn and Sivasithamparam (1987) found that herbicide applications (diquat + paraquat, glyphosate, or trifluralin) increased the incidence of Take-all disease in wheat, which they attributed to shifts observed in fungal and bacterial components of the microbial community. Indeed, subsequent research indicated shifts in community composition resulting from herbicide application, either through direct toxicity or alteration of root exudates, could have an impact on the pathogen load in soil and potential for disease development in crops. Kremer et al. (2005) found that exudates from glyphosate treated plants significantly increased the growth of two of four strains of *Fusarium* assayed. This study also demonstrated that some strains of *Fusarium* could utilize glyphosate as a nutrient source (Kremer et al. 2005). In a later study, Kremer and Means (2009) found glyphosate-resistant (GR) soybeans treated with glyphosate had higher levels of *Fusarium* spp. colonization of the root system. In addition to stimulating fungal growth, GR soybeans and glyphosate application also decreased levels of fluorescent Pseudomonads that serve as antagonists to fungal pathogens in soil (Kremer and Means 2009). A similar trend of higher *Fusarium* root colonization and lower Pseudomonad population was observed in the rhizosphere of glyphosate-treated plants (Zobiole et al. 2010). However, this increase in *Fusarium* is not universal and does not always result in higher colonization. Njiti et al. (2003) found that *Fusarium* colonization and sudden death syndrome (SDS) did not significantly increase following Roundup™ application in soybean, although they noted differences in cultivars with varying susceptibility to SDS. Liu et al. (1991) reported that changes in root exudates following glyphosate treatment could stimulate the germination and growth of *Pythium* spp. pathogens, possibly contributing to increases in infection seen in treated plants. Similarly, Meriles et al. (2006) found levels of *Fusarium* and *Pythium* in soil increased in proportion to glyphosate application rate. Alternatively, some herbicides can act against some plant pathogens, lowering crop disease. In one such instance, glyphosate exhibited anti-fungal activity against *Phakopsora pachyrhizi*, the causative agent of Asian soybean rust, and delayed onset of this disease in field trials (Feng et al. 2008). However, the benefit of glyphosate against natural infections was variable (Feng et al. 2008).

Weed residues left behind following herbicide application can also impact soil ecology, microbial community composition, and disease development. Cereal crops planted 1-3 d after a herbicide application experienced an increase in infection with *Pythium*. This was attributed to dead weed tissue that served to enhance the plant pathogen (Pittaway 1995). Descalzo et al. (1998) also reported dead root tissue from glyphosate application resulted in short-term increase of *Phythium* in soil.

Impact of Soil Characteristics

Trends reported for inhibition or stimulation of microbial activity at high application rates and a lack of significant responses at field rates suggest specific soil conditions might influence the degree to which a herbicide application will impact soil ecosystem function. Multiple studies have reported differing responses of soil microbial communities to herbicides when applied to soils with different textures and chemical characteristics. Banks et al. (2014) found soil texture, pH, and site history influenced the microbial community's response to treatments (glyphosate, atrazine and bentazon), with larger differences between treatments noted in a silt loam soil than a silt clay loam, particularly when looking at bacterial components of the community. Soil organic matter can also influence herbicide interactions in soil, with the effects of 2,4-D application on microbial community composition being greater in soils with low organic matter content (Zhang et al. 2010). The impact of different herbicides on nutrient cycling in soil can be strongly influenced by soil texture and application rate. Martens and Bremner (1993) found that six (amitrole, chlorpropham, 2,4-D amine, dinoseb, propham, and propanil) of 28 herbicides tested reduced nitrification rates in coarse but not fine texture soils when applied at 5 mg kg^{-1}.

At higher application rates (50 mg kg^{-1} soil), the influence of soil texture was more apparent. While all 28 herbicides inhibited nitrification in at least one of four soil textures, only four (propanil, propham, amitrole, and dinoseb) inhibited nitrification in all four soil types assayed (Martens and Bremner 1993). In fact, in many reported cases, the recommended application rates of herbicides are not sufficiently high enough to inhibit enzyme activities in most soil types. For example, Dzantor and Felsot (1991) found that alachlor applied with atrazine, metolachlor, and trifluralin did not significantly impact microbial communities when applied at field rates while esterase activities were inhibited when microbes were exposed to 10,000 mg kg^{-1}.

Influence of Transgenic Crops on Soil Systems

Transgenic herbicide resistant crops can alter soil conditions and release transgenic material into soil. One non-herbicide example of this is the release of the transgenic Cry1Ab protein in the soil of Bt-corn plots (Andersen et al. 2007). It has also been suggested that herbicide resistant crops can cause changes in the microbial soil community through horizontal gene transfer of transgenic genes upon decomposition of dead crop material (Dunfield and Germida 2004), and studies have found that some of these recombinant genes, including a CP4 5-enol-pyruvyl-shikimate-3-phosphate synthase (cp4 epsps) resistant to glyphosate, can persist in soil, although gene copy numbers were higher in larger soil aggregates (>2,000 μm) which contains undegraded plant material (Levy-Booth et al. 2009). Differences due to transgenic crops alone can have an impact on the soil ecosystem. Dunfield and Germida (2003) found the microbial community in the rhizosphere of transgenic canola differed significantly from a non-transgenic variety over the growing season, although these differences were transient and were not observed in the subsequent growing season.

New Formulations

Herbicides are frequently applied in combination in order to achieve greater weed control. As such, it is important to consider the cumulative or synergistic impact of multiple herbicide formulations on soil ecosystems. New formulations containing glyphosate in combination with dicamba or 2,4-D have been recently introduced. Given these formulations are new to the market, few studies are available on the combined effects of these active ingredients. Lupwayi et al. (2009) reported that application of 2,4-D plus glyphosate resulted in increases in microbial biomass carbon compared to application of glyphosate alone but only in one year of a two-year study. When glyphosate and 2,4-D are applied together, no significant impacts on substrate induced respiration, dehydrogenase activity, or bacterial community structure were observed at either low or high concentrations (Zabaloy et al. 2016). Similarly, Nandula and Tyler (2016) reported 2,4-D plus glyphosate application had insignificant and transient effects on phosphatase, β-glucosidase, N-acetylglucosaminidase, and cellobiohydrolase activities in the soybean rhizosphere. More experiments looking at long term, repeated exposure under different conditions are needed to better understand any potential impacts that these formulations might have on soil function.

Herbicide Degradation by Soil Microorganisms

Degradation of herbicides by microorganisms in soil is known to occur with most classes of herbicides. When applied at field rates, many herbicides, including alachlor, atrazine, metolachlor, and trifluralin, are degraded within a year of application (Dzantor and Felsot 1991). Much attention has been given to soil microorganisms that can degrade herbicides for several different reasons. Degradation of herbicides in soil can reduce the period of time the herbicide is effective in the field. Additionally, enhanced herbicide degradation can be a pivotal mechanism for minimizing the amounts of herbicides transported into downstream ecosystems. Lastly, application of herbicides may promote the proliferation of soil microorganisms capable of mineralizing herbicides by utilizing them as a carbon or nitrogen source. As a result, microbial herbicide degradation could minimize the amount of carry-over residues from year to year.

2,4-D can dissipate in soil within 30 days of application in rice paddies (Durga Devi et al. 2008), and its degradation is greatly influenced by soil moisture (Cattaneo et al. 1997), with decreasing moisture resulting in decreases in 2,4-D degradation and in the number and abundance of degrading bacteria (Han and New 1994). These results show how environmental conditions can alter herbicide interactions in soil. Culturing experiments have demonstrated that soil bacteria are capable of breaking down 2,4-D (Fournier 1980). The degradation of 2,4-D is further linked to the soil microbial community by the observation that it is degraded more rapidly in soils with higher levels of microbial biomass (Voos and Groffman 1997).

Herbicide degradation in soil can have a variety of effects on the soil ecosystem. It can shorten the effective period after application (resulting in increased application frequency), as well as alleviate toxicity to non-target organisms. Degradation of 2,4-D by bacteria in the rhizosphere can mitigate toxicity against sensitive plant species (Kraiser et al. 2013). The 2,4-D degrading bacterium, *Burkholderia* sp. YK-2 can overcome this herbicide's toxicity by increasing expression of stress shock proteins, although this mechanism is not as effective with higher exposure concentrations (Cho et al. 2000). Similar mechanisms may also be utilized by other herbicide-degrading bacteria and might explain why degradation rates for herbicides are negligible at higher concentrations where soil bacteria cannot overcome the toxic effects of these chemicals.

The degradation and fate of glyphosate in soil is more thoroughly reviewed elsewhere (Duke et al. 2012, Borggaard and Gimsing 2008) and only a brief overview is described here in the context of its impact on soil ecosystems of crop fields. Overall, glyphosate appears to persist less in the environment and has a lower toxicity compared to the herbicides it has replaced; however, this benefit may be counteracted by the accumulation of the degradation product, aminomethylphosphonic acid, in soils under repeated glyphosate application (Mamy et al. 2005, Mamy et al. 2010). Glyphosate can be used as a carbon, nitrogen, or phosphorus source by soil microorganisms, causing shifts in the heterotrophic populations in soils with a history of glyphosate application (Partoazar et al. 2011, Lancaster et al. 2010). This, in turn, could alter soil microbial functions. The ability to degrade glyphosate is common in members of the Rhizobiaceae (Liu et al. 1991), potentially impacting nodulation by *Rhizobium* and *Bradyrhizobium* in leguminous crops, since strains gaining a competitive advantage due to glyphosate degradation may not be the most efficient at plant colonization or nitrogen fixation.

In the case of dicamba, bacteria can utilize this herbicide as a sole carbon source (Krueger et al. 1991). The speed at which dicamba is broken down is more rapid in soils with higher microbial biomass (Voos and Groffman 1997). Thus, its application might influence community composition and activity of plant-associated bacteria. However, little research has been done on dicamba's impact in the rhizosphere of crops.

Herbicide degradation rates in soil can be highly dependent upon site history, with prior herbicide exposure priming the soil, allowing for the competitive advantage of bacteria that can utilize them as carbon or nitrogen sources. Barriuso and Houot (1996) found atrazine degradation was higher in corn fields receiving annual applications of this herbicide than in fields with no history of atrazine treatment. Over the years, microbial soil communities have been found to adapt to repeated applications of triazine herbicides, such as atrazine, resulting in accelerated degradation rates in soil [reviewed in Krutz et al. (2010)]. The rate of atrazine degradation increases with repeated applications and increased application frequency (Fang et al. 2015).

Concluding Remarks

The soil is a complex ecosystem of living organisms interacting with nonliving components to perform a variety of ecosystem functions deemed favorable to agricultural production.

In this chapter, we have discussed the potential impacts of some common methods of weed management on the various components of the soil ecosystem. Conservation practices including reduced tillage and use of cover crops generally tend to stimulate soil biological activities and increase organic matter levels, while reduced tillage can also lead to greater soil compaction. Effects of herbicide application are more complex, and appear to vary based on local soil conditions; reports of inhibitory, stimulatory, or no effect on biological activities and community composition in soil are conflicting. More work needs to be done to determine how long-term, repeated herbicide applications, as well as herbicide use in combination with conservation methods, influence soil ecosystems. Such integrated approaches combining multiple methods is the key to minimizing any detrimental effects of weed control on soil biology and promoting healthy, functioning soils in agroecosystems.

REFERENCES

Acosta-Martínez, V., M.M. Mikha and M.F. Vigil. 2007. Microbial communities and enzyme activities in soils under alternative crop rotations compared to wheat-fallow for the Central Great Plains. Applied Soil Ecology 37(1-2): 41–52. doi: 10. 1016/j.apsoil.2007.03.009.

Acosta-Martínez, V., T.M. Zobeck, T.E. Gill and A.C. Kennedy. 2003. Enzyme activities and microbial community structure in semiarid agricultural soils. Biology and Fertility of Soils 38(4): 216–227. doi: 10.1007/s00374-003-0626-1.

Ahmed, S., C. Piggin, A. Haddad, S. Kumar, Y. Khalil and B. Geletu. 2012. Nematode and fungal diseases of food legumes under conservation cropping systems in northern Syria. Soil and Tillage Research 121: 68–73. doi: 10.1016/j.still.2012.01.19.

Al-Kaisi, M.M., X. Yin and M.A. Licht. 2005. Soil carbon and nitrogen changes as influenced by tillage and cropping systems in some Iowa soils. Agriculture, Ecosystems & Environment 105(4): 635–647. doi: 10.1016/j. agee.2004.08.002.

Alletto, L.,Y. Coquet, P. Benoit, D. Heddadj and E. Barriuso. 2010. Tillage management effects on pesticide fate in soils. A review. Agronomy for Sustainable Development 30(2): 367–400. doi: 10.1051/agro/2009018.

Almeida, Á.M.R., L. Amorim, A. Bergamin Filho, E. Torres, J.R.B. Farias, L.C. Benato, M.C. Pinto and N. Valentim. 2003. Progress of soybean charcoal rot under tillage and no-tillage systems in Brazil. Fitopatologia Brasileira 28: 131–135.

Andersen, M.N., C. Sausse, B. Lacroix, S. Caul and A. Messéan. 2007. Agricultural studies of GM maize and the field experimental infrastructure of ECOGEN. Pedobiologia 51(3): 175–184. doi: 10.1016/j.pedobi.2007.03.005.

Arango, L., K. Buddrus-Schiemann, K. Opelt, T. Lueders, F. Haesler, M. Schmid, D. Ernst and A. Hartmann. 2014. Effects of glyphosate on the bacterial community associated with roots of transgenic Roundup Ready® soybean. European Journal of Soil Biology 63: 41–48

Araújo, A.S.F., R.T.R. Monteiro and R.B. Abarkeli. 2003. Effect of glyphosate on the microbial activity of two Brazilian soils. Chemosphere 52(5): 799–804. doi:10.1016/S0045-6535(03)00266-2.

Babujia, L.C., A.P. Silva, A.S. Nakatani, M.E. Cantão, A.T.R. Vasconcelos, J.V. Visentainer and M. Hungria. 2016. Impact of long-term cropping of glyphosate-resistant transgenic soybean [Glycine max (L.) Merr.] on soil microbiome. Transgenic Res. 25(4): 425–440. doi: 10.1007/s11248-016-9938-4.

Banks, M.L., A.C. Kennedy, R.J. Kremer and F. Eivazi. 2014. Soil microbial community response to surfactants and herbicides in two soils. Applied Soil Ecology 74: 12–20. doi: 10.1016/j. apsoil.2013.08.018.

Barriuso, E. and S. Houot. 1996. Rapid mineralization of the s-triazine ring of atrazine in soils in relation to soil management. Soil Biology and Biochemistry 28(10-11): 1341–1348.

Barriuso, J., S. Marin and R.P. Mellado. 2010. Effect of the herbicide glyphosate on glyphosate-tolerant maize rhizobacterial communities: a comparison with pre-emergency applied herbicide consisting of a combination of acetochlor and terbuthylazine. Environ Microbiol. 12(4): 1021–1030. doi: 10.1111/j.1462-2920.2009.02146.x.

Barriuso, J., S. Marin and R.P. Mellado. 2011. Potential accumulative effect of the herbicide glyphosate on glyphosate-tolerant maize rhizobacterial communities over a three-year cultivation period. PLoS One 6(11): e27558. doi: 10.1371/journal. pone.0027558.

Barriuso, J. and R.P. Mellado. 2012. Glyphosate affects the rhizobacterial communities in glyphosate-tolerant cotton. Applied Soil Ecology 55: 20–26. doi: 10.1016/j. apsoil. 2011.12. 010.

Barthès, B. and E. Roose. 2002. Aggregate stability as an indicator of soil susceptibility to runoff and erosion, validation at several levels. CATENA 47(2): 133–149. doi: 10.1016/S0341-8162(01)00180-1.

Bécaert, V., R. Samson and L. Deschênes. 2006. Effect of 2, 4-D contamination on soil functional stability evaluated using the relative soil stability index (RSSI). Chemosphere 64(10): 1713–1721. doi: 10.1016/j. chemosphere.2006.01.008.

Benjamin, J.G., M. Mikha, D.C. Nielsen, M.F. Vigil, F. Calderón and W.B. Henry. 2007. Cropping intensity effects on physical properties of a no-till silt loam. Soil Science Society of America Journal 71(4): 1160–1165. doi: 10.2136/sssaj2006.0363.

Blanco-Canqui, B., M.M. Mikha, J.G. Benjamin, L.R. Stone, A.J. Schlegel, D.J. Lyon, M.F. Vigil and P.W. Stahlman. 2009a. Regional study of no-till impacts on near-surface aggregate properties that influence soil erodibility. Soil Science Society of America Journal 73(4): 1361–1368. doi: 10.2136/sssaj2008. 0401.

Blanco-Canqui, H., M.M. Claassen and L.R. Stone. 2010. Controlled traffic impacts on physical and hydraulic properties in an intensively cropped no-till soil. Soil Science Society of America Journal 74(6): 2142–2150. doi:10.2136/sssaj2010.0061.

Blanco-Canqui, H. and R. Lal. 2008. No-tillage and soil-profile carbon sequestration: an on-farm assessment. Soil Science Society of America Journal 72(3): 693–701. doi:10. 2136/sssaj2007.0233.

Blanco-Canqui, H., L.R. Stone, A.J. Schlegel, D.J. Lyon, M.F. Vigil, M.M. Mikha, P.W. Stahlman and C.W. Rice. 2009b. No-till induced increase in organic carbon reduces maximum bulk density of soils. Soil Science Society of America Journal 73(6): 1871-1879. doi: 10.2136/sssaj2008.0353.

Blouin, M., M.E. Hodson, E.A. Delgado, G. Baker, L. Brussaard, K.R. Butt, J. Dai, L. Dendooven, G. Peres, J.E. Tondoh, D. Cluzeau and J.J. Brun. 2013. A review of earthworm impact on soil function and ecosystem services. European Journal of Soil Science 64(2): 161–182. doi: 10.1111/ejss.12025.

Borggaard, O.K. and A.L. Gimsing. 2008. Fate of glyphosate in soil and the possibility of leaching to ground and surface waters: a review. Pest Management Science 64(4): 441–456. doi: 10.1002/ps.1512.

Carlisle, S.M. and J.T Trevors. 1986. Effect of the herbicide glyphosate on nitrification, denitrification, and acetylene reduction in soil. Water, Air, and Soil Pollution 29(2): 189–203. doi:10.1007/bf00208408.

Castellanos-Navarrete, A., C. Rodríguez-Aragonés, R.G.M. de Goede, M.J. Kooistra, K.D. Sayre, L. Brussaard and M.M. Pulleman. 2012. Earthworm activity and soil structural changes under conservation agriculture in central Mexico. Soil and Tillage Research 123: 61–70. doi: 10.1016/j. still.2012.03.011.

Cattaneo, M.V., C. Masson, and C.W. Greer. 1997. The influence of moisture on microbial transport, survival and 2,4-D biodegradation with a genetically marked *Burkholderia cepacia* in unsaturated soil columns. Biodegradation 8(2): 87–96. doi: 10.1023/A:1008236401342.

Chan, K.Y. 2001. An overview of some tillage impacts on earthworm population abundance and diversity—implications for functioning in soils. Soil and Tillage Research 57(4): 179–191. doi: 10.1016/S0167-1987(00)00173-2.

Cherni, A.E., D. Trabelsi, S. Chebil, F. Barhoumi, I.D. Rodríguez-Llorente and K. Zribi. 2015. Effect of glyphosate on enzymatic activities, rhizobiaceae and total bacterial communities in an agricultural tunisian soil. Water, Air, and Soil Pollution 226(5): 1–11. doi: 10.1007/s11270-014-2263-8.

Cho, Y.S., S.H. Park, C.K. Kim and K.H. Oh. 2000. Induction of stress shock proteins DnaK and GroEL by phenoxyherbicide 2,4-D in *Burkholderia* sp. YK-2 isolated from rice field. Current Microbiology 41(1): 33–38.

Coupland, D. and J.C. Caseley. 1979. Presence of ^{14}C activity in root exudates and guttation fluid from *Agropyron repens* treated with ^{14}C-labelled glyphosate. New Phytol. 83(1): 17–22. doi: 10.1111/j.1469-8137.1979.tb00721.x.

Crittenden, S.J., T. Eswaramurthy, R.G.M. de Goede, L. Brussaard and M.M. Pulleman. 2014. Effect of tillage on earthworms over short- and medium-term in conventional and organic farming. Applied Soil Ecology 83: 140–148. doi: 10.1016/j.apsoil.2014.03.001.

Crouzet, O., I. Batisson, P. Besse-Hoggan, F. Bonnemoy, C. Bardot, F. Poly, J. Bohatier and C. Mallet. 2010. Response of soil microbial communities to the herbicide mesotrione: A dose-effect microcosm approach. Soil Biology and Biochemistry 42(2): 193–202. doi: 10.1016/j.soilbio.2009.10.016.

Dao, T.H. 1996. Tillage system and crop residue effects on surface compaction of a Paleustoll. Agronomy Journal 88(2): 141–148. doi: 10.2134/agronj1996.00021962008800020005x.

Descalzo, R.C., Z.K. Punja, C.A. Lévesque and J.E. Rahe. 1998. Glyphosate treatment of bean seedlings causes short-term increases in Pythium populations and damping off potential in soils. Applied Soil Ecology 8(1-3): 25–33.

Ding, G., J. Novak, D. Amarasiriwardena, P. Hunt and B. Xing. 2002. Soil organic matter characteristics as affected by tillage management. Soil Science Society of America Journal 66(2): 421–429.

Dominati, E., M. Patterson and A. Mackay. 2010. A framework for classifying and quantifying the natural capital and ecosystem services of soils. Ecological Economics 69(9): 1858–1868. doi: 10.1016/j.ecolecon.2010.05.002.

Doran, G., P. Eberbach and S. Helliwell. 2006. The sorption and degradation of the rice pesticides fipronil and thiobencarb on two Australian rice soils. Aust J Soil Res. 44(6): 599–610. doi: 10.1071/sr05173.

Doube, B.M., J.C. Buckerfield and J.A. Kirkegaard. 1994. Short-term effects of tillage and stubble management on earthworm populations in cropping systems in southern New South Wales. Australian Journal of Agricultural Research 45(7): 1587–1600.

Drijber, R.A., J.W. Doran, A.M. Parkhurst and D.J. Lyon. 2000. Changes in soil microbial community structure with tillage under long-term wheat-fallow management. Soil Biology and Biochemistry 32(10): 1419–1430. doi: 10.1016/S0038-0717(00)00060-2.

Druille, M., M.N. Cabello, M. Omacini and R.A. Golluscio. 2013. Glyphosate reduces spore viability and root colonization of arbuscular mycorrhizal fungi. Applied Soil Ecology 64: 99–103. doi: 10.1016/j.apsoil.2012.10.007.

Duke, S.O., J. Lydon, W.C. Koskinen, T.B. Moorman, R.L. Chaney and R. Hammerschmidt. 2012. Glyphosate effects on plant mineral nutrition, crop rhizosphere microbiota, and plant disease in glyphosate-resistant crops. Journal of Agricultural and Food Chemistry 60(42): 10375–10397. doi: 10.1021/jf302436u.

Dunfield, K.E. and J.J. Germida. 2003. Seasonal changes in the rhizosphere microbial communities associated with field-grown genetically modified canola (*Brassica napus*). Applied and Environmental Microbiology 69(12): 7310–7318. doi: 10.1128/AEM.69.12.7310-7318.2003.

Dunfield, K.E. and J.J. Germida. 2004. Impact of genetically modified crops on soil- and plant-associated microbial communities. Journal of Environmental Quality 33(3): 806–815.

Durga Devi, K.M., S. Beena and C.T. Abraham. 2008. Effect of 2,4-D residues on soil microflora. J Trop Agric. 46(1-2): 64–66.

Dzantor, E.K. and A.S. Felsot. 1991. Microbial responses to large concentrations of herbicides in soil. Environmental Toxicology and Chemistry 10(5): 649–655.

Emmerling, C. 2001. Response of earthworm communities to different types of soil tillage. Applied Soil Ecology 17(1): 91–96. doi: http://doi.org/10.1016/S0929-1393(00)00132-3.

Ernst, G. and C. Emmerling. 2009. Impact of five different tillage systems on soil organic carbon content and the density, biomass, and community composition of earthworms after a ten year period. European Journal of Soil Biology 45(3): 247–251.

Errouissi, F., S. Ben Moussa-Machraoui, M. Ben-Hammouda and S. Nouira. 2011. Soil invertebrates in durum wheat (*Triticum durum* L.) cropping system under Mediterranean semi arid conditions: a comparison between conventional and no-tillage management. Soil and Tillage Research 112(2): 122–132. doi: http://doi. org/10.1016/j. still.2010.12.004.

Fabra, A., R. Duffard and A. Evangelista De Duffard. 1997. Toxicity of 2,4- dichlorophenoxyacetic acid to *Rhizobium* sp in pure culture. Bulletin of Environmental Contamination and Toxicology 59(4): 645–652. doi: 10.1007/s001289900528.

Fang, H., J. Lian, H. Wang, L. Cai and Y. Yu. 2015. Exploring bacterial community structure and function associated with atrazine biodegradation in repeatedly treated soils. Journal of Hazardous Materials 286: 457–465. doi: 10.1016/j. jhazmat. 2015. 01.006

Feng, P.C.C., C. Clark, G.C. Andrade, M.C. Balbi and P. Caldwell. 2008. The control of Asian rust by glyphosate in glyphosate-resistant soybeans. Pest Management Science 64(4): 353–359. doi: 10.1002/ps.1498.

Feng, Y., A.C. Motta, D.W. Reeves, C.H. Burmester, E. van Santen and J.A. Osborne. 2003. Soil microbial communities under conventional-till and no-till continuous cotton systems. Soil Biology and Biochemistry 35(12): 1693–1703. doi: http://doi. org/10. 1016/j.soilbio.2003.08.016.

Fernandez, A.L., C.C. Sheaffer, D.L. Wyse, C. Staley, T.J. Gould and M.J. Sadowsky. 2016. Associations between soil bacterial community structure and nutrient cycling functions in long-term organic farm soils following cover crop and organic fertiliser amendment. Science of the Total Environment 566–567: 949–959. doi: 10.1016/j. scitotenv.2016.05.073

Fischer, R.S., A. Berry, C.G. Gaines and R.A. Jensen. 1986. Comparative action of glyphosate as a trigger of energy drain in Eubacteria. Journal of Bacteriology 168(3): 1147–1154.

Fontana, M., A. Berner, P. Mäder, F. Lamy and P. Boivin. 2015. Soil organic carbon and soil bio-physicochemical properties as co-influenced by tillage treatment. Soil Science Society of America Journal 79(5): 1435–1445. doi: 10.2136/sssaj2014.07.0288.

Fournier, J.C. 1980. Enumeration of the soil micro-organisms able to degrade 2,4-D by metabolism or co-metabolism. Chemosphere 9(3): 169–174. doi: 10.1016/0045-6535(80)90089-2.

Frey, S.D., E.T. Elliott and K. Paustian. 1999. Bacterial and fungal abundance and biomass in conventional and no-tillage agroecosystems along two climatic gradients. Soil Biology and Biochemistry 31(4): 573–585. doi: 10.1016/S0038-0717(98)00161-8.

Gál, A., T.J. Vyn, E. Michéli, E.J. Kladivko and W.W. McFee. 2007. Soil carbon and nitrogen accumulation with long-term no-till versus moldboard plowing overestimated with tilled-zone sampling depths. Soil and Tillage Research 96(1-2): 42–51. doi: 10.1016/j. still.2007.02.007.

Gallaher, R.N., D.W. Dickson, J.F. Corella and T.E. Hewlett. 1988. Tillage and multiple cropping systems and population-dynamics of phytoparasitic nematodes. Annals of Applied Nematology 2: 90–94.

García-Orenes, F., C. Guerrero, A. Roldán, J. Mataix-Solera, A. Cerdà, M. Campoy, R. Zornoza, G. Bárcenas and F. Caravaca. 2010. Soil microbial biomass and activity under different agricultural management systems in a semiarid Mediterranean agroecosystem. Soil and Tillage Research 109(2): 110–115. doi: 10.1016/j.still.2010. 05.005.

Gavassoni, W.L., G.L. Tylka and G.P. Munkvold. 2001. Relationships between tillage and spatial patterns of *Heterodera glycines*. Phytopathology 91(6): 534–545. doi: 10.1094/PHYTO.2001.91.6.534.

George, T.S., A.E. Richardson, S.S. Li, P.J. Gregory and T.J. Daniell. 2009. Extracellular release of a heterologous phytase from roots of transgenic plants: does manipulation of rhizosphere biochemistry impact microbial community structure? FEMS Microbiology Ecology 70(3): 433–445. doi: 10.1111/j.1574-6941.2009.00762.x.

Ghani, A., D.A. Wardle, A. Rahman and D.R. Lauren. 1996. Interactions between ^{14}C-labelled atrazine and the soil microbial biomass in relation to herbicide degradation. Biology and Fertility of Soils 21(1–2): 17–22.

Gomez, E., L. Ferreras, L. Lovotti and E. Fernandez. 2009. Impact of glyphosate application on microbial biomass and metabolic activity in a Vertic Argiudoll from Argentina. European Journal of Soil Biology 45(2): 163–167. doi: 10.1016/j.ejsobi. 2008.10.001.

Gonod, L.V., F. Martin-Laurent and C. Chenu. 2006. 2,4-D impact on bacterial communities, and the activity and genetic potential of 2,4-D degrading communities in soil. FEMS Microbiology Ecology 58(3): 529–537. doi: 10.1111/j.1574-6941.2006. 00159.x.

Gonzalez-Lopez, J., J. Moreno, T. De la Rubia, M.V. Martinez-Toledo and A. Ramos-Cormenzana. 1984. Toxicity of dicamba (2-methoxy-3,6-dichlorobenzoic acid) to *Azotobacter vinelandii*. Folia Microbiologica 29(2): 127–130.

Gossen, B.D. and D.A. Derksen. 2003. Impact of tillage and crop rotation on ascochyta blight (*Ascochyta lentis*) of lentil. Can J Plant Sci. 83(2): 411–415. doi: 10.4141/P02-088.

Govaerts, B., M. Fuentes, M. Mezzalama, J.M. Nicol, J. Deckers, J.D. Etchevers, B. Figueroa-Sandoval and K.D. Sayre. 2007. Infiltration, soil moisture, root rot and nematode populations after 12 years of different tillage, residue and crop rotation managements. Soil and Tillage Research 94(1): 209–219. doi: 10.1016/j.still.2006.07.013.

Green, J.M. 2014. Current state of herbicides in herbicide-resistant crops. Pest Management Science 70(9): 1351–1357. doi: 10.1002/ps.3727.

Gupta, D. and N. Joshi. 2009. Changes in microbial biomass and phosphatase activity exposed to 2,4-D and glyphosate. Journal of Environmental Research and Development 3(3): 663–669.

Halpern, M.T., J.K. Whalen and C.A. Madramootoo. 2010. Long-term tillage and residue management influences soil carbon and nitrogen dynamics. Soil Science Society of America Journal 74(4): 1211–1217. doi: 10.2136/sssaj2009.0406.

Han, S.O. and P.B. New. 1994. Effect of water availability on degradation of 2, 4-dichlorophenoxyacetic acid (2,4-d) by soil microorganisms. Soil Biology and Biochemistry 26(12): 1689–1697. doi: 10.1016/0038-0717(94)90322-0.

Haney, R.L., S.A. Senseman and F.M. Hons. 2002. Effect of roundup ultra on microbial activity and biomass from selected soils. Journal of Environmental Quality 31(3): 730–735.

Haney, R.L., S.A. Senseman, F.M. Hons and D.A. Zuberer. 2000. Effect of glyphosate on soil microbial activity and biomass. Weed Science 48(1): 89–93.

Hart, M.M., J.R. Powell, R.H. Gulden, K.E. Dunfield, K. Peter Pauls, C.J. Swanton, J.N. Klironomos, P.M. Antunes, A.M. Koch and J.T. Trevors. 2009. Separating the effect of crop from herbicide on soil microbial communities in glyphosate-resistant corn. Pedobiologia. 52(4): 253–262. doi: 10.1016/j.pedobi.2008.10.005.

Helgason, B.L., F.L. Walley, J.J. Germida. 2009. Fungal and bacterial abundance in long-term no-till and intensive-till soils of the Northern Great Plains. Soil Science Society of America Journal 73(1): 120–127. doi: 10.2136/sssaj2007.0392.

Helgason, B.L., F.L. Walley and J.J. Germida. 2010. No-till soil management increases microbial biomass and alters community profiles in soil aggregates. Applied Soil Ecology 46(3): 390–397. doi: 10.1016/j.apsoil.2010.10.002.

Hill, R.L. and M. Meza-Montalvo. 1990. Long-term wheel traffic effects on soil physical properties under different tillage systems. Soil Science Society of America Journal 54(3): 865–870. doi: 10.2136/sssaj1990.03615995005400030042x.

Hou, R., Z. Ouyang, Y. Li, D.D. Tyler, F. Li and G.V. Wilson. 2012. Effects of tillage and residue management on soil organic carbon and total nitrogen in the north China plain. Soil Science Society of America Journal 76(1): 230–240. doi: 10.2136/sssaj2011.0107.

Hubbard, V.C., D. Jordan and J.A. Stecker. 1999. Earthworm response to rotation and tillage in a Missouri claypan soil. Biology and Fertility of Soils 29(4): 343–347. doi: 10.1007/s003740050563.

Jenkins, M.B., M.A. Locke, K.N. Reddy, D.S. McChesney and R.W. Steinriede. 2017. Impact of glyphosate-resistant corn, glyphosate applications and tillage on soil nutrient ratios, exoenzyme activities and nutrient acquisition ratios. Pest Management Science 73(1): 78–86. doi: 10.1002/ps.4413.

Jia, S., X. Zhang, X. Chen, N.B. McLaughlin, S. Zhang, S. Wei, B. Sun and A. Liang. 2016. Long-term conservation tillage influences the soil microbial community and its contribution to soil CO_2 emissions in a Mollisol in Northeast China. Journal of Soils and Sediments 16(1): 1–12. doi: 10.1007/s11368-015-1158-7.

Johnson, W.M. 1963. The seventh approximation: a symposium. Soil Science Society of America Journal 27: 212–215.

Joschko, M., R. Gebbers, D. Barkusky, J. Rogasik, W. Höhn, W. Hierold, C.A. Fox and J. Timmer. 2009. Location-dependency of earthworm response to reduced tillage on sandy soil. Soil and Tillage Research 102(1): 55–66. doi: http://doi. org/10. 1016/j.still.2008.07.023.

Kladivko, E.J. 2001. Tillage systems and soil ecology. Soil and Tillage Research 61(1–2): 61–76. doi: http://doi. org/10.1016/S0167-1987(01)00179-9.

Kladivko, E.J., N.M. Akhouri and G. Weesies. 1997. Earthworm populations and species distributions under no-till and conventional tillage in Indiana and Illinois. Soil Biology and Biochemistry 29(3–4): 613–615. doi: 10.1016/S0038-0717(96)00187-3.

Kochsiek, A.E., J.M.H. Knops, C.E. Brassil and T.J. Arkebauer. 2013. Maize and soybean litter-carbon pool dynamics in three no-till systems. Soil Science Society of America Journal 77(1): 226–236. doi: 10.2136/sssaj2012.0175.

Kraiser, T., M. Stuardo, M. Manzano, T. Ledger and B. González. 2013. Simultaneous assessment of the effects of an herbicide on the triad: rhizobacterial community, an herbicide degrading soil bacterium and their plant host. Plant and Soil 366(1–2): 377–388. doi: 10.1007/s11104-012-1444-8.

Kremer, R.J. and N.E. Means. 2009. Glyphosate and glyphosate-resistant crop interactions with rhizosphere microorganisms. European Journal of Agronomy 31(3): 153–161. doi: 10.1016/j. eja.2009.06.004.

Kremer, R.J., N.E. Means and S. Kim. 2005. Glyphosate affects soybean root exudation and rhizosphere micro-organisms. International Journal of Environmental Analytical Chemistry 85(15): 1165–1174. doi: 10. 1080/03067310500273146.

Krueger, J.P., R.G. Butz and D.J. Cork. 1991. Use of dicamba-degrading microorganisms to protect dicamba susceptible plant species. Journal of Agricultural and Food Chemistry 39(5): 1000–1003.

Krutz, L.J., M.A. Locke and R.W. Steinriede Jr. 2009. Interactions of tillage and cover crop on water, sediment, and pre-emergence herbicide loss in glyphosate-resistant cotton: implications for the control of glyphosate-resistant weed biotypes. Journal of Environmental Quality 38(3): 1240–1247.

Krutz, L.J., D.L. Shaner, M.A. Weaver, R.M.T. Webb, R.M. Zablotowicz, K.N. Reddy, Y. Huang and S.J. Thomson. 2010. Agronomic and environmental implications of enhanced s-triazine degradation. Pest Management Science 66(5): 461–481. doi: 10.1002/ps.1909.

Kumar, S., A. Kadono, R. Lal and W. Dick. 2012. Long-term no-till impacts on organic carbon and properties of two contrasting soils and corn yields in Ohio. Soil Science Society of America Journal 76(5): 1798–1809. doi: 10. 2136/sssaj2012.0055.

Laitinen, P., S. Rämö and K. Siimes. 2007. Glyphosate translocation from plants to soil—does this constitute a significant proportion of residues in soil? Plant and Soil 300(1–2): 51–60. doi: 10.1007/s11104-007-9387-1.

Lal, R. and J.M. Kimble. 1997. Conservation tillage for carbon sequestration. Nutrient Cycling in Agroecosystems 49(1): 243–253. doi: 10.1023/a:1009794514742.

Lancaster, S.H., R.L. Haney, S.A. Senseman, F.M. Hons and J.M. Chandler. 2006. Soil microbial activity is affected by Roundup WeatherMax and pesticides applied to cotton (*Gossypium hirsutum*). Journal of Agricultural and Food Chemistry 54(19): 7221–7226. doi: 10.1021/jf061673p.

Lancaster, S.H., E.B. Hollister, S.A. Senseman and T.J. Gentry. 2010. Effects of repeated glyphosate applications on soil microbial community composition and the mineralization of glyphosate. Pest Management Science 66(1): 59–64. doi: 10.1002/ps.1831.

Lane, M., N. Lorenz, J. Saxena, C. Ramsier and R.P. Dick. 2012. Microbial activity, community structure and potassium dynamics in rhizosphere soil of soybean plants treated with glyphosate. Pedobiologia 55(3): 153–159. doi: 10.1016/j.pedobi.2011. 12.005.

Ławniczak, Ł., A. Syguda, A. Borkowski, P. Cyplik, K. Marcinkowska, L. Wolko, T. Praczyk, L. Chrzanowski and J. Pernak. 2016. Influence of oligomeric herbicidal ionic liquids with MCPA and Dicamba anions on the community structure of autochthonic bacteria present in agricultural soil. Science of the Total Environment 563–564: 247–255. doi: 10.1016/j. scitotenv.2016.04.109.

Levanon, D., J.J. Meisinger, E.E. Codling and J.L. Starr. 1994. Impact of tillage on microbial activity and the fate of pesticides in the upper soil. Water, Air, and Soil Pollution 72(1): 179–189. doi: 10.1007/bf01257123.

Levy-Booth, D.J., R.H. Gulden, R.G. Campbell, J.R. Powell, J.N. Klironomos, K.P. Pauls, C.J. Swanton, J.T. Trevors and K.E. Dunfield. 2009. Roundup Ready® soybean gene concentrations in field soil aggregate size classes. FEMS Microbiology Letters 291(2): 175–179. doi: 10.1111/j.1574-6968.2008.01449.x.

Li, L., F.J. Larney, D.A. Angers, D.C. Pearson and R.E. Blackshaw. 2015. Surface soil quality attributes following 12 years of conventional and conservation management on irrigated rotations in southern Alberta. Soil Science Society of America Journal 79(3): 930–942. doi: 10.2136/sssaj2015.02.0051.

Liebig, M.A., A.J. Jones, L.N. Mielke and J.W. Doran. 1993. Controlled wheel traffic effects on soil properties in ridge tillage. Soil Science Society of America Journal 57(4): 1061–1066. doi: 10. 2136/sssaj1993.03615995005700040030x.

Liebig, M.A., D.L. Tanaka and B.J. Wienhold. 2004. Tillage and cropping effects on soil quality indicators in the northern Great Plains. Soil and Tillage Research 78(2): 131–141. doi: 10.1016/j. still.2004.02.002.

Liphadzi, K.B., K. Al-Khatib, C.N. Bensch, P.W. Stahlman, J.A. Dille, T. Todd, C.W. Rice, M.J. Horak and G. Head. 2005. Soil microbial and nematode communities as affected by glyphosate and tillage practices in a glyphosate-resistant cropping system. Weed Science 53(4): 536–545. doi: 10.1614/WS-04-129R1.

Liu, C.M., P.A. McLean, C.C. Sookdeo and F.C. Cannon. 1991. Degradation of the herbicide glyphosate by members of the family Rhizobiaceae. Applied and Environmental Microbiology 57(6): 1799–1804.

Locke, M.A. and C.T. Bryson. 1997. Herbicide-soil interactions in reduced tillage and plant residue management systems. Weed Science 45(2): 307–320.

Locke, M.A., L.A. Gaston and R.M. Zablotowicz. 1996. Alachlor Biotransformation and Sorption in Soil from Two Soybean Tillage Systems. Journal of Agricultural and Food Chemistry 44(4): 1128–1134. doi: 10.1021/jf950466e.

Locke, M.A., L.J. Krutz, R.W. Steinriede Jr. and S. Testa III. 2015. Conservation management improves runoff water quality: implications for environmental sustainability in a glyphosate-resistant cotton production system. Soil Science Society of America Journal 79(2): 660–671. doi: 10.2136/sssaj2014.09.0389.

Locke, M.A., K.N. Reddy and R.M. Zablotowicz. 2002. Weed management in conservation crop production systems. Weed Biol Manag. 2: 123–132.

Locke, M.A., D.D. Tyler and L.A. Gaston. 2010. Soil and water conservation in the Mid-South United States: lessons learned and a look to the future. pp. 201–236. *In:* Zobeck, T.M. and Schillinger, W.F. (Eds.) Soil and Water Conservation Advances in the United States, SSSA Special Publication 60. Soil Science Society of America, Madison, WI, USA.

Locke, M.A. and R.M. Zablotowicz. (2004). Pesticides in soil: benefits and limitations to soil health. pp. 239-260. *In:* Schjonning, P., Elmholt, S. and Christensen, B.T. (Eds.) Managing Soil Quality: Challenges in Modern Agriculture. CAB International.

Locke, M.A., R.M. Zablotowicz, P.J. Bauer, R.W. Steinriede and L.A. Gaston. 2005. Conservation cotton production in the southern United States: Herbicide dissipation in soil and cover crops. Weed Science 53(5): 717–727. doi: 10.1614/WS-04-174R1.1.

Locke, M.A., R.M. Zablotowicz and K.N. Reddy. 2008. Integrating soil conservation practices and glyphosate-resistant crops: Impacts on soil. Pest Management Science 64(4): 457–469.

Locke, M.A., R.M. Zablotowicz, R.W. Steinriede Jr, S. Testa and K.N. Reddy. 2013. Conservation management in cotton production: long-term soil biological, chemical, and physical changes. Soil Science Society of America Journal 77(3): 974–984.

Locke, M.A., R.M. Zablotowicz and M.A. Weaver. 2006. Herbicide fate under conservation tillage, cover crop, and edge-of-field management practices. pp. 373–392. *In:* H.P. Sigh, D.R. Batish and R.K. Kohli. (Eds.) Handbook of Sustainable Weed Management. Haworth Press, Binghamton, New York.

López-Fando, C. and A. Bello. 1995. Variability in soil nematode populations due to tillage and crop rotation in semi-arid Mediterranean agrosystems. Soil and Tillage Research 36(1): 59–72. doi: 10.1016/0167-1987(95)00496-3.

Lupwayi, N.Z., G.W. Clayton, J.T. O'Donovan, K.N. Harker, T.K. Turkington and Y.K. Soon. 2006. Soil nutrient stratification and uptake by wheat after seven years of conventional and zero tillage in the Northern Grain belt of Canada. Canadian Journal of Soil Science 86(5): 767–778.

Lupwayi, N.Z., K.N. Harker, G.W. Clayton, J.T. O'Donovan and R.E. Blackshaw. 2009. Soil microbial response to herbicides applied to glyphosate-resistant canola. Agriculture, Ecosystems and Environment 129(1–3): 171–176. doi: 10.1016/j.agee. 2008.08.007.

Lupwayi, N.Z., K.N. Harker, G.W. Clayton, T.K. Turkington, W.A. Rice and J.T. O'Donovan. 2004. Soil microbial biomass and diversity after herbicide application. Can J Plant Sci. 84(2): 677–685.

Lupwayi, N.Z., F.J. Larney, R.E. Blackshaw, D.A. Kanashiro and D.C. Pearson. 2017. Phospholipid fatty acid biomarkers show positive soil microbial community responses to conservation soil management of irrigated crop rotations. Soil and Tillage Research 168: 1–10. doi: 10.1016/j.still.2016.12.003.

Macur, R.E., J.T. Wheeler, M.D. Burr and W.P. Inskeep. 2007. Impacts of 2,4-D application on soil microbial community structure and on populations associated with 2,4-D degradation. Microbiological Research 162(1): 37–45. doi: http://dx.doi.org/10. 1016/j.micres.2006.05.007.

Mahía, J., A. Cabaneiro, T. Carballa and M. Díaz-Raviña. 2008. Microbial biomass and C mineralization in agricultural soils as affected by atrazine addition. Biology and Fertility of Soils 45(1): 99–105. doi: 10.1007/s00374-008-0318-y.

Mamy, L., E. Barriuso and B. Gabrielle. 2005. Environmental fate of herbicides trifluralin, metazachlor, metamitron and sulcotrione compared with that of glyphosate, a substitute broad spectrum herbicide for different glyphosate-resistant crops. Pest Management Science 61(9): 905–916. doi: 10.1002/ps.1108.

Mamy, L., B. Gabrielle and E. Barriuso. 2010. Comparative environmental impacts of glyphosate and conventional herbicides when used with glyphosate-tolerant and non-tolerant crops. Environ Pollut. 158(10): 3172–3178. doi: 10.1016/j.envpol.2010.06.036.

Martens, D.A. and J.M. Bremner. 1993. Influence of herbicides on transformations of urea nitrogen in soil. Journal of Environmental Science and Health – Part B. Pesticides, Food Contaminants, and Agricultural Wastes 28(4): 377–395.

Martínez, I., A. Chervet, P. Weisskopf, W.G. Sturny, A. Etana, M. Stettler, J. Forkman and T. Keller. 2016. Two decades of no-till in the Oberacker long-term field experiment: Part I. Crop yield, soil organic carbon and nutrient distribution in the soil profile. Soil and Tillage Research 163: 141–151. doi: http://doi.org/10.1016/j.still.2016.05.021.

Martínez, J.M., J.A. Galantini, M.E. Duval and F.M. López. 2017. Tillage effects on labile pools of soil organic nitrogen in a semi-humid climate of Argentina: a long-term field study. Soil and Tillage Research 169: 71–80. doi: http://doi.org/10.1016/j.still.2017.02.001.

Masvaya, E.N., J. Nyamangara, K. Descheemaeker and K.E. Giller. 2017. Tillage, mulch and fertiliser impacts on soil nitrogen availability and maize production in semi-arid Zimbabwe. Soil and Tillage Research 168: 125–132.

Matson, P.A., W.J. Parton, A.G. Power and M.J. Swift. 1997. Agricultural intensification and ecosystem properties. Science 277(5325): 504–509. doi: 10.1126/science.277.5325. 504.

Mbuthia, L.W., V. Acosta-Martínez, J. DeBryun, S. Schaeffer, D. Tyler, E. Odoi, M. Mpheshea, F. Walker and N. Eash. 2015. Long term tillage, cover crop, and fertilization effects on microbial community structure, activity: implications for soil quality. Soil Biology and Biochemistry 89: 24–34. doi: 10.1016/j.soilbio.2015.06.016.

Mekwatanakarn, P. and K. Sivasithamparam. 1987. Effect of certain herbicides on soil microbial populations and their influence on saprophytic growth in soil and pathogenicity of take-all fungus. Biology and Fertility of Soils 5(2): 175–180. doi: 10.1007/BF00257655.

Meriles, J.M., S. Vargas Gil, R.J. Haro, G.J. March and C.A. Guzman. 2006. Glyphosate and previous crop residue effect on deleterious and beneficial soil-borne fungi from a peanut-corn-soybean rotations. J Phytopathol. 154(5): 309–316. doi: 10.1111/j.1439-0434.2006.01098.x.

Mijangos, I., J.M. Becerril, I. Albizu, L. Epelde and C. Garbisu. 2009. Effects of glyphosate on rhizosphere soil microbial communities under two different plant compositions by cultivation-dependent and -independent methodologies. Soil Biology and Biochemistry 41(3): 505–513. doi: 10.1016/j.soilbio.2008.12.009.

Mikha, M.M. and C.W. Rice. 2004. Tillage and Manure Effects on Soil and Aggregate-Associated Carbon and Nitrogen Contribution. Soil Science Society of America Journal 68(3): 809–816. doi: 10.2136/sssaj2004.8090.

Miller, T.W. (2016). Integrated strategies for management of perennial weeds. Invasive Plant Sci Manage. 9(2): 148–158. doi: 10.1614/IPSM-D-15-00037.1.

Mina, B.L., S. Saha, N. Kumar, A.K. Srivastva and H.S. Gupta. 2008. Changes in soil nutrient content and enzymatic activity under conventional and zero-tillage practices in an Indian sandy clay loam soil. Nutrient Cycling in Agroecosystems 82(3): 273–281. doi: 10.1007/s10705-008-9189-8.

Mishra, U., D.A.N. Ussiri and R. Lal. 2010. Tillage effects on soil organic carbon storage and dynamics in Corn Belt of Ohio USA. Soil and Tillage Research 107(2): 88–96. doi: http://doi.org/10.1016/j.still.2010.02.005.

Mitchell, J.P., A. Shrestha, K. Mathesius, K.M. Scow, R.J. Southard, R.L. Haney, R. Schmidt, D.S. Munk and W.R. Horwath. 2017. Cover cropping and no-tillage improve soil health in an arid irrigated cropping system in California's San Joaquin Valley, USA. Soil and Tillage Research 165: 325–335. doi: 10.1016/j.still.2016.09.001.

Moore, E.B., M.H. Wiedenhoeft, T.C. Kaspar and C.A. Cambardella. 2014. Rye cover crop effects on soil quality in no-till corn silage–soybean cropping systems. Soil Science Society of America Journal 78(3): 968–976. doi: 10.2136/sssaj2013.09.0401.

Moorman, T.B. 1989. A review of pesticide effects on microorganisms and microbial processes related to soil fertility. Journal of Production Agriculture 2(1): 14–23. doi: 10.2134/jpa1989.0014.

Moorman, T.B., J.M. Becerril, J. Lydon and S.O. Duke. 1992. Production of hydroxybenzoic acids by *Bradyrhizobium japonicum* strains after treatment with glyphosate. Journal of Agricultural and Food Chemistry 40(2): 289–293.

Moreno, J.L., A. Aliaga, S. Navarro, T. Hernández and C. García. 2007. Effects of atrazine on microbial activity in semiarid soil. Applied Soil Ecology 35(1): 120–127. doi: 10.1016/j.apsoil.2006.05.002.

Muñoz-Romero, V., R.J. Lopez-Bellido, P. Fernandez-Garcia, R. Redondo, S. Murillo and L. Lopez-Bellido. 2017. Effects of tillage, crop rotation and N application rate on labile and recalcitrant soil carbon in a Mediterranean Vertisol. Soil and Tillage Research 169: 118–123. doi: http://doi.org/10.1016/j.still.2017.02.004.

Muruganandam, S., D.W. Israel and W.P. Robarge. 2009. Activities of nitrogen-mineralization enzymes associated with soil aggregate size fractions of three tillage systems. Soil Science Society of America Journal 73(3): 751–759. doi: 10.2136/sssaj2008.0231.

Nakatani, A.S., M.F. Fernandes, R.A. De Souza, A.P. Da Silva, F.B. Dos Reis-Junior, I.C. Mendes and M. Hungria. 2014. Effects of the glyphosate-resistance gene and of herbicides applied to the soybean crop on soil microbial biomass and enzymes. Field Crops Research 162: 20–29.

Nandula, V.K. and H.L. Tyler. 2016. Effect of new auxin herbicide formulations on control of herbicide resistant weeds and on microbial activities in the rhizosphere. American Journal of Plant Sciences 7: 2429–2439.

Nawaz, A., M. Farooq, R. Lal, A. Rehman and R. Hafeezur. 2017. Comparison of conventional and conservation rice-wheat systems in Punjab, Pakistan. Soil and Tillage Research 169: 35–43. doi: http://doi.org/10.1016/j.still.2017.01.012.

Neumann, G., S. Kohls, E. Landsberg, K. Stock-Oliveira Souza, T. Yamada and V. Römheld. 2006. Relevance of glyphosate transfer to non-target plants via the rhizosphere. J Plant Dis Prot (20): 963–969.

Nguyen, D.B., M.T. Rose, T.J. Rose, S.G. Morris and L. van Zwieten. 2016. Impact of glyphosate on soil microbial biomass and respiration: a meta-analysis. Soil Biology and Biochemistry 92: 50–57. doi: 10.1016/j.soilbio.2015.09.014.

Njiti, V.N., O. Myers Jr, D. Schroeder and D.A. Lightfoot. 2003. Roundup Ready soybean: glyphosate effects on *Fusarium solani* root colonization and sudden death syndrome. Agronomy Journal 95(5): 1140–1145.

Oleszczuk, P., I. Jośko, B. Futa, S. Pasieczna-Patkowska, E. Pałys and P. Kraska. 2014. Effect of pesticides on microorganisms, enzymatic activity and plant in biochar-amended soil. Geoderma. 214–215: 10–18. doi: 10.1016/j.geoderma.2013.10.010.

Olson, B.M. and C.W. Lindwall. 1991. Soil microbial activity under chemical fallow conditions: effects of 2,4-D and glyphosate. Soil Biology and Biochemistry 23(11): 1071–1075. doi: 10.1016/0038-0717(91)90046-M.

Overstreet, L.F. and G.D. Hoyt. 2008. Effects of strip tillage and production inputs on soil biology across a spatial gradient. Soil Science Society of America Journal 72(5): 1454–1463. doi: 10.2136/sssaj2007.0143.

Pankhurst, C., C. Kirkby, B. Hawke and B. Harch. 2002. Impact of a change in tillage and crop residue management practice on soil chemical and microbiological properties in a cereal-producing red duplex soil in NSW, Australia. Biology and Fertility of Soils 35(3): 189–196. doi: 10.1007/s00374-002-0459-3.

Partoazar, M., M. Hoodaji and A. Tahmourespour. 2011. The effect of glyphosate application on soil microbial activities in agricultural land. African Journal of Biotechnology 10(83): 19419-19424. doi: 10.5897/AJB11.2440.

Perucci, P., S. Dumontet, S.A. Bufo, A. Mazzatura and C. Casucci. 2000. Effects of organic amendment and herbicide treatment on soil microbial biomass. Biology and Fertility of Soils 32(1): 17–23. doi: 10.1007/s003740000207.

Petchey, O.L. and K.J. Gaston. 2006. Functional diversity: back to basics and looking forward. Ecology Letters 9(6): 741–758. doi: 10. 1111/j.1461-0248.2006.00924.x.

Peters, R.D., A.V. Sturz, M.R. Carter and J.B. Sanderson. 2003. Developing disease-suppressive soils through crop rotation and tillage management practices. Soil and Tillage Research 72(2): 181–192. 10.1016/S0167-1987(03)00087-4.

Pikul, J.L., R.C. Schwartz, J.G. Benjamin, R.L. Baumhardt and S. Merrill. 2006. Cropping system influences on soil physical properties in the Great Plains. Renewable Agriculture and Food Systems 21(1): 15–25. doi: 10.1079/RAF2005122.

Pittaway, P.A. 1995. Opportunistic association between *Pythium* species and weed residues causing seedling emergence failure in cereals. Australian Journal of Agricultural Research 46(3): 655–662. doi: 10.071/AR9950655.

Raczkowski, C.W., J.P. Mueller, W.J. Busscher, M.C. Bell and M.L. McGraw. 2012. Soil physical properties of agricultural systems in a large-scale study. Soil and Tillage Research 119: 50–59. doi: http://doi.org/10.1016/j.still.2011.12.006.

Rahman, L., K.Y. Chan and D.P. Heenan. 2007. Impact of tillage, stubble management and crop rotation on nematode populations in a long-term field experiment. Soil and Tillage Research 95(1–2): 110–119. doi: http://doi.org/10.1016/j.still.2006.11.008.

Ratnayake, M. and L.J. Audus. 1978. Studies on the effects of herbicides on soil nitrification. Pesticide Biochemistry and Physiology 8(2): 170–185.

Reddy, K.N. and M.A. Locke. 1996. Imazaquin spray retention, foliar washoff and runoff losses under simulated rainfall. PESTIC. SCI. 48(2): 179–187. doi: 10.1002/(SICI)1096-9063(199610)48:2<179::AID-PS457>3.0. CO,2-M.

Reddy, K.N., M.A. Locke and K.D. Howard. 1995. Bentazon Spray Retention, Activity, and Foliar Washoff in Weed Species. Weed Technol. 9(4): 773–778.

Reddy, K.N., R.M. Zablotowicz, M.A. Locke and C.H. Koger. 2003. Cover crop, tillage, and herbicide effects on weeds, soil properties, microbial populations, and soybean yield. Weed Science 51(6): 987–994.

Rivarola, V., A. Fabra, G. Mori and H. Balegno. 1992. In vitro protein synthesis is affected by the herbicide 2,4-dichlorophenoxyacetic acid in *Azospirillum brasilense*. Toxicology 73(1): 71–79. doi: 10.1016/0300-483X(92)90171-A.

Ros, M., M. Goberna, J.L. Moreno, T. Hernandez, C. García, H. Insam and J.A. Pascual. 2006. Molecular and physiological bacterial diversity of a semi-arid soil contaminated with different levels of formulated atrazine. Applied Soil Ecology 34(2–3): 93–102. doi: 10.1016/j.apsoil.2006.03.010.

Rovira, A.D., K.R.J. Smettem and K.E. Lee. 1987. Effect of rotation and conservation tillage of earthworms in a red-brown earth under wheat. Australian Journal of Agricultural Research 38(5): 829–834. doi: 10.1071/AR9870829.

Sainju, U.M., T. Caesar-TonThat and J.D. Jabro. 2009. Carbon and nitrogen fractions in dryland soil aggregates affected by long-term tillage and cropping sequence. Soil Science Society of America Journal 73(5): 1488–1495. doi: 10.2136/sssaj2008.0405.

Sainju, U.M., W.B. Stevens and T. Caesar-TonThat. 2014. Soil carbon and crop yields affected by irrigation, tillage, cropping system, and nitrogen fertilization. Soil Science Society of America Journal 78(3): 936–948. doi: 10.2136/sssaj2013.12.0514.

Sannino, F. and L. Gianfreda. 2001. Pesticide influence on soil enzymatic activities. Chemosphere 45(4–5): 417–425.

Sanyal, D. and A. Shrestha. 2008. Direct effect of herbicides on plant pathogens and disease development in various cropping systems. Weed Science 56(1): 155–160. doi: 10.1614/WS-07-081.1.

Schafer, J.R., S.G. Hallett and W.G. Johnson. 2014. Rhizosphere microbial community dynamics in glyphosate-treated susceptible and resistant biotypes of giant ragweed (*Ambrosia trifida*). Weed Science 62(2): 370–381. doi: 10.1614/ws-d-13-00164.1.

Sebiomo, A., V.W. Ogundero and S.A. Bankole. 2011. Effect of four herbicides on microbial population, soil organic matter and dehydrogenase activity. African Journal of Biotechnology 10(5): 770–778.

Shaver, T.M., G.A. Peterson and L.A. Sherrod. 2003. Cropping intensification in dryland systems improves soil physical properties: regression relations. Geoderma. 116(1–2): 149–164. doi: http://doi.org/10.1016/S0016-7061(03)00099-5.

Shipitalo, M.J., L.B. Owens, J.V. Bonta and W.M. Edwards. 2013. Effect of no-till and extended rotation on nutrient losses in surface runoff. Soil Science Society of America Journal 77(4): 1329–1337. doi: 10.2136/sssaj2013.01.0045.

Singh, P. and N. Ghoshal. 2010. Variation in total biological productivity and soil microbial biomass in rainfed agroecosystems: impact of application of herbicide and soil amendments. Agriculture, Ecosystems and Environment 137(3–4): 241–250. doi: 10.1016/j.agee.2010.02.009.

Six, J., E.T. Elliott and K. Paustian. 2000. Soil macroaggregate turnover and microaggregate formation: a mechanism for C sequestration under no-tillage agriculture. Soil Biology and Biochemistry 32(14): 2099–2103. doi: http://doi.org/10.1016/S0038-0717(00)00179-6.

Sofo, A., A. Scopa, S. Dumontet, A. Mazzatura and V. Pasquale. 2012. Toxic effects of four sulphonylureas herbicides on soil microbial biomass. Journal of Environmental Science and Health – Part B. Pesticides, Food Contaminants, and Agricultural Wastes 47(7): 653–659. doi: 10.1080/03601234.2012.669205.

Soon, Y.K. and M.A. Arshad. 2004. Tillage, crop residue and crop sequence effects on nitrogen availability in a legume-based cropping system. Canadian Journal of Soil Science 84(4): 421–430. doi: 10.4141/s04-023.

Spargo, J.T., M.A. Cavigelli, M.M. Alley, J.E. Maul, J.S. Buyer, C.H. Sequeira and R.F. Follett. 2012. Changes in soil organic carbon and nitrogen fractions with duration of no-tillage management. Soil Science Society of America Journal 76(5): 1624–1633. doi: 10.2136/sssaj2011.0337.

Steenwerth, K. and K.M. Belina. 2008a. Cover crops and cultivation: impacts on soil N dynamics and microbiological function in a Mediterranean vineyard agroecosystem. Applied Soil Ecology 40(2): 370–380. doi: 10.1016/j.apsoil.2008.06.004.

Steenwerth, K. and K.M. Belina. 2008b. Cover crops enhance soil organic matter, carbon dynamics and microbiological function in a vineyard agroecosystem. Applied Soil Ecology 40(2): 359–369. doi: 10.1016/j.apsoil.2008.06.006.

Tanney, J.B. and L.J. Hutchison. 2010. The effects of glyphosate on the in vitro linear growth of selected microfungi from a boreal forest soil. Canadian Journal of Microbiology 56(2): 138–144. doi: 10.1139/W09-122.

Teasdale, J.R., C.E. Beste and W.E. Potts. 1991. Response of weeds to tillage and cover crop residue. Weed Science 39(2): 195–199.

Teasdale, J.R., L.O. Brandsæedter, A. Calegari and F. Skora Neto. 2007. Cover crops and weed management. pp. 49–64. *In:* Non-Chemical Weed Management: Principles, Concepts and Technology. CABI Publishing.

Timmons, F.L. 1970. A history of weed control in the United States and Canada. Weed Science 18(2): 294–307.

Tu, C.M. 1994. Effects of herbicides and fumigants on microbial activities in soil. Bulletin of Environmental Contamination and Toxicology 53(1): 12–17.

Umiker, K.J., J.L. Johnson-Maynard, T.D. Hatten, S.D. Eigenbrode and N.A. Bosque-Pérez. 2009. Soil carbon, nitrogen, pH, and earthworm density as influenced by cropping practices in the Inland Pacific Northwest. Soil and Tillage Research 105(2): 184–191. doi: 10.1016/j. still.2009.09.001.

Voos, G. and P.M. Groffman. 1997. Relationships between microbial biomass and dissipation of 2,4-D and dicamba in soil. Biology and Fertility of Soils 24(1): 106–110.

Wardle, D.A. 1992. A comparative assessment of factors which influence microbial biomass carbon and nitrogen levels in soil. Biological Reviews 67(3): 321–358. doi: 10.1111/j.1469-185X.1992. tb00728.x.

Wardle, D.A. and D. Parkinson. 1990. Effects of three herbicides on soil microbial biomass and activity. Plant and Soil 122(1): 21–28. doi: 10.1007/BF02851906.

Weaver, M.A., L.J. Krutz, R.M. Zablotowicz and K.N. Reddy. 2007. Effects of glyphosate on soil microbial communities and its mineralization in a Mississippi soil. Pest Management Science 63(4): 388–393. doi: 10.1002/ps.1351.

White, P.M. and C.W. Rice. 2009. Tillage effects on microbial and carbon dynamics during plant residue decomposition. Soil Science Society of America Journal 73(1): 138–145. doi: 10.2136/ sssaj2007.0384.

Wolińska, A. and Z. Stępniewska. 2012. Dehydrogenase Activity in the Soil Environment. pp. 183–210. *In:* Canuto, R.A. (Ed.) Dehydrogenases. InTech. doi: 10.5772/48294.

Yeomans, J.C. and J.M. Bremner. 1985. Denitrification in soil: effects of herbicides. Soil Biology and Biochemistry 17(4): 447–452.

Yu, Y., H. Zhang and Q. Zhou. 2011. Using soil available P and activities of soil dehydrogenase and phosphatase as indicators for biodegradation of organophosphorus pesticide methamidophos and glyphosate. Soil and Sediment Contamination 20(6): 688–701.

Zabaloy, M.C., I. Carné, R. Viassolo, M.A. Gómez and E. Gomez. 2016. Soil ecotoxicity assessment of glyphosate use under field conditions: microbial activity and community structure of Eubacteria and ammonia-oxidising bacteria. Pest Management Science 72(4): 684–691.

Zabaloy, M.C., J.L. Garland and M.A. Gómez. 2008. An integrated approach to evaluate the impacts of the herbicides glyphosate, 2,4-D and metsulfuron-methyl on soil microbial communities in the Pampas region, Argentina. Applied Soil Ecology 40(1): 1–12. doi: 10.1016/j.apsoil.2008.02.004.

Zabaloy, M.C. and M.A. Gómez. 2005. Diversity of rhizobia isolated from an agricultural soil in Argentina based on carbon utilization and effects of herbicides on growth. Biology and Fertility of Soils 42(2): 83–88. doi: 10.1007/s00374-005-0012-2.

Zabaloy, M.C. and M.A. Gómez. 2008. Microbial respiration in soils of the Argentine Pampas after metsulfuron methyl, 2,4-D, and glyphosate treatments. Communications in Soil Science and Plant Analysis 39(3–4): 370–385. doi: 10.1080/00103620701826506.

Zablotowicz, R.M., M.A. Locke and L.A. Gaston. 2007. Tillage and cover effects on soil microbial properties and fluometuron degradation. Biology and Fertility of Soils 44(1): 27–35.

Zablotowicz, R.M., M.A. Locke, L.A. Gaston and C.T. Bryson. 2000. Interactions of tillage and soil depth on fluometuron degradation in a Dundee silt loam soil. Soil and Tillage Research 57(1–2): 61–68. doi: 10.1016/S0167-1987(00)00150-1.

Zablotowicz, R.M., K.N. Reddy, M.A. Weaver, A. Mengistu, L.J. Krutz, R.E. Gordon and N. Bellaloui. 2010. Cover crops, tillage, and glyphosate effects on chemical and biological properties of a lower Mississippi Delta soil and soybean yield. Environmental Research Journal 4(3/4): 227–251.

Zhang, C., X. Liu, F. Dong, J. Xu, Y. Zheng and J. Li. 2010. Soil microbial communities response to herbicide 2,4-dichlorophenoxyacetic acid butyl ester. European Journal of Soil Biology 46(2): 175–180. doi: 10.1016/j.ejsobi.2009.12.005.

Zobiole, L.H.S., R.J. Kremer, R.S. Oliveira and J. Constantin. 2010. Glyphosate affects micro-organisms in rhizospheres of glyphosate-resistant soybeans. Journal of Applied Microbiology 110(1): 118–127.

Effects of Herbicides on Freshwater Ecosystems

Gertie Arts[*1] and Mark Hanson[2]

[1] Wageningen Environmental Research (Wageningen University and Research),
P.O. Box 47, 6700 AA Wageningen,
The Netherlands, Europe
[2] University of Manitoba, Winnipeg, Manitoba, Canada

Introduction

Freshwater ecosystems are important to society as they provide essential services, such as drinking water, food resources, nutrient flow, enhance water quality, modulate climate, and provide aesthetic and cultural value (Körner 2002, Van Donk and Van de Bund 2002, Wetzel 2001). In addition, they are the habitat and a resource for many organisms that are not strictly aquatic, such as waterfowl and mammals (Soininen et al. 2015). The value of these services has been long recognized; however, this has not prevented freshwater systems from being threatened by a wide array of stressors (e.g., eutrophication, invasive species, habitat loss, over use through irrigation, salinization). In agricultural settings, plant protection products (PPPs, e.g., pesticides) are of concern as they can enter freshwater ecosystems by spray drift, drainage, and run-off from intensively cropped areas, as well as direct application in special cases, such as invasive species control (Schäfer et al. 2016). In the context of these complex issues in water quality, the specific objective of this chapter is to discuss potential environmental issues related to herbicides and their use as well as to briefly review the possible effects of herbicides on freshwater ecosystems. While the focus is on herbicides, the risk presented by other chemicals that can affect primary producers will also be discussed, specifically plant growth regulators and fungicides that exhibit herbicidal activity.

PPPs are substances or mixtures of substances that protect crop plants by eliminating, modifying, or limiting the growth of pests, weeds, or undesired plants (as defined by the European and Mediterranean Plant Protection Organization, 2004). In contrast to other groups of chemical contaminants, the mode of action and resulting impact of most PPPs is relatively specific: they affect those organisms that are most closely related to target pest organisms. Herbicides are a PPP that specifically affect primary producers (Hutson 1998) with a range of modes of action to target undesirable organisms, typically weeds (Rashid et al. 2010). Examples are photosynthetic inhibitors, lipid synthesis inhibitors, amino acid synthesis inhibitors, growth regulators, metabolism (i.e., nitrogen) disruptors, pigment synthesis inhibitors, cell membrane disruptors and root growth inhibitors. As a result herbicides might exert direct effects on sensitive non-target primary producers in freshwater systems. Primary producers are distinguished

---

*Corresponding author: Gertie.Arts@wur.nl

from other groups of organisms by their ability to perform photosynthesis and can be divided into microscopic algae, including photosynthetic bacteria, and aquatic macrophytes. Aquatic macrophytes are a diverse assemblage of plants that have become adapted for life wholly or partially in water, and large enough to be seen with the naked eye (Maltby et al. 2010). These non-target direct effects might lead to indirect effects on other populations in the food web via a number of mechanisms, which will be discussed later in this chapter (Fleeger et al. 2003).

Ecological Risk Assessment (ERA) of Herbicides

Since aquatic primary producers fulfil important roles in freshwater systems they are explicitly considered in the risk assessment of pesticides all over the world (EC 2000, 2009, 2013, EFSA 2013, EPA 2000). Risk assessment can be either prospective or retrospective from a methodological point of view, and in Europe and North America both approaches are laid down in different legislation (EC 2000, 2009, EPA 2000). In Europe, PPPs need authorisation prior to placing them in the market. A dual system is in place, under which the European Food Safety Authority (EFSA) evaluates active substances used in plant protection products and Member States evaluate and authorize the products at the national level (EC 2009). The risk assessment for herbicides in Europe requires three standard tests, i.e., a standard green alga test and a second standard test with an alga species from a different taxonomic group, i.e., either a diatom or a blue-green alga, and a test with a higher plant, i.e., *Lemna* sp. In the United States, the Environmental Protection Agency (EPA) administers the Federal Insecticide, Fungicide, and Rodenticide Act (FIFRA), which dictates federal control on the sale, distribution, and use of herbicides and pesticides. FIFRA requires that four algal species are tested for pesticides registration in the United States (i.e., *Pseudokirchneriella subcapitata, Anabaena flos-aquae,* the diatom *Navicula pelliculosa* and *Skeletonema costatum*), as well as one vascular plant (i.e., *Lemna* spp.) (EPA 2000). Similar to the US and Europe, Canada requires data on the effects of herbicides to non-target aquatic plants. There the approval of herbicides is overseen by the Pest Management Regulatory Agency (PMRA), which is under the auspices of Health Canada. For Latin America, the international framework for risk assessments (protection goals, effects, and exposure assessments, risk characterization, and risk mitigation) is broadly applicable (Carriquiriborde 2014) however, it requires further refinement when used in the region.

In general, and using Europe as an example, prospective risk assessments are based on data from simple laboratory tests, laboratory tests with either additional test species or including modified exposure conditions, population- and ecosystem-level controlled tests in microcosm and mesocosm studies mimicking aquatic ecosystems, and finally modelling approaches under a tiered risk assessment framework. In the case of herbicides, the simple laboratory tests in the first tier include standard tests with *Lemna* and algae (OECD 2006a,b). When either *Lemna* and algae are not sensitive, or tests with terrestrial plants show a greater sensitivity towards dicot species or the plant is exposed via the sediment, a second macrophyte test is needed with a rooted macrophyte. In this case the standard rooted test species is the dicot *Myriophyllum spicatum* (OECD 2014, EFSA 2013). At the higher tiers, a range of algae and macrophyte growth forms might be used in more representative systems (e.g., micro/mesocosms) for the purposes of refined risk assessment for herbicides. In all cases a prospective assessment is based on specific uses of PPPs in certain crops (Brock et al. 2006). In Europe, the retrospective risk assessment, which is the Water Framework Directive (EC 2000), follows a general approach and largely makes use of monitoring data in the exposure assessments (EC 2000). For the effect assessment the WFD addresses the potential ecological risks of all toxic chemicals (including PPPs) and is, therefore, much wider than PPPs only. These other chemicals comprise priority substances, priority hazardous substances, or substances of concern in specific river basins. For the pesticide risk assessment, the WFD (EC 2000) and the Pesticide Directive (EC 2009) make use of the same data. However, they apply different assessment factors and the WFD does not follow a tiered approach (Smit et al. 2013).

Problems with ERA are the proper linkage with the protection goals, which are essential to protect the ecosystems in the field. In this context, the Ecosystem Services Approach has been developed by EFSA (Nienstedt et al. 2012). Also, lower tiers need to be protective for effects in mesocosm studies and the field (Van Wijngaarden and Arts 2018). This validation of the tiered risk assessment is needed in order to calibrate the standard first-tier effect assessment: if we miss sensitive species here, higher tiers might turn out to be more sensitive than the first-tier risk assessment.

Action of Herbicides in Algae and Plants

Until recently photosynthesis inhibitors were by far the most studied herbicides in aquatic ecotoxicology (Brock et al. 2000). Only over the last decade have more ecotoxicological data from non-photosynthesis inhibitors become available in the open literature (Giddings et al. 2013). Herbicides typically act on plant specific pathways by blocking photosynthesis, carotenoid synthesis, or aromatic and branched chain amino acid synthesis that are essential in plants but not in mammals (Casida 2009). Table 4.1 gives an overview of different modes of action for herbicides. Besides herbicides, certain fungicides can exert a herbicidal mode of action of which chlorothalonil and fluazinam are examples. Photosynthesis inhibitors act immediately (Brock et al. 2000), and once plants are surrounded by clean water and exposure to the herbicide has stopped, recovery to original growth rates is typically rapid (Kersting and van Wijngaarden 1999, Brain et al. 2012a,b, Baxter et al. 2014). It is very specific to this group of herbicides that the algae and macrophytes are generally equally sensitive. This is not the case for the ALS-inhibitors (acetolactate synthase inhibitors, i.e., sulfonylureas, Table 4.1), which represent a different mode of action (Ferenc 2001). This group of herbicides selectively inhibit ALS and biosynthesis of essential branched-chain amino acids. The pesticides of this group are weak acids and form a special chemical group of herbicides (Ferenc 2001). Weak acids are pH-sensitive (Ferenc 2001). Comparison of the sensitivity of the macrophyte *Lemna minor* and the alga *Pseudokirchneriella subcapitata* revealed a sensitivity up to 1,000-folds greater for the macrophyte compared to the alga (Cedergreen and Streibig 2005), for which the pH of the apoplast of multicellular plants might be an explanatory factor. Besides the ALS-inhibitors, glyphosate is also under the amino acid inhibitor class that blocks amino acid synthesis (EPSP synthase inhibitor, Table 4.1). Synthetic auxins form a second group of herbicides that act specifically on macrophytes but not, or to a lesser extent, on algae. This is caused by the fact that synthetic auxins are plant hormones non-toxic to algae (Cedergreen and Streibig 2005). Many other herbicides act on plant specific pathways in both algae and macrophytes (Casida 2009). In those cases sensitivity might be more related to growth rate than to mode of action. Examples are included in Table 4.1 and comprise pigment inhibitors which inhibit the enzyme 4-hydroxyphenylpyruvate dioxygenase (HPPD) (Mitchell et al. 2001), lipid synthesis inhibitors which inhibit the enzyme acetyl-CoA carboxylase (ACCase), which catalyses the first step in fatty acid synthesis and is important for membrane synthesis, cell membrane inhibitors like PPO (Protoporphyrinogen oxidase) inhibitors (http://herbicidesymptoms. ipm. ucanr. edu/ MOA/PPO_inhibitors/) or dinitroanilines which disrupt microtubule production in roots.

Herbicide efficacy trials in terrestrial plants might show differences in mode of action and sensitivity between monocotyledonous species (monocots) and dicotyledonous species (dicots); these differences are not always observed in aquatic plants (Arts et al. 2008). Within aquatic macrophytes, differences in growth forms—including emergent, submerged, free-floating and sediment-rooted and floating growth-forms—seem to be more important in explaining differences in sensitivity among aquatic macrophytes than differences between monocotyledonous and dicotyledonous species; however, differences between monocots and dicots might play a role for certain herbicidal modes of action. Ecophysiological differences between terrestrial and aquatic plants are certainly important. While terrestrial plants have a cuticle that protects the plant against desiccation, this protection is not needed in submerged aquatic plants. Submerged aquatic plants are composed of thin leaves of a few cell layers thick while a cuticula and stomata are missing and solidity structures are not needed when surrounded by water. These differences in structure and texture might enhance the permeability of submerged leaves and stems of aquatic macrophytes for intrusion of herbicides.

Table 4.1. Toxic mode of action of different groups of herbicides (effects on plant growth), the site of the action and examples of each mode of action and active ingredient (after https://ag. purdue.edu/ btny/weedscience/Documents/Herbicide_MOA CornSoy_12_2012%5B1%5D. pdf)

Mode of action	Site of action	Example chemical family	Example active ingredient
Lipid Synthesis Inhibitors	ACCase Inhibitors	Aryloxyphenoxy propionate	Fenoxaprop
Amino Acid Synthesis Inhibitors	ALS Inhibitor	Sulfonylurea	Prosulfuron
Amino Acid Synthesis Inhibitors	EPSP Synthase Inhibitor	None accepted	Glyphosate
Growth Regulators	Site Unknown	Phenoxy	2,4-D
Growth Regulators	Auxin Transport	Semicarbazone	Diflufenzopyr
Photosynthesis Inhibitors	Photosystem II Inhibitors	Triazine	Atrazine
Photosynthesis Inhibitors	Photosystem II Inhibitors	Nitrile	Bromoxynil
Photosynthesis Inhibitors	Photosystem II Inhibitors	Ureas	Linuron
Nitrogen Metabolism	Glutamine Synthesis Inhibitor	None accepted	Glufosinate
Pigment Inhibitors	Diterpene Synthesis Inhibitor	Isoxazolidinone	Domazone
Pigment Inhibitors	HPPD Inhibitor	Isoxazole	Isoxaflutole
Cell Membrane Disruptors	PPO Inhibitors	N-phenylphthalimide	Flumioxazin
Cell Membrane Disruptors	Photosystem I Electron Diverter	Bipyridilium	Paraquat
Seedling Root Growth Inhibitors	Microtubule Inhibitors	Dinitroaniline	Trifluralin
Seedling Shoot Growth Inhibitors	Lipid Synthesis Inhibitors (not ACCase)	Thiocarbamate	Butylate
Seedling Shoot Growth Inhibitors	Long-chain Fatty Acid Inhibitor	Chloroacetamide	Metolachlor

Response Variables in Herbicide-exposed Primary Producers

Aquatic macrophytes and algae can show a wide variety of sub-lethal effects as a result of exposure to herbicides (Arts et al. 2008). Risk assessments with aquatic plants have to include response variables that are relevant to protection goals and ecology as well as sensitive to the compound to be tested or to a series of compounds to be tested. In plant ecotoxicology, a response variable is called an 'endpoint' and can be defined as a variable reflecting plant performance and development during and after exposure to a toxic compound (Arts et al. 2008). These endpoints are variables measured in experiments or tests, e.g., cell density for algae, frond number of *Lemna*, number of leaves, shoot length, fresh weight, dry weight, and corresponding growth rates for rooted macrophytes (Arts et al. 2008). Also, physiological endpoints can be included as measurement endpoints, e.g., photosynthetic performance by the use of pulse amplitude modulated (PAM) fluorometry (Pedersen et al. 2013). Not all these endpoints are appropriate, as toxicological sensitivity, variance, and ecological relevance are important criteria to consider in the evaluation of the suitability of potential endpoints in plant toxicity experiments and risk assessment (Arts et al. 2008). Moreover, endpoints need to be linked to the specific protection goals (Nienstedt et al. 2012). For plants these specific protection goals are defined as short-term effects or an absence of effects on biomass of functional groups and keystone species as well as conservation of biodiversity at the watershed and/or landscape level (Nienstedt et al. 2012, Hommen et al. 2016).

Assessment endpoints are defined as standard endpoints used in OECD standard macrophyte tests, i.e., *Lemna* sp. and *Myriophyllum spicatum* tests performed for regulatory purposes (OECD 2006b, 2014). Plant length and biomass endpoints are characterised by low coefficients of variation (Knauer et al. 2006, 2008) and are therefore included in the risk assessment for PPPs (EFSA 2013). As plant biomass can be measured in experimental and field studies and predicted by models, this endpoint forms a link from standard toxicity tests to populations and communities in the field (Hommen et al. 2016). At the field- and landscape-level, surrogate measures for plant biomass might be assessed as well, e.g., plant abundance. Root endpoints are sensitive, however, they show high intrinsic variability (Hanson et al. 2003, Turgut and Fomin 2002, Arts et al. 2008) and are therefore not considered in the current risk assessment. However, sediment-exposure might require the evaluation of belowground macrophyte endpoints.

EFSA (2013) prefers the use of growth rate over yield (for discussion see van Wijngaarden and Arts 2018). While growth rate is based on a natural logarithmic function, yield is not. The latter is only a quotient comparing biomass or length over the experimental period. Intrinsically, yield is dependent on test duration, as longer tests will end up in lower effect concentrations if the incipient has not been reached and unless any recovery has occurred. From a scientific point of view, growth rate is preferred over yield, but as yield is often the lowest endpoint in the risk assessment, endpoint selection is still part of a scientific discourse within the regulatory and scientific community in Europe.

Sensitivity of Primary Producers in Freshwater Ecosystems

Aquatic macrophytes are often classified by their growth habit, the four categories being—emergent and sediment-rooted, rooted and floating-leaved, free-floating, and sediment-rooted and submerged (Maltby et al. 2010). In aquatic macrophyte risk assessment it has become standard to include a range of morphologically and taxonomically different macrophytes in species sensitivity distributions (SSDs) (EFSA 2013). Where feasible, endpoints should be based on a common measurement and duration of exposure for all species (Maltby et al. 2010, EFSA 2013). Data on single-species macrophyte toxicity were analysed to assess the relative sensitivity of *L. gibba* and *M. spicatum* (Giddings et al. 2013). Specifically, 11 herbicides and three fungicides SSDs were plotted and the position of *L. gibba* and *M. spicatum* and the sensitivity of the standard algae test species for pesticide registration in the United States under FIFRA were evaluated. The duckweed *L. gibba* was among the most sensitive species for approximately 50% of the chemicals evaluated, while *M. spicatum* was among the most sensitive macrophyte species for approximately 25% of the chemicals evaluated. In most cases the lowest FIFRA algal endpoint was less than the most sensitive macrophyte species. For risk assessment, a consistent result was that a combination of *L. gibba* and the FIFRA algae was protective of effects in 12 of the 14 herbicides, while in the other two *M. spicatum* was the most sensitive species (Giddings et al. 2013). This suggests that standard test species plus *Myriophyllum spicatum* are generally protective for effects on primary producers. For aquatic plants, the authors concluded that no single species was consistently the most sensitive, but current risk assessment approaches are protective. This has been confirmed by other studies (e.g., Cedergreen et al. 2004a, 2004b) and is consistent with terrestrial plant work (Boutin et al. 2004). Further research is needed to explore the relationship between the mode of action of the compound and sensitivity of macrophyte growth form and/or taxonomy.

Indirect Effects

In an ecological context, changes in primary productivity, whether increases or decreases, can have profound impacts on higher trophic levels (see Figure 4.1). The changes observed in higher trophic levels are not always a result of a direct action on the species themselves—e.g., by direct effects of insecticides on predators—rather it can also be an ecologically-mediated response, or indirect effect, as a result of a direct change of e.g., herbicides in the status of organisms with which they are interacting. Contaminants acting on sensitive species in the food web might trigger a cascade of indirect effects both bottom–up as well as top–down (Fleeger et al. 2003, Figure 4.1). The principles and discussion around top–down and bottom–up theories in food chains have been described by Vadas (1989) in terms of biomass and turnover patterns. Here, we consider bottom–up and top–down mechanisms solely in the context of the influence of contaminants and propagation of indirect effects of contaminants in the food chain. Bottom–up effects take place via the detritivorous and primary production food chain and the autochthonous organic carbon cycle. Also herbicides, fungicides with an herbicidal mode of action, bactericides or antibiotics can exert bottom–up effects by affecting aquatic macrophytes or algae. A specific type of bottom–up effects is the disruption of the autochthonous organic carbon cycle by toxic effects on microbes and bacterivorous nanoplankton and microplankton (Reynolds 2008). These bottom–up effects are best known through studies examining eutrophication and the effects in the food web, specifically with nutrients, such as phosphorous (Schindler et al. 1977, 1993, Johnson et al. 2007). Bottom–up effects can mimic the effects of nutrient addition: it results in elevated primary productivity, followed shortly thereafter by increases

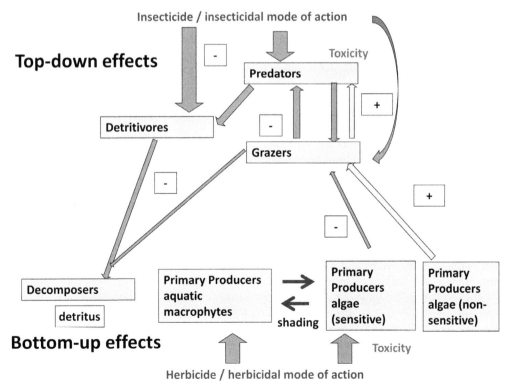

Figure 4.1. A fictive example of a contaminant-induced effects cascade through a simplified aquatic food web initiated by top-down or bottom-up toxic effects. - represent decrease of populations. + represent increase of populations.

in secondary productivity (typically zooplankton and other invertebrates) in response to the greater abundance of food resources (typically in the form of phytoplankton and periphyton). If sustained for a sufficient duration, this increase in secondary production can then enhance higher tropic levels that feed upon the invertebrates, including fish. Overall, the rate of energy flow through the system would increase. In contrast, herbicides could also act in the opposite direction to a nutrient addition. In that case, declines in primary productivity, if sustained, could then result in a decrease in secondary consumers, which then in turn might result in a decrease in fish and other higher trophic levels. While the links driving these indirect effects are conceptual sound, eutrophication including long-term, pronounced effects have only been reported when the threshold levels for effects of the PPPs under consideration were exceeded by more than 10 times (Arts and De Lange 2008). This situation might not be very common. This is in part due to a number of factors. First, not all primary producers will respond in a similar manner to the same toxicant, e.g., there is differential sensitivity to herbicides across species. Due to functional redundancy in freshwater ecosystems, it takes a significant input of herbicide to impair the community of primary producers sufficiently to then reduce secondary consumers via a lack of food (or the appropriate food or the quality of food). In natural systems, primary production has been observed to be sustained even following the loss of a significant number of species as a result of functional redundancy (Lawton and Brown 1994). The ecological risk assessment process discussed earlier aims to minimize the likelihood of this occurring through normal use of PPPs. Secondly, unlike nutrients, such as phosphorous, which can be entrained for long durations (years) in a freshwater ecosystem, especially smaller isolated ones, such as lentic wetlands, herbicidal effects are more transient due to natural degradation and loss of the active molecule. In flowing (lotic) systems, the compound is moved out of the system rapidly, again reducing the likelihood of a prolonged response. Essentially, any impairment is temporary, and the time needed to cause significant impairment at higher levels (zooplankton or fish), is not sufficient. Finally, should significant impacts on primary production occur, recovery of the algae community is swift due to their life history strategies, e.g., rapid growth and replacement (Brain et al. 2012a). Impacts on the macrophyte community might be more pronounced, as these organisms have other life cycle strategies and do not recover as quickly as phytoplankton, but recovery does occur (Brain et al. 2012b). Recovery is also likely as many herbicides are not algicidal in nature, but rather algistatic, in that they inhibit growth rather than outright kill cells and plants. Work with stream mesocosms and atrazine have demonstrated the rapid recovery of primary production following pulsed exposures typical of field exposures in agricultural settings and a lack of impacts on fish (King et al. 2016), further limiting the likelihood of a cascade of effects to higher trophic levels.

In their review, Fleeger et al. (2003) summarized studies that could capture the indirect effects of contaminants, including herbicides, following a direct change on a set of target populations. Overall, there was no consistent response in higher trophic levels following the introduction of a herbicide. Shifts in the phytoplankton community were reported as resulting in no change, a decrease, or an increase in zooplankton. This is not surprising, as even the direct effects of a herbicide will be more complicated than a simple decrease overall in abundance, diversity and primary productivity, as noted in the examples above. Essentially, each herbicide will have a different possible response depending on its mode of action in the target species, the exposure concentration, and the duration of exposure. Experimentally, examples of this bottom–up effect of herbicides in freshwater ecosystems have been demonstrated through the use of mesocosms, which allow for more natural assemblages of phytoplankton, periphyton, zooplankton and other invertebrates, and sometimes fish. Early work by DeNoyelles et al. (1982) and Kettle et al. (1987) with atrazine in outdoor mesocosms documented the direct effects of atrazine on the phytoplankton and macrophyte communities, and subsequent indirect effects on benthic invertebrates and fish (sunfish: *Lepomis macrochirus*, channel catfish *Ictalurus punctatus*, and gizzard shad *Dorosoma cepedianum*). After 136 days of exposure to atrazine at 20 and 500 µg/L there was no difference in fish survivorship from controls for any species, but sunfish reproduction was significantly reduced in both treatments by greater than 95%.

This was attributed to atrazine's direct reduction in macrophyte cover in the mesocosms (60% and 90% coverage, respectively), which subsequently reduced the habitat for benthic invertebrates, the preferred food source for the sunfish, leading indirectly to reduced fecundity. Still, even this study is equivocal due to the inclusion of grass carp, which will decimate macrophytes in their own right and the challenge of isolating this effect. Further complicating the situation are examples where introduction of an herbicide can result in an overall increase in primary productivity. For example, in an 11-day mesocosm study examining the response of phytoplankton communities in response to glyphosate significant increases (by a factor of two) in primary production were observed due to the more sensitive micro- and nano-plankton being replaced by less sensitive picocyanobacteria (Perez et al. 2007). This response is further complicated by glyphosate being able to act as a fertilizer due to the phosphorous moiety in the glyphosate that was rapidly released through degradation, which in this study was equivalent to 840 and 1680 μg/L P. It should be noted that in all the above examples, the concentrations and durations of exposure employed in the studies that produced the relevant effect are well above those values typically seen in freshwater ecosystems, or even allowed as a result of the ecological risk assessment process (Arts and De Lange 2008). Regardless, post-registration monitoring is typical, as well as regular re-evaluations, and these can potentially lead to a change in the regulation of compounds.

An interesting, but rare, example of the potential of herbicides to act indirectly and impair other higher level organisms in freshwater ecosystems is around symbionts. The green algae *Oophila amblystomatis* colonizes the eggs of the yellow-spotted salamander (*Ambystoma maculatum*), from which it derives protection from grazing and nitrogen compounds, while providing the developing embryo with dissolved oxygen and removing waste compounds (Kerney et al. 2011). When the growth and development of the algae are impaired, this can reduce the developmental success of the salamander, presenting a plausible mechanism by which a herbicide might impair a vertebrate, albeit indirectly. Baxter et al. (2014) were successful in isolating the algae from the egg and performing standard 96-hour toxicity tests with atrazine, reporting EC50s greater than those typically seen for standard test species. In a follow-up laboratory study with atrazine and whole egg masses exposed through embryo to hatch (>60 days), no impact on the salamander or algae were observed, again at concentrations significantly greater than those observed in surface waters (Baxter et al. 2015). A cautionary tale from this work is the need to properly identify the species one is working with, especially with algae, which in many cases have unresolved taxonomic issues. In the method development phase of this work, genetic analyses were used to confirm the identity of the isolated algae, which was found to be a new species of symbiont from those reported from other geographical locations for the salamander (Rodriquez-Gil et al. 2014).

Herbicide Interaction with Nutrients, Pesticide Mixtures, and Other Ecologically Relevant Scenarios in Freshwater Ecosystems

As with many stressors in natural systems, herbicides rarely occur alone, and the complex mixtures (chemical, physical, and biological) pose challenges for our ability to interpret possibly ecologically significant impairment. Consistent with their use patterns and the mechanisms by which many herbicides are transported off-field following application (e.g., run-off), the most commonly co-occurring stressor with herbicides tends to be nutrients. Fertilizers are applied widely in agriculture, and regardless of the form, the movement of excess phosphorous and nitrogen has consequences for water quality as noted previously. In streams, this relationship between nutrients and the herbicide atrazine was demonstrated in a monitoring study of field run-off in the Mid-Western United States (Andrus et al. 2013). When herbicides and nutrients co-occur at sufficient concentrations to cause a change in primary productivity, they tend to counter-act each other, as has been observed experimentally. For example, a mesocosm study examining the interaction between atrazine and phosphorous found that the addition of phosphorous masked reductions in dissolved oxygen with little to no difference in biomass, even at greatly elevated atrazine concentrations (Baxter et al. 2013). At the field level, work examining the relationship between various water quality parameters (including atrazine and nutrients) and the macrophyte community in streams and rivers from agricultural catchments found nutrients to be the driver for observed declines in community structure (Dalton et al. 2015). Floristic quality decreased as nitrate increased in the system and from this they concluded that nutrient enrichment has adverse effects on riparian and aquatic plant communities that override any possible effects of atrazine. The role of nutrients in driving periphyton community structure in agricultural streams, as opposed to atrazine was also noted in the Mid-Western United States (Andrus et al. 2013).

The complexity of the nutrient-herbicide relationship is highlighted by concerns around amphibian parasitism. It has been hypothesized that herbicides could increase parasitism by trematodes, which promote limb deformity in frogs, make them more susceptible to predation. The trematodes rely on snails as an intermediary host, who in turn prefer periphyton as a food source. The concern is that herbicides could shift primary production from phytoplankton to periphyton in agricultural settings due to increased water clarity via the loss of algae (Rohr et al. 2008). With the greater abundance of food, snail numbers might increase with excessive herbicide inputs, assuming that the periphyton is not impaired by the herbicide either. First it must be noted that the concentrations required to cause a sufficient decline in phytoplankton communities to cause a shift to periphyton are simply not seen in surface waters (Solomon et al. 1996, 2008). A more parsimonious explanation for the observation of increased parasitism in agriculture settings is due to increased nutrient loading, which tends to enhance primary productivity overall, to the benefit of the snails, and subsequently the trematodes. Johnson et al. (2007) demonstrated that by simply adding phosphorous to an experimental aquatic system, periphyton increased, snail abundance and biomass increased, with an overall increase in the rate of infection for snails by trematode cercariae. Further work at the landscape level by Hartson et al. (2011) found that phosphorous concentrations in ponds positively influenced amphibian parasite distributions.

In agricultural systems, herbicides can co-occur not just with nutrients, but also with other pesticides, leading to concerns about possible synergy of effects (i.e., greater than predicted, based on the individual components). Still, work has shown that most mixtures of pesticides with an herbicide component at environmentally relevant concentrations are unlikely to cause any significant impairment (Cedergreen 2014). Hartgers et al. (1998) examined a mixture of the herbicides atrazine, diuron, and metolachlor over a range of concentrations with continuous exposure for four weeks, followed by another four weeks of monitoring with natural dissipation. Effects were only observed at concentrations well above environmental relevance and they concluded that current risk assessment practices would be protective of this mixture at realistic environmental concentrations. Van den Brink et al. (2009) conducted a chronic microcosm study where they examined the response of phytoplankton and zooplankton to both atrazine and the insecticide lindane. Even after greater than 60 days exposure to atrazine at concentrations upwards of 250 µg/L, there was little to no response observed in the phytoplankton or periphyton communities. A recent review summarized the outcomes of studies reported in the peer-reviewed literature for mixtures of herbicides and herbicides with other chemical contaminants (Cedergreen 2014). It concluded that herbicides as mixtures can be synergistic at times, but that these responses are typically not observed at environmental concentrations.

In the available scientific literature, most publications studying herbicides focus on the phytoplankton and periphyton community which is a consequence of the fact that, for many modes of action, algae are among the most sensitive organisms (Giddings et al. 2013). In previous paragraphs, it was discussed that ALS-inhibitors and synthetic auxins primarily act selectively on macrophytes, potentially shifting the equilibrium between algae and submerged macrophytes in freshwater ecosystems in the direction of algae dominance. These changes can potentially pose a long-term risk to macrophyte-dominated communities and ecosystems (Wendt-Rasch et al. 2004) at sufficient exposure. In freshwater systems, the competition between phytoplankton, periphyton, (submerged) macrophytes and filamentous algae is a delicate equilibrium. Herbicides might disturb this equilibrium by primarily acting on one of these primary producer groups, e.g., on filamentous algae in the case of metazachlor (Mohr et al. 2007), on algae (many examples discussed already with atrazine), or rooted macrophytes (Wendt-Rasch et al. 2003). These disturbances can have potential indirect effects on the other primary producer groups, resulting in complex and non-linear indirect effects.

Agricultural systems might be characterized by a range of different exposure regimes, ranging from chronic to pulsed. In order to study the effects of the herbicide linuron, a photosynthesis II inhibiting herbicide, under different regimes experimental ditch studies were performed (Cuppen et al. 1997, Van den Brink et al. 1997). The effects of the herbicide were studied via different exposure regimes in experimental ditches: a chronic exposure regime and a pulsed exposure regime. In the chronic studies it was found that at the highest concentrations of 50 and 150 µg/L the plant biomass of the submerged macrophyte *Elodea nuttallii* significantly decreased and also the oxygen levels and pH levels decreased when compared to controls. These decreases were followed by an increase of Chlorophyll-a levels in periphyton and phytoplankton (Cuppen et al. 1997, Van den Brink et al. 1997). In contrast to the effects of chronic exposure, pulsed-dose experiments with the same herbicide linuron did not result in a consistent decline of the submerged macrophytes and an increase in Chlorophyll-a (Kersting et al. 1999, Van Geest et al. 1999). These authors only found effects that were less severe. Apparently chronic levels of this herbicide were a prerequisite for observing these shifts in primary producer dominance.

Gaps and Uncertainties

There are a number of recognized gaps in our understanding of the potential impacts of herbicides on freshwater ecosystems. As noted previously, a significant challenge with phytoplankton, and especially periphyton, is taxonomic confirmation of the species studied, even in the laboratory (Baxter et al. 2015). This is especially true in mesocosm and field studies where taxonomy can be even more difficult due to the shear diversity of species needing identification (Culverhouse et al. 2003). This poor taxonomy can have significant consequences for interpreting data, leading to wrong conclusions (Bortolus 2008). Any new species designed to screen for herbicidal effects should be confirmed through genetic sequencing. New techniques in genetic sequencing and identification of environmental DNA might be beneficial in assessing herbicidal effects on algae communities at the population and ecosystem level. In addition to taxonomic challenges the baseline dynamics of an ecosystem needs further research in order to increase our understanding and to collect data for calibration and validation of models.

Another gap is a formal methodological and assessment framework to characterize recovery in primary producers. In algae and duckweed, recovery following exposure to herbicides can be assessed relatively easily (Brain et al. 2012a,b, Baxter et al. 2014). For sediment-rooted macrophytes, assessment of recovery is more complicated. As opposed to algae and free-floating species like *Lemna*, their growth is much slower and recovery following exposure is therefore less quick. Recovery in growth rate might be achieved within a short timeframe. However, recovery of biomass can only be expected within a longer time frame, even in the next growing season (Hommen et al. 2016). What are needed are standard methods for the assessment of recovery of primary producers, as well as accepted definitions of recovery for use in ecological risk assessment.

Concluding Remarks

It can be reasonably concluded that potential effects of herbicides on freshwater ecosystems mainly propagate through the foodweb via bottom–up effects. While direct effects on primary producers can often be predicted from the mode of action of the herbicide, indirect effects are dependent on the foodweb in the specific freshwater system under consideration as shown in this chapter. It should be noted that in all the examples above, the concentrations employed in the studies that produced the reported relevant effects are well above the exposures typically seen in freshwater ecosystems, or even allowed as a result of the ecological risk assessment process. Herbicides do not act alone, but can interact with other compounds and nutrients in freshwater ecosystems. New developments in assessment of environmental DNA might be helpful for assessment of herbicidal effects on the algae community. Methods for the assessment of recovery are needed for all primary producer groups. Accepted definitions of recovery are most urgent in the risk assessment for rooted aquatic macrophytes. Here matrix- and individual-based population models and ecosystem models (e.g., CASM and AQUATOX) might help as they are useful tools to help understand and predict effects.

REFERENCES

Andrus, J.M., D. Winter, M. Scanlan, S. Sullivan, W. Bollman, J.B. Waggoner, A.J. Hosmer and R.A. Brain. 2013. Seasonal synchronicity of algal assemblages in three Midwestern streams exhibiting varying levels of atrazine and other chemicals from agricultural field inputs. Science of the Total Environment 458–460: 125–139.

Arts, Gertie H.P., J. Dick M. Belgers, Conny H. Hoekzema and Jac T.N.M. Thissen. 2008. Sensitivity of submersed freshwater macrophytes and endpoints in laboratory toxicity tests. Environmental Pollution 153: 199–206.

Arts, G.H.P. and H.J. de Lange. 2008. Can loading of aquatic systems with plant protection products boost eutrophication? Wageningen, Alterra, Alterra-rapport 1747. 35 blz., 6. Figure, 2 tab., 57 ref. (in Dutch).

Baxter, L.R., P.K. Sibley, K.R. Solomon and M.L. Hanson. 2013. Interactions between atrazine and phosphorus in aquatic systems: effects on phytoplankton and periphyton. Chemosphere 90(3): 1069–1076.

Baxter, L., R. Brain, J.L. Rodriguez-Gil, A. Hosmer, K. Solomon and M. Hanson. 2014. Response of the green alga *Oophila* sp., a salamander endosymbiont, to a PSII-inhibitor under laboratory conditions. Environmental Toxicology and Chemistry 33(8): 1858–1864.

Baxter, L., R.A. Brain, A.J. Hosmer, M. Nema, K.M. Müller, K.R. Solomon and M.L. Hanson. 2015. Effects of atrazine on egg masses of the yellow-spotted salamander (*Ambystoma maculatum*) and its endosymbiotic alga (*Oophila amblystomatis*). Environmental Pollution 206: 324–331.

Bortolus, A. 2008. Error cascades in the biological sciences: the unwanted consequences of using bad taxonomy in ecology. AMBIO: A Journal of the Human Environment 37(2): 114–118.

Boutin, C., N. Elmegaard. and C. Kjaer. 2004. Toxicity testing of fifteen non-crop plant species with six herbicides in a greenhouse experiment: implications for risk assessment. Ecotoxicology 13(4): 349–369.

Brain, R.A., J.R. Arnie, J.R. Porch. and A.J. Hosmer. 2012a. Recovery of photosynthesis and growth rate in green, blue–green, and diatom algae after exposure to atrazine. Environmental Toxicology and Chemistry 31(11): 2572–2581.

Brain, R.A., A.J. Hosmer, D. Desjardins, T.Z. Kendall, H.O. Krueger and S.B. Wall. 2012b. Recovery of duckweed from time-varying exposure to atrazine. Environmental Toxicology and Chemistry 31(5): 1121–1128.

Brock, T.C.M., G.H.P. Arts, L. Maltby and P.J. Van den Brink. 2006. Aquatic Risks of Pesticides, Ecological Protection Goals and Common Aims in European Union Legislation. Integrated Environmental Assessment and Management 2(4): 20–46.

Brock, T.C.M., J. Lahr and P.J. Van den Brink. 2000. Ecological risks of pesticides in freshwater systems: Part 1: herbicides. Wageningen, Alterra report 088. 124 pp.

Carriquiriborde, P., P. Mirabella, A. Waichman, K. Solomon, P.J. Van den Brink and S. Maund. 2014. Aquatic risk assessment of pesticides in Latin America. Integrated Environmental Assessment and Management 1561. 10(4): 539–542.

Casida, J.E. 2009. Pesticide toxicology: the primary mechanism of pesticide action. Chemical Research in Toxicology. 22(4): 609–619.

Cedergreen, N., N.H. Spliid and J.C. Streibig. 2004a. Species-specific sensitivity of aquatic macrophytes towards two herbicide. Ecotoxicology and Environmental Safety 58(3): 314–323.

Cedergreen, N., J.C. Streibig and N.H. Spliid. 2004b. Sensitivity of aquatic plants to the herbicide metsulfuron-methyl. Ecotoxicology and Environmental Safety 57(2): 153–161.

Cedergreen, N. and J.C. Streibig. 2005. The toxicity of herbicides to non-target aquatic plants and algae: assessment of predictive factors and hazard. Pest Management Science 61(12): 1152–1160.

Cedergreen, N. 2014. Quantifying synergy: a systematic review of mixture toxicity studies within environmental toxicology. PloS One 9(5): 96580.

Culverhouse, P.F., R. Williams, B. Reguera, V. Herry and S. González-Gil. 2003. Do experts make mistakes? A comparison of human and machine identification of dinoflagellates. Marine Ecology Progress Series 247: 17–25.

Cuppen, J.G.M., P.J. Van den Brink, E. Camps, K.F. Uil and T.C.M. Brock. 2000. Impact of the fungicide carbendazim in microcosms. I. Water quality, breakdown of particulate matter and responses of macroinvertebrates. Aquat. Toxicol. 48: 233–250.

Dalton, R.L., C. Boutin and F.R. Pick. 2015. Nutrients override atrazine effects on riparian and aquatic plant community structure in a North American agricultural catchment. Freshwater Biology 60(7): 1292–1307.

DeNoyelles, F., W.D. Kettle and D.E. Sinn. 1982. The responses of plankton communities in experimental ponds to atrazine, the most heavily used pesticide in the United States. Ecology 63(5): 1285–1293.

EC. 2009. Regulation (EC) No 1107/2009 of the European parliament and the council of 21 October 2009 concerning the placing of plant protection products on the market and repealing Council Directives 79/117/EEC and 91/414/EEC. Official Journal of the European Union L 309: 1–50.

EC. 2000. Directive 2000/60/EC of the European parliament and of the Council of 23 October 2000 establishing a framework for Community action in the field of water policy. Information available via http://ec. europa. eu/environment/water/water-framework/index_en. html.

EC. 2013. Commission Regulation (EU) No 283/2013 of 1 March 2013 setting out the data requirements for active substances, in accordance with Regulation (EC) No. 1107/2009 of the European Parliament and of the Council concerning the placing of plant protection products on the market. Official Journal of the European Union.

EFSA PPR Panel (EFSA Panel on Plant Protection Products and their Residues), 2013. Guidance on tiered risk assessment for plant protection products for aquatic organisms in edge-of-field surface waters. EFSA Journal 11(7): 3290, 268 pp. doi: 10.2903/j.efsa.2013.3290.

Ferenc, S.A. (Ed.) 2001. Impacts of Low-Dose, High-Potency Herbicides on Nontarget and Unintended Plant Species. SETAC technical publications series. 177 pp.

Fleeger, J.W., K.R. Carman and R.M. Nisbet. 2003. Indirect effects in aquatic ecosystems. The Science of the Total Environment 317: 207–233.

Giddings, J., G. Arts and U. Hommen. 2013. The Relative Sensitivity of Macrophyte and Algal Species to Herbicides and Fungicides: an analysis using species sensitivity distributions. Integrated Environmental Assessment and Management 9(2): 308–318.

Hanson, M.L., H. Sanderson and K.R. Solomon. 2003. Variation, replication, and power analysis of *Myriophyllum* spp. microcosm toxicity data. Environmental Toxicology and Chemistry 22: 1318–1329.

Hartgers, E.M., G.R. Aalderink, P.J. Van den Brink, R. Gylstra, J.W.F. Wiegman and T.C. Brock. 1998. Ecotoxicological threshold levels of a mixture of herbicides (atrazine, diuron and metolachlor) in freshwater microcosms. Aquatic Ecology 32(2): 135–152.

Hartson, R.B., S.A. Orlofske, V.E. Melin, R.T. Dillon and P.T. Johnson. 2011. Land use and wetland spatial position jointly determine amphibian parasite communities. EcoHealth 8(4): 485–500.

Health Canada Pest Management Regulatory Agency (PMRA). 2002. Pest control products act. https://www.canada.ca/en/health-canada/services/consumer-product-safety/pesticides-pest-management/public/protecting-your-health-environment/pest-control-products-acts-and-regulations-en.html

Hommen, U., W. Schmitt, S. Heine, T.C.M. Brock, S. Duquesne, P. Manson, G. Meregalli, H. Ochoa-Acuña, P. van Vliet and G.H.P. Arts. 2016. How TK-TD and population models for aquatic macrophytes could support the risk assessment for plant protection products. Integrated Environmemtal Assessment and Management 12(1): 82–95.

Hutson, David Herd. 1998. Metabolic Pathways of Agrochemicals: herbicides and plant growth regulators. Vol. 1. Royal Society of Chemistry.

Johnson, P.T., J.M. Chase, K.L. Dosch, R.B. Hartson, J.A. Gross, D.J. Larson, D.R. Sutherland and S.R. Carpenter. 2007. Aquatic eutrophication promotes pathogenic infection in amphibians. Proceedings of the National Academy of Sciences 104(40): 15781–15786.

Kerney, R., E. Kim, R.P. Hangarter, A.A. Heiss, C.D. Bishop and B.K. Hall. 2011. Intracellular invasion of green algae in a salamander host. Proc Nat Acad Sci USA 108: 6497–6502.

Kersting, K. and R.P.A. Van Wijngaarden. 1999. Effects of a pulsed treatment with the herbicide afalon (active ingredient linuron) on macrophyte-dominated mesocosms. I. Responses of ecosystem metabolism. Environmental Toxicology and Chemistry 18(12): 2859–2865.

Kettle, W.D., B.D. Heacock and A.M. Kadoum. 1987. Diet and reproductive success of bluegill recovered from experimental ponds treated with atrazine. Bulletin of Environmental Contamination and Toxicology 38(1): 47–52.

King, R.S., R.A. Brain, J.A. Back, C. Becker, M.V. Wright, V. Toteu Djomte, W.C. Scott, S.R. Virgil, B.W. Brooks, A.J. Hosmer and C.K. Chambliss. 2016. Effects of pulsed atrazine exposures on autotrophic community structure, biomass, and production in field-based stream mesocosms. Environmental Toxicology and Chemistry 35(3): 660–675.

Knauer, K., S. Mohr and U. Feiler. 2008. Comparing growth development of *Myriophyllum* spp. in laboratory and field experiments for ecotoxicological testing. Environ. Sci. Pollut. Res. 15: 322–331.

Knauer, K., M. Vervliet-Schneebaum, R.J. Dark and S.J. Maund. 2006. Methods for assessing the toxicity of herbicides to submersed aquatic plants. Pest. Manage. Sci. 62: 715–722.

Körner, S. 2002. Submerged macrophytes—an important tool for lake restoration in Germany? Wasser und Boden 54: 38–41.

Lawton, J.H. and V.K. Brown. 1994. Redundancy in ecosystems. pp. 255–270. *In:* Biodiversity and Ecosystem Function. Springer Berlin Heidelberg.

Maltby, L., D. Arnold, G. Arts, J. Davies, F. Heimbach, C. Pickl and V. Poulsen. 2010. Aquatic Macrophyte Risk Assessment for Pesticides. 135. SETAC Europe Workshop AMRAP, Wageningen, Netherlands. SETAC Press & CRC Press, Taylor and Francis Group, Boca Raton, London, New York.

Mitchell, G., D.W. Bartlett, T.E. Fraser, T.R. Hawkes, D.C. Holt, J.K. Townson and R.A. Wichert. 2001. Mesotrione: a new selective herbicide for use in maize. Pest Management Science 57(2): 120–128.

Mohr, S., R. Berghahn,. M. Feibicke, S. Meinecke, T. Ottenströer, I. Schmiedling, R. Schmiediche and R. Schmidt. 2007. Effects of the herbicide metazachlor on macrophytes and ecosystem function in freshwater pond and stream mesocosms. Aquatic Toxicology 82: 73–84.

Nienstedt, K.M., T.C.M. Brock, J. van Wensem, M. Montforts, A. Hart, A. Aagaard, A. Alix, J. Boesten, S.K. Bopp, C. Brown, E. Capri, V. Forbes, H. Kopp, M. Liess, R. Luttik, L. Maltby, J.P. Sousa, F. Streissl and A.R. Hardy. 2012. Development of a framework based on an ecosystem services approach for deriving specific protection goals for environmental risk assessment of pesticides. Science of the Total Environment 415: Special Issue: SI, 31–38.

OECD. 2006a. Freshwater Algae and Cyanobacteria Growth Inhibition Test. OECD Guidelines for the Testing of Chemicals. 25 pp.

OECD. 2006b. *Lemna* sp. Growth Inhibition Test. OECD Guidelines for the Testing of Chemicals. 22 pp.

OECD. 2014. Water-sediment *Myriophyllum spicatum* toxicity test. OECD Guidelines for the Testing of Chemicals. 18 pp.

Pedersen, O., T. Colmer and K. Sand-Jensen. 2013. Underwater Photosynthesis of Submerged Plants – Recent Advances and Methods. Frontiers in Plant Science 4: 140.

Pérez, G.L., A. Torremorell, H. Mugni, P. Rodríguez, M.S. Vera, M.D. Nascimento, L. Allende, J. Bustingorry, R. Escaray, M. Ferraro and I. Izaguirre. 2007. Effects of the herbicide Roundup on freshwater microbial communities: a mesocosm study. Ecological Applications 17(8): 2310–2322.

Rashid, B., T. Husnain and S. Riazuddin. 2010. Herbicides and Pesticides as Potential Pollutants: A Global Problem. pp. 427–447. *In:* Plant Adaptation and Phytoremediation. Springer Netherlands. DOI: 10.1007/978-90-481-9370-7_19

Reynolds, C.S. 2008. A Changing Paradigm of Pelagic Food Webs. Hydrobiology 93(4–5): 517–531.

Rodríguez-Gil, J.L., R. Brain, L. Baxter, S. Ruffell, B. McConkey, K. Solomon and M. Hanson. 2014. Optimization of culturing conditions for toxicity testing with the alga *Oophila* sp. (Chlorophyceae), an amphibian endosymbiont. Environmental Toxicology and Chemistry 33(11): 2566–2575.

Rohr, J.R., A.M. Schotthoefer, T.R. Raffel, H.J. Carrick, N. Halstead, J.T. Hoverman, C.M. Johnson, L.B. Johnson, C. Lieske, M.D. Piwoni and P.K. Schoff. 2008. Agrochemicals increase trematode infections in a declining amphibian species. Nature 455(7217): 1235–1239.

Schäfer, R.B., B. Kühn, E. Malaj, A. König and R. Gergs. 2016. Contribution of organic toxicants to multiple stress in river ecosystems. Freshwater Biology 61: 2116–2128.

Schindler, D.W. 1977. Evolution of phosphorus limitation in lakes. Science 195(4275): 260–262.

Schindler, D.E., J.F. Kitchell, X.I. He, S.R. Carpenter, J.R. Hodgson and K.L. Cottingham. 1993. Food web structure and phosphorus cycling in lakes. Transactions of the American Fisheries Society 122(5): 756–772.

Smit, C.E., G.H.P. Arts, T.C.M. Brock, T.E.M. ten Hulscher, R. Luttik and P.J.M. vanVliet. 2013. Aquatic effect and risk assessment for plant protection products: evaluation of the Dutch 2011 proposal. Alterra Wageningen UR, 2013 (Alterra-rapport 2463), 146 p.

Soininen, J., P. Bartels, J. Heino, M. Luoto and H. Hillebrand. 2015. Toward more integrated ecosystem research in aquatic and terrestrial environments. Bioscience 65(2): 174–182.

Solomon, K.R., D.B. Baker, R.P. Richards, K.R. Dixon, S.J. Klaine, T.W. La Point, R.J. Kendall, C.P. Weisskopf, J.M. Giddings, J.P. Giesy and L.W. Hall. 1996. Ecological risk assessment of atrazine in North American surface waters. Environmental Toxicology and Chemistry 15(1): 31–76.

Solomon, K.R., J.A. Carr, L.H. Du Preez, J.P. Giesy, R.J. Kendall, E.E. Smith and G.J. Van Der Kraak. 2008. Effects of atrazine on fish, amphibians, and aquatic reptiles: a critical review. Critical Reviews in Toxicology 38(9): 721–772.

Turgut, C. and A. Fomin. 2002. The ability of *Myriophyllum aquaticum* (Vell.) Verdcourt in the uptake and the translocation of pesticides via roots with a view to using the plants in sediment toxicity testing. Journal of Applied Botany 76: 62–65.

USEPA (US Environmental Protection Agency). 2000. Federal Insecticide, Fungicide, and Rodenticide Act (FIFRA).

Vadas Jr, R.L. 1989. Food web patterns in ecosystems: a reply to Fretwell and Oksanen. Oikos. 56(3): 339–343.

Van den Brink, P.J., E.M. Hartgers, U. Fettweis, S.J.H. Crum, E. Van Donk, T.C.M. Brock. 1997. Sensitivity of macrophyte-dominated freshwater microcosms to chronic levels of the herbicide linuron. I. Primary producers. Ecotoxicol. Environ. Saf. 38: 13–24.

Van den Brink, P.J., S.J. Crum, R. Gylstra, F. Bransen, J.G. Cuppen and T.C. Brock. 2009. Effects of a herbicide–insecticide mixture in freshwater microcosms: risk assessment and ecological effect chain. Environmental Pollution 157(1): 237–249.

Van Donk, E. and W.J. Van de Bund. 2002. Impact of submerged macrophytes including charophytes on phyto- and zooplankton communities: allelopathy versus other mechanisms. Aquat. Bot. 72: 261–274.

Van Geest, G.J., N.G. Zwaardemaker, R.P.A. Van Wijngaarden and J.G.M. Cuppen. 1999. Effects of a pulsed treatment with the herbicide afalon (active ingredient linuron) on macrophyte-dominated mesocosms. I. Structural responses. Environmental Toxicology and Chemistry 18(12): 2866–2874.

Van Wijngaarden, P.A. René and Gertie H.P. Arts. 2018. Is the Tier-1 effect assessment for herbicides protective for aquatic algae and vascular plant communities? Environmental Toxicology and Chemistry 37(1): 175–183.

Wendt-Rasch, L., P. Pirzadeh and P. Woin. 2003. Effects of metsulfuron-methyl and cypermethrin exposure on freshwater model ecosystems. Aquatic Toxicology 63: 243–256.

Wetzel, R.G. 2001. Limnology: lake and river ecosystems. Monograph. Academic Press.

Direct and Indirect Effects of Herbicides on Insects

John L. Capinera

Entomology & Nematology Department, University of Florida, Gainesville, FL 32611, USA
E-mail: Capinera@ufl.edu

Introduction

Increased agricultural productivity is due to many factors, including selective breeding, better fertilization and irrigation practices, and importantly – pest management. Chemical pesticides are an integral part of most plant production systems, and are the principal tools by which weeds, insects (and a few other arthropods), and plant pathogens (mostly fungi) are managed. In many crops, and certainly on most acreage, herbicides are used more than insecticides and fungicides. Use varies among crops, however. For example, although herbicide use on wheat grown in the USA greatly exceeds insecticide and fungicide use, in potatoes the amounts of herbicides, insecticides, and fungicides applied are equivalent. Also, there are shifts in use due to the advent of new technologies, occurrence of new pests, and changes in pesticide costs and in the economics of crop production. Although the nominal cost of pesticides has increased over the past 50 years, the pesticide cost share (pesticide costs relative to other inputs) has diminished (Fernandez-Cornejo et al. 2014).

Historically, herbicides have been considered to selectively kill weeds, whereas insecticides kill insects, and fungicides kill fungi. Although substantially true, there are many situations wherein there are direct or indirect effects of herbicides on nontarget organisms. Thus, herbicides can directly affect some insects, or indirectly affect them by modifying the plant community available to the insects, or the plant pathogens available to be transmitted by insects that vector pathogens. Here I review some of the interrelationships of herbicides and insects, focusing on: (1) physiological effects of herbicides on insects; (2) modification of the plant community by herbicides, and the consequences for insect abundance and diversity; and (3) effects of herbicides on occurrence of plant pathogens available for insects to vector to crops. The integration of herbicide application and insect herbivory is suggested as an important area for future research.

Physiological Effects of Herbicides on Insects

Most herbicides have been designed to take advantage of biochemical pathways that are unique to plants, and even pathways that are unique to certain plants, thereby achieving selectivity and sparing crop plants from injury. Consequently, they tend to have lower levels of toxicity to animals than insecticides, and they are not normally thought of as affecting insects.

Indeed, most herbicides have not been shown to have direct effects on arthropods (Norris and Kogan 2005).

However, the physiology of insects can be significantly disrupted by some herbicides, sometimes leading to death. For example, when thiobencarb and endothall herbicides were fed to cabbage looper (*Trichoplusia ni* [Hübner]: Lepidoptera: Noctuidae) larvae, thiobencarb inhibited cuticular hydrocarbon deposition, and was as toxic as many insecticides, displaying an LD_{50} value of < 300 mg/kg. Endothall was slightly less toxic, but caused prolonged development time of larvae and produced pupae that were smaller in size (Brown 1987). Dicamba was reported by Bohnenblust et al. (2013) to not affect survival of corn earworm, *Helicoverpa zea* (Boddie) (Lepidopera: Noctuidae), or painted lady, *Vanessa cardui* (Linnaeus) (Lepidoptera: Nymphalidae), but larval and pupal mass of *V. cardui* were negatively affected by higher concentrations of this herbicide. Ladybird beetle larvae (Coleoptera: Coccinellidae) have long been known to be poisoned by 2,4-D, and development time of survivors prolonged (Adams 1960). Subsequently, Adams and Drew (1965) postulated that aphid populations in grain fields could increase due to destruction of their coccinellid natural enemies by herbicides. Sotherton (1982) and Trumble and Kok (1979) also reported the toxicity of 2,4-D to Coleoptera, in this case *Gastrophysa polygoni* (Linnaeus) (Chrysomelidae) and *Rhinocyllus conicus* (Frölich) (Curculionidae). Toxicity of 2,4-D to coccinellids was recently confirmed by Michaud and Vargas (2010), who reported that it caused mortality to two species, though two other herbicide mixtures (Dupont Ally™ and Syngenta Rave™) had no measurable effect on these beneficial predators. Similarly, the susceptibilities of the weed-feeding moth *Leucoptera spartifoliella* Hübner (Lepidoptera: Lyonetiidae), seed beetle *Bruchidius villosus* F. (Coleoptera: Chrysomelidae), and psyllid *Arytainilla spartiophila* (Förster) (Hemiptera: Psyllidae) were determined for three herbicides and two surfactants at field rates of application by Affeld et al. (2004). These authors reported that the insects varied in susceptibility (*A. spartifoliella* was most susceptible), and that the herbicides (triclopyr, triclopyr + picloram, glyphosate) varied in toxicity (glyphosate tended to be less toxic than the other herbicides). Also, the surfactants (polydimethylsiloxane, dimethicone copolyol) tested alone were toxic to all insects. Herbicides were also evaluated on young honey bees, *Apis mellifera* L. (Hymenoptera: Apidae) (Morton et al. 1972). This assessment of 17 herbicides showed that although most were relatively non-toxic, several were toxic when ingested, namely paraquat, MAA, MSMA, DSMA, hexaflurate, and cacodylic acid. Of these, paraquat, MAA, and cacodylic acid were most hazardous to bees. The herbicide 2,4-DB was found to cause very little ($< 7\%$) mortality to two species of insect natural enemies: *Hippodamia convergens* Guérin-Méneville (Coleoptera: Coccinellidae) and *Chrysoperla carnea* Stephens (Neuroptera: Chrysopidae) (Wilkinson et al. 1975). In contrast to some earlier studies, glyphosate-based herbicide was found to have negative effects on rose-grain aphid, *Metopolophium dirhodum* (Walker) (Hemiptera: Aphididae), aphid mortality increased with increasing concentration of the herbicide, development rate was increased, and fecundity decreased, indicating significant physiological effects (Saska et al. 2016). Using a life table analysis to project population changes, the authors of this glyphosate study estimated an aphid population of nearly nine million in 60 days in the glyphosate-free population, in contrast to a population of less than one million aphids where aphids were treated at a high concentration of glyphosate. In some cases, it is not the herbicide per se that negatively affects insects, but the surfactant used in conjunction with the herbicide (Affeld et al. 2004).

The disruptive effects of some herbicides are not limited to terrestrial invertebrates. For example, Crosby and Tucker (1966) reported that diquat and dichlobenil could cause immobilization, disruption of development, and toxicity to *Daphnia magna* Saussure (Cladocera: Daphniidae), depending on the concentration and duration of exposure. However, using these same herbicides, Wilson and Bond (1969) reported that although the amphipod, *Hyalella azteca* Saussure (Amphipoda: Dogielinotidae) was highly sensitive to herbicides, several types of aquatic insects were less affected. On the other hand, Kreutzweiser et al. (1992) reported that trichlopyr ester affected the survival of three aquatic insect species: *Simulium* sp. (Diptera: Chironomidae), *Isogenoides* sp. (Ephemeroptera: Perlodidae), and *Dolophilodes distinctus* (Trichoptera: Pilopotamidae). However, several other test species were not affected by trichlopyr ester, and mortality was not increased in any test species when hexazinone was applied. Interestingly, there also were behavioral responses noted in the stream-dwelling insects, some species displayed a greater tendency to drift when exposed to herbicides. This could result in increased exposure to mortality, but it is difficult to assess.

Some effects of herbicides are difficult to discern. Although herbicides tend not to be potent neurotoxins (paraquat is an exception [Costa et al. 2008]) they sometimes increase the toxicity of insecticides. For example, Lichtenstein et al. (1973) evaluated the addition of four herbicides (atrazine, simazine, monuron 2,4-D) to two insecticides (parathion, DDT), on three species of flies (Diptera) and found that all of the herbicides increased the toxicity of the insecticides. Atrazine was most effective, increasing toxicity about 5-fold. They also evaluated the addition of atrazine to 12 insecticides, including representatives of the organophosphate, carbamate, and chlorinated hydrocarbon classes of insecticides. Again, the insecticide toxicity was synergized, with toxicity increasing from 2.2 to 8.6-fold, depending on the insecticide. In a more recent study, Pape-Lindstrom and Lydy (1997) reported synergistic activity of atrazine with several organophosphate insecticides; however, the toxicity of methoxychlor (a chlorinated hydrocarbon) when combined with atrazine was less than additive.

In typical agricultural environments, where both herbicides and insecticides are being applied, enhanced insect suppression might be overlooked and mortality attributed solely to insecticides. If only pest insects are present, this could also be viewed as a useful outcome. However, if there are desirable insects in the field, as in the case of many predators and parasites of pests, or if they are important pollinators, then the increase in toxicity due to the addition of herbicides to insecticides might be detrimental rather than beneficial.

In some cases, herbicides have also been shown to stimulate or benefit arthropods. Some herbicides function as plant growth regulators. If the growth and physiological processes of plants are affected by endogenous application of a chemical, it should not be surprising that insects might respond to the change in the chemistry of their host plants, with some responses being positive. Maxwell and Harwood (1960) reported that the reproductive rates of pea aphid, *Acyrthosiphon pisum* (Harris) (Hemiptera: Aphididae) and cabbage aphid, *Brevicoryne brassicae* (L.) (Hemiptera: Aphididae) were markedly increased following application of 2,4-D, which was attributed to nutritional factors (increased nitrogen), though grasshoppers (Orthoptera: Acrididae) and caterpillar spp. (Lepidoptera) were not affected. Similarly, Oka and Pimentel (1974) reported stimulation of corn leaf aphid, *Rhopalosiphum maidis* (Fitch) (Hemiptera: Aphididae) by application of 2,4-D. Also, Wu et al. (2001) found that four (butachlor, metolachlor, oxadiazon, bentazone) of 11 herbicides tested on the brown planthopper, *Nilaparvata lugens* (Stål) (Hemiptera: Delphacidae), increased the growth rate and reproduction of the insect, and decreased the resistance (measured by the degree of plant damage) of the rice plants to planthopper feeding. Overall, this positive type of response is less common, or perhaps less well documented, or restricted to insects that feed directly on the vascular system of plants, but further demonstrates the variable, seemingly unpredictable nature of the physiological response of insects to herbicides.

Modification of the Plant Community by Herbicides, and Consequences for Insects

Biological diversity impinges on many essential aspects of ecosystem maintenance, including nutrient cycling, decomposition, soil formation, water availability, fire ecology, pest impacts, and wildlife abundance. In arable and pastoral habitats, weeds often comprise a major component of the producer community. Most insects evolved with, and continue to be associated with, non-cultivated plants. An impressively large number of different insects can be associated with weeds. For example, Table 5. 1 contains a list of selected weeds and the number of insect species found in association with selected weeds in Great Britain. Note that there are numerous insect species associated with each weed species, and that there is a mixture of host-specific insect

Table 5.1. The numbers of phytophagous insects associated with selected weed species in Great Britain (adapted from Marshall et al. 2002)

Weed species	Plant family	No. of insect families	No. of insect species	No. of host-specific species
Capsella bursa-pastoris	Brassicaceae	5	13	2
Chenopodium album	Amaranthaceae	15	31	2
Cirsium arvense	Asteraceae	19	50	5
Galium aparine	Rubiaceae	13	30	4
Lamium purpureum	Lamiaceae	8	18	2
Poa annua	Poaceae	15	53	7
Polygonum aviculare	Polygonaceae	15	61	4
Rumex obtusifolius	Polygonaceae	15	79	4
Senecio vulgaris	Asteraceae	10	46	4
Solanum nigrum	Solanaceae	3	7	0
Sonchus oleraceus	Asteraceae	14	28	1
Stellaria media	Carophyllaceae	12	71	4

species and generalists feeding on weeds. Disturbance to the weed flora, whether due to grazing, tillage, burning, or herbicide application, can significantly affect insect communities, though most insects are not host specific. However, insect–weed–crop relationships are potentially very resilient. Loss of weeds may reduce plant diversity, but at least some of the entomofauna may move to nearby crops or weeds. New (2005) provides an excellent overview of insect diversity in the context of agricultural practices.

Traditionally, crop selection and insecticide use have been the principal determinants of insect abundance and diversity in arable lands. However, another important determinant of insect abundance and diversity is herbicide use (Freemark and Boutin 1995). Abundance of broad-leaved weeds is reduced by dicotyledon (broad-leaf plants)-specific herbicide use, grass weeds are suppressed by monocot (grass)-specific herbicides, and both types of weeds by broad-spectrum herbicides. Herbicide use has also resulted in declining seedbanks in some soils (Marshall et al. 2003). Thus, herbicides can significantly modify plant communities. Removing any major form of plant life can clearly affect not only the herbivore community, but also the natural enemies that feed on the herbivores, and the detritivores that consume remnants of plants and animals. For example, in a major study involving several crops, Hawes et al. (2003) reported that the abundance of herbivores, detritivores, pollinators, and natural enemies all tracked the abundance of their food resources, and that they were sensitive to changes in weed communities.

Nectar is an important food resource for many insects, particularly some beneficial species such as ants (Hymenoptera: Formicidae), bees (Hymenoptera: Apidae), and parasitic wasps (Hymenoptera: many families), but also many pest insects such as flies (Diptera), and butterflies and moths (Lepidoptera). Nectar is widely produced by dicots, but only infrequently by monocots (Mizell and Mizell 2008). Floral nectaries are assumed to have evolved to enhance pollination and reproduction of plants, though some dicots can be both insect- and wind-pollinated. Although floral nectaries are most apparent and generally more important ecologically, extraflora nectaries also occur in many plants. Extrafloral nectaries seem to function mostly to attract predatory insects that protect plants from herbivorous insects, and these plant structures can be quite important to the plants bearing them, especially in the tropics. Extrafloral nectaries also occur on the plant early in its development, before floral structures are formed. Thus, predatory insects are attracted to plants and provide protection from herbivory over a protracted time, not just at the time of plant reproduction (Röse et al. 2006). Nectar supply is clearly an important determinant of insect abundance, but also indicative of plant health. Because grasses generally are wind pollinated rather than insect pollinated, the herbicides that affect dicots, or both monocots and dicots, are more disruptive to insect populations than are the monocot-specific herbicides because monocots usually lack both floral and extrafloral nectaries. Herbivorous insects often feed on plants within the same plant family (Capinera 2005). Herbicides also can display plant family-level selectivity. Thus, due to herbicide selectivity, chemical weed control often fosters survival of weeds related to the crop. It is not surprising, therefore, that among the problem weeds in corn, *Zea mays* L. (Poaceae), are grass weeds, such as giant foxtail, *Setaria faberi* R.A. Herrm (Poaceae), and large crabgrass, *Digitaria sanguinalis* (L.) Scop. (Poaceae) (Chege et al. 2009). These plants all support larval growth of the most serious pest of corn, the western corn rootworm, *Diabrotica virgifera* LeConte (Coleoptera: Chrysomelidae). Similarly, forage grasses, such as western wheatgrass, *Pascopyrum smithii* (Rydb.), pubescent wheatgrass, *Elytrigia intermedia* (Host), and side-oats grama, *Bouteloua curtipendula* Michx. (all Poaceae), also support the growth of rootworm larvae (Oyediran et al. 2004). Thus, the occurrence of crop pests is attributable to much more than just the crop plants, and destruction of other grasses by herbicides can influence abundance of the rootworm beetles.

Another interesting example of the influence of alternate hosts involves the biology of bean aphid, *Aphis fabae* Scopoli (Hemiptera: Aphididae) and its crop host sugar beet, *Beta vulgaris* L. (Amaranthaceae). Lambsquarters, *Chenopodium album* (Amaranthaceae), a weed in the same family as sugar beet, serves as an alternate host during the period of the year when sugar beet is growing. The difficulty of controlling lambsquarters in sugar beet fields, which could serve as a host for aphids, can be offset by applying insecticide to the crop and weeds contained therein. However, not only does insecticide application for this purpose impose additional costs to producers and the environment, but lambsquarters also tends to invade other disturbed areas, such as field edges and irrigation ditches. Thus, management of nearby non-crop areas also is necessary. Also interesting is the tendency of bean aphids to migrate, often for several kilometers, to a winter host. Many aphids, including bean aphid, disperse away from the crop as it matures, seeking a suitable food source for the adult aphids, and a safer, more permanent oviposition site where eggs will survive the winter. In this case, the eggs often survive the winter on perennial shrubs such as burning bush, *Euonymus* spp. (Celastroideae), and snowball, *Viburnum* spp. (Adoxaceae). These are deciduous shrubs that are not related to the summer hosts, so the occurrence of this insect is dependent on more than just crops and related weeds. However, we can also view the deciduous shrubs as weeds, and in some areas the overwintering hosts of aphids are removed to minimize egg survival and subsequent aphid problems. On the other hand, if biodiversity is to be maintained, there are benefits to having distant (out of the crop fields) refugia for insects.

Weeds can also serve as important hosts for generalist herbivores, or those insects that do not display a strong affinity for a particular taxon of plants. Probably foremost among these are certain species of grasshoppers (Orthoptera: Acrididae or Romaleidae). In North America, the most important crop-feeding species are two-striped grasshopper, *Melanoplus bivittatus* (Say), differential grasshopper, *M. differentialis* (Thomas), red-legged grasshopper, *M. femurrubrum* (De Geer), migratory grasshopper, *M. sanguinipes* (F.), Packard's grasshopper, *M. packardi* Scudder, eastern lubber grasshopper, *Romalea microptera* (Beauvois), and American grasshopper, *Schistocerca americana* (Drury) (Capinera 2005, 2008). Grasshopper problems almost always originate outside the crop field, with crop loss typically resulting from dispersal of hoppers into crops from nearby weedy areas. Weedy areas supporting grasshopper populations are typically fencerows, irrigation ditches, roadsides, fallow fields, and senescent or abandoned cropland. Occasionally, pastures or rangeland are a source of crop-feeding grasshoppers, but more typically it is areas where the soil was disturbed, allowing weeds to flourish. The weeds support grasshopper feeding, but in addition the weedy areas may provide an undisturbed site for egg deposition, and areas that typically are not high priority for treatment with insecticides. After hatching and gaining size and increased mobility, the grasshoppers disperse in search of food, which unfortunately often is found in cropland. An example of this can be seen in Figure

Figure 5.1. Wheat stubble and weed-infested strips of cropland (a) harbor grasshoppers that disperse to young wheat (b) to feed. The darker area to the right of this image (c) has taller wheat seedlings because it is not yet damaged by the grasshoppers dispersing from the source of insects (a).

5.1, which shows the effects on invasion by grasshoppers on the margin of a newly sprouted wheat field. In arid-land wheat production, one strategy to optimize production is to grow crops in alternate years. In years when wheat is not growing, wheat stubble is left in place to intercept blowing snow and allow moisture to accumulate in the soil. The strips of cropland that do not have actively growing wheat (Figure 5.1, part a) often produce robust weed populations, which is supportive of grasshopper populations. When newly planted wheat emerges, however, the cultivated crop is very attractive to the grasshoppers, which disperse from the wheat stubble to feed on nearby areas of the wheat crop (Figure 5.1, part b). The effect of this is clearly visible by comparing the area where grasshoppers have been feeding (Figure 5.1, part b) to the area where the wheat remains largely undisturbed (Figure 5.1, part c). This problem can be alleviated by timely application of herbicide to germinating weeds in the wheat stubble.

Because the abundance of many insect species is linked to the occurrence of weeds (or other alternate hosts), herbicide use can reduce the abundance of some weed and crop-feeding species by depriving them of a host during part of their development, or their entire life cycle. When herbicide-treated and untreated fields have been compared, herbicides have been shown to suppress not only weed density and diversity, but invertebrate populations as well (Moreby and Southway 1999, Buckelew et al. 2000, Norris and Kogan 2005, Egan et al. 2014). Some of this decrease in insect abundance can be attributed to physiological disruption of the insects (as discussed previously), but certainly more of it is due to loss of weed hosts. If weeds are diverse and not closely related to the crop plant, there is great likelihood of increased diversity and increased abundance of the entomofauna. Research has repeatedly demonstrated that insect populations are higher in weedy fields. For example, populations of Japanese beetle, *Popillia japonica* Newman (Coleoptera: Scarabaeidae), were higher in weedy nurseries and fields of soybean, *Glycine max* (L.) Merrill (Hammond and Stinner 1987), northern corn rootworm, *Diabrotica barberi* (Smith & Lawrence), and western corn rootworm, *D. virgifera* (both Coleoptera: Chrysomelidae), population densities were higher in weedy corn fields (Pavuk and Stinner 1994), and overall insect population densities were greater in soybean where weeds were more plentiful (Buckelew et al. 2000). It is perhaps worth noting that although herbicides kill certain weeds, they also induce species replacement. As one weed species is eliminated, another proliferates (Freemark and Boutin 1995), with the replacement usually representing a different family and attracting its unique complement of insects. The shift in plant community may well affect herbivore species diversity in a field. However, reduction in weed abundance does not assure reduction in insect abundance, as weeds differ in palatability to herbivores and attractiveness to pollinators.

Though elimination of weeds might appear to be overwhelmingly beneficial because we normally think of the insects associated with weeds as being herbivorous, and therefore plant pests, this is not always the case. Herbicide use can also affect the abundance of pollinator, predatory, and parasitic insects, which are beneficial organisms. Alarmingly, Bohnenblust et al. (2016) reported that drift of dicamba onto weeds not only suppressed flowering by the weeds, but reduced visitation by pollinators to flowers that were produced on the weeds. Pollen and nectar (both floral and extrafloral) produced by weeds is frequently said to enhance survival and reproduction of beneficial insects, though most of the data are observational, and unsupported by experimental data (Norris and Kogan 2005). Frequency of visitation data showing that flowering plants are attractive to predatory and parasitic insects are plentiful (e.g., Al-Doghairi and Cranshaw 1999, Carreck and Williams 2002, Ambrosino et al. 2006, Campbell et al. 2016) but this is not exactly the same as enhanced reproduction.

Nevertheless, some experimental data support the benefits of having flowering plants present amongst crops. For example, Nicholls et al. (2000) demonstrated reductions in western grape leafhopper, *Erythronura elegantula* Osborne (Hemiptera: Cicadellidae), and western flower thrips, *Frankliniella occidentalis* (Pergande) (Thysanoptera: Thripidae) abundance where flowering ground cover was present. Similarly, margins of wheat fields where herbicides were not applied had not only greater numbers of weeds and greater diversity of weeds, but also hosted more predatory insects, particularly generalist predators (Chiverton and Sotherton 1991). Consistent with this, strips of flowering plants (including weeds) are often sown in European crop fields to foster the occurrence of beneficial insects (Abivardi 2008, Holland et al. 2016).

It is important to recognize that the effects of herbicide use are not limited to the crop, the weeds in the crop, or even to the herbivores that feed on the crop and weeds. The next layer of consumers, often principally birdlife, also can be affected because they depend on insects and weed seeds as critical food resources (Capinera 2010). Taylor et al. (2006), for example, documented that bromxynil and imazamethabenz herbicides could indirectly affect populations of ringnecked pheasant, *Phasianus colchicus* (L.) and gray partridge, *Perdix perdix* (L.), by modifying weed populations and the arthropod community the plants supported. The weight of chick-food (ground-dwelling) insects was much higher in weedy plots than where herbicide was applied. The importance of insects to birds cannot be overstated. Indeed, assessment of North American bird feeding behavior shows that 61% are primarily insectivorous, 28% are partially insectivorous, and only 11% are not insectivorous. In Central Europe, 89% of birds are insectivorous when in their breeding period. Although it comes as no surprise that certain birds (e.g., woodpeckers, cuckoos, purple martins, nighthawks, swifts, swallows, flycatchers) feed extensively or almost exclusively on insects, it is not widely recognized that many other birds feed heavily on insects, especially during the breeding period when rapidly growing nestlings need a high-protein diet. The overall proportion of the diet (based on stomach contents) that consists of insect material for selected birds is 68.0% for eastern bluebird, *Sialia sialis* (L.), 81.9% for western bluebird, *Sialia mexicana* (Swainson), 59.6% for wood thrush, *Hylocichla mustelina* (Gmelin), 53.0% for rusty blackbird, *Euphagus carolinus* (Muller), 83.4% for Baltimore oriole, *Icterus galbula* (L.), 29.0% for northern cardinal, *Cardinalis cardinalis* (L.), and 66.6% for tufted titmouse, *Baeolophus bicolor* (L.). Examples of marked seasonal shifts in feeding behavior (% of stomach volume consisting of insects during winter and summer, respectively) include increase from 43% to 99% in northern flicker, *Colaptes cafer* (L.), 49% to 96% in eastern meadowlark, *Sturnella magna* (L.), 45% to 89% in Carolina chickadee, *Poecile carolinensis* (Audubon), 2% to 59% in chipping sparrow, *Spizella passerine* (Bechstein), 5% to 80% in bobolink, *Dolichonyx oryzivorus* (L.) and 3% to 60% in redwing blackbird, *Agelaius phoeniceus* (L.) (Martin et al. 1961). Without ready access to insects for food, brood size and survival are greatly diminished in most birds. This phenomenon is not limited to birds, of course. Among the small mammals that feed heavily on insects are bats, shrews, moles, mice, chipmunks, armadillos, opossums, skunks, raccoons, foxes, and badgers. Also, it is not simply the abundance of insects that influences vertebrate feeding behavior, insect diversity is also important. Most vertebrate wildlife will not eat just any insect—they have specific preferences and needs. Thus, to maintain diversity in wildlife populations, we must maintain diversity in insect populations. The importance of wildlife extends beyond their ecological significance, as some wildlife are important in both the culture and economics of farming and ranching. Note that it is not necessary to have weedy crop fields to support robust wildlife populations, but it is beneficial to have field margins, fence rows, irrigation ditches, and pastureland that are relatively free of herbicide treatment.

Effects of Herbicides on Occurrence of Microbial Plant Pathogens Available for Insects to Vector to Crops

Microbial plant pathogens may be found in diverse environments, including living and dead plants, in both crop plants and weeds, in association with aboveground and belowground plant tissue, or existing free in the soil and water. Also, for some pathogens, insects play a major role in acquisition, harboring, and transmission of the pathogens. Plant pathogens may be transmitted on the legs, mouthparts, and bodies of their insect vectors. Fungi and bacteria often are transmitted externally on insect bodies, whereas viruses, mollicutes (phytoplasmas), xylem- and phoem-inhabiting fungi and bacteria, and nematodes, are typically carried by insects internally, with feeding required to transmit the latter pathogen to the plant. Some pathogens (certain viruses) even replicate within the cells of their insect vectors as well as their host plants. The association of plant pathogens with insects is significant; for example, Agrios (2008) estimated that 30–40% of damage and losses caused by plant diseases is due to insect involvement.

Plant disease often results from the movement and feeding of insects within a crop field. However, most crop plants initially are free of disease-causing pathogens, as growers typically use seed or transplants that are free or nearly free of pathogens. The source of many plant pathogens is weeds (reservoirs) that are growing in or near crop fields, and it is the feeding by insects on these weeds (acquisition) that, after dispersal, introduces the disease-causing agents into crops (transmission). Indeed, sometimes weeds infected with plant disease are more preferred by insects relative to non-infected plants, as are weeds over crop plants. This is the case with green peach aphid, *Myzus persicae* (Sulzer) (Hemiptera: Aphididae) in its feeding on hairy nightshade, *Solanum sarrachoides*, and potato *Solanum tuberosum* (both Solanaceae), which can be hosts of potato leafroll virus (Srinivasan et al. 2006). Often, only a low proportion of plants succumb initially to insect-transmitted disease, but as the population of insects increases in abundance over the growing season, and as the insects move from plant to plant to feed, the proportion of infected plants increases. Thus, insects are important in transmission of pathogens from weeds to crop plants (inoculation into fields), and also from crop plants to crop plants (inoculation within the fields) (Wisler and Norris 2005). Some insects can move long distances, either by active flight or more passively in association with weather events (Thresh 1983, Stinner et al. 1983).

One of the most damaging insect-transmitted plant diseases is due to the bacterium *Xylella fastidiosa* (Hopkins and Purcell 2002). This plant pathogen has several strains and causes diseases in several important plants. The strains are somewhat host-specific, but, if taken together, this bacterium has quite a wide host range. The bacteria inhabit the xylem of the host plant and are transmitted by several insects. The most important of the insect vectors are several species of xylem-feeding insects, especially sharpshooter leafhoppers and spittlebugs (Hemiptera: Cicadellidae and Cercopidae, respectively) (Redak et al. 2004). The importance of the vectors varies regionally; however, the most important generally in North America is the glassy-winged sharpshooter, *Homalodisca vitripennis* (Germar) (Hemiptera: Cicadellidae). The name of the disease caused by *X. fastidiosa* differs among the host plants. In grape, it is known as Pierce's disease of grape, but it also is known as alfalfa dwarf disease, citrus variegated chlorosis, phony peach, plum leaf scald, almond leaf scorch, bacterial leaf scorch of coffee, oak leaf scorch, olive quick decline, and leaf scorch of oleander, pear, maple, mulberry, elm, sycamore, and others. The plant host range for glassy-winged leafhopper numbers in the 100s, including common weeds such as lambsquarter, *Chenopodium album* L. (Amaranthaceae), sunflower, *Helianthus annuus* L. (Asteraceae), amaranth, *Amaranthus* spp. (Amaranthaceae), pokeweed, *Phytolacca* spp. (Phytolaccaceae), goldenrod, *Solidago* spp. (Asteraceae), as well as woody plants, such as sumac, *Rhus* spp. (Anacardiaceae), mountain ash, *Sorbus* spp. (Rosaceae), choke cherry, *Aronia* spp. (Rosaceae), and many others. In relatively xeric environments, such as California, it can be beneficial to eliminate non-cultivated plants so as to deprive the insects of feeding and breeding sites, and sites from which to acquire *X. fastidiosa*. Purcell and Frazier (1985) suggested weed control and irrigation as a means to accomplish this goal. In mesic environments, it is nearly impossible to manage the disease this way due to the wide host range and abundance of the bacterium, hosts, and insect vectors.

An impressively wide array of weeds can serve as reservoirs of plant disease. For example, Table 5.2 shows some examples of weeds serving as plant virus reservoirs in Florida, and the plant viruses they harbor. Note that this represents only a few of the plants that function as virus reservoirs, and that additional virus hosts are constantly being discovered. From this list, it is evident that a large number of different viruses are found in weeds, many different plant families are represented, and that some viruses occur in more than one weed species. Plant viruses are not the only plant pathogens found in weeds, but they are probably the most important with respect to insect transmission. Virus transmission by insects to crop and ornamental plants has increased in significance in recent years (Whitfield et al. 2015) due to the invasion of effective disease vectors (especially whiteflies and thrips) to new locations throughout the world, movement of plant viruses either in their vector or plant material, and the evolution of insecticide resistance in these insects.

Tomato spotted wilt virus, a tospovirus, is one of the most damaging plant diseases in the southern United States. It has been shown to infect over 1,000 different species of plants, including 15 families of monocots and 69 species of dicots (Groves et al. 2002, Parella et al. 2003, Momol et al. 2004). Included in its host range are very important food crops such as peanut, *Arachis hypogaea* L. (Fabaceae), pepper, *Capsicum annuum* L. (Solanaceae), tobacco, *Nicotiana tabacum* L. (Solanaceae), tomato, *Lycopersicon esculentum* Miller (Solanaceae), lettuce, *Lactuca sativa* L. (Asteraceae), bean, *Phaseolus vulgaris* L. (Fabaceae), and cucumber, *Cucumis sativa* L. (Cucurbitaceae). Numerous ornamental crops are also affected, including amaryllis, *Amaryllis belladonna* L. (Amaryllidaceae), begonia, *Begonia obliqua* L. (Begoniaceae), chrysanthemum, *Chrysanthemum indicum* L. (Asteraceae), zinnia, *Chrysogonium peruvianum* L. (Asteraceae), cosmos, *Cosmos* spp. (Asteraceae), gerbera, *Gerbera* spp. (Asteraceae), gladiolus, *Gladiolus communis* L. (Iridaceae), and peony, *Paeonia* spp. (Paeoniaceae). Many other important plant families contain hosts of this virus, including Amaranthaceae, Caryophyllaceae, Chenopodiaceae, Convolvulaceae, Cruciferae, Malvaceae, Plumbaginaceae, Portulacaceae, Tropaeolaceae, and Verbenaceae. Winter weeds and perennial weeds are critical in the maintenance of tomato spotted wilt virus (Groves et al. 2001, Chatzvivassiliou et al. 2007). Thrips (Thysanoptera: Thripidae) serve as the vectors of the virus from weeds to crop plants. In the southeastern USA, tobacco thrips, *Frankliniella fusca* (Hinds), serves to inoculate crops in the spring, whereas western flower thrips, *F. occidentalis* (Pergande), is the major vector in the autumn. Acquisition of the virus occurs only in the first instar larval stage, so the weed must be a suitable host for at least a full generation to function as a virus inoculum host (Srinivasan et al. 2014). Not all disease hosts support thrips reproduction, so some weeds serve as 'dead-end hosts' for the virus. In North Carolina, the weeds that supported the largest numbers of tobacco thrips were prickly sow-thistle, *Sonchus asper* (L.) (Asteraceae), common chickweed, *Stellari media* (L.) (Carophyllaceae), and common dandelion, *Taraxacum officianale* (L.) (Asteraceae) (Groves et al. 2002). Thus, weed management should be integrated with insect management for successful suppression of this virus disease (Duffus 1971, Costa 1976).

The final example of plant disease transmission by insects involving 'weeds' involves a fungus, *Raffaellea lauricola* (and possibly other species), that is transmitted by redbay ambrosia beetle, *Xyleborus glabratus* Eichhoff (Coleoptera: Curculionidae) (Harrington et al. 2010), and

Table 5.2. Some common weeds found in Florida vegetable fields, and the insect-transmitted virus diseases they harbor (adapted from Goyal et al. 2012).

Weed common name	Weed scientific name	Weed family	Disease name
Alyceclover	*Alysicarpus vaginalis* (L.)	Fabaceae	Watermelon mosaic virus
American burnweed	*Erechtites hieraciifolius* (L.)	Asteraceae	Bidens mottle virus
American pokeweed	*Phytolacca americana* L.	Phytolaccaceae	Cucumber mosaic virus
Balsam apple	*Momordica charantia* L.	Cucurbitaceae	Cucurbit leaf crumple virus
Balsam apple	*Momordica charantia* L.	Cucurbitaceae	Papaya ringspot virus Type W
Balsam apple	*Momordica charantia* L.	Cucurbitaceae	Squash vein yellowing virus
Balsam apple	*Momordica charantia* L.	Cucurbitaceae	Zucchini yellow mosaic virus
Beggarticks	*Bidens alba* (L.)	Asteraceae	Bidens mottle virus
Beggarticks	*Bidens alba* (L.)	Asteraceae	Tomato spotted wilt virus
Big chickweed	*Cerastium fontanum* Baumg.	Caryophyllaceae	Tomato spotted wilt virus
Bull thistle	*Cirsium vulgare* (Savi)	Asteraceae	Lettuce mosaic virus
Burr clover	*Trifolium fucatum* Lindl.	Fabaceae	Lettuce mosaic virus
Butterweed	*Packera glabella* (Poir.)	Asteraceae	Bidens mottle virus
Canadian horseweed	*Erigeron canadensis* (L.)	Asteraceae	Bidens mottle virus
Canadian toadflax	*Nuttallanthus canadensis* L.	Scrophilariaceae	Tomato spotted wilt virus
Carolina cranesbill	*Geranium carolinianum* L.	Geraniaceae	Tomato spotted wilt virus
Carolina desert chicory	*Pyrrhopappus carolinianus* (Walter)	Asteraceae	Tomato spotted wilt virus
Cheeseweed mallow	*Malva parviflora* L.	Malvaceae	Lettuce mosaic virus
Cheeseweed mallow	*Malva parviflora* L.	Malvaceae	Tomato spotted wilt virus
Chicory	*Chicorium intybus* L.	Asteraceae	Lettuce mosaic virus
Citron	*Citrus medica* L.	Rutaceae	Watermelon mosaic virus
Clasping Venus' looking-glass	*Triodanis perfoliata* (L.)	Campanulaceae	Tomato spotted wilt virus
Cocklebur	*Xanthium strumarium* L.	Asteraceae	Tobacco rattle virus
Common chickweed	*Stellaria media* L.	Carophyllaceae	Lettuce mosaic virus
Common chickweed	*Stellaria media* L.	Carophyllaceae	Tomato spotted wilt virus
Common groundsel	*Senecio vulgaris* L.	Asteraceae	Lettuce mosaic virus
Common plantain	*Plantago major* L.	Plantaginaceae	Tobacco mosaic virus
Common sowthistle	*Sonchus oleraceus* L.	Asteraceae	Tobacco etch virus
Creeping cucumber	*Melothria pendula* L.	Cucurbitaceae	Papaya ringspot virus type W
Creeping cucumber	*Melothria pendula* L.	Cucurbitaceae	Squash vein yellowing virus
Creeping cucumber	*Melothria pendula* L.	Cucurbitaceae	Zucchini yellow mosaic virus
Curlytop knotweed	*Persicaria lathifolia* L.	Polygonaceae	Tobacco rattle virus

possibly other species. This ambrosia beetle has a symbiotic relationship with the fungus that it vectors; both are recent invaders of the USA. The beetles bore into the sapwood of the host tree, forming tunnels called galleries that are inoculated with fungal spores harbored in mycangia found near the insect's mandibles (Brar et al. 2013). The fungal spores germinate and infect living tissues of the host tree. As the fungus grows in the galleries and adjacent sapwood, it disrupts the flow of water and nutrients in the tree, resulting in death of the plant. Both the redbay ambrosia beetle adults and larvae feed on the fungus. Thus, the beetle benefits from having this fungal food source, but the species must continually inoculate trees if it and its progeny are to survive.

The disease resulting from inoculation of trees by redbay ambrosia beetles is known as laurel wilt, and it affects several trees in the family Lauraceae found in the southeastern USA, including redbay, *Persea borbonia* (L.), swampbay, *P. palustris* (Raf.), sassafras, *Sassafras albidum* (Nutt), camphor tree, *Cinnamomum camphora* (L.), pondspice, *Litsea aestivalis* (L.), and avocado, *P. americana* (Mill). Some of these host plants support vigorous reproduction of the insect vector, particularly redbay and swampbay (Pisani et al. 2015). From an economic perspective, avocado is the most important host of the disease, and the only cultivated host, with the other host plants essentially functioning as weeds. Avocado is not a good reproductive host for the ambrosia beetle, but avocado plantings are inoculated by *X. glabratus* beetles originating outside the avocado groves, and then secondarily by other ambrosia beetles within the groves (Carrillo et al. 2013). As is usually the case with insect-vectored plant disease, elimination of weed hosts (e.g., swampbay) in the vicinity of the avocado plantings is recommended. Because wood containing the fungus and beetles may be saved and transported for use as firewood, which will enhance spread of the disease agent and its vector, sanitation by destruction of dead and dying trees is also recommended.

The pattern of weeds serving as reservoirs of plant pathogens is a consistent problem with respect to disease-crop relationships. Regardless of the taxon of the plant pathogen, insects often serve as effective vectors for transmission of the disease-causing agent. The effectiveness of pathogen transmission and induction of pathogenesis vary among species of vectors, species and strains of pathogens, and plant species. Nevertheless, the role of insects in vectoring plant disease-causing agents from weeds to crops is epidemiologically and economically important in crop production systems. Disruption of disease transmission can be accomplished by various means (e.g., elimination of weeds, elimination of vectors, use of disease-resistant crops) but typically involves herbicides for weed suppression or insecticides for vector suppression.

Prediction of Herbicide–Insect Relationships

The impacts of herbicides on insects can be direct, particularly due to toxicity and disruption of insect development by herbicidal chemicals. However, sometimes the direct effects are subtle, such as enhanced toxicity of insecticides, and sometimes surprising, such as stimulation of reproduction by insects. This topic has received considerable attention, but the results are inconsistent among chemicals and insects, so it is difficult to predict the direct consequences of herbicide application on insects without specific research on this subject. Except in a few cases, the direct impacts of herbicides on insects seem to be minimal.

There also are numerous indirect effects of herbicides on insects, though central to most indirect effects is destruction of weeds. This is largely because insects, at several trophic levels, are so dependent on weeds as primary or supplemental sources of nutrition and habitation. The suitabilities of weeds for insect food and habitation, though not always known, are often predictable (Capinera 2005) or at least easily measurable. Also, we can reliably predict that insect diversity, and usually insect abundance, will be favored by an abundance of weeds. In contrast, indirect effects that involve plant disease transmission are somewhat less predictable because although we can assess the ability of weeds to serve as disease reservoirs, they do not always do so. Thus, we can increasingly predict what weed species have the potential to favor plant disease transmission, but unless we have a good fix on the abundance of vectors and the incidence of plant pathogens in weed populations, our predictive abilities are limited.

Some of the interrelationships of herbicides with weeds and insects, and also with predators of insects, plant pathogens transmitted by insects, and seeds produced by plants, can be portrayed diagrammatically by a food web (Figure 5.2). Insects largely coevolved with angiosperms, so the relationship of insects with plants is not only important, but complex (Southwood 1972). The most important relationship is the role of insects as primary consumers. Weed seed as well as the vegetative portions of the weeds support primary consumers, but weed seed is also important in the diet of carabid beetles, which are ordinarily considered to be predators (secondary consumers). Also shown in the food web is the role of insects as vectors of plant pathogens. Note that only certain insects, generally those with piercing-sucking mouthparts (order Hemiptera) are effective vectors. Finally, we must mention the important role of insects in pollination of (some) weeds, as indeed the weed flora would be markedly different if there were only wind-pollinated plants. Both nectar and pollen from weeds are important food resources for some insects, and it is these insects that are most important in pollination. In this diagram (Figure 5.2), pollinators are represented by a moth (order Lepidoptera), but pollination services are provided by several taxa, most notably bees (order Hymenoptera) and flies (order Diptera).

The application of herbicides to plants has its greatest effects on the plants, but secondarily on herbivorous insects (primary consumers) and to a lesser degree on carnivorous insects (secondary consumers). The declining importance of herbicides as we move up the food web is

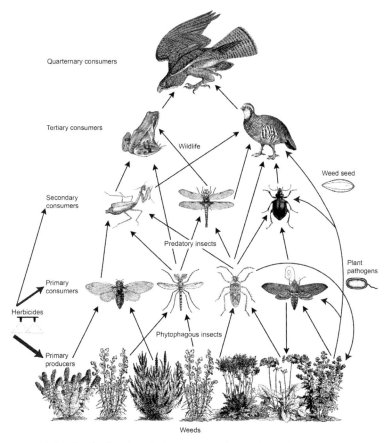

Figure 5.2. A food web showing the important relationships of weeds with insects and wildlife, and the level of impact (expressed by the width of the arrows) caused by herbicides at different trophic levels.

reflected in this diagram by the narrowing of the arrows connecting herbicides to the different trophic levels. The absence of arrows connecting herbicides to the higher tropic levels indicates a decline in effect, though not a complete absence. Certainly, wildlife is affected by herbicides, even if it is rather indirectly, but the higher trophic levels are less assuredly affected due to the greater vagility of organisms in the higher levels.

A major challenge facing us in the future is to predict the outcomes of coupling herbivory of weeds by insects with herbicide application. Certainly, both weed herbivory and herbicide application can be beneficial to crop protection and production, but in order to optimize weed control we need to learn to integrate these factors. Integration of herbicides and insects has long been discussed (e.g. , Andres 1982, Messersmith and Adkins 1995, DiTomaso et al. 2016). The Messersmith and Adkins (1995) publication, in particular, discusses the compatibility and incompatibility issues associated with these two approaches. Not a great deal of research has addressed integration, but there are good examples of successful integration, and DiTomaso (2008) provided a useful, though short, list of herbicides that are compatible with insect biocontrol agents. For example, Paynter (2002), reported that an insect biological control agent, the stem miner *Neurostrotata gunniella* Busck (Lepidoptera: Gracillariidae), can survive on young *Mimosa pigra* L. (Mimosaceae) seedlings if herbicide application is delayed for two weeks after appearance of the leaf mines. Lym and Nelson (2002) showed that leafy spurge, *Euphorbia esula* L. (Euphorbiaceae), densities could be reduced faster when *Aphthona* spp. flea beetles (Coleoptera: Chrysomelidae) and 2,4-D herbicide were used in concert than when either was used alone. Previously, it had been demonstrated that the hawk moth *Hyles euphorbiae* L. (Lepidoptera: *Sphingidae*) and the gall midge *Spurgia esulae* Gagné (Diptera: Cecidomyiidae) could be integrated with herbicide applications to leafy spurge (Lym and Carlson 1994, Rees and Fay 1989). The limited research that demonstrates effective integration tends to stress modification of the timing of herbicide application to avoid affecting insect herbivore populations, or leaving untreated patches of the weed as refugia for the herbivorous insects. These practices are not always achievable, in large measure because many weeds are invaders from another continent, and their herbivorous insects have not been imported to feed on them, and local insect herbivores are not adapted to feed on them to any great extent. However, where reliable levels of herbivory occur and herbicides are also used, research on integration of these two approaches should be high priority.

REFERENCES

Abivardi, C. 2008. Flower strips as ecological compensation areas for pest management. pp. 1489–1494. *In:* J.L. Capinera (Ed.) Encyclopedia of Entomology (2nd ed.). Springer, Dordrecht, Netherlands.

Adams, J.B. 1960. Effects of spraying 2,4-D amine on coccinellid larvae. Can. J. Zool. 38: 285–288.

Adams, J.B. and M.E. Drew. 1965. Grain aphids in New Brunswick. III. Aphid populations in herbicide-treated oat fields. Can J. Zool. 43: 789–794.

Affeld, K., K. Hill, L.A. Smith and P. Syrett. 2004. Toxicity of herbicides and surfactants to three insect biological control agents for *Cytisus scoparius* (Scotch broom). pp. 375–380. *In:* Cullen, J.M., Briese, D.T., Kriticos, D.J., Lonsdale, W.M., Morin, L. and Scott, J.K. (Eds.) Proceedings of the 11th International Symposium on Biological Control of Weeds, Canberra, Australia, CSIRO.

Agrios, G.N. 2008. Transmission of plant diseases by insects. pp. 3853–3885. *In:* J.L. Capinera (Eds.) Encyclopedia of Entomology (2nd ed.). Springer, Dordrecht, Netherlands.

Al-Doghairi, M.A. and W.S. Cranshaw. 1999. Surveys on visitation of flowering landscape plants by common biological control agents in Colorado. J. Kansas Entomol. Soc. 72: 190–196.

Ambrosino, M.D., J.M. Luna, P.C. Jepson and S.D. Wratten. 2006. Relative frequencies of visits to selected insectary plants by predatory hoverflies (Diptera: Syrphidae), other beneficial insects, and herbivores. Environ. Entomol. 35: 394–400.

Andres, L.A. 1982. Integrating weed biological control agents into a pest-management program. Weed Sci. 30 (Suppl. 1): 25–30.

Bohnenblust, E. , J.F. Egan, D. Mortensen and J. Tooker. 2013. Direct and indirect effects of the synthetic-auxin herbicide dicamba on two lepidopteran species. Environ. Entomol. 42: 586–594.

Bohnenblust, E.W., A.D. Vaudo, J.F. Egan, D.A. Mortensen and J.F. Tooker. 2016. Effects of the herbicide dicamba on nontarget plants and pollinator visitation. Environ. Toxicol. Chem. 35: 144–151.

Brar, G.S., J.L. Capinera, P.E. Kendra, S. McLean and J.E. Peña. 2013. Life cycle, development, and culture of *Xyleborus glabratus* (Coleoptera: Curculionidae: Scolytinae). Fla. Entomol. 96: 1158–1167.

Brown, J.J. 1987. Toxicity of herbicides thiobencarb and endothall when fed to laboratory-reared *Trichoplusia ni* (Lepidoptera: Noctuidae). Pest. Biochem. Physiol. 27: 97–100.

Buckelew, L.D., L.P. Pedigo, H.M. Mero, M.D.K Owen and G.L. Tylka. 2000. Effects of weed management systems on canopy insects in herbicide-resistant soybeans. J. Econ. Entomol. 93: 1437–1443.

Campbell, J.W., A. Irvin, H. Irvin, C. Stanley-Stahr and J.D. Ellis. 2016. Insect visitors to flowering buckwheat, *Fagopyrum esculentum* (Polygonales: Polyonaceae), in north-central Florida. Fla. Entomol. 99: 264–268.

Capinera, J.L. 2005. Relationships between insect pests and weeds: an evolutionary perspective. Weed Sci. 53: 892–901.

Capinera, J.L. 2008. Grasshopper pests in North America. pp. 1682–1689. *In:* J.L. Capinera (Ed.) Encyclopedia of Entomology (2nd ed.). Springer, Dordrecht, Netherlands.

Capinera, J.L. 2010. Insects and Wildlife: Arthropods and their Relationships with Wild Vertebrate Animals. Wiley-Blackwell, Chichester, West Sussex, UK.

Carreck, N.L. and I.H. Williams. 2002. Food for insect pollinators on farmland: insect visits to flowers of annual seed mixtures. J. Insect Conserv. 6: 13–23.

Carrillo, D., R.E. Duncan, J.N. Ploetz, A.F. Campbell, R.C. Ploetz and J.E. Peña. 2014. Lateral transfer of a phytopathogenic symbiont among native and exotic ambrosia beetles. Plant Pathol. 63: 54–62.

Chatzivassiliou, E.K., D. Peters and N.I. Katis. 2007. The role of weeds in the spread of *tomato spotted wild virus* by *Thrips tabaci* (Thysanoptera: Thripidae) in tobacco crops. J. Phytopath. 155: 699–705.

Chenge, P.C., T.L. Clark and B.E. Hibbard. 2009. Initial larval feeding on an alternate host enhances western corn rootworm (Coleoptera: Chrysomelidae) beetle emergence on Cry3Bb1-expressing maize. J. Kansas Entomol. Soc. 82: 63–75.

Chiverton, A.A. and N.W. Sotherton. 1991. The effects on beneficial arthropods of the exclusion of herbicides from cereal crop edges. J. Appl. Ecol. 28: 1027–1039.

Costa, A.S. 1976. Whitefly-transmitted plant diseases. Annu. Rev. Phytopath. 14: 429–449.

Costa, L.G., G. Giordano, M. Guizzetti and A. Vitalone. 2008. Neurotoxicity of pesticides: a brief review. Front. Biosci. 13: 1240–1249.

Crosby, D.G. and R.K. Tucker. 1966. Toxicity of aquatic herbicides to *Daphnia magna*. Science 154: 289–290.

Di Tomaso, J.M. 2008. Integration of biological control into weed management strategies. pp. 649–654. *In:* Julien, M.H., Sforza, R., Bon, M.C., Evans, H.C., Hatcher, P.E., Hinz, H.L. and Rector, B.G. (Eds.) Proceedings of the 12th International Symposium on Biological Control of Weeds. Wallingford, UK. CAB International.

Di Tomaso, J.M., R.A. Van Steenwyk, R.M. Nowierski, J.L. Vollmer, E. Lane, E. Chilton, P.L. Burch, P.E. Cowan, K. Zimmerman and C.P. Dionigi. 2016. Enhancing the effectiveness of biological control programs on invasive species through a more comprehensive pest management approach. Pest Manage. Sci. doi: 10.1002/ps.4347

Duffus, J.E. 1971. Role of weeds in the incidence of virus diseases. Annu. Rev. Phytopath. 9: 319–340.

Egan, J.F., E. Bohnenblust, S. Goslee, J.F. Tooker and D.A. Mortensen. 2014. Effects of simulated dicamba herbicide drift on the structure and floral resource provisioning of field edge and old field plant communities. Agric. Ecosyst. Environ. 185: 77–87.

Fernandez-Cornejo, J., R. Nehring, C. Osteen, S. Wechsler, A. Martin and A. Vialou. 2014. Pesticide Use in US Agriculture: 21 Selected Crops, 1960–2008, EIB-124, US Department of Agriculture, Economic Research Service.

Freemark, K. and C. Boutin. 1995. Impacts of agricultural herbicide use on terrestrial wildlife in temperate landscapes: a review with special reference to North America. Agric. Ecosyst. Environ. 52: 67–91.

Goyal, G., H.K. Gill and R. McSorley. 2012. Common weed hosts of insect-transmitted viruses of Florida vegetable crops. University of Florida, ENY-863: 1–12.

Groves, R.L., J.F. Walgenbach, J.W. Moyer and G.G. Kennedy. 2001. Overwintering of *Frankliniella fusca* (Thysanoptera: Thripidae) in winter annual weeds infected with *Tomato spotted wilt virus* and patterns of virus movement between susceptible weed hosts. Phytopathology 91: 891–899.

Groves, R.L., J.F. Walgenbach, J.W. Moyer and G.G. Kennedy. 2002. The role of weed hosts and tobacco thrips, *Frankliniella fusca*, in the epidemiology of *Tomato spotted wild virus*. Plant Dis. 86: 573–582.

Harrington, T.C., D.N. Aghayeve and S.W. Fraedrich. 2010. New combinations in *Raffaelea, Ambrosiella,* and *Hyalorhinoccladiella,* and four new species from the redbay ambrosia beetle, *Xyleborus glabratus.* Mycotaxon 111: 337–361.

Hammond, R.B. and B.R. Stinner. 1987. Soybean foliage insects in conservation tillage systems: effects of tillage, previous cropping history, and soil insecticide application. Environ. Entomol. 16: 524–531.

Hawes, C., A.J. Haughton, J.L. Osborne, D.B. Roy, S.J. Clark, J.N. Perry, P. Rotherly, D.A. Bohan, D.R. Brooks, G.T. Champion, A.M. Dewar, M.S. Heard, I.P. Woiwod, R.E. Daniels, M.W. Young, A.M. Parish, R.J. Scott, L.G. Firbank and G.R. Squire. 2003. Responses of plan and invertebrate trophic groups to contrasting herbicide regimes in the Farm Scale Evaluation of genetically modified herbicide-tolerant crops. Phil. Trans. R. Soc. Lond. B 358: 1899–1913.

Hopkins, D.L. and A.H. Purcell. 2002. *Xylella fastidiosa*: cause of Pierce's disease of grapes and other emergent diseases. Plant Dis. 86: 1056–1066.

Holland, J.M. , F.J.J.A. Bianchi, M.H. Entling, A-C. Moonen, B.M. Smith and P. Jeanneret. 2016. Structure, function and management of semi-natural habitats for conservation biological control: a review of European studies. Pest Management Science 72: 1638–1651.

Kreutzweiser, D.P., S.B. Holmes and D.J. Behmer. 1992. Effects of herbicides hexazinone and triclopyr ester on aquatic insects. Ecotox. Environ. Safe. 23: 364–374.

Lichtenstein, E.P., T.T. Liang and B.N. Anderegg. 1973. Synergism of insecticides by herbicides. Science 181: 847–849.

Lym, R.G. and R.B. Carlson. 1994. Effect of herbicide treatment on leafy spurge gall midge (*Spurgia esulae*) population. Weed Technol. 8: 285–288.

Marshall, E.J.P., V.K. Brown, N.D. Boatman, P.J.W. Lutman, G.R. Squire and L.K. Ward. 2003. The role of weeds in supporting biological diversity within crop fields. Weed Res. 43: 77–89.

Martin, A.C., H.S. Zim and A.L. Nelson. 1961. American Wildlife and Plants. A Guide to Wildlife Food Habits. Dover Publications, New York.

Maxwell, R.C. and R.F. Harwood. 1960. Increased reproduction of pea aphids on broad beans treated with 2,4-D. Ann. Entomol. Soc. Amer. 53: 199–205.

Messersmith, C.G. and S.W. Adkins 1995. Integrating weed-feeding insects and herbicides for weed control. Weed. Technol. 9: 199–208.

Michaud, J.P. and G. Vargas. 2010. Relative toxicity of three wheat herbicides to two species of Coccinellidae. Insect Sci. 17: 434–438.

Mizell, III, R.F. and P.A. Mizell 2008. Plant extrafloral nectaries. pp. 2914–2921. *In:* J.L. Capinera (Ed.) Encyclopedia of Entomology (2nd ed.). Springer, Dordrecht, Netherlands.

Momol, M.T., S.M. Olson, J.E. Funderburk, J. Stavinsky and J.J. Marois. 2004. Integrated management of tomato spotted wilt on field-grown tomatoes. Plant Dis. 88: 882–890.

Moreby, S.J. and S.E. Southway. 1999. Influence of autumn applied herbicides on summer and autumn food availability to birds in winter wheat fields in southern England. Agric. Ecosys. Environ. 72: 285–297.

Morton, H.L., J.O. Moffett and R.L. Macdonald. 1972. Toxicity of herbicides to newly emerged honey bees. Environ. Entomol. 1: 102–104.

New, T.R. 2005. Invertebrate Conservation and Agricultural Ecosystems. Cambridge University Press, Cambridge, UK. 354 p.

Nicholls, C.I., M.P. Parrella and M.A. Altieri. 2000. Reducing the abundance of leafhoppers and thrips in a northern California organic vineyard through maintenance of full season floral diversity with summer cover crops. Agric. Forest Entomol. 2: 107–113.

Norris, R.F. and M. Kogan. 2005. Ecology of interactions between weeds and arthropods. Annu. Rev. Entomol. 50: 479–503.

Oyediran, I.O., B.E. Hibbard and T.L. Clark. 2004. Prairie grasses as hosts of the western corn rootworm (Coleoptera: Chrysomelidae). Environ. Entomol. 33: 740–747.

Oka, I.N. and D. Pimentel. 1974. Corn susceptibility to corn leaf aphids and common corn smut after herbicide treatment. Environ. Entomol. 3: 911–915.

Pape-Lindstrom, A.A. and M.J. Lyndy. 1997. Synergistic toxicity of atrazine and organophosphate insecticides contravenes the response addition mixture model. Environ. Toxicol. Chem. 11: 2415–2420.

Parella, G., P. Gognalons, K. Gebre-Selassiè, C. Vovlas and G. Marchoux. 2003. An update on the host range of *tomato spotted wilt virus*. Plant Pathol. 85 (special issue): 227–264.

Pavuk, D.M. and B.R. Stinner. 1994. Influence of weeds within *Zea mays* crop plantings on populations of adult *Diabrotica barberi* and *Diabrotica virgifera virgifera*. Agric. Ecosys. Environ. 50: 165–175.

Pisani, C., R.C. Ploetz, E. Stover, M.A. Ritenour and B. Scully. 2015. Laurel wilt in avocado: review of an emerging disease. Int. J. Plant Biol. Res. **3: 1043.** 7 pp.

Purcell, A.H. and N.W. Frazier. 1985. Habitats and dispersal of the leafhopper vectors of Pierce's disease in the San Joaquin Valley. Hilgardia 53: 1–32.

Redak, R.A., A.H. Purcell, J.R.S. Lopes, M.J. Blua, R.F Mizell III and P.C. Andersen. 2004. The biology of xylem fluid-feeding insect vectors of *Xylella fastidiosa* and their relation to disease epidemiology. Annu. Rev. Entomol. 49: 243–270.

Rees, N.E. and P.K. Fay. 1989. Survival of leafy spurge hawk moths when larvae are exposed to 2,4-D or picloram. Weed Technol. 3: 429–431.

Röse, U.S.R., J. Lewis and J.H. Tumlinson. 2006. Extrafloral nectar from cotton (*Gossypium hirsutum*) as a food source for parasitic wasps. Funct. Ecol. 20: 67–74.

Saska, P., J. Shuhrovec, J. Lukás, H. Chi, S.-J. Tuan and A. Honek. 2016. Treatment by glyphosate-based herbicide alters life history parameters of the rose-grain aphid *Metopolophium dirhodum*. Sci. Repts. 6: 27801. doi: 10. 1038/srep27801

Sotherton, N.W. 1982. The effects of herbicides on the chrysomelid beetle *Gastrophysa polygoni* (L.) in the laboratory and field. Z. Angew. Entomol. 94: 446–451.

Southwood, T.R.E. 1972. The insect/plant relationship—an evolutionary perspective. pp. 3–30. *In:* Southwood, T.R.E. (Ed.) Insect/Plant Relationships. Symposium 6 of the Royal Entomological Society, Oxford, UK. Blackwell Scientific Publications.

Srinivasan, R., J.M. Alvarez, S.D. Eigenbrode and N.A. Bosque-Pérez. 2006. Influence of hairy nightshade *Solanum sarrachoides* (Sendtner) and potato leafroll virus (Luteoviridae: *Polerovirus*) on the host preference of *Myzus persicae* (Sulzer) (Homoptera: Aphididae). Environ. Entomol. 35: 546–553.

Srinivasan, R., D. Riley, S. Diffie, A. Shrestha and A. Culbreath. 2014. Winter weeds as inoculum sources of tomato spotted wilt virus and as reservoirs for its vector, *Frankliniella fusca* (Thysanoptera: Thripidae) in farmscapes of Georgia. Environ. Entomol. 43: 410–420.

Stinner, R.E., C.S. Barfield, J.L. Stimac and L. Dohse. 1983. Dispersal and movement of insect pests. Annu. Rev. Entomol. 23: 19–35.

Taylor, R.L., B.D. Maxwell and R.J. Boik. 2006. Indirect effects of herbicides on bird food resources and beneficial arthropods. Agric. Ecosys. Environ. 116: 157–164.

Thresh, J.M. 1983. The long-range dispersal of plant viruses by arthropod vectors. Phil. Trans. R. Soc. London B. 302: 497–528.

Trumble, J.T. and L.T. Kok. 1979. Compatibility of *Rhinocyllus conicus* and 2,4-D (LVA) for musk thistle control. Environ. Entomol. 8: 421–422.

Whitfield, A.E., B.W. Falk and D. Rotenberg. 2015. Insect vector-mediated transmission of plant viruses. Virology 479–480: 278–289.

Wilkinson, J.D., K.D. Biever and C.M. Ignoffo. 1975. Contact toxicity of some chemical and biological pesticides to several insect parasitoids and predators. Entomophaga 20: 113–120.

Wisler, G.C. and R.F. Norris. 2005. Interactions between weeds and cultivated plants as related to management of plant pathogens. Weed. Sci. 53: 914–917.

Wu, J.C. , J.X. Xu, J.L. Liu, S.Z. Yuan, J.A. Cheng and K.L. Heong. 2001. Effects of herbicides on rice resistance and on multiplication and feeding of brown planthopper (BPH), *Nilaparvata lugens* (Stål). Inter. J. Pest Manage. 47: 153–159.

Evolution of Herbicide-Resistant Weeds

Nilda Roma-Burgos*[1], Ian M. Heap[2], Christopher E. Rouse[1] and Amy L. Lawton-Rauh[3]

[1] Department of Crop, Soil, and Environmental Sciences, University of Arkansas, 1366 W. Altheimer Drive, Fayetteville, AR 72704, USA
[2] International Survey of Herbicide-Resistant Weeds, PO Box 1365, Corvallis, OR 97339, USA
[3] Department of Genetics and Biochemistry, Clemson University, 105 Collings Drive, Clemson SC 29634 USA

Introduction

Herbicides are still the most efficient and cost-effective means of weed control in the major agronomic crops of the world. Since the introduction of modern synthetic herbicides in the 1940s herbicides have dominated weed management. As predicted by Harper (1954), the evolution of herbicide-resistant (HR) weeds is an inevitable outcome of intensive herbicide use. Herbicide-resistant weeds have increased steadily since the mid-1970s, but was easily taken care of by the vibrant agricultural chemical industry of the 1970s and 1980s through steady introduction of new herbicide sites of action (SOAs). There was little incentive for growers to manage for herbicide resistance proactively because they could rely upon new herbicide SOAs to manage HR weeds (Llewellyn et al. 2002). The intensifying environmental regulations during the 1980s and 1990s led to a substantial increase in the cost to develop and register new herbicides. Combined with the introduction of glyphosate-resistant (GR) crops, which moved the economics of weed control from herbicide fees to trait technology fees, the increasing cost of herbicide development and registration greatly reduced the potential economic benefit of bringing a new herbicide to market. Thus, the introduction of new herbicide SOAs ceased (Figure 6.1).

At the time of introduction of GR crops there were no reports of GR weeds despite more than 20 years of glyphosate usage, leading some in the industry to claim that the evolution of GR weeds was unlikely. However, herbicide resistance is a numbers game and as vast hectareage of crops were treated with glyphosate, it quickly became apparent that GR weeds would become a major issue. We are currently at a point in weed science where the focus on herbicides for weed control over the past 70 years has left us vulnerable to their loss of efficacy due to weed resistance (Figure 6.2). As a stop-gap measure, companies are creating new traits in crops and stacking these traits so that growers can use existing herbicides in new situations, and provide a new herbicide SOA for some crops.

Examples of these are resistance traits to: glyphosate + glufosinate (soybean, corn, cotton), glyphosate + ALS inhibitors (soybean, corn, canola), glyphosate + glufosinate + 2,4-D (soybean,

---

*Corresponding author: nburgos@uark.edu

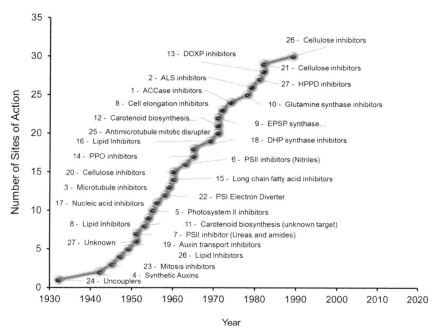

Figure 6.1. Introduction time of new herbicide sites of action (WSSA codes).

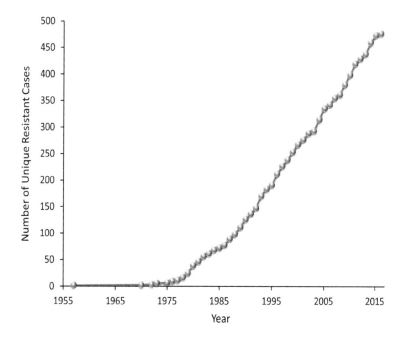

Figure 6.2. The chronological increase in unique (species by site of action) herbicide-resistant weed cases globally.

cotton), glyphosate + glufosinate + dicamba (soybean, corn, cotton), glyphosate + glufosinate + HPPD inhibitors (soybean and cotton), glyphosate + glufosinate + 2,4-D + ACCase inhibitors (corn), and glufosinate + dicamba (wheat) (Green 2016). Among the trade names associated with these traits are GlyTol Liberty Link® (glyphosate + glufosinate), Xtend Flex® (glyphosate +

glufosinate + dicamba), Roundup Ready®Xtend (glyphosate + dicamba), Enlist™ (glyphosate + 2,4-D) and Enlist E3™ (glyphosate + glufosinate + 2,4-D). The utility of auxinic herbicides in these stacks is primarily to control GR broadleaf weeds, especially GR *Amaranthus palmeri* S. Wats., *A. tuberculatus* (Moq.) Sauer, and *Kochia scoparia* (L.) Schrad. These technologies are certainly welcome, but will only buy us about 10 to 15 years before their widespread use results in multiple resistance.

To continue to provide effective economic weed control in the long term it will be necessary to incorporate a variety of weed control practices aimed at thwarting resistance evolution. A relatively recent innovation is the Harrington Seed Destructor (HSD) in Australia (Walsh et al. 2012). This unit intercepts chaff during crop harvest, and with a grinder as its central component, destroys weed seeds that are harvested by combining. The HSD has been reported to destroy >95% of weed seeds in the chaff of wheat (*Triticum aestivum* L.), barley (*Hordeum vulgare* L.), and lupin (*Lupinus angustifolius* L.) (Walsh et al. 2012). The major weeds tested in these crops include *Avena fatua* L., *Bromus* spp., *Lolium rigidum* Gaudin, and *Raphanus raphanistrum* L. The HSD is now being tested in some states in the United States and Canada. Research in Western Canada showed that the HSD controlled a variety of small- to large-seeded weed species 97.7 – 99.8% including *A. fatua*, *Brassica napus* L., *Galium spurium* L., *K. scoparia* and *Setaria viridis* L. Beauv. (Tidemann et al. 2017). The seed destructor is proving to be an excellent non-chemical tool for integrated weed management. Producers may be tempted to switch to HSD as their main weed management tool because of its consistently high efficacy; however, weeds adapt to any control measure, whether cultural, mechanical, or chemical. The HSD may facilitate selection for species, or ecotypes of species, that mature earlier than the crop or resume development after crop harvest, those of short stature in order to escape harvest (i.e., weeds in lawns), or highly shattering ecotypes.

Herbicides will most likely continue to be a major part of agronomic crop weed management in the next 50 years, particularly as new herbicide SOAs are introduced, and more stacked herbicide traits are used to manage resistant weeds. Cultural practices and new technologies will become even more important than they have been in the past 60 years. Researchers are very active in the areas of robotics, reliant on computer-aided sensors for weed detection and building upon earlier work by several groups such as that of Peña et al. (2013), mechanical inter-row weeders (Fennimore et al. 2016), and nanotechnology applications for herbicide formulations, delivery, and detection (Burgos et al. 2017, Pereira et al. 2014).

Resistance to Herbicides

Herbicide-resistant weeds are a predictable result of natural selection. Herbicides are chemicals that cause the suppression and death of weeds by interfering with their biology. Many herbicides achieve this by entering the leaf surface, moving within either the phloem or xylem and then into cells, where they bind and inactivate key biochemical enzymes or disrupt the photosynthetic apparatus. Weed populations, which are genetically diverse, contain rare genetic variants that allow the weed to survive herbicide treatments. Tolerance or resistance is achieved by either preventing the herbicide from reaching its site of action (SOA), metabolising or changing the herbicide so it is no longer active, or by altering the SOA such that the herbicide can no longer bind to the catalytic site and inhibit the function of its target (Délye et al. 2013, Jang et al. 2013, Sammons and Gaines 2014, Yu and Powles 2014). Alteration of the target site is referred to as target-site resistance (TSR). These rare variants are selected only after repeated use of the same herbicide SOA as the most susceptible individuals in the seedbank and the progeny are removed by the herbicide. Tolerance of the population to the herbicide increases when the first selected target-site mutant reproduces, and gradually increases the frequency of resistant alleles in the population (Salas et al. 2016), to a level where it is deemed resistant to the commercial rate of a herbicide.

Population-level tolerance to the herbicide increases gradually with cycles of exposure as some individuals survive inadvertent sublethal doses and accumulate herbicide-stress-adaptation genes, leading to non-target site resistance (NTSR). Selection for resistance by low-dose glyphosate applications have been documented in *Lolium* spp. in Australia (Busi et al. 2012) and Italy (Collavo and Sattin 2014). A number of plants in the field may receive a sublethal dose of the herbicide due to various reasons including environmental factors (rain event, drought stress), application factors (canopy shielding, edge of field obstructions, skips, wheel tracking), and biological factors (differential emergence and plant size, emergence after herbicide application, clumping distribution). Low-dose selection can impart high or low-level resistance in as few as three cycles of selection depending on the breeding behavior, ploidy level and other biological factors. For example, low-dose selection of diploid, cross-pollinated, rigid ryegrass (*L. rigidum* Gaud.) with diclofop resulted in a 40-fold increase in LD_{50} after three cycles, but only three-fold increase in self-pollinated, hexaploid wild oat (*Avena fatua* L.) (Busi et al. 2016). The resulting resistance level from low-dose selection could be modified further by the nature of the herbicide SOA as indicated by only a 2.8-fold increase in Palmer amaranth (*Amaranthus palmeri* S. Wats.) tolerance to dicamba after three cycles of selection (Tehranchian et al. 2017). Like rigid ryegrass, Palmer amaranth is also a cross-pollinated, genetically diverse diploid species. It appears that resistance evolution is relatively slow in species with high ploidy level, because of the dilution effect of multiple genomes, and in species where the target site is highly conserved, highly critical for plant survival, or physiologically complex.

Usually, the grower will notice the change in weed population response only when the herbicide efficacy falls below their criteria of acceptable weed control, which is when the population has evolved resistance to the recommended field dose of the herbicide. The time it takes for field-scale resistance to occur depends on many factors and their interactions. Among the most important factors affecting resistance evolution are population size, standing genetic variation, and the herbicide site of action.

The Effect of Population Size, Standing Variation, and Herbicide Site of Action

The adaptation and evolutionary persistence of resistance to herbicides are a numbers game. Smaller population sizes (small seed bank) lead to lower risks for adaptations to survive and spread through a population. This is because random chance, or genetic drift, plays a larger role in persistence of alleles between generations than selection. The strength of selection has to be very high to compensate for random allele frequency changes due to genetic drift. Populations, or species, with high standing genetic variation will have a high propensity for selection of genetic variants carrying a herbicide resistance trait. The nature of the target site also affects the rate of successful selection of resistant individuals. For example, given the same population size, selection leading to an increase in the number of ALS inhibitor-resistant individuals would take fewer generations than most other targets because the target enzyme, ALS (acetolactate synthase), maintains function with more non-synonymous mutations (DNA mutations that change the amino acid sequence) at various loci involved in herbicide binding. Thus, the use of ALS herbicides has resulted in the fastest documented evolution of resistant genotypes illustrated by the detection of sulfonylurea-resistant *Alopecurus myosuroides* Huds. in the UK in 1982 when this ALS inhibitor was commercialised for cereal production barely two years (Brown and Cotterman 1994, Heap 2017, Moss and Cussans 1985). In contrast, the binding domain of EPSPS (5-enol-pyruvyl shikimate phosphate synthase), the target site for glyphosate, is highly conserved; the enzyme loses function when non-synonymous mutations occur in the protein-coding regions that fold into its binding pocket. The first documented resistance to glyphosate occurred in 1996 (Powles et al. 1998), about 23 years after its commercialisation in 1974 (Duke 2017). Apparently, the *EPSPS* gene accumulates fewer non-synonymous mutations because mutations at the binding domain are generally lethal. The resistance-conferring mutations in GR weeds endow only low-level resistance. Therefore, if a field with low weed density (e.g., 1,000 to 10,000 individuals per hectare) is treated with an ALS inhibitor, the risk for, or time to, resistance evolution would be similar to that of a field with a high density of weeds (e.g., 1,000,000 to 10,000,000 individuals per hectare) treated with glyphosate.

As mentioned above, weeds can become resistant in many ways.The likelihood of any particular type of resistance mechanism being selected is largely determined by the herbicide SOA and the weed species involved. Some herbicide SOAs are prone to target site resistance because although the target site is modified by single amino acid mutations that prevent the herbicide from binding, the herbicide target is still able to function. Acetolactate synthase-, ACCase-, and photosynthesis inhibitors fall into this category, and are particularly prone to target site resistance. Whilst there are some target site resistance to glyphosate, the EPSPS enzyme is highly conserved (genetic code for the enzyme is very similar) across a wide array of plant species, indicating that most enzyme variants over time have been detrimental to the survival of an individual. Most single point mutations in the EPSPS enzyme result in a non-functioning or an inefficient fitness-compromised enzyme. As such, glyphosate is not prone to triggering target-site resistance, a major reason that glyphosate in general is a low-risk herbicide when compared to ALS-, ACCase-, and photosynthesis inhibitors.

It cannot be stressed enough that whilst non-herbicidal weed control techniques (cultural controls, mechanical controls, etc.) alone may not provide acceptable weed control, they can reduce population size significantly, thereby curtailing resistance evolution by reducing the effectiveness of selection due to the increased impact of genetic drift. In particular, if a weed has evolved resistance to one herbicide site of action and has built up high population numbers in a field, the aim should be to bring the population size down through non-herbicidal means first before using the next herbicide SOA, otherwise, rapid evolution of multiple resistance will ensue.

Resistance Terminologies and Biological Principles

Resistance is the inherited ability of a plant to survive and reproduce following exposure to a dose of herbicide normally lethal to the wild type. In a plant, resistance may occur naturally or may be induced by techniques, such as genetic engineering or selection of variants produced by tissue culture or mutagenesis (WSSA 1998). Resistance evolution would occur faster in a population with high standing (natural) genetic variation than in one with low genetic variation.

Target-site Resistance (TSR) results from an amino acid substitution, which alters the 3-D conformation of the substrate-binding domain and reduces the affinity of the herbicide molecule to the target site. In general, TSR is conferred by dominant or semi-dominant single, nuclear genes (Délye et al. 2013). Exceptions are the recessive resistance to microtubule assembly inhibitors (Délye et al. 2004) and the cytoplasmic inheritance of resistance to PSII inhibitors (Preston and Mallory-Smith 2001). The nature of binding-site mutation × herbicide interaction results in different levels of resistance and different cross-resistance or multiple-resistance patterns.

Nontarget-site Resistance (NTSR) is a multigene, complex resistance mechanism that occurs outside the herbicide-binding domain. The following processes endow NTSR: i) reduced absorption of the herbicide into the plant tissue, ii) reduced translocation to the target site or compartmentalisation/ sequestration of the herbicide away from the target, iii) enhanced herbicide metabolism, iv) protection from the damaging elements of herbicide action, and v) overproduction of herbicide target to compensate for those inhibited by the herbicide (Délye et al. 2013). Unlike TSR, NTSR is generally associated with abiotic stress-coping mechanisms in plants. Thus, NTSR may be endowed by the same genes that endow protection from strong oxidants or tolerance to drought, flooding, cold, or heat stress. Consequently, NTSR mechanisms also result in unpredictable resistance patterns and may endow broad resistance to unrelated herbicide chemistries. The number of NTSR cases is increasing.

Cross-Resistance is resistance to two or more herbicides in one HR plant, which is conferred by a single mechanism (Beckie and Tardif 2012). This resistance pattern can be due to target site modification, which compromises the binding affinity of multiple herbicides that are most structurally similar. Thus, cross-resistance occurs most often to herbicides from the same chemical family (i.e. mutations at Pro_{197}, conferring resistance to sulfonylureas only) and occasionally to herbicides from different chemical families (i.e., $Asp_{376}Glu$, endowing resistance to imidazolinones, sulfonylureas, and triazolopyrimidines) (see ALS mutation table by Tranel and Wright 2017 and review by Yu and Powles 2014). Cross-resistance may also ensue from NTSR mechanisms, such as increased production of a cytochrome P450 enzyme, which can metabolise herbicides with particular functional groups, resulting in resistance to multiple herbicides from the same or different families. For example, CYP76B1 can metabolise phenylureas chlorotoluron and linuron (Robineau et al. 1998) and CYP81A6 can metabolise bentazon and sulfonylurea herbicides (Zhang et al. 2007), the overexpression of these CytP450s would result in cross-resistance.

Multiple-Resistance is resistance to two or more herbicides in one HR plant, which is conferred by different mechanisms (Beckie and Tardif 2012). It results from the accumulation of multiple genes endowing different resistance mechanisms in one plant. The accumulation of multiple resistance mechanisms can result from successive selection with different herbicide modes of action, such as observed with *Echinochloa* spp. multiple resistance in rice production (Rouse et al. 2017, Talbert and Burgos 2007). Multiple resistance also results from hybridisation between plants within a population or species and across genetically compatible species. As we switch from one herbicide to another, or stack multiple HR traits in crops to manage HR weeds, cases of multiple resistance is continuing to increase (Heap 2017).

Distribution

The global distribution of HR weeds is primarily determined by the years and intensity of herbicide use and the level of integration of non-herbicidal weed control methods. Herbicides first became widely used in the developed world, initially in North America and Europe. It is not surprising that weed resistance first became a major issue in cropping systems in these regions (Figure 6.3) a few decades after the commercialisation of herbicides. Today, herbicide-resistant weeds have been reported in 69 countries (Heap 2017). Initially this was driven by the widespread use of triazine herbicides in corn production, and the 1970s saw an explosion of atrazine resistance identification and research. The introduction of ALS inhibitors and ACCase inhibitors in the late 1980s helped to control triazine-resistant weeds, but shortly after led to the rapid appearance of weeds with resistance to these herbicide SOAs. Resistance did not appear often in the developing world where weed control relied upon cultural methods and, in particular, handweeding. As labour became more expensive in Asian countries (South Korea, Thailand, China, etc.), these countries transitioned from handweeding to herbicides, and as a result, we see a corresponding rise in the appearance of HR weeds. The number of unique HR weed cases (species by SOA) globally is increasing steadily in a predictable manner. Although the first cases of herbicide resistance were documented in 1957, the steady increase in resistance cases started in the late 1970s, fuelled by triazine-resistant weeds in corn production (Figure 6.4). Unless otherwise stated, the data used in tables and graphs in this chapter come from the International Survey of Herbicide-Resistant Weeds' website at http://www.weedscience.org.

Weeds had evolved resistance to 23 of the 26 herbicide SOAs (Table 6.1). The herbicide SOAs that had selected the greatest number of resistant weed species are the ALS inhibitors (159), followed by the PSII-, ACCase-, and EPSP synthase inhibitors, and synthetic auxins. These numbers are only partial metrics as these do not present the area infested and seriousness of each case of resistance. While there are no good data to account for the area infested with HR weeds it can be stated that the largest area infested was initially with PSII-inhibitor-resistant weeds in the 1970s to 1980s.

Resistance to PSII inhibitors were then surpassed in severity by ALS inhibitor- and ACCase inhibitor-resistant weeds in the 1990s (Figure 6.4). A more troubling scenario is the geometric rise in the number of weed species evolving resistance to glyphosate since the first report in 1997 in *L. rigidum* in Australia (Powles et al. 1998). In 20 years, we observed an average of two

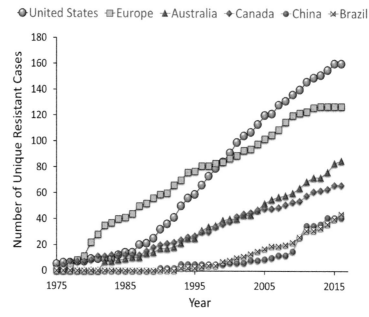

Figure 6.3. The chronological increase in unique herbicide-resistant cases for selected countries and Europe.

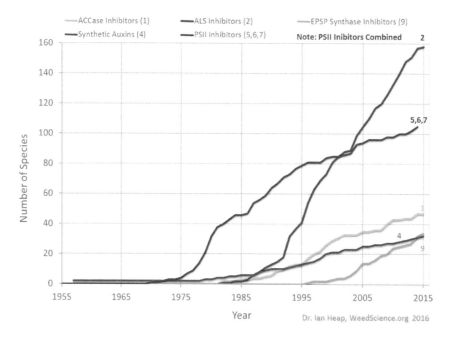

Figure 6.4. The chronological increase in the number of herbicide-resistant weed species to five major herbicide sites of action.Numbers in brackets are Weed Science of America (WSSA) codes for herbicide classification by site of action.

new GR weed species per year (Figure 6.5). Of the 39 GR weeds, only seven are of serious economic concern, however, this situation is changing rapidly due to the widespread adoption of GR crops in North and South America. The synthetic auxins have been used longer than any other herbicide SOA, and arguably on a greater area, yet of the 32 weed species that have

Table 6.1. Global summary of unique resistance cases by site of action[1] as of December 2017

Herbicide group	HRAC group	Example herbicide	Dicots	Monocots	Total
ALS inhibitors	B	Chlorsulfuron	97	62	159
Photosystem II inhibitors	C1	Atrazine	50	23	73
ACCase inhibitors	A	Sethoxydim	0	48	48
EPSP synthase inhibitors	G	Glyphosate	21	18	39
Synthetic auxins	O	2,4-D	28	8	36
PSI electron diverter	D	Paraquat	22	10	32
PSII inhibitor (Ureas and amides)	C2	Chlorotoluron	10	18	28
PPO inhibitors	E	Fomesafen	10	3	13
Microtubule inhibitors	K1	Trifluralin	2	10	12
Lipid inhibitors	N	Triallate	0	10	10
Carotenoid biosynthesis (unknown target)	F3	Amitrole	1	5	6
Long chain fatty acid inhibitors	K3	Butachlor	0	5	5
PSII inhibitors (Nitriles)	C3	Bromoxynil	3	1	4
Carotenoid biosynthesis inhibitors	F1	Diflufenican	3	1	4
Glutamine synthase inhibitors	H	Glufosinate-ammonium	0	3	3
Cellulose inhibitors	L	Dichlobenil	0	3	3
Antimicrotubule mitotic disrupter	Z	Flamprop-methyl	0	3	3
HPPD inhibitors	F2	Isoxaflutole	2	0	2
DOXP inhibitors	F4	Clomazone	0	2	2
Mitosis inhibitors	K2	Propham	0	1	1
Unknown	Z	Endothall	0	1	1
Cell elongation inhibitors	Z	Difenzoquat	0	1	1
Nucleic acid inhibitors	Z	MSMA	1	0	1
Total			*250*	*236*	*486*

[1]Herbicide Resistance Action Committee (HRAC) herbicide group codes.

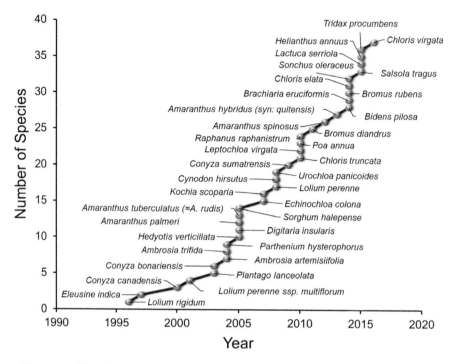

Figure 6.5. The chronological increase in glyphosate-resistant weeds worldwide.

evolved resistance to the synthetic auxins only a handful have become significant economic problems.

The primary factors influencing the selection of HR weeds are population size (the number of individuals exposed to the selector), the site of action itself as discussed previously, and time (cycles of selection). The number of ALS inhibitor-resistant weeds increased at around 10 years of commercialisation and increased very rapidly compared to any other herbicide SOA (Figure 6.4). Resistance to ACCase inhibitors became prominent at around 13 years of usage and continued to increase; however, this trend will level off as the remaining key grass weed species that are yet to evolve resistance to ACCase inhibitors are few. Glyphosate was used for 22 years prior to the first documented case of glyphosate resistance; however, the rate of new glyphosate resistance cases is increasing at a rate similar to that of the ACCase inhibitors. Glyphosate is a non-selective, highly effective herbicide that was used only in plantation crops and non-crop areas during the first 22 years of its commercial use. It was expensive then. Upon the commercialisation of GR crops in the mid-1990s, glyphosate use expanded to major agronomic crops (corn, cotton, soybean) covering millions of hectares. The price of glyphosate also dropped, making it one of the cheapest herbicides, and fuelling its massive use. These factors resulted in a sharp rise in the GR weeds in about 30 years from first commercialisation. Monitoring of resistant weeds intensified, primarily due to the importance and dire implications of the occurrence of HR weeds in general on GR crops and the upcoming transgenic crops with stacked HR traits.

Although HR weeds are reported in only 69 countries globally, weed resistance probably occurs wherever there is prolonged use of herbicides. North America and Europe have been using herbicides longer than any other region, and these countries were first to encounter HR weeds. Initially, in the late 1970s to the early 1990s Europe had more cases of herbicide resistance (predominantly triazine-resistant weeds) than the United States; however, the United States surpassed Europe in the mid-1990s due to the greater use of ALS inhibitors and glyphosate in the USA than in Europe.

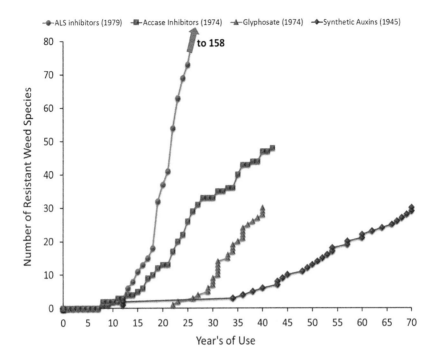

Figure 6.6. The number of herbicide-resistant weed species to four herbicide sites-of-action group in relation to the number of years that the herbicide group has been used. Years in brackets indicate the year that the group was first used commercially.

Australia and Canada are both large producers of wheat and small grains and have virtually identical rates of increase in HR weeds (Figure 6.3). Resistance problems in these countries are driven by economically important resistant grasses in cereals (i.e. *A. myosuroides*, *Avena fatua* L., *Lolium* spp.), which do not have much options for control. Canada also has a number of HR weeds in its small soybean-producing region in Southern Ontario. Recently, new soybean varieties have allowed soybean production in Western Canada and it is likely we will see GR weeds increasing in these regions. The production of rice, corn, and soybean has been increasing steadily with the greatest increase being in soybean – from about 25 million ha in 1960 to almost 150 million ha in 2015.

Resistant Weed Problems in Selected Major Crops

The top five agronomic crops in the world based on area of production in decreasing order are wheat, rice, corn, soybean, and cotton (Figure 6.7). Since 1960, wheat production had remained above 200,000,000 ha. Cotton production remained relatively flat, at around 30 million ha. During the same period, the productivity of all these crops increased (Figure 6.8). The development and use of improved varieties or hybrids and the use of herbicides for efficient and effective weed control are major factors contributing to increasing crop productivity. Areas planted with these crops are managed primarily with herbicides; not surprisingly, the largest numbers of weed resistance cases to herbicides are associated with these crops topped by wheat with almost 140 cases (Figure 6.9). Considering that not all unique cases of resistance are reported and the extent of research on HR weeds differ across countries, it is certain that these numbers represent, but underestimate, the occurrence of resistance in crop (and non-crop) fields. The use of sustainable farming practices is increasing gradually, but we have yet to attain broad-scale sustainability in our management of soil health, use of irrigation water, and use of pesticides among other agriculture-related pursuits.

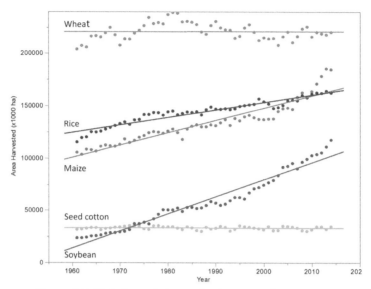

Figure 6.7. Global trends in production area of major crops.

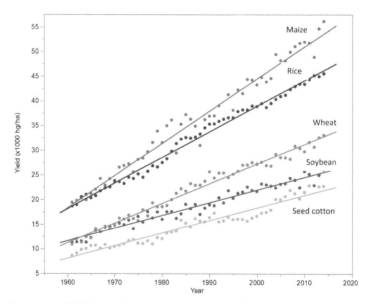

Figure 6.8. Global trends in average yield per unit area of major crops.

Rice: A Case Study in Resistance Evolution

Rice ranks third among cereal grains produced worldwide in 2016/17, followed by corn and wheat (https://www.statista.com/statistics/263977/world-grain-production-by-type/). Whereas corn is used for food, feed, and biofuel, rice is used exclusively for human consumption and is the main food for more than half of the world's population. The major obstacles that threaten rice production are water, emergence of new insect pest and disease problems with subsequent adaptations to insecticides and fungicides, resistance to herbicides and climate change (Singh et al. 2017).

The greatest biological constraint in rice production are weeds (Naylor 1994). At the dawn of rice production, farmers managed weeds with work animals and handweeding; thus, farm sizes were generally small. Rice was transplanted, which facilitates inter-row cultivation with

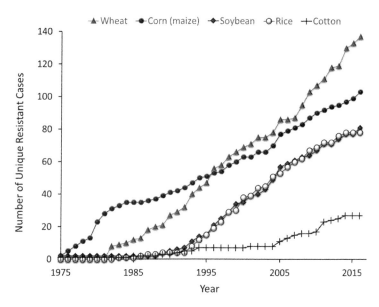

Figure 6.9. The chronological increase in unique herbicide-resistant weed cases for selected crops.

small implements for weed control. Additional weed control is afforded by maintaining a permanent flood. In traditional rice systems, relatively few herbicides are used because soil puddling, transplanting, and ponding water are effective weed control measures (GRiSP 2013). Farm mechanization allowed cultivation of larger tracts of land. The green revolution ushered in high-yielding, semi-dwarf varieties that require high fertilizer inputs to reach its yield potential. The combination of semi-dwarf rice and high fertilizer application favored more weed growth, which was solved with the use of herbicides. These conditions, and the development of herbicide technology, allowed further expansion of rice production areas and increased farm sizes. The efficiency of herbicides compared to handweeding or mechanical weeding inevitably made herbicide the main tool for weed control. Economic and social factors also made dependence on herbicides inevitable. Historically, rice production in many regions of the world was driven by human labor. Rising labor costs made manual transplanting and handweeding uneconomical. Also, the rural demographics had changed across the past decades. In underdeveloped and developing countries, women and children were the traditional workers in rice fields, but economic and educational development fueled a steady migration of the younger population to seek employment in urban areas, creating severe labor shortage at the farm. Thus, rice production systems shifted to direct seeding starting in the early 1990s, which favors weed infestation, hence more herbicide use (Chauhan et al. 2017, GRiSP 2013, Naylor 1994). Thus, a large proportion of rice production fields worldwide have been managed with herbicides for decades. In the USA, herbicides comprise 55–95% of chemical inputs in rice production (Scott et al. 2013, Arkansas Crop Enterprise Budgets at https://www.uaex.edu/farm-ranch/economics-marketing/farm-planning/budgets/crop-budgets.aspx).

Diversity in Herbicide SOAs Alone does not Quell Resistance Evolution

Rice herbicides represent at least 10 SOAs, including ACCase-, ALS-, and PSII inhibitors (Table 6.2). Some of these herbicides, such as the non-selective glyphosate or the auxin herbicide 2,4-D, are applied for pre-plant desiccation of vegetation. Today, weeds in rice evolved resistance to nine of these groups of herbicides. Following the overall global trend in resistance evolution, the largest number of resistance cases (79) in rice production systems involve ALS inhibitors (Table 6.3). It is a fact that ALS contains multiple conserved regions where the five chemical families

Table 6.2. Herbicide sites of action (SOAs) used in rice production

Site of action classification[1]		Some herbicide examples
A/1	Acetyl-coA carboxylase (ACCase) inhibitor	Cyhalofop, fenoxaprop
B/2	Acetolactate synthase (ALS) inhibitor	Bensulfuron-methyl, bispyribac-sodium, halosulfuron, imidazolinone, imazapic, imazapyr, imazosulfuron, orthosulfamuron, penoxsulam, pyrazosulfuron
C3/6	Photosystem II (PSII) inhibitor	Bentazon
C2/7	Photosystem II (PSII) inhibitor	Propanil
E/14	Protoporphyrinogen oxidase (PPO) inhibitor	Acifluorfen, carfentrazone, oxadiazon, saflufenacil
G/9	Enolpyruvyl shikimate 3-phosphate synthase (EPSPS) inhibitor	Glyphosate
H/10	Glutamine synthetase	Glufosinate
K1/3	Microtubule inhibitor	Pendimethalin
K3/15	Very long chain fatty acid synthesis inhibitor	Butachlor, pretilachlor
N/8	Lipid synthesis inhibitor	Molinate, thiobencarb
O/4	Synthetic auxin	2,4-D; quinclorac; triclopyr

[1] Letter codes are classification by the Herbicide Resistance Action Committee (HRAC). Numeric codes are by the Weed Science Society of America (WSSA) classification.

of ALS inhibitors bind and mutations at eight sites endow numerous cross resistance patterns to ALS inhibitors (McCourt et al. 2006; Tranel et al. 2017; Yu and Powles 2014). Several of these herbicides are sulfonylureas (bensulfuron, halosulfuron, imazosulfuron, etc.). One sulfonylurea herbicide that is used intensively is halosulfuron because of its high activity on many species of sedges. Some ALS inhibitors (e.g. penoxsulam) are active on grasses, others (i.e. bensulfuron) are active on aquatic broadleaf weeds, while others have species-specific activity on both. Therefore, a great diversity of species (42 total) had evolved resistance to ALS inhibitors.

The second highest number of cases involve a PSII inhibitor, propanil (Table 6.3). This was the first, most intensively used rice herbicide globally, akin to the role of atrazine in corn production. Propanil is a broad-spectrum herbicide with no soil activity. In North America, it was the backbone of chemical weed control in rice for three decades before resistance evolved among *Echinochloa* populations (Talbert and Burgos 2007). In spite of this, it is still used today with several other herbicide SOAs because of its broad-spectrum activity. Resistance to selective grass herbicides (i.e., cyhalofop, fenoxaprop) is also widespread (18 cases). Grasses are dominant weed problems in rice, which necessitate regular use of ACCase inhibitor herbicides.

Resistance to synthetic auxins (13 cases) primarily involves quinclorac, which is used worldwide to control grass weeds in rice (Table 6.3). In this respect, quinclorac differs from the other auxinic herbicides such as benzoic acids (e.g. dicamba) or phenoxies (e.g. 2,4-D), which are effective only on broadleaf weeds. Quinclorac-resistant grasses are primarily *Echinochloa* species.

A few items are noteworthy. Inhibitors of very long-chain fatty acid synthesis (e.g. acetochlor, alachlor, dimethenamid, *S*-metolachlor, pyroxasulfone) are important herbicides

Table 6.3. Resistance to rice herbicides by site of action (SOA) classification

Resistance type	Herbicide target	HRAC/WSSA	Dicot	First year	Mono-cot	First year	Total
Single	ALS	B/2	19	1994	60	1993	79
	PSII	C2/7			20	1987	20
	ACCase	A/1			18	1994	18
	Synthetic auxin	O/4			13	1983	13
	Lipid synthesis	N/8			3	1993	3
	DOX-P synthase	F4/13			1	2008	1
	EPSPS	G/9	1	2003			1
	Very long chain fatty acid synthesis	K3/15			1	1993	1
Multiple, 2	ACCase, ALS	A/1; B/2			7	2006	7
	ALS, synthetic auxin	B/2; O/4	1	2002	3	1998	4
	Synthetic auxin; ALS	O/4; B/2			4	1998	4
	PSII; synthetic auxin	C2/7; O/4			3	1999	3
	PSII; ALS	C2/7; B/2			3	2009	3
	ACCase, EPSPS	A/1; G/9			1	2008	1
Multiple, 3	ACCase, ALS, PSII	A/1; B/2; C2/7			3	1998	3
	ALS, PSII, synthetic auxin	B/2; C2/7; O/4			2	2011	2
	ACCase, ALS, synthetic auxin	A/1; B/2; O/4			1	2015	1
Multiple, 4	ACCase, PSII, synthetic auxin, glutamine synthetase	A/1; C2/7; O/4; H/10			1	2010	1
	ACCase, ALS, PSII, synthetic auxin	A/1; B/2; C2/7; O/4			1	2011	1
Multiple, 6	ACCase, ALS, DOX-P synthase, lipid synthesis, synthetic auxin, PSII	A/1; B/2; F4/13; N/8; O/4; C2/7			1		1
Total			*21*		*146*		*167*

for agronomic crops because of their strong residual activity on annual grasses and small-seeded broadleaf weeds. These compounds, except pyroxasulfone, had been used extensively on agronomic crops for more than three decades, yet no cases of resistance have been reported involving these type of herbicides. However, the major grass weeds in rice (*E. crus-galli*, *E. oryzoides*, *E. phyllopogon*) had evolved resistance to butachlor or pretilachlor (Table 6.3).

A more worrisome development in rice production is the evolution of multiple resistance (Table 6.3). This occurred primarily because of successive selection with one herbicide SOA after another, such as the case with *Echinochloa* resistance in North America (Fischer et al. 2000, Rouse et al. 2017, Talbert and Burgos 2007, Yasour et al. 2008, 2009, 2011). Thus, the availability of several herbicide SOAs does not guarantee mitigation of resistance evolution if the driver weed (such as *Echinochloa* spp. in the case of rice) is managed continually with the same herbicide SOA over long periods. Integration of control measures has to be done including cultural (changing tillage approaches, water management, planting dates, crop rotation), biological (competitive or weed-suppressive varieties), post-harvest/pre-plant weed management, and preventive measures.

Resistance Evolution is not Related to Area of Production

The first case of HR weed reported in rice production system was *Sphenoclea zeylanica* Gaertn. in the Philippines in 1983 (Mercado et al. 1990). Since then, new unique cases of resistant weeds have been reported every year. The USA leads the world in the total number of resistance cases, at 39 (Figure 6.10). Japan ranks second with 21 unique cases; followed by South Korea with 14; Brazil, China, and Italy with 10; Malaysia, 9; and Venezuela, 8. The frequency of resistance occurrence is independent of production area; rather, it is strongly linked to the intensity of herbicide use. Japan harvested 1.57 million ha of rice in 2016, which was only slightly larger than the 1.25 million ha harvested in the USA in the same year. The majority of rice in Japan are machine-transplanted. Like all other rice-producing countries, labor availability for rice farming in Japan has declined drastically and weeds are managed primarily with herbicides (Matsunaka 2001). Between 1950 and 1990 Japan had the largest increase in herbicide use in Asia; South Korea was third. In the early 1990s it was estimated that 75–100% of rice fields in Korea were sprayed with herbicides (Naylor 1994), most likely multiple times in a year. Japan and South Korea lead the Asian rice-producing countries with herbicide-resistant weed problems. Among the countries that reported resistant weeds in rice, the top three rice producers are China, Indonesia, and Thailand with 30.16, 12.16, and 10.08 million ha, respectively. These three countries reported a combined 15 cases of HR weeds in rice while the USA, with only 1.25 million ha, had 39 cases. India, and Vietnam, which are among the leading rice producers in the world, so far, have not reported cases of resistance. This does not necessarily mean that there are no HR weeds in Vietnam nor India. It could be that resistance monitoring and testing have not been done in these countries. The absence of HR weed report from India may also reflect the fact that the adoption of chemical weed control in India was slow as can be deduced from the review of herbicide use in Asia by Naylor (1994). Historically, farmers in India had relied on manual labor for weeding crop fields until recently when labor is becoming scarce and expensive, just like in other countries. Brazil has 1.6 times more rice hectares than the USA, yet Brazilian rice farmers spend less per hectare on chemical inputs, which are mostly herbicides, than USA farmers (Singh et al. 2017). Coincidentally, Brazil has only 10 cases of reported resistant weeds in rice, almost four times less than those of the USA.

Herbicide-resistant Weeds in Rice

At least 167 total unique cases of resistance, including those with different patterns of multiple resistance, have been reported in rice production systems (Table 6.3). Two groups of species are noteworthy. Of 62 HR weed species in rice, 14 belong to the genus *Echinochloa*, comprising almost one-half (46%) of the reported cases. Resistance to four SOAs in *E. colona* and *E. crus-galli* has been reported in the southern USA (Heap 2017) while resistance to six SOAs has been reported in *E. phyllopogon* in California, USA (Fischer et al. 2000, Yasour et al. 2008, 2009, 2011).

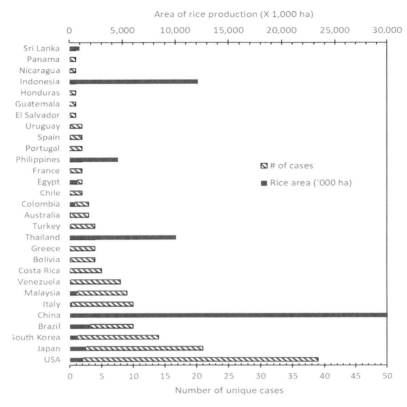

Figure 6.10. Unique cases of resistance and area of rice production, global summary.

The most common resistance problem in this genus are to propanil, quinclorac, and ACCase inhibitors. The high level of adoption of the ALS-resistant (Clearfield™) rice technology accompanied by the use of imidazolinone herbicides, plus the use of other ALS-inhibitor herbicides, resulted in the evolution of ALS-resistant *Echinochloa* spp. Although the genus is primarily self-pollinated it has high ploidy level (6×) and has shown broad adaptability to diverse agroecological environments. This, and its dominance in rice fields (large population size), provide the foundation for rampant cases of resistance relative to other grass species. The second largest group of resistant species are sedges, specifically, in the genus *Cyperus, Sagittaria,* and *Schoenoplectus.* There are 15 resistant species in this group, which collectively account for 37 (22%) of the reported cases. Of the species in this group, small umbrella sedge (*Cyperus difformis*) has shown the most propensity to evolve resistance (10 cases). Because these species are controlled with an ALS-inhibitor herbicide, all cases except two, involve resistance to ALS inhibitors. The exceptions are one case each of *C. difformis* and *Schoenoplectus mucronatus* with multiple resistance to ALS- and PSII inhibitors.

Resistance to Herbicides in Wheat Production

Wheat is the cereal food grain that is produced in the largest land area (Figure 6.7) and the most volume because it is adaptable to a wide range of environments and is second to rice as a source of calories for humans (Awika et al. 2011). In general, wheat is more adaptable to a wide range of growth conditions than other major cereal crops, and is thus the most widely cultivated food crop in the world. China is the world's largest wheat producer followed by India, Russia, USA, and France (https://www.worldatlas.com/articles/top-wheat-producing-countries.html). Like rice, there are no commercial transgenic HR traits in wheat. Selective control of grasses in wheat is problematic as grass weeds often have the ability to mimic the mechanisms of wheat that allow it to tolerate grass herbicides. Herbicides in wheat represent 13 modes of action (Table 6.4). The non-selective ones (glyphosate, glufosinate, and paraquat) are used for pre-plant or post-harvest weed management although glyphosate is the most commonly used globally among these. Growers mix non-selective herbicides with other herbicide modes of action (i.e. auxinic herbicides) to improve control of difficult species or mix soil-active herbicides (i.e. flumioxazin, oxyfluorfen) to provide residual weed control. The primary herbicides in-crop for wheat are ACCase-, ALS-, and PSII-, inhibitors, which are highly effective on the dominant grass weed problems.

Resistance to herbicides in wheat was first reported in 1982, involving *A. myosuroides* resistant to ACCase inhibitors (A/1) in the UK (Heap 2017). In this year also, *L. rigidum* resistant to three SOAs [chlorsulfuron (B/2), diclofop-methyl (A/1), trifluralin (K1/3)] was reported in Australia. The most problematic species are *A. fatua, L. rigidum, L. multiflorum, A. myosuroides, Phalaris* spp., and *Setaria* spp. (Table 6.5). These species have evolved target-site and metabolic resistance to ACCase- and ALS-inhibitors used to control them. *A. fatua*, with multiple herbicide resistance, is widespread in North America, especially in Canada (Beckie et al. 1999). It is also problematic in Europe, the Middle East, and to a lesser extent in Australia. *L. rigidum* is the most serious resistant weed in wheat production in Australia (Owen et al. 2007). While *L. rigidum* is a problem in parts of southern Europe, *L. multiflorum* is the one which is of particular concern in wheat production in much of the USA and Europe (Heap 2017, Rauch et al. 2010, Salas et al. 2013).

A. myosuroides is the most serious weed of wheat in Europe, being the one with the most cases of resistance reported (Heap 2017). It is interesting that *A. myosuroides* has not become a resistant weed in parts of the USA (for instance in Oregon where it occurs as a weed of wheat). *Phalaris canariensis* L. and *Phalaris minor* Retz. are major HR weeds of wheat in warmer wheat

Table 6.4. Herbicide sites of action (SOAs) used pre-plant or in-crop in wheat production

Site of action classification[1]		Some herbicide examples
A/1	Acetyl-coA carboxylase (ACCase) inhibitor	Clodinafop, diclofop, pinoxaden, tralkoxydim
B/2	Acetolactate synthase (ALS) inhibitor	Chlorsulfuron, flufenacet, mesosulfuron, pyroxsulam, sulfosulfuron
C1/5	Photosystem II (PSII) inhibitor	Metribuzin
C3/6	Photosystem II (PSII) inhibitor	Bromoxynil
C2/7	Photosystem II (PSII) inhibitor	Chlorotoluron, isoproturon
D/22	Photosystem I (PS I) inhibitor	Paraquat
E/14	Protoporphyrinogen oxidase (PPO) inhibitor	Carfentrazone, flumioxazin, oxyfluorfen, saflufenacil
F2/27	4-hydroxyphenyl-pyruvate-dioxygenase (4-HPPD) inhibitor	Pyrasulfotole
G/9	Enolpyruvyl shikimate 3-phosphate synthase (EPSPS) inhibitor	Glyphosate
H/10	Glutamine synthetase inhibitor	Glufosinate
K1/3	Microtubule inhibitor	Pendimethalin, trifluralin
K3/15	Very long chain fatty acid synthesis inhibitor	Pyroxasulfone
N/8	Lipid synthesis inhibitor	Prosulfocarb, triallate
O/4	Synthetic auxin	2,4-D; clopyralid; dicamba; MCPA

[1] Letter codes are classification by the Herbicide Resistance Action Committee (HRAC). Numeric codes are by the Weed Science Society of America (WSSA) classification.

growing regions, such as India and Mexico (Singh 2007). Resistance to ACCase inhibitors has evolved in 15 grass species.

Overall, weeds in wheat had evolved resistance to 10 of 13 herbicide SOAs used for weed control (Table 6.5). There were 313 cases reported, involving 74 species (Heap 2017). The largest resistance problem is to ALS inhibitors involving *K. scoparia* and *Stellaria media* (L.) Vill. and 58 other species. Twenty-one species evolved resistance to two SOAs; this is mostly to ACCase- and ALS inhibitors. Seven species evolved resistance to three SOAs, with A and B groups plus a third SOA. The largest cases of multiple resistance reported involved *Lolium* species, followed by *A. myosuroides* as a distant second. *L. rigidum* in Australia evolved resistance to seven SOAs.

Table 6.5. Global summary of reported herbicide-resistance cases in wheat by site of action (SOA)

SOA[1]	Target site	# Cases	Total # species	Top species	# Cases
A/1	Acetyl-coA carboxylase (ACCase) inhibitor	71	15	*Avena* spp.	29
				Lolium spp.	18
				Phalaris spp.	11
B/2	Acetolactate synthase (ALS) inhibitor	136	60	*Kochia scoparia*	15
				Stellaria media	11
C2/7	Photosystem II (PSII) inhibitor	10	7	*Alopecurus myosuroides*	4
G/9	Enolpyruvyl shikimate 3-phosphate synthase (EPSPS) inhibitor	10	9		
K1/3	Microtubule inhibitor	7	4	*Setaria viridis*	4
O/4	Synthetic auxin	12	8		
MR/2 SOA	A/1 + (B/2, K1/3, C2/7, G9); B/2 + (K1/3, C1/5, C2/7, O/4, G/9); G/9 + O/4	49	21	*Lolium* spp.	22
				A. myosuroides	10
MR/3 SOA	A/1 + B2 + (K1/3, N/8, C2/7, F2/27, K3/15)	14	7	*Avena* spp.	6
				Lolium spp.	6
MR/4 SOA	A/1, B/2, G/9, K1/3	1	1	*L. rigidum*	
MR/7 SOA	A/1, B/2, F4/13, K1/3, K2/23, K3/15, N/8	1	1	*L. rigidum*	
Total		*313*	*74[2]*		

[1] Letter codes are classification by the Herbicide Resistance Action Committee (HRAC). Numeric codes are by the Weed Science Society of America (WSSA) classification.
[2] Several species are resistant to more than one herbicide SOA.

Resistance to Herbicides in Corn Production

There are 14 herbicide SOAs used in corn (Table 6.6). The herbicide used for the longest period is 2,4-D and the one that has been used over the largest crop area globally is atrazine. It is no surprise that the first cases of any HR weeds that were of significant economic concern were triazine-resistant weeds in corn. The first of these was atrazine-resistant *Amaranthus hybridus* L., identified in 1972 in corn fields in Maryland (Heap 2017). To date, the largest group of HR weeds in corn were selected by PSII inhibitors (C1/5) across 42 weed species (Table 6.7). In fact, the first 41 HR weed species identified in corn globally (mainly US, Europe, and Canada from 1972 to 1990) were resistant to PSII inhibitors, selected primarily by atrazine, and to a lesser extent by simazine, metribuzin and cyanazine. The most common and widespread cases of triazine-resistant weeds in corn involve *Amaranthus* spp. (in order of severity or occurrence: *A. retroflexus* L., *A. hybridus*, *A. tuberculatus*, *A. palmeri*, *A. powellii* S. Wats., *A. blitoides* S. Wats., *A. blitum* L., *A. cruentus* L., and *A. viridis* L.), *Chenopodium album* L., *K. scoparia*, *Solanum nigrum* L., *Senecio vulgaris* L., and *Conyza canadensis* (L.) Cronquist. Triazine-resistant grasses were less prevalent, but the most important were *Setaria* spp. (*S. faberi* Herrm., *S. pumila* (Poir.) Roem. & Schult., *S. viridis* and *S. verticillata* (L.) P. Beauv.), as well as *Echinochloa crus-galli* (L.) P. Beauv., and *Digitaria sanguinalis* (L.) Scop. Out of the 73 weeds in all crops/situations known to have evolved resistance to PSII inhibitors, 50 were found in corn production.

While triazine-resistant weeds are still dominant in corn production in North America and Europe, they declined in importance after the introduction of ALS inhibitors in the late 1980s and then again by the introduction of glyphosate in Roundup Ready® corn in the late 1990s. The second wave of HR weeds in corn was to the ALS inhibitors, which selected resistance faster than any other herbicide SOA. A total of 19 weed species had evolved single resistance to ALS inhibitors in corn production. The most problematic ALS inhibitor-resistant broadleaf weeds in corn are *Amaranthus* spp. (*A. tuberculatus*, *A. palmeri*, *A. hybridus*, *A. retroflexus*, *A. powellii*), *Ambrosia* spp. (*A. artemisiifolia* L., *A. trifida* L.), *C. canadensis*, and *Solanum ptycanthum* Dunal. The worst ALS inhibitor-resistant grasses in corn production include *Sorghum* spp. (*S. halepense* [L.] Pers. and *S. bicolor* [L.] Moench), *E. crus-galli*, *S. faberi*, *S. viridis*, and *D. sanguinalis*.

In the mid-1990s many of these species, particularly the broadleaves, had evolved multiple resistance to ALS- and PSII inhibitors. Growers in Europe did not use ALS inhibitors to the extent that US growers did; thus, European farmers did not experience the severity of ALS-inhibitor-resistance problems as did the US farmers. In North America, Roundup Ready® corn and soybean technologies arrived in perfect timing as the best solution to the mounting resistance problem with ALS- and PSII-inhibitors in these crops. Roundup Ready® soybean was released in 1996, shortly followed by Roundup Ready® corn and others.

Growers adopted the Roundup Ready® technology rapidly, primarily due to their multiple-herbicide-resistance predicament in corn and soybean. *C. canadensis* was the first glyphosate-resistant (GR) weed in Roundup Ready® corn, but was quickly superseded in importance and abundance by GR *A. palmeri* and *A. tuberculatus* (Table 6.7). Successive selection with different herbicide SOAs led to the current struggle with GR weeds in corn with multiple resistance to various combinations of ALS-, PSII-, or protoporhyrinogen oxidase (PPO) inhibitors, but usually with ALS inhibitors.

Herbicide-resistant Weeds in Soybean Production

Soybean is one of the four crops grown in the largest area worldwide, occupying a little over 120 million ha (Figure 6.7). The US leads global soybean production at 34% (34.4 million ha), closely followed by Brazil at 30% (29 million ha), and Argentina as distant third at 18% (20.3 million ha) (FAOStat. http://www.fao.org/faostat/en/#home). Farm sizes in major soybean-producing countries are in the hundreds of hectares. Production is highly mechanized, utilizing the most advanced technologies, to maximize efficiency. Besides modern machineries, herbicides enable soybean production in vast tracts of land. Growers have a large array of herbicides and herbicide mixtures to use for weed management pre-plant, in-crop, and post-

Table 6.6. Herbicide sites of action (SOAs) used pre-plant, in-crop, or post-harvest in corn production

SOA[4]	Target site	Some herbicide examples
A/1	Acetyl-coA carboxylase (ACCase)	Clethodim[1]
B/2	Acetolactate synthase (ALS)	Halosulfuron, nicosulfuron, flumetsulam, primisulfuron, rimsulfuron, thiencarbazone
C1/5	Photosystem II (PSII)	Metribuzin[1,3]
C3/6	Photosystem II (PSII)	Bentazon, bromoxynil
C2/7	Photosystem II (PSII)	Diuron[1]
D/22	Photosystem I (PS I)	Paraquat
E/14	Protoporphyrinogen oxidase (PPO)	Carfentrazone[2], flumioxazin[3], fluthiacet-methyl, saflufenacil
F2/27	4-hydroxyphenyl-pyruvate-dioxygenase (4-HPPD)	Bicyclopyrone, isoxaflutole, tembotrione, topramezone
G/9	Enolpyruvyl shikimate 3-phosphate synthase (EPSPS)	Glyphosate
H/10	Glutamine synthetase	Glufosinate
K1/3	Microtubule	Pendimethalin
K3/15	Very long chain fatty acid synthesis	Dimethenamid-P, *S*-metolachlor, pyroxasulfone
N/8	Lipid synthesis	
O/4	Synthetic auxin	2,4-D; clopyralid; dicamba

[1] Pre-plant
[2] Desiccant
[3] Post-harvest
[4] Letter codes are classification by the Herbicide Resistance Action Committee (HRAC). Numeric codes are by the Weed Science Society of America (WSSA) classification.

harvest. Labelled herbicides comprise 12 SOAs (Table 6.8) and there are hundreds of formulated premixes from which to choose. There are several HR soybean cultivars, primarily with glyphosate resistance trait (Roundup Ready®) but also those with resistance to sulfonylurea herbicides (STS® beans) or glufosinate (Liberty Link® beans). Among the most widely used herbicide chemistries historically outside of the HR crop technologies are the chloroacetamides (K3/15), dinitroanilines (K1/3), diphenylethers (E/14), imidazolinones (B/2), and selective grass herbicides (A/1). Soybean is also highly amenable to being grown in rotation with other crops, especially corn, cotton, grain sorghum, or wheat. In spite of this, the intensity of selection pressure exerted by herbicides had selected for resistant weeds as early as in the mid-70s.

The first reported HR weed in soybean was *Eleusine indica*, resistant to dinitroaniline herbicides (DNAs), pendimethalin and trifluralin (microtubule inhibitors), in 1974 in the USA (Mudge et al. 1984). At the time, DNAs had been used for about 25 years, and had been used in soybean alone or in mixtures with other pre-emergence herbicides. At about the same time, *Chenopodium album* resistant to triazinone, metribuzin (PSII inhibitor), was also reported in the USA in 1975 (Heap 2017). Metribuzin has not been used as much as the other pre-emergence herbicides in soybean in the USA. This population was cross-resistant to atrazine and was most likely selected by the latter in corn because corn-soybean rotation is the principal cropping system in the US cornbelt.

Table 6.7. Global summary of reported herbicide-resistance cases in corn by site of action (SOA)

SOA[1]	Target site	# Cases	Total # species	Top species	# Cases
A/1	Acetyl-coA carboxylase (ACCase)	1	1	*Setariafaberi*	1
B/2	Acetolactate synthase (ALS)	58	19	*Sorghum halepense*	10
				S. bicolor	8
C1/5	photosystem II (PSII)	155	42	*Chenopodium album*	34
				Amaranthus retroflexus	15
				A. hybridus	14
C2/7	Photosystem II (PSII)	1	1	*Euphorbia heterophylla*	1
C3/6	Photosystem II (PSII)	1	1	*A. hybridus*	1
D/22	Photosystem I (PS I) electron diverter	1	1	*Eleusine indica*	1
F2/27	4-hydroxyphenyl-pyruvate-dioxygenase (4-HPPD)	2	2	*Amaranthus palmeri*	1
				A. tuberculatus	1
G/9	Enolpyruvyl shikimate 3-phosphate synthase (EPSPS)	92	14	*A. palmeri*	9
				Conyza canadensis	8
				Kochia scoparia	7
O/4	Synthetic auxin	3	2	*K. scoparia*	2
MR/2 SOA	A/1+ (B/2, G/9); B/2 + (C1/5, E/14, F2/27, G/9); C1/5 + (C2/7, F2/27); G/9 + (D/22, E/14, O/4)	27	13	*A. palmeri*	6
				A. tuberculatus	4
				Lolium perenne var. *multiflorum*	3
MR/3 SOA	B/2 + C1/5 + (E/14, F2/27, G/9); B/2 + E/14 + G/9	5	3	*A. palmeri*	2
				A. tuberculatus	2
MR/4 SOA	B/2 + C1/5 + G/9 + (E/14, F2/27, O/4)	3	2	*A. tuberculatus*	2
MR/5 SOA	B/2, C1/5, E/14, F2/27, O/4	1	1	*A. tuberculatus*	1
Total		304	61[2]		

[1] Letter codes are classification by the Herbicide Resistance Action Committee (HRAC). Numeric codes are by the Weed Science Society of America (WSSA) classification.

[2] Several species are resistant to multiple herbicide SOAs.

Table 6.8. Herbicide sites of action (SOAs) used pre-plant, in-crop,
or post-harvest in soybean production

SOA code[5]	Target site	Some herbicide examples
A/1	Acetyl-coA carboxylase (ACCase)	Clethodim, sethoxydim, fluazifop, fenoxaprop, quizalofop
B/2	Acetolactate synthase (ALS)	Chloransulam-methyl, chlorimuron, flumetsulam, imazaquin, imazethapyr, thifensulfuron,
C1/5	Photosystem II (PSII)	Metribuzin
C3/6	Photosystem II (PSII)	Bentazon
C2/7	Photosystem II (PSII)	Diuron[1]
D/22	Photosystem I (PS I)	Paraquat[1,2,3]
E/14	Protoporphyrinogen oxidase (PPO)	Acifluorfen, carfentrazone[2], fomesafen, flumiclorac, flumioxazin, fluthiacet-methyl, lactofen, sulfentrazone
G/9	Enolpyruvyl shikimate 3-phosphate synthase (EPSPS)	Glyphosate[4]
H/10	Glutamine synthetase	Glufosinate
K1/3	Microtubule	Pendimethalin, trifluralin
K3/15	Very long chain fatty acid synthesis	Acetochlor, dimethenamid-P, pyroxasulfone, *S*-metolachlor
O/4	Synthetic auxin	2,4-DB

[1] Pre-plant
[2] Post-directed
[3] Desiccant
[4] Pre-plant, in-crop (GR soybean), or pos-harvest
[5] Letter codes are classification by the Herbicide Resistance Action Committee (HRAC). Numeric codes are by the Weed Science Society of America (WSSA) classification.

To date, 297 cases of resistance have been reported in soybean encompassing 46 species (Table 6.9). The largest proportion of these reports originated from the US (74%). The second and third largest soybean-producing countries, Brazil and Argentina, reported 9% and 6%, respectively (Heap 2017). Collectively, 90% of reported HR weeds in soybean originated from these countries.

Resistance to glyphosate is now the most predominant problem in soybean production, occurring across 17 species (Table 6.9), reflecting the sharp increase in resistance to glyphosate soon after the commercialization of GR crops in the mid-90s. The single, most abundant GR species was *C. canadensis*. Including *C. bonariensis* and *C. sumatrensis*, this genus comprised 23% of reported GR cases. The *Ambrosia* genus (*A. artimisiifolia* and *A. trifida*) also comprised 23% of GR cases. These were surpassed only by the *Amaranthus* genus, led by *A. palmeri* and *A. tuberculatus*, which comprised 33% of reported cases.

Although resistance to glyphosate was reported most often, resistance to ALS inhibitors occurred in almost twice as many species as that of glyphosate (Table 6.9). This demonstrates the relatively high frequency of resistance-conferring alleles with respect to the ALS-inhibitor target compared to that of the glyphosate target, EPSPS. *Amaranthus tuberculatus* was the most common ALS-resistant species. All *Amaranthus* species comprised 36% of reported ALS-resistant weeds in soybean.

The most recent, economically important, resistance evolution in soybean production is to the PPO inhibitors (E/14) in the USA. This was first found in *A. tuberculatus* in Kansas, USA in 2000 (Shoup et al. 2003) and a year after, was also found in Illinois (Patzoldt et al. 2005). As with glyphosate, this was particularly alarming because the PPO inhibitors were among the most reliable herbicides in soybean with broad applications (pre-plant, pre-emergence, or post-emergence). The PPO inhibitors have been used since the early 1970s; thus far, only 13 species had evolved resistance to this group of herbicides and only in this decade, about 30 years from first commercialisation of this type of chemistry. These herbicides are the primary

Table 6.9. Global summary of reported herbicide-resistance cases in soybean by site of action (SOA)

SOA code[1]	Target site description	# Cases	Total # species	Top species	# Cases
A/1	Acetyl-coA carboxylase (ACCase)	20	13	*Sorghum halepense*	5
B/2	Acetolactate synthase (ALS)	93	27	*Amaranthus tuberculatus*	13
C1/5	Photosystem II inhibitors (PSII)	5	4	*Chenopodium album*	2
C2/7	Photosystem II inhibitors (PSII)	1	1	*Euphorbia heterophylla*	1
D/22	Photosystem I (PS I) electron diverter	3	2	*Conyza canadensis*	2
E/14	Protoporphyrinogen IX oxidase (PPO)	7	5	*A. tuberculatus*	3
G/9	Enolpyruvyl shikimate 3-phosphate synthase (EPSPS)	107	17	*Conyza canadensis*	22
K1/3	Microtubule inhibitors	3	2	*Amaranthus palmeri*	2
O/4	Synthetic auxins	2	2	*Amaranthus hybridus*	1
				Daucus carota	1
MR/2 SOA	B/2 + (C1/5, E/14, G/9, A/1); A/1 + G/9; D/22 + G/9; E14 + G/9; C1/5 + G/9	47	15	*Amaranthus palmeri*	11
MR/3 SOA	B/2 + E/14 + G/9; B/2 + C1/5 + E/14	6	3	*Ambrosia artemisiifolia*	3
MR/4 SOA	B/2 + C1/5 + G/9 + (E/14, F2/27)	2	1	*A. tuberculatus*	2
MR/5 SOA	B/2 + C1/5 + E/14 + F2/27 + O/4	1	1	*A. tuberculatus*	1
Total		297	46[2]		

[1] Letter codes are classification by the Herbicide Resistance Action Committee (HRAC). Numeric codes are by the Weed Science Society of America (WSSA) classification.
[2] Several species are resistant to multiple herbicide SOAs.

options for the management of ALS-, PSII-, and EPSPS-inhibitor-resistant *Amaranthus* species. PPO-resistant *A. palmeri* was confirmed lately in soybean and cotton fields (Salas et al. 2016) and had been detected quasi-simultaneously across several states in the southern USA. Resistance to PPO inhibitors occurred mostly in the USA (*Amaranthus* spp. and *Ambrosia* spp.), but has also been reported in Brazil, Bolivia, China, Israel, and Spain (Heap 2017).

Multiple resistance were reported in 15 species, occurring most often in *A. palmeri* (Table 6.9). Double resistance generally involved glyphosate plus any other SOA, or an ALS inhibitor plus any other SOA. Resistance to three SOAs among *A. palmeri* populations is now common in the US mid-south, at almost 50% of populations tested in Arkansas (Salas et al. 2017). This involved resistance to ALS-, EPSPS-, and PPO inhibitors. *A. tuberculatus* has evolved resistance to four and five herbicide SOAs.

Herbicide-Resistant Weeds in Cotton

The global cotton production area had never risen above 40 million ha since crop production data were archived starting in the 1960s (Figure 6.7). Instead, the global cotton production area is on a slight downward trend. This is contrary to other major cereal and agronomic crops, which are increasing in production to meet the ever-increasing demand. Two factors might have contributed largely to this pattern. First, the global profitability for cotton has been historically lower compared to that of other major crops because of the high input costs required for growing cotton. Second, the synthetic alternatives to cotton are gaining more market share, resulting in reduced demand for cotton fiber and the inevitable reduction in cotton price to stay competitive. The top three cotton-producing countries are China, India, and the USA (FAOStat. http://www.fao.org/faostat/en/#home). Pakistan and Brazil make up the top five. Unlike the major cereals and soybean, cotton is a long-season crop. It is generally grown on wide rows to promote more boll production and needs up to four herbicide applications during the growing season. It is common knowledge that cotton farmers in North America, South America (Argentina, Brazil), and Australia use high volumes of herbicides to manage weeds.

Cotton herbicides represent 13 SOAs (Table 6.10). Prior to the commercialisation of GR cotton, farmers used a variety of soil-applied herbicides (diuron, fluometuron, norflurazon, pendimethalin, trifluralin, *S*-metolachlor). Later on, the PPO inhibitors (flumioxazin and fomesafen) were also registered for pre-plant or post-directed use in cotton. The ALS-inhibitor herbicides were used to supplement post-emergence control of broadleaf weeds. The advent of GR cotton shifted herbicide use heavily towards glyphosate and less of the other SOAs.

The majority (81%) of resistance reports in cotton originated from the USA and 10% was from Brazil (Table 6.11). The rest were from Australia, China, Greece, Israel, and Paraguay. India and Pakistan, which are among the top five largest cotton producers, did not report any resistant weeds in cotton. Most likely this was because cotton farmers in these countries have smaller farm sizes, use more cultivation and manual interrow weeding, and use less herbicides.

The first HR weed reported in cotton was *E. indica* resistant to pendimethalin and trifluralin (microtubule inhibitors) in 1973 in the USA (Mudge et al. 1984), at least 20 years before GR cotton was commercialised. This resistance event did not spread far nor did it escalate, primarily because the trait is recessive (Zeng and Baird 1997) unlike all other resistance traits, which are dominant or partially dominant. Eventually, resistance evolved to the ALS inhibitors in 1994 in *A. palmeri* and *A. tuberculatus* (Heap 2017). Today, almost all *A. palmeri* and *A. tuberculatus* populations are ALS-resistant. The largest weed resistance problem in cotton is to glyphosate, a consequence of the extremely high adoption of GR cotton. The most predominant GR weed is *A. palmeri*. The explosion of resistance to ALS-inhibitors and glyphosate has compelled the affected farmers to plant glufosinate-resistant (Liberty Link®) varieties. At the same time, the use of PPO inhibitors in cotton production increased. Thus, resistance to PPO inhibitors also evolved among *A. palmeri* populations starting in the US mid-south in 2014 in both cotton and soybean fields (Salas et al. 2017).

Table 6.10. Herbicide sites of action (SOAs) used preplant, in-crop, or postharvest in cotton production

SOA code[5]	Target site	Some herbicide examples
A/1	Acetyl-coA carboxylase (ACCase)	Clethodim, fluazifop, quizalofop, sethoxydim
B/2	Acetolactate synthase (ALS)	Pyrithiobac, trifloxysulfuron
C1/5	Photosystem II (PSII)	Prometryn
C2/7	Photosystem II (PSII)	Diuron, fluometuron, linuron
D/22	Photosystem I (PS I)	Paraquat[1,2,3]
E/14	Protoporphyrinogen oxidase (PPO)	Carfentrazone[1], fomesafen[1,2], flumioxazin[1,2], sulfentrazone[1]
F/1	Phytoene desaturase	Fluoridone, norflurazon
G/9	Enolpyruvyl shikimate 3-phosphate synthase (EPSPS)	Glyphosate
H/10	Glutamine synthetase	Glufosinate
K1/3	Microtubule	Pendimethalin
K3/15	Very long chain fatty acid synthesis	Pyroxasulfone, S-metolachlor
O/4	Synthetic auxin	2,4-D[1,4], dicamba[1,4]
Z/17	Unknown	MSMA[2]

[1] Pre-plant
[2] Post-directed
[3] Desiccant
[4] Post-harvest
[5] Letter codes are classification by the Herbicide Resistance Action Committee (HRAC). Numeric codes are by the Weed Science Society of America (WSSA) classification.

Herbicide-Resistant Weeds in Orchards

Among perennial crops, orchards have the most number of reported HR weeds. This is probably because orchards generally are managed to maintain a weed-free strip on both sides of the trees, small fruits, or vines. Once the crop is established, this clean strip cannot be tilled so as to not damage the crop roots. Therefore, herbicides are applied multiple times in a year to keep this band weed-free. To date, there are 140 cases of HR weeds reported in orchards, involving 46 species (Table 6.12). The USA had the most number (25 cases) of HR weeds reported. Spain ranked second with 12 reported cases. There may be other countries with high incidences of weed resistance in orchards, but researchers and resources for research in this crop category might be lacking, resulting in limited detection and underreporting of resistance cases.

The largest category of reported HR weeds were to triazines (47 cases) and to glyphosate (46 cases). The most prevalent resistant species to triazines (C1/5) were *Senecio vulgaris* and *Poa annua* (Table 6.12). Both of these species are cool-season weeds and are among the principal weed species exposed to the application of residual herbicides (such as the triazines) in the fall to early spring. In fact, the first species that evolved resistance to triazines in orchards was *S. vulgaris* in 1970 in the USA (Heap 2017). Before the advent of Roundup Ready® crops, glyphosate was used globally for weed control in perennial crops. It was not until 1997 when resistance to glyphosate was reported in *L. rigidum* in New South Wales, Australia. The other

Table 6.11. Global summary of reported herbicide-resistance cases in cotton
by site of action (SOA)

SOA code[1]	Target site description	# Cases	Total # species	Top species	# Cases
A/1	Acetyl-coA carboxylase (ACCase)	6	2	*Sorghum halepense*	5
B/2	Acetolactate synthase (ALS)	8	6	*Amaranthus palmeri*	3
E/14	Protoporphyrinogen IX oxidase (PPO)	2	2	*Amaranthus retroflexus A. palmeri*	
G/9	Enolpyruvyl shikimate 3-phosphate synthase (EPSPS)	35	12	*A. palmeri*	11
K1/3	Microtubule	10	3	*Eleusine indica*	7
Z/17	Nucleic acid	7	1	*Xanthium strumarium*	7
MR/2 SOA	B/2 + C1/5; B/2 + G/9; A/1 + G/9; B/2 + G/9	10	5		
Total		*78*	*18[2]*		

[1] Letter codes are classification by the Herbicide Resistance Action Committee (HRAC). Numeric codes are by the Weed Science Society of America (WSSA classification.
[2] Several species are resistant to multiple herbicide SOAs.

non-selective herbicide widely used in orchards is paraquat. Paraquat is used extensively to manage vegetation around perennial crops because of its contact activity. Paraquat drift on green tissue is not detrimental compared to drift with the systemic, non-selective herbicide glyphosate.

Thus, it is also in this situation where we observed the most number of resistance cases to paraquat, a PS I inhibitor, compared to all other crop or non-crop situations. Resistance to paraquat is the third most common in orchards and was first reported in 1988 in *C. canadensis*. To date, *Conyza* spp. are still the most dominant paraquat-resistant weeds in orchards.

Herbicide-resistant Weeds in Non-crop Areas

Non-crop situations include roadsides, railways, industrial sites, and fence lines (Table 6.13). In many cases these sites are treated with long residual herbicides with the aim of controlling/ suppressing weeds over long periods, applying strong selection pressures for resistance. In this situation, selection for TSR and NTSR can occur simultaneously as the highly effective herbicides quickly remove the most susceptible genotypes from the soil seedbank, resulting in a high possibility of selecting for a genotype with resistance-conferring mutation at the target site. Concurrently, the slow dissipation rate of herbicide molecules in the soil exposes several cohorts of germinating seeds to a continuously declining concentration of herbicide in the soil-water phase, effecting a sustained low-dose selection. This condition selects for multiple, low-effect genes that protect plants from abiotic stress, including herbicide stress, potentially resulting in NTSR.

Table 6.12. Global summary of reported herbicide-resistance cases in orchards
by site of action (SOA)

SOA[1]	Target site	# Cases	Total # species	Top species	# Cases
B/2	Acetolactate synthase (ALS)	4	4		
C1/5	Photosystem II (PSII)	47	21	*Senecio vulgaris*	9
				Poa annua	7
C2/7	Photosystem II (PSII)	2	1	*Conyza canadensis*	
D/22	Photosystem I (PS I) electron diverter	19	15	*Conyza* spp.	6
F3/11	4-hydroxyphenyl-pyruvate-dioxygenase (4-HPPD)	3	3	*Agrostis stolonifera, Poa annua, Polygonum aviculare*	
G/9	Enolpyruvyl shikimate 3-phosphate synthase (EPSPS)	46	16	*Conyza* spp.	18
				Lolium spp.	17
H/10	Glutamine synthetase	1	1	*Lolium perenne*	
K1/3	Microtubules	1	1	*Echinochloa crus-galli*	
MR/2 SOA	A/1 + G/9; B/2 + G/9; F3/11 + G/9; C1/5 + C2/7; D/22 + G/9; E/14 + G/9; G/9 + H/10	12	8	More or less evenly distributed	
MR/3 SOA	F3/11 + G/9 + H/10; A/1 + D/22 + G/9	3	3	*L. rigidum, L. perenne, L. perenne* ssp. *multiflorum*	
MR/4 SOA	A/1 + D/22 + G/9 + (B/2, H/10)	2	2	*Eleusine indica, L. perenne* ssp. *multiflorum*	
Total		140	27[2]		

[1] Letter codes are classification by the Herbicide Resistance Action Committee (HRAC). Numeric codes are by the Weed Science Society of America (WSSA) classification.
[2] Some species are resistant to more than one herbicide SOA.

These non-crop sites (except for industrial sites) are usually long and thin, which allow significant immigration of weed seeds and pollen from adjacent land. In addition, roadsides and railways come with extremely effective weed-propagule dispersal agents via the vehicles and trains travelling on them. These factors combine to make non-crop situations particularly vulnerable to the selection of HR weeds. Even though monitoring of resistance in these areas is generally sparse, about 150 cases of resistance have been reported across 60 species (Table 6.13). Sixty-six (44%) of these cases were from roadsides, 34 (23%) were from railways, and 22 (15%) were from pastures. This resistance pattern generally follows the relative intensities of herbicide use in these areas as is observed in crop production fields. Forty-seven (32%) of resistance cases reported were to PSII inhibitors (C1/5) involving 23 species, the top two being triazine-

Table 6.13. Global summary of reported herbicide-resistance cases in non-crop areas by site of action (SOA)[1]

SOA[2]	Target site description	# Cases	Total # species	Top species	# Cases
A/1	Acetyl-coA carboxylase (ACCase)	4	4		
B/2	Acetolactate synthase (ALS)	26	15	*Kochia scoparia*	8
C1/5	Photosystem II (PSII)	47	23	*Conyza canadensis*	8
				Kochia scoparia	5
C2/7	Photosystem II (PSII)	2	2	*Alopecurus myosuroides, Chloris barbata*	
D/22	Photosystem I (PS I) electron diverter	10	6	*Conyza* spp.	6
F3/11	4-hydroxyphenyl-pyruvate-dioxygenase (4-HPPD)	2	1	*Lolium rigidum*	2
G/9	Enolpyruvyl shikimate 3-phosphate synthase (EPSPS)	24	8	*Conyza bonariensis*	6
				L. rigidum	6
				C. canadensis	5
N/26	Lipid synthesis, not ACCase	3	3		
O/4	Synthetic auxin	9	8		
MR/2 SOA	A/1 + (B/2, D/22); B/2 + (C1/5, O/4); C1/5 + C2/7; D/22 + G/9	15	11		
MR/3 SOA	A/1 + B/2 + K1/3; B/2 + C1/5 + O/4;	2	2	*L. rigidum*	1
				A. tuberculatus	1
MR/5 SOA	A/1 + B/2 + C1/5 + D/22 + G/9	1	1	*L. rigidum*	1
Total		*149*	*60*[3]		

[1] Non-crop areas include bushland reserve, forests, industrial areas, pasture, pasture seed, railways, rangeland, and roadsides.
[2] Letter codes are classification by the Herbicide Resistance Action Committee (HRAC). Numeric codes are by the Weed Science Society of America (WSSA) classification.
[3] Some species are resistant to multiple herbicide SOAs.

resistant *C. canadensis* and *K. scoparia*. The primary selector of these species was atrazine. The second highest incidence of weed resistance was to ALS inhibitors (16% of cases reported) and glyphosate. Unlike in crop production areas, resistance to glyphosate (15% of cases) in non-crop areas was as predominant as resistance to ALS inhibitors, reflecting the fact that glyphosate is the single, most commonly used foliar herbicide in non-crop areas. Glyphosate is applied extensively on roadsides. Weeds will continue to evolve resistance to chemical weed control at an accelerating pace. The increasing occurrence of NTSR mechanisms involving protection genes (i.e. cytochrome P450s and glutathione transferases), which are constitutively expressed in resistant populations, means that these selected weed populations are primed to withstand higher levels of herbicide stress. We need to implement more integrated weed management methods because our current practices are not sustainable.

Multiple Resistance

Even more concerning than the increasing cases of resistance to various SOAs across different weed species is the upward trend in stacking of multiple resistance traits within one population and, even worse, in one plant. Resistance to two herbicide SOAs had evolved within populations of nearly 100 species and about 50 species were reported resistant to three herbicide SOAs (Figure 6.11). Fewer cases were confirmed to have stacked resistance within one plant, but the accumulation of resistance traits in one plant is an inevitable consequence of sequential selection with different herbicide SOAs and gene flow. Stacking of resistance traits also means stacking of multiple resistance mechanisms – TSR + one or more NTSR mechanisms. Leading among these multiple-resistant species is *L. rigidum* (Owen et al. 2014).

The Most Problematic Weed Species

Resistance to herbicides in primary weedy species of major global crops, have major economic and ecological consequences. Herbicide-resistant weeds drive the majority of our agricultural pursuits today. Years of weed resistance data have shown accelerated evolution of weed populations across species in managed ecosystems across the globe. We can extract invaluable lessons from this to guide our scientific investigations and innovations to meet the universal challenges of managing crops and weeds sustainably. The poaceae plant family is most problematic, with a total of 80 HR species (Figure 6.12). These are infesting primarily the cool-season cereal crops, such as *A. myosuroides* (Délye et al. 2010) and *Lolium* spp. (Owen et al. 2014, Salas et al. 2013) in wheat.

The singular specie that is emerging to be most problematic is *L. rigidum*, which has evolved resistance to 12 SOAs (Figure 6.13, Appendix Table 1). Its close relative, *L. perenne* ssp. *multiflorum*, is catching up fast, with resistance to eight SOAs. *Lolium* spp. are major weeds in cool-season cereals, especially wheat. *A. fatua, P. annua,* and *A. myosuroides* are also among the array of highly problematic weeds in cereal crops. New herbicides for cereals must control one or more of these primary cool-season grass problems. However, we learned from the progression of HR weed cases that just rolling out another herbicide does not stop weed resistance evolution; instead, we are observing increasing multiple resistance evolution.

E. crus-galli and *E. colona* are the primary grass weeds of rice, which had evolved resistance to 10 and 7 herbicide SOAs, respectively (Figure 6.13). In terms of potential global economic impact, these species are more formidable because these infest not only rice, but also many other crops. Furthermore, three other species in this genus (*E. crus-galli* var. *formosensis, E. oryzoides,* and *E. phyllopogon*) had evolved resistance to herbicides. This genus is composed of over 60 species with high ploidy level ($4\times – 6\times$) (Thakur et al. 1999, Yabuno 1962) and wide adaptability to a broad range of environments. *Echinochloa* spp. has the potential to surpass *Lolium* spp. as the worst herbicide-resistant grass weed as plants with multiple genomes can stack TSR and NTSR traits across its genome sets (Yu et al., 2013), resulting in more complex and unpredictable resistance profiles than in diploid species.

Because NTSR mechanisms are multi-genic and are geared toward adaptation to abiotic (herbicide) stress, species harboring these mechanisms can acquire resistance to herbicide SOAs yet to be used in certain crop or non-crop situations.

The Asteraceae family would compose the second largest group of HR species (Figure 6.12), which would include *Ambrosia* spp., *Conyza* spp., *Xanthium strumarium,* etc. (Appendix Table 1). However, the one family that could rival *Echinochloa* and *Lolium* in resistance evolution

Appendix Table 1. Global summary of unique resistance cases by site of action and weed species as of November, 2017. International Weed Survey (Heap 2017).

Species	Total	A	B	C1	C2	D	G	K1	O	F2	Other
Lolium rigidum	12	1	1	1	1	1	1	1			5
Echinochloa crus-galli var. *crus-galli*	9	1	1	1	1			1	1		3
Poa annua	9		1	1	1	1	1	1			3
Avena fatua	8	1	1					1			5
Eleusine indica	8	1	1	1		1	1	1			2
Lolium perenne ssp. *multiflorum*	8	1	1		1	1	1				3
Alopecurus myosuroides	7	1	1	1	1			1			2
Echinochloa colona	7	1	1	1	1		1		1		1
Amaranthus hybridus	6		1	1			1		1		2
Amaranthus palmeri	6		1	1			1	1		1	1
Amaranthus tuberculatus	6		1	1			1		1	1	1
Amaranthus retroflexus	5		1	1	1						2
Ambrosia artemisiifolia	5		1	1	1		1				1
Conyza canadensis	5		1	1	1	1	1				0
Kochia scoparia	5		1	1	1		1		1		0
Lolium perenne	5	1	1				1				2
Raphanus raphanistrum	5		1	1			1		1		1
Senecio vernalis	5		1	1	1						2
Alopecurus japonicus	4	1	1		1	1					0
Bidens pilosa	4		1	1		1	1				0
Bromus tectorum	4	1	1	1	1						0
Chenopodium album	4		1	1	1				1		0
Conyza bonariensis	4		1	1		1	1				0
Conyza sumatrensis	4		1			1	1				1
Ischaemum rugosum	4	1	1		1	1					0
Senecio vulgaris	4		1	1	1						1
Setaria viridis	4	1	1	1				1			0
Sisymbrium orientale	4		1	1					1		1
Sorghum halepense	4	1	1			1	1				0
Alopecurus aequalis	3	1	1					1			0
Amaranthus blitum (ssp. *oleraceus)*	3		1	1		1					0
Amaranthus powellii	3		1	1	1						0
Apera spica-venti	3	1	1		1						0
Avena sterilis	3	1	1								1
Avena sterilis ssp. *ludoviciana*	3	1	1								1

(Contd.)

Appendix Table 1. (*Contd.*)

Species	Total	A	B	C1	C2	D	G	K1	O	F2	Other
Beckmannia syzigachne	3	1			1			1			0
Bromus diandrus	3	1	1				1				0
Descurainia sophia	3		1						1		1
Digitaria sanguinalis	3	1	1	1							0
Echinochloa oryzoides	3	1	1								1
Echinochloa phyllopogon	3	1	1								1
Euphorbia heterophylla	3		1		1						1
Hordeum murinum ssp. *glaucum*	3	1	1			1					0
Lactuca serriola	3		1				1		1		0
Phalaris minor	3	1	1		1						0
Phalaris paradoxa	3	1	1	1							0
Setaria faberi	3	1	1	1							0
Setaria viridis var. *major*	3	1	1	1							0
Sinapis arvensis	3		1	1					1		0
Solanum ptycanthum	3		1	1	1						0
Sonchus oleraceus	3		1				1		1		0
Stellaria media	3		1	1					1		0
Amaranthus blitoides	2		1	1							0
Amaranthus spinosus	2		1				1				0
Amaranthus viridis	2		1	1							0
Ambrosia trifida	2		1				1				0
Arctotheca calendula	2					1			1		0
Bidens subalternans	2		1	1							0
Brachypodium distachyon	2	1		1							0
Bromus diandrus ssp. *rigidus*	2	1	1								0
Bromus sterilis	2	1	1								0
Capsella bursa-pastoris	2		1	1							0
Centaurea cyanus	2		1						1		0
Chloris barbata	2			1	1						0
Cynosurus echinatus	2	1	1								0
Cyperus difformis	2		1		1						0
Digitaria insularis	2	1					1				0
Digitaria ischaemum	2	1							1		0
Echinochloa crus-galli var. *formosensis*	2	1	1								0
Echinochloa erecta	2				1						1
Epilobium ciliatum	2			1		1					0

(*Contd.*)

Fimbristylis miliacea	2		1						1		0
Galeopsis tetrahit	2		1						1		0
Galium aparine	2		1						1		0
Galium spurium	2		1						1		0
Hedyotis verticillata	2					1	1				0
Helianthus annuus	2		1				1				0
Hirschfeldia incana	2		1						1		0
Hordeum murinum ssp. leporinum	2	1				1					0
Leptochloa chinensis	2	1		1							0
Limnocharis flava	2		1						1		0
Limnophila erecta	2		1						1		0
Papaver rhoeas	2		1						1		0
Parthenium hysterophorus	2		1				1				0
Phalaris brachystachys	2	1	1								0
Polygonum aviculare	2			1							1
Polygonum convolvulus	2		1	1							0
Polygonum lapathifolium	2		1	1							0
Polygonum persicaria	2		1	1							0
Portulaca oleracea	2			1	1						0
Ranunculus acris	2		1						1		0
Rottboellia cochinchinensis	2	1	1								0
Sagittaria montevidensis	2		1								1
Salsola tragus	2		1				1				0
Schoenoplectus mucronatus	2		1		1						0
Setaria pumila	2		1	1							0
Solanum nigrum	2			1		1					0
Sonchus asper	2		1	1							0
Urochloa panicoides	2			1			1				0
Vulpia bromoides	2			1	1						0
Xanthium strumarium	2		1								1
Others	152	10	74	25	1	13	11	1	11		6
TOTAL	484	48	159	73	28	32	39	12	36	2	55

[1]A = acetyl Co-A carboxylase (ACCase) inhibitors; B = acetolactate synthase (ALS) inhibitors; C1, C2 = photosystem II inhibitors; D = photosystem I electron diverter; G = EPSP synthase inhibitor; K1 = microtubule assembly inhibitors; O = synthetic auxins; F2 = 4-hydrophenyl-pyruvate-dioxygenase (HPPD) inhibitors

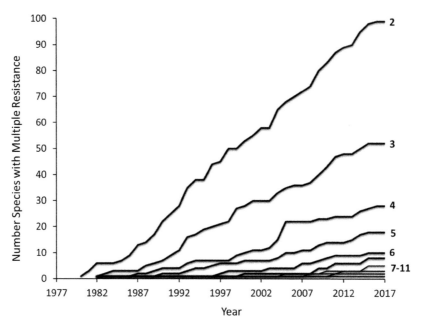

Figure 6.11. The chronological increase in the number of weed species that have evolved resistance to two or more herbicide sites of action.

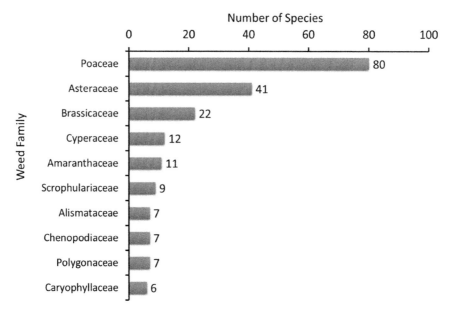

Figure 6.12. The number of weed species that have evolved herbicide-resistance by family (top 10).

and economic impact is Amaranthaceae. Seven species in the genus *Amaranthus* had evolved resistance to herbicides and four species (*A. hybridus, A. palmeri, A. tuberculatis, A. retroflexus*) had evolved resistance to 5–6 herbicide SOAs (Figure 6.13). Like *Echinochloa* spp., *Amaranthus* spp. are global weeds with a large number of weedy species. Unlike *Echinochloa*, *Amaranthus* spp. are diploids (Costea et al. 2004); however, *A. palmeri* and *A. tuberculatus* are dioecious and have

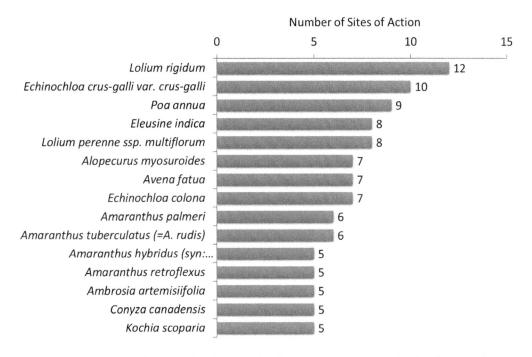

Figure 6.13. Top 15 weed species that have evolved resistance to multiple herbicide sites of action.

high genetic mutation rates or high standing variation within populations (Chandi et al. 2013). Moreover, *Amaranthus* spp. (especially *A. palmeri* and *A. tuberculatus*) produce copious, tiny seeds that can be transported over long distances via farm machinery, crop grain transportation, wind, and water. Its pollen also can travel over long distances (Sosnoskie et al. 2012). Therefore, selection for resistance in *Amaranthus* has occurred, and spread, independently or via gene flow at a high rate. Being an obligate outcrosser and having the adaptation traits for long-distance gene flow allows rapid accumulation of TSR and NTSR genes similar to what is possible with polyploid species.

The Cost of Resistance

The simplicity, affordability, and unparalleled high efficacy of glyphosate has become its demise. A case study in Georgia, USA, on the management of GR *A. palmeri* showed that following the confirmation of resistance to glyphosate in 2005, cotton growers had resorted to hoe-weeding 52% of their cotton fields to remove remaining *A. palmeri* after herbicide applications in-season, costing them $54 ha^{-1} (Sosnoskie and Culpepper 2014). This is in addition to using residual herbicides pre-plant and various combinations of residual and foliar herbicides in-crop. The cotton growers also tilled 20% of their fields as they resorted to using the old, microtubule-inhibitor herbicide trifluralin, which needs to be soil-incorporated. Mechanical cultivation was used in 44% of their fields, and 19% of cotton fields were deep-turned after harvest to bury *A. palmeri* seeds deep into the soil and prevent emergence. Cultivation and mechanical/chemical-intensive practices did not increase weed management costs (Lambert et al. 2017), but these practices bear negative environmental impacts. Increased tillage and cultivation means that soil-conservation-friendly farming has to be abandoned for a certain period to mitigate the HR weed problem. Intensified herbicide use means applying more herbicide SOAs at more times during the crop-growing season. This nullifies the effort of reducing the pesticide load in the environment. Depending on the crop management changes adopted, farmers spend between

$85 and $138 ha^{-1} to combat weed resistance (Lambert et al. 2017). There are other peripheral costs associated with HR weeds including manpower and financial resources to research, plan, and implement stricter and comprehensive regulations for herbicide registrations, promote the adoption of best management practices, and educate agricultural practitioners about resistance and resistance management. In fact, the complex tools for weed management that are forthcoming require intensive education for Extension Agents, Crop Consultants, and farmers to use successfully. The success of cotton farmers in Georgia, USA in managing *A. palmeri* post-resistance evolution to glyphosate was attributed to intense educational and outreach efforts (Sosnoskie and Culpepper 2014).

The Future of Weed Control

We have yet to see a drop in the total volume of herbicides used globally. Except for a seemingly uncharacteristic drop in total herbicide volume used in 2015 (Figure 6.14), we predict that herbicide use will remain high in the next several decades as herbicides remain to be the primary tool for weed management. In fact, a new generation of crops with stacked HR traits are entering the market, a short-term answer to the management of multiple-resistant weeds. Among these are soybean, cotton, and corn with stacked resistance traits including resistance to the auxinic herbicide 2,4-D (Wright et al. 2010) named Enlist®, or those stacked with resistance to dicamba (Behrens et al., 2007) named Engenia® or Xtend®. Following these are crops that would have up to four stacked HR traits to include resistance to HPPD-inhibitor herbicides. The common base of these stacked traits are the non-selective herbicides glyphosate and glufosinate. The driver species for these recent stacks are the rapidly expanding occurrence of *A. palmeri* and *A. tuberculatus* resistant to glyphosate, ALS inhibitors, and PSII inhibitors. Concurrently, we are witnessing the sharp rise in GR weeds following the high, global adoption of GR crops. In fact, the highest number of evolved resistant weeds occurred in GR crops (Figure 6.15). Of note also is that although *de novo* resistance alleles to auxinic herbicides are rare, the list of weeds with resistance to auxinic herbicides has increased sharply since the 1980s (Figure 6.16). *A. tuberculatus* resistant to 2,4-D (Bernards et al. 2012) is one of the most recent additions to this list. Many weed species are now at the cusp of being selected for resistance to various herbicide SOAs being used.

Transitioning from herbicides to other weed control practices is a little like moving from fossil fuels to renewable energy; the speed at which it happens is determined by the need for change. Because herbicides are so effective and, in many cases, are the cheapest and easiest form of crop weed control, there is little incentive to pour resources into alternatives until herbicides fail. This is short sighted. It is clear that the impact of HR weeds is increasing far quicker than the supply of new herbicide (SOA) solutions, and research into alternatives for herbicides needs greater funding now, rather than when most herbicides are failing. There will be no single solution to replace herbicides and it is unlikely that the solutions will be as effective and economical as herbicides (at least in the short term), but that should not stop the allocation of necessary resources to secure the future of weed management. Advances in the field of genetic engineering is likely to provide new weed control technologies, not only in the existing form of creating HR crops, but in more creative ways. These would range from designer biological control agents, to endowing crops with potent phytotoxins (allelopathy) (Duke et al. 2002), to crops with competitive abilities and resistance to abiotic stresses, and designer weed control tools utilising gene silencing technologies (Burgos et al. 2017). Advances in next generation sequencing (NGS) and bioinformatics technologies will complement the OMICS (transcriptomics, proteomics, metabolomics) and biotechnology fields of research. Collectively, these technologies will lead to breakthroughs in developing climate-resilient, weed-competitive, high-yielding crops with broad herbicide tolerance. Breaking the ceiling of crop improvement will ease the burden on herbicides for weed management and will result in more effective and efficient integrated weed management programs.

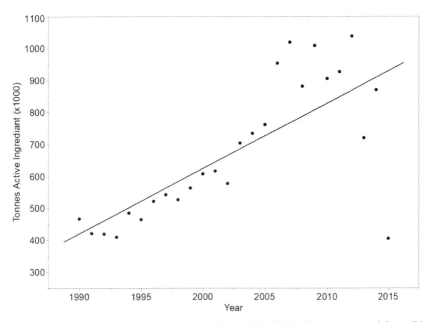

Figure 6.14. Global trend in total herbicide use from 1990–2015. (Data extracted from FAOStat. http://www.fao.org/faostat/en/#home . Accessed November 28, 2017)

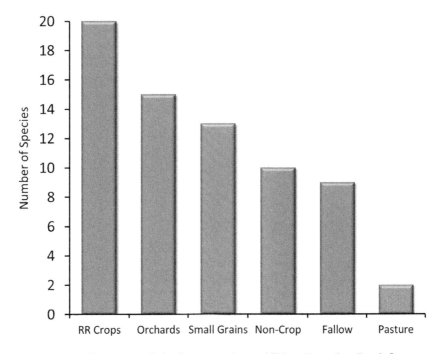

Figure 6.15. The impact of glyphosate-resistant (GR) or Roundup Ready® crops (RR) technology on selection for resistance.

The field of computer science, robotics, and sensing technology is advancing at an exponential rate and is likely to be a rich source of new weed control methods. Robotic weeders are in their infancy, starting with innovations like the one developed for intra-row weeding and thinning of vegetable crops (Fennimore et al. 2016). As the efficiency of these robotic weeders

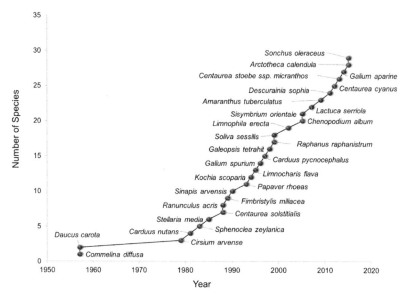

Figure 6.16. The chronological increase in synthetic auxin-resistant weeds.

rises and their price drops, it is likely we will see them as common weed control solutions in orchards, vineyards, and some row crops within 15 years. Remote sensing technology applications is invaluable in big agriculture to achieve precise and more efficient applications of crop production inputs (fertiliser, pesticides, water) and scouting for weeds. The greater challenge is correct identification of weed species, especially at the young stage targeted for herbicide application. Progress in this area is being made, albiet slowly (Andújar et al. 2011, Midtiby et al. 2011, Peña et al. 2013). We will also see more applications of nanotechnology for weed management, specifically in improved herbicide formulations (Grillo et al. 2012; Periera et al. 2014) and herbicide detection in the plant or the environment (Jia et al. 2016; Rahemi et al. 2015).

The complexity of weed resistance patterns, and the cost of weed resistance, will demand complex weed management tools. The situation demands that crop production becomes even more science-based and technologically advanced. The challenge for farmers will be both economical and educational.

REFERENCES

Andújar, D., A. Ribeiro, C. Fernández-Quintanilla and J. Dorado. 2011. Accuracy and feasibility of optoelectronic sensors for weed mapping in wide row crops. Sensors 11: 2304–2318.

Awika, J. 2011. Major cereal grains production and use around the world. Advances in Cereal Science: Implications to Food Processing and Health Promotion ACS Symposium Series; American Chemical Society: Washington, DC, 2011. pp. 1-13 Available at: http://pubs.acs.org/doi/pdf/10.1021/bk-2011-1089.ch001. [Accessed on December 30, 2017].

Beckie, H.J. and F.J. Tardif. 2012. Herbicide cross resistance in weeds. Crop Prot. 35: 15–28.

Beckie, H.J., A.G. Thomas, A. Légère, D.J. Kelner, R.C. Van Acker and S. Meers. 1999. Nature, occurrence, and cost of herbicide-resistant wild oat (*Avena fatua*) in small-grain production areas. Weed Technol. 13: 612–625.

Behrens, M.R., N. Mutlu, S. Chakraborty, R. Dumitru, W.Z. Jiang et al. 2007. Dicamba resistance: enlarging and preserving biotechnology-based weed management strategies. Science 316: 1185–1188.

Bernards, M.L., C.J. Roberto, G.R. Kruger, R. Gaussoin and P.J. Tranel. 2012. A waterhemp (*Amaranthus tuberculatus*) population resistant to 2,4-D. Weed Sci. 60: 379–384.

Brown, H.M. and J.C. Cotterman. 1994. Recent advances in sulfonylurea herbicides. *In:* Stetter, J. (Ed.) Herbicides inhibiting branched-chain amino acid biosynthesis. Chem. Plant Prot. vol. 10. Springer, Berlin, Heidelberg.

Burgos, N.R., C.E. Rouse, V. Singh, R.A. Salas-Perez and M. Bagavathiannan. 2017. Technological advances for weed management. pp. 71–86. *In:* N. Chandrasena and A. Narayana Rao (Eds.) Commemorating 50 Years (1967–2017). 50th Anniversary Celebratory Volume. Asian-Pacific Weed Sci Soc (APWSS); Indian Soc Weed Sci (ISWS), and Weed Sci Soc of Japan (WSSJ).

Busi, R., P. Neve and S. Powles. 2012. Evolved polygenic herbicide resistance in *Lolium rigidum* by low-dose herbicide selection within standing genetic variation. Evol Applic. 6: 231–241.

Buis, R., M. Girotto and S. Powles. 2016. Response to low-dose herbicide selection in self-pollinated *Avena fatua*. Pest Manag Sci. 72: 603–608.

Chandi, A., S.R. Milla-Lewis, D.L. Jordan, A.C. York, J.D. Burton, M.C. Zuleta, J.R. Whitaker and A.S. Culpepper. 2013. Use of AFLP markers to assess genetic diversity in Palmer amaranth (*Amaranthus palmeri*) populations from North Carolina and Georgia. Weed Sci. 61: 136–145.

Collavo, A. and M. Sattin. 2014. First glyphosate-resistant *Lolium* spp. biotypes found in a European annual arable cropping system also affected by ACCase and ALS resistance. Weed Res. 54: 325–334 doi: 10.1111/wre.12082

Costea, M., S.E. Weaver and F.J. Tardif. 2004. The biology of Canadian weeds *Amaranthus retroflexus* L., *A. powellii* S. Watson and *A. hybridus* L. Canadian J. Plant Sci. 84: 631–668.

Dauer, J.T., D.A. Mortensen and M.J. Vangessel. 2007. Temporal and spatial dynamics of long-distance *Conyza canadensis* seed dispersal. J Applied Ecol. 44: 105–114.

Délye, C., M. Jasieniuk and V. Le Corre. 2013. Deciphering the evolution of herbicide resistance in weeds. Trends in Genetics 29: 649–658. doi: 10.1016/j.tig.2013.06.001

Délye, C., M. Séverine, A. Bérard, B. Chauvel, D. Brunel, J.P. Guillemin, F. Dessaint and V. Le Corre. 2010. Geographical variation in resistance to acetyl-coenzyme A carboxylase-inhibiting herbicides across the range of the arable weed *Alopecurus myosuroides* (black-grass). New Phytol. 186: 1005–1017. doi: 10.1111/j.1469-8137.2010.03233.x

Duke, S.O., B.E. Scheffler, F.E. Dayan and W.E. Dyer. 2002. Genetic engineering crops for improved weed management traits. pp. 52–66. *In:* K. Rajasekaran, T.J. Jacks and J.W. Finley (Eds.) ACS Symposium Series No. 829, Crop Biotechnology, ACS Books, Washington, DC.

Fennimore, S.A., D.C. Slaughter, M.C. Siemens, R.G. Leon and M.N. Saber. 2016. Technology for automation of weed control in specialty crops. Weed Technol. 30: 823–837.

Fischer, A.J., C.M. Ateh, D.E. Bayer and J.E. Hill. 2000. Herbicide-resistant early (*Echinochloa oryzoides*) and late (*E. phyllopogon*) watergrass in California rice fields. Weed Sci. 48: 225–230.

Green, J.M. 2016. Current state of herbicides in herbicide-resistant crops. Pest Manag Sci. 70: 1351–1357.

GRiSP (Global Rice Science Partnership). 2013. Rice Almanac, 4th edition. Los Baños (Philippines): International Rice Research Institute. 283 p.

Grillo, R., N.Z.P. dos Santos, C.R. Maruyama, A.H. Rosa and R. de Lima. 2012. Poly (ε-caprolactone) nanocapsules as carrier systems for herbicides: physicochemical characterization and genotoxicity evaluation. J. Hazardous Materials 231: 1–9.

Heap, I.M. 2017. The International Survey of Herbicide Resistant Weeds. Available at: www.weedscience.com (Accessed on March 25, 2017).

Iwakami, S., A. Uchino, Y. Kataoka, H. Shibaike, H. Watanabe and T. Inamura. 2014. Cytochrome P450 genes induced by bispyribac-sodium treatment in a multiple-herbicide-resistant biotype of *Echinochloa phyllopogon*. Pest Manag. Sci. 70: 549–558. doi:10.1002/ps.3572

Jang, S.R., J. Marjanovic and P. Gornicki. 2013. Resistance to herbicides caused by single amino acid mutations in acetyl-CoA carboxylase in resistant populations of grassy weeds. New Phytol. 197: 1110–1116. doi: 10.1111/nph.12117

Jasieniuk, M. et al. 1996. The evolution and genetics of herbicide resistance in weeds. Weed Sci. 44: 176–193.

Jia, J.L., X.Y. Jin, Q.L. Liu, W.L. Liang, M.S. Lin et al. (2016). Preparation, characterization, and intracellular imaging of 2,4-Dichlorophenoxyacetic acid conjugated gold nanorods. J. Nanosci. Nanotechnol. 16: 4936–4942.

Karim, R.S.M., A.B. Man and I.B. Sahid. 2004. Weed problems and their management in rice fields of Malaysia: an overview. Weed Biol Manag. 4: 177–186.

Lambert, D.M., J.A. Larson, R.K. Roberts, B.C. English, X. Zhoua, L.L. Falconerb, R.J. Hogan Jr., J.L. Johnson and J.M. Reeves. 2017. "Resistance is futile": estimating the costs of managing herbicide resistance as a first-order Markov process and the case of U.S. upland cotton producers. Agric. Econ. 48: 387–396.

Llewellyn, R.S., R.K. Lindner, D.J. Pannell and S.B. Powles. 2002. Resistance and the herbicide resource: perceptions of Western Australian grain growers. Crop Prot. 21: 1067–1075.

Matsunaka, S. 1976. Diphenylethers. pp. 709–739. *In:* Kearney, P.C. and Kaufman D.D. (Eds.) Herbicides: Chemistry, Degradation, and Mode of Action. 2. Marcel Dekker, New York.

Matsunaka, S. 2001. Historical review of rice herbicides in Japan. Weed Biol Manag. 1: 10–14.

McCourt, J.A., S.S. Pang, J. King-Scott, L.W. Guddat and R.G. Duggleby. 2006. Herbicide-binding sites revealed in the structure of plant acetohydroxyacid synthase. Proc. Natl. Acad. Sci. USA 103: 569–573.

Mercado, B.L., S.K. De Datta, T.R. Migo and A.M. Baltazar. 1990. Growth behavior and leaf morphology of Philippine strains of *Sphenoclea zeylanica* showing differential response to 2,4-D. Weed Res. 30: 245–250.

Midtiby, H.S., S.K. Mathiassen, K.J. Andersson and R.N. Jørgensen. 2011. Performance evaluation of a crop/weed discriminating microsprayer. Computers and Electronics in Agric. 77: 35–40.

Moss, S.R. and G.W. Cussans. 1985. Variability in the susceptibility of *Alopecurus myosuroides* (black-grass) to chlortoluron and isoproturon. Aspects Appl. Biol. 9: 91–98.

Mudge, L.C., B.J. Gossett and T.R. Murphy. 1984 Resistance of goosegrass (*Eleusine indica*) to dinitroaniline herbicides. Weed Sci. 32: 591–594.

Naylor, R. 1994. Herbicide use in Asian rice production. World Development. 22: 55–70.

Owen, M.J., M.J. Walsh, R.S. Llewellyn and S.B. Powles. 2007. Widespread occurrence of multiple herbicide resistance in Western Australian annual ryegrass (*Lolium rigidum*) populations. Australian J Agric. Res. 58: 711–718.

Patzoldt, W.L., P.J. Tranel and A.G. Hager. 2005. A waterhemp (*Amaranthus tuberculatus*) biotype with multiple resistance across three herbicide sites of action. Weed Sci. 53: 30–36.

Peña, J.M., J. Torres-Sánchez, A.I. de Castro, M. Kelly and F. López-Granados. 2013. Weed mapping in early-season maize fields using object-based analysis of unmanned aerial vehicle (UAV) images. PLoS ONE 8(10): e77151. doi: 10.1371/journal.pone.0077151

Pereira, A.E.S., R. Grillo and N.F.S. Mello. 2014. Application of poly (epsilon-caprolactone) nanoparticles containing atrazine herbicide as an alternative technique to control weeds and reduce damage to the environment. J. Hazardous Matter 268: 207–215.

Powles, S.B., D.F. Lorraine-Colwill, J.J. Dellow and C. Preston. 1998. Evolved resistance to glyphosate in rigid ryegrass (*Lolium rigidum*) in Australia. Weed Sci. 46: 604–607.

Preston, C. and C.A. Mallory-Smith. 2001. Biochemical mechanisms, inheritance, and molecular genetics of herbicide resistance in weeds. pp. 23–60. *In:* S.B. Powles and D.L. Shaner (Eds.) Herbicide Resistance and World Grains. CRC, Boca Raton, FL.

Rahemi, V., J.M.P.J. Garrido, F. Borges, C.M.A. Brett and E.M.P.J. Garrido. 2015. Electrochemical sensor for simultaneous determination of herbicide MCPA and its metabolite 4-chloro-2-methyphenol. Application to photodegradation environmental monitoring. Environ. Sci. Pollution Res. 22: 4491–4499.

Rahman, M.M., M.A. Islam, M. Sofian-Azirun, N.B. Amru and S. Ismail. 2014. Control measures of sprangletop (*Leptochloa chinensis*) resistant biotype using propanil, quinclorac and cyhalofop-butyl. Int. J. Agric. Biol. 16: 801–806.

Robineau, T., Y. Batard, S. Nedelkina, F. Cabello-Hurtado, M. LeRet, O. Sorokine, L. Didierjean and D. Werek-Reichhart. 1998. The chemically inducible plant cytochrome P450 CYP76B1 actively metabolizes phenylureas and other xenobiotics. Plant Physiol. 118: 1049–1056.

Rauch, T.A., D.C. Thill, S.A. Gersdorf and W.J. Price. 2010. Widespread occurrence of herbicide-resistant Italian ryegrass (*Lolium multiflorum*) in Northern Idaho and Eastern Washington. Weed Technol. 24(3): 281–288. doi:.org/10.1614/WT-D-09-00059

Rouse, C.E., N.R. Burgos, N.R. Steppig, Z.T. Hill and J.L. Gentry. 2016. Characterization and biology of a new Arkansas rice weed: *Schoenoplectus mucronatus*. Proc. Southern Weed Sci. Soc. Meeting 69: 219.

Rouse, C.E., N.R. Burgos, J.K. Norsworthy, T.M. Tseng, C.E. Starkey and R.C. Scott. 2017. *Echinochloa* resistance to herbicides continues to increase in Arkansas rice fields. Weed Technol. (In press).

Salas, R.A., N.R. Burgos, A. Mauromoustakos, R.B. Lassiter, R.C. Scott and E.A. Alcober. 2013. Resistance to ACCase and ALS inhibitors in *Lolium perenne* ssp. *multiflorum* in the United States. J. Crop and Weed 9: 168–183.

Salas, R.A., N.R. Burgos, P.J. Tranel, S. Singh, L. Glasgow, R.C. Scott, R.L. Nichols. 2016. Resistance to PPO-inhibiting herbicide in Palmer amaranth from Arkansas. Pest Manag Sci. doi: 10.1002/ps.4241

Salas-Perez, R.A., N.R. Burgos, G. Rangani, S. Singh, J.P. Refatti, L. Piveta, P.J. Tranel, A. Mauromoustakos and R.C. Scott. 2017. Frequency of Gly^{210} deletion mutation among PPO-inhibitor-resistant Palmer amaranth (*Amaranthus palmeri*) populations. Weed Sci. 65: 718–731. doi: 10.1017/wsc.2017.41

Sammons, R.D. and T.A. Gaines. 2014. Glyphosate resistance: state of knowledge. Pest Manag. Sci. 70: 1367–1377. doi: 10.1002/ps.3743

Scott, R.C., J.K. Norsworthy, T. Barber and J. Hardke. 2013. Rice weed control. pp. 53–62. *In:* J.T. Hardke (Ed.) Arkansas Rice Production Handbook. University of Arkansas Division of Agriculture Cooperative Extension Service Publications MP192.

Singh, S. 2007. Role of management practices in control of isoproturon-resistant littleseed canarygrass (*Phalaris minor*) in India. Weed Technol. 21: 339–346.

Singh, V., S. Zhou, Z. Ganie, B. Valverde, L. Avila, E. Marchesan, A. Merotto, G. Zorrilla, N. Burgos, J. Norsworthy and M. Bagavathiannan. 2017. Rice Production in the Americas. pp. 137–168. *In:* B.S. Chauhan (Ed.) Rice Production Worldwide . Springer, New York, USA.

Shoup, D.E., K. Al-Khatib and D.E. Peterson. 2003. Common waterhemp (*Amaranthus rudis*) resistance to protoporphyrinogen oxidase inhibiting herbicides. Weed Sci. 51: 145–150.

Sosnoskie, S.M. and S.A. Culpepper. 2014. Glyphosate-resistant Palmer amaranth (*Amaranthus palmeri*) increases herbicide use, tillage, and hand-weeding in Georgia cotton. Weed Sci. 62: 393–402.

Sosnoskie, L.M., T.M. Webster, J.M. Kichler, A.W. MacRae, T.L. Grey and A.S. Culpepper. 2012. Pollen-mediated dispersal of glyphosate-resistance in Palmer Amaranth under field conditions. Weed Sci. 60: 366-373.

Talbert, R.E. and N.R. Burgos. 2007. History and management of herbicide-resistant barnyardgrass (*Echinochloa crus-galli*) in Arkansas rice. Weed Technol. 21: 324–331.

Tehranchian, P., J.K. Norsworthy, S. Powles, M.T. Bararpour, M.V. Bagavathiannan, T. Barber and R.C. Scott. 2017. Recurrent sublethal-dose selection for reduced susceptibility of Palmer amaranth (*Amaranthus palmeri*) to dicamba. Weed Sci. 65: 206–212.

Tidemann, B.D., L.M. Hall, K.N. Harker and H.J. Beckie. 2017. Factors affecting weed seed devitalization with the Harrington Seed Destructor. Weed Sci. 65: 650–658.

Thakur, M.J., I.N. Mishra and S. Jha. 1999. Panlynological studies of selected species of the tribes Andropogoneae and Paniceae. Environ. and Ecol. 17: 22–25.

Tranel, P.J., T.R. Wright and I.M. Heap. Mutations in herbicide-resistant weeds to ALS inhibitors. Online. Available at: http://www.weedscience.com. (Accessed on December 5, 2017).

Walsh, M.J., R.B. Harrington and S.B. Powles. 2012. Harrington seed destructor: a new nonchemical weed control tool for global grain crops. Crop Sci. 52: 1343–1347. doi: 10.2135/cropsci2011.11.0608

Wright, T.R., G. Shan, T.A. Walsh, J.M. Lira, C. Cui et al. 2010. Robust crop resistance to broadleaf and grass herbicides provided by aryloxyalkanoate dioxygenase transgenes. Proc Nat'l. Acad Sci, USA 107: 20240–20245.

WSSA. 1998. Herbicide resistance and herbicide tolerance definitions. Available at: http://wssa.net/wssa/weed/resistance/ (Accessed on December 26, 2017).

Yabuno, T. 1962. Cytotaxonomic studies on the two cultivated species and the wild relatives of genus *Echinochloa*. Cytologia 27: 296-305.

Yasuor, H., P.L. Ten Brook, R.S. Tjeerdema and A.J. Fischer. 2008. Responses to clomazone and 5-ketoclomazone by Echinochloa phyllopogon resistant to multiple herbicides in Californian rice fields. Pest Manag. Sci. 64: 1031–1039.

Yasuor, H., M.D. Osuna, A. Ortiz, N. Saldain, J.W. Eckert and A.J. Fischer. 2009. Mechanism of resistance to penoxsulam in late watergrass (*Echinochloa phyllopogon* [Stapf] Koss.). J. Agric. Food Chem. 57: 3653–3660. doi: 10.1021/jf8039999

Yasuor, H., M. Milan, J.W. Eckert and A.J. Fischer. 2011. Quinclorac resistance: a concerted hormonal and enzymatic effort in *Echinochloa phyllopogon*. Pest Manag. Sci. doi: 10.1002/ps.2230

Yu, Q., M.S. Ahmad-Hamdani, H. Han, M.J. Christoffers and S.B. Powles. 2013. Herbicide resistance-endowing ACCase gene mutations in hexaploid wild oat (*Avena fatua*): insights into resistance evolution in hexaploid species. Heredity 110: 220–231.

Yu, Q. and S.B. Powles. 2014. Resistance to AHAS inhibitor herbicides: current understanding. Pest Manag. Sci. doi: 10.1002/ps.3710

Zeng, L. and W.V. Baird. 1997. Genetic basis of dinitroaniline herbicide resistance in a highly resistant biotype of goosegrass (*Eleusine indica*). J Heredity 88: 427–432.

Zhang, L., Q. Lu, H. Chen, G. Pan, S. Xiao, Y. Dai, Q. Li, J. Zhang, X. Wu, J. Wu, J. Tu and K. Liu. 2007. Identification of a cytochrome P450 hydroxylase, CYP81A6, as the candidate for the bentazon and sulfonylurea herbicide resistance gene, *Bel*, in rice. Mol Breeding 19: 59–68. doi: 10.1007/s11032-006-9044-z.

Microbial Herbicides

Alan K. Watson

Department of Plant Sciences, Macdonald Campus of McGill University,
21,111 Lakeshore Road, Ste-Anne-de-Bellevue, H9X 2V9, QC, Canada
E-mail: alan.watson@mcgill.ca

Introduction

Weeds are the major constraint to crop production but weeds are also prone to disease and there are several ways these natural enemies may be used to suppress weeds. To begin this chapter, the following terms are defined: biological weed control, biopesticides, bioherbicide, microbial herbicide, and biochemical herbicide. Biological control of weeds is defined "as the use of an agent, a complex of agents, or biological processes to bring about weed suppression. All forms of macrobial and microbial organisms are considered as biological control agents. Examples of biological control agents include, but are not limited to, arthropods (insects and mites), plant pathogens (fungi, bacteria, viruses, and nematodes), fish, birds, and other animals" (http://wssa. net/wssa/weed/biological-control/). The United States Environmental Protection Agency (EPA) defines biopesticides as "naturally occurring substances that control pests (biochemical pesticides), microorganisms that control pests (microbial pesticides), and pesticidal substances produced by plants containing added genetic material (plant-incorporated protectants) or PIPs" (www.epa.gov/pesticides/biopesticides). Bioherbicides are living phytopathogenic microorganisms (microbial herbicides) or microbial phytotoxins (biochemical herbicides) that are field applied in ways like conventional chemical herbicides. Microbial herbicides are presented in this chapter and biochemical herbicides have been discussed in Chapter 8.

Biological Weed Control

Biological weed control can be realized by two main strategies: classical biocontrol and bioherbicide strategies (Templeton et al. 1979, Yandoc et al. 2000, Evans 2013). Classical biological weed control targets exotic, non-native weed species that have arrived from another part of the world without their natural enemies and have become dominant in their new habitats (Wapshere 1974, Watson 1991b). These invasive weeds often infest large areas of marginal lands, such as pastures and rangelands. Host specific natural enemies (mostly insects, but occasionally plant pathogens) are obtained from a target weed's native range, and host specificity and impact are carefully evaluated. Founding populations of the host specific biocontrol agent are released (inoculated) into the weed infested areas of the invaded country. After release, the biocontrol agent populations are monitored to evaluate establishment, spread and level of weed control. Classical biocontrol (CBC) is an ecological approach that can provide sustainable, long-term control.

The bioherbicide strategy or inundative (IBC) approach involves the use of local endemic phytopathogenic microorganisms (fungi, bacteria, viruses, and nematodes) to control a native or naturalized weed. These weeds are already in dynamic equilibrium with their natural enemies and can be controlled by manipulation of existing weed-natural enemy relationships. Large inundative populations of an existing natural enemy are mass produced, formulated and applied on the crop fields like a chemical herbicide, following extensive testing to ensure that non-target species are not negatively affected. Weed control is rendered, but the biocontrol agent normally does not survive in high numbers requiring retreatment, and the control is not sustained (like a chemical herbicide). Bioherbicides are a technological approach, like chemical herbicides, and deliver transient, non-sustained weed control.

Biological weed control can also follow a third strategy, herbivory. Fish, tadpole shrimps, ducks and goats provide weed control in Asian rice fields (de Datta and Baltazar 1996, Shibayama 2001). Rice-fish-duck systems often have significantly higher weed control than other farming systems tested (Men et al. 1999, Zhang et al. 2010). In North America, leafy spurge (*Euphorbia esula* L.) is effectively controlled with grazing of sheep and goats (Landgraf et al. 1984, Sedivec et al. 1996).

Classical Biological Weed Control with Phytopathogens

Much of the effort and success of classical biological weed control has been dominated by insect biocontrol agents (Wapshere 1982, Waterhouse 1994, Julien et al. 2012, Winston et al. 2014, Day and. Winston, 2016) but many biopesticide papers reviewed for this chapter encompassed reference to obligate fungal pathogens that have been introduced as classical biocontrol (CBC) agents into various countries of the world (Barreto 2007, Barreto et al. 2012, Barton 2004, 2005, Bruckart 2005, Burdon et al. 2002, Cullen 1985, Cullen et al. 1973, Evans 1995, 2013, Evans et al. 2001, Hershenhorn et al. 2016, Watson 1991b, Winston et al. 2014). Weed species targeted with classical biocontrol are primarily weeds of aquatic systems and rangelands; nonetheless success with plant pathogenic fungi has been recorded in several grassland systems (Table 7.1).

Host specificity is the most important factor when evaluating microorganism for CBC (Wapshere 1982, Watson 1985, Evans 2013). Extensive life-history studies are conducted in the microbe's native range and before release of the microbe, a full life history risk assessment is presented to authorities in a receiving country (Barreto et al. 2012, Evans 2013). This forms the basis of the pest risk assessment, of which the principal objective is to demonstrate specificity to the target weed. CBC safety record and success rate have been very good (Evans 2013).

Microbial Herbicides

Microbial herbicides comprise living plant pathogenic organisms including fungi (mycoherbicides), bacteria (bacto-herbicides), viruses (viral-herbicides) and nematodes (nematoda-herbicides). Most bioherbicide research activity has been with fungal plant pathogens and much less effort with plant pathogenic bacteria, viruses or nematodes.

Mycoherbicides

Mycoherbicides "are simply plant-pathogenic fungi developed and used in the inundative strategy to control weeds the way chemical herbicides are used" (TeBeest and Templeton 1985). Interest in mycoherbicide research began with knowledge of the Lubao mycoherbicide in China. Lubao No. 1, a formulated suspension of *Colletotrichum gloeosporioides* (Penz.) Penz. & Sacc. f. sp. *cuscutae* for the control of dodder (*Cuscutae australis* R. Br.), a weed in soybean fields (Gao and Yu 1992). Lubao was discovered in 1963 and by the late 1970s was applied to 670,000 ha of soybean (Zhang et al. 2011).

Table 7.1. Successful classical biological weed control projects for cropland weeds (adapted from Barreto et al. 2012 and Winston et al. 2014)

Weed target	Pathogen	Country of introduction	Control status	References
Acacia saligna (Labill.) H. L. Wendl. (Port Jackson willow)	*Uromycladium tepperianum* (Sacc.) McAlpine	South Africa	Significant	Morris 1987, 1997
Ageratina riparia (Regel) R.M. King & H. Rob (mistflower)	*Entyloma ageratinae* R.W. Barreto & H.C. Evans	USA, New Zealand, South Africa	Significant	Morin et al. 1997, Fröhlich et al. 1999
Carduus nutans L. (nodding thistle)	*Puccinia carduorum* Jacky	USA (continental)	Significant	Politis et al. 1984, Bruckart 2005.
C. pycnocephalus L. (Italian thistle) and *C. tenuiflorus* Curtis (slender-flower thistle)	*Puccinia cardui-pycnocephali* P. Syd. & Syd.	Australia	Significant	Burdon et al. 2000
Chondrilla juncea L. (skeleton weed)	*Puccinia chondrillina* (Bubak & Syd.)	Australia, USA (continental), Canada	Significant/ Partial	Cullen et al. 1973, Cullen 1985
Clidemia hirta (L.) D. Don (Koster's curse)	*Colletotrichum gloeosporioides* f. sp. *clidemiae* E.E. Trujillo, Latterell & A.E. Rossi	USA (Hawaii)	Partial	Trujillo 2005
Cryptostegia grandiflora (Roxb. ex R. Br.) R. Br. (rubber vine)	*Maravalia cryptostegiae* (Cummins) Y. Ono	Australia	Significant	Tomley and Evans 2004
Passiflora tarminiana Coppens & V.E. Barney (banana poka)	*Septoria passiflorae* Louw	USA (Hawaii)	Significant	Trujillo 2005
Rubus constrictus P.J. Müll. & Lefèvre, *R. ulmifolius* Schott (wild blackberry)	*Phragmidium violaceum* (Schultz) G. Winter	Chile	Significant	Oehrens and Gonzales 1977
R. fruticosus aggregate (shrubby blackberry)	*Phragmidium violaceum*	Australia	Partial	Evans et al. 2004

In 1973, reports of the biological control of milkweed vine (*Morrenia odorata* Lindl.) with a race of *Phytophthora citrophthora* (Butler) Butler (Burnett et al. 1973, 1974) and the biological control of northern jointvetch [*Aeschynomene virginica* (L.) B.S.P.] in rice with *Colletotrichum gloeosporioides* Penz. Sacc. f. sp. *aeschynome* (Daniel et al. 1973) were published in the scientific literature. Subsequently, *Phytophthora citrophthora* was registered as a microbial pest control product by the United States Environmental Protection Agency (EPA) as DeVine® for control of stranglervine (*Morrenia odorata*) in Florida citrus groves in 1981 (Ridings 1986) and the next year, *Colletotrichum gloeosporioides* f. sp. *aeschynome* was registered as Collego® for the control of northern jointvetch (*Aeschynomene virginica*) in fields of rice and soybeans in Arkansas, Louisiana and Mississippi (TeBeest and Templeton 1985, Bowers 1986, Smith 1986).

Templeton (1982a) outlined the discovery, development, and deployment phases involved in forming a biological herbicide. The discover phase involves the collection, isolation, identification, and culture maintenance of a weed pathogen. The development phase includes inoculum production, culture conditions, disease etiology, field trials, and host range determination. Product formulation, mass production scale-up, intellectual property protection, patents, government registration approval and commercialization complete the deployment phase. Later, Bailey et al. (2009) and Bailey and Falk (2011) suggested different approaches were needed to evaluate scientific and commercial potential of a bioherbicide organism because "commercialization is the ultimate goal, then the science must consider factors deemed important to the industry" (Bailey et al. 2009).

Early success of Luboa, Collego and DeVine in the late 1970s and early 1980s was followed by relatively well-funded research in many countries (Templeton 1982b, Charudattan 1991). Hundreds of weeds were targeted with fungi, bacteria, viruses and nematodes resulting in numerous manuscripts and patents being fashioned. Considerable basic knowledge was acquired but success was limited, if measured in number of registrations and commercially viable products. Loss of virulence, limited market size, small specialist market, plus control persistence were factors responsible for the demise of Luboa, Collego and DeVine mycoherbicides, all single weed target-restricted products.

There have been numerous comprehensive regional and world reviews on the progress and listings of microbial herbicide research and development projects (Templeton et al. 1979, Templeton 1982b, TeBeest and Templeton 1985, Charudattan 1991, Morris 1991, Watson 1991a, Yoo 1991, TeBeest et al. 1992, Watson 1994, Evans 1995, Cother 1996, Fujimori 1999, Morris et al. 1999, Rosskopf et al. 1999, Watson 1999, Auld 2000, Charudattan and Dinoor 2000, Müller-Schärer et al. 2000, Evans et al. 2001, Li et al. 2003, Boyetchko and Peng 2004, Barton 2005, Charudattan 2005, Trujillo 2005, Chutia et al. 2007, Vurro and Evans 2007, Ash 2010, Bailey et al. 2010, Ash 2011, Barreto et al. 2012, Stubbs and Kennedy 2012, Aneja et al. 2013, Evans 2013, Bailey 2014, Winston et al. 2014, Patel and Patel 2015, Harding and Raizada 2015, Pacanoski 2015, Cordeau et al. 2016, Hershenhorn et al. 2016, Cai and Gu 2016, Gaddeyya et al. 2017, Watson 2017). After DeVine and Collego were marketed, very few commercial microbial herbicide products were registered during the past 35 years (Table 7.2). A commercial product must perform under field condition, be economically produced and formulated to retain sufficient shelf-life during commercial distribution (Zorner et al. 1993, Charudattan 1991, Bailey and Falk 2011). Mycoherbicides have not met earlier expectations and have contributed little to weed management in cropping systems (Zhang et al. 2011, Barreto et al. 2012, Evans 2013, Hershenhorn et al. 2016).

Questions arise after examining the number of potential microbial herbicides cited in the literature, and realizing that very few microbial herbicides have made it to the market place. Why the limited success? Was the wrong target selected? Would a different pathogen be preferred? What factors instigated inconsistent field results? Disease development involves interplay between the host, the pathogen and the environment, a susceptible host, a virulent pathogen and favorable moisture and temperature conditions are essential for disease to occur. Constraints to bioherbicide development including biological (low virulence), environmental (temperature and dew period requirements), technological (mass production, formulation issues), and commercial (patent, registration, market analysis) factors have been considered over the past 25 years to explain the lack of commercial success (Watson and Wymore 1990, Auld and Morin 1995, Mortensen 1998, Pacanoski 2015).

Considerable efforts to overcome these constraints were instigated in many research groups. Several reviews covering improvements in adjuvants, formulation and application techniques were published (Boyette et al. 1991, 1996, Green et al. 1998). Innovative approaches including vegetable oil suspensions (Auld 1993), sodium alginate granules (Walker and Connick 1988), pesta-like formulation (Connick et al. 1991, Elzein et al. 2008), and invert emulsions (Womack et al. 1996) were used to overcome dew period requirements.

Table 7.2. World list of registered microbial herbicides for cropland weeds

Product	Microorganism	Weed target	Reference	Present status
Luboa	*Colletotrichum gloeosporioides* f. sp. *cuscutae* T.Y. Zhang,	*Cuscuta* spp. (dodder)	Gao and Yu 1992	Unknown, maybe local cottage industry
DeVine®	*Phytophthora palmivora* (E.J. Butler) E.J. Butler	*Morrenia odorata* (Hook. & Arn.) Lindl. (strangler vine)	Ridings 1986	No longer available
Collego® (reregistered as Lockdown™)	*Colletotrichum gloeosporioides* f. sp. *aeschynomene = C. aeschynomenes* B. Weir & P.R. Johnst (ATCC 20358)	*Aeschynomene virginica* (L.) B.S.P (Northern jointvetch)	Bowers 1986, TeBeest et al. 1992, Cartwright et al. 2010	May still be available, but small market
Casst™	*Alternaria cassiae* Jurair & A. Khan (NRRL #12553)	*Cassia obtusifolia* L. (sicklepod), *C. occidentalis* L. (coffee senna), *Crotalaria spectabilis* Roth. (showy crotalaria)	Bannon 1988, Walker and Riley 1982	Never commercialized
BioMal™	*Colletotrichum gloeosporioides* (Penzig) Penzig & Saccardo f. sp. *malvae* Mortensen	*Malva pusilla* Sm. (roundleaf mallow)	Boyetchko et al. 2007, Mortensen 1988	Not marketed. Unable to be economically mass produced
Dr. BioSedge®	*Puccinia canaliculata* (Schwein.) Lagerh.	*Cyperus esculentus* L. (yellow nutsedge)	Phatak et al. 1983	Never marketed, no production
Biochon™	*Chondrostereum purpureum* (Pers.) Pouzar	*Prunus serotina* Ehrh. (black cherry)	De Jong et al. 1990	Removed from the market
StumpOut™ (registered)	*Cylindrobasidium laeve* (Pers.) Chamuris	*Acacia mearnsii* De Wild. (black wattle)	Morris et al. 1999	Seldom produced
Hakatak (not registered)	*Colletotrichum acutatum* J. H. Simmonds	*Hakea sericea* Schrad. & J.C. Wendl. (silky hakea)	Morris 1989	Occasionally produced on request
Camperico™ (JT-P482)	*Xanthomonas campestris pv. poae* (Pammel 1895) Dowson 1939 emend. Vauterin et al. 1995	*Poa annua* L. (annual bluegrass)	Fujimori 1999, Nishino & Tateno 2000	Not available, difficult to produce
Chontrol™	*Chondrostereum purpureum* (PFC2139)	*Alnus rubra* Bong. and *A. sinuata* (Regel) Rydb. (red and sitka alders)	Becker et al. 2005, Hintz 2007	Mycologic Inc. Product available (?)
Myco-Tech Paste™	*Chondrostereum purpureum* (HQ1).	Brush weeds in rights-of-ways and forest plantations	Bailey 2014	Company closed

(*Contd.*)

Table 7.2. (*Contd.*)

Product	Microorganism	Weed target	Reference	Present status
Woad Warrior®	*Puccinia thlaspeos* Ficinus & C. Schub. 1823 (strain woad)	*Isatis tinctoria* L. (Dyer's woad)	Kropp et al. 1996, 2002	Not marketed
Smolder™	*Alternaria destruens* E.G. Simmons	*Cuscuta* spp. (dodder)	Bewick et al. 2000	Project terminated
Sarritor™	*Sclerotinia minor* Jagger	*Taraxacum officinale* (L.) Weber ex F.H. Wigg (dandelion) and other broadleaf weeds	Abu-Dieyeh & Watson 2007, Health Canada 2010	Company restructuring
Wilson Lawn Bio-Phoma, Premier Tech	*Phoma macrostoma* Montagne	Broadleaf turf weeds	Bailey & Falk 2011, Bailey 2014	Market pending, manufacturing difficulties
SolviNix™ Strain U2	*Tobamovirus*, Group IV ((+) ss RNA), Virgaviridae, Tobacco Mild Green Mosaic Tobamovirus	*Solanum viarum* Dunal (tropical soda apple)	Charudattan and Hiebert 2007, Charudattan 2016	Commercial product available

Some mycoherbicide failures have been attributed to poor virulence of the pathogen. Weed pathogens have coevolved with their plant host, they are in biological balance and if the necrotrophic fungal pathogen was hypervirulent it would lead to self-extinction (Gressel 2001). Efforts to increase virulence of bioherbicide candidate pathogens have included interfering with host plant defense mechanisms (Sharon et al. 1992, Ahn et al. 2005), using synergies to enhance virulence (Hodgson et al. 1988, Wymore et al. 1987, Gressel 2010), genetically enhancing virulence (Tiourebaev et al. 2001, Thompson et al. 2007, Nzioki et al. 2016) and engineering hypervirulence (Sharon et al. 2001, Amsellem et al. 2002, Cohen et al. 2002a, Gressel et al. 2007, Meir et al. 2009). These, and other advances in molecular biology, may help resolve bioherbicide deficiencies and the realisation of functional, commercial microbial herbicides (Gressel et al. 2007, Ash 2011).

Many, well-funded bioherbicide search and development research projects were carried out in many countries in the world from the 1970s, 1980s to early 2000s, but few have matured into commercial products. George Templeton (1992a, b) coined the term 'orphaned mycoherbicides' for ones that provided effective weed control of their target weed but did not become commercialized due to low market potential, mass production difficulties or other concerns (Table 7.3).

Microbial herbicide success has occurred with virulent, broad host range pathogens *Chondrosterum purpureum* and *Sclerotinia* spp. Chontrol™ (*Chondrosterum purpureum*) is one of the few microbial herbicides available today and is used to control re-sprouting of hard-wood species (Hintz 2007). *Sclerotinia minor* has been developed as the Sarritor™ microbial herbicide for control of dandelion, broadleaved plantain (*Plantago major* L.) and other broadleaved weeds (Abu-Dieyeh and Watson 2007, Health Canada 2010, Watson and Bailey 2013). Government pesticide restriction and bans in Canada expedited the research and commercialization of Chontrol™ and Sarritor™.

Sclerotinia sclerotiorum is a voracious, virulent pathogen, an ideal microbial herbicide for broadleaf weed control, but its broad host range is expanded due to the sporogenic (ascospores) phase endangering broadleaf crops (Watson 2007). In New Zealand, *S. sclerotiorum* is being used to control Canada thistle (*Cirsium arvense* L.) in pastures and risk analysis simulates the dispersal of ascospores (de Jong et al. 2002). Safety zones for susceptible horticultural crops away from a *S. sclerotiorum*-based mycoherbicide treatment have been determined regarding variations in regional and yearly climate (Bourdôt et al. 2006).

Charles Wilson's (1970) 'commencement' paper on 'Plant pathogens in weed control' mentioned *Sclerotium rolfsii* Sacc., another virulent, aggressive, broad host range crop pathogen, could be considered as a biocontrol agent. Several groups have reported bioherbicide research interest with *Sclerotium rolfsii*: Mishra et al. (1995) for the control of *Parthenium* in India, Tang et al. (2011) for broadleaf weed control in dry, direct-seeded rice fields and Gibson et al. (2014) for control of swallowworts (*Vincetoxicum* spp.) in eastern North America. In Australia, three virulent, broad host range, destructive fungi *Lasiodiplodia pseudotheobromae*, *Neoscytalidium novaehollandiae* and *Macrophomina phaseolina* are combined in capsule that is injected into *Parkinsonia* shrub trunks (Cripps 2017). Weed control should be very good with these pathogens, but regulatory acceptance may be challenging. This Australian effort is reminiscent of reports of a local research foundation providing farmers with spores of the fungus, *Acremonium* (*Cephalosporium*) *diospyri* (Crand.) W. Gams, for application to cut stumps for control of common persimmon (*Diospyros virginiana* L.), an invasive weed in Oklahoma grasslands (Wilson 1965, Griffith 1970).

Early on, the bioherbicide industry indicated major efforts were needed in mass production, formulation and delivery technologies before commercialisation of additional bioherbicides could occur (Zorner et al. 1993). A "bioherbicide innovation chain" was proposed by Bailey et al. (2009) to assist researchers and industry to work together to increase microbial herbicide product commercialisation. The nine-step process, from discovery to technology adoption, requires involvement of scientists, market experts and a solid industrial partner to link research activities with business models. Future microbial herbicide projects are encouraged to follow the bioherbicide innovation chain proposed and tested by Bailey et al. (2009).

Table 7.3. Orphaned microbial herbicide candidates targeting crop weeds

Code	Pathogen agents	Weed targets	References
VELGO	*Colletotrichum coccodes* (Wallr.) S. Hughes	*Abutilon theophrasti* Medik (velvetleaf)	Wymore et al. 1988, DiTommaso and Watson 1995
IMI 48942	*Colletotrichum orbiculare* (Berk. and Mont.) v. Arx	*Xanthium spinosum* L. (Bathurst burr).	Auld and Say 1999, Chittick and Auld 2001
NRRL 13737	*Colletotrichum truncatum* (Schw.) Andrus et Moore	*Sesbania exaltata* (Raf.) Rydb. ex A.W. Hill. (hemp sesbania)	Jackson and Bothast 1990, Jackson and Schisler 1995, Boyette et al. 2007
Myco-herb®	*Lewia chlamidosporiformans* B.S. Vieira & R.W. Barreto	*Euphorbia heterophylla* L. (wild poinsettia)	Vieira et al. 2008, Vieira and Barreto 2010
MTB-951	*Drechslera monoceras* (Drechsler) Subram. et Jain (=*Exserohilum monoceras* [Drechsler] Leonard et Suggs)	*Echinochloa crus-galli* (L.) P. Beauv. (barnyardgrass)	Fujimori 1999, Hirase et al. 2004, 2006
JTB-808	*Exserohilum monoceras* (Drechsler) K. J. Leonard & Suggs	*Echinochloa crus-galli* (barnyardgrass)	Tsukamoto et al. 1997, 1998, 2001
QZ-2000	*Curvularia eragrostidis* (Henn.) J.A. Mey.	*Digitaria sanguinalis* (L.) Scop. (large crabgrass)	Zhu and Qiang 2004, Wang et al. 2013
FOXY 2, M12-4A, PSM197	*Fusarium oxysporum* Schlecht. emend. Snyder & Hansen f. sp. *strigae* Elzein & Thines	*Striga hermonthica* (Del.) Benth. (witchweed)	Ciotola et al. 1995, Marley et al. 1999, Elzein and Kroschel. 2004, Venne et al. 2009, Watson 2013
FOG	*Fusarium oxysporum* Schltdt.	*Phelipanche ramosa* (L.) Pomel (branched broomrape).	Müller-Stöver et al. 2009a, Kohlschmid et al. 2009
FT2	*Fusarium oxysporum*	*P. ramosa*	Boari and Vurro 2004, Cipriani et al. 2009
FOXY	*Fusarium oxysporum*	*Phelipanche aegyptiaca* (Pers.) Pomel (Egyptian broomrape), *P. ramosa, Orobanche cernua* Loefl. (nodding broomrape)	Amsellem et al. 2001a, b, Cohen et al. 2002b
FARTH	*Fusarium arthrosporioides* Sherb.	*P. aegyptiaca, P. ramosa, O. cernua*	Amsellem et al. 2001a, b, Cohen et al. 2002b
FOO	*Fusarium oxysporum* Schlecht. f. sp. *orthoceras* (Appel & Wollenw) Bilay	*Orobanche cumana* Wallr (sunflower broomrape), *O. cernua, P. aegyptiaca*	Thomas et al. 1998, Müller-Stöver et al. 2004, 2009b

Bacto-Herbicides

Interest in using soil-borne phytopathogenic bacteria for biological weed control has been strong (Johnson . et al. 1996, Kremer and Kennedy 1996), but commercial success of a bacterial herbicide has not been realised (Barreto et al. 2012). Several deleterious rhizobacteria (DRB), *Pseudomonas fluorescens* strain D7, *Pseudomonas fluorescens* strain BRG100, *Pseudomonas fluorescens* strain G2-11 and *Pseudomonas trivialis* X33d are being evaluated for weed control. DRB suppress weed seed germination and early growth of the weed and function as plant growth promoting rhizobacteria (PGPR) favoring crop plant development to the detriment of weed growth. *Pseudomonas fluorescens* strain D7 was shown to suppress downy brome (*Bromus tectorum* L.) infesting winter wheat crops in the Pacific Northwest U.S. (Kennedy et al. 1991, 2001). *Pseudomonas fluorescens* strain BRG100 adversely affects germination and root growth of green foxtail (*Setaria viridis* (L.) P. Beauv. and wild oat (*Avena fatua* L.) (Caldwell et al. 2012). Economic and technical analyses of large scale production of pre-emergent *Pseudomonas fluorescens* microbial bioherbicide for green foxtail and wild oat control were conducted in Canada to support bioherbicide research and development investment and commercialization strategies (Mupondwa et al. 2015). *Pseudomonas fluorescens* strain G2-11 was isolated from roots of giant foxtail (*Setaria faberi* Herrm.) and herbicidal performance was affected by formulation and soil properties (Zdor et al. 2005). A semolina-kaolin granular formulation (Pesta) improved *Pseudomonas trivialis* X33 biocontrol of ripgut brome (*Bromus diandrus* Roth) in durum wheat (Mejri et al. 2012).

One plant pathogenic bacteria, *Xanthomonas campestris* pv. *poae* P-482, Camperico®, was commercialized by Japan Tobacco Ltd as a microbial bioherbicide for the control of the turfgrass weed, annual bluegrass (*Poa annua* L.) in Japan (Fujimori 1999, Imaizumi et al. 1997, Nishino and Tateno 2000). Research with *Xanthomonas campestris* pv. *poannua* was also conducted in the U.S. (Zhou and Neal 1995), but Camperico® is not available in Japan nor elsewhere. Even though, another isolate of *Xanthomonas campestris* has recently been studied in the United States for the control of horseweed (*Conyza canadensis* (L.) Cronquist) as a major glyphosate herbicide-resistant weed in limited and no tillage cropping systems (Boyette and Hoagland 2015).

Viral-Herbicides

Plant viruses have seldom been evaluated as potential microbial herbicides as they are not ideal candidates for biocontrol agents, due to their general broad host ranges and their need for vectors like surface abrasion or injection/transmission of viral particles by insects, fungi or nematodes into host plant cells. Nevertheless, one plant virus, tobacco mild green mosaic tobamovirus (TMGMV), *Tobamovirus,* Group IV ((+) ss RNA), Virgaviridae, SolviNix™, was recently registered as a microbial herbicide for the control of tropical soda apple, *Solanum viarum* Dunal (Solanaceae) (EPA 2015, Charudattan 2016).

Solanum viarum is native to southeastern Brazil, northeastern Argentina, Paraguay, and Uruguay but recently arrived in the USA becoming a serious, invasive weed of rangeland in Florida (Medal et al. 2012). Once introduced, *S. viarum* rapidly invades cropland establishing large impenetrable, monotypic stands (Charudattan and Hiebert 2007). SolviNix is usually spot-spray applied with high-pressure sprayers, providing excellent control of tropical soda apple (Ferrell et al. 2008). SolviNix can be mixed with herbicides to control other weeds as well. The commercialization of SolviNix by BioProdex Inc. is another example of a university–private company collaboration for successful microbial herbicide development. BioProdex is partnering with a Brazilian company interested in applying SolviNix for native Solanaceae weeds, and expansion into other countries is planned.

Nematoda-Herbicides

Silverleaf Nightshade

Silverleaf nightshade (*Solanum elaeagnifolium* Cav.) is an important native perennial weed in western U.S. and this weed has also invaded Australia, India and South Africa (Parker 1991a). Silverleaf nightshade plants are commonly parasitized by a leaf and stem galling nematode (*Orrina phyllobia* (Thorne) Brezeski, Anguinidae: Nematoda). Adult nematodes and larvae infect and initiate gall formation in fresh juvenile leaves and stems (Notham and Orr 1982). The galled leaves and stems soon become dry and are abscised. Adults die but second generation infective larvae enter a state of anhydrobiosis and can remain in that state for several years. Dried galled plant debris can simply be collected and large numbers of infective larvae can be readily distributed to other silver nightshade populations. With the arrival of moisture, anhydrobiosis state is overcome and infective larvae search for nightshade shoots to invade. The nematode can reduce the biomass and density of silverleaf nightshade. A government funded pilot project developed and implemented the mass rearing of *O. phyllobia*, but when compared to other weed control options, mass rearing was not competitive (Parker 1991a, b). Perhaps with improved, cost-effective mass rearing technology, this augmentation tactic could contribute to an integrated weed management approach for silverleaf nightshade suppression.

Russian Knapweed

Russian knapweed, *Acroptilon repens* (L) DC [*Rhaponticum repens* (L.) Hidalgo], Asteraceae, is a deep-rooted, aggressive perennial, native to Eurasia, that rapidly spreads forming persistent dense monotypic stands degrading native range and crop land. Russian knapweed arrived in Canada as a contaminant of Turkestan alfalfa seed in the early 1900s, and was recorded in the U.S. in California in 1910. Russian knapweed infestations are common in western Canada, widespread in the western and central regions of the U.S., and problematic in Afghanistan, Argentina, Australia, India, Iran, Turkey, and South Africa (http://www.cabi. org/isc/datasheet/2946). Biological control research activities on Russian knapweed involved collaboration amongst Canadian and Soviet scientists evaluating several insects and one nematode, as potential biocontrol agents (Watson and Harris 1984). The stem-gall nematode, *Subanguina picridis* (Kirj.) Brezeski (Anguinidae: Nematoda), was shown to be host limited, damaging to Russian knapweed and was released on Russian knapweed infestations in Canada and in the United States (Watson and Harris 1984). The life cycle is almost indistinguishable from the silverleaf nightshade nematode, including the anhydrobiosis capabilities of the infective larvae, facilitating the collection of galls from infected plants and re-distribution to additional sites. Soviet scientist developed a process to extract larvae from field collected galls and prepared water suspensions for sprayer applications (Kovalev et al. 1973). In efforts to augment this biocontrol organism in the United States, nematodes extracted from galls were encapsulated in calcium alginate granules, oil coated, dried and frozen (−20°C) to provide nine months' shelf life (Caesar-Ton That et al. 1995). *Subanguina* fecundity and gall numbers were low, limiting distribution. To resolve this problem, an *in vitro* mass culture system on callus, excised roots, and shoot tissues of *Acroptilon repens* was developed (Ou and Watson 1992). Mass cultured *S. picridis* were virulent and rapidly increased in population size. In three months, the initial 50 larvae increased to 7,000–10,000 per petri dish, a 140- to 200-fold increase (Ou and Watson 1993). Mass rearing of *S. picridis* is achievable but there has been limited interest in commercialization.

Sustainability, Safety, Hazards and Risks of Microbial Herbicides

Webster's New World dictionary defines sustainability as the ability to 'keep in existence, keep up, maintain or prolong' (Neufeldt 1988). The goal of microbial herbicide research is development of commercially acceptable weed control products that effectively suppress weed growth and promote crop growth. Microbial herbicides must be economically produced, formulated with lengthy shelf life, and perform consistently under field conditions (Jorner et al. 1993). Microbial herbicides would seemingly be sustainable in terms of human health, environment pollution and social aspects, but not in economic terms presently as there has been no successful, widely marketed, microbial herbicide to date.

Plant pathogens used as microbial may cause risks to non-target organisms including plants, animals, microbes and humans that must be rigorously scrutinized (Hoagland et al. 2007, Saharan and Mehta 2008, Bailey 2013). All microorganisms proposed as microbial herbicides must be tested and registered for use following various national or regional guidelines for registration as microbial pest control products. Prior to marketing and using a microbial herbicide, Australian Pesticides and Veterinary Medicines Authority (APVMA), Health Canada's Pest Management Regulatory Agency (PMRA), United States Environmental Protection Agency (EPA), and Organisation for Economic Co-operation and Development (OECD 2003) agencies rigorously evaluate the proposed microbial herbicide to assure that its use will not pose unreasonable risks or harm to human health and the environment. Also see Kabaluk et al. (2010), to view the regulation of microbial pesticides in representative jurisdictions worldwide. Once released, microbial herbicides should be monitored for stability, impact and persistence. Genetic markers have been developed for risk assessment and monitoring persistence of both registered broad host range microbial herbicides, Chontrol (Hintz et al. 2001) and Sarritor (Pan et al. 2010).

Concluding Remarks

Classical biological weed control with fungal plant pathogens has provided good control of several dominant invasive weeds in grassland ecosystems. Success with microbial herbicides has generally been limited to non-agricultural systems and no bioherbicide product has been developed for a major crop weed. Many microorganisms have been studied, mass-produced, formulated, field tested and controlled target weeds, but were deemed commercially acceptable. Herbicide resistance, absence of new chemistries, government restrictions, public pressure and expansion of organic agriculture support the need for non-chemical weed control. Future business models, innovative ideas and collaborative efforts will expand the prospects for microbial herbicides, such as witchweed biocontrol on a toothpick (Nzioki et al. 2016).

REFERENCES

Abu-Dieyeh, M. and A.K. Watson. 2007. Efficacy of *Sclerotinia minor* for dandelion control: effect of dandelion accession, age and grass competition. Weed Res. 47: 63–72.

Ahn, B., T. Paulitz, S. Jabaji-Hare and A.K. Watson. 2005. Enhancement of *Colletotrichum coccodes* virulence by inhibitors of plant defence mechanisms. Biocontrol Sci. Techn. 15(3): 299–308.

Amsellem, Z., B.A. Cohen and J. Gressel. 2002. Engineering hypervirulence in a mycoherbicidal fungus for efficient weed control. Nature Biotech. 20: 1035–1039.

Amsellem, Z., S. Barghouthi, B. Cohen, Y. Goldwasser, J. Gressel, L. Hornok, Z. Kerenyi, Y. Kleifeld, O. Klein, J. Kroschel, J. Sauerborn, D. Müller-Stöver, H. Thomas, M. Vurro and M.C. Zonno. 2001a. Recent advances in the biocontrol of *Orobanche* (broomrape) species. Biocontrol 46: 211–228.

Amsellem, Z., Y. Kleifeld, Z. Kerenyi, L. Hornok, Y. Goldwasser and J. Gressel. 2001b. Isolation, identification and activity of mycoherbicidal pathogens from juvenile broomrape plants. Biol. Control 21: 274–284.

Aneja, K.R., V. Kumar, P. Jiloha, M. Kaur, C. Sharma, P. Surain, R. Dhiman and A. Aneja. 2013. Potential bioherbicides: Indian perspectives. pp. 197–215. *In:* R.K. Salar, S.K. Gahlawat, P. Siwach and J.S. Duhan (Eds.) Biotechnology: Prospects and Applications. © Springer India, New Delhi.

Ash, G.J. 2010. The science, art and business of successful bioherbicides. Biol. Control 52: 230–240.

Ash, G.J. 2011. Biological control of weeds with mycoherbicides in the age of genomics. Pest Technol. 5 (Special Issue 1): 41-47. © 2011 Global Science Books.

Auld, B.A. 1993. Vegetable oil suspension emulsion reduces dew dependence of a mycoherbicide. Crop Prot. 12: 477–479.

Auld, B.A. 2000. Success in biological control of weeds by pathogens, including bioherbicides. pp. 323–340. *In:* G. Gurr and S. Wratten (Eds.) Biological Control: Measures of Success. Kluwer Academic Publishers. Dordrecht, Netherlands.

Auld, B.A. and L. Morin. 1995. Constraints in the development of bioherbicides. Weed Technol. 9(3): 638–652.

Auld, B.A. and M.M. Say. 1999. Comparison of isolates of *Colletotrichum orbiculare* from Argentina and Australia as potential bioherbicides for *Xanthium spinosum* in Aust. Agric. Ecosyst. Environ. 72: 53–58.

Bailey, K.L. 2010. Canadian innovations in microbial biopesticides. Can. J. Plant Pathol. 32: 113–121.

Bailey, K.L. 2014. The Bioherbicide Approach to Weed Control Using Plant Pathogens. pp. 245–266. *In:* D.P. Abrol (Ed.) Integrated Pest Management: Current Concepts and Ecological Perspective, © 2013 Elsevier Inc. London, UK.

Bailey, K.L. and S. Falk. 2011. A Phoma story. Pest Technol. 5 (Special Issue 1): 73–79. Global Science Books.

Bailey, K.L., S. Boyetchko and T. Langle. 2010. Social and economic drivers shaping the future of biological control: a Canadian perspective on the factors affecting the development and use of microbial biopesticides. Biol. Control 52(3): 221–229.

Bailey, K.L., S.M. Boyetchko, G. Peng, R.K. Hynes, W.G. Taylor and W.M. Pitt. 2009. Developing weed control technologies with fungi. pp. 1–44. *In:* R.K. Mahendrah (Ed.) Current Advances in Fungal Technology. I.K. International, New Delhi, India.

Bannon, J.S. 1988. CASST™ herbicide (*Alternaria cassiae*), A case history of a mycoherbicide. Am. J. Alt. Agr. 3(2–3): 73–76.

Barreto, R.W. 2007. Latin American weed biological control science at the crossroads. pp. 109–121. *In:* Julien, M.H., Sforza, R., Bon, M.C., Evans, H.C., Hatcher, P.E., Hinz, H. and Rector, B.G. (Eds.). Proceedings XII International Symposium on Biological Control of Weeds. La Grande Motte, France. CABI.

Barreto, R.W., C.A. Ellison, M.K. Seier and H.C. Evans. 2012. Biological control of weeds with plant pathogens: Four decades on. pp. 299–350. *In:* D.P. Abrol and U. Shankar (Eds.) Integrated Pest Management: Principles and Practice. CABI, Wallingford, Oxfordshire, UK.

Barton, J. 2004. How good are we at predicting the field host-range of fungal pathogens used for classical biological control of weeds? Biol. Control 31: 99–122.

Barton, J. 2005. Bioherbicides: All in a day's work . . . for a superhero. pp. 4–6 *In:* What's New in Biological Control of Weeds? Manaaki Whenua Issue 34 November 2005, Landcare Research, Lincoln, New Zealand.

Becker, E., S.F. Shamoun and W.E. Hintz. 2005. Efficacy and environmental fate of *Chondrostereum purpureum* used as a biological control for red alder (*Alnus rubra*). Biol. Control 33: 269–277.

Bewick, T.A., J.C. Porter and R.C. Ostrowski. 2000. Smolder™: A bioherbicide for suppression of dodder (*Cuscuta* spp.). Proc. Southern Weed Sci. Soc. Abstracts 53: 152.

Boari, A. and M. Vurro. 2004. Evaluation of *Fusarium* spp. and other fungi as biological control agents of broomrape (*Orobanche ramosa*). Biol. Control 30: 212–219.

Bourdôt, G.W., D. Baird, G. Hurrell and M.D. de Jong. 2006. Safety zones for a *Sclerotinia sclerotiorum*-based mycoherbicide: Accounting for regional and yearly variation in climate. Biocontrol Sci. Techn. 16: 345–358.

Bowers, R.C. 1986. Commercialization of Collego™ – an industrialist's view. Weed Sci. (Suppl.) 34: 24–25.

Boyetchko, S.M. and G. Peng. 2004. Challenges and strategies for development of mycoherbicides. pp. 111–121. *In:* D.K. Arora (Ed.) Fungal Biotechnology in Agricultural, Food, and Environmental Applications. Marcel Dekker, New York, USA.

Boyetchko, S.M., K.L. Bailey, R.K. Hynes and G. Peng. 2007. Development of Bio Mal. pp. 274–283. *In:* C. Vincent, M.S. Goettel and G. Lazarovits (Eds.) Biological Control: A Global Perspective. CABI, Wallingford, UK.

Boyette, C.D. and R.E. Hoagland. 2015. Bioherbicidal potential of *Xanthomonas campestris* for controlling *Conyza canadensis*. Biocontrol Sci. Techn. 25: 229–237.

Boyette, C.D., R.E. Hoagland and M.A. Weaver. 2007. Biocontrol efficacy of *Colletotrichum truncatum* for hemp sesbania (*Sesbania exaltata*) is enhanced with unrefined corn oil and surfactant. Weed Biol. Mange. 7(1): 70–76.

Boyette, C.D., P.C. Jr. Quimby, A.J. Caesar, J.L. Birdsall, W.J. Jr. Connick, D.J. Daigle, M.A. Jackson, G.E. Eagley and H.K. Abbas. 1996. Adjuvants, formulations and spraying systems for improvement of mycoherbicides. Weed Technol. 10: 637–644.

Boyette, C.D., P.C. Jr. Quimby, W.J. Jr. Connick, D.J. Daigle and F.E. Fulgham. 1991. Progress in the production, formulation and application of mycoherbicides. pp. 209–222. *In:* D.O. TeBeest. (Ed.) Microbial Control of Weeds. Chapman & Hall. New York, USA.

Bruckart, W.L. 2005. Supplemental risk evaluations and status of *Puccinia carduorum* for biological control of musk thistle. Biol. Control 32: 348–355.

Burdon, J.J., P.H. Thrall, P.H. Groves and P. Chaboudez. 2002. Biological control of *Carduus pycnocephalus* and *C. tenuiflorus* using the rust fungus *Puccinia cardui-pycnocephali*. Plant Prot. Quart. 15: 14–17.

Burnett, H.C., D.P.H. Tucker and W.H. Ridings. 1974. *Phytophthora* root and stem rot of milkweed vine. Plant Dis. Rep. 58: 355–357.

Burnett, H.C., D.P.H. Tucker, M.E. Patterson and W. H. Ridings. 1973. Biological control of milkweed vine with a race of *Phytophthora citrophthora*. Proc. FL. State Hortic. Soc. 85: 111–115.

Caesar-Ton That, T.C., W.E. Dyer, P.C. Quimby and S.S. Rosenthal. 1995. Formulation of an endoparasitic nematode *Subanguina picridis* Brezeski, a biocontrol agent for Russian knapweed [*Acroptilon repens* (L.) DC.]. Biol. Control 5(2): 262–266.

Cai, X. and M. Gu. 2016. Bioherbicides in organic horticulture. Horticulturae 2: 1–10.

Caldwell, C.J., R.K. Hynes, S.M. Boyetchko and D.R. Korber. 2012. Colonization and bioherbicidal activity on green foxtail by *Pseudomonas fluorescens* BRG100 in a pesta formulation. Can J Microbiol. 58(1): 1–9.

Cartwright, K., D. Boyette and M. Roberts. 2010. Lockdown: Collego bioherbicide gets a second act. Phytopathology 100: S162.

Charudattan, R. 1991. The mycoherbicide approach with plant pathogens. pp. 24–57. *In:* D.O. TeBeest (Ed.) Microbial Control of Weeds. Chapman & Hall, New York, USA.

Charudattan, R. 2005. Ecological, practical, and political inputs into selection of weed targets: what makes a good biological control target? Biol. Control 5(3): 183–196.

Charudattan, R. 2016. SolviNix LC, the first registered bioherbicide containing a plant virus as the active ingredient (Abstract 99). Proceedings 7th International Weed Science Congress, Prague.

Charudattan, R. and A. Dinoor. 2000. Biological control of weeds using plant pathogens: accomplishments and limitations. Crop Prot. 19(8–10): 691–695.

Charudattan, R. and E. Hiebert. 2007. A plant virus as a bioherbicide for tropical soda apple, *Solanum viarum*. Outlooks Pest Manage. 18: 167–171.

Chittick, A.T. and B.A. Auld. 2001. Polymers in bioherbicide formulation: *Xanthium spinosum* and *Colletotrichum orbiculare* as a model system. Biocontrol Sci. Techn. 11(6): 691–702.

Chutia, M., J.J. Mahanta, N. Bhattacharyya, M. Bhuyan, P. Boruah and T.C. Sarma. 2007. Microbial herbicides for weed management: prospects, progress and constraints. Plant Pathol. J. 6(3): 210–218.

Ciotola, M., A.K. Watson and S.G. Hallett. 1995. Discovery of an isolate of *Fusarium oxysporum* with potential to control *Striga hermonthica* in Africa. Weed Res. 35: 303–309.

Cipriani, M.G., G. Stea, A. Moretti, C. Altomare, G. Mulè and M. Vurro. 2009. Development of a PCR-based assay for the detection of *Fusarium oxysporum* strain FT2, a potential mycoherbicide of *Orobanche ramosa*. Biol. Control 50: 78–84.

Cohen, B., Z. Amsellem, R. Maor, A. Sharon and J. Gressel. 2002a. Transgenically enhanced expression of indole-3-acetic acid confers hypervirulence to plant pathogens. Phytopathology 92(6): 590–596.

Cohen, B.A., Z. Amsellem, S. Lev-Yadun and J. Gressel. 2002b. Infection of tubercles of the parasitic weed *Orobanche aegyptiaca* by mycoherbicidal *Fusarium* species. Annals Botany 90: 567–578.

Connick, W.J. Jr., C.D. Boyette and J.R. McAlpine. 1991. Formulations of mycoherbicides using a pesta like process. Biol. Control 1: 281–287.

Cordeau, S., M. Triolet, S. Wayman, C. Steinberg and J.P. Guillemin. 2016. Bioherbicides: dead in the water? A review of the existing products for integrated weed management. Crop Prot. 87: 44–49.

Cother, E.J. 1996. Bioherbicides and weed management in Asian rice fields. pp. 183–200. *In:* R. Naylor (Ed.) Herbicides in Asian rice: transitions in weed management. Palo Alto, CA: Institute for International Studies, Stanford University and Manila (Philippines): International Rice Research Institute. 270 pp.

Cripps, S. 2017. New parkinsonia bioherbicide control demonstrated on Aramac property. Available at: http://www.northqueenslandregister.com.au/story/595596/graziers-gunning-for-parkinsonia

Cullen, J.M. 1985. Bringing the cost benefits of analysis of biological control of *Chondrilla juncea* up to date. pp. 145–152. *In:* Delfosse, E.S. (Ed.) Proceedings IV Symposium on Biological Control of Weeds. Vancouver. Agriculture Canada.

Cullen, J.M., P.F. Kable and M. Catt. 1973. Epidemic spread of a rust imported for biological control. Nature 244: 462–464.

Daniel, J.T., G.E. Templeton, R.J. Jr. Smith and W.T. Fox. 1973. Biological control of northern jointvetch in rice with an endemic fungal disease. Weed Sci. 21: 303–307.

Day, M.D. and R.L. Winston. 2016. Biological control of weeds in the 22 Pacific island countries and territories: current status and future prospects. Neo Biota. 30: 167–192.

De Datta, S.K. and A.M. Baltazar. 1996. Integrated weed management in rice in Asia. pp. 145–165 *In:* R. Naylor (Ed.) Herbicides in Asian Rice: Transitions in Weed Management. Institute for International Studies, Stanford University and Manila (Philippines): International Rice Research Institute. 270 pp.

De Jong, M.D., P.C. Scheepens and J.C. Zadoks. 1990. Risk analysis for biological control: a Dutch case study in biocontrol of *Prunus serotina* by the fungus *Chondrostereum purpureum*. Plant Dis. 74: 189–194.

De Jong, M.D., G.W. Bourdôt, G.A. Hurrell and D.J. Saville. 2002. Risk analysis for biological weed control – simulating dispersal of *Sclerotinia sclerotiorum* (Lib.) de Bary ascospores from a pasture after biological control of *Cirsium arvense* (L.) Scop. Aerobiologia 18: 211–222.

DiTommaso, A. and A.K. Watson. 1995. Impact of a fungal pathogen, *Colletotrichum coccodes* on growth and competitive ability of *Abutilon theophrasti*. New Phytol. 131: 51–60.

EPA (Environmental Protection Agency). 2015. Biopesticides Registration Action Document: Tobacco Mild Green Mosaic Tobamovirus Strain U2. PC Code: 056705. United States Environmental Protection Agency.

Elzein, A. and J. Kroschel. 2004. *Fusarium oxysporum* Foxy 2 shows potential to control both *Striga hermonthica* and *S. asiatica*. Weed Res. 44: 433–438.

Elzein, A., J. Kroschel and G. Cadisch. 2008. Efficacy of Pesta granular formulation of Striga-mycoherbicide *Fusarium oxysporumf.* sp. *strigae* Foxy 2 after 5-year of storage. J. Plant Dis. Prot. 115(6): 259–262.

Evans, H.C. 1995. Fungi as biocontrol agents of weeds: a tropical perspective. Can. J. Bot. 73(Suppl. 1): S58–S64.

Evans, H.C. 2013. Biological control of weeds with fungi. pp. 145–172. *In:* F. Kempken. (Ed.), Agricultural Applications, 2nd Edition, The Mycota XI, Springer Verlag Berlin Heidelberg.

Evans, H.C., M.P. Greaves and A.K. Watson. 2001. Fungal biocontrol of weeds. pp. 169–192. *In:* T.M. Butt, C.W. Jackson and N. Magan (Eds.) Fungi as Biocontrol Agents: Progress, Problems and Potential. CABI Publishing, CABI, Wallingford, Oxfordshire, UK.

Evans, K.J., L. Morin, E. Bruzzese and R.T. Roush. 2004. Overcoming limits on rust epidemics in Australian infestations of European blackberry. pp. 514. *In:* Cullen, J.M., Briese, D.T., Kriticos, D.J., Lonsdale, W.M., Morin, L. and Scott, J.K. (Eds.) The Proceedings of 9th International Symposium on Biological Control of Weeds. CSIRO, Canberra, Australia.

Ferrell, J., R. Charudattan, M. Elliott and E. Hiebert. 2008. Effects of selected herbicides on the efficacy of *Tobacco mild green mosaic virus* to control tropical soda apple (*Solanum viarum*). Weed Sci. 56: 128–132.

Frohlich, J., S.V. Fowler, A. Gianotti, R.L. Hill, E. Killgore, L. Morin, L. Sugiyama and C. Winks. 1999. Biological control of mist flower (*Ageratina riparia*, Asteraceae) in New Zealand. pp. 6–11. *In:* O'Callaghan, M. (Ed.) The Proceedings of 52nd N.Z. Plant Protection Conference, The New Zealand Plant Protection Society Inc.

Fujimori, T. 1999. New developments in plant pathology in Japan. Australas. Plant Pathol. 28: 292–297.

Gaddeyya, G., G. Easteru Rani, B. Susmitha and K. Subhashini. 2017. Microbial technology in weed management: a special reference of biological control of horse purslane weed. Int. J. Current Adv. Res. 6(7): 4978–4991.

Gao, Z.Y. and J.E. Yu. 1992. Biological control of *Cuscuta* spp. with Lubao #1, a product formulated from *Colletotrichum* sp. Chin. J. Biocontrol. 8(4): 173–175.

Gibson, D.M., R.H. Vaughan, J. Biazzo and L.R. Milbrath. 2014. Exploring the feasibility of *Sclerotium rolfsii* VrNY as a potential bioherbicide for control of swallowworts (*Vincetoxicum* spp.). Invasive Plant Sci. Manage. 7(2): 320–327.

Green, S., S.M. Stewart-Wade, G.J. Boland, M.P. Teshler and S.H. Liu. 1998. Formulating microorganisms for biological control of weeds. pp. 249–281. *In:* G.J. Boland and L.D. Kuykendall (Eds.) Plant-Microbe Interactions and Biological Control. Marcel Dekker, New York.

Gressel, J. 2001. Potential fail safe mechanisms against the spread and introgression of transgenic hypervirulent biocontrol fungi. Trends Biotechnol. 19: 149–154.

Gressel, J. 2010. Herbicides as synergists for mycoherbicides, and vice versa. Weed Sci. 58: 324–328.

Gressel, J., S. Meir, Y. Herschkovitz, H. Al-Ahmad, I. Greenspoon, O. Babalola and Z. Amsellem. 2007. Approaches to and successes in developing transgenically enhanced mycoherbicides. pp. 297–305. *In:* M. Vurro and J. Gressel (Eds.) Novel Biotechnologies for Biocontrol Agent Enhancement and Management. Springer, Dordrecht.

Griffith, C.A. 1970. Persimmon wilt research. Annual report 1960–1970. Noble Foundation Agricultural Division, Ardmore, OK.

Harding, D.P. and M.H. Raizada. 2015. Controlling weeds with fungi, bacteria and viruses: a review. Front. Plant Sci. 6: 659. Published online 2015 August 28. doi: 10. 3389/fpls. 2015. 00659.

Health Canada. 2010. *Sclerotinia minor* strain IMI 344141. Registration decision RD 2010-08. 22 September 2010. Pest Management Regulatory Agency. Health Canada HC Pub: 100362. IBBN: 978-1-100-16450-2 (978-1-100-16451-9). 6 p.

Hershenhorn, J., F. Casella and M. Vurro. 2016. Weed biocontrol with fungi: past, present and future. Biocontrol Sci. Techn. 26(10): 1313–1328.

Hintz, W.E. 2007. Development of *Chondrostereum purpureum* as a mycoherbicide for deciduous brush control. pp. 284–290. *In:* C. Vincent, M.S. Goettel and G. Lazarovits (Eds.) Biological Control: A Global Perspective. CAB International, Wallingford, UK. 432 p.

Hintz, W.E., E.M. Becker and S.F. Shamoun. 2001. Development of genetic markers for risk assessment of biological control agents. Can. J. Plant Pathol. 23(1): 13–18.

Hirase, K., M. Nishida and T. Shinmi. 2006. Effects of lodging of *Echinochloa crusgalli* L. on the herbicidal efficacy of MTB-951, a mycoherbicide using *Drechslera monoceras* (Drechsler) Subram. et Jain (=*Exserohilum monoceras* [Drechsler] Leonard et Suggs). Weed Biol. Manage. 6(1): 30–34.

Hirase, K., S. Yoshigai, M. Nishida, Z. Takanaka and T. Shinmi. 2004. Influence of water management, application timing and temperature on efficacy of MTB-951, a mycoherbicide using *Drechslera monoceras* to control *Echinochloa crus-galli* L. Weed Biol. Manage. 4(2): 71–74.

Hoagland, R., C. Boyette, M. Weaver and H. Abbas. 2007. Bioherbicides: research and risks. Toxin Rev. 26: 313–342.

Hodgson, R.H., L.A. Wymore, A.K. Watson, R. Snyder and A. Collette. 1988. Efficacy of *Colletotrichum coccodes* and thidiazuron for velvetleaf (*Abutilon theophrasti*) control in soybean (*Glycine max*). Weed Technol. 2: 473–480.

Imaizumi, S., T. Nishino, K. Miyabe, T.M. Fujimori and M. Yamada. 1997. Biological control of annual bluegrass (*Poa annua* L.) with a Japanese isolate of *Xanthomonas campestris* pv. *poae* (JT-P482). Biol. Control 8: 7–14.

Jackson, M.A. and D.A. Schisler. 1995. Liquid culture production of microsclerotia of *Colletotrichum truncatum* for use as bioherbicidal propagules. Mycol. Res. 99(7): 879–884.

Jackson, M.A. and R.J. Bothast. 1990. Carbon concentration and carbon-to-nitrogen ratio influence submerged-culture conidiation by the potential bioherbicide *Colletotrichum truncatum* NRRL 13737. Appl. Environ. Microbiol. 56(11): 3435–3438.

Johnson, D.R., D.L. Wyse and K.L. Jones. 1996. Controlling weeds with phytopathogenic bacteria. Weed Technol. 10: 621–624.

Julien, M., R. McFadyen and J. Cullen (Eds.) 2012. Biological Control of Weeds in Australia. CSIRO Publishing, Melbourne, Australia. 620 p.

Kabaluk, J.T., A.M. Svircev, M.S. Goettel and S.G. Woo (Eds.) 2010. The Use and Regulation of Microbial Pesticides in Representative Jurisdictions Worldwide. IOBC Global. pp. 99.

Kennedy, A.C., B.N. Johnson and T.L. Stubbs. 2001. Host range of a deleterious rhizobacterium for biological control of downy brome. Weed Sci. 49: 792–797.

Kennedy, A.C., L.F. Elliott, F.L. Young and C.L. Douglas. 1991. Rhizobacteria suppressive to the weed downy brome. Soil Sci. Soc. Am. J. 55: 722–727.

Kovalev, O.V., L.G. Danilov and T.S. Ivanova. 1973. Method of controlling Russian knapweed. Opisanie Izobreteniia Kavtorskomu Svidetel'stvu Byull. 38: 2. (Translation – Translation Bureau, Canada Department of Secretary of State, No. 619707, Ottawa).

Kremer, R.J. and A.C. Kennedy. 1996. Rhizobacteria as biocontrol agents of weeds. Weed Technol. 10: 601–609.

Kropp, B.R., D.R. Hansen and S.V. Thomson. 2002. Establishment and dispersal of *Puccinia thlaspeos* in field populations of Dyer's woad. Plant Dis. 86(3): 241–246.

Kropp, B.R., D.R. Hansen, K.M. Flint and S.V. Thomson. 1996. Artificial inoculation and colonization of Dyer's woad (*Isatis tinctoria*) by the systemic rust fungus *Puccinia thlaspeos*. Phytopathology 86: 891–896.

Landgraf, B.K., P.K. Fay and K.M. Havstad. 1984. Utilization of leafy spurge (*Euphorbia esula* L.) by sheep. Weed Sci. 32(3): 348–352.

Li, Y., Z. Sun, X. Zhuang, L. Xu, S. Chen and M. Li. 2003. Research progress on microbial herbicides. Crop Prot. 22(2): 247–252.

Medal, J., W. Overholt, R. Charudattan, J. Mullahey, R. Gaskalla, R. Díaz and J. Cuda (Eds.) 2012. Tropical Soda Apple Management Plan, University of Florida-IFAS. Gainesville, Florida Department of Agriculture and Consumer Services-DPI, https://plants.ifas.ufl.edu/plant-directory/solanum-viarum/

Marley, P.S., S.M. Ahmed, J.A.Y. Shebayan and S.T.O. Lagoke. 1999. Isolation of *Fusarium oxysporum* with potential for biocontrol of the witch weed (*Striga hermonthica*) in the Nigerian savanna. Biocontrol Sci. Techn. 9: 159–163.

Meir, S., A. Amsellem, H. Al-Ahmad, E. Safran and J. Gressel. 2009. Transforming a NEP1 toxin gene into two *Fusarium* spp. to enhance mycoherbicide activity on *Orobanche* – failure and success. Pest Manage. Sci. 65: 588–595.

Mejri, D., E. Gamalero and T. Souissi. 2012. Formulation development of the deleterious rhizobacterium *Pseudomonas trivialis* X33d for biocontrol of brome (*Bromus diandrus*) in durum wheat. J. Appl. Microbiol. 114(1): 219–228.

Men, B.X., T.K. Tinh, T.R. Preston, B. Ogle and J.E. Lindberg. 1999. Use of local ducklings to control insect pests and weeds in the growing rice field. Livestock Res. Rural Dev. 11: 8.

Mishra, J., A.K. Pandey and S.K. Hasija. 1995. Evaluation of *Sclerotium rolfsii* Sacc. as mycoherbicide for *Parthenium*: factors affecting viability and virulence. Indian Phytopath. 48: 476–479.

Morin, L., R.L. Hill and S. Matayoshi. 1997. Hawaii's successful biological control strategy for mist flower (*Ageratina riparia*) – can it be transferred to New Zealand? Biocontrol News Inform. 18: 77N–88N.

Morris, M.J. 1987. Biology of the Acacia gall rust, *Uromycladium tepperianum*. Plant Path. 36(1): 100–106.

Morris, M.J. 1989. A method for controlling *Hakea sericea* Shrad. seedlings using the fungus *Colletotrichum gloeosporioides* (Penz.) Sacc. Weed Res. 29: 449–454.

Morris, M.J. 1997. Impact of the gall-forming rust fungus *Uromycladium tepperianum* on the invasive tree *Acacia saligna* in South Africa. Biol. Control 10: 75–82.

Morris, M.J., A.R. Wood and A. den Breeÿen. 1999. Plant pathogens and biological control of weeds in South Africa: a review of projects and progress during the last decade. pp. 129–137. *In:* T. Olckers and M.P. Hill (Eds.) Biological control of weeds in South Africa (1990-1998). Entomological Society of Southern Africa.

Mortensen, K. 1988. The potential of an endemic fungus, *Colletotrichum gloeosporioides*, for biological control of round-leaved mallow (*Malva pusilla*) and velvetleaf (*Abutilon theophrasti*). Weed Sci. 36: 473–478.

Müller-Schärer, H., P.C. Scheepens and M.P. Greaves. 2000. Biological control of weeds in European crops: recent achievements and future work. Weed Res. 40: 83–98.

Müller-Stöver, D., E. Kohlschmid and J. Sauerborn. 2009a. A novel strain of *Fusarium oxysporum* from Germany and its potential for biocontrol of *Orobanche ramosa*. Weed Res. 49(Suppl. 1): 175–182.

Müller-Stöver, D., H. Thomas, J. Sauerborn and J. Kroschel. 2004. Two granular formulations of *Fusarium oxysporum* f. sp. *orthoceras* to mitigate sunflower broomrape *Orobanche cumana*. Bio. Control 49(5): 595–602.

Müller-Stöver, D., R. Batchvarova, E. Kohlschmid and J. Sauerborn. 2009b. Mycoherbicidal management of *Orobanche cumana*: observations from three years of field experiments. pp. 86. *In:* Rubiales, D., Westwood, J. and Uledag, A. (Eds.) The Proceedings of the 10th International World Congress of Parasitic Plants, Kusadasi, Turkey.

Mupondwa, E., X. Li, S. Boyetchko, R. Hynes and J. Geissler. 2015. Technoeconomic analysis of large scale production of pre-emergent *Pseudomonas fluorescens* microbial bioherbicide in Canada. Bioresour. Technol. 175: 517–528.

Neufeldt, E. (Ed.) 1988. Webster's New World Dictionary. Third College Edition. Simon & Schuster, New York, USA.

Nishino, J. and A. Tateno. 2000. Camperico® – a new bioherbicide for annual bluegrass in turf. Agrochemicals Japan 77: 13–16.

Notham, F.E. and E.C. Orr. 1982. Effects of a nematode on biomass and density of silverleaf nightshade. J. Range Manag. 35(4): 536–537.

Nzioki, H.S., F. Oyosi, C.E. Morris, E. Kaya, A.L. Pilgeram, C.S. Baker and D.C. Sands. 2016. *Striga* biocontrol on a toothpick: a readily deployable and inexpensive method for smallholder farmers. Front. Plant Sci. 08. doi. :10.3389/fpls.2016.01121

Oehrens, E. and S. Gonzales. 1977. Dispersion, ciclo biologico y danos causados por *Phragmidium violaceum* (Schulz) Winter en zarzamora (*Rubus constrictus* Lef. et M. y *R. ulmifolius* Schott.) en laszonas centro-sur de Chile. Agro Sur. 5: 73–85.

OECD, 2003. Guidance for Registration Requirements for Microbial Pesticides. OECD Series on Pesticides. 18: 51. http://www.oecd.org/officialdocuments/publicdisplaydocumentpdf/?cote= env/jm/mono(2003)5&doclanguage=en

Ou, X. and A.K. Watson. 1992. *In vitro* culture of *Subanguina picridis* in *Acroptilon repens* callus, excised roots, and shoot tissues. J. Nematol. 24: 199–204.

Ou, X. and A.K. Watson. 1993. Mass culture of *Subanguina picridis* and its bioherbicidal efficacy on *Acroptilon repens*. J. Nematol. 25: 89–94.

Pacanoski, Z. 2015. Bioherbicides. pp. 153–274. *In:* A. Price, J. Kelton and L. Sarunaite (Eds.) Bioherbicides, Herbicides, Physiology of Action, and Safety, In Tech. doi: 10. 5772/61528.

Pan, L., G.J. Ash, B. Ahn and A.K. Watson. 2010. Development of strain specific molecular markers for the *Sclerotinia minor* bioherbicide strain IMI 344141. Biocontrol Sci. Techn. 20(9): 939–959.

Parker, P.E. 1991a. Nematodes as biological control agents of weeds. pp. 58–68. *In:* D.O. TeBeest (Ed.) Microbial Control of Weeds. Chapman & Hall, New York, USA.

Parker, P.E. 1991b. Nematode control of silverleaf nightshade, a biological control pilot project. Weed Sci. 34(Suppl. 1): 33–34.

Patel, R. and D.R. Patel. 2015. Biological control of weeds with pathogens: current status and future trends. Int. J. Pharm. Life Sci. 6(6): 4531–4550.

Phatak, S.C., D.R. Summer, H.D. Wells, D.K. Bell and N.C. Glaze. 1983. Biological control of yellow nutsedge with the indigenous rust fungus *Puccinia canaliculata*. Science 219(4591): 1446–1447.

Politis, D.J., A.K. Watson and W.L. Bruckart. 1984. Susceptibility of musk thistle and related composites to *Puccinia carduorum*. Phytopathology 74: 687–691.

Ridings, W.H. 1986. Biological control of strangler vine in citrus – a researcher's view. Weed Sci. 34(Suppl. 1): 31–32.

Rosskopf, E.N., R. Charudattan and J.B. Kadir. 1999. Use of plant pathogens in weed control. pp. 891–918. *In:* T.W. Fisher, T.S. Bellows, L.E. Caltagirone, D.L. Dahlsten, C.B. Huffaker and G. Gordh (Eds.) Academic Press. San Diego, USA.

Saharan, G.S. and N. Mehta. 2008. *Sclerotinia* as Mycoherbicide. pp. 377–381. *In: Sclerotinia* Diseases of Crop Plants: Biology, Ecology and Disease Management. Springer, Dordrecht. 486 p.

Sedivec, K., T. Hanson and C. Heiser. 1995. Controlling leafy spurge with goats and sheep. EXT NDSU Extension Service, North Dakota State University, Fargo. https://library. ndsu. edu/ repository/bitstream/handle/10365/17574/R-1093-1995. pdf?sequence=2

Sharon, A., S. Barhoom and R. Maor. 2001. Genetic engineering of *Collectotrichum gloeosporioides* f. sp. *aeschynomene*. pp. 240–247. *In:* M. Vurro et al. (Eds.) Enhancing Biocontrol Agents and Handling Risks, IOS Press, Washington, DC.

Sharon, A., Z. Amsellem and J. Gressel. 1992. Glyphosate suppression of an elicited defense response: increased susceptibility of *Cassia obtusifolia* to a mycoherbicide. Plant Physiol. 98: 654–659.

Shibayama, H. 2001. Weeds and weed management in rice production in Japan. Weed Biol. Manage. 1(1): 53–60.

Smith, R.J. Jr. 1986. Biological control of northern jointvetch in rice and soybeans – a researcher's view. Weed Sci. 34(Suppl. 1): 17–23.

Stubbs, T.L. and A.C. Kennedy. 2012. Microbial Weed Control and Microbial Herbicides. pp. 135–166. *In:* R. Alvarez-Fernandez (Ed.) Herbicides – Environmental Impact Studies and Management. Intech Open, DOI: 10.5772/32705.

Suzuki, H. 1991. Biological control of a paddy weed, water chestnut, with a fungal pathogen. pp. 78-86. *In:* The Biological Control of Plant Diseases. Food and Fertiliser Technology Center Book Series No. 42. Taipei, Taiwan.

Tang, W., Y-Z. Zhu, H-Q. He, S. Qiang and B.A. Auld. 2011. Field evaluation of *Sclerotium rolfsii*, a biological control agent for broadleaf weeds in dry, direct-seeded rice. Crop Prot. 30(10): 1315–1320.

TeBeest, D.O. and G.E. Templeton. 1985. Mycoherbicides: progress in the biological control of weeds. Plant Dis. 69: 6–10.

TeBeest, D.O., X.B. Yang and C.R. Cisar. 1992. The Status of Biological Control of Weeds with Fungal Pathogens. Ann. Rev. Phytopathol. 30: 637–657.

Templeton, G.E. 1982a. Biological herbicides: discovery, development, deployment. Weed Sci. 30: 430–433.

Templeton, G.E. 1982b. Status of weed control with plant pathogens. pp. 29–44. *In:* R. Charudattan and H.L. Walker (Eds.) Biological Control of Weeds with Plant Pathogens. John Wiley and Sons, New York, USA.

Templeton, G.E. 1992a. Potential for developing and marketing mycoherbicides pp. 264–268. *In:* Combellack, J.H., Levick, K.J., Parsons, J. and Richardson, R.G. (Eds.) The Proceedings of the. 1st International Weed Control Congress, Melbourne, Australia. Weed Sci. Soc. Victoria Inc.

Templeton, G.E. 1992b. Some "orphaned" mycoherbicides and their potential for development. Plant Prot. Q. 7: 149–150.

Templeton, G.E., D.O. TeBeest and R.J. Jr. Smith. 1979. Biological control of weeds with mycoherbicides. Ann. Rev. Phytopathol. 17: 301–310.

Thomas, H., J. Sauerborn, D. Müller-Stöver, A. Ziegler, J.S. Bedi and J. Kroschel. 1998. The potential of *Fusarium oxysporum* f. sp. *orthoceras* as a biological control agent for *Orobanche cumana* in sunflower. Biol. Control 13: 41–48.

Thompson, B.M., M.M. Kirkpatrick, D.C. Sands and A.L. Pilgeram. 2007. Genetically enhancing the efficacy of plant pathogens for control of weeds. pp. 267-275. *In:* M. Vurro and J. Gressel (Eds.) Novel Biotechnologies for Biocontrol Agent Enhancement and Management. Springer, Dordrecht.

Tiourebaev, K., G. Semenchenko, M. Dolgovskaya, M. McCarthy, T. Anderson, L. Carsten, A. Pilgeram and D. Sands. 2001. Biological control of infestations of ditchweed (*Cannabis sativa*) with *Fusarium oxysporum* f. sp. *cannabis* in Kazakhstan. Biocontrol Sci. Techn. 11(4): 535–540.

Tomley, A.J. and H.C. Evans. 2004. Establishment of, and preliminary impact studies on, the rust, *Maravalia cryptostegiae*, of the invasive alien weed, *Cryptostegia grandiflora* in Queensland, Australia. Plant Path. 53: 475–494.

Trujillo, E.E. 2005. History and success of plant pathogens for biological control of introduced weeds in Hawaii. Biol. Control 33: 113–122.

Tsukamoto, H., M. Gohbara, M. Tsuda and T. Fujimori. 1997. Evaluation of pathogenic fungi for biological control of *Echinochloa* species. Ann. Phytopathol. Soc. Jpn. 63(5): 366–372.

Tsukamoto, H., M. Tsuda, M. Gohbara and T. Fujimori. 1998. Effect of water management on mycoherbicidal activity of against *Echinochloa oryzicola* Ann. Phytopathol. Soc. Jpn. 64: 526–531.

Tsukamoto, H., M. Takabayashi, T. Hieda and M. Gohbara. 2001. Strains belonging to *Exserohilum monoceras*, and uses thereof. US Patent # 6,313,069 B1.

Venne, J., F. Beed, A. Avocanh and A.K. Watson. 2009. Integrating *Fusarium oxysporum* f. sp. *strigae* into cereal cropping systems in Africa. Pest Manage. Sci. 65(5): 572–580.

Vieira, B.S. and R.W. Barreto. 2010. Liquid culture production of chlamydospores of *Lewia chlamidosporiformans* (Ascomycota: Pleosporales), a mycoherbicide candidate for wild poinsettia. Australas. Plant Pathol. 39(2): 154–160.

Vieira, B.S., K.L. Nechet and R.W. Barreto. 2008. *Lewia chlamidosporiformans*, a mycoherbicide for control of *Euphoria heterophylla*: isolate selection and mass production. pp. 221–226. *In:* Julien, M.H., Sforza, R., Bon, M.C., Evans, H.C., Hatcher, P.E., Hinz, H.L. and Rector, B.G. (Eds.) The Proceedings of the 12th International Symposium on Biological Control of Weeds, La Grande Motte, France. © CABI.

Vurro, M. and H.C. Evans. 2007. Opportunities and constraints for the biological control of weeds in Europe. pp. 455–462. *In:* Julien, M.H., Sforza, R., Bon, M.C., Evans, H.C., Hatcher, P.E., Hinz, H. and Rector, B.G. (Eds.) The Proceedings of the 12th International Symposium on Biological Control of Weeds, La Grande Motte, France. CABI.

Walker, H.L. and J.A. Riley. 1982. Evaluation of *Alternaria cassiae* for the biocontrol of sicklepod (*Cassia obtusifolia*). Weed Sci. 30(6): 651–654.

Walker, H.L. and W.J. Jr. Connick. 1983. Sodium alginate for production and formulation of mycoherbicides. Weed Sci. 31: 333–338.

Wang, J., X. Wang, B. Yuan and S. Qiang. 2013. Differential gene expression for *Curvularia eragrostidis* pathogenic incidence in crabgrass (*Digitaria sanguinalis*) revealed by cDNA-AFLP analysis. PLoS ONE 8(10): e75430. doi: 10.1371/journal.pone.0075430.

Wapshere, A.J. 1982. Biological Control of Weeds. pp. 47–**56**. *In:* W. Holzner and N. Numata (Eds.) Biology and Ecology of Weeds. Junk Publishers, The Hague.

Waterhouse, D.F. 1994. Biological control of weeds: Southeast Asian prospects. ACIAR Monograph. 26: 302.

Watson, A.K. 1985. Host specificity of plant pathogens in biological weed control. pp. 577–586. *In:* Delfosse, E.S. (Ed.) The Proceedings of the 6th International Symposium on Biological Control of Weeds, Vancouver. Ottawa. Agriculture Canada.

Watson, A.K. 1986. The biology of *Subanguina picridis*, a potential biological control agent of Russian knapweed. J. Nematol. 18: 149–154.

Watson, A.K. 1991a. Prospects for bioherbicide development in Southeast Asia. pp. 65–73. *In:* Swarbrick, J.T., Nishimoto, R.K. and Soerjani, M. (Eds.) The Proceedings of the 13th Asian Pacific Weed Science Society Conference, Jakarta, Indonesia. Asian Pacific Weed Science Society and Weed Science Society of Indonesia.

Watson, A.K. 1991b. The classical approach with plants pathogens. pp. 3–23. *In:* D.O. TeBeest (Ed.) Microbial Control of Weeds. Chapman & Hall, New York, USA.

Watson, A.K. 1994. Current status of bioherbicide development and prospects for rice in Asia. pp. 195–201. *In:* H. Shibayama, K. Kiritani and J. Bay-Petersen (Eds.) Integrated Management of Paddy and Aquatic Weeds in Asia. FFTC Book Series No. 45. Food and Fertiliser Technology Center for the Asian and Pacific Region, Taipei, Taiwan.

Watson, A.K. 1999. Can viable weed control be attainable with microorganisms. pp. 59–63. *In:* Hong, L.W., Sastroutomo, S.S., Caunter, I.G., Ali, J., Yeang, L.K., Vijaysegaran, S. and Sen, Y.H. (Eds.) Biological Control in the Tropics: Towards Efficient Biodiversity and Bioresource Management for Effective Biological Control. Serdang, Malaysia. CABI. Oxford University Press.

Watson, A.K. 2007. *Sclerotinia minor* – Biocontrol target or agent? pp. 205–211. *In:* M. Vurro and J. Gressel (Eds.) Novel Biotechnologies for Biocontrol Agent Enhancement and Management. Springer, Dordrecht.

Watson, A.K. 2013. Biocontrol. pp. 469–497. *In:* D.M. Joel, J. Gressel and L.J. Musselman (Eds.) Parasitic Orobanchaceae, Parasitic Mechanisms and Control Strategies. Springer, Heidelberg.

Watson, A.K. 2017. Biocontrol and Weed Management in Rice of Asian Pacific Region. pp. 113–134. *In:* A. Rao and H. Matsumoto (Eds.) Weed Management in Rice of Asian-Pacific Region. Pacific Weed Science Society. @ APWSS 2017.

Watson, A.K. and K. Bailey. 2013. *Taraxacum officinale* Weber, Dandelion (Asteraceae). pp. 383–391. *In:* P. Mason and D. Gillespie (Eds.) Biological Control Programmes in Canada 2001-2012. CABI Publishing, Wallingford, UK.

Watson, A.K. and L.A. Wymore. 1990. Identifying limiting factors in the biocontrol of weeds. pp. 305–316. *In:* R. Baker and P. Dunn (Eds.) New Directions in Biological Control, UCLA Symp. Mol. Cell. Biol., New Series 112, Alan R. Liss, New York.

Watson, A.K. and P. Harris. 1984. *Acroptilon repens* (L.) DC, Russian knapweed (Compositae). pp. 105–110. *In:* J.S. Kelleher and M.A. Hulme (Eds.) Biological Control Programmes against Insects and Weeds in Canada 1969-1980. Farnham Royal: Commonwealth Agricultural Bureaux, Farnham Royal, Slough, England.

Wilson, C.L. 1965. Consideration of the use of persimmon wilt as a silvicide for weed Persimmons. Plant Dis. Rep. 49(9): 789–791.

Wilson, C.L. 1970. Use of plant pathogens in weed control. PANS 16(3): 482–487.

Winston, R.L., M. Schwarzländer, H.L. Hinz, M.D. Day, M.J.W. Cock and M.H. Julien (Eds.). 2014. Biological Control of Weeds: A World Catalogue of Agents and Their Target Weeds, 5th edition. USDA Forest Service, Forest Health Technology Enterprise Team, Morgantown, West Virginia. FHTET-2014-04. 838 p.

Womack, J.G., G.M. Eccleston and M.N. Burge. 1996. A vegetable oil based invert emulsion for mycoherbicides delivery. Biol. Control 6: 23–28.

Wymore, L.A., A.K. Watson and A.R. Gotlieb. 1987. Interaction between *Colletotrichum coccodes* and thidiazuron for control of velvetleaf (*Abutilon theophrasti*). Weed Sci. 35: 377–383.

Wymore, L.A., C. Poirier, A.K. Watson and A.R. Gotlieb. 1988. *Colletotrichum coccodes*, a potential bioherbicide for control of velvetleaf (*Abutilon theophrasti*). Plant Dis. 72: 534–538.

Yandoc-Ables, C.B., E.N. Rosskopf and R. Charudattan. 2006. Plant Pathogens at Work: Progress and Possibilities for Weed Biocontrol. Part 1: Classical vs. Bioherbicidal Approach. Available at: http://www.apsnet.org/publications/apsnetfeatures/Pages/WeedBiocontrolPart1.aspx

Yoo, I.D. 1991. Status and perspectives of microbial herbicide development. Korean J. Weed Sci. 11: 47–55.

Zdor, R., C. Alexander and R. Kremer. 2007. Weed suppression by deleterious rhizobacteria is affected by formulation and soil properties. Comm. Soil Sci. Plant Anal. 36(9–10): 1289–1299.

Zhang, D., Q.W. Min, S.K. Cheng, H.L. Yang, L. He, W.J. Jiao et al. 2010. Effects of different rice farming systems on paddy field weed community. [Article in Chinese] Chin. J. Appl. Ecol. 21(6): 1603–1608.

Zhang, J., S. Yang, Z. Zhou and L. Yu. 2011. Development of bioherbicides for control of barnyard grass in China. Pest Technol. 5 (Special Issue 1): 56–60.

Zhou, T. and J.C. Neal. 1995. Annual bluegrass (*Poa annua*) control with *Xanthomonas campestris* pv. *poannua* in New York State. Weed Technol. 9: 173–177.

Zhu, Y. and S. Qiang. 2004. Isolation, pathogenicity and safety of *Curvularia eragrostidis* isolate QZ-2000 as a bioherbicide agent for large crabgrass (*Digitaria sanguinalis*). Biocontrol Sci. Techn. 14(8): 769–782.

Zorner, P.S., S.L. Evans and S.D. Savage. 1993. Perspectives on providing a realistic technical foundation for the commercialization of bioherbicides. pp. 79–86 *In:* S.O. Duke, J.J. Menn and J.R. Plimmer (Eds.) Pest Control with Enhanced Environmental Safety, ACS Symposium Series, Vol. 524.

8

Natural Product-Based Chemical Herbicides

Stephen O. Duke[*1], **Daniel K. Owens**[2] **and Franck E. Dayan**[3]

[1] USDA, ARS, Natural Products Utilization Research Unit, Thad Cochran Research Center, School of Pharmacy, University of Mississippi, University, MS 38667, USA
[2] Molecular Biosciences and Bioengineering, University of Hawaii at Manoa, Honolulu, HI 96822, USA
[3] Bioagricultural Sciences and Pest Management, Colorado State University, Ft. Collins, CO 80523, USA

Introduction

There is great interest in using natural compounds as herbicides. This interest is fueled by the growing adoption of organic agriculture as well as the desire in conventional agriculture for pesticides with a 'softer' toxicological and environmental profile than many synthetic herbicides (Dayan and Duke 2014, Duke et al. 2014, Glare et al. 2012, Marrone 2014, Seiber et al. 2014). In surveys of organic farmers, weed management issues are often listed as their biggest production problem. Compared to insect and plant pathogen management products for organic agriculture, there are relatively few natural products, or even natural product-based, herbicides available for weed management.

Another driver of the trend toward natural product-based herbicides is rapidly-increasing evolution of resistance to commercial synthetic herbicides (Heap 2017). This problem is exacerbated by the fact that there has not been a new herbicide mode of action introduced in over 30 years (Duke 2012). Natural phytotoxins (toxins that kill plants) often have modes of action that are not represented by the available synthetic herbicides (Dayan and Duke 2014, Duke et al. 2000, Duke and Dayan 2015), so it is unlikely that existing herbicide-resistant weeds will be resistant to them. This aspect of natural phytotoxins has led the agrochemical industry to focus more on natural compounds as leads for new herbicides.

Between 1997 and 2010, natural products accounted for only 6.4% of new active ingredients registered by the United Stated Environmental Protection Agency (USEPA), and almost 15% of registrations were synthetic compounds derived from natural compounds (Cantrell et al. 2012). However, specifically for weed management, only 8% of registrations were synthetic, natural compound-derived products, and there were no natural compounds registered. Thus, even though weed management is the primary pest management problem for organic farmers, relatively little is being successfully done to solve this issue with natural chemicals.

Not all natural product-based herbicides are considered biopesticides by the USEPA. Biochemical biopesticides are defined by the USEPA as "naturally occurring substances that control pests by non-toxic mechanisms". Conventional pesticides, by contrast, are generally "synthetic materials that directly kill or inactivate the pest". They further state that "because it

*Corresponding author: stephen.duke@ars.usda.gov

is sometimes difficult to determine whether a substance meets the criteria for classification as a biochemical pesticide, the EPA has established a special committee to make such decisions". Thus, by their definition biochemical pesticides cannot kill by a toxic mechanism, but to manage weeds in a field during a growing season requires that the weeds must be killed rapidly. Therefore, the special committee has obviously made some exceptions to the criterion that biochemical biopesticides must control weeds by non-toxic mechanisms. If the USEPA deems a product a biochemical biopesticide, the regulatory requirements for approval are much less stringent than for synthetic, conventional pesticides, significantly lowering the cost of getting the product to the market.

This chapter will update previous reviews (Copping and Duke 2007, Dayan and Duke 2010) that have discussed available natural products for weed management. It will also discuss the prospects for natural products that have been proposed as herbicides, but are not yet on the market.

Current Natural Product-Based Herbicides

Copping and Duke (2007) listed all natural product and natural product-based herbicides that had been or were available on the pesticide market up through the time of that publication. Table 8.1 lists some of the natural product herbicide ingredients that are currently or have been commercially available in some part of the world. It does not include two USEPA approved biochemical bioherbicides, iron HEDTA [(2-hydroxyethyl)ethylenediaminetetraacetic acid] and sodium ferric EDTA (ethylenediaminetetraacetic acid), because neither HEDTA nor EDTA are natural products. The natural product 9,10-anthraquinone is listed by the USEPA as a biopesticide for use as a bird repellent; however it is patented, but not approved, as a highly selective blue green algicide as well (Nanayakkara and Schrader 2008).

Some of the natural products for weed management have very small markets, so there are products that may have been missed in Table 8.1. Products that are no longer available have left the marketplace mostly because of poor performance. In other words, their efficacy in the field did not justify the cost of their use. Most of these products are crude mixtures, such as essential oils or mixtures of organic and fatty acids. New commercial products are often generated by simply mixing new proportions and/or combinations of existing active ingredients. Finding the exact ingredients of some of these products is not easy. The Organic Materials Review Institute (OMRI) lists pest management products (https://www.omri.org/omri-lists/download) and generic materials (https://www.omri.org/purchase-generic-materials-list) that they deem acceptable for organic agriculture in the USA. Their listing of acceptable herbicides is much shorter than that for acceptable insecticides and fungicides for plant protection. Furthermore, there is little difference between some of the products that are listed by OMRI for weed management. Not all active ingredients listed in Table 8.1 are OMRI listed. Moreover, not all products listed by OMRI for vegetation management are listed in Table 8.1, as inorganic chemicals, such as copper sulfate are listed by OMRI as acceptable for algae management, and only organic materials are covered in this review. The USDA National Organic Program allows a wider range of products for weed management than OMRI. The USDA National Organic Program requires that all ingredients in organic pesticide preparations be certified organic by a fifteen-member board (National Organic Standards Board). Materials produced by genetic engineering and chemically-synthesized compounds are excluded. There are a few exceptions for chemically-synthesized compounds, as long as they do not contaminate the crop, soil or water or have adverse effects on the crop or human health. For example, synthesized soaps are allowed to control weeds in some situations. Natural substances are allowed unless specifically prohibited (e.g. , strychnine, a natural compound, is specifically prohibited). Natural organic and inorganic products that are acceptable for weed management in organic agriculture vary between countries. For example, OMRI has different listings for Canada and the USA.

Although microbial bioherbicides (another USEPA category of bioherbicides reviewed by Harding and Raizada 2015 and Cordeau et al. 2016) are not covered in this review, there

Table 8.1. Examples of natural products commercialized for weed management

Commercial products	Main component	Mode of action or molecular target	Concentration
Essential oils			
Avenger® Weed Killer	Citrus oil	Unknown	17.5%
Weed Zap®	Clove/Cinnamon oil	Unknown	45%/45%
Weed Blitz®	Pine oil	Unknown	13.6%
EcoExempt HC®	2-phenethyl propionate/clove oil	Membrane disruption	24%/24%
Crude botanicals			
AgraLawn CrabGrass Killer®	Cinnamon bark	Unknown	0.95%
Concern Weed Prevention Plus®	Corn gluten (small peptides)	Unknown	100%
Microbial broth			
Bialaphos	L-alanyl-L-alanylphosphinothricin	Glutamine synthetase	
Organic acids			
Nature's Way Organic Weed Spray ®	Acetic acid and clove oil	Unknown	24%/8%
Scythe® herbicide	Pelargonic acid	Membrane disruption	57%

is strong evidence that much of the activity of microbial bioherbicides is due to phytotoxins produced by the microbe, either in fermentation before the microbe is used or *in planta*, after the weed is inoculated with a live pathogenic microbe. For example, the USEPA approved the use of thaxtomins (4-nitroindol-3-yl-containing 2,5-dioxopiperazines) as a microbial bioherbicide to be applied as a killed, non-viable *Streptomyces acidiscabies* (potato scab) preparation in 2013. Thaxtomin has a unique molecular site of action in cellulose synthesis (Wolf-Rüdiger et al. 2003). Although approved for sale and by OMRI, this product has not been commercialized at the time of this writing.

In general, the cost of weed management with the natural product herbicides available is significantly greater than with commercial herbicides. For example, in a study of weed management along roadsides, Young (2004) found glyphosate to give much better weed control than acetic acid or pine oil and to cost (including application cost) 50 to 80 times less than these two products.

Shrestha et al. (2012) found a D-limonene bioherbicide product to give very poor weed control compared to steam or flame treatment, and the cost was five- and 10-fold higher, respectively. Few such economic studies comparing natural product herbicides with conventional and/or alternative weed management methods have been made. The lack of good efficacy and the relatively high cost of the existing biochemical herbicide products support the view that economical and efficacious natural herbicides are greatly needed.

Another problem with many of these products is the relatively high application rates that are needed. For example, in the study by Young (2004), approximately 100 and 130 L ha^{-1} of acetic acid and pine oil, respectively, were used, whereas glyphosate was used less than 1 kg ha^{-1}. Despite more than 100-fold difference in use rate, glyphosate efficacy was essential 100% at

21 days after spraying, whereas, acetic acid efficacy ranged from 35% to 73%, depending on the weed species, and pine oil efficacy ranged from 24% to 63%. Such high usage rates may reflect non-specific modes of action, such as general effects on membranes or pH effects, rather than having specific molecular targets for which relatively little chemical is needed. This is especially true for products consisting of a mixture of compounds, such as essential oils or fatty acids. Also, these products are generally 'burn down' contact herbicides that kill the foliage to which they are applied at high concentration, with no translocation of the phytotoxic chemical to portions of the plant shoot that do not directly contact the herbicide. This allows for re-growth from protected meristems.

There are only two conventional herbicides or herbicide classes that are clearly derived from natural products (Figure 8.1). Glufosinate is a racemic mixture of synthetic L-phosphinothricin {4-[hydroxy(methyl)phosphinyl]-L-homoalanine} and its D-enantiomer (reviewed by Lydon and Duke 1999) (Figure 8.1). The L-form is a naturally-occurring compound produced by some *Streptomyces* species. The unnatural D-enantiomer is inactive as a herbicide, because it does not inhibit glutamine synthetase (GS), the molecular target site of L-phosphinothricin. Because glufosinate is a synthetic form of a natural product, and also contains a synthetic, inactive enantiomer, organic farmers are not allowed to use it. Bialaphos is a *Strepmomyces*-produced tripeptide that breaks down to L-phosphinothricin in the plant, thereby killing the plant by inhibition of GS. A small market for bialaphos exists in Japan, where it is produced by fermentation. Considering the extreme need by organic farmers in the US and Europe for more efficacious natural product-based herbicides, it is perplexing that this product is not marketed outside Japan. Numerous other natural product inhibitors of GS exist (e.g., phosalacine) and should be sufficiently active to be used as herbicides, even though they are not as active as L-phosphinothricin or GS inhibitors (Lydon and Duke 1999). Cost and/or intellectual property issues (see below) may have prevented development of some or all of these potential products.

Glufosinate is a successful herbicide that has been made more successful through use of the *bar* (bialaphos resistance) or *pat* (phosphinothricin acetyl-transferase) gene from *Streptomyces* spp. to make crops resistant to L-phosphinothricin (Duke 2014). Since glufosinate is a non-selective herbicide, this makes the herbicide/transgene combination useful for control of most weeds in the crop. The market penetration of this herbicide/transgene combination was initially poor because of the superiority of the glyphosate/glyphosate-resistant (GR) crop combination (Duke and Powles 2008). However, glufosinate use in the U.S. is now growing rapidly due to the increased value of glufosinate-resistant crops as a result of the rapidly increasing spread of GR weeds in GR crops (Heap 2014).

The other significant natural product-derived herbicide group is the triketone inhibitors of *p*-hydroxyphenylpyruvate dioxygenase (HPPD) (Lee et al. 1997) (Figure 8.1). Inhibition of HPPD stops plastoquinone production, a required co-factor for phytoene desaturase, which is required for carotenoid production. As with herbicides that inhibit carotenoid synthesis more directly, white foliage is the telltale symptom of HPPD inhibitors. HPPD inhibition is the last mode of action of synthetic herbicides to be introduced (Duke 2012). These highly successful herbicides were at least initially based on leptospermone—a natural HPPD inhibitor produced

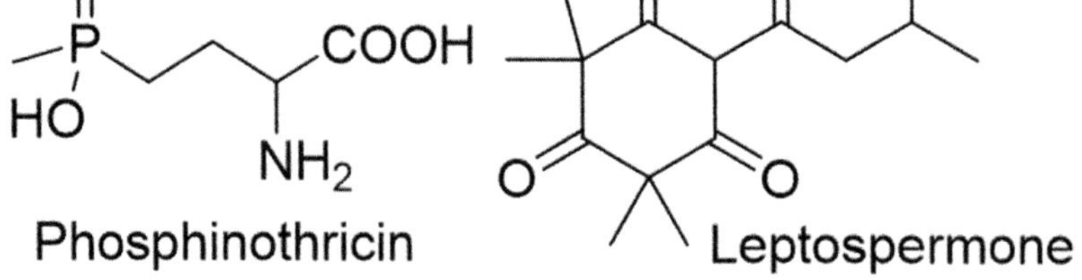

Figure 8.1. Chemical structures of highly phytotoxic natural compounds that have been the basis for commercial, synthetic herbicides.

by several woody plants found in Australia and New Zealand. Manuka oil distilled from the manuka tree (*Leptospermum scoparium*) is composed of approximately 18% natural triketones (Dayan et al. 2007). This oil has excellent pre-emergence herbicidal activity on large crabgrass (*Digitaria sanguinalis*) when applied to soil as a 1% (v/v) aqueous solution (Dayan et al. 2011). The half-life in soil of leptospermone, the dominant triketone, as a manuka oil constituent was more than two weeks. This is unique, as few natural compounds have good pre-emergence activity. Three of the four triketones of manuka oil are HPPD inhibitors, and one of these (grandiflorone) has HPPD inhibitory activity almost as strong as that of sulcotrione, a commercial triketone herbicide (Dayan et al. 2007). Although there has been interest, to our knowledge, no one has made an effort to develop natural triketones as biopesticides. The activity of manuka oil as a herbicide was improved by formulation with bioherbicide essential oil or acetic acid products (O'Sullivan et al. 2015).

When compared to synthetic herbicides, there are no good (inexpensive and efficacious) natural product herbicides available. Thus, the products that are available are used by those who can afford or are willing to use such products to fulfill organic agriculture requirements, to reduce residues of synthetic pesticides in or on their products, and/or to attempt to reduce the environmental impact of weed management. In many cases, the use of non-chemical methods of weed management (e.g., flaming, hand weeding, automated weeding or mulching) is preferable to the weak natural product herbicides that are available (Fennimore et al. 2016, Young et al. 2014). Furthermore, compared to insecticides and fungicides, natural products have not been the inspiration for many commercial, synthetic herbicides (Gerwick and Sparks 2014, Sparks et al. 2017). Nevertheless, there are many natural phytotoxic compounds that have potential as bioherbicides or as leads for new herbicides.

Potential Herbicides from Natural Products

Many natural products have been patented as herbicides, and many more have been found to be phytotoxic and proposed to be useful as herbicides, but not patented. Examples of the latter case are many of the phytotoxins reported to be involved in allelopathy (chemical warfare between plants). Ironically, a frequent rationale for much of the work on allelopathy has been to discover new herbicides. However, the compounds reported from plants are rarely sufficiently active as phytotoxins to be used as herbicides, and, in many cases, the compounds are even unlikely to actually be involved in allelopathy because of insufficient quantity produced by the plant, weak phytotoxicity, and/or little or no phytotoxicity in soil (Dayan and Duke 2009, Duke 2010, 2015, Duke et al. 2009). Much of the literature on the phytotoxicity of natural compounds should be viewed critically, as in many cases only petri dish or leaf prick bioassays are reported, and these bioassays are not good indicators of pre-emergence or post-emergence herbicidal activity on plants grown in soil. For example, Heisey and Putnam (1986) reported the microbial phytotoxins geldanamyin and nigericin (Figure 8.2) as promising herbicides in their initial studies using petri dish bioassay without soil, but neither compound performed well as a herbicide on pot-grown plants (Heisey and Putnam 1990). However, in this paper and many others involving scientists not affiliated with the pesticide industry, little effort has been made to optimize activity by formulation. So, many of these papers may underestimate the potential of these products.

Many of the more potent natural phytotoxins are produced by plant pathogens which apparently use them as virulence factors. Relatively little effort has been made to discover the phytotoxins of pathogens that infect weeds with the exceptions of those of Evidente, Strobel and Sugawara (Cimmino et al. 2015, Strobel et al. 1991, Kenfield et al. 1988).

Nevertheless, some natural products that are highly potent phytotoxins are not used as herbicides, despite the fact that they often have novel modes of action and/or a good weed spectrum. There are three main reasons that potent natural compounds have not been commercialized as herbicides: i) toxicity, ii) cost, and iii) intellectual property issues. We give examples of each of these below.

That 'natural' signifies 'safe' is a common misconception. Some of the most toxic substances known are natural chemicals (e.g., botulinum toxin and aflatoxin). So, we should not be surprised that a significant number of natural phytotoxins are too toxic to humans to be considered for development of a herbicide. A good example of this is AAL-toxin (Figure 8.2), which is an extremely potent phytotoxin (Abbas et al. 1994, 1995a, b). However, it is a close chemical analog of the fumonisins (Figure 8.2), which are potent mycotoxins associated with cancer, encephalitis, and other ailments (Abbas et al. 2002, Nelson et al. 1993). Fumonisins are also quite phytotoxic (Abbas et al. 1992, 2002, Tanaka et al. 1993). Both fumonisins and AAL-toxin act by inhibiting ceramide synthase in plants, a target site that is also present in animals (Abbas et al. 1994, 1995a, 2002). This would be a new mode of action for a herbicide if it were not for the extreme animal toxicity of all known compounds that inhibit ceramide synthase. There are many other natural phytotoxins with good herbicidal activity that also have known or suspected toxicology profiles that would preclude them from use as a commercial herbicide.

Many natural compounds are much more structurally complex than synthetic herbicides, often with more than one chiral carbon. The same problem exists for natural product pharmaceuticals. When obtaining commercial supplies from the natural sources is difficult and expensive, chemical synthesis is an alternative. However, the complexity of many natural pharmaceuticals has made chemical synthesis prohibitive. Examples of this problem include the anti-cancer drug taxol from the yew (*Taxus* spp.) tree and the anti-malarial drug artemisinin from *Artemisia annua* (Figure 8.2). Both taxol and artemisinin are highly phytotoxic (Duke et al.

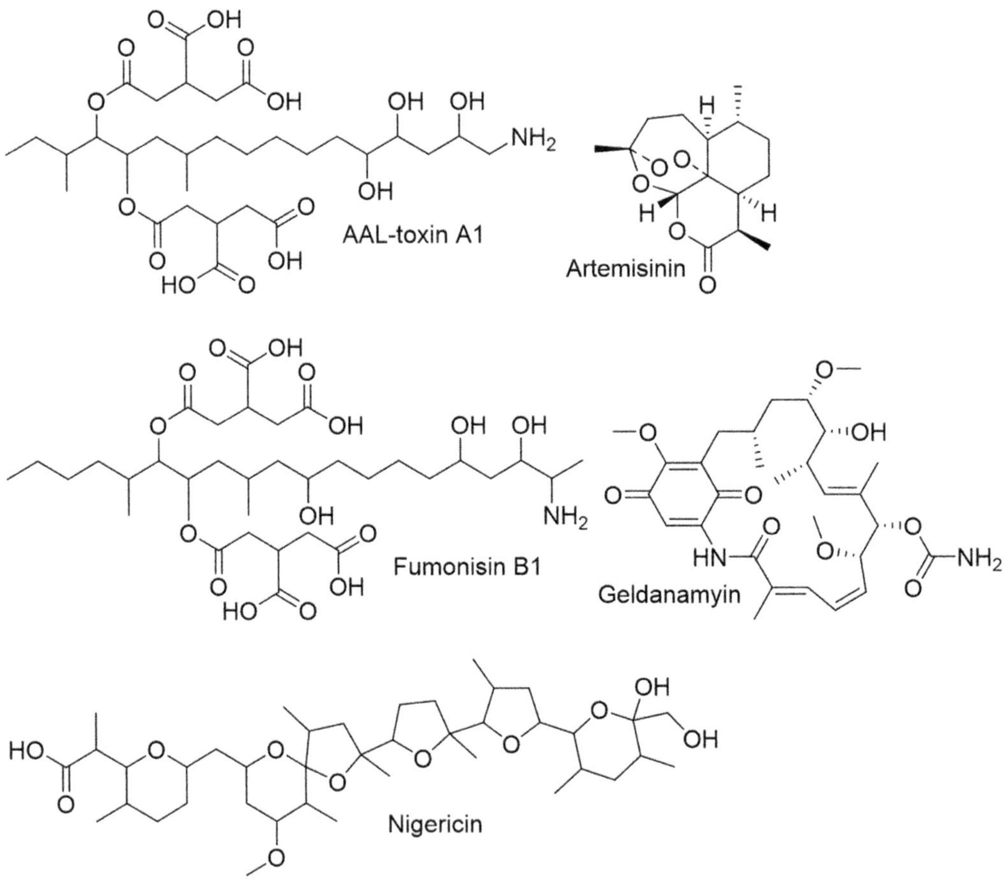

Figure 8.2. Chemical structures of highly phytotoxic natural compounds that have not led to new herbicides.

1987, Ramalakshmi and Muthuchelian 2013, Vaughan and Vaughn 1988), but the cost of these compounds would prohibit their use as herbicides.

Lastly, intellectual property barriers have limited the availability of natural herbicides. The phytotoxicity of many natural compounds has been discovered in academic laboratories that either have no interest in their use as a herbicide or have no understanding of the need for intellectual property (a patent) or the need to protect intellectual property by not publicly divulging results (orally, electronically, or in print) before a patent is filed. Without the protection from competition afforded by patent rights, most companies will not take the costly risk of investment in research and development to bring a product to market. However, new methods of production of already known natural phytotoxins can be patented, just as such patents have been used to extend patent rights of synthetic pesticides. If the new production method reduces production costs, such a patent can make the natural product more attractive for commercialization. In addition to these issues, international agreements regarding intellectual property rights regarding compounds from indigenous species or discovered from indigenous knowledge of native people have discouraged discovery efforts for pharmaceuticals and pesticides from natural sources. In summary, without the protection of robust patents, companies are unlikely to risk the cost of development and marketing of products, even if they would be beneficial to the user, the public, and/or the environment.

There are a number of natural phytotoxins with moderate to strong activity that have potential for development as herbicides. Table 8.2 provides a few examples of such compounds. We will briefly discuss some of these examples.

Rhizobitoxine is highly phytotoxic, with herbicidal activity comparable to or better than the synthetic herbicide amitrole on some weed species (Owens 1973) (Figure 8.3). It irreversibly inhibits β-cystathionase, an enzyme required for methionine synthesis (Giovanni et al. 1973). This is a unique mode of action that is not used by commercial herbicides. At the time this compound and its herbicidal properties were discovered and patented (Owens 1972), there were many inexpensive, synthetic herbicides being introduced. Furthermore, at that time there was almost no demand for bioherbicides, natural product herbicides or herbicides with new molecular target sites. Furthermore, almost no resistance to any herbicide had been reported

Table 8.2. Examples of natural products that have desirable properties as herbicides but have not been developed

Natural compound	Mode of action or molecular target	Source	References
Ascaulitoxin	Unknown	*Ascochyta caulina*	Duke et al. 2011
Cornexistin	Aspartate aminotransferase	*Paecilomyces variotii*	Nakajima et al. 1989
Hydantocidin	Adenylosuccinate synthetase	*Steptomyces hygroscopicus*	Fonné-Pfister et al. 1996, Heim et al. 1995, Nakajima et al. 1991
Momilactone B	Unknown	*Oryza* spp.	Kato-Noguchi 2004
Rhizobitoxine	β-cystathionase	*Bradyrhizobium japonicum*	Owens 1973
Sarmentine	Membrane disruptor and PSII inhibitor	*Piper* spp.	Dayan et al. 2015
Sorgoleone	PSII inhibition	*Sorghum* spp.	Gonzalez et al. 1997, Einhellig et al. 1993
Tentoxin	CF$_1$ ATPase	*Alternaria alternata*	Groth 2002
Visnagin	Unknown	*Ammi visnaga*	Traviani et al. 2016

at the time of the patenting of rhizobitoxine. Thus, the need for such a product then was much less than now.

Hydantocidin, a microbial product, has herbicidal activity comparable to glyphosate and bialaphos (Nakajima et al. 1991) (Figure 8.3). Several patents from a number of companies exist on it and close structural analogs, indicating that at one time there was considerable interest in it as a herbicide. It has very low mammalian toxicity with an acute lethal dose value of more than 1 g kg^{-1} when given orally and more than 100 mg kg^{-1} when intravenously fed to mice (Nakajima et al. 1991). Hydantocidin must be phosphorylated *in planta* in order for it to inhibit adenylosuccinate synthetase, its target site (Fonné-Pfister et al. 1996, Heim et al. 1995). This would have added a new mode of action to those of conventional herbicides. The reason(s) why none of the companies working on this compound chose to develop it are publicly unknown, but these patents were issued close to the time that GR crops were introduced, an event that substantially reduced the value of the non-glyphosate herbicide market.

Tentoxin is a cyclic tetrapeptide produced by the plant pathogenic fungus *Alternaria alternata* (Figure 8.3) (Meyer et al. 1975). It is active on many plant species, including weed species, but is not active on maize or soybean (Duke 1986, Lax et al. 1988). Tentoxin is a potent inhibitor of chloroplastic (CF1) ATPase (Arntzen 1972, Meiss et al. 2008, Steele et al. 1976). Tentoxin is highly stable and has soil activity (Duke 1986). There are no published data on its mammalian toxicity, but its mode of action is plant specific. Thus, it appears to have all of the requisites for a good selective herbicide, but it is a structurally complicated molecule that is produced in low quantities by *A. alternata* (Lax et al. 1994). As with hydantocidin, a great amount of money and effort has been applied to the study and development of tentoxin as a natural herbicide. A major

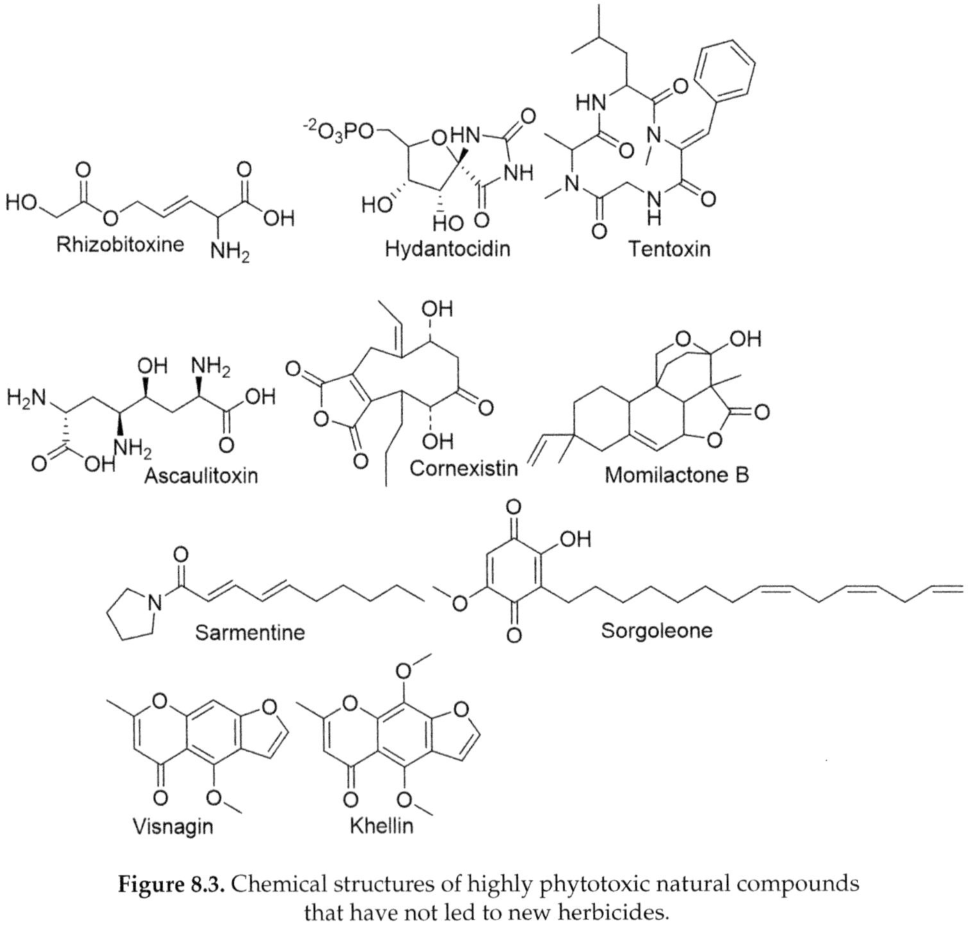

Figure 8.3. Chemical structures of highly phytotoxic natural compounds that have not led to new herbicides.

factor limiting the development of tentoxin as a commerical herbicide has been the difficulty in economically synthesizing this cyclic tetrapeptide with several chiral centers. Consequently, no commercial herbicides based on tentoxin have been developed that act on this target site nor have any successful herbicides been derived from the tentoxin backbone. Modern molecular methods could be employed to generate high-producing strains of the microbe. Tentoxin has been synthesized with high yield (Loiseau et al. 2002), so it might be more economical to chemically synthesize it rather than produce it with new fermentation technologies, but cost of synthesis may still be too high for a pesticide.

The aglycone of ascaulitoxin (Figure 8.3) is a potent phytotoxin produced by the phytopathogen *Ascochyta caulina* (Evidente et al. 1998). It inhibits plant growth at sub-micromolar concentrations. Structurally, ascaulitoxin is a fairly simple non-protein amino acid, but the presence of four chiral centers makes its stereospecific synthesis difficult and costly. Complementation studies determined that the toxic effect of ascaulitoxin could be reversed by most amino acids, whereas addition of sucrose slightly increased its activity (Duke et al. 2011). Its mode of action was investigated with metabolomics. While these studies suggested that ascaulitoxin interfers with amino acid metabolism, its target site in plants is still unknown (Duke et al. 2011).

Cornexistin (Figure 8.3) is an anhydride natural product produced by the fungal pathogen *Paecilomyces variotii* SANK 21086 with good post-emergence activity against young annual and perennial monocotyledonous and dicotyledonous plants with selectivity for corn (Nakajima et al. 1989). It is moderately active on aspartate aminotransferase (AAT) but its mechanism of action remains to be fully elucidated.

Rice produces a number of bioactive metabolites, including phenolic acids, fatty acids, phenylalkanoic acids, hydroxamic acids, terpenes, and indoles. Of these, the labdane-diterpenoid momilactones have allelochemical properties (Kato-Noguchi et al. 2010). They are released from rice roots into the soil and inhibit the growth of plant species growing nearby (Kato-Noguchi 2004). For example, momilactones inhibit the growth of barnyardgrass (*Echinochloa crus-galli*) to different degrees with momilactone B (Figure 8.3) being more than 20 times more potent than momilactone A. Genetic studies have identified the presence of a dedicated momilactone biosynthetic gene cluster in the rice genome (Kato-Noguchi and Peters 2013).

Sarmentine is an amide that has been isolated from *Piper longum* L. fruit with broad-spectrum, contact herbicidal activity which resembles the effects of herbicidal soaps, such as pelargonic acid (Huang et al. 2010) (Figure 8.3). Investigations into the mode of action of sarmentine demonstrated that it was able to destabilize the plasma membrane in a manner similar to pelargonic acid, but sarmentine was 10 to 30 times more active, and the effect is enhanced by light. Binding competition assays against ^{14}C-labeled atrazine showed that sarmentine was able to disrupt photosystem II by interacting with the Q_B plastoquinone binding site. However, as sarmentine was also able to disrupt membranes in darkness, it was apparent that the compound likely has multiple sites of action. Subsequently, its inhibitory activity on enoyl-ACP reductase, a critical enzyme in fatty acid synthesis, was found to contribute to membrane instability. Therefore, although similar to a herbicidal soap in terms of phenotypic response, sarmentine's herbicidal activity appears to result from a combination of different mechanisms (Dayan et al. 2015). It is particularly advantageous for a natural product to have more than one mode of action. As mentioned in the introduction, weeds have evolved resistance to a great number of synthetic herbicides. These compounds have unique, single target sites, and the intense selection pressure they impart can lead to the selection of single-point mutations to the molecular target site to impart resistance. Evolution of resistance to natural products with more than one mechanism of action, thus requiring two target mutations, would be much less likely.

Sorgoleone, an allelochemical exuded from the root hairs of *Sorghum bicolor* species, has been extensively studied (Głab et al. 2017) (Figure 8.3). It is produced exclusively in root hairs that contain the entire biosynthesis machinery required for the production and release of this phytotoxin in the rhizosphere (Dayan et al. 2009). Sorgoleone is a potent inhibitor of photosynthesis but is too lipophilic to translocate in the transpiration stream of the xylem

(Dayan et al. 2009). Consequently, it is most active in small-seeded plants growing around sorghum plants, and it controls their growth early in their development. Sorgoleone also inhibits root ATPase (Hejl and Koster 2004) and HPPD (Meazza et al. 2002). Thus, evolution of target site resistance to sorgoleone is less likely than for commercial herbicides that have only one molecular target.

There are a great number of other natural products with interesting phytotoxicity. However, an exhaustive list of these many natural products is beyond the scope of this review. For example, the furanochromenes, khellin and visnagin (Figure 8.3) are moderately active phytotoxic phytochemicals (Traviani et al. 2016), for which a patent has been filed.

A patent was filed before publishing the greenhouse activity of these two compounds. Even though the other phytotoxins listed in Table 8.2 have all been reported in the literature, thus precluding a patent, or have been patented as herbicides in patents that have expired, patent protection might still be obtained on the novel ways of production of these compounds or on chemical analogs that might have improved activity or better physicochemical properties.

Concluding Remarks

Although there are two examples (glufosinate and the triketone HPPD inhibitors) of natural products leading to highly successful commercial herbicides, the success of natural products or natural product-derived products in weed management is weak compared to that for insecticides and fungicides (Gerwick and Sparks 2014, Sparks et al. 2017). The biochemical bioherbicides, available for organic farmers and those who wish to reduce synthetic herbicide use, are ineffective and costly to use. In conventional agriculture, there has been no new herbicide mode of action introduced in about 30 years, even though the need for new modes of action is tremendous because of evolved herbicide resistance in hundreds of weed species. Several highly potent natural phytotoxins have novel modes of action. Thus, the growing need for new biochemical bioherbicides for organic agriculture and herbicides with new herbicide modes of action for conventional agriculture are both drivers for intensification of research into natural products as herbicides or templates for synthetic herbicides.

REFERENCES

Abbas, H.K., R.N. Paul, C.D. Boyette, S.O. Duke and R.F. Vesonder. 1992. Physiological and ultrastructural effects of fumonisin on jimsonweed leaves. Can. J. Bot. 70: 1824–1833.

Abbas, H.K., T. Tanaka, S.O. Duke, J.K. Porter, E.M. Wray, L. Hodges, A.E. Sessions, E. Wang, A.H. Merrill and R.T. Riley. 1994. Fumonisin and AAL-toxin-induced disruption of sphingolipid metabolism with accumulation of free sphingoid bases: involvement in plant disease. Plant Physiol. 106: 1085–1093.

Abbas, H.K., S.O. Duke, R.N. Paul, R.T. Riley and T. Tanaka. 1995a. AAL-toxin, a potent natural herbicide disrupts sphingolipid metabolism in plants. Pestic. Sci. 43: 181–187.

Abbas, H.K., T. Tanaka, S.O. Duke and C.D. Boyette. 1995b. Susceptibility of various crop and weed species to AAL-toxin, a natural herbicide. Weed Technol. 9: 125–130.

Abbas, H.K., S.O. Duke, W.T. Shier and M.V. Duke. 2002. Inhibition of ceramide synthesis in plants by phytotoxins. pp. 211–229. *In:* R.K. Upadhyay (Ed.) Advances in Microbial Toxin Research and Its Biotechnological Exploitation. Kluwer Academic/Plenum Publ., London.

Arntzen, C.J. 1972. Inhibition of photophosphorylation by tentoxin, a cyclic terapeptide. Biochim. Biophys. Acta, Bioenergetics 283: 539–542.

Cimmino, A., M. Masi, M. Evidente, S. Superchi and A. Evidente. 2015. Fungal phytotoxins with potential herbicidal activity: chemical and biological characterization. Nat. Prod. Rep. 32: 1629–1653.

Cantrell, C.L., F.E. Dayan and S.O. Duke. 2012. Natural products as sources for new pesticides. J. Nat. Prod. 75: 1231–1242.

Copping, L.G. and S.O. Duke. 2007. Natural products that have been used commercially as crop protection agents – a review. Pest Manag. Sci. 63: 524–554.

Cordeau, S., M. Triolet, S. Wayman, C. Steinberg and J.-P. Guillemin. 2016. Bioherbicides: Dead in the water? A review of the existing products for integrated weed management. Crop. Protect. 87: 44–49.

Dayan, F.E. and S.O. Duke. 2009. Biological activity of allelochemicals. pp. 361–384. *In*: A. Osbourn and V. Lanzotti (Eds.) Plant-Derived Natural Products – Synthesis, Function and Application. Springer, Dordrecht, Germany.

Dayan, F.E. and S.O. Duke. 2010. Natural products for weed management in organic farming in the USA. Outlooks Pest Manag. 21: 156–160.

Dayan, F.E. and S.O. Duke. 2014. Natural compounds as next generation herbicides. Plant Physiol. 166: 1090–1105.

Dayan, F.E. and S.B. Watson. 2011. Plant cell membrane as a marker for light-dependent and light-independent herbicide modes of action. Pestic. Biochem. Physiol. 101: 182-190.

Dayan, F.E., S.O. Duke, A. Sauldubois, N. Singh, C. McCurdy and C.L. Cantrell. 2007. *p*-Hydroxyphenylpyruvate dioxygenase is a target site for β-triketones from *Leptospermum scoparium*. Phytochemistry 68: 2004–2014.

Dayan, F.E., J.L. Howell and J.D. Weidenhamer. 2009. Dynamic root exudation of sorgoleone and its *in planta* mechanism of action. J. Exp. Bot. 60: 2107–2117.

Dayan, F.E., J. Howell, J.P. Marais, D. Ferreira and M. Koivunen. 2011. Manuka oil, a natural herbicide with preemergence activity. Weed Sci. 59: 464–469.

Dayan, F.E., D.K. Owens, S.B. Watson, R.N. Asolkar and L.G. Boddy. 2015. Sarmentine, a natural herbicide from *Piper* species with multiple herbicide mechanisms of action. Front. Plant Sci. 6: 222. doi: 10. 3389/fpls. 2015. 00222.

Duke, S.O. 1986. Microbial phytotoxins as herbicides – a perspective. pp. 287–304. *In:* A.R. Putnam and C.S. Tang (Eds.) The Science of Allelopathy, John Wiley, New York, USA.

Duke, S.O. 2010. Allelopathy: current status of research and future of the discipline: a commentary. Allelopathy J. 25: 17–30.

Duke, S.O. 2012. Why have no new herbicide modes of action appeared in recent years? Pest Manag. Sci. 68: 505–512.

Duke, S.O. 2014. Biotechnology: Herbicide-Resistant Crops. pp. 94–116. *In:* N. Van Alfen (Ed.) Encyclopedia of Agriculture and Food Systems. Vol. 2. Elsevier, San Diego, CA, USA.

Duke, S.O. 2015. Proving allelopathy in crop-weed interactions. Weed Sci. 63: 121–132.

Duke, S.O. and F.E. Dayan. 2015. Discovery of new herbicide modes of action with natural phytotoxins. Amer. Chem. Soc. Symp. Ser. 1204: 79–92.

Duke, S.O. and S.B. Powles. 2008. Glyphosate: A once in a century herbicide. Pest Manag. Sci. 64: 319–325.

Duke, S.O., A.C. Blair, F.E. Dayan, R.D. Johnson, K.M. Meepagala, D. Cook and J. Bajsa. 2009. Is (–)-catechin a novel weapon of spotted knapweed (*Centaurea stoebe*)? J. Chem. Ecol. 35: 141–153.

Duke, S.O., A. Evidente, M. Fiore, A.M. Rimando, F.E. Dayan, M. Vurro, N. Christiansen, R. Looser, J. Hutzler and K. Grossmann. 2011. Effects of the aglycone of ascaulitoxin on amino acid metabolism in *Lemna paucicostata*. Pestic. Biochem. Physiol. 100: 41–50.

Duke, S.O., D.K. Owens and F.E. Dayan. 2014. The growing need for biochemical bioherbicides. Amer. Chem. Soc. Symp. Ser. 1172: 31–43.

Duke, S.O., J.G. Romagni and F.E. Dayan. 2000. Natural products as sources of new mechanisms of herbicidal action. Crop Protect. 19: 583–589.

Duke, S.O., K.C. Vaughn, E.M. Croom Jr. and H.N. Elsohly. 1987. Artemisinin, a constituent of annual wormwood (*Artemisia annua*), is a selective phytotoxin. Weed Sci. 35: 499–505.

Einhellig, F.A., J.A. Rasmussen, A.M. Hejl and I.F. Souza. 1993. Effects of root exudate sorgoleone on photosynthesis. J. Chem. Ecol. 19: 369-375.

Evidente, A., R. Capasso, A. Cutignano, O. Taglialatela-Scafati, M. Vurro, M.C. Zonno and A. Motta. 1998. Ascaulitoxin, a phytotoxic bis-amino acid *N*-glucoside from *Ascochyta caulina*. Phytochemistry 48: 1131–1137.

Fonné-Pfister, R., P. Chemla, E. Ward, M. Girardet, K.E. Kreutz, R.B. Honzatko, H.J. Fromm, H.-P. Schär, M.G. Grüter and S.W. Cowan-Jacob. 1996. The mode of action and the structure of a herbicide in complex with its target: binding of activated hydantocidin to the feedback regulation site of adenylosuccinate synthetase. Proc. Natl. Acad. Sci. 93: 9431–9436.

Fennimore, S.A., D.C. Slaughter, M.C. Siemens, R.G. Leon and M.N. Saber. 2016. Technology for automation of weed control in specialty crops. Weed Technol. 30: 823–837.

Gerwick, B.C. and T.C. Sparks. 2014. Natural products for pest control: an analysis of their role, value and future. Pest Manag. Sci. 70: 1169–1185.

Giovanelli, J., L.D. Owens and S.H. Mudd. 1973. β-Cystathionase: *In vivo* inactivation by rhizobitoxine, and role of the enzyme in methionine biosynthsis in corn seedlings. Plant Physiol. 51: 492–503.

Głab, L., J. Sowinski, R. Bough and F.E. Dayan. 2017. Allelopathic potential of *Sorghum bicolor* (L.) Moench in weed control: a comprehensive review. Adv. Agron. 145: 43–95.

Glare, T., J. Caradus, W. Gelernter, T. Jackson, N. Keyhani, J. Köhl, P. Marrone, L. Morin and A. Stewart. 2012. Have biopesticides come of age? Trends Biotechnol. 30: 250–258.

Gonzalez, V.M., J. Kazimir, C. Nimbal, L.A. Weston and G.M. Cheniae. 1997. Inhibition of a photosystem II electron transfer reaction by the natural product sorgoleone. J. Agric. Food Chem. 45: 1415–1421.

Groth, G. 2002. Structure of spinach chloroplast F1-ATPase complexed with the phytopathogenic inhibitor tentoxin. Proc. Natl. Acad. Sci. USA 99: 3464–3468.

Harding, D.P. and M.N. Raizada. 2015. Controlling weeds with fungi, bacteria and viruses: a review. Front. Plant Sci. 6: 669. doi: 10. 3389/fpls. 2015. 00659.

Heap, I. 2014. Global perspective of herbicide-resistant weeds. Pest Manag. Sci. 70: 1306–1315.

Heim, D.R., C. Cseke, B.C. Gerwick, M.G. Murdoch and S.B. Green. 1995. Hydantocidin: a possible proherbicides inhibiting purine biosynthesis at the site of adenylosuccinate synthase. Pestic. Biochem. Physiol. 53: 138–145.

Heisey, R.M. and A.R. Putnam. 1986. Herbicidal effects of geldanamycin and nigericin, antibiotics from *Streptomyces hygroscopicus*. J. Nat. Prod. 49: 859–865.

Heisey, R.M. and A.R. Putnam. 1990. Herbicidal activity of the antibiotics geldanamycin and nigericin. J. Plant Growth Regul. 9: 19–25.

Hejl, A.M. and K.L. Koster. 2004. The allelochemical sorgoleone inhibits root H^+-ATPase and water uptake. J. Chem. Ecol. 30: 2181–2191.

Huang, H., C.M. Morgan, R.N. Asolkar, M.E. Koivunen and P.G. Marrone. 2010. Phytotoxicity of sarmentine isolated from long pepper (*Piper longum*) fruit. J. Agric. Food Chem. 58: 9994–10000.

Kato-Noguchi, H. 2004. Allelopathic substances in rice root exudates: rediscovery of momilactone B as an allelochemical. J. Plant Physiol. 161: 271–276.

Kato-Noguchi, H., M. Hasegawa, T. Ino, K. Ota and H. Kujime. 2010. Contribution of momilactone A and B to rice allelopathy. J. Plant Physiol. 167: 787–791.

Kato-Noguchi, H. and R.J. Peters. 2013. The role of momilactones in rice allelopathy. J. Chem. Ecol. 39: 175–185.

Kenfield, D., G. Bunkers, G.A. Srobel and F. Sugawara. 1988. Potential new herbicides – phytotoxins from plant pathogens. Weed Technol. 2: 519–524.

Lax, A.R., H.S. Shepherd and J.V. Edwards. 1988. Tentoxin, a chlorosis-inducing toxin from *Alternaria* as a potential herbicide. Weed Technol. 2: 540–544.

Lax, A.R., J.M. Bland and H.S. Shepherd. 1994. Biorational control of weeds with fungi and peptides. Amer. Chem. Soc. Symp. Ser. 551: 268–277.

Lederer, B., T. Fujimori, Y. Tsujino, K. Wakabayahi and P. Böger, 2004. Phytotoxic activity of middle-chain fatty acids II: peroxidation and membrane effects. Pestic. Biochem. Physiol. 80: 151–156.

Lee, D.L., M.P. Prisbylla, T.H. Cromartie, D.P. Dagarin, S.W. Howard, W.M. Provan, M.K. Ellis, T. Fraser and L.C. Mutter. 1997. The discovery and structural requirements of inhibitors of *p*-hydroxyphenylpyruvate dioxygenase. Weed Sci. 45: 601–609.

Liu, D.L. and N.E. Christians. 1994. Isolation and identification of root-inhibiting compounds from corn gluten hydrolysate. J Plant Growth Regul. 13: 227–230.

Loiseau, N., F. Cavelier, J.P. Noel and J.M. Gomes. 2002. High yield synthesis of tentoxin, a cyclic tetrapeptide. J. Peptide Sci. 8: 335–346.

Lydon, J. and S.O. Duke. 1999. Inhibitors of glutamine biosynthesis. pp. 445–464. *In:* B.K. Singh (Ed.) Plant Amino Acids: Biochemistry and Biotechnology. Marcel Dekker, New York City, USA.

Marrone, P.G. 2014. The market and potential for biopesticides. Amer. Chem. Soc. Symp. Ser. 1172: 245–258.

Meazza, G., B.E. Scheffler, M.R. Tellez, A.M. Rimando, J.G. Romagni, S.O. Duke, D. Nanayakkara, I.A. Khan, E.A. Abourashed and F.E. Dayan. 2002. The inhibitory activity of natural products on plant *p*-hydroxyphenylpyruvate dioxygenase. Phytochemistry 60: 281-288.

Meiss, E., H. Konno, G. Groth and T. Hisabori. 2008 Molecular processes of inhibition and stimulation of ATP synthase caused by the phytotoxin tentoxin. J. Biol. Chem. 283: 24594–24599.

Meyer, W.L., G.E. Templeton, C.I. Grable, R. Jones, L.F. Kuyper, R.B. Lewis, C.W. Sigel and S.H. Woodhead. 1975. Use of ¹H nuclear magnetic resonance spectroscopy for sequence and configuration analysis of cyclic tetrapeptides: the structure of tentoxin. J. Amer. Chem. Soc. 97: 3802–3809.

Nakajima, M., K. Itoi, Y. Takamatsu, S. Sato, Y. Furukawa, K. Furuya, T. Honma, J. Kadotani, M. Kozasa and T. Haneishi. 1989. Cornexistin: a new fungal metabolite with herbicidal activity. J. Antibiot. 44: 1065–1072.

Nakajima, M., K. Itoi, Y. Takamatsu, T. Kinoshita, T. Okazaki, K. Kawakubo, M. Shindo, T. Honma, M. Tohjigamori and T. Haneishi. 1991. Hydantocidin: a new compound with herbicidal activity from *Streptomyces hyrgoscopis*. J. Antiobiot. 44: 293–300.

Nanayakkara, N.P.D. and K.K. Schrader. 2008. Synthesis of water-soluble 9,10-anthraquinone analogues with potent cyanobacterial activity toward the musty-odor cyanobacterium *Oscillatoria perornata*. J. Agric. Food Chem. 56: 1002–1007.

Nelson, P.E., A.E. Desjardins and R.D. Plattner. 1993. Fumonisins, mycotoxins produced by *Fusarium* species: biology, chemistry, and significance. Annu. Rev. Phytopathol. 31: 233–252.

O'Sullivan, J., R. Van Acker, R. Grohs and R. Riddle. 2015. Improved herbicide efficacy for organically grown vegetables. Org. Agric. 5: 315–322.

Owens, L.D. 1972. Rhizobitoxine as a post-emergent herbicide. US Patent No. 3,672,862, issued June 27, 1972.

Owens, L.D. 1973. Herbicidal potential of rhizobitoxine. Weed Sci. 21: 63–66.

Poignant, P. 1954. Chemical structure and herbicidal activity of a group of organic acids. Compt. Rend. 239: 822–824.

Ramalakshmi, S. and K. Muthuchelian. 2013. Studies on cytotoxic, phytotoxic and volatile profile of the bark extract of the medicinal plant, *Mallotus tetracoccus* (Roxb.) Kurz. African J. Biotechnol. 12: 6176–6184.

Seiber, J.N., J.R. Coats, S.O. Duke and A.D. Gross. 2014. Biopesticides: state of the art and future opportunities. J. Agric. Food Chem. 62: 11613–11619.

Shrestha, A., M. Moretti and N. Mourad. 2012. Evaluation of thermal implements and organic herbicides for weed control in a nonbearing almond (*Prunus dulcis*) orchard. Weed Technol. 26: 110–116.

Sparks, T.C., D.R. Hahn and N.V. Garizi. 2017. Natural products, their derivatives, mimics and synthetic equivalents: role in agrochemical discovery. Pest Manag. Sci. 73: 700–715.

Steele, J.A., T.F. Uchytil, R.D. Durbin, P. Bhatnagar and D.H. Rich. 1976. Chloroplast coupling factor 1: a species-specific receptor for tentoxin. Proc. Natl. Acad. Sci. 73: 2245–2248.

Strobel, G., D. Kenfield, G. Bunkers, F. Sugawara and J. Clardy. 1991. Phytotoxins as potential herbicides. Experientia 47: 819–826.

Tachibana, K. 2003. Bialaphos, a natural herbicide. Meiji Seika Kenkyu Nenpo 42: 44–57.

Tanaka, T., H.K. Abbas and S.O. Duke. 1993. Structure-dependent phytotoxicity of fumonisins and related compounds in a duckweed bioassay. Phytochemistry 33: 779–785.

Travaini, M.L., G.M. Sosa, E.A. Ceccarelli, H. Walter, C.L. Cantrell, N.J. Carrillo, F.E. Dayan, K.M. Meepagala and S.O. Duke. 2016. Khellin and visnagin, furanochromenes from *Ammi visnaga* (L.) Lam., as potential bioherbicides. J. Agric. Food Chem. 64: 9475–9487.

Vaughan, M.A. and K.C. Vaughn. 1988. Mitotic disrupters from higher plants and their potential uses as herbicides. Weed Technol. 4: 533–539.

Wolf-Rüdiger, S., B. Fry, A. Kochevenko, D. Schindelasch, L. Zimmerli, S. Somerville, R. Loria and C.R. Somerville. 2003. An Arabidopsis mutant resistant to thaxtomin A, a cellulose synthesis inhibitor from *Streptomyces* species. The Plant Cell 15: 1781–1784.

Young, S.L. 2004. Natural product herbicides for control of annual vegetation along roadsides. Weed Technol. 18: 580–587.

Young, S.L., F.J. Pierce and P. Nowak. 2014. Introduction: scope of the problem – rising costs and demand for environmental safety in weed control. pp. 1–8. *In:* S.L. Young and F. Pierce (Eds.) Automation: The Future of Weed Control in Cropping Systems. Springer Science and Business Media, Dordrecht, Germany.

Allelopathy for Sustainable Weed Management

Yoshiharu Fujii* and Kwame Sarpong Appiah

Department of International Environmental and Agricultural Science, Tokyo University of Agriculture and Technology, 3-5-8, Saiwai-cho, Fuchu, Tokyo 183-8509, Japan

Introduction

Studies about the potential uses of allelopathy in agriculture have been explored by many scientists (Rice 1984, Weston and Duke 2003, Weston 2005). Cheema et al. (2000) reported the use of extracts of allelopathic plants instead of herbicides. Studies on plants that prevent other plants from growing around them are gaining momentum, and there is the increasing possibility of using bioactive compounds from these plants as selective bio-herbicides (Duke 2010). Allelochemicals released by this category of plant species are viewed as a potentially new source for safer herbicides (Dayan et al. 2009). Plants with allelopathic potentials are seen as sustainable alternatives that can be utilized directly or indirectly to control weed growth and reduce the heavy reliance on herbicides to mitigate the problems that may arise from the excessive use of herbicides (Appiah et al. 2015a).

Several experiments have been conducted to standardize the experimental methodology for allelopathic research (Fujii et al. 1990a, 1990b). Similarly, bioassay experiments related to allelopathy have been developed using plant extracts, plant root exudates and plant leachates (Fujii et al. 2007). The Plant Box method, a specific bioassay method to analyze the allelopathic activity of exudates released by roots into the environment using non-nutrient agar as the growing medium was developed (Fujii et al. 2007). The Sandwich method is specifically used to test the allelopathic activity of leaf litter leachates (Fujii et al. 1991a, 2003, 2004, Morikawa et al. 2012a). The Dish Pack method is used to test for the effects of allelochemicals released through volatilization using filter paper as the growing medium for the test species (Fujii et al. 2005). These specific methods are important to differentiate between allelopathic activities and resource competition among other plant species (Fujii et al. 2007). In the course of screening for potential allelopathic plants, we found that plants of medicinal value have strong allelopathic activity (Fujii et al. 1991c, Itani et al. 1998, Fujii et al. 2003, Appiah et al. 2017). Plants of medicinal value are easier to screen, possibly due to the abundance of bioactive compounds compared to those in the other plants (Fujii et al. 2003). By these bioassays and field observation experiments, it was found that the most practical application of allelopathy for weed management was the utilization of allelopathic plants as cover crops/mulch. The most successful application

---

* Corresponding author: yfujii@cc.tuat.ac.jp

of velvet bean (*Mucuna pruriens* var. *utilis*) and hairy vetch (*Vicia villosa*) as allelopathic cover crops/mulch were explained (Fujii 2003).

Specific Bioassay Methods for Evaluating Allelopathy

The observation of germination and growth of test species in a petridish used to be the commonest method for the study of plant allelopathy. This approach has been used to evaluate the effect of allelopathic plants on other plant species (Albuquerque et al. 2011). There exist several bioassays for testing plant allelopathy. However, the elucidation of allelochemicals and their involvement in the phenomenon of allelopathy are challenging due to the lack of route-specific bioassay methodologies. From this viewpoint, we developed and standardised several bioassays and assessment methods to explore route-specific identification of allelochemicals under laboratory conditions. Consequently, we developed three route-specific bioassays named the Plant Box method, Sandwich method, and Dish Pack method. In this chapter, the laboratory-based assessment methods for plant allelopathy are explained.

Plant Box Method

The plant box method was developed based on the dose-response principle where the distance of the donor plant to the bioassay species (receiver plant) is used (Fujii et al. 1991b). It is related to the inhibitory phenomenon that occurs in the receiver plant due to the concentration of the root exudates in the growth medium. In this method, agar is used as the growth medium, and this allows the allelochemicals to move from the roots of the donor plant to that of the test species. The donor plant strongly inhibited the test species that were sown closely to the roots of the donor plant (Fujii et al. 2007). This method was used by Appiah et al. (2015a) to study the allelopathic activity of selected *Mucuna pruriens* genotypes to check the strongly inhibiting mucuna genotype. Also, this method had been used to identify the allelopathic activity of 19 medicinal plants from Pakistan (Syed et al. 2014). The root exudates of *Sarcococca saligna* exhibited the highest inhibition compared to the other 18 species by causing a reduction of 78% in the radicle growth of *Lactuca sativa* (Syed et al. 2014). This method was repeated about 12,000 times, and the allelopathic activities of about 2,000 plant species have been evaluated in the last 26 years. Some of the results were published (Fujii et al. 1991a, 1992a, 2007, Appiah et al. 2015a), but most of them are not yet published. Table 9.1 summarizes the selected species that showed strong allelopathic potential through root exudates and could be useful for weed management.

Sandwich Method

The sandwich method is used to evaluate the potential allelopathic effect of the leaf litter leachate of plants under laboratory conditions (Fujii et al. 2003). The Standard Deviation Variance (SDV) concept was adopted to categorize the allelopathic activity (Fujii et al. 2003, 2004). Under this concept, the mean and standard deviation were calculated, and the criterion of the SDV was evaluated to indicate the strongly inhibiting species. This approach is efficient, reliable, less time consuming, and effective in the screening for allelopathic plants on a large scale (Morikawa et al. 2012b, Fujii et al. 2003). About 10 mg or 50 mg of dried leaves (at 60°C for several hours) will be placed at the bottom of a six-well multi-dish. Around 5 mL of an autoclaved agar cooled to a temperature of 35°C–40°C is then pipetted into the six-well multi-dish. The same quantity of another layer of agar is added soon after the initial agar had gelatinized. This will create two layers of agar with the plant sample between them and hence the name Sandwich method.

Five lettuce seeds are vertically placed on the second layer of the solidified agar. The multi-well dishes are then labelled and covered with plastic tape and incubated at 22°C for 3 days. The hypocotyl and radicle lengths of lettuce were measured after 3-day incubation. Treatments are replicated three times and data presented as the mean of the three replicates. Agar with no plant material is used as the untreated control. Seeds of the test plant species are sown on the surface of the second layer of agar. This Sandwich bioassay had been used (Fujii et al. 2003, Morikawa et al. 2012a, Itani et al. 2013, Appiah et al. 2015, Mardani et al. 2015, Mishyna et al. 2015a, Ismail et al. 2016) to screen large quantities of plants. This method is useful in determining allelopathic activities by leachates of plant litter under laboratory conditions. Appiah et al. (2015b) used this

Table 9.1. Assessment of allelopathic activity by Plant Box method

Scientific name (English name)	Inhibition (%)[*1]
Mucuna pruriens var. *utilis* (Velvetbean, cv. Hassjo)	96
Abutilon theophrasti (Chinese Jute)	91
Symphytum peragrinum (Russian Comfrey)	89
Mucuna pruriens var. *utilis* (Velvetbean, cv. Florida)	88
Imperata cylindrica var. *koenigii* (Cogon grass)	88
Triticum polonicum (Polish Wheat)	87
Panicum miliaceum (Millet)	86
Ruta graveolens (Common Rue)	82
Vicia faba (Broad Bean)	81
Vicia villosa var. *dasycarpa* (Woolly Pod Vetch)	81
Vicia villosa var. *villosa* (Hairy vetch)	80
Linum usitatissimum (Flax)	80
Symphytum officinale (Comfrey)	79
Melilotus albus (White Sweet Clover)	77
Secalotricum ryedax (Triticale)	77
Portulaca oleracea (Purslane)	76
Vicia sativa (Common Vetch)	75
Canavalia ensiformis (Jack Bean)	72
Anthoxanthum odoratum (Sweet Vernalgrass)	72
Secale cereale (Rye)	71
Brassica oleracea var. *italica* (Broccoli)	71
Avena sativa (Oat)	70
Lavandula angustifolia (Lavender)	68
Vigna unguiculata (Cowpea)	67
Medicago sativa (Alfalfa, cv. dupy)	66
Eleusine coracana (Finger Millet)	66
Triticum aestivum (Wheat, cv. Nourin 61)	66
Trifolium incarnatum (Crimson Clover)	64
Hordeum vulgare (Barley)	62
Capsicum annuum (Red Pepper)	61
Crotaralia spectabilis (Sunn Hemp)	60
Panicum maximum (Guinea Grass)	60
Dolicos lablab (Lablab Bean)	59
Latyrus sativus (Grass Pea)	59
Brassica napus (Rape)	59
Pisum sativum (Pea)	58
Cicer arietinum (Chickpea)	56
Festuca arundinacea (Tall Fescue)	55
Sorghum sudanense (Sudan Grass)	55
Lycopercicum esculentum (Tomato)	54
Sorghum dochna (Sorghum)	53
Calopogonium mucunoides (Calopogonio)	51
Vigna unguiculata subsp. *sesquipedalis* (Asparagus Pea)	51
Brassica campestris (Turnip rape)	50
Arachis hypogaea (Peanut)	50

(Contd.)

Solanum melongena var. *esculentum* (Egg Plant)	49
Trifolium repens (White Clover)	47
Raphanus sativus (Radish)	47
Vicia angustifolia var. *segetalis* (*Karasu-no-endou*)	45
Zoysia japonica (Japanese Lawn Grass)	42
Cucumis sativus (Cucumber)	42
Mentha piperita (Peppermint)	42
Momordica charantia (Balsam Pear)	42
Astragalus sinicus (Chinese Milk Vetch)	41
Lupinus albus (White Lupine)	40
Poa pratensis (Kentucky Bluegrass)	40
Perilla frutescens (Perilla, Egoma)	40
Sesbania cannabina (Sesbania, cv. Densuke)	39
Cucumis melo (Melon)	39
Cucurbita pepo (Pumpkin)	38
Lolium perenne (Perennial Ryegrass)	36
Glycine max (Soybean, cv. Tachi-nagaha)	35
Allium cepa (Onion)	35
Zea mays (Corn, cv. Pioneer Dent)	34
Tagetes patula (French Marygold)	32
Trifolium pratense (Red Clover)	28
Mentha arvensis (Japanese Mint)	28
Lolium multiflorum (Italian Ryegrass)	26
Phalaris arundinacea (Reed Canary grass)	24
Helianthus annuus (Sunflower)	22
Allium fistulosum (Welsh Onion)	17
Mentha pulegium (Pennyroyal mint)	17
Vigna angularis (Adzuki Bean)	16
Spinacia oleracea (Spinach)	7
Phleum pratense (Timothy)	4

[*1] Inhibition (%) means the radicle growth inhibition percentage at the surface of the root zone separation tube compared to that of control.

Sandwich method to screen 251 plant species in the Sino-Japanese floristic region to evaluate their allelopathic activities. Currently, the allelopathic activities of about 4,000 plant species have been assessed by using this method. A portion of this data has been published, but the bulk of the data is yet to be reported. Table 9.2 summarizes the selected species that could be useful for weed management.

Dish Pack Method

This approach was first adopted by Fujii et al. (2005) to test for the presence of volatile allelochemicals from plant species. This technique is widely used (Amini et al. 2014, Appiah et al. 2015, Mardani et al. 2015) due to the rapid and efficient way of determining the presence of volatile allelochemicals in plants. Multi-well plastic dishes with six wells (36 mm×18 mm) are used with this method. The distances from the point where plant samples are placed (source well) to the center of other wells are 41, 58, 82, and 92 mm. Around 200 mg of oven-dried plant material is used to fill the source well, while filter papers moistened with 0.7 mL of distilled water are laid in the other wells. The control treatment does not contain any plant sample at the source well.

Table 9.2. Assessment of allelopathic activity by Sandwich method

Scientific name	Inhibition (%)[*1]	Scientific name	Inhibition (%)[*1]
Melilotus officinalis	95	*Portulaca oleracea*	58
Lycoris radiata	94	*Linum usitatissimum*	58
Fagopyrum esculentum	90	*Glycine max subsp. Soja*	57
Mucuna pruriens	89	*Ipomoea nil*	55
Vicia villosa	88	*Setaria italica*	55
Melilotus albus	87	*Vulpia myuros*	53
Vicia hirsute	85	*Oxalis articulata*	51
Oxalis tuberosa	82	*Trifolium incarnatum*	51
Canavalia ensiformis	77	*Houttuynia cordata*	50
Brachiaria decumbens	76	*Phalaris canariensis*	50
Brassica juncea	76	*Echinochloa crus-galli*	50
Vicia tetrasperma	75	*Panicum miliaceum*	47
Phalaris minor	74	*Clitoria ternatea*	46
Eragrostis tef	74	*Chloris gayana*	46
Festuca arundinacea	74	*Mentha pulegium*	45
Lathyrus latifolius	74	*Humulus lupulus*	44
Ipomoea aquatica	73	*Coronilla varia*	44
Tropaeolum tuberosum	72	*Phleum pratense*	43
Perilla frutescens	72	*Coix lacryma-jobi*	43
Dactylis glomerata	70	*Saccharum officinarum*	43
Lolium perenne	70	*Piper nigrum*	43
Cymbopogon citratus	70	*Ipomoea batatas*	42
Chloris gayana	69	*Trifolium pratense*	42
Trifolium dubium	69	*Petasites japonicus*	42
Vigna mungo	68	*Artemisia princeps*	41
Poa annua	68	*Trifolium repens*	39
Lotus corniculatus	68	*Urtica dioica*	37
Pennisetum americanum	68	*Thymus quinquecostatus*	36
Sorghum bicolor	66	*Astragalus sinicus*	34
Brassica juncea	66	*Panicum dichotomiflorum*	34
Capsicum annuum	65	*Vicia angustifolia*	32
Abutilon theophrasti	65	*Pachyrhizus erosus*	31
Gossypium arboreum	64	*Sorghum halepense*	27
Ullucus tuberosus	64	*Mentha arvensis*	27
Eleusine coracana	64	*Zea mays*	26
Berberis sieboldii	63	*Coix lacryma-jobi*	22
Sorghum sudanense	63	*Thymus vulgaris*	21
Vulpia bromoides	62	*Melissa officinalis*	21
Medicago sativa	62	*Commelina communis*	20
Sesamum indicum	62	*Ceratonia siliqua*	18
Cassia obtusifolia	62	*Mazus reptans*	17
Nicotiana tabacum	61	*Arachis hypogaea*	17
Avena fatua	60	*Polymnia sonchifolia*	9
Elymus repens	59	*Manihot esculenta*	8

[*1] Inhibition (%) means the radicle growth inhibition % compared to control.

Seven lettuce seeds (*Lactuca sativa* var. Great Lakes 366) are placed on the filter paper in each well. The multi-well dishes should be tightly sealed to avoid desiccation and the apparent loss of volatile compounds. The plates are then wrapped with aluminium foils and placed in an incubator at 22°C for 3 days. The radicle and hypocotyl lengths are measured after 3 days of incubation and compared to that of the control to calculate the degree of inhibition. The allelopathic activities of about 600 plant species have been evaluated by using this method. Through this technique, *Heracleum sosnowskyi* and *Crocus sativus* were identified as allelopathic species leading to the identification of octanal and safranal as volatile allelochemicals from these species, respectively (Mishyna et al. 2015b, Mardani et al. 2015). The bulk of this data on volatile allelopathy of plants is yet to be published. Table 9.3 summarises the results of selected species and volatile chemicals identified.

Sustainable Weed Control Using Allelopathic Approach

Application of Plant Residue with Allelopathic Activity

The residues of allelopathic plants can be used in weed management by incorporating the plant debris in different quantities into the soil. The effects of plant residues or chemicals released from decomposition on the target weed species were studied in the greenhouse or the field (Albuquerque et al. 2011). At the incorporation rate of 40 g/1,000 g soil, the residue of *Parthenium hysterophorus* decreased the size and dry weight of *Brassica oleracea*, *B. campestris*, and *B. rapa*. This inhibition was primarily due to the water-soluble phenolic compounds released by *P. hysterophorus* (Singh et al. 2005). Similarly, soil incorporated with *Chenopodium murale* inhibited the germination, nodulation, and macromolecule content of *Cicer arietinum* and *Pisum sativum* (Batish et al. 2007). This approach is useful from the standpoint of using natural resources to minimise the excessive application of herbicides. However, the process is labour-intensive and usually not readily adopted by most of the farmers.

Application of Allelopathic Plants as Mulching Materials

Plant residues, ground cover crops, and plant mulch are being used in weed control management activities as they can be obtained in large quantities from the field. Mulching can physically reduce the amount of sunlight, temperature, and moisture, which are very important for weed-seed germination (Davies et al. 2008). The effectiveness of using mulch as weed control strategy increases when the mulching materials have strong allelopathic activities. This was observed when *Ageratum conyzoides* was used as mulch at the rate of 2 t per ha in the paddy field. *A. conyzoides* inhibited approximately 86% of the weed population and 75% of the dry weight of weeds compared to that of the control. The inhibited weeds included *Graticola japonica*, *Lindernia pyxidaria*, *Echinochloa oryzicola*, *Eleocharis acicularis*, *Monochoria vaginalis,* and *Rotala indica* (Xuan et al. 2004). Although this approach is better than the application of plant residue, it is equally labor-intensive, and the effect is not consistent year by year.

Application of Cover Crops with Allelopathic Properties

Cover crops are plant species that are usually not the main crop but are introduced into the cropping system especially when the soil is left open and not cultivated. Cover crops are also used in uncultivated areas to prevent the erosion of soil and to conserve the moisture and nutrient content in the soil (Gallandt et al. 1999). Ground cover crops are mostly important in rotational cropping systems because they are fast-growing species that form a dense cover on the soil surface to prevent the germination and growth of weed species (Singh et al. 2001).

The use of allelopathic cover crops that can inhibit the growth of other plants has been suggested by Fujii (2003) as one of the effective ways of integrating the concept of allelopathy into weed control. Fujii (2001) tested 53 species of ground cover crops using the Plant Box method. The study indicated that certain cover crops have the potential to be used in weed control as cover crops. These crops included *Avena sativa*, *Hordeum vulgare*, *Secale cereal*, *Mucuna pruriens*, and *Vicia villosa*. *Vicia villosa* (hairy vetch) was able to control weeds in the paddy fields

Table 9.3. Assessment of allelopathic activity by Dish Pack method

Scientific name	Radicle[*1]	Hypocotyl[*1]	Major volatile chemicals identified
Cleome spinosa	100	100	Methyl Isothiocyanate
Papaver rhoeas	84	80	2-Hexenal
Hibiscus cannabinus	69	54	2-Hexenal, 3-Hexenal
Solidago altissima	68	61	α-Pinene, Limonene, Myrcene, Ocimene
Vicia villosa	66	58	2-Hexenal
Rosmarinus officinalis	64	73	α-Pinene, Camphor, 1,8-Cineole
Crotalaria agatiflora	61	49	2-Hexenal, *trans*-3-Hexenol
Artemisia princeps	59	60	β-Pinene, 1,8-Cineol, 2-Octenal
Vinca major	55	50	*cis*-3-Hexenyl acetate, *trans*-3-Hexenol
Ipomoea aquatica	46	42	2-Hexenal, 4-Pentenal
Mucuna pruriens	40	45	Hexenal, 2-Hexenal, 3-Hexenal
Fagopyrum esculentum	34	49	2-Hexenal, 3-Hexenal
Arctotheca calendula	34	10	β-Pinene, 2-Hexenal
Phlox subulata	32	34	Limonene
Phacelia tanacetifolia	29	20	Myrcene, Limonene, 2-Hexenal
Thymus serphyllum	26	29	Terpinen, Cymene, Isocaryophyllene
Oxalis articulata	25	22	3-Hexen-1-ol,acetate
Chamomilla nobilis	21	25	Ocimene, Cyclopropanecarboxylic acid
Festuca myuros	18	44	*cis*-3-Hexenyl acetate
Coreopsis tinctoria	17	21	Limonene, α-Phellandrene, α-Pinene
Mentha pulegium	13	8	Pulegone, Myrcene, Limonene
Cymbopogon citratus	8	25	Myrcene, Citral,
Houttuynia cordata	7	25	Myrcene, β-Pinene, Ocimene, Limonene
Lycoris radidata	6	13	2-Hexenal
Ocimum basilicum	3	13	Linalol, 1,8-Cineole

[*1] Inhibition (%) means the growth inhibition percentage compared to control.

by forming a thick cover on the paddy soil surface during the summer season when the fields were dry (Fujii 2001). This method is a traditional way of farming, and not labour consuming compared to the previous two methods. Velvet bean and hairy vetch are especially practical as cover crops.

Practical Application of Velvet Bean (*Mucuna pruriens* var. *utilis*)

Velvet bean (*Mucuna pruriens* (L.) DC. var. *utilis* or *Stizolobium deeringianum* Piper et Tracy) is a tropical legume, generally grown as green manure. It is recognized that velvet bean increases the yield of its companion graminaceous crops and also smothers the growth of other harmful weeds, such as nutsedge (*Cyperus* spp.) and alang-alang (*Imperata cylindrica*) (Taib et al. 1979, Lorenzi 1984). A series of experiments were performed to screen allelopathic plants with special emphasis on chemical interactions among them. The results indicated that velvet bean was the most promising candidate (Fujii et al. 1990a, 1991d). A field test also showed that velvet bean stands minimized weed population as compared to those of tomato, eggplant, upland rice, and fallow (Fujii 1991f, 1991g).

The genus *Mucuna* consists of about 100 species growing in the tropics and subtropics (Tateishi and Ohashi 1981, Wilmot-Dear 1983). There are two subgenera in *Mucuna*: one is *Mucuna* that is perennial and woody, and the other is *Stizolobium*, which is annual or biennial and herbaceous. The whole plant is utilized as green manure and cover crop, the leaves for fodder, the grains for food and seeds, and the stems for medicine in Africa and China (Watt and

Breyer-Brandwijk 1962). Grain yields reach as high as 1.5–2.0 t/ha, and fresh leaves and stems weigh 20–30 t/ha, indicating that velvet bean is one of the most productive crops in the world.

Survey of Allelopathic Plants

Seventy plant species were tested for their allelopathic activities following Richards' function, generalized logistic curve (Fujii et al. 1990a). To destroy the enzymes that degrade some chemical constituents of a plant, and to minimize the changes of the organic chemical constituents, the leaves, stems, and roots were dried at 60°C for 24 hours. Around 100 mg of the dried samples each was extracted with 10 mL water. The extraction mixtures were sonicated for 60 s to complete the extraction of chemicals. The extract was filtered with Whatman No. 4 filter papers. Ten lettuce seeds were placed in 4.5 cm diameter petri dishes containing 0.5 mL of test solution on Whatman No. 1 filter papers. The petri dishes were then incubated in the dark at 25°C.

Numbers of the germinated seeds were counted, and hypocotyl and radicle growth were measured on the fourth day. The parameters for germination tests were: onset of germination (Ts), germination rate (R), and final germination percentage (A). A simplex method was applied for the computer simulation of germination curves with the Richards' function. It was observed that the activity of velvet bean was distinctive (Table 9.4). Some other plants, such as *Artemisia princeps, Houttuynia cordata, Vicia angustifolia*, and *Colocasia esculenta* also showed inhibitory response.

Through this screening by the Plant Box method (Fujii and Shibuya 1991b), it was found that velvet bean showed the strongest inhibitory activity. Evaluation by the Sandwich method also showed strong inhibitory activity by velvet bean (Fujii and Shibuya 1990a).

Weed Appearance in the Fields with Velvet Bean Stands

Planting of velvet bean and some other plants were repeated for a period of two to three years (Fujii 1991d). Plants were grown in lysimeters (each size is 10 m²) with six replications, where the surface soils of 10 cm depth were replaced with uncultivated soils in the starting year. Each plot received a standard level of chemical fertilizers: N, P, K of 80, 80, 80 g/10 m² except for the fallow plot. Table 9.5 shows weed populations in spring in the continuous cropping fields grown in lysimeters. Weed populations in the spring in the continuous cropping fields grown in lysimeters revealed that the velvet bean plot showed a lower population of weeds, dominated by sticky chickweed (*Cerastium glomeratum*) than the other plots of eggplant, tomato plant, upland rice, and fallow did.

Mixed Culture of Velvet Bean by Allelopathy Discrimination Methods

Allelopathy of velvet bean in the field was confirmed using the Stairstep method (Fujii et al. 1991f) and the substitutive experiment (Fujii et al. 1991d, 1991g). The Stairstep experiment was designed according to the method of Bell and Koeppe (1972) with three replications within two mixed plants. Circulation of nutrients solution was about 600 to 800 ml/hr per pot. The half strength of Hoagland's solution was circulated. The substitutive experiment in this study was modified based on Fujii et al. (1991d, 1991g). The Stairstep method is a kind of sand culture with a nutrient solution re-circulating through a staircase bed. Through the use of this method, the presence of velvet bean reduced the growth of lettuce shoot to 70% of the control. This result indicates that velvet bean root exudates have allelopathic substances (Table 9.6).

Isolation and Identification of Allelochemicals

Some fractions were extracted from fully expanded leaves and roots of velvet bean with 80% ethanol. The acid fraction of the extract inhibited the growth of lettuce seedlings. This fraction was subjected to silica gel column chromatography and HPLC with an ODS column, and the major inhibitor was identical to L-3,4-dihydroxyphenylalanine (L-DOPA) (Figure 9.1) (Fujii et al. 1991e, Fujii 1994). The identification was confirmed by co-chromatography with an authentic sample using two HPLC column systems (silica gel and ODS) equipped with an electro-conductivity detector.

Table 9.4. Screening of allelopathic plants with lettuce germination/growth test

Plant (part[1])	Germination test					Growth test		Extraction ratio[8]
	A[2]	R[3]	Ts[4]	I[5]	T$_{50}$[6]	Hypocotyl[7]	Radicle[7]	
(Compositae)								
Ambrosia elatior (S)	94	74	2.1	34	1.6	139	54	10
Artemisia princeps (S) $$$	65	20	2.9	5	3.3	51	50	20
Carthamus tinctorius (W)	100	173	0.9	206	0.7	141	65	8
Helianthus annuus (S) $	86	38	1.2	27	1.5	102	33	10
Helianthus tuberosus (S)	91	96	1.4	62	1.3	104	67	25
Ixeris debilis (W)	85	96	1.3	71	1.6	114	63	10
Saussurea carthamoides (S)	97	74	2.1	34	1.7	139	63	10
(Poaceae)								
Avena sativa (L)	98	117	1.4	88	1.0	105	105	2.5
Hordeum vulgare (L)	100	102	0.9	114	1.0	144	65	6.3
Oryza sativa (L)	100	226	2.2	105	1.0	114	77	12.5
Sasa sinensis (S)	94	55	3.2	17	2.7	134	44	25
Secale cereale (L) $$	91	62	1.2	48	1.3	79	21	10
Sorghum sudanense (S) $	86	66	1.3	47	1.3	107	31	10
(Fabaceae)								
Arachis hypogaea (L) $	83	90	4.9	16	1.8	98	60	10
Glycine max (S)	96	44	0.6	70	1.4	117	41	10
Lupinus albus (S) $	95	98	2.8	33	1.6	100	37	12.5
Mucuna pruriens (L) $$$	96	82	9.3	9	4.6	79	**26**	25
Pisum sativum (S)	99	45	0.5	99	1.1	115	38	10
Vicia angustifolia (S) $	97	60	3.6	16	2.8	126	**22**	6.7
Vicia hirsuta (S) $	100	62	3.6	18	2.8	114	24	6.7

(Contd.)

(Chenopodiaceae)								
Beta vulgaris (S)	96	86	1.5	56	1.2	109	64	5
Chenopodium album (L)	98	43	1.0	44	1.9	90	48	10
Spinacia oleracea (L)	94	68	2.4	28	1.7	119	38	5
(Pologonaceae)								
Fagopyrum esculentum (S)	100	235	2.4	100	1.0	107	60	12.5
Polygonum blumei (S) $$	84	4	1.3	31	1.5	86	37	25
(Labiatae)								
Lamium amplexicaule (W) $	85	54	2.4	19	2.0	70	45	10
Melissa officinalis (L) $$	39	23	3.7	3	2.3	101	57	8
Mentha spicata (L)$	99	51	1.9	27	1.9	121	28	8
Salvia officinalis (L)	94	106	3.3	31	1.3	112	67	10
(Solanaceae)								
Lycopersicon esculentum (S)	96	136	5.9	23	1.9	135	37	10
Solanum melongena (S)	86	83	4.9	15	1.9	125	51	10
Solanum tuberosum (L)	99	75	1.3	127	1.3	127	62	6
(Cucurbitaceae)								
Citrullus lanatus (L)	95	102	3.7	26	1.3	133	69	6
Citrullus lanatus (Stem)	96	116	3.0	36	1.7	129	59	6
Cucumis sativus (S)	99	123	3.1	41	1.3	187	78	5
Cucurbita maxina (S)	93	153	4.8	30	1.8	119	50	12.5
(Other genus)								
Anaranthus tricolor (L)	92	66	4.0	15	2.4	93	81	6
Brassica campestris (L)	93	27	0.5	58	1.6	141	94	3
Brassica oleracea (L)$	76	97	5.6	14	1.4	146	88	5
Brassica juncea (S)	87	61	1.6	34	1.5	154	71	3

(Contd.)

Table 9.4. (*Contd.*)

Plant (part[*1])	Germination test					Growth test		Extraction ratio[*8]
	A[*2]	R[*3]	Ts[*4]	I[*5]	T$_{50}$[*6]	Hypocotyl[*7]	Radicle[*7]	
Brassica napus (S)	84	85	1.3	56	1.2	108	98	10
Calystegia hederacea (S)	96	66	2.4	27	1.9	94	60	10
Cerastium glomeratum (W) $	90	74	2.1	31	1.7	103	29	10
Garium spurium (W) $	92	65	2.1	29	1.8	85	58	10
***Houttuynia cordata* (S) $$$**	**98**	**33**	**3.6**	**9**	**3.4**	**62**	**26**	**5**
Impatiens balsamina (L)	93	101	3.3	28	1.9	117	64	6
Oenothera biennis (S)	84	48	1.3	31	1.5	105	39	25
Paederia scandens (L)	97	46	1.5	86	1.2	123	92	12.5
Paulovinia tomentosa (L)	100	53	1.2	45	1.5	119	61	12.5
Plantago major (L)	88	101	3.5	26	1.6	121	73	5
Portulaca oleracea (W)	90	117	4.8	22	1.9	119	49	3
Stellaria media (W)	97	69	1.4	51	1.4	99	67	5

[*1] Abbreviations of plant parts are as follows: S: Shoot, W: Whole plant (= S+R), L: Leaf

[*2] Germination percentage at the end of germination process speculated with cumulative germination curves fitted to Richards' function (% of control).

[*3] Germination Rate (% of germinated seeds per day, % of control)

[*4] Start of germination (a time spent until one seed germinate, ratio to control)

[*5] Germination Index (I = A × R/Ts)

[*6] 50 % germination time (a time spent until 50% of seed which can germinate, ratio to control)

[*7] % of control (control dish is cultured with water)

[*8] Extraction ratio [mg–D.W./ml]. Extraction ratio was determined in order that EC of the assay solution did not exceed 1 mS/cm.

[*9] Plant name with underline denotes strong inhibition in either of next parameters: hypocotyl elongation, radicle elongation, A (germination %), I (germination index).

$ marks after plant name shows the degree of inhibition. When each value exceed the criteria of average±σ, we judge the possibility of inhibition. The number of $ is the number of inhibition in four criteria of [*9].

Table 9.5. Weed population in continuous cropping fields

Crop	Treatment	Weed population (g dry weight per m²)	Weed species observed [6]
Upland Rice	3yr.c [1]	5.11 (49.4) [4]	1),3),5),6),7),8),9),10),11) [5]
Eggplant	3yr.C	16.82 (40.1)	1),2),3),5),6),7),8),9),10),11),12),13),14)
Tomato	3yr.c	4.92 (64.9)	1),5),6),9),12),13),17)
Velvet bean	2yr.c	0.00 (0.0)	No emergence
Velvet bean	1yr.c,1yr.f [2]	3.05 (74.8)	1),10),12),13),16),18)
Fallow	3yr.f [3]	0.97 (37.3)	1),2),6),10),12),13),15),16)

[1] Continuous cropping for 3 years.
[2] Cultivated for 1 year, followed by fallow next year (test year).
[3] Fallow for 3 years, without fertilizer.
[4] Numbers in parenthesis are percentages of chickweed, a dominant species.
[5] Species appeared in each plot: 1) Sticky chickweed (*Cerastium glomeratum*), 2) 'Miminagusa' (*Cerastium vulgatum* var. *augustifolium*), 3) Annual fleabane (*Erigeron annuus*), 4) Philadelphia fleabane (*Erigeron philadelphicus*), 5) Starwort (*Stellaria alsine* var. *undulata*), 6) Floating foxtail (*Alopecurus geniculatus*), 7) Narrowleaf vetch (*Vicia angustifolia*), 8) Flexuosa bittercress (*Cardamine flexuosa*), 9) 'Inugarashi' (*Rorippa atrovirens*), 10) Common dandelion (*Taraxacum officinale*), 11) Japanese mugwort (*Artemisia princeps*), 12) Danadian fleabane (*Erigeron canadensis*), 13) 'Hahakogusa' (*Gnaphalium affine*), 14) Blady grass (*Imperata cylindrica*), 15) Meadowgrass (*Poa annua*), 16) Creeping wood-sorrel (*Oxalis corniculata*), 17) Shepherd's-purse (*Capsella bursa-pastoris*), 18) Prickly sowthistle (*Sonchus asper*).
[6] Surveyed on 14 April 1988. Source: Fujii et al. (1991)11.

Table 9.6. Effect of mixed culture of velvet bean to the growth of lettuce and kidney bean by Stairstep method

Receiver plant	Donor plant	Leaf area (cm²)	Shoot dry weight (g)	Root dry weight (g)
Lettuce	Lettuce	30.4[b] (89)[1]	53.9[b] (96)	12[b] (101)
	Velvet bean	21.5[c] (63)	39.3[c] (70)	5.7[c] (48)
	None	34.2[a] (100)	56.3[a] (100)	11.9[a] (100)
Kidney bean	Kidney bean	87.9[a] (97)	343[a] (96)	148[b] (79)
	Velvet bean	81.4[a] (90)	344[a] (96)	153[b] (81)
	None	90.3[a] (100)	358[a] (100)	188[a] (100)

[1] Numbers in the parentheses are percentage of control. Means followed by the same letter within the same column are not significantly different at 1% level (Duncan's multiple range test).
Source: Fujii et al. (1991f)

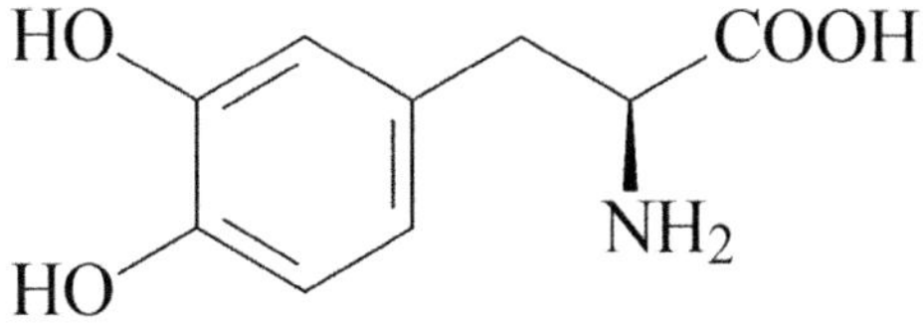

Figure 9.1. Chemical structure of L-DOPA (L-3,4-dihydroxyphenylalanine).

The active compound of velvet bean in restraining the growth of companion plants was confirmed to be L-DOPA. Velvet bean seeds are known to contain a high concentration of L-DOPA (6–9%) (Damodaran 1937, Rehr et al. 1973), which plays an essential role as a chemical barrier against insect attacks (Bell and Janzen 1971). In the mammalian brain, L-DOPA is the precursor of dopamine, a neurotransmitter, and also essential intermediates of alkaloids in plants. In animal hair, skin, feathers, fur and insect cuticle, L-DOPA is oxidized to dopaquinone and finally converted to melanin. Since L-DOPA is an intermediate species in such a biochemical pathway and is rapidly metabolized, normal tissues keep the low content of L-DOPA.

Fresh velvet bean leaves contain as much as 1% L-DOPA of their weight. It exudes from the roots, and its concentration reaches about 1 ppm in water-culture solution, and 50 ppm in the vicinity of roots. This concentration of L-DOPA is high enough to reduce the growth of surrounding species, and the inhibition of growth in a mixed culture is shown in agar-medium culture (Fujii and Shibuya 1991a, 1991b). L-DOPA also leaches out from leaves with raindrops or fog dew. Since velvet bean produces about 20–30 tons of fresh leaves and stems per hectare, approximately 200–300 kg of L-DOPA per hectare may be released into soils in a year.

Phytotoxic Effects and Mechanism of Action of L-DOPA

L-DOPA suppressed the radicle growth of chickweed and lettuce to 50% of the control at 50 ppm (0.2 mM) (Fujii et al. 1991e, Fujii 1994). It was, however, less effective on the hypocotyl growth and practically ineffective on germination. L-DOPA strongly inhibited the growth of *Cerastium glomeratum, Spergula arvensis* (both Caryophyllaceae), *Linum usitatissimum*, and *Lactuca sativa*, and moderately inhibited the growth of some Compositae, with very limited effect on Poaceae and Fabaceae. Such selective action is similar to other allelochemical candidates (Chou and Kuo 1986, Elacovitch and Stevens 1985).

The L-DOPA contained in fresh velvet bean leaves is entirely responsible for the plant growth inhibition through its crude extract. The result that L-DOPA actively suppressed the growth of chickweed agrees with weed inhibition exhibited by the velvet bean under field conditions. All these data suggest that L-DOPA functions as an allelopathic substance. In the case of older leaves, the content of dopamine increases, and L-DOPA and dopamine are presumably changed to catechol in the litter as in the case of L-mimosine (Figure 9.2).

The inhibitory activity of catechol on lettuce radicle growth is almost the same as L-DOPA, but catechol is more toxic to hypocotyl growth and germination of lettuce (Figure 9.2). Table 9.7 shows the inhibitory activities of L-DOPA, dopamine, and catechol on some test plants. In all the plants tested, dopamine showed no practical inhibition on radicle growth, but catechol showed stronger inhibition of other weeds than L-DOPA.

It is an earlier thought that velvet bean smothers weeds under the rapid and thick covering effect of the leaves (Bunch and Staff 1985, Soule 1997). However, the results above suggest that L-DOPA or its associated compounds accumulate to extremely high concentrations in some plants and function as allelochemicals in reducing weed population. The role of L-DOPA in velvet bean seeds was earlier regarded as a chemical barrier to insect attacks. However, it is now confirmed that it also plays another important role in its allelopathic activity in weed control.

Recommendation of Velvet Bean for Weed Control

Velvet bean has special important agronomic abilities, such as weed smothering (Fujii et al. 1991d), tolerance to pests (Bell and Janzen 1971, Hulugalle et al. 1986), suppression of nematode populations (Reddy et al. 1986, Tenente et al. 1980, Tenente et al. 1982), and soil improvement in its physical structure (Hulugalle et al. 1986). This plant could be widely used to reduce applications of synthetic herbicides to a sustainable level. Yields of velvet bean seeds are very high in the tropics. If the detrimental factors, such as L-DOPA and trypsin inhibitors could be eliminated through proper cooking (Ravindran and Ravindran 1988), it would also contribute to the alleviation of the food insecurity in some tropical countries.

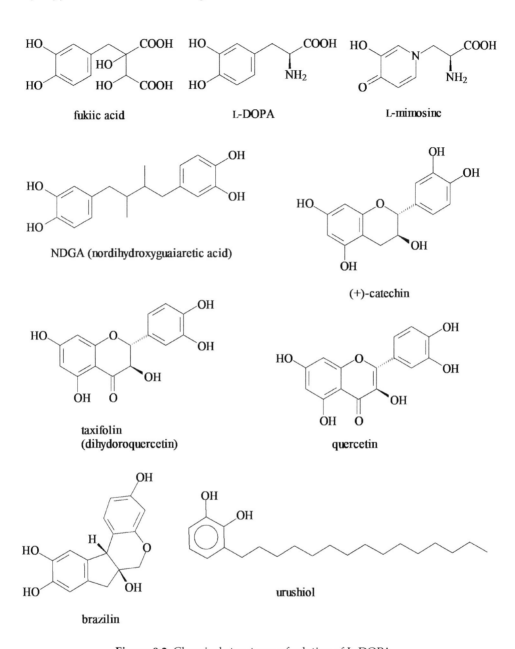

Figure 9.2. Chemical structures of relative of L-DOPA.

Practical Application of Hairy Vetch (*Vicia villosa*)

Screening for Allelopathic Cover Crops by the Plant Box Test

Primary selection of cover crops with allelopathic potentials was made using 'Plant Box' method, developed for the specific assessment of plant allelopathy through root exudates (Fujii, 1991b). Young plants were cultivated for one to two months in a sand culture containing a nutrient solution. The receptor species used for bioassay was lettuce (*Lactuca sativa*) because of its high sensitivity to bioactive substances.

The results of screening of candidates species for allelopathic cover crops by Plant Box method are shown in Table 9.1. In this table, radicle percentage means the percentages of the

Table 9.7. Effects of L-DOPA and related compounds in velvet bean on the
growth of radicles of lettuce and some weeds

Compounds	*Lactuca sativa* [3]	*Solidago altissima* [4]	*Taraxacum officinale* [5]	*Amaranthus lividus* [6]
		EC_{50} (mM) [1]		
L-DOPA	0.20	0.46	1.3	0.76
Dopamine	6.3	>3.2	1.6	>3.2
Catechol [2]	0.73	0.36	0.73	<0.27

Compound	*Miscanthus sinensis* [7]	*Setaria faberi* [8]	*Cerastium glomeratum* [9]	*Spergula arvensis* [10]
		EC_{50} (mM) [1]		
L-DOPA	0.86	2.0	0.10	0.20
Dopamine	>3.2	4.4	>3.2	1.6
Catechol [2]	0.73	2.7	0.55	1.4

[1] 50% inhibition concentration
[2] Pyrocatechol
[3] Lettuce
[4] Tall goldenrod
[5] Common dandelion
[6] Wild blite
[7] Japanese silver grass
[8] Giant foxtail.
[9] Sticky chickweed (mouse-ear)
[10] Corn spurrey

root radicle by the length of the young lettuce plants present in the root zone of each donor plant, based on the calculation of radicle length within the root zone controlled by donor plants. As shown in Table 9.1, leguminous cover crops, such as velvet bean, hairy vetch, yellow sweet clover, and white sweet clover showed strong allelopathic inhibitory activities. Among these legumes, velvet bean, *Crotalaria*, *Canavalia*, *Cajanus*, *Cicer*, *Vigna*, and *Glycine* are summer cover crops, and most of the others are winter cover crops. *Melilotus* and *Pueraria* are perennial crops. However, it is essential to know the characteristics of each cover crop for possible use in agricultural production.

Screening of Cover Crops for Weed Control in the Experimental Fields

Field experiments were conducted using the small-scale field to assess the weed suppression activity of these cover crops (Fujii and Shibuya 1992b). Most of the spring-sown cover crops were not promising as shown in Table 9.8. In Japanese weather conditions, there is a rainy season in June, and soon after the beginning of this season, there is the incidence of plant diseases and vigorous growth of weeds. Some of the cover crops, such as *Helianthus*, *Celosia*, and *Panicum* showed relatively high inhibitory activity on weeds followed by *Mucuna*, *Vigna*, and *Cassia*. In the field, competition for nutrients and light must be the most essential factors for plant growth. However, all of these plants have vigorous growth rate and relatively large leaves. Most of these cover crops were reported as allelopathic, and have relatively strong inhibitory activity in the Plant Box test (Table 9.1). For example, *Helianthus* and *Celosia* have been reported with allelopathic potentials (Rice 1984). However, 20% of weeds remaining on the fields look weedy and unacceptable. Considering these effects, spring-grown cover crops under these conditions are not acceptable to Japanese farmers (Fujii et al. 1994).

Table 9.9 shows the results from the trial for fall seeding. If cover crops are sown in fall, they tend to grow slowly but they grow steadily in winter to make enough biomass in spring and can eliminate the vigorous weeds. Table 9.9 shows a part of the result of fall-sown cover

Table 9.8. Effect of spring-sown cover crops on the weed growth in the field

Cover crop	W^{*1} (%)	Cover crop	W^{*1} (%)
Helianthus annuus	85	*Ricinus communis*	14
Celosia argentea	85	*Phaseolus vulgaris*	10
Panicum maximum	83	*Vicia hirsta*	9
Mucuna pruriens	48	*Medicago sativa*	9
Panicum milliaceum	46	*Luffa cylindrical*	9
Vigna angularis	41	*Glycine ussuriensis*	8
Cassia occidentalis	39	*Mormodica charantia*	3
Corchorus olitorius	36	*Canavalia ensiformis*	1
Corchorus capsularis	34	*Vigna angularis*	-2
Gossypium barbadense	33	*Vigna umbellate*	-4
Tephrosia candida	26	*Vigna radiate*	-5
Panicum ramosum	25	*Glycine max*	-12
Amaranthus tricolor	21	*Citrullus lanatus*	-16
Panicum virgotum	19	*Cajanus cajan*	-19
Medicago rugosa	17	*Carthamus tinctorius*	-23
Setaria italic	16	*Crotalaria juncea*	-33

Percentage of weed control. 100 % means complete inhibition.
Transplanting date: 23 May 1992, Sampling date: 25 August 1992.
No replication, each plot consists of 1 × 4 m (4 m²).

crops. As a primary experiment, there were no replications in this test. However, it is obvious that the dry weight of each cover crop is the most important factor to reduce the growth of weeds. This is true for rye, oat, wheat, woolly pod vetch, and Italian ryegrass. As a result of the overwhelming canopy of these crops, weeds have little space for rapid growth. However, when compared, the relationship between dry weight and weed suppression activity of *Brassica*, *Vicia*, and *Medicago*, biomass does not correlate well with weed suppression, and allelopathy may play a role in these cases. Table 9.10 shows the result of a more precise study with four replications. Hairy vetch, wheat, barley, oat, and rye showed significant inhibitory activity of weeds without weeding, but Chinese milk vetch, which is a traditional green manure in Japan and China, showed little weed suppression.

In conclusion, spring-sown cover crops are not practically promising in Japanese conditions, but fall-sown cover crops, such as rye, wheat, hairy vetch, barley, and oat have excellent inhibitory activity on weeds from spring to early summer. Their inhibitions of weeds are almost the same as that of traditional methods of weed control, such as herbicide application and rice straw mulch.

Application of Hairy Vetch to Abandoned Paddy Field in the Experimental Station

From the results of screening of cover crops, we focused on hairy vetch, with five field trials conducted on the experimental field in Shikoku National Agricultural Experiment Station. (1) A large scale application test of hairy vetch was conducted in comparison with Chinese milk vetch on a uniform paddy field of 1,000 m². This field was divided into 16 blocks, and four replications of four different cover crop trials were designed. Cover crops used in these designs were: (i) hairy vetch, (ii) hairy vetch and oat, (iii) Chinese milk vetch, and (iv) no treatment as a control. This experiment was started in 1992 and continued until 1994. Data from 1992 to 1994 was reported (Fujii, 2001). (2) Changing the seeding ratio from standard seeding to an increased four times was examined using a field of 500 m², and arranged with a Split-split-plot test with

Table 9.9. Effect of fall-sown cover crops on weed control in the field

Cover crop (English name)	W[*1] (%)	Dry weight (g m^{-2})
Control (No weeding)	0	–
Vicia villosa (Hairy Vetch)	100	1171
Secale cereale (Rye)	100	1671
Brassica napus (Rape)	100	998
Latyrus stivus (Grass Pea)	99	891
Avena sativa (Oat)	99	1426
Hordeum vulgare (Barley)	99	1173
Triticum aestivum (Wheat)	99	1751
Brassica campestris (Field Mustard)	97	834
Pisum sativum (Pea)	95	1359
Brassica alba (White Mustard)	95	416
Vicia dasycarpa (Woolly Pod Vetch)	91	2409
Lolium multiflorum (Italian Ryegrass)	89	1799
Vicia sativa (Common Vetch)	87	1016
Medicago sativa (Alfalfa)	77	384
Phleum pratense (Timothy)	69	445
Lolium perenne (Perennial Ryegrass)	66	516
Lupinus albus (Lupin)	45	341
Vicia angustifolia (Karasuni-endou)	44	473
Festuca arundinacea (Tall Fesque)	44	304
Astragalus sinicus (Chinese Milk Vetch)	31	167
Festuca eltior (Meadow Fesque)	31	372
Vicia hirsuta (Suzumeno-endou)	30	304
Trifolium repens (Ladino Clover)	22	426

[*1] Percentage of weed control. 100% means complete inhibition. Dry weight of weeds in the control plot is 381 g m^{-2} (=3810 kg ha^{-1}).
Dominant weeds were *Capsella bursa-pastoris* and *Lamium amplexicaule*.
Seeding date: November 5, 1992, Sampling date: April 20, 1993.
No replication, each plot consists of 2 × 2 m (4 m^2).

four replications. This experiment started from October 25, 1993 and weed and crop yield was measured on May 20, 1994. (3) Changing the seeding date from October to February was tested using the same field with three replications. This test was started from 1992 and ended in 1994. (4) Effect of soil-water contents on the growth of hairy vetch was tested in the lysimeter. This lysimeter was designed to know the suitable water ratio for crops, and it is possible to change the water level by Stair-step system of overflowing. Each block was about 10 m^2, and there were four stages with two replications. The soil type of the four tests above was sandy loam, paddy field converted to the upland condition. (5) The combination of leguminous cover crops and barley were examined on the slope land experimental field of Shikoku National Agricultural Experiment Station on the Oo-asa mountain with a total area of about 800 m^2. The slope angle of this field was 9 degree. Seeding date was November 6, 1993, and crop and weed yield were measured on May 30, 1994.

The results from the large-scale applications of hairy vetch, Chinese milk vetch, and mixed planting of vetch and oats are shown in Tables 9.11 and 9.12. These data were taken between 1992 and 1994, and in each year four replications were taken (Hanano et al. 1998). From these data, it is evident that hairy vetch almost completely inhibited the growth of weeds in spring. On the contrary, Chinese milk vetch, traditionally used in Japanese paddy field as green manure, inhibit up to 80% of the weed biomass. However, leaving 20% of weeds to grow

Table 9.10. Effect of selected fall-sown cover crops on the weed control

Cover crop (English name)	W[*1] (%)	Dry weight (g m[-2])
Control (No weeding)	0 a[*2]	-
Secale cereale (Rye)	99 d	693
Triticum aestivum (Wheat)	99 d	1751
Avena sativa (Oat)	99 cd	994
Hordeum vulgare (Barley)	99 cd	1173
Vicia villosa (Hairy Vetch)	98 cd	816
Brassica campestris (Field Mustard)	97 cd	834
Trifolium repens (White Clover)	78 cd	356
Medicago sativa (Alfalfa)	77 cd	384
Lupinus albus (Lupin)	49 b	341
Astragalus sinicus (Chinese Milk Vetch)	36 b	167
Melilotus albus (White Sweet Clover)	-16 a	30
Rice straw mulch (10,000 kg ha[-1])	87 cd	(1,000)
[Herbicide (Benthiocarb + Prometrin, 40 kg ha[-1])]	91 cd	-

[*1] Percentage of weed control. 100% means complete control. Dry weight of weeds in the control plot is 381 g m[-2].
The same letter means not significantly different by Duncan's Multiple Range Test (P>0. 01).
Dry weight of each cover crop.
Dominant weeds species in this field were *Capsella bursa-pastoris* and *Lamium amplexicaule*.
Seeding date: November 5, 1992, Sampling date: April 20, 1993.
Four replications by complete randomized block design, and each plot consists of 2 × 2 m (4 m²) block.

Table 9.11. Effect of cover crops on weed control in abandoned paddy field, first trial on the experimental station

Cover crop (English name)	W (%)[*1]	Crop yield (g m[-2])
First sampling (May 7)		
Control (No weeding)	0 a[*2]	-
Astragalus sinicus (Chinese Milk Vetch)	82 b	431 a
Vicia villosa (Hairy Vetch)	99 b	584 ab
A. sativa + *V. villosa* (Oat and Hairy Vetch Mix.)	99 b	730 b
Second sampling (June 10)		
Control (No weeding)	0 a	-
Astragalus sinicus (Chinese Milk Vetch)	59 b	135 a
Vicia villosa (Hairy Vetch)	100 c	147 b
A. sativa + *V. villosa* (Oat and Hairy Vetch mix)	100 c	137 a

[*1] Percentage of weed control. 100% means complete control. Dry weight of weeds in the control plot are 281 (First) and 155 (Second) g m[-2].
The same letter means not significantly different by Duncan's Multiple Range Test (P>0. 05).
Dominant weeds species in this field are *Alopecurus aequalis* and *Lamium amplexicaule*.
Seeding date: 28 October 1992, Sampling date: 7 May (First), 10 June (Second), 1993.
Four replications, each plot is 5 × 10 m (500 m²).
Place: SNAES Experimental Field P8 (Zentsuji, Kagawa), abandoned paddy field.

will make this field weedy the following year, and will be abandoned in two or three years without weeding. These results correspond with the observation of farmers that continuous use of Chinese milk vetch will cause a severe infestation of weeds. The addition of oat to hairy vetch was aimed to increase the weed suppression ability. In both years, the addition of oat decreased the population of weeds, to nearly complete inhibition of weed until fall (Table 9.12). Mixed planting of hairy vetch and oat has companionship, and both yields, per acre, increased. However, if no care was taken, the outlook of the field from spring to summer of the mixed cover field was not beautiful because of the remaining stems of oat. Hairy vetch however kept the stand height of maximum 50 cm, and the outlook of this field was uniform and flat and free from weeds. Hairy vetch died itself when the maximum temperature reached 30°C. In our experimental field, hairy vetch made a straw-like mulch without any work, and this mulch protected the field from weed invasion after the death of mother plants. Then, we concluded that using hairy vetch alone is a better recommendation to the farmers because of the simplicity of sowing and minimal labor requirement.

The combination of leguminous cover crops and barley were examined on the slope land experimental field of Shikoku National Agricultural Experiment Station. This experiment aimed to use barley as a cover crop in slope land. Barley, also known as 'Hadaka-mugi,' a naked barley, is the traditional cultivar and suitable for Shikoku and southeast area of Japan. Hairy vetch slightly reduced the growth of barley, but the weed suppression by hairy vetch and the combination of hairy vetch and barley were the best followed by the combination of red clover and barley. About optimum seeding rate, standard seeding rate was enough for the weed control. However, increasing the seed volume made no difference on biomass production and weed suppression. The cost per 1,000 m² (Standard Japanese unit of farming) is about JP¥ 2500 ($20 by current Japanese market price), hence increasing the seeding rate is not cost-effective. As for optimum seeding date for weed suppression in Japan, late seeding tends to grow more weeds, and it was concluded that early planting no later than the first week of November is recommended.

Table 9.12. Effect of cover crops on weed control in abandoned paddy field, second trial on the experimental station

Cover crop (English name)	W (%)[*1]	Crop yield (g m⁻²)
First sampling (June 2)		
Control (No weeding)	0 a [*2]	----
Astragalus sinicus (Chinese Milk Vetch)	83 b	220 a
Vicia villosa (Hairy Vetch)	100 b	619 ab
A. sativa + V. villosa (Oat and Hairy Vetch mix.)	100 b	629 + 367 b
Second sampling (October 7)		
Control (No weeding)	0 a	----
Astragalus sinicus (Chinese Milk Vetch)	77 b	----
Vicia villosa (Hairy Vetch)	90 c	----
A. sativa + V. villosa (Oat and Hairy Vetch mix.)	97 c	----

[*1] Percentage of weed control. 100% means complete control. Dry weight of weeds in the control plot are 156 (First) and 247 (Second) g m⁻².
[*2] The same letter means not significantly different by Duncan's Multiple Range Test ($P>0.05$).
Dominant weeds species in this field are *Alopecurus aequalis* and *Lamium amplexicaule*.
Seeding date: 10 November 1993.
Four replications, each plot is 5×10 m (500 m²).
Place: SNAES Experimental Field P8 (Zentsuji, Kagawa), abandoned paddy field.

Direct Application Test of Hairy Vetch on Farmers' Field

Direct application and exhibition trial of hairy vetch were done using the field of cooperative farmers by courtesy of each district counsellors for farmers (in Japan, there are counsellors for farmers in each county. These counsellors are public service officers belonging to the Ministry of Agriculture). Six different farming systems were chosen including: (1) Paddy field in Man'nou town with an area of about 800 m². This paddy field was in a slightly mountainous area, with slightly dry condition. Rice production on this field had just come to a halt. (2) Paddy field in Marugame city with an area of 1,000 m². This field was close to the road and in wet condition. This field was abandoned two years before the start of these field applications. (3) Paddy field in Zentsuji city with an area of about 900 m². This field was abandoned for 4 or 5 years, and they used Chinese milk vetch as green cover crop, but suffered from severe weed infection. (4) Grassland in Tyu-nan town land size of 600 m². This grassland was in a slope land and used as an exhibition. (5) Orchard for Kaki, Japanese persimmon in Kounan town, Oka village. The area used for cover crop trial in the first year was about 5,000 m² in 1992, and extended to 10,000 m² in the following year and continues until now. (6) Pear Orchard in Toyohama town. The area used for the trial for hairy vetch was about 80,000 m².

Direct application and exhibition trial of hairy vetch were done using the field of cooperative farmers. In most of the cases, the weed suppression by hairy vetch was enough, and the impressions of farmers were excellent. Before these trials, there was no custom of using hairy vetch as cover crops to control the weeds in Japan. After our recommendation of hairy vetch in Japan, many farmers started using hairy vetch. Now, about 25 years after the first introduction, hairy vetch covers about 10,000 ha as a cover crop and is now the No. 2 green manure crop next to Chinese milk vetch in Japan.

Isolation and Identification of Allelopathic Substances from Hairy Vetch

The crude extract of fresh leaves and stems of hairy vetch showed inhibitory activity on hypocotyl and radicle elongation of lettuce. The crude extract from 0.1 g F.W. of hairy vetch inhibited hypocotyl elongation of lettuce by 50%. Bioassay of fractions isolated from the plant was made during all the fractionation procedures. The amounts of samples used for the bioassay were calculated on the basis of the fresh weight of the extracted hairy vetch. The fraction showing the strongest inhibition was further fractionated chromatographically. Finally, it gave a major compound for plant growth inhibition.

This compound was identified as cyanamide (Figure 9.3) from the IR and 1H NMR spectra. This assignment was confirmed by comparing these spectra with those of authentic cyanamide.

The quantitative analysis confirmed that the hairy vetch used for the isolation of a plant growth inhibitor contained 130 mg of cyanamide per gram of fresh plant body (Kamo et al. 2005).

The crude extract and authentic cyanamide were compared with regard to their ability to inhibit the growth of lettuce hypocotyls on the basis of their cyanamide concentration. In all the range of concentrations examined, the growth inhibition of the crude extract on the lettuce hypocotyls was well explained by the action of cyanamide contained in the extract. The growth inhibition on lettuce radicle, however, did not well correspond with that of authentic cyanamide, especially in the region of lower concentration. At higher concentration, the crude extract inhibited radicle growth and was well explained by the action of cyanamide. Other lesser-contributing phytotoxic compounds in the crude extract could contribute to the total inhibitory activity on the elongation of lettuce radicle together with cyanamide. These results indicate that cyanamide is a major allelochemical in hairy vetch (Kamo et al. 2003).

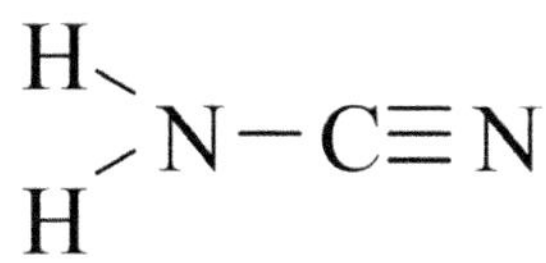

$$H_2N-C\equiv N$$

Figure **9.3.** Chemical structure of cyanamide.

The content of cyanamide in hairy vetch was 0.13 ± 0.04 g (± SD) per seed. To clarify whether cyanamide is biosynthesized *in vivo* or not, hairy vetch was grown without nutrients in an illuminated growth chamber for nine days, and cyanamide content in each plant was determined. Cyanamide existed mainly in the shoot part, but was also found in fewer amounts in the endosperm covered with the seed coat and in the root parts. The total amount of cyanamide was 5.1 ±2.8 g (± SD) per the whole seedling. This indicates that the amount of cyanamide increased after the germination to approximately 40 times greater than the initial amount.

Cyanamide has been produced industrially and utilized for drugs and agrochemicals, but was considered to be absent in natural products. It was not certain at the early stage of this study whether the cyanamide isolated was from a natural source, since unexpected contamination by artificial cyanamide might be possible in the field. However, it was confirmed that cyanamide is obviously biosynthesized in hairy vetch. L-Cyanoalanine, known as a neurotoxic compound, has been reported to occur in the seeds of common vetch, *V. sativa*, and other *Vicia* spp. (Odriozola et al. 1990). However, the relationship between cyanamide and L-beta-cyanoalanine is unknown.

Cyanamide was identified as a natural product, although it has been synthesized for over 100 years for industrial and agricultural purposes. The distribution of natural cyanamide appears to be limited, as indicated by our previous investigation of 101 weed species (Kamo et al. 2008). In the present study, to evaluate the distribution of natural cyanamide in *Vicia* species, we monitored the cyanamide contents in *V. villosa* subsp. *varia*, *V. cracca*, and *V. amoena* during their pre-flowering and flowering seasons. It was confirmed that *V. cracca* was superior to *V. villosa* subsp. *varia* in accumulating natural cyanamide. However, *V. amoena* was unable to biosynthesize this compound. The localization of cyanamide in the leaves of *V. villosa* subsp. *varia* seedlings were also clarified. In a screening study to find cyanamide-biosynthesizing plants, only *Robinia pseudo-acacia* was found to contain cyanamide among the 452 species of higher plants tested. Among the 553 plant species investigated, only three species including *V. villosa* subsp. *varia*, *V. cracca* and *R. pseudo-acacia* had the ability to biosynthesize cyanamide (Kamo et al. 2008).

Recommendation of Hairy Vetch as an Allelopathic Ground Cover Crop

Hairy vetch is a well-known green manure and cover crop in the United States and Europe. The origin of hairy vetch is considered to be in the area extending from West Asia to the eastern Mediterranean coast. It was cultivated in England and Germany in the early nineteenth century and then introduced to the USA in the middle of the nineteenth century. Hairy vetch had a good reputation from USDA recommendations and is now widely distributed in the southern part of the USA. Hairy vetch was introduced to Japan in the early twentieth century and showed good results at Agricultural Experimental Stations, but was not distributed until the present.

There are some reports on allelopathy and weed control by hairy vetch. Lazauskas and Balinevichiute (1972) tested the inhibitory activity of seed extracts on barley and found that hairy vetch showed the strongest activity. White et al. (1989) reported that the incorporation of the residue of hairy vetch and crimson clover reduced the emergence of *Solanaceae* weeds to about 60–80%, and water extract of hairy vetch had the strongest inhibitory activity. Johnson et al. (1993) reported that the mulch made from hairy vetch or rye completely inhibited the weed in non-tillage systems. Teasdale and Daughtry (1993) reported that the living mulch of hairy vetch showed better inhibitory activity than the desiccated one. Abdul-Baki and Teasdale (1993) reported a unique system using hairy vetch mulch to compensate for the vinyl plastic film mulch in tomato production. There are many reports and field observation about the weed suppression ability of hairy vetch, but the contribution of allelopathy and its allelochemicals were unknown.

We isolated and identified cyanamide as the main allelochemical responsible for the plant growth inhibition activity of hairy vetch (Kamo et al. 2008). Hairy vetch has many merits other than weed control in the field. Some of the other advantages of hairy vetch include:, nitrogen fixation to reduce chemical fertilizer, organic materials to reduce chemical fertilizer or soil conditioner, soil erosion control by acting as surface cover, promotion of soil porosity through its deep root system, thick cover palliate the microclimate to reduce maximum temperature and increase minimum temperature, induction of carnivorous ladybug to reduce the population harmful insects.

After these series of experiments, it was concluded that hairy vetch is the most promising allelopathic cover crop for the control of weeds in abandoned fields, grassland, and orchard in the central and southern parts of Japan.

Concluding Remarks

The application of the allelopathic approach as one of the strategies in sustainable weed control provides an alternative methodology for underdeveloped and developing countries to establish a sustainable and environment-friendly agricultural system. However, in order to ensure that the allelopathic approach is successful, it should be practically simple and economically viable. This can be achieved with the use of widespread and easily accessible local plants with allelopathic properties. Among these plants, velvet bean and hairy vetch are the most promising allelopathic cover crops. Velvet bean is now widely spreading as a crop in South America, with wide distribution in Southeast Asia and African countries. Hairy vetch is gradually developing as a cover crop in Japan and is now being planted on an area of about 10,000 ha, making it the No. 2 cover crop in Japan after Chinese milk vetch, a traditional cover crop. However, Chinese milkvetch is non-allelopathic and is recently getting increasingly challenging to grow due to invasive insects. These recent advances in the utilization of allelopathic plants are aided by the ability to isolate and identify allelochemicals. Further research is needed to develop and make these compounds useful in weed control. However, a multidisciplinary approach is needed to assess the allelopathic influence and plant interactions in integrated weed management strategy.

REFERENCES

Abdul-Baki, A.A. and J.R. Teasdale. 1993. A no-tillage tomato production system using hairy vetch and subterranean clover mulches. HortScience 28: 106–108.

Albuquerque, M.B., R.C. Santos, L.M. Lima, P.A.M. Filho, R.J.M.C. Nogueira, C.A.G. Câmara and A.R. Ramos. 2011. Allelopathy, an alternative tool to improve cropping systems. A review. Ag. Sustain. Devel. 31(2): 379–395.

Appiah, K.S., C.A. Amoatey and Y. Fujii. 2015a. Allelopathic activities of selected *Mucuna pruriens* on the germination and initial growth of lettuce. International Journal of Basic and Applied Science 4(4): 475–481.

Appiah, K.S., Z. Li, R. Zeng, S. Luo, Y. Oikawa and Y. Fujii. 2017b. Determination of allelopathic potentials in plant species in Sino-Japanese floristic region by sandwich method and dish pack method. International Journal of Basic and Applied Science 4(4): 381–394.

Appiah, K.S., H.K. Mardani, A. Osivand, S. Kpabitey, C.A. Amoatey, Y. Oikawa and Y. Fujii. 2017. Exploring alternative use of medicinal plants for sustainable weed management. Sustainability 9: 1468. doi: 10. 3390/su9081468.

Batish, D.R., K. Lavanya, H.P. Singh and R.K. Kohli. 2007. Phenolic allelochemicals released by Chenopodium murale affect the growth, nodulation and macromolecule content in chickpea and pea. Plant Growth Regulation 51: 119–128.

Bell, E.A. and D.H. Janzen. 1971. Medical and ecological considerations of L-Dopa and 5-HTP in seeds. Nature 229: 136–137.

Bell, D.T. and D.E. Koeppe. 1972. Noncompetitive effects of giant foxtail on the growth of corn. Agron. J. 64: 321–325.

Bunch, R. and E. Staff. 1985. Green Manure Crops. ECHO Technical Note.

Cheema, Z.A. and A. Khaliq. 2000. Use of sorghum allelopathic properties to control weeds in irrigated wheat in a semi-arid region of Punjab. Agriculture, Ecosystems and Environment 79: 105–112.

Chou, C-H. and Y-L. Kuo. 1986. Allelopathic research of subtropical vegetation in Taiwan. III. Allelopathic exclusion of understory by *Leucaena leucocephala* (Lam.) de Wit. J. Chem. Ecol. 12: 1431–1448.

Damodaran, M. and R. Ramaswamy. 1937. Isolation of L-3,4-dihydroxyphenylalanine from the seeds of *Mucuna pruriens*. Biochem. J. 31: 2149–2152.

Davies, G., B. Turner and B. Bond. 2008. Weed Management for Organic Farmers, Growers and Smallholders: A Complete Guide. The Crowood Press Ltd, Wiltshire.

Dayan, F.E., C.L. Cantrell and S.O. Duke. 2009. Natural products in crop protection. Bioorganic and Medicinal Chemistry 17(12): 4022–4034.

Duke, S.O. 2010. Allelopathy: current status of research and future of the discipline: a commentary. Allelopathy Journal 2(1): 17–30.

Elacovitch, S.D. and K.L. Stevens. 1985. Phytotoxic properties of nordihydroguaiaretic acid, a lignan from *Larrea tridentata* (Creosote bush). J. Chem. Ecol. 11: 27–33.

Fujii, Y., T. Shibuya and T. Yasuda. 1990a. Method for screening allelopathic activities by using the logistic function (Richards' function) fitted to lettuce seed germination and growth curves. Weed Res. Japan 35: 353–361.

Fujii, Y., T. Shibuya and T. Yasuda. 1990b. Survey of Japanese weeds and crops for the detection of water-extractable allelopathic chemicals using Richards' function fitted to lettuce germination test. Weed Res. Japan 35: 362–370.

Fujii, Y. and T. Shibuya. 1991a. A new bioassay for allelopathy with agar medium I. Assessment of allelopathy from litter leachate by sandwich method. Weed Res. Japan 36(Suppl): 150–151.

Fujii, Y. and T. Shibuya. 1991b. A new bioassay for allelopathy with agar medium. II. Assessment of allelopathy from root exudates. Weed Res. Japan 36(Suppl): 152–153.

Fujii, Y., M. Furukawa, Y. Hayakawa, K. Sugawara and T. Shibuya. 1991c. Survey of Japanese medicinal plants for the detection of allelopathic properties. Weed Research Japan 36: 36–42.

Fujii, Y., T. Shibuya and Y. Usami. 1991d. Allelopathic effect of *Mucuna pruriens* on the appearance of weeds. Weed Res. Japan 36: 43–49.

Fujii, Y., T. Shibuya and T. Yasuda. 1991e. L-3,4-Dihydroxyphenylalanine as an allelochemical candidate from *Mucuna pruriens* (L.) DC. var. *utilis*. Agr. Biol. Chem. 55: 617–618.

Fujii, Y., T. Shibuya and T. Yasuda. 1991f. Discrimination of allelopathy of velvetbean (*Mucuna pruriens*) with stairstep experiment and rotary greenhouse experiments. Jap. J. Soil Sci. Plant Nutri. 62: 258–264.

Fujii, Y., T. Shibuya and T. Yasuda. 1991g. Intercropping of velvetbean (*Mucuna pruriens*) by substitutive experiments: suggestion of companion plants with corn and kidney bean. Jap. J. Soil Sci. Plant Nutri. 62: 363–370.

Fujii, Y. and T. Shibuya. 1992a. Establishment of a new bioassay for allelopathy: assessment of allelopathic activity from root exudates by using Plant Box and agar medium. Weed Res. Japan 36(Suppl): 152–153.

Fujii, Y. and T. Shibuya. 1992b. Allelopathy of hairy vetch: assessment of allelopathic activity and weed control in the field. Weed Res. Japan 37(Suppl): 160–161.

Fujii, Y. 1994. Screening of allelopathic candidates by new specific discrimination and assessment methods for allelopathy, and the identification of L-DOPA as the allelopathic substance from the most promising velvetbean. The Bulletin of the National Institute of Agro-Environmental Sciences 10: 115–218.

Fujii, Y., S. Ono and K. Sato. 1994. Weed suppression by winter cover crops such as hairy vetch, oats, rye, barley and wheat, and its relation to Plant Box Test. Weed Res. Japan 39(Suppl): 258–259.

Fujii, Y. 2001. Screening and future exploitation of allelopathic plants as alternative herbicides with special reference to hairy vetch. Journal of Crop Production 4: 257–275.

Fujii, Y. 2003. Allelopathy in the natural and agricultural ecosystems and isolation of potent allelochemicals from velvet bean (*Mucuna pruriens*) and hairy vetch (*Vicia villosa*). Biological Sciences in Space 17: 6–13.

Fujii, Y., S. Parvez, M.M. Parvez, Y. Ohmae and O. Iida. 2003. Screening of 239 medicinal plant species for allelopathic activity using the sandwich method. Weed Biology and Management 3: 233–241.

Fujii Y., T. Shibuya, K. Nakatani and T. Yasuda. 2004. Assessment method for allelopathic effect from leaf litter leachates. Weed Biology and Management 4: 19–23.

Fujii, Y., M. Matsuyama, S. Hiradate and H. Shimozawa. 2005. Dish pack method: a new bioassay for volatile allelopathy. pp. 493–497. *In:* Harper, J.D.I., An, M., Wu, H. and Kent, J.H. (Eds.) The Proceedings of the 4th World Congress, Allelopathy, "Establishing the Scientific Base". Wagga Wagga, New South Wales, Australia.

Fujii, Y., D. Pariasca, T. Shibuya, T. Yasuda, B. Kahn and G.R. Waller. 2007. Plant-box method: a specific bioassay to evaluate allelopathy through root exudates. pp. 39–56. *In:* Y. Fujii and S. Hiradate (Eds.) Allelopathy: New Concepts and Methodology. Science Publishers, Enfield.

Gallandt, E.R., M. Liebman and D.R. Huggins. 1999. Improving soil quality: implications for weed management. Journal of Crop Production 2(1): 95–121.

Hanano, Y., Y. Fujii, K. Satoh, S. Osozawa and S. Fujihara. 1998. Bulletin of the Shikoku National Agricultural Experiment Station 62: 45–70.

Hulugalle, N.R., R. Lal and C.H.H. Terkuile. 1986. Amelioration of soil physical properties by *Mucuna* after mechanized land clearing of a tropical rain forest. Soil Science 141: 219–224.

Ismail, B.S., H. Syamimi, J.A. Wan Juliana and Y. Nornasuha. 2016. Allelopathic potential of the leaf and seed of *Pueraria javanica* Benth. on the germination and growth of three selected weed species. Sains Malaysiana 45(4): 517–521.

Itani T., K. Hirai, Y. Fujii, H. Kanda and M. Tamaki. 1998. Screening for allelopathic activity among weeds and medicinal plants using the 'sandwich method'. Weed Sci. Techn. 43: 258–266.

Itani, T., Y. Nakahata and H. Kato-Noguchi. 2013. Allelopathic activity of some herb plant species. Int J Agric Biol. 15: 1359–1362.

Johnson, G.A., M.S. Defelice and Z.R. Helsel. 1993. Cover crop management and weed control in Corn. Weed Technology 7: 425–430.

Kamo, T., S. Hiradate and Y. Fujii. 2003. First isolation of natural cyanamide as a possible allelochemical from hairy vetch *Vicia vilosa*. J. Chem. Ecol. 29: 275–283.

Kamo, T., M. Endo, M. Sato, R. Kasahara, H. Yamaya, S. Hiradate, Y. Fujii, N. Hirai and M. Hirota. 2008. Limited distribution of natural cyanamide in higher plants: occurrence in *Vicia villosa* subsp. *varia*, *V. cracca*, and *Robinia pseudo-acacia*. Phytochemistry 69: 1166–1172.

Lazauskas, P. and Z. Balinevichiute. 1972. Influence of the excretions from *Vicia villosa* seeds on germination and primary growth of some crops and weeds. pp. 76–79. *In:* Physiological-Biochemical Basis of Plant Interactions in Phytocenoses. Naukova Dumka, Kiev.

Lorenzi, H. 1984. Consideracoes sobre plantas daninhas no plantio direto. pp. 24–35. *In:* Torrado, P.V. and Aloisi, R.R. (Eds.) Plant Direto no Brasil. Fundacao Cargill, Campinas.

Mardani, H., T. Sekine, M. Azizi, M. Mishyna and Y. Fujii. 2015. Identification of Safranal as the Main Allelochemical from Saffron (*Crocus sativus*). Natural Product Communications. 10: 1–3.

Mishyna, M., N. Laman, V. Prokhorov and Y. Fujii. 2015a. Angelicin as the Principal Allelochemical in *Heracleum sosnowskyi* Fruit. Natural Product Communications 10(5): 767–770.

Mishyna, M., N. Laman, V. Prokhorov, J.S. Maninang and Y. Fujii. 2015b. Identification of Octanal as plant growth inhibitory volatile compound released from *Heracleum sosnowskyi* Fruit. Natural Product Communications 10(5): 771–774.

Morikawa, C.I.O., R. Miyaura, Y. Tapia, M.D.L. Figueroa, E.L.R. Salgado and Y. Fujii. 2012a. Screening of 170 Peruvian plant species for allelopathic activity by using the sandwich method. Weed Biology and Management 12: 111.

Morikawa, C.I.O., R. Miyaura, G.V. Segovia, E.L.R. Salgado and Y. Fujii. 2012b. Evaluation of allelopathic activity from Peruvian plant species by sandwich method. Pakistan Journal of Weed Science Research 18: 829–834.

Odriozola, E., E. Paloma, T. Lopez and C. Campero. 1990. An outbreak of *Vicia villosa* (hairy vetch) poisoning in grazing Aberdeen Angus bulls in Argentina. Vet. Hum. Toxicol. 33: 278–280.

Premchand. 1981. Presence of feeding deterrent in velvetbean (Mucuna cochinchinensis). Indian J. Entomology 43: 217–219.

Ravindran, G. and G. Ravindran 1988. Nutritional and anti-nutritional characteristics of mucuna (*Mucuna utilis*) bean seeds. J. Sci. Food Agric. 46: 71–79.

Reddy, K.C., A.R. Soffes, G.M. Prine and R.A. Dunn. 1986. Tropical legumes for green manure. II. Nematode populations and their effects on succeeding crop yields. Agron. J. 78: 5–10.

Rehr, S.S., D.H. Janzen and P.P. Feeny. 1973. L-Dopa in legume seeds: a chemical barrier to insect attack. Science 181: 81–82.

Rice, E.L. 1984. Manipulated ecosystems: roles of allelopathy in agriculture. pp. 8–12. *In:* Allelopathy 2nd ed. Academic Press. London.

Singh, H.P., D.R. Batish and R.K. Kohli. 2001. Allelopathy in agroecosystems: an overview. pp. 1–41. *In:* R.K. Kohli, H.P. Singh and D.R. Batish (Eds.) Allelopathy in Agroecosystems. Hawthorn Press, New York, USA.

Singh, H.P., D.R. Batish, J.K. Pandher and R.K. Kohli. 2005. Phytotoxic effects of *Parthenium hysterophorus* residues on three *Brassica* species. Weed Biology and Management 5: 105–109.

Soule, M.J. 1997. Farmer assessment of velvetbean as a green manure in Veracruz, Mexico: experimentation and expected profits. NRG paper 97–02. Mexico, DF: CIMMYT.

Syed, S., M.I. Al-Haq, Z.I. Ahmed, A. Razzaq and M. Akmal. 2014. Root exudates and leaf leachates of 19 medicinal plants of Pakistan exhibit allelopathic potential. Pakistan Journal of Botany 46(5): 1693–1701.

Taib, I.M., L. Sin and A.F. Alif. 1979. Chemical weed control in legume management. pp. 375–391. *In:* The Proceedings of the Rubber Research Institute of Malaysia Planters' Conference.

Tateishi, Y. and H. Ohashi. 1981. Eastern Asiatic species of *Mucuna* (Leguminosae). Bot. Mag. Tokyo 94: 91–105.

Teasdale, J.R. and C.S.T. Daughtry. 1993. Weed suppression by live and desiccated hairy vetch. Weed Science 41: 207–212.

Tenente, R.C.V. and L.G.E. Lordello. 1980. Influence of *Stizolobium aterrimum* on the life-cycle of *Meloidogyne incognita*. Sociedade Brasileira de Nematologia. 213–215.

Tenente, R.C.V., L.G.E. Lordello and J.F.S. Dias. 1982. A study of the effect of root exudates of *Stizolobium aterrimum* on the hatching, penetration and development of *Meloidogyne incognita* race 4. Sociedade Brasileira de Nematologia. 271–284.

Watt, J.M. and M.G. Breyer-Brandwijk. 1962. Medicinal and poisonous plants of southern and eastern Africa. 2nd ed. pp. 631–634. E.&S. Livingstone, Edinburgh and London.

Weston, L.A. and S.O. Duke. 2003. Weed and Crop Allelopathy. CRC Crit Rev Plant Sci. 22: 367–389.

Weston, L.A. 2005. History and current trends in the use of allelopathy for weed management. Hortechnology 15: 529–534.

White, R.H., A.D. Worsham and U. Blum. 1989. Allelopathic potential of legume debris and aqueous extracts. Weed Science 37: 674–679.

Wilmot-Dear, C.M. 1983. A revision of *Mucuna* (Legminosae-Phaseolae) in China and Japan. Kew Bulletin 39: 23–65.

Xuan, T.D., S. Tawata, N.H. Hong, T.D. Khanh and C.I. Min. 2004b. Assessment of phytotoxic action of *Ageratum conyzoides* L. (billy goat weed) on weeds. Crop Protection 23(10): 915–922.

Genetically Engineered Herbicide Tolerant Crops and Sustainable Weed Management

Mahima Krishnan* and Christopher Preston

School of Agriculture, Food and Wine, University of Adelaide, Australia

Introduction

Weeds have major impacts on productivity of crops through competition for resources and therefore there is a strong desire to control them to reduce their impact. While there are a range of control measures that can be used to control weeds, herbicides have become the tool of choice due to being highly effective, easy to use and their relatively low cost (Gianessi and Reigner 2007, Gianessi 2013). However, herbicides are unable to control all weeds in cropping systems. For example, weeds that are closely related to the crop cannot be controlled with crop-selective herbicides (Mazur and Falco1989). In addition, herbicides can cause damage to the crop or to the following crops (Duke 2005, Hollaway et al. 2006). These issues could be solved through the production of crops with novel tolerance to herbicides (Mazur and Falco 1989).

GM-HT crops have been widely adopted in many cropping systems where they have been introduced. GM-HT brought considerable benefits to farmers in terms of better weed control and simplified weed management systems particularly in the Americas (Duke 2015). While there are examples of HT crops that were not produced by GM methods—most notably crops resistant to the imidazolinone herbicides, this chapter will not consider those in any great detail. However, many of the points made about GM-HT crops apply to these crops as well. The distinction of GM-HT crops is that they require additional layers of regulatory scrutiny prior to their introduction (Leyser 2014). In addition, there has been well-organised opposition to GM crops by a range of interest groups (Aerni 2013, Twardowski and Małyska 2015). Both of these have tended to delay the introduction of GM-HT crops and even stop their introduction in some areas like the EU (Smyth and Phillips 2014, Gleim et al. 2016). This has had ramifications for the sustainable use of GM-HT crops. The regulatory and market hurdles have likely reduced investment and delayed the introduction of new traits.

In this chapter, we will review the GM-HT traits available and describe their adoption in several markets. We will discuss the positive and negative impacts of GM-HT crops on farming systems and weed management and consider their role in sustainable weed management. Finally, we will discuss the options for using these crops more sustainably for weed management and consider why this has often not occurred to date.

*Corresponding author: mahima.krishnan@adelaide.edu.au

History and Adoption of GM-HT Crops

There have been a number of GM-HT traits commercialised. These are listed in Table 10.1 by crop species and countries where approval for environmental release have occurred. This list is dominated by traits providing resistance to glyphosate. However, resistance to other herbicides, such as bromoxynil, glufosinate, 2,4-D and dicamba have also been approved. Initially, approvals were for crops with tolerance to a single herbicide; however, in more recent years crops with stacked HT have become more common.

Glyphosate is the world's most widely used herbicide, largely due to its relatively low environmental toxicity and its broad-spectrum control of almost any weed (Duke and Powles 2008, Giesy et al. 2000, Smith and Oehme 1992). One of the problems with a broad-spectrum herbicide like glyphosate is that it cannot be used to control weeds in crop. Therefore, there was considerable interest over a long time in developing crops with tolerance to glyphosate (Kishore et al. 1992).

The advent of GM technology and the identification of EPSPS, the target for glyphosate, in a strain of agrobacterium insensitive to glyphosate led to the development of the revolutionary Roundup Ready crops. The CP4-EPSPS, the first glyphosate tolerance gene, was transformed into soybean and subsequently commercialized as the first Roundup Ready crop in 1996 (Huang et al. 2015, Barry et al. 1992). Following soy, Roundup Ready canola (1996), cotton (1996), sugar beet (2003) and alfalfa (2005) were commercially released containing the same transgene. Roundup Ready canola, in addition to CP4-EPSPS, also had glyphosate oxidase (GOX), a glyphosate detoxification gene, as part of the transgenic cassette. The CP4-EPSPS transgene is insensitive to glyphosate as a result of a unique structural conformation of EPSPS, which allows the binding of glyphosate but, is not inhibited by it (Funke et al. 2006).

Another glyphosate insensitive EPSPS was identified through site-directed mutagenesis in maize which had two amino acid substitutions, T102I and P106S or TIPS, which reduced the binding ability of glyphosate without affecting enzyme functionality. The 2mepsps or TIPS-EPSPS was used to confer glyphosate tolerance to the first glyphosate tolerant corn (event GA21) in 1998 (Huang et al. 2015, Lebrun et al. 2003) and subsequently used in other crops (Table 10.1).

Other Roundup Ready crops contained, along with the CP4-EPSPS gene, glyphosate oxidase (GOX), a glyphosate detoxification gene (Pedotti et al. 2009). Another glyphosate deactivation gene, glyphosate acetyletransferase (GAT), was also used to transform crops to make them glyphosate tolerant (Castle et al. 2004).

Improvement of glyphosate tolerant traits didn't stop with just the introduction of the aforementioned transgenes. Efforts have been made to optimise the expression of the transgenic EPSPS and glyphosate deactivation/detoxification genes through promoter engineering. The second generation of glyphosate tolerant plants contain promoters which facilitate gene expression in all tissues at risk of glyphosate damage (McElroy et al. 1990, Wendy et al. 2002, Heck et al. 2005, Huang et al. 2015).

After glyphosate tolerance, glufosinate tolerance is the next most popular GM trait introduced into crops. One of the major contributing factors that is responsible for the prevalence of glufosinate tolerance transgene in crops is its use as a selection marker to screen for plants which contain other important transgenic traits like insect resistance in the same cassette (Huang et al. 2015). The bialaphos resistance (bar) gene which encodes for a phosphinothricin acetyltransferase (PAT) was isolated from *Streptomyces hygroscopicus* and *Streptomyces viridochromogenes* (Block et al. 1987, Thompson et al. 1987). This gene was used to develop the transgenic Liberty Link crops (Huang et al. 2015).

Other introduced GM-HT traits include bxn gene in cotton and canola which confers tolerance to oxynil herbicides (Stalker et al. 1988). Sulfonylurea tolerant traits have been used in some crops but, haven't been a popular trait as the risk of development of resistance in weeds is high (Preston and Powles 2002, ISAAA 2016).

Table 10.1. GM herbicide tolerance traits approved for commercial production and countries where they have been approved[a] (Adapted from ILSI Research Foundation, 2016)

Crop species	Herbicide	Trait	Country	Year approved for environmental release[b]
Glycine max (Soybean)	Glyphosate	CP4 EPSPS	USA	1993
			Canada	1995
			Argentina	1996
			Uruguay	1996
			Brazil	1998
			Paraguay	2004
			Bolivia	2005
	Glufosinate	Phosphinothricin N-acetyltransferase	USA	1996
			Canada	1999
			Brazil	2010
			Argentina	2011
			Uruguay	2012
	Glyphosate + chlorsulphuron	Glyphosate N-acetyltransferase + zm-hra	USA	2007
			Canada	2009
	Imazapic + Imazapyr	Csr1-2	Brazil	2009
			Canada	2012
			Argentina	2013
			USA	2014
			Uruguay	2014
			Paraguay	2014
	Glyphosate + Isoxaflutole	2mepsps + hppdPF W336	Canada	2012
			USA	2013
			Brazil	2015
	Glufosinate + Mesotrione	Phosphinothricin N-acetyltransferase + avhppd-03	Canada	2014
			USA	2014
	Glufosinate + 2,4-D	Phosphinothricin N-acetyltransferase + aryloxyalkanoate dioxygenase	Canada	2012
			USA	2014
			Brazil	2015
	Glyphosate + Glufosinate + 2,4-D	2mepsps + Phosphinothricin N-acetyltransferase + aryloxyalkanoate dioxygenase	Canada	2013
			USA	2014
			Brazil	2015
			Argentina	2015
	Glyphosate + Glufosinate + 2,4-D	CP4 EPSPS + Phosphinothricin N-acetyltransferase + aryloxyalkanoate dioxygenase	Canada	2013
	Glyphosate + Dicamba	CP4 EPSPS + dicamba monooxygenase	Colombia	2012
			Canada	2012
			USA	2015

(Contd.)

Table 10.1. (*Contd.*)

Crop species	Herbicide	Trait	Country	Year approved for environmental release[b]
Gossypium hirsutum (cotton)	Glyphosate	CP4 EPSPS	USA	1995
			Argentina	1999
			Australia	2000
			Colombia	2004
			Brazil	2008
			Paraguay	2013
	Glyphosate	2mepsps	USA	2009
			Brazil	2010
			Argentina	2012
	Glufosinate	Phosphinothricin N-acetyltransferase	USA	2003
			Australia	2006
			Brazil	2008
			Colombia	2010
	Bromoxynil	Nitrilase	USA	1994
	Glufosinate + 2,4-D	Phosphinothricin N-acetyltransferase+ Aryloxyalkanoate dioxygenase	USA	2015
	Glufosinate + Dicamba	Phosphinothricin N-acetyltransferase + Dicamba monooxygenase	USA	2015
	Glyphosate + Dicamba + Glufosinate	CP4 EPSPS + dicamba monooxygenase + Phosphinothricin N-acetyltransferase	Australia	2016
Brassica napus (Canola)	Glyphosate	CP4 EPSPS + Glyphosate oxidase	Canada	1995
			USA	1999
			Australia	2003
	Glyphosate	CP4 EPSPS	Canada	2012
			USA	2013
			Australia	2014
	Glyphosate	Glyphosate N-acetyltransferase	Canada	2012
			USA	2013
			Australia	2016
	Glufosinate	Phosphinothricin N-acetyltransferase	Canada	1996
			USA	1999
			Australia	2013
	Bromoxynil	Nitrilase	Canada	1997
	Glufosinate + Glyphosate	Phosphinothricin N-acetyltransferase + CP4 EPSPS	Australia	2016
Zea mays (Maize or Corn)	Glufosinate	Phosphinothricin N-acetyltransferase	USA	1995
			Canada	1996
			Argentina	1998

(*Contd.*)

		South Africa	2003
		Uruguay	2004
		Brazil	2007
		Colombia	2008
Glyphosate	CP4 EPSPS	USA	1996
		Canada	1996
		South Africa	2002
		Argentina	2004
		The Philippines	2005
		Colombia	2007
		Brazil	2008
		Uruguay	2011
Glyphosate	2mepsps	USA	1997
		Canada	1998
		Argentina	1998
		Brazil	2008
		Colombia	2008
		The Philippines	2009
		South Africa	2010
		Uruguay	2011
		Paraguay	2015
Glyphosate	EPSPS grg23ace5	USA	2013
		Canada	2014
Glyphosate + Glufosinate	CP4 EPSPS + Phosphinothricin N-acetyltransferase	Canada	2005
		Colombia	2008
		USA	2009
		Argentina	2009
		Brazil	2009
		Uruguay	2011
		Paraguay	2015
Glyphosate + chlorsulfuron	Glyphosate N-acetyltransferase + zm-hra	Canada	2009
		USA	2009
		Argentina	2011
2,4-D	aryloxyalkanoate dioxygenase	Canada	2012
		Brazil	2015
		USA	2014
Glyphosate + 2,4-D	CP4 EPSPS + aryloxyalkanoate dioxygenase	Canada	2013
		Brazil	2015
Glyphosate + Glufosinate + 2,4-D	CP4 EPSPS + Phosphinothricin N-acetyltransferase + aryloxyalkanoate dioxygenase	Canada	2014
		Brazil	2016

(Contd.)

Table 10.1. (*Contd.*)

Crop species	Herbicide	Trait	Country	Year approved for environmental release[b]
Brassica rapa (Canola)	Glyphosate	CP4 EPSPS	Canada	1997
	Glufosinate	Phosphinothricin N-acetyltransferase	Canada	1998
Solanum tuberosum (Potato)	Glyphosate	CP4 EPSPS	USA	1999
			Canada	2001
Oryza sativa (Rice)	Glufosinate	Phosphinothricin N-acetyltransferase	USA	1999
Medicago sativa (Alfalfa or Lucerne)	Glyphosate	CP4 EPSPS	USA	2005
Beta vulgaris (Sugar Beet)	Glyphosate	CP4 EPSPS	USA	2005
			Canada	2005
	Glufosinate	Phosphinothricin N-acetyltransferase	USA	1998
			Canada	2001

[a] Does not include traits used solely for selection purposes, countries, such as Japan, where approval for environmental release does not equate to approval for commercial production or situations where approval is for breeding and seed production purposes only.

[b] Despite approvals for environmental release, commercial production may be delayed for a variety of reasons, so this date is not the same as the first year of commercial production.

Some new herbicide tolerant traits that have been introduced into crops are auxin-mimics, dicamba and 2,4-D resistance. Dicamba mono-oxygenase which has been transformed into cotton, soy and maize deactivates the dicamba, which enables the use of the herbicide both pre- and in-crop to control weeds (Behrens et al. 2007). Crops, namely maize, soy and cotton, have also been engineered to have 2,4-D tolerance through introduction of aad-1 and aad-12 and are being released as Enlist and Enlist Duo, when combined with glyphosate tolerance, by Dow AgroSciences (Wright et al. 2010).

A more recent GM-HT trait is the HPPD gene from maize and oats have been introduced into soy to confer herbicide tolerance to mesotrione and isoxaflutole herbicides, respectively (Siehl et al. 2014). This trait is stacked with other more GM herbicide tolerance traits like glyphosate or glufosinate tolerance (ISAAA 2016).

It is clear from Table 10.1 that for most species early introductions of HT traits were for single traits. However, over the past decade there have been an increasing number of stacked GM-HT traits approved and in the past few years several triple stacks have been approved. Stacked GM-HT traits offer potential advantages in allowing multiple herbicides to be used. However, there can be disadvantages. Stacked traits can trigger greater regulatory review that slows their development (Que et al. 2010). In some countries, this problem can be overcome by using breeding stacks of existing approved traits; however, then the problem of keeping the stack together may increase costs of seed. As the regulatory environment for stacked GM traits across the world is not harmonized, data requirements for regulatory agencies vary. In some countries, such as the EU, there are additional data requirements for breeding stacks (Kramer et al. 2016). As crops need to be approved for import of product regardless of whether they are going to be grown in that country, lack of regulatory harmonization is slowing the introduction and adoption of GM-HT stacked traits.

Adoption of GM-HT Crops

USA

Farms in the USA were the first to grow glyphosate tolerant crops, namely the Roundup Ready soybean which was released in 1996. With the promise of easier weed management through in-crop application of glyphosate, reduced to no-till farming, a favorable regulatory environment and a drop in glyphosate prices in 2000, cultivation of RR soybean went from 17% of the soybean area in 1997 to 93% in 2012 (Huang et al. 2015, Bonny 2016). GM-HT cotton was 10% of total cotton acreage in 1997 and increased to 80% in 2012 and GM-HT corn was at 73% in 2012. In the first year of GM-HT crop cultivation cost savings, as a result of reduced/no-till and consequent labor savings, were projected to be between $17–$30 per hectare in the US as a whole (Traxler 2006). Fuel savings alone were projected to be in the vicinity of 53L/ha as a result of reducing the trips from reduced tillage on the fields (Schwember 2008). Thus, the main incentive for GM-HT crops was the economic benefits of reduced farming inputs in order to achieve similar yields as non-GM counterparts.

Argentina

As in the USA, glyphosate tolerant soybeans were the first GM crops to be cultivated in 1996 in Argentina. Uptake of GM soybean was more rapid in Argentina compared to US, from 1% of total GM soybean acreage in 1996/97 to over 90% of soybean acreage being GM in 2001/02 (Trigo and Cap 2003). One of the contributing factors to the rapid deployment of GM crops in Argentina was the streamlined regulatory mechanisms that were in place prior to the commercialisation of GM soy which took place in 1996. The establishment of the National Seeds Institute, INASE, and the National Advisory Agricultural Biotechnology Commission (CONABIA) in 1991 helped with the rapid evaluation and approval of the new technology (Trigo and Cap 2003). GM seeds were also cheaper to acquire in Argentina compared to the USA as the relative lack of IP protection in Argentina meant that Monsanto did not have the monopoly on the GM seeds. Other companies like Nidera were able to develop and sell Roundup Ready technology without paying royalties. With multiple companies competing for the same market share, seed prices were pushed down and quickly adopted by farmers (Traxler 2006).

Canada

Although Canada only has 6.4% of the global GM acreage, approximately 95% of canola, 80% of grain corn, 60% of soy and nearly 100% of sugarbeet in Canada is GM containing either one or a combination of two main herbicide tolerance traits: glyphosate or glufosinate tolerance (CBAN 2015). As with the US, a favorable regulatory environment combined with the upholding of IP pertaining to the herbicide tolerant technologies developed by companies like Monsanto facilitated the spread of the GM crops (Smyth 2014). The rapid uptake of the GM technologies in Canada, as in the US, has been a result of economic benefits to the farmers.

Brazil

Brazil is the second largest grower of GM-HT crops, globally with a contribution of 23.3% of global GM acreage (CBAN 2015). Despite being second to US in terms of GM crop production, Brazil was slower to adopt GM-HT crops. While the Roundup Ready technology was approved in 1998, it was only cultivated legally in 2003 (Mendonça-Hagler et al. 2008). Despite farmer enthusiasm, international NGOs and environmental groups stalled its release for five years after approval (Paarlberg 2001, Mendonça-Hagler et al. 2008). Brazil, however, had the advantage of looking to other countries that had taken up the technology many years earlier and learn from various surveys conducted in farms across the Americas. Release of Roundup Ready soy was approved only after performing trials on sites that were representative of the future cultivation sites (Mendonça-Hagler et al. 2008). Cultivation of GM-HT corn was also heavily regulated in order to prevent contamination of the commercial non-GM landraces in Brazil and adoption has, as a result, been slower than neighboring countries. Due to the large market share Brazil has in soy, canola and corn globally, companies like Monsanto and Dow still have a presence in the country and the upholding of IP rights of the HT traits developed by the respective companies has maintained high interest levels in the country by the multinational companies (Mendonça-Hagler et al. 2008).

Australia

Australia has only 0.3% of the global share in GM acreage but is ranked as 13th in the world as a GM crop cultivator (CBAN 2015). However, cotton in Australia is almost 100% GM with most of it containing an insect resistance gene, Bt. Roundup Ready cotton was introduced in 2000 and was rapidly adopted as there were associated economic and environmental benefits (Werth et al. 2008). In contrast to the rapid adoption of GM cotton in Australia, the political climate wasn't ideal for growing 'food crops' like canola. Approvals for Roundup Ready and InVigor varieties occurred in 2004 but, due to moratoria placed on cultivation of GM canola in several states, commercial-scale cultivation didn't occur until 2008 (Hudson and Richards 2014).

Impacts of GM-HT Crops on Weed Management and Farming Systems

Positive Impacts

A key benefit of GM-HT crops has been the ability to control weeds using herbicide within the crop, particularly in situations where weed species are closely related to the growing crop, such as brassica weeds in canola (Senior and Dale 2002) and red rice in rice (Steele et al. 2002). More broadly, the adoption of GM-HT crops has led to simpler and less expensive weed control. Brookes and Barfoot (2016) estimate the total benefit of GM-HT soybeans in the US as $12.9 billion due to cost savings alone.

The availability of new herbicide resistance in crops has had other positive benefits. Tillage of cropping areas was reduced and in many cases no-till systems were adopted as a direct result of growing GM-HT crops. This has several advantages like the preservation of soil structure, increased water infiltration, maintenance of microbial activity, increase in soil carbon and reduction in soil erosion as a result of reduced tillage (Beckie et al. 2006, Cerdeira and Duke 2006, Huang et al. 2015). No-till also meant fewer passes across each field, saving on labor and farm operations (Cerdeira and Duke 2006). In some circumstances, no-till also improves yield (Beckie et al. 2006, Traxler 2006). Importantly, in areas where water is limited, such as most dryland farming operations, no-till allows timely sowing of the crop, maximising utilization of moisture and light during the growing season (Beckie et al. 2006, Farooq et al. 2011). An example of this is canola production in Canada which typically relied on soil incorporated pre-emergent herbicides, like ethalfluralin, for control of broadleaf weeds. This often meant that an application of herbicide had to follow the spring thaw prior to sowing the crop. Depending on the ability to get equipment across the land, this could delay sowing by some weeks (Beckie et al. 2006). The ability to control all weeds post-emergent with broad-spectrum herbicides, such as glyphosate and glufosinate, allowed sowing canola immediately after the spring thaw, maximising the growing season for canola (Beckie et al. 2006). In Argentina, the adoption of GM-HT soybeans allowed sowing of the crop immediately following harvest of the previous winter crop, providing the opportunity for double cropping in some areas, with a concomitant increase in productivity (Brookes and Barfoot 2005, Trigo and Cap 2006).

Where the advantages of GM-HT crops have been less obvious, adoption has been slower. For example, the adoption of GM-HT corn in the US, lagged adoption of GM-HT soybeans and canola (Dill et al. 2008). For corn there were already several effective post-emergent herbicides available, meaning that the weed control advantages were not as great and the extra seed costs made the economics of GM-HT crops not as attractive as other crops (Owen 2000). This situation changed with the introduction of stacked GM-HT and IT cultivars and approvals for import into Europe. The value of the Bt trait for the control of rootworm and European stalk borer enhanced adoption (Dill et al. 2008).

Adoption of GM-HT canola in Australia has also been relatively slow. Following the end of a moratorium on production of GM canola in 2008, adoption of glyphosate tolerant was slow (Hudson and Richards 2014) and in 2016 represented 23% of the canola area sown (Agricultural Biotechnology Council of Australia 2016). In Australia triazine-tolerant canola had already been widely adopted and provided acceptable control of grass weeds (Salisbury et al. 2016). The extra costs associated with the production of glyphosate-tolerant canola (technology fee, extra transport costs for delivery and price discounts on grain) have not made the crop attractive enough for growers in the Eastern states. In Western Australia, GM-HT canola has been more widely adopted as the herbicide provides better control of the weed spectrum (Hudson and Richards 2014).

Negative Impacts of GM-HT Crop Adoption on Farming Systems

The negative impacts of GM-HT crop adoption on farming systems are relatively few in number. The two major negative impacts have been weed species' shifts and the evolution of herbicide-resistant weeds. Both of these negative impacts occurred as a result of the choice by farmers to rely heavily or exclusively on a single herbicide for control of weeds in GM-HT crops. The ability to use a single broad spectrum herbicide to control all or most weeds within the crop and to simplify weed management was too tempting (Duke and Powles 2009).

Weed species shifts were predicted from prior to the introduction of GM-HT crops (Duke 1996, Shaner 2000). There was already a considerable amount of literature examining the relative tolerance of various weed species to glyphosate, in particular, and predictions were made for which weeds were likely to become more common in glyphosate-tolerant cropping systems (Reddy 2004). In areas where rotations consisting mostly or completely of glyphosate tolerant crops, species shifts were indeed found (Webster and Sosnoski 2010, Werth et al. 2013). The invasion of *Commelina benghalensis* in cotton production in the southern USA was an early example (Culpepper 2006, Webster and Sosnoski 2010).

Of much greater impact than species shift was the evolution of glyphosate-resistant weeds. Prior to the introduction of GM-HT crops, it was thought that weeds species would find it exceptionally difficult to evolve resistance to glyphosate (Laura et al. 1997). However, the first glyphosate resistant weed population was identified just prior to the introduction of glyphosate resistant crops. *Lolium rigidum* evolved resistance to glyphosate in no-till grain production and orchards in Australia in 1996 (Powles et al. 1998, Pratley et al. 1999), demonstrating the potential for glyphosate resistance evolution.

By 2016, a total of 36 weeds species evolved resistance to glyphosate in 27 countries (Table 10.2). Glyphosate use in GM-HT crops has played a major role in the selection for glyphosate resistance in 14 of these weed species in at least one country and a more minor role for a further seven species. However, glyphosate resistance does evolve in the absence of GM crops with resistance appearing in countries that grow no GM crops, such as Japan, France and Switzerland (Table 10.2) (Heap 2017). There are examples of weed species that have evolved resistance in one country driven by glyphosate use in GM-HT crops, and in a second country where GM-HT crops have played no role in resistance evolution. Glyphosate resistance is likely to occur in all situations where glyphosate is used intensively, not just in GM-HT crops.

However, the subsequent very intensive use of glyphosate in glyphosate-tolerant crops quickly resulted in the evolution of glyphosate-resistant weeds. Firstly in *Conyza canadensis* in soybean production in the US (Van Gessel 2001). Resistance in *C. canadensis* spread rapidly across soybean production areas, partly aided by the rapid dispersal of *C. canadensis* seed (Dauer et al. 2006), but more by the practices of soybean farmers, who tended to grow continuous soybeans and use glyphosate both for burn down and in crop uses (Koger et al. 2004, Owen and Zeleya 2005, Davis et al. 2008). Glyphosate resistant *C. canadensis* was relatively easily managed by the incorporation of 2,4-D or other herbicides into the burn-down herbicide application (Everitt and Keeling 2007, Duke and Powles 2009, Owen et al. 2011).

A far more difficult issue arose with the evolution of glyphosate resistant *Amaranthus palmeri* in cotton production in the southern US (Culpepper et al. 2006). This is an aggressive weed species that produces large amounts of seed, is dioecious and germinates throughout the season. Once again, glyphosate resistant *A. palmeri* spread rapidly across cotton and soybean production areas in the southern US and even expanded into soybean and corn production areas (Duke and Powles 2009). Shortly afterwards, glyphosate resistance evolved in the related species *Amaranthus rudis* in soybean and corn production regions throughout the midwest of the US (Legleiter and Bradley 2008). While seed and pollen movement of these weed species may have played a role in the rapid spread of resistance, in this case the similarity of practices used by farmers was again the main culprit (Culpepper et al. 2008). Unlike the situation with *C. Canadensis*, however, there was no simple management practice that could be easily introduced for these weed species. This had very significant impacts on productivity in fields infested with *A. palmeri*, particularly in the relatively less competitive cotton crops. In response to glyphosate resistant *A. palmeri*, farmers adopted expensive and intensive tactics, including multiple applications of soil applied herbicides and the reintroduction of hand weeding. Tillage was also reintroduced in some circumstances (Price et al. 2011, Shaw et al. 2012, Sosnoski and Culpepper 2014).

Other glyphosate resistant weeds have also evolved in response to the adoption of GM-HT glyphosate resistant crops. In the Missouri River valley and through southern Ontario in Canada, glyphosate-resistant *Ambrosia trifida* and *Ambrosia artemisiifolia* occur (Norsworthy et al. 2011, Vink et al. 2012, Heap 2017). In the Great Plains of the US and prairies of Canada, glyphosate resistance has evolved in *Kochia scoparia* (Beckie et al. 2013, Hall et al. 2014, Wiersma et al. 2015). However, in this case, glyphosate use in chemical fallows has played a significant role (Beckie et al. 2013). The evolution of glyphosate resistant weeds has had a significant impact on corn yield and weed control costs in the US (Wechsler et al. 2017). Argentina has seen the evolution of glyphosate resistant *Sorghum halepense* in glyphosate resistant soybean production (Vila-Aiub et al. 2007). Likewise, Brazil has seen the evolution of glyphosate-resistant *Digitaria insularis* in glyphosate-resistant soybeans (de Carvalho et al. 2012) (Table 10.2). In all of these cases, the intensive use of glyphosate has led to the evolution of glyphosate-resistant weeds (Heap 2017).

The Strange Case of Glufosinate

The other main broad spectrum herbicide used in GM-HT crops in contrast has seen relatively little evolution of glufosinate resistance. This is not because resistance to glufosinate is particularly difficult for plants to evolve. Glufosinate resistance has evolved in several weed species but, significantly, none of those events has been related to GM-HT crop production (Heap 2017).

The lack of resistance to glufosinate associated with GM-HT crops is related to the use patterns of the herbicide. The major GM-HT crop with glufosinate resistance is canola grown in Canada and the US, where it makes up almost 50% of the canola sown (Beckie et al. 2011). In these areas there is the alternative option of glyphosate tolerant canola, which means that glufosinate may not be used every time canola is grown. More important is the fact that canola in these areas is rotated with other crops, rather than being sown every year. In the non-canola production years, glufosinate is not used. This means that glufosinate will be used at most once every two years (Beckie et al. 2011). This has taken the selection pressure off the herbicide and delayed the evolution of resistance.

Table 10.2. Weed species that have evolved resistance to the herbicide by country and the role of GM-HT crops in the selection for resistance (Heap 2017)

Weed species	Country	Year first detected	Role of GM crops in selection[a]
Amaranthus hybridus	Argentina	2013	Major
Amaranthus palmeri	USA	2005	Major
	Argentina	2015	Major
	Brazil	2015	Major
Amaranthus spinosus	USA	2012	Major
Amaranthus tuberculatus	USA	2005	Major
	Canada	2014	Major
Ambrosia artemisiifolia	USA	2004	Major
	Canada	2012	Major
Ambrosia trifida	USA	2004	Major
	Canada	2008	Major
Bidens pilosa	Mexico	2014	None
Brachiaria eruciformis	Australia	2014	None
Bromus diandrus	Australia	2011	None
Bromus rubens	Australia	2014	None
Chloris elata	Brazil	2014	Major
Chloris truncata	Australia	2010	None
Chloris virgata	Australia	2015	None
Conyza bonariensis	South Africa	2003	None
	Spain	2004	None
	Brazil	2005	None
	Israel	2005	None
	Colombia	2006	None
	USA	2007	None
	Australia	2010	Minor
	Greece	2010	None
	Portugal	2010	None
Conyza canadensis	USA	2000	Major
	Brazil	2005	Minor
	China	2006	None
	Spain	2006	None
	Czech Republic	2007	None
	Canada	2010	Major
	Poland	2010	None
	Italy	2011	None
	Portugal	2011	None
	Greece	2012	None

(Contd.)

Table 10.2. (*Contd.*)

Weed species	Country	Year first detected	Role of GM crops in selection[a]
	Japan	2014	None
Conyza sumatrensis	Spain	2009	None
	Brazil	2010	Major
	France	2010	None
	Greece	2012	None
Cynodon hirsutus	Argentina	2008	Major
Digitaria insularis	Paraguay	2005	Major
	Brazil	2008	Major
	Argentina	2014	Major
Echinochloa colona	Australia	2007	Minor
	USA	2008	Minor
	Venezuela	2008	None
	Argentina	2009	Major
Eleusine indica	Malaysia	1997	None
	Colombia	2006	None
	Bolivia	2007	Major
	China	2010	None
	Costa Rica	2010	None
	USA	2011	Major
	Argentina	2012	Major
	Indonesia	2012	None
	Japan	2013	None
	Brazil	2016	Major
Hedyotis verticillata	Malaysia	2005	None
Kochia scoparia	USA	2007	Minor
	Canada	2012	Minor
Lactuca serriola	Australia	2015	None
Leptochloa virgata	Mexico	2010	None
Lolium perenne	Argentina	2008	Minor
	New Zealand	2012	None
	Portugal	2013	None
Lolium multiflorum	Chile	2001	None
	Brazil	2003	Minor
	USA	2004	Minor
	Spain	2006	None

(*Contd.*)

	Argentina	2007	Minor
	Italy	2008	None
	Japan	2011	None
	Switzerland	2011	None
	New Zealand	2012	None
Lolium rigidum	Australia	1996	Minor
	USA	1998	None
	South Africa	2001	None
	France	2005	None
	Spain	2006	None
	Israel	2007	None
	Italy	2007	None
Parthenium hysterphorus	Colombia	2004	None
	USA	2014	None
Plantago lanceolata	South Africa	2003	None
Poa annua	USA	2010	None
Raphanus raphanistrum	Australia	2010	Minor
Salsola tragus	USA	2015	None
Sonchus oleraceus	Australia	2014	Minor
Sorghum halepense	Argentina	2005	Major
	USA	2007	Major
Tridax procumbens	Australia	2016	None
Urochloa panicoides	Australia	2008	None

[a] Major: Most or all resistant populations have occurred in GM-HT crop fields following use of glyphosate, Minor: Most resistant populations have occurred outside GM-HT crop fields, None: All resistant populations occurred in cropping systems or in phases of the rotation where GM-HT crops were not grown.

Impacts of GM-HT Crops on the Environment

In addition to their impacts in production systems, GM-HT crops can also have impacts on the environment. These impacts can be both positive and negative. The main impacts of GM-HT crops considered have been: changes to the environmental impacts of herbicides used, changes in tillage and the impact on CO_2 emissions, out-crossing to related plant species, and impacts on non-target species.

A number of studies have attempted to assess the environmental impact of GM-HT crops by using environmental impact (EI) based on the environmental impact quotient (EIQ) of pesticides. The EIQ has three components: an applicator component, a consumer component and an ecological component that are averaged to create a value for the EIQ (Kovach et al. 1992). Brookes and Barfoot (2005, 2011, 2017) have made several examinations of the EI of GM-HT crops. Up to 2004, GM-HT crops had reduced the EI of herbicides by 3% for maize, 19% for soybeans, 21% for canola and 22% for cotton (Brookes and Barfoot 2005). This was largely due to the substitution of glyphosate for more toxic herbicides. By 2011, the reduction in EI was 10% for cotton, 13% for maize, 14% for soybeans, 30% for canola, and 1% for sugar beets (Brookes and Barfoot 2017). A review collating 13 years of data on GM US maize and soybeans found that while herbicide use in some fields actually went up as a result of increasing weed resistance, the EIQ scores for farms containing GM-HT crops actually reduced or were the same compared to the conventional systems (Perry et al. 2016).

Another environmental advantage of GM-HT crops has been the reduction in tillage associated with the adoption of some of these crops (Cerdeira and Duke 2006, Smyth et al. 2011). Brookes and Barfoot (2017) estimate that by 2011 the reduction in tillage in GM-HT crops had reduced CO_2 emissions by 2.5 million kg just through reductions in fuel use. Smyth et al. (2011) estimated that in Canada alone one million tonnes of carbon had been sequestered by changes in tillage practices by the adoption of HT canola.

There has been considerable interest in the potential impacts of gene flow from GM crops to other plant species (Ellstrand et al. 1999, Snow 2002, Lu and Snow 2005). While gene flow has occurred from HT crop plants to wild relatives in the field (Reiger et al. 2001, Gealy et al. 2003, Simard et al. 2007), to date there is no evidence this is having a detrimental environmental impact. Likewise, GM-HT crops can escape cultivation and appear in disturbed habitats, such as road sides (Knipsel and McLaughlan 2009, Nishizawa et al. 2009). However, again there is no evidence this is having a significant environmental impact.

Impacts on non-target species can occur with GM-HT crops either through changes in the crop species mix in the landscape or through changes in herbicide practices. The farm scale evaluation of GM crops in the UK attempted to address the non-target impacts of changing herbicide practice. It showed a reduction in weed seeds present in fields when more effective herbicides were used (Heard et al. 2003). This resulted in a reduction in some insect species (Hawes et al. 2003). There was an increase in detritivores in the GM-HT crops, possibly caused by the decaying biomass from weeds that were treated with herbicide later than those in conventional crops (Brooks et al. 2003). There were flow on effects leading to reductions in abundance of birds in GM-HT crops, typically granivores and species relying on insects attracted to weed flowers (Chamberlain et al. 2007).

Increased herbicide efficacy can also have an impact on insect species by depriving them of essential hosts. The increased efficacy of glyphosate in GM-HT corn and soybean fields in the US has resulted in a large reduction of milkweed (*Asclepias syriaca*) in crops (Hartzler 2010). Even though milkweed is less common in crop fields compared with roadsides (Hartzler and Buhler 2000), this reduction in milkweed in crops has been correlated with a reduction in monarch butterfly (*Danaus plexippus*) populations in the US (Pleasants and Oberhauser 2013).

Role of GM-HT Crops in Sustainable Weed Management

GM-HT crops by providing additional herbicide options for weed control have a major role to play in sustainable weed management. In part they have fulfilled this role in some situations where existing herbicide resistance was a constraint to production, such as soybean production in the US. However, for the most part this promise has been unfulfilled. The main reason for this has been the behavior of farmers. There has been a preference for the use of GM-HT crops to simplify and improve farming systems over implementation of more sustainable weed management (Duke and Powles 2008).

Often the response to the introduction of GM-HT crops is to use the over the top broad spectrum herbicide exclusively for weed management. This has simplified weed management decisions and made farm management easier, particularly where a single post-emergent herbicide has replaced multiple herbicides and tillage. Even where crop rotation has been practiced, multiple crops with the same HT trait has resulted in the same herbicide being used in every crop (Duke 2015). This has been the principal reaction of farmers who have adopted GM-HT corn, soybeans or cotton wherever they have been commercialised. The contrast has been the adoption of GM-HT canola in Canada. This is the most successful example of sustainable weed management using GM-HT crops; however, this has been more inadvertent than deliberate. In the case of GM-HT canola, two different traits were made available at about the same time. In contrast to some other examples, such as cotton in the US, both traits were adopted by growers (Beckie et al. 2006). This allowed the rotation of glyphosate and glufosinate for weed control where individual farmers opted to use both traits (Beckie et al. 2011). This was aided by the fact that canola is typically rotated with at least one crop between canola sowings. The ability to rotate herbicide modes of action has played a significant role in delaying the onset of glyphosate and glufosinate resistant weeds in Canadian canola production (Beckie et al. 2011).

However, even the availability of multiple GM-HT traits may not be sufficient to encourage the rotation of herbicides and other practices. For example, when both bromoxynil and glyphosate tolerance in cotton were available in the US, growers tended to use one trait or the other. Bromoxynil tolerant cotton was principally grown in Texas, whereas it was not grown at all in the southeastern cotton production areas. Likewise, the introduction of bromoxynil tolerant canola in Canada in 2000 was not widely adopted by growers and had disappeared by 2003 (Beckie et al. 2011).

The marketing of additional HT traits is not sufficient to encourage improved weed management sustainability. Herbicides that are insufficiently active against key weed species, or that have impacts on rotational crops will struggle to be adopted where a better alternative exists (Bryan 2006).

Future HT Traits and their Role in Weed Management

Dicamba and 2,4-D

The evolution of glyphosate resistant weeds, particularly *Amaranthus* spp. in the US precipitated interest in new herbicide tolerance traits in crops. In the US, much of the interest has been in developing crops that offer new alternatives for the control of broadleaf weeds, such as the *Amaranthus* spp. in cotton and soybeans. Two traits have recently been approved in the US. These are dicamba resistance and 2,4-D resistance. Both of these traits will come stacked with glyphosate resistance. Dicamba and glyphosate resistance will be further stacked with glufosinate resistance for some crops. The availability of stacked HT traits will allow choice of herbicide options for growers and facilitate rotation of herbicide use (Meyer et al. 2015, Schulz and Segobye 2016). Unfortunately, it is likely that these crops will be most widely adopted in areas where resistance to glyphosate in broadleaf weeds is already widespread. This means that at least for these glyphosate resistant weed species, the majority of the selection pressure will occur on dicamba or 2,4-D. Like glyphosate, resistance to the hormone herbicides took a long time to evolve (Schulz and Segobye 2016, Heap 2017). However, there has been an increasing number of reports of resistance to this mode of action in recent years. Where resistance consists of complex traits, involving several genes, resistance evolution can be slowed due to the need to select each of the resistance alleles (Jasieniuk et al. 1996). Where the inheritance of resistance to hormone herbicides has been examined, there are examples of multi-gene inheritance (Weinberg et al. 2006) and recessive gene inheritance (Sabba et al. 2003), both of which would tend to delay the evolution of resistance. However, in other species, resistance is endowed by a single largely dominant allele that will mean resistance can be selected more easily (Preston et al. 2009, Preston and Malone 2015).

It is likely that the relatively slow evolution of resistance to the hormone herbicides was the result in part of low intensity of selection. Examples of resistance occurred in situations like the control of thistles in permanent pastures (Bourdôt et al. 2007). The increased use of these herbicides where resistance to alternative herbicides has occurred has resulted in an increasing frequency of resistance to this mode of action. It is likely that the introduction of crops with resistance to these herbicides will accelerate the evolution of resistance in weeds.

Other Modes of Action

Additional HT traits are being developed, but these remain some years away. These include resistance to ACCase and HPPD inhibitors. None of these herbicides are immune to resistance evolution in weeds. Indeed for both modes of action, resistance has evolved more quickly after the introduction of the herbicides than it did with glyphosate (Green 2012, Heap 2017). Therefore, the introduction of future HT traits might lead to rapid and extensive resistance if the herbicides are already being widely used and resistance in weeds occurs frequently.

Concluding Remarks – A New Paradigm Regarding HT Crops in IWM

The experience of the introduction of GM-HT crops has been wide adoption where the associated herbicides provided benefits to growers, particularly in the control of difficult weeds or for simplicity of weed management. While these crops had the potential to increase the sustainability of weed management through the introduction of alternative herbicide practices, this has rarely occurred. Instead in most locations, farmers have relied on a single herbicide for the bulk of their weed control and this has inevitably led to herbicide resistance in weeds.

The one exception to this experience is the adoption of GM-HT canola in Canada. In that case, two herbicide resistance traits were introduced at approximately the same time and were adopted in roughly equal amounts (Beckie et al. 2011). However, in other areas despite the availability of more than one HT trait, growers selectively used a single trait. This was often due to the lack of performance of one of the GM-HT introductions.

Given previous experience, how could GM-HT crops be better used to improve sustainability of weed control? Clearly, the solution is to create systems where a single herbicide is not used exclusively for weed management. There are considerable difficulties in achieving this and it will require a multi-faceted approach.

One option would be a regulatory framework that limited the amount of area that could be planted to a particular trait. Such an option was successfully employed with the introduction of BT cotton in Australia (Fitt 2000). Such regulatory approaches have difficulties with equity, who gets to use the crops, and with resistance selection occurring regardless if the herbicides are being used elsewhere in the environment.

Another approach would involve farmer education of the risks of resistance and the need to rotate herbicide modes of action and to introduce other tactics to delay the onset of herbicide resistance in weeds. To achieve a sustainable outcome from such a course of action, useful alternatives must be available (Asmus and Schroeder 2016). The historical approach that mostly led to a single HT trait being introduced resulted in growers using that trait exclusively. Even in situations where more than one trait was introduced, often a single trait was preferred. There are several reasons why growers would choose such a strategy. If one herbicide was considerably more efficacious than the other or because one trait occurred in better yielding varieties of the crop. The situation was made worse by existing uses of the same herbicide. Where a HT trait introduced was for a herbicide already widely used elsewhere in the cropping system the risks were greater (Green 2012, 2014).

In the early stage of the introduction of HT crops, there was considerable concern about the potential introduction of stacked HT crops. Much of this concern revolved around the ability to control volunteers of the crop if it was resistant to multiple herbicides. In hindsight, introducing stacked traits at the beginning would have provided growers with more ability to rotate herbicide modes of action or to use a second herbicide to control survivors of the first herbicide. There are several difficulties in implementing such a plan. The first is having sufficient effective HT traits available. A second difficulty is that stacked traits are likely to suffer greater regulatory difficulties in approvals. Finally, even if stacked traits were available, there is no guarantee that growers would use them wisely.

Clearly to achieve a more sustainable use of GM-HT crops in agriculture involves more sustainable use of the herbicide products. Offering stacked HT traits allows growers to use one, more or all of the herbicides to improve weed management. This could encourage more rotation of herbicide modes of action. However, other practices must be used to sustain the herbicides for the longer term. Finding ways to ensure growers use these additional tactics for weed control will be the difficult task. Regulation could be counter-productive, however, incentivising growers and a targeted education program may be a better approach. Ultimately, it is the HT trait providers and the growers who have the most to lose from the loss of the herbicides to resistance and both should invest in their more sustainable use.

REFERENCES

Aerni, P. 2013. Resistance to agricultural biotechnology: the importance of distinguishing between weak and strong public attitudes. Biotechnology Journal 8: 1129–1132.

Agricultural Biotechnology Council of Australia. 2016. GM canola growth in Australia. Available at: https://www. abca. com. au/wp-content/uploads/2016/08/2016-GM-canola-growth-in-Australia. pdf (Accessed 30 December 2016).

Asmus, A. and J. Schroeder. 2016. Rethinking outreach: collaboration is key for herbicide-resistance management. Weed Science 64: 655–660.

Barry, G.F., G.M. Kishore and S.R. Padgette. 1992. Glyphosate tolerant 5-enolpyruvylshikimate-3-phosphate synthases. International patent number # WO 92/04449.

Beckie, H.J., K.N. Harker, L.M. Hall, S.I. Warwick, A. Légère, P.H. Sikkema, G.W. Clayton, A.G. Thomas, J.Y. Leeson, G. Séguin-Swartz and M.J. Simard. 2006. A decade of herbicide-resistant crops in Canada. Canadian Journal of Plant Science 86: 1243–1264.

Beckie, H.J., K.N. Harker, A. Legere, M.J. Morrison, G. Seguin-Swartz and K.C. Falk. 2011. GM canola: the Canadian experience. Farm Policy Journal 8: 43–49.

Beckie, H.J., R.E. Blackshaw, R. Low, L.M. Hall, C.A. Sauder, S. Martin, R.N. Brandt and S.W. Shirriff. 2013. Glyphosate- and acetolactate synthase inhibitor-resistant kochia (*Kochia scoparia*) in Western Canada. Weed Science 61: 310–318.

Behrens, M.R., N. Mutlu, S. Chakraborty, R. Dumitru, W.Z. Jiang, B.J. Lavallee, P.L. Herman, T.E. Clemente and D.P. Weeks. 2007. Dicamba resistance: enlarging and preserving biotechnology-based weed management strategies. Science 5828: 1185–1188.

Block, M.D., J. Botterman, M. Vandewiele, J. Dockx, C. Thoen, V. Gosselé, N.R. Movva, C. Thompson, M.V. Montagu and J. Leemans. 1987. Engineering herbicide resistance in plants by expression of a detoxifying enzyme. The EMBO Journal 6: 2513–2518.

Bonny, S. 2016. Genetically modified herbicide-tolerant crops, weeds, and herbicides: overview and impact. Environmental Management 57: 31–48.

Bourdôt, G.W., S.V. Fowler, G.R. Edwards, D.J. Kriticos, J.M. Kean, A. Rahman and A.J. Parsons. 2007. Pastoral weeds in New Zealand: status and potential solutions. New Zealand Journal of Agricultural Research 50: 139–161.

Brookes, G. and P. Barfoot. 2005. GM crops: the global economic and environmental impact – the first nine years 1996–2004. AgBioForum 8: 187–196.

Brookes, G. and P. Barfoot. 2011. Global impact of biotech crops: environmental effects 1996–2009. GM Crops 2: 34–49.

Brookes, G. and P. Barfoot 2016. Global income and production impacts of using GM crop technology 1996–2014. GM Crops & Food 7: 38–77.

Brookes, G. and P. Barfoot. 2017. Environmental impacts of genetically modified (GM) crop use 1996–2015: impacts on pesticide use and carbon emissions. GM Crops & Food 8: 117–147.

Brooks, D.R., D.A. Bohan, G.T. Champion, A.J. Haughton, C. Hawes, M.S. Heard, S.J. Clark, A.M. Dewar, L.G. Firbank, J.N. Perry and P. Rothery. 2003. Invertebrate responses to the management of genetically modified herbicide–tolerant and conventional spring crops. I. Soil-surface-active invertebrates. Philosophical Transactions of the Royal Society of London B: Biological Sciences 358: 1847–1862.

Bryan, G.Y. 2006. Changes in herbicide use patterns and production practices resulting from glyphosate-resistant crops. Weed Technology 20: 301–307.

Castle, L.A., D.L. Siehl, R. Gorton, P.A. Patten, Y.H. Chen, S. Bertain, H.J. Cho, N. Duck, J. Wong, D. Liu and M.W. Lassner. 2004. Discovery and directed evolution of a glyphosate tolerance gene. Science 304: 1151–1154.

CBAN. 2015. Where in the world are GM crops and foods? GMO Inquiry 2015. Ontario, Canada. CBAN.

Cerdeira, A.L. and S.O. Duke. 2006. The current status and environmental impacts of glyphosate-resistant crops. Journal of Environmental Quality 35: 1633–1658.

Chamberlain, D.E., S.N. Freeman and J.A. Vickery. 2007. The effects of GMHT crops on bird abundance in arable fields in the UK. Agriculture, Ecosystems & Environment 118: 350–356.

Culpepper, A.S. 2006. Glyphosate-induced weed shifts. Weed Technology 20: 277–281.

Dauer, J.T., D.A. Mortensen and R. Humston. 2006. Controlled experiments to predict horseweed (*Conyza canadensis*) dispersal distances. Weed Science 54: 484–489.

de Carvalho, L.B., H. Cruz-Hipolito, F. González-Torralva, P.L. da Costa Aguiar Alves, P.J. Christoffoleti and R. De Prado. 2011. Detection of sourgrass (*Digitaria insularis*) biotypes resistant to glyphosate in Brazil. Weed Science 59: 171–176.

Dill, G.M., C.A. Cajacob and S.R. Padgette. 2008. Glyphosate-resistant crops: adoption, use and future considerations. Pest Management Science 64: 326–331.

Duke, S.O. 2005. Taking stock of herbicide-resistant crops ten years after introduction. Pest Management Science 61: 211–218.

Duke, S.O. 2015. Perspectives on transgenic, herbicide-resistant crops in the United States almost 20 years after introduction. Pest Management Science 71: 652–657.

Duke, S.O. and S.B. Powles. 2008. Glyphosate: a once-in-a-century herbicide. Pest Management Science 64: 319–325.

Duke, S.O. and S.B. Powles. 2009. Glyphosate-resistant crops and weeds: now and in the future. AgBioForum 12: 346–357.

Ellstrand, N.C., H.C. Prentice and J.F. Hancock. 1999. Gene flow and introgression from domesticated plants into their wild relatives. Annual Review of Ecology and Systematics 30: 539–563.

Farooq, M., K.C. Flower, K. Jabran, A. Wahid and K.H.M. Siddique. 2011. Crop yield and weed management in rainfed conservation agriculture. Soil and Tillage Research 117: 172–183.

Fitt, G.P. 2000. An Australian approach to IPM in cotton: integrating new technologies to minimise insecticide dependence. Crop Protection 19: 793–800.

Funke, T., H. Han, M.L. Healy-Fried, M. Fischer and E. Schönbrunn. 2006. Molecular basis for the herbicide resistance of Roundup Ready crops. Proceedings of the National Academy of Sciences 103: 13010–13015.

Gealy, D.R., D.H. Mitten and J.N. Rutger. 2003. Gene flow between red rice (*Oryza sativa*) and herbicide-resistant rice (*O. sativa*): implications for weed management 1. Weed Technology 17: 627–645.

Gianessi, L.P. 2013. The increasing importance of herbicides in worldwide crop production. Pest Management Science 69: 1099–1105.

Gianessi, L.P. and N.P. Reigner. 2007. The Value of Herbicides in US Crop Production. Weed Technology 21: 559–566.

Giesy, J.P., S. Dobson and K.R. Solomon. 2000. Ecotoxicological Risk Assessment for Roundup® Herbicide. pp. 35–120. *In:* Ware, G.W. (Ed.) Reviews of Environmental Contamination and Toxicology: Continuation of Residue Reviews. Springer, New York.

Gleim, S., S.J. Smyth and P.W. Phillips. 2016. Regulatory System Impacts on Global GM Crop Adoption Patterns. Estey Centre Journal of International Law and Trade Policy 17.

Green, J.M. 2012. The benefits of herbicide-resistant crops. Pest Management Science 68: 1323–1331.

Green, J.M. 2014. Current state of herbicides in herbicide-resistant crops. Pest Management Science 70: 1351–1357.

Hall, L.M., H.J. Beckie, R. Low, S.W. Shirriff, R.E. Blackshaw, N. Kimmel and C. Neeser. 2013. Survey of glyphosate-resistant kochia (*Kochia scoparia* L. Schrad.) in Alberta. Canadian Journal of Plant Science 94: 127–130.

Hartzler, R.G. 2010. Reduction in common milkweed (*Asclepias syriaca*) occurrence in Iowa cropland from 1999 to 2009. Crop Protection 29: 1542–1544.

Hartzler, R.G. and D.D. Buhler. 2000. Occurrence of common milkweed (*Asclepias syriaca*) in cropland and adjacent areas. Crop Protection 19: 363–366.

Hawes, C., A.J. Haughton, J.L. Osborne, D.B. Roy, S.J. Clark, J.N. Perry, P. Rothery, D.A. Bohan, D.R. Brooks, G.T. Champion and A.M. Dewar. 2003. Responses of plants and invertebrate trophic groups to contrasting herbicide regimes in the farm scale evaluations of genetically modified herbicide–tolerant crops. Philosophical Transactions of the Royal Society B. Biological Sciences 358: 1899–1913.

Heap, I. 2017. The International Survey of Herbicide Resistant Weeds [Online]. Available at: www. weedscience. org. (Accessed on February 21, 2017).

Heard, M.S., C. Hawes, G.T. Champion, S.J. Clark, L.G. Firbank, A.J. Haughton, A.M. Parish, J.N. Perry, P. Rothery, R.J. Scott and M.P. Skellern. 2003. Weeds in fields with contrasting conventional and genetically modified herbicide–tolerant crops. I. Effects on abundance and diversity. Philosophical Transactions of the Royal Society of London B. Biological Sciences 358: 1819–1832.

Heck, G.R., C.L. Armstrong, J.D. Astwood, C.F. Behr, J.T. Bookout, S.M. Brown, T.A. Cavato, D.L. Deboer, Y.M. Deng, C. George, J.R. Hillyard, C.M. Hironaka, A.R. Howe, E.H. Jakse, B.E. Ledesma, T.C. Lee, R.P. Lirette, M.L. Mangano, J.N. Mutz, Y. Qi, R.E. Rodriguez, S.R. Sidhu, A. Silvanovich, M.A. Stoecker, R.A. Yingling and J. You. 2005. Development and characterization of a CP4 EPSPS-based, glyphosate-tolerant corn event. Crop Science 45: 329–339.

Hollaway, K.L., R.S. Kookana, D.M. Noy, J.G. Smith and N. Wilhelm. 2006. Crop damage caused by residual acetolactate synthase herbicides in the soils of south-eastern Australia. Australian Journal of Experimental Agriculture 46: 1323–1331.

Huang, J., C. Ellis, B. Hauge, Y. Qi and M. Varagona. 2015. Herbicide tolerance. pp. 213–237. *In:* K. Azhakanandam, A. Silverstone, H. Daniell and M.R. Davey (Eds.) Recent Advancements in Gene Expression and Enabling Technologies in Crop Plants. Springer New York, New York, USA.

Hudson, D. and R. Richards. 2014. Evaluation of the agronomic, environmental, economic, and coexistence impacts following the introduction of GM Canola to Australia (2008–2010). AgBioForum 17: 1–12.

ISAAA. 2016. Commercial GM Trait: Herbicide Tolerance [Online]. Available at: http://www. isaaa. org/gmapprovaldatabase/commercialtrait/default. asp?TraitTypeID=1andTraitType=Herbici de%20Tolerance.

Jasieniuk, M., A.L. Brûlé-Babel and I.N. Morrison 1996. The evolution and genetics of herbicide resistance in weeds. Weed Science 44: 176–193.

Kishore, G.M., Padgette, S.R. and Fraley, R.T. 1992. History of herbicide-tolerant crops, methods of development and current state of the art: emphasis on glyphosate tolerance. Weed Technology 6: 626–634.

Knispel, A.L. and S.M. McLachlan. 2010. Landscape-scale distribution and persistence of genetically modified oilseed rape (*Brassica napus*) in Manitoba, Canada. Environmental Science and Pollution Research 17: 13–25.

Kovach J., C. Petzoldt, J. Degni and J. Tette. 1992. A method to measure the environmental impact of pesticides. New York Food and Life Sciences Bulletin 139: 8.

Kramer, C., P. Brune, J. McDonald, M. Nesbitt, A. Sauve and S. Storck-Weyhermueller. 2016. Evolution of risk assessment strategies for food and feed uses of stacked GM events. Plant Biotechnology Journal 14: 1899–1913.

Laura, D.B., S.R. Padgette, S.L. Kimball and H.W. Barbara. 1997. Perspectives on glyphosate resistance. Weed Technology 11: 189–198.

Lebrun, M., A. Sailland, G. Freyssinet and E. Degryse. 2003. Mutated 5-enolpyruvylshikimate-3-phosphate synthase, gene coding for said protein and transformed plants containing said gene. U.S. Patent # 6, 566, 587.

Leyser, O. 2014. Moving beyond the GM debate. PLOS Biology 12: e1001887.

Lu, B.R. and A.A. Snow. 2005. Gene flow from genetically modified rice and its environmental consequences. AIBS Bulletin 55: 669–678.

Mazur, B.J. and S.C. Falco. 1989. The development of herbicide resistant crops. Annual Review of Plant Biology 40: 441–470.

McElroy, D., W. Zhang, J. Cao and R. Wu. 1990. Isolation of an efficient actin promoter for use in rice transformation. The Plant Cell 2: 163–171.

Mendonça-Hagler, L., L. Souza, L. Aleixo and L. Oda. 2008. Trends in biotechnology and biosafety in Brazil. Environmental Biosafety Research 7: 115–121.

Meyer, C.J., J.K. Norsworthy, B.G. Young, L.E. Steckel, K.W. Bradley, W.G. Johnson, M.M. Loux, V.M. Davis, G.R. Kruger, M.T. Bararpour, J.T. Ikley, D.J. Spaunhorst and T.R. Butts. 2015. Herbicide program approaches for managing glyphosate-resistant Palmer Amaranth (*Amaranthus palmeri*) and Waterhemp (*Amaranthus tuberculatus* and *Amaranthus rudis*) in future soybean-trait technologies. Weed Technology 29: 716–729.

Nishizawa, T., N. Nakajima, M. Aono, M. Tamaoki, A. Kubo and H. Saji. 2009. Monitoring the occurrence of genetically modified oilseed rape growing along a Japanese roadside: 3-year observations. Environmental Biosafety Research 8: 33–44.

Norsworthy, J.K., D. Riar, P. Jha and R.C. Scott. 2011. Confirmation, control, and physiology of glyphosate-resistant giant ragweed (*Ambrosia trifida*) in Arkansas. Weed Technology 25: 430–435.

Owen, M.D.K. 2000. Current use of transgenic herbicide-resistant soybean and corn in the USA. Crop Protection 19: 765–771.

Paarlberg, R.L. 2001. The politics of precaution: genetically modified crops in developing countries. International Food Policy Research Institute (IFPRI). Washington D.C. USA.

Pedotti, M., E. Rosini, G. Molla, T. Moschetti, C. Savino, B. Vallone and L. Pollegioni. 2009. Glyphosate resistance by engineering the Flavoenzyme Glycine Oxidase. Journal of Biological Chemistry 284: 36415–36423.

Perry, E.D., F. Ciliberto, D.A. Hennessy and G. Moschini. 2016. Genetically engineered crops and pesticide use in US maize and soybeans. Science Advances 2: p. e1600850. doi: 10.1126/sciadv. 1600850.

Pleasants, J.M. and K.S. Oberhauser. 2013. Milkweed loss in agricultural fields because of herbicide use: effect on the monarch butterfly population. Insect Conservation and Diversity 6: 135–144.

Powles, S.B., D.F. Lorraine-Colwill, J.J. Dellow and C. Preston. 1998. Evolved resistance to glyphosate in rigid ryegrass (*Lolium rigidum*) in Australia. Weed Science 46: 604–607.

Powles, S.B., C. Preston, I.B. Bryan and A.R. Jutsum. 1996. Herbicide Resistance: Impact and Management. pp. 57–93. *In:* L.S. Donald (Ed.) Advances in Agronomy. Academic Press, California, USA.

Pratley, J., N. Urwin, R. Stanton, P. Baines, J. Broster, K. Cullis, D. Schafer, J. Bohn and R. Krueger. 1999. Resistance to Glyphosate in *Lolium rigidum*. I. Bioevaluation. Weed Science 47: 405–411.

Preston, C. and J.M. Malone. 2015. Inheritance of resistance to 2,4-D and chlorsulfuron in a multiple-resistant population of *Sisymbrium orientale*. Pest Management Science 71: 1523–1528.

Preston, C. and S.B. Powles. 2002. Evolution of herbicide resistance in weeds: initial frequency of target site-based resistance to acetolactate synthase-inhibiting herbicides in *Lolium rigidum*. Heredity 88: 8–13.

Preston, C., D.S. Belles, P.H. Westra, S.J. Nissen and S.M. Ward. 2009. Inheritance of resistance to the auxinic herbicide dicamba in kochia (*Kochia scoparia*). Weed Science 57: 43–47.

Price, A.J., K.S. Balkcom, S.A. Culpepper, J.A. Kelton, R.L. Nichols and H. Schomberg. 2011. Glyphosate-resistant Palmer amaranth: a threat to conservation tillage. Journal of Soil and Water Conservation 66: 265–275.

Que, Q., M-D.M. Chilton, C.M. de Fontes, C. He, M. Nuccio, T. Zhu, Y. Wu, J.S. Chen and L. Shi. 2010. Trait stacking in transgenic crops: challenges and opportunities. GM Crops 1: 220–229.

Reddy, K.N. 2004. Weed control and species shift in Bromoxynil- and Glyphosate-Resistant Cotton (*Gossypium hirsutum*) Rotation Systems. Weed Technology 18: 131–139.

Rieger, M.A., T.D. Potter, C. Preston and S.B. Powles. 2001. Hybridisation between *Brassica napus* L. and *Raphanus raphanistrum* L. under agronomic field conditions. Theoretical and Applied Genetics 103: 555–560.

Sabba, R.P., I.M. Ray, N. Lownds and T.M. Sterling. 2003. Inheritance of resistance to clopyralid and picloram in yellow starthistle (*Centaurea solstitialis* L.) is controlled by a single nuclear recessive gene. Journal of Heredity 94: 523–527.

Salisbury, P.A., W.A. Cowling and T.D. Potter. 2016. Continuing innovation in Australian canola breeding. Crop and Pasture Science 67: 266–272.

Schulz, B. and K. Segobye. 2016. 2,4-D transport and herbicide resistance in weeds. Journal of Experimental Botany 67: 3177–3179.

Schwember, A.R. 2008. An update on genetically modified crops. Ciencia e investigación agraria 35: 231–250.

Senior, I.J. and P.J. Dale. 2002. Herbicide-tolerant crops in agriculture: oilseed rape as a case study. Plant Breeding 121: 97–107.

Shaner, D.L. 2000. The impact of glyphosate-tolerant crops on the use of other herbicides and on resistance management. Pest Management Science 56: 320–326.

Shaw, D., S. Culpepper, M. Owen, A. Price and R. Wilson. 2012. Herbicide-resistant weeds threaten soil conservation gains: finding a balance for soil and farm sustainability. Cast Issue Paper 49: 1–16.

Siehl, D.L., Y. Tao, H. Albert, Y. Dong, M. Heckert, A. Madrigal, B. Lincoln-Cabatu, J. Lu, T. Fenwick, E. Bermudez, M. Sandoval, C. Horn, J.M. Green, T. Hale, P. Pagano, J. Clark, I.A. Udranszky, N. Rizzo, T. Bourett, R.J. Howard, D.H. Johnson, M. Vogt, G. Akinsola and L.A. Castle. 2014. Broad 4-hydroxyphenylpyruvate dioxygenase inhibitor herbicide tolerance in soybean with an optimized enzyme and expression cassette. Plant Physiology 166: 1162–1176.

Simard, M.J., A. Légère and S.I. Warwick. 2006. Transgenic *Brassica napus* fields and *Brassica rapa* weeds in Quebec: sympatry and weed-crop in situ hybridization. Canadian Journal of Botany 84: 1842–1851.

Smith, E. and F. Oehme. 1992. The biological activity of glyphosate to plants and animals: a literature review. Veterinary and Human Toxicology 34: 531–543.

Smyth, S.J. 2014. The state of genetically modified crop regulation in Canada. GM Crops and Food 5: 195–203.

Smyth, S.J. and P.W.B. Phillips. 2014. Risk, regulation and biotechnology: the case of GM crops. GM Crops and Food 5: 170–177.

Smyth, S.J., M. Gusta, K. Belcher, P.W. Phillips and D. Castle. 2011. Environmental impacts from herbicide tolerant canola production in Western Canada. Agricultural Systems 104: 403–410.

Snow, A.A. 2002. Transgenic crops – why gene flow matters. Nature Biotechnology 20: 542.

Stalker, D.M., K.E. McBride and L.D. Malyj. 1988. Herbicide resistance in transgenic plants expressing a bacterial detoxification gene. Science 242: 419–423.

Steele, G.L., J.M. Chandler and G.N. McCauley. 2002. Control of red rice (*Oryza sativa*) in imidazolinone-tolerant rice (*O. sativa*). Weed Technology 16: 627–630.

Thompson, C.J., N.R. Movva, R. Tizard, R. Crameri, J.E. Davies, M. Lauwereys and J. Botterman. 1987. Characterization of the herbicide-resistance gene bar from *Streptomyces hygroscopicus*. The EMBO Journal 6: 2519–2523.

Traxler, G. 2006. The GMO experience in North and South America. International Journal of Technology and Globalisation 2: 46–64.

Trigo, E.J. and E.J. Cap. 2006. Ten years of genetically modified crops in Argentine agriculture. Argenbio. Argentina.

Trigo, E.J. and E.J. Cap. 2003. The impact of the introduction of transgenic crops in Argentinean agriculture. AgBioForum 6: 87–94.

Twardowski, T. and A. Małyska. 2015. Uninformed and disinformed society and the GMO market. Trends in Biotechnology 33: 1–3.

VanGessel, M.J. 2001. Glyphosate-resistant horseweed from Delaware. Weed Science 49: 703–705.

Vila-Aiub, M.M., M.C. Balbi, P.E. Gundel, C.M. Ghersa and S.B. Powles. 2007. Evolution of glyphosate-resistant Johnsongrass (*Sorghum halepense*) in glyphosate-resistant soybean. Weed Science 55: 566–571.

Vink, J.P., N. Soltani, D.E. Robinson, F.J. Tardif, M.B. Lawton and P.H. Sikkema. 2012. Occurrence and distribution of glyphosate-resistant giant ragweed (*Ambrosia trifida* L.) in southwestern Ontario. Canadian Journal of Plant Science 92: 533–539.

Webster, T.M. and L.M. Sosnoskie. 2010. Loss of glyphosate efficacy: a changing weed spectrum in Georgia cotton. Weed Science 58: 73–79.

Weinberg, T., G.R. Stephenson, M.D. McLean and J.C. Hall. 2006. MCPA (4-chloro-2-ethylphenoxyacetate) resistance in hemp-nettle (*Galeopsis tetrahit* L.). Journal of Agricultural and Food Chemistry 54: 9126–9134.

Wendy, A.P., V. Ryan, W.W. John, L.E. Keith, J. Thomas and W. Randy. 2002. Reproductive abnormalities in glyphosate-resistant cotton caused by lower CP4-EPSPS levels in the male reproductive tissue. Weed Science 50: 438–447.

Werth, J.A., C. Preston, I.N. Taylor, G.W. Charles, G.N. Roberts and J. Baker. 2008. Managing the risk of glyphosate resistance in Australian glyphosate-resistant cotton production systems. Pest Management Science 64: 417–421.

Wechsler, S.J., J.R. McFadden and D.J. Smith. 2017. What do farmers' weed-control decisions imply about glyphosate resistance? Evidence from surveys of US corn fields. Pest Management Science (in press).

Wiersma, A.T., T.A. Gaines, C. Preston, J.P. Hamilton, D. Giacomini, C. Robin Buell, J.E. Leach and P. Westra. 2015. Gene amplification of 5-enol-pyruvylshikimate-3-phosphate synthase in glyphosate-resistant *Kochia scoparia*. Planta 241: 463–474.

Wright, T.R., G. Shan, T.A. Walsh, J.M. Lira, C. Cui, P. Song, M. Zhuang, N.L. Arnold, G. Lin, K. Yau, S.M. Russell, R.M. Cicchillo, M.A. Peterson, D.M. Simpson, N. Zhou, J. Ponsamuel and Z. Zhang. 2010. Robust crop resistance to broadleaf and grass herbicides provided by aryloxyalkanoate dioxygenase transgenes. Proceedings of the National Academy of Sciences 107: 20240–20245.

SECTION II

Sustainable Weed Control in Crops and Cropping Systems

CHAPTER

11

Sustainable Weed Control in Small Grain Cereals (Wheat/Barley)

Alistair J. Murdoch

School of Agriculture, Policy and Development, University of Reading,
Earley Gate, PO Box 237, Reading RG6 6AR, U.K.
E-mail: a.j.murdoch@reading.ac.uk

Importance of Small Grain Cereal Cropping Systems

Small grain cereals are important food and feed crops, and in 2014 were grown on 289 million ha (mi ha) resulting in 928 million tonnes (mi t) grain (Table 11.1). This chapter focuses on the two main crops—wheat and barley—though oats, rye and triticale are locally important (Table 11.1).

Table 11.1. Global production of small-grained cereals in 2014. Compiled and calculated from data in FAOSTAT (2017)

Crop	Area, mi ha	Production, mi t	Yield, t/ha
Wheat	220.42	729.0	3.31
Barley	49.43	144.5	2.92
Oats	9.59	22.7	2.37
Rye	5.31	15.2	2.87
Triticale	4.14	17.0	4.10

Wheat (*Triticum* spp.) grows most successfully at latitudes of 30° to 60°N and 27° to 40°S (Nuttonson 1955), but it is found in the tropics at higher altitudes and even within the Arctic. Percival (1921) reported wheat was grown in Tibet at altitudes up to 4,570 metres above sea level. It is the world's most important small-grained temperate cereal crop with production of 729 mi t in 2014 (Table 11.1), production having been similar since then (736 mi t in 2016, FAO 2017).

Barley is also an important temperate cereal, being fourth in global importance among cereals after wheat, rice and maize. Like wheat, it is commonly grown in temperate cereal systems, but also at a wide range of latitudes and altitudes. It is particularly favoured in more hostile, drier environments (Crop Trust 2017).

Impact of Weeds on Small Grain Cereals

Potential crop losses due to weeds were estimated at 32% (range 26–40%) exceeding those of pests (18%) and pathogens (15%) (Royal Society 2009). Yield loss is, however, only part of the story; the social consequences and the opportunity costs of the other economic activities people could do if they did not have to weed their crops are often ignored especially for small-scale, resource-poor farmers. For example, Holm (1971) argued that, "more energy is expended for the weeding of man's crops than for any other single human task". The need for weed control in wheat is shown by the potential losses due to weeds being greater than for other crop protection problems (Table 11.2). The success of the efforts expended on weeding in wheat is illustrated first of all by the estimate that potential losses due to weeds are nearly a half of the total for all crop protection problems (23% out of 49.8%) whereas they are close to a quarter of actual losses (7.7% out of 28.2%) and secondly, by the efficacy of control being much greater for weeds of wheat than for pests and diseases (Table 11.2).

Table 11.2. Estimated global potential and actual yield losses of wheat attributable to weeds, pests and diseases, together with efficacy of control. Estimates assume global wheat production of 785 mi t. in 2001–03. Ranges were estimated across 19 regions. Adapted and calculated from Oerke (2006)

	Potential yield losses, %		Actual yield losses, %		† Efficacy of control, %	
	Mean	Range	Mean	Range	Mean	Range
Weeds	23.0	18–29	7.7	3–13	67	55–83
Animal pests	8.7	7–10	7.9	5–10	9	0–29
Pathogens	15.6	12–20	10.2	5–14	35	30–58
Viruses	2.5	2–3	2.4	2–4	4	– 33–0
Total	49.8	44–54	28.2	14–40	43	26–68

† Efficacy calculated as (1-[actual yield loss]/[potential yield loss])*100. Ranges of efficacy are of limited accuracy due to low values and method of calculation.

Major Weeds

Black-grass (*Alopecurus myosuroides* Huds.) is native to Eurasia and is widespread in Europe. It has become a significant, invasive weed of rotations including winter cereals in Western Europe. Changes in farming practice, such as the widespread adoption of minimum tillage instead of ploughing and a decrease in spring cropping, have encouraged its spread (CABI 2017b). It is a major challenge to cereal growers in England, France, Germany, Belgium and the Netherlands. It is also spreading northwards in the UK and increasing in Denmark, southern Sweden and Poland (Moss 2013). For conventional (non-organic) systems, its importance is exacerbated by the evolution of herbicide resistant populations.

Using data in the 2015 survey of weeds in the United States and Canada (Van Wychen 2016), the most troublesome species in spring cereals were *Avena fatua* L. and *Kochia scoparia* (L.) Schrad with four of the top six, broad-leaved weeds (Table 11.3). In winter grains, the top four were grass weeds but the broad-leaved weed, *K. scoparia*, was the only species appearing in the top six for both winter and spring cereals (Table 11.3).

Wild-oat (*A. fatua*) probably has its centre of origin in Central Asia, but occurs globally in crops within arable rotations. As just noted, it is a particular problem in spring cereals, although it frequently infests winter cereals also. It competes particularly successfully with small grain cereals, such as wheat and barley in part due to its greater height (up to 120 cm) compared to modern semi-dwarf cultivars (Holm et al. 1977, CABI 2017a).

Weeds in arable fields tend to reflect the soil seed bank of the weeds and, for example, the median soil seed bank of 64 arable fields comprised 4,360 viable seeds per square metre with a

Table 11.3. Relative importance[†] of of six most troublesome weed species of spring and winter cereals in the United States and Canada calculated from responses in the 2015 Survey of Weeds (Van Wychen 2016)

Weeds of spring cereals		Weeds of winter cereals	
Avena fatua	57%	*Lolium perenne* L. ssp. *multiflorum* (Lam.) Husnot	35%
Kochia scoparia	24%	*Bromus tectorum* L.	30%
Galium spp.	10%	*Secale cereale* L.	17%
Cirsium arvense (L.) Scop.	8%	*Aegilops cylindrica* Host	13%
Setaria viridis (L.) Beauv.	7%	*Kochia scoparia*	10%
Polygonum convolvulus L.	7%	*Stellaria media* (L.) Vill.	9%

† Calculation of relative importance: Respondents listed the five most troublesome weeds in their area. Taking the first three, the species a respondent ranked as the most troublesome was scored 3, the second, 2, and the third, 1. The weighted scores were summed for each species and expressed as a percentage of the maximum score ($3n$) where n is the number of respondents giving a valid response. $n = 30$ and 34 for spring and winter cereals, respectively.

range of 1,500 to 67,000 (Roberts and Chancellor 1986). The most prevalent weeds were present in a majority of fields, with *Poa annua* L. present in all 64 fields assessed and >625 viable seeds m^{-2} in 35 of them (Table 11.4). These are some of the common annual weeds of arable fields.

Table 11.4. Prevalence of annual weeds in the soil seed banks of 64 arable fields in Midlands of England in 1976–77 (Data from Roberts and Chancellor 1986)

Weed species	Present in fields, % (n=64)	Number of fields containing >625 viable seeds m^{-2}
Poa annua L.	100	35
Polygonum aviculare L.	92	17
Stellaria media (L.) Vill.	90	20
Fallopia convolvulus (L.) Á. Löve*	70	2
Aethusa cynapium L.	68	7
Veronica persica Poir.	67	19
Alopecurus myosuroides Huds.	67	18
Chenopodium album L.	66	7

* Synonym: *Bilderdykia convolvulus* (L.) Dumort.

The incidence of these weeds does not, however, reflect their economic importance. For example, Wilson and Wright (1990) reported 2% yield losses of cereals occurred with populations of 0.5, 1.6, 5.4, 8.3, and 39 plants m^{-2} for *Avena fatua*, *Galium aparine* L., *Poa trivialis* L., *Alopecurus myosuroides* and *Veronica hederifolia* L., respectively.

Much higher numbers of seeds were found in 37 Danish cereal fields by Jensen (1969). He reported an average of 62,700 seeds m^{-2} in 'heavily infested' fields. It is probably possible to account for the large seed banks by high numbers of *Juncus bufonius* L. var. *bufonius* seeds in some fields and methodological differences (Murdoch 2006).

Understanding and quantifying the life cycles and ecology of weeds is a key element in devising more rational and integrated methods of weed control. This knowledge may assist in exploring options for sustainable intensification of cereal growing designed to minimise unnecessary use of herbicides and to mitigate the potential for development of herbicide resistance. Using *Avena fatua* as a case study, it is clear that although perhaps half of the potential seeds produced may be inviable or empty, 10% may contaminate the harvested grain (Figure 11.1). Losses in the soil may be considerable (90% of seeds >one year old may be depleted per annum, shown as a proportion of 0. 1 in Figure 11.1) but only if there is a high level of available nitrate in the soil (Murdoch and Roberts 1982). Integrated weed management strategies may exploit the latter in order to increase the rate of depletion of the soil seed bank. Conversely, inappropriate agronomy may mean smaller losses or greater seed production than suggested (Figure 11.1).

Weed Strengths and Weaknesses

Life cycle diagrams (Figure 11.1) linked to an understanding of driving variables can be a powerful tool for a systematic consideration of the strengths and weaknesses of different weed species and to identify vulnerabilities which may be exploited for Integrated Weed Management (IWM).

Seed production is frequently a defining trait in annual arable weeds, characterised both by fecundity and plasticity, such that large numbers of seeds may be produced in favourable conditions, but even in adverse conditions, such as in a highly competitive crop, at least some seeds are produced (Table 11.5). These large numbers give rise to the old adage: "One year's seeding: seven years' weeding", a statement which is well-supported by experimental evidence of the longevity of arable weed seeds in cultivated soil (Murdoch 2006).

This fecundity gives rise to another strength of many but not all arable weeds, namely the size of the soil seed bank as discussed above.

Contrary to notion that a weed is 'a plant out of place' (Blatchley 1912), a further strength of weeds is that the dormancy of many, especially small-seeded weed species facilitates their survival in the soil seed bank of arable fields for many years and then to germinate in the right place at the right time, when conditions are most favourable to their establishment (Murdoch 2013). Arable weeds are arguably, therefore, very much in their 'place'! Examples of this adaptation are numerous. For example seeds of *Stellaria media* L. failed to germinate at constant temperature in darkness on paper moistened with water. Even exposure to an alternating temperatures only gave 6% germination. However, exposure to light produced about 50% while combining light and nitrate increased germination to 100% (Figure 11.2). *Chenopodium album* L. shows greater adaptation to alternating temperatures provided light and nitrate are available, with a preference for longer periods each day at the upper temperature (Figure 11.3) with the highest germination in a regime of approximately 3/20°C (8h/16h). Soil disturbance can only serve to increase the probability that such species will germinate and emerge, an inference of considerable relevance especially when shallow or inter-row tillage is used to control weeds of cereals.

It is equally important and somewhat surprising to learn that some weeds of cereal crops have very little seed dormancy, an example being *Bromus sterilis* L. (syn. *Anisantha sterilis* (L.) Nevski). While the seeds have sufficient dormancy to prevent precocious germination on the mother plant, and while secondary dormancy may be induced through exposure to light, most seeds are non-dormant and seeds can be eliminated by a combination of burial after shedding and delayed drilling, such as spring cropping (Peters et al. 1993, Andersson et al. 2002). Burial at a depth from which they will not emerge by total inversion ploughing is a totally effective option for recently-shed seeds. Unfortunately in the UK, adoption of earlier autumn drilling of winter cereals in combination with dry autumns led to this weed causing significant infestations in some fields (Peters et al. 1993).

Other grass weeds show more primary dormancy (Murdoch 2013) but requirements for relief of this dormancy means they will not respond to creation of a false seedbed. For example, *Avena fatua* is better adapted to germinate under cool conditions without exposure to either

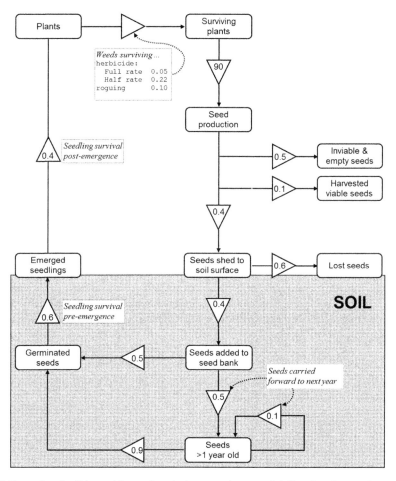

Figure 11.1. Life cycle of wild-oat (*Avena fatua*). Arrows show multiplication factors for each stage of the life cycle based on assumptions in Murdoch (1988) but values range widely. Depletion of buried seeds by loss of viability is not shown.

light or fluctuating temperatures (Figure 11.4). Such seeds will remain largely unaffected by creating a false seedbed, and indeed their longevity will be enhanced since burial of the seeds by the creation of the false seedbed will remove them from seed-eating birds and exposure to the environmental fluctuations of the soil surface (Wilson and Cussans 1975).

To complete the picture, characteristics of treatments which relieve primary dormancy of freshly-harvested or dry-stored weed seeds can only give a rough indication of what might happen to seeds in the soil. Buried seeds generally exhibit an annual dormancy cycle in which secondary dormancy is induced and relieved over the course of a year. Such cycles were first clearly described for a common weed of cereal crops, namely *Polygonum aviculare* L. (Courtney 1968). Many other examples occur (Murdoch and Roberts 1998) and the seasons in which low dormancy occurs is linked to periodicity of seedling emergence (Baskin and Baskin 1985). For example, a spring-germinating summer annual like *P. aviculare* loses dormancy in winter and dormancy is induced in late spring (Courtney 1968). The dormancy observed in the seed population is a combination of any residual innate dormancy plus induced dormancy—the two types of dormancy being generally indistinguishable after burial. Seeds germinate when times of low dormancy coincide with environmental conditions suitable for germination (Murdoch and Roberts 1996) resulting in the periodicity of seedling emergence which characterises many species (Roberts 1986) provided moisture is available (Roberts and Potter 1980).

Table 11.5. Seed outputs per unit area and per plant per year of selected species

Species	Seeds per plant	Seeds m⁻²
Alopecurus myosuroides Huds.	43	2,500
Avena fatua L.	22 (range: 16-184)	1,000 (range: 393-4784)
Chenopodium album L.	3,000	-
Papaver rhoeas L.	17,000 (1,300 per fruit)	-
Stellaria media (L.) Vill.	2,500 (5-16 seeds per capsule)	-

Sources: Salisbury (1961), Sagar and Mortimer (1976)
- Not available

In *A. fatua*, for example, the dormancy of seeds retrieved over four years followed an annual cycle (Murdoch and Roberts 1996). Induction of dormancy in the late spring was especially associated with increasing daily maximum soil temperatures above 20 °C provided the soil water potential at the soil surface was between field capacity (–10 kPa) and *c.* –100 kPa. Even lower water potentials during summer led to a slight loss of dormancy presumably due to dry after-ripening, but the main decline in dormancy, when seeds regained their responsiveness to low temperatures and nitrate, occurred in autumn and early winter when the soil was again at field capacity and the daily maximum soil temperature was below 20 °C (Figure 11.4). These responses also reflect seed-to-seed variation in dormancy. Thus *in situ* germination occurred late in winter when dormancy was least and some retrieved seeds had lost sufficient dormancy so that they would germinate given water and air. Other seeds from exactly the same seed population retained a measure of dormancy such that they still required darkness and/or nitrate and/or low temperatures to germinate. Understanding that there is seed-to-seed variation in dormancy even within seed populations originating from the same field and at the same time in the same year is essential to the development of more rational weed management strategies which are designed to exploit weaknesses in the life cycles of weeds (compare Figure 11.1). This variation in relative dormancy in an ostensibly homogeneous population of seeds of *Avena fatua* can be illustrated clearly by the population response to temperatures between 3 and 20 °C (Figure 11.4). This seed-to-seed variation in dormancy is a common feature of all germination studies and quantifying this variation is a key component in modelling germination and dormancy (Murdoch 2013, Murdoch and Kebreab 2013).

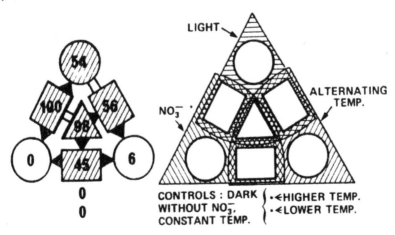

Figure 11.2. Germination of *Stellaria media* seeds. Germination in factorial combinations of constant temperature (25 °C) compared to alternating (3/25 °C, 16h/8h), in water compared to 0. 01 M KNO₃, and with or without exposure to light. Germination at both control constant temperatures is shown underneath the triangular diagram. Cross-hatching indicates germination is significantly higher than in the control. The solid greater than (>) symbols indicate statistically significant differences. The equals (=) symbols indicate no significant difference (Roberts 1973).

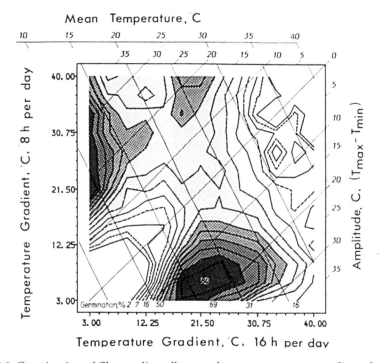

Figure 11.3. Germination of *Chenopodium album* seeds on a temperature gradient plate with gradients operating from left to right for 16 hours per day and from bottom to top for 8 hours. The darker the shading, the higher the germination, plotted as contours on a scale of normal deviates, back-transformed values being given at the bottom of the diagram (Murdoch et al. 1989).

Vleeshowers (1997) also distinguished between the seasonal cycle of dormancy and residual dormancy calling the latter a germination requirement. The results in Figures 11.2, 11.3 and 11.4 emphasize that the practical outworking of dormancy is not an identical characteristic of all seeds in the seed population. A germination test estimates what can be thought of as the mean level of dormancy.

Thus dormancy of buried seeds of *A. fatua* was relieved from October through to March and secondary dormancy appeared to be induced after the end of March (Figure 11.4). The induction of dormancy is only partly accounted for by the loss of relatively non-dormant seeds by *in situ* germination leaving a residual population of more dormant seeds.

Fecundity, the annual dormancy cycle and periodicity of seedling emergence are major traits conferring weediness to weeds of small-grained cereals and also make elimination of these weeds problematic due to the longevity of their seeds in the soil. Circumstantial evidence for the longevity of seeds is supported by classical seed burial trials commenced by Beal in 1879 in Michigan and Duvel in 1902 at Arlington, Virginia as well as more recent examples (Murdoch and Ellis 2000). Shorter-term studies are more useful from a farming perspective since they put reports of record-breaking, extreme individuals into the context of the overall seed population and allow the probabilities of persistence from year-to-year to be calculated and compared between species and environments.

Without further seed introductions, that is with 100% weed control, persistent soil seed banks approximately follow a negative exponential decay model on a year-to-year basis although decay in the first year may differ from that in subsequent years (as implied for buried seeds of *Avena fatua* in Figure 11.1). Annual probabilities of decline vary greatly both with species and environment, the frequency of tillage, soil type and fertility being significant factors (Roberts 1970, 1981, Murdoch and Ellis 2000, Lutman et al. 2002). For example, while the annual rates of depletion of some weeds typical of small-grained cereal crops declined rapidly (e.g.,

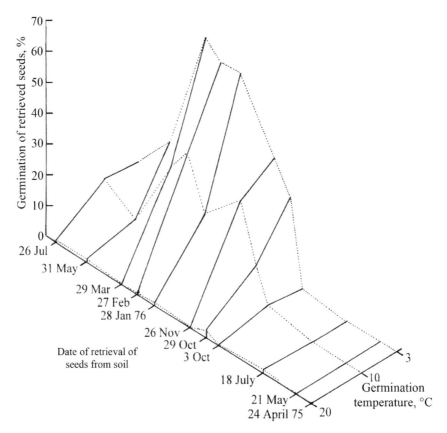

Figure 11.4. Germination of seeds of *Avena fatua* after retrieval on various dates from a 23 cm burial depth. Seeds were buried in December 1975, retrieved on the dates shown and then processed in a dark room equipped with a green safe light. Germination was carried out in incubators at temperatures of 3 °C, 10 °C and 20 °C in 0. 01 M KNO$_3$ and in darkness (Original data, A J Murdoch).

Galium aparine—58% decline pa) and others very slowly (e.g., *Papaver rhoeas* L. —9% decline pa), most declined at 20%–40% pa.

Predicting germination and emergence of weeds is needed to optimize timings of post-emergence herbicides and also of non-chemical methods, such as inter-row tillage. Various models have been developed and compared (e.g., Chantre et al. 2013), but these need to be combined with models of dormancy release (e.g., Blanco et al. 2014). Parameterising such models for different agro-ecosystems and different ages and cohorts of seeds in the soil seed bank is probably unrealistic especially as the influence of the crop is not accounted for. Simpler approaches may be all that is needed. For example, many years ago, Roberts at the then National Vegetable Research Station in the UK, showed the association of flushes of weed seed germination of annual broad-leaved weeds in conjunction with cultivation (seed-bed preparation) and rainfall (Roberts and Ricketts 1979).

Current Weed Control Practices

Prevention

Seed influx from sources of seeds outside a given field are unlikely to cause a significant infestation in the year of their introduction. Such introductions are however highly significant as a factor to consider in IWM, even though the numbers may seem insignificant and indeed, they usually are. The risk, however, is a new weed species or a new and perhaps herbicide-resistant biotype, may be introduced. For example, standard certified cereal seed in the UK must not contain more than 20 weed seeds kg⁻¹, equivalent to 0. 3 seeds m⁻² for a seed rate of 150 kg ha⁻¹. A similar argument applies to irrigation water, but contamination of manure and compost can be more significant (Fenner and Thompson 2005). Avoiding the introduction of new species should be a key element of IWM and so if there is a risk of such introductions, then monitoring is essential. The method of monitoring should also be evaluated. Isolated introductions are likely to be missed if weed scouting is dependent on images captured by anything other than proximal sensors since the spatial resolution of images captured by remote sensing by Remotely Piloted Aircraft Systems (RPAS, popularly known as UAVs or drones), aircraft or satellite are currently only sufficient to detect fairly dense patches of weeds (Murdoch et al. 2014). Biosecurity should, therefore, be considered and the IWM message is: look out for new weed species and herbicide-tolerant biotypes, since these may in time become serious problems.

Mechanical/Physical Weed Control

Soil disturbance is one of the oldest methods of weed control and may be utilised for this purpose before drilling as part of both primary and secondary cultivations (e.g., ploughing and harrowing). In a meta-analysis of 25 experiments, Lutman et al. (2013) found that, relative to non-inversion tine tillage, direct drilling led to a (non-significant) 16% increase in the number of *A. myosuroides* plants in the following crop; total inversion tillage with a mouldboard plough reduced the infestation by an average of 69%. Scherner et al. (2016) similarly showed that the annual grass weeds, *Apera spica-venti* L. and *Vulpia myuros* L., which like *A. myosuroides* have relatively short-lived soil seed banks, have become more widespread with the adoption of non-inversion tillage when preparing seedbeds for winter cereals in Europe. Interestingly, Scherner et al. (2017) found that direct drilling (zero tillage) increased the thermal time for emergence of *A. spica-venti* and *V. myuros* meaning they were more likely to escape early autumn herbicide treatments.

It is, therefore, important to stress that, unless integrated with other approaches, total inversion tillage is often essential to ensure seeds are buried at depths from which they cannot emerge. Moreover, for ploughing to be effective where there is a persistent soil seed bank, rotational tillage should be practised to avoid restoring surviving seeds to the soil surface in the next season! The variability of the responses to tillage is perhaps more important to a farmer than the average effect. Thus, although direct drilling had no significant effect on final emergence, in Lutman et al.'s (2013) meta-analysis, over half the experiments (13) had increases in infestation—up to 344% in the worst case, while six showed a decrease of up to 78%.

Post-emergence weed control can be carried out with shallow tines although incurring some damage to the crop (Welsh et al. 1996) but the net effect of the weed control achieved appears to be positive (Melander et al. 2005). Inter-row cultivations can be carried out without crop damage but a row spacing wider than the usual 12 cm is desirable for small-grained cereals even with a vision guided hoe to reduce the risk of crop damage. While wider row spacing up to 22 cm was claimed not to incur a yield penalty (Tillett et al. 1999), further studies in Italy have suggested the contrary as discussed below under seed rates. For non-chemical weed control, the yield losses are a cost, but for conventional farming systems using herbicides, the risks of wider row spacing may not be acceptable.

Shallow tillage can also be used to exploit the stimulatory effects of light, nitrate and fluctuating temperatures on the germination of some weed seeds (see above). No-till cereal systems can exploit this trait to suppress germination, since the soil remains undisturbed, or alternatively, shallow tillage can be used to promote germination prior to crop drilling by creating a 'stale' or 'false' seedbed in order to 'fool' the weed seeds into germinating. They may then be controlled by harrowing before the crop is drilled or by spraying before it emerges.

It should also be noted that tillage will almost always lead to a flush of weed seedlings and so it is important to note that inter-row tillage systems mentioned must be designed with this probability in mind. A good rule of thumb is that given moisture, shallow tillage will stimulate 3–6% of the viable seeds in the soil seedbank to germinate (compare Roberts and Ricketts 1979). So systems designed to use inter-row tillage must ensure that the crop is kept weed-free for the duration of the critical weed-free period of the crop. As a minimum, tillage is needed by the start of the weed-free period and is not needed after the end of it. For winter wheat in the UK, yield losses in excess of 5% were predicted if the crop was not kept weed-free at thermal times between 500 °C and 1,000 °C days after sowing (October to January) (Welsh et al. 1999).

Cultural Weed Control

Seed Rates and Seed Quality

Andrew and Storkey (2017) simulated yield losses of winter wheat caused by an infestation of 80 *A. myosuroides* plants m^{-2} over 10 years and the average yield loss increased from 9.4% with a crop plant density of 300 plants m^{-2} to 15% as crop density decreased to 150 plants m^{-2}. Korres and Froud-Williams (2002) studied weed suppression of a natural weed infestation comprising mainly annual broad-leaved weeds and *Poa annua* by six winter wheat cultivars. Averaged across all cultivars, weed dry weight in late June, approximately eight months after sowing, was reduced by more than 50% by approximately doubling crop plant density (125 compared to 270 wheat plants m^{-2}). Although a yet higher density of 380 plants m^{-2} failed to affect weed dry weight, the numbers of weed reproductive structures were approximately halved relative to 270 plants m^{-2} (1,387 compared to 2,736 m^{-2}). Sowing the crop 30 days later (in late October rather than late September) also reduced percentage yield loss from 19% to 5%.

Many reports confirm that increasing seed rate may enhance crop competitiveness and/or weed suppression although close examination shows some results are more equivocal. Thus in Korres and Froud-Williams (2002), crop plant density did not significantly affect weed dry weights or reproductive structures when assessed 70 days after sowing, whereas differences became apparent later during the growing season.

One problem with increasing seed rate is that intraspecific competition among crop plants may increase due to the increase in rectangularity of the crop. For example, using 12 cm rows, the distance between wheat plants within each row decreases from 5.6 cm to 2.8 cm if crop density is doubled from the 150 to 300 m^{-2}. Each plant is therefore 'allocated' a rectangle of 2.8 × 12 cm or 33.3 cm^2 (Table 11.6). Planting 'on the square' with 5.8 cm between rows and the same distance between plants within the row, would make a lot more sense for the crop and increase its potential competitiveness. The argument becomes even more compelling at a crop density of 450 plants m^{-2} (Table 11.6). In a study of the weed competition on yield and quality of durum wheat (*Triticum durum* Desf.) in Italy, de Vita et al. (2017) planted the crop at rates of 190, 380 and 570 seeds m^{-2} and row spacings of 5, 15 and 25 cm. Interestingly, in this experiment, the seed rate did not affect weed dry biomass when assessed at the end of tillering (presumably

Table 11.6. Crop architecture and rectangularity for different seed rates at row spacings of 12 and 25 cm

Crop density, plants m^{-2}	Crop density per linear metre	Land area per plant, cm^2	Gap between rows, cm, (A)	Gap between plants within rows, cm, (B)	Rectangularity, (A)/(B)	Optimum distance between plants, cm
150	18	66.7	12	5.6	2.2	8.2
300	36	33.3	12	2.8	4.3	5.8
450	54	22.2	12	1.8	6.5	4.7
150	37.5	66.7	25	2.7	9.4	8.2
300	75	33.3	25	1.3	18.8	5.8
450	112	22.2	25	0.9	28.1	4.7

Growth Stage (GS) 30, Tottman 1987) which may explain the conflict with Korres and Froud-Williams (2002) who found very large differences with seed rate after GS69. Reduced inter-row distance, however, at about GS30, reduced the weed dry biomass of mostly broad-leaved weeds from approx. 110 to 70 and 22 g m^{-2} for the semi-dwarf cv. PR22D89 at 25, 15 and 5 cm row spacings, respectively. Wheat yields and crop nitrogen uptake were also higher for the narrow row-spacing even in the weed-free controls. The benefits of less rectangular planting arrangements are supported by other studies including simulation modelling (e.g., Evers and Bastiaans 2016, Renton and Chauhan 2017).

A somewhat neglected aspect of weed suppression is that of seed quality even though high vigour seeds not only emerge more rapidly but also give higher emergence in stressful environments (Khah et al. 1986, 1989). These effects of crop seed vigour on emergence are given added importance since crop yield losses are greatly affected by the relative times of emergence of crops and weeds (O'Donovan et al. 1985, Cousens et al. 1987). Combined effects of seed vigour and seed rate can, therefore, be highly significant and, without other weed control interventions, may make the difference between some yield and no yield (Figure 11.5).

In Figure 11.6, relationships are shown for weed seedlings emerging three days before, at the same time (with) and three days after crop emergence. Curves are calculated from parameter values in Cousens et al. (1987). Advancing crop emergence by two or three days can have a significant effect on yield losses and weed competitiveness is affected by the relative time of emergence (RTE) of weeds relative to the emergence of the crop (Figure 11.6). It is likely that the impact of RTE on crop yields is insufficiently recognised and, as a result, factors likely to shift this parameter in ways designed to enhance the crop's competitiveness are not given due weight.

Competitive Crops

The Green Revolution has been of particular importance from the perspective of weed control, since many current commercial wheat cultivars are now semi-dwarf, shorter-stemmed cultivars due to expression of reduced height genes (*Rht*) (Addisu et al. 2008, Gooding et al. 2012).

This observation is important since agronomic cultivar traits are evaluated in weed-free conditions. Indeed, unlike other biotic constraints, such as pests and diseases, competitiveness against weeds is not typically a criterion for plant breeding (Seefeldt et al. 1999) and nor is it generally quantified as a trait to help farmers choose varieties (e.g., AHDB 2017). The main exception to these generalisations is for organic farming where the effect of dwarfing genes in

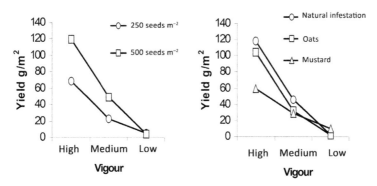

Figure 11.5. Impacts of seed rate and seed quality (vigour) on grain yield of spring wheat cultivar Chablis in a field experiment near Reading, Berkshire, UK, during the 2002 growing season. Seed vigour levels were achieved by ageing subsamples of the same seed lot for 38 hours (medium vigour, germination 87%) and 48 hours (low vigour, germination 77%), the high vigour being the untreated control (germination 98%). Treatments comprised two seed rates (recommended - 250 seeds m^{-2} - and twice the recommended rate, SED = 11, DF = 17) and three weed treatments (SED = 13): (i) naturally occurring weeds or as model weeds, (ii) *Avena sativa* var. Firth (oats) and (iii) *Sinapis alba* L. (mustard) (Al Allagi and Murdoch 2003)

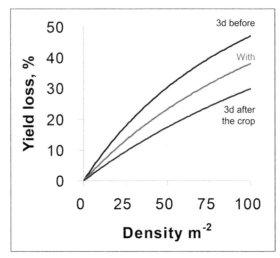

Figure 11.6. Predicted yield loss of spring wheat as a function of the density of *Avena fatua* and its time of emergence relative to that of the crop. Curves are calculated from parameter values in Cousens et al. (1987).

wheat is recognised to lead to increased weed infestations (Cosser. et al. 1996, 1997) such that taller and traditional cultivars tend to be more weed suppressive (Hoad et al. 2008, Wolfe et al. 2008). More generally, crop height may be more important than relative time of emergence in mitigating yield losses from weeds in winter wheat (Harris 2011).

In evaluating varietal traits, it is difficult to isolate effects of single traits like plant height on weed suppression. Using Near Isogenic Lines (NILs) in a common genetic background can help to overcome the problem of comparing different cultivars where various traits may be influencing competitiveness. Thus, using NILs in the genetic background of the wheat cultivar, Mercia, Kumuthini et al. (2010) compared percentage yield losses due to weed infestations for the 'tall' 90 cm NIL (*rht*) with a dwarf NIL (<40 cm) containing *Rht12*. The taller line had 67% and 23% yield losses in 2007/8 and 2008/9, respectively, compared to 95% and 63% for the dwarf one in experiments where the weeds emerged at the same time as the crop.

In addition to cultivar height at harvest, other traits associated with greater weed suppression by wheat and barley cultivars include early ground cover and tillering, leaf angle and canopy structure, early vigour and allelopathy (Worthington and Reberg-Horton 2013). However, not only are varietal characteristics important, cereal species themselves differ qualitatively in competitiveness, due to differences in these traits.

In experiments with *Avena sativa* L. and *Sinapis alba* L. as model weeds, spring barley and triticale were better able to maintain their yield under severe weed pressure, showing 27.6% and 39.2% yield loss compared to the weed-free control, whereas the most competitive wheat cultivars tested (Axona) had a 46% yield loss while and the least competitive (cv. Status) sustained a 75.1% yield loss (Table 11.7).

The weed suppressive ability of a crop tends to be a corollary of the ability to maintain yields in a weedy environment. Thus, barley and triticale suppressed weeds by 59% and 15%, respectively, more than the average for wheat cultivars. Among wheat cultivars, Axona and Paragon had 50% less weed dry matter compared to the poor competitor, Status (Al-Allagi 2007, compare with yield maintenance in Table 11.7). Using Principal Components Analysis, Al-Allagi (2007) showed that the importance of traits linked to the ability to maintain their yields under weed pressure could be ranked in order of importance as follows: (i) maintaining fertile tillers under weed pressure, (ii) ground cover at Growth Stage (GS) 15-16, (iii) height at GS 83, (iv) leaf area at GS 15-16, (v) dry biomass at GS 15-16, (vi) height at GS 15-16, and (vii) mean time for emergence. Variations in plant architecture especially during early growth thus appear to be of value in enhancing crop competitive ability. In a more recent

Table 11.7. Grain yields (t/ha) of spring cereal genotypes grown near Reading, U.K., in weed-free plots and the percentage grain yield loss (GYL) in plots sown with either *Avena sativa* or *Sinapis alba* as model weeds. Genotypes are placed in order of their ability to maintain the yield relative to the weed-free yield in the absence of any other weed control. (DF for SED: 44). Adapted from Al-Allagi (2007)

Genotype	Weed-free yield, t/ha	GYL due to Avena, %	GYL due to Sinapis, %	Mean GYL, %
Barley	7.44	27.2	28.1	27.6
Triticale	5.39	25.3	53	39.2
Wheat cv. Axona	4.74	36.3	55.6	46.0
Wheat cv. Paragon	6.01	53.3	63.8	58.6
Wheat cv. Alder	4.83	53.6	73.8	63.7
Wheat cv. Status	6.19	72.4	77.8	75.1
SED Genotypes	0.34	5.46		
SED Weeds		3.15		
SED Genotype x Weed		9.45		

study with winter wheat, barley and oats, Andrew et al. (2014, 2015) identified similar traits in suppression of *A. myosuroides* and *S. media*, although their significance varied between experiments.

Rotations

Rotations have been a key element of non-chemical weed control strategies since they give the opportunity to break the life cycle of problem weeds and to use different control methods. Adapting the life cycle model of *A. fatua* (Figure 11.1), the potential impacts of 'cleaning' crops for weed management in an arable rotation are apparent since full-rate herbicide was required in only six out of 21 seasons or in eight for half-rate treatment (Figure 11.7, Murdoch et al. 2003).

Sowing Date

A further critical factor which can reduce the impacts of weeds on small-grained cereal crops is sowing date. Where drilling can be delayed without an unacceptable yield penalty and/or risk of an adverse sowing environment, weed control can be enhanced especially if the weeds emerge and can be controlled prior to crop drilling. Moreover, the efficacy of the pre- and post-emergence herbicides may increase significantly with later drilling (Hull et al. 2014).

Chemical Weed Control

Availability of herbicides is subject to commercial constraints, societal and political pressures, legislation and resistance in target organisms. Most countries seek to impose some restrictions on the use of pesticides. In the European Union for example, active substances are given EU-wide approval, but are subject to various directives governing their sustainable use and the need to keep contamination of waterways below certain levels. In addition, approved chemicals must only be applied to crops for which the approval has been granted and according to 'label'. The label specifies measures to minimise the risks to people, non-target organisms, the environment and, for food crops, there are also maximum permissible residue limits. Under these rules, some chemicals which are important in arable rotations have lost approval and Clark (2014) considered that others of importance to European cereal growers are at risk including, 2,4-D, glyphosate and mecoprop. In addition some products of importance in crop rotations involving cereals and which can be applied to control weeds, such as *Alopecurus myosuroides* in crops other than cereals, are at risk of losing their approved status (for example, carbetamide, clopyralid and propyzamide,

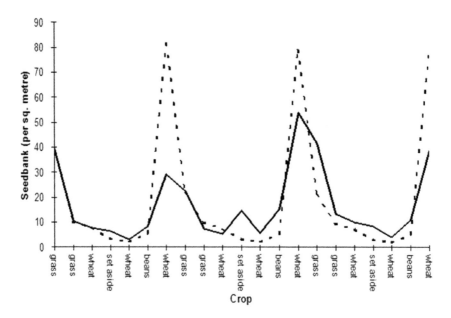

Figure 11.7. Predicted soil seed bank of *Avena fatua* in a seven-year arable farming rotation, which included a two-year grass ley. Herbicide use for *A. fatua* was modelled on the basis that herbicides were applied to infestations of 1 plant m^{-2} requiring either six full rate (- - - -) or eight half rate (⸺) herbicide applications, which were assumed to achieve 95% or 78% control of seed production, respectively. Adapted from Murdoch et al. (2003).

which are used in oilseed rape/canola). Nevertheless, a range of pre-, peri- and post-emergence herbicides are available and those most commonly applied to wheat and barley crops are tabulated for broadleaved and grass wheat weeds in small-grain cereals (Table 11.8). Out of 15,087 t of all pesticides applied to all arable crops in the UK, over half was applied to wheat (8,296 t) and nearly one-fifth of all UK pesticide applications to arable crops were the herbicides applied to the wheat crop (2,925 t) (Garthwaite et al. 2015).

Herbicide Resistance

Heap (2017) lists 315 records of herbicide resistance in wheat. Of these, 62 reported multiple resistance to two (43/62) or to three or more (19/62) modes of action. The most extreme case was a population of the grass weed, *Lolium rigidum* Gaudin in South Australia, where it occurs in spring barley and wheat and has evolved multiple resistance to chlorpropham, chlorsulfuron, clomazone, diclofop-methyl, ethalfluralin, fluazifop-P-butyl, imazapyr, metolachlor, metsulfuron-methyl, quizalofop-P-ethyl, sethoxydim, tralkoxydim, triallate, triasulfuron, and trifluralin and there may be cross-resistance to yet other herbicides with the same modes of action. This population is also the only reported instance of a weed infesting wheat with resistance to mitosis inhibitors and DOXP inhibitors [Report by Preston (2012) in Heap (2017)].

Resistance to ALS inhibiting herbicides is the most frequent in weeds of wheat (192/315 reports in Heap 2017) and the second most common is ACC-ase inhibitors (116/315 reports) and both of these are represented among the most common herbicides used in wheat and barley in the UK (Table 11.8). Resistance in weeds of small-grained cereals is widespread and increasing (Hull et al. 2014). Trends towards the use of herbicide tolerant crops and, where these are not used, to shorter rotations and earlier drilling as is the case for winter cereals in the UK for example, has increased the likelihood of herbicide resistance. Resistance is 'seen' by

Table 11.8. The top five active ingredients applied to wheat and winter barley in the UK in the 2013–14 growing season. Numerical information adapted from Garthwaite et al. (2015)

Crop/Herbicide active ingredient (a.i.)	Action/inhibitory mode of action (HRAC class*)	Weeds	Timing	Area treated, 1000s ha†	Amount used in UK, t a.i.	Proportion of full rate
Wheat						
Iodosulfuron-methyl-sodium/ mesosulfuron-methyl (Atlantis)	ALS inhibitor (branched chain amino acid synthesis) (B)	G	Post	825	12	0.95
Glyphosate	EPSP synthase (G)	All	Predrill or pre-harvest	818	601	0.47
Diflufenican/flufenacet (Liberator)	Pigment synthesis (F_1)/ Cell division (K_3)	G+B	Pre	748	173	0.84
Fluroxypyr (e.g. Starane)	Auxin (O)	B	Post	478	64	0.38
Flufenacet/pendimethalin	As above/Microtubule assembly (K_1)	G+B	Pre to GS23	460	505	0.76
Winter barley						
Glyphosate	As above	All	Predrill or pre-harvest	187	132	0.46
Diflufenican/Flufenacet (Liberator)	As above	G+B	Pre	184	41	0.80
Pinoxaden	ACCase (lipid synthesis)	G	Post to GS41	166	6	0.61
Chlorotoluron/Diflufenican	Photosynthesis at PS II (C2)/As above	G+B	Pre-Post	91	127	0.47
Pendimethalin/Picolinafen	As above/Pigment synthesis (F_1)	G+B	Pre	89	67	0.69

G: grass weeds, B: broad-leaved weeds, Pre: pre-emergence, Post: post-emergence

† Total areas: 1. 94 mi ha (wheat), 0. 429 mi ha (winter barley). * (HRAC 2017)

farmers in that doses, which were once effective, no longer appear to be, even where application conditions are ideal and the product is applied correctly. Pragmatically, the dose required for efficacy exceeds the permitted 'label' rate and indeed resistance may appear to be total in many plants within weed populations such that no dose appears to be sufficient to control them. This is well-illustrated by comparing the dose response curves of susceptible and resistant *A. myosuroides* populations to the post-emergence graminicide, Atlantis (Iodosulfuron-methyl-sodium/ mesosulfuron-methyl). In 2014, this herbicide was applied to a larger area (c. 43%) of the UK wheat crop than any other herbicide including glyphosate (Table 11.8). In glasshouse trials, the 'Rothamsted Susceptible' biotype was controlled at 1/8th of the recommended (approved) dose rate whereas the two resistant lines tolerated 4x that dose. Despite the use of a herbicide mixture, the tolerant genotypes were resistant to both herbicides, exhibiting cross-resistance (Moss 2010).

Integrating Weed Control Methods to Achieve Sustainability

Integrated weed control should enhance the efficacy of weed control but is unlikely to attract high rates of adoption among farmers unless gross margins are maintained or improved; it is not sufficient to minimise the application costs. Integration at its simplest may simply combine pre- and post-emergence herbicides as well as herbicide mixtures (Pannacci and Onofri 2016). The next step is integration with non-chemical weed control methods (Pannacci and Tei 2014) and this is clearly needed to minimise risks of herbicide resistance. An important goal of integrating methods is to contribute to the sustainability of crop production by balancing crop productivity with the use of finite resources, while always seeking to minimise adverse environmental impacts (Dorwin and Lesch 2010).

Integrated and sustainable weed control for small-grained cereals must first minimise the risk of introducing new weeds at every management level. Legislation is appropriate at national and regional levels. But the ultimate responsibility has to be the farmer's. At farm and field scales and perhaps within fields, standard operating procedures are needed to reduce the risk of introducing new weed problems or spreading them to clean areas during agricultural operations.

Where a weed is present in a field or farm, the goals of weed control need to be agronomically sustainable. Simplistically, irrespective of the short-term measures needed to control a current infestation, the long-term goals of weed control must have a sound basis in weed population dynamics. Should the goal be to contain the infestation and minimise yield losses in each crop or would a long-term goal of elimination be applicable? Modelling weed populations helps to answer the question. For example, given a soil seed bank of 500 viable weed seeds m^{-2} and an infestation of only 10 weeds m^{-2} (assuming only 2% emergence from the soil seedbank into the crop), would elimination be feasible? Assuming a negative exponential decline of the soil seed bank with 33% depletion of the soil seedbank each year and 100% effective weed control so that no new seeds are produced, 23 years are predicted to be needed to reach 10 weeds ha^{-1} (Murdoch 1988). Clearly, 10 weeds ha^{-1} is not elimination and since 100% efficacy is unlikely, the goal for most annual weeds of small-grained cereals has to be containment. Only in exceptional cases where depletion rate exceeds 90% per annum, is elimination likely to become a realistic goal. Moreover because of the fecundity of most weeds (Table 11.5), the efficacy of weed control required to contain an infestation is very high. While it is clear that crop rotation is a key contribution to long-term sustainability (Figure 11.7), the problem of occasional crops where weeds fail to be controlled must be addressed (Figure 11.7). Using results in Lutman et al. (2013), Moss (unpublished) suggested it may be possible to 'stack' several methods to control *A. myosuroides* even in a crop like winter wheat. Thus, although non-chemical approaches do not offer silver bullets, the individual efficacies of control by ploughing, delayed drilling, an increased seed rate and a more competitive cultivar are on average, respectively, 69%, 31%, 26% and 22%. Assuming these are independent effects and can be stacked, their combined effect would result in 88% control of *A. myosuroides* in winter wheat. The remaining 12% of uncontrolled plants may still affect yield and will certainly produce seeds and so a crop rotation would be needed for an organic system or a herbicide could be added for the conventional. Assuming 90% control was achieved with the herbicide, the overall efficacy of the stacked, integrated strategy would be close to 99%.

Sustaining the Use of Herbicides

Should some land be devoted to intensive cereal production, while other zones are kept for environmental benefits and ecosystem services? This question has an underpinning premise: that intensive cereal production, especially, where it involves the use of agro-chemicals cannot be sustainable. Use of many herbicides can, however, be sustainable if their application rates are optimised and if risks to people and the environment are minimised, mitigated and monitored. The trade-off between these dual requirements is less than might be imagined. First of all, the efficacy of herbicides can be enhanced and so the amounts required can be reduced by more efficient application methods (Butler Ellis et al. 2006, Jensen 2010). Moreover, efficacy is only achievable if the herbicide actually hits the weeds and run-off is avoided, for example, arable weeds with erectiform or needle-like leaves are more difficult to target than those with planiform leaves.

Given a goal of reducing the amount of herbicide used without compromising efficacy, it is self-evident that efficacy will be greater if herbicides are applied when weeds are most susceptible and accessible, such as at early growth stages before canopy closure when weeds are less likely to be protected by the crop. More precise targeting is also possible for weeds which occur in patches, when patch spraying minimises unnecessary applications to the soil and crop. The trade-off is more difficult to avoid where adverse environmental impacts are likely to vulnerable zones, such as headlands and field margins close to waterways or hedges. Finally, post-emergence, non-residual herbicides usually have less environmental impact and where available and effective, their use is preferable.

Sustainable use of chemicals is also much more likely if integrated with non-chemical weed and crop management. Not only does this reduce risks of herbicide resistance, but dose rates required or the frequency of treatment will also be lowered. For the farmers, the real trade-off is an economic one such that the decision to control weeds must be based on the probability that a profit will be achieved through their control—whether by chemical or non-chemical means or by an integration of several approaches. This profit may be considered on a year-to-year or crop-to-crop basis, but the biology of weeds and the longevity of their soil seed banks means that long-term implications of allowing seed shedding on subsequent crops must be considered. Similarly, as noted already, crop hygiene must be observed to prevent new introductions.

Where herbicides are to be used, an environmental risk assessment of spray drift or of potential exposure and damage to non-target organisms is desirable. Options designed to mitigate risks in small-grained cereal crops include operating nozzles at lower pressures, using air assistance and directing spray nozzles downwards, and, for the last swath, either leaving it untreated or using lower drift nozzles and perhaps only applying sprayer washings.

For post-emergence herbicides, the dose to optimise efficacy needs to be determined for common weed species (Pannacci 2016). Optimising herbicide dose rates and use of improved weed emergence models may improve the targetting and efficacy of post-emergence weed control (Masin et al. 2014).

Understanding how spatial variation in abiotic constraints affects cereal yields, weed infestations and herbicide efficacy, may also help in optimisation. The approaches and technologies of precision agriculture within individual fields are then likely to yield more sustainable and integrated approaches to weed and indeed cereal crop management.

Even in developing countries, farmers are well aware of spatial variability, at least at the level of farms and villages (Samake et al. 2005). Within farmers' fields, however, spatial variation of the key biotic constraints including weeds and their association with spatial variation of abiotic constraints, such as soil moisture and fertility is seldom known or understood. Most advice to farmers, therefore, follows a 'one-size-fits-all' approach, ignoring the typically patchy distribution of many weeds in cereal fields. New technologies, such as RPAS allow weed mapping, but often 'after the horse has bolted', that is when it is too too late for treatment and there may also be too little information if weed identification needs to be verified by field walking.

Patchy distributions of weeds in fields are frequently noted (Murdoch et al. 2014). In such cases, should the whole field be treated uniformly or the patches differently? Treating the whole field uniformly is likely to waste resources and money and may risk an adverse environmental impact. However, if only patches are sprayed, is that sufficient to control weeds and maintain yields across the whole field? A related question is: how stable are these patches? When should and how easy is it for weeds to be mapped? For example, mapping might seem relatively simple with an RPAS or proximal sensor when the weeds are flowering or fruiting, but that is too late to apply anything other than the most drastic control measures, such as destroying the patches. Indeed such patches reflect failure of control earlier in the growing seasons and so some have sought to eliminate such patches where they have problems with herbicide resistance.

In general adoption of site specific weed management has been limited because of the uncertainty and research is needed to quantify probabilities of outcomes so that farmers' different risk and time preferences can be accommodated in decision support systems. A recent report (EIP-Agri 2015) emphasised that one of the "challenges for innovation in the coming years [is that] a change in research attitudes is needed encouraging researchers to take into account farmers' opinions and advice. "

More precise application technologies are challenging for small grain cereals due to narrow row spacing. For example, Midtiby et al. (2011) described a microspraying system, which worked well between rows but incurred unacceptable risks of crop damage within rows due to the tendency of spray droplets to drift. Similarly, Miller et al.'s (2010) spot sprayer was only suitable for large weeds. Christensen et al. (2009) described a droplet applicator rather than a sprayer. Here the risk of drift is lower, due to use of gravity-fed droplets applied via a field robot (see Klose et al. 2008). Leaf-specific control with a 'point and shoot' ejector emitting individually metered and targetted droplets is an advance on this but is currently most applicable for field vegetables (Murdoch et al. 2017) and its use for small grain cereals is unlikely to be achieved for several years.

Concluding Remarks

The concept of sustainability for farmers must connect with the maintenance of their livelihoods as well as their self-interest in wanting to preserve the land for their posterity. With increasing problems of herbicide resistance, and because the efficacy of weed control from non-chemical methods is seldom sufficient to contain infestations, integrated weed management for small grain cereals is essential especially in temperate latitudes. For policy makers, drivers are more complex but are driven by the need to preserve food and nutritional security in the long-term. These get translated into various legal frameworks and various Directives and Regulations relate to the use of pesticides in general and herbicides in particular. For example in the EU, concerns about pesticides in waterways are such that the Water Framework Directive (2000/60/EC) allows maxima per litre of drinking water of 1 µg for any one pesticide and a total of 5 µg for all pesticides. The Sustainable Use Directive (SUD 2009/128/EC) explicitly promotes integrated weed management (IWM) designed to reduce or even eliminate the need for herbicides and the "risks and impacts of [their] use on human health and the environment".

The issue of loss of approvals of active ingredients and the numerous hurdles and dossiers of information required to develop new actives are a strong disincentive to manufacturers. Nevertheless unless a range of products is available, the risks of losing actives due to herbicide resistance is also greater since farmers will have to resort to using chemicals with the same mode of action. Moreover, the "greater the use of one active, over a large area, the more likely it is to appear in water" (Clark et al. 2009).

Overall, chemical weed control must be integrated with non-chemical approaches to be sustainable. The "many little hammers" (Liebmann and Gallandt 1997) approach of stacking treatments is also vital for non-chemical methods, as their efficacy is lower and there are no silver bullets. Another reason for stacking is that weeds are just as capable of developing "resistance" to non-chemical methods as they are to chemical ones (Harker 2013). Adaptation is a key trait for weed survival for a weed is ultimately the right plant in the right place at the right time. Integrated and sustainable weed management in small grain cereal crops must therefore be flexible and varied to limit the likelihood of such adaptation.

Acknowledgment

The author would like to thank Dr. Peter Lutman for reviewing the chapter.

REFERENCES

Addisu, M., J.W. Snape, J.R. Simmond and M.J. Gooding. 2009. Reduced height (Rht) and photoperiod insensitivity (Ppd) allele associations with establishment and early growth of wheat in contrasting production systems. Euphytica 166: 249–267.

AHDB. 2017 online. AHDB Recommended Lists for cereals and oilseeds for 2017/18. Summer edition 2017. https://cereals.ahdb.org.uk/media/800462/ahdb-recommended-list-web.pdf. (Accessed August 2017).

Al-Allagi, M. 2007. Integrated weed management in spring cereals. The contributions of crop genotype and seed quality. PhD Thesis, University of Reading, U.K.

Al Allagi, M.D. and A.J. Murdoch. 2003. Suppressing weed competition: the interaction of seed quality and seed rate in spring wheat. 1: pp. 667–670. *In*: Proceedings of the 2003 Crop Science & Technology Conference, Glasgow.

Andersson, L., P. Milberg, W. Schütz and O. Steinmetz. 2002. Germination characteristics and emergence time of annual Bromus species of differing weediness in Sweden. Weed Research 42: 135–147.

Andrew, I.K.S., J. Storkey and D.L. Sparkes. 2015. A review of the potential for competitive cereal cultivars as a tool in integrated weed management. Weed Research 55: 239–248.

Andrew, I., J. Storkey and D. Sparkes. 2014. Identifying the traits in cereals that confer greater competitive ability against *Alopecurus myosuroides*. Aspects of Applied Biology 127, Crop Production in Southern Britain: Precision Decisions for Profitable Cropping. pp. 165–171. Association of Applied Biologists, Wellesbourne, U.K.

Baskin, J.M. and C.C. Baskin. 1985. The annual dormancy cycle in buried weed seeds: a continuum. Bioscience 35: 492–498.

Blatchley, W.S. 1912. The Indiana weed book. Nature Publishing Company.

Bond, W. and A.C. Grundy. 2001. Non-chemical weed management in organic farming systems. Weed Research 41: 383–405.

Butler Ellis, M.C., S. Knight, P.C.H. Miller. 2006. The effects of low application volumes on the performance of a T2 fungicide application. Aspects of Applied Biology 77: 433–438.

CABI. 2017a online. *Avena fatua*. *In*: Invasive Species Compendium. Wallingford, UK: CAB International. Available at: www. cabi. org/isc. (Accessed in August 2017).

CABI. 2017b online. *Alopecurus myosuroides*. *In*: Invasive Species Compendium. Wallingford, UK: CAB International. Available at: www. cabi. org/isc. (Accessed in August 2017).

Christensen, S., H.T. Søgaard, P. Kudsk, M. Nørremark, I. Lund, E.S. Nadimi and R. Jørgensen. 2009. Site-specific weed control technologies. Weed Research 49: 233–241.

Clarke, J., S. Wynn, S. Twining, P. Berry, S. Cook, S. Ellis and P. Gladders. 2009. Pesticide availability for cereals and oilseeds following revision of Directive 91/414/EEC, effects of losses and new research priorities. HGCA Research Review 70: 131.

Cosser, N.D., M.J. Gooding, W.P. Davies and R.J. Froud-Williams. 1996. Effects of wheat dwarfing genes on grain yield and quality of wheat in competition with *Alopecurus myosuroides*. pp. 1089–1094. *In*: The Proceedings of the 2nd International Weed Control Congress, Copenhagen, Denmark.

Cosser, N.D., M.J. Gooding, A.J. Thompson and R.J. Froud-Williams. 1997. Competitive ability and tolerance of organically grown wheat cultivars to natural weed infestations. Annals of Applied Biology 130: 523–535.

Courtney, A.D. 1968. Seed dormancy and field emergence in *Polygonum aviculare*. Journal of Applied Ecology 5: 675–684.

Cousens, R., P. Brain, J.T. O'Donovan and P.A. O'Sullivan. 1987. The use of biologically realistic equations to describe the effects of weed density and relative time of emergence on crop yield. Weed Science 35: 720–725.

Crop Trust. 2017. Barley. Available at: www. croptrust. org/crop/barley (Accessed on July 17, 2017).

Curtis, B.C. undated - online. Wheat in the world. Available at: http://www.fao.org/docrep/006/y4011e/y4011e04.htm (Accessed in August 2017).

De Vita, P., S.A. Colecchia, I. Pecorella and S. Saia. 2017. Reduced inter-row distance improves yield and competition against weeds in a semi-dwarf durum wheat variety. European Journal of Agronomy 85: 69–77.

EIP-AGRI. 2015. Final Report of Focus Group on Precision Farming. pp. 44. Available at: https://ec.europa.eu/eip/agriculture/sites/agri-eip/files/eip-agri_focus_group_on_precision_farming_final_report_2015.pdf. (Accessed in August 2017).

Evers, J.B. and L. Bastiaans. 2016. Quantifying the effect of crop spatial arrangement on weed suppression using functional-structural plant modelling. Journal of Plant Research 129: 339–351.

Fenner, M. and K. Thompson. 2005. The ecology of seeds. Cambridge University Press.

FAO (Food and Agriculture Organization). 2017. World Food Situation. Available at: http://www.fao.org/worldfoodsituation/csdb/en/ (Accessed on July 12, 2017).

FAOSTAT. 2017. Food and agriculture data. Available at: http://www.fao.org/faostat/en/#home (Accessed on July 12, 2017).

Garthwaite, D., I. Barker, R. Laybourn, A. Huntly, G.P. Parrish, S. Hudson and H. Thygesen. 2015. Pesticide Usage Survey Report 263 – Arable Farm Crops in the United Kingdom, 2014. York: FERA.

Gooding, M.J., R.K. Uppal, M. Addisu, K.D. Harris, C. Uauy, J.R. Simmonds and A.J. Murdoch. 2012. Reduced height alleles (Rht) and Hagberg falling number of wheat. Journal of Cereal Science 55: 305–311.

Harker, K.N. 2013. Slowing weed evolution with integrated weed management. Canadian Journal of Plant Science 93: 759–764.

Harris, K.D., A.J. Murdoch and M.J. Gooding. 2010. Effect of plant height and relative time of weed emergence in relation to yield maintenance and weed suppression by Rht lines of winter wheat. Abstract A0126. *In*: The Proceedings of the 15th EWRS Symposium, Kaposvar, Hungary.

Harris, K.D. 2011. Effect of crop height, relative time of weed emergence, seed vigour and dormancy on yield maintenance and weed suppression by Rht lines of winter wheat. PhD thesis, University of Reading.

Heap, I. 2017. The international survey of herbicide resistant weeds. Available at: www.weedscience.org. (Accessed in August 2017).

Hoad, S.P., C.F.E. Topp and D.H.K. Davies. 2008. Selection of cereals for weed suppression in organic agriculture: a method based on cultivar sensitivity to weed growth. Euphytica. 163: 355–366.

Holm, L. 1971. The role of weeds in human affairs. Weed Science 19: 485–490.

Holm, L.G., D.L. Plucknett, J.V. Pancho and J.P. Herberger. 1977. The World's Worst Weeds: Distribution and Biology. University Press of Hawaii Honolulu, USA.

HRAC. 2017 online. Global herbicide classification. Available at: http://hracglobal.com/tools/world-of-herbicides-map (Accessed in August 2017).

Hull, R., L.V. Tatnell, S.K. Cook, R. Beffa and S.R. Moss. 2014. Current status of herbicide-resistant weeds in the UK. Aspects of Applied Biology 127. pp. 261–272. *In*: Crop Production in Southern Britain: Precision Decisions for Profitable Cropping. Association of Applied Biologists, Wellesbourne, U.K.

Jensen, H.A. 1969. Content of buried seeds in arable soil in Denmark and its relation to the weed population. Dansk Botanisk Arkiv 27: 1–56.

Jensen, P.K. 2010. Improved control of annual grass weeds with foliar-acting herbicides using angled applications. Aspects of Applied Biology 99: 81–88.

Khah, E.M., R.H. Ellis and E.H. Roberts. 1986. Effects of laboratory germination, soil temperature and moisture content on the emergence of spring wheat. The Journal of Agricultural Science 107: 431–438.

Khah, E.M., E.H. Roberts and R.H. Ellis. 1989. Effects of seed ageing on growth and yield of spring wheat at different plant-population densities. Field Crops Research 20: 175–190.

Klose, R., A. Ruckelshausen, M. Thiel and J. Marquering. 2008. Weedy – a sensor fusion based autonomous field robot for selective weed control pp. 167–172. *In*: Proceedings 66th International Conference Agricultural Engineering/AgEng, Stuttgart-Hohenheim, VDI-Verlag.

Korres, N.E. and R.J. Froud Williams. 2002. Effects of winter wheat cultivars and seed rate on the biological characteristics of naturally occurring weed flora. Weed Research 42: 417–428.

Liebman, M. and E.R. Gallandt. 1997. Many little hammers: ecological management of crop-weed interactions. pp. 291–343. *In*: Jackson, L.E. (Ed.) Ecology in Agriculture, Academic Press, San Diego, CA.

Lutman, P.J.W., S.R. Moss, S. Cook and S.J. Welham. 2013. A review of the effects of crop agronomy on the management of *Alopecurus myosuroides*. Weed Research 53: 299–313.

Masin, R., D. Loddo, V. Gasparini, S. Otto and G. Zanin. 2014. Evaluation of weed emergence model. Alert Inf for maize in soybean. Weed Science 62: 360–369.

Melander, B., I.A. Rasmussen and P. Bàrberi. 2005. Integrating physical and cultural methods of weed control—examples from European research. Weed Science 53: 369–381.

Midtiby, H.S., S.K. Mathiassen, K.J. Andersson and R.N. Jørgensen. 2011. Performance evaluation of a crop/weed discriminating microsprayer. Computers and Electronics in Agriculture 77: 35–40.

Miller, P., A. Lane, N. Tillett and T. Hague. 2010. Minimising the environmental impact of weed control in vegetables by weed detection and spot herbicide application. Final Report R270. AHDB/HDB, Stoneleigh.

Moss, S. 2013 online. Black-grass (*Alopecurus myosuroides*). Available at: https://cereals. ahdb. org. uk/media/1108866/Black-grass-everything-you-really-wanted-to-know-WRAG-May-2013-. pdf. (Accessed in August 2017).

Moss, S.R., L.V. Tatnell, R. Hull, J.H. Clarke, S. Wynn and R. Marshall. 2010. Integrated management of herbicide resistance. HGCA Project Report No. 466. 115 pp.

Murdoch, A.J. 1988. Long-term profit from weed control. *In*: Weed Control in Cereals and the Impact of Legislation on Pesticide Application. Aspects of Applied Biology 18: 91–98.

Murdoch, A.J. 1998. Dormancy cycles of weed seeds in soil. *In*: Weed Seedbanks: Determination, Dynamics and Manipulation. Aspects of Applied Biology 51: 119–126.

Murdoch, A.J. 2006. Soil seed banks. pp. 501–520, *In*: Basra, A. (Ed.) Handbook of Seed Science and Technology. Haworth's Food Products Press, Binghamton, New York.

Murdoch, A.J. 2013. Seed dormancy. pp. 151–177. *In*: Gallagher, R.S. (Ed.) Seeds: The Ecology of Regeneration and Plant Communities, 3rd edition, CAB International, Wallingford, Oxon UK.

Murdoch, A.J. and R.H. Ellis. 2000. Dormancy, viability and longevity. pp. 183–214. *In*: Fenner, M. (Ed.) Seeds: The Ecology of Regeneration and Plant Communities, 2nd edition. CAB International, Wallingford, Oxon UK.

Murdoch, A.J. and E. Kebreab. 2013. Germination ecophysiology. pp. 195–219. *In*: Parasitic Orobanchaceae. Springer Berlin Heidelberg.

Murdoch, A.J. and E.H. Roberts. 1982. Biological and financial criteria of long-term control strategies for annual weeds. pp. 741–748. *In*: Proceedings of the 1982 British Crop Protection Conference – Weeds.

Murdoch, A.J. and E.H. Roberts. 1996. Dormancy cycle of *Avena fatua* seeds in soil. pp. 147–152. *In*: Proceedings Second International Weed Control Congress Copenhagen, Denmark.

Murdoch, A.J., C. Flint, R. Pilgrim, P. De La Warr, J. Camp, B. Knight, P. Lutman, B. Magri, P. Miller, T. Robinson, S. Sanford and N. Walters. 2014. eyeWeed: automating mapping of black-grass (*Alopecurus myosuroides*) for more precise applications of pre- and post-emergence herbicides and detecting potential herbicide resistance. Aspects of Applied Biology 127. pp. 151–158. *In*: Crop Production in Southern Britain: Precision Decisions for Profitable Cropping. Association of Applied Biologists, Wellesbourne, U.K.

Murdoch, A.J., N. Koukiasas, R.A. Pilgrim, S. Sanford, P. De La Warr and F. Price-Jones. 2017. Precision targeting of herbicide droplets potentially reduces herbicide inputs by at least 90%. pp. 39–44. *In*: Grove, I. and Kennedy, R. (Eds.) Aspects of Applied Biology 135. Precision Systems in Agricultural and Horticultural Production. Association of Applied Biologists, Wellesbourne, U.K.

Murdoch, A.J., E.H. Roberts and C.O. Goedert. 1989. A model for germination responses to alternating temperatures. Annals of Botany 63: 97–111.

Murdoch, A.J., S.J. Watson and J.R. Park. 2003. Modelling the soil seed bank as an aid to crop management in Integrated Arable Farming Systems. pp. 521–526. *In*: Proceedings of the 2003 Crop Science & Technology Conference, Glasgow.

Nuttonson, M.Y. 1955. Wheat-climatic relationships and the use of phenology in ascertaining the thermal and photothermal requirements of wheat. Washington, D.C., American Institute of Crop Ecology. Cited by Curtis (undated).

O'Donovan, J.T., E.A.D.S. Remy, P.A. O'Sullivan, D.A. Dew and A.K. Sharma. 1985. Influence of the relative time of emergence of wild oat (*Avena fatua*) on yield loss of barley (*Hordeum vulgare*) and wheat (*Triticum aestivum*). Weed Science 33: 498–503.

Oerke, E-C. 2006. Crop losses to pests. Journal of Agricultural Science 144: 31–43.

Olsen, J., L. Kristensen, J. Weiner and H.W. Griepentrog. 2005. Increased density and spatial uniformity increase weed suppression by spring wheat. Weed Research 45: 316–321.

Pannacci, E. 2016. Optimization of foramsulfuron doses for post-emergence weed control in maize (*Zea mays* L.). Spanish Journal of Agricultural Research 14: 1005.

Percival, J. 1921. The wheat plant. A monograph. New York, NY, USA, E.P. Dutton & Company.

Peters, N.C.B., R.J. Froud-Williams and J.H. Orson. 1993. The rise of barren brome (*Bromus sterilis*) in UK cereal crops. pp. 773–780. *In*: Proceedings Brighton Crop Protection Conference, Weeds, Brighton, UK. 22–25 November 1993.

Preston, C. 2012 online. Multiple resistant rigid ryegrass. Available at: http://www.weedscience.org/Details/Case.aspx?ResistID=389 (Accessed in August 2017).

Renton, M. and B.S. Chauhan. 2017. Modelling crop-weed competition: why, what, how and what lies ahead? Crop Protection 95: 101–108.

Roberts, E.H. 1973. Oxidative processes and the control of seed germination. pp. 189–218. *In*: W. Heydecker (Ed.) Seed Ecology. Pennsylvania State University Press. London: Buttterworth.

Roberts, H.A. 1970. Report National Vegetable Research Station, Wellesbourne for 1969: 25.

Roberts, H.A. 1981. Seed banks in soils. Advances in Applied Biology 6: 1–55.

Roberts, H.A. 1986. Seed persistence in soil and seasonal emergence in plant species from different habitats. Journal of Applied Ecology 23: 639–656.

Roberts, H.A. and R.J. Chancellor. 1986. Seed banks of some arable soils in the English midlands. Weed Research 26: 251–257.

Roberts, H.A. and M.E. Potter. 1980. Emergence patterns of weed seedlings in relation to cultivation and rainfall. Weed Research 20: 377–386.

Roberts, H.A. and M.E. Ricketts. 1979. Quantitative relationships between the weed flora after cultivation and the seed population in the soil. Weed Research 19: 269–275.

Royal Society. 2009. Reaping the benefits: science and the sustainable intensification of global agriculture. The Royal Society, London. 86 pp.

Sagar, G.R. and A.M. Mortimer. 1976. An approach to the study of the population dynamics of plants with special reference to weeds. Applied Biology 1: 1–47.

Salisbury, E. 1961. Weeds & Aliens. Collins, London.

Samaké, O., E.M.A. Smaling, M.J. Kropff, T.J. Stomph and A. Kodio. 2005. Effects of cultivation practices on spatial variation of soil fertility and millet yields in Mali. Agriculture, Ecosystems and Environment 109: 335–345.

Scherner, A., B. Melander and P. Kudsk. 2016. Vertical distribution and composition of weed seeds within the plough layer after eleven years of contrasting crop rotation and tillage schemes. Soil and Tillage Research 161: 135–142.

Scherner, A., B. Melander, P.K. Jensen, P. Kudsk and L.A. Avila. 2017. Reducing tillage intensity affects the cumulative emergence dynamics of annual grass weeds in winter cereals. Weed Research 57: 314–322.

Seefeldt, S.S., A.G. Ogg and Y. Hou. 1999. Near-isogenic lines for *Triticum aestivum* height and crop competitiveness. Weed Science 47: 316–320.

Tillett, N.D., T. Hague, A.M. Blair, P.A. Jones, R. Ingle and J.H. Orson. 1999. Precision inter-row weeding in winter wheat. pp. 975–980. *In*: Proceedings 1999 Brighton Crop Protection Conference—Weeds.

Tottman, D.R. 1987. The decimal code for the growth stages of cereals, with illustrations. Annals of Applied Biology 110: 441–454.

Van Wychen, L. 2016. 2015 Survey of the most common and troublesome weeds in the United States and Canada. Weed Science Society of America National Weed Survey Dataset. Available at: http://wssa.net/wp-content/uploads/2015-Weed-Survey_final.xlsx (Accessed in November 2017).

Vleeshouwers, L.M. 1997. Modelling weed emergence patterns. Ph.D. Thesis, Wageningen Agricultural University, The Netherlands.

Watson, S.J., A.J. Murdoch and J. Park. 1999. Seed bank depletion of wild oat and cleavers in integrated arable farming systems. pp. 569–570. *In*: Proceedings of the Brighton Crop Protection Conference—Weeds, 2.

Welsh, J.P., H.A.J. Bulson, C.E. Stopes, R.J. Froud-Williams and A.J. Murdoch. 1996. Weed control in organic winter wheat using a spring-time weeder. pp. 1127–1132. *In*: Proceedings Second International Weed Control Congress. Copenhagen, Denmark.

Welsh, J.P., H.A.J. Bulson, C.E. Stopes, A.J. Murdoch and R.J. Froud-Williams. 1999. The critical weed-free period in organically-grown winter wheat. Annals of Applied Biology 134: 315–320.

Wilson, B.J. and G.W. Cussans. 1975. A study of the population dynamics of *Avena fatua* L. as influenced by straw burning, seed shedding and cultivations. Weed Research 15: 249–258.

Wilson, B.J. and K.J. Wright. 1990. Predicting the growth and competitive effects of annual weeds in wheat. Weed Research 30: 201–211.

Wolfe, M.S., J.P. Baresel, D. Desclaux, I. Goldringer, S. Hoad, G. Kovacs, F. Löschenberger, T. Miedaner, H. Østergård and E.T. Lammerts van Bueren. 2008. Developments in breeding cereals for organic agriculture. Euphytica 163: 323–346.

Worthington, M. and C. Reberg-Horton. 2013. Breeding cereal crops for enhanced weed suppression: optimizing allelopathy and competitive ability. Journal of Chemical Ecology 39: 213–231.

12

Sustainable Weed Control in Maize

Per Kudsk[1]*, Vasileios P. Vasileiadis[2] and Maurizio Sattin[2]

[1] Aarhus University, Department of Agroecology, DK-4200 Slagelse, Denmark, Europe
[2] National Research Council of Italy (CNR), Institute of Agro-Environmental and Forest Biology, 35020 Legnaro (PD), Italy, Europe

Introduction

Maize (*Zea mays* L.) is grown throughout the world and is ranking third among the world's cereal crops after wheat and rice. Maize is a very versatile crop grown for human consumption, as an animal feed, for energy and for industrial purposes. In 2014, maize was grown on an area of ca. 184 million hectare according to FAO (http://www. fao. org/faostat/en/#data/QC). Nearly 1/3rd of the total area was cropped in Asia, of which more than 50% was grown in China, followed by Africa and North America, each with around 1/6 of the area, while the areas in South America and Europe constituted ca. 1/8th and 1/10th of the total area. Total maize grain production in 2014 was around 1,000 Mt of which 35% was produced in the US reflecting that the average yields were significantly higher in North America (11.8 tons/ha) compared to the other regions, for example, Asia with 5.5 tons/ha and Africa (2.3 tons/ha) (http://www.fao.org/faostat/en/#data/QC). Average grain yields in Europe were 7.6 tons/ha, i.e. also considerably lower than in North America. A distinct feature of European maize production is that within the 28 Member States of the European Union approximately 40% of the area is harvested as green maize for silage (mainly Northern Europe) or energy use (mainly Germany) (http://ec.europa.eu/eurostat/web/agriculture/data/main-tables).

Maize is normally grown either in short rotations or in monocultures. In the North-Central region of the United States, the main maize-growing area in North America and often referred to as the 'Corn Belt', maize is typically grown in either a two-year rotation with soybean or in monoculture. In Europe, crop rotations tend to be more diverse with cereals, soybean, potato and grass being the most common rotational crops but maize monoculture is also common particularly in regions with intensive dairy farming (Vasileiadis et al. 2011). In China, intercropping, mixed cropping and relay cropping are still practiced but the trend is towards more simplified cropping systems including maize monoculture (Zhang et al. 2015).

The most significant change to maize cultivation in recent years has been the introduction of herbicide-tolerant maize cultivars and most notably glyphosate tolerant cultivars that have had a dramatic effect on weed management practices and, as a result of this, the challenges that farmers growing herbicide-tolerant cultivars are facing nowadays (Duke and Powles 2009). In the US, genetically modified glyphosate-tolerant cultivars constitute ca. 90% of the total maize area (https://www.ers.usda.gov/data-products/adoption-of-genetically-engineered-crops-in-the-us. aspx) but genetically modified maize, including Bt maize, only constitutes 29% of the maize area grown worldwide which is significantly lower than for soybean and cotton (James 2015).

Weed Flora and Impact

The weed flora of maize fields varies between continents. Forcella et al. (1997), in a weed seedbank emergence study comprising 22 locations across the maize-growing area of the United States, identified the following 15 weed species as the most important: *Abutilon theophrasti, Amaranthus* spp., *Ambrosia artemisiifolia, Chenopodium album, Helianthus annuus, Kochia scoparia, Panicum miliaceum, Polygonum aviculare, Polygonum convolvulus, Polygonum pensylvanicum, Portulaca oleracea* L, *Setaria faberi, Setaria pumila, Setaria viridis,* and *Solanum physalifolium*. Since this survey was done maize cultivation in the United States has changed and is now dominated by genetically modified glyphosate-tolerant varieties (https://www.ers.usda.gov/data-products/adoption-of-genetically-engineered-crops-in-the-us.aspx). Webster and Nichols (2012) found no significant changes in the weed flora of maize fields between 1994 and 2009, in contrast to soybean and cotton. This difference was attributed to the much faster adoption by US farmers of glyphosate-tolerant varieties in soybean and cotton compared to maize. With the recent increase in the cultivation of herbicide tolerant maize varieties in the US (James 2015) it is likely that also the weed flora composition in maize will have changed but no studies have been done so far that can confirm this assumption.

Recently, two expert-based surveys on weed distribution in maize in Europe were published. Meissle et al. (2010), covering 11 maize producing areas in eight countries, listed *C. album* as the most important dicotyledonous species and *Echinochloa crus-galli* (L.) Beauv. and *S. viridis* as the most important monocotyledonous weed species. Jensen et al. (2011), surveying weed species distribution in 11 countries, listed 203 weed species of which 61 annual and perennial weed species were considered to be 'very common' in at least one country but only one weed species, *C. album*, was classified as being 'very common' in all 11 countries. In a weed survey in the Czech Republic, maize was the crop with the lowest number of weed species (Pysek et al. 2005).

The surveys done in the United States and Europe have shown that the weed flora is dominated by annual weed species reflecting that maize is an annual crop. However, both Meissle et al. (2010) and Jensen et al. (2011) mentioned perennial species like *Elymus repens, Cirsium arvense,* and *Convolvolus arvensis* L., but maybe more interestingly Meissle et al. (2011) reported that species like *Sorghum halepense* (L.) Pers, *Calystegia sepium* (L.) R. Br. and *Fallopia convolvulus* (L.) were an increasing problem in some maize producing regions. Lower susceptibility to the most widely used herbicides and adoption of non-inversion tillage systems were suggested as most likely causes for this shift. In glyphosate-tolerant maize, perennial weeds are less likely to pose problems due to the high efficacy of glyphosate on most perennial weeds.

In sub-Saharan Africa up to 50 million hectares of farmland is infested by the parasitic weed species *Striga asiatica* and *S. hermonthica* adversely affecting the livelihood of up to 300 million people (Nzioki et al. 2016). *Striga* spp. are major constraints to crop production that can lead to complete crop failure and some of the most severe effects are observed in maize (Atera et al. 2013).

Weeds are a major constraint to maize production due to the wide row spacing that makes the crop susceptible to weed competition (Cerrudo et al. 2012). Several studies have attempted to determine the critical period for weed control, i.e., when the crop is most susceptible to weed competition and yield losses are maximum (Zimdahl 1988). The critical period for weed control depends on cropping conditions, maize type, weed flora and, not at least, the level of acceptable yield loss. The majority of the studies have shown that it begins at the very early growth stages and can last until the 10- to 12-leaf stage (e.g., Isik 2006, Williams II, 2006, Page et al. 2012, Tursun et al. 2016), i.e., weeds have to be effectively controlled for a relatively long period. Oerke (2006) concluded that weeds are the most important pest group in maize and that potential yield losses are ca. 30% exceeding the sum of losses caused by the other pest groups. The actual average yield losses are ca. 10%, but with large regional differences ranging from only 5% in Europe to 19% in West Africa. These differences primarily reflect the differences in farmer's access to effective weed control measures but also the severity of weed infestation. For example, under African conditions *Striga* ssp can cause crop failure and 100% yield loss (Atera et al. 2013).

Weed Management in Maize

Maize is grown in all parts of the world in high-productive and very intensively managed cropping systems with high level of inputs of fertilizers and pesticides, as found in North America and Europe, as well as in extensively managed systems as exemplified by resource-poor smallholder farmers in sub-Saharan Africa. Weed management practices vary accordingly from totally relying on herbicides to doing weed management by hand. In the following sections the various weed management practices and their application in maize will be presented.

Preventive and Cultural Weed Management

The main objectives of preventive and cultural weed management are to avoid the introduction of new weed species into a field, reduce weed density to minimise competition with the crop and to reduce the impact on crop yields by favouring crop growth and hampering weed growth. Cultural weed management encompasses a suite of weed control tactics but in this chapter only the most widely adopted in maize cultivation will be addressed. For more details on the application of cultural weed management in maize see Mhlanga et al. (2016a).

Preventive Measures

In modern farming, field machinery is one of the main sources of spreading weeds from field to field and proper cleaning of machinery before moving to the next field should be a standard procedure. On farms with livestock the animals or the manure may serve as a vector spreading weeds between fields and farms. With herbicide resistance being a perpetual increasing problem, avoiding the spread of weeds has become even more important (Vencill et al. 2012). For weed species where seed dispersal is wind-assisted, the spread from one field to another is difficult to control and scouting is the only option to detect resistant individuals. *Conyza canadensis* represents such a case with its wind-dispersed seeds that are able to enter the planetary boundary level and be exposed to long distance dispersal (Shields et al. 2006).

Crop Rotation

Crop rotation is the backbone of sustainable weed management (Schreiber 1992). The more diverse the crop rotation, in terms of life spans (annual, biannual or perennial) and cropping season (winter versus summer crops), the more diverse the weed flora will be, which may allow for a reduction in herbicide inputs (Liebman et al. 2014, Simic et al. 2016). A more diverse crop rotation prevents the build-up of large populations of adapted weed species, as well as allowing for more herbicide diversification with respect to herbicide sites of action reducing the risk of evolution of herbicide resistance (Beckie and Harker 2017). The importance of herbicide diversity and preventing the build-up of large weed populations dominated by one or a few weed species was once again manifested by the rapid occurrence of glyphosate-resistant weed biotypes following the adoption of glyphosate tolerant crops in North and South America (Heap 2017).

Despite the long-term benefits of crop rotation maize is often grown in monoculture. The reasons are many, including lower profitability of crop rotation in the short-term, specialization and lack of market opportunities for other crops (Lamichhane et al. 2016). The latter can be an important hindrance to crop protection in developing countries (Thierfelder et al. 2013). Recently, crop rotation was stimulated in the EU through changes to the Common Agricultural Policy subsidy schemes.

Seed Rates and Row Spacing

Crop competitiveness against weeds can be enhanced by increasing crop density. Higher crop densities will accelerate canopy closure and reduce the amount of light penetrating the canopy and reaching the soil surface and the weeds growing under the crop canopy. Increasing crop density has been shown to reduce biomass of *Cyperus esculentus* (Ghafar and Watson 1983) and *A. theophrasti* (Teasdale 1998). Similarly, Saberali et al. (2008) found that increasing crop density reduced weed biomass of *C. album* and that two-row planting, i.e., planting on both sides of the ridge, enhanced the suppressive effect of the crop. In other studies increasing seed rates were combined with narrower row spacing. Narrowing row distance will result in a more uniform plant stand which, in theory, should improve crop competitiveness. Anderson (2000) found a 60% reduction in the abundance of *Setaria italica* by halving row distance and increasing seed rate by 27%, while Teasdale (1995) found a 36% reduction in weed cover by halving row spacing and doubling seed rate. Marin and Weiner (2014) compared row spacing to a spatial uniform grid pattern with the same intra- and inter-row distance between plants and found up to 75% lower biomass of the invasive weed species *Bracharia brizantha* with a uniform grid pattern (Figure 12.1).

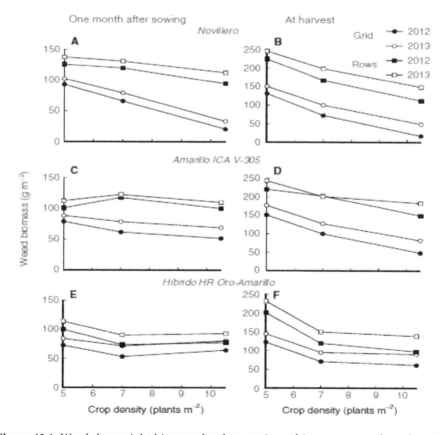

Figure 12.1. Weed dry weight biomass for three maize cultivars sown at three densities either in rows or in a grid with the same intra- and inter-row distance between the maize plants. Data from two growing seasons (Marin and Weiner 2014).

Improved crop competitiveness through a higher seed rate and/or reduced row spacing should allow for a reduction in herbicide input. Teasdale (1995) found that the efficacy of 25% of the recommended herbicide dose in combination with a high seed rate and narrow row spacing was comparable to that of the full dose applied in plots with standard seed rate and row spacing. In contrast, Dalley et al. (2004) found no significant differences in weed biomass between 38 and 76 cm row distance following application of glyphosate as a single or a sequential treatment at different timings. Similarly, Johnson et al. (1998) found no interactions between the performance of eight herbicide treatments and row distance (51 and 76 cm) on *S. faberi* and *A. artemisiifolia* and in a follow-up study no interactions were found between row spacing and herbicide timing (Johnson and Hoverstad 2002). Johnson and Hoverstad (2002) speculated that a higher spray interception by the crop at the narrow row spacing may offset a positive effect of narrow row spacing on crop-weed competition.

Competitive Cultivars

Competitive cultivars are considered a key element of an integrated weed management strategy and particularly for small grain cereals like wheat and barley competitive cultivars have received significant attention (see review by Andrew et al. 2015). In the case of maize, competitive cultivars have received less attention. Intuitively, cultivars with planophile leaves would be assumed to be more competitive than cultivars with erectophile leaves because canopy light extinction is increased. This will, however, reduce photosynthesis in the leaves nearest to the ear. As these leaves are known to be the source of the majority of assimilates transported to the developing ear leaf erectness can be expected to be positively correlated with maize yield (Hammer et al. 2009). Maybe the scarcity of studies is caused by a lack of commercially available competitive varieties because of a more pronounced negative correlation between competitiveness and yield potential than observed with small grain cereals.

Ford and Pleasant (1994) examined the weed suppressive effect of six maize hybrids by comparing their performance at four weed management strategies (from no control to broadcast herbicide application) and found significant interactions suggesting that some hybrids were more competitive than others. Studies by Lindquist et al. (1998) and Marin and Weiner (2014), comparing modern maize cultivars to traditional cultivars, suggested that genotype x environment interactions were very important, i.e., cropping system and choice of cultivar should go hand-in-hand. In other studies, rapid early growth has been found to confer higher crop competitiveness (Travlos 2011a, 2011b).

Competitiveness in small grain cereals has often been linked to allelopathic properties but even though maize contains and can exudate benzoxazinoids, a group of chemical compounds known to possess allelopathic properties (Kato-Noguchi et al. 2000), this potentially beneficial property of maize has not been studied in detail. In contrast, many studies have looked into the potential allelopathic effects of weeds on maize growth or the use of extracts of allelopathic plant species for weed control in maize but this topic is considered to be outside the scope of the present review.

Considering *Striga* spp. resistance or tolerance rather than competiveness is the desirable attribute. While resistance is defined as the ability to prevent infection and/or reproduction, tolerance is the capacity to withstand an infection without or with a minimum loss of yield (Rodenburg and Bastiaans 2011). Midega et al. (2016) reported that the local Kenyan landraces tolerated *S. hermonthica* infections better than modern hybrids and produced higher yields and that the increase in the *S. hermonthica* seedbank following three years of cultivation was higher for the hybrid cultivars than for two of the six landraces.

Sowing Date

Sowing time can influence weed flora composition. For example, Vidotto et al. (2016) found higher infestation levels of *C. album*, *A. theophrasti*, and *F. convolvulus* at early compared to normal sowing time but lower frequencies of *Panicum dichotomiflorum*, *S. halepense*, *A. retroflexus*, and *P. oleracea*.

The most obvious benefit of delayed sowing, with respect to weed management, is that it opens up for establishing a false or stale seedbed, i.e., to initiate weed germination prior to maize sowing and thereby reducing weed infestation in the crop (Buhler 2002). Besides a lower weed infestation a stale seedbed can delay the emergence of the weeds that germinate after sowing thus contributing to enhanced crop competitiveness (Fahad et al. 2014, Travlos 2011a, Travlos 2011b).

In some countries the trend has been the opposite, i.e., sowing is earlier than it used to be. This is the case in Italy where late winter sowing is seen as an approach to avoid flowering during the hottest time of the year (Otto et al. 2009) and to reduce problems related to mycotoxins. However, this will increase the problems with late winter and early spring germinating weeds, postpone the critical period of weed control and reduce the benefits of applying pre-emergence herbicides (Otto et al. 2009).

Intercropping

Intercropping is common in Africa, South America and some parts of Asia. The underlying concept of intercropping is to increase yields by making better use of light, water, and nutrients (Vandermeer 1989). A potential benefit of intercropping is a better suppression of weeds while, on the other hand, intercropping may render the use of physical and chemical weed control methods impossible or very difficult. As maize is often grown at wide row spacing it is very suitable for intercropping and many studies have examined maize intercropping, in particular, with legume crops. The results reported by Jamshidia et al. (2013) illustrate the potential benefits of intercropping. In their study maize was intercropped with cowpea and the inclusion of cowpea reduced weed biomass by ca. 40% compared to maize alone. Under weed-free conditions intercropping with cowpea had a negligible effect on maize yield (up by 4. 2%) but under weed-infested conditions maize yield was increased by up to 32%.

In Africa, one of the main drivers for intercropping is the management of *S. asiatica* and *S. hermonthica,* and of the stem borer (Woomer et al. 2008). Several legumes have been studied (e.g., Atera et al. 2013) but the most successful in terms of both *Striga* control and net profit are fodder legumes belonging to the genus *Desmodium* that are used in a so-called 'push-pull' strategy (Khan et al. 2007, Midega et al. 2014). The 'push-pull' strategy is a chemical-based strategy on the exudation of compounds preventing attachment of *Striga* to the maize roots and subsequent emergence (Khan et al. 2016).

Fertilizer Management

Nitrogen fertilization can affect crop-weed competition. In field experiments Wortman et al. (2011) found that the interference of *A. theophrasti* in maize decreased as nitrogen rate was increased. Likewise, Evans et al. (2003a) found that increasing nitrogen rates benefitted the crop more than the weeds, improving crop competitiveness and reducing the length of the critical period for weed control (Evans et al. 2003b). Crop competitiveness can be improved by making fertilizers easily accessible to the maize crop and difficult to access for the weeds. This strategy could be designated 'feeding the crop and starving the weeds' (DiTomasso 1995). Maqbool et al. (2016) compared fertilizer placement 5 cm from the maize row and 5 cm below the maize seeds but no significant differences in weed biomass or maize yield were observed between the two fertilizer treatments although the below seed treatment tended to produce lower weed biomass and higher yields.

Physical Weed Management

The term physical weed management is an umbrella term for a variety of weed management tactics ranging from hand weeding over harrowing and flaming to advanced sensor-based robotic systems. Physical weed management practices have received an increased attention in recent years, even in conventional farming systems, partly due to a declining performance of many herbicides due to the evolution of herbicide resistance and partly due to innovations in sensor technology and the appearance of novel implements for physical weed control.

Hand Weeding

Hand weeding is still a common weed control practice in developing countries. Hand weeding is time consuming and labour constraints for weeding and land preparation are believed to have led to a reduction in the area under cultivation in sub-Saharan Africa. A recent study by Nyamangara et al. (2014), studying the uptake of a conservation agricultural practice with the potential to increase maize yields by smallholder farmers in Zimbabwe, provides some insight into this issue. Despite higher crop yields, a higher gross margin and higher returns to investment conservation agriculture has not been widely adopted in the region. A survey on 50 farms revealed that labour demand under conservation agriculture was twice that of the conventional mouldboard ploughing system (84.7 versus 48.1 man day ha^{-1}). This was partly due a higher weed pressure and, consequently, more weeding operations. The authors concluded that the study highlighted the need to explore other weed control methods including herbicides as part of a novel conservation agriculture package for resource-poor smallholder farmers to overcome the constraints imposed by lack of labour (Table 12.1).

Table 12.1. Returns on investment for maize fields under hand-hoe based reduced tillage (conservation agriculture) and conventional tillage. Results from a survey of 50 farms in Zimbabwe. Adapted from Nyamangara et al. (2014)

	Item	Reduced tillage		Conventional tillage	
		Quantity	Cost	Item	Cost
Income		1603 kg/ha	440.83 US\$ ha^{-1}	991 kg/ha	272.53 US\$ ha^{-1}
Variable costs	Seed, fertiliser and other inputs		206.14 US\$ ha^{-1}		185.91 US\$ ha^{-1}
	Labour	84,73 h	162.68 US\$ ha^{-1}	38,60 h	74.11 US\$ ha^{-1}
Returns	Gross margin		72.00 US\$ ha^{-1}		12.50 US\$ ha^{-1}
	Cost per kg maize		0.23 US\$ kg^{-1}		0.26 US\$ kg^{-1}
	Returns to labour		0.44 US\$ day^{-1}		0.17 US\$ day^{-1}

Hand weeding is traditionally the responsibility of women and this was reflected in a recent study examining the attitude of small-holder farmers in the Kwa Zulu Natal province of South Africa to genetically modified crops including herbicide-tolerant maize. Whereas higher yields was mentioned as the main reason for adoption by male farmers labour-saving was the main driver for adoption by female farmers (Gouse et al. 2016).

Mowing

Mowing has been listed as one of the most important weed management tactics in no tillage organic farming in Brazil (Lemos et al. 2013) but mowing has also been studied in conventional maize farming. Combining various herbicide strategies with inter-row mowing provided full weed control and reduced herbicide inputs by ca. 50% (Donald 2006, 2007). Mowing has also been used as the only tool for vegetation management in maize grown together with a living mulch of white clover (Deguchi et al. 2014). White clover was grown as a source of nitrogen but did also suppress weed growth.

Soil Tillage Including Weed Harrowing and Inter-Row Cultivation

Primary soil tillage has a significant impact on weed seed distribution in the soil profile and, thus, the potential number of weeds emerging in a field. In no-tillage systems weed seeds tend to be located in the upper soil layer while mouldboard ploughing results in a more uniform distribution in the soil profile (Buhler et al. 1997, Scherner et al. 2016). Secondary tillage will influence the ratio of the potentially germinable weed seeds that will eventually emerge. For example, a stale or false seed bed will provoke germination of weeds that can be killed prior to sowing maize (Shaw 1996).

Following crop emergence mechanical weed control methods can be applied. Being a row crop, two methods are available in maize: weed harrowing and inter-row cultivation (also referred to as hoeing). Weed harrowing can be carried out either before and after crop emergence. Pre-emergence weed harrowing can be very effective and shows little discrimination between weed species (Rasmussen 1996). In contrast, post-emergence weed harrowing is more effective against small weed species with a prostrate growth habit like *Stellaria media* and *Veronica* spp. than against larger weed species with an erect growth habit like *Sinapis arvensis* and *C. album* (Rasmussen and Svenningsen 1995). The benefit of post-emergence weed harrowing is a trade-off between the effect on the weeds and the damage to the crop due to soil cover and physical damage (Rasmussen et al. 2008). Pannacci and Tei (2014) found that weed harrowing as a stand-alone method was inferior to other mechanical weed control methods but efficacy could be improved when combined with other mechanical methods. Recently, Rueda-Ayala et al. (2015) examined the performance of an ultrasonic sensor-guided weed harrowing system for online weed control. Weed harrowing intensity was adjusted according to weed density with aggressive harrowing and potentially severe crop damage at high weed densities, gentle harrowing at low weed densities and no harrowing at very low weed densities. Average level of weed control of two passes with the weed harrow was only 51% but without significant crop damage. It is envisaged that this simple and relatively cheap system could be further optimized.

Inter-row cultivation is a more intensive form of soil cultivation compared to weed harrowing. It has been postulated that inter-row cultivation may have positive effects on crop growth as a result of nitrogen mineralization and moisture conservation as well as negative effects due to root damage. The effect of inter-row cultivation on the yield of weed-free maize was studied in 13 field experiments in the Netherlands (Vanderwerf et al. 1991). No effects were observed in 10 of the 13 experiments while yield increased in two and decreased in one experiment. In a follow-up study Vanderwerf and Tollenaar (1993) found a good correlation between root damage, measured as xylem exudation, and crop response. In contrast to post-emergence weed harrowing inter-row cultivation can provide effective control of most annual weed species. Eadie et al. (1992) and Murphy et al. (1996) studied the effect of the duration of inter-row cultivation on weed control and maize yield and found that extending inter-row cultivation until late into the season had no positive effect on yields. This was attributed to late germinating weeds having no significant effects on maize yields. Inter-row cultivation in maize was found to be a very effective weed control method particularly when combined with ridging in the intra-row (Panacci and Tei 2014). Similarly, Vasileiadis et al. (2015, 2016) found that the performance of inter-row cultivation combined with an early post-emergence herbicide as a band application was comparable to that of standard herbicide broadcast application.

More recently various intra-row weeding tools, such as finger and torsion weeders have attracted attention but these so-called 'intelligent weeding tools' have primarily been studied in transplanted vegetable crops and other high-value crops (van der Weide et al. 2008, Melander et al. 2015). Although they are suitable for weed control in maize, the high costs of investment and low working capacity is prohibitive for their use in low-value crops like maize.

Thermal Weed Control

Thermal weed control can be flaming, steaming or application of hot water. Of these, mainly flaming has been studied in maize. Flaming is an attractive non-chemical weed control method as it does not disturb the soil, unlike weed harrowing and inter-row cultivation, and therefore does not initiate new flushes of weed seed germination. Flaming is, however, expensive and compared with weed harrowing and inter-row cultivation energy consumption is high, albeit lower than for steaming and hot water application (Ascard et al. 2007).

Maize is a relatively heat-tolerant crop but with maize, as with other agronomic crops, the key to crop tolerance is avoiding targeting the growing points (Datta and Knezevic 2013). Effect and selectivity of flaming in maize has been studied in detail by Ulloa and colleagues. They concluded that broadleaf weeds were more susceptible than grass weeds and that weeds at early growth stages were more easily controlled than weeds on later growth stages (Ulloa et al. 2010a, 2010b). Concerning crop tolerance field maize was found to be more tolerant than other maize types (Knezevic et al. 2009) and flaming at the 5 or 7 leaf-stage of maize caused less visual damage and lower yield reduction than flaming at the 2 leaf-stage (Ulloa et al. 2010c, 2010d, 2011). Effects on weeds and the maize crop were higher when flaming was done in the afternoon compared to earlier in the day and it was suggested that this could be due to lower relative water content in the plants (Ulloa et al. 2012). Although selectivity between crop and weed was not improved, flaming in the afternoon was recommended because propane consumption was lower. High level of weed control was also achieved when broadcast or in-row flaming was combined with inter-row cultivation (Stepanovic et al. 2016).

In summary, flaming is an alternative to other weed control methods because maize, in contrast to most other arable and horticultural crops, not only tolerates pre-emergence but also post-emergence flaming as it was also recently confirmed by Martelloni et al. (2016).

Cover Crops

The use of cover crops as a weed management tool has received increased attention in recent years. Cover crops are an integrated part of conservation agriculture practices and also widely adopted in organic farming, but less so in conventional farming. Traditionally cover crops have been grown as a source of nutrients and as a mean to reduce the loss of nutrients from the soil. Thus the main focus has been on nitrogen-fixation legumes, but also non-legumes have been studied primarily for their putative effects on pests including weeds (Cherr et al. 2006). Cover crops can suppress weeds prior to sowing of the main crop through competition for water, nutrients and light and the release of chemical compounds, known as allelochemicals, that can reduce weed germination in the subsequent crop when incorporated into the soil or left on the soil surface. Besides nutrient management and effects on pests cover crops are also used to reduce soil erosion.

Among the legumes, the ones receiving most attention as cover crops for weed suppression in maize has been lupin (*Lupinus sativa*), common vetch (*Vicia sativa*) and various clover species while rye (*Secale cereal*), ryegrass (*Lolium* spp.) and various Brassicas are the most studied non-legumes (e.g., Martins et al. 2016, Cutti et al. 2016, Dorn et al. 2015, Brust et al. 2014, Gavazzi et al. 2010). The results reported in the literature are variable and reflects that the growth of cover crops is very much influenced by environmental conditions and that year-to-year and site-to-site variations are to be expected.

Under Brazilian conditions Martins et al. (2016) found that a cover crop of black oat (*Avena strigosa*) and lupin was more effective in suppressing weed growth in maize than forage radish (*Raphanus sativa*) while Cutti et al. (2016) found common vetch and rye to be more effective than Italian ryegrass (*Lolium multiflorum*). Kaefer et al. (2012) reported that the shorter the period between management of a cover crop of black oat and maize sowing, the higher the effect on weeds in the succeeding maize crop. Dorn et al. (2015) compared the effect of various cover crops on weed growth in maize and sunflower in conventional and organic farming and found higher effects in conventional (96–100%) than in organic farming (19–87%). No explanation was provided but the fact that cover crops were terminated with glyphosate and left on the soil surface in conventional farming systems while incorporated on the organic farms may explain the observed differences. Similar differences have previously been observed with rye (Teasdale et al. 2012). In studies in Zimbabwe soil coverage of the cover crops tended to overrule choice of crop species on the effect on *C. esculentus* in the succeeding maize crop (Mhlanga et al. 2016b).

Several studies have shown effects of rye as a cover crop on weed occurrence in maize. Moonen and Bàrberi (2004) found a reduction in the number of seeds in the weed seedbank of 25% after seven years of a rye cover crop compared to a cropping system with no cover crops. Gavazzi et al. (2010) found 61% and 96% reduction in grass and broadleaf weed density in maize (no-till and conventional till) following a rye cover crop and similar results were reported by Tabaglio et al. (2013) for no-till maize.

Under temperate conditions like in northern European conditions the short growing season in the autumn prior to maize establishment in the spring is a challenge and to be successful with cover crops rapid germination and early growth is needed. Brust et al. (2014) screened a range of potential cover crops and found Tartary buckwheat (*Fagopyrum esculentum*) and forage radish to be the most suitable crops under German conditions.

In most cover crops studies no adverse effects on maize establishment, growth and yield were reported but in some cases cover crops not only affected weed germination and weed growth but also crop growth (e.g., Bezuidenhout et al. 2012).

Chemical Weed Control

Chemical weed control with synthetic herbicides is the most widely adopted method of weed control in maize. Herbicides can be grouped according to site of action, site of uptake and mobility in the plants (Kudsk 2017). Site of action is an important parameter considering herbicide resistance whereas site of uptake and mobility in plants are more relevant considering optimum use of herbicides and the impact of biotic and abiotic parameters on herbicide performance.

Herbicides can be classified as either *soil-active* or *foliar-active* herbicides. Soil-active herbicides, also referred to as *residual herbicides* are usually applied to the soil and sometimes also incorporated into the soil before sowing (pre-planting) or applied before or shortly after weed emergence (pre-emergence) and the below-ground plant parts, i.e., roots and the emerging shoot, are the main sites of uptake. Foliar-active herbicides are applied after crop and weed emergence (post-emergence) reflecting that leaves and stems are the main sites of uptake. Many herbicides are both soil- and foliar-active.

Compared to many other crops the number of herbicides available to maize farmers is high representing a broad range of modes of action. Herbicides vary in their efficacy against individual weed species, i.e., a reduced rate may be sufficient to control some weed species whereas other will survive. This information is often not provided by the agrochemical distributors but is essential for farmers for making decisions on the correct herbicide(s) and doses to use (Kudsk and Moss 2017). An example of the kind of information on herbicide efficacy that is decisive for optimizing herbicide use was provided by Pannacci and Covarelli (2009) and Pannacci (2016) who sprayed a dose range of mesotrione and foramsulfuron and estimated the field doses giving 95% effect against a broad range of weed species at various growth stages (Table 12.2). Their results revealed consistent differences between weed species but only minor differences between years (with the exception of foramsulfuron in 2013) and thus provide the farmers with a solid basis for making decisions on herbicide and herbicide dose.

Pre-Plant and Pre-Emergence Application

Pre-plant herbicide application can be used to control weeds surviving from the preceding crop and is therefore a common practice in no-tillage systems. Glyphosate is the preferred herbicide for this purpose but other foliar-active herbicides like 2,4-D, dicamba, diquat, paraquat and some sulfonylurea herbicides can also be applied. Application of the other foliar-active herbicides is common if glyphosate-tolerant maize is cultivated. A foliar-active herbicide can be applied in mixture with a soil-active herbicide to also provide some long-term control.

Table 12.2. Estimated ED_{95} doses for mesotrione and foramsulfuron applied over several growing seasons to field-grown maize at the 4-5 (mesotrione) and 5-6 (foramsulfuron) leaf stage of the maize crop. Standard errors are in parentheses. Adapted from Pannacci and Covarelli (2009) and Pannacci (2016)

Weed species	Mesotrione ED_{95} (g ha^{-1})				Foramsulfuron ED_{95} (g ha^{-1})		
	2000	2002	2004	2011	2012	2013	2014
Abutilon theophrasti			21.6 (0.3)	29.4 (0.7)	25.6 (0.1)		24.7 (0.5)
Amaranthus retroflexus	45.8 (5.7)	38.5 (4.9)	39.6 (1.9)	15.0 (0.4)	11.7 (0.9)	19.7 (0.01)	16.0 (0.3)
Chenopodium album	24.1 (0.1)	26.4 (0.1)	28.4 (0.1)	24.6 (0.1)	23.6 (0.1)	52.6 (7.2)	
Echinocloa crus-galli			89.0 (7.2)	18.1 (0.1)	20.7 (0.05)	48.1 (6.9)	34.7 (2.0)
Polygonum lapathifolium				56. 6 (7.8)	43.6 (10.1)	>61	>61
Polygonum persicaria			42.2 (2.9)				
Portulaca oleracea		>100	>150				
Setaria viridis					15.5 (0.5)		
Sinapis arvensis						12.7 (0.9)	
Solanum nigrum	16.6 (0.6)		30.7 (0.1)	15.4 (0.4)	26.4 (1.3)		14.7 (0.7)
Xanthium strumarium	25.7 (0.2)						

This practice is often recommended where low soil moisture at the time of sowing could be an issue. Atrazine, which can also be used pre- and post-emergence, has been widely used for this purpose. Atrazine has, however, been banned in the European Union since 2007 and has come under scrutiny in other countries, e.g., the United States. Other residual herbicides used pre-plant are acetochlor, alachlor, dimethenamid-P, isoxafluatole, metoloachlor-s, pendimethalin, pyroxasulfone, saflufenacil and thiencarbazone-methyl. Otherwise pre-plant application of residual herbicides is recommended for herbicides that have to be incorporated into the soil to prevent loss due to volatilisation or photodegradation like trifluralin.

Pre-emergence applications can be done from the time of sowing and until just before crop emergence. In regions where rainfall is low or erratic the efficacy of pre-emergence herbicides can be improved by a shallow incorporation into the soil but this may also increase the risk of crop damage in case of excessive rainfall. The herbicides available for pre-emergence application are very much the same as those used for pre-plant application but, in addition, soil-applied herbicides with shorter residual activity, such as clopyralid and mesotrione are also recommended for pre-emergence applications.

The adsorption of residual herbicides to the soil colloids, and thus availability to plants, is determined by the physico-chemical properties of the herbicide and the soil texture. For many residual herbicides adsorption is low on soils with a low organic matter and to prevent crop damage the recommended doses are often lower on such soil conditions. Conversely, some residual herbicides are ineffective on soils with a high organic matter content and therefore not recommended for use on these soils.

Pre-emergence herbicides can be applied as a broadcast application or only to the maize rows and combined with inter-row cultivation or inter-row flaming either alone or in combination with a reduced input of post-emergence herbicides (Vasileiadis et al. 2015).

Post-emergence Application

Many of the herbicides authorised for pre-plant and pre-emergence use can also be applied post-emergence, for example, 2,4-D, dicamba, acetochlor, atrazine, clopyralid and mesotrione and in glyphosate-tolerant maize also glyphosate. Other herbicides like the sulfonylureas floramsulfuron, nicosulfuron and rimsulfuron as well as bromoxynil, carfentrazone-ethyl, temborione and topramezone are only recommended for post-emergence use.

The performance of post-emergence herbicides is affected by biotic factors such as weed growth stage and the level of crop competition as well as abiotic factors such as the environmental conditions around the time of application (Kudsk 2017). Spray adjuvants for maize herbicides have attracted a lot of interest (e.g., Gitsopoulos et al. 2010, Knezevic et al. 2010, Idziak and Woznica 2013). They can have a significant influence on the performance of post-emergence herbicides, allowing for the use of lower herbicide doses (Panacci 2016), but can also increase the risk of crop damage.

Glyphosate tolerant maize was introduced in the US in 1998 (Duke and Powles 2009). Before that US farmers have had access to sethoxydim tolerant (Dotray et al. 1993) and imidazolinone tolerant maize cultivars (Tan et al. 2005), which, in contrast to the glyphosate-tolerant varieties, were developed using mutagenesis and selection, i.e., not transgenic crops. Around the time glyphosate tolerant varieties were introduced also glufosinate tolerant varieties were marketed but they never gained the same currency as the glyphosate tolerant cultivars due to their narrow weed spectrum compared to glyphosate. The uptake of glyphosate tolerant maize in the US has been much slower than the uptake of glyphosate-tolerant soybean and cotton that was introduced around the same time (https://www. ers. usda. gov/data-products/adoption-of-genetically-engineered-crops-in-the-us/recent-trends-in-ge-adoption/). Following the introduction in the US, herbicide-tolerant maize varieties have been approved in a number of countries and are now grown on large scale in Argentina, Brazil and some other Latin-American countries (James 2015).

An important benefit of herbicide tolerant varieties is that weed control programmes could be simplified using only one foliar-applied broad-spectrum herbicides rather than a suite of pre- and post-emergence herbicides with complementary weed spectrums. Another advantage has been that it enabled a shift from conventional to no and reduced tillage, which is considered to be more sustainable than inversion tillage (Duke and Powles 2009). Finally, the acreage that a farmer can manage increased significantly allowing farmers to grow more land and increase their income. The downside of the wide-spread uptake of herbicide tolerant crops has been the emergence of resistant weed biotypes (Heap 2017). Imidazolinone and glyphosate tolerant maize cultivars also provide a highly effective technology for controlling parasitic weeds like *S. asiatica* and *S. hermonthica* (Ransom et al. 2012).

The expectation was that growing glyphosate-tolerant crops would reduce herbicide use thus also provide environmental benefits. Benbrook (2012) reported that this had not been the case for either glyphosate-tolerant maize or other glyphosate-tolerant crops. The conclusions were, however, challenged by Brookes et al. (2012) who criticised Benbrook for relying on data from the USDA that did not disaggregate pesticide use by trait. Recently, a study by Perry et al. (2016), based on pesticide use data collected from 1998–2011 on farms that had adopted or not adopted genetically engineered crops, showed that farmers growing glyphosate-tolerant maize used less herbicide active ingredients but that the difference has become less during the study period and was just 1.2% in 2011 (Figure 12.2). This is due to an increase in the use of other herbicides in glyphosat tolerant maize as only 19% of the land planted with glyphosate tolerant maize was treated exclusively with glyphosate in 2011 compared to more than 60% in 2000). Weighted by the environmental impact quotient (EIQ) the difference in herbicide use was 9.8%. These results contradict the conclusions of Benbrook (2012) but, nonetheless, the current recommendation to combine the use of glyphosate with residual herbicide to control glyphosate resistant weed biotypes will gradually eliminate the benefits of a lower environmental impact with glyphosate tolerant maze. A similar effect can be expected from the ongoing introduction of maize cultivars with tolerance to both glyphosate and an auxin herbicide (Li et al. 2013, Zhou et al. 2016).

Herbicide Resistance

The first case of herbicide resistance in maize in Delaware, US dates back to the 1970s and was in *Amaranthus hybridus* that evolved resistance to triazine herbicides (Schnappinger et al. 1979). In

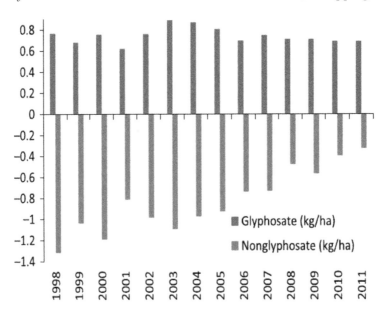

Figure 12.2. Difference in herbicide use between farmers who have adopted glyphosate tolerant maize and those cultivating conventional cultivars (Perry et al. 2016).

1991 the first case of ACCase resistance was found in *Setaria faberi* (Heap 2017) and a few years later the first cases of ALS resistance was reported in *Amaranthus rudis* and *Amaranthus palmeri* (Horak and Peterson 1995). Later resistance to other photosystem II inhibitors, synthetic auxins, PPO inhibitors and glyphosate was reported and recently many cases of multiple resistance have been found.

In total 302 unique cases of herbicide resistance have been found in maize covering 60 weed species (Heap 2017). With the exception of one case in South Africa and one in Indonesia all other cases of herbicide resistance in maize originates from the Americas and Europe with the US leading in number of unique cases. The major resistance issues in North and South America are glyphosate resistance and multiple resistance (Schultz et al. 2015, Varanasi et al. 2015, Bagavathiannan and Norsworthy 2016), while in Europe resistance to ALS inhibitors is the main challenge (Panozzo et al. 2015).

Managing the evolution of herbicide resistance has become a main focus in those regions where herbicides are the mainstay of weed management in maize. The key message to farmers is 'diversification' not only in terms of herbicide sites of action but also weed management tools. The preventive measures that will delay the evolution of herbicide resistance are basically the same as the ones recommended to farmers for switching to more integrated weed management practices (Barzman et al. 2015).

Integrated Weed Management (IWM)

As shown in the previous sections several weed management tools are available to maize growers. Besides the preventive and cultural weed management tools that are available for most annual crops also inter-row cultivation, one of the most effective mechanical weed control measures, is an option because maize is grown in rows. Furthermore, as maize tolerates heat flaming, even in the row, this is also an option. Nonetheless weed management in the developed part of the world is almost solely based on the use of herbicides. In this section the experiences with integrated weed management (IWM) strategies in maize will be summarized.

In the US where glyphosate-tolerant cultivars covers the majority of the maize area, IWM has focused on making the cultivation of glyphosate-tolerant maize more sustainable (Owen et al. 2015). The farmers' willingness to adopt IWM is often limited to herbicide rotation whereas other IWM tools are not widely adopted (Owen et al. 2015). In this context the series of papers presenting the outcome of the so-called benchmark study is interesting. The purpose of the benchmark study was to compare grower practices with academic recommendations, side by side, in 156 fields of which 106 had maize in the crop rotation (Shaw et al. 2010). Some of the main findings from this study were that herbicide diversification reduced weed population densities and so did crop rotation whether glyphosate-resistant or conventional cultivars were cultivated (Wilson et al. 2011). Although the inclusion of more herbicide sites of actions were more costly, the economic returns were not negatively affected because crop yields often increased (Edwards et al. 2013). Herbicide diversification can be achieved either by mixing two or more herbicide sites of action or by sequential application of herbicides with different sites of action during the growing season. In a retrospective study, using the information from more than 500 sites, Evans et al. (2015) found that mixing herbicides with different sites of action was more effective in preventing the evolution of glyphosate resistance in *Amaranthus tuberculatus* than sequential application.

In the early era of glyphosate-tolerant maize, Norsworthy and Frederick (2005) published results from a study comparing IWM that included the use of cover crops, narrow row spacing and surface tillage in maize treated either with atrazine and metolachlor pre-emergence or glyphosate post-emergence. Cover crops of rye and wheat in the glyphosate treated plots and surface tillage in the plots with no cover crops had no effect on weed control but narrow row distance (38 vs 76 cm) improved the control of *Senna obtusifolia* and *Sida rhombifolia*.

In the near future maize cultivars with stacked tolerance to glyphosate and either 2,4-D or dicamba will be introduced in the US (Zhou et al. 2016). This technology will allow farmers to apply 2,4-D and dicamba for the control glyphosate resistant biotypes of broadleaved weeds. Besides the potential crop damage by spray drift widespread use of these two auxin herbicides may cause to auxin-susceptible crops (Egan et al. 2014), which has been sought minimized through the development of formulations with low potential for off-target movement (Li et al. 2013) and extensive educational programmes for farmers, the sustainability of this technology, as a tool to minimize problems with herbicide resistance, has also been questioned (Mortensen et al. 2012). If tolerance to auxins is introduced in a region where resistance to glyphosate is widespread, the risk of evolution of resistance to both herbicides is comparable to the risk of resistance evolving to the auxin applied alone. Even in situation where resistance has not yet evolved to any of the two herbicides, combining tolerance to glyphosate and auxin is not an ideal combination because of the lack of overlap in weed spectrum with glyphosate providing broad spectrum weed control and auxins only controlling broadleaf weeds (Gressel et al. 2017). Thus, stacking of herbicide tolerance traits is not a substitute for truly integrated solutions although herbicide tolerant cultivars have the potential to become a valuable component of IWM strategies if used wisely (Lamichhane et al. 2016).

In Europe and other parts of the world, where herbicide-tolerant maize is not available, research in IWM tends to have more focus on integrating non-chemical and chemical control options. Consultations with farmers, extension services, academia, etc., revealed not only an interest in innovative technologies and integrated pest management (IPM) including novel technologies for weed management but also that some of these technologies were on the brink of being implemented (Vasileiadis et al. 2011). A multi-criteria assessment of potential IPM scenarios revealed environmental benefits and the same economic sustainability as the current practice (Vasileiadis et al. 2013). Later, the IWM strategies that emerged from the abovementioned consultations were evaluated on-farm in different European countries. As the IWM strategies were adapted to the local conditions they varied between locations. One study, including three countries over two years, compared broadcast pre- and/or post-emergence herbicide applications to early post-emergence band application of herbicides in combination with inter-row cultivations (Southern Germany), early post-emergence herbicide broadcast application according to a predictive model of weed emergence followed by inter-row cultivation (Northern Italy) and tine-harrowing followed by a low dose post-emergence herbicide application (Slovenia) (Vasileiadis et al. 2015). In all three locations the IWM strategies provided sufficient weed control with a reduced input of herbicides and, importantly, the same yield and economic return as the conventional system. In another on-farm evaluation over two seasons broadcast herbicide applications were compared to early post-emergence band application of herbicides followed by one inter-row cultivations (Hungary and Slovenia), two inter-row cultivations (Southern Germany) or pre-emergence band application followed by inter-row cultivation (Northern Italy)(Vasileiadis et al. 2016). The IWM strategies provided effective weed control in the maize rows but only partial control between the rows, however the economic returns were the same for the conventional and IWM systems (Figure 12.3).

These findings are in line with the results of Pannacci and Tei (2014), who found a very high level of weed control by combining band application of herbicides with inter-row cultivation and inter-row cultivation with ridging but without herbicides was found to be almost as effective. In that study other non-chemical methods provided insufficient weed control.

In a recent study various low-input maize cultivation systems were compared to a traditional maize monoculture (Giuliano et al. 2016). A reduction in the input of herbicides was just one of several objectives of the study, hence it is difficult to draw conclusions specifically on weed management. However, in one of the low input systems the major change was to introduce mouldboard ploughing prior to sowing in combination with a reduction in herbicide use by 50% by combining chemical and mechanical weed control measures. Overall, no differences in weed abundancy were found between the two systems when assessments were made after weed control measures had been applied.

In developing countries shortage of labour for hand weeding is a major impediment and reduces maize yields (Weber et al. 1995). Hence, in contrast to regions with intensive farming the

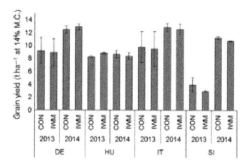 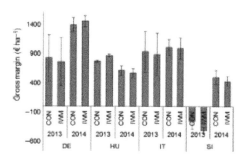

Figure 12.3. Grain yield (t ha⁻¹) and gross margin (Euro ha⁻¹) following conventional weed management (CON) and integrated weed management (IWM) tested in Germany (DE), Hungary (HU), Italy (IT) and Slovenia (SI) in 2013 and 2014. Columns indicate means and bars standard errors (Vasileiadis et al. 2016).

objective of IWM is economic rather than environmental sustainability (Akobundu, 1996). An example is the study by Chikoye et al. (2004) who compared treatments combining herbicides, a *Mucuna cochinchinensis* cover crop, different intensities/timing of manual weeding and crop densities. The farmer's weed management choice was the reference treatment. The highest weed biomass and the lowest yields were observed with farmer's choice and the authors recommended a combination of early herbicide treatment, cover crop and three times manual weeding as an optimal IWM strategy. The authors stressed that early herbicide applications should be part of an IWM strategy because subsistence farmers in Africa typically delay hand weeding until all crops are sown leading to significant yield losses in the early sown crops.

A severe weed problem in Sub-Saharan maize cultivation is *Striga* spp. Due to underground parasitism IWM strategies targeting other weeds are normally not effective against *Striga* spp. An integrated approach comprising the use of non-contaminated seeds, trap crops/inter-cropping, resistant/tolerant cultivars, improved soil fertility and direct control measures such as herbicides or hand pulling is needed to achieve satisfactory control (Ransom 1996). The reason is that except for the use of seed dressings of imazapyr in combination with imidazolinone-tolerant maize varieties (Ransom et al. 2012) and post-emergence applications of glyphosate in glyphosate-tolerant maize cultivars direct control measures by farmers are not very effective. In a comprehensive review published recently Fernandez-Aparici et al. (2016) discussed the tools available i) to reduce seed bank viability, ii) to reduce the ability of *Striga* spp. to timely detect the host, iii) to minimize penetration into the vascular system and iv) to increase the tolerance of the crop to parasitism, which should be elements of novel integrated management approaches in conventional maize cultivars.

Concluding Remarks

In developed agriculture weed management in maize is almost solely based on the use of herbicides and this pattern has been reinforced in those parts of the world where transgenic herbicide tolerant maize cultivars are widely cultivated. This has led to a push for changes by the public opinion and legislators and also many farmers are now realizing that the current strategy is not sustainable and that the focus of weed management should be on IWM and agronomic, economic and environmental sustainability. This trend is more notable in Europe than in the US where the introduction of glyphosate-tolerant maize provided the farmers with relief and postponed the decision on the inevitable changes but the current problems with glyphosate-resistant weed biotypes have clearly shown that this was only a stay of execution. Transgenic maize cultivars with stacked resistance genes will only be a success if they are considered in a truly IWM context.

In contrast, in resource-poor countries weed management in maize is manual and despite farmers spend up to 50–70% of their time on weed control (Chikoye et al. 2004) yield losses are significant and therefore economic sustainability of weed management in maize should be the main focus to ensure food security in these parts of the world. Very likely this implies an increased use of herbicides, at least temporary, until other weed control tools have been developed, adapted to the local conditions and validated but it is important that herbicides are considered one of several IWM tools and not the tool.

The benefits of weed control may not be limited to the effect on weeds. Recently, Reboud et al. (2016) reported that omitting the use of herbicides increased the content of mycotoxins in the grain and, unexpectedly, resulted in a higher increase in mycotoxin contents than omitting insecticide use. The authors concluded that the results support the hypothesis that some weed species can harbour *Fusarium* species. Integrating weed management and management of other pests, as well as optimizing the ecosystem services that weeds can provide, should be the long-term goal of IWM. The 'push-pull' strategy aiming at controlling both *Striga* spp. and Lepidopteran stem borers is an example on this from developing countries but very few examples of such holistic approaches are known from regions with high input crop production systems. Recently, however, an example of a holistic approach was published involving the weed species *Asclepias syriaca* in maize. The underlying concept is not to completely kill off the weed *A. syriaca* with glyphosate in glyphosate-tolerant maize but maintain it at low densities (< 1 stem m^{-2}) as it harbours aphids that provides a food source for a parasitoid wasp attacking the eggs of the European corn borer and also is the host for the monarch butterfly, a non-target organism of high conservation value (DiTomasso et al. 2016).

Future approaches ask for a better integration of weed management with other pests and diseases management. This will require a better understanding of the potential interactions between the various pest groups and therefore may not always be possible but should be exploited.

REFERENCES

Akobundu, I.O. 1996. Principles and prospects for integrated weed management in developing countries. pp. 591–600. *In:* Proceedings of the Second International Weed Control Congress, Copenhagen, Denmark.

Anderson, R.L. 2000. Cultural systems to aid weed management in semiarid corn (*Zea mays*). Weed Technol. 14: 630–634.

Andrew, I.S.K., J. Storkey and D.L. Sparkes. 2015. A review of the potential for competitive cereal cultivars as a tool in integrated weed management. Weed Res. 55: 239–248.

Ascard, J., P.E. Hatcher, B. Melander and M.K. Upadhyaya. 2007. Thermal weed control. pp. 155–175. *In:* Upadhyaya, M.K. and Blackshaw, R.E. (Eds.) Non-chemical Weed Management. CAB International, UK.

Atera, A.E., T. Ishii, J.C. Onyango, K. Itoh and T. Azuma. 2013. *Striga* infestation in Kenya: status, distribution and management options. Sustain. Agric. Res. 2: 99–108.

Atera, A.E., F. Kundu and K. Itoh. 2013. Evaluation of intercropping and permaculture farming systems for control of *Striga asiatica* in maize, Central Malawi. Trop. Agric. Develop. 57: 114–119.

Bagavathiannan, M.V. and J.K. Norsworthy. 2016. Multiple-herbicide resistance is widespread in roadside palmer amaranth populations. PLos ONE 11: e0148748. doi:10.1371/journal.pone.0148748.

Barzman, M., P. Bàrberi, A.N.E. Birch, P. Boonekamp, S. Dachbrodt-Saaydeh, B. Graf, B. Hommel, J.E. Jensen, J. Kiss, P. Kudsk, J.R. Lamichhane, A. Messéan, A-C. Moonen, A. Ratnadass, P. Ricci, J-L. Sarah and M. Sattin. 2015. Eight principles of integrated pest management. Agron. Sustain. Dev. 35. doi: 10. 1007/s13593-015-0327-9

Beckie, H.J. and K.N. Harker. 2017. Our top 10 herbicide-resistant weed management practices. Pest. Manag. Sci. 73: 1045–1052. doi: 10.1002/ps.4543

Benbrook, C.M. 2012. Impacts of genetically engineered crops on pesticide use in the U.S. – the first 16 years. Env. Sci. Eur. 24: 24.

Bezuidenhout, S.R., C.F. Reinhardt and M.I. Whitwell. 2012. Cover crops of oats, stooling rye and three annual ryegrass cultivars influence maize and *Cyperus esculentus* growth. Weed Res. 52: 153–160.

Brookes, G., J. Carpenter and A. McHughen. 2012. A review and assessment of 'Impact of genetically engineered crops on pesticide use in the US – the first sixteen years: Benbrook, C. (2012). Available at: www.ask-force.org/web/Pesticides/Brookes-Carpenter-Hughes-Rebuttal-Benbrook-2012 pdf (Accessed on January 4, 2017).

Brust, J., W. Claupein and R. Gerhards. 2014. Growth and weed suppression ability of common and new cover crops in Germany. Crop Prot 63: 1–8.

Buhler, D.D. 2002. Challenges and opportunities for integrated weed management. Weed Sci. 50: 273–280.

Buhler, D.D., R.G. Hartzler and F. Forcella. 1997. Implications of weed seedbank dynamics to weed management. Weed Sci. 45: 329–336.

Cerrudo, D., E.R. Page, M. Tollenaar, G. Stewart and C.J. Swanton. 2012. Mechanisms of yield loss in maize caused by weed competition. Weed Sci. 60: 225–232.

Cherr, C.M., J.M.S. Scholberg and R. McSorley. 2006. Green manure approaches to crop production: a synthesis. Agron. J. 98: 302–319.

Chikoye, D., V.M. Manyong, R.J. Carsky, G. Gbehounou and A. Ahanchede. 2002. Response of speargrass (*Imperata cylindrica*) to cover crops integrated with handweeding, and chemical control in maize and cassava. Crop Prot. 21: 145–156.

Chikoye, D., S. Schulz and F. Ekeleme. 2004. Evaluation of integrated weed management practices for maize in the northern Guinea savanna of Nigeria. Crop Prot. 23: 895–900.

Cutti, L., F.P. Lamego, A.C. Morales de Aguiar, T.E. Kaspary and C.A.G. Rigon. 2016. Winter cover crops on weed infestation and maize yield. Revista Caatinga 29: 885–891.

Dalley, C.D., J.J. Kells and K.A. Renner. 2004. Effect of glyphosate application timing and row spacing on weed growth in corn (*Zea mays*) and soybean (*Glycine max*). Weed Technol. 18: 177–182.

Datta, A. and S.Z. Knezevic. 2013. Flaming as an alternative weed control method for conventional and organic agronomic crop production systems: a review. Adv. Agron. 118: 339–428.

Deguchi, S., S. Uozumi, E. Touno, M. Kaneko and K. Tawaraya. 2014. White clover living mulch controlled only by mowing supplies nitrogen to corn. Soil Sci. Plant Nutrit. 60: 183–187.

DiTomasso, A., K.M. Averill, M.P. Hoffman, J.R. Fuchsberg and J.E. Losey. 2016. Integrating insect resistance and floral resource management in weed control decision-making. Weed Sci. 64: 743–756.

DiTomasso, J.M. 1995. Approaches for improving crop competitiveness through the manipulation of fertilization strategies. Weed Sci. 43: 491–497.

Donald, W.W. 2006. Preemergence banded herbicides followed by only one between-row mowing controls weeds in corn. Weed Technol. 20: 143–149.

Donald. W.W. 2007. Control of both winter annual and summer annual weeds in no-till corn with between-row mowing systems. Weed Technol. 21: 591–601.

Dorn, B., W. Jossi and G.A. van der Heijden. 2015. Weed suppression by cover crops: comparative on-farm experiments under integrated and organic conservation tillage. Weed Res. 55: 586–597.

Dotray, P.A., L.C. Marshall, W.B. Parker, D.L. Wyse, D.A. Somers and B.G. Gengenbach. 1993. Herbicide tolerance and weed control in sethoxydim-tolerant corn (*Zea mays*). Weed Sci. 41: 213–217.

Duke, S.O. and S.B. Powles. 2009. Glyphosate-resistant crops and weeds: now and in the future. Ag Bioforum 12: 346–357.

Eadie, A.G., C.J. Swanton, J.E. Shaw and G.W. Anderson. 1992. Banded herbicide applications and cultivation in a modified no-till corn (*Zea mays*) system. Weed Technol. 6: 535–542.

Edwards, C.B., D.L. Jordan, M.D.K. Owen, P.M. Dixon, B.G. Young, R.G. Wilson, S.C. Weller and D.R. Shaw. 2014. Benchmark study on glyphosate-resistant crop systems in the United States. Economics of herbicide resistance management practices in a 5 year field-scale study. Pest Manag. Sci. 70: 1924–1929.

Egan, J.F., K.M. Barlow and D.A. Mortensen. 2014. A meta-analysis on the effects of 2,4-D and dicamba drift on soybean and cotton. Weed Sci. 62: 193–206.

Evans, J.A., P.J. Tranel, A.G. Hager, B. Schutte, C. Wu, L.A. Chatham and A.S. Davis. 2015. Managing the evolution of herbicide resistance. Pest Manag. Sci. 72: 74–80.

Evans, S.P., S.Z. Knezevic, J.L. Lindquist and C.A. Shapiro. 2003a. Influence of nitrogen and duration of weed interference on corn growth and development. Weed Sci. 51: 546–556.

Evans, S.P., S.Z. Knezevic, J.L. Lindquist, C.A. Shapiro and E.E. Blankenship. 2003b. Nitrogen application influences the critical period for weed control in corn. Weed Sci. 51: 408–417.

Fahad, S., S. Hussain, S. Saud, S. Hassan, H. Muhammad, D. Shan, C. Chen, C. Wu, D. Xiong, S.B. Khan, A. Jan, K. Cui and J. Huang. 2014. Consequences of narrow crop row spacing and delayed *Echinocloa colona* and *Trianthema portulacastrum* emergence for weed growth and crop yield loss in maize. Weed Res. 54: 475–483.

Fernández-Aparicio, M., X. Reboud and S. Gibot-Leclerc. 2016. Broomrape weeds. Underground mechanisms of parasitism and associated strategies for their control: a review. Front. Plant Sci. 7: 135.

Forcella, F., R.G. Wilson, J. Dekker, R.J. Kremer, J. Cardina, R.L. Anderson, D. Alm, K.A. Renner, R.G. Harvey, S. Clay and D.D. Buhler. 1997. Weed seed bank emergence across the Corn Belt. Weed Sci. 40: 67–76.

Ford, G.T. and J.M. Pleasant. 1995. Competitive ability of six corn (*Zea mays* L.) hybrids with four weed control practices. Weed Technol. 8: 124–128.

Gavazzi, C., M. Schulz, A. Marocco and V. Tabaglio. 2010. Sustainable weed control by allelochemicals from rye cover crops: from the greenhouse to field evidence. Allelopathy J. 25: 259–273.

Ghafar, Z. and A.K. Watson. 1983. Effect of corn (*Zea mays*) population on the growth of yellow nutsedge (*Cyperus esculentus*). Weed Sci. 31: 588–592.

Gitsopoulos, T.K., V. Melidis and G. Evgenidis. 2010. Response of maize (*Zea mays* L.) to post-emergence applications of topramezone. Crop Prot. 29: 1091–1093.

Giuliano, S., M.R. Ryan, G. Véricel, G. Rametti, F. Perdrieux, E. Justes and L. Alletto. 2016. Low-input cropping systems to reduce input dependency and environmental impacts in maize production: a multi-criteria assessment. Eur. J. Agron. 76: 160–175.

Gouse, M., D. Sengupta, P. Zambrano and J.F. Zepeda. 2016. Genetically modified maize: less drudgery for her, more maize for him? Evidence from smallholder maize farmers in South Africa. World Dev. 83: 27–38.

Gressel, J., A. Gassmann and M.D.K. Owen. 2017. How well will stacked transgenic pest/herbicide resistances delay pests from evolving resistance? Pest Manag. Sci. 73: 22–34.

Hammer, G.L., Z. Dong, G. McLean, A. Doherty, C. Messina, J. Schussler, C. Zinselmeier, S. Paszkiewicz and M. Cooper. 2009. Can changes in canopy and/or root system architecture explain historical maize yield trends in the US Corn Belt? Crop Sci. 49: 299–312.

Heap, I. 2017. International survey of herbicide resistant weeds. Accessed January 3, 2017. Available from: http://www.weedscience.org/In.asp (Accessed on January 3, 2017)

Horak, M.J. and D.E. Peterson. 1995. Biotypes of palmer amaranth (*Amaranthus palmeri*) and common waterhemp (*Amaranthus rudis*) are resistant to imazethapyr and thifensulfuron. Weed Technol. 9: 192–195.

Idziak, R. and Z. Woznica. 2013. Effect of nitrogen fertilisers and oil adjuvants on nicosulfuron efficacy. Turkish J. Filed Crops 18: 174–178.

Isik, D., H. Mennan, B. Bekir, A. Oz and M. Ngouajio. 2006. The critical period for weed control in corn in Turkey. Weed Technol. 20: 867–872.

James, C. 2015. 20th Anniversary (1996 to 2015) of the Global Commercialization of Biotech Crops and Biotech Crop Highlights in 2015. *ISAAA Brief* No. 51 (Executive Summary). ISAAA: Ithaca, NY. Available at: http://isaaa.org/resources/publications/briefs/51/executivesummary/default.asp

Jamshidia, K., A.R. Yousefia and M. Oveisib. 2013. Effect of cowpea (*Vigna unguiculata*) intercropping on weed biomass and maize (*Zea mays*) yield. New Zealand J. Crop Hort. Sci. 41: 180–188. doi: 10.1080/01140671.2013.807853

Jensen, P.K., V. Bibard, E. Czembor, S. Dumitru, G. Foucart, R.J. Froud-Williams, J.E. Jensen, M. Saavedra, M. Sattin, J. Soukup, A.T. Palou, J.-B. Thibord, W. Voegler and P. Kudsk. 2011. Survey of weeds in maize crops in Europe. DJF Report Agricultural Science 149: 44.

Johnson, G.A. and T.R. Hoverstad. 2002. Effect of row spacing and herbicide application timing on weed control and grain yield in corn (*Zea mays*). Weed Technol. 16: 548–553.

Johnson, G.A., T.R. Hoverstad and R.E. Greenwald. 1998. Integrated weed management using narrow row spacing, herbicides, and cultivation. Agron. J. 90: 40–46.

Kaefer, J.E., V.F. Guimaraes, A. Richart, R. Campagnolo and T.A. Wendling. 2012. Influence of black-oats chemical management periods on the incidence of weeds and productive performance of maize. Semini-Ciencas Agrarias 33: 481–489.

Kato-Noguchi, H., Y. Sakata, K. Takenokuchi, S. Kosemura and S. Yamamura. 2000. Allelopathy in maize: isolation and identification of allelochemicals in maize seedlings. Plant Prod. Sci. 3: 43–46.

Khan, Z., C.A.O. Midega, A. Hooper and J. Pickett. 2016. Push-pull: chemical ecology-based integrated pest management technology. J. Chem. Ecol. 42: 689–697.

Khan, Z., C.A.O. Midega, A. Hassanali, J.A. Pickett and L.J. Wadhams. 2007. Assessment of different legumes for the control of *Striga hermonthica* in maize and sorghum. Crop Sci. 47: 730–734.

Knezevic, S.Z., A. Datta, J. Scott and L.D. Charvat. 2010. Application timing and adjuvant type affected saflufenacil efficacy on selected broadleaf weeds. Crop Prot. 29: 94–99.

Knezevic, S.Z., C.M. Costa, S.M. Ulloa and A. Datta. 2009. Response of corn (*Zea mays* L.) types to broadcast flaming. pp. 92–97. *In:* Proceedings of the 8th European Weed Research Society Workshop on Physical and Cultural Weed Control. Zaragoza, Spain.

Kudsk, P. 2017. Optimising herbicide performance. pp. 149–180. *In:* Hatcher, P.E. and Froud-Williams, R.J. (Eds.) Weed Research: Expanding Horizons. Wiley, Oxford, UK.

Kudsk, P. and S. Moss. 2017. Herbicide dose: what is a low dose. pp. 15–24. *In:* Duke, S., Kudsk, P. and Solomon, K. (Eds.) Pesticide Dose: Effects on the Environment and Target and Non-Target Organisms. American Chemical Society, Washinghton DC, USA.

Lamichhane, J.R., Y. Devos, H.J. Beckie, M.K.D. Owen, P. Tillie, A. Messean and P. Kudsk. 2016. Integrated weed management systems with herbicide-tolerant crops in the European Union: lessons learnt from home and abroad. Critical Reviews in Biotechnology 37: 459–475. DOI: 10.1080/07388551.2016.1180588

Lemos, J.P., J.C.C. Galvão, A.A. Silva, A. Fontanetti, P.R. Cecon and L.M.C. Lemos. 2013. Management of *Bidens pilosa* and *Commelina benghalensis* in organic corn cultivation under no-tillage. Planta Daninha 31: 351–357.

Li, M., H. Tank, A. Kennedy, H. Zhang, B. Downer, D. Ouse and L. Liu. 2013. Enlist duo herbicide: a novel 2,4-D plus glyphosate premix formulation with low potential for off-target movement. pp. 3–14. *In:* Pesticide Formulations and Delivery Systems: Innovative Legacy Products for New Uses, American Society for Testing and Materials Special Technical Publications, Book Series Vol. 1558, Tampa, Florida.

Liebman, M., Z.J. Miller, C.L. Williams, P.R. Westerman, P.M. Dixon, A. Heggenstaller, A.S. Davis, F.D. Menalled and D.N. Sundberg. 2014. Fates of *Setaria faberi* and *Abutilon theophrasti* seeds in three crop rotation systems. Weed Res. 54: 293–306.

Lindquist, J.L., D.A. Mortensen and B.E. Johnson, 1998. Mechanisms of corn tolerance and velvetleaf suppressive ability. Agron. J. 90: 787–792.

Maqbool, M.M., A. Tanveer, A. Ali, M.N. Abbas, M. Imran, M. Ahmad and A.A. Abid. 2016. Growth and yield response of maize (*Zea mays*) to inter and intra-row weed competition under different fertiliser application methods. Planta Daninha 34: 47–56.

Marìn, C. and J. Weiner. 2014. Effects of density and sowing pattern on weed suppression and grain yield in three varieties of maize under high weed pressure. Weed Res. 54: 467–474.

Martelloni, L., M. Fontanelli, C. Frasconi, M. Raffaelli and A. Pruzzi. 2016. Cross-flaming application for intra-row weed control in maize. Appl. Engineer. Agric. 32: 569–578.

Martins, D., C.G. Goncalves and A.C. da Silva Junior. 2016. Winter mulches and chemical control of weeds in maize. Revista Ciencia Agronomica 47: 649–657.

Meissle, M., P. Mouron, T. Mus, F. Bigler, X. Pons, V.P. Vasileiadis, S. Otto, D. Antichi, J. Kiss, Z. Pálinkás, Z. Dorner, R. van der Weide, J. Groten, E. Czembor, J. Adamczyk, J.-B. Thibord, B. Melander, G. Cordsen Nielsen, R.T. Poulsen, O. Zimmermann, A. Verschwele and E. Oldenburg. 2010. Pests, pesticide use and alternative options in European maize production: current status and future prospects. J. Appl. Entomol. 134: 357–375.

Melander, B., B. Lattanzi and E. Pannacci. 2015. Intelligent versus non-intelligent mechanical intra-row weed control in transplanted onion and cabbage. Crop Prot. 72: 1–8.

Mhlanga, B., S. Cheesman, B.S. Chauhan and C. Theirfelder. 2016b. Weed emergence as affected by maize (*Zea mays* L.) – cover crop rotations in contrasting arable soils of Zimbabwe under conservation agriculture. Crop Prot. 81: 47–56.

Mhlanga, B., B.S. Chaudan and C. Theirfelder. 2016a. Weed management in maize using crop competition: a review. Crop Prot. 88: 28–36.

Midega, C.A.O., J. Pickett, A. Hooper, J. Pittchar. and Z.R. Khan 2016. Maize landraces are less affected by *Striga hermonthica* relative to hybrids in western Kenya. Weed Technol. 30: 21–28.

Midega, C.A.O., D. Salifu, T.J. Bruce, J. Pittchar, J.A. Pickett and Z.R. Khan. 2014. Cumulative effects and economic benefits of intercropping maize with food legumes on *Striga hermonthica* infestations. Field Crops Res. 155: 144–152.

Moonen, A.C. and P. Bàrberi 2004. Size and composition of the weed seedbank after 7 years of different cover-crop-maize management systems. Weed Res. 44: 163–177.

Mortensen, D.A., J.F. Egan, B.D. Maxwell, M.R. Ryan and R. Smith. 2012. Navigating a critical juncture for sustainable weed management. BioSci. 62: 75–84.

Murphy, S.D., Y. Yakubu, S.F. Weise and C.J. Swanton. 1996. Effect of planning patterns and inter-row cultivation on competition between corn (*Zea mays*) and later emerging weeds. Weed Sci. 44: 856–870.

Norsworthy, J.K. and J.R. Frederick. 2005. Integrated weed management strategies for maize (*Zea mays*) production on the southeastern coastal plains of North America. Crop Prot. 24: 119–126.

Nyamangara, J., N. Mashingaidze, E.N. Masvaya, K. Nyegerai, M. Kunzekweguta, R. Tirivavi and K. Mazvimadvi. 2014. Weed growth and labor demand under hand-hoe based reduced tillage in smallholder farmer's fields in Zimbabwe. Agric Ecosyst. Environm. 187: 146–154.

Nzioki, H.S., F. Oyosi, C.E. Morris, E. Kaya, A.L. Pilgeram, C.S. Baker and D.C. Sands. 2016. *Striga* biocontrol on a toothpick: a readily deployable and inexpensive method for smallholder farmers. Front. Plant Sci. 7: 1121.doi. org/10.3389/fpls.2016.01121

Oerke, E.C. 2006. Crop losses to pests. J. Agric. Sci. 144: 31–43.

Otto, S., R. Masin, G. Casari and G. Zanin. 2009. Weed–corn competition parameters in late-winter sowing in Northern Italy. Weed Sci. 57: 194–201.

Owen, M.D.K., H.J. Beckie, J.Y. Leeson, J.K. Norsworthy and L.E. Steckel. 2015. Integrated pest management and weed management in the United States and Canada. Pest Manag. Sci. 71: 357–376.

Page, E.R., D. Cerrudo, P. Westra, M. Loux, K. Smith, C. Foresman, H. Wright and C.J. Swanton. 2012. Why early season weed control is important in maize. Weed Sci. 60: 423–430.

Pannacci, E. 2016. Optimization of foramsulfuron doses for post-emergence weed control in maize (*Zea mays*). Spanish J. Agric. Res. 14: e1005.

Pannacci, E. and G. Covarelli. 2009. Efficacy of mesotrione used at reduced doses for post-emergence weed control in maize (*Zea mays* L.). Crop Prot. 28: 57–61.

Pannacci, E. and F. Tei. 2014. Effects of mechanical and chemical methods on weed control, weed seed rain and crop yield in maize, sunflower and soybean. Crop Prot. 64: 51–59.

Panozzo, S., L. Scarabel, A. Balogh, J. Heini, I. Dancza and M. Sattin. 2015. First European cases of *Sorghum halepense* (L.) resistant to ALS inhibitors: resistance patterns and mechanisms. p. 97. *In:* Proceedings of 17th European Weed Research Society Symposium "Weed Management in Changing Environments", Wageningen, The Netherlands.

Perry, E.D., F. Ciliberto, D.A. Hennessy and G. Moschini. 2016. Genetically engineered crops and pesticide use in US maize and soybeans. Sci. Adv. 2: e1600850.

Pysek, P., V. Jarosik, Z. Kropac, M. Chytry, J. Wild and L. Tichy. 2005. Effects of abiotic factors on species richness and cover in Central European weed communities. Agric. Ecosyst. Environ. 109: 1–8.

Ransom, J.K. 1996. Integrated management of Striga spp. in the agriculture of sub-Saharan Africa. pp. 623–628. *In:* Proceedings of the Second International Weed Control Congress, Copenhagen, Denmark.

Ransom, J., F. Kanampiu, J. Gressel, H. de Groote, M. Burnet and G. Odhiambo. 2012. Herbicide applied to imidazolinone resistant-maize seed as a *Striga* control option for small-scale African farmers. Weed Sci. 60: 283–289.

Rasmussen, J. 1996. Mechanical weed management. pp. 943–948. *In:* Proceedings of the Second International Weed Control Congress, Copenhagen, Denmark.

Rasmussen, J., B.M. Bibby and A.P. Schou. 2008. Investigating the selectivity of weed harrowing with new methods. Weed Res. 48: 523–532.

Rasmussen, J. and T. Svenningsen. 1995. Selective weed harrowing in cereals. Biol. Agric. Hort. 12: 29–46.

Reboud, X., N. Eychenne, M. Délos and L. Folcher. 2016. Withdrawal of maize protection by herbicides and insecticides increases mycotoxins contamination near maximum thresholds. Agron. Sustain. Dev. 36: 43. DOI 10. 1007/s13593-016-0376-8

Rodenburg, J. and L. Bastiaans. 2011. Host-plant defence against *Striga* spp.: reconsidering the role of tolerance. Weed Res. 51: 438–441.

Rueda-Ayala, V., G. Peteinatos, R. Gerhards and D. Andujar. 2015. A non-chemical system for online weed control. Sensors 15: 7691–7707.

Saberali, S.F., M.A. Baghestani and E. Zand. 2008. Influence of corn density and planting pattern on the growth of common lambsquarters (*Chenopodium album* L.). Weed Biol. Manag. 8: 54–63.

Scherner, A., B. Melander and P. Kudsk. 2016. Vertical distribution and composition of weed seeds within the plough layer after eleven years of contrasting crop rotations and tillage schemes. Soil and Tillage Res. 161: 135–142.

Schnappinger, M.G., J.V. Parochetti, T.C. Harris and S.W. Pruss. 1979. Triazine resistant redroot pigweed control in field corn in Maryland. pp. 8–9. *In:* Proc. Northeast. Weed Sci. Soc. 33.

Schreiber, M.M. 1992. Influence of tillage, crop-rotation, and weed management on giant foxtail (*Setaria faberi*) population dynamics and corn yield. Weed Sci. 40: 645–653.

Schultz, J.L., L.A. Chatham, C.W. Riggins, P.J. Tranel and K.W. Bradley. 2015. Distribution of herbicide resistances in Missouri waterhemp (*Amaranthus rudis* Sauer) populations. Weed Sci. 63: 336–345.

Shaw, D.R. 1996. Development of stale seedbed weed control programs for southern row crops. Weed Sci. 44: 413–416.

Shaw, D.R., M.D.K. Owen, P.M. Dixon, S.C. Weller, B. Young, R.G. Wilson and D.L. Jordan. 2010. Benchmark study on glyphosate-resistant cropping systems in the United States. Part 1: Introduction to 2006–2008. Pest Manag. Sci. 67: 741–746.

Shields, E.J., J.T. Dauer, M.J. VanGessel and G. Neumann. 2006. Horseweed (*Conyza canadensis*) seed collected in the planetary boundary layer. Weed Sci. 54: 1063–1067.

Simic, M., I. Spasojevic, D. Kovacevic, M. Brankov and V. Dragievic. 2016. Crop rotation influence on annual and perennial weed control and maize productivity. Romanian Agric. Res. 33: 125–132.

Stepanovic, S. and A. Datta, B. Neilson, C. Bruening, A. Shapiro, G. Gogos and Z. Knezevic. 2016. Effectiveness of flame weeding and cultivation for weed control in organic maize. Biol. Agric. Hort. 32: 47–62.

Tabaglio, V., A. Marocco and M. Schulz. 2013. Allelopathic cover crop of rye for integrated weed control in sustainable agroecosystems. Ital. J. Agron. 8: 35–40.

Tan, S., R.R. Evans, M.L. Dahmer, B.K. Singh and D.L. Shaner. 2005. Imidazolinone-tolerant crops: history, current status and future. Pest Manag. Sci. 61: 245–257.

Teasdale, J.R. 1995. Influence of narrow row high population corn (*Zea mays*) on weed-control and light transmittance. Weed Technol. 9: 113–118.

Teasdale, J.R. 1998. Influence of corn (*Zea mays*) population and row spacing on corn and velvetleaf (*Abutilon theophrasti*) yield. Weed Sci. 46: 447–453.

Teasdale, J.R., C.P. Rice, G. Cai and R.W. Mangum. 2012. Expression of allelopathy in the soil environment: soil concentration and activity of benzoxazinoid compounds released by rye cover crop residue. Plant Ecol. 213: 1893–1905.

Thierfelder, C., S. Cheesman and L. Rusinamhodzi. 2013. Benefits and challenges of crop rotation in maize-based conservation agriculture (CA) cropping systems of southern Africa. Int. J. Agric. Sustain. 11: 108–124.

Travlos, I.S., G. Economou and P.J. Kanatas. 2011a. Corn and barnyardgrass competition as influenced by relative time of weed emergence and corn hybrid. Agron. J. 103: 1–6.

Travlos, I.S., P.J. Kanatas, G. Economou, V.E. Kotoulas, D. Chachalis and S. Tsioris. 2011b. Evaluation of velvetleaf interference with maize hybrids as influenced by relative time of emergence. Expl. Agric. 48: 122–137.

Tursun, N., A. Datta, M.S. Sakinmaz, Z. Kantarci, S.Z. Knezevic and B.S. Chauhan. 2016. The critical period for weed control in three corn (*Zea mays*) types. Crop Prot. 90: 59–65.

Ulloa, S.M., A. Datta and S.Z. Knezevic. 2010a. Growth stage influenced differential response of foxtail and pigweed species to broadcast flaming. Weed Technol. 24: 319–325.

Ulloa, S.M., A. Datta and S.Z. Knezevic. 2010b. Tolerance of selected weed species to broadcast flaming at different growth stages. Crop Prot. 29: 1381–1388.

Ulloa, S.M., A. Datta, G. Malidza, R. Leskovsek and S.Z. Knezevic. 2010c. Timing and propane dose of broadcast flaming to control weed population influenced yield of sweet maize (*Zea mays* L. var. rugosa). Field Crops Res. 118: 282–288.

Ulloa, S.M., A. Datta, S.D. Cavalieri, M. Lesnik and S.Z. Knezevic. 2010d. Popcorn (*Zea mays* L. var. everta) yield and yield components as influenced by the timing of broadcast flaming. Crop Prot. 29: 1496–1501.

Ulloa, S.M., A. Datta, C. Bruening, B. Neilson, J. Miller, G. Gogos and S.Z. Knezevic. 2011. Maize response to broadcast flaming at different growth stages: effects on growth, yield and yield components. Eur. J. Agron. 34: 10–19.

Ulloa, S.M., A. Datta, C. Bruening, G. Gogos, T.J. Arkebauer and S.Z. Knezevic. 2012. Weed control and crop tolerance to propane flaming as influenced by the time of day. Crop Prot. 31: 1–7.

van der Weide, R., P.O. Bleeker, V.T.J.M. Acthen, L.A.P. Lotz, F. Fogelfors and B. Melander. 2008. Innovation in mechanical weed control in crop rows. Weed Res. 48: 215–224.

Vandermeer, J. 1989. The Ecology of Intercropping. Cambridge University Press, Cambridge, UK.

Vanderwerf, H.M.G., J.J. Klooster, D.A. Vanderschans, F.R. Boone and B.W. Veenm. 1991. The effect of inter-row cultivation on yield of weed-free maize. J Agron. Crop Sci. 166: 249–258

Vanderwerf, H.M.G. and M. Tollenaar. 1993. The effect of damage to the root-system caused by inter-row cultivation on growth of maize. J Agron. Crop Sci. 171: 31–35.

Varanasi, V.K., A.S. Godar, R.S. Currie, A.J. Dille, C.R. Thompson, P.W. Stahlman and M. Jugulam. 2015. Field-evolved resistance to four modes of action of herbicides in a single kochia (*Kochia scoparia* L. Schrad.) population. Pest Manag. Sci. 71: 1207–1212.

Vasileiadis, V.P., M. Sattin, S. Otto, A. Veres, Z. Pálinkás, R. Ban, X. Pons, P. Kudsk, R. van der Weide, E. Czembor, A.C. Moonen and J. Kiss. 2011. Crop protection in European maize-based cropping systems: current practices and recommendations for innovative Integrated Pest Management. Agr. Syst. 104: 533–540.

Vasileiadis, V.P., A.C. Moonen, M. Sattin, S. Otto, X. Pons, P. Kudsk, A. Veres, Z. Dorner, R. van der Weide, E. Marraccini, E. Pelzer, F. Angevin and J. Kiss. 2013. Sustainability of European maize-based cropping systems: economic, environmental and social assessment of current and proposed innovative IPM-based systems. Eur. J. Agron. 48: 1–11.

Vasileiadis, V.P., S. Otto, W. van Dijk, G. Urek, R. Leskovšek, A. Verschwele, L. Furlan and M. Sattin. 2015. On-farm evaluation of integrated weed management tools for maize production in three different agro-environments in Europe: agronomic efficacy, herbicide use reduction and economic sustainability. Eur. J. Agron. 63: 71–78.

Vasileiadis, V.P., W. van Dijk, A. Verschwele, I.J. Holb, A. Vamos, G. Urek, R. Leskovšek, L. Furlan and M. Sattin. 2016. Farm-scale evaluation of herbicide band application integrated with inter-row mechanical weeding for maize production in four European regions. Weed Res. 56: 313–322.

Vencill, W.K., R.L. Nichols, T.M. Webster, J.K. Soteres, C. Mallory-Smith, N.R. Burgos, W.G. Johnson and M.R. McClelland. 2012. Herbicide resistance: toward an understanding of resistance development and the impact of herbicide-resistant crops. Weed Sci. 60 (Special Issue): 2–30.

Vidotto, F., S. Fogliatto., M. Milan and A. Ferrero. 2016. Weed communities in Italian maize fields as affected by pedo-climatic traits and sowing time. Eur. J. Agron. 74: 38–46.

Webster, T.M. and R.L. Nichols. 2012. Changes in the prevalence of weed species in the major agronomic crops of the Southern United States: 1994/1995 to 2008/2009. Weed Sci. 60: 145–157.

Williams, M.M. II. 2006. Planting date influences critical period of weed control in sweet corn. Weed Science 54: 928–933.

Wilson, R.G., B.G. Young, J.L. Matthews, S.C. Weller, W.G. Johnson, D.L. Jordan, M.D.K. Owen, P.M. Dixon and D.R. Shaw. 2011. Benchmark study on glyphosate-resistant cropping systems in the United States. Part 4: Weed management practices and effects on weed populations and soil seedbanks. Pest Manag. Sci. 67: 771–780.

Woomer, P.L., M. Bokanga and G.D. Odhiambo. 2008. *Striga* management and the African farmer. Outlook on Agriculture 37: 277–282.

Wortman, S.E., A.S. Davis, B.J. Schutte and J.L. Lindquist. 2011. Integrating management of soil nitrogen and weeds. Weed Sci. 59: 162–170.

Zhang, W., C. Cheng, Z. Song, A. Deng and Z. He. 2015. Farming systems in China: Innovations for sustainable crop production. pp. 43–54. *In:* Sadras, V.O. and Caldrini, D. (Eds.) Crop Physiology: Applications for Genetic Improvement and Agronomy. Elsevier, Oxford, UK.

Zhou, X., S.L. Rotondaro, M. Ma, S.W. Rosser, E.L. Olberding, B.M. Wendelburg, Y.A. Adelfinskaya, J.L. Balcer, T.C. Blewett and B. Clemenst. 2016. Metabolism and residues of 2,4-Dichlorophenoxyacetic acid in DAS-40278-) maize (*Zea mays*) transformed with aryloxyalkanoate dioxygenase-1 gene. J. Agric. Food Chem. 64: 7438–7444.

Zimdahl, R.L. 1988. The concept and application of the critical weed-free period. pp. 145–155. *In:* Altieri, M.A. and Liebmann, M. (Eds.) Weed Management in Agroecosystems: Ecological Approaches. CRC Press, Boca Raton, Florida.

Sustainable Weed Control in Grain Sorghum

Lauren M. Schwartz-Lazaro*[1] and Karla L. Gage[2]

[1] School of Plant, Environmental, and Soil Sciences, Louisiana State University AgCenter,
Baton Rouge, LA, USA
[2] Department of Plant, Soil, and Agricultural Systems and Plant Biology,
Southern Illinois University, Carbondale, IL, USA

Introduction

Grain sorghum is an important cultivated crop, planted on 16.5 hectares and producing 11.7 million metric tons or 462 million bushels in the US in 2016 (USDA-NASS 2016). The US is currently the world's top sorghum producer and exporter in 2016. The leading states in grain sorghum production, in 2016, were Kansas, Texas, Arkansas, Oklahoma, and Colorado. According to the US Grains Council, grain sorghum is the third most important cereal crop grown in the US and the fifth most important crop worldwide. Sorghum is a staple cereal crop for millions of people in the marginal, semi-arid environments of Africa and South Asia. Its unique and advanced ability to grow in regions of low and variable rainfall highlight its potential to impact agricultural productivity in widespread water-limited environments (Mann et al. 1983). Originating and evolving across the diverse environmental landscape of Africa, morphological and physiological adaptation strategies have advanced sorghum as a naturally heat and drought-tolerant warm season C_4 grass that is more efficient at utilizing water, nitrogen, and energy resources with respect to other major crops, including maize (*Zea mays* subsp. *mays*) and wheat (*Triticum* spp.) (de Vries et al. 2010). Cultivated in diverse climates and environmental conditions, the challenges of increasing performance and yield on marginal lands and cooler climates remains at the forefront of sorghum improvement efforts worldwide (Shoemaker and Bransby 2010).

Botanical Description

The systematics, origin, and evolution of sorghum has been extensively discussed (de Wet and Harlan 1971, de Wet and Huckabay 1967, Harlan 1975, Harlan and de Wet 1972, Snowden 1936). Sorghum is currently classified under the genus *Sorghum* (Clayton and Renvoize 1986) and is broken down into three subspecies: *S. bicolor* subsp. *bicolor, S. bicolor* subsp. *drummondii,* and *S. bicolor* subsp. *verticilliflorum.* Cultivated sorghum, *S. bicolor* subsp. *bicolor,* are represented by agronomic cultivars such as grain sorghum, sweet sorghum, sudangrass and broomcorn (Berenji

---

*Corresponding author: llazaro@agcenter.lsu.edu

and Dahlberg 2004). Additionally, there are at least two weedy sorghums that are widespread and are extremely problematic: johnsongrass (*S. halepense* [L.] Pers.) and shattercane (*S. bicolor* [L.] Moench subsp. *arundinaceum*). Johnsongrass is known as one of the world's worst weeds. It reproduces through both seed and rhizome production and is considered a major perennial weed in many of the crops worldwide. Control of this weedy sorghum is both time consuming and difficult.

Importance

Sorghum grain is used primarily for livestock feed in the US. Recently, the US has started using sorghum as a renewable fuel and has introduced this species to the gluten-free food market. In Africa and India, it is an important part of the diet in the form of unleavened bread, boiled porridge or gruel, and specialty foods, such as popped grain and beer. Grain sorghum is a potential field crop in Europe for cattle feed (Berenji and Dahlberg 2004). Sorghum has a very high nutrient content and in developing countries the addition of sorghum into added food and beverage products is an important driver for economic development (Taylor et al. 2006).

Traditionally, sorghum has been used in unfermented and fermented breads, porridges, couscous, rice-like products, snacks, and malted alcoholic and non-alcoholic beverages in the diets of many African and Asian countries. There are also several groups working on unique health properties associated with sorghum grain that could have an impact on its use in the health food industry. Ciacci et al. (2007) reported on the in vitro and in vivo safety of sorghum food products and found that sorghum did not show toxicity for celiac patients and can be considered safe for use by those with celiac disease. Schober et al. (2005) developed several gluten-free sorghum products and studied the effects of different sorghum hybrids on food characteristics. Groups such as Lloyd Rooney at Texas A&M University, Scott Bean, US Department of Agriculture, Agricultural Research Service (USDA-ARS), Manhattan, Kansas, and Ron Prior, USDA-ARS, Little Rock, Arkansas are exploring the antioxidant activities of some unique sorghum cultivars and other such nutritional aspects of sorghum relevant to its use as human food. There is growing evidence that some of these sorghums have high anti-inflammatory and anti-colon cancer activities (Dykes and Rooney 2006).

Grain sorghum has been successfully used as cattle feed (Berenji and Kunc 1995). Sweet sorghum and sudangrass in Europe is used for cattle feed, similar to the use of silage maize. Proper cultivar choice and production technology will completely eliminate prussic acid (HCN) problems sometimes associated with fresh sweet sorghum or sudangrass used for feed (Kunc et al. 1995). Improvements in feed technology already common in the US are highlighted by various feeding guides produced by the United Sorghum Checkoff Program and could enhance the feed use of grain sorghum in Europe.

Sorghum is an excellent crop for production of renewable fuels (Berenji 1994, Kisgeci et al. 1983). Sorghum is an example of an annual crop that could be both a short- and long term solution as a renewable, sustainable biomass feedstock. Sorghum is unique among the potential renewable energy feedstock crops in that it can be used in all the various processes being considered for biofuel production. The ethanol market is one of the fastest growing segments of the sorghum industry in US representing the single-largest value-added market for grain sorghum producers in the US. There are currently eight ethanol plants in USA that use about 15 to 20% of the US grain sorghum crop each year. Equal quantities of ethanol are produced from the same amount of grain sorghum as from maize. Similar experimental results have been achieved in Europe (Kisgeci and Pekie 1983). Research is underway to evaluate the use of sweet sorghum in processes similar to what is currently available in sugarcane for the production of ethanol. Recently, Dahlberg et al. (2011) published research on compositional and agronomic evaluation of sorghum biomass. They reported that sorghum forages could produce high biomass yields over several years and that, using theoretical estimates for ethanol productions, these forages could average 6,146 L ha^{-1} of renewable fuels with a maximum production of 8,422 L ha^{-1} from the top ranked forage hybrids. These findings and sorghum's diversity as a feedstock for renewable fuels production has potential for Europe as it attempts to formulate alternative energy production strategies.

Other varieties of sorghum also have unique uses (Dahlberg et al. 2011). For example, sweet sorghum has a high dry matter yield and is typically used for sugar production. In the US, these sorghums are also processed to make a sweet syrup, similar to molasses. Sudangrass tends to grow rapidly and is best used for repeated cutting, animal grazing or baling for hay. Broomcorn is a specialty sorghum recognizable by long panicles (heads) which are composed of long, fine, elastic branches called fibres with seeds on their tip used for manufacture of corn brooms (Berenji and Kisgeci 1996). This widespread utilization of sorghum is based upon its diverse genetic background.

Weed Impact

Major Weeds

Bridges (1992) listed more than 40 common or troublesome weeds of grain sorghum production areas in the US. Generally, grass weeds have the greatest impact on sorghum production because grasses are difficult to control after crop emergence. In a 2012 survey of weed scientists in seven southern grain sorghum producing states (Alabama, Arkansas, Florida, Georgia, Mississippi, Missouri, and Texas) there were 25 unique species or species complexes listed as top ten most troublesome: barnyardgrass (*Echinochloa crus-gali* (L.) P. Beauv.), bermudagrass (*Cynodon dactylon* (L.) Pers.), broadleaf signalgrass (*Urochloa platyphylla* (Munro ex C. Wright) R.D. Webster), browntop millet (*Urochloa ramosa* (L.) Nguyen), common cocklebur (*Xanthium strumarium* L.), crabgrasses (*Digitaria* spp.), fall panicum (*Panicum dichotimiflorum* Michx.), field bindweed (*Convolvulus arvensis* L.), Florida pusely (*Richardia scabra* L.), goosegrass (*Eleusine indica* (L.) Gaertn.), Italian ryegrass (*Lolium perenne* L. subsp. *multiflorum* (Lam.) Husnot), Johnsongrass (*S. halepense* (L.) Pers.), kochia (*Bassia scoparia* (L.) A.J. Scott), marestail (*Conyza canadensis* (L.) Cronquist), morningglories (*Ipomoea* spp.), nutsedges (*Cyperus* spp.), pigweeds (*Amaranthus* spp.), Russian thistle (*Salsola kali* L.), shattercane (*S. bicolor* (L.) Moench subsp. *arundinaceum*), sicklepod (*Senna obtusifolia* (L.) Irwin & Barneby), silverleaf nightshade (*Solanum eleagnifolium* Cav.), sunflower (*Helianthus annuus* L.), Texas millet (*Urochloa texana* (Buckley) R. Webster), wild poinsettia (*Euphorbia cyathophora* Murray), and wild radish (*Raphanus raphanistrum* L.). While less than half of these species are grasses, six of seven states listed johnsongrass as first or second most-troublesome. Shattercane was number one in Missouri but was not on the top ten list of any of the other six states (Webster 2012).

In Europe there are at least two distinct areas (SE Hungary and NE Serbia) where shattercane has been reported. Shattercane is problematic because of the seed dispersal mechanism, as well as the high likelihood of out-crossing between cultivated grain sorghum and johnsongrass. Horizontal gene flow between cultivated and weedy sorghums presents a unique problem in sorghum (Sikora and Berenji 2008).

The hemi-parasitic weed, Striga (*Striga* spp.), may significantly impact production. Sorghum is a major crop host for the following species of Striga: *S. hermonthica* (Africa), and *S. asiatica* (Africa, India, China, Indonesia, Philippines, US (N and S Carolina)), *S. densiflora* (India), *S. aspera* (Africa), *S. euphrasiodes* (India). The most economically damaging species are *S. hermonthica* and *S. asiatica* (Musselman 1980). In corn, millet (*Pennisetum glaucum* (L.) R. Br.), and sorghum in Africa, *Striga hermothica* is one of greatest causes of crop loss (Doggett 1965, Oswald 2005), commonly causing estimated declines of 50% or more (House and Vasudeva Rao 1982). However, actual losses are difficult to determine due to lack of data. A regression model approach estimates that *S. asiatica* causes average sorghum losses of 18 and 25% in India, depending upon environmental conditions, but losses may reach 98% in some years (Rao et al. 1989). Sorghum cultivars with high levels of tolerance and resistance to Striga infection may be planted where production is at risk (Rodenburg et al. 2005).

Weed Competition

Weeds compete with grain sorghum for light, nutrients, and soil water, resulting in reduced yields, lower grain quality, and increased production costs. Crop-weed competition for moisture has received the most attention, perhaps because it frequently is the most limited resource in semi-arid environments. Water consumption by weeds reduces the amount of soil water available to support crop growth, thereby contributing to crop water stress and directly influencing the duration of critical weed-free period for crops. The amount of water needed to produce a kilogram of dry matter is a measure of water use efficiency (WUE) (Stahlman and Wicks 2000). Water use efficiencies of most C4 plants are considerably greater than those of C3 plants. Grain sorghum and several important weeds of grain sorghum like Palmer amaranth are C4 plants. Plants that produce the most dry matter with the least amount of water are the most efficient, weed species having high WUE typically are highly competitive with crops (Stahlman and Wicks 2000). Grain sorghum yield reductions of up to 85% have been reported in the presence of severe weed competition (Okafor and Zitta 1991). With many weeds being problematic in grain sorghum, it is crucial that an effective weed control program is implemented.

Impact on Yield Loss

Research indicates that the percentage of grain sorghum yield lost from weed competition exceeds that of most other grain crops. Yield losses to weeds generally range from 30 to 50% (Stahlman and Wicks 2000), but weed competition can interact with other environmental variables to cause losses between 15 and 97% (Peerzada et al. 2017). Stahlman and Wicks (2000) reviewed and compiled the results of grain sorghum yield reductions due to weed interference from 27 different sources spanning from 1954 to 1998. Grain sorghum yield reductions were reported as low as 4–18% (Burnside and Wicks 1969) and as high as 26 to 100% (Vencill and Banks 1994).

Many of the weeds listed above contributed to these documented yields losses, but 12 out of these 27 sources reported that a 'mixed population' of weed species caused these yield losses. Many weeds can interfere with grain sorghum establishment and harvest but if weeds are controlled within the first four weeks after crop emergence, yield loss from later emerging weeds is minimal (Moore et al. 2004). In a study of the impact of redroot pigweed (*Amaranthus retroflexus* L.) competition on yield, losses were documented only when pigweed emerged prior to the 5.5-leaf stage (Knezevic et al. 1997).

Weed Control

Mechanical Weed Control

Soil Cultivation/Tillage

Typically, a combination of cultivation and chemical weed control is implemented and is most effective in grain sorghum. Cultivation, however, can prune roots and cause stress to the crop if plows are used too close to established plants. In a comparison of various mechanical and chemical weed control practices, rotary hoed twice, cultivated once, cultivated twice, harrowed twice, pre-emergence herbicides only, or post-emergence application of 2,4-D only, sorghum yields were 99, 98, 97, 93, 85, and 83% of the weed-free control, respectively (Wiese et al. 1964).

Sorghum can be grown under no-till, conservation tillage, reduced tillage, or stubble mulch tillage conditions. Stubble mulch tillage helps control soil erosion and was originally developed under extreme conditions, such as the extreme drought and associated severe wind erosion that plagued the US Great Plains and Canadian Prairie Provinces in the 1930s. This type of tillage replaced clean tillage for dryland (non-irrigated) small grain production, primarily winter wheat and dryland grain sorghum (Allen and Fenster 1986). Adequate crop residues are generally retained on the soil surface to control wind and water erosion (McCalla and Army 1961). Water conservation and erosion control is also improved which is highly important for dryland crop production in semi-arid regions. Producers in these regions are adopting reduced tillage and no-tillage (NT) production methods that retain more residues on the surface. Use of these methods led to greater water conservation, which, along with such factors as improved weed control, cultivars, and fertilizer practices, resulted in a shift towards crop rotation of sorghum following wheat in either a two-year, one crop or a no-till double crop system. While mechanical or manual soil cultivation is perhaps the most common form of physical weed control, between-row mowing in no-till sorghum has been shown to be very effective at reducing weed competition (Donald 2007).

Irrigation

Irrigation has been shown to initiate the germination of weeds. This initial germination allows for the weeds to be controlled at an early stage in crop development. The total amount of water that a grain sorghum crop needs during the growing season can vary from 16 to 25 inches. In most seasons, the amount of water required is about 20 inches, this is dependent on rainfall and soil moisture (Musik et al. 1963, Tacker et al. 2004). Maximum water usage occurs from the boot to bloom stages where the water needs ranges from 0.2 to 0.3 inches per day. Moisture stress anytime during the growing season can affect plant development and overall yield.

Cover Crops

Cover crops have been extensively studied in a variety of cropping systems and they can provide soil protection from rain or runoff, add organic matter to the soil, can fix nitrogen, suppress soil diseases and pests, and can help suppress weeds by providing ground cover through the fall, winter, and early spring. Incorporating cover crops into crop rotations allows for weeds to be inhibited in two primary ways: 1. in the fall, cover crops can prevent growth and development of weeds through direct competition. Cover crops fill any open gaps in cropping systems that would otherwise be occupied by weeds (Liebman and Staver 2001), and 2. in the spring, cover crop residues can be incorporated into the soil which can suppress weed emergence and growth (Al Khatib 1997).

Cultural Weed Control

Seed Rates

Seeding rate of sorghum may vary, depending upon the rainfall and prevailing growth conditions (Vanderlip et al. 1998). Ottman and Olsen (2009) stated that the optimum seeding rate for grain sorghum is 4.5 kg ha^{-1} with the goal to achieve a population of 250,000 plants ha^{-1}. In Kansas, populations of 24,000 and 100,000 plants ha^{-1} are recommended for low rainfall (< 50 cm) and irrigated conditions, respectively (Shroyer et al. 1998). However, seeding rates can be increased by 20% in cases of reduced row spacing (narrower than 75 cm) with a short season hybrid or if planting in double rows. Too low of a plant population ultimately increases the risk of high weed densities, affects light interception, and decreases the maximum yield potential of the crop. Therefore, ideal plant populations are essential in order to reach the maximum yield potential in sorghum (Linneman 2011).

Stahlman and Wicks (2000) stated that the ability to adjust the tillering and head size allows grain sorghum to produce similar yields over a wide range of seeding rates. However, an increased number of heads per plant has been observed when sorghum is planted at lower densities as compared to recommended densities (Lafarge and Hammer 2002). Nevertheless, crops with low seeding rates and late emerging tillers are often less competitive to weeds and usually delay grain harvest (Stahlman and Wicks 2000). Al-Bedairy et al. (2013) recorded reduced weed population (21–42%) and dry biomass (88–99%) at 6.6, 13.3, and 26.6 plants m^{-2} densities. However, Gholami et al. (2013) reported no significant reduction in weed density and weed biomass when sorghum was planted at different densities (190,000 and 266,000 plants ha^{-1}). Combinations of competitive cultivars with reduced narrow spacing may create a favorable environment for crops with a detrimental impact on weed growth and seed production due to increased competition and reduced light interception.

Competitive Crops/Cultivars

Field crops usually vary in their ability to tolerate and compete for resources against weed pressure in a cropping system (Hoffman and Buhler 2002). Within a single crop, different cultivars possess diverse competitive responses against divergent weed populations. In general, the competitive potential of a crop usually depends upon its ability to access and utilize resources like light, moisture, nutrients, and space. Selecting a competitive cultivar is one way to potentially suppress the weed growth and seed production without the risk of sacrificing crop yield (Fric 2000). Specific characteristics that enhance the crop competitiveness may include rapid emergence, rapid biomass accumulation, leaf characteristics, canopy structure, tillering capacity, and height (Hoffman and Buhler 2002). In addition, the allelopathic potential of a crop cultivar can contribute to negatively affecting the growth and densities of weed population in a farming system (Bhadoria 2011).

Light is considered to be the most limiting factor to plant growth, when moisture and nutrients are sufficient (Poorter and Nagel 2000). Increased weed density and diverse morphological characteristics affect the light interception in the canopy and absorption of photosynthetic active radiation (PAR) by the crop, resulting in reduced crop leaf area index and chlorophyll content (Tollenaar et al. 1994). Graham et al. (1988) observed decreased leaf area and light absorption with increased weed density, exerting a large negative impact on crop yield. Stahlman and Wicks (2000) stated that light quality becomes increasingly important in determining the final yield as the sorghum crop reaches anthesis and maturity. Therefore, selection of competitive sorghum cultivars and the use of increased plant densities negatively influences light interception by weeds (Gholami et al. 2013). These cultivars restrict light penetration to weeds by absorbing the light in the canopy, resulting in reduced weed dry matter production (Mishra et al. 2015).

In addition, sorghum is reported as one of the most allelopathic crops used extensively as cover and smother crops and is also incorporated in the soil for weed suppression (Alsaadawi et al. 1986, Putnam 1990). In the Southern US, growers customarily use sorghum hybrids as a smother crop in order to reduce the weed infestation through allelopathic effect in the succeeding years (Weston 1996). Similarly, several studies have documented sorghum allelopathy, defined as the ability to provide short-term weed suppression due to the release of phenolic acid from decomposed parts (Alsaadawi et al. 2007). Weed population density and dry biomass decrease with increased competitive ability of sorghum in combination with strong allelopathic effects through root exudation (Al-Bedairy et al. 2013). Historically, cultivated *Sorghum* species have been used in weed management programs as a smother crop among other crops for its competitive suppression of weed species growing during the same time (Overland 1966). Weston et al. (2013) observed strong allelopathic effects of living sorghum plants and residues in both monoculture and multiple cropping systems, inhibiting the growth of several competing weeds.

Rotations

Crop rotation allows for a diverse set of weed control practices that include various cultural (i.e., row spacing, planting date), mechanical (i.e., tillage), and chemical weed control in the different crops (Liebman and Dyck 1993). This type of management limits the opportunity for herbicide resistance to evolve. In addition, it is important to review herbicide labels and their plant-back restrictions to grain sorghum to prevent carryover injury. In the Great Plains and Midwest region of the US, grain sorghum is typically rotated with winter wheat. In areas with sufficient rainfall to allow continuous cropping, sorghum often is planted following crops other than wheat, such as corn, cotton, or soybean. Corn can be used as a rotational crop in areas with sufficient moisture. However, the high use of ALS-inhibiting herbicides can cause grain sorghum crop injury from carryover. Using sorghum in a rotation with broadleaf crops, such as cotton and soybean, breaks up disease, insect, and weed life cycles prevalent in these crops.

Research shows typically higher cotton and soybean yields when following sorghum. Weed resistance to herbicides is becoming a major issue in cotton and soybeans. Planting sorghum in rotation with these crops allows for the use of alternative herbicides for control of resistant biotypes. Sorghum residue has allelopathic properties and may be used to suppress weed species in various rotational crops, such as corn, rice, wheat, beans, and crops in the Brassica genus (Cheema et al. 2004).

Furthermore, it is essential that weeds be controlled in the rotation crop to maintain the land, minimize moisture loss, and reduce the weed seed bank. Any uncontrolled weeds can produce seed which will be a direct input back to the soil seedbank (Schwartz et al. 2016). This ultimately leads to weed problems in the future.

Row Spacing

In grain sorghum, row spacing may range from 15 to 100 cm with a variety of configuration, generally planted on beds or on flats (Vanderlip et al. 1998, Ottman and Olsen 2009). However, row spacing varying from 25 to 102 cm are usually adopted in modern grain sorghum production systems, with 76-cm row spacing being most common (Hewitt 2015). Grain sorghum grown in narrower rows more effectively competes against weeds as compared to wider rows providing increased grain production (Smith et al. 1990). Although Vanderlip et al. (1998) stated that, while narrower row spacing have not consistently yielded better than sorghum grown on 45-cm spaced rows, narrower rows are likely to shade the soil faster, improve weed control, and reduce soil erosion.

Several studies published on grain sorghum have reported increased crop yield when planted in narrower rows. For example, Bishnoi et al. (1990) reported that narrow row spacing of 45-cm resulted in increased grain sorghum yields and reduced weed population (25–54%) as compared to sorghum grown in 60- and 90-cm row spacing. Staggenborg et al. (1999) found that 25-cm spaced rows reduced weed emergence by 24 and 45% as compared to 51-cm and 76-cm spaced rows, respectively). Narrow-row sorghum systems offer more advantage over wide-row from early canopy closure, limiting the light interception and results into reduced weed population during early growing season (Staggenborg et al. 1999).

Shading of weeds due to high plant population densities and uniform crop distribution directly affects the light interception by weeds (Forcella et al. 1992). Reduced amount and quality of solar radiation reaching the soil surface negatively influences weed seed germination (Locke et al. 2002). Wider row spacing (76–107 cm) in grain sorghum results in slower canopy formation and provides unshaded conditions for weeds, allowing weeds to compete with the crop (Stahlman and Wicks 2000). Before the establishment of grassy weed species, a quick canopy closure over the inter-rows were reported in high density sorghum crops (Smith et al. 1990). Everaarts (1993) concluded that narrower row spacing gives grain sorghum a competitive advantage over weeds and crop row spacing of <76 cm would increase grain yield in areas with higher yield potential.

Chemical Weed Control

Chemical weed control options in sorghum are limited, as compared to other crops (Table 13.1). Pre-emergence herbicide applications are used for broad spectrum grass and broadleaf control, but control failures are risked when excessive or inadequate activating rainfall is received. Once the crop has emerged there are few post-emergence grass control options, and directed applications may be required to minimize crop injury. Traditionally, the sorghum cropping system has been heavily reliant upon atrazine, and atrazine continues to be an important herbicide in sorghum production (Thompson et al. 2017). Atrazine in premixes will extend the pre-emergence control of grasses. However, use of safened seed is necessary with chloracetamide herbicide applications. Although new chemistries are unlikely to be developed for use in sorghum systems, crop plant resistance to ALS- and ACCase-inhibiting herbicides may be exploited to provide more options for post-emergence control of weeds (Kerschner

Table 13.1. List of single and premixes of active ingredients and WSSA Site of Action (SOA) groups labeled for use in grain sorghum production Adapted from http://www.agrian.com/labelcenter/results.cfm

Preplant/preemerge	SOA group	Postemerge	SOA group	Postplant incorporated	SOA group
Acetochlor + atrazine	15 + 5	2,4-D	4	Pendimethalin	3
Alachlor	15	Atrazine	5	Pendimethalin + atrazine	3 + 5
Atrazine	5	Bentazon	6	Trifluralin	3
Bromoxynil	6	Bentazon + atrazine	6 + 5		
Dicamba + nicosulfuron	4 + 2	Bromoxynil	6		
Dimethenamid-P	15	Carfentrazone	14		
Flumioxazin	14	Dicamba	4		
Glyphosate + s-metolachlor	9 + 15	Dicamba + atrazine	4 + 5		
Glyphosate + s-metolachlor + atrazine	9 + 15 + 5	Dicamba + halosulfuron	4 + 2		
Halosulfuron	2	Dimethenamid-P	15		
Linuron	7	Diuron (DIRECTED)	7		
Mesotrione	27	Fluroxypyr	4		
Mesotrione + s-metolachlor + atrazine	27 + 15 + 5	Fluroxypyr + bromoxynil	4 + 6		
Metolachlor	15	Glyphosate (DIRECTED)	9		
Metolachlor + atrazine	15 + 5	Halosulfuron	2		
Nicosulfuron	2	Linuron	7		
Paraquat	22	Mesotrione (DIRECTED)	27		
Propazine	5	S-metolachlor	15		
Prosulfuron	2	S-metolachlor + atrazine	15 + 5		
Quinclorac	4	Metsulfuron + 2,4-D	2 + 4		
Quinclorac + atrazine	4 + 5	Prosulfuron	2		
Saflufenacil	14	Prosulfuron + atrazine	2 + 5		
Saflufenacil + dimethenamid-P	14 + 15	Pyrasulfotole + bromoxynil	27 + 6		
S-metolachlor	15	Quinclorac	4		
S-metolachlor + atrazine	15 + 5				
Tribenuron + thifensulfuron	2				

et al. 2012, Werle et al. 2014). Non-GMO lines of ALS-inhibitor-tolerant grain sorghum have been developed by crossing ALS-inhibitor-resistant shattercane with grain sorghum, allowing broadcast post-emergence applications of rimsulfuron and nicosulfuron and providing the only post-emergence grass control system in grain sorghum.

In the utilization of new herbicide tolerant technologies in grain sorghum, gene flow between weedy relatives is an important consideration. Shattercane and johnsongrass can outcross with grain sorghum (Schmidt et al. 2013, Smeda et al. 2000), and therefore, provide the greatest threat to the sustainability of an herbicide tolerant system. Shattercane × grain sorghum crosses may occur frequently, since both species are diploids, however, johnsongrass is a tetraploid, and therefore, successful crosses with grain sorghum are predicted to be less frequent than with shattercane. Strict stewardship guidelines should be followed to preserve herbicide-tolerant grain sorghum technology.

In addition to the above considerations for chemical control, postemergence applications of some herbicides, such as 2,4-D must be carefully timed by crop stage to avoid injury. Applications during emergence to the four-leaf stage or from 30 cm to soft-dough stage (when the seed begins to rapidly gain starch and protein) may cause crop injury. Schweizer et al. (1978) found that two postemergence applications of 2,4-D and dicamba, alone or mixed, increased the chances of crop injury. However, there may be differences in hybrid sensitivity to 2,4-D.

Active Ingredients

Herbicide Resistance: Herbicide resistant weeds may threaten the effectiveness of the already limited chemical options in sorghum production. There is already occurrence of resistance to ALS-inhibitors, ACCase-inhibitors, glyphosate, and microtubule inhibitors in weedy *Sorghum* spp. (Heap 2017). Herbicide resistant shattercane has been documented in eight states within the US (Illinois, Indiana, Iowa, Kansas, Nebraska, Ohio, Pennsylvania, and Virginia) (Heap 2017). Six out of eight cases were in maize cropping systems, and two of eight were in maize-soybean rotation. All cases of resistance were to ALS-inhibiting herbicides, sulfonureas (nicosulfuron, oxasulfuron, and primsulfuron-methyl) and imidazolinones (imazamox, imazapyr, and imazethapyr) (Heap 2017). Herbicide resistant johnsongrass has been documented in nine states within the US (Arkansas, Indiana, Kentucky, Louisiana, Mississippi, Tennessee, Texas, West Virginia, and Virginia) in corn, cotton, and soybean cropping systems, as well as in nine countries (Argentina, Chile, Greece, Israel, Italy, Mexico, Serbia, Spain, and Venezuela) in corn, cotton, soybeans, and tomatoes (Heap 2017). Like shattercane, johnsongrass has evolved resistance to ALS-inhibiting herbicides, as well as ACCase-inhibitors, EPSPS inhibitors (glyphosate), and one case of microtubule inhibitor resistance (pendimethalin) in Mississippi.

Herbicide resistant broadleaves are also a management challenge. Atrazine-resistant pigweeds (Palmer amaranth, *Amaranthus palmeri* S. Watson, waterhemp, *Amaranthus tuberculatus* Moq. Sauer) may limit the usefulness of atrazine premixes in preemergence weed control, allowing *Amaranthus* spp. to establish early and impact yield. Of all the herbicide resistance cases documented in grain sorghum cropping systems, three *Amaranthus* spp. (*A. palmeri, A. tuberculatus*, and *A. retroflexus*) make up 8 of the 11 unique cases of resistance (Heap 2017). All of these cases of resistance in *Amaranthus* spp. are documented in Kansas and Texas, with resistance to ALS inhibitors, photosystem II inhibitors (atrazine), and glyphosate. There is one population of *A. palmeri* in Kansas with multiple 3-way resistance to ALS inhibitors, photosystem II inhibitors, and HPPD inhibitors. Additionally, multiple two-way resistant *Kochia scoparia*, with resistance to glyphosate and synthetic auxins (dicamba), was discovered in grain sorghum in Kansas in 2013.

There are two other cases of documented herbicide resistance in sorghum cropping systems. Both are grass species in Australia, *Echinochloa colona* and *Urochloa panicoides*, resistant to photosystem II inhibitors (atrazine) and EPSPS inhibitors (glyphosate), respectively (Heap 2017).

Integrated Weed Control

One of the foundational papers on Integrated Weed Management (IWM) in the US, Thill et al. (1991), cited weed biology, crop manipulation, biological control, herbicide technology, and education as important IWM topics following a WSSA symposium on IWM in crop production. The foundation of all IWM program approaches is based, to some degree, in weed ecology and biology. Knowledge and identification of the weed species present in the field is critical to management timing and methodology. Appropriate timing of management must not only be considered for the life history characteristics of the weed, but also during the correct management period for the crop; generally, the first 4–6 weeks of crop growth are critical to maintain weed-free for maximum yield, and for sorghum, this is generally prior to the 5.5-leaf stage (Knezevic et al. 1997).

Environmental resources (light, soil moisture, temperature, nutrients, and space) all interact with weed-crop biological and ecological factors to further influence management techniques. For example, mechanical weed control through physical handweeding or the use of machinery is best accomplished with minimal damage to the crop when weeds are small but there is some moisture present to soften the soil (Vijayakumar et al. 2014). The integration of mechanical and chemical methods increases the diversity of management and lowers the selection pressure for weed evolution, but management timing and soil moisture availability remain important considerations for chemical control as well as mechanical control. While effective preemergence herbicide programs may contribute more to a weed control program than mechanical practices alone, the right amount of rainfall must be received to activate the herbicide but not enough to leach the herbicide out of the zone of seed germination. Post-emergence options for grass control remain limited in sorghum, relative to other major crops, even with the development of herbicide-tolerant crop lines. Therefore, these limitations create a high level of importance in targeting grass species ecology and biology. Best management practices include the rotation of herbicide sites of action to minimize the likelihood of the evolution of resistance, as well as rotation with other crops that allow herbicide applications that effectively control the grass species.

Cultural practices, such as selecting competitive cultivars, proper row spacing, utilizing crop rotations, cover crops, and managing inputs, complement successful mechanical and chemical programs, but again, the efficacy of these practices also relies upon environmental factors. As an example, Wiese et al. (1964) showed that soil moisture availability shifted the crop-weed dynamics of competition. Rainfall immediately after planting often leads to a flush of weeds, which increased the level of competition with the emerging crop. If postemergence control strategies fail, weed species will decrease crop yield and produce seeds. The ultimate success of an integrated approach to weed management will be measured by the seasonal contribution to the weed seedbank, as growers strive to target control strategies to decrease seed production of competitive fall-reproducing grass and broadleaf species, as well as winter annual seed production in April and early May.

Concluding Remarks

Sorghum is a hardy crop that thrives in environmentally diverse systems, thus it is used in many parts of the world. Throughout the world, the species of weeds affecting this crop and their appropriate management plans vary greatly and depend upon region. An integrated weed management approach, tailored to local needs, should be implemented to protect crop yield and improve resource use efficiency.

Acknowledgements

The authors would like to thank Ronald Krausz for his timely edits on this chapter. We could not have done this without him.

REFERENCES

Al-Bedairy, N.R., I.S. Alsaadawi and R.K. Shati. 2013. Combining effect of allelopathic *Sorghum bicolor* L. (Moench) cultivars with planting densities on companion weeds. Archives of Agronomy and Soil Science 59: 955–961.

Al Khatib, K. 1997. Weed suppression with *Brassica* green manure crops in green pea. Weed Science 45: 439–445.

Allen, R.R. and C.R. Fenster. 1986. Stubble-mulch equipment for soil and water conservation in the Great Plains. Journal of Soil Water Conservation 41: 11–16.

Alsaadawi, I.S., J.K. Al-Uqaili, A.J. Alrubeaa and S.M. Al-Hadithy. 1986. Allelopathic suppression of weed and nitrification by selected cultivars of *Sorghum bicolor* (L.) Moench. Journal of Chemical Ecology 12: 209–219.

Bhadoria, P.B.S. 2011. Allelopathy: a natural way towards weed management. American Journal of Experimental Agriculture 1: 7–20.

Bishnoi, U.R., D.A. Mays and M.T. Fabasso. 1990. Response of no-till and conventionally planted grain sorghum to weed control method and row spacing. Plant and Soil 129: 117–120.

Berenji, J. 1994. Bioalkohol od sirka i oiooke. Revija agronomska saznanja 4: 19–21.

Berenji, J. and J. Dahlberg. 2004. Perspectives of sorghum in Europe. Journal of Agronomic Crop Science 1905: 332–338.

Berenji, J. and J. Kisgeci. 1996. Broomcorn-classical example of industrial use of sorghum. pp. 43–48. *In:* Proceedings of 1st European Seminar on Sorghum for Energy and Industry. Toulouse, France.

Berenji, J.B. and V. Kunc. 1995. Prinos i kvalitet sirka za zrno. Zbornik radova Naunog instituta za ratarstvo I povrtarstvo. Novi Sad 23: 309–318.

Bridges, D.C. 1992. Crop losses due to weeds in the United States. p. 403. *In:* Proceedings of the Weed Science Society of America.

Burnside, O.C. and G.A. Wicks. 1969. Influence of weed competition on sorghum growth. Weed Science 17: 332–334.

Cheema, Z.A., A. Khaliq, and S. Saeed. 2004. Weed control in maize (*Zea mays* L.) through sorghum allelopathy. Journal of Sustainable Agriculture 23: 73–86.

Ciacci, C., L. Maiuri, N. Caporaso, C. Bucci, L. Del Giudice, D.R. Massardo, P. Pontieri, N. Di Fonzo, S.R. Bean, Ban Loerger and M. Londei. 2007. Celiac disease: *in vitro* and *in vivo* safety and palatability of wheat free sorghum food products. Clinical Nutrition 26: 799–805.

Clayton, W.D. and S.A. Renvoize. 1986. Genera Graminum grasses of the world. Kew Bulletin Additional Series XIII, pp. 338–345. Royal Botanic.

Dahlberg, J.A., E. Wolfrum, B. Bean and L. Rooney. 2011. Compositional and agronomic evaluation of sorghum biomass as a potential feedstock for renewable fuels. Journal of Biobased Materials and Bioenergy 5: 507–513.

de Vries, S.C., G.W.J van de Ven, M.K. van Ittersum and K.E. Giller. 2010. Resource use efficiency and environmental performance of nine major biofuel crops, processed by first-generation conversion techniques. Biomass and Bioenergy 34: 588–601.

de Wet, J.M.J. and J.R. Harlan. 1971. The origin and domestication of *Sorghum bicolor*. Economic Botany 25: 128–135.

de Wet, J.M.J. and J.P. Huckabay. 1967. The origin of *Sorghum bicolor*. II. Distribution and domestication. Evolution 21: 787–802.

Doggett, H. 1965. Striga hermonthica on sorghum in East Africa. The Journal of Agricultural Science 65: 183–194.

Donald, W.W. 2007. Between-row mowing systems control summer annual weeds in no-till grain sorghum. Weed Technology 21: 511–517.

Dykes, L. and L.W. Rooney. 2006. Sorghum and millet phenols and antioxidants. Journal of Cereal Science 44: 236–251.

Everaarts, A.P. 1993. Effects of competition with weeds on the growth, development and yield of sorghum. The Journal of Agricultural Science 120: 187–196.

Forcella, F., M.E. Westgate and D.D. Warnes. 1992. Effect of row width on herbicide and cultivation requirements in row crops. American Journal of Alternative Agriculture 7: 161–167.

Fric, B. 2000. Back to the Basics: A Manual for Weed Management on Organic Farms. Organic Producers Association of Manitoba, Virden, MB. pp. 3–30.

Gholami, S., M. Minbashi, E. Zand and G. Noormohammadi. 2013. Non-chemical management of weeds effects on forage sorghum production. International Journal of Advances Biological Biomedical Research 1: 614–623.

Graham, P.L., J.L. Steiner and A.F. Wiese. 1988. Light absorption and competition in mixed sorghum-pigweed communities. Agronomy Journal 80: 415–418.

Harlan, J.R. 1975. Crops and man. American Society of Agronomy, Madison, Wisconsin.

Harlan, J.R. and J.M.J. de Wet. 1972. A simplified classification of cultivated sorghum. Crop Science 12: 172–176.

Heap, I. 2017. The International Survey of Herbicide Resistant Weeds. Available at: http://www.weedscience.com.

Hewitt, C.A. 2015. Effect of row spacing and seeding rate on grain sorghum tolerance of weeds. Doctoral dissertation Kansas State University, Manhattan, Kansas.

Hoffman, M.L. and D.D. Buhler. 2002. Utilizing *Sorghum* as a functional model of crop-weed competition. I. Establishing a competitive hierarchy. Weed Science 50: 466–472.

House, L.R. and M.J. Vasudeva Rao. 1982. Striga–problems and prospects. Proceedings of the ICRISAT AICSIP (ICAR) Working Group Meeting on Striga Control.

Kershner, K.S., K. Al-Khatib, K. Krothapalli and M.R. Tuinstra. 2012. Genetic resistance to acetyl-coenzyme a carboxylase-inhibiting herbicides in grain sorghum. Crop Science 52: 64–73.

Kisgeci, J., A. Mijavec and J. Berenji. 1983. Growing sorghum on marginal lands as raw material for the production of fuel alcohol. CNRE Bulletin 2: 26–29.

Kisgeci, J. and B. Peki. 1983. Sorghum growing in Vojvodina in order to obtain alcohol as a blend with gasoline for motor fuel. Bilten za hmelj, sirak i lekovito bilje 15: 1–165.

Knezevic, S.Z., J.H.J. Michael and L.R. Vanderlip. 1997. Relative time of redroot pigweed (*Amaranthus retroflexus* L.) emergence is critical in pigweed-sorghum (*Sorghum bicolor* [L.] Moench) competition. Weed Science 45: 502–508.

Kunc, V., M. Latkovska, M. Krajinovie and J. Berenji. 1995. HCN content of forage sorghums and sudangrasses. Zbornik Matice Srpske za prirodne nauke 89: 53–61.

Lafarge, T.A. and G.L. Hammer. 2002. Tillering in grain sorghum over a wide range of population densities: modelling dynamics of tiller fertility. Annals of Botany 90: 99–110.

Liebman, M. and E. Dyck. 1993. Crop rotation and intercropping strategies for weed management. Ecol. Appl. 3: 92–122.

Liebman, M. and C.P. Staver. 2001. Crop diversification for weed management. pp. 322–374. *In:* Liebman, M., C.L. Mohler and C.P. Staver (Eds.) Ecological Management of Agricultural Weeds . Cambridge University Press, Cambridge, UK.

Linneman, J.W. 2011. Developing row spacing and planting density recommendations for sweet sorghum production in the Southern Great Plains. Doctoral dissertation Oklahoma State University, Oklahoma, US.

Locke, M.A., K.N. Reddy and R.M. Zablotowicz. 2002. Weed management in conservation crop production systems. Weed Biology and Management 2: 123–132.

Mann, J.A., C.T. Kimber and F.R. Miller. 1983. The origin and early cultivation of sorghums in Africa. Texas Agricultural Experiment Station Bulletin 1454.

Marín, C. and J. Weiner. 2014. Effects of density and sowing pattern on weed suppression and grain yield in three varieties of maize under high weed pressure. Weed Research 54: 467–74.

McCalla, T.M. and T.J. Army. 1961. Stubble mulch farming. Advances in Agronomy 13: 125–196.

Mishra, J., S. Rao and J. Patil. 2015. Response of grain sorghum (*Sorghum bicolor*) cultivars to weed competition in semi-arid tropical India. Indian Journal of Agricultural Science 85: 688–694.

Moore, J.W., D.S. Murray and R.B. Westerman. 2004. Palmer amaranth (*Amaranthus palmeri*) effects on the harvest and yield of grain sorghum (*Sorghum bicolor*). Weed Technology 18: 23–29.

Musik, J.T., J.W. Grimes and G.M. Harron. 1963. Irrigation water management and nitrogen fertilization of grain sorghum. Agronomy Journal 55: 295–298.

Musselman, L.J. 1980. The biology of Striga, Orobanche, and other root-parasitic weeds. Annual Review of Phytopathology 18: 463–489.

Okafar, L.I. and C. Zitta. 1991. The influence of nitrogen on sorghum-weed competition in the tropics. Tropical Pest Management 37: 138–143.

Oswald, A. 2005. Striga control—technologies and their dissemination. Crop Protection 24: 33–342.

Ottman, M. and M. Olsen. 2009. Growing Grain Sorghum in Arizona. Arizona Cooperative Extension. University of Arizona, College of Agricultural Life Sciences, Tucson, Arizona.

Overland, L. 1966. The role of allelopathic substances in the 'Smother Crop' barley. American Journal of Botany 53: 423–432.

Peerzada, A.M., H.H. Ali and B.S. Chauhan. 2017. Weed management in sorghum (*Sorghum bicolor* [L.] Moench) using crop competition: a review. Crop Protection 95: 74–80.

Poorter, H. and O. Nagel. 2000. The role of biomass allocation in the growth response of plants to different levels of light, CO_2, nutrients and water: a quantitative review. Functional Plant Biology 27: 595–607.

Putnam, A.R. 1990. Vegetable weed control with minimal inputs. Horticulture Science 25: 155–158.

Rao, M.J., V.L. Vasudeva, Chidley and L.R. House. 1989. Estimates of grain yield losses caused in sorghum (*Sorghum bicolor* [L.] Moench) by *Striga asiatica* (L.) Kuntze obtained using the regression approach. Agriculture, Ecosystems & Environment 25: 139–149.

Rodenburg, J., L. Bastiaans, E. Weltzien and D.E. Hess. 2005. How can field selection for Striga resistance and tolerance in sorghum be improved? Field Crops Research 93: 34–50.

Schober, T.J., M. Messerschmidt, S.R. Bean, S.H. Park and E.K. Arendt. 2005. Gluten-free bread from sorghum: quality differences among hybrids. Cereal Chemistry 82: 394–404.

Schmidt, J.J., J.F. Pedersen, M.L. Bernards and J.L. Lindquist. 2013. Rate of shattercane × sorghum hybridization in situ. Crop Science 53: 1677–1685.

Schwartz, L.M., J.K. Norsworthy, B.G. Young, K.W. Bradley, G.R. Kruger, V.M. Davis, L.E. Steckel and M.J. Walsh. 2016. Waterhemp (*Amaranthus tuberculatus*) and Palmer amaranth (*Amaranthus palmeri*) seed production retention at soybean (*Glycine max*) harvest. Weed Technology 30: 284–290.

Schweizer, E.E., J.F. Swink and P.E. Heikes. 1978. Filed bindweed (*Convolvulus arvensis*) control in corn (*Zea mays*) and sorghum (*Sorghum bicolor*) with dicamba and 2,4-D. Weed Science 26: 665–668.

Shoemaker, C.E. and D.I. Bransby. 2010. The role of sorghum as a bioenergy feedstock. *In:* R. Bruan, D. Karlen, D. Johnson (Eds.) Sustainable alternative fuel feedstock opportunities, challenges and roadmaps for six US regions. Proceedings of the Sustainable Feedstocks for Advance Biofuels Workshop.

Shroyer, J., H. Kok and D. Fjell. 1998. Seedbed preparation and planting practices. Grain Sorghum Production Handbook, Kansas Coop. Extn. Serv., Manhattan, KS C-687 Revised Sikora, V. and Berenji, J. 2008. Interspecies hibridizacija u okviru roda Sorghum. Bilten za hmelj, sirak I lekovito bilje 4: 17–22.

Smeda, R.J., R.S. Currie and J.H. Rippee. 2000. Fluazifop-P Resistance Expressed as a Dominant Trait in Sorghum (*Sorghum bicolor*). Weed Technology 14: 397–401.

Smith, B.S., D.S. Murray, J.D. Green, W.M. Wanyahaya and D.L. Weeks. 1990. Interference of three annual grasses with grain sorghum (*Sorghum bicolor*). Weed Technology 4: 245–249.

Snowden, J.D. 1936. The cultivated races of sorghum. Ad lard and Son Ltd London.

Staggenborg, S.A., D.L. Fjell, D.L. Devlin, W.B. Gordon and B.H. Marsh. 1999. Grain sorghum response to row spacings and seeding rates in Kansas. Journal of Production Agriculture 12: 390–395.

Stahlman, P.W. and G.A. Wicks. 2000. Weeds and their control in grain sorghum. pp. 535–582. *In:* C.W. Smith and R.A. Frederikson (Eds.) Sorghum: Origin, History, Technology, and Production, Wiley Publishers, New York, USA.

Tacker, P., E. Vories and G. Huitink. 2004. Drainage and irrigation. pp. 11–20. *In:* Arkansas Grain Sorghum Production Handbook. Arkansas Cooperative Extension Service Miscellaneous Publications 297. Little Rock, AR: University of Arkansas.

Taylor, J.R.N., T.J. Schober and S.R. Bean. 2006. Novel food and non-food uses for sorghum and millets. Journal of Cereal Science 44: 252–271.

Thill, D.C., J.M. Lish, R.H. Callihan and E.J. Bechinski. 1991. Integrated weed management: a component of integrated pest management – a critical review. Weed Technology 5: 648–656.

Thompson, C.R., J.A. Dille and D.E. Peterson. 2017. Weed Competition and Management in Sorghum. *In:* I. Ciampitti and V. Prasad (Eds.) Sorghum: State of the Art and Future Perspectives, Agronomy Monograph 58. ASA and CSSA, Madison, WI. doi: 10.2134/agronmonogr58.2014.0071

Tollenaar, M., A.A. Dibo, A. Aguilara, S.F. Weise and C.J. Swanton. 1994. Effect of crop density on weed interference in maize. Agronomy Journal 86: 591–595.

United States Department of Agriculture National Agricultural Statistics Service. 2016. USDA NASS Quick Stats. Available at: http://quickstats.nass.usda.gov/

Vanderlip, R., K. Roozeboom, D. Fjell, J. Hickman, H. Kok, J. Shroyer, D. Regehr, D. Whitney, R. Black, D.H. Rodgers and D. Jardine. 1998. Grain Sorghum Production Handbook, 3, Kansas Cooperative Extension Services, Manhattan, KS.

Vencill, W.K. and P.A. Banks. 1994. Effects of tillage systems and weed management on weed populations in grain sorghum (*Sorghum bicolor*). Weed Science 42: 541–547.

Vijayakumar, M., C. Jayanthi, R. Kalpana and D. Ravisankar. 2014. Integrated weed management in sorghum (*Sorghum bicolor* [L.] Moench) – a review. Agricultural Reviews 35: 79–91.

Webster, T.M. 2012. Weed survey-grass crops subsection. Proceedings of South Weed Science Society. 65: 267–288.

Werle, R., J.J. Schmidt, J. Laborde, A. Tran, C.F. Creech and J.L. Lindquist. 2014. Shattercane X ALS-tolerant sorghum F1 hybrid and shattercane interference in ALS-tolerant sorghum. Journal of Agricultural Science 6: 159–165.

Weston, L.A. 1996. Utilization of allelopathy for weed management in agroecosystems. Agronomy Journal 88: 860–866.

Weston, L.A., I.S. Alsaadawi and S.R. Baerson. 2013. Sorghum allelopathy from ecosystem to molecule. Journal of Chemical Ecology 39: 142–153.

Wiese, A.F., J.W. Collier, L.E. Clark and U.D. Havelka. 1964. Effect of weeds and cultural practices on sorghum yields. Weeds 12: 209–211.

14

Sustainable Weed Control in Rice

Khawar Jabran*[1], Ahmet Uludag[2] and Bhagirath S. Chauhan[3]

[1] Department of Plant Protection, Faculty of Agriculture and Natural Sciences, Duzce University, Duzce, Turkey

[2] Plant Protection Department, Faculty of Agriculture, Canakkale Onsekiz Mart University, Canakkale, Turkey

[3] Centre for Plant Science, Queensland Alliance for Agriculture and Food Innovation (QAAFI), The University of Queensland, Gatton, Queensland 4343, Australia

Introduction

Rice is included among the top cereals of the world owing to a large area cultivated under this crop and huge grain production. Billions of people obtain their calories from rice grains on daily basis. The way of consuming rice is not limited to eating of boiled rice grains. The rice dishes, such as *'Biryani'* and Chinese rice are famous in several parts of the world. 'Rice milk' also provides an important way of consuming rice. A large number of sweet dishes are also made using rice grain as an ingredient in various parts of the world. Other than this, a large number of by-products are made from the rice grains and vegetative parts.

Currently, rice is grown over an area of more than 160 million ha while the global rice production exceeds 740 million tons (FAO 2014) (Figure 14.1).

Asia is undoubtedly the top rice producing region in the world followed by Americas, Africa and Europe, respectively (FAO 2014) (Figure 14.2).

Top five rice producing countries (keeping in regard the quantity of paddy produced) of the world are: Republic of China, India, Indonesia, Bangladesh and Vietnam (FAO 2014) (Figure 14.3).

Rice is a member of Poaceae family and has two cultivated species, i.e., (i) *Oryza sativa* L. and (ii) *Oryza glaberrima* Steud. The species *O. sativa* is grown in most parts of the world while *O. glaberrima* is limited to African countries. Hence, *O. glaberrima* is usually named as African rice. *O. sativa* is further divided into two subspecies, i.e., (i) *indica*, and (ii) *japonica*. *O. glaberrima* is believed to have an origin in Africa, while *O. sativa* is supposedly originated from the foot plains of Himalayas.

Rice may be produced under various production systems using different planting methods (Ehsanullah et al. 2007). Two distinct systems may include irrigated and rainfed rice. Irrigated rice may have various types, of which flooded rice is most important. Flooded rice may be grown as transplanted or direct-seeded. Similarly, irrigated rice may be grown as transplanted or direct-seeded. Direct-seeded rice sown on dry soil and irrigated occasionally is known as aerobic rice.

*Corresponding author: khawarjabran@gmail.com

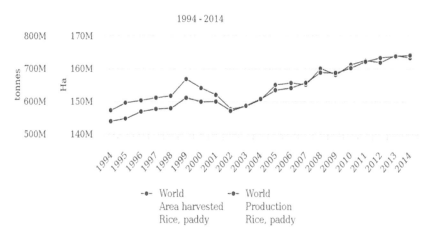

Figure 14.1. Global paddy rice production.
(Source: http://fenix.fao.org/faostat/beta/en/#data/QC/visualize)

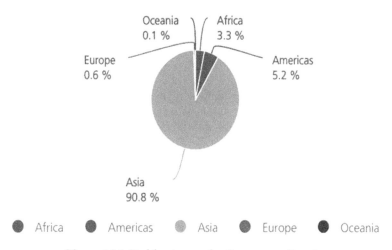

Figure 14.2. Paddy rice production per continent.
(Source: http://fenix.fao.org/faostat/beta/en/#data/QC/visualize)

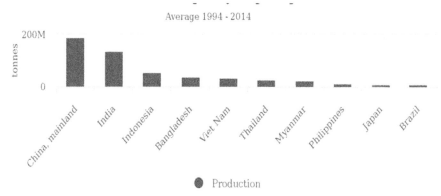

Figure 14.3. Top paddy rice production countries.
(Source: http://fenix.fao.org/faostat/beta/en/#data/QC/visualize)

Such rice sowing is aimed at saving of water. In this article, we have considered aerobic rice and dry direct-seeded rice as anonymous terms. As aerobic rice is irrigated occasionally, moisture conservation techniques may be required for effective water management. For instance, use of a soil cover (e.g., a straw or plastic mulch) has been suggested in the aerobic rice system in order to improve its water retention, water productivity, paddy yield, grain quality and economic returns (Jabran et al. 2016, Jabran et al. 2015b, Jabran et al. 2015d).

A number of factors can be listed that act as serious production constraints in rice cropping. The most important of these may include poor crop management (particularly low plant population and poor fertilizer management), heat stress, drought stress, insect pests, diseases and weeds. Other pathogens such as viruses, snails, nematodes etc. may have an occasional likeliness of infesting the rice fields and damaging rice crop (Babatola 1984, Hibino 1990). Nevertheless, weeds are among the major pests of rice, the losses caused by weeds to rice usually surpass those caused by insect pests and diseases. Under certain cases, weeds have been reported to cause a major loss in rice productivity or a total crop failure (23-100%) (Jabran and Chauhan 2015).

Results from research indicate that there are several potential methods for controlling weeds in rice. However, sustainability of such weed control has been questionable. Importantly, the sustainability of weed management is challenged by factors, such as evolution of herbicide resistance in weeds, water scarcity and climate change. Evidence from recent scientific literature and personnel observations indicates that weed control in rice could be sustainable if it is based on multiple (integrated) and innovative strategies. Multiple weed control strategies in rice should make an integrated use of techniques, such as classical cultural control, herbicides along with innovative methods (e.g., use of allelopathy or crop competition for weed control) (Farooq et al. 2011b, Jabran et al. 2015a, Sardana et al. 2016).

The objectives of this book chapter include reviewing the important weeds infesting rice fields in various parts of the world, the damages and yield losses caused by weeds in rice, the opportunities for managing weeds in rice, and how the sustainability of weed control in rice may be improved.

Impact of Weeds on Rice Crop

Field observations, research data, and recent literature from across the world indicate that rice is infested by a complex weed flora. All kind of weeds, such as narrow-leaves, broad-leaves and sedges severely infest the rice crop. However, a difference in weed flora may be noted in different rice production systems or different rice cropping systems. Kraehmer et al. (2016) reviewed the distribution of weeds in rice growing areas of the world. They found that several weeds were dominant in rice fields throughout the world independent of the soil, environment, and management practices. Important of these were the weeds (*Echinochloa* spp., *Cyperus* spp., *Fimbristylis* spp., *Leptochloa* spp. and *Scirpus* spp.) that were dominating the rice fields over the decades without distinction of climate or geographical areas. Other than this, it was noted that among the sedges (*Cyperus* spp.), *C. difformis* L., *C. rotundus* L. and *C. iria* L. were the respectively important sedges infesting rice fields in different rice cropping areas of the world (Haefele et al. 2000) (Figure 14.4).

Echinochloa crus-galli (L.) P. Beauv. with other *Echinochloa* species is considered as the most important narrow-leaved weed infesting rice fields followed by weeds such as *E. colona* (L.) Link (Bajwa et al. 2015, Haefele et al. 2000, Peerzada et al. 2016). Several of the other *Echinochloa* species also infest the rice fields, however, generally these do not have a major importance (Jabran and Chauhan, 2015). Other than this, weedy rice is becoming an important production constraint in several of the rice growing regions of the world (Kraehmer et al. 2016). In a study from West Africa, the weeds infesting rice fields were *Bolboschoenus maritimus* L., *Schoenoplectus senegalensis* (Steudel) Raynal, *C. iria*, *Diplachne fusca* (L.) P. Beauv. ex Stapf, *C. difformis* L. *E. colona*, and *Sphenoclea zeylanica* Gaertner (Haefele et al. 2000). Along with others, weedy rice

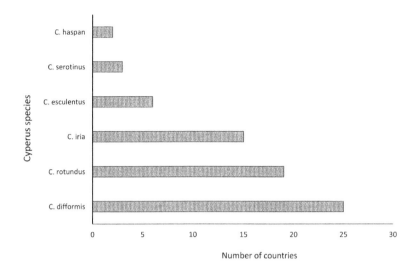

Figure 14.4. Number of rice growing countries where *Cyperus* species have been reported to infest rice fields. Adapted from Kraehmer et al. (2016)

(*Oryza sativa* L.) is also attaining importance as a dangerous rice weed in several of rice growing areas in the world (Olofsdotter et al. 2000, Chauhan et al. 2014, Dai et al. 20014).

Water management and production system (e.g., aerobic or flooded rice) impacts the composition and intensity of weeds prevailing in rice fields (Kraehmer et al. 2016). For instance, the work of Jabran et al. (2015c) indicated that *Trianthema portulacastarum* (L.) and *Cynodon dactylon* (L.) Pers. were the weeds infesting aerobic rice fields but absent in flooded rice. In contrast, *E. crus-galli* was prevailing in flooded rice but absent in aerobic rice (Jabran et al. 2015c). However, the results of other studies indicated that *E. crus-galli* was infesting aerobic rice too (Akbar et al. 2011, Jabran et al. 2012a, Jabran et al. 2012b). A review by Jabran and Chauhan (2015) indicated that 90 weed species could infest aerobic rice fields in different parts of the world, and yield losses caused by weeds in aerobic rice were in a range of 23-100%. Yield losses caused by weeds have been noted in all the rice growing regions of the world. A yield loss of more than half ton of paddy grains was noted with 10% of relative biomass of weeds in rice (Haefele et al. 2000). A study from Pakistan compared non-treated control plots and weed-free rice plots and found more than 70% decrease in paddy yield caused by the free growing weeds in aerobic rice fields (Jabran et al. 2012b). More than 35% decline in rice yield may be noted if weeds were not subjected to management (Oerke 2006).

Weed Control Strategies in Rice

Preventive Measures

Different rice production systems may have a different choice of preventive measures as weed control methods. The classical preventive measures for weed management should be integrated with other management practices to improve the efficiency of an integrated weed control plan. Clean cultivation, the application of weed seed-free water, flooding, the use of clean crop seeds which are free of weed seeds (both for nursery and direct sowing), the application of clean fertilizer, and the use of weed seed-free equipment are generally included in classical preventive weed control measures in rice (Jabran and Chauhan 2018). Additionally, practices, such as proper crop residue management and zero-level seed production can provide an important contribution as preventive weed control methods in rice. More or less, all of these classical weed preventive measures are true for reducing weed pressure in different rice production systems (Jabran et al. 2014). However, the probable order representing the importance of these preventive measures is: irrigation with weed seed-free water > use of weed seed-free rice seed > clean cultivation > use of sterilized equipment. An important strategy includes the cutting of weed species manually immediately after the seed set, and before seed ripening and dispersal. This is especially true for *Echinochloa* spp. (*E. crus-galli*, in particular) whose height is generally greater than the height of rice plants. These weeds are generally difficult to be separated from rice plants at the seedling and vegetative stages but are easily identifiable at the seed set stage. Cleaning of rice field from weeds before shattering of seeds in the field can help in attaining long-term and sustainable reduction of weeds in the rice fields (Jabran et al. 2014).

Rice crop requires more number of irrigations than the other cereals. This frequent irrigation may lead to the addition of weed seeds to the rice fields, particularly if the source of irrigation is canal water. In such cases, weeds may grow in several repeated flushes. Clean cultivation will reduce the probability of addition of weed seeds to irrigation water, while a hindrance (screen or sieve) maintained at the water entry point of the field to avoid the weed seed entry to the rice fields will effectively reduce weed intensity in aerobic rice systems.

Clean cultivation, the use of sterilized equipment, and the use of weed seed-free rice seeds are among the basic principles of weed prevention (Rao et al. 2007). Clean cultivation and the use of sterilized equipment can avoid the transfer of the propagules of weeds, such as *C. dactylon* and *Cyperus* spp. Generally, during seedling transplanting, a number of weeds from the nursery can be easily avoided to be transplanted in rice fields. However, in the case of direct seed drilling, weeds have a higher probability of establishing in rice fields due to the presence of weed seeds in rice seeds. Therefore, the use of certified seeds may be helpful to avoid weed seed entry caused by direct drilling (Rao et al. 2007).

In addition to ample water supply to rice plant, the flooding (in rice) provides several ecological benefits. Weed suppression is most important among these benefits (Chauhan 2012). Conventionally flooded rice (CFR) is kept inundated during the major part of its growing season. This flooding in the form of a layer of water blocks sunlight reaching the weed seeds, and hence inhibit the germination of weeds. Moreover, the germinated weeds also get a suppressive effective owing to layer of standing water. Flooding has been among the classical methods of weed control in rice (Rao et al. 2007, Chauhan 2012). This method benefits the weed control process directly and indirectly. In the indirect use, the layer of water improve the retention and efficacy of several herbicides. Severe water scarcity across the world hinders the flooding of rice. Many of the farmers have turned to aerobic rice cultivation that require lower water input than CFR. This has decreased the chances for use of flooding for controlling weeds. Importantly, farmers can make an occasional use of flooding rather than continuous inundation of rice fields keeping in regard the emergence of a weed flush.

In conclusion, adopting preventive weed management practices can importantly reduce weed intensity in aerobic rice fields. This would be helpful in implementing more reliable and durable weed control for improved rice productivity.

Cultural Weed Control

Sowing Date

Adjusting sowing of rice according to germination timing of weed may be useful to avoid weeds. For instance, *E. crus-galli* germinating 30 days later than rice crop may not cause a major decline in rice productivity (Gibson et al. 2002). Knowledge of ecology of weeds growing in rice is required to avoid competition of weeds with rice.

Competitive Rice Cultivars

Rice cultivars with superior morphological trait can be preferred over the rice cultivars with weaker morphological traits (Dingkuhn et al. 1999). Superior morphological traits make the crop cultivars competitive against weeds and help them in suppressing weeds. Recently, researchers have strongly advised using crop cultivars with a strong competitive potential against weeds (Sardana et al. 2016). Improving the grain yield is usually an important criterion for breeding new varieties, however, breeders may need to revise this criterion and need to consider the weed suppressing ability of genotypes in their breeding programs. The problems, such as high weed intensity in rice, high costs of weed control and evolution of herbicide resistance in weeds support the idea to have weed suppressing rice cultivars. Rice cultivars having a fast early development and constructing a good canopy may take advantage of their growth to suppress weeds (Mennan et al. 2012). Similarly, rice cultivars with a greater height and tillering, and high leaf area and dry matter accumulation are expected to possess a suppressive effect on rice weeds. Dingkuhn et al. (1999) reported that high tillering and specific leaf area were the characters that could provide rice an advantage over weeds.

Seeding Rates and Row Spacing

Use of a rice seed rate higher than normal (or transplanting more number of seedlings than recommended) can provide the rice crop a competitive advantage over the weeds because of more number of rice tillers per unit area (Chauhan 2012). This will leave less space and other resources for weeds (Sardana et al. 2016). In a similar manner, sowing of rice in narrow rows may also leave less space for weed plants (Sardana et al. 2016, Chauhan and Johnson, 2010, 2011). Subsequently, the rice plants will be able to achieve a canopy coverage earlier than usual and shade the weeds. This shading of rice on weed plants will let the rice plants absorb more of solar radiation than weeds.

Fertilizer Management

Fertilizer application is purposed to provide nutrition to crop plants. However, weeds equally compete with crop plants for the uptake of nutrients applied in the form of fertilizers. Rice crop faces strongly competing weeds such as *Echinochloa* spp., *Cyperus* spp., etc. (Jabran et al. 2012b, Kraehmer et al. 2016). Weeds growing in rice may absorb nutrients in high quantities, even higher than crops (Moody 1981). It is important to manage fertilizer in an appropriate way so that most of it is consumed by rice plants rather than competing weeds (Jabran and Chauhan 2015). For instance, fertilizer may not be applied in rice when weeds are growing actively. Fertilizer application in weed-free rice fields will be most advantageous for the rice plants. Combining appropriate fertilizer management with weed control had synergistic effect on grain yield of rice grown in West Africa (Haefele et al. 2000). In a study from Pakistan, fertilizer (nitrogen and phosphorus) application in rice fields that were cleaned from weeds through hoeing provided the highest grain yields (3 t/ha), while the fields applied with the same quantity of fertilizer without weed control had a yield of 1.9 t/ha, i.e., 37% lower than the plots where fertilizer was applied after weed control (Ullah et al. 2009). Ineffective weed control combined with fertilizer application lead to poor yield than the effective weed control and fertilizer application practice (Mahajan and Timsina 2011).

Rotations

Crop rotation can break the weed cycle, and has been found effective in weed control in rice (Chauhan 2012). Effect of rotation may be improved through inclusion of an allelopathic crop in rotation (Farooq et al. 2011b, Jabran et al. 2015a).

Mechanical/Physical Weed Control

Stale Seedbed, Puddling and Soil Solarization

Stale seedbed is a perfect method of non-chemical weed control that has been practiced over the decades as a successful weed control technique in the rice crop. The plan of weed control through stale seedbed is simple and highly effective. This includes irrigating rice fields well before the sowing season. This is followed by a soil cultivation operation that brings the soil conditions at a level favoring the germination and growth of weeds. Most of the weeds in the top soil germinate, the germinating weeds are allowed to grow for nearly a couple of weeks or little more. The germinated weeds are then killed through use of herbicides (such as glyphosate or paraquat) or through mechanical weed control. Stale seedbed has been successful for control in all kinds of rice production systems provided that there is enough time to carry out this practice and the environmental conditions and climate is favorable for it (Jabran et al. 2014).

Puddling is an important agronomic practice in case the rice crop is grown through the conventionally flooded method. Puddling includes an intensive tillage in inundated fields that causes an uprooting of all the weeds existing in the field. Further, these weeds are buried in the soil through the process of puddling that closes the chance of their regrowth. A thin layer of standing water is an important characteristic of puddled rice fields. This layer of water inhibits the germination of weeds and also the growth of germinated weeds.

Soil solarization is usually used in vegetable crops, however, keeping in view the solar radiations received in rice growing areas (that mostly include the sub-tropical and tropical), solarization may help in considerable weed suppression in the rice crop (Jabran and Chauhan, 2015).

Handweeding, Hand Hoeing and Cutting

Hand weeding can be done both in aerobic and flooded rice while hand hoeing is convenient in aerobic rice. Hand weeding has been among the old methods of weed control in rice where farmers used to move all across the rice fields and pull the plants growing in the field other than rice. This practice has been popular in many Asian countries. Farmers with small holding can particularly benefit from this technique. However, a single spell of hand weeding may not be enough to take out all weeds from the field. The weeds will emerge again (although with a lower intensity), hence, the field may require a second, third or even fourth hand weeding. Hand hoeing also provides effective weed control in rice (Akbar et al. 2011). Like hand pulling, the single spell of hand hoeing may not be enough to control all weeds (Akbar et al. 2011). Rice requires a higher number of irrigations than other crops, hence, weeds emerge in multiple flushes, each irrigation may stimulate more weeds to germinate.

Soil Cultivation, Tillage, Modern Robots

Tillage in flooded rice aims at puddling the flooded field, while in aerobic rice, it usually aims at preparing a fine seedbed (Jabran et al. 2015c, d). Either puddling or tillage, both help in a strong suppression of weeds through their uprooting and burial. Puddling has a strong suppressive effect on rice weeds. Particularly, the well-grown, emerging and germinating weeds are uprooted and buried in the soil, hence, leaving a rare chance for re-establishment of these weeds (Hussain et al. 2018). Although impossible in flooded fields, aerobic rice provides an opportunity for tilling the soils while the rice is at its vegetative stage. Both the tractor drawn tools and the intelligent weeders may be used to control weeds in rice.

Mulching

Mulching has been included among the principal components of conservation agriculture and can provide weed suppression under field conditions (Farooq et al. 2011a). Various kinds of mulches may be applied to control weeds in rice fields (Chauhan 2012). Most important of these may include straw and plastic mulches. Mulches will block the sunlight to weeds, raise the soil temperature and put a physical pressure on germinating weeds. Such factors lead to weed suppression under field conditions. For instances, the role of plastic mulch as a weed control technique has been observed in direct-seeded rice in Pakistan (personnel observation, data not published). This may provide some positive results if the plastic mulching is done before weed emergence. Straw mulch (for example, a mulch from wheat residues) has also been effective in controlling weeds in the rice crop (Singh et al. 2007).

Allelopathy

Chemical communications among the plants (and also microorganisms many times) is called allelopathy (Farooq et al. 2011b, Jabran and Farooq 2013, Jabran et al. 2015a). Rice has been observed to excrete several phytotoxic substances (allelochemicals) that can damage the rice weeds (Jabran and Farooq 2013, Jabran 2017a, b). Importantly, allelopathy can be used to control weeds in the rice crop in several ways (Jabran 2017a,b). Most important among these may be the growing rice cultivars with the weed suppressive ability through their allelopathic potential (Masum et al. 2016, Olofsdotter et al. 2002, Olofsdotter et al. 1999, Thi et al. 2014). Another important way of utilising the phenomenon of allelopathy to control weeds in rice include the use allelopathic mulches (Wathugala and Ranagalage 2015). Other than this, use of plant aqueous extracts with an allelopathic activity has also been mentioned to hold promising results for weed control in rice crop.

Weed scientists in many of the rice growing countries are focussing on selection and breeding of rice cultivars that hold an allelopathic potential and can suppress weeds under field conditions (Farooq et al. 2008, Gealy et al. 2013, Seal and Pratley 2010, Jabran 2017a, b). A large number of allelopathic rice genotypes have been known to suppress the noxious rice weed, i.e., *E. crus-galli* (Lee et al. 2004, Mennan et al. 2011). In China, Huagan-1 and Huagan-3 were available as allelopathic cultivars, and could suppress several of the noxious rice weeds under field conditions (Kong et al. 2011). There has been efforts to determine the genes linked to the expression of allelopathy in rice (Fang et al. 2010). In conclusion, the allelopathic phenomenon holds the merit for utilization to control weeds in rice.

Chemical Weed Control

Use of herbicides becomes inevitable when rice fields are heavily infested with noxious weeds. Reliance on herbicides to control weeds has been increased after the rice production has witnessed a shift from CFR to DSR. Herbicides for weed control may be selected depending on the intensity and nature of the weeds. Usually, more than one herbicide may be required to control weeds in aerobic rice (Chauhan et al. 2015). Phenoxy and sulfonylurea were effective in suppressing the broadleaved and sedge weeds in rice grown in India (Mahajan and Chauhan 2013). Ethoxysulfuron, oxadiazon and fenoxaprop were a few among the herbicides that could be used either alone or in combination with other herbicides for controlling weeds in rice (Chauhan et al. 2015). Table 14.1 provides an overview of herbicides being used for weed control in rice. Although several herbicides may provide effective weed control in rice, in the wake of herbicide resistance evolution in weeds and a high infestation of weeds in rice crops, the weed scientists always suggest that herbicide application should be integrated with other weed control methods for witnessing effective and sustainable weed management. Reliance on herbicides as a sole method of weed control in rice may intensify the problem of herbicide resistance in weeds.

Table 14.1. Herbicides for weed control in rice

Herbicides	Dose (g a.i./ha)	Decrease in weed intensity (%)	Region	References
Pendimethalin	825	50-76	Pakistan	Jabran et al. 2012a, b
Bispyribac-sodium	25	90-94	Pakistan	Jabran et al. 2012a
		78	India	Kabdal et al. 2014
Penoxsulam	15	57-82	Pakistan	Jabran et al. 2012a, b
Penoxsulam	22.5	66	India	Kabdal et al. 2014
Butachlor	1800	74-81	Pakistan	Akbar et al. 2011
Pretilachlor	1250	87	Pakistan	Akbar et al. 2011
Pretilachlor	1000	55	India	Kabdal et al. 2014
Pyrazosulfuron	20	40	India	Kabdal et al. 2014
2,4-D-amine	-	-	Mauritania	Haefele et al. 2001
Propanil	-	-	Mauritania	Haefele et al. 2001

Integrated Weed Control

Achieving a sustainable weed control in rice is not possible if different weed control methods are not practiced in a proper combination (Chauhan et al. 2015). Under the field conditions, the farmers, in general, use multiple management methods to control weeds in rice. In many parts of the world, a manual or mechanical control method is combined with herbicide application in order to achieve satisfactory weed suppression. In the wake of problems, such as climate change and herbicide resistance evolution in weeds, it is required to educate the farming community to integrate various conventional and non-conventional weed control methods for effectively suppressing weeds. For instance, weed competitive rice cultivars or the rice cultivars possessing an allelopathic potential may be grown to suppress the weeds in rice and supplemented with other methods such as manual, mechanical and chemical weed control (Jabran and Chauhan 2015). It is particularly true for the rice crop that the weed control methods such as stale-seedbed or soil solarization may be practiced prior to the sowing/transplanting of rice followed by the use of competitive/allelopathic cultivars and an application of herbicides (Jabran and Chauhan 2015). This sequential use of various strategies will help to achieve sustainable weed control in rice. Keeping in regard the local conditions and available facilities, other cultural and physical weed control methods can be combined appropriately with herbicide application. Strategies, such as seed rate, planting density, row direction, etc., will always have a room to be adjusted with the frame of integrated weed management (IWM) in rice. Most suitable among the preventive measures may be chosen to combine with other methods of weed control being practice in rice. This means a few of the preventive strategies will always be a part of IWM in rice. Similarly, appropriate use of fertilizers will also be a part of the package of IWM all the times i.e. fertilizers should nourish the crop and not the weeds thereby applying fertilizer when the crop is free of weeds. Other than preventive measures and wise-fertilizer application, growing of a competitive/allelopathic cultivar should always be a part of IWM in rice. For cases such as DSR where the crop is heavily infested with weeds, more than one herbicide may be combined with the cultural or mechanical means of weed control. This means a pre-emergence and an early post-emergence or post-emergence herbicide will be applied in addition to cultural practices, such as stale-seedbed preparation. In the CFR production systems, flooding and puddling of the fields combined with a pre-emergence herbicide and occasionally a post-emergence herbicide as well will help to provide an effective weed control in transplanted rice. Many farmers also use to uproot some of the weeds that were left uncontrolled even after the application of these methods. This presents a good example of IWM in the CFR.

Evaluation of Weed Control Sustainability

Currently, the evolution of herbicide resistance in weeds may be considered as a major challenge to the sustainability of weed management in rice. According to Heap (2014), more than 30% herbicide-resistant weeds in rice are those that have evolved resistance against the ALS-inhibitor herbicides, while the most important of these weeds are *E. crus-galli*, *Sagittaria montevidensis* Cham. & Schltdl., *C. difformis*, and *Alisma plantago-aquatica* L. Some examples of rice weeds that have evolved resistance against the ACCase-inhibitor herbicides may include *E. crus-galli*, *C. difformis*, and *Leptochloa chinensis* (L.) Nees (Maneechote et al. 2005, Pornprom et al. 2006, Huan et al. 2013, Mennan et al. 2013).

A heavy infestation of weeds in non-conventionally grown rice (i.e., DSR or aerobic rice) is the other important challenge to the sustainability of weed management in rice. A diversity in the practiced weed control methods will be required to meet these challenges. For instance, use of IWM (as explained in the section 4.6 of this chapter) may help to suppress many of the weeds that are resistant to herbicides. Similarly, the intensity of weed infestation in DSR can be reduced if the herbicide application is integrated with the practices, such as soil solarization or stale-seedbed preparation. Occasionally, farmers may also need to practice the classical agronomic techniques, such as crop rotation. This will help to break the cycles of established weeds. DSR is taking the place of CFR in several countries—the farmers can even rotate these methods each year within their fields to disturb the weed selection and break the weed cycles. Most importantly, the weeds that were not suppressed by any of the control methods will produce seeds to increase weed infestations in the coming years. Such weeds should not be allowed to produce and propagate their seeds in the rice fields—although, this may increase the expenditures to remove these weeds from the fields.

Concluding Remarks

Evolution of herbicide resistance in weeds is an important challenge that can impact the sustainability of weed management in rice. Weed control in rice should start with quarantine before farmers care their fields and crops. Growers need to focus on multiple weed control practices and apply this in integration for achieving a sustainable weed control in rice. High weed infestations noted in dry seeded rice is the other challenge that is impacting the sustainability of weed control practices in rice production. Combining cultural, mechanical and other weed control practices (such allelopathy) with the herbicide application may help to control weeds effectively under dry-seeded rice.

REFERENCES

Akbar, N., Ehsanullah, Jabran, K. and M.A. Ali. 2011. Weed management improves yield and quality of direct seeded rice. Aus. J. Crop Sci. 5: 688.

Babatola, J. 1984. Rice nematode problems in Nigeria: their occurrence, distribution and pathogenesis. Inter. J. Pest Manag. 30: 256–265.

Bajwa, A.A., K. Jabran, M. Shahid, H.H. Ali and B.S. Chauhan. 2015. Eco-biology and management of Echinochloa crus-galli. Crop Prot. 75: 151–162.

Chauhan, B.S. 2012. Weed ecology and weed management strategies for dry-seeded rice in Asia. Weed Technol. 26: 1–13.

Chauhan, B.S., S. Ahmed, T.H. Awan, K. Jabran and S. Manalil. 2015. Integrated weed management approach to improve weed control efficiencies for sustainable rice production in dry-seeded systems. Crop Prot. 71: 19–24.

Chauhan, B.S. and D.E. Johnson. 2011. Row spacing and weed control timing affect yield of aerobic rice. Field Crops Res. 121: 226–231.

Chauhan, B.S. and D.E. Johnson. 2010. Implications of narrow crop row spacing and delayed Echinochloa colona and Echinochloa crus-galli emergence for weed growth and crop yield loss in aerobic rice. Field Crops Res. 117: 177–182.

Chauhan, B.S., A.S. Abeysekera, M.S. Wickramarathe, S.D. Kulatunga and U.B. Wickrama. 2014. Effect of rice establishment methods on weedy rice (*Oryza sativa* L.) infestation and grain yield of cultivated rice (O. sativa L.) in Sri Lanka. Crop Prot. 55: 42–49.

Dai, L., W. Dai, X. Song, B. Lu and S. Qiang 2014. A comparative study of competitiveness between different genotypes of weedy rice (*Oryza sativa* L.) and cultivated rice. Pest Manag. Sci 70: 113–122.

Dingkuhn, M., D.E. Johnson, A. Sow and A.Y. Audebert. 1999. Relationships between upland rice canopy characteristics and weed competitiveness. Field Crops Res. 61: 79–95.

Ehsanullah, N.A., K. Jabran and M. Tahir. 2007. Comparison of different planting methods for optimization of plant population of fine rice (*Oryza sativa* L.) in Punjab (Pakistan). Pak. J. Agric. Sci. 44: 597–599.

Fang, C.-X., H.-B. He, Q.-S. Wang, L. Qiu, H.-B. Wang, Y.-E. Zhuang, J. Xiong and W.-X. Lin. 2010. Genomic analysis of allelopathic response to low nitrogen and barnyard grass competition in rice (*Oryza sativa* L.). Plant Growth Regul. 61: 277–286.

FAO. 2014. Food and Agriculture Organization of United Nations. Available at: http://www.fao.org/faostat/en/#data/QC/visualize. (Accessed on January 09, 2017).

Farooq, M., K. Flower, K. Jabran, A. Wahid and K.H. Siddique. 2011a. Crop yield and weed management in rainfed conservation agriculture. Soil Til. Res. 117: 172–183.

Farooq, M., K. Jabran, Z.A. Cheema, A. Wahid and K.H. Siddique. 2011b. The role of allelopathy in agricultural pest management. Pest Manag. Sci. 67: 493–506.

Farooq, M., K. Jabran, H. Rehman and M. Hussain. 2008. Allelopathic effects of rice on seedling development in wheat, oat, barley and berseem. Allelopathy J. 22: 385–390.

Gealy, D.R., K.A. Moldenhauer and M.H. Jia. 2013. Field performance of STG06L-35-061, a new genetic resource developed from crosses between weed-suppressive Indica rice and commercial southern US long-grains. Plant Soil 370: 277–293.

Gibson, K., A. Fischer, T. Foin and J. Hill. 2002. Implications of delayed Echinochloa spp. germination and duration of competition for integrated weed management in water-seeded rice. Weed Res. 42: 351–358.

Haefele, S., D. Johnson, S. Diallo, M. Wopereis and I. Janin. 2000. Improved soil fertility and weed management is profitable for irrigated rice farmers in Sahelian West Africa. Field Crops Res. 66: 101–113.

Haefele, S., M. Wopereis, C. Donovan and J. Maubuisson. 2001. Improving the productivity and profitability of irrigated rice production in Mauritania. Europ. J. Agron. 14: 181–196.

Heap, I. 2014. Global perspective of herbicide-resistant weeds. Pest Manag. Sci. 70: 1306–1315.

Hibino, H. 1990. Insect-borne viruses of rice, pp. 209–241. *In:* Advances in Disease Vector Research. Springer.

Huan, Z., Z. Xu, D. Lv and J. Wang. 2013. Determination of AC Case Sensitivity and Gene Expression in Quizalofop–Ethyl-Resistant and Susceptible Barnyard grass (*Echinochloa crus-galli*) Biotypes. Weed Sci. 61: 537–542.

Hussain, M., S. Farooq, C. Merfield and K. Jabran. 2018. Mechanical Weed Control. *In:* Jabran, K. and Chauhan. B.S. (Eds.) Non-Chemical Weed Control. Sciencedirect, Academic Press, USA.

Jabran, K. and B.S. Chauhan. 2018. Non-Chemical Weed Control (1st Edition). Elsevier, Academic Press, USA.

Jabran, K., M. Farooq and M. Hussain. 2014. Need for integrated weed management in fine grained dry direct seeded rice. pp. 556–562. *In:* Proceedings of the Fifth International Scientific Agricultural Symposium. "Agrosym 2014", Jahorina, Bosnia and Herzegovina, October 23–26, 2014. University of East Sarajevo, Faculty of Agriculture.

Jabran, K. 2017a. Manipulation of Allelopathic Crops for Weed Control. Springer International Publishing AG, Gewerbestrasse 11, 6330 Cham, Switzerland. doi: 10.1007/978-3-319-53186-1

Jabran, K. 2017b. Rice Allelopathy for Weed Control. *In:* Manipulation of Allelopathic Crops for Weed Control. Springer International Publishing AG, Gewerbestrasse 11, 6330 Cham, Switzerland. doi: 10.1007/978-3-319-53186-1_5.

Jabran, K. and B.S. Chauhan. 2015. Weed management in aerobic rice systems. Crop Prot. 78: 151–163.

Jabran, K. and M. Farooq. 2013. Implications of potential allelopathic crops in agricultural systems. pp. 349–385. *In:* Allelopathy. Springer, Berlin Heidelberg.

Jabran, K., M. Farooq, M. Hussain, M. Khan, M. Shahid and L. DongJin. 2012a. Efficient weeds control with penoxsulam application ensures higher productivity and economic returns of direct seeded rice. Inter. J. Agric. Biol. 14: 901–907.

Jabran, K., M. Hussain, S. Fahad, M. Farooq, A.A. Bajwa, H. Alharrby and W. Nasim. 2016. Economic assessment of different mulches in conventional and water-saving rice production systems. Environ. Sci. Pol. Res. 23: 9156–9163.

Jabran, K., M. Hussain, M. Farooq, M. Babar, M.N. Doğan and D.J. Lee. 2012b. Application of bispyribac-sodium provides effective weed control in direct-planted rice on a sandy loam soil. Weed Biol. Manag. 12: 136–145.

Jabran, K., G. Mahajan, V. Sardana and B.S. Chauhan. 2015a. Allelopathy for weed control in agricultural systems. Crop Prot. 72: 57–65.

Jabran, K., E. Ullah and N. Akbar. 2015b. Mulching Improves crop growth, grain length, head rice and milling recovery of basmati rice grown in water-saving production systems. Inter. J. Agric. Biol. 17: 920–928.

Jabran, K., E. Ullah, M. Hussain, M. Farooq, N. Haider and B.S. Chauhan. 2015c. Water saving, water productivity and yield outputs of fine-grain rice cultivars under conventional and water-saving rice production systems. Exp. Agric. 51: 567–581.

Jabran, K., E. Ullah, M. Hussain, M. Farooq, U. Zaman, M. Yaseen and B.S. Chauhan. 2015d. Mulching improves water productivity, yield and quality of fine rice under water-saving rice production systems. J. Agron. Crop Sci. 201: 389–400.

Kabdal, P., T. Pratap, V. Singh, R. Singh and S. Singh. 2014. Control of complex weed flora in transplanted rice with herbicide mixture. Indian J. Weed Sci. 46: 377–379.

Kong, C.H., X.H. Chen, F. Hu and S.Z. Zhang. 2011. Breeding of commercially acceptable allelopathic rice cultivars in China. Pest Manag. Sci. 67: 1100–1106.

Kraehmer, H., K. Jabran, H. Mennan and B.S. Chauhan. 2016. Global distribution of rice weeds – a review. Crop Prot. 80: 73–86.

Lee, S., Y. Ku, K. Kim, S. Hahn and I. Chung. 2004. Allelopathic potential of rice germplasm against barnyard grass. Allelopathy J. 13: 17–28.

Mahajan, G. and J. Timsina. 2011. Effect of nitrogen rates and weed control methods on weeds abundance and yield of direct-seeded rice. Arch. Agron. Soil Sci. 57: 239–250.

Mahajan, G. and B.S. Chauhan. 2013. Herbicide options for weed control in dry-seeded aromatic rice in India. Weed Technol. 27: 682–689.

Maneechote, C., S. Samanwong, Zhang Xiao Qi and S.B. Powles. 2005. Resistance to ACCase-inhibiting herbicides in sprangletop (Leptochloa chinensis). Weed Sci. 53: 290–295.

Masum, S.M., M.A. Hossain, H. Akamine, J.I. Sakagami and P.C. Bhowmik. 2016. Allelopathic potential of indigenous Bangladeshi rice varieties. Weed Biol.Manag. 16: 119–131.

Mennan, H., M. Ngouajio, M. Sahín and D. Isik. 2011. Allelopathic potentials of rice (*Oryza sativa* L.) cultivars leaves, straw and hull extracts on seed germination of barnyard grass (*Echinochloa crus-galli* L.). Allelopathy J. 28: 167–178.

Mennan, H., M. Ngouajio, M. Sahin, D. Isik and E.K. Altop. 2012. Competitiveness of rice (*Oryza sativa* L.) cultivars against *Echinochloa crus-galli* (L.) Beauv. in water-seeded production systems. Crop Prot. 41: 1–9.

Mennan H., E. Kaya-Altop, S. Rasa, J.C. Streibig, D. Yatmaz, U. Budak, D. Sariaslan. 2013. Resistance to ACCase and ALS inhibiting herbicides in cereals in Turkey, What have we learned? EWRS 16th SYMPOSIUM, Samsun, Turkey.

Moody, K. 1981. Weed-fertiliser interactions in rice. International Rice Research Institute, Los Banos, Laguna,Philippines.

Oerke, E.-C. 2006. Crop losses to pests. J. Agric. Sci. 144: 31–43.

Olofsdotter, M., L.B. Jensen and B. Courtois. 2002. Improving crop competitive ability using allelopathy—an example from rice. Plant Breeding 121: 1–9.

Olofsdotter, M., D. Navarez, M. Rebulanan and J. Streibig. 1999. Weed suppressing rice cultivars – does allelopathy play a role? Weed Res. 39: 441–454.

Olofsdotter, M., B.E. Valverde and K.H. Valverde. 2000. Herbicide resistant rice (*Oryza sativa* L.): global implications for weedy rice and weed management. Annals Appl. Biol. 137: 279–295.

Peerzada, A.M., A.A. Bajwa, H.H. Ali and B.S. Chauhan. 2016. Biology, impact, and management of Echinochloa colona (L.) Link. Crop Prot. 83: 56–66.

Pornprom, T., P. Mahatamnuchoke and K. Usui. 2006. The role of altered acetyl-CoA carboxylase in conferring resistance to fenoxaprop-P-ethyl in Chinese sprangletop (Leptochloa chinensis (L.) Nees). Pest Manag. Sci. 62: 1109–1115.

Rao, A., D. Johnson, B. Sivaprasad, J. Ladha and A. Mortimer. 2007. Weed management in direct-seeded rice. Adv. Agron. 93: 153–255.

Sardana, V., G. Mahajan, K. Jabran and B.S. Chauhan. 2016. Role of competition in managing weeds: an introduction to the special issue. Crop Prot. 95: 1–7.

Seal, A. and J. Pratley. 2010. The specificity of allelopathy in rice (Oryza sativa). Weed Res. 50: 303–311.

Singh, S., J. Ladha, R. Gupta, L. Bhushan, A. Rao, B. Sivaprasad and P. Singh. 2007. Evaluation of mulching, intercropping with Sesbania and herbicide use for weed management in dry-seeded rice (Oryza sativa L.). Crop Prot. 26: 518–524.

Thi, H.L., C.H. Lin, R.J. Smeda and F.B. Fritschi. 2014. Isolation and purification of growth-inhibitors from Vietnamese rice cultivars. Weed Biol. Manag. 14: 221–231.

Ullah, E., A. Ur-Rehman, Q. Arshad and S. Shah. 2009. Yield response of fine rice to NP fertiliser and weed management practices. Pak. J. Bot. 41: 1351–1357.

Wathugala, D. and A. Ranagalage. 2015. Effect of incorporating the residues of Sri Lankan improved rice (*Oryza sativa* L.) varieties on germination and growth of barnyard grass (*Echinochloa crus-galli*). J. Nat. Sci. Found. Sri Lanka 43: 57–64.

CHAPTER

15

Sustainable Weed Control in Soybean

Nicholas E. Korres*[1], Krishna N. Reddy[2], Christopher Rouse[1] and Andy C. King[1]

[1] Crop, Soil and Environmental Sciences, University of Arkansas, Fayetteville, 72704 AR, USA
[2] USDA-ARS, Crop Production Systems Research Unit, PO Box 350, Stoneville, MS 38776, USA

Importance of Soybean

Soybean (*Glycine max* [L.] Merr.) is one of the most important food crops globally (Datta et al. 2017). This crop can be classified as leguminous, oil-seed, vegetable or even fuel source, depending upon its usage. A series of traits have typified the plant as one of the most attractive crops globally (Grau 2005). Soybean exhibits low water content (12–14%), high nutritive value and the final product can be used as a source for the production of a variety of derivatives, i.e., for human food, animal feed, oil or industrial products (Anonymous 2017). This has resulted in an increasing demand for soybean worldwide with subsequent large investments in research and development and the widespread use of relatively newly developed soybean transgenic cultivars. In the US, 94% of the total soybean acreage in 2016 was planted with transgenic herbicide-resistant cultivars (NASS 2017). These transgenic cultivars were resistant to glyphosate and glufosinate, 5-enolpryuvylshikimate-3-phosphate synthase (EPSPS) and glutamine synthase inhibitor type herbicides. Advantages of the transgenic cultivars include increased yields, reduced costs due to reduced herbicide use, and increased range of appropriate planting environments (Kaimowitz and Smith 2001).

Though originating in Asia, seven of the top ten soybean producers today are found in the Americas (Figure 15.1) (FAO 2017). The contribution of soybean to these economies is significant. In 2015, for example, 33,811,817 ha of soybeans were harvested in the United States (NASS 2015) with a national average yield of approximately 3 tons/ha and total value of $40.9 billion (Anonymous 2017).

In 2013, soybean exports earned $620 million for Bolivia, over $1 billion for Canada, $1.89 billion for Uruguay, $2.41 billion for Paraguay, $2.7 billion for India, $10.7 billion for Argentina and $23 billion for Brazil (FAO 2017).

Weeds and Weed Impact on Soybean

Infestations of crop weeds are a universal annual threat to productivity, especially in the major field crops including soybean (Walsh et al. 2013). Prolonged weed interference can significantly reduce soybean yield and yield quality, thus, early-season weed management is required for the

---

*Corresponding author: nkorres@yahoo.co.uk; korres@uark.edu

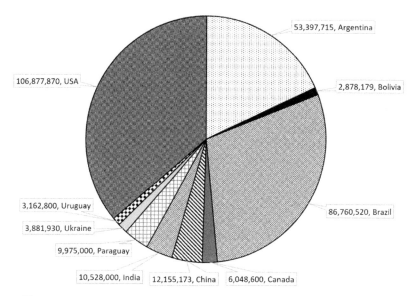

Figure 15.1. Tonnage soybean production of the top ten soybean producer countries in 2014 (FAO 2017).

achievement of economically acceptable yields (Knezevic et al. 2003, Hock et al. 2005, Korres and Norsworthy 2015a). Weed species with differing competitive ability commonly found in soybean include *Amaranthus palmeri* S. Wats. (Palmer amaranth), *Amaranthus retroflexus* L. (redroot pigweed), *Amaranthus tuberculatus* Moq. Sauer (tall waterhemp), *Chenopodium album* L. (common lambsquarters), *Ambrosia* spp. (ragweed), *Conyza* spp. (horseweed), *Xanthium strumarium* L. (cocklebur), *Abutilon theophrasti* Medik. (velvetleaf), *Ipomoea* spp. (morning-glories), *Senna obtusifolia* L.H.S. Irwin & Barneby (sicklepod), *Sida spinosa* L. (prickly sida), *Sorghum halepense* L. (johnsongrass), *Digitaria sanguinalis* (L.) Scop. (large crabgrass), *Urochloa platyphylla* (Nash) R.D. Webster (broadleaf signalgrass), *Eleusine indica* (L.) Gaertn. (goosegrass), and *Lolium* spp. (ryegrass) (Johnson et al. 1998, Zimdahl 2004, Korres et al. 2015a, b).

Potential yield loss resulting from weed infestations is estimated to be 13 to 27% across the US, and up to 9% in Canada (Chandler et al. 1984). Weed competition in soybean can reduce yield from 8 to 55% in the US, and 10% in Ontario, Canada (Van Acker et al. 1993, Swanton et al. 1993). Intervention strategies using best management practices (BMP) have the potential to successfully mitigate weed competition but may result in unacceptable levels of yield loss. For soybean, a reported 2 to 20% of yield loss may occur when BMP with an herbicide component are used, but, as much as 15 to 65% yield loss is expected in corn when the herbicide is excluded (Bridges 1992). The resulting yield loss directly impacts the production value of the crop with an estimated annual loss in value of $16 billion across the US states, ranging from $18.7 million in Delaware in the Northeast to $2.7 billion in Illinois in the Midwest (Soltani et al. 2017).

Non-chemical Weed Control

In recent years, non-chemical weed management approaches have gained a renewed interest due to public awareness of health issues, environmental pollution concerns and food production cost (Korres 2017). It has been estimated that U.S. farmers spend over $3.5 billion annually on chemical weed control and over $2.5 billion for non-chemical weed control (Cahoon et al. 2016). Using 2003 dollar values, the loss in food and fiber without the use of herbicides and the likely substitution for alternatives (for example, non-chemical control methods) is worth $13.3 billion. The following sections discuss various non-chemical weed control methods in soybean in an attempt to highlight their advantages and disadvantages.

Mechanical Weed Control

Mechanical weed control may involve weeding the whole crop, or it may be limited to selective inter-row weeding through the use of implements that have been designed to control weeds within the crop row by directing soil along the crop row to cover small weeds (Klooster 1982). A considerable diversity of mechanical weeders exists ranging from basic hand tools to sophisticated tractor driven devices including cultivating tools (e.g., hoes, harrows, tines and brush weeders), cutting tools (e.g., mowers and strimmers) and dual purpose implements like thistle-bars that operate either as cultivating or as cutting tools (Bond and Grundy 2001). As stated by Kunz et al. (2015) mechanical weed control and particularly weed hoeing is a promising alternative to chemical weed control which can be applied between row (inter-row hoeing) and within crop rows (intra-row hoeing). For soybean in Europe, in-row hoeing occupied around 20% of production fields whereas intra-row hoeing is used less than 1% (Kunz et al. 2015). Hoeing can control both larger weeds and grass-weeds, which are difficult to remove by flexible tine harrows and the risk of crop damage for inter-row hoeing is usually lower compared to harrowing (Lotjonen and Mikkola 2000, Rasmussen and Ascard 1995). In general, the effects of mechanical weed control strongly depend on soil conditions, weed species, growth stage of weed species, and also mechanical methods itself. Highest efficacies, for example, were achieved when crops at the time of hoeing were taller than the weeds (Bowman 1997, Van der Weide et al. 2008). In addition, Korres et al. (2015b) found that mowing exerts no effects on naturally occurring weed flora at roadsides in Eastern Arkansas Mississippi River Delta area and suggested as an alternative, and economically feasible solution, the use of herbicide control where appropriate.

Weber et al. (2016) reported a high weed control efficiency with preemergence (PRE) herbicides and a combination of PRE fb post-emergence (POST) herbicide applications. On the contrary, harrowing resulted in less efficient weed control as it differed among locations and years. However, weeds were highly controlled in the inter-row area by hoeing, although its efficiency was reduced compared to hand weeding. Mechanized weed control by cultivators pulled either by animals or tractors, is widely accepted in various agricultural economies including Brazilian agriculture (Silva et al. 2007), being one of the main methods of weed control especially on small planted areas. The main limitations of this method are the difficulty of controlling weeds in the crop rows, low efficiency when performed in wet conditions and it is also inefficient to control weeds that reproduce by vegetative parts (Silva et al. 2007).

Cover Crops

Cover crops, crops grown between cropping seasons in arable farming systems or between row middles of orchard trees, are an important multipurpose agronomic tool that enhance plant growth by improving soil chemical, biological, and physical properties in various cropping systems (Alberts and Neibling 1994, Dabney et al. 2001, Korres 2005, Price and Norsworthy, 2013). In addition, the integration of cover crops into cropping systems has proved to be an effective management strategy for weed suppression (Liebl et al. 1992, Skroch et al. 1992, Chauhan and Abugho 2013, Korres 2017) through weed biomass reduction (Korres and Norsworthy, 2015), and consequently, reductions in weed reproductive capacity (Korres and Froud-Williams 2002). Crops, such as *Vicia villosa* Roth (hairy vetch) or *Secale cereale* L. (winter rye) can provide uniform and dense ground cover when properly managed, while crops like *Coronilla varia* L. (crown vetch) can provide long-term soil management (Korres 2005). Other crops that could be used as cover crops include *Trifolium incarnatum* L. (crimson clover), *Trifolium pratense* L. (red clover), *Trifolium repens* L. (white clover), *Pisum* spp. (dry peas), *Coronilla varia, Lotus corniculatus* L. (bird's-foot trefoil), *Avena sativa* L. (oat), *Lolium* spp. (ryegrass), *Festuca* spp. (fescues), *Poa* spp. (bluegrass), *Bromus inermis Leyss.* (smooth brome), *Phleum pratense* L. (Timothy grass) and *Dactylis glomerata* L. (cock's-foot) (Korres 2005). Winter cereals offer many benefits as cover crops because they produce high amounts of biomass, are easily established, easily terminated and they provide excellent groundcover during winter period (Brown et al. 1985, Schomberg et al. 2006). Winter rye (*S. cereale*), for example, decreased weed seed germination and delayed weed seedling emergence as a result of a dense ground cover due to high amounts of biomass produced (Schomberg et al. 2006). This results in high weed seedling physical suppression and reduced light transmission to the soil surface (Akemo et al. 2000, Teasdale and Mohler 2000). Cover crops can also buffer the temperature fluctuations that some weeds use as a signal for germination in the spring (Teasdale and Mohler 1993) because the mulch created by the cover crop blocks light transmittance to the soil surface and also maintains soil moisture.

Cover crops like *S. cereale* produce allelopathic compounds, for example, benzoxazinoid compounds such as 2,4-dihydroxy-1,4, (2H) - benzoxazin-3-one, and benzoxazolin-2(3H)-one (Barnes and Putnam, 1987, Barnes et al. 1987, Chase et al. 1991, Schulz et al. 2013), which mainly affect the germination and growth of small seeded weeds (Hartwig and Ammon 2002, Kruidhof et al. 2011).

In soybean cropping systems, the use of cover crops has been proved a useful management approach for weed control. A *S. cereale* cover crop provided 90% control of weeds whereas the application of herbicides into the soybean-cover crop system did not increase crop yield significantly (Liebl et al. 1992). Robinson and Dunham (1954) found that soybean yields were increased and weeds suppressed when wheat or rye was used as cover crop. On the contrary, *Medicago sativa* (alfalfa), *Vicia* spp., (vetch), *Trifolium* spp. (clovers), *Bromus* spp. (bromegrass) and *Phleum pratense* (timothy grass) exhibited unsatisfactory weed control and *Pisum sativum* (pea) caused lodging of the soybean crop. Norsworthy et al. (2016), investigating various at harvest and autumn management systems for the suppression of glyphosate resistant *Amaranthus palmeri*, found *S. cereale* as cover crop one of the most effective methods to suppress the weed population and seed production.

Winter cover crops, aside from numerous benefits on soil, can also be useful tools to suppress or replace winter annual weed species (Reddy 2001). The long growing season in the lower Mississippi River Delta region permits the use of winter cover crops in row crop production (Reddy 2001). Cover crop residues provide early-season suppression of certain weeds, but do not provide full-season weed control (Koger et al. 2002, Reddy 2001, 2003). Thus cover crops can eliminate pre-emergence herbicide and late-season weeds can be managed with post-emergence herbicides on an as-needed basis. However, it should be noted that in cover crop systems, input costs are often higher because of the additional cost of seed, planting, and cover crop desiccation. For example, the additional cost resulted in a lower net return with the rye cover crop ($29/ha) compared with the no-cover crop ($84/ha) system, even though soybean yield in the rye cover crop system was comparable to that from the no-cover crop system (Reddy 2003). Nevertheless, a rye cover crop-based soybean production could be a desirable agricultural system for those producers who need ground cover on highly erodible land to prevent soil erosion, reduce nutrient and pesticide movement, reduce herbicide selection pressure, and augment sequestration of atmospheric CO_2 into soil.

Cultural Weed Control

Cultural practices, such as seeding rate, row spacing, competitive cultivars, fertilization management and others can significantly impact weed control (Anderson 1996, Grichar et al. 2004, O'Donovan et al. 2001, Korres 2017). The evolution of many weed species with resistance to multiple sites of action will require producers to diversify their production systems and integrate cultural and herbicidal control methods (Bradley 2013, Heap 2017, Norsworthy et al. 2012, Schultz et al. 2015).

Seeding Rate

Crop competitiveness against weeds can be enhanced by increasing crop density, an easily manipulative cultural practice (Buehring et al. 2002, Korres and Froud-Williams 2002). The canopy closure at high crop densities is accelerated causing reduction in the amount of light transmitted to the soil surface and to the weeds growing beneath the crop canopy (Korres and Norsworthy 2017). This results in decline of weed population, lower weed biomass and seed production (Korres and Norsworthy 2017). Harder et al. (2007) reported that soybean densities of 124,000 to 198,000 plants ha[-1] exerted no significant differences on weed biomass at various row widths. According to the same authors, soybean densities of 300,000 to 445,000 plants ha[-1] were significantly suppressive to weed biomass especially at narrow row widths. Korres and Norsworthy (2015b) showed that increased seed rates in drill-seeded soybean from 125,000 to 400,000 seeds/ha resulted in decrease of *Amaranthus palmeri* biomass and seed production by 3 and 10-fold, respectively, compared to these observed in the absence of crop competition (Figure 15.2).

Buehring et al. (2002) found that a soybean population of approximately 690,000 plants ha[-1] resulted in 92% control of *Senna obtusifolia* (sicklepod) compared to 29% control when crop was established at densities of 270,000 plants ha[-1]. Norsworthy and Oliver (2002) observed greater *Sesbania exaltata* (bigpod sesbania) biomass reduction with soybean populations of 520,000 plants ha[-1] compared to 217,000 and 370,000 plants ha[-1]. Nevertheless, compensatory effects in combination with environmental conditions (Benbella and Paulsen 1998) impose a wide range of crop yield responses to crop density manipulation. Korres and Norsworthy (2015b), investigating the effects of three soybean seeding rates (i.e., 125,000, 250,000 and 400,000 seeds/ha) in a drill-seeded soybean cropping system, found no significant yield increases between 250,000 and 400,000 seeds/ha. Norsworthy and Oliver (2001) stated that the seed cost associated with high seeding rates (i.e. >450,000 plants ha[-1]) can exceed the benefit for higher weed control.

Row Spacing

The adoption of narrow-row spacing (i.e., 19 or 38 cm in narrow-row spacing as opposed to 76 or 92 cm in wide-row) has primarily been driven by the potential for higher yields in the narrow-compared to the wide-row production systems (Bradley 2006). Moreover, narrow-row spacing can have a significant impact on weed populations and enhance the efficiency of weed management systems. This is due to early season space capture within and between the rows (Harder et al. 2007), hence rapid canopy closure compared to wide-row soybean, which results in reduced light amount that reaches the soil surface (Puricelli et al. 2003, Steckel and Sprague 2004) (Figure 15.3).

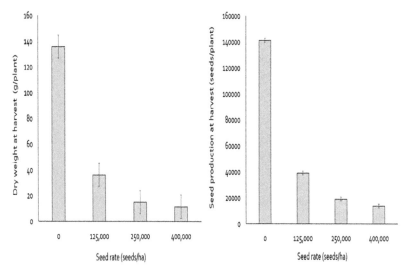

Figure 15.2. Effects of drill-seeded soybean seed rate on Palmer amaranth dry weight (left) and seed production (right). Vertical bars represent the standard error of the mean (Adapted from Korres and Norsworthy 2015b).

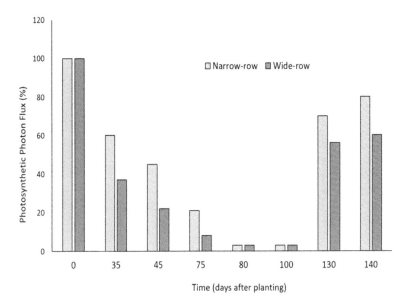

Figure 15.3. Percentage radiation interception [measured as photosynthetic photon flux (PPF)] in relation to days after planting in narrow- and wide-row soybean at soil surface (based on Puricelli et al. 2003).

The reductions in light penetration and time to canopy closure have a profound influence on the likelihood of weed emergence later in the growing season (Yelverton and Coble 1991). In addition, the critical period of weed control in soybean can be affected by narrow-row spacing. The critical period of weed control is an interval of time in the growth of a crop during which it is essential to control weeds in order to prevent unacceptable yield losses (Knezevic et al. 2002). This indicates the significance of manipulating soybean row spacing since planting in wide rows will require implementation of weed removal practices much earlier than in narrow rows. Narrow-row spacing can also influence light quality by altering the red: far red light that reaches to the soil surface. As reported by Graming and Stoltenberg (2009), *Chenopodium album* L. (common lambsquarters) responded to light quality alterations through elongation of the main stem, reduced leaf area and seed production.

Korres and Norsworthy (2015b) reported that inter-row distance in wide-row soybean affected *A. palmeri* height, dry weight and, consequently, seed production. The greater the distance from the crop, the lesser the competition effects on *A. palmeri*, which resulted in higher Palmer amaranth biomass and subsequent seed production (Figure 15.4) indicating the importance of row spacing manipulation for the control of this weed.

Harder et al. (2007) reported reductions in weed density and biomass production following an effective POST herbicide application in 19 compared to 76 cm soybean rows. Buehring et al. (2002) observed 29% higher control of *Senna obtusifolia* (sicklepod) in 19 cm compared to 76 cm soybean row whereas Steckel and Sprague (2004) reported a 57% *Amaranthus tuberculatus* (waterhemp) biomass reduction in 19 cm in comparison with 76 cm soybean row at V2 to V3 soybean growth stage. However, Johnson et al. (1998) found that *Setaria faberi* (giantfoxtail) was able to successfully complete its life cycle in narrow row cropping systems.

Sowing Date

Sowing date can influence the type and the degree of weed infestation along with the composition of the weed flora during the growing season. Vidotto et al. (2016), for example, found higher infestation levels of *Chenopodium album* (common lambsquarters), *Abutilon theophrasti (velvetleaf)*, and *Falopia convolvulus* (black-bindweed**)** at early compared to

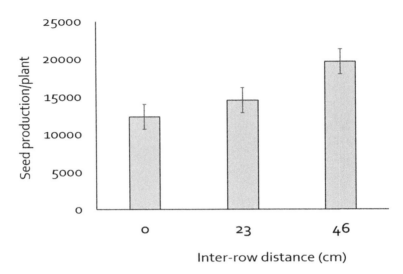

Figure 15.4 Effects of inter-row distance from the crop on *Amaranthus palmeri* seed production in wide-row soybean. Vertical bars represent the standard error of the mean (adapted from Korres and Norsworthy 2015).

conventional sowing time in maize. They also stated that early sowing time resulted in lower frequency of *Panicum dichotomiflorum* (fall panicgrass), *Sorghum halepense* (Johnsongrass), *Amaranthus retroflexus* and *Portulaca oleracea* (common purslane). The early soybean production system (ESPS) which entails early planting of short-season varieties (Heatherly and Spurlock 1999) has become the normal soybean production practice in the Mid-South US. These earlier-maturing varieties would enter and conclude critical reproductive stages before the onset of non-optimal soil moisture, thus enhancing the possibility of increased production. Furthermore, the use of early-maturing cultivars planted at higher crop densities for earlier canopy closure, hence greater light interception (Seversike and Purcell 2006), is a cultural method that could possibly enhance crop competitiveness. However, Bennet and Shaw (2000) reported that late-maturing soybean cultivars depressed weed seed production and seed weight of both *Ipomoea lacunosa* L. (pitted morningglory) and *Sesbania exaltata* presumably due to their ability to maintain vegetative growth longer (Nordby et al. 2002). Crops sown at the optimum time with adequate soil moisture, particularly in dry areas, and temperature will always be more vigorous and suppressive against weeds than those sown in less optimum conditions. If not carefully managed, early sowing might pose the risk of cool and wet soils, frost and a greater possibility of plant disease and herbicide injury. Weed species ecology in conjunction with the potential severity of weed infestation may be used as a guide for the determination of the most appropriate time for sowing knowing that early-emerging weeds usually interfere with crops more than late-emerging weeds (Korres 2005).

Use of Competitive Cultivars

The response of a crop to weed competition can be considered as a) the ability of the crop to tolerate weed competition (i.e., the ability of the crop to maintain high yield under weedy conditions) or b) the ability of the crop to suppress the growth of weeds (i.e., it is usually determined by comparing different biological characteristics in mixtures to that in pure stands) (Callaway 1992, Korres and Froud-Williams 2004, Andrews et al. 2015).

Crop tolerance to weed competition varies widely over seasons and locations (Cousens and Mokhtari 1998, Olesen et al. 2004), hence, the ability of a crop to suppress weeds is considered in this chapter. It has been stated by various authors that developing crop cultivars with an enhanced ability to suppress weeds would be a sustainable contribution towards improved weed management including soybean cropping systems (Bennet and Shaw 2000, Vollmann et al. 2010). Despite possible benefits, the selection of competitive crops has been explored to a limited extent as a late season opportunity for weed control. Competitive soybeans were studied in the late 1990's and there were a few soybean varieties with weed-suppressive traits. Traits that convey crop competitiveness include increased leaf area, height, leaf area expansion rate and plant canopy (Pester et al. 1999, Bussan et al. 1997). These traits mainly were associated with the manipulation of light interception by crop canopy during the growing season. In addition, cultivar maturity has also been studied as a potential characteristic that could enhance the competitive advantage in soybean. Nevertheless, it is still unclear whether early or late maturing soybeans are more competitive with weeds (Yelverton and Coble 1991, Vollmann et al. 2010, Nordby et al. 2007).

Crop Rotations

Continuous cultivation of a single crop or crops having similar management practices allows weed species to become dominant in the system, and over time, these weed species become hard to control (Chauhan et al. 2012). One of the main cultural practices is crop rotation, the benefits of which depend on the selection of crops and their sequence in the rotation system. If the main goal, for example, is weed control through crop rotation, the choice of the rotating crop should be based on crops with contrasting growth habits and cultural needs (Korres 2005, 2017).

According to Silva et al. (2013) crop rotation disturbs the life cycle of weeds by preventing them from becoming established and dominant. The interference of an established weed flora with the crop is significantly increased if the same cultural techniques are applied in sequence for long time periods in the same field (Silva et al. 2013). Kelley et al. (2003) reported that soybean production was improved by using crop rotation as a management practice. Other studies have found yield decreases when soybean was grown continuously in monoculture than when rotated with another crop (Crookston et al. 1991, Meese et al. 1991, West et al. 1996).

Fertiliser Management

Nitrogen (N), phosphorus (P), and potassium (K), among other nutrients, are the most influential macronutrients for plant growth, development, and establishment. Korres et al. (2017a), for example, reported the importance of total soil N and extractable soil phosphorus on the occurrence of *A. palmeri* at field margins in the Mississippi River Delta area in eastern Arkansas. The growth of weeds under increasing rates of nitrogen is species dependent (Blackshaw et al. 2003) and amaranths have been found to thrive in nitrogen-enriched environments (Korres et al. 2017a, b). Nitrogen accumulation and growth, for example, was greater in redroot pigweed compared with the crop when grown in a nitrogen-enriched environment (Teyker et al. 1991). The importance of fertilizer application timing on weed growth has been discussed by Liebman and Davis (2000) with respect to the ability of the weed to absorb nutrients earlier and more rapidly than the crop, and suggested that delayed fertilization is likely to be most useful for managing small-seeded weeds, e.g. *A. palmeri* in large-seeded crops, such as soybean, hence, fertilizer management, particularly for K and P in soybean could be planned accordingly. Alternatively, the use of nitrogen-demanding winter cereal cover crops (Dabney et al. 2001, Kaspar et al. 2008) prior to soybean planting can be an invaluable tool in integrated *A. palmeri* control methods.

Preventive Measures

Preventing weed seed inputs to the soil seedbank is an effective means of reducing the impact of weeds on subsequent crops while prolonging the efficacy of herbicide-based weed management programs (Walsh et al. 2013).

Intensive tillage, stubble burning, crop desiccation, windrowing, weed seed collection at harvest using chaff carts or baling systems, cover crops, and herbicide programs are some of the methods used for preventing the influx of weed seeds back into the soil seedbank (Chauhan and Abugho 2013, Devenish and Leaver 2000, Fogelfors 1982, Storrie 2014, Walsh and Powles 2004).

The at-harvest management of weed seed bearing crop residues has been reported as an effective method in reducing seed return into the weed seedbank as in the case of *Lolium* spp. (ryegrass), *Raphanus raphanistrum* L. (wild radish), *Avena fatua* (wild oat), *Bromus* spp. (bromegrass) (Walsh et al. 2013, Shirtliffe and Entz 2005, Walsh and Powles 2004, Walsh and Newman 2007). A recent work by Norsworthy et al. (2016) on integration of herbicide programs with crop residues management (i.e., crop residues removal or narrow windrow burning or incorporation of crop residues into the soil during bed formation) or the use of cover crops in wide-row soybean highlighted the effectiveness of these preventive approaches for controlling the population density and seed production of glyphosate resistant *A. palmeri.*

Chemical Weed Control

Prior to the initial introduction of glyphosate-resistant (GR) cultivars, weeds in soybean were controlled with herbicides applied as burndown (i.e., before planting of the crop), pre-emergence (PRE) or post-emergence (POST). In general, prior to GR soybean, weed control programs were dominated by imidazolinones and dinitroaniline herbicides (Cantwell et al. 1989, Givens et al. 2009), such as imazaquin, imazethapyr, pendimethalin, and trifluralin. An herbicide program typically utilised a pre-emergence application for grass control which was usually followed by (fb) a two-pass of acetolactate synthase (ALS) inhibitors applied POST. This has proved to be an efficient weed control herbicide program for most weed species except for large-seeded weeds including common cocklebur (*X. strumarium*), morningglories (*Ipomoea* spp.) and sicklepod (*S. obtusifolia*). Weeds with high rates of reproduction, such as *Amaranthus* spp., were also difficult to control (Price et al. 2011). Tillage was included for managing weeds because of its appreciable reductions of weed biomass between the rows (Snipes and Muller 1992) and the control of biennial weeds (Brown and Whitwell 1988). Nevertheless, as reported by Weber et al. (2016) and Gehring et al. (2014), the use of PRE fb POST herbicides in European cropping systems including soybean lead to significant control of a diversified weed flora including *Matricaria chamomilla* and *Avena fatua* particularly when metribuzin was included into the program.

The adoption of GR soybean, which has enabled growers to reduce tillage while achieving season-long weed control through multiple applications (Wilcut et al. 1995), was rapid (Dill et al. 2008). By 2005, for example, 87% of the total U.S. soybean acreage included the GR varieties (Fernandez-Cronejo and Caswell 2006). As stated by Riar et al. (2013), the release and subsequence widespread adoption of GR soybeans have resulted in the decreased use of residual herbicides and led to the substitution of commonly used postemergence products, such these mentioned earlier, with glyphosate (Young 2006).

Herbicides, other than glyphosate, used in the autumn regardless of crop rotation are 2,4-D, chlorimuron, dicamba, paraquat or flumioxazin (Givens et al. 2009). These herbicides are often applied in the autumn to control weeds that would otherwise be difficult to manage in the spring, prior to soybean planting (Wicks et al. 2000).

The majority of growers in the US use a spring pre-plant burndown application, mainly glyphosate, 2,4-D, although paraquat can be also used in soybean rotations with corn or cotton (Givens et al. 2009). According to Scott (2011) a tank-mix of paraquat plus flumioxazin or glyphosate plus flumioxazin is a recommended strategy for herbicide resistance reduction.

Soil-applied residual or PRE-herbicides are an effective weed management tool for controlling *A. palmeri* and many other weeds early in the cropping season, allowing for the inclusion of an additional herbicide mode of action to reduce glyphosate-resistance pressure before crop canopy formation (Bell et al. 2015). Pre-emergence herbicides can also delay the time of postemergence herbicide application, which allows the postemergence herbicide to manage weeds that have delayed-emergence and/or multiple germination flushes during the growing season. Studies show incorporation of a residual herbicide with glyphosate can manage glyphosate resistant weed populations by reducing weed population, hence seed production and, consequently, additional inputs in soil seedbank (Benbrook 2012). Whitaker et al. (2010) reported that in a conventional soybean production system, a PRE-application of *S*-metolachlor or pendimethalin, in addition to either flumioxazin, fomesafen, or metribuzin plus chlorimuron, increased control of *A. palmeri* by 27, 29, and 22%, respectively, when the first POST herbicide application was applied to 10 to 15 cm tall *A. palmeri*, compared to the non-treated control.

After crop planting, an early POST application can enhance weed control in addition to any PRE-applied residual herbicide. Tank-mixtures, such as glyphosate + S-metolachlor or glufosinate + S-metolachlor could be an option for mid-season weed control in glyphosate resistant soybeans or in LibertyLink® soybeans (i.e., glufosinate resistant soybean). Another option for small (<5 cm) or recently emerged weeds (10 to 14 days) is fomesafen or lactofen, which is an effective POST herbicide up to canopy row closure (Scott et al. 2015).

In general, the POST herbicide applications in glyphosate resistant soybean include ideally two (Gonzini et al. 1999, Swanton et al. 2000) or more glyphosate applications in a continuous glyphosate resistant soybean or soybean rotational systems including cotton. Nevertheless, the number of glyphosate applications depends on soybean row width, planting date, soybean maturity period, duration of crop growth. In addition, a relatively lower percentage of growers use non-glyphosate herbicides prior to GR soybean rotations such as diuron, fluometuron, pendimethalin, s-metolachlor and trifluralin (Young 2006).

Applications of glufosinate are similar to that of glyphosate. The use of two applications are more effective than one when the product is applied at the recommended rate. The first application is recommended to be made 7–10 days after soybean emergence, when weeds are 5 to 7.5 cm tall, followed by a second application 10–14 days later (Scott et al. 2015). Applications of glufosinate are not recommended after reproductive stage 1 (R1 soybean growth stage).

Herbicide Resistance

Herbicide resistance is not new, but the rapid spread of herbicide resistant weeds in recent years is mainly the result of three practices. The use of monocultures, the overreliance on a particular class or family of herbicides and the negligence of other non-herbicide control measures are the major reasons for the rapid evolution of herbicide resistance (Mortensen et al. 2012). This outbreak has widely spread; more than 500 herbicide resistant cases have been reported (Heap 2017) and millions hectares of cropland have been infested with herbicide resistant weeds (Mortensen et al. 2012). These weeds have evolved resistance to various herbicides with different mechanisms of action (Table 15.1). Glyphosate, was an effective solution against weeds resistant to ALS, PS-II, and ACCase inhibitors, but the overreliance on it as a primary or only herbicide used in many systems, is one of the major factors for the evolution of GR weeds (Conner et al. 2003, Firbank and Forcella 2000, Powles 2003, Shaner 2000, Watkinson et al. 2000, Young 2006).

The occurred changes within weed communities are not surprising considering the level of selection pressure that glyphosate resistance technology imparted on cropping systems (Owen 2008). Resistance mechanisms to herbicides in weeds include reduced cellular transport to active meristematic tissues and insensitive altered EPSPS (Powles and Preston 2006), enzymatic insensitivity (Burgos et al. 2001, Sprague et al. 1997), point mutation (Hirschberg et al. 1987, Tranel and Wright 2002) or gene amplification (Gaines et al. 2009).

Integrated Weed Control

Weed management approaches that involve no herbicide use are gaining a renewed interest because of public awareness of health issues, environmental pollution concerns and cost of food production (Korres 2017). The rapid spread of herbicide-resistant weeds coupled with the decline in the development of new herbicide mechanisms of action (Strek 2014) necessitates the

Table 15.1.* Most important herbicide resistant weeds in soybean

Weed species	Location	Resistance**
Amaranthus palmeri	Argentina, USA,	ALS, EPSP, Microtubule inhibitors, PPO, PS II
Amaranthus tuberculatus	Canada, USA	ALS, PS II, EPSP, PPO, HPPD
Ambrosia artemisifolia	Canada, USA	ALS, EPSP, PPO,
Ambrosia trifida	Canada, USA	ALS, EPSP
Conyza bonariensis	Brazil	EPSP
Conyza canadensis	Brazil, Canada, USA	ALS, EPSP, PS I
Conyza sumatrensis	Brazil	ALS, EPSP
Cynodon hirsutus	Argentina	EPSP
Digitaria insularis	Argentina, Brazil, Paraguay	ACCase, EPSP
Echinochloa colona	Argentina, Bolivia	ACCase, EPSP
Eleusine indica	Argentina, Bolivia, Brazil, USA	ACCase, ALS, EPSP, Microtubule
Kochia scoparia	Canada, USA	ALS, EPSP
Lolium multiflorum	Brazil, USA	ALS, EPSP, ACCase
Lolium perenne	Argentina	EPSP
Parthenium hysterophorus	Brazil	ALS
Sorghum halepense	Argentina, Italy, USA	ACCase, ALS, EPSP

* Based on Heap 2018, **ALS = Acetolactate synthase inhibiting herbicides, EPSP = Enolpyruvyl Shikimate-3-Phosphate Synthase Inhibitors, ACCase = Acetyl-CoA carboxylase inhibiting herbicides, PPO = Protoporphyrinogen oxidase, HPPD = 4-Hydroxyphenylpyruvate dioxygenase, PS I, PS II = Photosystem I and photosystem II inhibitors.

utilization of all available weed management options such as cultural, physical/mechanical, biological and preventative tactics in conjunction with herbicides. Most experts agree that weed management systems in the days ahead will be more complicated (Thompson 2012). Mortensen et al. (2012) stated that if herbicide-resistant-weed problems were addressed only with a single mode of action, herbicide resistance evolution will most likely prevail. Many stakeholders have started realizing that weed management systems should rely less on the panacea of glyphosate and few other herbicide products with a single mechanism of action. Integrated weed management system concerns incorporating various tactics to enhance the efficacy of weed control and reduce the rate by which weeds evolve resistance have been reported (Booth and Swanton, 2002, Norsworthy et al. 2012). These tactics could possibly include, amongst others, a combination of crop rotation, crop competitiveness, row spacing, cover crops, tillage, fertilizer placement, weed thresholds, the use of herbicide products with more than one mode of action or the use of various prevention methods. (Thill et al. 1991, Norsworthy et al. 2012, Scott 2011). The incorporation of rice into the soybean cropping system to control various herbicide resistant weed species, provided that adequate weed control in levees can be achieved, seems an interesting option (Scott et al. 2015). The use of cover crops, followed by a narrow-row soybean with the incorporation of a POST herbicide application the first year fb crop rotation the second year, could suppress weed flora effectively. Korres et al. (2016), Norsworthy et al. (2016) reported that the use of various herbicide programs and several at harvest and post-harvest methods (i.e., harvest weed seed control through crop residual collection, cover crop or soybean residue management) achieved significant control of glyphosate resistant *A. palmeri* population and seed production (Figure 15.5).

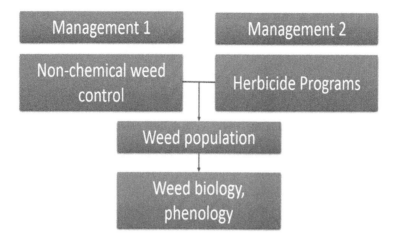

Figure 15.5. An integrated weed management scheme for the control of glyphosate resistant *Amaranthus palmeri*.

Davis and Johnson. (2008) reported the control of glyphosate resistant *Conyza canadensis* (L.) Cronquist (horseweed) in no-till soybean by integrating cover crops and soil-applied residual herbicides. Despite the evidence that indicate the efficiency of integrated weed management approaches, more research is required by combining various tactics for the determination of the one with the greatest impact on weed emergence, growth, and weed seed production (Barberi and Mazzoncini 2001, Liebman and Dyck 1993). Some sets of management practices are complimentary while others are not, hence multi-tactical weed management should employ combinations of weed management practices that result in interactions that are synergistic or additive (Korres and Norsworthy 2014).

Concluding Remarks

The challenge of agricultural sustainability requires solving the trade-off between producing satisfying levels of agricultural products, both in terms of quantity and quality, and reducing the environmental impacts and preserving non-renewable resources. Weed management is a key issue, because herbicides are the most applied pesticides around the world and they are most often detected substances in the surface and ground waters. Therefore, it is necessary to adopt correct strategies for weed management. Simple measures like choosing the correct cultivar, adopting correct tillage practices, using cover crops and crop rotation are responsible for decreasing the use of herbicides and, consequently, contributing to environmental sustainability.

REFERENCES

Akemo, M.C., E.E. Regnier, M.A. Bennett. 2000. Weed suppression in spring-sown rye (*Secale cereale*) pea (*Pisum sativum*) cover crop mixes. Weed Technol. 14: 545–549.

Alberts, E.E. and W.H. Neiblin. 1994. Influence of crop residues on water erosion. pp. 19–39. *In:* Unger, P.W. (Ed.) Managing Agricultural Residues. Lewis Publishing, Ann Arbor, MI.

Anderson, W.P. 1996. Weed Science Principles and Applications. 3rd edition. West Publishing Company, St. Paul, MN. 388 p.

Andrews, I.K.S., J. Storkey and D.L. Sparkes. 2015. A review of the potential for competitive cereal cultivars as a tool in integrated weed management. Weed Res. 55: 239–248 .

Anonymous. 2017. Soystats. A reference guide to important soybean facts and figures. The American Soybean Association. Available at: www.soystats.com (Accessed in April 2017).

Barberi, P. and M. Mazzoncini. 2001. Changes in weed community composition as influenced by cover crop and management system in continuous corn. Weed Sci. 49: 491–499.

Barnes, J.P., A.R. Putman, B.A. Burke and A.J. Aasen. 1987. Isolation and characterization of allelochemicals in rye herbage. Phytochemistry 26: 1385–1390.

Barnes, J.P. and A.R. Putnam. 1987. Role of benzoxazinones in allelopathy by rye (*Secale cereale* L.). J. Chem. Ecol. 13: 889–906.

Bell, H.D., J.K. Norsworthy and R.C. Scott. 2015. Effect of drill-seeded soybean density and residual herbicide on Palmer amaranth (*Amaranthus palmeri*) emergence. Weed Technol. 29: 697–706.

Benbella, M. and G.M. Paulsen. 1998. Efficacy of treatments for delaying senescence of wheat leaves: II. Senescence and grain yield under field conditions. Agron. J. 90: 332–338.

Benbrook, C.M. 2012. Impacts of genetically engineered crops on pesticide use in the U.S. – the first sixteen years. Environ. Sci. Eur. 24: 1–13.

Bennett, A.C. and D.R. Shaw. 2000. Effect of Glycine max cultivars and weed control weed seed characteristic. Weed Sci. 48: 431–435.

Blackshaw, R.E., R.N. Brandt, H.H. Jazen, T. Entz, C.A. Grant and D.A. Derksen. 2003. Differential response of weed species to added nitrogen. Weed Sci. 52, 532–539.

Bond, W. and A.C. Grundy. 2001. Non-chemical weed management in organic farming systems. Weed Res. 41: 383–405.

Booth, B.D. and C.J. Swanton. 2002. Assembly theory applied to weed communities. Weed Sci. 50: 2–13.

Bowman, G. 1997. Steel in the Field: A Farmer's Guide to Weed Management Tools, Greg Bowmann: Beltsville, MD, USA.

Bradley, K.W. 2013. Herbicide-resistance in the Midwest: current status and impacts. Weed Sci. Soc. Am. Abstr. 53: 271.

Bradley, K.W. 2006. A review of the effects of row spacing on weed management in corn and soybean. Online. Crop Manag doi: 10.1094/CM-2006–0227-02-RV.

Bridges, D.C. 1992. Crop losses due to weeds in the United States – 1992. WSSA special publication, Champaign, IL.

Brown, S.M., T. Whitwell, J.T. Touchton and C.H. Burmester. 1985. Conservation tillage systems for cotton production (USA). Soil Sci. Soc. Am. J. 49: 1256–1260.

Brown, S.M. and T. Whitewell. 1988. Influence of tillage on horseweed (*Conyza canadensis*). Weed technol. 2: 269–270.

Buehring, N.W., D.R. Shaw and G.R.W. Nice. 2002 Sicklepod (*Senna obtusifolia*) control and soybean (*Glycine max*) response to soybean row spacing and population in three weed management systems. Weed Technol. 16: 131–141.

Burgos, N.R., Y. Kuk and R.E. Talbert. 2001. *Amaranthus palmeri* resistance and differential tolerance of *Amaranthus palmeri* and *Amaranthus hybridus* to ALS inhibitor herbicides. Pest Manag. Sci. 57: 449–457.

Bussan, A.J., O.C. Burnside, J.H. Orf and K.J. Puettmann. 1997. Field evaluation of soybean (*Glycine max*) genotypes for weed competitiveness. Weed Sci. 45: 31–37.

Cahoon, C.W., M.L. Flessner, R.L. Ritter, B.A. Majek, W.S. Curran, R. Chandran and M. VanGessel. 2016. Weed control in field crops. Introduction to weed management. pp. 5–3. *In:* Pest management guide, field crops 2016 and 2017. Virginia Cooperative Extension, Virginia Tech. Available at: https://pubs.ext.vt.edu/456/456–016/456–016.html (Accessed in April 2017).

Callaway, M.B. 1992. A compendium of crop varietal tolerance to weeds. Am. J. Alternative Agr. 7: 169–180.

Cantwell, J.R., R.A. Liebl and F.W. Slife. 1989. Imazethapyr for Weed Control in Soybean (*Glycine max*) Weed Technol. 3: 596–601.

Chandler, J.M., A.S. Hamill and A.G. Thomas. 1984. Crop Losses Due to Weeds in Canada and the United States. WSSA special publication, Champaign, IL.

Chase, W.R., M.G. Nair and A.R. Putman. 1991. 2, 2'-oxo-1, 1'-azobenzene-selective toxicity of rye (*Secale cereale* L.) allelochemicals to weed and crop species II. J. Chem. Ecol. 19: 9–19.

Chauhan, B.S. and S.B. Abugho. 2013. Effect of crop residue on seedling emergence and growth of selected weed species in a sprinkler-irrigated zero-till dry-seeded rice system. Weed Sci. 61: 403–409.

Chauhan, B.S., R.G. Singh and G. Mahajan. 2012. Ecology and management of weeds under conservation agriculture. A. Rev. Crop. Pr. 38: 57–65.

Conner, A.J., T.R. Glare and J.P. Nap. 2003. The release of genetically modified crops into the environment. Part II. Overview of ecological risk assessment. Plant J. 33: 19–46.

Cousens, R.D. and S. Mokhtari. 1998. Seasonal and site variability in the tolerance of wheat cultivars to interference from *Lolium rigidum*. Weed Res. 38: 301–307.

Crookston, R.K., J.E. Kurle, P.J. Copeland, J.H. Ford and W.E. Lueschen. 1991. Rotational cropping sequence affects yield of corn and soybean. Agron. J. 83: 108–113.

Dabney, S.M., J.A. Delgado and D.W. Reeves. 2001. Using winter cover crops to improve soil and water quality. Commun. Soil Sci. Plant. Anal. 32: 1221–1250.

Datta, A., H. Ullah, N. Tursun, T. Pornprom, S.Z. Knezevic and B.S. Chauhan. 2017. Managing weeds using crop competition in soybean (*Glycine max* [L.] Merr.). Crop Prot. 95: 60–68.

Davis, V.M. and W.G. Johnson. 2008. Glyphosate-resistant Horseweed (*Conyza canadensis*) Emergence, Survival, and Fecundity in No-till Soybean. Weed Sci. 56: 231–236.

Devenish, K.L. and L.J. Leaver. 2000. The fate of ryegrass seed when sheep graze chaff cart heaps. Asian-Aus J Anim Sci 13 (Suppl.): 484. Available at: http://www.asap.asn.au/livestocklibrary/2000/Devenish_0631.pdf (Accessed in March 2017).

Dill, G.M., C.A. Jacob and S.E. Padgette. 2008. Glyphosate-resistant crops: adoption, use and future considerations. Pest Manag. Sci. 64: 326–331.

FAO (Food and Agriculture Organization of the United Nations). 2017. FAO Stat. Available at: http://www.fao.org/faostat/en/#data/QC (Accessed on November 7, 2017).

Fernandez-Cornejo, J. and M.F. Caswell. 2006. The first decade of genetically engineered crops in the United States. Washington, DC: US Department of Agriculture, Economic Research Service, Economic Information Bulletin No. 11.

Firbank, L.G. and F. Forcella. 2000. Genetically modified crops and farmland diversity. Science 289: 1481–1482 .

Fogelfors, H. 1982. Collection of chaff, awns and straw when combining and its influence on the seed bank and the composition of the weed flora. pp. 339–345. *In:* Weeds and Weed Control: 23rd Swedish Weed Conference. Uppsala, Sweden: Department of Plant Husbandry and Research Information Centre, Swedish University of Agricultural Science.

Gaines, T.A., W. Zhang, D. Wang, B. Bukun, S.T. Chisholm, D.L. Shaner, S.J. Nissen, W.L. Patzoldt, P.J. Tranel, A.S. Culpepper, T.L. Grey, T.M. Webster, W.K. Vencill, R.D. Sammons, J. Jiang, C. Preston, J.E. Leach and P. Westra. 2009. Gene amplification confers resistance in *Amaranthus palmeri*. Proc. Nat. Acad. Sci. 107: 1029–1034.

Gehring, K., T. Festner, R. Gerhards, K. Husgen and S. Thyssen. 2014. Chemical weed control in soybean (*Glycine max*, L.). Julius-Kuhn-Archiv 443: 701–708.

Givens, W.A., D.R. Shaw, W.G. Johnson, S.C. Weller, B.G. Young, R.G. Wilson, M.D.K. Owen and D. Jordan. 2009. A grower survey of herbicide use patterns in glyphosate-resistant cropping systems. Weed Technol. 23: 156–161.

Gonzini, L.C., S.E. Hart and L.M. Wax. 1999. Herbicide combinations for weed management in glyphosate-resistant soybean (*Glycine max*). Weed Technol. 13: 354–360.

Gramig, G.G. and D.E. Stoltenberg. 2009. Adaptive responses of field grown common lambsquarters (*Chenopodium album*) to variable light quality and quantity environments. Weed Sci. 57: 271–280.

Grau, R.H. 2005. Globalization and soybean expansion into semiarid ecosystems of Argentina. Ambio 34: 265–266.

Grichar, W.J., B.A. Bessler and K.D. Brewer. 2004. Effect of row spacing and herbicide dose on weed control and grain sorghum yield. Crop Prot. 23: 263–267.

Harder, D.B., K.A. Renner and C.L. Sprague. 2007. Effect of soybean row width and population on weeds, crop yield and economic return. Weed Technol. 21: 744–752.

Hartwig, N.L. and H.U. Ammon. 2002. Cover crops and living mulches. Weed Science 50: 688–699.

Heap, I. 2017. The International Survey of Herbicide Resistant Weeds. Available at: http://www.weedscience.org/summary/home.aspx. (Accessed in December 2017).

Heatherly, L. and S. Spurlock. 1999. Yield and economics of traditional and early soybean production system (ESPS) seedings in the midsouthern United States. Field Crops Res. 63: 35–45.

Hirschberg, J., A.B. Yehuda, I. Pecker and N. Ohad. 1987. Mutations resistant to Photosystem II herbicides. pp. 357–366. *In:* von Wettstein, D., Chua, N.-M. (Eds.) Plant Molecular Biology. Plenum Press, New York.

Hock, S.M., S.Z. Knezevic, A.R. Martin and J.L. Lindquist. 2005. Influence of soybean row width and velvetleaf emergence time on velvetleaf (*Abutilon theophrasti*). Weed Sci. 53: 160–165.

Johnson, G.A., T.R. Hoverstad and R.E. Greenwald. 1998. Integrated weed management using narrow corn row spacing, herbicides, and cultivation. Agron J 90: 40–46.

Kaimowitz, D. and J. Smith. 2001. Soybean technology and the loss of natural vegetation in Brazil and Bolivia. pp. 195–211. *In:* Angelsen, A. and Kaimowitz, D. (Eds.) Agricultural Technologies and Tropical Deforestation. CABI Publishing, Wallingford, UK.

Kaspar, T.C., E.J. Kladivko, J.W. Singer, S. Morse and D.R. Mutch. 2008. Potential and limitations of cover crops, living mulches, and perennials to reduce nutrient losses to water sources from agricultural fields in the Upper Mississippi River Basin. Gulf Hypoxia and Local Water Quality Concerns Workshop. St. Joseph, Michigan: ASABE. pp. 127–148.

Kelley, K.W., J.H. Long Jr and T.C. Todd. 2003 Long-term crop rotations affect soybean yield, seed weight, and soil chemical properties. Field Crops Res. 83: 41–50.

Klooster, J.J. 1982. The role of soil tillage in weed control. pp. 256–261. *In:* Proceedings of the 9th Conference of the Soil Tillage Research Organization. Institution of Agricultural Engineering, Wageningen.

Knezevic, S.Z., S.P. Evans, E.E. Blankenship, R.C. Van Acker and J.L. Lindquist. 2002. Critical period of weed control: The concept and data analysis. Weed Sci. 50: 773–786.

Knezevic, S.Z., S.P Evans and M. Mainz. 2003. Row spacing influences the critical timing for weed removal in soybean (*Glycine max*). Weed Technol. 17: 666–673.

Koger, C.H., K.N. Reddy and D.R. Shaw. 2002. Effects of rye cover crop residue and herbicides on weed control in narrow and wide row soybean planting systems. Weed Biology and Management. 2: 216–224.

Korres, N.E. 2017. Agronomic weed control: A trustworthy approach for sustainable weed management systems. pp. 97–114. *In:* Jabran K. and Chauhan B.S. (Eds.) Non-chemical Weed Control. Pub. Elsevier.

Korres, N.E. and J.K. Norsworthy. 2017. Palmer amaranth (*Amaranthus palmeri*) demographic and biological characteristics in wide-row soybean. Weed Sci. 65: 491–503.

Korres, N.E., J.K. Norsworthy, K. Brye, V. Skinner Jr and A. Mauromoustakos. 2017a. Relationships between soil properties and the occurrence of the most agronomically important weed species in the field margins of eastern Arkansas – implications for weed management in field margins. Weed Res. 57: 159–171.

Korres, N.E., J.K. Norsworthy, T. Fitz Simons, T.L. Roberts and D.M. Oosterhuis. 2017b. Differential response of Palmer amaranth (*Amaranthus palmeri*) gender to abiotic stress. Weed Sci. 65(2): 213–227.

Korres, N.E., J.K. Norsworthy and R.C. Scott. 2016. Fall management practices and herbicide programs for controlling Palmer amaranth population and seed production in soybean. pp. 122–125. *In:* Ross J. (Ed.) Arkansas soybean research studies 2014, University of Arkansas, Research and Extension, Research Series 631.

Korres, N.E. and J.K. Norsworthy. 2015a. Influence of a rye cover crop on the critical period for weed control in cotton. Weed Sci. 63: 346–352.

Korres, N.E. and J.K. Norsworthy. 2015b. Influence of Palmer amaranth density and emergence date on seed production in wide row and drill-seeded soybean. Proceedings of the Weed Science Society of America Annual Meeting, Lexington, KN.

Korres, N.E., J.K. Norsworthy, M.V. Bagavathiannan and A. Mauromoustakos. 2015a. Distribution of arable weed populations along Eastern Arkansas Mississippi Delta roadsides: occurrence, distribution, and favoured growth habitats. Weed Technol. 29: 587–595.

Korres, N.E., J.K. Norsworthy, M.V. Bagavathiannan and A. Mauromoustakos. 2015b. Distribution of arable weed populations along Eastern Arkansas Mississippi Delta roadsides: factors affecting weed occurrence. Weed Technol. 29: 596–604.

Korres, N.E. 2005. Encyclopaedic dictionary of weed science. Theory and digest. Andover, UK: Intercept, Paris: Lavoisier Publishing. pp. 177–178.

Korres, N.E. and R.J. Froud-Williams. 2004. The interrelationships of winter wheat cultivars, crop density and competition of naturally occurring weed flora. Biol. Agric. Hortic. 22(1): 1–20.

Korres, N.E. and R.J. Froud-Williams. 2002. Effects of winter wheat cultivars and seed rate on the biological characteristics of naturally occurring weed flora. Weed Res. 42: 417–428.

Kruidhof, H.M., E.R. Gallandt, E.R. Haramoto and L. Bastiaans. 2011. Selective weed suppression by cover crop residues: effects of seed mass and timing of species' sensitivity. Weed Res. 51: 177–186.

Kunz, C., F.J. Weber, R. Gerhards. 2015. Benefits of precision farming technologies for mechanical weed control in soybean and sugar beet-comparison of precision hoeing with conventional mechanical weed control. Agronomy 5: 130–142.

Liebl, R., F.W. Simmons, L.M. Wax and E.W. Stoller. 1992. Effects of rye (*Secale cereale*) mulch on weed control and soil moisture in soybean (*Glycine max*). Weed Technol. 6: 838–846.

Liebman, M. and A.S. Davis. 2000. Integration of soil, crop and weed management in low-external-input farming systems. Weed Res. 40: 27–47.

Liebman, M. and E. Dyck. 1993. Crop rotation and intercropping strategies for weed management. Ecol. Appl. 3: 92–122.

Liebman, M. and A.S. Davis. 2000. Integration of soil, crop and weed management in low-external-input farming systems. Weed Res. 40: 27–47.

Lotjonen, T. and H.J. Mikkola. 2000. Three mechanical weed control techniques in spring cereals. Agric. Food Sci. Finland 9: 269–278.

Meese, B.G., P.R. Carter, E.S. Oplinger and J.W. Pendleton. 1991. Corn/soybean rotation effect as influenced by tillage, nitrogen, and hybrid/cultivar. J Prod. Agric. 4: 74–80.

Mortensen, D.A., J.F. Egan, B.D. Maxwell, M.R. Ryan and R.G. Smith. 2012. Navigating a Critical Juncture for Sustainable Weed Management. Bio Science 62: 75–84.

NASS (National Agricultural Statistics Service). 2017. United States Department of Agriculture (USDA). Statistics by Subject. Available at: http://www.nass.usda.gov/Statistics_by_Subject/result.php?74123334-32CB-374E-8F2A-151E45AE60D6&sector=CROPS&group=FIELD%20CROPS&comm=SOYBEANS (Accessed in December 2017).

NASS (National Agricultural Statistics Service). 2015. United States Department of Agriculture (USDA). Quick Stats (Crops). Available at: http://quickstats.nass.usda.gov. (Accessed in September 2015).

Nordby, D.E., D.L. Alderks and E.D. Nafziger. 2002. Competitiveness with weeds of soybean cultivars with different maturity and canopy width characteristics. Weed Technol. 21: 1082–1088.

Norsworthy, J.K., N.E. Korres, M.J. Walsh, and S.B. Powles. 2016. Integrating herbicide programs with harvest weed seed control and other fall management practices for the control of glyphosate-resistant Palmer amaranth. Weed Sci. 64: 540–550.

Norsworthy, J.K. and L.R. Oliver. 2001. Effect of seeding rate of drilled glyphosate-resistant soybean (*Glycine max*) on seed yield and gross profit margin. Weed Technol. 15: 284–292.

Norsworthy, J.K. and L.R. Oliver. 2002. Hemp sesbania interference in drill-seeded glyphosate-resistant soybean. Weed Sci. 50: 34–41.

Norsworthy, J.K., S.M. Ward, D.R. Shaw, R.S. Llwellyn, R.L. Nichols, T.M. Webster, K.W. Bradley, G. Frisvold, S.B. Powles, N.R. Burgos, W.W. Witt and M. Barrett. 2012. Reducing the Risks of Herbicide Resistance: Best Management Practices and Recommendations. Weed Sci. (Special Issue) 60: 31–62.

O'Donovan, J.T., K.N. Harker, G.W. Clayton, J.C. Newman, D. Robinson and L.M. Hall. 2001. Barley seeding rate influence the effects of variable herbicide rates on wild oat. Weed Sci. 49: 746–754.

Olesen, J.E., P.K. Hansen, J. Berntsen and S. Christensen. 2004. Simulation of above-ground suppression of competing species and competition tolerance in winter wheat varieties. Field Crops Res. 89: 263–280.

Owen, M.D. 2008. Weed species shifts in glyphosate-resistant crops. Pest Manag. Sci. 64: 377–387.

Pester, T.A., O.C. Burnside and J.H. Orf. 1999. Increasing crop competitiveness to weeds through crop breeding. J. Crop. Prod. 2: 31–58.

Powles, S.B. 2003. My view. Weed Sci. 51: 471.

Powles, S.B. and C. Preston. 2006. Evolved glyphosate resistance in plants: biochemical and genetic basis of resistance. Weed Technol. 20: 282–289.

Price, A.J. and J.K. Norsworthy. 2013. Cover crops for weed management in southern reduced tillage vegetable cropping systems. Weed Technol. 27: 212–217.

Price, A.J., K.S. Balkcom, S.A. Culpepper, J.A. Kelton, R.L. Nichols and H. Schomberg. 2011. Glyphosate-resistant Palmer amaranth: a threat to conservation tillage. J. Soil and Water Conserv. 66: 265–275.

Puricelli, E.C., D.E. Faccini, G.A. Orioli and M.R. Sabbatini. 2003. Spurred anoda (*Anoda cristata*) competition in narrow- and wide-row soybean (*Glycine max*). Weed Technol: 17: 446–451.

Rasmussen, J. and J. Ascard. 1995. Weed control in organic farming systems. pp. 49–67. *In:* Glen, D.M., Greaves, M.P. and Anderson, H.M. (Eds.) Ecology and Integrated Farming Systems: Proceedings of the 13th Long Ashton International Symposium on Arable Ecosystems for the 21st Century. John Wiley & Sons: Chichester, UK.

Reddy, K.N. 2001. Effects of cereal and legume cover crop residues on weeds, yield, and net return in soybean (*Glycine max*). Weed Technol. 15: 660–668.

Reddy, K.N. 2003. Impact of rye cover crop and herbicides on weeds, yield, and net return in narrow-row transgenic and conventional soybean (*Glycine max*). Weed Technology 17: 28–35.

Riar, D.S., J.K. Norsworthy, L.E. Steckel, D.O. Stephenson, T.W. Eubank and R.C. Scott. 2013. Assessment of weed management practices and problem weeds in the midsouth United States – soybean: a consultant's perspective. Weed Technol. 27: 612–622.

Robinson, R.G. and R.S. Dunham. 1954. Companion crops for weed control in soybeans. Agron J 46: 278–281.

Schomberg, H.H., R.G. McDaniel, E. Mallard, D.M. Endale, D.S. Fisher and M.L. Cabrera. 2006. Conservation tillage and cover crop influences on cotton production on a southeastern US coastal plain soil. Agron J 98: 1247–1256.

Schultz, J.L., L.A. Chatham, C.W. Riggins, P.J. Tranel and K.W. Bradley. 2015. Distribution of herbicide resistance and molecular mechanisms conferring resistance in Missouri waterhemp (*Amaranthus rudis Sauer*) populations. Weed Sci 63: 336–345.

Schulz, M., A. Marocco, V. Tabaglio, F.A. Macias and J.M.G. Molinillo. 2013. Benzoxazinoids in rye allelopathy – from discovery to application in sustainable weed control and organic farming. J Chem Ecol 39: 154–174.

Scott, R.C. 2011. Prevention and Control of Glyphosate-Resistant Pigweed in Soybean and Cotton. Arkansas Cooperative Extension Service Publications FSA2152. University of Arkansas, Little Rock, AR.

Scott, R.C., L.T. Barber, J.W. Boyd, G. Selden, J.K. Norsworthy and N. Burgos. 2015. Recommended Chemicals for Weed and Brush Control. University of Arkansas Division of Agriculture, Cooperative Extension Service. MP44.

Seversike, T.M. and L.C. Purcell. 2006. Multifoliate soybean: breeding and management strategies for ultra-early production systems in Arkansas. Page 4 in Arkansas Crop Protection Association. Abstracts Research Conference. Vol. 10. Clarion Inn, Fayetteville, Arkansas, November 27–28, 2006.

Shaner, D.L. 2000. The impact of glyphosate-tolerant crops on the use of other herbicides and on resistance management. Pest Man. Sci. 56: 320–326.

Shirtliffe, S.J. and M.H. Entz. 2005. Chaff collection reduces seed dispersal of wild oat (Avena fatua) by a combine harvester. Weed Sci. 53: 465–470.

Silva, A.F., L. Galon, I. Aspiazu, E.A. Ferreira, G. Concenco, E.U, Ramos Jr. and P.R.R. Rocha. 2013. Weed Management in the soybean crop. pp. 85–112. *In:* El-Shemy H.A. (Ed.) Soybean-Pest-Resistance. Pub. InTech, Rijeka, Croatia.

Silva, A.A., F.A. Ferreira, L.R. Ferreira and J.B. Santos. 2007. Metodos de controle de plantas daninhas. pp. 64–82. *In:* Silva, A.A. and Silva, J.F. (Eds.) Topicos em manejo de plantas daninhas. Viçosa: Universidade Federal de Vicosa.

Skroch, W.A., M.A. Powell, T.E. Bilderback and P.H. Henry. 1992. Mulches: durability, aesthetic value, weed control and temperature. J Enviorn Hort 10: 43–45.

Snipes, C.E. and T.C. Mueller. 1992. Cotton (*Gossypium hirsutum*) yield response to mechanical and chemical weed control systems. Weed Sci. 42: 249–254.

Soltani, N., J.A. Dille, I.C. Burke, W.J. Everman, M.J. VanGessel, V.M. Davis and P.H. Sikkema. 2017. Perspectives on Potential Soybean Yield Losses from Weeds in North America. Weed Technol 31: 148–154.

Sprague, C.L., E.W. Stoller, L.M. Wax and M.J. Horak. 1997. Palmer amaranth (*Amaranthus palmeri*) and common waterhemp (*Amaranthus rudis*) resistance to selected ALS inhibiting herbicides. Weed Sci. 45: 192–197.

Steckel, L.E. and C.L. Sprague. 2004. Late-season common waterhemp (*Amaranthus rudis*) interference in narrow- and wide-row soybean. Weed Technol. 18: 947–952.

Storrie, A.M. (Ed.). 2014. Integrated weed management in Australian cropping systems. Canberra, Australia: Grains Research and Development Corporation. 386 p.

Strek, H.J. 2014. Herbicide resistance. What have we learned from other disciplines? J. Chem. Biol. 7: 129–132.

Swanton, C.J., K.N. Harker and R.L. Anderson. 1993. Crop losses due to weeds in Canada. Weed Technol. 7: 537–542.

Swanton, C.J., A. Shrestha, K. Chandler and W. Deen. 2000. An economic assessment of weed control strategies in no-till glyphosate-resistant soybean (*Glycine max*). Weed Technol. 14: 755–763.

Teasdale, J.R. and C.L. Mohler. 1993. Light transmittance, soil temperature, and soil moisture under residue of hairy vetch and rye. Agron. J. 85: 673–680.

Teyker, R.H., H.D. Hoelzer and R.A. Liebl. 1991. Maize and pigweed response to nitrogen supply and form. Plant Soil 135: 287–292.

Thill, D.C., J.M. Lish, R.H. Callihan and E.J. Bechinski. 1991. Integrated weed management: a component of integrated pest management – a critical review. Weed Technol. 5: 648–656.

Thompson, H. 2012. War on weeds loses ground. Nature 485 doi: 10.1038/485430a.

Tranel, P.J. and T.R. Wright. 2002. Resistance of weeds to ALS-inhibiting herbicides: what have we learned? Weed Sci. 50: 700–712.

Van Acker, R.C., S.F. Wiese and C.J. Swanton. 1993. Influence of interference from a mixed weed species stand on soybean (*Glycine max* [L.] Merr.) growth. Can. J. Plant Sci. 73: 1293–1304.

Van der Weide, R.Y., P.O. Bleeker, V.T.J.M Achten, L.A.P Lotz, F. Fogelberg and B. Melander. 2008. Innovation in mechanical weed control in crop rows. Weed Res. 48: 215–224.

Vidotto, F., S. Fogliatto, M. Milan. and A. Ferrero. 2016. Weed communities in Italian maize fields as affected by pedo-climatic traits and sowing time. Eur. J. Agron. 74: 38–46.

Vollmann, J., H. Wagentristl and W. Hartl. 2010. The effects of simulated weed pressure on early maturity soybeans. Eur. J. Agron. 32: 243–248.

Walsh, M.J. and P. Newman. 2007. Burning narrow windrows for weed seed destruction. Field Crops Res. 104: 24–30.

Walsh, M.J., P. Newman and S.B. Powles. 2013. Targeting weed seeds in-crop: a new weed control paradigm for global agriculture. Weed Technol. 27: 431–436.

Walsh, M.J. and S.B. Powles. 2004. Herbicide resistance: an imperative for smarter crop weed management. pp. 1–6. *In:* Proceedings of the 4th International Crop Science Congress. Brisbane, Australia: The Regional Institute Ltd.

Watkinson, A.R., R.P. Freckleton, R.A. Robinson and W.J. Sutherland. 2000. Predictions of biodiversity response to genetically modified herbicide-tolerant crops. Science 289: 1554–1557.

Weber, J.F., C. Kunz and R. Gerhards. 2016. Chemical and mechanical weed control in soybean (*Glycine max*). Julius-Kuhn-Archiv 452: 171–176.

West, T.D., D.R. Griffith, G.C. Steinhardt, E.J. Kladivko and S.D. Parsons. 1996. Effect of tillage and rotation on agronomic performance of corn and soybean: twenty-year study on dark silty clay loam soil. J Prod Agric 9: 241–248.

Whitaker, J.R., A.C. York, D.L. Jordan and A.S. Culpepper. 2010. Palmer amaranth (*Amaranthus palmeri*) control in soybean with glyphosate and conventional herbicide systems. Weed Technol. 24: 403–410.

Wicks, G.A., G.W. Mahnken and G.E. Hanson. 2000. Effect of herbicides applied in winter wheat (*Triticum aestivum*) stubble on weed management in corn (*Zea mays*). Weed Technol. 14: 705–712.

Wilcut, J.W., A.C. York, W.J. Grichar and G.R. Wehtje. 1995. The biology and management of weeds in peanut (*Arachis hypogaea*). pp. 207–244. *In:* H.E. Pattee and H.T. Stalker (Eds.) Advances in Peanut Science. Am. Peanut Res. and Educ. Soc., Stillwater, OK.

Yelverton, F.H. and H.D. Coble. 1991. Narrow row spacing and canopy formation reduces weed resurgence in soybeans (*Glycine max*). Weed Technol. 5: 169–174.

Young, B.G. 2006. Changes in herbicide use patterns and production practices resulting from glyphosate-resistant. Weed Technol. 20: 301–307.

Zimdahl, R.L. 2004. Weed-crop Competition: A Review, 2nd ed. Blackwell Publishing, Ames, Iowa, USA.

Sustainable Weed Control in Cotton

Krishna N. Reddy* and William T. Molin

USDA-ARS, Crop Production Systems Research Unit, PO Box 350, Stoneville, MS 38776, USA

Introduction

Cotton is a major cash crop grown throughout the world for fiber and oil seed. It is grown in more than 70 countries under different environmental conditions such as tropical, sub-tropical, and temperate climates. Cotton is perennial in nature and may be grown as a ratoon crop, but is generally grown as an annual. Globally, cotton was raised on 29.7 million hectares with a production of 106.7 million bales of lint cotton in 2016–17 (USDA 2017). The top three cotton producing countries are India, China, and the United States. The United States ranks third in production and produces cotton in 17 southern states from Virginia to California covering 4 million hectares (USDA 2016a).

Cotton is sensitive to competition from weeds, adverse environmental conditions, and insect and disease pressures. Weeds reduce cotton yield and fiber quality, impede production practices, lower machinery efficiency, and increase production costs. Weeds also serve as hosts and habitats for pests and as such pose a major threat to maximizing cotton production. The magnitude of economic losses from weeds depends largely on the degree of infestation. Although crop production practices exert selection pressure on weed communities, these practices also create niches that favor or disfavor various species. Considering that weeds can exploit a variety of soils, climatic conditions, and cultural practices, effective control of weeds requires the use of site-specific weed management strategies.

Cotton weed control technology has progressively changed from very labor-intensive hand hoeing and animal drawn tillage to mechanical tillage and use of herbicides. In 1997, genetically engineered cotton was introduced initiating the era of herbicide-resistant biotechnology (Duke and Powles 2009). McWhorter and Abernathy (1992) have published a book titled *Weeds of Cotton: Characterization and Control* which consists of 15 chapters from leading scientists on various aspects of weeds and methods used to control them in cotton. This book provides a treasure of information from historical times through the early 1990s prior to the commercialization of herbicide-resistant (HR) cotton. Two HR (bromoxynil and glyphosate) cotton cultivars were introduced in the mid-1990s and another (glufosinate) appeared in the mid-2000s drastically changing cotton production (Reddy and Nandula 2012). Most of today's commercial cotton farms use a combination of best management practices for weed control methods in combination with HR varieties (Hurley et al. 2009). This chapter provides a summary of weed management

*Corresponding author: krishna.reddy@ars.usda.gov

tactics used in cotton production, expands on the impact of modern HR cropping systems and HR weeds, and discusses upcoming technologies.

Weeds in Cotton

Cotton growth, from planting to harvest, generally occurs over a five to six-month period. During the first 11 to 13 weeks, growth is relatively slow compared to competing weed species and other crops such as corn (*Zea mays* L.) and soybean (*Glycine max* [L.] Merr). This pre-canopy closure period provides opportunities for weeds to become established and compete for resources intended for optimal cotton growth, such as water, nutrients and light. Both cotton and weed establishment, as well as their growth and yield, are driven by environmental factors such as favorable and unfavorable temperatures, adequate and inadequate moisture, and soil nutrient status.

The level of competition with cotton depends upon the species present, their densities and proximities to the crop as well as time of emergence relative to cotton and growth rates. Many of the weeds commonly found in cotton are likely to exert competitive stress on cotton because of differences in growth rate. Keeley and Thullen (1993) found the relative growth rate of cotton based on plant height to be much less than johnsongrass (*Sorghum halepense* [L.] Pers.), morningglories (*Ipomoea* spp.), barnyardgrass (*Echinochloa crus-galli* [L.] Beauv.) and Palmer amaranth (*Amaranthus palmeri* [S.] Wats.) at three weeks after planting (WAP), and these weeds continued to outgrow cotton for the first 9 WAP. This timeframe would be well into the critical weed free period (CWFP). Other weeds, such as bermudagrass (*Cynodon dactylon* [L.] Pers.), yellow nutsedge (*Cyperus esculentus* [L.]) and black nightshade (*Solanum nigrum* [L.]), had relative growth rates less than cotton and by 9 WAP cotton was taller.

The primary goal in cotton production is development of a healthy plant capable of sustaining vigorous growth with the initiation, retention and development of squares into mature bolls. Squares begin to develop at about 40 days after planting (DAP, 6 to 8 leaf stage) and continue to form until about 120 DAP. Weeds can severely impact square retention and development for the first 5 to 6 weeks after squaring is initiated, and longer periods if weeds are exceptionally tall. CWFP has been established for each cotton growing region to identify the intervals during which weeds exert the greatest negative impact on cotton growth and yield (Bukun 2004, Papamichail et al. 2002). The CWFPs are often location dependent, and are influenced by cultural practices and the weed species present (Bukun 2004, Coble and Byrd 1992, Papamichail et al. 2002). In order to effectively use CWFP in cotton, a knowledge of the weed species common to a region, their rates of growth and germination window, and estimated degree of competitiveness, is required. CWFPs cannot be precisely defined because of variability between years but generally fall in a range of 4 to 10 weeks after emergence. Some CWFP were determined for single important weeds and others for weed complexes. For example, the CWFP for cocklebur (*Xanthium pensylvanicum*) was 8 to 10 weeks (Snipes et al. 1982) and for coastal bermudagrass was 4 to 7 WAP (Vencill et al. 1993). Bukun (2004) determined CWFP for a weed complex and estimated a CWFP between 1 or 2 and 11 or 12 WAP. A general conclusion from CWFP studies is that weed control practices should be initiated before onset of CWFP and continue until yield is no longer impacted. With this information in hand, a weed control program can be designed to take advantage of the current management tools to control most weeds.

Weeds present during the first three weeks after planting may not irreversibly affect cotton growth and yield provided they are removed at the end of this time period. Papamichail et al. (2002) in a two-year study reported that although there was no reduction in cotton height where weeds were present for the first 3 to 5 weeks after crop emergence, an 8% to 13% yield reduction was observed when weeds were in competition with cotton for the first three weeks. Cardoso et al. (2011) reported an 83% reduction in yield when weeds were present throughout the growing season. Cardoso et al. (2011) determined a CWFP of 31 to 74 days after emergence and weeds controlled during this period resulted in only a 5% reduction in yield. Cardoso et al. (2011) and

Papamichail et al. (2002) also stated that initiating control efforts at the beginning of the CWFP were more influential on yield than the ending and concluded that defining the correct moment to start weed control was more important to yield than determining when to cease it.

Weeds may also reduce cotton yield and fiber quality at harvest. For example, morningglories may grow through a canopy forming dense entanglements across the top of the cotton which hamper harvest efficiency (Authors personal observation). Morningglories produce numerous seed capsules which shatter upon harvest replenishing the soil seed bank. Browntop millet (*Urochloa ramosa* [L.] Nguyen) culms, many of which exceed 2 metres in height, also form entanglements and are captured during harvest adding grassy trash to the lint (Authors personal observation). Both of these weeds can collect in spindle heads and break off brackets supporting moisture pads in picker heads. More robust weeds such as mature pigweeds (e.g., Palmer amaranth) can cause mechanical damage by obstructing picker heads (Smith et al. 2000). Therefore, scouting and removing these weeds from fields before harvest reduce equipment breakdowns.

Tillage Choices and Seedbed Preparation

The objective of weed control is not only to maximize crop yields but also to prevent or minimize recurrence weeds. Generally, some combinations of chemical, mechanical, and cultural methods are used to achieve effective weed control. Cotton may be produced with or without preseason tillage. If tillage is performed tillage may be in the form of reduced tillage (RT) which may include bedding, cultivations and subsoiling, or conventional tillage (CT) which also includes deep tillage with a chisel and moldboard plow and combinations of both. The choice of production system is clearly up to the farmer and is based on a knowledge of the land, yearly fluctuations in weather patterns, weed problems, and farm capabilities.

The following three comparative studies demonstrate some of the various production parameters considered in cotton production. An economic study with CT and no-tillage (NT) cotton found no differences between these systems (Varner et al. 2011). However, machinery labor and average machinery investment costs were approximately 50% less for NT than for CT systems while herbicide costs were 40% greater for the CT system. The net return for the NT system was $78.53/ha whereas the CT system was –$9.31/ha. Over the six years of the study, only two years resulted in positive returns in either of the systems (Varner et al. 2011). Martin and Hanks (2008) compared CT, NT, and RT with and without subsoiling and/or a wheat (*Triticum aestivum* [L.]) cover crop. Results indicated that the highest returns and lowest relative risk were obtained from a NT system among the systems studied. Cover crops did not increase yield enough to offset the expenses associated with their establishment. Subsoiling also did not increase returns enough to overcome the added expense and may have even reduced yields. The CT system had relatively high returns, but was found to be among the riskiest (highest variance) of the treatments analyzed. Blaise (2006) compared the effects of CT to two levels of RT on weed populations. The CT program consisted of moldboard plowing and four inter-row cultivations, and the RT program consisted of application of preemergence (PRE) herbicide with or without two cultivations. Weed density was significantly lower in RT than in CT largely because of the addition of herbicide to the production plan. One drawback of tillage is that it may stimulate weed germination and emergence (Taylor et al. 2005). Awned canarygrass (*Phalaris paradoxa* [L.]) seeds emerged in May following tillage but emerged in July without tillage. Tillage altered the periodicity of weed seed emergence and accelerated the decline of seeds for this species. Mohler et al. (2006) studied vertical movement of seeds in soil following moldboard, chisel, disk and rotary tiller soil treatments. They found that moldboard plowing brought deeply buried seed to the soil surface which may further exacerbate weed control problems.

These studies show that cotton can be grown with minimal soil disturbance using preplant herbicides to kill existing weeds and PRE herbicides to kill weeds emerging with the crop. However, if there is insufficient activating rainfall, the crop may suffer from increased early season weed competition requiring the use of follow-up treatments of post-emergence (POST) herbicides. Seedbed imperfections may cause soil moisture and drying irregularities, and unwanted exposure to herbicides, which may hinder early seedling growth. In addition, due to the length of time to canopy closure in cotton, POST herbicides are often needed to control new flushes of weeds (Buchanan 1992, Wilcut et al. 1993). Before the introduction of herbicide-resistant cotton, POST broadleaf herbicides, such as fluometuron and MSMA were mainly applied POST-directed because of their effects on delaying cotton maturity (Culpepper and York 1997). When bromoxynil-resistant cotton was introduced, a bromoxynil plus pyrithiobac POST mixture offered cotton producers another POST weed control option (Paulsgrove et al. 2005).

Tilled and bedded fields provide several advantages but increase production costs. Cotton production has traditionally relied on extensive tillage prior to planting due to the fact that early season cotton growth is extremely sensitive to weed competition. Fall and spring tillage operations kill existing weed populations and incorporate crop and weed residues into the soil. In addition, in areas with the capacity to furrow irrigate, cotton is often planted on well-prepared raised seedbeds. These beds allow the soil to warm and dry faster, and they facilitate runoff following rainfall. Beds established in the fall or spring can be smoothed just prior to planting to expose warm, moist soil underneath the ridge cap. Planting into good soil moisture at optimum depth across the field and closing the seed furrows contributes to uniform crop emergence and improves crop competitiveness. A uniform seedbed also allows for an even distribution of soil applied products and lessens chances of herbicide injury to the crop.

St. John et al. (2011) published guidelines for preparation of seed beds which are applicable to most crops and are summarized as follows. Soil in a well-prepared seedbed provides a bed with a tilth that ensures good contact between seed and the soil. This facilitates moisture transfer, promotes germination, and maximizes conditions for seedling establishment. The soil below the seed needs to be firm but with a soil strength sufficiently low to allow radical and hypocotyl vertical extension. The soil above the seed, if of high strength, may impede upward thrust and if too loose or shallow may prevent shedding of the seed coat. A thin crust over loose soil may cause serpentine growth and broken hypocotyls. Soil moisture in the planting zone must increase with depth and be sufficient to complete germination during the first wetting cycle or the seed may dry and die. What is essential is that cotton seeds do not undergo repeated cycles of wetting and drying. The advantages of bedded cotton was demonstrated by Harrison et al. (2009) in a seven-year study to determine whether NT cotton maintained yields equivalent to a one-pass raised bed system. The raised bed systems were found to be essential for consistently high cotton yields and showed a 14% higher lint yield than NT. The factors that promote cotton seed germination such as favorable temperatures, moisture, and soil contact also provide suitable conditions for weed growth.

Cultivation

Cultivation of row middles (inter-row cultivation) during cotton growing season can complement chemical weed control. Weeds in the cotton row, intra-row weeds, are unaffected by mechanical cultivation and must be controlled by herbicides, hand hoeing, or both. Cultivation kills weeds by cutting shoots from roots, uprooting plants, and burying plants. Cultivation is effective when weeds are small, as larger weeds are difficult to bury in the soil. Generally, annual weeds are more easily managed by cultivation than perennial weeds. Cultivation is effective on seedlings of perennial weeds, but older plants with extensive underground root system can survive the disruption. Cultivation during dry periods promotes desiccation. Rainfall immediately following cultivation often results in incomplete control of weeds due to regrowth. Notably, in the United States, cultivation for weed control in cotton has decreased from 89% of planted acres in 1996 to 39% of planted acres in 2007 (Table 16.1), a time frame that coincides with the introduction of HR cotton.

Table 16.1. Weed management practices for cotton in the USA, 1996-2007 (USDA 2016c)

Control tactics	Percent of planted acres						
	1996	1997	1998	1999	2000	2003	2007
Field scouted for weeds	71	74	75	80	82	85	93
Area treated with herbicides	92	96	94	96	94	97	96
Herbicide-resistant seed used	NA	11	33	40	58	67	90
Non-herbicide-resistant seed used	NA	NA	NA	NA	NA	NA	10
Preplant herbicides used	7	7	7	10	25	57	43
Preemergence herbicides used	90	95	89	89	79	71	73
Postemergence herbicides used	62	73	69	70	76	79	89
Cultivation for weed control	89	87	72	77	63	54	39

Cultural Weed Control

Traditionally, in the United States, cotton is grown in rows spaced 76 to 102 cm apart. Ultra-narrow row (UNR) cotton production, which uses row spacings of 25 cm or less, has also been explored (Reddy 2001a). The weed species encountered in wide-row, UNR, and narrow-row (NR, 38-cm wide) cotton are similar, however, there are fewer late-season options to control weeds that escape early-season control in UNR and NR cotton because mechanical cultivation and hoeing are not an option. Weed control in UNR and NR cotton is dependent on broadcast application of PRE and POST herbicides as inter-row cultivation is not possible (Reddy 2001a). The UNR and NR cotton systems can provide early canopy closure compared to wide-row cotton and thus increase shading on weeds escaping herbicide treatments. Development of planters capable of UNR planting patterns and harvesting equipment (finger stripper) and variable-row system spindle-type pickers capable of picking cotton in 38-cm rows has rejuvenated interest in UNR and NR cotton production systems (Nichols et al. 2004, Reddy et al. 2009, Willcutt et al. 2006).

Increasing plant density and reducing row spacing may suppress weeds in cotton, but the extent of weed suppression is dependent on weed architecture (Manalil et al. 2016). Tursun et al. (2016) found that NR spacings (50 cm) had greater competitiveness against weeds compared to wider row spacings (70 to 90 cm).

Tall weeds, such as hemp sesbania (*Sesbania herbacea* [P. Mill.] McVaugh), sicklepod (*Senna obtusifolia* [L.] H.S. Irwin & Barneby), *Amaranthus* spp., Johnsongrass (*Sorghum halepense* [L.] Pers.), and spurred anoda (*Anoda cristata* [L.] Schlecht.) can grow through cotton canopy as cotton does not provide much competition. In a comparison of wide (WR, 101 cm row width) and UNR (25 cm row width) cotton, yield loss at eight spurred anoda plants/m exceeded 55% (Molin et al. 2006). The decrease in seed cotton yield in WR was due to a decrease in bolls per plant, cotton density remained constant at 9 plants/m across all spurred anoda densities. In UNR, as spurred anoda density increased from 0 to 8 plants/m, cotton density decreased from 25 to 14 plants/m (Molin et al. 2006). It is clear that row spacing affected weed growth differentially.

For weeds with less biomass during early growth, cotton growth may exceed that of weeds. Hyssop spurge (*Chamaesyce hyssopifolia* [L.] Small) and prickly sida (*Sida spinosa* [L.]), both having less biomass than spurred anoda, were more affected by UNR than by WR spacing (Molin et al. 2004). Row spacing did not affect weed height in either case, but under UNR, main stem dry weight and seed capsule production and primary, secondary and tertiary branching of prickly sida was significantly reduced (Molin et al. 2004). Total dry weight and branching was also significantly reduced in hyssop spurge in UNR cotton, indicating a reduction in carbon allocation to branches. Light penetration through the canopy was reduced 47% in WR and 85% in UNR for 60 cm tall cotton which indicates compaction of cotton canopy. For 100 cm tall cotton, light penetration was reduced by 87% and 97%, respectively. In addition, late season weed cover established from a second flush of weeds (e.g., browntop millet) was reduced 60% to 85% in plots in which a tall, spreading cultivar was grown compared to shorter cultivars in two of three years (Stetina et al. 2010). Reductions in browntop millet were observed with CT compared to minimum tillage and in late planted compared to early planted cotton (Stetina et al. 2010). Taken together, these results support the need for further research in using plant architecture to create a competitive advantage for cotton in weed control to develop cultivar based weed control strategies.

Both UNR and NR cotton systems were never widely adopted due to economic concerns. Narrow row systems had higher seed costs due to increased plant densities. There were also potentially higher ginning penalties due to increased wear on gins because of excessive trash levels associated with UNR cotton and reduced fiber quality manifested as higher neps associated with finger-stripper picked cotton. There was less producer interest in variable-row system spindle-type pickers that would allow picking cotton planted at different row widths. Also, there was the loss in ability to perform inter-row cultivation (Brown et al. 1998, Valco et al. 2001). As such, the dominance of conventional wide-row cotton production systems prevailed.

One area for improved weed control in cotton that has not been well developed is selection of cultivars based on weed suppression attributes. Cultivars might be selected for faster germination and seedling establishment, and faster or greater canopy development in addition to yield and favorable fiber properties. If cultivars with more competitive attributes could be developed, comparative testing would need to be examined in different soils, environments, and against different weed populations to verify their weed control potential. Development of a cotton cultivar with competitive advantages would likely include the studies on the contributions of seedling vigor, early canopy closure, leaf orientation, leaf area development, branching pattern and plant height (Eslami 2015) of both cotton and competitive weeds. Some commercial cotton seed companies indicate maturity (early, mid, full), seedling vigor, plant height (short, medium, medium tall, tall),canopy architecture (compact, columnar), leaf type (normal, okra), and disease resistance characteristics for their varieties so there is awareness of canopy architecture. Cotton cultivars with tall canopy architectures have been shown to have greater effects on mid and late season browntop millet populations than cultivars with more columnar architectures (Stetina et al. 2010).

Cotton traditionally was produced on the same ground year after year in the Mid-South. However, cotton rotated with corn has been shown to have agronomic benefits (Martin and Hanks 2009). From a weed control standpoint, rotating crops is beneficial as a cultural practice because it allows application of herbicide with different modes of action. Cotton rotated with corn in Arkansas produced 12% higher yield than monocrop cotton (Paxton et al. 1995). In a six year cotton-corn rotation study in Mississippi, it was demonstrated that yields of cotton and corn increased every year following rotation with each other. Weed shifts towards yellow nutsedge (*Cyperus esculentus* [L.]) in cotton was apparent and was mitigated by rotating with corn (Reddy et al. 2006). In a separate four year study, purple nutsedge (*Cyperus rotundus* L.) populations greatly decreased when cotton was rotated with soybean compared with continuous cotton (Bryson et al. 2003). Evidently, rotating cotton with other crops has the potential to mitigate specific weed problems and often result in increased crop yields.

Winter cover crops can be used as part of the cultural approach to weed control. In addition to reducing soil erosion and improving soil fertility and crop performance, some cover crops exhibit allelopathic effects on weeds. Winter cover crops offer the potential to slow developmentof weed problems in the spring which otherwise might become unmanageable. For example, CT cotton production is threatened by widespread glyphosate-resistant (GR) Palmer amaranth. Cover crop residues, shown to provide early-season weed control, can be used in cotton along with other weed control tactics to manage GR Palmer amaranth growth (Price et al. 2012). Winter cereal (barley, *Hordeum vulgare* L., triticale, X *Triticosecale* Wittmack, and rye, *Secale cereale* L.) mulches have suppressed germination of annual grass weeds (barnyardgrass,*Echinochloa crus-galli* [L.] Beauv., bristly foxtail, *Sateria verticillata* [L.] Beauv., and large crabgrass, *Digitaria sanguinalis* [L.] Scop). in cotton compared to no mulch. However, a combination with inter-row cultivation and herbicides was essential to maximize cotton yields (Vasilakoglou et al. 2006). Integration of cereal rye cover crop into a conservation cotton production system reduced Palmer amaranth density and increased yield compared to no cover crop. However, integration of cover crop requires use of effective herbicide programs (Price et al. 2012). Although early-season weed control is possible with cover crops, season-long control requires the use of herbicides (Reddy 2001b, Vasilakoglou et al. 2006). Season-long weed control for Palmer amaranth is especially necessary due to its extended germination period.

Weed Shifts and Stability Among the Most Troublesome Weeds in Cotton

The Southern Weed Science Society (SWSS) has periodically conducted a weed survey of the most common, and the most troublesome, weeds in cotton production. These surveys appear every four years and from 2001 onward are available at the SWSS website. Table 16.2 shows the changes in the ten most troublesome weeds since 1974. Results of similar surveys were reported by Bryson and Keeley (1992) and Webster and Coble (1997). These surveys contained lists of weeds as ranked by research and extension personnel before the release of the first transgenic herbicide-resistant (HR) crops. Changes in the weed flora were found although morningglories, nutsedges, bermudagrass, and pigweeds (*Amaranthus* spp.) are present at each interval, and compared to 1987, morningglories and nutsedges were among the top five most troublesome weeds in 1974, 1983, 1995, 2009 and 2013.

Palmer amaranth (*Amaranthus palmeri* [S.] Wats.), pigweeds, and Florida pusley (*Richardia scabra* [L.]) were also among the five most troublesome species in 2009. The weeds with the largest increases in importance in cotton were common ragweed (*Ambrosia artemisiifolia* [L.]) and two species with tolerance to glyphosate, Benghal dayflower (*Commelina benghalensis* [L.]) and Florida pusley (Webster and Nichols 2012). Shaner (2000) prognosticated that primary reliance on glyphosate for weed control, following introduction of HR cotton and soybean varieties, would result in a shift in the weed spectrum toward more tolerant weeds species. He noted that certain weeds, such as morningglories and hemp sesbania have a higher tolerance to glyphosate and are not controlled at rates generally applied in HR crops.

In the 2009 weed survey (SWSS 2009), some states grouped all pigweeds together as 'pigweed species' whereas other states separated the pigweeds into different categories. Georgia listed GR and acetolactate synthase-resistant (ALSR) Palmer amaranth as the #1 most troublesome weed, GR Palmer amaranth as #2 and 'pigweed species' as #4. Six of the ten states reporting in 2009 considered Palmer amaranth as the #1 most troublesome weed. In the 2013 survey (SWSS 2013), Florida, Alabama and Georgia separated GR Palmer amaranth from other pigweeds species and seven of eight states considered GR Palmer as the #1 most troublesome weed. This separation indicates the importance and severity of the GR Palmer amaranth problem in cotton.

Noted changes in the most troublesome weeds was a decrease in Texas millet (*Urochloa texana* [Buckl.] R. Webster) from five of ten states reporting to one of eight states reporting (#9 in Florida), and in bermudagrass from eight states reporting in 2005, five states reporting in 2009 to three states in 2013 (SWSS 2009, SWSS 2013). Another change was specifically naming yellow nutsedge rather than combining nutsedges in one group. This may have be due to the greater activity of glyphosate on purple nutsedge. The surveys also list the ten most common weeds in addition the ten most troublesome weeds. Morningglories species and nutsedge species are common to both weed lists (SWSS 2009, SWSS 2013).

Herbicide Usage and New Products

Prior to herbicides, weeds were controlled mainly by inter-row cultivation and hand hoeing. The use of herbicides in row crops began in the 1940s with the commercialization of 2,4-D

Table 16.2. Ten most troublesome weeds in cotton that appeared in annual weed surveys conducted by the Southern Weed Science Society. The information was extracted from annual weed surveys published in the Proceedings of Southern Weed Science Society and other publications (Buchanan 1974, Bryson and Keeley 1992, Webster and Coble 1997, Webster 2013, Webster and Nichols 2012)

Rank	1974	1983	1987	1995	2009	2013
1.	Prickly sida	Johnsongrass	Morningglories	Morningglories	Palmer amaranth	Palmer amaranth[a]
2.	Johnsongrass	Prickly sida	Cocklebur	Nutsedges	Morningglories	Morningglories
3.	Nutsedges	Nutsedges	Johnsongrass	Sicklepod	Nutsedges	Nutsedges[b]
4.	Morningglories	Morningglories	Purple nutsedge	Bermudagrass	Pigweeds	Horseweed
5.	Crotons	Common cocklebur	Prickly sida	Common cocklebur	Florida pusley	Florida pusley
6.	Pigweeds	Bermudagrass	Yellow nutsedge	Prickly sida	Bermudagrass	Prickly sida
7.	Ragweeds	Spurges	Pigweed	Spurges	Bengal dayflower	Goosegrass
8.	Goosegrass	Sicklepod	Spurges	Velvetleaf	Sicklepod	Spurges
9.	Crabgrasses	Silverleaf nightshade	Bermudagrass	Coffee senna	Texas millet	Johnsongrass
10.	Common cocklebur	Goosegrass	Sicklepod	Pigweeds and Silverleaf nightshade	Cocklebur and Velvetleaf	Bermudagrass

[a] Glyphosate resistant Palmer amaranth specifically separated from Palmer amaranth and pigweeds in 4 of the 8 states.

[b] Yellow nutsedge specifically separated from nutsedges in general in 2 of 8 states.

(Timmons 2005). At present, cotton production relies heavily upon herbicides to control weeds because they are effective and economical tools. Applications of two or more herbicides as preplant, PRE, and POST is not uncommon. Approximately 5% of cotton acres were treated with herbicides in 1952 and by 1980, herbicide use had reached 90% of U.S. cotton acres planted. After 1980, herbicide use stabilized and was about 92% in 2015 as most cotton area was already treated with herbicides (Fernandez-Cornejo et al. 2014, USDA 2016b). Notably, the top four most heavily used herbicides in 2015 were glyphosate (various formulations), trifluralin, diuron, and fomesafen (USDA 2016b). Of the treated acres in 2007, 43%, 73%, and 89% of acres were treated with preplant, PRE, and POST herbicides, respectively (Table 16.1) (USDA 2016c).

Most of the herbicides available for use in cotton in 1994 can still be used in 2014 in both HR and non-HR cotton (MSU 2016). These include glyphosate, benzoate, dinitroaniline, diphenylethers, organic arsenical, triazines, ureas, chloroacetamides, and POST grass herbicides and these are still available for use. Prior to 1996, glyphosate was largely used for preplant burndown treatments. Bugle (fenoxaprop) was discontinued in cotton and several new herbicides were added. Two herbicides with natural selectivity in cotton, both ALS inhibitors (pyrithiobac and trifloxysulfuron), were labeled for POST over-the-top use in cotton. Pyrithiobac was labeled in the United States in 1996 and trifloxysulfuron in 2004. Flumioxazin was labeled for use in cotton in 2004 for winter weed control, burndown, preplant, PRE, layby, and in-season cotton use. Aacetochlor was labeled for PRE use in cotton in 2012.

Currently, there are about 31 herbicides with different modes of action used in cotton, and many of them are limited to specific situations. Herbicides for weed control are applied in seven different application methods: i) postharvest (fallow seedbed), ii) preplant foliar (burndown), iii) preplant incorporated, iv) PRE, v) POST over-the-top, vi) directed-POST, and vii) spot treatments. Weed control guidelines published by the respective State Cooperative Extension Services (e.g., *Weed Control Guidelines for Mississippi*, a yearly publication, MSU 2016), the *Herbicide Handbook* (WSSA 2007), and manufacturers' specimen labels are available from their websites. Whereas herbicide and crop rotation programs provide means to control existing weeds, they may also cause shifts in weed populations to those more resilient to the current control strategy.

Cotton area planted and herbicide usage pattern in Mississippi during 1990 to 2016 is shown in Figure 16.1 (USDA 2016e). The decrease in glyphosate usage from 2005 to 2010 probably reflects the rapid decrease in cotton acreage with simultaneous loss in cotton infrastructure (gins, storage, and transport). Many of the PRE-herbicides were used far less following the introduction of HR cotton. Metolachlor use was not reported each year but may have averaged approximately 68,000 kg/year (USDA 2016e). In 2005 and 2010, 26,700 and 62,100 kg/year were applied, respectively (USDA 2016e). Data for lactofen, clomazone and pyrithiobac and clethodim were mostly less than 13,600 kg and not included in the figure (USDA 2016e).

A recent trend in Mississippi is the increased use of preplant foliar herbicides. Combinations include dicamba, fomesafen, glyphosate, glyphosate plus 2,4-dichlorophenoxyacetic acid (2,4-D), glyphosate plus carfentrazone, glyphosate plus dicamba, glyphosate plus dicamba plus 2,4-D, glyphosate plus saflufenacil, flumioxazin, oxyfluorfen, glufosinate, paraquat and prometryn. The added herbicides for use as postharvest, fallowbed and preplant foliar weed control are likely the result of increased conservation tillage. An effective control option in no-till cotton is an early preplant application of flumioxazin followed by a PRE application of fomesafen (Cahoon et al. 2014). This combination could promote conservation tillage efforts because it provides excellent control under NT conditions but places greater reliance on protoporphyrinogen IX oxidase inhibiting herbicides.

With the introduction of HR cotton came the POST over-the-top uses of glyphosate, glufosinate, and bromoxynil. Bromoxynil-resistant cotton was released in 1995, but it was short lived. Shortly thereafter, glyphosate-resistant cotton was released in 1996. Registration of Liberty (glufosinate), a glutamine synthetase inhibitor, for use in Liberty Link cotton came in 2004 (Reddy and Nandula 2012). LibertyLink (glufosinate-resistant) systems came after this time period and in 2010 only 6,800 kg of glufosinate were applied. As glufosinate-resistant

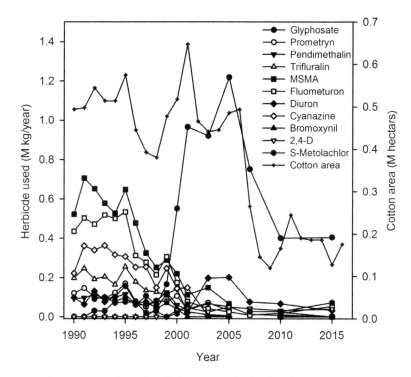

Figure 16.1. Cotton area planted and changes in the herbicide use pattern in Mississippi from 1990 to 2016 (USDA 2016e).

cotton varieties were introduced, especially those stacked with the GR trait, one would expect a rise in glufosinate use.

The phenomenal success of GR cotton has greatly impacted herbicide use patterns. The use of PRE and POST herbicides declined from 1996, the year GR cotton was commercialized compared to 2007 (Table 16.3). In the United States, glyphosate use has increased significantly with a concomitant decrease in use of other herbicides (Figure 16.1). Glyphosate use in cotton, based on a total active ingredient basis, has increased from 0.6 million kg per year in 1996 to 7.5 million kg per year in 2005 (USDA 2016d). This represents a 13-fold increased use of glyphosate since commercialization of GR cotton. Webster and Nichols (2012) also presented the change in treated cotton area from 1990 to 2007 which showed a rise in dinitroaniline and photosystem II inhibitor use from 1990 to 1995 followed by a decrease in use from 1995 to 2007 to less than half of that applied in 1995. They also showed a gradual increase in glyphosate use from 1990 to 1997 followed by a four to five-fold increase in use above 1997 levels from 1997 to 2000. These changes in herbicide use pattern to essentially one herbicide has contributed to the rise in herbicide resistance. Now that trend is being reversed.

Changes in Weed Management Practices

The most notable change in weed management practices in cotton in the past 30 years has been the development and rapid adoption of HR cotton varieties which permit the use of effective, broad spectrum, POST herbicides like glyphosate, glufosinate and bromoxynil (Cerdeira and Duke 2006). Bromoxynil had limited activity on grasses but controlled a broad spectrum of broadleaf weeds (Culpepper and York 1997, Corbett et al. 2002). In 1996, GR cotton was commercialized and quickly gained wide spread acceptance. In the United States, the cotton area planted with GR cultivars has increased from 4% in 1997 to 89% in 2016 (USDA 2016a). Glyphosate provided broad spectrum control of narrowleaf and broadleaf weeds, and eventually could be applied

Table 16.3. Major chemical family of herbicides used for weed control in cotton in the USA, 1996-2007 (USDA 2016c)

Herbicide chemical family	Percent of herbicide acre treatments						
	1996	1997	1998	1999	2000	2003	2007
Benzoates	3.2	7.1	5.6	5.0	5.0	3.2	1.3
Dinitroanilines	25.6	23.6	25.1	24.3	20.2	15.4	13.7
Organic arsenicals	11.8	12.0	8.5	7.4	5.9	2.5	0.8
Phosphonic acid	3.5	5.1	13.8	17.2	30.8	59.8	59.7
Triazines	12.9	12.4	10.6	9.4	8.3	2.4	2.2
Ureas	20.3	19.2	18.0	17.5	14.4	8.0	5.6
Other herbicides	22.6	20.5	18.3	19.1	15.4	8.7	16.6

over the top multiple times per season. The rapid adoption was also driven by inconsistencies with PRE herbicide systems resulting from a lack of activation rainfall, narrow weed control spectra and lesser crop selectivity of available products (Dill et al. 2008).

Weed management systems in cotton changed following the introduction of GR cotton (Culpepper and York 1998). Glyphosate applied POST over the top three to four weeks after planting followed by glyphosate or cyanazine plus MSMA POST directed six to seven weeks after planting provided the same control as the standard system of trifluralin and fluometuron, PPI and PRE, followed by fluometuron and MSMA early POST followed by cyanazine and MSMA mid-POST (Culpepper and York 1998). They also showed three applications of glyphosate were no more effective than two, and that trifluralin and fluometuron were of no benefit in systems with glyphosate applied twice or glyphosate followed by cyanazine plus MSMA. Glyphosate was an effective replacement for PPI and PRE herbicides, and could substitute for a treatment of cyanazine plus MSMA. Pyrithiobac applied POST over the top could substitute for a glyphosate application. Glyphosate was also less expensive than combinations of cyanazine plus MSMA, trifluralin and fluometuron, or fluometuron plus MSMA, or pyrithiobac, with the end result that there were fewer herbicide applications, less total herbicide, less pesticide storage, and easier handling. Glyphosate applied twice provided control of all species equal to that of the aforementioned standard system. Culpepper and York (1998) found yields and net returns were equivalent between glyphosate and standard systems. The authors also note that glyphosate systems give the added benefit of crop rotational flexibility. This research was prophetic as there has been substantial change in the herbicides used with the loss of cyanazine in the cotton market and substantial decreases in use of fluometuron, trifluralin, pendimethalin, MSMA, prometyrn and diuron (Figure 16.1). However, Benbrook (2009), in a report on pesticide use, reported that herbicide use on HR crops has veered sharply upward and crop years 2007 and 2008 accounted for 46% of the increase. This increase probably resulted from increases in HR weed populations necessitating a return to use of PRE herbicides.

Stacked herbicide resistance traits, such as glyphosate and glufosinate resistance, in combination with aggressive use of incentives to growers to utilize an array of herbicides with different modes of action and cultural practices, should be effective additions to the weed management tools used. Dicamba-resistant cotton and soybean were approved for sale and stacked varieties with resistance to dicamba and glyphosate are planned.

Herbicide-resistant cotton with resistance to both glyphosate and glufosinate was introduced in 2013. These products, dubbed second and third generation varieties, are directed toward addressing weed management issues, such as weed shifts and weed resistance, and for combating GR weeds and other weeds poorly controlled by glyphosate. Third generation cotton will combine dicamba resistance with glyphosate and glufosinate resistance. These enhancements will provide growers with choices and flexibility in their management of GR and other tough-to-control weeds. The three-way glyphosate, dicamba, glufosinate combinations are Monsanto's Roundup Ready Xtend Crop System (Bollman 2013). Bayer CropScience announced several FiberMax® and Stoneville® cotton varieties with stacked GlyTol®

LibertyLink® technology for a wide window of over-the-top applications of glufosinate and glyphosate. Fibermax/Stoneville introduced GlyTol/Liberty Link stacked varieties.

The Enlist™ weed control system with Colex-D technology from Dow AgroSciences, consists of glyphosate and the choline salt of 2,4-D and targets hard-to-control and resistant weeds (Siebert 2013). The new choline formulation will provide growers key benefits such as ultra-low-volatility, minimized potential for physical drift, decreased odor and improved handling characteristics (Siebert 2013). Single gene technology will continue to be available from Bayer CropScience under the label The LibertyLink program (Light 2013). Liberty® herbicide contains the active ingredient glufosinate-ammonium and provides control of other herbicide resistant weed species as well as non-selective control of a wide range of broadleaf and grass weed species in a matter of days as compared to weeks. The weed control strengths of this herbicide include morningglories, common cocklebur, and glyphosate-resistant weed species. There are continued efforts to look for traits with additional modes of action for tolerance.

Cotton with resistance to glyphosate, 2,4-D and glufosinate (Merchant et al. 2014) and glyphosate, dicamba and glufosinate (Cahoon et al. 2015) have been developed and evaluated for control of GR Palmer amaranth. These systems provide greater control of Palmer amaranth and improve its management which in turn increases grower flexibility to control this weed without injury to cotton.

As a result of the emergence of HR cotton, the potential for reducing machinery, fuel and labor costs in cotton production was realized which led to increases in CT practices (Cerdiera and Duke 2006). Conservation tillage is a practice that minimizes soil disturbance, reduces passes through the field, and leaves a minimum of 30% of the soil surface covered by the previous year's crop residue. However, the residue remaining after cotton stalks are shred may be insufficient to cover 30% of the soil surface and may not provide weed control advantages. There may be greater reliance placed on herbicides for weed control, and perennial and biennial weeds might increase under such conditions. Hence, to be compliant with CT programs, a winter cover crop may be needed to achieve the targeted residue coverage. Establishment of a cover crop in the fall may necessitate the termination of the cover crop in advance of planting cotton the following spring. Despite the increased availability and use of GR cotton, only 21% of the cotton acres were reported to be conservation tillage in the USA (CTIC 2016). This may reflect the historical use of more rigorous tillage operations to suppress weeds prior to planting cotton. Banerjee et al. (2007), in studies on adoption of CT practices and HR cotton, found no relationship between these management decisions. Rather, prior experience with cotton under NT conditions was the driving force on adoption of these practices.

Conservation tillage systems were explored as a means to reduce costs in cotton production well before introduction of herbicide resistant varieties (Keeling et al. 1989). Toler et al. (2002), in studies conducted in 1994 and 1995, evaluated weed and cotton response to several herbicide programs and found PRE and early POST herbicides to be an essential component of reduced tillage systems. Optimum cotton production was achieved with broadcast applications of PRE and early POST herbicides for competitive weeds in cotton including sicklepod, Palmer amaranth, goosegrass (*Eleusine indica* [L.] Gaertn.), and southern crabgrass. Without an aggressive standard program, levels of these weeds were unacceptably high. When GR cotton became available, the growers gained an effective POST, broad spectrum weed control tool. A survey of tillage trends following the adoption of glyphosate resistant crops reported a shift from more tillage-intense systems to reduced or no-till systems after including GR crops into their production system (Givens et al. 2009). This survey showed that cotton producers willingly adopted conservation tillage as a production system when effective POST weed control was available.

Herbicide-Resistant Weeds

Continuous cotton, cotton grown on the same land year after year, was common in the South especially when irrigation and rotation were not possible. Following introduction of GR crops, land which may have been rotated between cotton, corn and soybean was now rotated between GR cotton, GR corn, and GR soybean. As a result, weeds have been exposed to repeated selection pressures from the same herbicide. Resistance has evolved to both PRE and POST herbicides and includes several of the most troublesome weeds in cotton most notably horseweed (*Conyza canadensis* [L.] Cronq.), Italian ryegrass [*Lolium perenne* L. ssp. *multiflorum* (Lam.) Husnot] and Palmer amaranth (Table 16.2). GR horseweed populations were reported in cotton in Mississippi, and another in Mississippi had resistance to glyphosate and photosystem I inhibitors (Eubank et al. 2012). Resistant forms were also present in corn, rice and soybeans. Horseweed is now listed as one of the ten most troublesome weeds (Table 16.2) in the 2013 SWSS weed survey. GR Italian ryegrass was confirmed in Mississippi (Nandula et al. 2007) and has spread into in to cotton, corn and soybean. GR Italian ryegrass is rapidly spreading through the south and may soon become a candidate for the ten most troublesome weeds list.

Several Palmer amaranths have also evolved resistance to both glyphosate and pyrithiobac, an ALS inhibitor (Nandula et al. 2012). These multiple resistant Palmer amaranths have been found in several cotton-producing states (Heap 2017). GR Palmer amaranth is competitive in cotton, corn and soybean, and was listed as the most troublesome weed in cotton in 2009 and 2013 (Webster and Nicholls 2012, Webster 2013). Similar levels of resistance have appeared in other *Amaranthus* species including tall waterhemp (*Amaranthus tuberculatus* [Moq.] Sauer) and spiny amaranth (*Amaranthus spinosus* [L.]), both of which are present in cotton, corn and soybeans. The resistance in spiny amaranth was the result of hybridization between GR Palmer amaranth and spiny amaranth (Nandula et al. 2014).

Although there are no reports of herbicide resistance in barnyardgrass, sedges or junglerice (*Echinochloa colona* [L.] Link) in cotton, resistance in these weeds to ALS inhibitors has appeared in rice (Heap 2017). In barnyardgrass, resistance to synthetic auxins, and photosystem II, ACCase, lipid, cellulose, and DOXP inhibitors also been identified (Heap 2017). In Mississippi, a population with multiple resistances to four sites of action ACCase, ALS, photosystem II and cellulose inhibitors has been found. This case is interesting in that resistances to each of the four modes of action have not been reported on an individual basis. The breadth of resistances in barnyardgrass and the proximity of these populations to one another and to cotton production areas may have potential problems in cotton in the future. Glyphosate resistant goosegrass was reported in cotton in 2010 in Mississippi (Molin et al. 2013) but this has not become a serious issue.

To meet the challenge of resistant weed management, new control strategies based on preplant, PRE, and POST herbicides have been forthcoming. Extensive use of PPO inhibitors may speed development of new resistances. Hager (2013) addressed this issue showing waterhemp differences in sensitivity to flumioxazin at rates below the field scale application rate. As Hager (2013) pointed out, as degradation occurs, eventually a field rate is reached that will control sensitive plants but resistant plants survive providing a selective advantage for resistant plants. Cahoon et al. (2014) encourage use of diuron or fluometuron PRE with fomesafen to help prevent selection for resistance to PPO inhibitors. Seed companies have responded to the threat of resistant weeds by developing second and third generation transgenic lines having resistance to multiple herbicides, such as glyphosate, glufosinate, dicamba and 2,4-D, which allows application of herbicides with different mechanisms of action over the top of crops to control resistant weeds.

Integrated Weed Management

Integrated weed management (IWM) is an approach where any or all available chemical and non-chemical control tactics are employed to manage specific weeds regardless of herbicide resistance. The chemical methods include: residual herbicides, full-labelled rate applications, sequences, tank mixtures, application timing, and herbicide rotations. The non-chemical methods include: mechanical disturbance to separate roots from soil (preplant tillage, in-crop cultivation, hand hoeing, and postharvest tillage) and cultural practices to increase competition and shading (related to row spacing, plant population, competitive cultivars, crop rotation, cover crops). IWM strategies must be effective in disrupting establishment and spread of weeds to reduce seed production to prevent replenishment of the weed seedbank. GR weeds, especially GR Palmer amaranth, are a major problem in cotton production under CT system in highly erodible soils (Price et al. 2011). As a result, some growers are re-thinking their conservation tillage strategy and returning to high-intensity tillage practices to manage GR weeds (Hollis 2015). All weed control tactics must be considered without exclusion of one for the other. It should be noted that what works on one farm may not work on another—one program seldom fits all scenarios. Diversified approaches must be developed to match farm, county, state, and region-specific weed problems. Norsworthy et al. (2012) have published a document with several best management practices for HR weed management.

HR weed management in cotton requires diversification of control tactics. The predominant change in weed management practices from 1990s to current day, was the rapid adaption of glyphosate in herbicide-resistant cultivars of corn, cotton, and soybean resulting in changes in the weed flora of crops and the prevalence of glyphosate resistant weeds (Webster and Nichols 2012). Since, herbicides remain the most economical means of weed management, any diversification will increase production costs. In post HR cotton era, typical weed management systems in cotton include: soil-applied, POST (multiple herbicide options with single and stacked-trait cotton), POST-directed herbicides, several cultivations, and hand hoeing (as a last resort) to kill weeds, prevent seed production, and deplete the soil seedbank. To manage GR Palmer amaranth in cotton, growers are rotating herbicide chemistries, applying residual herbicides throughout the cropping season, and integrating with hand weeding and preplant, in-crop, and postharvest tillage (Sosnoskie and Culpepper 2014, Riar et al. 2013). The authors reported that these diverse, complex, and relatively expensive systems were effective in controlling GR Palmer amaranth in GR cotton.

Communication through the Internet

Access to current information regarding updates on herbicide-resistant varieties, new and old herbicides and combinations, and resistance management programs, is greatly needed. Fortunately, another striking change in agriculture has been scripting of up-to-date procedures and recommendations for managing cotton by the extension services in each of the states where it is grown, and development of easy access to this information through the internet (Riar et al. 2013). Procedures and recommendations are now available from every cotton growing region across the United States. Development and use of the internet in agriculture along with wireless connection capabilities has become the standard method for accessing the latest information on cotton weed management. The internet allows farmers to identify weeds, gather information on weed development, and obtain the latest information on control strategies. Also, they can link directly to businesses to place orders for agrochemicals from their trucks while in the field. Direct access to many of the weed journals and popular press newspapers, such as *Farm Press* is also possible. Agricultural extension services have placed most of their publications and recommendations on university web sites. Included are emails, addresses and phone numbers of local extension specialists. For example, North Carolina Cooperative Extension Service published a 208-page document entitled '2013 Cotton Information' which covers every aspect of cotton production with strong emphasis on weed management and provides sound recommendations for avoiding yield and grade reductions in the final product. It includes: lists of extension agents, tillage methods, variety comparisons, nutritional requirements and fertilizer recommendations, weed, disease and insect control recommendations, methods to follow the progress of the crop, models of budgets and profitability expectations and many more detailed items. These databases have and continue to provide instantaneous access to solutions to production problems.

Future Outlook

Current changes in herbicide usage trends will likely continue for some time which will result in an ever increasing number of species evolving resistance. There will also be an increase in the cases of multiple resistances adding to the complexity of making management decisions. For example, multiple resistance in Palmer amaranth to glyphosate and ALS inhibiting herbicides, which have been identified in Mississippi Palmer amaranth populations (Molin et al. 2016), will likely be further complicated by development of resistance to burn down herbicides such as flumioxazin or oxyfluorfen and related herbicides. Instances of hybridization between GR Palmer amaranth and other Amaranth species will continue to increase providing new platforms for genetic introgression and increasing species diversity (Nandula et al. 2014). The number and diversity of herbicide-resistant alleles in the seed bank may increase and fuel introgression of resistance mechanisms in future generations. Without proper stewardship, new herbicides may be used to exhaustion contributing to their early failure. These failures may force the use of more costly eradication methods, such as the use of hoe crews, resulting in increased food and fiber costs.

As technology increases, there may be several new methods of weed control technology on the horizon. High frequency electromagnetic waves (microwaves) (Brodie et al. 2012), high energy electric currents (Diprose et al. 1984, Vigneault and Benoît. 2001) and modernized flaming devices will be advanced to the marketplace. Spraying technology will likely continue to be improved. Drones with GPS devices, software and cameras, will be used for weed recognition, herbicide injury to crops and weeds and to direct applications to weed infested areas rather than make applications to the whole farm (Lin et al. 2017, Huang et al. 2016, Reddy et al. 2014). Furthermore, new herbicide combinations and HR cotton technologies will be forthcoming, re-establishing the predominance of POST applications which will drastically reduce the pesticide load on the environment and lead to reduced instances of herbicide-laden runoff into streams and rivers.

REFERENCES

Banerjee, S., S.W. Martin, R.K. Roberts, J.A. Larson, R. Hogan, Jr., J.L. Johnson, K.W. Paxton and J.M. Reeves. 2007. Adoption of Conservation-Tillage Practices in Cotton Production. Selected Paper prepared for presentation at the Southern Agricultural Economics Association Annual Meetings Mobile, Alabama, February 4-7, 2007.

Benbrook, C. 2009. Impacts of genetically engineered crops on pesticide use in the United States: the first thirteen years. The Organic Center. Available at: www.organic-center.org. (Accessed on October 18 , 2016).

Blaise, D. 2006. Effect of tillage systems on weed control, yield and fiber quality of upland (*Gossypium hirsutum* L.) and Asiatictree cotton (*G. arboreum* L.). Soil & Tillage Research 91: 207–216.

Bollman, S. 2013. Roundup Ready® Xtend Crop System for cotton. Proc. Beltwide Cotton Conferences, San Antonio, Texas, January 7–10, 2013. 393 p.

Brodie, G., C. Ryan and C. Lancaster. 2012. Microwave Technologies as Part of an Integrated Weed Management Strategy: A Review. International Journal of Agronomy. Vol. 2012, Article ID 636905, 14 p. doi: 10.1155/2012/636905.

Brown, A.B., T.L. Cole and J. Alphin. 1998. Ultra narrow row cotton: economic evaluation of 1996 BASF field Plots. 1998 Proc. Beltwide Cotton Conf. pp. 88–91.

Bryson, C.T. and P.E. Keeley. 1992. Reduced tillage systems. pp. 323–363. *In:* C.G. McWhorter and J.R. Abernathy (Eds.) Weeds of Cotton: Characterization and Control. The Cotton Foundation, Memphis, TN.

Bryson, C.T., K.N. Reddy and W.T. Molin. 2003. Purple nutsedge (*Cyperus rotundus*) population dynamics in narrow row transgenic cotton (*Gossypium hirsutum*) and soybean (*Glycine max*) rotation. Weed Technology 17: 805–810.

Buchanan, G.A. 1974. Weed survey—southern states. South. Weed Sci. Soc. Res. Rep. 27: 215–249.

Buchanan, G.A. 1992. Trends in weed control methods. pp. 47–72. *In:* C.G. McWhorter and J.R. Abernathy (Eds.) Weeds of Cotton: Characterization and Control. The Cotton Foundation, Memphis, TN.

Bukun, B. 2004. Critical periods for weed control in cotton in Turkey. Weed Research 44: 404–412.

Cahoon, C.W., A.C. York, D.L. Jordan, W.J. Everman and R.W. Seagroves. 2014. An alternative to multiple protoporphyrinogen oxidase inhibitor applications in no-till cotton. Weed Technol. 28(1): 58–71.

Cahoon, C.W., A.C. York, D.L. Jordan, W.J. Everman, R.W. Seagroves, A.S. Culpepper and P.M. Eure. 2015. Palmer amaranth (*Amaranthus palmeri*) management in dicamba-resistant cotton. Weed Technol. 29: 758–770.

Cardoso, G.D., P.L.C.A. Alves, L.S. Severino and L.S. Vale. 2011. Critical periods of weed control in naturally green colored cotton BRS Verde. Industrial Crops and Products 34: 1198–1202.

Cerdeira, A.L. and S.O. Duke. 2006. The current status and environmental impacts of glyphosate-resistant crops: a review. J. Environ. Qual. 35: 1633–1658.

Coble, H.D. and J.D. Byrd. 1992. Interference of weeds with cotton. pp. 73–84. *In:* C.G. McWhorter and J.R. Abernathy (Eds.) Weeds of Cotton: Characterization and Control. The Cotton Foundation, Memphis, TN.

Corbett, J.L., S.D. Askew, D. Porterfield and J. Wilcut. 2002. Bromoxynil, prometryn, pyrithiobac and MSMA weed management systems for bromoxynil-resistant cotton (*Gossypium hirsutum*). Weed Tech. 16: 712–718.

CTIC (Conservation Technology Information Center). 2016. 2008 Amendment to the national crop residue management survey summary. Available at: http://ctic.org/crm/. (Accessed on October 17, 2016).

Culpepper, A.S. and A.C. York. 1997. Weed management in no-tillage bromoxynil-tolerant cotton (*Gossypium hirsutum*). Weed Tech. 11: 335–345.

Culpepper, A.S. and A.C. York. 1998. Weed management in glyphosate-tolerant cotton. J. Cotton Sci. 2: 174–185 [Online]. Available at: http://www.cotton.org/journal/1998-02/4/174.cfm

Dill, G.M., C.A. CaJacob and S.R. Padgette. 2008. Glyphosate-resistant crops: adoption, use and future considerations. Pest Manag. Sci. 64: 326–331.

Diprose, M.F., F.A. Benson and A.J. Willis. 1984. The effect of externally applied electrostatic fields, microwave radiation and electric currents on plants and other organisms, with special reference to weed control. Botanical Rev. (Apr.-Jun., 1984) 50(2): 171–223.

Duke, S.O. and S.B. Powles. 2009. Glyphosate-resistant crops and weeds: now and in the future. AgBioForum 12: 346–357.

Eslami, S.V. 2015. Weed management in conservation agricultural systems. pp. 87–124. *In:* Chauhan, B.S., Mahajan, G. (Eds.) Recent Advances in Weed Management. Springer.

Fernandez-Cornejo, J., R. Nehring, C. Osteen, S. Wechsler, A. Martin and A. Vialou. 2014. Pesticide use in US agriculture: 21 selected crops, 1960–2008. United States Department of Agriculture, Economic Research Service, Economic Information Bulletin Number 124. pp. 80. May 2014.

Givens, W.A., D.R. Shaw, G.R. Kruger, W.G. Johnson, S.C. Weller, B.G. Young, R.G. Wilson, M.D.K. Owen and D. Jordan. 2009. Survey of tillage trends following the adoption of glyphosate-resistant crops. Weed Tech. 23: 150–155.

Hager, A. 2013. Herbicide resistance: are soil applied herbicides immune? Available at: http://bulletin.ipm.illinois.edu/?p=290. (Accessed on October 17, 2016).

Harrison, M.P., N.W. Buehring and R.R. Dobbs. 2009. Long term tillage effect cotton growth and yield. pp. 1255–1259. *In:* Proceedings, Beltwide Cotton Conference, CD-ROM.

Heap, I. 2017. The International Survey of Herbicide Resistant Weeds. Online. Internet. 1 October 2017. Available at: www.weedscience.org.

Hollis P. 2015 Conservation tillage systems threatened by herbicide-resistant weeds. Available at: http://www.gmwatch.org/en/news/archive/2015-articles/15996-conservation-tillage-threatened-by-herbicide-resistant-weeds.

Hurley, T.M., P.D. Mitchell and G.B. Frisvold. 2009. Weed management costs, weed best management practices, and the Roundup Ready® weed management program. AgBioForum 12: 281–290.

Keeley, P.E. and R.J. Thullen. 1993. Weeds in cotton: their biology, ecology, and control. U.S. Department of Agriculture, Agricultural Research Service, Technical Bulletin 1810, 35 p.

Keeling, J.W., E. Segarra and J.R. Abernathy. 1989. Evaluation of conservation tillage cropping systems for cotton on the Texas southern High Plains. J. Prod. Agric 2: 269–273.

Light, G.G. 2013. The LibertyLink® System and Future solutions. Proc. Beltwide Cotton Conferences, San Antonio, Texas, January 7–10, 2013. pp. 390–391.

Manalil, S., O. Coast, J. Werth and B.S. Chauhan. 2016. Weed management in cotton (*Gossypium hirsutum* [L.]) through weed-crop competition: a review. Crop Protection (Online, accessed on January 19, 2016). doi: 10.1016/j.cropro.2016.08.008

Martin, S.W. and J. Hanks. 2008. Partial returns from cotton conservation tillage practices in the Mississippi Delta. Mississippi Agricultural and Forestry Experiment Station. Bulletin 1171.

Martin, S.W. and J. Hanks. 2009. Economic analysis of no tillage and minimum tillage cotton-corn rotations in the Mississippi Delta. Soil and Tillage Research 102: 135–137.

Merchant, R.M., A.S. Culpepper, P.M. Eure, J.S. Richburg and L.B. Braxton. 2014. Controlling glyphosate resistant Palmer amaranth (*Amaranthus palmeri*) in cotton with resistance to glyphosate, 2,4-D, and glufosinate. Weed Technol. 28: 291–297.

McWhorter, C.G. and J.R. Abernathy. 1992. Weeds of Cotton: Characterization and Control. The Cotton Foundation, Memphis, Tennessee, USA.

Mohler, C.L., J.C. Frisch and C.E. McCulloch. 2006. Vertical movement of weed seed surrogates by tillage implements and natural processes. Soil Till. Res. 86: 110–122.

Molin, W.T., D. Boykin, J.A. Hugie, H.H. Ratnayaka and T.M. Sterling. 2006. Spurred anoda (*Anoda cristata*) interference in wide row and ultra narrow row cotton. Weed Science 54: 651–657.

Molin, W.T., J.A. Hugie and K. Hirase. 2004. Prickly sida (*Sida spinosa* [L.]) and spurge (*Euphorbia hyssopifolia* [L.]) response to wide row and ultra narrow row cotton (*Gossypium hirsutum* [L.]) management systems. Weed Biology and Management 4: 222–229.

Molin, W.T., V.K. Nandula, A.A. Wright and J.A. Bond. 2016. Transfer and Expression of ALS Inhibitor Resistance from Palmer Amaranth (*Amaranthus palmeri*) to an *A. spinosus* × *A. palmeri* Hybrid. Weed Sci. 64(2): 240–247.

Molin,W.T., A.A. Wright and V.K. Nandula. 2013. Glyphosate-resistant goosegrass from Mississippi. Agronomy 3: 474–487. doi: 10.3390/agronomy3020474

MSU (Mississippi State University). 2016. Weed control guidelines for Mississippi. Mississippi Agricultural and Forestry Experiment Station and Mississippi Cooperative Extension Service, Mississippi State University, Mississippi State, Mississippi. Available at: https://extension.msstate.edu/sites/default/files/publications/publications/p1532_1.pdf. (Accessed on October 17, 2016).

Nandula, V.K., K.N. Reddy, C.H. Koger, D.H. Poston, A.M. Rimando, S.O. Duke, J.A. Bond and D.N. Ribeiro. 2012. Multiple resistance to glyphosate and pyrithiobac in Palmer amaranth (*Amaranthus palmeri*) from Mississippi and response to flumiclorac. Weed Science 60: 179–188.

Nandula, V.K., D.H. Poston, T.W. Eubank, C.H. Koger and K.N. Reddy. 2007.Differential response to glyphosate in Italian ryegrass (*Lolium multiflorum*) populations from Mississippi. Weed Technology 21: 477–482.

Nandula, V.K., A.A. Wright, J.A. Bond, J.D. Ray, T.W. Eubank and W.T. Molin. 2014. EPSPS amplification in glyphosate-resistant spiny amaranth (*Amaranthus spinosus*): a case of gene transfer via interspecific hybridization from glyphosate-resistant Palmer amaranth (*Amaranthus palmeri*). Pest Manage. Sci. 70: 1902–1909. doi: 10.1002/ps.3754

Nichols, S.P., C.E. Snipes and M.A. Jones. 2004. Cotton growth, lint yield, and fiber quality as affected by row spacing and cultivar (Online). J. Cotton Sci. 8: 1–12. Available at: http://www.cotton.org/journal/2004-08/1/ (Verified on June 17, 2008).

Norsworthy, J.K., S. Ward, D. Shaw, R. Llewellyn, R. Nichols, T. Webster, K. Bradley, G. Frisvold, S. Powles, N. Burgos, W. Witt and M. Barrett. 2012. Reducing the risks of herbicide resistance: best management practices and recommendations. Weed Sci. 60 (Special Issue): 31–62.

Papamichail, D., I. Elefttherohorinos, R. Froud-Williams and F. Gravanis. 2002. Critical periods of weed competition in Greece. Phytoparasitica 30: 105–111.

Paulsgrove, M.D., W.L. Barker and J.W. Wilcut. 2005. Bromoxynil-resistant cotton and selected weed response to mixtures of bromoxynil and pyrithiobac. Weed Technology 19: 753–761.

Paxton, K.W., A. Jana and D.J. Boquet. 1995. Economics of cotton production within alternative crop rotation systems. pp. 379–381. *In:* Proc. Beltwide Cotton Conf. Memphis, TN: National Cotton Council of America.

Price, A.J., K.S. Balkcom, S.A. Culpepper, J.A. Kelton, R.L. Nichols and H. Schomberg. 2011. Glyphosate-resistant palmer amaranth: a threat to conservation tillage. Journal of Soil and Water Conservation 66: 265–275.

Price, A.J., K.S. Balkcom, L.M. Duzy and J.A. Kelton. 2012. Herbicide and cover crop residue integration for *Amaranthus* control in conservation agriculture cotton and implications for resistance management. Weed Technology 26: 490–498.

Reddy, K.N. 2001a. Broadleaf weed control in ultra narrow row bromoxynil-resistant cotton (*Gossypium hirsutum*). Weed Technology 15: 497–504.

Reddy, K.N. 2001b. Effects of cereal and legume cover crop residues on weeds, yield, and net return in soybean (*Glycine max*). Weed Technology 15: 660–668.

Reddy, K.N., I.C. Burke, J.C. Boykin and J.R. Williford. 2009. Narrow-row cotton production under irrigated and non-irrigated environment: plant population and lint yield. Journal of Cotton Science 13: 48–55.

Reddy, K.N., M.L. Locke, C.H. Koger, R.M. Zablotowicz and L.J. Krutz. 2006. Cotton and corn rotation under reduced tillage management: impacts on soil properties, weed control, yield, and net return. Weed Science 54: 768–774.

Reddy, K.N. and V.K. Nandula. 2012. Herbicide resistant crops: history, development and current technologies. Indian Journal of Agronomy 57: 1–7.

Riar, D.S., J.K. Norsworthy, L.E. Steckel, D.O. Stephenson, IV, T.W. Eubank, J. Bond, and R.C. Scott. 2013. Adoption of best management practices for herbicide-resistant weeds in Midsouthern United States cotton, rice, and soybean. Weed Technology 27: 788–797. doi: 10.1614/WT-D-13-00087.1

Shaner, D. 2000. The impact of glyphosate-tolerant crops on the use of other herbicides and on resistance management. Pest Manag. Sci. 56: 320–326.

Siebert, J.D. 2013. The Enlist™ Weed Control System – Technologies and Stewardship to Meet Diverse Agricultural Needs. Proc. Beltwide Cotton Conferences, San Antonio, Texas, January 7–10, 2013, 392 p.

Smith, D.T., R.V. Baker and G.L. Steele. 2000. Palmer amaranth (*Amaranthus palmeri*) impacts on yield, harvesting, and ginning in dryland cotton (*Gossypium hirsutum*). Weed Technol. 14: 122–126.

Snipes, C.E., G.A. Buchanan, J.E. Street and J.A. McGuire. 1982. Competition of common cocklebur (*Xanthium pensylvanicum*) with cotton (*Gossypium hirsutum*). Weed Sci. 30: 553–556.

Sosnoskie, L.M. and A.S. Culpepper. 2014. Glyphosate-resistant palmer amaranth (*Amaranthus palmeri*) increases herbicide use, tillage, and hand-weeding in Georgia cotton. Weed Sci. 62: 393–402.

Stetina, S.R., W.T. Molin and W.T. Pettigrew. 2010. Effects of varying planting dates and tillage systems on reniform nematode and browntop millet populations in cotton. Online. Plant Health Progress. doi: 10.1094/PHP-2010-1227-01-RS.

St. John, L., D.J. Tilley, D. Ogle, J. Jacobs, L. Holzworth and L. Wiesner. 2011.Principles of seedbed preparation for conservation seedings. Available at: https://www.nrcs.usda.gov/Internet/FSE_PLANTMATERIALS/publications/idpmctn10748.pdf

SWSS. 2009. Proceedings 2009 Proceedings, Southern Weed Science Society, 590 pp. Vol. 62 (Proceedings of the Southern Weed Science Society 66th Annual Meeting Royal Sonesta Hotel Houston, TX28-30 January 2013) 300 pp.

SWSS 2013. Proceedings 2009 Proceedings, Southern Weed Science Society, 590 pp. Volume 66 (Proceedings of the Southern Weed Science Society 66th Annual Meeting Royal Sonesta Hotel Houston, TX 28-30 January 2013) 300 pp.

Taylor, I.N., S.R. Walker and S.W. Adkins. 2005. Burial depth and cultivation influence emergence and persistence of *Phalaris paradoxa* seed in an Australian sub-tropical environment. Weed Research 45: 33–40.

Timmons, F.L. 2005. A history of weed control in the United States and Canada. Weed Science 53: 748–761.

Toler J.E., E.C. Murdock and A. Keeton. 2002. Weed management systems for cotton (*Gossypium hirsutum*) with reduced tillage. Weed Tech. 16: 773–780.

Tursun, N., A. Datta, S. Budak, Z. Kantarci and S.Z. Knezevic. 2016. Row spacing impacts the critical period for weed control in cotton (*Gossypium hirsutum*). Phytoparasitica 44: 139–149.

USDA (United States Department of Agriculture, National Agricultural Statistics Service). 2016a. Economics, Statistics and Market Information System. Acreage. Available at: http://usda.mannlib.cornell.edu/MannUsda/view DocumentInfo.do?documentID=1000. (Accessed on August 24, 2016).

USDA (United States Department of Agriculture, National Agricultural Statistics Service). 2016b. 2015 Agricultural Chemical use survey. Available at: https://www.nass.usda.gov/Surveys/Guide_to_NASS_Surveys/Chemical_Use/2015_Cotton_Oats_Soybeans_Wheat_Highlights/ChemUseHighlights_Cotton_2015.pdf (Accessed on October 19, 2016).

USDA (United States Department of Agriculture, Economic Research Service). 2016c. Crop production practices database. Available at: http://www.ers.usda.gov/data-products/arms-farm-financial-and-crop-production-practices/arms-data.aspx. (Accessed on October 21, 2016).

USDA (United States Department of Agriculture). 2016d. National Agricultural Statistics Service. Agricultural Chemical database (Online). Available at: http://www.pestmanagement.info/nass/. (Accessed on October 22, 2016).

USDA (United States Department of Agriculture). 2016e. National Agricultural Statistics Service. Quick Stats (Online). Available at: https://quickstats.nass.usda.gov/. (Accessed on March 24, 2017).

USDA (United States Department of Agriculture). 2017. Foreign Agricultural Service. Market and Trade data PSD Online. Available at: https://apps.fas.usda.gov/psdonline/app/index.html#/app/compositeViz. (Accessed on October 2, 2017).

Valco, T.D., W.S. Stanley and D.D. McAllister, III. 2001. Ultra narrow row cotton ginning and textile performance results. 2001 Proc. Beltwide Cotton Conf. pp. 355–357.

Varner, B.T., F.M. Epplin and G.L. Strickland. 2011. Economics of no-till versus tilled dryland cotton, grain sorghum, and wheat. Agronomy Journal 103: 1329–1338.

Vasilakoglou, I., K. Dhima, I. Eleftherohorinos and A. Lithourgidis. 2006. Winter cereal cover crop mulches and inter-row cultivation effects on cotton development and grass weed suppression. Agronomy Journal 98: 1290–1297.

Vencill, W.K., L.J. Giraudo and G.W. Langdale. 1993. Soil moisture relations and critical period of *Cynodon dactylon* (L.) Pers. (Coastal bermudagrass) competition in conservation tillage cotton (*Gossypium hirsutum* L.). Weed Res. 33: 89–96.

Vigneault, C. and D.L. Benoît. 2001. Electrical weed control: theory and applications. pp. 174–188. *In:* C. Vincent, B. Panneton and F. Fleurat-Lessard. (Eds.) Physical Control Methods in Plant Protection. Springer.

Webster, T.M. 2013. Weed Survey – Southern States. pp. 275–287. *In:* Proceedings, Southern Weed Science Society. Vol. 66.

Webster, T.M. and H.D. Coble. 1997. Changes in the weed species composition of the United States: 1974 to 1995. Weed Technology 11: 308–317.

Webster, T.M. and R.L. Nichols. 2012. Changes in the prevalence of weed species in the major agronomic crops of the southern United States: 1994/1995 to 2008/2009. Weed Sci. 60: 145–157.

WSSA (Weed Science Society of America). 2007. Herbicide Handbook. 9th edition. Weed Science Society of America, Champaign, Illinois. 458 p.

Wilcut, J.W., A.C. York and D.L. Jordan. 1993. Weed management for reduced-tillage southeastern cotton. pp. 29–35. *In:* M.R. McClelland, T.D. Valco and R.E. Frans. (Eds.) Conservation-Tillage Systems for Cotton: A Review of Research and Demonstration Results from Across the Cotton Belt. Arkansas Agric. Exp. Sta. Special Report 160.

Willcutt, M.H., E.P. Columbus, N.W. Buehring, R.R. Dobbs and M.P. Harrison. 2006. Evaluation of a 15-inch spindle harvester in various row patterns, three years progress. 2006 Proc. Beltwide Cotton Conf. pp. 531–547.

Sustainable Weed Control in Oilseed Rape

Peter J.W. Lutman

26 Singlets Lane, Flamstead, St Albans AL3 8EP, UK
E-mail: peter.lutman@btinternet.com

Introduction

Oilseed rape (also called canola), primarily *Brassica napus* L., was the seventh most widely grown crop in the world in 2014 (35.8 Mha) (Faostat 2016). Although most production is of *B. napus*, some *B. campestris* L. (syn *B. rapa* L.) (turnip rape) is also grown in some countries, as it is more frost resistant. The major producers, worldwide, are shown in Table 17.1, with Canada and China, followed by the nations of the European Union (including UK) and India being the major growers. However, production methods and crop types grown vary greatly. Crops grown in Europe are primarily autumn-sown meaning that the crop remains in the ground for virtually 12 months, whilst in Canada, Australia and the USA, because of the more severe climate (temperature, rainfall limitations) the crop is 'spring' sown, meaning it is present in the ground for only around six months. These differences have a considerable impact on yields and on weed management. Yields in spring crop producing countries are in the region of 1-2 t/ha, whilst in autumn rape yields are much higher (3-4 t/ha).

Not only does the type of oilseed rape grown vary across the world, the degree of mechanisation employed in production varies greatly. For example, in China and India much of the agronomy is done by hand with hand planting and hand harvesting. Mechanisation of production systems is accelerating but still much of the production is small-scale, small-farmer based, with much weed control based on hand weeding. In China, much of the crop is transplanted (Hu et al. 2017), like rice, which poses very different weed management issues to 'conventional' seed based production systems. In contrast, production in Europe and North America is fully mechanised, with increasingly sophisticated management and highly technological equipment.

In much of the world oilseed rape is grown primarily in rotation with cereal crops in arable production systems. Consequently, the weed flora tends to reflect that of the dominating cereals. Indeed, the problems associated with managing weeds in the cereal crops are having an increasing effect on the weed flora and its management in oilseed rape. Cultivation practices, in seed-based production systems vary, though in many countries non-inversion tillage and direct drilling dominate and have largely replaced the more traditional plough based tillage. This too has had (is having) impacts on the weed flora and its management. A further issue that is particularly significant for oilseed rape, because of its small seeds and their potential to persist, is the problem caused to subsequent arable crops by volunteer oilseed rape. Harvesting

Table 17.1. Major worldwide producers of oilseed rape: area grown and yields (average 2010–14) (Source: Faostat 2016)

Country	Area grown (Mha)	Yield t/ha
Australia	2.42	1.28
Canada	7.76	1.95
China	7.24	1.83
France	1.51	3.37
Germany	1.39	3.78
India	6.30	1.18
Poland	0.87	2.71
Russian Federation	0.93	1.20
Ukraine	0.82	2.11
United Kingdom	0.70	3.48
USA	0.57	1.78
(European Union)	6.69	3.12)
World	35.8	

(mechanized and hand harvesting) can leave substantial numbers of seeds in the field, which then produce plants that emerge in subsequent crops.

Pesticides are the main tools for crop protection in oilseed rape production in Europe, North America and Australia. As with all intensively grown arable crops, pest, disease and weed control is of prime importance. Fungicides, insecticides and herbicides are widely used by most growers, though environmental concerns and pest resistance are reducing pesticide dependence and alternative cultural practices are being developed. Weed control has been, and continues, to be based mainly around herbicides. But there is a major dichotomy between Europe and N. America, as the former has not approved the cultivation of the genetically modified herbicide tolerant (GMHT) cultivars that dominate production in N. America. This has major implications for weed control and will be discussed in greater detail later in the chapter. Some rape is produced organically, but the market is small, though the crops produced by the small-scale producers in Asia are often 'organically' grown, with little or no use of pesticides.

Weeds of Oilseed Rape and their Impact

The weeds that do occur in oilseed rape vary somewhat around the world but historically 'European' weeds tend to dominate arable crops in N. America and Australia as well as in Europe, though the relative importance varies. As mentioned in the introduction, the weeds of the cereal crops are often the primary weeds of oilseed rape but the two crop types do create differing problems and provide an opportunity to manage the problems of the other.

Major Weeds

Surveys of weeds are not very common and those that have been done vary in their scope and methodology, but they do provide some indication of the major problem weeds. In this chapter I have endeavoured to seek out information from the main oilseed rape growing areas of the world that is reasonably up-to-date. The data for the UK comes from a report by Cook et al. (2015) which includes a survey of growers' views of major weeds in 2009 plus some earlier more objective survey data. In contrast, the French data from Fried et al. (2015) are based on an extensive survey in France with weed counts carried out in autumn and spring in 2002–10. Similarly, the Canadian data from Leeson (2016) are based on counts of weeds in summer on a large number of fields in Sakatchewan (representative of much of Canadian production in the prairie states) in 2014–15. The information from Germany (Hanzlik and Gerowitt 2012) is the result of surveys in 2005–07 of untreated areas of 1,463 fields in the autumn. The Australian data (Llewellyn et al. 2016) are based on opinions of the major weeds from an extensive farmer survey in 2014. The method of surveying does influence the results as objective surveys tend to record smaller, less aggressive, species more highly than surveys based on farmer opinion, and in contrast, farmer surveys tend to over record vigorous species that are hard to control. Thus, *Viola arvensis* and *Veronica* spp. feature highly in the French and German surveys but are not mentioned in the UK farmer survey despite their relative abundance in the UK (Lutman et al. 2009).Despite these differences, the surveys do identify the main weeds of concern in oilseed rape.

The important weeds to occur in oilseed rape can be split into several groups for comparison: dicotyledonous species (excluding Brassicaceae), dicotyledonous species belonging to the Brassicaceae (Cruciferae), annual grasses (including volunteer cereals) and perennials. Table 17.2 lists the 28 most significant weed species reported in recent papers in Europe, Canada and Australia and includes some indication of relative importance in those countries. Annual grasses cause major problems worldwide, partly because they tend to be quite abundant but also because they have become increasingly resistant to the herbicides used for their control both in oilseed rape and in cereals. *Alopecurus myosuroides* (L.) is the problem in the UK and France, whilst *Setaria viridis* (L.) P. Beauv. and *Avena* spp. pose the problem in Canada and *Lolium rigidum* (Gaudin) and *Avena* spp. are the key weeds in Australia. *Apera spica-venti* (L.) P. Beauv. is commoner than *A. myosuroides* in Germany (although not included in above list) and fills the same niche. All these grasses exhibit rapid germination post crop planting, high competitive ability and abundant seed return. The problematic dicotyledonous species of main concern can be classified in two ways, by their periodicity of emergence and their taxonomic family. Weed species in the same family as oilseed rape (Brassicaceae) are particularly difficult to manage as they tend not to be susceptible to most herbicides that can be used in rape, *Raphanus raphanistrum* posing appreciable problems in Australia. However, where herbicide tolerant (HT) cultivars are grown (e.g., glyphosate and glufosinate resistant) the weeds in the Brassicaceae can be controlled more easily. In Table 17.2 none of these weeds feature as major problems in Canada, although they are present, as this is where HT cultivars have been grown most widely. Spring-sown rape has a different flora to autumn-sown rape. In the spring crop, spring emerging weeds, such as *Chenopodium album* (L.) and *Fallopia convolvulus* (L.) Love cause problems in many countries. In contrast, the autumn germinating species, such as *Papaver rhoeas* (L.), *Veronica* spp. and *Galium aparine* (L.) are of more importance where winter rape crops are grown. The frequent appearance of *Geranium* spp. as major weeds in European countries reflects their intrinsic resistance to many of the major broad-leaved weed herbicides available to European growers. Perennial weeds are not of major concern, though *Elymus repens* (L.) Gould remains problematic in some crops in the UK and *Taraxacum* is increasing in Canada, perhaps a reflection of the widespread adoption of no-till production systems. Thistles, both *Sonchus* and *Cirsium* species, both annual and perennial, are also of concern in Europe and Canada. This may be for two reasons: i) they tend to emerge after the main herbicide treatments have been applied and ii) they are not very susceptible to standard herbicides (unless HT cultivars are grown).

Weed Effects on Crop Production

The primary effect of weeds in oilseed rape is their competitive effect on crop growth and seed yield, as it is in most arable crops. Weeds can also adversely affect crop quality, by contaminating the harvested seeds and can lower harvesting efficiency.

Weed Competition

Although oilseed rape is a tall and aggressive plant, once established, and hence can compete strongly with weeds, it is very vulnerable to weed competition at its early stages. The impact

Table 17.2. Most important weeds of oilseed rape in Europe, Canada and Australia

Category	Species	UK[1]	Germany[2]	France[3]	Canada[4]	Australia[5]
d	Arctotheca calendula (L.) Levyns					*
d	Chenopodium album L.		*	*	*	
d	Cirsium/Sonchus spp ¶	*		*	**+	**
d	Fallopia convolvulus (L.) A.Love				***	*
d	Galium spp	***	*	*	**+	
d	Geranium spp	**	*+	**+		
d	Kochia scoparia (L.) Schrad.				*	
d	Lamium spp		*			
d	Matricaria/Tripleurospermum spp	**	**			
d	Papaver rhoeas L.	***	*			
d	Stellaria media (L.) Vill.	*	**	*	*	
d	Veronica spp		*	**		
d	Vicia sativa L.					**
d	Viola arvensis Murray		**	***		
db	Brassica rapa L.					**
db	Capsella bursa pastoris (L.) Medik	**	**	**+	*	
db	Raphanus raphanistrum L.	(*)		*		***
db	Sinapis arvensis L.	*		**	*-	**
g	Alopecurus myosuroides Huds	***		***		
g	Avena spp				***	***
g	Bromus/Anisantha spp	(*)				**

(Contd.)

	Species			
g	*Echinochloa crus galli* (L.) P. Beauv.			*+
g	*Lolium* spp		*	***
g	*Setaria viridis* (L.) P. Beauv.			***
g	volunteer cereals	*		**
gp	*Elymus repens* (L.) Gould.	(*)		
dp	*Taraxacum* spp			*+

g = Annual grass
gp = Perennial grass
d = Dicot
db = Dicot Brassicaceae
dp = Perennial dicot

+ = Increasing - = Decreasing
*** = Most common
* = Abundant but less common
(*) = Not highly frequent but significant
¶ = Includes both annual and perennial spp.

[1] Cook et al. 2015
[2] Hanzlik and Gerowitt 2012
[3] Fried et al. 2015
[4] Leeson 2016
[5] Llewellyn et al. 2016

of weeds on the crop depends on the weed species, their abundance and the relative vigour of the crop (density and plant size). A further complicating factor is whether the crop is spring-sown, at a time of increasing temperatures or autumn-sown when temperatures are declining. Temperatures influence the relative growth rate of the rape and weeds, as the base temperature for oilseed rape (5°C) (Morrison et al. 1989) tends to be higher than for many weeds, that germinate in autumn (base temperatures for green area accumulation: *A. myosuroides* 0.4°C, *Stellaria media* [L.] Vill.–1.7°C, *G. aparine* 1.9°C [Storkey and Cussans 2000]). However, this does not apply to spring-sown rape where spring-germinating weeds can have higher base temperatures (base temperatures for germination: *Amaranthus retroflexus* L. 8.9°C, *C. album* 5.9°C, *Solanum nigrum* L. 11.6°C [Guillemin et al. 2013]). Consequently, weeds in autumn-sown rape crops are relatively more competitive when temperatures are low. The key to minimising the effects of weeds on this crop is to ensure that the crop establishes well and grows vigorously after emergence. In both autumn- and spring-sown crops two factors control this early vigour, temperature and rainfall. In autumn-sown crops it is critical to establish the crop whilst autumn temperatures are high and this in Europe means sowing the crop in August or early in September. Crops sown later tend to grow less well because of lower temperatures. It is true that 'climate change' has been influencing rape establishment, as in the 21st Century later sown rape has been more successful. Early sowing, whilst good for vigor, exposes the seeds to potential inadequate moisture, as soils in late summer tend to be drier than those in autumn, so it is important that the crop is sown into moist seedbeds. In spring-sown crops the critical issues are again moisture and temperature. If the crop is sown too early when temperatures are low it may not establish well. Similarly, if the seedbed is too dry then establishment will be poor and weeds will take advantage of the spaces.

Non-edaphic factors can also impact on the vigour of young oilseed rape plants. Pest and disease attacks can reduce crop growth, thus making the crop more vulnerable to weeds. For example, the recent withdrawal of the neonicotinoid herbicides for use in oilseed rape from many European countries has resulted in the re-appearance of significant attacks from *Psylliodes chrysocephala* L. (cabbage-stem flea beetle).The larvae of this pest can severely reduce the vigour of surviving plants, thus exposing them to greater weed competition.

The response of oilseed rape to weed competition can be very variable. This is partly due to crop vigour but also to the density and species composition of the weed flora. Harker (2001) reported ten experiments in Canada in spring rape investigating yield loss from weeds and only found detectable responses in four of them. However, where yield losses were found the crop yield was reduced by a mean of 40%. Research in the UK in winter rape, whilst often detecting yield losses from annual grasses and volunteer cereals often failed to record impacts on yields from broad-leaved species (Davies et al. 1989, Lutman et al. 2000). It has been possible to rank weed species as to their respective risks to yield but not to clearly identify precise responses. For example, *G. aparine* has the ability to be far more competitive than *S. media* which in turn is potentially more damaging than *V. arvensis* (Lutman et al. 1995). Canadian research also indicates differing competitive abilities of their major weeds, for example, Blackshaw et al. (1987) showing that *S. arvensis* was much more competitive than *C. album*. French studies have categorised weeds according to their periodicity of emergence and their relative height (compared to rape) to achieve a ranking of their likely impact (Primot et al. 2006).

Influence of Time of Weed Removal on Yields. One component of the impact of weeds on crop yield relates to the time of control. A lot of research has been done into the 'critical periods' of weed removal in a range of crops: how long can weeds persist in the crop before they affect yields: for how long must the weeds be removed to ensure later emerging plants do not affect yields (see Knezevic et al. 2002). Research in spring rape in Canada reported by Martin et al. (2001) showed that the crop needs to be kept weed-free until the four-leaf stage. In most situations subsequently emerging weeds did not merit control. In another study in Canada delaying weed control until the six-leaf stage did not significantly reduce yields compared to removal at the two- and four-leaf stages (Harker et al. 2003). Studies in the UK compared the impact of relative time of removal of individual species on yields of autumn-sown rape and showed that volunteer barley should be controlled in early autumn, whilst *S. media* could often be left until late autumn whilst control of *G. aparine* could be left until the spring (Freeman and Lutman 2004).Thus, the importance of time of weed removal depends on the type of crop (spring versus autumn-sown rape) and the abundance and species of weeds present. The opportunity to delay control also depends on the availability of suitable control techniques to remove larger weeds at the later times. This is covered in a subsequent section.

Effects of Weeds on Crop Quality and on Harvesting

Weeds remaining at harvest time can adversely affect crop harvesting and seed quality. The presence of large amounts of green material in the crop prior to harvest can obstruct harvesters and increase seed moisture levels. A critical component of the effect of *G. aparine* on the crop relates to the biomass of the weed late in the growing season. Other late growing weeds such as *Sonchus* and *Cirsium* spp. can have a similar effect. The pre-harvest application of glyphosate, approved in some countries to aid harvesting, will reduce the effect of these late growing weeds. Weeds in the Brassicaceae can affect the rape seed's oil quality, as seeds of, for example, *Sinapis arvensis* (L.) and *R. raphanistrum* have a much higher erucic acid and glucosinolate content than oilseed rape and thus admixture of seeds of these weeds in the rape seed will adversely affect oil and protein meal quality. The round seeds of G. aparine can also contaminate the harvested rape seeds, in extreme cases resulting in crop rejection. Llewellyn et al. (2016) reported that 19% of Australian growers surveyed had had to clean their seed and 12% had incurred a financial penalty because of weed seed contamination (*R. raphanistrum* being a particularly common weed in Australia (Table 17.2). Thus, the impact of weeds is wider than simply their competitive effects on crop growth and yields.

Weed Management

Non-chemical Weed Control

Pre-crop Sowing

The management of weeds in oilseed rape is initiated by the cultivation practices employed prior to sowing. The primary tillage choices have evolved in recent years for a range of reasons. Traditionally, arable crops including oilseed rape were established following ploughing or deep non-inversion cultivation either with tines and/or discs. For financial and environmental reasons, the intensity of tillage prior to sowing oilseed rape has declined such that shallow tillage to incorporate trash from the previous crop and direct drilling with no cultivation is now common in Europe (Melander et al. 2013) and N. America. A survey of Canadian growers in 2012 showed that only 12% used 'traditional' tillage, 30% used minimum tillage and 57% established their rape crops by direct/zero tillage (Canola Council of Canada 2013). There has also been a dramatic increase in no-till and reduction in tillage passes over the past 15 years in Australia (Llewelyn et al. 2016). Less intensive tillage is faster and cheaper than conventional tillage and also is believed to result in less soil erosion, maintained or increased soil organic matter, increased soil microbial and faunal populations and increased soil moisture storage. However, these less intensive cultivations prior to drilling do tend to result in a greater number of weeds being present post sowing. Often such tillage goes hand-in-hand with the use of the herbicide glyphosate at drilling and/or post drilling to control the surviving weeds (see section on herbicides below). This approach uses shallow cultivation post-harvest of the previous crop to stimulate weed germination followed by the use of glyphosate to destroy the emerged weeds. If time and appropriate weather permits this can be done more than once. Reduced tillage also promotes a specialised weed flora adapted to the conditions created: annual grass weeds becoming a particular problem (Melander et al. 2013). There is some debate as to whether in-crop weed infestations are potentially higher or lower in no till/minimum till situations, compared to more intensive cultivation systems. It is true that in minimum tillage systems far more of the seeds shed in the previous crop remain on or near the soil surface where they can germinate. But, they are also more exposed to seed predation. Additionally, in more intensive cultivated soils, especially ploughed ones, weed seeds become buried such that fewer will germinate in the next crop but will persist for longer in the soil seedbank to infest crops in later years. Whether an increased (or decreased) weed flora matters to the emerging oilseed rape crop also depends on the tools (often herbicides) available to control them. If control is difficult any cultural tool that will reduce populations is helpful, but if control is simple then population size is less important.

Cover Crops and Oilseed Rape

There has been some interest in sowing another crop either prior to sowing the oilseed rape or between the rows, on the basis that this can be killed at a later stage, will suppress weed germination and can have other benefits. But, the introduction of a pre-sowing cover crop increases production costs and its destruction prior to establishing the next crop can be challenging. According to Owen et al. (2015) adoption of cover cropping in N. America has been 'low', as other Integrated Pest Management (IPM) tactics are more attractive. An alternative approach to cover crops is the sowing of an intercrop between the rows of oilseed rape. For example, Cadoux et al. (2015), in France, report promising results from sowing frost sensitive legumes between the rows of autumn-sown oilseed rape, as yields were not reduced by the inter-crops and weed growth was less. However, a study exploring the effect of inter-cropping rape and buckwheat (*Fagopyrum esculentum* Moench) in Germany showed no beneficial effect from the buckwheat on the weeds and rape yields were reduced (Stumm et al. 2009). Most interest in Europe relating to cover crops is associated with their establishment prior to spring-sown crops when these are planted as an alternative to autumn-sown crops, to reduce problems with annual grass weeds and to reduce nitrate leaching over the winter. Little research has been done on spring rape. Most of the emphasis has been on spring cereals.

Enhancement of Crop Growth

The approaches to optimising the effect of the crop on competing weeds are many and varied. Clearly, as outlined in the previous sections, the vigour of the crop is a critical factor in determining the impact of weeds. Thus, timing of drilling (in combination with prevailing weather) can be of great significance. Crops sown at the optimum time with adequate soil moisture will always be more vigorous and suppressive than those sown in less optimum conditions. This is particularly critical with oilseed rape, as plant vigour in for example cereals is not so dramatically affected by adverse conditions. Other factors, such as crop density and row spacing and differences in vigor between cultivars, and indeed seed size have all been explored by researchers. It is clear from Canadian research that higher seed rates and larger seeds will increase crop biomass, which helps weed suppression, but this does not necessarily result in higher seed yields (in the absence of weeds) (Harker et al. 2015).Much research has shown that oilseed rape yields are not very responsive to changes in seed rate (e.g., Kutcher et al. 2013). As crop seed is expensive there is pressure on farmers to reduce seed rates but this may not be a good choice if it reduces the crop's ability to suppress weeds. There is also evidence from Canadian research that hybrid cultivars are more vigorous and thus compete more strongly against weeds (Zand and Beckie 2002, O'Donovan et al. 2007).This can result in higher yields. Other research has also confirmed that cultivars do differ in their competitive abilities (Galon et al. 2015, Lemerle et al. 2011) and Lemerle et al. (2014) suggests that appropriate cultivar choice could be a low-cost tactic for integrated weed management. A further agronomic option to manipulate the crop's competitive ability is to adjust row spacings. It is fairly clear that widening row distances from a 'standard' 12 or 20 cm to 50 cm or more can reduce yields and reduce crop competition with weeds between the rows (Kutcher et al. 2013), even if seed density m^{-2} is maintained. Plasticity of oilseed rape means that this yield loss does not always arise (Cook et al. 2015), however, widening the row distances can open up opportunities for alternative weed management practices (see below). Overall, research shows that increasing seed rate and growing more competitive cultivars (especially hybrids) will increase the crops ability to suppress weeds, even if it does not raise yields. The overriding factor controlling the competitive ability of the crop is how well (and quickly) it establishes. This is determined by the weather (rainfall, temperature) which is not under the control of the grower but even so needs to be given careful consideration.

Direct Physical Weed Control

A range of tools are available to the grower to endeavour to kill weeds without killing the crop plants: tine weeders, rotary hoes, weed harrows, etc. The prime difficulty with overall mechanical weeding is that the more vigorous the disturbance the better the weed control but the greater the damage to the crop. There is an extensive literature on physical weed control in a range of arable and horticultural crops (van der Weide et al. 2008, Melander et al. 2013). Clearly, mechanical control is particularly important to growers of organic rape crops but is becoming increasingly relevant to 'conventional' growers as herbicide-based control becomes more difficult. Research in France (Lieven et al. 2008, Lieven and Lucas 2009) has explored the efficacy of a rotary hoe and a tine weeder, both in terms of safety to the rape crop and their performance on the weeds. Good control is possible with the mechanical systems but appropriate timing both in relation to crop and weed growth stage was critical for success, as also was a high level of operator skill. Soil type also influences the success of mechanical control and interacts with the type of hoe, tine weeders being more suitable for clay-loams and rotary hoes more suited to silts (Lieven et al. 2008). It is not easy to achieve good control as can be seen by the range of efficacies recorded in French research studying hoe performance (Table 17.3). Weed size at the time of hoeing is critical, as performance is much better on smaller plants (cotyledons–2 leaves). Grass weeds are less easy to control mechanically than broad-leaved weeds (Table 17.3).

Table 17.3. Efficacy of hoes on grass and broad-leaved weeds in oilseed rape in France

Weed type	Area cultivated	Mean % control	Range of response and number of trials
Grass weeds	Inter-row	73	(34-100) 10
	In row	34	(0-72) 10
	Both areas	54	(4-100) 13
Broad-leaved weeds	Inter-row	82	(64-96) 37
	In row	31	(0-81) 37
	Both areas	63	(30-100) 72

Terres Inovia 2016a

The rooting characteristics of the weeds also impacts on their sensitivity to mechanical control as tap rooted weeds (e.g., *S. arvensis, Sonchus* spp.) are more difficult to control than more fine rooted ones (e.g., *V.arvensis*). The success of the type of mechanical cultivation depends on the machine used as some are more appropriate for use on very small crop plants whilst others should only be used on larger ones (Table 17.4).

Table 17.4. Selectivity of hoes according to oilseed rape growth stage

Hoe type	Pre-em	Cotyledons	2 leaves	3 leaves	4 leaves	5 leaves
Tine weeder	+++	---	---	+++	+++	+++
Rotary hoe	+++	+++	+++	+++	-	--
In-row hoe	---	---	--	+++	+++	+++

+++ = Possible use, good selectivity, - = Possible use, the selectivity is reduced, -- = Possible use but the selectivity starts to decrease, --- = Not recommended for use, the selectivity is insufficient (from Lieven et al. 2008)

Other studies on mechanical weed control have tended to focus on inter-row techniques, in association with widening row widths, as the potential vigour of oilseed rape can effectively minimise the competition from in-row weeds (Melander et al. 2013). Such an approach could also include a combination of inter-row hoeing and intra-row herbicide treatment (discussed below).

Chemical Weed Control

The control of weeds in oilseed rape in Europe, N. America, and Australasia has been for the last 40 years (or since the crop was introduced) based mainly on herbicides. Weed control in the increasingly mechanised production in China and India is also being more and more based on herbicides. Although herbicide development in rape has not been as intensive as it has been on cereals, rice, soyabeans and maize, over the decades the chemical industry has developed a range of effective products for the crop. However, in recent years the flow of new herbicides has declined and few new products have emerged. In N. America since the mid-1990s there has been a switch from traditional selective herbicides to rape crops bred to be resistant to herbicides (HT rape).The glyphosate and glufosinate resistant cultivars have been created using genetic modification (GMHT), whilst imidazolinone (and triazine) resistant cultivars have been created by conventional breeding. Now over 90% of the crop in Canada is resistant to the herbicides glyphosate, glufosinate or those in the imidazolinone group (Smyth et al. 2011). In Europe, where HT rape resistant to glyphosate and glufosinate has not been approved, increasingly strict regulations have resulted in a lot of older herbicides being withdrawn and others having their uses restricted (limited doses, more restricted timings). At the same time, intensive herbicide use has resulted in the development of increasing numbers of herbicide resistant weed biotypes. This applies to traditional selective herbicides and to products used in HT rape. This scenario is common to all arable crops worldwide but there are specific aspects that are especially relevant to oilseed rape.

Conventional Herbicide-based Weed Control

Conventional weed control after sowing oilseed rape can be split into two classes: pre-emergence herbicides mainly focused on broad-leaved weed control (this could include very early post-emergence treatments) and post-emergence treatments aimed primarily at grass weed control. For many years the control of grass weeds has been achieved with a range of aryloxyphenoxy or cyclohexanedione herbicides (colloquially called 'fops' and 'dims' or ACCase herbicides: HRAC resistance group A: WSSA resistance group 1) but over the last 20 years herbicide resistance in the target weeds has become widespread. As conventional weed control in N. America and Australia has been largely replaced by HT rape (both GMHT and conventional bred HT cultivars) this section focusses on weed management practices in Europe.

Table 17.5 presents data on herbicide use in the UK in 2014, but this is reasonably representative of products available across Europe though the spectrum available changes according to more or less stringent regulatory regimes. In the UK, most crops received 3–4 herbicides, generally a combination of pre-emergence/early post-emergence products followed by one or two post-emergence ones for grass weed control. In Denmark, the number of herbicides available is more restricted: only seven are approved (Ørum 2014) compared to more than 16 in the UK (Table 17.5). Most of the Danish approved products are for grass weed control and few pre-emergence products are approved. More than 17 herbicides are approved in France (Terres Inovia 2016c). Additionally, there is variation in the dose that can be applied, because of climatic and cultural differences, and differences in the regulations across the EU.

Pre-emergence Weed Control

In many European countries, a number of pre-emergence herbicides can be used either alone or in mixtures to control broad-leaved weeds. Metazachlor is a key component of these mixtures accounting in the UK for more than 500,000 treated ha (c. 25% of treated area). Specific weeds drive the precise choice of products. For example, to control *Geranium* spp. dimethenamid or bifenox should be used, whilst the addition of quinmerac to metazachlor improves the control of *P. rhoeas* and *G. aparine*. For a number of weed species failure to apply a suitable pre-emergence herbicide will result in no control, as the approved post-emergence products do not control a very wide spectrum of broad-leaved weeds. A particular problem for oilseed rape is the control of other weeds in the Brassicaceae, such as *S. arvensis* and *R. raphanistrum*. None of the most widely used herbicides will control these weeds. In recent years rape has been bred to be resistant to the imidazolinone herbicides (using conventional breeding techniques) and these cultivars now have a niche in Europe, especially where Brassica weeds are a particular problem. The imazamox mentioned below Table 17.5 is for this purpose.

Table 17.5. Herbicide use in oilseed rape in the UK 2014 (sown area 674,000 ha (98% winter rape)

Herbicide	Target timing/weed type	Area treated (sprayed hectares) × 1000
Metazachlor (alone or with quinmerac)	Pre-em blw*	286
Clomazone	Pre-em blw	141
Dimethenamid (with metazachlor & quinmerac)	Pre-em blw	264
Bifenox	Pre-em blw	87
Propyzamide (alone or with aminopyralid)	Post-em grass & blw	415
Carbetamide	Post-em grass & blw	75
Clopyralid (alone or with picloram)	Post-em blw	121
Clethodim	Post-em grass	133
Cycloxydim	Post-em grass	34
Fluazifop p-butyl	Post-em grass	101
Propaquizafop	Post-em grass	331
Quizalofop (p-ethyl & p-tefuryl)	Post-em grass	131
Tepraloxydim	Post-em grass	68
Total area treated (major products)		2,187

* blw = Broad-leaved weeds (from Garthwaite et al. 2015)
Other herbicides also used on smaller areas of oilseed rape: dimethachlor, napropamide, imazamox.

Post-emergence Weed Control

As mentioned above, the broad-leaved weed control spectrum of the limited number of post-emergence herbicides is restricted. Clopyralid has a particular use in controlling *Sonchus* spp. and aminopyralid has the potential to control *P. rhoeas*. Post-emergence treatments are primarily targeted at grass weeds and volunteer cereals from the previous crop. The latter are fairly easy to control but as the volunteers can be aggressive competitors with rape they sometimes merit a specific early autumn treatment. However, their main purpose is to control annual grasses such as *A. myosuroides, Bromus/Anisantha* spp., *Avena* spp. and *Lolium* spp. Historically, these have been well controlled by the aryloxyphenoxy and cyclohexanedione products but resistance to these 'ACCase' herbicides is now extremely widespread in Europe and elsewhere in the world rendering many of these products increasingly ineffective. However, their use is still very widespread in Europe. In the UK, nearly 800,000 ha (over 30% of the treated area) was treated with these herbicides for control of volunteer cereals and annual grasses. Where target-site resistance has occurred some ACCase herbicides are less vulnerable to resistance than others (see below).The only alternatives for grass weed control are propyzamide and carbetamide.

Both are soil acting products and perform better when temperatures are relatively low, so tend to be applied late in the autumn. Their performance tends to be more vulnerable to environmental issues (rainfall, temperature) than the ACCase herbicides.

Problems of Herbicide-based Weed Control

Weed Resistance

Herbicide resistance in weeds is a common problem for developed agriculture worldwide and can affect virtually all herbicides and many, many weed species, both broad-leaved and grass species. World-wide c. 25 weed species in oilseed rape have been recorded as exhibiting resistance to herbicides, the vast majority being grass weeds resistant to ACCase herbicides (Heap 2016). In other crops, the control of annual grasses is affected by resistance to other herbicide groups, such as sulfonylureas but this is not an issue for oilseed rape (but see below for discussion of imidazolinone resistant rape). Resistance can be due to both target site mutations and/or enhanced metabolism. In surveys of the frequency of resistance to ACCase herbicides major resistance occurrence has been recorded for example in Canada (*A. fatua*) (Beckie et al. 2013), Australia (*L. rigidum*) (Boutsalis et al. 2012) (*A. fatua*) (Owen and Powles 2009), France (*A. myosuroides*) (Delye et al. 2007), UK (*A. myosuroides*) (Hull et al. 2014) and in China (*Alopecurus japonicas* Steud) (Yang et al. 2007). In virtually all cases most tested samples exhibited resistance to more than one of these herbicides. The extreme difficulty of controlling these resistant grass weeds (with herbicides) is frequently highlighted by authors, for example Saini et al. (2016) report the problems of managing clethodim resistant *L. rigidum* in canola in South Australia, concluding that all their options failed to give acceptable control of seed production for continuous cropping systems. Different ACCase herbicides have differing resistance characters and several researchers report that clethodim is less vulnerable to target site resistance than other 'fop' and 'dim' herbicides, Moss et al. (2012) concluding that the efficacy of clethodim and tepraloxydim, on *A. myosuroides* was much less affected by resistance than cycloxydim and propaquizafop. Beckie and Tardif (2012) showed that only 2 of 11 target resistance mutations present in annual grasses confer resistance to clethodim. Thus, in the absence of clear data on herbicide resistance from specific tests it would seem sensible (at the time of writing) to prefer clethodim to other ACCase herbicides for grass weed control in rape. But as the Saini et al. (2016) paper highlights, control of grass weeds even with this herbicide can fail. An alternative strategy is to use propyzamide as a non-ACCase tool to control grass weeds. This can be effective and is quite widely used (Table 17.5) in winter rape though it does have some drawbacks, as its performance is better at low temperatures and so in many European countries application should not be made until November. Carbetamide occupies a similar niche though is not so frequently used. A further problem with both these herbicides is that they are vulnerable to leaching into drainage systems, a particular issue as the optimum time of treatment is often a time of high rainfall.

Herbicide Leaching and Contamination of Water Supplies

A number of key oilseed rape herbicides have the potential to be leached into ground and surface water and hence become transported to sources of drinking water. Within the European Union there is a limit of 0.1 µg/l for any pesticide in drinking water. Where arable agriculture is a major component of water catchments, oilseed rape herbicides can greatly exceed this limit for short periods during the autumn/winter, especially when heavy rain occurs soon after application (Figure 17.1).

The key affected rape herbicides are metazachlor, propyzamide, carbetamide and clopyralid. EU regulation is endeavouring to reduce this risk by encouraging better herbicide stewardship so that for example application is avoided when a rainfall event is predicted. A practical difficulty, mentioned above, is that the late autumn application window for propyzamide and carbetamide, in particular, coincides with winter rainfall and wet soils which cause more drain flow and hence more herbicide leaching. The spikes in propyzamide seen in Figure 17.1 clearly emphasize the difficulty for farmers to meet the drinking water standard. The Cherwell data

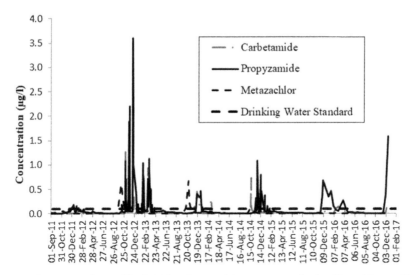

Figure 17.1. Concentrations of herbicides detected in raw water in the River Cherwell (UK) 2011-2017, compared with the EU Drinking Water Standard (Data provided by the Voluntary Initiative, 2017; Source: Water UK).

was collected from a catchment that includes a substantial area of undulating arable farm land. These issues are of primary concern in Europe, as elsewhere in the world the affected herbicides are not widely used and/or the climate is not so wet and thus leaching is less of an issue. Additionally, the EU's restrictive limit for pesticides in water of 0.1 µg/l, has not been adopted elsewhere in the world and so the presence of herbicides in water is not such a significant issue for farmers (and regulators). In Europe, more stringent control of the use of herbicides breaching the EU water standard may be required. In the UK, the farming industry has set up voluntary initiatives aimed at encouraging users to apply the products that can impact on water quality in the safest possible ways. Guidance is provided to farmers advising them on when and how to use these products (http://www.voluntaryinitiative.org.uk/schemes/ stewardship/metazachlor-matters/).

Herbicide Tolerant (HT) Oilseed Rape

In the 1990s commercial herbicide tolerant (HT) oilseed rape was developed by plant breeders. Initial cultivars were based on triazine herbicides which had been created in the 1980s in Canada by conventional breeding methods. Uptake was initially limited. By 1997 both glyphosate and glufosinate resistant cultivars created by genetic modification techniques had also been developed. This was followed by a further conventional breeding programme that created imidazolinone tolerant rape. These cultivars were rapidly adopted in N. America and by 1999, 76% of the Canadian canola crop was sown with the three new herbicide tolerant cultivars (Harker et al. 2000) and by 2006, 95% of the crop was sown with these cultivars (48% glyphosate, 37% glufosinate and 10% imidazolinone) (Smyth et al. 2011). Similar adoption of herbicide tolerant rape has occurred in the USA with 95% of the crop now being tolerant to glyphosate or glufosinate (Fernandez-Cornejo et al. 2016). The uptake of HT rape in Australia has been slightly different, as triazine tolerant cultivars, which were useful for the control of herbicide resistant *L. rigidum* and for the control of weeds in the Brassicaceae, especially *R. raphanistrum*, have been extensively adopted. In 2015 50–60% of the crop was triazine tolerant (Oliver et al. 2016). Imidazolinone tolerant cultivars are also grown in Australia but as the preferred herbicides imazamox, imazapyr and imazapic are all in the sulfonyl urea group of herbicides they are vulnerable to the development of resistant weeds, a major issue in Australia. Glyphosate resistant cultivars are also grown on a minority of farms, whilst glufosinate tolerant cultivars, although in development, are not yet sold to farmers (in 2015).

Europe has not approved the cultivation of genetically modified herbicide tolerant (GMHT) crops. This decision is partly related to concerns about geneflow between GM and non-GM rape and the lack of agreement on isolation/purity conditions, and partly to socio-political arguments about GM crops (Parisi et al. 2016, Schenkelaars and Wesseler 2016). However, the conventionally bred imidazolinone tolerant cultivars are approved in Europe and now occupy a small percentage of the area sown, especially where weeds in the Brassicaceae (e.g., *S. arvensis*, *R. raphanistrum*) are a particular problem. The suggested herbicide mixture of imazamox and metazachlor whilst being effective on the target weeds in the Brassicaceae can be weak on other common weeds including annual grasses. They are also adversely affected by the common occurrence of sulphonyl urea resistance in annual grasses (selected in cereals and other crops).

The adoption of HT rape has provided farmers, at least initially, with a suit of new tools with which to manage intractable weeds in oilseed rape. The attraction of triazine resistant rape to manage weeds in the Brassicaceae is one example. HT rape is also invaluable to control grass weeds resistant to the ACCase herbicides. The availability of effective post-emergence herbicides also opened the door to the potential for crops to be established by minimum tillage or by direct drilling, with considerable cost savings and environmental benefits. The abandonment of intensive tillage for direct drilling coupled with the use of glyphosate (and glufosinate) has major benefits, partly because of the avoidance of environmental damage caused by tillage and partly because glyphosate is more environmentally benign than the 'standard' rape herbicides it replaces (Duke and Powles 2008, Smyth et al. 2011).

However, the widespread adoption of HT crops, especially in N. America, with its associated very limited range of herbicides has resulted in the rapid development of herbicide resistant weeds. Heap (2016) reports over 268 incidents/species resistant to glyphosate, mainly linked to the planting of glyphosate tolerant crops. In general, these are not linked directly to herbicide use in oilseed rape but to maize soyabean and cotton, but some glyphosate resistance, arising in these other crops does affect weeds in oilseed rape. Beckie et al. (2013) in their review of herbicide resistant weeds in Canada failed to find any glyphosate or glufosinate resistant weeds, though appreciable numbers of species were resistant to sulphonyl ureas (imidazolinone resistant rape) and to ACCase herbicides. However, subsequently Beckie et al. (2015) report the presence of glyphosate resistant *Kochia scoparia* in rape in Canada. The mixed cropping commonly adopted in Canada and use of the three different HT systems seems to have minimized the appearance of resistance in weeds to glyphosate (and glufosinate). This contrasts with the monocultures often grown in the USA where glyphosate resistance is far more widespread. In general, glufosinate resistance has been quite rare, Heap (2016) only reporting six incidents., mainly in *Lolium perenne* L. and *Lolium rigidum*. Some other weeds in Australia do show resistance to glyphosate (Heap 2016) but this is probably not linked to HT rape, as the area sown to rape is still quite limited.

The widespread appearance of glyphosate resistance in N. America has led the plant breeders to create double resistant types, e.g. cultivars resistant to glyphosate and dicamba. As yet these have not been included in oilseed rape cultivars but in Australia triazine and glyphosate resistance has been bred into their rape cultivars (Oliver et al. 2016). Whether this is a good idea in the long-term is debatable as weeds with double resistance will probably arise and all it will achieve is a deferment in the time when production systems have to become less dependent on herbicides.

A further concern linked to HT rape is the management of volunteer rape, arising from the copious numbers of seeds shed at harvest. Some rape seeds can persist for 10 years (Lutman et al. 2003).Volunteer rape can be a serious weed in other crops and if they have been bred to be resistant to glyphosate for example then their management is that much more difficult (Australian Oilseeds Federation 2014). The role of HT rape volunteers has been much discussed in Europe in the context of isolation of GM and conventional rape and the risks of gene flow between different rape types (Lutman et al. 2005, Colbach et al. 2008). It is also a relevant issue in relation to the cultivation of non-GMHT cultivars, such as those resistant to imidazolinone herbicides (Krato and Petersen 2012). The imidazolinone resistant rape is also resistant to the widely used sulphonyl urea herbicides which are often key components of weed control in cereal crops. So, control of volunteer rape with this resistance character is more difficult, though alternatives are available.

Integrating Cultural and Chemical Weed Control

The problems of achieving reliable, cost-effective weed control in rape with the continued rise of herbicide resistant weeds outlined in the previous paragraphs, together with, at least in Europe, the presence of herbicides in water, is forcing farmers to re-evaluate their approach to weed control. This does not only apply to oilseed rape but to all the other major crops. A new approach is needed. An initial scenario is to introduce a greater diversity of crops and to reduce reliance on herbicides by combining cultural and chemical control. A return to more cultural based management is not attractive to growers as in the past it has been more complex, time consuming and less reliable than the totally herbicide-based approaches adopted over the last 30 years. But, if herbicides don't work how else does one suppress weeds adequately to ensure profitable yields?

Oilseed rape can be highly suppressive of weeds as discussed in the section on biology, so ensuring optimum establishment, which may mean a move away from direct drilling to more tillage-based establishment, is vital, despite its drawbacks. Adequate moisture in the seedbed and warm soil conditions are critical. Similarly, raising seed rates from a minimum is probably a sound strategy, despite increasing the cost of seeds, even though in ideal conditions very low densities (< 20 plants m^{-2}) can make a successful crop. The use of stale seedbeds to eliminate the weeds prior to sowing, either with a cultivator and/or with herbicides can also help to reduce the pressure on the in-crop herbicide treatments.

One strategy to reduce dependence on herbicides is to increase row width to 25 cm, or wider, to facilitate inter-row cultivation or spatially selective herbicide use between the rows (though very wide widths of 50 cm or more distorts the plant spatial distribution with possible negative effects on yields). Such combinations can be successful as shown in Figure 17.2 (Terres Inovia 2016b). A combination of hoeing/tine weeding between the rows and herbicides in the row can give high levels of control and also reduce the environmental load from the herbicides (Lieven et al. 2008, Nilsson et al. 2014). Such tactics are being enhanced by the involvement of GPS navigation systems and other computer-based technologies to accurately control the position of the interrow treatment, allowing much faster treatment and hoeing much closer to

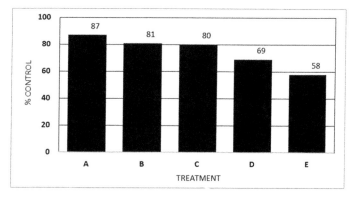

Figure 17.2. Percentage weed control achieved with herbicides and/or mechanical hoes (in France) (Terres Inovia 2016b).

A – Pre-emergence herbicide in row: hoeing between row
B – Overall pre-emergence herbicide
C – Reduced rate pre-emergence herbicide overall: hoeing in row
D – Reduced rate pre-emergence herbicide in row: hoeing between row
E – Hoeing between row only

the crop plants. Such approaches are being driven by horticultural growers, as weed control in their high value crops is even more difficult, but are equally applicable to oilseed rape (Melander et al. 2013, Fennimore et al. 2016).

These technologies can also be exploited even in totally herbicide-based systems by using shielded sprayers and GPS Navigation to apply a non-selective herbicide between the rows and a selective rape herbicide in the row. The use of glyphosate between rows and metazachlor/trifluralin/propyzamide based mixtures in the row has been explored in research programmes with some success (Cook et al. 2015) but although the within-crop application of glyphosate is permitted in EU vegetable crops it is not, as yet approved for oilseed rape. Such approaches reduce the exposure of water systems to individual herbicides as each product is only applied to part (c. 50%) of the field.

The Future

The way forward, as 'traditional' herbicide-based weed control becomes less tenable as the sole approach to weed control, is to consider the introduction of more integrated approaches. Integrated Weed Management (IWM) is relevant to the majority of major arable crops although there are specific issues with oilseed rape that are less relevant to cereals or maize, for example. Oilseed rape has specific problems with the control of other Brassica weeds and is particularly vulnerable to weeds when it grows poorly, when temperatures and moisture levels are less than optimum. In contrast, its potential rapid growth in ideal conditions does give it the ability to be very effective in suppressing weeds (see weed biology section). Firstly, growers have to look at their rotations to ensure that the planned rape crop is not being exposed to a massive weed seedbank. Introduction of a greater proportion of spring crops in the rotation can help reduce the pressure from annual grass weeds, in particular, when autumn-sown rape crops are grown. Indeed, the cropping of spring rape itself has a similar benefit but the low profitability of this crop in climatic areas suited for autumn-sown rape means that it is not an attractive solution for farmers. Clearly, in countries, such as Canada, where spring rape is often the only option for climatic reasons then the ability to switch between autumn- and spring-sown crops is limited.

Weed control in the previous crop is particularly critical if less intensive tillage is to be used to establish the rape. Ploughing may be expensive and more environmentally detrimental than minimum tillage but it is extremely effective (when done properly) at reducing potential seedling emergence in the crop, by burying seeds. So, the tillage practices used to establish the crop also need to be considered when endeavouring to minimize weed infestations. Sound post-harvest weed control is essential, prior to sowing rape crops, to ensure minimal weed seedling presence at the time of drilling. Attention to detail is critical, irrespective of the method of control being used (cultural and/or chemical). The next issue is to ensure that the seeds are sown when moisture and temperatures are optimum. If that can be achieved, and it is not easy as to-date our ability to precisely predict future weather is limited, then the crop will grow rapidly and will much more effectively smother weeds, reducing the need for weed control. This applies to both autumn and spring-sown crops. However, long-term control of weeds is more critical in autumn-sown crops where the plants can be present for 11 months, than it is for spring-sown crops that only grow for about five months. For the majority of growers some form of herbicide use is likely to be chosen, at least for the immediate future. However, it should be a support for the more agronomic approaches suggested above and should not be the main tool to control weeds. Some forms of in-crop mechanical weed control may be appropriate, as outlined above, either as a replacement for, or adjunct to, herbicide application(s). High standard agronomic management is more complex and the risks of failure are greater, than the traditional herbicide based approaches, but the continuing problems with herbicide-based control due to resistance means that in my view there is little choice but to reinvent agronomy based weed control using tools of the 21st Century to support the best cultural practices.

REFERENCES

Australian Oilseeds Federation. 2014. Canola volunteer control. pp. 4. Available at http://www.australianoilseeds.com/__data/assets/pdf_file/0018/9261/Canola_volunteer_control_guide_-_2014.pdf (Accessed in Januar 2017).

Beckie, H.J., R.H. Gulden, N. Shaikh, E.N. Johnson, C.J. Willenborg, C.A. Brenzil, S.W. Shirriff, C. Lozinski and G. Ford. 2015. Glyphosate-resistant kochia (*Kochia scoparia* L. Schrad.) in Saskatchewan and Manitoba. Canadian Journal of Plant Science 95: 345–349.

Beckie, H.J., C. Lozinski, S. Shirriff and C.A. Brenzil. 2013. Herbicide-resistant weeds in the canadian prairies: 2007 to 2011. Weed Technol. 27: 171–183.

Beckie, H.J. and F.J. Tardif. 2012. Herbicide cross resistance in weeds. Crop Protection 35: 15–28.

Blackshaw, R.E., G.W. Anderson and J. Dekker. 1987. Interference of *Sinapis arvenis* and *Chenopodium album* in spring rapeseed (*Brassica napus*). Weed Res. 27: 207–214.

Boutsalis, P., G.S. Gill and C. Preston. 2012. Incidence of Herbicide Resistance in Rigid Ryegrass (*Lolium rigidum*) across Southeastern Australia. Weed Technology 26: 391–398.

Cadoux, S., G. Sauzet, M. Valantin-Morison, C. Pontet, L. Champolivier, C. Robert, J. Lieven, F. Flénet, O. Mangenot, P. Fauvin and N. Landé. 2015. Intercropping frost-sensitive legume crops with winter oilseed rape reduces weed competition, insect damage, and improves nitrogen use efficiency. OCL – Oilseeds and fats, Crops and Lipids 22(3): 11, D302.

Canola Council of Canada. 2013. Canola Encyclopaedia: Tillage. Available at: http://www.canolacouncil.org/canola-encyclopedia/field-characteristics/tillage/ (Accessed in November 2016).

Colbach, N., C. Durr, S. Gruber and C. Pekrun. 2008. Modelling the seed bank evolution and emergence of oilseed rape volunteers for managing co-existence between GM and non-GM varieties. European Journal of Agronomy 28: 19–32.

Cook, S.K., M. Ballingall, R. Stobart, T. Doring, P. Berry and D. Ginsburg. 2015. New approaches to weed control in oilseed rape. UK Agriculture and Horticulture Development Board: Project Report No. 530. pp. 196.

Davies, D.H.K., K.C. Walker and G.P. Whytock. 1989. Herbicide use and yield response in winter oilseed rape in Scotland. Aspects of Applied Biology 24, Production and Protection of Oilseed Rape and other Brassica Crops. pp. 227–235.

Delye, C., Y. Menchari, J-P. Guillemin, A. Matejicek, S. Michel, C. Camilleri and B. Chauvel. 2007. Status of black grass (*Alopecurus myosuroides*) resistance to acetyl-coenzyme – A carboxylase inhibitors in France. Weed Research 47: 95–105.

Duke, S.O. and S.B. Powles. 2008. Mini-review Glyphosate: a once-in-a-century herbicide. Pest Management Science 64: 319–325.

Faostat. 2016. Data on oilseed rape areas and production (yield). Available at: http://www.fao.org/faostat/en/#data/QC (Accessed in December 2016).

Fennimore, S.A., D.C. Slaughter, M.C. Siemens, R.G. Leon and M.N. Saber. 2016. Technology for Automation of Weed Control in Specialty Crops. Weed Technology 30: 823–837.

Fernandez-Cornejo, J., S. Wechsler and D. Milkove. 2016. The Adoption of Genetically Engineered Alfalfa, Canola, and Sugarbeets in the United States. USDA Economic Research Service, Economic Information Bulletin Number 163: 22.

Freeman, S. and P.J.W. Lutman. 2004. The effects of timing of control of weeds on the yield of winter oilseed rape (*Brassica napus*), in the context of the potential commercialisation of herbicide-tolerant winter rape. Journal of Agricultural Science 146: 263–272.

Fried, G., B. Chauvel and X. Reboud. 2015. Weed flora shifts and specialisation in winter oilseed rape in France. Weed Res. 55: 514–524.

Galon, L., L.R. Agazzi, L. Vargas, F. Nonemacher, F.J.M. Basso, G.F. Perin, F.F. Fernandez, C.T. Forte, A.A. Roche and R. Trevisot. 2015. Competitive ability of canola hybrids with weeds. Planta Daninha 33: 413–423.

Garthwaite, D., I. Barker and R. Laybourn. 2015. Pesticide Usage Survey 263, Arable Crops in the United Kingdom 2014. 87 p. Available at: https://secure.fera.defra.gov.uk/pusstats/surveys/documents/arable2014v2.pdf (Accessed in November 2016).

Guillemin, J.P., A. Gardarin, S. Granger, C. Reibel, N. Munier-Jolain and N. Colbach. 2013. Assessing potential germination period of weeds with base temperatures and base water potentials. Weed Research 53: 76–87.

Hanzlik, K. and B. Gerowitt. 2012. Occurrence and distribution of important weed species in German winter oilseed rape fields. Journal of Plant Diseases and Protection 119: 107–120.

Harker, K.N. 2001. Survey of yield losses due to weeds in central Alberta. Canadian Journal of Plant Science 81: 339–342.

Harker, K.N., R.E. Blackshaw, K.J. Kirkland, D.A. Derksen and D. Wall. 2000. Herbicide-tolerant canola: weed control and yield comparisons in western Canada. Canadian Journal of Plant Science 80: 647–654.

Harker, K.N., G.W. Clayton, R.E. Blackshaw, J.T. O'Donovan and F.C. Stevenson. 2003. Seeding rate, herbicide timing and competitive hybrids contribute to integrated weed management in canola (*Brassica napus*). Canadian Journal of Plant Science 83: 433–440.

Harker, K.N., J.T. O'Donovan, E.G. Smith, E.N. Johnson, G. Peng, C.J. Willenborg, R.H. Gulden, R. Mohr, K.S. Gill and L.A. Grenkow. 2015. Seed size and seeding rate effects on canola emergence, development, yield and seed weight. Canadian Journal of Plant Science 95: 1–8.

Heap, I. 2016. International Survey of herbicide resistant weeds. Available at: http://www.weedscience.org (Accessed in November 2016).

Hu, Q., W. Hua, Y. Yin, X. Zhang, L. Liu, Jiaqin Shi, Y. Zhao, L. Qin, C. Chen and H. Wang. 2017. Rapeseed research and production in China. The Crop Journal 5: 127–135.

Hull, R., L.V. Tatnell, S.K. Cook, R. Beffa and S.R. Moss. 2014. Current status of herbicide-resistant weeds in the UK. Aspects of Applied Biology 127, Crop Production in Southern Britain: Precision Decisions for Profitable Cropping. pp. 261–272.

Knezevic, S.Z., S.P. Evans, E.E. Blankenship, R.C. Van Acker and J.L. Lindquist. 2002. Critical period for weed control: the concept and data analysis. Weed Sci. 50: 773–786.

Krato, C. and J. Petersen. 2012. Gene flow between imidazolinone-tolerant and susceptible winter rape varieties. Weed Res. 52: 187–196.

Kutcher, H.R., T.K. Turkington, G.W. Clayton and K.N. Harker. 2013. Response of herbicide-tolerant canola (*Brassica napus* L.) cultivars to four row spacings and three seeding rates in a no-till production system. Canadian Journal of Plant Science 93: 1229–1236.

Leeson, J.Y. 2016. Saskatchewan Weed Survey of Cereal, Oilseed and Pulse Crops in 2014 and 2015. Agriculture and Agri-Food Canada, Saskatoon Research Centre, Weed Survey Series Publication 16-1: 356.

Lemerle, D., P. Lockley, E. Koetz, D. Luckett and H.W. Wu. 2011. Manipulating canola agronomy for weed suppression. Proceedings 17th Australian Research Assembly on Brassicas (ARAB). Wagga Wagga. pp. 181–183.

Lemerle, D., D.J. Luckett, P. Lockley, E. Koetz and H.W. Wu. 2014. Competitive ability of Australian canola (*Brassica napus*) genotypes for weed management. Crop and Pasture Science 65: 1300–1310.

Lieven, J., L. Quere and J.L. Lucas. 2008. Oilseed rape weed integrated management: concern of mechanical weed control. ENDURE International Conference 2008 Diversifying crop protection, La Grande-Motte, France, O-29 pp 4.

Lieven, J. and J.L. Lucas. 2009. Contribution des méthodes de lutte mécanique au désherbage intégré des grandes cultures oléagineuses. OCL – Oilseeds and Fats, Crops and Lipids 16: 169–181.

Llewellyn R., D. Ronning, M. Clarke, A. Mayfield, S. Walker and J. Ouzman. 2016. Impact of weeds on Australian grain production: the cost of weeds to Australian grain growers and the adoption of weed management and tillage practices. Report for Grains Research and Development Corporation. CSIRO, Australia. 109 p.

Lutman, P., K. Berry, R. Payne, J. Sweet, E. Simpson, J. Law, K. Walker, P. Wightman, G. Champion, M. May and M. Lainsbury. 2005. Persistence of seeds from crops of conventional and genetically modified herbicide tolerant oilseed rape (*Brassica napus*). Proceedings of the Royal Society B, 272: 1909–1915.

Lutman, P.J.W., P. Bowerman, G.M. Palmer and G.P. Whytock. 1995. A comparison of the competitive effects of eleven weed species on the growth and yield of oilseed rape. Proceedings 1995 Brighton Crop Protection Conference (Weeds), pp. 877–882.

Lutman, P.J.W., P. Bowerman, G.M. Palmer and G.P. Whytock. 2000. Response of oilseed rape to interference from Stellaria media. Weed Res. 40: 255–270.

Lutman, P.J.W., S.E. Freeman and C. Pekrun, 2003. The long-term persistence of seeds of oilseed rape (*Brassica napus*) in arable fields. Journal of Agricultural Science 141: 231–240.

Lutman, P., J. Storkey, H. Martin and J. Holland. 2009. Abundance of Weeds in Arable Fields in Southern England in 2007/08. Aspects of Applied Biology 91, Crop Protection in Southern Britain. pp. 163–168.

Martin, S.G., R.C. Van Acker and L.F. Friesen. 2001. Critical period of weed control in spring canola. Weed Sci. 49: 326–333.

Melander, B., N. Munier-Jolain, R. Charles, J. Wirth, J. Schwarz, R. van der Weide, L. Bonin, P.K. Jensen and P. Kudsk. 2013. European perspectives on the adoption of nonchemical weed management in reduced tillage systems for arable crops.Weed Technol. 27: 231–240.

Morrison, M.J., P.B.E. McVetty and C.F. Shaykewich. 1989. The determination and verification of a baseline temperature for the growth of Westar summer rape. Canadian Journal of Plant Science 69: 455–464.

Moss, S.R., C. Riches and D. Stormonth. 2012. Clethodim: it's potential to combat herbicide-resistant *Alopecurus myosuroides* (black-grass). Aspects of Applied Biology 117, Crop Protection in Southern Britain. pp. 39–45.

Nilsson, A.T.S., A. Lundkvist, T. Verwijst, M. Gilbertsson, P-A. Algerbo, D. Hansson, A. Andersson, P. Ståhl and M. Stenberg. 2014. Integrated control of annual weeds by inter-row hoeing and intra-row herbicide treatment in spring oilseed rape. 26. Deutsche Arbeitsbesprechung über Fragen der Unkrautbiologie und -bekämpfung, 11-13. März 2014 in Braunschweig.Julius-Kühn-Archiv 443: 746–750.

O'Donovan, J.T., R.E. Blackshaw, K.N. Harker, G.W. Clayton, J.R. Mayo, L.M. Dosdall, D.C. Maurice and T.K. Turkington. 2007. Integrated approaches to managing weeds in spring-sown crops in western Canada. Crop Protection 26: 390–398.

Oliver, D.P., R.S. Kookana, R.B. Miller and R.L. Correll. 2016. Comparative environmental impact assessment of herbicides used on genetically modified and non-genetically modified herbicide-tolerant canola crops using two risk indicators. Science of the Total Environ. 557–558: 754–763.

Ørum, J.E. 2014. Bekæmpelsesmiddel-statistik 2014. Publikationer: Miljø-og Fødevareministeriet, Denmark. pp. 92.

Owen, M.D.K., H.J. Beckie, J.Y. Leeson, J.K. Norsworthy and L.E. Steckel. 2015.Integrated pest management and weed management in the United States and Canada. Pest Management Science 71: 357–376.

Owen, M. and S.B. Powles. 2009. Distribution and frequency of herbicide resistant wild oat (*Avena* spp.) across the Western Australian grain belt. Crop and Pasture Sci. 60: 25–31.

Parisi, C., P. Tillie and E. Rodríguez-Cerezo. 2016. The global pipeline of GM crops out to 2020. Nature Biotech. 34: 31–36.

Primot, S., M. Valantin-Morison and D. Makowski. 2006. Predicting the risk of weed infestation in winter oilseed rape crops. Weed Res. 46: 22–33.

Saini, R.K., S.G.L. Kleeman, C. Preston and G.S. Gill. 2016. Alternative herbicides for the control of clethodim resistant rigid ryegrass (*Loliun rigidum*) in Clearfield canola in Southern Australia. Weed Technol. 30: 423–430.

Schenkelaars, P. and J. Wesseler. 2016. Farm-level GM coexistence policies in the EU: context, concepts and developments. Eurochoices 15(1): 5–10.

Smyth, S.J., M. Gusta, K. Belcher, P.W.B. Phillips and D. Castle. 2011. Changes in Herbicide Use after Adoption of HR Canola in Western Canada. Weed Technology 25: 492–500.

Storkey, J. and J.W. Cussans. 2000. Relationship between temperature and the early growth of *Triticum aestivum* and three weed species. Weed Science 48: 467–473.

Stumm, C., M. Berg and U. Köpke. 2009. Anbau und Düngung von Winterraps (*Brassica napus* L.) im Ökologischen Landbau. Pflanzenbau: Tagungsbandes der 10. Wissenschaftstagung Ökologischer Landbau. (erschienen Mayer, J., Alföldi, T., Leiber, F., Dubois, D., Fried, P., Heckendorn, F. et al. Zurich Feb. 2009) Band 1: 193–196.

Terres Inovia. 2016a. Efficacy of hoes on grass and broad-leaved weeds in oilseed rape in France. Available at: http://www.terresinovia.fr/colza/cultiver-du-colza/desherbage/lutte-mecanique /efficacite-des-outils-mecaniques/ (Accessed in November 2016).

Terres Inovia. 2016b. Mechanical weed control in oilseed rape. Available at: http://www.terresinovia. fr/colza/cultiver-du-colza/desherbage/lutte-mecanique/(Accessed Nov 16).

Terres Inovia. 2016c. Les stratégies herbicides: Efficacité des programmes + caractéristiques des produits. Available at: http://www.terresinovia.fr/fileadmin/cetiom/Cultures/Colza/desherbage/tableau2016_colza_varietes_Terres_Inovia.pdf

van der Weide, R.Y., P.O. Bleeker, V.T.J.M. Achten, L.A.P. Lotz, F. Fogelberg and B. Melander. 2008. Innovation in mechanical weed control in crop rows. Weed Res. 48: 215–224.

Voluntary Initiative. 2017. Voluntary Initiative: Promoting Responsible Pesticide Use. Available at: http://www.voluntaryinitiative.org.uk/

Yang, C., L. Dong, J. Li and S.R. Moss. 2007. Identification of Japanese foxtail (*Alopecurus japonicus*) resistant to haloxyfop using three different assay techniques. Weed Sci. 55: 537–540.

Zand, E. and H.J. Beckie. 2002. Competitive ability of hybrid and open-pollinated canola (*Brassica napus*) with wild oat (*Avena fatua*). Canadian Journal of Plant Science 82: 473–480.

CHAPTER
18

Sustainable Weed Management in Peanut

Ramon G. Leon*[1], David L. Jordan[2], Grace Bolfrey-Arku[3] and Israel Dzomeku[4]

[1] 4402C Williams Hall, Department of Crop and Soil Sciences, North Carolina State University, Raleigh, North Carolina, 27695, USA
[2] 4207 Williams Hall, Department of Crop and Soil Sciences, North Carolina State University, Raleigh, North Carolina, 27695, USA
[3] Council for Scientific and Industrial Research, Crops Research Institute, P.O. Box 3785, Kumasi, Ghana
[4] Department of Agronomy, Faculty of Agriculture, University of Development Studies, Box TI1350, Tamale, Ghana

Introduction

Peanut is an important crop for human nutrition because of the high oil and protein content present in the seed (Davis and Dean 2016). Its biochemical characteristics give this crop considerable value and flexibility for preparation, processing, and consumption. In developed countries peanut is usually processed to make products, such as butter and cooking oil, but a large proportion of the production is used for roasting and incorporation into other foods, such as cookies, snacks, and candies (American Peanut Council 2017). In many developing countries, peanut plays a very important role for both food security and income generation, not only due to its nutritional value but also due to its adaptability to a wide range of climatic and soil conditions (Valentine 2016). For example, in the Sub-Saharan Africa peanut is the most consumed legume after cowpea (*Vigna unguiculata* [L.] Walpers), and this region comprises 40% of the world's peanut harvested area, contributing to 26% of the world's production (Fletcher and Shi 2016) with West Africa responsible for almost two-thirds of this production (Debrah and Waliyar 1996). From 2010 to 2013, 9.3%, 26.1%, and 63.7% of the world's production of peanut was found in the Americas, Africa, and Asia, respectively (Fletcher and Shi 2016). Peanut yields in Argentina, India, and USA during this period of time were 2.94, 1.01, and 4.22 MT/ha, respectively (Fletcher and Shi 2016). Average yield of peanut in Nigeria and Senegal, important West African peanut production countries, was 1.19 and 0.91 MT/ha, respectively (Fletcher and Shi 2016, NAERLS 2010, NPAFS 2010).

The characteristics of peanut-based cropping systems vary depending on ecological conditions and technology availability. However, in most cases peanut is grown in rotation or association with other crops, but crop rotation criteria are mainly based on economic and nutritional needs, not on weed management strategies. In the USA, peanut is commonly rotated with crops, such as cotton (*Gossypium hirsutum* L.), corn (*Zea mays* L.), grain sorghum (*Sorghum bicolor* [L.] Moench. subsp. bicolor), wheat (*Triticum aestivum* L.), and less frequently, due to disease issues, with soybean (*Glycine max* [L.] Merr.) (Johnson et al. 2001, Jordan et al. 2008, Lamb et al. 1993), although there are certain areas in which peanut is produced

*Corresponding author: rleon@ncsu.edu

predominantly as a monoculture (Berger et al. 2015). Rotation with other crops can reduce soil parasitic nematode populations that can negatively impact peanut yield (Barker and Olthof 1976, Culbreath et al. 1992, Rodriguez-Kabana et al. 1987, Rodriguez-Kabana and Touchton 1984). In low-input systems with little or no fertilizer and herbicide availability, especially in developing countries, peanut is grown by smallholder farmers with farm size of one to four hectares (Angelucci and Bazzucchi 2013, Bolfrey-Arku et al. 2006b) either as a monoculture or in association with corn, cassava (*Manihot esculenta* Crantz) through intercropping, and other crops that might complement nutrition requirements.

Peanut yield and quality can be affected by both abiotic and biotic factors. Pests, including weeds, insects, diseases, and nematodes, can reduce peanut yield substantially (Lynch and Mack 1995, Sherwood et al. 1995, Wilcut et al. 1995). In developing countries, competition from weeds is one of the major constraints for optimal peanut production at levels even more limiting than in the USA (Akobundu 1987). For example, in West Africa, cogongrass (*Imperata cylindrica* [L.] Raeusch.), which is present in about 97% of peanut fields, is a threat to the livelihood of more than 200 million people (Anoka 1995, Bolfrey-Arku et al. 2006b, Chikoye et al. 1999, Jagtap 1995). Under these conditions, peanut yield loss due to untimely or ineffective weed control could range from 21 to 80% (Arthur et al. 2016, Bolfrey-Arku et al. 2006a, Carson 1979). To minimize weed interference, peanut farmers must incur high weed management costs because of the extensive hand-labor necessary to control weeds in absence of herbicides. Managing weed populations may consume 40–50% of the total available farm labor pool in Sub-Saharan Africa. Although most farmers do not fully appreciate interactions between weeds and other pests, unmanaged weed infestations have been shown to reduce peanut production by favoring pathogen infection. For instance, rhizomes of cogongrass can pierce peanut kernels creating entry points for disease pathogens including *Aspergillus* spp., while blue boneset (*Chromolaena odorata* [L.] King and H. Rob) harbors several aphids (*Aphis spp.*) that transmit the peanut rosette virus disease, which also has Benghal dayflower (*Commelina benghalensis* L.) as potential host. Furthermore, there are multiple reports of weed species including bristly starbur (*Acanthospermum hispidum* DC.), tropical whiteweed (*Ageratum conyzoides* L.), wild poinsettia (*Euphorbia heterophylla* L.), purple fleabane (*Vernonia cinerea* Less.), and Benghal dayflower that are alternative hosts for nematodes, such as *Meloidogyne*, *Pratylenchus*, and *Paratrichodorus* species that infect peanut (Agostinho et al. 2006, Osei et al. 2005).

Weed Communities and their Importance on Peanut Production

Weed species and distribution can be considered at local, regional, national, and international levels, but although weed community composition varies between farms and regions, there are several weed species that are consistently associated with peanut production across different latitudes and environments. Under subtropical and temperate conditions, weed communities associated with peanut production comprise multiple species, but annual dicotyledonous and annual and perennial monocotyledonous weeds have the largest impact on production. Although grass weed species can cause yield reductions from 7% to more than 60% as a result of season–long interference (Everman et al. 2008a, Johnson and Mullinix 2005, York and Coble 1977), these weeds are frequently considered to be minor problems in peanut in high-input systems because of the availability of pre-emergence herbicides, such as pendimethalin, *S*-metolachlor, and flumioxazin and selective post-emergence systemic herbicides including clethodim, sethoxydim, and fluazifop-P-butyl, that provide effective control of the most common and frequent grass weed species including large crabgrass (*Digitaria sanguinalis* [L.] Scop.), goosegrass (*Eleusine indica* [L.] Gaertn.), crowfootgrass (*Dactyloctenium aegyptium* [L.] Wild.), browntop millet (*Urochloa ramosa* [L.] Nguyen), fall panicum (*Panicum dichotomiflorum* Michx.) and Texas panicum (*Panicum texanum* Buckl.) among others (Burke et al. 2004, Grichar 1991, Grichar and Boswell 1986, Johnson and Mullinix 2005). In contrast, the monocotyledonous species purple nutsedge (*Cyperus rotundus* L.), yellow nutsedge (*Cyperus esculentus* L.), and Benghal dayflower significantly interfere with peanut growth, and their control can be challenging depending on availability of herbicides and production practices (Webster and MacDonald 2001, Webster et al. 2005, Webster et al. 2007).

There are multiple dicotyledonous species that have been reported as problematic in weed communities associated with peanut. Among them, species such as Palmer amaranth (*Amaranthus palmeri* [S.] Wats.), sicklepod (*Senna obtusifolia* [L.] H.S. Irwin and Barneby), Florida beggarweed (*Desmodium tortuosum* [Sw.] DC., and morningglory species (*Ipomoea* spp. and *Jacquemontia tamnifolia* [L.] Griseb.) have been for many years not only frequently found in peanut fields, but also difficult to control (Barbour and Bridges 1995, Cardina and Brecke 1991, Webster and MacDonald 2001). Their success in peanut cropping systems depends on traits, such as high seed production (e.g., Palmer amaranth and sicklepod), seeds with hard seedcoats ensuring a persistent seed bank, and large seeds that can germinate from deeper within the soil (e.g., sicklepod, morningglory, and Benghal dayflower). Also, evolution of resistance to herbicides commonly used in peanut and in rotating crops has favoured Palmer amaranth populations (Berger et al. 2015, Poirier et al. 2014, Wise et al. 2009). It is worth noting that there are reports of weed species, such as bristly starbur, common cocklebur (*Xanthium strumarium* L.), common ragweed (*Ambrosia artemisiifolia* L.), crownbeard (*Verbesina encelioides* [Cav. Benth. & Hook. F. ex Gray]), horsenettle (*Solanum carolinense* L.), silver nightshade (*Solanum elaeagnnifolium* Cav.), and tropic croton (*Croton glandulosus* var. *septentrionalis* Muewll.-Ag.) that can be problematic in peanut under specific situations (Clewis et al. 2001, Farris and Murray 2006, Hackett et al. 1987a, 1987b, Royal et al. 1997, Walker et al. 1989, Webster and MacDonald 2001), but they are not necessarily a wide–spread problem (Webster 2013).

In peanut producing areas closer to the Equator, weed communities tend to be more diverse than in subtropical and temperate regions, but interestingly many of the most troublesome weeds are found in both tropical and temperate regions (Table 18.1). Thus, species such as purple nutsedge and Benghal dayflower, which can reduce productivity by 76% in commercial peanut fields (Agostinho et al. 2006), are weeds of economic importance in Africa, Asia, and USA (Table 18.1). However, in Africa a grass species, cogongrass, is one of the most challenging weed species to manage by peanut farmers, while in the USA, grass weed control is commonly a minor problem in this crop. It is worth noting that Sub-Saharan Africa growers assess the importance of weed species in peanut production based on: i) difficulty of manual or chemical control, ii) interference with peanut, iii) rapid growth, iv) labour demand, v) profuse seed production, and vi) impact on yield. In the case of purple nutsedge, cogongrass and Benghal dayflower, which are listed among the world's worst weeds (Holm 1977, Holm et al. 1977, Webster et al. 2005), the difficulty for their control in low-input systems is their vegetative propagation by tubers, rhizomes, and stem fragments, respectively (Budd et al. 1979, Riar et al. 2016). Wild poinsettia because of its profuse growth (25 to 110 plants/m^2) is very competitive during peanut establishment and a problem for manual and chemical control in peanut cropping systems. Follow-up chemical weed control is also limited by the availability of acceptable and appropriate herbicides that suit the system.

Weed Management Strategies and Tools

The Southern Region IPM (Integrated Pest Management) Center (cIPM) in the USA has developed the prevention, avoidance, monitoring, and suppression (PAMS) approach to IPM, which are the parts of successful IPM programs for all pests in peanut including weeds. In this chapter, we will describe components found within PAMS that influence weed populations and their management in peanut.

PAMS relies on the principle of developing proactive approaches to manage weed populations. Thus, rather than just killing established weeds to minimize weed interference and yield loss, actions must be taken to prevent introduction, establishment, and population growth of new and existing weed species. This requires growers to visualize weed management in- and off-peanut growing season and within and out of fields where peanut is grown. In this way, the ultimate goal is to have a system that creates an unfavorable environment for

Table 18.1. Examples of ten weed species frequently associated with peanut production in Argentina, Ghana, India, and the USA. List is organized alphabetically and the order does not represent any ranking based on frequency or importance on peanut production

Country	Species	Family	Life cycle	Class
Argentina	*Amaranthus hybridus* L.	Amaranthaceae	Annual	Dicot
	Amaranthus palmeri S. Wats.	Amaranthaceae	Annual	Dicot
	Eleusine indica L.	Poaceae	Annual	Monocot
	Digitaria sanguinalis (L.) Scop.	Poaceae	Annual	Monocot
	Sorghum halepense (L.) Pers.	Poaceae	Perennial	Monocot
	Gomphrena pulchella Mart.	Amaranthaceae	Annual	Dicot
	Cynodom hirsutus Stent	Poaceae	Perennial	Monocot
	Chloris spp.	Poaceae	Annual	Monocot
	Euphorbia spp.	Euphorbiaceae	Annual	Dicot
	Salsola kali L.	Chenopodiaceae	Annual	Dicot
Ghana	*Acanthospernum hispidum* DC	Asteraceae	Annual	Dicot
	Ageratum conyzoides L.	Asteraceae	Annual	Dicot
	Boerhavia diffusa L.	Nyctaginaceae	Annual	Dicot
	Brachiaria spp.	Poaceae	Annual/Perennial	Monocot
	Commelina benghalensis L.	Commelinaceae	Perennial	Monocot
	Cyperus rotundus L.	Cyperaceae	Perennial	Monocot
	Digitaria horizontalis Willd.	Poaceae	Annual	Monocot
	Euphorbia heterophylla L.	Euphorbiaceae	Annual	Dicot
	Imperata cylindrica (L.) Beauv.	Poaceae	Perennial	Monocot
	Synedrella nodiflora (L.) Gaertn.	Asteraceae	Annual/Perennial	Dicot
India	*Achyranthes aspera* L.	Amaranthaceae	Annual/Perennial	Dicot
	Amaranthus spinosus L.	Amaranthaceae	Annual	Dicot
	Brachiaria spp.	Poaceae	Annual/Perennial	Monocot
	Celosia argentea L.	Amaranthaceae	Annual	Dicot
	Commelina sp.	Commelinaceae	Annual/Perennial	Monocot
	Cynodon dactylon (L.) Pers.	Poaceae	Perennial	Monocot
	Cyperus rotundus L.	Cyperaceae	Perennial	Monocot
	Dacytloctenium aegyptium (L.) Wild	Poaceae	Annual	Monocot
	Eleusine indica L.	Poaceae	Annual	Monocot
	Phyllanthus niruri L.	Euphorbiaceae	Annual	Dicot
USA	*Amaranthus palmeri* S. Wats.	Amaranthaceae	Annual	Dicot
	Ambrosia artemisiifolia L.	Asteraceae	Annual	Dicot
	Commelina benghalensis L.	Commelinaceae	Perennial	Monocot
	Cyperus rotundus L.	Cyperaceae	Perennial	Monocot
	Dacytloctenium aegyptium (L.) Wild	Poaceae	Annual	Monocot
	Desmodium tortuosum (Sw.) DC	Fabaceae	Annual	Dicot
	Digitaria sanguinalis (L.) Scop.	Poaceae	Annual	Monocot
	Eleusine indica L.	Poaceae	Annual	Monocot
	Ipomoea spp.	Convolvulaceae	Annual	Dicot
	Senna obtusifolia (L.) H.S. Irwin & Barneby	Fabaceae	Annual	Dicot

Source: Adapted from Arthur et al. 2016, Bolfrey et al. 2006b, Dzomeku et al. 2009, Jat et al. 2011, Morichetti personal communication, Webster 2013.

weeds before, during, and after the growing season. Furthermore, using PAMS increases the importance of cultural practices as part of an integrated weed management (IWM) program and helps decrease the weight that herbicides have on overall weed control also reducing the risk of herbicide resistance evolution. Peanut farmers consider weed management as one component of successful peanut production, and balancing weed management strategies with strategies for other pests in a comprehensive IPM approach is critical.

Prevention of Weed Infestations

Similar to other pests, preventing weeds from entering fields is an important first step in minimising interference of weeds with peanut. Sanitation, including preventing weed seed or reproductive propagules from entering fields from relative short or vast distances is an easy and extremely valuable way to reduce weed management cost in the long run. Sanitation practices tend to represent a minor cost when compared with managing established weed populations (Anderson 2007). In the southern USA weeds, such as Palmer amaranth and sicklepod, most likely increased their range due to movement by equipment used either in peanut or crops rotated with peanut. Yellow and purple nutsedge tubers are difficult to remove after harvest due to size similarity to peanut kernels, so they are frequently transported to new fields in non-certified peanut seed. Likewise, movement of rhizomes and stolons from perennial weeds, such as Benghal dayflower and bermudagrass (*Cynodon dactylon* L.), respectively, from field borders into the field itself can increase infestation and requirements for management. Many of the weed seeds introduced into fields will become entrenched for years to come because of dormancy factors that ensure long-term survival. For example, tubers from nutsedge and seed from sicklepod can survive in soil for over four years (Warren and Coble 1999). In West Africa, dispersal of weeds does not often occur by equipment but can result from movement in and through domesticated livestock and wild animals and from transport of halum and storage for livestock consumption.

Interestingly, weeds can play important roles in minimizing infestation of some arthropods. In the southern USA, spider mites (*Tetranychus urticae* Koch Acari: Tetranychidae) often infest fields more slowly or less frequently when weeds in field borders are left undisturbed during hot and dry periods of the growing season (Boykin et al. 1984). However, the practice of not disturbing adjacent vegetation can allow weeds to produce seed and perennial weeds to develop a greater number of vegetative propagules that can spread into fields.

Avoiding Weeds

Crop rotation is considered an essential element of sustainable peanut production through decreased incidence of disease and reductions in plant parasitic nematode populations (Johnson et al. 2001, Jordan et al. 2008, Lamb et al. 1993, Rodriguez-Kabana et al. 1994). Rotating peanut with cotton, corn, or other crops that are not suitable hosts for pathogens can result in greater peanut yield than shorter rotations with more frequent plantings of peanut (Jordan et al. 2008). Peanut plants are often more competitive with weeds when they are not experiencing disease or injury from nematodes. However, ultimately the economic value of the rotational crop often determines the cropping system (Johnson et al. 2001). Crop rotation does allow effective rotation of herbicides and in many cases herbicide mechanisms of action. This can enable farmers to use herbicides for several years prior to peanut plantings that may be more effective than herbicides applied in the peanut crop. For example, sicklepod, common ragweed, and eclipta (*Eclipta prostrata* L.) can be difficult to control in peanut once they become relatively large. Populations of these weeds can be controlled relatively well in glyphosate-resistant crops, such as corn, cotton, and soybean. Also, in some cases bermudagrass and sedges can be controlled better over time in other crops than in peanut.

While cultivar resistance has proven to be an excellent avoidance strategy in peanut for disease, tolerance of weed interference in peanut has been much less effective. Despite its short stature compared to crops, such as corn, when grown at the right planting density and arrangement, peanut can form a dense canopy that limits the amount of solar radiation reaching the soil reducing soil temperature and weed seed germination. Unfortunately, although variation in growth habit and canopy architecture has been documented among peanut botanical types and cultivars, these differences have little effect on weed suppression and weed tolerance (Agostinho et al. 2006, Leon et al. 2016, Place et al. 2012). Furthermore, unlike diseases, cultivars with maturity dates that may allow earlier harvest have not proven effective in peanut. Therefore, breeding efforts to increase competitive and weed suppression in peanut cultivars have not been pursued.

Monitoring Weeds in Peanut

Accurate and timely identification of weeds is important in making decisions on management. Most growers' concerns about weed-crop interactions are generally limited to the risk of yield loss, so their control efforts tend to be biased towards weeds that they perceive as a threat because of their size or control difficulty. This narrow view of weed management makes growers, regardless of technological level, miss opportunities for weed management during land preparation and after yield has been set. Thus, weed populations should be characterized every year during and off-season to develop a clear understanding of weed community structure and population densities and distribution. Only with this information growers can truly assess whether their management programs are solving weed issues. However, for a grower to be successful yield losses must be prevented, and to achieve this goal weeds must be controlled at the most effective timing. Because of its canopy architecture, growth habit and planting arrangements, peanut exhibit clear critical periods of competition (CPC). The duration of the CPC can vary depending on weed species, but most research indicates that weed-free conditions are needed from 3 to 9 weeks after planting (WAP) (Hauser and Buchanan 1981). By controlling weeds during the CPC yield reductions can be avoided or maintained below 10%. Although weeds emerging after the CPC would typically not impact yield, not controlling them might increase weed seed banks making weed management more challenging in future growing seasons. The digging requirement in mechanized systems from relatively low populations of weeds, most notably monocotyledonus species, further reduces yield. In fields where peanut pods are gleaned this is not a major factor. Therefore, assessing weed survival before peanut canopy closure is key in determining the need for control actions not to avoid yield losses in the current season but to maintain weed populations at manageable levels in the near future. Everman et al. (2008a, 2008b) conducted a set of very detailed field experiments to determine CPC in peanut. In most cases, these researchers found that CPC lasted approximately five weeks, but this period can occur earlier or later depending on weed community. When annual grasses were the predominant weeds, CPC extended from 4.3 to 9 WAP, but when annual broadleaf species were the main problem CPC occurred earlier from 2.6 to 8 WAP. When the weed community comprised both grass and broadleaf species, CPC behave similarly to the broadleaf community and occurred from 3 to 8 WAP clearly indicating that broadleaf weed characteristics are driving forces for weed interference in peanut. In many situations weed control actions during CPC are conducted once, twice or thrice depending on crop variety, location/season, timing of activity, labor availability, weed type and level of infestation. However, in systems relying on labor, research recommends hand weeding twice at two and four weeks after planting (Naim et al. 2010). Growing early maturing and branched peanut cultivars may eliminate the need for a third weeding if land preparation is thorough. For fields dominated with difficult to control weeds, three weeding events may be necessary if sole manual control is employed. It is worth noting that hand weeding should not be employed at start of pegging to avoid disturbing the flowers or growing kernels. Despite the well-defined CPC, season-long weed control is frequently needed because of the short stature of peanut throughout the growing season and the need to dig peanut from the ground and in mechanized systems invert vines for harvest. Tall weeds and dense weed populations of tall or short weeds can complicate and even prevent peanut harvest. Furthermore, weeds can also interfere with other pest management operations, such as fungicides sprays for disease management.

Determining the impact of diverse and non-uniformly distributed weeds on peanut yield and strategies for their management can be challenging. Bennet et al. (2003) ranked the competitiveness of major weeds in peanut to develop threshold-based weed management decision systems for post-emergence herbicides, referred to as Herbicide Application Decision Support System (WebHADSS). While this approach to weed management has not been adopted by growers, the information does assist educational institutions describe relationship of crop yield and weed species and density and how distribution of weeds can affect herbicide selection (Jordan et al. 2003b, Robinson et al. 2007). Certainly, evolved resistance changed approaches to managing sub-economic threshold population of weeds. However, there is potential and a clear need of decision support systems especially for weed control timing. New computing technologies have made possible to use not only computers but also smartphones to give direct access to growers to weather information and automated detection systems (Pongnumkul et al. 2015, Shaw 2005). These technologies are applicable and have great potential to improve weed detection and to support the decision-making process of peanut growers.

Cultural Weed Management in Peanut

When prevention and avoidance tactics are not completely effective, a range of tactics can be employed to minimize interference of weeds with peanut. Minimizing injury from insects, such as thrips (*Franklinellia* spp.) and ensuring uniform stand establishment through effective early season disease management programs increases the rate at which the peanut canopy closes (Chahal et al. 2014, Drake et al. 2009). Furthermore, rapid canopy closure is one of the most important processes to increase the ability of peanut to compete more effectively with weeds. Weed establishment before canopy closure is the period that represents the most susceptible growing phase in which weeds can impact peanut yield and determine weed seed bank replenishment at the end of the growing season. Furthermore, because of the short stature of the peanut canopy, most weeds that successfully establish before canopy closure will grow taller than peanut plants and will likely outcompete the crop impacting yield and complicating harvest (Burke et al. 2007, Cardina and Brecke 1991, Hauser et al. 1975). Therefore, the use of planting densities and arrangements that minimize the time to canopy closure are key cultural practices of effective integrated weed management programs for peanut (Colvin et al. 1985a, Hauser and Buchanan 1981). Twin-rows have been a preferred planting arrangement to favor faster canopy closure and to optimize seeding rates (Brecke and Stephenson 2006, Colvin et al. 1985b). However, not all cultivars benefit the same from twin-row planting arrangements. Jordan et al. (2010) showed that a bunch type growing habit cultivar responded more positively to twin-row planting than a cultivar with prostrate growth habit. Place et al. (2010) and Lanier et al. (2004) reported greater weed control in twin-row planting patterns compared with single row plantings, although the increase in weed control was not adequate to eliminate the need for herbicides to protect yield. In a similar study Johnson et al. (2005) found that narrow rows did not reduce weed density, but peanut yields were higher than planting in wide rows, suggesting that growth suppression of emerged weeds was an important addition to the overall weed management. Although increasing peanut density by using twin row planting patterns and higher seeding rates can reduce weed infestation and subsequent interference as well as reducing incidence of tomato spotted wilt of peanut (Johnson et al. 2005), stem rot disease (caused by *Sclerotium rolfsii*) can increase under high population plantings (Sconyers et al. 2007, Wehtje et al. 1994). Thus, achieving a balance between weed suppression, yield maximization, and disease prevention must be a central goal when deciding peanut planting patterns.

Tillage prior to planting and during the early period of the season can be effective in minimizing weeds in peanut (Johnson and Mullinix 1995). However, in some regions reduced tillage systems have become common and weeds in those systems must be controlled with effective pre-plant or burndown herbicide programs. Also, cultivation after planting can be effective in controlling weeds between rows. However, in-row weeds that are not affected by cultivation during the season must be controlled with alternative measures and cultivation can bring non-treated soil from earlier herbicides with residual activity to the surface and result in additional emergence of weeds. Additionally, soil-borne pathogens can increase more quickly as pathogens that cause disease, such as stem rot and tomato spotted wilt can infect plants more quickly (Garren 1961, 1964, Johnson et al. 2001).

While most peanut production is in conventional tillage, in reduced tillage systems cover crops have been effective in protecting soil resources prior to planting peanut and can suppress weed populations (Lassiter et al. 2011, Price et al. 2007). Biomass-intensive cover crop systems can control weeds effectively in peanut, but challenges with pegging, digging and inverting peanut vines and presence of foreign matter in the farmer stock peanut can limit the use of this approach to manage weeds or for weed management. Furthermore, even when the cover crop leaves high residues on the ground, weed suppression is partial, and other control actions are required. Thus, in conservation-tillage systems the limitation in the use of cultivation make necessary to use herbicide programs (Tubbs and Gallaher 2005). However, integrating herbicide programs with high-residue cover crops allows reducing the number of herbicide applications while maintaining weed control and peanut yield (Price et al. 2007). In contrast, when the cover crop does not produce enough biomass to leave a dense and abundant layer of residues on the ground for weed suppression, the benefits of this cultural practice are minimal and effective herbicide programs are required (Dobrow et al. 2011).

In low-input systems, soil cultivation for land preparation is either manual or mechanical with some minimum chemical application (Dzomeku et al. 2009). In West Africa, the major land preparation methods are slash and burn, plough and minimum tillage, but slashing may be mechanical or manual with planting on the flat, mounds, or ridges. The slash-burn-ridge combination is more frequently employed on farmer-owned or family lands while tractor is used for tilling mostly rented properties or large areas. Mechanical tillage with tractors and animals are successfully used for early-season weed control in the tropics under varying conditions. In Ghana, tractor slashing and plowing are more suitable and prevalent in the forest-savannah-transition and the savanna zones where large tracts of land are cultivated. Difficult to control weeds or fields with thick vegetation are double ploughed after slashing.

Hand weeding, a very labor-intensive practice, is unavoidable in peanut production particularly in the tropics where it is grown as part of multi-cropping system and could be eco-friendly. Hand weeding consist of hand-slashing and hoeing with labor for weeding either hired or by family and might require a total of 10–68 man-days/ha depending on weed type (Arthur et al. 2016, Bolfrey-Arku et al. 2006a, Chikoye et al. 2000, Gianessi 2009). These labor-intensive practices are more easily implemented and more practical in terms of time, monitoring and crop safety, when conducted on fields planted in rows relative to uniform/random planting.

In low input-systems and small farms, hand-slashing is the major pre-plant vegetation control method in the tropical forest zones due to the fragile soils often pre-disposed to erosion. The machete (diverse forms) is the main tool for slashing or clearing bush from fallow or fields previously cropped, leaving the underground parts of weeds to minimizing soil erosion. The slashed vegetation from the fallow are burned, ploughed in or allowed to regrow for pre-plant herbicide application in minimum tillage systems in the forest or savanna zones, either way, planting is on the flat, hoe ridged or mounded. Hand-slashing is also used for the control of annual weeds as well as a coping strategy for perennial plant regrowth in peanut cropping systems. Manual slashing can be used in wet weather or on wet soils, and can be employed as over the top weeding of tall escaped weeds. However, yield reduction due to accidental mechanical damage of the crop is common. Hoe-weeding is applicable to both annual and perennial weeds and a widely used manual control method particularly in the tropics but is dependent on rainfall and soil moisture conditions. Hoeing is particularly useful for uprooting plants that propagate by vegetative parts though caution must be exercised to prevent the spread of vegetative propagules. The heavy short or long handled hoes are used in the humid and savanna zones to prepare ridges or mounds as well as for weed control. Hoeing is implemented after weed emergence but before pegging usually within six weeks after planting. Loosening the soil is an important activity to encourage peanut pegging, and penetration of young pods into the soil. Thus, hoeing is commonly used at the time of the second weeding with the dual goal of removing weeds during CPC and favor peanut pegging.

The height differential offered by peanut relative to weeds, such as Palmer amaranth and common ragweed can be partially controlled by mowing (Prostko 2011). However, this approach is often considered a salvage tactic to facilitate more efficient digging and inversion of vines during harvest. Also, traffic through fields to mow weeds earlier in the season can result in vine damage and spread of disease and spider mites in some instances.

Chemical Weed Management in Peanut

Herbicides are important components of IPM programs. As discussed earlier, applications should be made after the economic benefit of use is considered. The limited competitive ability of peanut, especially with tall weeds, and the need to dig and invert vines often makes herbicide applications essential. While a significant number of herbicides are available for peanut, herbicides with the least negative impact should be selected whenever possible. Other factors that should be considered when applying herbicides include developing an effective resistance management strategy and including non-treated vegetated borders located near sensitive waterways.

Herbicides are the most widely used weed control tool in high-input systems. Conversely, low-input systems relying on labor for weed control use herbicides only for pre-plant vegetation weed control, and in some cases pre-emergence and post-emergence herbicides are used on fields that have difficult to control weeds or where labor cost and availability are limiting (Ahmed et al. 2008).

Weed control using herbicides requires an integration of pre-emergence and post-emergence herbicides to minimize the number of applications and maximize weed control throughout the growing season (Brecke and Stephenson 2006, Clewis et al. 2007, Colvin et al. 1985a,1985b, Wehtje et al. 2000). Most peanut cultivars have adequate tolerance to herbicides from multiple mechanisms of action including: acetolactate-synthase–(ALS), acetyl-coA-carboxylase (ACCase), protoporphyrinogen oxidase–(PPO), carotenoid biosynthesis–, mitosis–, fatty acid and lipid biosynthesis–inhibitors, and photosystem I (PSI)–electron diverters. Additionally, peanut has some tolerance to synthetic auxins, but it is particularly tolerant to 2,4-DB (Baughman et al. 2002), which makes possible the selective control of important dicotyledonous species with post-emergence applications (Chahal et al. 2011). Peanut is highly susceptible to photosystem II–inhibitors (e.g., metribuzin, atrazine), enolpyruvyl shikimate-3-phosphate synthase (EPSPS)–inhibitors (e.g., glyphosate), and glutamine synthase–inhibitors (e.g., glufosinate), which are herbicides commonly used in crops in rotation with peanut. In fact, metribuzin, glyphosate, and glufosinate are used in soybean and cotton to control volunteer peanut plants when grown in rotation. Pre-emergence control in peanut is predominantly achieved with mitosis inhibitors from the chemical families dinitroaniline (e.g., pendimethalin and trifluralin) and chloroacetamide (e.g., acetochlor, alachlor dimethenamid-p, and S-metolachlor). These herbicides should be incorporated into the soil to avoid losses due to volatility and photodegradation and to maximize control efficacy. Additionally, the imidazolinone (ALS–inhibitors) herbicides imazethapyr and imazapic are commonly used as early post-emergence herbicides due to their broad spectrum of control, considerable soil residual activity, and high safety on peanut (Richburg et al. 1995, 1996, Wilcut et al. 1996).

Pendimethalin and trifluralin are commonly applied in pre-emergence immediately after planting. However, applications at pre-planting that are incorporated mechanically or with rainfall or irrigation tend to provide more consistent weed control. These two herbicides control grassy weeds and small seeded dicotyledonous species, such as common lambsquarters

(*Chenopodium album* L.) and pigweeds (*Amaranthus* spp.), but their activity on other important weed species is limited. Therefore, alachlor, *S*-metolachlor, acetochlor, dimethenamid-p, and imazethapyr applied in pre-emergence soon after peanut planting can greatly help control species, such as morningglory, Benghal dayflower, and sicklepod. *S*-metolachlor is particularly effective for Bengal dayflower (Morichetti et al. 2012) and yellow nutsedge (Grichar et al. 1996) control, so in fields with high infestations of these weeds herbicide programs that lack *S*-metolachlor tend to fail in providing adequate control. Consequently, *S*-metolachlor is currently the most widely used chloroacetamide herbicide in peanut. *S*-metolachlor can cause peanut stunting when applied in pre-emergence especially under cold and wet conditions (Cardina and Swann 1988, Wehtje et al. 1988). For this reason, applications at cracking are recommended (Grichar et al. 1996). *S*-metolachlor is usually more toxic to seedlings shortly after germination and before emergence. By applying *S*-metolachlor once the peanut plant has initiated emergence through the soil, exposure of the peanut hypocotyl and radicle to the herbicide is reduced increasing the safety of the application.

Recently, the use of the PPO-inhibitor flumioxazin for pre-emergence control in peanut has increased considerably. Flumioxazin provides very high control of a broad spectrum of species, but it is particularly effective against dicotyledonous species that are hard to control in peanut (e.g., Florida beggarweed and Benghal dayflower) (Askew et al. 1999, Grey et al. 2003, Grichar and Colburn 1996, Morichetti et al. 2012, Tredaway-Ducar et al. 2009). Furthermore, it provides effective control of species—that have evolved resistance to ALS-inhibitors including Palmer amaranth (Morichetti et al. 2012). Flumioxazin usually causes only minor injury to peanut when applied in pre-emergence (Askew et al. 1999, Burke et al. 2002), but if it is applied after germination initiation, the risk of direct contact with the plant increases, and flumioxazin is highly injurious on emerged peanut plants (Leon and Tillman 2015). Despite its pre-emergence activity, flumioxazin has a low-soil persistence making it an excellent herbicide for rotational systems to minimize carryover issues (Grey et al. 2002).

Early emerging weeds play a major role in peanut yield and seed bank replenishment. Weeds that escape pre-emergence control and get established with the crop are the ones that have the highest potential for causing peanut yield loss and are more likely to survive post-emergence control actions due to their larger size when these actions are implemented. For these reasons, early emerging weeds are also more likely to produce large amounts of seed at the end of the season maintaining or increasing the weed seed bank. The ability of peanut plants to withstand paraquat (PSI electron diverter) injury at early growth stages enables growers to broadcast this herbicide at cracking (i.e., from peanut hypocotyl emergence to the 3- to 5-leaf stage) to control early emerging weeds that survived pre-emergence control (Wilcut and Swann 1990). After paraquat applications, peanut exhibits significant signs of injury for a short time, but the plants quickly recover and by the end of the season, yield loss due to paraquat injury is negligible in most cases, and when it occurs it is considerably lower than yield losses caused by weed interference (Carley et al. 2009, Wehtje et al. 1991, Wilcut et al. 1994). However, paraquat injury can interact with other stresses losing crop safety. For example, when paraquat is applied to peanut with significant damage from thrips, yield is often lower than when peanut experiences either injury from paraquat or from thrips feeding (Brecke et al. 1996, Drake et al. 2009). Depending on weed community composition, growers might combine paraquat with bentazon and/or S-metolachlor at cracking to increase control spectrum and residual activity (Carley et al. 2009, Wehtje et al. 1992).

When weed emergence is delayed in relation to peanut emergence or escapes from early control occur, control actions are needed at the end of the CPC to protect peanut yield and minimise weed seed bank growth (Bauer and Mortensen 1992, Gallandt 2006, Jones and Medd 2000). Timing of these post-emergence control actions is a critical component of successful integrated management programs because optimum timing ensures efficacy and reduces the need of follow up applications.

Post-emergence applications must be done with the objective of eliminating the highest number of weeds favoring weed-free conditions until peanut canopy closure is reached. However, early post-emergence control actions are preferred over late actions because the former tend to be more effective than the latter (Everman et al. 2006, Wilcut 1991). Post-emergence herbicides for peanut include acifluorfen and lactofen (PPO–inhibitors), 2,4–DB (synthetic auxin), chlorimuron, diclosulam, imazethapyr, and imazapic (ALS–inhibitors). Acifluorfen and lactofen have contact activity, so applications must be done when the weeds are small and preferably between groundcracking and the following two to three weeks to maximize weed control (Jordan et al. 1993). These two herbicides are frequently applied in combination with 2,4–DB and crop oil concentrate (Burke et al. 2002, Ferrell et al. 2013, Grichar 1997). The addition of 2,4–DB increases control efficacy by providing systemic activity. The use of acifluorfen and lactofen combined with 2,4–DB has improved control of ALS–resistant weeds in peanut fields. However, these herbicide treatments can cause visible peanut injury for 2–4 weeks after application, although yield penalties are rarely observed (Ferrell et al. 2013, Grey et al. 2000).

The development of ALS–inhibitors, such as chlorimuron, diclosulam, imazethapyr, and imazapic in the 1990s, dramatically changed weed control in peanut production. These herbicides provided not only effective control of both monocotyledonous and dicotyledonous species, but also both foliar absorption for control of emerged weeds and soil residual activity for preventing new seedling recruitment (Grey et al. 2003, Jordan et al. 2009b). Besides the versatile uses these ALS–inhibiting herbicides allow, they exhibit very consistent safety on peanut (Grey and Wehtje 2005). These characteristics partially explain the high reliance peanut growers have had on these herbicides during the last 20 years, which in fact favored the evolution of resistance in important and problematic species including Palmer amaranth.

Imazapic is perhaps the preferred post-emergence herbicide by most peanut growers in USA. Since its introduction in the late 1990s, imazapic has been extensively used because it effectively controls most weed species of economic importance in peanut fields without the need of tank-mixing other herbicides. Additionally, it has one of the longest residual activities of all ALS–inhibiting herbicides registered for use in peanut. Diclosulam is another ALS-inhibiting herbicide that is commonly used in peanut providing effective weed control when it is used in pre-emergence (Bailey et al. 1999, Grey et al. 2001, Grichar et al. 1999) especially when combined with herbicides, such as *S*-metolachlor, but it can also be used in early post-emergence applications (Everman et al. 2006).

Chlorimuron is particularly effective for the control of Florida beggarweed at later stages (Johnson et al. 1992b). Therefore, in situations with dense populations of this weed species, and where early post-emergence applications failed, chlorimuron can be applied at canopy closure or later to mitigate escapes and prevent yield loss, harvest interference, and weed seed bank increases. Chlorimuron has been reported causing reductions in peanut growth, and after chlorimuron applications, runner market types tend to form a more compact canopy that might not close completely between rows (Johnson et al. 1992a, b). Despite its slight effect on peanut growth, no consistent reductions in peanut yield have been observed after chlorimuron applications (Prostko et al. 2009). Nonetheless, greater incidence of tomato spotted wilt (caused by a *Tospovirus*) has been observed when chlorimuron is applied.

A current trend is to co-apply as many pesticides as possible to increase pest control, reduce the number of application trips over the field saving time, fuel, and making possible to cover a larger area when application conditions are favorable (Jordan et al. 2011, Lancaster et al. 2007, 2008). However, this practice can affect weed control when the tank-mix contains pesticides that are not compatible (Chahal et al. 2012b, 2013, Jordan et al. 2011). For example, efficacy of clethodim and sethoxydim on grass weed species, and 2,4-DB and imazethapyr on broadleaf species, such as entire leaf morningglory and smooth pigweed (*Amaranthus hybridus* L.) can be reduced when these herbicides are tank-mixed with copper-based fungicides (Jordan et al. 2003a, Lancaster et al. 2005c, 2005d). Conversely, in a similar study Lancaster et al. (2005a) found that tank-mixing 2,4-DB with different fungicides did not affect sicklepod control compared to 2,4-DB alone. This type of specific interactions between herbicides and other pesticides highlights the importance of paying attention to aspects, such as crop safety, and weed and pest control efficacy when trying new tank-mixtures. In some cases, increasing the herbicide rate can overcome the problem and increase weed control efficacy (Lancaster et al. 2005b), but in most cases certain tank-mixes should be avoided. Herbicide safety on peanut can also be affected by damage caused by other pests. The ability of peanut plants to recover from early season injury caused by tobacco thrips [*Frankliniella fusca* (Hinds)] can be reduced by post-emergence herbicide applications potentially delaying maturity and yield, although this type of combined stress response is not consistent and depends on environmental conditions (Brecke et al. 1996, Chahal et al. 2014, Drake et al. 2009, Funderburk et al. 1998).

In a recent survey (McClean et al. 2017), growers indicated that 4–5 products applied simultaneously to control weeds, insects, and pathogens that cause disease in peanut was not uncommon. Chahal et al. (2012a, 2012c, 2012d) attempted to address this issue relative to insect, disease, and weed management. Although response was variable across pests, pesticides, and other component mixtures, often one component affected efficacy, and including more components to the mixture did not result in a complete loss of control (Chahal et al. 2012a, 2012c, 2012d). In many instances weed control did not differ by more than 15% for most combinations compared with the herbicide applied alone. These researchers also observed little to no effect of herbicides on efficacy of fungicides and insecticides. In other research (Jordan et al. 2006, 2012), herbicides did not adversely affect boron and manganese accumulation in peanut leaves or efficacy of fungicides. However, the adjuvant system used for the herbicide did affect boron and manganese accumulation in tissue. Defining interaction of pesticides and other products applied to peanut continues to be a challenge because of the number of products available for the major pest management disciplines and the overlap of pests in the field (Jordan et al. 2011).

Herbicide Resistance Issues

Herbicide resistance issues in fields are a consequence of repeated use of herbicides with the same mechanisms of action not only within peanut fields but also in rotational crops. The majority of post-emergence herbicides used in peanut, especially those that have both systemic and soil residual activity, are ALS–inhibitors. Currently, weed species resistant to ALS–inhibiting herbicides are a serious threat to peanut production (Berger et al. 2015, Wise et al. 2009). Without these herbicides, there are few post-emergence alternatives. PPO-inhibitors have provided effective means to manage ALS-resistant weeds at pre-emergence and post-emergence, but the risk exists for overusing this mechanism of action, which could result in resistance evolution. This risk is present not only in fields growing peanut as a monoculture, but also in other phases of the crop rotation. PPO-inhibitors, such as acifluorfen and flumioxazin are currently important tools to manage glyphosate- and ALS-resistant weed species in cotton and soybean, crops that are commonly rotated with peanut (Culpepper et al. 2006, Wise et al. 2009). Therefore, if growers do not implement a diverse and properly designed IWM program, reliance on PPO-inhibitors in all phases of the rotation will likely favour PPO-resistance evolution. 2,4-DB has also been an important tool to control ALS-resistant weeds in peanut. The recent release of cotton and soybean resistant to 2,4-D and dicamba will increase the use of these herbicides in rotations including these crops and peanut. In a scenario in which weeds evolve resistance to PPO-inhibitors and synthetic auxins, post-emergence options to control dicotyledonous weed species in peanut will be extremely limited, and intensive mechanical control will be needed to avoid yield reductions.

Challenges for the Sustainability of Weed Management in Peanut

Africa offers a good illustration about how weed issues are not adequately addressed in peanut production because of lack of awareness at the farmer and governmental levels. Availability of resources and technology to effectively manage weeds also contributes to challenges in managing weeds for smallholder farmers. Identification of the most limiting factors must drive how potential approaches to solve these issues are prioritized (Dankyi et al. 2005). Peanut is important in diet of people in resource-poor communities and command a high demand in the market, yet farmers have few if any price incentives to motivate investment in improved technologies for weed control. Impact of weeds is underestimated, and government spending in Africa for this crop pest is virtually non-existent. Consequently, seed systems are poorly developed and marketing along the value chain is disjointed with insufficient government intervention leaving the farmers vulnerable to market fluctuation and access to markets. Within the African context, at least in Ghana, the choice of land preparation method, a key activity towards effective weed control is closely related to the land tenure system, which needs a more robust policy to drive change. Weed management practices must be linked to socio-cultural systems while the knowledge gap of chemical weed control on human health and the environment must be improved across all participants. Determining the relationships among weed management and other production practices and managing weeds through integrated crop and pest management strategies will be the most effective way for sustainable weed control, but it requires major efforts in training of farmers.

In low-input systems, it is imperative to reduce labor requirements, which are the principal challenge to effectively manage weeds in peanut. Crop rotation and sequence and appropriate and effective cultural practices could help minimize or prevent contributions of weed seed to the soil seed bank and subsequently reduce the time devoted to hand-removal of weeds in peanut. The key to effective weed control is to have appropriate tools for timely removal of weeds, which is elusive to most African peanut producers due to the drudgery and amount of time employed in hand weeding. Research indicates that peanut requires 378 hours of hand weeding per ha, which limits the area under cultivation (Akobundu 1987, 1991). Efforts to reduce this amount of labor can vary from modification of tools, such as hoes to make them user-friendly and more effective to introduction of large mechanical cultivators or increase access to herbicides. For example, the major handweeding tool, the hoe, can be improved and small mechanized weeders can be introduced to minimize the stooping position by use of the hoe that results in backache and certain cases of permanent spinal deformation (Nwuba and Kaul 1986, Oyedemi and Olajide 2002). Improvement of this manual weed management practices would make peanut production more attractive to the youth and reduction in urban migration.

Herbicides could greatly simplify weed management, but their use in peanut production by smallholder farmers in Africa is less than 5% (Bolfrey-Arku et al. 2006b, Mavudzi et al. 2001, Overfield et al. 2001). Although positive extensive studies have been conducted on the economics, yield, costs and time of herbicide use (Ayeni 1997, Benson 1982, Chikoye et al. 2007), yet the dissemination to smallholder farmers has been slow. Peanut is basically grown as a cash crop for all categories of farmers, thus for herbicides to be successfully and sustainably applied growers must be assisted to access credits and technical support on identifiable knowledge gap on human health and environment. Herbicides must be tailored to specific cropping systems and available in quantities suitable for use. Interactions of pest management practices including herbicides and other pesticides also need to be defined in the African context (Abudulai et al. 2017, Arthur et al. 2016). The knowledge base of extension services and actors in the supply chain must be upgraded through the various technology transfer methods. This will enable all participants to be positioned for sustainable weed management practices and technologies for enhanced productivity.

In contrast to low-input systems with resource constraints, in high-input systems a major threat to sustainable weed management is the evolution of herbicide resistant weeds, resulting from herbicide over-use, lack of rotation of mechanisms of action, and inadequate integration with other forms of weed control (e.g., cultural and mechanical). ALS resistance in common ragweed and Palmer amaranth has been confirmed in peanut fields in southeastern USA (Berger et al. 2015, Chandi et al. 2012). Because herbicides that inhibit ALS are widely used in peanut, sustainable weed management in the long run is in question. Research has been conducted to find alternative weed control programs, especially relying on other herbicide mechanisms of action. For example, Chandi et al. (2012) developed management strategies to address ALS resistance in corn, cotton, peanut, and soybean. While a greater number of options are available in corn, cotton, and soybean to manage ALS-resistant common ragweed and Palmer amaranth, in peanut the primary resistance management tool is use of PPO-inhibiting herbicides including acifluorfen, flumioxazin, and lactofen (Chandi et al. 2012, Ferrell et al. 2013). While paraquat controls these weeds when they are small, it can be applied only within the first 28 days after peanut emerges (Wilcut et al. 1995, Wilcut and Swann 1990). Other herbicides applied at planting or postemergence are only marginally effective on these weeds. The fact that resistance to PPO-inhibitors evolved recently in Palmer amaranth in soybean fields (Heap 2017) indicates that current strategies to manage ALS-resistant populations of this weed in peanut might not last too long. Therefore, peanut growers in high-input systems will likely be forced to incorporate non-chemical practices into their production systems. This will not be easy in part because the requirement of pegging and the need to dig and invert vines prior to harvest and because the efficacy of weed control is limited when using only mechanical means in the absence of labor (Johnson et al. 2013, Johnson and Davis 2015).

Under the assumption that proper training and resources are allocated to support growers, the sustainability of peanut weed management will likely depend on the ability of growers to rotate with other crops and how this rotation and sequence of crops might allow diversifying the tools that are used to control weeds. This need is not unique to weed management. Management of pathogens that cause economically-damaging diseases and nematodes in peanut also rely heavily on crop rotation to reduce inoculum (Jordan et al. 2008, Rodriguez-Kabana et al. 1994). Crop diversity can influence herbicide selection and in some cases, allows growers to rotate herbicide mechanisms of action and implement more freely mechanical control. Weed suppressive crops that are not rotated frequently with peanut to reduce the weed seed bank and subsequent infestation are more feasible in cropping systems with greater diversity (Liebman and Dyck 1993). However, some cropping sequences can limit herbicide options in peanut because of the possible negative impact of herbicides remaining in the soil on subsequent crops (Matocha et al. 2003, Wiatrak et al. 2009). A major challenge for this approach is to change financial and social incentives to support production at longer time scales and move from a year-to-year basis to a multi-year rotational system. Otherwise, it would be difficult for growers to devote off-season resources to manage weed problems affecting peanut. In low input systems, alternative crops with high oil and protein content should be used as part of the rotation to maintain a balanced nutritional supply to the communities served by peanut growers.

Concluding remarks

Weed management is a key component for the sustainability of peanut production. Weeds are a problem that is always present and impacts not only productivity and profitability but also the implementation of other agronomic practices and more importantly it largely determines labor requirements especially in low-input systems. Despite the relatively high efficacy of current techniques for weed control, evolution of herbicide resistant weeds and the lack of herbicide alternatives in high-input systems, and the increasing cost of labor and lack of herbicides and mechanical control alternatives in low-input systems, threaten weed management sustainability in peanut. These contrasting challenges should be addressed with integrated management approaches, such as PAMS, to reduce the reliance on a single weed control tool (e.g., herbicides and labor in high- and low-input systems, respectively) and create management programs that are diverse, versatile and applicable in both high- and low-input systems. Crop rotation, establishing adequate stands in appropriate planting configurations, achieving adequate levels of soil fertility and plant health, and protection of peanut from pests other than weeds will be key components of sustainable peanut-based cropping systems. Developing comprehensive weed management strategies that enable farmers to adopt conservation tillage systems successfully will also affect environmental stewardship, especially in areas where soil resources have been depleted and water availability is limited. The future of weed management in peanut will likely depend on how weed management is achieved not only in peanut but also in its rotational partner-crops.

REFERENCES

Abudulai, M., J. Naab, S.S. Seini, I. Dzomeku, K. Boote, R. Brandenburg and D. Jordan. 2018. Peanut (*Arachis hypogaea*) response to weed and disease management in northern Ghana. Int. J. Pest Management 64: 204–209.

Agostinho, F.H., R. Gravena, P.L.C.A. Alves, T.P. Salgado and E.D. Mattos. 2006. The effect of cultivar on critical periods of weed control in peanuts. Peanut Sci. 33: 29–35.

Ahmed, Y.M., A.S. Mostafa, L.A. Reda, A.M. Khozimy and Y.Y. Mosleh. 2008. Efficacy of the selected herbicides in controlling weeds and their side effects on peanut. J. Plant Prot. Res. 48: 355–363.

Akobundu, I.O. 1987. Weed control in other food crops. pp. 522. *In:* Weed Science in the Tropics. Principles and practices. A John Wiley and Sons Publication, Chichester.

Akobundu, I.O. 1991. Weeds in human affairs in sub-saharan Africa: implications for sustainable food production. Weed Technol. 5: 680–690.

American Peanut Council, USA. 2017. The peanut industry. Available at: http://www.peanutsusa. com/about-peanuts/the-peanut-industry3.html. [Accessed on April 3, 2017].

Anderson, R.L. 2007. Managing weeds with a dualistic approach of prevention and control. A review. Agron. Sustain. Dev. 27: 13–18.

Angelucci, F. and A. Bazzucchi. 2013. Analysis of incentives and disincentives for groundnuts in Ghana. Technical notes series, MAFAP, FAO, Rome.

Anoka, A. 1995. Phenology of speargrass [*Imperata cylindrica* (L.) Räeuschel variety Africana (Andeass) C.E. Hubbard], and the contributions of bush-fire, cultivation and nitrogen fertiliser to its persistence in arable lands. PhD. dissertation, University of Reading, UK. 187 p.

Arthur, S., G. Bolfrey-Arku, M.B. Mochiah, J. Sarkodie-Addo, W.O. Appaw, D.L. Jordan and R.L. Brandenburg. 2016. Influence of herbicides and fungicides on peanut production and quality in Ghana. pp. 100. *In:* 48th Proceedings of the American Peanut Research and Education Society, Inc. Annual meeting, July 12–14, Hilton Clearwater Beach, Clearwater, Florida.

Askew, S.D., J.W. Wilcut and J.R. Cranmer. 1999. Weed management in peanut (*Arachis hypogaea*) with flumioxazin preemergence. Weed Technol. 13: 594–598.

Ayeni, A.O. 1997. Use and optimisation of imidazolinone herbicides in legume production in Nigeria. The Brighton Crop Protection Conference, Weeds. pp. 693–698.

Bailey, W.A., J.W. Wilcut, D.L. Jordan, C.W. Swann and V.B. Langston. 1999. Weed management in peanut (*Arachis hypogaea*) with diclosulam preemergence. Weed Technol. 13: 450–456.

Barbour, J.C. and D.C. Bridges. 1995. A model of competition for light between peanut (*Arachis hypogaea*) and broadleaf weeds. Weed Sci. 43: 247–257.

Barker, K.R. and T.H.A. Olthof. 1976. Relationships between nematode population densities and crop responses. Annu. Rev. Phytopathol. 14: 327–353.

Bauer, T.A. and D.A. Mortensen. 1992. A comparison of economic and economic optimum thresholds for two annual weeds in soybeans. Weed Technol. 6: 228–235.

Baughman, T.A., W.J. Grichar and D.L. Jordan. 2002. Tolerance of Virginia-type peanut to different application timings of 2,4-DB. Peanut Sci. 29: 126–128.

Bennett, A.C., A.J. Price, M.C. Sturgill, G.S. Buol and G.G. Wilkerson. 2003. HADSSTM, Pocket HERBTM and Web HADSSTM: Decision Aids for Field Crops. Weed Technol. 17: 412–420.

Benson, J.M. 1982. Weeds in tropical crops: review of abstracts on constraints in production caused by weeds in maize, rice, sorghum-millet, groundnuts and cassava. FAO Plant Production and Protection Paper. 32(1).

Berger, S.T., J.A. Ferrell, P.J. Dittmar and R. Leon. 2015. Survey of glyphosate- and imazapic-resistant Palmer amaranth (*Amaranthus palmeri*) in Florida. Crop, Forage & Turfgrass Management. doi: 10.2134/cftm2015.0122

Bolfrey-Arku, G.E.K., O.U. Onokpise, A.G. Carson, D.G. Shilling and C.C. Coultas. 2006a. The speargrass (*Imperata cylindrica* (L.) Beauv.) menace in Ghana: incidence, farmer perceptions and control practices in the forest and forest-savanna transition agro-ecological zones of Ghana. West African J. Appl. Ecol. 10: 177–188.

Bolfrey-Arku, G., M. Owusu-Akyaw, J.V.K. Afun, J. Adu-Mensah, F.O. Anno-Nyako, E. Moses, K. Osei, S. Osei-Yeboah, M.B. Mochiah, I. Adama, A.A. Dankyi, R.L. Brandenburg and D.L. Jordan. 2006b. Survey of weed management in peanut (*Arachis hypogaea* L.) fields in southern Ghana, West Africa. Peanut Sci. 33: 90–96.

Boykin, L.S., W.V. Campbell and L.A. Nelson. 1984. Effect of barren soil borders and weed border treatments on movement of the two-spotted spider mite into peanut fields. Peanut Sci. 11: 52–55.

Brecke, B.J., J.E. Funderburk, I.D. Teare and D.W. Gorbet. 1996. Interaction of early-season herbicide injury, tobacco thrips injury and cultivar on peanut. Agron. J. 88: 14–18.

Brecke, B.J. and D.O. Stephenson IV. 2006. Weed management in single- vs. twin-row peanut (*Arachis hypogaea*). Weed Technol. 20: 368–376.

Budd, G.D., P.E.L. Thomas, J.C.S. Allison. 1979. Vegetative regeneration, depth of germination and seed dormancy in *Commelina benghalensis* L. Rhod. J. Agr. Res. 17: 151–153.

Burke, I.C., S.D. Askew, J.W. Wilcut. 2002. Flumioxazin systems for weed management in North Carolina peanut (*Arachis hypogaea*). Weed Technol. 16: 743–748.

Burke, I.C., A.J. Price, J.W. Wilcut, D.L. Jordan, A.S. Culpepper and J. Tredaway-Ducar. 2004. Annual grass control in peanut (*Arachis hypogaea*) with clethodim and impazapic. Weed Technol. 18: 88–92.

Burke, I.C., M. Schroeder, W.E. Thomas and J.W. Wilcut. 2007. Palmer amaranth interference and seed production in peanut. Weed Technol. 21: 367–371.

Cardina, J. and B.J. Brecke. 1991. Florida beggarweed (*Desmodium tortuosum*) growth and development in peanuts (*Arachis hypogaea*). Weed Technol. 5: 147–153.

Cardina, J. and C.W. Swann. 1988. Metolachlor effects on peanut growth and development. Peanut Sci. 15: 57–60.

Carley, D.S., D.L. Jordan, R.L. Brandenburg and L.C. Dharmasri. 2009. Factors influencing response of Virginia market type peanut (*Arachis hypogaea*) to paraquat under weed-free conditions. Peanut Sci. 36: 180–189.

Carson, A.G. 1979. Weed competition and control in peanut (*Arachis hypogaea* L.). Ghana J. Agric. Sci. 9: 169–173.

Chahal, G.S., D.L. Jordan, F.L. Brandenburg, B.B. Shew, J.D. Burton, D. Danehower and A.C. York. 2012a. Interactions of agrochemicals applied to peanut, Part 3: Effects on insecticides and prohexadione calcium. Crop Prot. 41: 150–157.

Chahal, G.S., D.L. Jordan, P.M. Eure and R.L. Brandenburg. 2014. Compatibility of acephate with herbicides applied postemergence in peanut. Peanut Sci. 4: 58–64.

Chahal, G.S., D.L. Jordan, B.B. Shew, R.L. Brandenburg, J.D. Burton, D. Danehower and P.M. Eure. 2012b. Influence of selected fungicides on efficacy of clethodim and 2,4-DB. Peanut Sci. 39: 121–126.

Chahal, G.S., D.L. Jordan, B.B. Shew, R.L. Brandenburg, J.D. Burton, D. Danehower and A.C. York. 2012c. Interactions of agrochemicals applied to peanut, Part 2: Effects on fungicides. Crop Prot. 41: 143–149.

Chahal, G.S., D.L. Jordan, B.B. Shew, R.L. Brandenburg, A.C. York, J.D. Burton and D. Danehower. 2012d. Interactions of agrochemicals applied to peanut, Part 1: Effects on herbicides. Crop Prot. 41: 134–142.

Chahal, G.S., D.L. Jordan, A.C. York, R.L. Brandenburg, B.B. Shew, J.D. Burton and D. Danehower. 2013. Interactions of clethodim and sethoxydim with other pesticides. Peanut Sci. 40: 127–134.

Chahal, G.S., D.L. Jordan, A.C. York and E.P. Prostko. 2011. Palmer amaranth control with combinations of 2,4-DB and diphenylether herbicides. Crop Manag. doi: 10.1094/CM-2011-0802-01-RS.

Chandi, A., D.L. Jordan, A.C. York and B.R. Lassiter. 2012. Confirmation and management of common ragweed resistant to diclosulam. Weed Technol. 26: 29–36.

Chikoye, D., F. Ekeleme and J.T. Ambe. 1999. Survey of distribution and farmers' perceptions of speargrass (*Imperata cylindrica* [L.] Raeuschel) in cassava-based systems in West Africa. Int. J. Pest Manage. 45: 305–311.

Chikoye, D., V.M. Manyong and F. Ekeleme. 2000. Cogongrass suppression by intercropping cover crops in corn/cassava systems. Weed Sci. 49: 658–667.

Chikoye, D., U.E. Udensi, A.F. Lum and F. Ekeleme. 2007. Rimsulfuron for postemergence weed control in corn in humid tropical environments of Nigeria. Weed Technol. 21: 977–981.

Clewis, S.B., S.D. Askew and J.W. Wilcut. 2001. Common ragweed interference in peanut. Weed Sci. 49: 768–772.

Clewis, S.B., W.J. Everman, D.L. Jordan and J.W. Wilcut. 2007. Weed management in North Carolina peanut (*Arachis hypogaea*) with *S*-metolachlor, diclosulam, flumioxazin, and sulfentrazone systems. Weed Technol. 21: 629–635.

Colvin, D.L., G.R. Wehtje, M. Patterson and R.H. Walker. 1985a. Weed management in minimum-tillage peanuts (*Arachis hypogaea*) as influenced by cultivar, row, and herbicides. Weed Sci. 33: 233–237.

Colvin, D.L., R.H. Walker, M.G. Patterson, G. Wehtje and J.A. McGuire. 1985b. Row pattern and weed management effects on peanut production. Peanut Sci. 12: 22–27.

Culbreath, A.K., M.K. Beute, B.B. Shew and K.R. Barker. 1992. Effects of *Meloidogyne hapla* and *M. arenaria* on black rot severity in new *Cylindrocladium*-resistant peanut genotypes. Plant Dis. 76: 352–357.

Culpepper, A.S., T.L. Grey, W.K. Vencill, J.M. Kichler, T.M. Webster, S.M. Brown, A.C. York, J.W. Davis and W.W. Hana. 2006. Glyphosate-resistant Palmer amaranth (*Amaranthus palmeri*) confirmed in Georgia. Weed Sci. 54: 620–626.

Dankyi, A.A., M. Owusu-Akyaw, V.M. Anchirinah, J. Adu-Mensah, M.B. Mochiah, E. Moses, J.F.K. Afun, G. Bolfrey-Arku, K. Osei, S. Osei-Yeboah, I. Adama, R.L. Brandenburg, J.E. Bailey and D.L. Jordan. 2005. Survey of production and pest management practices for peanut (*Arachis hypogaea* L.) in selected villages in Ghana, West Africa. Peanut Sci. 32: 91–97.

Davis, J.P. and L.L. Dean. 2016. Peanut consumption, flavor and nutrition. pp. 289–346. *In:* Stalker, H.T., Wilson, R.F. (Eds.) Peanuts: Genetics, Processing, and Utilization. AOCS Monograph Series, AOCS Press, Elsevier, 478 pages.

Debrah, S.K. and F. Waliyar. 1996. Groundnut production and utilization in Africa: past trends, projections and opportunities for increased production. 5th Regional groundnut workshop for West Africa. No. 18–21, 1996. Accra, Ghana.

Dobrow Jr, M.H., J.A. Ferrell, W.H. Faircloth, G.E. MacDonald, B.J. Brecke and J.E. Erickson. 2011. Effect of cover crop management and preemergence herbicides on the control of ALS-resistant Palmer amaranth (*Amarantus palmeri*) in peanut. Peanut Sci. 38: 73–77.

Drake, W.L., D.L. Jordan, R.L. Brandenburg, B.R. Lassiter, P.D. Johnson and B.M. Royals. 2009. Peanut cultivar response to damage from tobacco thrips and paraquat. Agron. J. 101: 1388–1393.

Dzomeku, I.K., M. Abudulai, R.L. Brandenburg and D.L. Jordan. 2009. Survey of weeds and management practices in peanut (*Arachis hypogaea* L.) in the savanna ecology of Ghana. Peanut Sci. 36: 165–173.

Everman, W.J., I.C. Burke, S.B. Clewis, W.E. Thomas and J.W. Wilcut. 2008a. Critical period of grass vs. broadleaf weed interference in peanut. Weed Technol. 22: 68–73.

Everman, W.J., S.B. Clewis, W.E. Thomas, I.C. Burke and J.W. Wilcut. 2008b. Critical period of weed interference in peanut. Weed Technol. 22: 63–67.

Everman, W.J., S.B. Clewis, Z.G. Taylor and J.W. Wilcut. 2006. Influence of diclosulam postemergence application timing on weed control and peanut tolerance. Weed Technol. 20: 651–657.

Farris, R.L. and D.S. Murray. 2006. Influence of crownbeard (*Verbesina encelioides*) densities on peanut (*Arachis hypogaea*) yield. Weed Technol. 20: 627–632.

Ferrell, J.A., R.G. Leon, B. Sellers, D. Rowland and B. Brecke. 2013. Influence of lactofen and 2,4-DB combinations on peanut injury and yield. Peanut Sci. 40: 62–65.

Fletcher, S.M. and Z. Shi. 2016. An overview of world peanut markets. pp. 267–288. *In:* Stalker, H.T., Wilson, R.F. (Eds.) Peanuts: Genetics, Processing, and Utilization. AOCS Monograph Series, AOCS Press, Elsevier, 478 p.

Funderburk, J.E., D.W. Gorbet, I.D. Teare and J. Stavisky. 1998. Thrips injury can reduce peanut yield and quality under conditions of multiple stress. Agron. J. 90: 563–566.

Gallandt, E.R. 2006. How can we target the weed seedbank? Weed Sci. 54: 588–596.

Garren, K.H. 1961. Control of *Sclerotium rolfsii* through cultural practices. Phytopathology 51: 124–128.

Garren, K.H. 1964. Inoculum potential and differences among peanuts in susceptibillity to *Sclerotium rolfsii*. Phytopathology 54: 279–281.

Gianessi, L. 2009. Solving Africa's weed problem: increasing crop production and improving lives of women. Crop protection Research Institute. Available at: https://croplifefoundation.org/wp-content/uploads/2015/12/solving-africas-weed-problem-report 1.pdf. (Accessed May 11, 2017).

Grey, T.L., D.C. Bridges and B.J. Brecke. 2000. Response of seven peanut (*Arachis hypogaea*) cultivars to sulfentrazone. Weed Technol. 14: 51–56.

Grey, T.L., D.C. Bridges and E.F. Eastin. 2001. Influence of application rate and timing of diclosulam on weed control in peanut (*Arachis hypogaea* L.). Peanut Sci. 28: 13–19.

Grey, T.L., D.C. Bridges, E.F. Easting and G.E. MacDonald. 2002. Influence of flumioxazin rate and herbicide combinations on weed control in peanut (*Arachis hypogaea* L.). Peanut Sci. 29: 24–29.

Grey, T.L., D.C. Bridges, E.P. Prostko, E.F. Eastin, W.C. Johnson III, W.K. Vencill, B.J. Brecke, G.E. MacDonald, J.A. Tredaway-Ducar, J.W. Everest, G.R. Wehtje and J.W. Wilcut. 2003. Residual weed control with imazapic, diclosulam, and flumioxazin in southeastern peanut (*Arachis hypogaea*). Peanut Sci. 30: 27–34.

Grey, T.L. and G.R. Wehtje. 2005. Residual herbicide weed control systems in peanut. Weed Technol. 19: 560–567.

Grichar, W.J. 1991. Sethoxydim and broadleaf herbicide interaction effects on annual grass control in peanuts (*Arachis hypogaea*). Weed Technol. 5: 321–324.

Grichar, W.J. 1997. Influence of herbicides and timing of application on broadleaf weed control in peanut (*Arachis hypogaea*). Weed Technol. 11: 708–713.

Grichar, W.J. and T.E. Boswell. 1986. Postemergence grass control in peanut (*Arachis hypogaea*). Weed Sci. 34: 587–590.

Grichar, W.J. and A.E. Colburn. 1996. Flumioxazin for weed control in Texas peanut (*Arachis hypogaea* L.). Peanut Sci. 23: 30–36.

Grichar, W.J., A.E. Colburn and P.A. Baumann. 1996. Yellow nutsedge (*Cyperus esculentus*) control in peanut (*Arachis hypogaea*) as influenced by method of metolachlor application. Weed Technol. 10: 278–281.

Grichar, W.J., P.A. Dotray and D.C. Sestak. 1999. Diclosulam for weed control in Texas peanut. Peanut Sci. 26: 23–28.

Hackett, N.M., D.S. Murray and D.L. Weeks. 1987a. Interference of horsenettle (*Solanum carolinense*) with peanuts (*Arachis hypogaea*). Weed Sci. 35: 780–784.

Hackett, N.M., D.S. Murray and D.L. Weeks. 1987b. Interference of silver nightshade (*Solanum elaeagnifolium*) on Spanish peanuts (*Arachis hypogaea*). Peanut Sci. 14: 39–41.

Hauser, E.W. and G.A. Buchanan. 1981. Influence of row spacing, seeding rates and herbicide systems on the competitiveness and yield of peanuts. Peanut Sci. 8: 74–81.

Hauser, E.W., G.A. Buchanan and W.J. Ethredge. 1975. Competition of Florida beggarweed and sicklepod with peanuts. I. Weed-free maintenance and weed competition. Weed Sci. 23: 368–372.

Heap, I. 2017. The international survey of herbicide resistant weeds. Available at: http://weedscience.org. (Accessed on May 11, 2017).

Holm, L.R., D.L. Pluncknett, J.V. Pancho and J.P. Herberger. 1977. *Imperata cylindrica* (L.) Beauv. pp. 62–71. *In:* The World's Worst Weeds: Distribution and Biology. University Press of Hawaii, Honolulu, USA.

Holm, L.G. 1977. The World's Worst Weeds, Honolulu: University Press of Hawaii, pp. 225–235.

Jagtap, S.S. 1995. Environmental characterization of the moist lowland savanna of Africa. Proceedings of the International Workshop "Moist Savannas of Africa: Potentials and Constraints for Crop Production. Cotonou, Benin, IITA. pp. 9–30.

Jat, R.S., H.N. Meena, A.L. Singh, J.N. Surya and J.B. Misra. 2011. Weed management in groundnut (*Arachys hypogaea* L.) in India – a review. Agric. Rev. 32: 155–171.

Johnson III, W.C., T.B. Brenneman, S.H. Baker, A.W. Johnson, D.R. Sumner and B.G. Mullinix Jr. 2001. Tillage and pest management considerations in a peanut–cotton rotation in the southeastern Coastal Plain. Agron. J. 93: 570–576.

Johnson III, W.C., M.A. Boudreau and J.W. Davis. 2013. Combinations of corn gluten meal, clove oil, and sweep cultivation are ineffective for weed control in organic peanut production. Weed Technol. 27: 417–421.

Johnson III, W.C. and J.W. Davis. 2015. Perpendicular cultivation for improved in-row weed control in organic peanut production. Weed Technol. 29: 128–134.

Johnson III, W.C., C.C. Holbrook, B.G. Mullinix Jr and J. Cardina. 1992a. Response of eight genetically diverse peanut genotypes to chlorimuron. Peanut Sci. 19: 111–115.

Johnson III, W.C. and B.G. Mullinix Jr. 2005. Texas panicum (*Panicum texanum*) inteference in peanut (*Arachis hypogaea*) and implications for treatment decision. Peanut Sci. 32: 68–72.

Johnson III, W.C., B.G. Mullinix Jr and S.M. Brown. 1992b. Phytotoxicity of chlorimuron and tank mixtures on peanut (*Arachis hypogaea*). Weed Technol. 6: 404–408.

Johnson III, W.C., E.P. Prostko and B.G. Mullinix Jr. 2005. Improving the management of dicot weeds in peanut narrow row spacings and residual herbicides. Agron. J. 97: 85–88.

Johnson III, W.C. and B.G. Mullinix Jr. 1995. Weed management in peanut using stale seedbed techniques. Weed Sci. 43: 293–297.

Jones, R.E. and R.W. Medd. 2000. Economic thresholds and the case for longer term approaches to population management decisions. Weed Technol. 14: 337–350.

Jordan, D.L., G.S. Chahal, S.H. Lancaster, J.B. Beam and A.C. York. 2011. Defining interactions of herbicides with other agrochemicals applied to peanut. pp. 73–92. *In:* Soloneski, S., Larramendy, M. (Eds.) Herbicides, theory and Applications. ISBN: 978-953-307-975-2, InTech, Rijeka, Croatia. 610 p.

Jordan, D.L., A.S. Culpepper, W.J. Grichar, J. Tredaway-Ducar, B.J. Brecke and A.C. York. 2003a. Weed control with combinations of selected fungicides and herbicides applied postemergence to peanut (*Arachis hypogaea*). Peanut Sci. 30: 1–7.

Jordan, D.L., S.H. Lancaster, J.E. Lanier, P.D. Johnson, J.B. Beam, A.C. York and R.L. Brandenburg. 2012. Influence of application variables on efficacy of manganese-containing fertilisers applied to peanut (*Arachis hypogaea* L.). Peanut Sci. 39: 1–8.

Jordan, D.L., S.H. Lancaster, J.E. Lanier, P.D. Johnson, J.B. Beam, A.C. York, R.L. Brandenburg, F.R. Walls, S. Casteel and C. Hudak. 2006. Influence of application variables on efficacy of boron-containing fertilisers applied to peanut (*Arachis hypogaea* L.). Peanut Sci. 33: 104–111.

Jordan, D.L., S.H. Lancaster, J.E. Lanier, B.R. Lassiter and P.D. Johnson. 2009a. Peanut (*Arachis hypogaea*) and eclipta response to flumioxazin. Weed Technol. 22: 231–235.

Jordan, D.L., S.H. Lancaster, J.E. Lanier, B.R. Lassiter and P.D. Johnson. 2009b. Weed management in peanut (*Arachis hypogaea*) with herbicide combinations containing imazapic and other pesticides. Weed Technol. 23: 6–10.

Jordan, D.L., G. Place, R.L. Brandenburg, J.E. Lanier and D.L. Carley. 2010. Response of Virginia market type peanut to planting pattern and herbicide program. Crop Manage. doi: 10.1094/CM-2010-0430-01-RS.

Jordan, D.L., G.G. Wilkerson and D.W. Krueger. 2003b. Evaluation of scouting methods in peanut (*Arachis hypogaea*) using theoretical net returns from HADSS™. Weed Technol. 17: 358–365.

Jordan, D.L., J.W. Wilcut and C.W. Swann. 1993. Application timing of lactofen for broadleaf weed control in peanut (*Arachis hypogaea*). Peanut Sci. 20: 129–131.

Jordan, D.L., B.B. Shew, J.S. Barnes, T. Corbett, J. Alston, P.D. Johnson, W. Ye and R.L. Brandenburg. 2008. Pest reaction, yield, and economic return of peanut cropping systems in the North Carolina Coastal Plain. Crop Manage. doi: 10.1094/CM-2008-1008-01-RS.

Lamb, M.C., J.I. Davidson and C.L. Butts. 1993. Peanut yield decline in the southeast and economically reasonable solutions. Peanut Sci. 20: 36–40.

Lancaster, S.H., J.B. Beam, J.E. Lanier, D.L. Jordan and P.D. Johnson. 2007. Compatibility of diclosulam with postemergence herbicides and fungicides. Weed Technol. 21: 869–872.

Lancaster, S.H., D.L. Jordan and P.D. Johnson. 2008. Influence of graminicide formulation on compatibility with other pesticides. Weed Technol. 22: 580–583.

Lancaster, S.H., D.L. Jordan, J.F. Spears, A.C. York, J.W. Wilcut, D.W. Monks, R.B. Batts and R.L. Brandenburg. 2005a. Sicklepod (*Senna obtusifolia*) control and seed production following 2,4-DB applied alone and with fungicides or insecticides. Weed Technol. 19: 451–455.

Lancaster, S.H., D.L. Jordan, A.C. York, I.C. Burke, F.T. Corbin, Y.S. Sheldon, J.W. Wilcut and D.W. Monks. 2005b. Influence of selected fungicides on efficacy of clethodim and sethoxydim. Weed Technol. 19: 397–403.

Lancaster, S.H., D.L. Jordan, A.C. York, J.W. Wilcut, R.L. Brandenburg and D.W. Monks. 2005c. Interactions of late-season morningglory (*Ipomoea* spp.) management practices in peanut (*Arachis hypogaea*). Weed Technol. 19: 803–808.

Lancaster, S.H., D.L. Jordan, A.C. York, J.W. Wilcut, D.W. Monks and R.L. Brandenburg. 2005d. Interactions of clethodim and sethoxydim with selected agrichemicals applied to peanut. Weed Technol. 19: 456–461.

Lanier, J.E., S.H. Lancaster, D.L. Jordan, P.D. Johnson, J.F. Spears, R. Wells, C.A. Hurt and R.L. Brandenburg. 2004. Sicklepod control in peanut seeded in single and twin row planting patterns. Peanut Sci. 31: 36–40.

Lassiter, B.R., D.L. Jordan, G.G. Wilkerson, B.B. Shew and R.L. Brandenburg. 2011. Influence of cover crops on weed management in strip tillage peanut. Weed Technol. 25: 568–573.

Leon, R.G., M.J. Mulvaney and B.J. Tillman. 2016. Peanut cultivars differing in growth habit and canopy architecture respond similarly to weed interference. Peanut Sci. 43: 133–140.

Leon, R.G. and B.J. Tillman. 2015. Postemergence herbicide tolerance variation in peanut germplasm. Weed Sci. 63: 546–554.

Liebman, M. and E. Dyck. 1993. Crop rotation and intercropping strategies for weed management. Ecol. Appl. 3: 92–122.

Lynch, R.E. and T.P. Mack. 1995. Biological and biotechnical advances for insect management in peanut. pp. 95–159. *In:* Pattee, H.E., Stalker, H.T. (Eds.)Advances in Peanut Science, American Peanut Research and Education Society, Stillwater, OK, 614 p.

Matocha, M.A., W.J. Grichar, S.A. Senseman, C.A. Gerngross, B.J. Brecke and W.K. Vencill. 2003. The persistence of imazapic in peanut (*Arachis hypogaea*) crop rotations. Weed Technol. 17: 325–329

Mavudzi, Z., A.B. Mashingaidze, O.A. Chivinge, J. Ellis-Jones and C. Riches. 2001. Improving weed management in a cotton-maize system in the Zambezi valley Zimbabwe. Brighton Crop Protection Conference, Weeds. pp. 169–174.

McClean, B., B. Sandlin, B. Barrow, J. Hurry, M. Leary, M. Shaw, M. Carroll, T. Adams, A. Bradley, P. Smith, R. Thagard, A. Whitehead, B. Parish, J. Holland, T. Britton, J. Morgan, A. Cochran, C. Ellison, M. Huffman, M. Seitz, D. Lilley, L. Grimes, M. Malloy, D. King, R. Wood, A. Williams, T. Whaley, N. Harrell, D.L. Jordan, B.B. Shew, R.L. Brandenburg, D.J. Anco, D.J. Croft, A. Warner, P. Dehond, H. Mikell, J. Varn, J. Crouch, M. Balota, H. Mehl, S.V. Taylor, J. Spencer, J. Reiter and L. Preisser. 2017. Results from surveys on application variables associated with production and pest management in peanut in North Carolina, South Carolina, and Virginia. Proc. Am. Peanut Res. Educ. Soc. 49: 59.

Morichetti, S., J. Ferrell, G. MacDonald, B. Sellers and D. Rowland. 2012. Weed management and peanut response from applications of saflufenacil. Weed Technol. 26: 261–266.

Naim, A.M., M.A. Eldouna and A.E. Abdalla. 2010. Effect of weeding frequencies and plant density on the vegetative growth characteristic in groundnut (*Arachis hypogea* L.) in North kordofan of Sudan. Int. J. Appl. Biol. Pharm. 1: 1188–1192.

NAERLS (National Agricultural Extension and Research Liaison Services). 2010. Agricultural performance survey of 2010 wet season in Nigeria, 182 pp.

NPAFS (National Programme for Agriculture and Food Security). 2010. Report of the 2009 agricultural production survey. Abuja, Nigeria, 86 pp.

Nwuba, E.I.U. and R.N. Kaul. 1986. The effect of working posture on the Nigerian Hoe farmer. J. Agri. Eng. Res. 33: 179–185.

Osei, K., M. Owusu-Akyaw, J.K. Twumasi, J.V.K. Afun, F.K. Anno-Nyako, J. Adu-Mensah, E. Moses, G. Bolfrey-Arku, S. Osei-Yeboah, M.B. Mochiah, I. Adama, R.L. Brandenburg, J.E. Bailey and D.L. Jordan. 2005. Incidence and potential host-plant resistance of peanut (*Arachis hypogaea* L.) to plant parasitic nematodes in southern Ghana. Peanut Sci. 32: 91–97.

Overfield, D., F.M. Murithi, J.N. Muthamia, J.O. Ouma, K.F. Birungi, J.M. Maina, G.N. Kibata, F.J. Musembi, G. Nyanyu, M. Kamidi, L.O. Mose, M. Odendo, J. Ndungu, G. Kamau, J. Kikafunda and P.J. Terry. 2001. Analysis of the constraints to adoption of herbicides by smallholder maize growers in Kenya and Uganda. The BCPC Conference Weeds. pp. 907–912.

Oyedemi, T. and A. Olajide. 2002. Ergonomic evaluation of an indigenous tillage tool employed in Nigerian agriculture. ASAE Annual Meeting. Paper Number 028001. American Society of Agricultural and Biological Engineers.

Place, G.T., S.C. Reberg-Horton and D.L. Jordan. 2010. Interaction of cultivar, planting pattern, and weed management tactics in peanut. Weed Sci. 58: 442–448.

Place, G.T., S.C. Reberg-Horton, D.L. Jordan, T.G. Isleib and G.G. Wilkerson. 2012. Influence of Virginia market type genotype on peanut response to weed interference. Peanut Sci. 39: 22–29.

Pongnumkul, S., P. Chaovalit and N. Surasvadi. 2015. Applications of smartphone-based sensors in agriculture: a systematic review of research. J. Sensors. doi: 10.1155/2015/195308.

Poirier, A.H., A.C. York, D.L. Jordan, A. Chandi, W.J. Everman and J.R. Whitaker. 2014. Distribution of glyphosate- and thifensulfuron-resistant Palmer amaranth (*Amaranthus palmeri*) in North Carolina. Int. J. Agron. doi: 10.1155/2014/747810.

Price, A.J., D.W. Reeves, M.G. Patterson, B.E. Gamble, K.S. Balkcom, F.J. Arriaga and C.D. Monks. 2007. Weed control in peanut grown in a high-residue conservation-tillage system. Peanut Sci. 34: 59–64.

Prostko, E.P. 2011. Non-selective applicators for the control of Palmer amaranth. Proc. South. Weed Sci. Soc. 64: 1.

Prostko, E.P., R.C. Kemmerait, P.H. Jost, W.C. Johnson III, S.N. Brown and T.M. Webster. 2009. The influence of cultivar and chlorimuron application timing on spotted wilt disease and peanut yield. Peanut Sci. 36: 92–95.

Riar, M.K., D.S. Carley, C. Zhang, M.S. Schroeder-Moreno, D.L. Jordan, T.M. Webster and T.W. Rufty. 2016. Environmental influences on growth and reproduction of invasive *Commelina benghalensis*. Int. J. Agron. doi: 10.1155/2016/5679249.

Richburg III, J.S., J.W. Wilcut, D.L. Colvin and G.R. Wiley. 1996. Weed management in southeastern peanut (*Arachis hypogaea*) with AC 263,222. Weed Technol. 10: 145–152.

Richburg III, J.S., J.W. Wilcut, A.K. Culbreath and C.K. Kvien. 1995. Response of eight peanut (*Arachis hypogaea*) cultivars to the herbicide AC 263,222. Peanut Sci. 22: 76–80.

Robinson, B.L., J.M. Moffitt, G.G. Wilkerson and D.L. Jordan. 2007. Economics and effectiveness of alternative weed scouting methods in peanut. Weed Technol. 21: 88–96.

Rodriguez-Kabana, R., N. Kokalis-Burelle, D.G. Robertson, P.S. King and L.W. Wells. 1994. Rotations with coastal bermudagrass, cotton, and bahiagrass for management of *Meloidogyne arenaria* and sothern blight in peanut. J. Nematol. 26: 665–668.

Rodriguez-Kabana, R., H. Ivey and P.A. Backman. 1987. Peanut-cotton rotations for management of *Meloidogyne arenaria*. J. Nematol. 19: 484–487.

Rodriguez-Kabana, R. and J.T. Touchton. 1984. Corn and sorghum rotations for management of *Meloidogyne arenaria* in peanut. Nematropica 14: 26–36.

Royal, S.S., B.J. Brecke and D.L. Colvin. 1997. Common cocklebur (*Xanthium strumarium*) interference with peanut (*Arachis hypogaea*). Weed Sci. 45: 38–43.

Sconyers, L.E., T.B. Brenneman, K.L. Stevenson and B.G. Mullinix. 2007. Effects of row pattern, seeding rate, and inoculation date on fungicide efficacy and development of peanut stem rot. Plant Dis. 91: 273–278.

Shaw, D. 2005. Translation of remote sensing data into weed management decisions. Weed Sci. 53: 264–273.

Sherwood, J.L., M.K. Beute, D.W. Dickson, V.J. Ellitt, R.S. Nelson, C.H. Opperman and B.B. Shew. 1995. Biological and biotechnological control advances in *Arachis* hypogaea. pp. 160–206. *In:* Pattee, H.E., Stalker, H.T. (Eds). Advances in Peanut Science, American Peanut Research and Education Society, Stillwater, OK, 614 p.

Tredaway-Ducar, J., J.B. Clewis, J.W. Wilcut, D.L. Jordan, B.J. Breckey, W.J. Grichar, W.C. Johnson III and G.R. Wehtje. 2009. Weed management using reduced rate combinations of diclosulam, flumioxazin, and imazapic in peanut. Weed Technol. 23: 236–242.

Tubbs, R.S. and R.N. Gallaher. 2005. Conservation tillage and herbicide management for two peanut cultivars. Agron. J. 97: 500–504.

Valentine, H. 2016. The role of peanuts in global food security. pp. 447–462. *In:* Stalker, H.T., Wilson, R.F. (Eds.) Peanuts: Genetics, Processing, and Utilization.AOCS Monograph Series, AOCS Press, Elsevier, 478 p.

Walker, R.H., L.W. Wells and J.A. McGuire. 1989. Bristly starbur (*Acanthospermum hispidum*) interference in peanut (*Arachis hypogaea*). Weed Sci. 37: 196–200.

Warren Jr., L.S. and H.D. Coble. 1999. Managing purple nutsedge (*Cyperus rotundus*) populations utilizing herbicide strategies and crop rotation sequences. Weed Technol. 13: 494–503.

Webster, T.M. 2013. Weed survey – Southern States 2013. Proc. South. Weed Sci. Soc. 66: 275–287.

Webster, T.M., M.G. Burton, A.S. Culpepper, A.C. York and E.P. Prostko. 2005. Tropical spiderwort (*Commelina benghalensis*): a tropical invader threatens agroecosystems of the southern United States. Weed Technol. 19: 501–508.

Webster, T.M., W.H. Faircloth, J.T. Flanders, E.P. Prostko and T.L. Grey. 2007. The critical period of Benghal dayflower (*Commelina benghalensis*) control in peanut. Weed Sci. 55: 359–364.

Webster, T.M. and G.E. MacDonald. 2001. A survey of weeds in various crops in Georgia. Weed Technol. 15: 771–790.

Wehtje, G., B.J. Brecke and N.R. Martin. 2000. Performance and economic benefit of herbicides used for broadleaf weed control in peanut. Peanut Sci. 27: 11–16.

Wehtje, G., R. Weeks, M. West, L. Wells and P. Pace. 1994. Influence of planter type and seeding rate on yield and disease incidence in peanut. Peanut Sci. 21: 16–19.

Wehtje, G., J.W. Wilcut, T.V. Hicks and J. McGuire. 1988. Relative tolerance of peanuts to alachlor and metolachlor. Peanut Sci. 15: 53–56.

Wehtje, G., J.W. Wilcut and J.A. McGuire. 1992. Influence of bentazon on the phytotoxicity of paraquat to peanuts (*Arachis hypogaea*) and associated weeds. Weed Sci. 40: 90–95.

Wehtje, G., J.W. Wilcut and J.A. McGuire. 1991. Foliar penetration and phytotoxicity of paraquat as influenced by peanut cultivar. Peanut Sci. 18: 67–71.

Wiatrack, P.J., D.L. Wright and J.J. Marois. 2009. Influence of imazapic herbicide simulated carryover on cotton growth, yields, and lint quality. Crop Manag. doi: 10.1094/CM-2009-0720-01-RS

Wilcut, J.W. 1991. Economic yield response of peanut (*Arachis hypogaea*) to postemergence herbicides. Weed Technol. 5: 416–420.

Wilcut, J.W., A.C. York, W.J. Grichar and G.R. Wehtje. 1995. The biology and management of weeds in peanut (*Arachis hypogaea*), pp. 160–216. In: Pattee, H.E., Stalker, H.T. (Eds.) Advances in Peanut Science, American Peanut Research and Education Society, Stillwater, OK, 614 p.

Wilcut, J.W., J.S. Richburg III, E.F. Eastin, G.R. Wiley, F.R. Walls Jr and S. Newell. 1994. Imazethapyr and paraquat systems for weed management in peanut (*Arachis hypogaea*). Weed Sci. 42: 601–607.

Wilcut, J.W., J.S. Richburg III, G.R. Wiley, F.R. Walls Jr. 1996. Postemergence AC 263,222 systems for weed control in peanut (*Arachis hypogaea*). Weed Sci. 44: 615–621.

Wilcut, J.W. and C.W. Swann. 1990. Timing of paraquat applications for weed control in Virginia-type peanuts (*Arachis hypogaea*). Weed Sci. 38: 558–562.

Wise, A.M., T.L. Grey, E.P. Prostko, W.K. Vencill and T.M. Webster. 2009. Establishing the geographical distribution and level of acetolactate synthase resistance of Palmer amaranth (*Amaranthus palmeri*) accesions in Georgia. Weed Technol. 23: 214–220.

York, A.C. and H.D. Coble. 1977. Fall panicum interference in peanut. Weed Sci. 25: 43–47.

Sustainable Weed Control in Tobacco

Matthew C. Vann*[1], Loren R. Fisher[2] and Matthew D. Inman[3]

[1] Department of Crop and Soil Sciences, North Carolina State University, 4212 Williams Hall, Raleigh, North Carolina 27695
[2] Department of Crop and Soil Sciences, North Carolina State University, 4216 Williams Hall, Raleigh, North Carolina 27695
[3] Department of Crop and Soil Sciences, North Carolina State University, 4222B Williams Hall, Raleigh, North Carolina 27695

Introduction

Indigenous to the Americas, tobacco has been commercially cultivated for over 400 years and is believed to have been consumed by native people for as many as 2,000 years ago (Collins and Hawks 2013a, Sykes 2008). Since 1612, the production and trade have expanded outside of the Americas to every continent except Antarctica, thus creating the infrastructure and demand for tobacco products around the world. The major production regions are currently found in both North (United States) and South America (Brazil), Africa (Malawi and Zimbabwe), and Asia (People's Republic of China, Indonesia, and the Philippines) (Anonymous 2016, Brown and Snell 2016). Accounting for an estimated total of 328.2 million kilograms of leaf in 2013 (Honig 2016a), the United States is the fourth largest tobacco producing country, behind the People's Republic of China, India, and Brazil (FAO 2012). In the same year, the price per pound of cured leaf for all tobacco types produced in the U.S. averaged $USD 4.84 kg^{-1}, accounting for a total farm gate value in excess of $USD 1.57 billion (Honig 2016b).

Despite great variation in production practices, all tobacco types are susceptible to disease and insect pests; however, weeds are also a major focus in management systems (Bailey 2013, Collins and Hawks 2013b, Fisher 2016, Pearce et al. 2015a, Vann et al. 2016). Weeds provide direct competition for sunlight, water, oxygen, carbon dioxide, and other essential plant nutrients. If left uncontrolled weed competition can greatly reduce leaf yield and quality. Additionally, select species, such as common ragweed (*Ambrosia artemisiifolia*) (Vann et al. 2016) and horsenettle (*Solanum carolinense*) (Lucas 1975), can serve as alternative hosts for major tobacco pathogens. Uncontrolled weeds may also interfere with hand and mechanical harvesting efforts, thus reducing harvest efficiency and increasing production costs. Lastly, weeds or weed seed inadvertently collected during harvest can remain intermingled with tobacco for the duration of curing and processing unless removed. The presence of organic and inorganic foreign material in unprocessed tobacco and manufactured products is prohibited.

The husbandry of tobacco, in modern times, is very much different from what is commonly practiced in other agronomic crops grown in industrialized regions of the world; however, many

*Corresponding author: matthew_vann@ncsu.edu

of the basic principles of production can be of great benefit to weed control. First, tobacco is a transplanted crop. Tobacco seedlings, more commonly referred to as transplants, are produced in outdoor plant beds or greenhouse float beds and, as a result, plants are 10 to 15 cm in total height when they are introduced for field production. This advanced stage of growth allows tobacco to be very competitive with weeds at the onset of the growing season. Second, tobacco leaves, not seed or fiber, are harvested, cured, and sold for use in products, such as cigarettes, cigars, moist and dry snuff, loose leaf chewing tobacco, and snus. To produce maximum yield, plants are managed to provide maximum ground cover which, in turn, can shade smaller weeds and inhibit growth.Third, tobacco requires large quantities of physical labor to produce. Sykes (2008) reports that approximately 1,228 man-hours were required in 1957 to produce one hectare of flue-cured tobacco in the United States. In modern times, this estimate has been reduced to an average of 124 man-hours per hectare thanks, in part, to technological improvements in mechanization, bulk leaf curing, and the adoption of various Crop Protection Agents (CPAs) that are used to manage tobacco pests (Sykes 2008). However, despite such dramatic advances, the fact remains that hand-labor is required for various production practices, therefore, it is plausible that it may also be utilized for weed control through hand-removal.

Weed Impact

Weed species and densities are variable among global tobacco producing regions; therefore, it is difficult to list each problematic weed that a producer or agronomist might encounter. However, Bailey (2013) has stated that from a global perspective the five most common and troublesome weed genera in tobacco are: *Amaranthus, Cyperus, Digitaria, Chenopodium,* and *Ipomoea*. More specifically, Bailey adds that the six most common and troublesome broadleaf weeds in tobacco are redroot pigweed (*Amaranthus retroflexus*), yellow nutsedge (*Cyperus esculentus*), ivyleaf morningglory (*Ipomoea hederacea*), common lambsquarters (*Chenopodium album*), common ragweed (*Ambrosia artemisiifolia*), and horsenettle (*Solanum carolinense*) (Bailey 2013). Bailey also states that the five most common and troublesome grass weed species in tobacco are large crabgrass (*Digitaria sanguinalis*), goosegrass (*Eleusine indica*), fall panicum (*Panicum dichotomiflorum*), giant foxtail (*Setaria faberi*), and johnsongrass (*Sorghum halepense*) (Bailey 2013). Webster (2013) offers further insight with the 10 most common and troublesome weeds provided by weed surveys from the U.S. tobacco producing states of Florida, Georgia, Kentucky, and North Carolina (Table 19.1). Based on Webster's survey results, there are 25 different weed species that are common and/or troublesome in U.S. tobacco production, 16 of which are present in each of the four surveyed states (Webster 2013) (Table 19.1).

In a very general sense, the most common weeds encountered by tobacco producers are warm-season annuals (Table 19.1). However, there are a few perennial species that can be found in minimum-tillage production sites or regions where perennial forages are included in long-term cropping rotations that include tobacco. The most troublesome perennial weeds in these systems are typically Johnsongrass (*Sorghum halepense*), honeyvine milkweed (*Cynanchum laeve*), and yellow nutsedge (*Cyperus erythrorhizos*) (Green et al. 2015). Both annual and perennial weed species are of primary concern at the onset of production and are typically less concerning later in the season as tobacco plants become more competitive. Some grass and *Amaranthus* weed species can have multiple growing cycles within a single season of flue-cured tobacco production. Flue-cured tobacco is harvested in multiple passes from the lower stalk into the upper stalk positions as senescence is initiated, and the duration of harvest may last longer than 45 days. It is during this extended phase of harvest that sunlight penetration through the crop canopy can increase, resulting in additional weed seed germination. During the latter portion of the growing season secondary tillage cannot be completed due to plant interference, furthermore, there are no herbicides labelled for application at this advanced crop stage.

Weed interference reduces yield, quality, and value of tobacco through direct competition for sunlight, water, and nutrients (Peedin 1999). In general, a transplanted tobacco seedling is innately more competitive with weed populations when compared to a direct seeded crop,

Table 19.1. The Southern States – 10 most common and troublesome weeds in tobacco (Webster 2013)

Ranking		States		
	Florida	Georgia	Kentucky	North Carolina
Ten Most Common				
1	Digitaria spp.	Digitaria spp.	Amaranthus hybridus L.	Cyperus spp.
2	Richardia scabra L.	Richardia scabra L.	Setaria spp.	Digitaria sanguinalis L.
3	Cyperus spp.	Ipomoea spp.	Digitaria sanguinalis L.	Amaranthus spp.
4	Senna obtusifolia	Cyperus spp.	Ipomoea spp.	Senna obtusifolia
5	Desmodium tortuosum	Amaranthus spp.	Chenopodium album L.	Urochloa platyphylla
6	Cynodon dactylon L.	Chenopodium album L.	Ambrosia artemisiifolia L.	Ambrosia artemisiifolia L.
7	Ipomoea spp.	Desmodium tortuosum	Sorghum halepense L.	Chenopodium album L.
8	Acanthospermum hispidum	Cynodon dactylon L.	Cyperus esculentus L.	Ipomoea spp.
9	Urochloa texana	Acanthospermum hispidum	Solanum carolinense L.	Eleusine indica L.
10	Amaranthus spp.	Senna obtusifolia	Galinsoga quadriradiata	Cynodon dactylon L.
Ten Most Troublesome				
1	Cyperus spp.	Cyperus spp.	Galinsoga quadriradiata	Ipomoea spp.
2	Cynodon dactylon L.	Ipomoea spp.	Cyperus esculentus L.	Amaranthus spp.
3	Ipomoea spp.	Cynodon dactylon L.	Solanum carolinense L.	Cyperus spp.
4	Commelina benghalensis L.	Amaranthus spp.	Ipomoea spp.	Digitaria sanguinalis L.
5	Senna obtusifolia	Richardia scabra L.	Cynanchum laeve	Senna obtusifolia
6	Desmodium tortuosum Sw.	Desmodium tortuosum Sw.	Sorghum halepense L.	Ambrosia artemisiifolia L.
7	Acanthospermum hispidum	Acanthospermum hispidum	Amaranthus hybridus L.	Xanthium strumarium L.
8	Amaranthus spp.	Commelina benghalensis L.	Ambrosia artemisiifolia L.	Solanum carolinense L.
9	Indigofera hirsuta	Senna obtusifolia	Chenopodium album L.	Chenopodium album L.
10	Euphorbia cyathophora	Xanthium strumarium L.	Physalis longifolia	Urochloa platyphylla

however, this initial advantage is eventually overcome, as weed presence can greatly reduce leaf yield and quality when weed control efforts are minimal. Wilson (1995) reports a 77% reduction in flue-cured tobacco yield and a 10% reduction in leaf quality when weed growth and development was not controlled in North Carolina. Results from Hauser and Miles (1975) demonstrate a 26% reduction in flue-cured yield under the same weedy conditions in the U.S. State of Georgia. Similar results have been documented in dark, burley, and Oriental tobacco production where long-term weed presence reduced yield 28 to 40, 95, and 50%, respectively, when compared to season-long weed-free treatments (Bailey 2013, Lolas 1986). Ultimately, there is a great variation in tobacco yield and quality losses that are experienced by weed competition. These losses are based upon tobacco type, weed species present, and the relative density of weed species, however, the point remains that weeds negatively impact tobacco yield, quality, and value.

For most crops the critical weed-free period is four to six weeks after crop emergence; however, it has been hypothesized that the transplanting aspect of tobacco production might reduce the weed-free period for tobacco by one to two weeks (Bailey 2013). This estimate appears to be relatively accurate for specific tobacco types in specific production regions. Medlen (1978) reports a decrease in tobacco yield and value of 88.90 kg ha^{-1} and \$USD 254.11 ha^{-1}, respectively, for every week that common ragweed (*Ambrosia artemisiifolia* L.) is present under field conditions. Medlen ultimately concludes that yield of flue-cured tobacco did not significantly decline when maintained free of common ragweed for two weeks following transplanting (Medlen 1978). Similarly, Lolas (1986) observed that burley and Oriental tobacco yield were not significantly affected by competition unless weeds remained longer than three weeks after transplanting. In contrast, the critical weed-free period for flue-cured tobacco produced in Zimbabwe has been proposed to occur between four and nine weeks after transplanting (Ian et al. 2013, Mashayamombe et al. 2013). Ultimately, the dramatic reduction in yield, quality, and value demonstrates the point that the competitive advantage offered by advanced seedling growth at the onset of production is eventually overcome by weed competition if early-season control measures are not implemented.

Weed Control

Weed management is achieved through various combinations of differing management practices. What follows is a brief outline of those management practices commonly utilized by tobacco producers: Mechanical and Physical, Cultural, Preventative, Chemical, and Integrated Weed Control.

Mechanical and Physical Weed Control

In regions where topography and soil type will allow, tobacco is grown in a ridged (bedded) culture using conventional tillage practices. In this system, producers will employ the use of a moldboard plow, chisel plow, or disc/harrow type implement to adequately prepare the soil for ridging. These practices are the primary means by which weeds and vegetative residues are destroyed prior to ridging (Bailey 2013). Following primary tillage, field cultivators and, sometimes, mechanical rotary tillers are used as finishing tools (Bailey 2013). The combination of these practices should result in soil conditions absent of vegetation, thus allowing for uniform application of CPAs and, ultimately, ridging.

Conservation tillage is utilized in regions where soil type and topography are prone to wind and/or water erosion. In the United States, conservation tillage practices (no-tillage and strip-tillage) are more commonly implemented in the burley, dark-air, and dark-fire production areas of Kentucky, Tennessee, and Virginia. Initial no-till evaluations in North Carolina report an 18% decline in burley tobacco yield due to lacking tillage and poor weed control (Fisher 2004). Later research demonstrates that the yield of no-till burley could be improved beyond that of conventional tillage when herbicides providing residual weed suppression were utilized

(Fisher 2004). Ultimately, tobacco production systems utilizing conservation tillage practices still rely heavily upon herbicides to adequately kill cover crops and to provide residual, in-season weed suppression (Parker et al. 2007, Pearce et al. 2015b). In North Carolina, the largest tobacco producing state in the U.S., conservation tillage accounts for less than one percent of all flue-cured production (L. Fisher, personal communication). The low adoption rates documented in flue-cured production are not attributed to current weed control or technological limitations, but rather environmental factors that are much different from those found in burley and dark tobacco producing areas (Fisher 2004). Where conservation tillage practices are employed but herbicide use is not available or prohibited, weed suppression is accomplished by physical removal of weed escapes (Parker et al. 2007).

Following transplanting, secondary cultivation in conventional systems is typically employed two to four times per season, depending upon growing conditions (Collins and Hawks 2013b). Tobacco is cultivated for three major reasons: i) weed control, ii) building of row ridges to promote drainage, and iii) reduction of soil crusting to promote drainage and gaseous exchange (Collins and Hawks 2013b). In addition to mechanical weed control, a light cultivation may aid in herbicide activation when rainfall and/or irrigation are lacking (Vann et al. 2016). In tobacco systems, where herbicides are not used, secondary cultivation can occur as many as seven times per season with hand-removal occurring once or twice (Klingman 1967). Hawks and Collins (1970) report that the number of post-transplanting cultivations can be reduced from five to two per season without reducing leaf yield when chemical weed control is utilized. In the same study it was also reported that treatments receiving no secondary cultivation produced the lowest yield and value despite receiving a herbicide application (Hawks and Collins 1970). Secondary cultivation is employed in nearly all global production regions, however, the practice does not provide sufficient control when weed density is extremely high. Vann (2015) reports a reduction of leaf yield by 2,182 kg ha^{-1}, leaf quality by 16%, and crop value by \$USD 1,847 ha^{-1} where cultivation was the single method employed for broadleaf weed control. These results demonstrate the value of combining cultivation and herbicide application to tobacco yield, quality, and value. The final cultivation after transplanting is referred to as 'layby' which typically occurs four to six weeks after transplanting when plants are approximately 30–35 cm in total height (Peedin 1999).

Handweeding may also be employed as a final attempt to remove weed escapes and is typically employed to prevent contribution of weed seed to the soil weed seed bank. Research conducted in North Carolina concluded that handweeding can be a profitable weed control practice where utilized in conjunction with a recommended herbicide program (Vann 2015). Vann reports that hand-removal of Palmer amaranth (*Amaranthus palmeri* S. Wats) in tobacco was accomplished in 172 min. ha^{-1} in treatments containing the herbicide sulfentrazone, alternatively, where sulfentrazone was not applied, hand-removal required 814 min. ha^{-1} (Vann 2015). In two subsequent years of this study where cotton was planted following tobacco, treatments with hand-weeding events in the previous seasons consistently contained a lower Palmer amaranth density than did counterpart treatments that were not previously hand-weeded (Vann 2015). Handweeding has also been reported to reduce weed density in flue-cured tobacco by 85% in Pakistan (Yousafzai et al. 2007). Producers considering handweeding for weed control are encouraged to remove weeds prior to seed set in order to prevent contamination of leaf material and to transport weeds from fields to prevent regrowth.

Alternatives to cultivation and hand-weeding for physical and/or mechanical weed control have been evaluated with varying degrees of success. Plastic film mulching is utilized in 98% of the tobacco area under cultivation in Japan, primarily as a means to prevent soil erosion and weed growth as well as to maintain soil moisture (Chida 2015). Plastic mulch is not used in the United States due to added cost of production for installation and removal as well as cropping rotations that do not allow for multi-crop use. Mowing is a specific practice employed in areas with high adoption rates of mechanized harvest, as it can prevent intermingling of weeds with harvested leaf and machinery, thus reducing the presence of organic foreign material in cured leaf. Most commonly, it is field borders and skip rows within a field setting that are maintained with a mower. Herbicide use in these areas is prohibited out of concern for crop injury and pesticide residues that can result from chemical drift. At present, steam or flame cultivation of tobacco are not recommended practices as there is great potential for leaf injury to occur during their implementation. However, with the recent demand for organically produced tobacco where herbicides are not permitted some growers in Georgia and South Carolina are finding that flame cultivation can be useful and an effective means of weed control after layby and particularly after harvest is initiated.

Cultural Weed Control

Tobacco is not directly seeded into the soil as other row crops, such as corn, cotton, peanut, or soybean, but rather are transplanted as young seedlings. The size and uniformity of transplants is thought to allow tobacco to be more competitive against weeds earlier in the growing season compared to seeded crops (Bailey 2013). In addition, tobacco is often transplanted into a clean, tilled bed earlier in the growing season compared to other crops and can become established before certain troublesome weed species germinate. In general, tobacco is tall (100–150 cm) and has large leaves that can quickly cover row middles in a relatively short duration of time, and as such, shading the soil and decreasing light penetration thereby reducing weed emergence and growth.

Fertility programs are strictly managed in tobacco production. Too little, too much, or the wrong application timing of a particular nutrient can have detrimental effects to yield, quality, and leaf chemistry, resulting in undesirable characteristics of the cured leaf (Peedin 1999, Tso 1990). Nutrients, specifically nitrogen, must be micromanaged throughout the life of the crop (Vann and Inman 2016). The strict management, placement, and timing of nutrients allow the crop to maximize uptake and growth, potentially further increasing crop competition against various weed species.

Including tobacco as a component of a cropping rotation is often viewed as a valuable resource from a weed management perspective, specifically in the United States where producers often grow a variety of crops on the same farm. The early season growth advantage and morphology of a tobacco plant, differing herbicide chemistry (although limited relative to other agronomic crops), cultivation, and intense management practices such as handweeding all contribute to the reduction of the soil seedbank. The various weed disturbance methods utilized make tobacco an effective rotation crop (Buhler et al. 1997, Liebman and Dyck 1993). One specific example of the rotational benefits offered by tobacco management is put forth by Vann (2015). In a three-year study in North Carolina, Vann (2015) concludes that density of Palmer amaranth (*Amaranthus palmeri* S. Wats) was reduced in cotton stands by as much as 78% where weed populations were managed with appropriate herbicide application and physical weed removal in tobacco in year one of the study. The same study also suggests the inclusion of sulfentrazone reduced Palmer amaranth density by 56% relative to treatments that did not include sulfentrazone. Herbicides used in tobacco production are not commonly used in most other crops, thereby reducing the exposure of those particular chemistries and potentially reducing the risk of evolution of herbicide resistance. In contrast, limited herbicide diversity and associated weed complexities are reasons why crop rotation is important.

Preventative Measures

The primary means of preventative weed control is to ensure that equipment and machinery are thoroughly cleaned prior to use. As farm operations become larger and equipment is traveling greater distances, the reality of preventing weed seed movement is becoming increasingly important (Buhler et al. 1997). It is recommended to clean and/or sanitize tillage, sprayer, and harvest equipment as operations move from field to field, especially if difficult to manage or herbicide-resistant weed species are present. Weed seed contamination of cured leaf is a major issue for tobacco sellers and buyers, especially troublesome species, such as Palmer amaranth (*Amaranthus palmeri* S. Wats). Weed seed most likely enters the supply chain by way of mechanical harvesters and makes its way into the cured product.

At harvest and post-harvest weed management is important to prevent local, regional, and global weed seed movement in exported leaf. Weed populations present at harvest are commonly handweeded and physically transported from the field. In general, tobacco harvest is completed earlier in the season than most other agronomic crops in tobacco producing regions. This can allow sufficient time for weeds to continue to germinate and/or grow, potentially contributing seed to the soil seedbank. Post-harvest stalk and root destruction is a normal operation. While this helps reduce disease and other pest buildup, a timely operation can mechanically suppress present weed communities and prevent further seed production.

Maintenance of field borders and ditches are always encouraged in order to prevent or reduce the introduction and movement of weed species and weed seed into a field, as well as the potential for outcrossing with resistant species and biotypes. Typically, growers often use a field cultivator to discard vegetation in field borders or unplanted areas of the field.

Chemical Weed Control

Herbicides account for approximately 10.4% of all pesticide use in the production of tobacco (Bailey 2013). However, in industrialized production regions, their use can account for a significant portion of a weed control program. Peedin (1999) estimates that chemical weed control is used on 80% of the tobacco area in the United States. Despite such a high rate of adoption, it is important for producers to understand that herbicides are only a single component of a tobacco weed control program and should not be used as the sole source of weed suppression.

At present there are only seven herbicide active ingredients federally approved for use in United States tobacco production: carfentrazone, clomazone, pebulate, pendimethalin, napropamide, sethoxydim, and sulfentrazone (Vann et al. 2016). It has been suggested that alachlor (Bailey 2013), dimethenamid (Masukwedza 2016), halosulfuron (A. Scholtz, personal communication), and S-metolachlor (Masukwedza 2016) are labeled in areas outside of the United States. Specific modes of action, chemical families, application rates, application timings, and targeted weed species for these herbicides can be found in Table 19.2. The expected control of specific monocot and dicot weed species offered by select herbicides is summarized in Tables 19.3 and 19.4, respectively. Additional information for each active ingredient is summarized by Bailey (2013). Additional active ingredients, such as fomesafen (Bridges and Stephenson 1991), imazaquin (Dhanapal et al. 1998), glufosinate (Whaley et al. 2016), glyphosate (Dhanapal et al. 1998), mesotrione (Whaley et al. 2016), and trifloxysulfuron (Bailey 2007, Porterfield et al. 2005), have been evaluated at differing application rates, methods, and timings but are not labeled at present.

Herbicide application for tobacco production typically occurs in one of four methods: pre-transplanting incorporated (PTI), pre-transplanting without incorporation (PRE-T), post-transplanting overtop (POST-OT), and post-transplanting directed (POST-Directed). Application method is dependent upon herbicide chemistry, tobacco tolerance, weed presence, and risk of herbicide residue detection after curing. A brief description and considerations for each application method are below.

Pre-transplanting Incorporated (PTI)

Pre-transplanting incorporated herbicides are uniformly applied to the soil surface and mechanically incorporated using a disc, harrow, field cultivator, or rotary-tiller type implements. Following incorporation, raised row ridges are formed and transplanting can be performed. Incorporation offers several advantages to producers. First, where approved, herbicides can be tank-mixed with other chemicals to reduce the number of applications a producer must make. Second, incorporation can sufficiently activate herbicides, thus reducing

Table 19.2. Mode of action (MOA), chemical family, application timings, application rates, and target weed species of selected herbicides used for tobacco production[a]

Herbicide	MOA[b]	Chemical family[b]	Application timing[c]	Application rate	Target weed species
				kg ai ha^{-1}	
Alachlor	Mitosis Inhibitor	Chloroacetamide	PTI and PRE-T	2.2 to 3.4	Grasses, sedges
Carfentrazone	Protox Inhibitor	Triazolinone	POST-Directed	0.03	Broadleaves, sedges
Cycloxydim	ACCase Inhibitor	Cyclohexanedione	POST-Directed	0.08 to 0.40	Grasses
Clomazone	DOXP Synthase Inhibitor	Isoxazolidinone	PTI, PRE-T, POST-OT	0.84 to 1.12	Grasses
Dimethenamid	Mitosis Inhibitor	Chloroacetamide	POST-Directed	0.50 to 0.86	Grasses, broadleaves
Fluazifop-P	ACCase Inhibitor	Aryloxyphenoxy-propionate	PRE-T and POST-OT	0.09 to 1.00	Grasses
Halosulfuron	ALS Inhibitor	Sulfonylurea	POST-Directed	0.04	Broadleaves, sedges
Propaquizafop	ACCase Inhibitor	Aryloxyphenoxy-propionate	POST-OT	0.05 to 0.15	Grasses
Napropamide	Mitosis Inhibitor	Acetamide	PTI, POST-OT, POST-Directed	1.12 to 2.24	Broadleaves, grasses
Pebulate	Lipid Synthesis Inhibitor	Thiocarbamate	PTI	4.48	Broadleaves, grasses
Pendimethalin	Microtubule Assembly Inhibitor	Dinitroaniline	PTI and POST-Directed	0.80 to 1.60	Broadleaves, grasses
S-Metoloachlor	Mitosis Inhibitor	Chloroacetamide	PTI and PRE-T	0.48 to 2.21	Grasses, sedges
Sethoxydim	ACCase Inhibitor	Cyclohexanedione	POST-OT	0.21 to 0.32	Grasses
Sulfentrazone	Protox Inhibitor	Triazolinone	PTI and PRE-T	0.28 to 0.42	Broadleaves, sedges

[a] Table adapted from Bailey 2013, Masukwedza 2016, Vann et al. 2016, and A. Scholtz, personal communication.

[b] Senseman, 2016

[c] PTI = Pre-Transplanting Incorporated, PRE-T = Pre-Transplanting without Incorporation, POST-OT = Post-Transplanting Overtop, POST-Directed = Post-Transplanting Directed to Row Middle.

Table 19.3. Expected control of selected monocot weed species from herbicide active ingredients labeled for use in United States tobacco production[a,b] (Vann et al. 2016)

Monocot species	Clomazone	Napropamide	Sethoxydim	Pendimethalin	Sulfentrazone	Pebulate	Carfentrazone
Echinochloa crusgalli	E	GE	E	GE	F	GE	N
Cynodon dactylon	PF	P	FG	P	P	P	N
Brachiaria platyphylla	E	G	E	G	F	P	N
Digitaria sanguinalis	E	E	GE	E	F	E	N
Dactyloctenium aegyptium	E	E	FG	E	F	E	N
Panicum dichotomiflorum	E	G	E	GE	-	G	N
Hordeum murinum	E	E	E	E	F	E	N
Elusine indica	E	E	GE	E	F	G	N
Sorghum halepense	G	F	E	G	-	G	N
Cenchrus echinatus	G	-	FG	G	-	G	P
Urochloa texana	G	-	E	G	F	P	N
Cyperus esculentus	P	P	N	P	E	FG	N
Cyperus rotundus	P	P	N	P	E	FG	N

[a] Ratings are based on average to good soil and weather conditions for herbicide performance and on proper application rate, technique, and timing.

[b] E = Excellent control (>90%), G = Good control (80-90%), F = Fair control (60-80%), P = Poor control (1-59%), N = No control, '-' = No data.

Table 19.4. Expected control of selected dicot weed species from herbicide active ingredients labeled for use in United States tobacco production[a,b] (Vann et al. 2016)

Dicot species	Clomazone	Napropamide	Sethoxydim	Pendimethalin	Sulfentrazone	Pebulate	Carfentrazone
Xanthium strumarium	F	P	N	P	FG	P	G
Portulaca oleracea	FG	E	N	P	G	P	P
Galinsoga quadriradiata	G	PF	N	P	G	P	P
Datura stramonium	G	P	N	P	-	P	G
Chenopodium album	G	G	N	G	E	G	G
Ipomoea spp.	P	P	N	P	E	P	E
Amaranthus spp.	P	G	N	G	E	G	E
Sida spinosa	E	P	N	P	G	P	P
Ambrosia artemisiifolia	G	F	N	P	P	P	N
Ambrosia trifida	PF	PF	N	P	-	P	N
Senna obtusifolia	P	P	N	P	-	P	N
Polygonum spp.	G	P	N	P	E	P	G

[a] Ratings are based on average to good soil and weather conditions for herbicide performance and on proper application rate, technique, and timing.
[b] E = Excellent control (>90%), G = Good control (80-90%), F = Fair control (60-80%), P = Poor control (1-59%), N = No control, '-' = No data.

the need for rainfall activation that is common to soil surface applied materials. Third, when poor field conditions postpone transplanting, PTI herbicides can help prevent weed growth that might be experienced with freshly tilled soil. The primary disadvantage of PTI herbicide applications is the potential for crop injury. Injury is greatest when herbicides are poorly incorporated, high application rates are utilized, or when multiple herbicides are tank-mixed (Vann et al. 2016). Furthermore, injury is more pronounced during cool, wet transplanting seasons when root growth is limited, resulting in prolonged exposure to herbicides. The herbicides most commonly applied PTI are alachlor, clomazone, pendimethalin, S-metolachlor, and sulfentrazone (Table 19.2).

Pre-transplanting without Incorporation (PRE-T)

Herbicides applied PRE-T are placed on top of raised row ridges. These materials are generally not activated without rainfall or mechanical incorporation. The delay in activation can inhibit weed control when rainfall is insufficient, therefore, it is recommended that producers utilize secondary cultivation to activate these materials post-transplanting (Vann et al. 2016). Prior to herbicide application it is recommended that producers remove the crest of the raised row ridge to the height at which transplanting will occur. Removal of the row ridge crest prior to application ensures that the soil is treated uniformly. If producers do not follow this practice treated soil is moved from the row ridge to the row middle during mechanical transplanting, thus leaving untreated bands of soil which can allow for weed growth. Although the PRE-T application method can require additional efforts on behalf of producers, this practice can significantly reduce the possibility of herbicide injury to plants.

For example, sulfentrazone injury to tobacco has been shown to range from 0 to 8% when applied PRE-T in comparison to 3 to 31% when applied PTI (Fisher 2003a, Fisher 2003b, Vann et al. 2016). Herbicides labeled for PTI application are commonly labeled PRE-T as well, the exception to this rule is fluazifop, which is not labeled PTI (Table 19.2).

Post-transplanting Overtop (POST-OT)

Currently, three active ingredients are labeled for POST-OT application. Sethoxydim can be applied up to 42 days prior to harvest, though residue concerns discourage late season use. Clomazone and napropamide are also labeled for this application timing, however only up to seven days after transplanting in the United States (Vann et al. 2016). If herbicide application is greatly delayed due to poor environmental conditions, weed growth can become too great for some materials to inhibit and non-chemical control methods must be utilized. If tobacco growth is rapid, soil coverage of these materials can be limited due to vegetative interception, being the major factor of the seven-day use window with clomazone and napropamide. Futhermore, clomazone and napropamide have pre-emergence activity and if weed emergence occurs between transplanting and application, weed suppression is reduced (Vann et al. 2016).

Post-transplanting Directed (POST-Directed)

Due to repeated cultivation and degradation of the herbicide, weed control may be reduced during the latter portion of the growing season. A layby herbicide application can be used to supplement cultivation disturbance. In most situations producers will complete the layby cultivation while simultaneously applying an herbicide behind the tillage equipment. This application prevents movement of the herbicide to the row ridge and allows for the herbicide to be directed to the row middle. Where this technique is not feasible, some growers will use drop lines to apply the herbicide immediately after the layby cultivation. Regardless of how the layby application occurs, it is recommended that producers apply these materials to the row middle in a band application to prevent them from coming into contact with tobacco plants. In the United States, carfentrazone, pendimethalin, and napropamide are the only herbicides labelled for this application (Table 19.2) (Vann et al. 2016). Carfentrazone is a contact herbicide, therefore, application at layby must be completed with a hooded or shielded sprayer to prevent crop injury. Carfentrazone may also be applied after the first harvest in flue-cured tobacco if it is directed underneath the crop canopy.

Herbicide Resistance

Heap (2016) reports that there are presently 471 unique cases of herbicide resistant weeds around the world. Specifically, resistance has been confirmed in 87 crops produced in 66 countries. While information specific to tobacco is not available, it can be noted that species resistance to herbicides used in tobacco production are present in at least 10 tobacco producing countries (Heap 2016). It is from this information one can assume that resistance is present in some tobacco production systems.

Fortunately for producers, there are generally a limited number of confirmed incidents of herbicide resistance to most of the active ingredients approved for use in tobacco (Heap 2016). For example, there are presently no reported cases of resistance to alachlor, dimethenamid, pebulate, napropamide, or S-metolachlor. In addition, confirmation of resistance to herbicides, such as clomazone, fluazifop, and pendimethalin has either occurred in regions where tobacco is not produced or in weed species that are not common in tobacco.

Despite the relatively low level of resistance reported in many tobacco herbicides, resistance is a large concern to producers, specifically in reference to those materials classified as protoporphyrinogen oxidase (PPO) inhibitors. Weed resistance to various families of PPO inhibitors has been confirmed in the tobacco producing countries of Brazil and the United States (Heap 2016). The single PPO inhibitor labeled for use in tobacco, sulfentrazone, offers exceptional suppression of weeds in *Amaranthus* and *Ipomoea* genera (Vann et al. 2016), some of which have been reported as the most common and most troublesome weeds in tobacco production (Bailey 2013, Webster 2013). Should that efficacy be reduced or, perhaps lost, producers will then be forced to utilize other non-chemical means of suppression.

Ultimately, herbicide resistance is at the forefront of grower consideration for weed management programs. Should resistance develop to the more efficacious chemistries, producers will lose some of the most effective options available for weed suppression. The strain placed upon the limited number of labeled materials would be greatly alleviated with the addition of active ingredients. However, given the hesitancy that is expressed by many pesticide manufacturers to label more herbicides for tobacco it is unlikely that acceptable alternatives will be approved for use in the near future. Despite such a bleak outlook, it has been hypothesized that tobacco producers are less likely to experience the severity of resistance related issues commonly associated with other agronomic crops. This is mainly due to the integrated nature of how tobacco is managed, the need for hand labor, and high value of the crop.

Integrated Weed Control

For the overwhelming majority of tobacco produced in the world, weed management occurs in very much the same manner as it has for decades. The exception to this is the option for chemical weed suppression in the industrialized regions of production. Despite the opportunity for herbicide application in numerous markets, tobacco producers continue to rely on a holistic approach to weed management. Many production practices are completed regardless of the need for weed control.

In brief, producers are encouraged to select field sites for tobacco production that have low weed densities. Producers will then repeatedly till the soil in preparation for herbicide application, row ridging, and transplanting. Though the number of herbicides labeled for tobacco production are far less than the number labeled for use in other agronomic crops, those that are labeled represent different chemical families and offer control of a wide range of weed species. Following transplanting, tobacco fields are cultivated and fertilized until plants are too large for these practices to occur mechanically. Most commonly it is four to six weeks after transplanting that cultivation ceases and, thus, plants are large enough to sufficiently outcompete weeds for essential inputs. As a last line of defense, producers physically remove weed escapes that are larger than tobacco plants. Physical labor is often required for 'topping' (removal of apical meristem), 'suckering' (removal of axillary shoots), and harvest, therefore, weed removal can be accomplished simultaneously.

Evaluation of Weed Control Sustainability

The sustainability of tobacco weed control is likely one of the most stable in the realm of production agriculture. As previously referenced, tobacco producers utilize mechanical/physical, cultural, preventative measures, and chemical weed control measures for crop management. Successful implementation of the outlined practices, along with factors that include favorable weather, sound agronomic management, and reliable sources of energy, producers have the opportunity to remain extremely profitable. For example, in the United States, high quality Virginia flue-cured tobacco with an average yield of 2,802 kg ha^{-1} can have a gross value of $US 12,350 ha^{-1}. Economic budgets from North Carolina State University imply that total cost per hectare for machine harvested leaf could be as low as $US 8,450 ha^{-1} (Brown and Snell 2016). With per hectare profitability as high as $US 3,900 ha^{-1}, it stands to reason that the economic opportunity and sustainability of tobacco production provide great incentive to U.S. farmers. The point of economic sustainability is even greater in developing countries where the "profitability of tobacco as a cash crop… is several times higher than that of any other competing commodity" according to a 2003 report from the World Health Organization (FAO 2003). Simply stated, tobacco production in many regions of the world allows for a cash influx that very well may not otherwise exist.

The two major concerns to the sustainability of tobacco weed management come in the form of soil erosion from repeated cultivation and the lacking number of suitable herbicides for chemical weed suppression relative to other crops. Current data is not available, however, in 1983 it was estimated that sheet and rill erosion accounted for 11.2 tons soil loss ha^{-1} in more than half of the U.S. tobacco production area (Larson et al. 1983). More specifically, Wood and Worsham (1986) put forth average annual soil losses in North Carolina that range from 33 to 40 tons ha^{-1}. Certainly these losses have been reduced in the decades following these reports, if not for any reason more than a declining area of tobacco production in the United States and North Carolina. Furthermore, the number of secondary cultivations employed per season has been reduced dramatically, in part, due to the use of herbicides. Historical estimates by Klingman (1967) place the number of secondary cultivations per season being as high as seven in the first third of the growing season. In modern times, two to four cultivations per season are more common.

The second point of concern to the sustainability of weed control in tobacco is the limited number of herbicide options available to producers. There are a total number of 14 tobacco herbicides representing seven Modes of Action (MOA) listed in Table 19.2. Of the chemicals referenced, three MOA's (mitosis inhibitor, protox inhibitor, and ACCase inhibitor) account for 10 of the herbicides. More specifically, in the United States, where herbicide application accounts for a large component of tobacco weed control programs, only six of the MOA's are represented (Table 19.2). Given that herbicidal options for weed suppression in large hectare crops, such as cotton, maize, and soybean can run into the hundreds, more options for tobacco producers are needed as well.

Concluding Remarks

The outlook for tobacco demand continues to remain strong, especially for the high quality, flavor styles of leaf that are produced in the United States, Brazil, Zimbabwe, and Malawi. If producers are to maintain the level of income associated with high yielding, high quality tobacco, sustainable weed control must continue to be emphasized. Tobacco has a long history of sustainable and integrated weed management, a history that is greater in uninterrupted duration more so than any other cash crop. Producers and researchers alike must continue to find alternative weed control measures in order to maintain the viability of their product and to address the growing concerns of herbicide resistant weeds as well as environmental degradation.

REFERENCES

Anonymous. 2016. Estimated leaf production by crop year. Published: Feb. 03, 2016. (Accessed on May 04, 2016). Retrieved from http://www.universalcorp.com/Resources/LeafProduction/World_Leaf_ Production_ February_2016.pdf.

Bailey, W.A. 2007. Dark tobacco (*Nicotiana tabacum*) tolerance to trifloxysulfuron and halosulfuron. Weed Technology 21(4): 1016–1022.

Bailey, W.A. 2013. Herbicides used in tobacco. pp. 175–199. *In:* A.J. Price and J.A. Kelton (Eds.) Herbicides – Current Research and Case Studies in Use. InTech, Rijeka, Croatia.

Bridges, D.C. and M.G. Stephenson. 1991. Weed control and tobacco (*Nicotiana tabacum*) tolerance with fomesafen. Weed Technol. 5(4): 868–872.

Brown, B. and W. Snell. 2016. US tobacco situation and outlook. pp. 7–16. *In:* L.R. Fisher (Ed.) Flue-Cured Tobacco Information (AG-187 Revised). North Carolina Cooperative Extension Service, Raleigh, North Carolina, USA.

Buhler, D.D., R.G. Hartzler and F. Forcella. 1997. Implications of weed seedbank dynamics to weed management. Weed Science 45: 329–336.

Chida, H. 2015. Mulching for tobacco cultivation in Japan. Abstract-AP13. 2015 CORESTA Agronomy/Phytopathology Conference, Izmir, Turkey.

Collins, W.K. and S.N. Hawks. 2013a. Origin and history. pp. 1–5. *In:* W.K. Collins and S.N. Hawks (Eds.) Principles of Flue-Cured Tobacco Production (2nd edition). Collins and Hawks, Raleigh, North Carolina, USA.

Collins, W.K. and S.N. Hawks. 2013b. Cultivation and weed management. pp. 67–74. *In:* W.K. Collins and S.N. Hawks (Eds.) Principles of Flue-Cured Tobacco Production (2nd ed.). Collins and Hawks, Raleigh, North Carolina, USA.

Dhanapal, G.N., S.J. Ter Borg and P.C. Struik. 1998. Postemergence chemical control of nodding broomrape (*Orobanche cernua*) in bidi tobacco (*Nicotiana tabacum*) in India. Weed Technol. 12(4): 652–659.

FAO (Food and Agriculture Organization of the United Nations). 2003. Projections of Tobacco Production, Consumption, and Trade to the Year 2010. FAO, Rome, Italy.

FAO (Food and Agriculture Organization of the United Nations). 2012. Top Production-Tobacco, Unmanufactured-2012. (Accessed on October 15, 2016). Retrieved from http://faostat.fao.org/site/339/default.aspx

Fisher, L.R. 2004. Potential for reduced tillage tobacco production in North Carolina. pp. 161–162. *In:* Proceedings of the 26th Southern Conservation Tillage Conference for Sustainable Agriculture, Raleigh, North Carolina, USA.

Fisher, L.R. (Ed.). 2016. 2016 Flue-Cured Tobacco Information (AG-187 Revised). North Carolina Cooperative Extension Service, Raleigh, North Carolina, USA.

Fisher, L.R., W.D. Smith and J.W. Wilcut. 2003a. Effect of sulfentrazone rate and application method on weed control and stunting in flue-cured tobacco. Tobacco Sci. 46: 12–16.

Fisher, L.R., W.D. Smith and J.W. Wilcut. 2003b. Effects of incorporation equipment, application method, and soil placement of sulfentrazone on injury to flue-cured tobacco. Tobacco Science 46: 1–4.

Green, J.D., N. Rhodes and C. Johnson. 2015. Weed management, pp. 54–56. *In:* B. Pearce and W.A. Bailey (Eds.) 2015–2016 Burley and Dark Tobacco Production Guide. University of Kentucky Cooperative Extension, Lexington, Kentucky, USA.

Hauser, E.W. and J.D. Miles. 1975. Flue-cured tobacco yield and quality as affected by weed control methods. Weed Res. 15: 211–215.

Hawks, S.N. and W.K. Collins. 1970. Effects of a herbicide and levels of cultivation on yield and value of flue-cured tobacco. Tobacco Science 14: 170–172.

Heap, I. 2016. The international survey of herbicide resistant weeds. Published: Jan. 31, 2017. (Accessed on January 31, 2017). Retrieved from http://www.weedscience.org/

Honig, L. 2016a. Crop production 2015 summary (January 2016). (Summary No. 1057-7823). USDA-NASS, Washington, DC, USA.

Honig, L. 2016b. Crop values 2015 summary (February 2016). (Summary No. 1949-0372). USDA-NASS, Washington, DC, USA.

Ian, M., R. Dzingai, M. Walter, and S. Ezekia. 2013. Impact of time of weeding on tobacco (*Nicotiana tabacum*) growth and yield. International Scholarly Research Notices 2013: 1–4.

Klingman, G.C. 1967. Weed control in flue-cured tobacco.Tobacco Science 11: 115–119.

Larson, W.E., F.J. Pierce and R.H. Dowdy. 1983. The threat of soil erosion to long-term crop production. Science 219: 458–465.

Liebman, M. and E. Dyck. 1993. Crop rotation and intercropping strategies for weed management. Ecological Applications 3(1): 92–122.

Lolas, P.C. 1986. Weed community interference in burley and oriental tobacco (*Nicotiana tabacum*). Weed Research 26: 1–7.

Lucas, G.B. 1975. Virus diseases-mosaic. pp. 427–456. *In:* G.B. Lucas (Ed.) Diseases of Tobacco (3rd edition.). Biological Consulting Associates, Raleigh, North Carolina, USA.

Mashayamombe, B.K., U. Mazarura and A. Chiteka. 2013. Effect of two formulations of sulfentrazone on weed control in tobacco (*Nicotiana tabacum* L.). Asian Journal of Agricultural and Rural Development 3(1): 1–6.

Masukwedza, R. 2016. Agrochemicals approved by the Tobacco Research Board: 12 October 2016. Published: Oct. 12, 2016. [Accessed on October 15, 2016]. Retrieved from http://www.kutsaga.co.zw/downloads/Agrochemicals12.10.16.pdf

Medlen, L.L. 1978. Common ragweed (*Ambrosia artemisiifolia* L.) interference in flue-cured tobacco (*Nicotiana tabacum* L.) and its effect on growth, yield, and quality. M.S. Thesis, North Carolina State University, Raleigh, North Carolina.

Parker, R.G., L.R. Fisher and D.S. Whitley. 2007. Weed management in conventional and no-till burley tobacco. pp. 51–70. *In:* L.R. Fisher (Ed.) Burley Tobacco Guide (AG-376 Revised). North Carolina Cooperative Extension Service, Raleigh, North Carolina, USA.

Pearce, B., W.A. Bailey and E. Walker (Eds.). 2015a. 2015–2016 Burley and Dark Tobacco Production Guide. University of Kentucky Cooperative Extension, Lexington, Kentucky, USA.

Pearce, B., E. Ritchey and T.D. Reed. 2015b. Field selection and soil preparation. pp. 22–26. *In:* B. Pearce and W.A. Bailey (Eds.) 2015–2016 Burley and Dark Tobacco Production Guide. University of Kentucky Cooperative Extension, Lexington, Kentucky, USA.

Peedin, G.F. 1999. Production practices: flue-cured tobacco. pp. 104–142. *In:* L.D. Davis and M.T. Nielson (Eds.) Tobacco: Production, Chemistry, and Technology. Blackwell Science, London, England.

Porterfield, D., L.R. Fisher, J.W. Wilcut and W.D. Smith. 2005. Tobacco response to residual and in-season treatments of CGA-362622. Weed Technol. 19(1): 1–5.

Senseman, S. 2016. Weed Science Society of America – Herbicide Mechanism of Action (MOA) Classification List. Published: Sept. 11, 2016. (Accessed on October 15, 2016). Retrieved from http://wssa.net/wp-content/uploads/WSSA-Herbicide-MOA-20160911.pdf

Sykes, L.M. 2008. Mechanization and labor reduction: a history of US flue-cured tobacco production, 1950-2008. Tobacco Sci. 1–83.

Tso, T.C. 1990. Mineral nutrition – primary elements. pp. 279–311. *In:* T.C. Tso (Ed.) Production, Physiology, and Biochemistry of Tobacco Plant. IDEALS, Inc., Beltsville, Maryland, USA.

Vann, M.C. 2015. Effects of soil tillage on flue-cured tobacco growth, weed control, and soil physical properties. PhD Dissertation, North Carolina State University, Raleigh, North Carolina.

Vann, M.C., L.R. Fisher, M.D. Inman, J.A. Priest and D.S. Whitley. 2016. Managing weeds. pp. 77–95. *In:* L.R. Fisher (Ed.) Flue-Cured Tobacco Information (AG-187 Revised). North Carolina Cooperative Extension, Raleigh, North Carolina, USA.

Vann, M.C. and M.D. Inman. 2016. Managing nutrients. pp. 61–76. *In:* L.R. Fisher (Ed.) Flue-Cured Tobacco Information (AG-187 Revised). North Carolina Cooperative Extension, Raleigh, North Carolina, USA.

Webster, T.M. 2013. Weed survey-southern states: broadleaf crops subsection. pp. 275–287. *In:* Proceedings of the Southern Weed Science Society. Houston, Texas, USA.

Whaley, W.T., L.R. Fisher and M.C. Vann. 2016. Evaluation of non-tobacco labeled herbicides for late season application. 2016 CORESTA Congress, Berlin, Germany.

Wilson, R.W. 1995. Effects of cultivation on growth of tobacco. (Technical Bulletin No. 116). Agricultural Experiment Station, Raleigh, North Carolina, USA.

Wood, S.D. and A.D. Worsham. 1986. Reducing soil erosion in tobacco fields with no-tillage transplanting. Journal of Soil and Water Conservation 41(3): 193–196.

Yousafzai, H.K., K.B. Marwat, M.A. Khan and G. Hassan. 2007. Efficacy of some pre and post emergence herbicides for controlling weeds of FCV tobacco (*Nicotiana tabacum* L.). *In:* Pakistan. African Crop Science Conference Proceedings 8: 1099–1103.

Sustainable Weed Control in Strawberry

Steven A. Fennimore*[1] and Nathan S. Boyd[2]

[1] University of California Davis, 1636 East Alisal St., Salinas, CA 93905
[2] University of Florida IFAS, Gulf Coast Research and Education Center, 14625 CR 672,
   Wimauma, FL 33598, USA

Introduction

Strawberry is an important horticultural crop in the United States with a total of 23,482 ha planted and a value of $2.2 billion. California and Florida are the largest strawberry fruit producers at 16,397 and 4,413 ha harvested, respectively (NASS 2016). Additionally, there are 2,676 commercial ha in eight other states. There are two phases of strawberry production: runner plant production and fruit production. California has both phases of production while Florida only produces fruit and purchases transplants from Canada, North Carolina, and California (Strand 2008). Strawberry is vegetatively propagated in field nurseries and as a result there is need for both runner plant and fruit production weed control programs. Weed management practices differ substantially between the nursery and fruit phases and as a result they will be described separately.

Nursery Production

Strawberry is vegetatively propagated and production of high quality nursery plants requires several years and occurs at multiple locations within California. The process begins with production of clean stock in a virus-free rearing facility (Kabir et al. 2005). Strawberry plants are then propagated during one or two eight-month production cycles at a low elevation nursery (LEN) in central California. A low elevation nursery is generally considered less than 150 m above sea level. The final season of plant production before the fruiting field takes place in high elevation nurseries (>1,000 m) (HEN) where the plants are exposed to low temperatures to stimulate fruit production and increase plant vigor (Larson and Shaw 2000). Plant harvest at the HEN generally takes place during September to November when plants are dug, sorted and packed for planting in fruiting fields (Kabir et al. 2005). California produces about one billion plants annually on 1,600 ha (UCANR 1999). Florida growers purchase bare-root transplants or plugs from nurseries predominately located in California, North Carolina and Canada. The use of plugs has increased in recent years in Florida due to more rapid establishment and earlier yield.

*Corresponding author: safennimore@ucdavis.edu

Nursery plants are grown on soils previously fumigated with methyl bromide plus chloropicrin (MB:Pic) under exemptions for quarantine and preshipment (QPS) as many SB plants are destined for export (USEPA 2016). Fumigation with MB:Pic forms the basis of the weed, nematodes and soilborne disease control program for strawberry nurseries. Methyl bromide fumigation does control weeds such as common lambsquarters (*Chenopodium album* L.), but it does not control weeds with hard seed coats that resist fumigant penetration like California burclover (*Medicago polymorpha* L.) and little mallow (*Malva parviflora* L.) (Fennimore et al. 2008). Weed control in the strawberry nursery is more difficult than in fruiting fields for several reasons: i) runner plants are grown on open ground without mulch to block the weeds, ii) herbicides are not used in strawberry nurseries because the daughter plants need to root without interference from herbicides, iii) mother plants are planted in rows, and inter-row cultivation is possible only in the early season before the field fills in with daughter plants making cultivation impossible, and iv) tolerance for weeds is very low due to the possibility of weed to strawberry disease transmission (Fennimore et al. 2008).

Strawberry Fruit Production

In colder regions of the US, strawberries are frequently grown as a perennial crop using the matted row system. Young plants are transplanted in the spring on open ground without plastic mulches. During the establishment year, the long summers promote extensive stolon formation which increases the strawberry plant population. Berries are typically harvested for two to three years after the establishment year. Weed competition during the first 1 to 2 months after planting has the greatest impact on subsequent fruit yield but dense weed populations during the production years can reduce yields, hinder harvest, function as hosts for pathogens, and discourage customers in u-pick operations (Pritts and Kelly, 2001). In many northern states, growers have switched to plasticulture production and day-neutral varieties to prolong their production season. This transformation is ongoing and will have a significant impact on weed management approaches in regions where matted row systems were historically the predominant production practice.

Weed control practices in matted row berries consists of establishment, renovation and winter dormant applications. In new plantings use of herbicides, cultivation and hand weeding are necessary to establish the plantings (Weber 2004). Herbicide applications are often timed to occur when the matted row strawberries are dormant, such as induced dormancy in summer after harvest or in fall when the dormant plants are less susceptible to herbicide induced injury (DeFrancesco 2016).

In Florida and California strawberries are grown as an annual crop using a plasticulture production system. In this type of production system the raised beds are formed, fumigated and covered with plastic mulch approximately 13 to 30 days prior to transplant. Bare-root or plug transplants are planted in double rows on the bed and irrigated with drip tape although in southern California four rows per bed are common. Preplant soil fumigation is primarily used for control of nematodes and soil-borne pathogens but also to suppress or control weeds. The most common methods of fumigation in California are drip chemigation following bed formation, drip tape and mulch installation (Figure 20.1A). However, broadcast fumigation treating 100% of the field area is also very common (Koike et al. 2013). Broadcast fumigants are applied prior to bed formation and include application of two sets of mulch films: fumigation tarps that are applied during fumigation, and mulch films applied to the finished raised beds before or after strawberry transplanting (Figure 20.1B). In Florida, fields are also fumigated using drip chemigation especially where a second strawberry crop is grown on the same raised bed in two successive years. However, for majority of the acreage, fumigants are applied with dual shanks during the bed formation process immediately prior to laying the plastic mulch. In some fields where nematodes cause persistent problems, fumigants are applied with deep shanks prior to the primary fumigation to control nematodes deep in the soil profile. It is also fairly common to inject metam potassium to terminate the crop in fields where nematodes or

Figure 20.1. (A) Fumigants applied by drip chemigation near Oxnard, CA (top). Photo by Husein Ajwa. (B) A typical broadcast fumigant application near Santa Maria, CA (bottom). Photo by Steve Fennimore. (C) Hand weeding in strawberry. (D) Nutsedge penetration of plastic mulch. Photo by Nathan Boyd

soilborne diseases are difficult to manage (Noling 2016). The total production costs in Florida per year average $67,000 per hectare making strawberry one of the most expensive crops to produce in the state (Mossler 2012).

Fruit production in California occurs on the central and southern coast (Stand 2008). Nearly all fruit is harvested from annual plantings made in the fall using plants obtained from high elevation nurseries described in the previous section or in mid-summer from 'frigo' plants which are pulled from cold storage. Fall plantings leave a gap in the production cycle during late fall and winter. The purpose of the mid-summer planting is to fill in gaps in the fruit production season that often occur in November and December. In Florida, 90% of the production occurs within Hillsborough County near Plant City in central Florida. Strawberry transplants obtained from California, Canada, and North Carolina are transplanted in September and October. Berry harvest begins in November with peak production in January and February. Berry harvests typically end sometime in March when berry quality and price begin to fall.

Weed control inputs consist of field selection and sanitation, crop rotation, cover crops, soil fumigation, herbicides, mechanical cultivation, mulches and hand weeding. In California, strawberries are frequently rotated with vegetable crops and cover crops on the central coast as the crop cycle is generally 14 months or more making continuous berry production on the same land impossible. On the south coast, such as near Oxnard, CA, continuous strawberry is more common due to the production cycle there that is less than 12 months (Strand 2008). In Florida, strawberries are a six-month crop and typically rotated with cover crops, such as sorghum-sudangrass or sunnhemp. In a portion of the hectarage strawberries are intercropped or rotated with crops, such as peppers, eggplant, cantaloupe, and watermelon. In many fields, fumigant tolerant weeds like California burclover and black medic (*Medicago lupulina* L.) build up over time and cause significant management problems. Nutsedge species (*Cyperus* spp.) can also build up over time especially in Florida and are difficult to manage due to the lack of registered herbicides and inconsistent fumigant efficacy. Weed management programs in strawberry are a multi-component system that utilizes physical and cultural weed control tools. Strawberry producers utilize a weed management system with redundancy in it that while expensive can be very effective. The lack of registered herbicides for use post-transplant especially on nutsedge species continues to be a problem. This is especially true in Florida where nutsedge emerging through the plastic and broadleaf and grass weeds emerging in the row middle can be very difficult to manage.

Weed Impact on Strawberry

The most common sedges in Florida strawberry fields include purple nutsedge (*Cyperus rotundus* L.), yellow nutsedge (*Cyperus esculentus* L.), and green kyllinga (*Kyllinga brevifolia* Rottb.). Nutsedge species are especially problematic because they are not adequately controlled by current fumigant programs and can puncture and damage the plastic mulch which can hinder its reuse in a second crop, such as cantaloupe. Green kyllinga only emerges in the row middles but it is very difficult to control due to the lack of effective herbicide options. The most common broadleaf weeds include black medic, Carolina geranium (*Geranium carolinianum* L.), American black nightshade (*Solanum americanum* P. Mill.), Florida pusley (*Richardia scabra* L.), cutleaf geranium (*Geranium dissectum* L.), common ragweed (*Ambrosia artemisiifolia* L.), horseweed (*Conyza canadensis* (L.) Cronq.), ragweed parthenium (*Parthenium hysterophorus* L.), dogfennel (*Eupatorium capillifolium* (Lam.) Small), carpetweed (*Mollugo verticillata* L.), and common purslane (*Portulaca oleracea* L.). Broadleaf weeds occur in the planting holes and in the role middles. A variety of grasses also occur with goosegrass being the most common. Yellow nutsedge is the most troublesome perennial weed in California. It can emerge from as deep in the soil as 45 cm (Stoller and Sweet 1987). Thus it is difficult to control the deepest tubers in the soil with fumigants. Among the most common weeds in California strawberry are annual bluegrass, little mallow, burclover, sweet clover, redstem filaree, annual sowthistle, hairy fleabane and horseweed (Strand 2008).

Limited research has examined the competitive interaction between strawberries and weeds. However, it is broadly acknowledged that competition with weeds can reduce berry yield and alter berry physico-chemical attributes (Jamwal and Wali 2014). Pritts and Kelly (2001) found that weed competition during the first 1 to 2 months after planting in a matted row system had the greatest impact on berry yield. Uncontrolled weed growth late in the season has limited effects on yield (Pritts and Kelly 1997). In a mature matted row system season-long uncontrolled weed growth can reduce productivity by 51% but yield losses are much less with lower weed densities (Pritts and Kelly 2004). The authors are not aware of any published research that examined the effects of weeds on strawberry yields in plasticulture production systems. However, Boyd and Reed (2016) found that purple nutsedge that emerged at the time of transplant at densities as high as 20 m^{-2} had no effect on berry yields. It is likely that the weed species, time of emergence, weed location in relationship to the crop, and weed density will all have an effect on the competitive interaction as has been observed in other crops grown in similar production systems (Morales-Payan et al. 1998, Motis et al. 2003, Gilreath and Santos, 2004).

A large number of weeds are hosts of nematodes, such as field bindweed (*Convulvulus arvensis* L.), shepherd's-purse (*Capsella bursa-pastoris* L.) Medik. little mallow and annual sowthistle (*Sonchus oleraceus* L.) (Strand 2008). In Florida, species, such as cudweed, dogfennel, ragweed and bidens (*Bidens alba*) are considered good hosts for sting nematode (Noling 2016) which is a serious pests in strawberry fields whereas weeds, such as Florida pusley and cutleaf primrose are considered poor to moderate nematode hosts (Rich et al. 2008). Very little is known about the effects of weed populations on nematode densities when the crop is growing. During fallow periods weeds that are not controlled can function as a food source. As a result, tillage or other weed management techniques can substantially decrease nematode populations during fallow periods (Johnson III et al. 2007) which should lead to decreased incidence during the cropping period. However, the relationship between weeds and nematodes is complex and weeds can protect nematodes from pesticides, protect them from non-favorable environmental conditions, suppress nematode numbers through antagonism, exert indirect effects via competition with other plants, or alter populations due to changes in the biotic and abiotic soil environment (Thomas et al. 2005). Certain cover crop species also function as hosts and it is widely believed, though not proven, that weeds growing within a non-host cover crop function as reservoirs for nematode populations. Much additional research is needed to examine the intricate relationship between weed and nematode population dynamics but there is little doubt that weed management in commercial strawberry fields especially during the fallow period will have some effect on nematode populations.

Even less information is known regarding the relationship between weeds and strawberry pathogens. Strawberry nurseries attempt to maintain weed-free production fields to avoid transmission of plant viruses and pathogens from weeds to strawberry runner plants (Fennimore et al. 2008). It is suspected that dense weed populations alter moisture dynamics and can alter strawberry foliar disease incidence. A variety of weeds are hosts for strawberry pathogens but almost no research has examined the effects of weeds on overall disease incidence in strawberry fields. We also know that many soil-borne pathogens can survive on the roots of weeds but little is known about the relationship between weeds, soil-borne pathogens, and disease incidence in strawberry.

The relationship between weeds and insects in strawberry fields is also complex and beyond the focus of this chapter. In brief, we know that weeds can be beneficial in some situations where flowering species attract pollinators and beneficial insects. At the same time, common insect pests can also feed or persist on weeds. For example, weeds along roadways and in ditches which are typically not controlled in Florida can function as hosts to a range of insect pests including spider mites (Mossler 2012). Various other insects including Western flower thrips have been observed on a range of weed species (Frantz and Mellinger 1990) but little research has been done to directly examine the effects of weed populations on insect pressure or crop damage.

Weed Management

Weed Prevention

Growers are encouraged to avoid fields with difficult to control weeds like field bindweed or nutsedge and to select fields with low weed populations whenever possible. If growers have a weedy field it is recommended to work in areas with low weed populations first and to clean equipment between fields. Annual weeds like little mallow, California burclover, black medic, and Carolina geranium are very tolerant of fumigants and fields infested with these weeds or perennial weeds should be controlled before rotation to strawberry (Strand 2008). Aggressive management, such as glyphosate application and/or tillage during fallow periods, can help remove or reduce weeds that are difficult to control in the strawberry crop. In California, weeds in ditches, field edges, and row middles are generally controlled to prevent weed seed production. In Florida, row middles are largely maintained weed-free but weeds generally are not controlled on field edges or in ditches. The different approaches are due in part to climate where heavy rains throughout the year in Florida can cause severe erosion where there is bare ground and the ample rainfall makes it very difficult to maintain weed-free areas.

Crop scouting and good field records are an integral part of good weed management practices. Fumigants and herbicides like flumioxazin and oxyfluorfen must be applied to raised beds under the plastic mulch as much as 30 days prior to strawberry transplanting and weed emergence. Consequently, the decision to apply herbicides is based on field history and weeds observed during scouting in previous seasons (Strand 2008). For example, little mallow is partially tolerant to soil fumigation, and if it is present in the field, plans should be made to apply flumioxazin or oxyfluorfen to the planting beds 30 days before transplanting (Samtani et al. 2012). Some weed species, such as nutsedges occur during fallow periods and during the cropping period. Scouting during the off-season can also help growers make weed management decisions for the cropping season.

Mechanical and Physical Weed Control

Handweeding

Strawberry is a labor-intensive crop. Berries are typically hand-picked twice a week and field crews are sent through the fields multiple times per season to remove runners. Handweeding frequently occurs at the same time as runner cutting but larger crews may be required if weed densities are high. It is difficult to provide an estimate of handweeding costs as the operation is frequently combined with other activities in Florida. A limited number of Florida growers interviewed informally by Nathan Boyd in 2017 estimated that pruning runners and handweeding done in conjunction cost them somewhere between \$494–\$1235 ha^{-1} in a single season even where herbicides were applied. Labor shortages in recent years combined with increasing labor costs have resulted in growers looking for ways to improve labor use efficiency in any way possible (Charlton and Taylor 2016). In runner plant nursery fields, the main forms of weed control are preplant fumigation, mechanical cultivation early in the season before runner initiation, and handweeding. Handweeding costs in MB:Pic fumigated plots ranged from \$101 to \$1,241 ha^{-1}, and in non-fumigated plots ranged from \$161 to \$2,257 ha^{-1} (Figure 20.1) (Fennimore et al. 2008). After the plants begin to set runners, the only way to weed the fields is by hand as mechanical cultivation is no longer an option due to the damage it would cause to runner plants.

Mulches

A mulch can consist of plastic tarps, fiber mats, or organic material, such as straw. Mulches serve multiple functions, such as warming or cooling the soil, protecting the fruit and foliage from soil-borne pathogens, enhancing moisture retention, repelling insects and weed control (Strand 2008). Clear tarps are most useful for warming the soil and encouraging early development of the crop, but they can enhance weed growth in non-fumigated soils and are not recommended for organic systems. Polyethylene mulches are available in black, blue, brown, green, red, white and yellow. Some of the colors (including white and yellow) are available on a black background that greatly improves weed control. Blue, red and clear plastic mulches give the poorest weed control because they permit the greatest amount of light to penetrate the plastic, often enough to allow weed germination and growth under the plastic (Johnson and Fennimore, 2005). Black, brown, and green mulches block light effectively and are recommended for use in organic systems. Growers should proceed with caution when using green tarps, since there is considerable variation in the amount of light intercepted by green mulches and that can result in variable levels of weed control. In Florida, almost all strawberries are grown with black plastic mulches. Colored mulches have not been adopted on a commercial scale and clear mulches are rarely if ever used. Low density polyethylene mulches were used historically but in recent years almost all growers have switched to barrier films which enhance fumigant retention and result in improved pest control. Growers that fumigate with dimethyl disulfide (DMDS) must use totally impermeable films (TIF) to reduce odor issues. As an added benefit, TIF mulches tend to reduce nutsedge density even when applied over non-fumigated soil (McAvoy and Freeman 2013).

    Mulches are an expensive but important component of the weed management program in fruiting fields. Polyethylene mulches retain fumigants in the soil for longer intervals and enhance herbicidal activity. They also inhibit broadleaf and grass emergence on the bed except in the planting holes. Organic strawberry growers do not have access to fumigants and soil active herbicides, and the result is very high hand weeding costs of $6,111 ha^{-1} (Bolda et al. 2014). Fumigants used in conventional fields reduce the cost of handweeding considerably compared to organic fields. Dara et al. (2011) estimated handweeding costs of $2,107 ha^{-1} in conventional fields near Santa Maria, CA. Unfortunately, research has not compared the costs and benefits in terms of weed control for strawberry growers in Florida. Despite the many benefits of mulches, disposal is a serious issue. Polyethylene mulch is the most common type of plastic mulch and most of it is either disposed of in landfills or burned (eXtension 2015). Costs of disposal have been estimated at $250 ha^{-1} and likely will be increasing (Shogren and Hochmuth 2004). If use of mulches is to be sustained long term, it will likely require development of efficient means of recycling of polyethylene mulch and/or development of a reliable and cost-effective biodegradable mulch that does not require disposal (eXtension 2015).

The use of plastic mulches also modifies herbicide efficacy and persistence. Grey et al. (2007) found that dissipation was more rapid for some herbicides on bareground versus under plastic mulch. The opposite trend was observed with other herbicides, such as sulfentrazone. Recent research conducted at the University of Florida examined the fate of fomesafen under plastic mulch in a range of different crops (Reed 2017). They found that persistence varied with the type of plastic mulch but not with the presence or absence of fumigation. In vegetable crops, fomesafen persisted under the mulch for much of the production season whereas persistence declined rapidly in strawberry crops. The rapid decline is attributed to the use of overhead irrigation in Florida strawberry production systems to aid with crop establishment. Herbicide persistence under mulches is likely to vary with herbicide chemistry, soil type, and water use patterns. Further research is needed to examine herbicide persistence under plastic mulches especially in situations where multiple crops are grown on the same bed.

Solarization

Soil solarization can be an effective option in the Central Valley and other warm areas of California (Elmore et al. 1997). However, in California most strawberries are grown in cooler locations along the coast, where results from soil solarization may be less satisfactory. In cooler areas, the heat from solarization does not penetrate very deep into the soil. However, because solarization kills most weed seeds in the upper layers of the soil, good weed control can still result (Gilbert et al. 2007, 2008, Samtani et al. 2012). The incorporation of broccoli residues into the soil immediately prior to solarization can enhance its effect (Stapleton, Elmore, and DeVay 2000). Soil solarization has also been evaluated for use in Florida but has not been adopted as a commercial practice. Chase et al. (1999) found that 92%–95% nutsedge control could be achieved with solarization if a thermal-infrared-retentive film was used. The use of solarization combined with anaerobic soil disinfestation (ASD) techniques has been widely evaluated but the requirement for extensive carbon inputs and water usage has limited its adoption.

Steam for soil disinfestation in strawberry is being evaluated in California where steam is applied with the objective of heating the soil to 70°C for 20 minutes. After the soil cools, strawberry can be transplanted. The main target of steam application is for control of soilborne diseases, however, steam is quite effective on weed seeds (Fennimore et al. 2014). The most likely use for steam is as a buffer zone treatment where fumigants cannot be applied (Fennimore and Goodhue, 2016). However, steam is not used commercially for strawberry and is currently under evaluation.

Cultural Weed Control

Crop rotation is an extremely important component of weed management programs. The long 14-month growing season on the central coast of California does not allow continuous strawberry production. Conventional growers can crop every other year, while organic growers typically place strawberry in the rotation only every four years (Koike et al. 2012). Typically, California growers rotate strawberries with vegetable crops, such as celery, cole crops, and lettuce (Strand 2008). Fumigants used in strawberry reduce incidence of vegetable diseases like *Sclerotinia minor* in lettuce while intensive cultivation and handweeding programs in vegetable crops reduce the incidence of weeds prior to strawberry planting. Intensive weed management in vegetable crops helps remove most weeds and minimize weed seed production. Cole crops like broccoli, also have the advantage that they provide some level of disease and weed control as a result of alleopathic compounds they produce (Koike et al. 2012). Strawberry nursery fields are generally used for strawberry no more than once every three years. At HEN fields, rotational crops are generally small grains.

Rotation of strawberry with vegetable crops and cover crops is common on the central coast of California and is an effective cultural practice. Crops like broccoli and lettuce have effective weed management programs and are integral to reducing weed seedbanks in fields rotated to strawberry (Strand 2008). On the south coast of California where continuous strawberry production occurs, annual use of fumigants and intensive management has reduced weed populations to clovers and little mallow, weeds with seed that are moderately resistant to control with fumigants (Fennimore et al. 2003). In Florida, management practices vary substantially between farms and even fields within a given farm. Significant portions of the industry grow strawberries every winter and cover crops during the spring and summer months. Diverse crop rotations are rarely adopted in California or Florida due to high land values and the expensive infrastructure needed to produce the crop. Instead, in Florida it is fairly common to grow vegetables on the same bed as the strawberries with the vegetables transplanted mid strawberry season or after the strawberries have been removed. This 'crop rotation' can provide many benefits in terms of economic return but can also complicate weed control as herbicide selection must take into account what types of crops will be grown. The crops are typically planted mid strawberry season (intercropping) or immediately after strawberry crop termination (multi-cropping). This technique has been shown to have the potential to increase overall yield per unit of land (Karlidag and Yildirim 2009). In Florida, strawberry plants and weeds are removed by hand at the end of the final strawberry harvest when intercropping. Herbicides or fumigants are used to terminate the strawberry crop and kill weeds when multi-cropping. Research has been conducted to identify optimal planting dates for vegetable crops when intercropping with strawberry (Santos et al. 2008) but very little research has examined the effects of intercropping or multi-cropping on weed population dynamics. One would anticipate that the long-term impacts of multi-cropping on weed populations would depend on the level of weed control achieved with the second crop.

It is also fairly common for Florida growers to leave the raised beds in place for two strawberry seasons. This practice relies on repeated applications of burn-down herbicides for weed control in the row middles and on the raised beds during the fallow period and herbicide resistant weeds, such as paraquat resistant goosegrass can be problematic. Other difficult to control weeds with the current registered herbicides, such as green kyllinga are also difficult to manage where repeated strawberry crops are grown on the same mulch. Rising fumigation and field preparation costs are driving the interest in finding alternative ways to produce multiple crops on the same raised beds. Finding alternative ways for growers to diversify their production system and adopt more diverse crop rotations that are economically viable is an ongoing challenge.

Cover crops are an important component of sustainable agro-ecosystems (Wang et al. 2005, Wyland et al. 1996). Properly managed, they improve soil quality, pest and disease management, nutrient cycling, decrease erosion, and can increase crop yields. Florida strawberry growers typically grow a cover crop during the summer months. The most common cover crop species are sorghum-sudangrass hybrids (*Sorghum bicolor* × *S. bicolor* var. *sudanese*) or sunn hemp (*Crotalaria juncea*). In recent years, sunn hemp has become the predominant species due to the extensive biomass produced even when grown on low fertility soils. It is considered a competitive crop and even relatively modest sunn hemp densities can substantially reduce weed biomass (Mosjidis and Wehtje 2011). In Florida, sunn hemp has been shown to produce more biomass than cowpea, velvetbean, or sorghum sudangrass (Wang et al. 2005). Several authors have reported suspected allelopathic properties of sunn hemp (Adler and Chase 2007) although it has never been proven in a field situation.

The management of weeds within the cover crop is a critical issue as weed seed production or vegetative growth of perennial weed species may increase weed problems and their associated management costs in subsequent cash crops (Boyd and Brennan 2006). A variety of techniques including increased seeding rates can enhance the cover crops' competitive ability with weeds (Brennan et al. 2009). Cover crops effectively reduce annual weed populations but may not have as significant an effect on perennial weeds, such as nutsedge species. The effect of the cover crop on weed population will vary with crop establishment, rate of cover crop growth, species, and cover crop biomass production.

Cover crops can be sown as a monoculture or as a mix of multiple species. Seed mixtures that contain legumes and cereals combine the nitrogen fixating capability of legumes with the nitrogen scavenging ability of cereal crops. Feasibility of a mixture depends upon the success of each component (Creamer et al. 1997) and cover crops composed of multiple species are often less competitive than monocultures. Research conducted in Salinas, California, found that rye monocultures were more competitive with weeds than rye-legume mixes (Brennan and Boyd, unpublished data). This can be attributed to the slow establishment and poor competitive ability of many legume species (Teasdale and Abdul-Baki 1998). Brennan et al. (2011) found that early season biomass production was greater with cereal monocultures than cereal-legume mixes and 40% cereal–60% legume mixes had greater early season biomass than those that only contained 10% cereal. This would suggest that mixes with greater cereal content would be more competitive with weeds. The legume component may be desirable for a variety of reasons including the increased yield often observed in the following cash crop when compared to cereal monocultures.

Considerable research has assessed the feasibility of using mustard cover crops to suppress soilborne diseases (such as *Verticillium dahliae*) and weeds in potatoes. Residues from a mustard cover crop can reduce the growth of weeds (Bialy et al. 1990). Glucosinolates in mustard cover crop residues are the major chemical components responsible for the fungicidal, herbicidal, and nematicidal activity of *Brassica* spp. (Fenwick, Heaney, and Mawson 1989). As the plant tissues breakdown in soil, glucosinolates are converted to isothiocyanates, thiocyanates, and other compounds that help with disease and weed control. Rapeseed foliage incorporated into the soil has been shown to control common lambsquarters and redroot pigweed to a degree nearly equal to that provided by a standard herbicide treatment (Boydston and Hang 1995). However, more modest control (i.e., 30–40%) of redroot pigweed and velvetleaf was observed in soybeans (Krishnan, Holshouser, and Nissen 1998). Recent results suggest that brassica seed meal does not control weeds, but that it does increase strawberry fruit yield in relation to the nontreated control (Fennimore et al. 2008, Fennimore et al. 2014). These results suggest potential benefits of planting mustard cover crops in organic strawberry fields, but further research and economic analysis are needed before we can accurately assess those benefits.

Chemical Weed Control

Soil Fumigation

Soil fumigants are applied as a preplant application to crops to control a wide range of soilborne pests. Fumigation is used extensively to control weeds, nematodes, and soil pathogens before planting strawberries (both nursery and field plantings). Fumigants are volatile organic chemicals that have a relatively high vapor pressure and low water solubility. Included are materials, such as Pic, MB and 1,3-dichloropropene (1,3-D). Other commercially important volatile soil pesticides not usually regarded as true soil fumigants include metam and dazomet (Table 20.1). All fumigants are similar in that when applied to the soil, volatilization takes place and the vapors diffuse through the soil to contact the organism to be controlled. Metam has a low vapor pressure and does not move easily through the soil as vapor, but it is more easily carried with water, either with sprinkler or drip irrigation. The overall effectiveness of any fumigant is determined by the dosage delivered to the pest and is a function of the concentration and time of exposure. This is referred to as the CT factor (concentration/time factor) (Ruzo 2006). Other important factors that bear heavily on the success or failure of fumigation is soil temperature, soil moisture, soil preparation, organic-matter content, application method, and surface seal.

Methyl bromide, metam, and dazomet have the best weed control activity of the products available in California. Methyl bromide cannot be used in Florida strawberry production. Growers in this region typically rely on various ratios of chloropicrin and 1,3-dichloropropene, metam potassium, or dimethyl disulfide. Application of emulsified formulations of fumigants by the drip irrigation system, i.e., chemigation, under plastic mulch has been shown to generally control weeds better than the traditional shank applications (Figure 20.1, Fennimore et al. 2003). Strawberries are typically grown in soils with low organic matter and high sand content in Florida. Consequently, although drip irrigation is desirable, it is difficult to get the fumigants to move to the edges of the beds (Candole et al. 2007, Jacoby 2012). The lack of lateral diffusion allows soilborne pathogens and weed seeds and propagules to survive on the non-treated soil along the bed edges.

Table 20.1. Soil applied chemicals used as fumigants

Chemical	Fungi	Nematodes	Weeds
1,3-dichloropropene	-	X	X*
Chloropicrin (Pic)	X	-	X*
Methyl bromide (MB)	X	X	X
Dazomet	X	X	X**
Metam	X	X	X**
MB + Pic	X	X	X
1,3-D + Pic	X	X	Shank - Drip X**

X = Acceptable control
* Control is not normally adequate from preplant treatments. Perennial weeds will be suppressed to some degree.
** Control is often variable because of application techniques relating to water and chemical movement in different texture soils.

Due to its effectiveness for control of soilborne fungi, nematodes, and weeds, MB was the most widely used pre-plant soil fumigant for production of vegetables, ornamentals, and strawberry for over 40 years (Rosskopf et al. 2005). Methyl bromide provided excellent pest control which the alternative fumigants have yet to match (Ruzo 2006). Methyl bromide use has declined in recent years due to its mandatory phase out as a result of its classification as a compound that depletes atmospheric ozone. Starting in 2017, no more MB will be used in California strawberry fruiting fields (USEPA 2016). Methyl bromide use has already been phased out in Florida and strawberry growers have transitioned to a variety of fumigant and management alternatives. Despite some success, pest pressures and production costs have continued to increase following the methyl bromide phase-out. Methyl bromide is still the predominant fumigant used in California strawberry nurseries due to the exemption for QPS. In fruiting fields, Pic or 1,3-D plus Pic combinations are the fumigants most commonly used in California and Florida strawberry (CADPR 2016). However, fumigant regulations in California, such as the township caps limit for 1,3-D use, as well as the ability to limit Pic dose near sensitive sites and prohibited in fumigant buffer zones (Carpenter et al. 2001) has resulted in reduced fumigant use. All other fumigants are under regulatory pressure to reduce fumigant emission and bystander exposure (Browne et al. 2013). The use of 'barrier films' that reduce or eliminate fumigant emissions by trapping fumigants in the soil where they kill soil pests and then degrade before tarp cutting are well established in California and Florida (Fennimore and Ajwa 2011, Qian et al. 2011).

Following the MB ban, extensive research was conducted to identify effective alternative fumigants that work as well as MB with several products registered, such as Pic, 1,3-D, and metam potassium that can be used alone, in combinations, or in sequence. In Florida, growers predominately rely on mixes of Pic and 1,3-D, metam potassium, or DMDS. Chloropicrin is considered effective on fungi and insects but not on nematodes and weeds (Hutchinson et al. 2000). However, Pic can be fairly effective on susceptible weeds like common chickweed (*Stellaria media* L.) and common purslane (Haar et al. 2003). 1,3-D is generally effective on nematodes and soilborne insects but less effective on pathogens and weeds (Noling and Becker, 1994). Metam potassium controls a broad spectrum of pests but is frequently applied for its herbicidal properties. Chloropicrin + 1,3-D programs when paired with an effective herbicide program can adequately control weeds in many crops. However, numerous edaphic, environmental, biological, and cultural factors influence soilborne pest populations, as well as the performance and consistency of all soil-applied fumigants (Gilreath et al. 2003b, Munnecke and Van Gundy 1979, Noling 2006).

Lower efficacy observed with the alternative fumigants is due in part to reduced volatility and ability to disperse in soil compared with methyl bromide (Ajwa et al. 2003). For example, metam potassium may move as little as 10 cm from the injection point. Lack of efficacy in some cases is not solely due to the products' inability to kill the pest, but also due to the fumigants not coming in contact with the pest. Efficacy may be improved with enhanced distribution or placement within the correct management zone where the pest is located. For example, supplemental Pic applied to soil beneath the edges of the plastic mulch where fumigants were not reaching when applied with standard shanks substantially reduced fusarium infection of tomato roots (Jacoby et al. 2015). Noling et al. (2015) also found that placement of fumigants beneath soil compaction zones in strawberry fields enhanced control of nematodes that occurred within this zone. The concept of management zones should also apply to weed control.

Most of the registered fumigants provide moderate to weak levels of nutsedge control. This is especially true of purple nutsedge which tends to be more tolerant of many fumigants (Culpepper and Langston 2004). The tolerance of nutsedge to current registered fumigants combined with the lack of herbicides for use in strawberry make purple nutsedge a critical issue in many fields. Florida growers with nutsedge frequently rely on combinations of Pic and DMDS to achieve control. When used in conjunction with TIF films adequately nutsedge control can be achieved (McAvoy and Freeman 2013).

Fumigation with MB, 1,3-D, Pic, and metam kills many weed seed and the reproductive structures of some perennials. Nearly all fumigant applications are either immediately covered with plastic mulch or are injected through the drip irrigation system under plastic mulch which helps maintain the fumigant concentration at levels that kill weeds. Drip injection of fumigants, such as 1,3-D or Pic often improves the weed control compared to shank fumigation of these same chemicals (Fennimore et al. 2003). However, it is important to thoroughly wet the planting bed during drip fumigant injection to ensure good weed control on the edges of the bed. Where drip fumigation is used, only the bed is treated, and the furrows are not fumigated meaning that weeds there will require control by other means, such as herbicides. Soil fumigants control weeds by killing both germinating seedlings and seeds that have not germinated. Methyl bromide, 1,3-D, Pic, and metam kill weed seedlings and seeds by respiration inhibition. However, to kill weed seeds, fumigants must be able to penetrate the seed coat and kill the seed embryo. It is more effective for fumigants to kill moistened seed, because the seed tissues swell with water and allow the fumigant to penetrate more thoroughly. Moist seeds also have higher respiration rates and are more susceptible to fumigants than low respiration dry seed. Seed of most species are susceptible to fumigants because the fumigant can penetrate the seed coat, but hard coated seed like little mallow and California burclover are difficult to control because the fumigant cannot rapidly penetrate the seed coat and kill the embryo. Frequently fumigated fields in southern California often build up populations of little mallow and California burclover over time because these weeds are not easily controlled by fumigants. The requirement for adequate soil moisture to wet weed seed means that proper irrigation before fumigation is one of the keys to effective weed control with all fumigants (Fennimore et al. 2003, Haar et al. 2003, Strand 2008).

For shank applications, good soil moisture is also necessary for proper fumigant diffusion. Too much water in the soil will retard diffusion while too little will allow the fumigant to diffuse too rapidly. Coarse-textured soils (sands) can be fumigated at higher moisture levels than fine-textured soils (clays). Soil moisture should be in the range of 15–75% of field capacity, depending on the soil type. Proper moisture content can be determined by squeezing a handful of soil into a ball. If it will not form a ball, it is too dry. If the ball will not break apart when touched with a finger, it is too wet. Soils to be fumigated should be in good tilth. Soil clods may shield weed seeds from adequate exposure to the fumigant or prevent a good seal. For drip applied fumigants it is necessary to apply the fumigant over several hours, faster for light soils, slower for heavy soils. Optimal soil moistures are needed so that the fumigant can disperse evenly across the bed and control the weed seed in the soil throughout the bed (Strand 2008).

Impermeable plastics or barrier films trap fumigants with the goal of increasing safety, reducing exposure and improving efficacy. Barrier films have been shown to dramatically improve the activity of fumigants whether applied by drip application or by shank application. Barrier films trap fumigants in the soil resulting in higher fumigant concentrations for a longer time than under traditional films. Higher fumigant concentrations are very favorable for weed control as the fumigant concentrations under barrier films are highest near the soil surface where shallow germinating weeds must be controlled (Fennimore and Ajwa 2011).

Shank-applied fumigants are applied through chisel injectors spaced 20 to 30 cm apart to a depth of 15 to 25 cm (Figure 20.1). Once placed in the soil, a seal must be formed on the soil surface to retain fumigant vapors. The type of sealing will vary with the volatility of the chemical being used. Usual methods include the use of a water seal, plastic film (most common), or packing the soil surface with rollers or drags. Due to its high volatility, MB must be sealed immediately with plastic covers. Plastic seals may also be used on the less volatile fumigants to increase herbicidal activity, especially if the fumigant is injected near the surface of the soil. The fumigant 1,3-D is primarily used to control nematodes and Pic is used to control fungal pathogens and when applied by shank application provide only limited weed control.

Metam can be applied either by soil injection or through sprinkler or drip irrigation water. If injected, the shanks must be about 15 cm apart because metam does not move well in the vapor phase. To be effective it must move to the site of the pest. This movement can be done by applying metam to the soil in enough water to form a drench. The material should be evenly distributed and then the soil packed and a water seal or polyethylene tarp applied. Best results can be achieved with a seal. After application, soil should not be disturbed for five days, then it should be cultivated. Three weeks or more should be allowed between treatment and planting, depending on soil and weather conditions, to avoid damage to the subsequent crop. Transplants are more susceptible to injury than direct-seeded plants.

Dazomet is a granular material that is applied to the soil surface and sealed with a water drench, or mechanically incorporated. Moisture must be reapplied to keep this water seal intact on the surface and to allow for conversion to the active ingredient methylisothiocyanate. Because it is not immediately water soluble, it cannot be applied as a drench or through irrigation systems. Following dazomet application, the field should be aerated for two weeks before planting or until the odor of the fumigant is no longer present. Low temperature will slow the degradation of the material and affect the control and replant time. If there is doubt about completeness of aeration of either metam or dazomet, a simple germination test of the soil can be used (Strand 2008).

Pre-emergence Herbicides for Plasticulture Production Systems

Strawberry growers in many regions of the United States apply pre-emergence herbicides under the plastic mulch for control of broadleaf weeds and grasses that emerge in the planting holes and nutsedge species that are capable of puncturing the mulch. The herbicides are typically applied immediately after the fumigant operation prior to laying the plastic mulch. In Florida, growers historically rely on oxyfluorfen plus napropamide for broadleaf weed and grass control due largely to research conducted by Gilreath et al. (2003) who found that napropamide plus oxyfluorfen had more fruit and increased yield compared to either product alone and also had the best grass and broadleaf weed control in Florida conditions. In California, Daugovish et al. (2008) found also that oxyfluorfen applied pre-transplant reduced broadleaf weeds, such as California burclover, hairy nightshade (*Solanum physalifolium* Rusby), little mallow, and shepherd's purse. In recent years, many growers in California and Florida have switched to pre-emergence applications of flumioxazin for weed control as it has proven to be safe on a variety of cultivars (Samtani et al. 2012). Strawberry tolerance to clopyralid, EPTC, fomesafen, metolachlor, napropamide, oxyfluorfen, pendimethalin, prodiamine, simazine, and terbacil applied to the bed top prior to laying the plastic mulch has also been shown (Boyd and Reed 2016, Stall et al. 1995).

Purple and yellow nutsedge are common weeds in commercial fields in Florida. Both species can penetrate the plastic mulch (Figure 20.1) and dense populations reduce yield and hinder harvest operations. There are currently no herbicides registered for use in strawberry that suppress or control either nutsedge species. Boyd and Reed (2016) found that fomesafen, S-metolachlor, and EPTC were safe on strawberry when applied under the plastic mulch but purple nutsedge control was inconsistent. Similar results have been observed in vegetable fields (Boyd 2015, Dittmar 2013) where suppression but not control is typically observed. Sulfentrazone was recently registered for pre-emergence use in strawberry. It controls a variety of broadleaf weeds, suppresses yellow nutsedge but purple nutsedge suppression is inconsistent (Boyd, unpublished results). Yellow and purple nutsedge remain a problem without a viable herbicide solution for strawberry growers. Results from vegetable trials suggest that post-transplant herbicide applied in conjunction with a pre-emergence herbicide may be needed to achieve season-long control (Adcock et al. 2008, Dittmar et al. 2012). Unfortunately, pre-emergence herbicides registered for strawberry only suppress nutsedge and there are no registered post-transplant herbicides with activity on nutsedge. To achieve satisfactory control levels, growers currently rely on fallow programs and fumigants for nutsedge control.

Weed management in the row middle (the bare ground between the raised beds) is typically the most difficult. Weeds between the beds reduce crop yields (Gilreath and Santos 2004), are a food source for nematodes (Rich et al. 2008), are an alternative host for pathogens (French-Monar 2006, Wisler and Norris 2005), and are an alternative hosts for insects (Bedford et al. 1998). Florida growers rely predominantly on herbicides as it is difficult to cultivate without causing mechanical damage to the plastic mulch. California growers, however, rely both on herbicides and mechanical cultivation to control weeds in the row middles (Strand 2008). Repeated herbicide applications are normally required to achieve season-long control. Growers often apply a pre-emergence herbicide tank-mixed with a burn-down herbicide if weeds have already emerged prior to crop transplant. Herbicide applications following transplant are more difficult due to the limited number of registered products and the risk associated with drift. The products applied and number of applications over a season varies according to grower preference and weed pressure. Boyd (2016) compared row middle herbicides in vegetable crops and found that paraquat tank-mixed with flumioxazin was one of the more consistent options. Flumioxazin tank-mixed with a burn-down herbicide, such as paraquat or carfentrazone has also been widely adopted by strawberry growers in Florida.

Post-emergence Herbicides for Plasticulture Production Systems

Grass and broadleaf weeds emerge in the planting holes in strawberry fields. Grasses are readily controlled with herbicides, such as sethoxydim and clethodim. Broadleaf weeds, such as black medic and Carolina geranium can be more difficult to control especially given the limited number of herbicides registered for post-emergence applications (Manning and Fennimore 2001). Clopyralid is the only post-emergence herbicide registered in Florida with efficacy on broadleaf weeds. Early research noted that clopyralid could cause minor damage and strawberry tolerance to clopyralid varied with rate in some varieties (Hunicutt et al. 2013a, b). More recent research has found that clopyralid is safe on the most common strawberry varieties grown in Florida over a wide range of rates and application times (Boyd and Dittmar 2015). Minor damage consisting of leaf curling may occur but this does not lead to a reduction in yield or berry quality. Strawberry tolerance to clopyralid was also noted in Ohio (Figueroa and Doohan 2006) and North Carolina (McMurray et al. 1996). Tolerance may be due at least in part to the limited translocation of the herbicide in the strawberry plant (Sharpe 2017). Research conducted in Florida found that clopyralid efficacy improved when black medic was sprayed when it was small (Sharpe et al. 2016). Herbicide applications early in the season before the crop canopy reaches full size and when the weeds are small may be the most viable management option (Sharpe 2017). Additional research is needed to identify other herbicides that can be safely applied after transplant.

Herbicides in Matted Row Systems

The type of herbicides registered for use in matted row and plasticulture production systems are fairly similar. The primary difference is that strawberries grown in the Southern U.S. using plasticulture production are almost exclusively grown as annuals with herbicide applications largely occurring prior to transplant. In matted row systems, strawberries are grown as perennials and herbicides can be applied at multiple time points. This may include pre-plant, at planting, late summer of the planting year, prior to mulching in the fall, during the fruiting year, and following renovation. Dormant periods when the crop is not growing facilitate the use of herbicides that cannot be used in plasticulture systems.

Weed management recommendations vary between regions but there are some consistent trends. A variety of herbicides, such as *S*-metolachlor and trifluralin can be applied pre-plant. The crop is most susceptible to weed competition shortly after transplant when there is a lot of bare soil. Products, such as terbacil, pendimethalin and napropamide can be used to control weed emergence during this period. Far fewer herbicides can be applied post-emergence during the planting year. Herbicide options are similar to plasticulture production and include clopyralid for broadleaf weeds and clethodim or sethoxydim for grasses. Herbicide applications in late fall prior to mulching with products, such as terbacil, napropamide and sulfentrazone can effectively control weeds through to the following spring.

During the fruiting years 2 and 3, the same herbicides can generally be used as during the planting year 1. Effective weed management in the late summer and fall following harvest when the field is semi-dormant is critical if growers plan to renovate the field. Herbicides that can be applied when the crop is semi-dormant include 2,4-D and clopyralid. This time period is especially important because 2,4-D is the only effective post-emergent herbicide that adequately controls perennial broadleaf weeds which can be problematic in a matted row system. Pre-emergence herbicides, such as terbacil, flumioxazin and oxyfluorfen can also be applied during this period as long as the maximum annual application rate is not exceeded. A note of caution that herbicide registrations vary between regions and there are a variety of application limitations based on crop variety, growth stage, soil type, and soil organic matter content. Read all labels carefully prior to herbicide application and follow local recommendations.

Weed Resistance to Fumigants and Herbicides

There are no known reports of weed resistance to fumigants in strawberry and very few incidences of herbicide resistance. Strawberry is a valuable crop with an integrated weed control program that has multiple redundancies built in, not the least of which is handweeding which removes weeds not controlled by other weed control tools (Strand 2008). Handweeding prevents most of the weeds from setting seed, and fumigants control many seeds in the soil seedbank. The result is that selection for resistant weeds in strawberry does not occur frequently. However, seeds of herbicide resistant weeds can move into strawberry fields from other locations and become established. For example, in Florida, paraquat resistant goosegrass and American black nightshade occur in the row middles in strawberry fields and can complicate weed management programs. Ragweed parthenium (*Parthenium hysterophorous*) is also a serious problem in some Florida strawberry fields. It is not controlled with glyphosate applications but it is not known for certain if it is a case of true resistance or tolerance. It is likely that increased reliance on a limited number of herbicide products combined with labour shortages will result in increased reports of herbicide resistance.

Sustainability of Weed Management Systems in Strawberry

Effects on Human Health and the Environment

Fumigants are volatile and move from the soil to the air at the application site and may move off site and produce adverse health effects in people from hours to days after application. Health effects of fumigants range from mild eye irritation to more severe effects, depending on the fumigant and the level of exposure (USEPA 2016b). Methods of reducing fumigant exposure include low permeable films that trap the chemical in the soil where it degrades, and in California and Florida, impermeable films are currently used to limit fumigant emissions (Fennimore and Ajwa 2011). Other measures to reduce fumigant exposure include buffer zones where fumigants cannot be applied due to the presence of sensitive sites like daycare centers, hospitals, prisons, and schools. Buffer zones of up to 410 m may mean nonfumigated areas where weeds can reproduce and infest the rest of the field (CADPR 2015, Goodhue et al. 2016). Increasingly strict regulation of fumigants will likely make them more difficult to use in the future, especially in California where public resistance to fumigant use runs high. Goodhue et al. (2016) analyzed the effects on land that can be fumigated and found that the effect of fumigant buffer zones on Pic application in Ventura County California ranged from 3% to 45% of the hectarage that cannot be fumigated. This means that under current regulations in California that buffer zones reduce the acreage available for production of fumigant dependent crops like strawberry. Therefore, either alternative methods of soil disinfestations, such as steam must be developed, or soil management systems developed that suppress soil pests (Fennimore and Goodhue 2016).

The essential role for fumigants is to control soilborne diseases that can kill strawberry plants. Development of disease resistant strawberry would greatly help in reducing dependency of strawberry producers on fumigants. However, elimination of fumigants from the strawberry weed management system would increase dependency on herbicides, mulches and handweeding. This would require considerable modification of the current weed management system for conventional strawberry.

Broadcast fumigation involves covering 100% of the field with film during the fumigation phase. The film is sent to the landfill after use for only a few days following fumigation, and a second set of plastic is installed on the raised beds for the duration of the season (Strand 2008). Given the difficulty of recycling 'dirty' mulch films and lack of a commercially viable biodegradable mulch film, discarding of mulch film in the landfill remains the standard industry practice. How much longer this plastic will be accepted at landfills is unknown. Technology to gather and prepare polyethylene mulch for recycling is clearly needed if use of this film is to be sustainable.

Benefits of the Current Weed Management System

Current weed management practices in strawberry rely on a diverse mixture of tools including handweeding, herbicides, fumigants, tillage, mechanical barriers, and cover crops. As a result, the development of herbicide resistance is not as great of an issue as it is in other crops. The use of plastic mulches significantly reduces weed pressure and weed management need focus on only row middles and planting holes with the exception of nutsedge species which can puncture the plastic mulch. Growers in Florida and California continue to modify their weed management practices following the loss of MB and management approaches are likely to continue to diversify and change over the next few years. The use of herbicides has increased in recent years due to the transition away from MB and the lack of reliable labor. The greatest challenge facing growers in Florida is the lack of registered herbicides that can be used following crop transplant. This shortfall has required growers to rely on more extensive management programs during fallow periods especially where nutsedge is a serious issue.

Strawberry has a very robust weed management system that utilizes multiple weed management tools integrated into a system that protects the crop from weed losses. Systems based on crop rotations, herbicides, soil fumigants, and physical control tools like plastic mulches and handweeding form an effective weed control system with redundancies. However, dependency of the system on fumigants, plastic mulches and abundant labor may not be sustainable in the long-term.

The strawberry weed management system is dependent on labor for handweeding as fumigants and herbicides are not adequate to provide complete control. Weeds that escape control from fumigation and herbicides, such as California burclover, can only be removed by hand where they are near the strawberry plant or under the plastic. Herbicides applied at higher rates would provide longer residual control and likely help weed control, but would likely increase the chance of injury to strawberry as well as yield loss. Novel methods of herbicide delivery through drip irrigation systems during the season or slow release herbicide formulations may be a way of extending weed control under plastic mulch without injuring strawberry.

Concluding Remarks

Strawberry has a well-developed weed management program with multiple levels of protection from weed loss. Crop rotations, fumigants, herbicides, mulches and handweeding effectively control weeds. However, there are considerable challenges to the sustainability of this system. Dependency of the system on high inputs of fumigants, plastic mulches and labor for hand weeding, all of which will be difficult to continue using at current levels.

It is likely that the strawberry production system in the future will be much more diverse than in the past. High end production systems will employ substrate production systems, such as used in Europe today. These systems do not use soil and weeds are not an issue. However, cost of substrate systems is very high in excess of $240,000 ha^{-1} vs. $89,572 ha^{-1} for conventional field growing costs in coastal California (Kubota 2015, Dara et al. 2011). These systems will likely only be used on a limited basis.

It is also very likely that recent trends favoring movement of fruit production to Mexico will continue as labor costs there are lower than the US and fumigant use restrictions are less strict than in California. Organic production will likely continue to grow as demand for organic fruit exceeds supply. However, with a four-year rotation, organic strawberry can only be grown once every four years. With limited land available, the hectarage of organic strawberry would be less than that of conventional strawberry. One possible means of expanding organic strawberry acreage would be to use steam for soil disinfestation which would allow a shorter rotation (Fennimore et al. 2014). Additionally, use of fumigants in parallel with steam for soil disinfestation are a strategy to maximize the use of land and labor where fumigant restrictions are high (Fennimore and Goodhue 2016).

REFERENCES

Adcock, C.W., W.G. Foshee III, G.R. Wehtje and C.H. Gilliam. 2008. Herbicide combinations in tomato to prevent nutsedge (*Cyperus esculentus*) punctures in plastic mulch for multi-cropping systems. Weed Technol. 22: 136–141.

Adler, M.J. and C.A. Chase. 2007. A comparative analysis of the allelopathic potential of leguminous summer cover crops: cowpea, sunn hemp and velvetbean. HortScience. 42: 289–293.

Bedford, I.D., A. Kelly, G.K. Banks, R.W. Briddon, J.L. Cenis and P.G. Markham. 1998. *Solanum nigrum*: an indigenous weed reservoir for a tomato yellow leaf curl gemini virus in southern Spain. Eur J Plant Path 104: 221–222.

Bolda, M.P., L. Tourte, K. Klonsky, R. Demoura and K.P. Tumbler. 2014. Sample costs to produce organic strawberries. University of California Cooperative Extension. Available at: http://coststudyfiles. ucdavis.edu/uploads/cs_public/94/4b/944b5aad-6660-4dcd-a449-d26361afcae2/strawberry-cc-organic-2014.pdf

Boyd, N.S. and E.B. Brennan. 2006. Weed management in a legume-cereal cover crop with the rotary hoe. Weed Technol. 20: 733–737.

Boyd, N.S., E.B. Brennan, R. Smith and R. Yokota. 2009. Effect of seeding rate and planting arrangement on rye cover crop and weed growth. Agron. J. 101: 1–5.

Boyd, N.S. and T. Reed. 2016. Strawberry tolerance to bed-top and drip-applied preemergence herbicides. Weed Technol. 30: 492–498.

Boyd, N.S., E.B. Brennan and S.A. Fennimore. 2006. Stale seedbed techniques for organic vegetable production. Weed Technol. 20: 1052–1057.

Boydston, R.A. and A. Hang. 1995. Rapeseed (*Brassica napus*) green manure suppresses weeds in potato (*Solanum tuberosum*). Weed Technol. 9: 669–675.

Brennan, E.B. and R.F. Smith. 2005. Winter cover crop growth and weed suppression on the central coast of California. Weed Technol. 19: 1017–1024.

Browne, G., S. Fennimore, A. Katten, K. Klonsky, R. Koda, D. Legard, P. Marrone, G. Obenauf, C. Shennan and J. Steggall. 2013. Nonfumigant Strawberry Production Working Group Action Plan. N. Gorder, M. Lee, M. Fossen, P. Verke and N. Davidson (Eds.) California Department of Pesticide Regulation. Available at: http://www.cdpr.ca.gov/docs/pestmgt/strawberry/work_group/action_plan.pdf

CADPR (California Department of Pesticide Regulation) 2015. Control measures for chloropicrin. Available at: http://www.cdpr.ca.gov/docs/whs/pdf/control_measures_chloropicrin summary.pdf

CADPR (California Department of Pesticide Regulation) 2016. Summary of pesticide use report data. Available at: http://www.cdpr.ca.gov/docs/pur/pur14rep/comrpt14.pdf

Candole, B.L., A.S. Csinos and D. Wang. 2007. Distribution and efficacy of drip-applied metam-sodium against the survival of Rhizoctonia solani and yellow nutsedge in plastic-mulched sandy soil beds. Pest Mgmt. Sci. 63: 468–475.

Carpenter J., L. Lynch and T. Trout. 2001. Township limits on 1,3-D will impact adjustment to methyl bromide phase-out. California Agriculture 55(3): 12–18. doi: 10.3733/ca.v055n03p12.

Charlton, D. and J.E. Taylor. 2016. A declining farm workforce: analysis of panel data from rural Mexico. Amer. J. Agr. Econ. doi: 10.1093/ajae/aaw018

Clay, D.V. and L. Andrews. 1984. The tolerance of strawberries to clopyralid: effect of crop age, herbicide dose, and application date. Asp. App. Biol. 8: 151–158.

CDMS (Crop Data Management Systems). 2008. Matran EC sample label. Available at: http://www.cdms.net/LDat/ld77P003.pdf

Creamer, N.G., M.A. Bennett and B.R. Stinner. 1997. Evaluation of cover crop mixtures for use in vegetable production systems. HortScience. 32: 866–870.

Culpepper, A.S. and D.L. Langston. 2004. Methyl bromide alternatives for nutsedge in pepper. Proc. South. Weed Sci. Soc. 57: 142.

Dara, S., K. Klonsky, R. Demoura and K.P. Tumbler. 2011. Sample costs to produce strawberries. University of California Cooperative Extension. Available at: http://coststudyfiles.ucdavis.edu/uploads/cs_public/87/d1/87d1dc5f-60ea-453c-a349-ef0a8ce3a851/strawberry_sc_smv2011.pdf

Daugovish, O. and S.A. Fennimore. 2008. Weeds. pp. 115–130 *In:* M.L. Flint (Ed.) UC integrated pest management for strawberries, Second Edition. Oakland: University of California Division of Agriculture and Natural Resouces, Publication 3351.

Daugovish, O., S.A. Fennimore and M.J. Mochizuki. 2008. Integration of Oxyfluorfen into Strawberry (*Fragaria* × *ananassa*) Weed Management Programs. Weed Technol. 22: 685–690.

De Francesco, J. 2016. Strawberries. Pacific Northwest Weed Management Handbook. Available at: https://pnwhandbooks.org/weed/horticultural/small-fruits/strawberries

Dilley, C.A., G.R. Nonnecke and N.E. Christians. 2002. Corn-based extracts to manage weeds and provide nitrogen in matted-row strawberry culture. HortScience. 37: 1053–1056.

Dittmar, P.J., D.W. Monks and K.M. Jennings. 2012. Effect of drip-applied herbicides on yellow nutsedge (*Cyperus esculentus*) in plasticulture. Weed Technol. 26: 243–247.

Dittmar, P.J. 2013. Weed control strategies in tomato. pp. 24. *In:* The Florida Tomato Proceedings. Available at: http://swfrec.ifas.ufl.edu/docs/pdf/veg-hort/tomato-institute/proceedings/ti13_proceedings.pdf.

Duniway, J.M. 2002. Status of chemical alternatives to methyl bromide for pre-plant fumigation of soil. Phytopathology 92: 1337–1343.

Elmore, C.L., J.J. Stapleton, C.E. Bell and J.E. DeVay. 1997. Soil solarisation: a nonpesticidal method for controlling diseases, nematodes, and weeds. Oakland: University of California Division of Agriculture and Natural Resources, Publication 21377.

eXtension. 2015. Current and future prospects for biodegradable plastic mulch in certified organic production systems. Available at: http://articles.extension.org/pages/67951/current-and-future-prospects-for-biodegradable-plastic-mulch-in-certified-organic-production-systems#.VBnhK_ldWSp

Figueroa, R.A. and D.J. Doohan. 2006. Selectivity and efficacy of clopyralid on strawberry (*Fragaria* × *ananassa*). Weed Technol. 20: 101–103.

Fennimore, S.A., M.J. Haar and H.A. Ajwa. 2003. Weed control in strawberry provided by shank- and drip-applied methyl bromide alternative fumigants. HortScience 38: 55–61.

Fennimore, S.A. and K.A. Roth. 2003. Predicting time to emergence, flowering, and seed set in burning nettle (*Urtica urens*). WSSA Abstracts 43: 23–24.

Fennimore, S.A., M.J. Haar, R.E. Goodhue and C.Q. Winterbottom. 2008. Weed Control in Strawberry Runner Plant Nurseries with Methyl Bromide Alternative Fumigants. HortScience. 43: 1495–1500.

Fennimore, S.A., H. Ajwa, K. Subbarao, F. Martin, G. Browne and S. Klose. 2008. Facilitating adoption of alternatives to methyl bromide in California strawberries. *In:* Annual International Research Conference on Methyl Bromide Alternatives and Emissions Reductions. Orlando, FL. Proceedings 11.

Fennimore, S.A. and R.E. Goodhue. 2016. Soil disinfestation with steam: a review of economics, engineering and soil pest control in California strawberry. International Journal of Fruit Production. doi: 10.1080/15538362.2016.1195312

Fenwick, G.R., R.K. Heaney and R. Mawson. 1989. Glucosinolates. pp. 97–141. *In:* P.R. Cheeke (Ed.) Toxicants of plant origin. Volume II: Glucosides. Boca Raton: CRC Press.

Forcella, F., S. Poppe, N. Hansen and E. Hoover. 2001. Weed management with bio-based mulches in transplanted strawberry. WSSA Abstracts 41: 106–107.

French-Monar, R.D., J.B. Jones and P.D. Roberts. 2006. Characterization of *Phytophthora capsici* associated with roots of weeds on Florida vegetable farms. Plant Dis. 90: 345–350.

Gilbert, C.A., S.A. Fennimore, K. Subbarao, R. Goodhue and J.B. Weber. 2007. Solarisation and steam heat for soil disinfestation in flower and strawberry. *In:* Annual International Research Conference on Methyl Bromide Alternatives and Emissions Reductions. San Diego, California. Proceedings 109.

Gilbert, C.A., S.A. Fennimore, K. Subbarao, R. Goodhue, J.B. Weber and J. Samtani. 2008. Soil disinfestation with steam and solarisation for flower and strawberry. *In:* Annual International Research Conference on Methyl Bromide Alternatives and Emissions Reductions. Orlando, Florida. Proceedings 68.

Gilreath, J.P., J. Jones, T. Motis, B. Santos, J. Noling and E. Rosskopf. 2003. Evaluation of various chemical treatments for potential as methyl bromide replacements for disinfestations of soilborne pests in polyethylene-mulched tomato. Proc. Fla. State Hort. Soc. 116: 151–158.

Gilreath, J.P., J.M. Mirusso, J.P. Jones, E.N. Rosskopf, J.W. Noling and P.R. Gilreath. 2002. Efficacy of broadcast Telone C-35 in tomato. Proc. Annual Int. Res. Conference on Methyl Bromide Alternatives and Emissions Reductions, MBAO. pp. 19-1 to 19-2.

Gilreath, J.P., J.W. Noling, J.P. Jones, S.J. Locascio and D.O. Chellemi. 2001. Three years of soil borne pest control in tomato with 1,3-D + Chloropicrin and solarisation. Proc. International Research Conference on Methyl Bromide Alternatives and Emissions Reductions. pp. 131–133.

Gilreath, J.P., B.M. Santos and T.N. Motis. 2003. Herbicide and mulch evaluations for weed management in west central Florida strawberries. Proc Fla Hort Soc. 116: 159–160.

Gilreath, J.P. and B.M. Santos. 2004. Efficacy of methyl bromide alternatives on purple nutsedge control (*Cyperus rotundus*) in tomato and pepper. Weed Technol. 18: 141–145.

Goodhue, R., M. Schweisguth and K. Klonsky. 2016. Revised chloropicrin use requirements impact strawberry growers unequally. California Agriculture 70: 116–123.

Grey, T.L., W.K. Vencill, N. Mantripagada and A.S. Culpepper. 2007. Residual Herbicide Dissipation from Soil Covered with Low-Density Polyethylene Mulch or Left Bare. Weed Sci. 55: 638–643.

Guthman, J. 2016. Strawberry growers wavered over methyl iodide, feared public backlash. California Agriculture 70: 124–129.

Haar, M.J., S.A. Fennimore, H.A. Ajwa and C.Q. Winterbottom. 2003. Chloropicrin effect on weed seed viability. Crop Prot. 22: 109–115.

Haar, M.J. and S.A. Fennimore. 2003. Evaluation of integrated practices for common purslane management in lettuce. Weed Technol. 17: 229–233.

Hartz, T.K., J.E. DeVay and C.L. Elmore. 1993. Solarizaton is an effective soil disinfestation technique for strawberry production. HortSci. 28: 104–106.

Hunnicutt, C.J., A.W. MacRae and V.M. Whitaker. 2013a. Response of four strawberry cultivars to clopyralid applied during the fruiting stage.HortTechnol 23: 301–305.

Hunnicutt, C.J., A.W. MacRae, P.J. Dittmar, J.W. Noling, J.A. Ferrell, C. Alves and T.P. Jacoby. 2013b. Annnual strawberry response to clopyralid applied during fruiting. Weed Technol. 27: 573–579.

Hutchinson, C.M., M.E. McGiffen Jr, H.D. Ohr, J.J. Sims and J.O. Becker. 2000. Efficacy of methyl iodide and synergy with chloropicrin for control of fungi. Pest Manag. Sci. 56: 413–418.

Jacoby, T. 2012. Improving the efficacy of methyl bromide alternatives for vegetable production in Florida. Ph.D. Thesis Dissertation. University of Florida.

Jamwal, K. and V.K. Wali. 2014. Impact of integrated weed management practices on physico-chemical attributes of strawberry (*Fragaria × ananassa* Duch.) cv. Chandler. Prog Hort 46: 31–33.

Johnson, M. 2003. Nonchemical weed control inputs for strawberry. MS Thesis. University of California, Davis.

Johnson III, W.C., R.F. Davis and B.G. Mullinix Jr. 2007. An integrated system of summer solarisation and fallow tillage for Cyperus esculentus and nematode management in the southeastern coastal plain. Crop Prot. 26: 1660–1666.

Johnson, M.S. and S.A. Fennimore. 2005. Weed and crop response to colored plastic mulches in strawberry production. HortScience 40: 1371–1375.

Kabir, Z., S.A. Fennimore, J.M. Duniway, F.N. Martin, G.T. Browne, C.Q. Winterbottom, H.A. Ajwa, B.B. Westerdahl, R.E. Goodhue and M.J. Haar. 2005. Alternatives to methyl bromide for strawberry runner plant production. HortScience 40: 1709–1715.

Karlidag, H. and E. Yildirim. 2009. Strawberry Intercropping with Vegetables for Proper Utilization of Space and Resources. J. Sust. Agric. 33: 107–116.

Koike, S.T., T.R. Gordon, O. Daugovish, H. Ajwa, M. Bolda and K. Subbarao. 2013. Recent developments on strawberry plant collapse problems in California caused by *Fusarium* and *Macrophomina*. Int. J. Fruit. Sci., 13: 76–83.

Krishnan, G., D.L. Holshouser and S. Nissen. 1998. Weed control in soybean (*Glycine max*) with green manure crops. Weed Technology 12: 97–102.

Kubota, C. 2015. Hydroponic strawberry costs and economics. Available at: http://cals.arizona.edu/strawberry/Hydroponic_Strawberry_Information_Website/Costs.html

Manning, G.R. and S.A. Fennimore. 2001. Evaluation of low-rate herbicides to supplement methyl bromide alternative fumigants to control weeds in strawberry. Hort Technology 11: 603–609.

McAvoy, T. and J.H. Freeman. 2013. Yellow Nutsedge (*Cyperus esculentus*) control with reduced rates of dimethyl disulfide in combination with totally impermeable film. Weed Technol. 27: 515–519.

McGuire, J.A. and J.A. Pitts. 1991. Broadleaf weed control in strawberries with postemergence-applied diphenyl ether herbicides. J. Amer. Soc. Hort. Sci. 116: 669–671.

McMurray, G.L., D.W. Monks and R.B. Leidy. 1996. Clopyralid use in strawberries (*Fragaria* × *ananassa* Duch.) grown on plastic mulch. Weed Sci 44: 350–354.

Morales-Payan, J.P., W.M. Stall, D.G. Shilling, R. Charudattan, J.A. Dusky and T.A. Bewick. 2003. Above- and belowground interference of purple and yellow nutsedge (*Cyperus* spp.) with tomato. Weed Sci. 51: 181–185.

Mosjidis, J.A. and G. Wehtje. 2011. Weed control in sunn hemp and its ability to suppress weed growth. Crop Prot. 30: 70–73.

Mossler, M.A. 2012. Florida Crop/Pest Management Profiles: Strawberry.Document CIR1239. Florida Cooperative Extension Service, Institute of Food and Agricultural Sciences, University of Florida. Available at: http://edis.ifas.ufl.edu/pdffiles/PI/PI03700.pdf

Motis, T.N., S.J. Locascio and J.P. Gilreath. 2001. Yellow nutsedge interference effects on fruit weight of polyethylene-mulched bell pepper. Proceedings of the Florida State Horticultural Society. 114: 268–271.

Munnecke, D.E. and S.D. Van Gundy. 1979. Movement of fumigants in soil, dosage response and differential effects. Annual Review Phytopathology 17: 405–429.

NASS. 2016. Vegetable 2015 Summary. Agricultural Statistics Board, NASS USDA, Washington, D.C. 75 pp.

Noling J.W. 2016.Nematode management in strawberries. Document ENY-031.Florida Cooperative Extension Service, Institute of Food and Agricultural Sciences, University of Florida. Available at: http://edis.ifas.ufl.edu/pdffiles/NG/NG03100.pdf

Noling, J.W. 2006. Identifying causes of pest control inconsistency with soil fumigation in Florida strawberry. Proc. International Research Conference on Methyl Bromide Alternatives and Emissions Reductions. pp. 471–475.

Noling, J.W. and J.O. Becker. 1994. The challenge of research and extension to define and implement alternatives to methyl bromide. Journal of Nematol. 26(4S): 573–586.

Pritts, M.P. and M.J. Kelly. 1997. Weed thresholds in strawberries. Proceedings of the Third International Strawberry Symposium. Acta Horticulture 2: 947–950.

Pritts, M.P. and M.J. Kelly. 2001. Early season weed competition reduces yield of newly planted matted row strawberries. HortScience 36: 729–731.

Pritts, M.P. and M.J. Kelly. 2004. Weed competition in a mature matted row strawberry planting. HortScience 39: 1050–1052.

Qian, Y., A. Kamel, C. Stafford, T. Nguyen, W.J. Chism, J. Dawson and C.W. Smith. 2011. Evaluation of the permeability of agricultural films to various fumigants.Environ. Sci. Technol. 45: 9711–9718.

Reed, T.V. 2017. Fomesafen persistence, movement, and efficacy for nutsedge control in Florida plasticulture production. Ph.D. Thesis Dissertation. University of Florida. pp. 124.

Rich, J.R., J.A. Brito, R. Kaur and J.A. Ferrell. 2008. Weed species as hosts of Meloidogyne: a review. Nematropica 39: 157–185.

Rosskopf, E.N., D.O. Chellemi, N. Kokalis-Burelle and G.T. Church. 2005. Alternatives to methyl bromide: a Florida perspective. Plant Health Progress. Available at: http://www.apsnet.org/online/feature/methylbromide/FloridaPerspective.pdf

Ruzo, L.O. 2006. Physical, chemical and environmental properties of selected chemical alternatives for the pre-plant use of methyl bromide as soil fumigant. Pest Manag. Sci. 62: 99–113.

Samtani, J.B., J.B. Weber and S.A. Fennimore. 2012. Tolerance of strawberry cultivars to oxyfluorfen and flumioxazin herbicides. HortScience 47: 848–851.

Sharpe, S., N.S. Boyd and P.J. Dittmar. 2016. Clopyralid dose response for two black medic (*Medicago lupulina*) growth stages. Weed Technol. 30: 717–724.

Sharpe, S. 2017. Use of clopyralid to control black medic (*Medicago lupulina*) in Florida strawberry (*Fragaria x ananasssa*) production. Ph.D. Thesis Dissertation.University of Florida. pp. 131.

Shem-Tov, S. and S. Fennimore. 2003. Seasonal changes in annual bluegrass (*Poa annua* L.) germinability and emergence in California. Weed Sci. 51: 690–695.

Shem-Tov, S., S.A. Fennimore and W.T. Lanini. 2006. Weed management in lettuce (*Lactuca sativa*) with pre-plant irrigation. Weed Technol. 20: 1058–1065.

Shogren, R.L. and R.C. Hochmuth. 2004. Field evaluation of watermelon growth on paper-polymerized vegetable oil mulches. HortScience 39: 1588–1591.

Stall, W.M., R.C. Hochmuth, J.P. Gilreath and T.E. Crocker. 1995. Tolerance of strawberries to preplant herbicides. Proc Fla Hort Soc 108: 245–248.

Stapleton, J.J., C.L. Elmore and J.E. DeVay. 2000. Solarisation and biofumigation help disinfest soil. California Agri. 54(6): 42–45.

Stoller, E.W. and R.D. Sweet. 1987. Biology and life cycle of purple and yellow nutsedges (*Cyperus rotundus* and *C. esculentus*). Weed Technol. 1: 66–73.

Strand, L. 2008. Integrated Pest Management for Strawberries, 2nd Ed. University of California Agricultural and Natural Resources Publication 3351.

Teasdale, J.R. and A.A. Abdul-Baki. 1998. Comparison of mixtures vs. monocultures of cover crops for fresh-market tomato production with and without herbicide. HortScience. 33: 1163–1166.

Thomasa, S.H., J. Schroederb and L.W. Murray. 2005. The role of weeds in nematode management. Weed Sci. 53: 923–928.

UCANR (University of California Agriculture and Natural Resources). 1999. Crop Profile for Strawberries in California. Available at: http://ucanr.edu/datastoreFiles/391-501.pdf

USEPA (United States Environmental Protection Agency). 2016a. Methyl bromide – allowed uses. Available at: https://www.epa.gov/ods-phaseout/methyl-bromide

USEPA (United States Environmental Protection Agency). 2016b. Soil fumigant toolbox. Available at: https://www.epa.gov/soil-fumigants

Wang, Q., W. Klassen, Y. Li., M. Codallo and A.A. Abdul-Baki. 2005. Influence of cover crops and irrigation rates on tomato yields and quality in a subtropical region. HortScience. 40: 2125–2131.

Weber, C. 2004. Weed management in matted row strawberry. The New York Berry News. Available at: http://www.fruit.cornell.edu/berry/ipm/ipmpdfs/strweedmgmt.pdf

Wisler, G.C. and R.F. Norris. 2005. Interactions between weeds and cultivated plants as related to management of plant pathogens. Weed Sci. 53: 914–917.

Wyland, L.J., L.E. Jackson, W.E. Chaney, K. Klonsky, S.T. Koike and B. Kimple. 1996. Winter cover crops in a vegetable cropping system: impacts on nitrate leaching, soil water, crop yields, pests and management costs. Agric. Ecos. Envir. 59: 1–17.

Sustainable Weed Control in Vegetables

Russell W. Wallace*[1], Timothy W. Miller[2] and Joseph G. Masabni[3]

[1] 1102 East FM 1294, Texas A&M AgriLife Research & Extension Center, Lubbock, TX 79403, USA
[2] 16650 State Route 536, Washington State University, Mount Vernon, WA 98273, USA
[3] 1710 N. FM 3053, Texas A&M AgriLife Research & Extension Center, Overton, TX 75684, USA

Introduction

Vegetable crop systems across the globe are very diverse and often are dependent on grower location and climate, growing preference, finances and potential market. There is an increasing movement towards small-scale farming which has driven an increased demand for organic and/or locally-grown produce sprayed with fewer pesticides (D'Souza and Ikerd 1996). Since most vegetable crops are slow growing, there is a significant need for integrated and sustainable approaches to weed control from planting to harvest (Tei and Pannacci 2017).

Weed control is essential for all successful and profitable vegetable production systems regardless of whether crops are produced on a large commercial scale or on small-acreage farms. In many cropping systems herbicides are required to effectively reduce weed competition and increase grower profitability. Non-chemical methods of controlling weeds have shown benefits, but growers often perceive these, among other reasons as too complex with more manual labor to manage, less effective than chemicals, and little evidence of success (Moss 2010). Vegetable weed control can be expensive and labor-intensive whether sustainable or conventional practices are used (Gianessi and Reigner 2007, Gnanavel 2015).

Herbicide use is a common method of weed control in both developed and developing countries; however, in many regions of the world, cultivation and handweeding may be the only option for successful weed control. Due to the lack of many adequate herbicides, handweeding continues to be a common and necessary practice in most vegetable production fields.

Conventional systems include crops grown in monocultures sprayed with pre- and/or post-emergence herbicides plus cultivation and handweeding as needed. Alternative cropping systems employed to control, reduce or suppress weeds include the use of plasticulture, modified field cultivators and killed or living mulches in no-till or reduced-tillage systems (Chen et al. 2017). Plastic mulches are known to significantly decrease weed populations around planted crops as well as increase soil temperatures and reduce soil moisture evaporation. Killed and living mulches aid in weed suppression through physical and allelopathic means, while

---

*Corresponding author: rwwallace@ag.tamu.edu

no-till and reduced-tillage systems, often combined with lower-rate herbicides, utilize plant residues to suppress weeds (Brennan 2017, Price and Norsworthy 2017). There is a recent rise in the number of acres of vegetables grown intensively in high tunnel systems; however, these systems generally require little or no herbicides as handweeding, plastic mulches or straw mulches are used to control weeds.

Although herbicides are generally effective, they can be influenced by soil type and climate and control may fluctuate in any given field depending on annual climatic conditions (Vollmer et al. 2017). Most vegetable herbicides do not control all weed species found in production fields often leading to the need for additional integrated and sustainable control techniques. When left uncontrolled or even at low population densities, weeds will compete for sunlight, space, soil moisture and essential crop nutrients. Even partial limitation of one or more of these critical growth factors may lead to significant reductions in vegetable yield and quality. Weedy fields also create environments conducive to increased pests and diseases. Weeds growing within or in areas surrounding production fields may harbor economically-damaging pests that act as vectors by infesting crops and reducing growth. Weeds may also reduce the visual and marketable quality of harvested vegetables. Visual appearance is as critical to overall vegetable marketability as yield.

A steady loss of herbicide active ingredients and their uses over the past decades has created a void for chemical weed control options (Fontanelli et al. 2015, Gianessi and Reigner 2007, Moss 2010, Tei and Pannacci 2017) and there has been a lack of new chemistries developed over the past three decades (Heap 2014). Herbicide manufacturers typically consider vegetables as high risk–low revenue cropping systems and are unlikely to target these crops unless there is a significant financial advantage. Therefore, integrating sustainable weed control options and developing new approaches is essential to the success of vegetable crop systems worldwide (Gnavanel 2015, Liebman et al. 2016, Pannacci et al. 2017).

In this chapter, practices and technology that benefit vegetable weed control and improve economic, environmental and farm sustainability will be reviewed.

Impact on Weeds in Vegetables

According to 2016 USDA statistics, US production of commercial vegetables and dry pulses totaled 127 billion pounds in 2015 with 35 billion pounds sold through fresh markets and the remainder sold to processors. To satisfy the higher demand for an increasingly diverse population, the US imported $11.9 billion of produce while exporting $6.8 billion of vegetables (Wells 2016). Therefore, the increased vegetable consumption signifies the need for increased farm sustainability including weed control options, not only in the US but worldwide.

Vegetable production is very diverse and complex techniques are often used for growing crops. Growers producing multiple crop species are often required to use different methods of weed control on the same field or farm. The number of vegetables marketed in the US alone is between 30 and 40 crops. Depending on markets, many growers will produce from 1–20 or more different vegetables in a given season. These are usually grown under widely diverse methods and require specific equipment and supplies to achieve success. Within each of these crops there is often a diversity of cultivars with unique growing habits and differential sensitivities to climate, weed competition, herbicides and production practices.

The impact of weeds is particularly damaging to vegetables when left uncontrolled or even at low population densities (Gnavanel 2015). Buckelew et al. (2006) reported that eastern black nightshade was competitive to tomatoes through 50 days after transplanting and resulted in up to a 20% loss of extra-large and jumbo tomatoes. However, Swanton et al. (2010) reported that the critical weed-free period for carrots was longer (930 Growing Degree Days [GDD]) for an early planted crops compared to carrots planted in later (444 GDD). Weeds and their competitiveness are therefore influenced by planting time and seasonal environmental conditions that can influence weed seed germination, emergence and weed growth during any vegetable crop production cycle.

The growth characteristics of weeds may also influence vegetable yield and quality. Depending on growth structure, even low weed populations may shade sunlight resulting in reduced crop growth. Weeds competing for space on vegetable planting beds may cause misshapen plants or delayed flowering, fruit set and harvest time. More importantly, weeds will compete for soil moisture and essential nutrients through root competition (Ugen et al. 2002). The partial limitation of one or more of these critical plant growth factors often leads to reduced vegetable yield and quality.

Commercial producers are particularly aware that their vegetable products may be rejected at any point down the marketing chain for reasons including sanitation (weed seeds found in packaged or canned products or seeds staining harvested produce), misshapen or small vegetables, nicks and cuts from weeds during harvest, the actual penetration of weed plant parts into underground or above-ground crops, and/or off-size or off-color fruit. In all processed vegetables there is a zero-limit to weed infestations and a near 100% control must be achieved in the field prior to harvest (Wallace et al. 2007). This often results in higher labor costs through the efforts of handweeding crews.

When combined with sustainable practices, herbicides play an important role in vegetable production. Many vegetable crops have short life cycles and herbicide residues are an important factor when planning field rotations. Carryover and herbicide residues from use in previous crops or the misapplication of herbicides often leads to reduced stands, delayed growth, lower yields or even crop injury and death (Greenland 2003). However, due to the low number of registered products, vegetable growers are often willing to risk herbicide injury to achieve adequate weed control. This indicates that there is significant need for more integrated and sustainable weed control options in order to achieve optimum control (Fennimore and Doohan 2008).

The expanding use of genetically-modified agronomic crops has also increased potential off-site herbicide drift to sensitive vegetable crops, and this remains a significant concern to specialty crop growers and researchers (Hatterman-Valenti et al. 2017). Herbicide drift to vegetables may cause multiple plant symptoms including leaf and stem malformations, root pruning, flower abscission, misshapen fruit or even crop death if doses are high. And, although no symptoms may be evident, where off-site drift has occurred the crop may be rejected due to illegal pesticide contamination.

The impact of herbicide-resistant weeds in vegetable production fields is not clear and future research is needed for assessment. Vegetable crop production systems are not likely a significant source of herbicide-resistant weeds. Many older herbicides used in vegetable fields can control to some degree the current herbicide-resistant weeds. However, management for resistance is still important and the use of all cultural, mechanical and herbicide options will be needed to prevent the spread of resistant weeds (Hatterman-Valenti et al. 2017, Norsworthy et al. 2012).

Troublesome weeds infesting vegetable production fields are diverse in number and species, as well as in their growth habit and life cycles. Although some weed species may be small or low-growing or population densities are low, competition for nutrients and moisture will influence yields, especially during the early- to mid-season. In many vegetable production fields there is found a diversity of annual and perennial grasses and broadleaf weeds, as well as sedges. Weeds have particularly environmental conditions required for germination and growth. However, many species may be present throughout both the early, mid and late planting seasons.

Due to the lack of systemic herbicides for organic production, perennial weeds are often difficult to control in organic systems. Knockdown herbicides, such as acetic acid (vinegar) or plant extract oils (clove, citrus, etc.) may benefit temporarily, but regrowth is likely. Perennial weeds may also spread within and between fields through transportation of reproductive parts.

Common and troublesome weeds found in vegetable fields include but are not limited to the following: annual broadleaves: *Acalypha, Amaranthus, Ambrosia, Capsella, Chamomilla, Chenopodium, Desmodium, Eclipta, Kochia, Lactuca, Lamium, Oenothera, Polygonum, Portulaca,*

Richardia, Salsola, Senecio, Senna, Sida, Sisymbrium, Solanum, Stellaria and *Xanthium*; perennial broadleaves: *Convolulus, Ipomoea* and *Solanum*; annual grasses: *Cenchrus, Digitaria, Echinochloa, Lolium, Panicum, Poa* and *Setaria*,; and perennial grasses: *Cynodon, Elytrigia* and *Sorghum* (Wallace and Miller, personal communication, 2017). A recent survey of researchers and the industry by Van Wychen (2016) reported many other important weed species.

Weed Control Options for Vegetables

There are diverse options for controlling weeds in vegetable crops, both in the US and worldwide. The production method used depends primarily on the crop produced and grower preference. The following are current non-chemical options used for weed control in vegetables. Generally, vegetable producers use integrated approaches of two or more methods to improve weed control or suppression and reduce hand labor. However, integrated weed control systems are not commonly practiced because they often fail to meet expectations (Young et al. 2017).

Mechanical/Physical Weed Control

Handweeding

Depending on the crop and production method used as well as the avenue for post-harvest sales, handweeding of production fields is more often than not essential. Handweeding has been employed longer than any other method and includes hand-pulling and/or hoeing (Blaxter and Robertson 1995). Handweeding is still widely used in the majority of vegetable fields around the world (Gianessi and Reigner 2007, Ross and Lembi 1985, Mohler 2001), and more so in organic systems. Most farmers use handweeding to augment other weed management options, including conventional herbicide systems where incomplete control often occurs due to the lack of adequate herbicides or environment factors. However, many developing countries are facing shortages of hand labor as the population becomes more urban, and therefore, herbicide use may continue to increase (Gianessi 2013).

Mowing

Mowing of weeds in vegetable production has seen limited use. Mowing usually occurs with tractor-pulled shredders or lawn mowers that fit between crop rows. Mowing is generally limited to the shredding cover crops prior to planting. While killing weeds by mowing between rows is beneficial, the effect is usually temporary and live weeds may still compete for moisture and nutrients. Donald (2000) reported that between-row mowing three times in soybeans was successful when banding over-the-row with herbicides was included. However, shading by the crop canopy following row closure also contributed to weed suppression. A grass monocrop, legume or a grass-legume mixture that was flail-mowed gave inadequate weed control in no-till green pepper (Chellemi and Rosskopf 2004, Díaz-Pérez et al. 2008).

Soil Cultivation and Tillage

Over the decades, many equipment types have been invented and used for soil tillage in horticultural crop production (Bellinder et al. 2000, Fennimore and Doohan 2008, Fennimore and Goodhue 2016, Ross and Lembi 1985). In its broadest term, tillage is soil cultivation that occurs prior to crop emergence or transplanting, or between rows of emerged crops. Tillage is generally conducted using tractors or animals (though some may occur with hand held equipment) to pull implements through the soil. It may be necessary to split tillage into primary and secondary operations. Primary tillage is defined as the initial breaking of soil which also buries plant residues and unwanted living vegetation. It often results in a rough, cloddy surface. Primary tillage implements include plows, such as moldboard, chisel, disk, and sweeps. Deep tillage may be useful to bury certain weeds to reduce their competitiveness.

Secondary tillage is used to break large soil clods leaving a smooth seedbed suitable for crop seeding or transplanting. Broadcast field cultivators, tine, disk harrows and rod weeders are widely-used secondary tillage implements (Van Der Weide et al. 2008). Further cultivation may include the bed-shaping process prior to seeding to prepare an appropriate seedbed or the laying of plastic mulch. Cultivation is often used following crop emergence where weed control is the principal objective (Bellinder et al. 2000). These implements include various between-row harrows, rotary hoes, rototillers, and sweep and even flex-tine and rolling cultivators. Often these implements will aid in fields where over-the-row banded herbicides are used.

Timing of tillage can impact weed species, weed community structure and subsequent weed influence on crops (Cordeau et al. 2017). They also reported that early tillage resulted in higher weed density by the end of the season; however, adjusting the timing of tillage and cropping practices may reduce ultimate weed interference with the crop.

Conservation-, reduced- and no-tillage production systems have gained favor in recent years as a means to reduce erosion and slow fertilizer and water runoff (Peigne et al. 2007, Price et al. 2011). No-tillage systems also increase soil organic matter content and may improve crop yield (Phatak et al. 2002, Price and Norsworthy 2013). Growers using no-till and reduced-tillage systems must rely on alternative methods other than cultivation for weed control. Reduced herbicide rates are commonly used with these systems and may be used in combination with crop residue for suppression (Bellinder et al. 2000, Bhullar et al. 2015).

Previous research conducted by Bellinder et al. (2000) compared in-row banded herbicide applications with and without flex-tine, rolling and shovel cultivators prior to potato hilling. It was reported that cultivation without banded herbicides showed higher weed densities compared to broadcast herbicides or plots with cultivation plus banded herbicides. In pumpkins, in-row banding of herbicides provided good control with no-tillage and strip-tillage systems, effectively suppressing weeds compared to conventional tillage (Rapp et al. 2004).

Few herbicide options, however, are available to organic and other low-input production systems (Fernandez et al. 2012, Mulvaney et al. 2011, Price and Norsworthy 2013, Walters and Young 2012). In such systems, reduced-tillage must be combined with adequate cover crop residues and mulches, or other alternative methods including soil solarization, flaming and/or use of biological control agents (Singh et al. 2005). Such weed control options will however, increase production costs.

Flaming and Steam

Propane flamers have been used for selective weed control in vegetable production since the early 1940s (Anderson 1977). In those systems, an open flame is directed toward the soil surface near the base of crop plants that are large enough to withstand a brief exposure to heat but also when weeds are young and more susceptible to plant cellular rupture and cuticle loss. Previously, flaming systems have historically been effective in cotton and corn, but less so in sensitive crops like soybeans (Anderson 1977). However, flaming for weed control has been more extensively researched in vegetable crops in Europe (Rasmussen et al. 2011).

Non-selective flaming can be employed using a stale seedbed approach where crops are seeded several days after seedbed preparation and initial weed emergence, but prior to crop emergence. This is similar to using nonselective herbicides or shallow cultivation in stale seedbeds. Organic vegetable growers have utilized this strategy for many years (Stopes and Millington 1991, Wookey 1985).

Successful use of flaming has been reported for small seeded, slow-to-germinate vegetable crops including carrot and parsnip, but less successfully on quick germinating crops like spinach (Cramer et al. 1991). Flaming and stale seedbed techniques reduced density and biomass of common purslane (*Portulaca oleracea*) and common chickweed (*Stellaria media*) in simulated vegetable seedings (Caldwell and Mohler 2001). In those studies, a single delayed flame treatment was usually as effective as multiple treatments.

Flaming is also useful for transplanted vegetables. A single flaming treatment four days after seedbed preparation and one day prior to transplanting lettuce reduced weed density by

62% (Balsari et al. 1994). Weed control data using flaming techniques continues to be lacking for most crops. Selective post-emergence flaming in annual vegetable crops is nearly absent from the literature.

Using steam for non-selective weed control prior to planting has been and continues to be researched but is expensive and not a common practice currently. Steam has been applied to greenhouse soils in Italy for high value ornamentals and vegetables (Gullino et al. 2003). Kerpauskas et al. (2006) reported the possibilities for use in onion, barley and maize and found significant reductions in weed biomass for each crop as well as 9%–22% yield increases. In another study steam reduced weed density and handweeding times compared to non-treated plots (Fennimore et al. 2014). Yields were equivalent to commercial chemigation using chloropicrin with 1,3-dichloropropene (Pic-Clor 60). Similar to flaming, research using steam as a weed control treatment is limited but advances in steam generator technology may permit more economical use in high value crops (Fennimore and Goodhue 2016).

Cover Crops and Mulching

Following plastic mulches, cover crops are perhaps the most widely used alternative method for sustainable weed control in vegetables. Another benefit of cover crops in sustainable systems is the addition of organic matter to soils (Brennan 2017).While most cover crops are planted off-season, they may be seeded anytime during the crop season. Cover crops are generally terminated with cultivation, mowing, rolling/crimping or an application of non-selective herbicides prior to seeding or transplanting vegetables. It is generally agreed that winter cover crops must be killed to avoid reducing yield in the following vegetable crop at least four or more weeks prior to planting (Zandstra et al. 1998).

Cover cropping also allows farmers to grow mulch in the same field where the vegetable crop is produced, thus eliminating transportation and handling costs that would result if the mulch were produced off-site and spread from an alternate location (Merwin et al. 1995). Producing cover crop mulches within the production field also eliminates the risk of introducing new weed populations or diseases from distant farms into new areas where those mulches will be utilized (Yordanova and Shaban, 2007).

Selected cover crops are frost sensitive and are reliably killed in locations with suitable low temperatures. Cover crop residues are normally left on the soil surface to shade the soil surface thereby reducing weed seed germination and growth. Living cover crop residues that are incorporated into the soil prior to crop seeding or transplanting are known as green manures or plow-down crops (Brennan 2017). Cover crops evaluated as green manures include cereals (rye, oat, wheat, triticale, and barley), legumes (pea, vetch, black medic, sweetclover, and clover) and others (buckwheat, rapeseed, and mustard). Excellent weed control is often achieved when cover crop foliage is present and killed residues remain on the soil surface. However, this system can lead to a modification in weed species composition in the field (Barberi and Mazzoncini 2001, Shrestha et al. 2002) as well as suppression of weed growth (Akemo et al. 2000, Caamal-Maldonado 2001, Herrero et al. 2001, Reddy 2001, Sainju and Singh 2001) among other potential benefits.

Ryegrass (*Lolium* spp.) mulch has been effectively used for weed suppression in tomato and pepper production (Edwards et al. 1995), and cereal rye reduced weed populations in reduced tillage potatoes (Wallace and Bellinder 1989). Hay mulch provided excellent weed control in lettuce (Kristiansen et al. 2008), while straw mulch was similarly effective in Chinese cabbage (Runham and Town 1995) and in broccoli (Yordanova and Shaban 2007). Even when applied several weeks after pepper transplanting, mulches can provide effective weed control (Law et al. 2006). Conversely, such mulches may not be appropriate for all crops as breakage of onion scapes has resulted during mulch application (Boyhan et al. 2006) and lower soil temperatures can reduce crop growth and productivity (Pedreros et al. 2008).

Weed control in crops grown after cover crop termination is less apparent unless cover crop residues release allelochemicals (Brennan and Smith 2005, Burgos and Talbert 2000, Khanh et al. 2005, Norsworthy et al. 2005, Price et al. 2008, Weir et al. 2004). Altering the soil microclimate through a reduction in light, temperature or moisture can also occur with crop residues (Creamer et al. 1996b, Masiunas et al. 1995). Cover crop residues left on the soil surface account for most of the weed suppression in rotationally-grown crops acting as a physical barrier to light and seedling growth (Teasdale and Mohler 2000). Allelochemicals, if present, have less impact compared to shading and mechanical resistance to seedling emergence.

Unfortunately, surface residues can also suppress vegetable growth causing a reduction in yield. Spring-sown and summer-incorporated buckwheat (*Fagopyrum esculentum* Moench), brown mustard (*Brassica juncea* L.), yellow mustard (*Sinapis alba* L.), and oat (*Avena sativa* L.) reduced seed production, emergence and growth of hairy galinsoga by 38% to 62%, but also reduced emergence and growth of seeded lettuce, Swiss chard, pea, and snap bean (Kumar et al. 2009). Further, when cover crop residues are inadequate to provide full weed control in the rotational crop, they must be augmented with other measures to achieve acceptable levels of control, such as high residue cultivators (Vollmer et al. 2010) or risk significantly increased pressure in the following crops (Eyre et al. 2011).

Mustard-family cover crops and/or amendments with mustard seed meal are sometimes used in vegetable production for disease, nematode or weed suppression (Meyer et al. 2015). These products have active ingredients including one or more allelochemicals. Glucosinolates in plant residues break down in the presence of myrosinase to form several allelochemicals including isothiocyanate (Brown et al. 1991). While reports detailing the effectiveness of these allelochemicals on weeds is limited, an application of 3% (w/w) *Brassica juncea* seed meal prior to seeding reduced early-season biomass of redroot pigweed, common lambsquarter and common chickweed by 74% to 99%. However, control varied by season and midseason weed biomass was not affected (Rice et al. 2007). In lettuce, mustard cover cropping reduced common purslane and hairy nightshade density, but other weeds were not affected (Bensen et al. 2009). Weed emergence in onion was reduced up to 91% by sequential applications of *Sinapis alba* seed meal (Boydston et al. 2011).

Recent research found positive results when incorporating cover cropping in reduced, no-tillage or inter-cropping vegetable systems including green pepper (Chellemi and Rosskopf 2004, Díaz-Pérez et al. 2008, Campiglia et al. 2012) and desert-grown pepper (Hutchinson and McGiffen 2000), tomatoes (Abdul-Baki et al. 1996) and zucchini squash (Walters et al. 2005) and onions (Vollmer et al. 2010). Winter rye reduced weed biomass in no-till sweet corn by an average 30% compared to plots seeded to hairy vetch or without a previous cover crop (Zotarelli et al. 2009). Lettuce transplanted into a dense mulch of senesced subterranean clover provided good weed control and yielded as well as cultivated lettuce (Stirzaker et al. 1993). A similar approach was effective in no-till tomato (Abdul-Baki and Teasdale 1993). Campiglia et al. (2015) reported the mulch strips of hairy vetch, phacelia, white mustard and barley showed some positive signs. Hairy vetch as a killed mulch was best at suppressing weeds while phacelia and white mustard were not. However, mixtures of cover crop species in no-till tomato were as effective as herbicides (Creamer et al. 1996a, Herrero et al. 2001).

Plasticulture and Polyethylene Mulches

Plastic mulches are commonly used in vegetable production worldwide, providing in-row weed suppression, improving crop growth and offering protection from certain insects (Bangrarwa et al. 2009, Locascio et al. 2005, Warnick et al. 2006). Polyethylene and biodegradable plastic films warm early-season soils (Dodds et al. 2003) and maintain higher soil moisture levels (Ham et al. 1991, Lamont 1993, Cowan et al. 2014) often leading to earlier harvests (Bonanno and Lamont 1987, Ibarra et al. 2001, Lamont 1993), improved yield (Brown et al. 1995, Leib et al. 2002) and cleaner fruit (Brown and Channell-Butcher 2001). Black polyethylene plastic is the most common color used in vegetables, although white on black mulch is often used in warmer climates where cooler soils are needed (Gordon et al. 2010). Plastic mulches come in a variety of colors to benefit specific crops; however, translucent mulch types may result in weeds growing beneath the plastic surface causing competition for moisture and nutrients. Dark-colored (black, blue, and red) plastic mulches increased early and total yield okra (*Abelmoschus esculentus*), while red has increased yield of tomato and bell pepper (Decoteau et al. 1989, Decoteau et al. 1990). In a comparison of wheat straw mulch versus plastic film, Díaz-Pérez et al. (2012) reported that both treatments reduced weed populations in both conventional and no-till broccoli systems, particularly when employed with cover crop residues.

However, not all weed species are controlled by plastic mulches. Yellow and purple nutsedge growing beneath black plastic mulch can penetrate the surface and reduce plastic film longevity while competing with the crop for nutrients and moisture (Wallace, personal experience, 2017). Collin et al. (2008) reported that pre-applied herbicides reduced nutsedge punctures in tomatoes grown on polyethylene mulch but not to the extent or duration that would allow sequential crops to receive the full benefit of non-punctured plastic.

Cultural Weed Control

Sustainable farming practices are aimed at growing a high quality and consistent vegetable crop at a reasonable profit, whether herbicides are used or not. Sustainable farming has an additional goal of maintaining environmental quality, both in the field as well as with products sold to consumers (Brennan 2017). Use of selected combinations of cultural practices is one of major importance and is generally the first approach used in sustainable crop production (Owen et al. 2015). Cultural practices influencing seasonal weed emergence and growth include field and in-field site selection, crop rotation, seeding time, properly-timed tillage and cultivation, cover cropping, water and fertilizer management, selecting appropriate and competitive varieties, pest control and weed management, amongst others. Cultural practices also include managing crop residues, using clean certified seed, and cleaning machinery between fields and after use. Use of weed-free compost for soil amendments is also critical. Controlling weeds around irrigation ditches, farm roads, or stockyards will aid in keeping production fields weed-free and are critical to preventing spread.

Seeding Rates

Increasing vegetable seeding rates can be expensive but may also benefit by shading out weeds. However, little research has been conducted on vegetables. Brennan et al. (2009) reported that as the seeding rate of cover crops increased weed biomass dry-matter decreased, which may be useful in both conventional and organic vegetable production systems. In Baby leaf spinach production in Texas, seeding rates have increased over the past decade. Baby leaf spinach can be planted at 3.5 million seeds per acre while whole leaf fresh-cut spinach is usually planted at 1.5 million seeds per acre. Along with seeding rates, changes in planting methods from two lines on a 40″ bed to up to 14 lines on 80″ beds appear to help suppress weeds (Ritchie 2016). However, handweeding high density spinach is difficult and has increased the need for selective post-emergence herbicides. Faster growing cultivars with higher seeding rates may also shade out selected weed seedlings (Ritchie 2016); however, research in this area is lacking.

Williams and Boydston (2013) reported that as sweet corn seeding levels increased the crop canopy became taller and thicker resulting in less wild-proso millet weed biomass, seed production, and germinability. However, effects on wild-proso millet growth and seed production were modest, at best, between corn populations used by growers and the higher population known to optimise yield of certain hybrids.

Competitive Crops, Intercropping and Living Mulches

As described previously, the effectiveness of living mulches and ground covers in vegetable production as a means of suppressing weeds is well known (Campiglia et al. 2012). Intercropping is defined as growing two or more crops simultaneously within a field during a single growing season (Liebman and Staver 2001). Cereal grains including rye, wheat and barley as well as broadleaf crops like vetches, peas and buckwheat have been extensively researched. Rye interseeded at planting in broccoli suppressed weeds and improved yields, but only compared to non-weeded controls (Brainard and Bellinder 2004). This result suggested that winter rye may be best integrated into broccoli production when sown at higher densities and in locations or seasons with lower initial temperatures and when combined with additional weed management tools.

In tomatoes, Gibson et al. (2011) found that living buckwheat mulch seeded at increasing rates after the critical period of competition (weed to crop) can be used to reduce the weed seed bank without reducing yields. Densely-growing living mulches can cover between crop soils reducing weed seed germination and subsequent weed growth. Charles et al. (2006) reported that oilseed radish consistently produced the greatest biomass and provided over 98% early-season weed biomass suppression in celery and that control was greater in the early season compared to late season. While living mulches may also compete with the production crop and possibly reduce its growth, this can be avoided when managed properly (Hartwig and Ammon 2002). To limit competition with a field crop, living mulch competition should be reduced through mowing, flaming, cultivation, or herbicides (Mohammadi 2012).

Even with good management practices living mulches may not be appropriate in all vegetable crop systems. For example, a sorghum Sudan grass cover crop persisted into late autumn and interfered with fall cabbage production through direct competition and possibly allelopathy (Finney et al. 2009). Similarly, winter rye living mulch reduced redroot pigweed biomass as much as an herbicide treatment but squash yield was reduced by allelopathy 20% to 50% during two years of study (Walters and Young 2008). Inter-cropping green pepper with cucurbits gave mixed results as pepper yield was not reduced with a melon crop but was with pumpkin. However, weed control was better with pumpkin (Akintoye and Adebayo 2013). Chase and Mbuya (2008) reported that of 12 winter cover crops, black oat, 'Wrens Abruzzi' rye and ryegrass competed with broccoli and lowered yields to that of the weedy control.

Crop Rotation

It is well known that crop rotation is critical to the reduction of weeds, as well as diseases and insects in vegetable fields and is an essential tool in all conventional, organic or sustainable systems (Price and Norsworthy 2013, Musser et al. 1985). Weed scientists and extension personnel recommend that growers rotate fields out of selected vegetable families for a minimum of three to four years. However, care in field selection is critical as crop injury may also occur to sensitive vegetables from herbicides sprayed in previously cropped fields. Soil type and other environmental factors can also influence the effect of residual herbicides (Greenland 2003). However, in many agronomic systems, growers often use herbicides to the exclusion of other integrated management strategies, including rotation (Owen et al. 2015).

Using herbicide-resistant crops the year prior to vegetable production to control perennial weeds can be a useful technique, but is only available in systems where chemicals are used. Tingle and Chandler (2004) reported that in three glyphosate-resistant corn-cotton rotations, weed control after three years was greater compared to a conventional herbicide program. However, good resistant weed management programs are also necessary (Heap 2014). Rotating herbicides with selected vegetable crops within the same field may also aid in controlling some species while reducing the potential for herbicide resistance in weeds.

Sowing Date

Tillage and vegetable planting dates are tools that can influence weed germination, emergence and subsequent growth and development. In reduced-tillage organic sweet corn, giant foxtail was the dominant species in earlier plantings while smooth pigweed was dominant at later times (Teasdale and Mirsky 2015). Williams (2006) reported that sweet corn planting date influenced the critical period for weed control and that weed biomass was significantly greater for early versus late planting dates. In spinach, weed diversity changes during early-season (fall) production compared to late-season (spring) in Texas (Ritchie 2016).

Fertilizer and Water Management

The rate of fertilizer (primarily nitrogen) and its application timing are known to influence weed growth. Cathcart et al. (2004) reported that green foxtail and redroot pigweed grown in a greenhouse under low nitrogen levels required much higher rates of selected herbicides to achieve a 50% reduction in weed growth and suggested that differences in field herbicide efficacy may be the result of varying level of soil nitrogen, which potentially could alter the weed community structure. This is especially true with the increasing organic and sustainable acreage where fresh manure and compost are integrated into the production system. Charles et al. (2006) reported that when integrated into celery production with low or half rate fertilizers, an oilseed radish cover crop gave good weed suppression while increasing yield. Finally, Sweeney et al. (2008) reported that spring nitrogen applications increased weed biomass but weed growth appeared to be dependent on the weed species, seed source, and environmental conditions. While producers making drastic reductions in fertilizer inputs may experience reductions in crop yield, results indicate that herbicide inputs could be reduced or eliminated periodically with no short-term yield loss in some crop systems. However, more information on weed responses to soil fertility and weed control is needed (Blackshaw and Brandt 2004).

Water management is critical to vegetable production but is also influential on weed emergence and growth. Properly managing water without economically damaging vegetable crops is critical to sustainable weed management. Preplant irrigation followed by a shallow cultivation prior to planting (stale seedbed technique) has been used by conventional and organic growers for decades to suppress early-season weeds. However, during production, low or high rates of water can influence weed growth and, therefore, subsequent decisions in weed management. Growers should have the necessary weed management tools available to control weeds under a variety of conditions. Weed species and density may change depending on soil conditions resulting in potential changes in weed management strategies. Soil conservation practices are critical to weed management and when not implemented, weed seeds and reproductive structures can be transported in excess runoff to other fields resulting in new infestations.

In vegetables, using plastic mulches coupled with drip irrigation conserves overall soil moisture for the crop while physically blocking in-row weeds and decreasing moisture-related weed seed germination between rows. Shem-Tov et al. (2006) found that the effective use of preplant irrigation and weed removal coupled with lower herbicide rates during lettuce production was sufficient to achieve good in-season weed control and crop yields. In nectarines, DaSilva et al. (2003) reported that the improved management and efficiency of irrigation may help to mitigate leaching of pesticides (herbicides) out of the treated zone, and into groundwater. Efficient irrigation improved the performance of simazine and that irrigation management should be considered when developing any weed control program.

Preventative Measures

Preventing seed distribution and subsequent emergence are common aspects of weed control in vegetable production fields. General preventative weed control methods include but are not limited to: i) obeying local, state, federal and international laws regulating seed distribution and transport; ii) using certified seed; using weed-free manures and hay in the production system; iii) proper sanitation of field and harvesting equipment (especially between fields); iv) eliminating weeds from nearby irrigation ditches or non-cultivated areas around production fields; and v) practicing zero-tolerance in production fields. Also, avoidance, weed monitoring and suppression are important components of weed prevention.

Crop Residue Management

Cover crops are widely used for weed prevention as well as soil and water conservation, increasing soil organic matter, improving soil fertility, improving field biodiversity (Price and

Norsworthy 2013). Winter cereals, legumes, and mustards are popular cover crops. Killing cover crops and managing their residues are critical to obtaining higher yielding vegetables. Prior to planting, cover crops may be rolled into a thick blanket of residue or chopped and left to dry or killed with an herbicide. Whether rolled, chopped, shredded or chemically-killed, cover crops form a preventative weed barrier between soil surfaces by physically blocking or shading weeds. Cooler soil temperatures may also reduce certain weed seed germination (Wallace and Bellinder 1989). Cover crops commonly used for winter or summer production include grasses: barley (*Hordeum vulgare*), cereal rye (*Secale cereale*), forage sorghum (*Sorghum bicolor*), millet (*Setaria italic*) and oats (*Avena strigosa*); and legumes clover (*Trifolium* spp.), cowpea (*Vigna unguiculata*), hairy vetch (*Vicia villosa*), Lablab (*Lablab purpureus*) and medics (*Medicago truncatula*). When managing cover crops efforts should be made to grow sufficient residue to cover the entire non-planted areas in order to be effective. Research shows that cover crop residues coupled with applications coupled with low-rate herbicide applications over the planted row is an effective alternative weed control option in vegetables (Wallace and Bellinder 1989, Brainard and Bellinder 2004).

Clean Seed

Purchasing and planting clean, certified seed is critical to preventing the dispersal of weed and introducing new species into vegetable production fields. Selecting a reliable seed source is important. Today, credible companies sell certified seeds guaranteeing a high percentage of genetic purity, a minimum level of germination quality and a minimum level of weed seeds. Certified seed may be more expensive, but it is well worth the cost to prevent infestations. While saving and sharing seed from one year to the next is common among small-acreage growers, extreme care should be taken to accurately remove all weed seed and debris.

Clean Machinery

Preventing field-to-field and within-field spread of weed seeds or their reproductive parts are critical to sustainable weed management (Norsworthy et al. 2012). Using uncleaned equipment (i.e., hand tools, plows, cultivators and harvesters) between fields can be problematic. Weed seeds clinging to soil debris or plant parts on field equipment, hand tools, work shoes and clothing can be transported between fields if not effectively cleaned. Growers managing weeds should adopt sustainable practices that include washing farm tools and equipment after each use, especially before entering other fields or farms. The simple act of washing equipment can reduce weed seed dispersal to new locations.

Clean Manure for Soil Amendments

Fresh manure and manure composts are common components of sustainable agriculture, especially in organic systems (Ozores-Hampton 2017). In cold climates, fresh manure is often spread in the fall and left to breakdown over the winter, then followed by soil incorporation during land preparation. In warmer climates where continuous cropping may occur, fresh manure is generally incorporated 90 days prior to planting to prevent crop injury. Composted manure has no time requirements between incorporation and crop harvest.

When using fresh or composted manure, care should be taken to prevent these materials being sources of weed seed dispersal (Ozores-Hampton 2017). While it is not possible to prevent air-transported weed seeds from infesting manure or compost piles, properly composted materials can kill certain weed species. Ruminants and other domesticated and non-domesticated animals feeding on weeds in pastures may infest usable manures with seed heads. In feedlots where manure is collected and spread to production fields, these may be a source of new infestations. Rupende et al. (1998) reported that 17 broadleaf and six grass weed species were found in manure heaped and left for one to five months but that heaping manure for at least three months significantly reduced weed seed viability. Cook et al. (2007) however reported that applications of dairy manure in corn did not increase weed populations nor require alternative weed management strategies.

Caution should be taken when using manure from grass-fed cattle where grassland herbicides like clopyralid, triclopyr and others in the pyridine carboxylic acid family are used. These herbicides can remain active in hay clippings and manure used for compost for extended periods (Davis et al. 2015). Symptoms of herbicide injury from composted manure include poor seed germination and seedling death, twisted, cupped or elongated leaves, misshapen fruit and reduced yields. It is important to know the source of compost prior to use.

Maintenance Around Irrigation Ditches and Roadsides

Controlling invasive weeds from outside sources is an essential management strategy in sustainable weed control. If left uncontrolled, irrigation ditches and roadside weed populations can be diverse communities of annual and perennial weeds that become sources of infestations into production fields. In one study, Clark et al. (2002) identified 95 weed species along Pennsylvania roadsides and that five species were prevalent in 50% of the locations surveyed. Mowing and herbicides are typically used to control weeds along roadsides, though some state laws and city ordinances restrict or prohibit their use. Irrigation ditches should be kept clean to prevent weed seed dispersal from running water or air movement into field production.

Chemical Weed Control

Active Ingredients

Herbicides and their selected active ingredients are important integrated tools for weed management in both organic and non-organic vegetable production systems. When used correctly, herbicides are a cost-effective means of safely controlling weeds in food crops (Gianessi 2013). Weed control in vegetables with herbicides can be very complex as there are over 40 vegetable species or more (and thousands of cultivars) grown on production farms worldwide. Across the US and globe, these production farms have a regional diversity in soil types and climates, and vegetables are often multi-cropped within the same field. There is no 'one size fits all' herbicide or active ingredient in vegetable production, and herbicide use at best can be risky with potential for crop injury. However, alternatives can be more costly and potentially injurious to the crops (Gianessi 2013).

In the US alone, there are over 50 chemically- and naturally-derived herbicides (active ingredients) registered for use on vegetables. Vegetable herbicides provide options for selectively controlling weeds including pre-plant, pre-emergence or post-emergence applications. However, there are many vegetable crops including herbs and other leafy greens where few, if any, herbicides are registered. Growers of these crops must, therefore, seek alternative, non-herbicidal solutions.

Since its inception over 50 years ago, the NIFA- and USDA-funded IR-4 Project has supported the regulatory approval of herbicides (and other pesticides) and their minor uses in specialty food crops (Kunkel et al. 2008). Through their efforts and supportive data, US-grown specialty crops remain some of the safest in the world. The United States Environmental Protection Agency (EPA) only allows the use of chemical products that have undergone rigorous food safety, environmental and human health risk assessments with strict adherence to modern safety standards (Baron et al. 2016). However, registering herbicide active ingredients and their potential uses requires multiple collaborative factors and input from manufacturers and industry, growers as well as approval from both the EPA and state agencies. Developing and registering a new active ingredient increased from $50 million in 1975 to over $200 million in 1995 (Ruegg et al. 2007). Due to their relative small acreage, such development costs are prohibitive for herbicides solely registered in vegetable crops.

Increased registrations and uses of glyphosate, 2,4-D and dicamba resistance for large-acreage agronomic crops has resulted in a significant decrease in herbicide development and registrations for vegetable crops. Although there are currently over 35 commonly-used herbicide active ingredients registered for use in US vegetable crops (see Table 20.1), the majority are products developed over 20–25 years earlier (Fontanelli et al. 2015, Gianessi and Reigner 2007, Moss 2010, Tei and Pannacci 2017). University researchers and the vegetable industry work in concert with the IR-4 Project to tirelessly register potential new uses on older chemistries to increase the herbicide management tools for growers (Kunkel et al. 2008).

Herbicide Resistance

Expanding development and the registered use of herbicide-resistant (glyphosate, 2,4-D and dicamba) agronomic crops across the US and globally has justifiably or unjustifiably increased small-acreage sustainable grower's concerns for the future of weed control (Owen et al. 2015). Negative publicity and a misinformed media have aided in this concern. According to Shaw (2016), however, "weed scientists for decades have conducted research and developed educational programs to prevent or mitigate evolution of herbicide resistance, yet resistance is more prevalent today than ever before". Thus, with every significant and positive development in herbicide-resistance strategies, there must be an accompanying need for risk assessment and impact on non-resistant crops, most importantly, food crops. If and when the need for resistance management in food crops increase due to resistant weed infestations, there will likely be an increase in food production costs.

While herbicide-resistant vegetables will not likely be developed nor accepted by the general public in the near future, it is critical for growers to employ resistance management tools to prevent potential infestations of resistant weeds in their respective fields. Using multiple herbicides, with differing modes of action, and employing alternative weed control options including both crop and herbicide rotations, will reduce the likelihood of developing herbicide resistant weeds in vegetable production fields. However, overusing and applying the same herbicides or herbicide families within crop fields, even when rotating crops may not necessarily lead to herbicide resistance, but will perhaps, lead to a buildup of non-controlled weed populations. This will also result in increased costs of handweeding or hoeing in those fields.

Past research has indicated that using lower herbicide rates in vegetable production controlled weeds effectively and helped reduce weed control costs with or without alternative strategies (Wallace and Bellinder 1990, Zhang et al. 2000). Multiple applications of low-rate herbicides have been evaluated for several decades in an effort to reduce crop injury, especially with post-emergence herbicides. Loken and Hatterman-Valenti (2010) reported that three microrate post-emergence herbicide applications provided greater and season-long weed control and higher onion yields compared with two applications across four herbicides and rates. Alternatively, Manalil et al. (2011) stated that using lower rates to control weeds was an "example of poor use of agrochemicals that can have potential adverse implications due to rapid herbicide resistance evolution". They concluded that diclofop resistance increased in herbicide-susceptible rigid ryegrass following lower rate applications and that herbicides should be applied according to recommended rates to ensure high weed mortality. Overall, small acreage food crops likely have little to no impact on weed selectivity to herbicide resistance.

Integrated Weed Control

Integrated weed management can be defined as an approach to weed control that combines different control measures to provide the crop with an advantage over in-field weeds (Harker and O'Donovan 2013). One of the main goals of integrating strategies is to suppress weeds while producing high quality crops and while reducing costs and impacts on the environment (Norsworthy et al. 2012, Pannacci et al. 2017). Sustainable farming has the additional goal of supplying consumer's demands for fewer pesticides in the food supply. Integrating two or more weed control practices should be one of the first approaches for controlling weeds in sustainable vegetable production.

Integrating on-farm weed management should include multiple, if not all of the following 12 strategies: i) regular scouting and record-keeping for each production field, ii) annual reviews of field histories, iii) zero weed tolerance for weeds in and around production fields, iv) herbicide-resistance management, v) rotating crop families and herbicides (including modes of action), vi) altering seeding time to avoid specific difficult-to-control weeds, vii) timing tillage and cultivation operations to prevent weed competition without injuring crop roots, viii) using cover crops and crop residues, ix) appropriate irrigation and fertilizer management, x) variety selection and using weed-free, certified seed, xi) cleaning and sanitizing farm equipment, hand tools and clothing to prevent spread, and xii) using clean compost and manure. While not necessarily guaranteeing excellent annual weed control, utilizing these important control strategies will aid in seasonal weed suppression and reduce spreading economically-damaging weeds to production fields.

As labor and handweeding costs continue to increase, alternative methods of selective chemical weed control are under development (Gianessi 2013, Fennimore et al. 2016). Using robotic technology which may improve sustainability, while reducing herbicide impacts on the environment is one such technology (Bawden et al. 2017, Young 2012). While potentially useful, robotic technology is limited by the detection and identification of weeds in the crop fields, especially when used under a wide range of environmental conditions (Slaughter et al. 2007) as well as weed density and row crop patterns (Fennimore et al. 2016). More research is needed to show the benefits of automated/robotic technologies before they can become standard in large-scale agriculture and in developing countries where such technology is cost-prohibited (Bawden et al. 2017, Onwude et al 2016, Rasmussen et al. 2012, Underwood et al. 2017).

Evaluation of Weed Control Sustainability

Perhaps no other sustainable management strategies are more important than those utilised in food crop production. When left uncontrolled, weeds significantly reduce vegetable yields and grower profitability, risking the sustainability of production. Positive weed control efforts in vegetable crops have improved production significantly since prior to the 1960s. Driven by increased consumer demands and the need for lower production costs, there has been increased opportunities for university research and extension funding to assist in evaluating and improving alternative and sustainable options for growers, and to provide higher quality and a safer food supply. Decades of sustainable research have provided vegetable growers with reasonable opportunities to achieve these demands. With a continued increase in educational opportunities, greater numbers of growers will likely be adopting sustainable practices.

Utilizing sustainable management strategies in theory reduces many of the competitive hazards of weeds to vegetable production and potential hazards of chemicals to the environment and consumers. Their integrated use and successful implementation should reduce negative effects associated with weeds. Crop and herbicide rotations are considered to be major components to integrated weed control. Alternatively, Garrison et al. (2014) reported that reducing crop rotation and increasing the number of consecutive plantings for a single crop actually decreased weed seed bank size by forcing weeds to compete with each other in similar environments for extended time periods. Their results suggested that while the same crop was planted in the same field for consecutive years, there were still traditional benefits of crop rotation with regards to weed control. Thus, sustainable weed control may be dependent on each individual crop system, as well as grower production preferences, and the appropriate management of all available sustainable tools.

Concluding Remarks

Vegetable crop systems across the globe are very diverse and dependent on grower location, regional climate, grower production preference, grower finances and the potential marketing and sales. Sustainable practices in vegetable crops production must include economic, environmental and social aspects to protect the environment and provides profitability for the farmer. Specific to weed control, integrated weed management strategies must be a part of this sustainable approach. There has been and will continue to be an increasing movement towards sustainable farming, in particular with small-scale farms. These have recently driven an increased demand for organic and/or locally-grown crops sprayed with fewer pesticides and grown non-conventionally. While much research has been conducted, there continues to be a significant need for alternative and improved sustainable approach to weed control in food production.

Weed control is an essential tool for successful and profitable vegetable production systems, whether produced on a large commercial or on small-acreage farms. The diverse cropping systems utilized in vegetable production create difficulty in recommending a 'one size fits all' approach to sustainable weed control. Whether producing vegetables organically or conventionally, implementing sustainable weed management strategies, including herbicides will reduce weed populations and crop competition. Incorporating sustainable strategies, whether for weed control or other pests, can be the cheapest option for weed control, but may not provide higher grower profits. However, implementing sustainable practices will provide the best opportunities to reduce the impact of weeds on vegetable crops, reduce the spread of noxious weeds while reducing the long-term effects of weeds on field ecological systems. Add to that the benefits of improving local and global environments as well as the safety of our food supply, sustainable practices, regardless of farm size and income, should be implemented by every vegetable producer worldwide.

REFERENCES

Abdul-Baki, A.A. and J.R. Teasdale. 1993. A no-tillage tomato production system using hairy vetch and subterranean clover mulches. HortSci. 28: 106–108.

Abdul-Baki, A.A., J.R. Teasdale, R. Korcak, D.J. Chitwood and R.N. Huettel. 1996. Fresh-market tomato production in a low-input alternative system using cover-crop mulch. HortSci. 31: 65–69.

Akemo, M.C., E.E. Regnier and M.A. Bennett. 2000. Weed suppression in spring-sown rye (*Secale cereale*)-pea (*Pisum sativum*) cover crop mix. Weed Technol. 14: 545–549.

Akintoye, H.A. and G.A. Adebayo. 2013. Growth and yield response of sweet pepper as influenced by live mulches. Acta Hortic. 1007: 543–548.

Anderson, W.P. 1977. Methods of weed control, burning. pp. 67–70. *In:* Weed Science Principles. West Publishing Company, St. Paul, MN.

Balsari, P., R. Berruto and A. Ferrero. 1994. Flame weed control in lettuce crop. Acta Hort. 372: 213–222.

Bangrarwa, S.K., J.K. Norsworthy and E.E. Gbur. 2009. Cover crop and herbicide combinations for weed control in polyethylene-mulched bell pepper. Hort Technol. 19: 405–410.

Barberi, P. and M. Mazzoncini. 2001. Changes in weed community composition as influenced by cover crop and management system in continuous corn. Weed Sci. 49: 491–499.

Baron, J.L., R. Holm, D. Kunkel, P.H. Schwartz and G. Markle. 2016. The IR-Project over 50 years of sustained success. pp. 10–25. *In:* Outlooks on Pest Management – February 2016. Research Information Ltd. http://ir4.rutgers.edu/Other/S4.pdf.

Bawden, O., J. Kulk, R. Russell, C. McCool, A. English, F. Dayoub, C. Lehnert and T. Perez. 2017. Robot for weed species plant-specific management. J. Field Robotics. 34: 1179–1199.

Bellinder, R.R., J.J. Kirkwyland, R.W. Wallace and J.B. Colquhoun. 2000. Weed control and potato (*Solanum tuberosum*) yield with banded herbicides and cultivation. Weed Technol. 14(1): 30–35.

Bensen, T.A., S.A. Fennimore, S. Shem-Tov, S.T. Koike, R.E. Smith and K.V. Subbarao. 2009. Mustard and other cover crop effects vary on lettuce drop caused by *Sclerotinia minor* and on weeds. Plant Dis. 93: 1019–1027.

Bhullar, M.S., S. Kaur, T. Kaur and A.J. Jhala. 2015. Integrated weed management in potato using straw mulch and atrazine. HortTechnology 25(3): 335–339.

Blackshaw, R.E., R.N. Brandt, H.H. Janzen and T. Entz. 2004. Weed species response to phosphorus fertilization. Weed Sci. 52(3): 406–412.

Blaxter, K. and N.R. Robertson. 1995. The control of weeds, pests and plant diseases. pp. 89–119. *In*: From Dearth to Plenty: The Modern Revolution in Food Production. Cambridge University Press, Cambridge, UK.

Bonanno, A.R. and W.J. Lamont. 1987. Effect of polyethylene mulches, irrigation method, and row cover on soil and air temperature and yield of muskmelon. J. Amer. Soc. Hort. Sci. 112: 735–738.

Boydston, R.A., M.J. Morra, V. Borek, L. Clayton and S.F. Vaughn. 2011. Onion and weed response to mustard (*Sinapis alba*) seed meal. Weed Sci. 59: 546–552.

Boyhan, G.E., R. Hicks and C.R. Hill. 2006. Natural mulches are not very effective for weed control in onions. HortTechnol. 16: 523–526.

Brainard, D.C. and R.R. Bellinder. 2004. Weed suppression in a broccoli–winter rye intercropping system. Weed Sci. 52(2): 281–290.

Brennan, E.B. 2017. Can we grow organic or conventional vegetables sustainably without cover crops? HortTechnology 27(2): 151–161.

Brennan, E.B., N.S. Boyd, R.F. Smith and P. Foster. 2009. Seeding rate and planting arrangement effects on growth and weed suppression of a legume-oat cover crop for organic vegetable systems. Agron. J. 101(4): 979–988.

Brennan, E.B. and R.F. Smith. 2005. Winter cover crop growth and weed suppression on the Central Coast of California. Weed Technol. 19: 1017–1024.

Brown, J.E. and C. Channell-Butcher. 2001. Black plastic mulch and drip irrigation affect growth and performance of bell pepper. J. Veg. Crop Prod. 7: 109–112.

Brown, J.E., S.P. Kovach, W.D. Goff, D.G. Himelrick, K.M. Tilt, W.S. Gazaway, L.M. Curtis and T.W. Tyson. 1995. Fumigation and mulch affect yield, weight, and quality of 'Pimiento' pepper (*Capsicum annuum* L.). J. Veg. Crop Prod. 1: 71–80.

Brown, P.C., M.J. Morra, J.P. McCaffrey, D.L. Auld and L. Williams III. 1991. Allelochemicals produced during glucosinolate degradation in soil. J. Chem. Ecol. 17: 2021–2034.

Buckelew, J.K., D.W. Monks, K.M. Jennings, G.D. Hoyt and R.F. Walls Jr. 2006. Eastern black nightshade (*Solanum ptycanthum*) reproduction and interference in transplanted plasticulture tomato. Weed Sci. 54(3): 490–495.

Burgos, N.R. and R.E. Talbert. 2000. Differential activity of allelochemicals from *Secale cereale* in seedling bioassays. Weed Sci. 48: 302–310.

Caamal-Maldonado, A., J.J.J. Nez-Osornio, A. Torres-Barraga and A.L. Anaya. 2001. The use of allelopathic legume cover and mulch species for weed control in cropping systems. Agron. J. 93: 27–36.

Caldwell, B. and C.L. Mohler. 2001. Stale seedbed practices for vegetable production. HortSci. 36: 703–705.

Campiglia, E., R. Mancinelli and E. Radicetti. 2012. Weed control strategies and yield response in a pepper crop (*Capsicum annuum* L.) mulched with hairy vetch (*Vicia villosa* Roth.) and oat (*Avena sativa* L.) residues. Crop Prot. 33: 65–73.

Campiglia, E., R. Mancinelli, E. Radicetti and F. Caporali. 2010. Hairy vetch (*Vicia villosa* Roth.) cover crop residue management for improving weed control and yield in no-tillage tomato (*Lycopersicon esculentum* Mill.) production. Eur. J. Agron. 33: 94–102.

Campiglia E., E. Radicetti and R. Mancinelli. 2015. Cover crops and mulches influence weed management and weed flora composition in strip-tilled tomato (*Solanum lycopersicum*). Weed Res. 55: 416–425.

Cathcart, R.J., K. Chandler and C.J. Swanton. 2004. Fertilizer nitrogen rate and the response of weeds to herbicides. Weed Sci. 52(2): 291–296.

Charles, K.S., M. Ngouajio, D.D. Warncke, K.L. Poff and M.K. Hausbeck. 2006. Integration of cover crops and fertilizer rates for weed management in celery. Weed Sci. 54(2): 326–334.

Chase, C.A. and O.S. Mbuya. 2008. Greater interference from living mulches than weeds in organic broccoli production. Weed Technol. 22: 280–285.

Chellemi, D.O. and E.R. Rosskopf. 2004. Yield potential and soil quality under alternative crop production practices for fresh market pepper. Renewable Agric. and Food Sys. 19(3): 168–175.

Chen, G., L. Kolb, A. Leslie and C. Hooks. 2017. Using reduced tillage and cover crop residue to manage weeds in organic vegetable production. Weed Technol. 31(4): 557–573.

Clark, B.J., W.S. Curran and M.W. Myers. 2002. Prevalence of weeds along Pennsylvania roadways: results from a ten-county survey. Proc. Northeast Weed Sci. Soc. 56: 99.

Collin, W. Adcock, G. Foshee III, G.R. Wehtje and C.H. Gilliam. 2008. Herbicide combinations in tomato to prevent nutsedge (*Cyperus esulentus*) punctures in plastic mulch for multi-cropping systems. Weed Technol. 22(1): 136–141.

Cook, A.R., J.L. Posner and J.O. Baldock. 2007. Effects of dairy manure and weed management on weed communities in corn on Wisconsin cash-grain farms. Weed Technol. 21(2): 389–395.

Cordeau, S., R.G. Smith, E.R. Gallandt and B. Brown. 2017. Timing of tillage as a driver of weed communities. Weed Sci. 65(4): 504–514.

Cowan, J.S., C.A. Miles, P.K. Andrews and D.A. Inglis. 2014. Biodegradable mulch performed comparably to polyethylene in high tunnel tomato (*Solanum lycopersicum* L.) production. J. Sci. Food Agric. 94: 1854–1864.

Cramer, C., G. Bowman, M. Brusko, K. Cicero, B. Hofstetter and C. Shirley. 1991. Direct seed and flame weed. pp. 74–76. *In:* Controlling Weeds with Fewer Chemicals. Rodale Institute, Emmaus, PA. USA.

Creamer, N.G., M.A. Bennett, B.R. Stinner and J. Cardina. 1996a. A comparison of four processing tomato production systems differing in cover crop and chemical inputs. J. Amer. Soc. Hort. Sci. 121: 559–568.

Creamer, N.G., M.A. Bennett, B.R. Stinner, J. Cardina and E.E. Regnier. 1996b. Mechanisms of weed suppression in cover crop-based production systems. HortSci. 31: 410–413.

Dasilva, A., C. Garretson, J. Troiano, G. Ritenour and C. Krauter. 2003. Relating simazine performance to irrigation management. Weed Technol. 17(2): 330–337.

Davis, J., S. Johnson and K. Jennings. 2015. Herbicide carryover in hay, manure, compost, and grass clippings. North Carolina Coop. Ext., NC State Univ. Pub. AG-72/W. 6 pages.

Decoteau, D.R., M.J. Kasperbauer and P.G. Hunt. 1989. Mulch surface color affects yield of fresh-market tomatoes. J. Amer. Soc. Hort. Sci. 114: 216–219.

Decoteau, D.R., M.J. Kasperbauer and P.G. Hunt. 1990. Bell pepper plant development over mulches of diverse colors. HortSci. 25: 460–462.

Díaz-Pérez, J.C., S.C. Phatak, J. Ruberson and R. Morse. 2012. Mulches increase yield and improve weed control in no-till organic broccoli (*Brassica oleracea* var. *botrytis*). Acta Hortic. 933: 337–342.

Díaz-Pérez, J.C., J. Silvoy, S.C. Phatak, J. Ruberson and R. Morse. 2008. Effect of winter cover crops and no-till on the yield of organically-grown bell pepper (*Capsicum annuum* L.). Acta Hortic. 767: 243–247.

Dodds, G.T., C.A. Madramootoo, D. Janik, E. Fava and A. Stewart. 2003. Factors affecting soil temperatures under plastic mulches. Trop. Agr. 80: 6–13.

Donald, W.W. 2000. Between-row mowing + in-row band-applied herbicide for weed control in *Glycine max*. Weed Sci. 48(4): 487–500.

D'Souza, G. and J. Ikerd. 1996. Small farms and sustainable development: is small more sustainable? J. Agric. and Appl. Econ. 28(1): 73–83.

Edwards, C.A., W.D. Shuster, M.F. Huelsman and E.N. Yardim. 1995. An economic comparison of chemical and lower-chemical input techniques for weed control in vegetables. Paper read at the Brighton Crop Prot. Conf. Weeds 20–23 November, Farnham, UK.

Eyre, M.D., S.J. Wilcockson, C. Leifert and C.N.R. Critchley. 2011. Crop sequence, crop protection and fertility management effects on weed cover in an organic/conventional farm management trial. Eur. J. Agron. 34: 153–162.

Fennimore, S.A. and D.J. Doohan. 2008. The challenges of specialty crop weed control, future directions. Weed Technol. 22(2): 364–372.

Fennimore, S.A. and R.E. Goodhue. 2016. Soil disinfestation with steam: a review of economics, engineering, and soil pest control in California strawberry. Inter. J. of Fruit Sci. Publ. Online. doi: 10.1080/15538362.2016.1195312

Fennimore, S.A., F.N. Martin, T.C. Miller, J.C. Broome, N. Dorn and I. Greene. 2014. Evaluation of a mobile steam applicator for soil disinfestation in California strawberry. HortSci. 49(12): 1542–1549.

Fernandez, A.L., C.C. Sheaffer, D.L. Wyse and T.E. Michaels. 2012. Yield and weed abundance in early- and late-sown field pea and lentil. Agron. J. 104: 1056–1064.

Finney, D.M., N.G. Creamer, J.R. Schultheis, M.G. Wagger and C. Brownie. 2009. Sorghum sudangrass as a summer cover and hay crop for organic fall cabbage production. Renew. Agr. Food Syst. 24: 225–233.

Fontanelli, M., L. Martelloni, M. Raffaelli, C. Frasconi, M. Ginanni and A. Peruzzi. 2015. Weed management in autumn fresh market spinach: a nonchemical alternative. HortTechnol. 25: 177–184.

Garrison, A.J., A.D. Miller, M.R. Ryan, S.H. Roxburgh and K. Shea. 2014. Stacked crop rotations exploit weed-weed competition for sustainable weed management. Weed Sci. 62(1): 166–176.

Gianessi, L.P. 2013. The increasing importance of herbicides in worldwide crop production. Pest. Manag. Sci. 69: 1099–1105.

Gianessi, L.P. and N.P. Reigner. 2007. The value of herbicides in U.S. crop production. Weed Technol. 21(2): 559–566.

Gibson, K.D., John McMillan, Stephen G. Hallett, Thomas Jordan and Stephen C. Weller. 2011. Effect of a living mulch on weed seed banks in tomato. Weed Technol. 25(2): 245–251.

Gnanavel, I. 2015. Eco-friendly weed control options for sustainable agriculture. Science International 3(2): 37–47.

Gordon, G.G., J.E. Brown, E.L. Vinson III, W.G. Foshee III and S.T. Reed. 2010. The effects of colored plastic mulches and row covers on the growth and yield of okra. HortTechnol. 20: 224–233.

Greenland, R.G. 2003. Injury to vegetable crops from herbicides applied in previous years. WeedTechnol. 17(1): 73–78.

Gullino, M.L., A. Camponogara, G. Gasparrini, V. Rizzo, C. Clini and A. Garibaldi. 2003. Replacing methyl bromide for soil disinfestation. Plant Dis. 87(9): 1012–1021.

Ham, J.M., G.J. Kluitenberg and W.J. Lamont. 1991. Potential impact of plastic mulches on the aboveground plant environment. Proc. Natl. Agr. Plastics Congr. 23: 63–69.

Harker, K.N. and J.T. O'Donovan. 2013. Recent weed control, weed management, and integrated weed management. Weed Technol. 27(1): 1–11.

Hartwig, N.L. and H.U. Ammon. 2002. Cover crops and living mulches. Weed Sci. 50: 688–699.

Hatterman-Valenti, H., G. Endres, B. Jenks, M. Ostlie, T. Reinhardt, A. Robinson, J. Stenger and R. Zollinger. 2017. Defining glyphosate and dicamba drift injury to dry edible pea, dry edible bean, and potato. HortTechol. 27: 502–509.

Heap, I. 2014. Global perspective of herbicide-resistant weeds. Pest. Manag. Sci. 70: 1306–1315.

Hutchinson, C.M. and M.E. McGiffen Jr. 2000. Cowpea cover crop mulch for weed control in desert pepper production. HortSci. 35: 196–198.

Herrero, E.V., J.P. Mitchell, W.T. Lanini, S.R. Temple, E.M. Miyao, R.D. Morse and E. Campiglia. 2001. Use of cover crop mulches in a no-till furrow-irrigated processing tomato production system. HortTechnol. 11: 43–48.

Ibarra, L., J. Flores and J.C. Díaz-Pérez. 2001. Growth and yield of muskmelon in response to plastic mulch and row covers. Scientia Hort. 87: 139–145.

Kerpauskas, P., A.P. Sirvydas, P. Lazauskas, R. Vasinauskiene and A. Tamosiunas. 2006. Possibilities of weed control by water steam. Agron. Res. 4(Special Issue): 221–226.

Khanh, T.D., M.I. Chung, T.D. Xuan and S. Tawata. 2005. The exploitation of crop allelopathy in sustainable agricultural productions. J. Agron. Crop Sci. 191: 172–184.

Kristiansen, P., B.M. Sindel and R.S. Jessop. 2008. Weed management in organic Echinacea (*Echinaceapurpurea*) and lettuce (*Lactuca sativa*) production. Renew. Agr. Food Syst. 23: 120–135.

Kumar, V., R.R. Bellinder and D.C. Brainard. 2009. Effects of spring-sown cover crops on establishment and growth of hairy galinsoga (*Galinsoga ciliata*) and four vegetable crops. HortSci. 44: 730–736.

Kunkel, D.L., F.P. Salzman, M. Arsenovic, J.J. Baron, M.P. Braverman and R.E. Holm. 2008. The role of IR-4 in the herbicide registration process for specialty food crops. Weed Technol. 22(2): 373–377.

Lamont, W.J. 1994. Opportunities for plasticulture in organic farming systems. Proc. Amer. Soc. Plasticulture 25: 34–35.

Lamont, W.J. 1993. Plastic mulch for the production of vegetable crops. HortTechnol. 3: 35–39.

Law, D.M., A.B. Rowell, J.C. Snyder and M.A. Williams. 2006. Weed control efficacy of organic mulches in two organically managed bell pepper production systems. HortTechnol. 16: 225–232.

Leib, B.G., A.R. Jarrett, M.D. Orzolek and R.O. Mumma. 2002. Drip chemigation of imidacloprid under plastic mulch increased yield and decreased leaching caused by rainfall. Trans. Amer. Soc. Agr. Eng. 43: 615–622.

Liebman, M. and C.P. Staver. 2001. Crop diversification for weed management. pp. 322–374. *In:* Ecological Management of Agricultural Weeds. Cambridge University Press, Cambridge, UK.

Liebman, M., B. Baraibar, Y. Buckley, D. Childs, S. Christensen, R. Cousens, H. Eizenberg, S. Heijting, D. Loddo, A. Merotto, M. Renton and M. Riemens. 2016. Ecologically sustainable weed management: how do we get from proof-of-concept to adoption? Ecol. Appl. 26: 1352–1369.

Locascio, S.J., J.P. Gilreath, S. Olson, C.M. Hutchinson and C.A. Chase. 2005. Red and black mulch color affects production of Florida strawberries. HortSci. 40: 69–71.

Loken, J.R. and H.M. Hatterman-Valenti. 2010. Multiple applications of reduced-rate herbicides for weed control in onion. Weed Technol. 24(2): 153–159.

Manalil, S., R. Busi, M. Renton and S.B. Powles. 2011. Rapid evolution of herbicide resistance by low herbicide dosages. Weed Sci. 59(2): 210–217.

Masiunas, J.B., L.A. West and S.C. Weller. 1995. The impact of rye cover crops on weed populations in a tomato cropping system. Weed Sci. 43: 318–323.

Merwin, I.A., D.A. Rosenberger, C.A. Engle, D.L. Rist and M. Fargione. 1995. Comparing mulches, herbicides, and cultivation as orchard groundcover management systems. HortTechnol. 5: 151–158.

Meyer, S.L.F., I.A. Zasada, S.M. Rupprecht, M.J. VanGessel, C.R.R. Hooks, M.J. Morra and K.L. Everts. 2015. Mustard seed meal for management of root-knot nematode and weeds in tomato production. HortTechnol. 25(2): 192–202.

Mohammadi, G.R. 2012. Living mulch as a tool to control weeds in agroecosystems: a review. *In:* A.J. Price (Ed.) Weed Control. Retrieved from http://cdn.intechopen.com/pdfs/29920.pdf (Accessed on December 22, 2016).

Mohler, C.L. 2001. Mechanical management of weeds. pp. 139–209. *In:* Ecological Management of Agricultural Weeds. Cambridge University Press, Cambridge, UK.

Moss, S.R. 2010. Non-chemical methods of weed control: benefits and limitations. Proc. 17th Annual Australasian Weeds Conference. pp. 14–19. New Zealand Plant Prot. Soc.

Mulvaney, M.J., A.J. Price and C.W. Wood. 2011. Cover crop residue and organic mulches provide weed control during limited-input no-till collard production. J. Sustain. Agr. 35: 312–328.

Musser, W.N., V.J. Alexander, B.V. Tew and D.A. Smittle. 1985. A mathematical programming model for vegetable rotations. J. Agric. and Appl. Econ. 17(1): 169–176.

Norsworthy, J.K., L. Brandenberger, N.R. Burgos and M. Riley. 2005. Weed suppression in *Vigna unguiculata* with a spring-seeded *Brassicaeae* green manure. Crop Prot. 24: 441–447.

Norsworthy, J.K., S.M. Ward, D.R. Shaw, R.S. Llewellyn, R.L. Nichols, T.M. Webster, K.W. Bradley, G. Powles, S.B. Frisvold, N.R. Burgos, W.W. Witt and M. Barrett. 2012. Reducing the risks of herbicide resistance: best management practices and recommendations. Weed Sci. 60: 31–62.

Onwude, D.I., R. Abdulstter, C. Gomes and N. Hashim. 2016. Mechanisation of large-scale agricultural fields in developing countries: a review. J. Sci. Food Agric. 96: 3969–3976.

Owen, M.D., H.J. Beckie, J.Y. Leeson, J.K. Norsworthy and L.E. Steckel. 2015. Integrated pest management and weed management in the United States and Canada. Pest. Manag. Sci. 71: 357–376.

Ozores-Hampton, M. 2017. Guidelines for assessing compost quality for safe and effective utilization in vegetable production. HortTechnol. 27(2): 162–165.

Pannacci, E., B. Lattanzi and F. Tei. 2017. Non-chemical weed management strategies in minor crops: a review. Crop Prot. 96: 44–58.

Pedreros, A., M.I. Gonzalez and V. Manosalva. 2008. Effect of organic mulching on growth and yield of raspberry cv. Heritage. Acta Hortic. 777: 473–475.

Peigné, J., B.C. Ball, J. Roger-Estrade and C. David. 2007. Is conservation tillage suitable for organic farming? A review. Soil Use and Management. 23: 129–144.

Phatak, S.C., J.R. Dozier, A.G. Bateman, K.E. Brunson and N.L. Martini. 2002. Cover crops and conservation tillage in sustainable vegetable production. pp. 401–403. *In:* E. van Santen (Ed.) Making Conservation Tillage Conventional: Building a Future on 25 Years of Research. Proc. 25th Ann. So. Conserv. Tillage Conf. for Sust. Agric.

Price, A.J., K.S. Balkcom, S.A. Culpepper, J.A. Kelton, R.L. Nichols and H. Schomberg. 2011. Glyphosate-resistant palmer amaranth: a threat to conservation tillage. J. Soil Water Conserv. 66: 265–275.

Price, A.J. and J.K. Norsworthy. 2013. Cover crops for weed management in southern reduced-tillage vegetable cropping systems. Weed Technol. 27: 212–217.

Price, A.J., M.E. Stoll, J.S. Bergtold, F.J. Arriaga, K.S. Balkcom, T.S. Kornecki and R.L. Raper. 2008. Effect of cover crop extracts on cotton and radish radicle elongation. Commun. Biometry Crop Sci. 3: 60–66.

Rapp, H.S., R.R. Bellinder, H.C. Wien and F.M. Vermeylen. 2004. Reduced tillage, rye residues, and herbicides influence weed suppression and yield of pumpkins. Weed Technol. 18(4): 953–961.

Rasmussen, J., H.W. Griepentrog, J. Nielsen and C.B. Henriksen. 2012. Automated intelligent rotor tine cultivation and punch planting to improve the selectivity of mechanical intra-row weed control. Weed Res. 52: 327–337.

Rasmussen, J., C.B. Henriksen, H.W. Griepentrog and J. Nielsen. 2011. Punch planting, flame weeding and delayed sowing to reduce intra-row weeds in row crops. Weed Res. 51: 489–498.

Reddy, K.N. 2001. Effects of cereal and legume cover crop residues on weeds, yield, and net return in soybean (*Glycine max*). Weed Technol. 15: 660–668.

Rupende, E., O.A. Chivinge and I.K. Mariga. 1998. Effect of storage time on weed seedling emergence and nutrient release in cattle manure. Exp. Agric. 34(3): 277–285.

Rice, A.R., J.L. Johnson-Maynard, D.C. Thill and M.J. Morra. 2007. Vegetable crop emergence and weed control following amendment with different *Brassicaceae* seed meals. Renewable Agr. and Food Syst. 22: 204–212.

Ritchie, E. 2016. Texas Wintergarden Spinach Producers Board, La Pryor. TX, USA. Pers. Comm.

Ross, M.A. and C.A. Lembi. 1985. Applied Weed Science. Prentice Hall, NJ, USA.

Ruegg, W.T., M. Quadranti and A. Zoschke. 2007. Herbicide research and development: challenges and opportunities. Weed Res. 47: 271–275.

Runham, S.R. and S.J. Town. 1995. An economic assessment of mulches in field scale vegetable crops. pp. 925–930. *In:* Brighton Crop Protection Conference: Weeds. British Crop Protection Council. Brighton, UK.

Sainju, U.M. and B.P. Singh. 2001. Tillage, cover crop, and kill-planting date effects on corn yield and soil nitrogen. Agron. J. 93: 878–886.

Shaw, D.R. 2016. The 'wicked' nature of the herbicide resistance problem. Weed Sci. 64(1): 552–558.

Shem-Tov, S., S.A. Fennimore and W.T. Lanini. 2006. Weed management in lettuce (*Lactuca sativa*) with preplant irrigation. Weed Technol. 20(4): 1058–1065.

Shrestha, A., S.Z. Knezevic, R.C. Roy, B. Ball-Coelho and C.J. Swanton. 2002. Effect of tillage, cover crop and crop rotation on the composition of weed flora in a sandy soil. Weed Res. 42: 76–87.

Singh, B.P., D.M. Granberry, W.T. Kelley, G. Boyhan, U.M. Sainju, S.C. Phatak, P.E. Sumner, M.J. Bader, T.M. Webster, A.S. Culpepper, D.G. Riley, D.B. Langston and G. Fonsah. 2005. Sustainable vegetable production. pp. 1–38. *In:* R. Dris (Ed.) Vegetables: Growing Environment and Mineral Nutrition. WFL Publisher, Helsinki, Finland.

Slaughter, D.C., D.K. Giles and D. Downey. 2007. Autonomous robotic weed control systems: a review. Computers and Electronics in Agric. 61(1): 63–78.

Stirzaker, R.J., B.G. Sutton and N. Collis-George. 1993. Soil management for irrigated vegetable production. I. The growth of processing tomatoes following soil preparation by cultivation, zero-tillage and an *in situ*-grown mulch. Aust. J. Agr. Res. 44: 817–829.

Stopes, C. and S. Millington. 1991. Weed control in organic farming systems. pp. 185–192. *In:* Brighton Crop Protection Conf. Weeds. Brit. Crop Prot. Council, Farnham, UK.

Swanton, C.J., J. O'Sullivan and D. Robinson. 2010. The critical weed-free period in carrot. Weed Sci. 58: 229–233.

Sweeney, A.E., K.A. Renner, C. Laboski and A. Davis. 2008. Effect of fertilizer nitrogen on weed emergence and growth. Weed Sci. 56(5): 714–721.

Teasdale, J.R. and S.B. Mirsky. 2015. Tillage and planting date effects on weed dormancy, emergence, and early growth in organic corn. Weed Sci. 63(2): 477–490.

Teasdale, J.R. and C.L. Mohler. 2000. The quantitative relationship between weed emergence and the physical properties of mulches. Weed Sci. 48: 385–392.

Tei, F. and E. Pannacci. 2017. Weed management systems in vegetables. *In:* P.E. Hatcher and R.J. Froud-Williams (Eds.) Weed Research: Expanding Horizons. John Wiley & Sons Ltd. Chichester, UK. doi: 10.1002/9781119380702.ch12

Tingle, C.H. and J.M. Chandler. 2004. The effect of herbicides and crop rotation on weed control in glyphosate-resistant crops. Weed Technol. 18(4): 940–946.

Underwood, J., A. Wendel, B. Schofield, L. McMurray and R. Kimber. 2017. Efficient in-field plant phenomics for row-crops with an autonomous ground vehicle. J. Field Robotics 34: 1061–1083.

Ugen, M.A., H.C. Wien and C.S. Wortmann. 2002. Dry bean competitiveness with annual weeds as affected by soil nutrient availability. Weed Sci. 50(4): 530–535.

Van Der Weide, R.Y., P.O. Bleeker, V.T.J.M. Achten, L.A.P. Lotz, F. Fogelberg and B. Melander. 2008. Innovation in mechanical weed control in crop rows. Weed Res. 48: 215–224.

Van Wychen, L. 2016. Survey of the most common and troublesome weeds in broadleaf crops, fruits and vegetables in the United States and Canada. Weed Sci. Soc. Amer. Nat. Survey Dataset. Available at: http://wssa.net/wp-content/uploads/2016_Weed_Survey_Final.xlsx (Accessed on January 24, 2016).

Vollmer, E.R., N. Creamer, C. Reberg-Horton and G. Hoyt. 2010. Evaluating cover crop mulches for no-till organic production of onions. HortSci. 45: 61–70.

Vollmer, K.M., M. VanGessel, Q. Johnson and B. Scott. 2017. Relative safety of pre-emergence corn herbicides applied to coarse-textured soil. Weed Technol. 31(3): 356–363.

Wallace, R.W. and R.B. Bellinder. 1989. Potato (*Solanum tuberosum*) yields and weed populations in conventional and reduced tillage systems. Weed Technol. 3: 590–595.

Wallace, R.W. and R.R. Belliner. 1990. Alternative tillage and herbicide options for successful weed control in vegetables. HortSci. 27(7): 745–749.

Wallace, R.W., A. Phillips and J.C. Hodges. 2007. Response of selected herbicides for weed control, crop injury and yield in processing spinach (*Spinacia oleracea* L.) production. Weed Technol. 21: 714–718.

Walters, S.A., S.A. Nolte and B.G. Young. 2005. Influence of winter rye and preemergence herbicides on weed control in no-tillage zucchini squash production. HortTechnol. 15: 238–243.

Walters, S.A. and B.G. Young. 2008. Utility of winter rye living mulch for weed management in zucchini squash production. Weed Technol. 22: 724–728.

Walters, S.A. and B.G. Young. 2012. Herbicide application timings on weed control and jack-o-lantern pumpkin yield. HortTechnol. 22: 201–206.

Warnick, J.P., C.A. Chase, E.N. Rosskopf, E.H. Simonne, J.M. Scholberg, R.L. Koenig and N.E. Roe. 2006. Weed suppression with hydramulch, a biodegradable liquid paper mulch in development. Renew. Agr. Food Syst. 21: 216–223.

Weir, T.L., S. Park and J.M. Vivanco. 2004. Biochemical and physiological mechanisms mediated by allelochemicals. Plant Biol. 7: 472–479.

Wells, H.F. 2016. Vegetables and Pulses Outlook. USDA-ERS. Available at: https://www.ers.usda.gov/webdocs/publications/vgs356/57264_vgs-356-revised.pdf?v=42516.

Williams, M.M. 2006. Planting date influences critical period of weed control in sweet corn. Weed Sci. 54: 928–933.

Williams II, M.M. and R.A. Boydston. 2013. Crop seeding level: implications for weed management in sweet corn. Weed Sci. 61(3): 437–442.

Wookey, C.B. 1985. Weed control practice on an organic farm. pp. 577–582. *In:* Brit. Crop Protection Conf. Weeds. Brit. Crop Protection Council, Croydon, UK.

Yordanova, M. and N. Shaban. 2007. Effect of mulching on weeds of fall broccoli. Bull. of Univ. of Agric. Sci. and Vet. Med. Cluj-Napoca 64: 99–102.

Young, S.L. 2012. True integrated weed management. Weed Res. 52: 107–111.

Young S.L., S.K. Pitla, F.K. Van Evert, J.K. Schueller and F.J. Pierce. 2017. Moving integrated weed management from low level to a truly integrated and highly specific weed management system using advanced technologies. Weed Res. 57: 1–5.

Zandstra, B.H., W.R. Chase and J.G. Masabni. 1998. Interplanted small grain cover crops in pickling cucumbers. HortTechnol. 8: 356–360.

Zhang, J., S.E. Weaver and A.S. Hamill. 2000. Risks and reliability of using herbicides at below-labeled rates. Weed Technol. 14(1): 106–115.

Zotarelli, L., L. Avila, J.M.S. Scholberg and B.J.R. Alves. 2009. Benefits of vetch and rye cover crops to sweet corn under no-tillage. Agron. J. 101: 252–260.

CHAPTER
22

Sustainable Weed Control in Coffee

Cláudio Pagotto Ronchi[*1] and Antonio Alberto da Silva[2]

[1] Federal University of Viçosa, Florestal Campus, Rodovia LMG, km 06, 35.690-000, Minas Gerais State, Brazil

[2] Federal University of Viçosa, Viçosa Campus, Avenida PH Rolfs s/n, 36.570-000, Minas Gerais State, Brazil

Introduction

Coffee is one of the most important commodities in international agricultural trade, generating over 90 billion dollars each year, with approximately 500 million people involved in its production, from cultivation to final consumption (DaMatta et al. 2010). Among more than a hundred species, only *C. arabica* L. (Arabica coffee) and *C. canephora* 'Pierre ex A. Froehner' (Robusta coffee) have a worldwide economic importance (DaMatta et al. 2010). In fact, mean data of the last four crop years (2013 to 2016) from the International Coffee Organization show that the world coffee production (Arabica + Robusta) is about 149, 4 million 60 kg bags, of which Arabica coffee accounts for about 59%, and Robusta coffee for the rest (Matiello et al. 2016).

Coffee is grown in tropical and subtropical regions in more than 60 countries spread over Africa, Asia and Oceania, Mexico and Central America, and South America (Matiello et al. 2016). Among all coffee-producing countries, seven of them produce more than five million bags. Brazil is the world's largest coffee producer (35% of total coffee production), followed by Vietnam, Colombia, Indonesia, Ethiopia, India, and Honduras. All together, these countries produced 80% of the world's green coffee in recent years. Arabica coffee is mainly produced in the American continent, whereas Robusta coffee is of chief importance in Africa and Asia (Matiello et al. 2016).

Although *C. canephora* also has a large economic importance worldwide, almost all published scientific articles that investigated weed control in coffee in the twenty-first century have considered weed control in *C. arabica*. Therefore, this chapter focusses specifically on weed control in Arabica coffee. Nonetheless, several weed control practices applied to Arabica coffee may also be recommended to Robusta coffee (Ronchi et al. 2016).

Coffee plants grow perennially, with a production cycle of up to 30 years. Although inter-rows spacing and coffee plant density may change significantly among coffee growing areas around the world, and even among areas into the same country, in Brazil, for example, coffee cropping systems consist mainly of large-spacing inter-rows (2.5 to 4.0 m), with an optimal planting density of about 5,000 trees ha^{-1} (DaMatta et al. 2010, Matiello et al. 2016). In Central

---

*Corresponding author: claudiopagotto@ufv.br

America (Costa Rica), a narrowed inter-rows spacing (~2.0 m) is used, but keeping the same planting density (Ramírez 2009) as in Brazil. In both cases, however, a high soil exposure is observed. In addition, coffee plants show a very low initial growth rate (DaMatta et al. 2010) that also impairs soil covering (Ronchi et al. 2001, Silva and Ronchi 2008). Thus, mainly during the juvenile phase (up to two years in fields), the coffee crop is highly sensitive to weed species competition (Ronchi and Silva 2006, Araújo et al. 2012). This results in a remarkable decrease in coffee growth and yield and weed control is one of the largest field management tasks, which can entail high costs (Allcântara and Ferreira 2000a, Silva and Ronchi 2008, Ramírez 2009).

Coffee is grown worldwide in different production-systems that have a direct impact on weed dynamics and hence its control strategy. For example, shading plantations with low soil disturbance is used in Central America (Aguilar et al. 2003, Ramíres 2009), whereas in Brazil, coffee is grown under full sunlight with constant soil disturbance, especially during harvesting (Matiello et al. 2016). In Brazil, the world's largest coffee producer, there are different primary coffee production regions, each utilizing particular cultural practices for crop management (Matiello et al. 2016). Therefore, adopted integrated weed management (IWM) practices will vary between coffee farms worldwide depending on the local characteristics. Actually, the adoption of site-specific IWM practices is the base for a sustainable weed management in any cropping system (Bajwa 2014).

Weed Impact

It is well known that weeds affect the coffee crop in several ways during its life cycle (Silva and Ronchi 2008). For example, it has been shown that young coffee plants suffer from competition with different weed species under both controlled conditions (Ronchi and Silva 2006, Ronchi et al. 2007, Fialho et al. 2010, 2011, Carvalho et al. 2013) and in field experiments (Lemes et al. 2010, Araújo et al. 2012, Magalhães et al. 2012). The reduction in plant growth correlated with decreasing photosynthetic efficiency (Matos et al. 2013) and nutrient accumulation by both coffee plant shoot (Ronchi et al. 2003, Carvalho et al. 2013) and root (Ronchi et al. 2007) systems. These studies also showed that the effect of weed competition on coffee plants was strongly dependent on both the weed species and its density, and on coffee plant age after transplanting. Therefore, an IWM in coffee must consider the characteristic of weed species individually, as well as their high potential of nutrient recycling.

The impaired crop growth due to weed competition soon after transplanting in the field will certainly cause irreversible losses in crop yield, as demonstrated by Lemes et al. (2010). These authors reported coffee yield loss of approximately 40% during the first three harvest cycles as a consequence of inadequate weed control practice that led to weed-crop competition after transplanting. In fact, yield losses due to weed competition varied from 24% (Moraima García et al. 2000) to 92% (Lemes et al. 2010) in different adult commercial coffee crops. Weeds can also hinder fertilizer application and coffee harvesting. Several other harmful effects of weed competition on the coffee crop are discussed elsewhere (Silva and Ronchi 2008).

Weed species biomass or composition significantly differ among coffee sites depending on several factors, including coffee cropping system—monocrop or intercropped (Concenço et al. 2014), use of shade trees and the level of shading (Silva et al. 2006), crop age, season, ground cover management (Aguilar et al. 2003), and the use of cover crops (Partelli et al. 2010, Moreira et al. 2013). Therefore, potential weed species-coffee interactions are high.

There are several weed species that occur in coffee plantations worldwide (Aguilar et al. 2003, Maciel et al. 2010, Partelli et al. 2010, Ferreira et al. 2011). Some examples are *Ageratum conyzoides, Baccharis trinervis, Borreria alata, Cyathula achirantoides,* and *Eleusine indica* in Venezuela (Sánchez and Gamboa 2004); *Cardamine flaccida, Commelina diffusa, Ipomoea purpurea, Spermacoce laevis,* and *Portulaca oleracea* in Costa Rica (Gómez 2005); *Amaranthus retroflexus, Cynodon dactylon, Cyperus rotundus, Oxalis latifolia,* and *Paspalum conjugatum* (Moreira et al. 2013), *Brachiaria brizanta, Panicum maximum, Commelina benghalensis, Sida cordifolia,* and *Cenchrus echinatus* (Silva et al. 2006), and *Acanthospermum australe, Urochloa decumbens, Urochloa plantaginea, Galinsoga parviflora,* and *Solanum americanum* (Araújo et al. 2012), all in Brazil. In the following paragraphs, emphasis is given to three important weed species occurring in Brazilian coffee growing areas: *Commelina* spp., which is a well-known weed in coffee plantations, and *Urochloa* spp. and *Ipomoea* spp., which have gained high importance in the newer-explored mechanized coffee crop areas.

Although the species of the genus *Commelina* are efficiently used as ground cover in shaded coffee plantations in Central America (Ramírez 2009), in Brazil, weed species of major importance in coffee plantations include *C. benghalensis* and *C. diffusa* (Ronchi et al. 2001). These weed species are resistant to water stress and grow exuberantly in coffee-induced shade environments. They also grow perennially and can easily propagate asexually under favorable soil moisture conditions. In the case of *C. benghalensis*, viable underground seed production is also observed (Kissmann 1997). Moreover, these weeds are tolerant to glyphosate, the most used non-selective post-emergent herbicide (Santos et al. 2002, Ronchi et al. 2002). Therefore, these weed species strongly compete with coffee plants, especially for water and nutrients, in addition to impairing coffee harvesting.

In several important Arabica coffee growing regions of Brazil, coffee plantations are commonly established on land previously used for grazing. Since *Urochloa decumbens* (Surinam grass) and other species of the genus *Urochloa* are the most used herbage in Brazil, it has become an important weed in some coffee crop areas (Dias et al. 2004, Souza et al. 2006, Araújo et al. 2012). In addition, *Urochloa* spp. is successfully used as ground cover to add biomass on the soil surface for soil moisture conservation and nutrient recycling (Alcântara and Silva 2010, Pedrosa 2013). In fact, the use of ground cover vegetation leads to a larger turnover of organic matter and nutrients, which is associated with increased soil water content, and improve the long-term soil fertility (Aguilar et al. 2003) and coffee plant nutrition (Pedrosa 2013). Although such practices may improve productivity, care must be taken to control Surinam grass, so that it does not grow too close or too tall, so as to compete with coffee and permanently impair coffee growth and yield, as demonstrated by Dias et al. (2004), Souza et al. (2006), and Araújo et al. (2012).

During the first months of coffee crop establishment in the field, the occurrence of high temperatures and abundant rainfall, associated with high irradiance levels, collectively promote the growth of *U. decumbens*, since this weed species shows a C4 photosynthetic metabolism. In contrast to the slow coffee plant growth just after transplanting (DaMatta et al. 2010), weed species, such as *U. decumbens* quickly develop large leaf area and height (a dense canopy), allowing weeds to better compete for light. This suggests that even before reaching pre-flowering stages, when slashing operations are commonly recommended for weed control in the coffee inter-rows (Silva and Ronchi 2008), weed shoots can shade both the lower and medium sections of the coffee plant canopy if they are growing near the crop line. This can interrupt the photosynthetic performance of coffee plants, which also affects the carbohydrate supply for crop growth.

Another weed species that has become increasingly important in the last years in Brazilian coffee areas is *Ipomoea* spp. (Morning glory). This weed species withstands common weed control methods (for example, mower operations, herbicide applications) applied in coffee plantations, since it is protected in the dense coffee rows when it germinates and emerges near the coffee trunk (Matiello and Santinato 2016). In addition to nutrient competition, it grows over the coffee plant canopy due to its twining climber habit (Lorenzi 2000), thereby strongly competing for light and impairing coffee harvesting. Morning glory shows high leaf area production that can completely cover the coffee plant canopy in areas where infestation is high. However, weed stems (or weed shoots) cannot be removed from the coffee hedgerows without significantly affecting the coffee plants (Matiello and Santinato 2016). Moreover, their seeds (which possess dormancy—Pazuch et al. 2015) are easily spread over the cropland by mechanized harvesting operations. Once Morning glory is able to seed, it can infect the entire coffee field resulting in high weed control costs.

Weed Control

Weed control occurs at all the steps of coffee production. It starts at an early stage when the land for coffee nurseries and both the soil and manure required to produce coffee seedlings have been identified. During the nursery period, coffee seedlings must also be kept weed-free. In the field, the area where coffee plants will be established must totally or at least partially cleared of weeds and natural vegetation, depending on the production systems. During the two years after transplanting, an appropriate weed control system is crucial for proper crop establishment minimizing any possibly harmful effects to the coffee environment (for example, high soil exposure). Finally, during the life of a productive plant, including the pruned ones, weeds should be managed using an appropriate (or sustainable) weed control practices in order to preserve yearly crop yields and shoot growth, as well as to ensure that coffee manage practices like fertilizer and pesticide applications, and coffee harvesting are efficiently applied to the crop.

In the sections below, regardless of the order, the primary weed control practices that are successfully used in each crop phase are discussed. Attention is paid to weed control in young coffee plantations, since this crop stage is considered a critical period for weed control in coffee, irrespective of the region where the crop is grown (Aguilar et al. 2003, Sánchez and Gamboa 2004, Sarno et al. 2004, Silva and Ronchi 2008, Ramírez 2009).

Soon after transplantation in the field, young coffee plants are highly sensitive to weed competition (Ronchi et al. 2003, 2007, Ronchi and Silva 2006, Fialho et al. 2010, 2011, 2012, Lemes et al. 2010, Araújo et al. 2012, Magalhães et al. 2012), with sensitivity to weed competition decreasing with age (Aguilar et al. 2003, Fialho et al. 2010). The weeds, irrespective of species types, are only harmful to young coffee plants when they grow near the coffee rows. For example, competition trials with young pot-grown coffee plants show that the distance between weeds and coffee plants affects coffee growth—the degree of weed competition increases as they get closer to the crop (Marcolini et al. 2009). Such negative interference occurs because proximity to neighboring plants leads to resource limitations for coffee plants, especially light and nutrients (Radosevich et al. 1996). This is because the main part of the root system of an Arabica coffee tree is generally concentrated in the first 0.2-m layer from the soil surface and is distributed near the coffee trunk (Ronchi et al. 2015), which aggravates the competition effect on coffee. Therefore, to preserve the initial growth of coffee plants, it is necessary to ensure that the area around coffee plants is kept bare, as evidenced in a study by Sarno et al. (2004) in Indonesia.

To prevent weed competition, farmers sometimes decide to eliminate all weeds from the crop fields, even in young and widely spaced (up to 3.8 m between rows) coffee plantations. However, such a drastic agronomic practice is not recommended, since it requires high inputs of energy for mechanical operations and, mainly, because soil exposure after weeding increases the likelihood of soil erosion, an unsustainable practice. Nonetheless, coffee is grown worldwide mainly on steep slopes where soil erosion is a severe problem (Sarno et al. 2004). In fact, several studies have demonstrated that natural soil coverage or the introduction of cover crops (discussed below) between coffee rows is effective for improving soil and crop characteristics (Aguilar 2003, Shivaprasad et al. 2005, Alcântara et al. 2009), especially in hilly areas (Sarno et al. 2004).

Adequate weed control in young coffee crops, and other perennial crops, is achieved by eliminating weed species only within a strip at both sides of the coffee row during the two years after transplanting. Weeds in the inter-rows may be managed by a desiccant non-selective herbicide, mower operations, and even by cover crops (Silva and Ronchi 2008). Managed this way, coffee can grow without weed competition and producers avoid unnecessary weeding. However, width of the weed control strip is randomly defined by growers, without any criteria. Generally, coffee genotype and crow architecture, plant spacing, soil texture, climatic conditions, and weed species density and diversity are factors that affect the width of a weed control strip. Age of the coffee plant also affects the width of the weed control strip (Araújo et al. 2012), since the diameter of the coffee plant canopy and the root system progressively increases after transplantation (DaMatta et al. 2010, Matiello et al. 2010, Araújo et al. 2012, Ronchi et al. 2015).

Even though coffee producers commonly keep young coffee rows permanently weeding, information regarding the criteria for weed control strip establishment in young coffee plantations is scarce in the published literature, which suggests that little importance has been given to this subject. In Brazil, adequate width of weed control strip at each side of coffee rows to prevent young coffee plants from weed competition ranges between 0.6 m (Dias et al. 2008) and 1.0 m (Souza et al. 2006). However, establishing a fixed weed control range irrespective of coffee plant age may not be an effective strategy as can be depicted from the work of Lemes et al. (2010). Indeed, Araújo et al. (2012) clearly demonstrated that integrated management of *U. decumbens* in young coffee crops must focus on weed control only in a minimum range along the coffee rows, which must increase with the coffee plant's age (Figure 22.1).

The weed strip control values reported by Araújo et al. (2012) for each crop age, however, are not standard fixed values for coffee plantations. Depending on both the occurring weed species and the adopted IWM methods applied in the inter-rows, as well as on several other crop aspects, weed control strips may increase or decrease. Consequently, each farm must be analyzed separately to properly define the best weed control strip. Over the past decades, some general agronomical recommendations to prevent weed competition focussed on controlling weeds only below the projected coffee plant canopy (Sánchez 1991, Silva and Ronchi 2008).

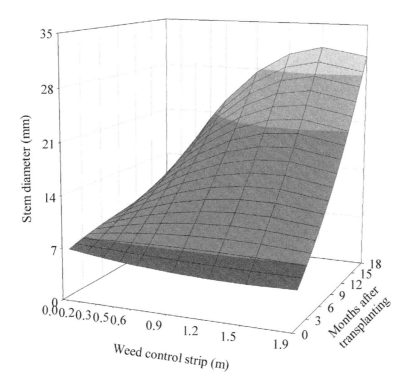

Figure 22.1. Coffee plant growth (represented by stem diameter) as a function of both the weed control strip and coffee plants' age in months after transplanting (MAT). In the first period after transplanting (within 2-3 MAT), weed competition exerts no effect or very little negative effect on plant growth; as the coffee plant ages, its growth is drastically reduced when weed-control practices are not employed. However, adequate coffee growth is ensured at approximately 18 MAT if a minimum weed strip control is applied to coffee rows. As the coffee plant ages, weed control strips must be progressively wider to prevent weed competition, and to enable the coffee plants to grow properly. This figure is adapted from Araújo et al. (2012).

Araújo et al. (2012) showed, however, that weed control must extend beyond the projected coffee plant canopy, as many Brazilian coffee growers are already applying. Recent research (Pedrosa 2013) has also shown that by controlling *U. decumbens* (that was sowed in the coffee inter-rows as ground cover—discussed above) in a 0.5-m control strip beyond the projected coffee plant canopy, nutrient (nitrogen) competition was reduced to only 1%, ensuring maximum coffee growth and yield.

Recommendation for controlling weeds beyond the projected coffee plant canopy does not mean that weeds should be eliminated in the entire inter-row spacing. As shown in Figure 22.1 (for detailed discussion see Araújo et al. (2012), maximum coffee plant growth for each coffee plant age was reached at a weed control strip smaller than the largest one that was possible; above such a critical weed control strip value, coffee growth was slightly reduced. This suggests that weed elimination in the inter-rows, but not the weed competition to coffee rows, may explain such a negative effect. Therefore, maintaining natural weed coverage in the center of coffee inter-rows may favor edaphoclimatic conditions for the improvement of coffee plant growth. In fact, there are many benefits of covering the inter-rows of coffee plantations, as will be discussed later. Moreover, the cost of weed control may be diminished if it is applied only to the minimum weed control strip in the coffee rows.

Mechanical/Physical Weed Control

Handweeding

Hand-weed control methods are of major importance in coffee farms, although they are slow and labor-intensive. During nursery stages, if preventive measures fail or selective herbicides are not used, weed plants that eventually emerge in the clods must be removed during the period of seedling formation and growth (Ronchi et al. 2001). Within two years after transplantation in the field, several hand hoeing operations are recommended to establish and maintain an adequate weed control strip along coffee rows, although herbicides can also judiciously be applied. In those coffee farms where selective pre-emergence herbicide application is the major method for weed control in the coffee rows, at least one hand hoeing operation is accomplished 2–3 weeks after transplanting prior to herbicide application to smooth the ground surface. In addition to controlling initial weed vegetation in the coffee rows, hand hoeing will ensure a high efficiency of pre-emergence herbicide application. In coffee inter-rows other weed control methods are recommended.

Some weed species are difficult to control and their seeds can be easily disseminated throughout the area. One example is *Digitaria insularis* (Sourgrass) an important role as a weed species in coffee plantations (Carvalho et al. 2013). If weed species are not controlled until dense infestations, further control operations may be highly expensive or ineffective. Hence, it is recommended that weed infestation be controlled early before it seeds (as a preventive measure), by mechanical methods, such as by a hoeing operation (Silva and Ronchi 2008).

Mowing

Mowing is one of the most widely used weed control practices in coffee plantations, especially in Brazil, to manage natural weed vegetation (or even planted cover crops) in the coffee inter-rows during the warm and wet season when they grow very fast. Mowing can be performed by hand-held or mechanized equipment. It is important to mow the weeds when they have developed maximum biomass, but before the seeds are viable. In addition, a regular mowing frequency that would prevent excessive shading to the bottom of the coffee plant canopy should be used. In addition to providing mulching, improving nutrient cycling, and controlling soil erosion (discussed below), successive mowing operations allow a constant weed re-growth and hence promotes the maintenance of soil organic carbon stocks (Cogo et al. 2013). Compared to other mechanical weed control methods, such as the use of disk harrow, the brush-cutter maintains a better state of soil particle aggregation (Siqueira et al. 2014a) and a lower soil bulk density (Siqueira et al. 2014b). In some coffee farms, mowing operations are commonly and successfully applied in alternating coffee inter-rows to conserve populations of the natural enemies of insect pests.

Soil Cultivation

The use of weed control practices comprising soil cultivation, especially those applied in the coffee inter-rows (such as the use of chisel plow, disk harrow, rotary hoe, and field cultivator), have drastically decreased in the last decades in coffee farms. This is in contrast to an increase in the use of desiccant herbicides and, mainly, mower operations. Nowadays, many coffee growers are conscious that soil disturbance must be reduced to prevent soil erosion and to ensure sustainable weed management. Although soil cultivation may have limitations (for example, it may not be efficient under high soil moisture content and for controlling asexually propagated weed species), and that its use may be partially restricted to both sloping and high planting density or intercropped areas, it is still successfully used.

In Brazil, for example, different types of disk harrow implements have been developed and used to till the soil along young coffee rows, before the coffee root system is fully developed. Moreover, in the coffee fields where sweeping and coffee gathering operations are necessary for the recovery of fallen fruit due to natural fall and mechanical harvesting, a brush shredder is commonly used to smooth the soil surface and to crush organic residues during the pre-harvesting period (Borba et al. 2016). In countries or coffee fields where there is no need to gather fallen coffee fruits (because they are completely harvested directly from coffee trees instead), such mechanic soil cultivation operations are not necessary.

Several important long-term field studies (Alcântara and Ferreira 2000b, Alcântara et al. 2007, 2009, Araujo-Junior et al. 2011a, b, Pais et al. 2011, 2013, Siqueira et al. 2014a, b, Pires et al. 2017a, b) have considered the impacts of mechanical (and many other) weed control methods, including rotary hoe, disk harrow, and brush shredder on soil physical and chemical properties as well as their effect on hydraulic attributes. The results of those studies were highly diverse and sometimes conflicting, showing large variations over the years and depending on both the field situations and sampled soil layers, which limits any generalization. For example, compared to the brush shredder, the disk harrow reduced the geometric mean diameter of soil aggregates, soil organic matter content, soil macroporosity, soil water retention capacity, and increased the soil bulk density (Siqueira et al. 2014a, b). However, Pais et al. (2011, 2013) observed that the brush shredder and disk harrow operations caused similar high level soil compaction. This brief analysis of those works shows that local soil conditions and weed-crop interactions must be known to precisely define the best soil cultivation method required for each weed-control situation, thereby preserving physical and chemical soil attributes. This is the basis of integrated weed control practices for sustainable agriculture.

Cover Crops/Mulching

As emphasized in the above sections, if weed vegetation is maintained at a sufficient distance from the coffee row (to prevent resource competition), there is no need to eliminate natural vegetation of the entire area (except during the harvesting period in some countries). Moreover, cover crops or green manure crops can be successfully intercropped with coffee. Both natural and introduced vegetation needs to be properly established and managed per local crop characteristics, ensuring the weed control practices benefits the environment, without affecting the coffee plants. Some cover crops, such as *Mucuna aterrima* (Fialho et al. 2011) and *U. decumbens* (Araújo et al. 2012) or possible inter-planted annual crops, such as *Gossypium hirsutum, Arachis hypogaea, Ricinus communis,* and *Zea mays* (Paulo et al. 2004) or *Phaseolus vulgaris* (Carvalho et al. 2007) may strongly affect coffee growth or yield if they are not appropriately established or managed. Therefore, much attention must be paid to the cover crops species, densities, planting-spacing, root system and crown architecture, and frequency of pruning or mowing/slashing operations, as well as on the coffee crop plant age and on the width of weed control strip.

Field studies carried out in different experimental conditions worldwide have demonstrated the high importance of green manure crops, mulching or even the maintenance of natural vegetation in the coffee inter-rows for both the crop and environment improvement (Aguilar et al. 2003, Sarno et al. 2004, Shivaprasad et al. 2005, Partelli et al. 2010, Pais et al. 2011, Araújo et al. 2012, Pedrosa 2013, Siqueira et al. 2014a, Martins et al. 2015). The benefits revealed in these studies include: prevention or reduction of soil exposure and erosion, improvement of physical, chemical and biological soil characteristics, maintenance of coffee yield potential, conserving soil moisture, improvement of soil organic carbon content, and weed suppression associated with allelopathic and physical effects or with direct resources competition. For example, Martins et al. (2015) found that the plots subjected to *M. deeringiana* cover crop in the inter-rows showed more than 90% reduction in weed density that were attributed to the allelopathic effects of such a cover crop. The other important effect of such sustainable practices is the reduction in soil temperature variation, preventing high soil temperatures from occurring. This is particularly important for coffee plants since their root systems are physiologically highly sensitive to high soil temperature (Franco 1958). In Central American countries, a selective ground cover management is carefully used in shaded coffee plantations to control weeds in patches, leaving uncontrolled species that are considered suitable to protect the soil from erosion and compaction, and to suppress more aggressive weed growth (Aguilar et al. 2003, Ramírez 2009).

However, the weed control methods and cover crops do not always alter some soil characteristics. For example, Martins et al. (2015, 2016a) assessed the effects of long-term different weed control methods and cover crops between coffee rows on the chemical soil characteristics, soil organic carbon and humic substances content, soil organic matter quality and weed diversity in Brazilian coffee plantations. They found that neither the different cover crops (*Arachis hypogeae*, *Mucuna deeringiana*, and natural vegetation—no weed control) nor the weed control methods (handweeding, portable mechanical mower and herbicide application) altered both the humic and fulvic acid carbon content at superficial soil layers (0 to 20 cm), and the organic matter humification degree. In addition, little or no changes were detected in chemical soil attributes among weed-control treatments. However, the top soil organic carbon content was up to 26% higher in the plots with no weed control as compared to other weed control methods, even to those possessing cover crops (Martins et al. 2015). Such increase in top soil organic carbon content was especially observed when mowed residues from coffee shrub pruning were left on the soil surface (Martins et al. 2016b).

Plant residues from other crops (if they are available in the farm without additional costs), from the coffee plants (leaves and stems), or from shade trees, especially after pruning operations, can be used as mulches. Actually, different types of organic materials, including coffee residues, such as coffee pulp, husk (Minassa 2014), and grounds (Yamane et al. 2014), have the potential to be used for controlling weeds through top dressing applications. For example, Yamane et al. (2014) recently demonstrated that top dressing application of coffee grounds at 16 kg m^{-2} resulted in significant weed control for half a year. Such an inhibition was a result of an allelopathic effect due the presence of caffeine, tannins, and polyphenols in the coffee grounds (Pandey et al. 2000). Regardless of the type of organic materials, their possible inhibitory effects on coffee plant growth and yield should also be evaluated before adding them to the coffee plantations to control weeds. Moreover, the studies mentioned above showed that adequate weed control is reached only if an optimum amount of top dressing organic material is used, otherwise it may stimulate weed growth.

Cultural Weed Control

Cultural weed control relies on the design and manipulation of cropping systems to reduce weed pressure. It is achieved through either reducing weed emergence or increasing crop competitive ability, or both, and can only be precisely established if the ecological characteristics of weeds and crops are known (Mohler 1996). Some traditional cultural weed control strategies cannot be applied in coffee (for example, crop rotations—coffee is perennial; exploring competitive ability of coffee cultivars—there is limited information about this strategy and, moreover, it is not a criteria for selecting coffee cultivars). The few cultural weed control methods that are particularly important and efficient in coffee plantations are discussed below.

Although coffee crop seedlings have an initial size advantage over weeds, coffee plants show a very low initial growth rate and low soil covering capacity as compared to annual crops. Therefore, high (or increasing) planting densities do not contribute to weed suppression during the critical period following transplanting. However, as the crop ages, high planting densities become effective in suppressing new weed growth compared to the conventional coffee planting density (Matiello et al. 2016). Such a positive effect on early weed suppression is almost the same as that induced by shading (Aguilar et al. 2003, Ramirez 2009). As coffee and shade trees age, the weed species composition also change (a reduction in the species less adapted to shade) and a reduction in weed biomass is commonly observed, which together significantly reduces the costs of weed control (Soto-Pinto et al. 2002, Aguilar et al. 2003, Silva et al. 2006, Ricci et al. 2008, Concenço et al. 2014). Weed suppression occurs because closer coffee plant canopies (as coffee ages and/or in high planting density) and over-story cover crop trees restrict light for weed growth, and probably also increase soil nutrient competition (Soto-Pinto et al. 2002, Aguilar et al. 2003). Thus, for cover crops, effective high planting density and shading trees must be established for each coffee plantation.

Other important cultural practices to improve crop competitive ability include choosing coffee cultivars adapted to each site, production of high-quality seedlings with a vigorous root system, transplanting only fully acclimated seedlings, establishing an equilibrated plant nutrition program, and controlling pest infestations and plant diseases. One practice that has been successfully used in coffee plantations, that minimizes weed-crop competition for nutrients, is fertilizer distribution using the drip fertigation system set up in the coffee row, near the trunk under the coffee plant canopy. Considering that roots of coffee plants, but not weeds, are concentrated in this soil region (Ronchi et al. 2015), nutrient uptake by the crop is improved. Long-term weed-crop competition trials conducted in modern Brazilian coffee plantations using such rigorous agronomic practices revealed a high tolerance of coffee plants to strong weed pressure (Ronchi C.P., unpublished data).

Preventive Weed Control Measures

Similar to cultural methods, preventive measures for weed suppression are less expensive and are advantageous to the coffee crop. There are very few, but relatively important preventive measures that should be applied in coffee production systems either to restrain the entrance or to diminish weed-seeds spreading over the coffee plantations.

The land where the coffee nursery will be established should be carefully selected. The nursery site should be free from hard-to-control weeds, such as *Cyperus rotundus* and *Cynodon dactylon,* or other weed species that are asexually propagated. Moreover, considering that most coffee seedlings are produced in polyethylene bags filled with a substrate composed of a mixture of soil and manure (clod-seedling), close attention should be paid to the origin of these substrate components. Although there are several physical and chemical methods potentially suitable for substrate sterilization (Miranda et al. 2007), it is not a common practice under field conditions. Therefore, the removal of topsoil layers (which represent the soil weed-seed bank—Roberts 1981) before soil withdrawing and the use of completely decomposed manure are important preventive measures that will drastically reduce the occurrence of weeds at the nursery stage (Ronchi et al. 2001). Other weed control practices, such as handweeding or pre-emergence herbicide application can be used during the nursery stage (see respective sections) to prevent the introduction of new weed species in new coffee areas.

In field-grown coffee plantations, preventive weed control practices basically include: i) cleaning manure for any soil amendments; ii) maintenance of weed-free farm roads by mowing operations or by desiccant herbicide applications that prevent soil erosion while keeping the soil covered; iii) cleaning machinery during or after any mechanized operation in the coffee farm;

iv) removing any new weed infestation before they become more dense; iv) controlling weed species up to flowering stage to prevent weed seeds spreading over the area by mechanical operations and animals, or to avoid increasing the soil weed-seed bank (Ronchi et al. 2014).

Weed species, such as *Bidens pilosa* and *Cenchrus echinatus*, for example, are commonly present in coffee areas and their seeds are easily dispersed by animals or even by the farm workers. Moreover, as mentioned earlier, in some important Brazilian coffee growing areas the mechanized coffee harvesting has contributed to the spread *Ipomoea* spp. seeds over the cropland (Matiello and Santinato 2016). By controlling this weed species during the early development stages or by cleaning harvesting machines frequently, the problem may diminish.

Chemical Weed Control

Chemical weed control is a component of IWM that is efficiently used in coffee plantations worldwide. The method is associated with several advantages, including reduced laboring and weed control costs (depending on labour costs), increased efficiency on perennial and asexually propagated weeds, improved mulching formation, and applicability during raining spells. However, if the chemicals are not correctly used, weed competition will not be reduced, coffee plants may become toxic (see discussion below), and health and environmental risks, such as soil contamination and natural selection of tolerant or resistant weed species can result (Ronchi et al. 2001). In addition, all situations that allow soil exposure, such as low plant densities, pruned crops, or early stages of adult coffee crops, benefit from herbicide application. Herbicides can be used in nurseries and in young or adult coffee plantations, either in the coffee rows or inter-rows. Herbicide mixture is an important common practice for increasing the spectrum of weed control in coffee plantations. Compared to the other crops, there are few effective herbicides formulations available for coffee.

Selective pre-emergent (for example, oxyfluorfen, alachlor, ametryn) or post-emergent (for example, Acetyl-CoA Carboxylase (ACCase) inhibitors, flumioxazin) herbicides are commonly used in young coffee plantations to establish weed control strips in the coffee rows (Rodrigues and Almeida 2011, Ronchi et al. 2014). These herbicides must be applied in a jet directed at the soil or weed seedlings, respectively, to prevent coffee poisoning (oxyfluorfen, for example, is not completely selective against Arabica—Magalhães et al. 2012—or Robusta—Yamashita et al. 2013—coffee) and to overcome umbrella effects of taller coffee plants. Application doses of these herbicides should be determined based mainly on the soil chemical characteristics for pre-emergent herbicides, and on the weed species and their development stage for post-emergent herbicides. Oxyfluorfen and ACCase inhibitors are also used in nurseries. Non-selective post-emergent herbicide (for example, glyphosate) application can also be recommended to control weeds in young coffee rows. However, they must be carefully applied to prevent coffee poisoning (see discussion below). If they are used when the weed vegetation is not too tall, lower doses might be applied, reducing the risk of coffee poisoning due to spray drift (Silva et al. 2014).

In adult coffee plants herbicides are mainly used in the inter-rows, but applications under the coffee plant canopy in the rows can be necessary (for example, for controlling *Ipomoea* spp.). In the inter-rows, they have been frequently used during the rainy season to control weed vegetation in a narrow strip beyond the projected coffee plant canopy, since they effectively prevent weed competition. Whether the entire or partial inter-row vegetation is desiccated, weed residues are retained on the soil surface as mulch, which contribute to soil and water conservation, nutrient cycling, and organic matter accumulation. In Brazil, pre-harvesting desiccation of the entire inter-row is necessary to allow high harvesting efficiency, particularly for gathering operations to recover fallen fruit from the soil surface (Ronchi et al. 2014). The commonly used post-emergent herbicides are carfentrazone-ethyl, 2,4-D, diquat, glufosinate-ammonium, metsulfuron methyl, paraquat, paraquat + diuron, saflufenacil, and sulfosate, while the commonly used pre-emergent ones are diuron, indaziflan, metribuzin, napropamide, oxyfluorfen, simazine, pendimentalin, and sulfentrazone (Aguilar et al. 2003, Sanches and Gamboa 2004, Gómez 2005, Rodrigues and Almeida 2011, Ronchi et al. 2014). Post-emergent herbicide applications likely cause lower impact on soil attributes compared to the pre-emergent ones (Alcântara and Ferreira 2000, Alcântara et al. 2007, 2009, Araujo-Junior et al. 2011b).

Herbicide Resistance

Glyphosate is the most used herbicide in coffee growing areas worldwide. It is used once during field preparation then several other times each year following crop establishment, as spray directed at the weeds in either young or adult coffee crops (Ronchi et al. 2014, 2017, Silva et al. 2015). Because of repeated applications in the same coffee area each year and over the years (considering coffee is perennial, and crop rotations are not possible), weed populations that have glyphosate tolerance, such as Benghal dayflower (*Commelina benhalensis* and *C. diffusa* - Santos et al. 2002) and Broadleaf buttonweed (*Spermacoce latifolia*) have spread throughout coffee crops, particularly in Brazil. Moreover, some weed biotypes of Hairy fleabane (*Conyza* spp.) and Sourgrass (*Digitaria insularis*) have shown high levels of glyphosate resistance (Christoffoleti et al. 2008, Carvalho et al. 2011). Therefore, the use of chemical weed control in coffee plantations has reduced effectiveness in these cases.

In order to control both the tolerant and resistant weed biotypes, as well as to prevent new biotype selection, integration of different weed control practices must be used. In the case of chemical control, herbicide associations are highly recommended (Silva and Silva 2013, Silva et al. 2015). One example is the use of 2,4-D associated to glyphosate for *Commelina* spp. and *Conyza* spp. control and for a broad spectrum of weed species, especially the broadleaf ones (Santos et al. 2002). Other herbicide mixtures that can be used efficiently to control Benghal dayflower species, under sequential applications, are paraquat+diuron and carfentrazone-ethyl+glyphosate (Ronchi et al. 2002), even though the use of paraquat have been strongly regulated worldwide. For controlling grass-resistant weed, post-emergence applications of ACCase inhibitors are recommended (Correia et al. 2015). However, caution must be exercised, because such grass-controlling herbicides show antagonism when they are associated with glyphosate (Barroso et al. 2014), and because indiscriminate use of 2,4-D affects the coffee plants (Ronchi et al. 2001, Ronchi et al. 2005).

Integrated Weed Control

Integrated weed management practices, which consider all plant growth factors, allow efficient utilization of environmental resources by the crops, provide increased yields and higher quality products (for example, no defensive residues in foods). Moreover, IWM aims to ensure production under environmental, economic, and social sustainability, increasing or at least maintaining the crop yields. For these reasons, IWM practices are increasingly used in many crops (Bajwa 2014). According to Silva and Silva (2013), if the IWM practices are precisely established, the negative effects of weeds on coffee plants (for example, crop yield reduction) and the environmental impacts of such practices are both diminished, in addition to a reduction in the production costs (for example, due to less use of herbicides and labor).

Integrated weed management in coffee is based on the rational combination of different weed control practices discussed herein (for example, preventive, cultural, mechanical, and chemical). Thus, no weed control practice is used in isolation (Bajwa 2014). For example, chemical weed control is not used as a sole weed control practice in coffee plantations. On the contrary, IWM associates herbicide use with other weed control practices, especially cultural practices, to ensure better conditions for coffee growth and development. Moreover, an IWM program must consider several aspects that include: potential benefits and harmful effects of each weed species, coffee crop characteristics, climatic conditions, costs of weed control, impacts of each practice on the environment, coffee crop and human health, labor supply, the availability of farm implements and machinery, critical periods of coffee-weed competition, and the expected effectiveness of the practice on controlling weeds. Several excellent examples of IWM in coffee have been reported by Sarno et al. (2004), Gómez (2005), Silva et al. (2006), Araújo et al. (2012), and Aguilar et al. (2013).

Evaluation of Sustainable Weed Control in Coffee

In agreement with the concept of a sustainable weed management in conservation agriculture presented by Bajwa (2014), a holistic analysis of all the scientific work cited herein clearly shows that the sustainability of weed management practices in coffee plantations has improved in the last decades worldwide. Coffee farmers, researchers, and technicians have changed from simply targeting to eradicate the weeds per se to the concept of sustainable and integrated weed management. Complete eradication of natural weed vegetation from coffee farms is now rarely used or recommended. Quite the opposite, special attention is paid to several advantages of maintaining and managing natural weed vegetation in coffee farms (discussed above), whether or not the crop is on a steep slope of a hilly area or on a gently sloping area. Thus, weed vegetation is eliminated only from areas where, and when it is necessary, reducing soil disturbance and erosion as well as costs of weed control. Moreover, the introduction of cover plants in coffee inter-rows has become a successful and common weed control practice in many coffee farms worldwide. Selective weed control has also been applied in some areas based on the ecological characteristics of each weed species, reducing their harmful effects on coffee plants while preserving their ground cover ability. Most likely, these practices have resulted in reduced labour and energy inputs in coffee plantations.

Although the sustainability of weed control has increased, coffee growers are challenged to choose the most suitable weed control practices for each field situation. Moreover, the growers must know how to apply each practice correctly to ensure satisfactory weed control without causing harmful effects on the environment or damaging crops. For example, although mechanical weed control methods can create environmental problems (soil compaction, hydric erosion, organic matter loss) and damage the coffee plants (young coffee plant stem and superficial root system) if they are not properly used, chemical control can also damage coffee plants. There are two major reasons to explain this. Inappropriate technologies for herbicide application can lead to herbicide spray drift, which affect the coffee plant, as shown in previous studies (França et al. 2013, Carvalho et al. 2014). Glyphosate drift occurring during the coffee fruit development stage can result in accumulation of herbicide residues in the harvested fruits, hampering coffee commercialization. Such cases of contamination have recently been detected in coffee beans from farms where post-emergence glyphosate was applied.

Secondly, coffee plants may also be affected by residues of herbicides that persist in soil for a long time after application. In addition to coffee plants and soil, water may also be contaminated. A small fraction of applied herbicide is absorbed by the weed vegetation, with a larger amount of the herbicide molecules being retained in soil. In the soil, herbicides undergo different processes, including retention, degradation, and transport that define their persistence level in soil. Thus, before using herbicides in coffee plantations, it is highly recommended to know the physical-chemical properties of both the soil and herbicides as well as their interactions with the environment in each site (Silva and Silva 2013). This is particularly important where herbicide associations are commonly used to broaden the weed control spectrum, since the interaction of herbicide molecules may affect soil processes of sorption, leaching, and persistence. Finally, understanding the interaction between the herbicide and the environment is vital when greater soil persistence herbicides (for example, indaziflan, oxyfluorfen, metsulfuron methyl, and ametryn) are used in coffee plantations.

Weed control practices are directly related to the economic sustainability of coffee production. Among the several factors that affect coffee crop profitability, the coffee yield (measured as 60 kg bags of green coffee per hectare) is of major importance, since profitability tends to increase with coffee yield. Similarly, reductions in production costs tend to increase the profits. Therefore, when weeds (and weed control) are considered in the production system they may decrease crop profitability by affecting both the coffee yield due the resource competition that impairs plant growth and the production costs because the coffee grower must spend money to control weeds. Thus, it can be deduced that the practices applied to control weeds in coffee must be properly balanced to reduce costs of weed control and to efficiently prevent weed competition to ensure high crop yield (Ronchi et al. 2014). For example, in a typical Brazilian Arabica coffee farm, where all the management practices, including weed control are mechanized, the costs of weed control is about 5% of the total input costs per year (Ronchi et al. 2014). Such weed control costs comprise costs of herbicides, diesel oil, labor, and equipment (springers, mowers, etc.) maintenance. Although weed control in coffee costs less, it should never be neglected otherwise the losses in crop yield due to weed competition will undoubtedly reduce the profitability of the crop (Ronchi et al. 2014).

Concluding Remarks

As discussed above, coffee growth and production may be strongly affected by weed competition. Integrating several weed control practices to control weeds is more effective than using a sole weed control method. Although there is less information about weed control in coffee compared to other crops, substantial advances have been made in the last decades to understand long-term effects of weed control practices on both the environment and the coffee crop worldwide. For instance, studies have shown that weed control in the entire coffee farm area is unnecessary, and that the beneficial effects of each weed species must also be considered. Therefore, selective weed control is crucial in some field situations.

It must be emphasized that the best weed control practices vary with local characteristics of each site or region of the world where coffee is grown. Generally, adequate weed control is achieved by eliminating weed species only within a strip at both sides of coffee rows, especially in the weed-sensitive younger crops. Weed vegetation in the inter-rows may be managed by a jet directed application of desiccant non-selective herbicides, mower operations, or by cover crops. Preventive and cultural methods are also of major importance, and must be based on the ecological characteristics of weeds and the crop. Finally, in a sustainable coffee production system the adoption of integrated weed management practices is, undoubtedly, key for reducing weed-coffee competition, weed control costs, risks of environmental and crop contamination, and cases of herbicide resistance, and to potentiate the beneficial characteristics of weed species, soil conservation, and coffee yield over years.

REFERENCES

Aguilar, V., C. Staver and P. Milberg. 2003. Weed vegetation response to chemical and manual selective ground cover management in a shaded coffee plantation. Weed Res. 43: 68–75.

Alcântara, E.N., J.C.A. Nóbrega and M.M. Ferreira. 2007. Métodos de controle de plantas invasoras na cultura do cafeeiro (*Coffea arabica* L.) e componentes da acidez do solo. Rev. Bras. Cienc. Solo 31: 1525–1533.

Alcântara, E.N. and M.M. Ferreira. 2000a. Efeito de diferentes métodos de controle de plantas daninhas sobre a produção de cafeeiros instalados em Latossolo Roxo distrófico. Ciênc. Agrotecnol. 24: 54–61.

Alcântara, E.N. and M.M. Ferreira. 2000b. Efeitos de métodos de controle de plantas daninhas na cultura do cafeeiro (*Coffea arabica* L.) sobre a qualidade física do solo. Rev. Bras. Cienc. Solo 24: 711–721.

Alcântara, E.N. and R.A. Silva. 2010. Manejo do mato em cafezais. pp. 519–572. *In:* P.R. Reis and R.L. Cunha (Eds.) Café Arábica do Plantio à Colheita. UR Epamig SM, Lavras, Brasil.

Alcântara, E.N., J.C.A. Nóbrega and M.M. Ferreira. 2009. Métodos de controle de plantas daninhas no cafeeiro afetam os atributos químicos do solo. Ciênc. Rural. 39: 749–757.

Araújo, F.C., C.P. Ronchi, W.L. Almeida, M.A.A. Silva, C.E.O. Magalhães and P.I.V. Good-God. 2012. Optimizing the width of strip weeding in arabica coffee in relation to crop age. Planta Daninha 30: 129–138.

Araujo-Junior, C.F., M.S. Dias Junior, P.T.G. Guimarães and E.N. Alcântara. 2011a. Sistema poroso e capacidade de retenção de água em Latossolo submetido a diferentes manejos de plantas invasoras em uma lavoura cafeeira. Planta Daninha 29: 499–513.

Araujo-Junior, C.F., P.T.G. Guimarães, M.S.D.E.N. Alcântara Junior and A.D.R. Mendes. 2011b. Alterações nos atributos químicos de um Latossolo pelo manejo de plantas invasoras em cafeeiros. Rev. Bras. Cienc. Solo 35: 2207–2217.

Bajwa, A.A. 2014. Sustainable weed management in conservation agriculture. Crop Prot. 65: 105–113.

Barroso, A.A.M., A.J.P. Albrecht, F.C. Reis and R.V. Filho. 2014. Interação entre herbicidas inibidores da ACCase e diferentes formulações de glyphosate no controle de capim-amargoso. Planta Daninha 32: 619–627.

Borba, M.A.P., T.O. Tavares, B.R. Oliveira, A.F. Santos and R.P. Silva. 2016. Perdas na varrição e recolhimento mecanizado do café em quatro manejos do solo. *In:* Proceedings of 45th Congresso Brasileiro de Engenharia Agrícola, Florianópolis, Brasil. SBEA.

Carvalho, A.J., M.J.B. Andrade and R.J. Guimarães. 2007. Sistemas de produção de feijão intercalado com cafeeiro adensado recém-plantado. Ciênc. Agrotecnol. 31: 133–139.

Carvalho, F.P., B.P. Souza, A.C. França, E.A. Ferreira, M.H.R. Franco, M.C.M. Kasuya and F.A. Ferreira. 2014. Glyphosate drift affects arbuscular mycorrhizal association in coffee. Planta Daninha 32: 783–789.

Carvalho, L.B., H. Cruz-Hipolito, F. Gonzalez-Torralva, P.L.C.A. Alves, P.J. Christoffoleti and R. Prado. 2011. Detection of Sourgrass (*Digitaria insularis*) biotypes resistant to glyphosate in Brazil. Weed Sci. 59: 171–177.

Carvalho, L.B., P.L.C.A. Alves and S. Bianco. 2013. Sourgrass densities affecting the initial growth and macronutrient content of coffee plants. Planta Daninha 31: 109–115.

Christoffoleti, P.J., A.J.B. Gali, S.J.P. Carvalho, M.S. Moreira, M. Nicolai, L.L. Foloni, B.A.B. Martins and D.N. Ribeiro. 2008. Glyphosate sustainability in South American cropping systems. Pest Manage. Sci. 64: 422–427.

Cogo, F.D., C.F. Araujo-Junior, Y.L. Zinn, M.S. Dias Junior, E.N. Alcântara and P.T.G. Guimarães. 2013. Estoques de carbono orgânico do solo em cafezais sob diferentes sistemas de controle de plantas invasoras. Semina: Cienc. Agrar. 34: 1089–1098.

Concenço, G., I.S. Motta, I.V.T. Correia, S.A. Santos, A. Mariani, R.F. Marques, W.G. Palharini and M.E.S. Alves. 2014. Infestation of weed species in monocrop coffee or intercropped with banana, under agroecological system. Planta Daninha 32: 665–674.

Correia, N.M., L.T. Acra and G. Balieiro. 2015. Chemical control of different *Digitaria insularis* populations and management of a glyphosate-resistant population. Planta Daninha 33: 93–101.

DaMatta, F.M., C.P. Ronchi, M. Maestri and R.S. Barros. 2010. Coffee: environment and crop physiology. pp. 181–216. *In:* F.M. DaMatta (Ed.) Ecophysiology of tropical tree crops. Nova Science Publishers, New York, USA.

Dias, G.F.S.D., P.L.C.A. Alves and T.C.S. Dias. 2004. *Urochloa decumbens* suppresses the initial growth of *Coffea arabica*. Sci. Agric. 61: 579–583.

Dias, T.C.S., P.L.C. Alves and L.N. Lemes. 2008. Faixas de controle de plantas daninhas e seus reflexos na produção do cafeeiro. Científica 36: 81–85.

Ferreira, E.A., A.C. França, R.F. Carvalho, J.B. Santos, D.V. Silva and E.A. Santos. 2011. Avaliação fitossociológica da comunidade infestante em áreas de transição para o café orgânico. Planta Daninha. 29: 565–576.

Fialho, C.M.T., A.C. França, S.P. Tironi, C.P. Ronchi and A.A. Silva. 2011. Interferência de plantas daninhas sobre o crescimento inicial de *Coffea arabica*. Planta Daninha 29: 137–147.

Fialho, C.M.T., G.R. Silva, M.A.M. Freitas, A.C. França, C.A.D. Melo and A.A. Silva. 2010. Competição de plantas daninhas com a cultura do café em duas épocas de infestação. Planta Daninha 28: 969–978.

França, A.C., F.P. Carvalho, C.M.T. Fialho, L. D'Antonino, A.A. Silva, J.B. Santos and L.R. Ferreira. 2013. Deriva simulada do glyphosate em cultivares de café Acaiá e Catucaí. Planta Daninha 31: 442–451.

Franco, C.M. 1958. Influence of temperature on growth of coffee plant. Bulletin No. 16. IBEC Research Institute, New York, USA.

Gómez, R. 2005. Efecto del control de malezas con paraquat y glifosato sobre La erosión y perdida de nutrimentos del suelo en cafeto. Agronomía Mesoamericana. 16: 55–87.

Kissmann, K.G. 1997. Plantas infestantes e nocivas. 2 ed. BASF Brasileira, São Paulo, Brasil.

Lemes, L.N., L.B. Carvalho, M.C. Souza and P.L.C.A. Alves. 2010. Weed interference on coffee fruit production during a four-year investigation after planting. Afr. J. Agric. Res. 5: 1128–1143.

Lorenzi, H.J. 2000. Plantas daninhas do Brasil. 3. ed. Inst. Plantarum, São Paulo, São Paulo.

Maciel, C.D.G., J. Poletine, A.M. Neto, N. Guerra and W. Justiniano. 2010. Levantamento fitossociológico de plantas daninhas em cafezal orgânico. Bragantia 69: 631–636.

Magalhães, C.E.O., C.P. Ronchi, R.A.A. Ruas, M.A.A. Silva, F.C. Araújo and W.L. Almeida. 2012. Seletividade e controle de plantas daninhas com oxyfluorfen e sulfentrazone na implantação de lavoura de café. Planta Daninha 30: 607–616.

Marcolini, L.W., P.L.C.A. Alves, T.C.S. Dias and M.C. Parreira. 2009. Effect of the density and of the distance of *Brachiaria decumbens* staff on the initial growth of *Coffea arabica* L. seedligns. Coffee Sci. 4: 11–15.

Martins, B.H., C.F. Araujo-Junior, M. Miyazawa and K.M. Vieira. 2016a. Humic substances and its distribution in coffee crop under cover crops and weed control methods. Sci. Agric. 73: 371–378.

Martins, B.H., C.F. Araujo-Junior, M. Miyazawa, K.M. Vieira, C.A. Hamanaka and A.S. Silva. 2016b. Weed control methods and coffee shrub residue effects on carbon stocks in a Latosol under conservation management practices. Agron. Sci. Biotechnol. 2: 66–78.

Martins, B.H., C.F. Araujo-Junior, M. Miyazawa, K.M. Vieira and D.M.B.P. Milori. 2015. Soil organic matter quality and weed diversity in coffee plantation area submitted to weed control and cover crops management. Soil Tillage Res. 153: 169–174.

Matiello, J.B. and F. Santinato. 2016. Corda-de-viola avança nos cafezais. Folha Técnica número 318. Fundação Procafé, Varginha, Brasil.

Matiello, J.B., R. Santinato, S.R. Almeida and A.L. Garcia. 2016. Cultura de café no Brasil: manual de recomendações, ed. 2015. Futurama Editora, São Paulo, Brasil.

Matos, C.C., C.M.T. Fialho, E.A. Ferreira, D.V. Silva, A.A. Silva, J.B. Santos, A.C. França and L. Galon. 2013. Características fisiológicas do cafeeiro em competição com plantas daninhas. Biosci. J. 29: 1111–1119.

Minassa, E.M.C. 2014. Efeito alelopático da palha de café (*Coffea canephora* L. e *Coffea arabica* L.) sobre plantas cultivadas e espontâneas. D.Sc. Thesis, Universidade Estadual do Norte Fluminense Darcy Ribeiro, Campos dos Goytacazes, Rio de Janeiro.

Miranda, G.R.B., R.J. Guimarães, V.P. Campos, E.P. Botrel, G.R.R. Almeida and R.G. Gonzalez. 2007. Métodos alternativos de desinfestação de plantas invasoras em substratos para formação de mudas de cafeeiro (*Coffea arabica* L.). Coffee Sci. 2: 168-174.

Mohler, C.L. 1996. Ecological bases for the cultural control of annual weeds. J. Prod. Agric. 9: 468-474.

Moraima Garcia, S., A. Canizares, F. Salcedo and L. Guillen. 2000. A contribution to determine critical levels of weed interference in coffee crops of Monagas state, Venezuela. Bioagro 12: 63–70.

Moreira, G.M., R.M. Oliveira, T.P. Barrella, A. Fontanétti, R.H.S. Santos and F.A. Ferreira. 2013. Fitossociologia de plantas daninhas do cafezal consorciado com leguminosas. Planta Daninha 31: 329–340.

Pais, P.S.M., M.S. Dias Junior, A.C. Dias, P. Iori, P.T.G. Guimarães and G.A. Santos. 2013. Load-bearing capacity of a Red-Yellow Latosol cultivated with coffee plants subjected to different weed managements. Cienc. Agrotecnol. 37: 145–151.

Pais, P.S.M., M.S.D. Junior, G.A. Santos, A.C. Dias, P.T.G. Guimarães and E.N. Alcântara. 2011. Compactação causada pelo manejo de plantas invasoras em Latossolo Vermelho-Amarelo cultivado com cafeeiros. Rev. Bras. Cienc. Solo 35: 1949–1957.

Pandey, A., C.R. Soccol, P. Nigam, D. Brand, R. Mohan and S. Roussos. 2000. Biotechnological potential of coffee pulp and coffee husk for bioprocesses. Biochem. Eng. J. 6: 153–162.

Partelli, F.L., H.D. Vieira, S.P. Freitas and J.A.A. Espindola. 2010. Aspectos fitossociológicos e manejo de plantas espontâneas utilizando espécies de cobertura em cafeeiro Conilon orgânico. Semina: Cienc. Agrar. 31: 605–618.

Paulo, E.M., R.S. Berton, J.C. Cavichioli and F.S. Kasai. 2004. Comportamento do cafeeiro Apoatã em consórcio com culturas anuais. Bragantia 63: 275–281.

Pazuch, D., M.M. Trezzi, F. Diesel, M.V.J. Barancelli, S.C. Batistel and R. Pasini. 2015. Superação de dormência em sementes de três espécies de *Ipomoea*. Cienc. Rural 45: 192–199.

Pedrosa, A.W. 2013. Eficiência da adubação nitrogenada no consórcio entre cafeeiro e *Brachiaria brizantha*. D.Sc. Thesis, Universidade de São Paulo, Piracicaba, São Paulo.

Pires, L.F., C.F. Araujo-Junior, A.C. Auler, N.M.P. Dias, M.S. Dias Junior and E.N. Alcântara. 2017a. Soil physico-hydrical properties changes induced by weed control methods in coffee plantation. Agric. Ecosyst. Environ. 246: 261–268.

Pires, L.F., C.F. Araujo-junior, N.M.P Dias, M.S. Dias Junior and E.N. Alcântara. 2017b. Weed control methods effect on the hydraulic atributes of a Latosol. Acta Sci. Agron 39: 119–128.

Radosevich, S.R., J. Holt and C. Ghersa. 1996. Physiological aspects of competition. pp. 217–301. *In:* Radosevich, S.R., J. Holt and C. Ghersa (Ed.) Weed ecology: Implication for Managements. John Wiley and Sons, Inc. New York, EUA.

Ramírez, J.E. 2009. Hacia la caficultura sostenible. C.R. ICAFE, San José, Costa Rica.

Ricci, M.S.F., E.M.V. Filho and J.R. Costa. 2008. Diversidade da comunidade de plantas invasoras em sistemas agroflorestais com café em Turrialba, Costa Rica. Pesqui. Agropecu. Bras. 43: 825–834.

Roberts, H.A. 1981. Seed banks in the soil. vol. 6. Advances in Applied Biology, Academic Press, Cambridge, UK.

Rodrigues, B.N. and F.S. Almeida. 2011. Guia de herbicidas. 6. ed. Edição dos Autores, Londrina, Brasil.

Ronchi, C.P., F.P. Carvalho and A.A. Silva. 2016. Manejo integrado de plantas daninhas. pp. 382–397. *In:* Ferrão, M.A., A.F.A. Fonseca, M.A. Ferrão and L.H. De Muner. Café conilon: 2^n edicão atualizada e ampliada. incaper, Espírito Santo, Brasil.

Ronchi, C.P. and A.A. Silva. 2003. Tolerância de mudas de café a herbicidas aplicados em pós emergência. Planta Daninha 21: 421–426.

Ronchi, C.P., A.A. Silva and L.R. Ferreira. 2001. Manejo de plantas daninhas em lavouras de café. Suprema Gráfica e Editora, Viçosa, Brasil.

Ronchi, C.P., A.A. Silva and L.R. Ferreira. 2001. Manejo de plantas daninhas em lavouras de café. UFV, Viçosa, Brasil.

Ronchi, C.P., A.A. Silva, A.A. Terra, G.V. Miranda and L.R. Ferreira. 2005. Effect of 2,4-dichlorophenoxyacetic acid applied as a herbicide on fruit shedding and coffee yield. Weed Res. 45: 41–47.

Ronchi, C.P., A.A. Silva, L.R. Ferreira, G.V. Miranda and A.A Terra. 2002. Mistura de herbicidas para o controle de plantas daninhas do gênero *Commelina*. Planta Daninha 20: 311–318.

Ronchi, C.P. and A.A. Silva. 2006. Effects of weed species competition on the growth of young coffee plants. Planta Daninha 24: 415–423.

Ronchi, C.P., A.A. Terra and A.A. Silva. 2007. Growth and nutrient concentration in coffee root system under weed species competition. Planta Daninha 25: 679–687.

Ronchi, C.P., A.A. Terra, A.A. Silva and L.R. Ferreira. 2003. Acúmulo de nutrientes pelo cafeeiro sob interferência de plantas daninhas. Planta Daninha 21: 219–227.

Ronchi, C.P., J.M. Souza Junior, W.L. Almeida, D.S. Souza, N.O. Silva, L.B. Oliveira, A.M.N.M. Guerra and P.A. Ferreira. 2015. Morfologia radicular de cultivares de café arábica submetidas a diferentes arranjos espaciais. Pesqui. Agropecu. Bras. 50: 187–195.

Ronchi, C.P., R.T. Ferreira and M.A.A. Silva. 2014. Manejo de plantas daninhas na cultura do café. pp. 132–154. *In:* Monquero, P.A. (Ed.) Manejo de plantas daninhas nas culturas agrícolas. RiMa Editora, São Carlos, Brasil.

Sánchez F.L. and E. Gamboa. 2004. Control de malezas con herbicidas y métodos mecánicos en plantaciones jóvenes de café. Bioagro 16: 1–4.

Sánchez, F.L.E. 1991. Lãs malezas em los cafetales: recomendaciones prácticas para su control. Fonaiap Divulga 9: 18–27.

Santos, I.C., F.A. Ferreira, A.A. Silva, G.V. Miranda and L.D.T. Santos. 2002. Eficiência do 2,4-D aplicado isoladamente e em mistura com glyphosate no controle da trapoeraba. Planta Daninha 20: 299–309.

Sarno, J. Lumbanraja, T. Adachi, Y. Oki, M. Senge and A. Watanabe. 2004. Effect of weed management in coffee plantation on soil chemical properties. Nutr. Cycling Agroecosyst. 69: 1–4.

Shivaprasad, P., I.B. Biradar, S.R. Salakinkop, Y. Raghuramulu, M.V. D'souza, N. Hariyappa, S.B. Hareesh, M.A. Murthy and Jayarama. 2005. Influence of soil cultivation methods in young coffee on soil moisture, weed suppression and organic matter. J. Coffee Res. 33: 1–14.

Silva, A.A., A.C. França, C.P. Ronchi and F.P. Carvalho. 2015. Manejo Integrado de Plantas Daninhas. pp. 104–128. *In:* Sakiyama, N., H. Martinez, M. Tomaz and A. Borém (Eds.) Café arábica: do plantio à colheita. UFV, Viçosa, Brasil.

Silva, A.A. and C.P. Ronchi. 2008. Manejo e controle de plantas daninhas em café. pp. 417–475. *In:* Vargas, L. and E.S. Roman (Eds.) Manual de manejo e controle de plantas daninhas. Embrapa Trigo, Passo Fundo, Brasil.

Silva, A.A. and J.F. Silva. 2013. Tópicos em manejo de plantas daninhas. UFV, Viçosa, Brasil.

Silva, S.O., S.N. Matsumoto, F.V. Bebé and A.R.S. José. 2006. Diversidade e frequência de plantas daninhas em associações entre cafeeiros e grevíleas. Coffee Sci. 1: 126–134.

Siqueira, R.H.S., M.M. Ferreira, E.N. Alcântara and R.C.S. Carvalho. 2014a. Agregação de um Latossolo Vermelho-Amarelo submetido a métodos de controle de plantas invasoras na cultura do café. Rev. Bras. Cienc. Solo. 38: 1128–1134.

Siqueira, S.H.R., M.M. Ferreira, E.N. Alcântara, B.M. Silva and R.C. Silva. 2014b. Water retention and S index of an Oxisol subjected to weed control methods in a coffee crop. Cienc. Agrotecnol. 38: 471–479.

Soto-Pinto, L., I. Perfecto and J. Caballero-Nieto. 2002. Shade over coffee: its effects on berry borer, leaf rust and spontaneous herbs in Chiapas, México. Agroforestry Syst. 55: 37–45.

Souza, L.S., P.H.L. Losasso, M. Oshiiwa, R.R. Garcia and L.A. Goes Filho. 2006. Efeitos das faixas de controle do capim-braquiária (*Brachiaria decumbens*) no desenvolvimento inicial e na produtividade do cafeeiro. (*Coffea arabica*). Planta Daninha 24: 715–720.

Yamane, K., M. Kono, T. Fukunaga, K. Iwai, R. Sekine, Y. Watanabe and M. Iijima. 2014. Field evaluation of coffee grounds application for crop growth enhancement, weed control, and soil improvement. Plant Prod. Sci. 17: 93–102.

Yamashita, O.M., J.V.N. Orsi, D.D. Resende, F.S. Mendonça, O.R. Campos, J.A. Massaroto and M.A.C. Carvalho. 2013. Deriva simulada de herbicidas em mudas de *Coffea canephora*. Sci. Agrar. Parana. 12: 148–156.

Advances in Weed Management in Tea

Probir Kumar Pal*, Sanatsujat Singh and Rakesh Kumar Sud

[1] Division of Agrotechnology of Medicinal, Aromatic and Commercially Important Plants, Council of Scientific and Industrial Research-Institute of Himalayan Bioresource Technology (CSIR-IHBT), Post Box No. 6, Palampur, Himachal Pradesh – 176061, HP, India

Introduction

Tea (*Camellia* spp.), a perennial woody evergreen plantation crop of Theaceae family and native to southern China, is commonly grown for non-alcoholic beverages. The commercial cultivation of tea is restricted only to certain regions of the World due to requirements of specific agro-climatic conditions, and it is very sensitive to the changes in growing conditions. Light and friable loam with porous sub-soil which permits a free percolation of water is best for tea cultivation. However, this plant prefers slightly acidic soil. The soil should be rich in humus. The temperature range between 21 °C to 29 °C is ideal for the production of tea. The high annual rainfall (150–250 cm) with even distribution is required for tea cultivation (Bose 2013). The ideal growing conditions for tea are only available in tropical and subtropical climates; however, some varieties can tolerate marine climates of British mainland and Washington area of the Unites States (FAO 2015). Thus, China, India, Sri Lanka are the major tea producing countries in the World. Among the plantation crops, tea is considered the most important crop as foreign exchange earner in Kenya, Sri Lanka, India and China. In India, tea is the second largest foreign exchange earner (Karmakar and Banerjee 2005). World tea production increased radically by 6 per cent in 2013, and this growth was owing to major increases in the key tea producing countries particularly China, India, Sri Lanka, Kenya and Indonesia (FAO 2015). Though China is the largest tea producing country, Kenya is the largest tea exporting country in the World market (FAO 2015). In India, the area under tea production was around 563.98 thousand hectares by the end of 2013. The maximum area is covered in Assam (304.40 thousand hectares) followed by West Bengal (140.44 thousand hectares), Tamil Nadu (69.62 thousand hectares) and Kerala (35.01 thousand hectares), respectively (IBEF 2017). During the financial year 2015–16, India recorded a total tea production of 1,233.14 million kg, which is the highest ever recorded so far (IBEF 2017).

The production as well as consumption of tea is increasing steadily. Nevertheless, the productivity and quality of tea is largely affected by the problematic weeds. In tea crop, weeds not only compete with resources but also hinder the plucking of buds and leaves. Moreover, weeds increase insect pests and disease invasion through creating micro-climate. The main reasons are plant architecture, ground exposure due to wider plant spacing and deep skiffing at regular intervals, heavy rainfall, and regular application of organic manure.

*Corresponding author: pkpal_agat@yahoo.in/palpk@ihbt.res.in

Weed Impact

Major Weeds and their Life Cycle

There are about 30,000 species of weeds in the world. Out of these about 18,000 are known as harmful weeds (Rodgers 1974) in the agricultural and non-agricultural systems. Weed problems in tea plantation differ very much from those in field crops and other plantation crops. This problem varies from one region to another and one farm to another. Moreover, each crop has a definite association with weed species. The major weeds associated with tea plantation all over the World are given in Table 23.1. However, about 130 common weeds in tea plantation have been reported by Dutta (1977). According to the duration of their life cycle, weeds are classified into three broad categories, that is, annual, biennial and perennial. There are both broadleaf weeds and grasses in each group. Annual weeds complete their life cycle within a year, and generally spread through seeds. Though the annual weeds are considered easy to control, they are very persistence due to production of large quantity of seeds in a short period. On the other hand, perennial weeds are very aggressive and competitive.

The majority of weeds in tea plantations are grasses followed by broad-leaf weeds (Singh et al. 1994). There is no serious problem with sedges in tea plantation (Singh et al. 1994). However, some ferns (Table 23.1) are also found in tea plantation. Since tea is grown under high humidity and limited sun light conditions, mosses tend to cover soil surface under the canopy and a large part of tea trunk and branches (Ronoprawiro 1976, 1981).

Beneficial Effect of Weeds inTea Plantation

Some beneficial effects of weeds in tea plantation are reported. In high rainfall areas, some non-competitive soft weeds reduce the risk of soil erosion due to run-off water (Manipura 1971; Sutidjo and Lubis 1971). Such type of weeds should not be removed from the field during selective weeding. Species, such as *Borreria latifolia* and *B. ocymoides* are less competitive at the early stage of growth; thus these species could be left in tea plantation for 6–8 weeks for covering the ground, which eventually suppress the growth of other weeds (Prematilake 1997). Mulch of grassy weeds also effectively controls weeds in the inter-row spaces of tea plantation under sloppy conditions. It had been reported that dry weight of tea root increased when treated with root-powder of *A. conyzoides* in combination with rock phosphate (Deori et al. 1997).

Impact on Yield Loss

Weeds generally compete with tea plants for nutrients, water, and sunlight. The competition between weed and crop plant is the most limiting factor in crop production. Intensity of this competition depends upon agro-climatic conditions, type of weed species, severity and duration of weed infestation, and competing capability of crop species. There are no ultimate statistics available regarding economics of weed control in tea; however, certain reports indicate that depending upon the intensity of infestation, weeds are estimated to cause yield losses between 15% and 40% (Rao 2000). It has been reported that uncontrolled weeds in tea plantations cause a loss of productivity to the extent of 50%–70% (Deka and Barua 2015, Hasselo and Sandanam 1965). Weeds also create favourable conditions in tea plantations for diseases through increasing the humidity around tea bushes (Hasselo and Sandanam 1965). In tea plantations, the period of active weed growth coincides with the period of active vegetative growth, which influences the deployment of labour for plucking.

Weed management in tea is the second costliest input after plucking (Sinha 1985). Weeds infestation in tea nursery is more severe because of favourable environmental conditions for plant growth. Thus, labour input cost is higher for nursery success. Weeds also grow abundantly from the time of plantation until the tea canopy covers the inter-row spaces sufficiently (Somaratne 1988). The critical period of weed competition with young tea is 8–16 weeks after planting, and tea growth is adversely affected when the weed infestation prevails for more than 12 weeks (Prematilake et al. 1999). Deka and Barua (2015) had reported that adverse effect

Table 23.1. Major weeds associated with tea plantation

S. No.	Major weeds	Category	Life cycle
1.	*Achyranthes aspera* L.	Broad leaf	Perennial
2.	*Ageratum conyzoides* L.	Broad leaf	Annual
3.	*Ageratum haustonianum* Milll.	Broad leaf	Annual
4.	*Artemisia vulgaris* L.	Broad leaf	Perennial
5.	*Arundinella bengalensis* (Spreng.) Druce	Grass	Perennial
6.	*Axonopus compressus* (Sw.) P. Beauv.	Grass	Perennial
7.	*Bidens pilosa* L.	Broad leaf	Annual
8.	*Borreria alata* (Aubl.) DC.	Broad leaf	Annual
9.	*Borreria hispida* Spruce ex K. Schum.	Broad leaf	Annual/Perennial
10.	*Capsella bursa-pastoris* L. Medik.	Broadleaf	Annual
11.	*Cassia tora* (L.) Roxb.	Broadleaf	Annual
12.	*Chromolaena adenophorum*	Broad leaf	Perennial
13.	*Chrysopogon aciculatus* (Retz.) Trin.	Grass	Perennial
14.	*Commelina benghalensis* L.	Broad leaf	Perennial
15.	*Cynodon dactylon* (L.) Pers.	Grass	Perennial
16.	*Drymaria cordata* (L.) Willd.	Broad leaf	Annual
17.	*Echinochloa colona* (L.) Link.	Grass	Annual
18.	*Erechtites valerianifolia* (Link ex Wolf) Less. ex DC	Broad leaf	Annual
19.	*Erigeron Canadensis* (L.) Cronquist	Broad leaf	Annual
20.	*Eupatorium odoratum* L.	Broad leaf	Perennial
21.	*Eupatorium riparium* Regel	Broad leaf	Perennial
22.	*Fragaria vesca* L.	Broad leaf	Perennial
23.	*Gnaphalium indicum* L.	Broad leaf	Annual
24.	*Hackelia uncinata* (Benth.)	Broad leaf	Perennial
25.	*Imperata cylindrica* (L.) P. Beau.	Grass	Perennial
26.	*Mikania cordata* (Burm.f.) B.L. Rob.	Broad leaf	Perennial
27.	*Mikania micrantha* Kunth.	Broad leaf	Perennial
28.	*Nephrodium* spp. L.	Fern	Perennial
29.	*Oxalis acetosella* L.	Broad leaf	Annual/Perennial
30.	*Panicum repens* L.	Grass	Perennial
31.	*Paspalum conjugatum* (sour paspalum).	Grass	Perennial
32.	*Pennisetum clandestinum* Hochst. ex Chiov.	Grass	Perennial
33.	*Polygonum alatum* Buch.-Ham. ex D. Don.	Broad leaf	Perennial
34.	*Polygonum chinense* L.	Broad leaf	Perennial
35.	*Polygonum perfoliatum* L.	Broad leaf	Perennial
36.	*Pteridium aquilinum* (L.) Kuhn	Fern	Perennial
37.	*Rubus spp.* L	Broad leaf	Perennial
38.	*Scoparia dulcis* L.	Broad leaf	Annual/Perennial

of weeds on tea more severe up to two years from plantation. It has been reported that the *Convolvulus arvensis*, a long slender and prostrate stems of field bind weed, reduced the quality of product by changing the colour, taste and smell of tea (Asghari and Mahmodi 1996).

Height of stem, length and total number of primary branches of two-year tea plants are adversely affected with infestation of *P. conjugatum*, *E. riparium* and *A. haustonianum* (Soedarsan et al. 1976). These effects were attributed to the intense root system and high regeneration rate (Singh 2013). The retarding effects of *I. cylindrica*, *A. vulgaris* and *P. repens* on tea plant become

been reported that deep hoeing during June and December in the Northeast India flat terrain provided higher average annual yield, whereas weed control by a sickle round the year gave the lowest yield (Sarkar et al. 1983). However, deep hoeing and scraping lead to devastating soil erosion on hill slope of high intensity rainfall areas, and repeatedly damage the roots of tea placed in upper 15 cm of soil (HPKV 1986). Bursulaya et al. (1990) had also reported that rotary cultivar deeper than 3–5 cm is not suggested in tea field because of shallow root formation. However, it had been reported that forking out of *Imperata cylindrica* rhizomes up to a depth of 45 cm gave long-lasting control (Sandanam and Jayasinghe 1977).

Mulching

Mulch acts as an effective physical hurdle to weed emergence. In tea plantation both organic and synthetic mulches are used. Organic mulching and thatching are the common practices in tea plantation to smother the weeds growth and conserve the soil moisture, and it enhances soil fertility status by addition of organic matter (Manipura et al. 1969). Black polyethylene mulch is very effective for satisfactory weeds control in nursery stage (Smale 1991). It had also been reported that mulching with black plastic, tea factory residues or peat improved weed control both in seed-borne as well as vegetatively propagated nurseries (Tabagari and Kopaliani 1984). For young as well as mature tea plantation, black low density polyethylene (LDPE) sheet is more effective to control the weeds (Tabagari and Kopaliani 1984, Korzun 1981, Zarnadze 1972; Pirtskhalaishvili and Kimutsadza 1972). Likewise, Singh et al. (1993) reported that LDPE mulch totally suppressed weeds in the inter-row spaces in young China hybrid tea when planted on slope.

In mature tea, rubber mulch was found more effective to check the weed growth and increased tea yield by about 39%, whereas peat mulch was not so useful (Zarnadze 1972). In Sri Lanka, weeds population were found to be significantly checked by the mulching with *Flemingia congesta* @1 kg dry matter per square meter (Prematilake et al. 1998). Mulching with Guatemala grass @45 t ha^{-1} can be effective in reducing weed growth in tea plantation (Sanusi 1977). However, in case of pruned tea, pruning litter provided best results to control weed in inter-rows spaces (Kogua 1975, Tabagari et al. 1988).

Cultural Method

Cultural method of weed control is a part of good agronomic practices in farming system. Thus, some of the cultural methods could be adopted without incurring additional cost of weed management. The important agronomic practices, which are suggested for effective suppression of weeds in plantation, are infilling and inter-planting, bush management, cover-cropping, and selective weeding.

Infilling and Inter-planting

The ground coverage by the crop is the most vital factor, which affects the growth of weeds. In tea plantation, the ground exposure is higher during young tea phase, i.e., particularly during first four years and during the first year after pruning. Thus, weed infestation is higher in the tea plantation having wider spacing and higher mortality. Under this situation, infilling the vacant areas with tea clone is important to cover the ground thereby to make a minimum room for weeds. Furthermore, suitable grasses should be planted until such time the vacancies get filled with tea (Prematilake 2003).

Bush Management

To check the weed population particularly in young tea plantation, healthy and well-spread nursery plants should be planted in the field so that plant could grow quickly. Good Agricultural Practices (GAP) like balanced and timely supply of manure, fertilizer and irrigation encourage the natural development of peripheral branches of the tea bush. In case of chinary or China hybrid tea bushes, the lower and thin shoots should be removed carefully immediately after pruning (Singh 2013).

Cover-crop and Selective Weeding

Inclusion of a creeping type cover crop during young stage of tea and during first year after pruning helps to reduce weed density in tea plantation. In tea plantation, the species, such as *Arachis pintoi, Desmodium ovalifolium,* and *Indigofera endocaphyla* could be established to cover the ground (Ekanayake 1996a). When the cover crops are used as dead mulch, weed suppression seems mostly to be the result of the physical effects of mulch, rather than to nutrient- or allelochemical-mediated effects (Teasdale and Mohler 2000). The dead mulch-cover suppresses weed germination and growth by preventing the light transmittance to soil surface. Some green manure crops, such as *Crotalaria juncea* and *Flemingia congesta* are also used in vacant areas of tea plantation to suppress the weed growth. It had also been reported that growing of white clover as intercrop and straw mulching were found to be effective ecological measures for weed control in tea plantations in China, (Xiao et al. 2008). On the other hand, clean weeding may lead to soil erosion in high rainfall areas. Thus, under such conditions, only noxious weeds should be controlled, but non-competitive soft weeds should be left on the field (Sutidjo and Lubis 1971). Some non-competitive weeds, such as *Borreria latifolia, B. ocymoides* can be left on the ground for 6–8 weeks to cover the ground and suppress the growth of other weeds (Prematilake 1997).

Chemical Method

Weed control in tea plantation is mostly done by physical, mechanical and cultural methods; however, these methods are time and labour consuming and thus very expensive. Chemical method of weed management is the most convenient and effective method among various weed management techniques for the tea plantation (Prematilake et al. 2004, Rajkhowa et al. 2005, Ilango et al. 2010, Mirghasemi et al. 2012) due to their efficiency, cost effectiveness and ease of operation. It had been reported that chemical weed control saved about 130 man-days ha[-1] year[-1] and about 6% increase in crop yield over manual weeding (Barbora 1971). A wide range of pre-emergent, contact, and translocated herbicides were tested against both dicot and monocot weeds in tea plantation (Barbora and Dutta 1972, Rahman et al. 1975, Rao 1981, Sharma et al. 1986). Moreover, chemical method also minimizes soil erosion and losses of plant nutrients from the tea field (Sivapalan 1983). The herbicides used in tea plantations are listed in Table 23.2 with classification; however, the most commonly useable are paraquat, glyphosate, simazine, 2,4-D sodium, 2,4-D amine, diuron, dalapon. Although, most of the approved herbicides are safe to tea, phytotoxicity in tea may occur due to application of herbicides at higher doses than the recommended doses, improper or non-targeted spraying, spray drift, leaching of pre-emergence herbicides by heavy rains, and age of the tea bush (Sinha and Borthakur 1992).

The pre-emergent herbicides like simazine, atrazine, fluchloralin, oxadiazon, methazole were recommended for chemical weed control in tea nursery (TRA 1978). The recommended dose of these herbicides is 2 kg ha[-1], and it should be applied three weeks before planting of clonal cuttings. It had also been reported that blend of simazine or atrazine with oxadiazon or fluchloralin were found more successful (Singh et al. 2014).

Weed growth in young tea (0–3 years old) is very vigorous particularly during the initial two years since the vacant area is not completely covered by the bush canopy (Rao 2000). The critical period of weed competition in young tea is between April and September (Rao and Singh

Table 23.2. The list of herbicides used in tea plantations

S. No	Classification	Name of herbicides
1.	Pre-emergent	Simazine, Atrazine, Atratone, Butachlor, Karmex, Prometryne, Oxadiazon, Oxyfluorfen, Pendimethalin, Methazole, Fluchloralin, Diuron, Dichlormate.
2.	Post-emergent	Gramoxone, Phenoxylene plus, Phordene, 2,4-D, Dalapon, Paraquat, Glyphosate, Dinoseb, Metribuzin, MSMA, Linuron.

1977). Thus the effective weed control measure should be adopted to make the ground weed-free during this period. However, the chemical weed control method for young tea is different from that for mature tea, because young tea plants are relatively more susceptible to herbicide and the weed flora is more diverse and intense. In young tea plantation, the application of oxyfluorfen at 0.125 kg ha^{-1} as pre-emergence during May followed by oxyfluorfen (0.06 kg ha^{-1}) + either paraquat (0.24 kg ha^{-1}) or 2,4-D (0.8 kg ha^{-1}) as post-emergence controlled most of the problematic weeds throughout the season (Ghosh and Ramakrishnan 1981). The bio-efficacy of pre-emergence herbicides, in general, is higher in the presence of moisture on the soil surface. Prematilake et al. (2004) reported that the application of oxyfluorfen @ 0.29 kg ha^{-1} + paraquat at 0.17 kg ha^{-1} or glyphosate at 0.99 kg ha^{-1} + kaolin at 3.42 kg ha^{-1} provided better weed control than handweeding in young tea in Sri Lanka. Simazine is also applied as pre-emergent on the clean soil of young tea plantation after the early rains between the end of March to end of April to keep the ground free from annual broad leaf weeds and grasses (Rao 2000). Dalapon and diuron are not recommended in tea younger than three years (CSIR 1985, 1990, Sinha and Borthakur 1992).

In mature tea, Oxyfluorfen, Simazine, Diuron, Imazapyr, Dithiopry, Thiazopry, and Norflurazon are the effective pre-emergent herbicides, which are used for controlling many broadleaf annual weeds and some annual grasses (Rao 2000). However, the effectiveness of diuron is higher than others. Oxyfluorfen at 0.25 kg ha^{-1} also provides good result for controlling the broad-leaved weeds without any phytotoxicity when it was applied to clean soil or to growing weeds (Singh et al. 2014). For suppressing seed-borne weeds, pendimethalin 0.75, oxyfluorfen 0.44, simazine 1.25, or atrazine 1.25 kg ha^{-1} were more effective pre-emergence treatments (60). Subsequent weed growth could be controlled with spot treatment of 2,4-D and/or paraquat (61).

Glyphosate is the most effective herbicide in tea plantations, and it controls many perennial weeds including *Setaria palmifolia*, *Imperata cylindrica* and *Cynodon dactylon* (Rahman et al. 1975, TRA 1976), when applied at 1.5 to 6.0 kg ha^{-1} (Rao et al. 1976, 1977, Awasthi and Rao 1979). Kabir et al. (1991) had also reported that glyphosate at 0.92 or 1.23 kg ha^{-1} was effective for the tea in Darjeeling area. Superiority of glyphosate in combination with diuron and other herbicides against weeds in tea had also been reported (Saikia et al. 1998, Mirghasemi et al. 2012). It has also been reported that glyphosate is an effective herbicide against deep-rooted broad-leaved weeds and hardy perennial grasses, and is not poisonous to tea bushes even when applied directly on the bushes at a rate of 1.68 kg a.i. ha^{-1} (UPASI 1978, Sharma and Satyanarayana 1976). However, studies at IHBT, Palampur indicated that use of glyphosate even at the rate of 1.03 kg ha^{-1} might cause phytotoxicity in seed-raised china hybrid tea plantations (Singh et al. 2014). On the other hand, Kumar et al. (2014) reported that the application of BCS AA 10717 –2% in combination with glyphosate 40–42% SC was quite effective without any phytotoxicity in tea. Intensive use of glyphosate can create new weed problems as less harmful species have 'shifted' to become more dominant and some species have developed resistance to glyphosate. Under this situation, using paraquat as an alternative non-selective herbicide is effective to avoid problems of weed shift and resistance. Another important herbicide for tea plantation is 2,4-D, which is extensively used for controlling the paraquat-resistant broad-leaved weeds at the rate of 1.12 kg ha^{-1} (Sharma 1975, 1977). Addition of paraquat at 0.3 kg ha^{-1} made it more efficient against grasses (TRA 1978, CSIR 1989).

Herbicide Resistance Build-up

The ability of a plant species to withstand the phytotoxicity of a chemical is known as resistance. The risk of developing resistance in weeds to any herbicide has become a serious issue (Prematilake 2003). *Erigeron sumatrensis* and *Crassocephalum crepidioides* are the weeds in the upcountry tea growing area, which developed resistance to paraquat (Marambe et al. 2002, 2003). Thus, different herbicides should be used in rotation with other weed management methods to avoid resistance development.

Biological Methods

There is an obvious lack of effort towards biological control of weeds deploying bio-herbicides or other bio-control agents in tea plantation. However, there are some reports on biological method for weed management in tea. The biological control for weeds in tea plantation is also called as Ecological control, since some plant species can suppress another weeds species by competing with growth factors (Prematilake 2003). For example, *Brachiaria brizantha* grass is used for controlling the *Panicum repens* (Ekanayake 1996b). Use of alke-strain of Tobacco mosaic virus was reported to control *Solanum carolinense*, an herbicide-resistant weed in tea (Izhevskii et al. 1981). It has also been reported that *Mikania micrantha* could be controlled by using the co-evolved rust fungus *Puccinia spegazzinii* (from Latin America) in tea in Assam, India (Ellison 2004). On the other hand, some information is available regarding the biology of a beetle, *Chabria sp.* (Chrysomelidae, Coleoptera), destroying *Borreria hispida*, a common weed of tea fields in Northeast India (Debnath 1989).

Integrated Weed Management

Integrated weed management (IWM) is a strategy for weed control, which considers the use of all available weed control methods (physical/mechanical, cultural, chemical, and biological) without relying on only one of these. When only one or few techniques of weed controls are used over a longer period, there is a possibility that weeds adapted and become tolerant to those techniques (Bhowmic 1997). Moreover, there are concerns about shifts in weed populations and developing resistance in weeds. Therefore, to avoid the problems of weed shift and development of resistance in weeds, there is a pressing need to control the weeds by using integrated weed-management strategies. In the current scenario, IWM is an environmental and social demand (Pannell 1990, Swanton and Weise 1991, Clements et al. 1994, Auld 2004). In case of tea weed management, these may include use of hoeing and scraping, slash weeding, hand pulling, mulching, infilling and inter-planting, bush management, cover cropping, selective weeding and judicious use of herbicides (Figure 23.2).

In tea plantation, IWM strategy is very effective in terms of yield, economical and environmental. For young tea plantation, a combination of inter-row mulching and oxyfluorfen and paraquat, followed by handweeding at every 6–8 weeks was considered the most effective weed management system (Prematilake et al. 2004). Though, chemical method of weed management is the most convenient and effective among various weed management techniques in tea plantation, there is a growing concern on phytotoxic effects on bush, existence of residues, development of resistance, and environmental pollution (Marambe et al. 2002, 2003, Ekanayake 1994). It is a mandatory to reduce the application of herbicides since tender leaves are the economical part of tea plant. Thus, IWM practices are recommended in tea plantation for sustaining productivity and reducing the use of environmental hazardous chemicals.

Concluding Remarks

In order to increase efficiency and cost effectiveness of the weed management in tea plantation with eco-friendly approach, research on biology of serious weed species, bio-herbicides, herbicides-resistant varieties, and integrated weed management become imperative. Research on weed ecology should be strengthened to understand the processes that regulate weed-crop interactions, weed population dynamics, adaptation and persistence under various management practices (Rao and Chauhan 2015). Herbicides-tolerant variety of tea may play a vital role to reduce the cost of weed management and to increase productivity of tea. Thus, there is a pressing need to develop such varieties. Research on bio-herbicides and biological agents for weeds management in tea plantation should be strengthened under integrated nutrient management system. The weed management in organic tea plantations is a new

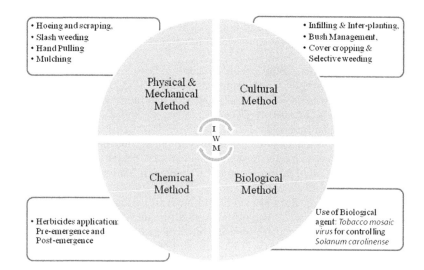

Figure 23.2. Component of integrated weed management (IWM) in tea.

challenge; however, exploiting of allelopathy and robotics might give useful solution for weed management in organic tea (Deka and Barua 2015).

There is an urgent need to develop decision support tool to increase the efficiency of weeds management technique in tea plantation. There is a huge scope to develop intelligent real-time system for regulating the usage of herbicides at the optimal level based on the prevalence of the existing weeds and apply herbicides automatically. Image processing technology should also be developed for identification and discrimination of weed types *viz.*, narrow and broad-leaf weeds in tea plantation. Thus, input cost of weeds management may be reduced through implementation of modern technique in the existing weeds management system.

Acknowledgements

Authors are thankful to Dr Sanjay Kumar, Director, CSIR-IHBT, Palampur for his constant encouragement for the work. Authors also acknowledge the Council of Scientific and Industrial Research, Government of India, for financial support.

REFERENCES

Asghari, J. and A. Mahmodi. 1996. The major weeds of croplands and rangelands of Iran (in Farsi). Guilan University, Rasht. Iran.

Auld, B.A. 2004. The persistence of weeds and their social impact. International Journal of Social Economics 31(9): 879–886.

Awasthi, R.C. and V.S. Rao. 1979. Economic evaluation of control of thatch grass (*Imperata*) and sour grass (*Paspalum*) weeds under experimental conditions in tea. Two and a Bud 26(1): 15–17.

Barbora, B.C. 1971. Economics of chemical weed control. *In:* Proc. 25th Conference, Tocklai Experimental Station. pp. 133–140.

Barbora, B.C. and S.K. Dutta. 1972. Planters Journal and Agriculturist. January 1972, pp. 16–19.

Bhowmic, P.C. 1997. Weed biology: importance to weed management. Weed Sci. 45: 349–356.

Bose, A. 2013. Climatic condition for growing tea. Retrieved from http://www.importantindia.com/4407/climatic-condition-for-growing-tea/

Bursulaya, T.T. 1990. Effect of different methods of inter-row soil management of tea root system development in the Kolkhida lowlands. Subtropicheskie Kul'tury 6: 22–26.

Clements, D.R., S.F. Weise and C.J. Swanton. 1994. Integrated weed management and weed species diversity. Phytoprotection 75: 1–18.

CSIR. 1985. Weed Control in Tea, CSIR Complex Palampur, Leaflet Tea 85/2: 6.

CSIR. 1989. Annual Report 1988-89, CSIR Complex Palampur (H.P.), India, 11.

CSIR. 1990. Weed Control in Tea, CSIR Complex Palampur Technical Folder Tea, 90/7.

Debnath, S. 1989. Natural occurrence of *Chabria* sp.: Chrysomelidae (Coleoptera), a beetle pest of Borria hispida. Two and a Bud 45(1): 24.

Deka, J. and I.C. Barua. 2015. Weeds of tea fields and their control. National Seminar on Plant Protection in Tea. Organised by Tea Research Association, Tocklai Tea Research Institute, Jorhat, Assam in association with Tea Improvement Consortium National Working Group on Plant Protection in Tea. pp. 26–27.

Deori, G., S.C. Barua and B.C. Borthakur. 1997. Growth of young tea plants in association with vesicular arbascular mycorrhiza. Ecological Agriculture and Sustainable Development. Vol. 1. *In:* Proceedings of International Conference on Ecological Agriculture: Towards Sustainable Development, Chandigarh, India, 15–17 November, 1997. pp. 417–423.

Dutta, A.C. 1977. Biology and control of Mikania. Two and a Bud 24(1): 17–20.

Ekanayake, P.B. 1994. Weed management in tea plantations, Weed management in the developing countries. FAO plant production and protection paper, 120 FAO, Rome. pp. 360–363.

Ekanayake, P.B. 1996a. Biological control of cooch grass. TRI update, A Newsletter. Tea Res. Inst. S.L. 1(3): 2.

Ekanayake, P.B. 1996b. A new cover crop. TRI update, A Newsletter. Tea Res. Inst. S.L. 1(l): 2.

Ellison, C.A. 2004. Biological control of weeds using fungal natural enemies: a new technology for weed management in tea? Int. J. Tea Sci. 3(1/2): 4–20.

FAO. 2015. World tea production and trade current and future development. Market and Policy Analyses of Raw Materials. Horticulture and Tropical (RAMHOT) Products Team, FAO Rome.

Ghosh, M.S. and L. Ramakrishnan. 1981. Study on economical weed management programme in young and pruned tea with oxyfluorfen. *In:* Proc. Eighth Asian-Pacific Weed Science Society Conference. pp. 119-125.

Hasselo, H.N. and S. Sandanam.1965. Chemical weed control in tea. The Tea Quarterly 36: 1.

HPKV. 1986. Consolidated Progress Report of Tea Experiment Station, 1962-85, HPKV, Palampur (H.P.), India. pp. 54–55.

IBEF. 2017. Brand India Plantations: Tea_Tea Statistics. (July 20, 2017). Retrieved from http://www.teacoffeespiceofindia.com/tea/tea-statistics.

Ilango, R.V.J., M. Saravanan, R. Parthibaraj and P.M. Kumar. 2010. Evaluation of Excel Mera-71 weed control in tea fields. Newsletter – UPASI Tea Research Foundation 20(1).

Izhevskii, S.S., A.E. Livshits, G.E. Murusidze and G.G. Gogoladze. 1981. Prospects of using Alke strain in integrated control of *Solanum carolinense* in tea plantations. Subtropicheskie Kul'tury 4: 60–65.

Kabir, S.E., T.C. Chaudhury and H.N. Ghosh. 1991. Evaluation of herbicides for weed control in Darjeeling tea. Indian Agric. 35(3): 179–185.

Karmakar, K.G. and G.D. Banerjee. 2005. The tea industry in India: a survey. Department of Economic Analysis and Research. National Bank for Agriculture and Rural Development, Mumbai. Occasional paper 39.

Kogua, M.B. 1975. The effect of tea prunings left in the inter rows on the soil water regime. Subtropicheskie Kul'tury 5: 27–30.

Korzun, B.V. 1981. The species composition of weeds on a tea plantation in Adygei. Subtropicheskie Kul'tury 166(1): 147–148.

Kumar, S., S.S. Rana, N.N. Angiras and Ramesh. 2014. Weed management in tea with herbicides mixture. Indian J. Weed Sci. 46(4): 353–357.

Manipura, W.B. 1971. Soil erosion and conservation in Ceylon with special reference to tea land. Tea Quarterly 42(4): 206–211.

Manipura, W.B., A. Somaratne and S.G. Jayasooriya. 1969. Some effects of mulching on the growth of young tea. Tea Quarterly 40: 153–159.

Marambe, B., J.M.A.K. Jayaweera and H.M.G.S.B. Hitinayake. 2003. Paraquat-resistant *Crassocepelum crepidiodes* in Sri Lanka. *In:* Proceedings of the 19th Asian-Pacific Weed Science Society (APWSS) conference held in Manila, Philillines on 17–21 March. pp. 782–787.

Marambe, B., S.P. Nissanka, L. de Silva, A. Anandacoomraswamy, M.G.D.L. Priyantha. 2002. Occurrence of paraquat-resistant *Erigeron sumatrensis* (Retz.) in up country tea lands of Sri Lanka. J. Plant Dis. Protect. XVIII: 973–978.

Mirghasemi, S.T., J. Daneshian and M.A. Baghestani. 2012. Investigating of increasing glyphosate herbicide efficiency with nitrogen in control of tea weeds. Int. J. Agri. Crop Sci. 4(24): 1817–1820.

Pannell, D.J. 1990. Responses to risk in weed control decisions under expected profit maximization. J. Agr. Econ. 41: 391–404.

Pirtskhalaishvili, S.K. and G.P. Kimutsadze. 1972. Comparative effectiveness of mulching tea interrows in Imeretiya. Subtropicheskie Kul'tury 4: 17–20.

Prematilake, K.G. 1997. Studies on weed management during early establishment of tea in low country of Sri Lanka, Ph. D Thesis, Dept. of Agric. Botany, School of Plant Sci. The Univ. of Reading, Reading, UK.

Prematilake, K.G. 2003. Weed management in tea in Sri Lanka. pp. 63–72. *In:* Twentieth Century Tea Research in Sri Lanka Ed Modder WWD.

Prematilake, K.G., R.J.F. Williams and P.B. Ekanayake. 1998. Potential of various green matters as mulches and their impact on weed suppression in tea lands. Trop. Agri. Res. 10: 309–323.

Prematilake, K.G., R.J.F. Williams and P.B. Ekanayake. 1999. Investigation of period threshold and critical period of weed competition in young tea. Brighton Crop Protection Conference: Weeds. *In:* Proceedings of an International Conference. 1: 363–368. Brighton, UK, 15–18 November.

Prematilake, K.G., R.J.F. Williams and P.B. Ekanayake. 2004. Weed infestation and tea growth under various weed management methods in a young tea (*Camellia sinensis* L. Kuntze) plantation. Weed Biol. Manag. 4(4): 239–248.

Rahman, F., A.K. Dutta, M.C. Saikia and B.C. Phukan. 1975. Studies on Roundup as a herbicide for tea. Two and a Bud. 22(1): 4–8.

Rajkhowa, D.J., R.P. Bhuyan and I.C. Barua. 2005. Evaluation of carfentrazone-ethyl 40 DF and glyphosate as tank mixture for weed control in tea. Indian J. Weed Sci. 37(1/2): 157–158.

Rao, A.N. and B.S. Chauhan. 2015. Weeds and weed management in india: a review. pp. 87–118. *In:* Weed Science in the Asian-Pacific Region.

Rao, V.D. and H.S. Singh. 1977. Effects of weed competition in young tea. pp. 15–19. *In:* Proc. Twenty-eight Conf., Tockli Experimental Station, Tea Research Association, Jorhat, Assam, India.

Rao, V.S. 2000. Principles of Weed Science. Oxford and IBH Publishing Co. Pvt. Ltd. pp. 497–498.

Rao, V.S., B. Kotoky, S.N. Sarmah. 1981. Perennial weed control in tea. pp. 259–299. *In:* Proceedings of the Eighth Asian-Pacific Weed Science Society Conference (Bangalore, 22–29 November 1981).

Rao, V.S., F. Rahman, H.S. Singh, A.K. Dutta, M.C. Saikia, S.N. Sharma and B.C. Phukan. 1976. Effective weed control in tea by glyphosate. Indian J. Weed Sci. 8(1): 1–14.

Rao, V.S., F. Rahman, H.S. Singh, A.K. Dutta, M.C. Saikia, S.N. Sharma and B.C. Phukan. 1977. Effective weed control in tea by glyphosate. Program and Abstracts of Papers, Weed Science Conference and Workshop in India, Paper No. 96, 58.

Rodgers, E.G. 1974. Weed prevention is the best control. Weeds Today 5(1): 8–22.

Ronoprawiro, S. 1976. Control of mosses in tea. pp. 365–369. *In:* Proc. Fifth Asian-Pacific Weed Science Society Conference. Tokyo, Japan.

Ronoprawiro, S. 1981. The possibility of using glyphosate to control mosses in tea. Ilmu Pertanian 3(1): 9–19.

Saikia, S., S. Baruah and A.C. Barbora. 1998. Inefficacy and economics of herbicidal combinations for control of *Polygonum chinense* (Linn.). Two and a Bud 45(2): 15–18.

Sandanam, S. and H.D. Jayasinghe. 1977. Manual and chemical control of *Imperata cylindrica* on tea land in Sri Lanka. PANS 23(4): 421–426.

Sanusi, M. 1977. Problems and control of weed on young tea in Indonesia. *In:* Proc. 16th Asian Pac. Weed Sci. Soc. Conf., Jakarta.

Sarkar, S.K., J. Chakravartee and S.D. Basu. 1983. Effects of soil stirring on yield of tea in heavy soil. Two and a Bud 30(1–2): 50–51.

Sharma, S.N., M.P. Sinha and K.C. Thakur. 1986. Effect of glyphosate on thatch grass (*Imperata cylindrica* L. Beuv) control. Two and a Bud 33: 53–55.

Sharma, V.S. 1975. Gramoxone-resistant weeds and their control. Planters Chronicle. 70(3): 59.

Sharma, V.S. and N. Satyanarayana. 1976. Recent developments in chemical weed control in tea fields. UPASI Tea Science Department Bulletin 33: 101–108.

Sharma, V.S. 1977. Chemical weed control in tea fields. Planter's Chronicle 72(2): 79–80.

Singh, R.D. 2013. Managing weeds in tea. pp. 301–316. *In:* P.S. Ahuja, A. Gulati, R.D. Singh (Eds.) Science of Tea Technology. Scientific Publishers, India.

Singh, R.D., B.K. Singha, R.K. Sud, M.B. Tamang and D.N. Chakrabarty. 1994. Weed flora in tea plantations of Himachal Pradesh. J. Econ. Taxon. Bot. 18(2): 399–418.

Singh, R.D., S.K. Sohani, B. Singh and D.N. Chakrabarty. 1993. Influence of long term mulching on yield of young tea and weeds. 3(S): pp. 34–36. *In:* Proc. Int. Symp. Indian Soc. Weed Sci. 18–20 November 1993, Hisar (Haryana), India.

Singh, R.D., R.K. Sud and P.K. Pal. 2014. Integrated weed management in plantation crops. pp. 255–280. *In:* B.S. Chauhan and G. Mahajan (Eds.) Recent Advances in Weed Management. Springer Science-Business Media, New York.

Sinha, M.P. 1985. A perspective weed control in tea. Two and a Bud 32(1–2): 35–39.

Sinha, M.P. and B. Borthakur. 1992. Important aspects of weed control in tea. pp. 134–139. *In:* Field Management in Tea (compilation of lectures), Tocklai Experimental Station, TRA.

Sivapalan, P. 1983. Minimizing soil erosion on tea estates in respect of manual weed control. Tea Q. 52: 81–83.

Smale, P.E. 1991. New Zealand has its own green tea industry. Horticulture in New Zealand 2(2): 6–9.

Soedarsan, A., Noormandias and H. Santika. 1976. Effects of some weed species on the growth of young tea. pp. 87–91. *In:* Proceedings of Fifth Asian-Pacific Weed Science Society Conference, Tokyo, Japan, 1975.

Somaratne, A. 1988. Weed management in tea plantation of Sri Lanka. *In:* Proceedings of the Regional Tea (Scientific) Conference (Ed. by Sivapalan, P. and Kathiravetpillai, A.). (Colombo, January 1988). S.L.J. of Tea Science Conference Issue.

Sutidjo, K. and J.R. Lubis. 1971. Some results of experiments and practical use of herbicides in tea estates in North Sumatra. Third Conference of the Asian-Pacific Weed Science Society. 7: 10.

Swanton, C.J. and S.F. Weise. 1991. Integrated weed management in Ontario: the rationale and approach. Weed Technol 5: 657–663.

Tabagari, L.G. and R.S. Kopaliani. 1984. Preliminary data on agrotechnical, measures in the interrows of young tea plantations, cv. Kolhida, on podzolic soil. Subtropicheskie Kul'tury 5: 67–75.

Tabagari, L.G., A.D. Mikeladze and I.D. Cheishvili. 1988. Results of preliminary studies on some agricultural measures for restoring neglected and unproductive tea plantations. Subtropicheskie Kul'tury 5: 67–75.

Teasdale, J.R. and C.L. Mohler. 2000. The quantitative relationship between weed emergence and the physical properties of mulches. Weed Sci. 48: 385–392.

TRA. 1976. Annual Scientific Report 1974-75, TRA, Jorhat, Assam, India. pp. 74.

TRA. 1978. Annual Scientific Report 1977-78, TRA, Jorhat, Assam, India. pp. 95.

UPASI. 1978. Annual Report 1976, United Planters' Association of Southern India. pp. 16–19.

Wettasinghe, D.T. 1969. Report of the low country station and the Kottawa substation for 1968. pp. 112–140. *In:* Fernando, L.H. (Ed.) Annual Report. Tea Research Institute, Talawakelle, Ceylon.

Wettasinghe, D.T. 1971a. Report of the Research Officer for the low country station and the Kottawa substation for 1969. pp. 145–167. *In:* Fernando, L.H. (Ed.) Annual Report. Tea Research Institute, Talawakelle, Ceylon.

Wettasinghe, D.T. 1971b. Report of the Research Officer for the low country station for 1970. pp. 126–142. *In:* Fernando, L.H. (Ed.) Annual Report. Tea Research Institute, Talawakelle, Ceylon.

Xiao, R., Z. Xiang, H. Xu, W.X. Shan, P. Chen, G.X. Wang and X. Cheng. 2008. Ecological effects of the weed community in tea garden with intercropping white clover and straw mulching. Trans. Chinese Soc. Agri. Engr. 4(11): 183–187.

Zernadze, D.N. 1972. The results of inter row mulching in a young tea plantation in the Kolkhida lowland. Subtropicheskie Kul'tury 1: 121–123.

24

Weed Management in Sugarcane

Ramon G. Leon[*1] and D. Calvin Odero[2]

[1] 4402C Williams Hall, Department of Crop and Soil Sciences, North Carolina State University, Raleigh, North Carolina, 27695, USA

[2] 3200 E Palm Beach Road, Everglades Research and Education Center, University of Florida, Belle Glade, Florida, 33430, USA

Introduction

Sugarcane (*Saccharum* spp. interspecific hybrids) is a globally important cash crop cultivated primarily for production of sucrose, commonly referred to as sugar. The crop accounts for 80% of global raw sugar production with the remainder supplied by sugarbeet (*Beta vulgaris* L.) (FAO 2015). Sugarcane, a member of the Poaceae family belongs to the Andropogoneae tribe and the genus *Saccharum*. Modern sugarcane cultivars are derived from complex interspecific crossings of the noble species, *Saccharum officinarum* L. and the wild species, *Saccharum spontaneum* L. The former species produces moderately tall, thick, and low fiber stalks with high sucrose while the latter species has high adaptability to drought, cold, disease, and poor growing conditions (Clements 1980). Sugarcane is a perennial crop propagated vegetatively using stem cuttings, also known as setts or seed-pieces. After the first crop, referred to as plant-cane, each succeeding crop or ratoon can last up to 15 years (Inman-Bamber 1994) depending on the prevailing environmental conditions, cultivar, yield, and pest pressure. Plant-cane requires 12 to 18 months or in some cropping systems up to 24 months before harvest. The maturity of ratoon cane is typically quicker than plant-cane, but will vary depending on the cultivar and environment. Among cultivated crops, sugarcane has one of the best efficiencies in converting atmospheric carbon dioxide and water in the presence of incident solar energy into chemical energy stored in stalks, which are crushed to extract juice for sugar production. Sugarcane is also used for production of bioenergy, feed, and fiber among other products. The by-product of sugarcane stalk crushing consisting of dry fibrous residue or bagasse is used for cogeneration of electricity to power sugar mills and supply excess electricity to the grid. Sugarcane molasses, another by-product of sugar production, is an important base component used to increase nutritional value and palatability of many livestock feeds. Molasses is also a promising dry-strength agent for replacement of cellulosic fibers in papermaking (Ashori et al. 2013). Still, there is potential for further diversification and utilization of sugarcane coproducts and byproducts using suitable technologies to ensure sustainability of the industry into the future. Such diversification programs can result in more value-added products and additional economic opportunities particularly in rural areas that make up the bulk of sugarcane cultivation.

*Corresponding author: rleon@ncsu.edu

Since its original cultivation for sugar production in India and southern China (Earle 1928), approximately 1.9 billion tons of sugarcane were produced on 27.1 million hectares with an average yield of 69.5 tons per hectare in 2014 in over 100 countries (Table 24.1). These represent 4.2-, 3.0-, and 1.4-fold increase in sugarcane tonnage, area under cultivation, and average yield, respectively, from 1961 to 2014. Sugarcane producing countries are mainly in the tropical region between latitude 23.5° North and 23.5° South, but production also occurs in sub-tropical and temperate environments, such as the United States, to a limit of approximately 35° North or South of the equator (Bakker 1999). Brazil is the leading sugarcane producer accounting for 39% of total global production (Table 24.1) while the first top three and top 15 countries account for 64% and 88% of production, respectively. The sugarcane industry is a key driver of rural development in many of these countries, particularly in the Caribbean, Latin America, Asia, and southern Africa, where it uses and supports different products in the agricultural, energy, and industrial sectors.

Cultivation of sugarcane has steadily increased in the past several decades to meet growing global sugar demand, making it the leading crop in terms of tonnage compared to major cereals and potatoes, which make up the majority of crop production and occupy several-fold a larger portion of the world's cultivated land (FAO 2015). The increased demand for sugarcane is attributed to socio-economic pressures, such as rising incomes and urbanization particularly in China and India that have resulted in shifts to diets higher in sugars, and the increased interest in renewable bioenergy derived from sugarcane to mitigate effects of greenhouse gas emission and fossil fuel dependency (FAO 2015). Bioethanol derived from sugarcane is now an important source of bioenergy in many countries particularly Brazil, where an Alcohol Program from sugarcane was launched in the mid-1970s to reduce the country's dependency on imported fossil fuel following the oil crisis in the 1970s (Moreira and Goldemberg 1999). Similar initiatives have been developed in other countries to spur research and investment in bioethanol production from sugarcane (Amores et al. 2013). Furthermore, the fibrous component of the crop conventionally used for electric energy generation is a promising feedstock for second-generation bioethanol (Dias et al. 2011). Cultivation of sugarcane will most likely increase throughout the 21st century in many production regions still characterized by poverty with diversification of more products and byproducts.

Table 24.1. Global sugarcane production data for top 15 countries in 2014[a]

Sugarcane producing country	Production area (1,000 hectares)	Average yield (tonnes per hectare)	Total production (1,000,000 tonnes)
Brazil	10,420	70.6	736.1
India	5,012	70.2	352.1
China	1,768	71.3	126.1
Thailand	1,353	76.6	103.7
Pakistan	1,140	55.1	62.8
Mexico	761	74.4	56.7
Indonesia	472	60.5	28.6
Cuba	450	40.0	18.0
Philippines	432	57.9	25.0
Colombia	401	91.0	36.5
Australia	375	81.3	30.5
Argentina	368	66.5	24.5
United States	351	78.5	27.6
Vietnam	305	65.0	19.8
South Africa	273	65.1	17.8
Total global production	27,124	69.5	1,884.2

[a] 2014 sugarcane production data from Food and Agriculture Organization of the United Nations, Statistics Division. Available at http://www.fao.org/faostat.

This chapter provides an overview of implications of weed management systems in sugarcane production emphasizing challenges resulting from managing diverse weed communities integrating multiple non-chemical weed control practices and a limited number of herbicide mechanisms of action.

Major Weeds and Impact

Weed Communities Associated with Sugarcane

Weed communities in sugarcane are diverse despite this crop being commonly grown as a monoculture. This high diversity has been reported in several regions throughout the world. For example, Rodríguez Cuevas and Romero Manzanares (1994) studied weed diversity in sugarcane fields in Holguín Province, Cuba, and they detected 123 plant species with high frequency of species, such as *Brachiaria fasciculata* (Sw.) Parodi, *Cyperus rotundus* L., *Chamaesyce hyssopifolia* (L.) Small, *Echinochloa colona* (L.) Link, *Euphorbia heterophylla* L., *Leptochloa panacea* (Retz.) Ohwi, *Panicum reptans* L., and *Rottboellia cochinchinensis* (Lour.) Clayton. Perdomo et al. (2004) conducted a weed diversity study in sugarcane fields in Tlaquiltenango, Morelos, Mexico, and reported 79 species with *Chamaesyce berteroana* (Balb. Ex Spreng.) Millsp. and *Leptochloa filiformis* (Pers.) P. Beauv. exhibiting the highest density and cover. Firehun and Tamado (2006) reported 180 taxa found in sugarcane fields in Ethiopia, with *Cyperus* spp., *Sorghum* spp., *Euphorbia hirta* L., *Rhyncosia malacophylla* (Spreng.) Bojer, and *Portulaca oleracea* L. as the most predominant species. In Brazil, several studies have reported that species richness within sugarcane fields (not including borders) ranged between 33 and 49 species, and that species, such as *Amaranthus* spp., *Cyperus* spp., *Ipomoea hederifolia* L., *P. oleracea*, *Chamaesyce* spp., *Euphorbia* spp., *Digitaria* spp., and *Eleusine indica* (L.) Gaertn. were the most important across different growing regions (Kuva et al. 2007, 2008, Monquero et al. 2008).

Leon et al. (2017a) studied weed diversity in sugarcane fields in Guanacaste, Costa Rica. They reported 120 species with similar compositions to those reported by Perdomo et al. (2004), Rodríguez Cuevas and Romero Manzanares (1994), Firehun and Tamado (2006), Kuva et al. (2007, 2008) and Monquero et al. (2008), with predominant species including *C. rotundus*, *R. cochinchinensis*, *E. colona*, *Leptochloa* spp., *Euphorbia* spp., *Chamaesyce* spp., *Rhynchosia* spp., and *P. oleracea*. Leon et al. (2017a) also determined that weed diversity is not evenly distributed in sugarcane fields and that beta-diversity explains most of the overall diversity in the system. They observed that areas within the sugarcane crop tend to have considerably lower diversity than borders and irrigation and drainage canals. Furthermore, they concluded that the number of key species affecting weed management decisions is relatively small, and these species are present not only inside the crop where weed diversity is low, but also outside where weed diversity is higher. The fact that six studies conducted in different ecosystems, albeit all were in tropical latitudes, had similar weed community composition suggests that crop management in sugarcane fields strongly determines which species can survive and thrive. This is an important observation because it opens the possibility that by identifying the conditions that favour the survival and growth of species, such as *R. cochinchinensis*, *C. rotundus*, and *L. filiformis*, it might be possible to develop cultural practices that will help reduce their success.

Weed Species of Economic Importance

Grass and nutsedge species are the most important weed groups complicating weed management and interfering with sugarcane production. Among grass weed species, johnsongrass (*Sorghum halepense* [L.] Pers.) has been considered not only one of the most important weed species in tropical and subtropical areas (Holm et al. 1977), but also a particularly challenging weed to control in sugarcane. Johnsongrass impact on sugarcane production varies depending on density, but there are multiple reports in which cane and sugar losses can reach up to 30% and >70% when johnsongrass densities are low and high, respectively (Ali et al. 1986, Millhollon 1980a, 1990, 1995). Although johnsongrass propagates by seeds, and control actions that prevent seedling recruitment help manage this weed, its creeping perennial growth and rapid vegetative propagation via rhizomes (Horowitz 1973, Monaghan 1979) makes this weed a persistent problem that is difficult to eliminate once mature plants are established. Johnsongrass interference is particularly serious in plant-cane (first year) because it can affect cane and sugar yields in the following two ratoons (second and third years) (Millhollon 1995). Therefore, aggressive control tactics during the plant year can simplify johnsongrass management in the following years. After establishment, overall sugarcane is a competitive crop, although there are differences in competition ability among sugarcane cultivars. Millhollon (1990) reported that johnsongrass interference depended on sugarcane cultivar, and the stalk number and size were the parameters that explained most of the differences among cultivars. Therefore, identifying cultivars that can tolerate johnsongrass interference and suppress its growth is an important component of sustainable weed management strategies for sugarcane.

Bermudagrass (*Cynodon dactylon* [L.] Pers.) is another perennial grass weed commonly found in sugarcane fields. However, unlike johnsongrass and other grass weeds with taller canopies, bermudagrass impact on cane and sugar yields tend to be considerably lower, with yield losses being less than 35% (Richard 1992, 1993, Richard and Dalley 2007). It is worth noting that although bermudagrass populations can increase over time within sugarcane fields (Richard 1992), its impact on cane and sugar yield is considerably higher during the plant than ratoon phases. For example, Richard and Dalley (2007) reported that bermudagrass caused up to 32% sugar yield loss in the plant-cane crop compared to less than 10% in the following two ratoons.

Itchgrass (*Rottboellia cochinchinensis*) is perhaps one of the most aggressive annual weed species in sugarcane. It is propagated by abundant production of seeds that can survive in the soil for several years (Thomas and Allison 1975) forming a persistent seed bank. Itchgrass seeds germinate predominantly from 0 to 2.5 cm depth, but some can germinate from up to 10 cm depth depending on soil type (Leon and Agüero 2001a), allowing seedling establishment in both conventional and reduced tillage systems. This weed species can reduce sugarcane yields over several years and increase its populations starting at a density of 1.8 plants m^{-2} (Millhollon 1992). Its competitive ability relies on its height reaching up to 3 m (Holm et al. 1977), high elongation rate (Millhollon 1965), and shade tolerance (Mercado 1978). This combination of growth traits allows itchgrass to survive and produce seed within sugarcane fields even after crop canopy closure. At the same time, itchgrass can exert continuous interference during the entire sugarcane growing season. Cane and sugar yield reductions resulting from season-long itchgrass interference have been reported to reach 43% to 72% (Lencse and Griffin 1991, Millhollon 1992). However, removal during the six weeks after itchgrass initial emergence can limit cane and sugar yield losses to less than 10% (Lencse and Griffin 1991). Similarly to johnsongrass, itchgrass interference predominantly reduces sugarcane stalk population although stalk height and weight can decrease especially under full-season competition (Millhollon 1992).

Purple nutsedge (*Cyperus rotundus*) and yellow nutsedge (*Cyperus esculentus* L.) are frequently found in sugarcane fields. Both of these species have short-statured canopies that are unlikely to compete for light with sugarcane. However, competition for soil moisture has been related to reductions in sugarcane yield that can be as high as 45% (Chapman 1966, Keeley 1987, Osgood et al. 1977). Furthermore, it is likely that allelopathic compounds released by tubers that these weed species produce underground (Jangaard et al. 1971; Sanchez-Tames et al. 1973, Stoller and Sweet 1987) can also affect sugarcane growth by affecting root growth. Studies have also showed that not only purple nutsedge tubers but also its leaves can release allelopathic compounds that limit the growth of other plants (Quayyum et al. 2000). Purple nutsedge adaptability allows it to colonize different management areas within sugarcane fields including row and furrows as well as drainage and irrigation canals (Leon et al. 2017a). Tillage and cultivation benefit purple nutsedge propagation and establishment by breaking the chains of tubers releasing apical dominance and promoting tuber sprouting. Passes with plows, disk cultivators, sweeps, and rippers are commonly done early during the growing season before the sugarcane can shade purple nutsedge plants. Therefore, tuber sprouting is promoted by cultivation precisely when the crop is less likely to suppress purple nutsedge growth and is more susceptible to interference.

Fall panicum (*Panicum dichotomiflorum* Michx.) has recently become a problem in sugarcane farms in subtropical conditions (Odero et al. 2016). This species has been successful because of its high seed production that ranges from 10,000 to up to 100,000 seeds per plant (Fausey and Renner 1997, Govinthasamy and Cavers 1995). Also, seeds can germinate from shallow (1 cm) to relatively deeper layers (5 cm) (Brecke and Duke 1980) in the soil, which makes it compatible with the different tillage and cultivation practices used in sugarcane production. Furthermore, fall panicum has exhibited tolerance to weed control actions commonly used in sugarcane (Odero et al. 2016). Fall panicum can reduce cane and sugar yield 20% to 60% when interference lasts for more than 20 weeks after sugarcane emergence, and as observed with the interference of other weed species, the reductions are mainly due to a decrease in the number of millable stalks (Odero et al. 2016).

Broadleaf weed species can also be problematic for sugarcane production. For example, morningglories (*Ipomoea* spp.) not only can reduce sugarcane yield from 21% to 36% if not controlled but also interfere with harvest operations by covering the crop (Bhullar et al. 2012). However, the availability of effective selective herbicides has made broadleaf weed control less challenging than grass and sedge control for sugarcane farmers.

Weed Interference and Weed Control Timing

Sugarcane is a very competitive crop because of its height and large leaf area index, which provide high levels of light interception (Keely 1987, Singels et al. 2005). However, weed infestations can reduce sugarcane yield especially when weeds are not controlled before canopy closure. After this point, sugarcane can effectively suppress weed growth. In most conditions, the critical period for weed control occurs between 3 and 12 weeks after crop planting/crop emergence (Odero et al. 2016, Yirefu et al. 2012). Before this period, impact on yield is limited, and after, weeds will be outcompeted by the crop and also the implementation of weed control practices is limited due to crop size. The duration of the critical period for weed control depends on sugarcane growth rate, timing to canopy closure, plant-cane vs. ratoon, and weed community composition and density. Although there is limited information, the critical period for weed control tends to be shorter in ratoon than plant-cane because shoot growth is faster and canopy closure occurs sooner in the former than the latter (Cheeroo-Nayamuth et al. 2000, Wallace et al. 1991). It must be pointed out that studies differ about how they measure this critical period. Some studies use time after sugarcane planting and others after sugarcane emergence (Odero et al. 2016, Yirefu et al. 2012). This could represent two to three weeks difference. This is important because weeds that emerge at the initiation of the critical period will be the ones that will impact yield the most if not controlled on time. Therefore, properly timing the first weed control actions will likely simplify management and reduce the risk of yield loss.

Weed interference is not uniform within the crop, and weeds growing within rows are more likely to impact yield than weeds between rows. Segura and Agüero (1997) reported that in plant-cane with high purple nutsedge populations (up to 400 shoots m^{-2}), controlling this weed only on a 0.5 m wide area on the planting row allowed the same cane and sucrose yield than controlling the entire area (i.e., 1.5 m row spacing). This approach considerably reduces herbicide use, but it is important to consider potential weed population growth between rows.

Besides direct competition with sugarcane plants, weeds can also affect sugarcane growth and production indirectly. For this reason, it is important to anticipate changes in weed communities that might favor the appearance of new weed problems. For example, as a result of glyphosate resistance evolution (Fernandez et al. 2015a, Odero 2012), Fernandez et al. (2015b) reported that ragweed parthenium (*Parthenium hysterophorus* L.), a species with potential to reduce crop yield and release of allelophatic compounds (Kohli et al. 2006, Reinhardt et al. 2004, Tamado and Milberg 2004), was increasing its populations in sugarcane fields during the fallow period preceding cane planting. In conditions where ragweed parthenium populations are too high, they can affect field preparation and cane planting.

Weed species have been documented as alternative host of important sugarcane insect pests, and their control might be necessary in order to prevent outbreaks of these pests. For example, larvae of the sugarcane root weevil (*Diaprepes abbreviataus* L., Coleoptera: Curculionidae) burrow through the soil until reaching sugarcane roots from which they feed causing root damage and sugarcane growth stunting (Cherry et al. 2011, Simpson et al. 1996). Several broadleaf weed species commonly found in sugarcane fields including spiny amaranth (*Amaranthus spinosus* L.), common purslane (*Portulaca oleracea*), and coffee senna (*Senna occidentalis* (L.) Link.) have been reported as alternative host (Odero et al. 2013), and their control has been proposed as an important component of integrated management strategies to control this sugarcane insect pest (Cherry et al. 2011, Odero et al. 2013).

Cultural Weed Control

Variation in morphology, growth rate, and emergence vigor in sugarcane germplasm should be considered to identify cultivars that have a higher weed suppression potential (Andrew et al. 2015, Jannink et al. 2000, Richard and Dalley 2007). Also, there might be variation for weed tolerance, which is the ability of the crop to minimize yield loss when suffering weed interference (Andrew et al. 2015, Leon et al. 2017b, Watson et al. 2006). Using cultivars with these traits is a low-cost component of integrated weed management programs that add a safety layer to minimize yield losses. Also, competitive cultivars help prevent weed seed bank increases especially when weeds escape control tools, such as herbicides and cultivation (Andrew et al. 2015, Segura and Agüero 1997). Richard and Dalley (2007) compared the competitive ability of several sugarcane cultivars that differed in emergence rate and stalk number against bermudagrass, but no clear differences in weed suppression were observed. These researchers proposed those traits, such as rapid emergence and high number of stalks might benefit sugarcane competitive ability. Firehun et al. (2012) studied how differences in leaf angle influenced weed suppressive ability among sugarcane cultivars. Their results demonstrated that there was an inverse relation between leaf angle and weed dry weight. Thus, a cultivar with an erect leaf pattern (i.e., 70–90° leaf angle) allowed up to 21% more weed biomass production than a cultivar with a sprawling pattern (i.e., 30–50° leaf angle). Additionally, the cultivar with erect leaf pattern required a longer weed-free period than the sprawling pattern cultivar to prevent yield loss due to weed interference. These findings highlight the importance of canopy architecture and rapid ground shading to maximize weed suppression.

Sugarcane is a very plastic crop that adjusts final stalk number depending on planting density and the number of established stalks by controlling tiller production (Bell and Garside 2005, Singels et al. 2005). For this reason, using high planting densities (buds per area) does not necessarily result in higher stalk numbers or denser canopies (Garside et al. 2002, Kanwar and Sharma 1974). However, row spacing determines canopy closure timing, which directly influences light interception and consequently weed suppression (Singels and Smit 2009). Kanwar and Sharma (1974) determined that, for plant-cane, row spacing between 60 and 120 cm maximized cane yield while wider row spacing tended to reduce yields. They also observed that narrow rows had higher tiller mortality than wider rows, which compensated for the higher planting density. Nevertheless, the importance of row spacing in first and second ratoons for stalk population was not so evident as in plant-cane. Similarly, Matherne (1974) found that row spacing between 0.9 and 1.0 m maximized yield while 1.8 m rows yielded the lowest. Richard et al. (1991) also reported that sugarcane yield was higher with row spacing of 0.9 and 1.2 m than 1.8 m, but the differences in yield were detected only for the plant-cane. Therefore, the use of narrow rows favors sugarcane production by increasing both yield and weed suppression potential. However, it is important balancing yield goals with the ability to conduct mechanized activities within the crop, such as cultivation and mechanical harvest. Very narrow rows might increase weed suppression, but they can also complicate weed management because of limitations on mechanization (Garside et al. 2009, Matherne 1974, Richard et al. 1991).

The fallow period between elimination of last ratoon and replanting represents an opportunity for implementing aggressive control actions to reduce weed populations, especially for perennial weed species in reduced- and no-tillage systems (Etheredge et al. 2009, Griffin et al. 2006, Richard 1995). Weed control during the fallow period can simplify weed management during the plant-cane year, which is the most susceptible to weed interference and the one that can determine weed pressure for following ratoon crops (Millhollon 1995). Griffin et al. (2006) explored including a short-lived crop, such as glyphosate resistant (GR) soybean during the summer fallow to offset the cost of an effective herbicide program to reduce johnsongrass populations before sugarcane replanting. They determined that growing GR soybeans for grain and using glyphosate as the main systemic herbicide for johnsongrass control resulted in positive net returns of up to $100 ha^{-1} while only applying glyphosate without growing soybean generated negative returns of at least $155 ha^{-1}. Interestingly, with both strategies johnsongrass control was similar, so taking advantage of a short rotation before replanting allows growers to cover the cost of reducing weed pressure while the sugarcane is not present.

Irrigation plays a major role in weed-crop interactions especially in dry areas because irrigation placement determines not only weed density but also where weed emergence is promoted (Grattan et al. 1988, Shresta et al. 2007, Shrivastava et al. 1994). Drip irrigation has been more widely used in vegetable and fruit production (Camp 1998). However, several studies have demonstrated that when drip line arrangements and irrigation strategies are properly designed to match crop water demand and rainfall patterns, drip irrigation promotes more efficient water use and similar sugarcane productivity when compared with surface irrigation systems (Batchelor et al. 1990, Hodnett et al. 1990, Surendran et al. 2016, Wiedenfeld 2004). Drip and subsurface irrigation can help decrease weed interference in sugarcane fields by reducing moisture on the soil surface, which is where most weed seed germination and successful seedling emergence occurs (Bullied et al. 2012, 2014, Forcella et al. 2000).

The development of mechanized green harvest has made it possible to avoid burning sugarcane fields before harvest. This practice has gained acceptance because it reduces greenhouse gas emissions and the crop residues left on the ground help maintain soil moisture reducing irrigation requirements (Leal et al. 2013). Sugarcane residues resulting from green harvest can reduce weed establishment by forming a barrier to weed seedling emergence (Leon and Agüero 2001b, Leal et al. 2013, Martins et al. 1999). In general, there is a positive relationship between the amount of sugarcane residues/straw and weed seedling emergence inhibition, but weed species respond differently to this inhibitory effect, which can generate weed community shifts (Martins et al. 1999, Silva Junior et al. 2016). Additionally, sugarcane straw has been documented releasing phytochemicals that inhibit root growth of annual weed species although their impact on perennial weeds is limited (Sampietro et al. 2007, Sampietro and Vattuone 2006a, 2006b). Viator et al. (2006) showed that water soluble chemicals present in sugarcane residues after harvest were able to reduce the germination, radicle length, and seedling dry weight of tall morningglory (*Ipomoea hederacea* Jacq.) in up to 6-, 10-, and 2-fold, respectively, compared to non-treated plants. However, these allelopathic effects were evident only in a silt loam soil, while they were absent in a clay soil. Therefore, it is likely that differences in soil texture and organic matter content might affect the value of sugarcane allelopathic suppression of weed populations.

Mechanical Weed Control

Due to the perennial or semi-perennial nature of sugarcane production, mechanical weed control is mainly conducted before planting during field preparation (Braunack and McGarry 2006, Silva-Olaya et al. 2013). However, cultivation between rows is a very important tool especially where pre-emergence herbicide use is limited. Cultivation after planting in plant-cane or after harvest in ratoon-cane is implemented to achieve several objectives: i) remove emerged weed seedlings in the furrow, ii) shape furrows to facilitate irrigation, and iii) raise beds to protect sugarcane roots and rhizomes and to bury weed seedlings within sugarcane rows (Mrini et al. 2001). Although all these goals are important, cultivation timing is usually determined as a function of irrigation needs and is limited by sugarcane size. Cultivation can only be done as long as sugarcane shoots are small enough that the risk of breaking them with tractors is minimal. Segura and Agüero (1997) showed that in plant-cane, two sweep-cultivations between rows at 24 and 52 days after planting, controlled purple nutsedge similarly to programs including both pre-emergence and postemergence herbicides. However, relying only on cultivation might favor purple nutsedge propagation, so it is important to monitor changes in population dynamics in the short- and medium-term to timely identify the need for changes in control practices.

During the last few years there has been an increase in the use of green harvest and reduced- and no-tillage to minimize soil compaction and greenhouse gas emissions (De Figueiredo and La Scala 2011, Wood 1991). However, in those production systems mechanical weed control is limited because of the abundant crop residue that lies on the soil surface, and other control tools, such as herbicides and handweeding gain importance for weed management (Grange et al. 2005, Judice et al. 2006).

Chemical Weed Control

The use of chemicals for weed control has become an integral component of sugarcane production in many countries. Chemical weed control in sugarcane began with the experimental use of sodium arsenite in Hawaii in 1913 followed by development of contact herbicide formulas between 1944 and 1961 based on sodium pentachlorophenate, diesel oil, aromatic oils, and pentachlorophenol (Hanson 1962). However, it was the discovery of phenoxy herbicides particularly 2,4-D that selectively killed broadleaf weeds in graminaceous crops in the early 1940s that revolutionized chemical weed control in many crops including sugarcane. The use of the phenoxy herbicide 2,4-D in sugarcane increased after its commercialization and is still a major herbicide used for broadleaf weed control in the crop. The benefits of herbicides including superior weed control and cost effectiveness compared to manual and mechanical weed control were observed from the early days of herbicide use in sugarcane. For example, 0.075% aqueous solution of the ammonium salt of 2,4-D at 0.20 $/ha provided control of broadleaf weeds in a sugarcane plantation in Puerto Rico within two weeks after application (Overbeek 1947). However, application of the 2,4-D solution to a dense stand of tall, mature weeds did not provide control (Overbeek 1947), illustrating the importance of appropriate timing of application on herbicide efficacy from the early days of commercialization of herbicides.

Rapid development of herbicides occurred after World War II with advances in chemistry resulting in production of several herbicides with different modes of action for selective control of broadleaf and grass weeds in sugarcane. These herbicides are applied preplant or prior to planting, pre-emergence to the crop, weed or both, and post-emergence after the crop, weed, or both have emerged to maximize weed control and herbicide selectivity. Sugarcane herbicides are used alone or in combination to broaden weed control spectrum and provide a weed-free environment for sugarcane growth and development especially early in the season before canopy closure when sugarcane is most vulnerable to weed competition. The common herbicides used for selective weed control in sugarcane include members of triazine, phenoxy, dinitroaniline, chloroacetamide, carbamate, and sulfonyl urea families.

Triazine herbicides atrazine, ametryn, simazine, metribuzin, and hexazinone are most commonly used in sugarcane for pre-emergence and post-emergence weed control. Atrazine is the most widely used because of its affordability, crop safety, and ability to provide consistent residual, broad spectrum weed control, and flexibility of application and tank-mixing (Smith et al. 2002). However, bacterial adaptations that enable enhanced degradation of atrazine and similar s-triazines have reduced its recalcitrant ability to provide persistent, season-long residual weed control (Krutz et al. 2010). Enhanced atrazine degradation and cross-adaptation with ametryn, a s-triazine and not the non-symmetrical metribuzin has been reported in Florida and Hawaii sugarcane soils with previous atrazine use history (Shaner et al. 2010). Rapid degradation of atrazine compared to metribuzin occurred under field conditions in Florida, indicating that metribuzin is a better option for weed control in sugarcane on soils exhibiting enhanced atrazine degradation (Odero and Shaner 2014b). Atrazine can be tank-mixed with other triazines (metribuzin and ametryn) or pendimethalin, S-metolachlor, alachlor, diuron, terbacil, or sulfentrazone to broaden weed control spectrum and provide longer residual control (Anonymous 1989, 2014, Jones and Griffin 2009, Odero and Dusky 2014, Orgeron 2016, Viator et al. 2002).

Ametryn is mainly used post-emergence in sugarcane to provide excellent control of small-seeded broadleaf and small grass weeds because of its limited pre-emergence activity. However, phytotoxicity can occur on sugarcane compared to atrazine especially at higher rates under relatively warm or hot conditions (Anonymous 1989, Odero and Dusky 2014, Smith et al. 2002). Similar to atrazine, the activity of ametryn is enhanced when tank-mixed with other triazines and other herbicides. Metribuzin provides pre-emergence and post-emergence control of small-seeded broadleaf and certain grass weeds. Metribuzin is mainly combined with pendimethalin when applied pre-emergence but can also be tank-mixed with hexazinone, terbacil, and clomazone particularly for control of problematic perennial grasses (Millhollon 1993, Orgeron 2016, Richard 1993). However, metribuzin is not used on mineral soils with low organic matter because its mobility and phytotoxicity are inversely correlated to soil organic matter content (Sharom and Stephenson 1976). Hexazinone provides control of many broadleaf and seedling grass weeds. The activity of hexazinone is greater when combined with diuron than when used alone (Anonymous 2014, Clement et al. 1989, Fadayomi 1988, Orgeron 2016). However, hexazinone is severely phytotoxic on sugarcane in light textured soils where injury is exacerbated with excessive rainfall (Millhollon 1980b). Simazine is another triazine herbicide labelled for control of small-seeded broadleaf and grass weeds in sugarcane presently used only in a limited number of countries where it is still registered for use (Smith et al. 2002).

Phenoxy herbicides 2,4-D and MCPA, and the benzoic acid herbicide dicamba are used in sugarcane to control many broadleaf weeds. Pre-plant or pre-emergence application of these herbicides before sugarcane emergence are made in combination with glyphosate or paraquat to broaden weed control spectrum. Use of a combination of 2,4-D and dicamba are common along with the combination of the two herbicides with atrazine, metribuzin or diuron (Griffin and Judice 2009). Many growers make layby application of these herbicides to ensure weed-free conditions until sugarcane harvest. Weed escapes especially *Ipomea* species that climb and wrap around sugarcane stalks, thereby impeding harvesting are controlled late in the season with over the canopy application of 2,4-D alone or in combination with dicamba (Siebert et al. 2004).

Dinitroaniline and chloroacetamide herbicides are used pre-emergence to provide residual control of many seedling grasses in sugarcane. Pendimethalin and trifluralin, dinitroaniline herbicides used in sugarcane, require incorporation into the soil mechanically or by irrigation or rainfall to improve their efficacy and minimize losses from volatilization and photodecomposition (Clement et al. 1989, Millhollon 1993, Odero and Shaner 2014a, Weber 1990). Chloroacetamides used in sugarcane include alachlor, acetochlor, S-metolachlor, and metazachlor. These dinitroaniline and chloroacetamide herbicides can be applied in combination with triazines, diuron, mesotrione, and terbuthylazine as tank-mixes or premixes to broaden weed control spectrum (Anonymous 2014, Chedzey and Findlay 1986). Other herbicides used pre-emergence in sugarcane alone or in combination with other herbicides for control of broadleaf and grass weeds include terbacil, diuron, isoxaflutole, mesotrione, amicarbazone, sulfentrazone, flumioxazin, EPTC, and tebuthiuron (Anonymous 2014, Odero and Dusky 2014, Orgeron 2016).

Post-emergence annual grass control in sugarcane is achieved using the carbamate herbicide, asulam and the sulfonyl urea herbicide, trifloxysulfuron. These herbicides are usually applied in combination to enhance efficacy especially on perennial grasses (Dalley and Richard 2008, Richard 1990). Grass control with post-directed application underneath the sugarcane canopy can be achieved using isoxaflutole and MSMA (Anonymous 2014). Sulfonyl urea herbicides halosulfuron and trifloxysulfuron, and the aryl triazinone, sulfentrazone are used to control *Cyperus* species in sugarcane (Etheredge et al. 2010). The sulfonyl urea herbicides can be applied post-emergence over-the-top of sugarcane while sulfentrazone can only be applied pre-emergence or be post-directed up to sugarcane layby.

Although the sugarcane crop can last for many years resulting in repeated application of the same herbicides season after season, there have been no reports of herbicide resistant weeds in the crop. Care must be taken when planning herbicide programs in sugarcane to manage against herbicide resistance because of the limited number of herbicides used in the crop. Use of herbicide tank-mixes, integrating other control methods particularly mechanical cultivation, rotational crops, and the fallow period provides ability to mitigate evolution of herbicide resistance in sugarcane.

Challenges for Sustainable Weed Control

The trend to reduce tillage and increase green harvest in sugarcane production represents an important challenge to sustainable weed control. Although those practices generate important benefits on aspects, such as soil health and quality, reduction in greenhouse gas emissions, reduction in fuel consumption, and increase in soil moisture retention, they increase the reliance on herbicides for effective weed control. Due to the limited number of selective mechanisms of action that can be used on sugarcane, a more intensive use of herbicides will favor herbicide resistance evolution, which ultimately jeopardize the sustainability of the sugarcane production system. In order to prevent this from happening, efforts should be made to maintain a strong strategy that emphasizes cultural weed management and diversification of weed control tools (Owen et al. 2015).

The lack of crop rotation is a driving factor promoting weed communities that are successful within sugarcane fields. Perennial species are particularly favored by the longer life cycle of this crop that prevents more dynamic changes in soil preparation, vegetation removal and diversity in weed control tools and crop canopy characteristics (i.e., shape, height, growing season). In order to avoid weed adaptations to current management practices, taking steps to diversify weed control and production practices is imperative. Potential options for diversification include crop rotation, intercropping, and rotation between conventional and green harvest and reduced tillage with cultivation. This type of dynamic change in production activities will counter weed adaptations at the species and community levels, limiting the possibilities of a few species becoming predominant as described in different sugarcane producing regions (Harker 2013, Owen et al. 2015).

Concluding Remarks

Weed management in sugarcane relies on diverse weed control among which, cultural practices, such as cultivar selection, irrigation and fertilization strategies, and weed control during fallow periods play a major role to maintain weed populations under manageable levels. Weed community composition in sugarcane fields seems to be highly determined by crop management practices, but research is needed to identify how to modify those practices to prevent population growth of frequent and problematic weed species. Finally, the increase in the use of green harvest and reduced-tillage in sugarcane fields will likely favor herbicide use increasing selection pressure for herbicide resistance evolution. Despite being a semi-perennial crop, sugarcane production allows enough diversification of weed control actions making possible the development of strong integrated programs that ultimately will ensure sustainable weed management.

REFERENCES

Ali, A.D., E. Reagan, L.M. Kitchen and J.L. Flynn. 1986. Effects of johnsongrass (*Sorghum halepense*) density on sugarcane (*Saccharum officinarum*) yield. Weed Sci. 34: 381–383.

Amores, M.J., F.D. Mele, L. Jiménez and F. Castells. 2013. Life cycle assessment of fuel ethanol from sugarcane in Argentina. Int. J. Life Cycle Assess. 18: 1344–1357.

Andrew, I.K.S., J. Storkey and D.L. Sparkes. 2015. A review of the potential for competitive cereal cultivars as a tool in integrated weed management. Weed Res. 55: 239–248.

Anonymous. 1989. Weeds in Australian Cane Fields. Part B: A Guide to the Selection of Chemical Control Measures. BSES Bulletin No. 28 October 1989.

Anonymous. 2014. Herbicide Guide, SASRI.

Ashori, A., M. Marashi, A. Ghasemian and E. Afra. 2013. Utilization of sugarcane molasses as a dry-strength additive for old corrugated container recycled paper. Composites Part B 45: 1595–1600.

Bakker, H. 1999. Sugarcane Cultivation and Management. Kluwer Academic/Plenum Publishers, New York. 679 p.

Batchelor, C.H., G.C. Soopramanien, J.P. Bell, R. Nayamuth and M.G. Hodnett. 1990. Importance of irrigation regime, dripline placement, and row spacing in the drip irrigation of sugar cane. Agric. Water Manag. 17: 75–94.

Bell, M.J. and A.L. Garside. 2005. Shoot and stalk dynamics and the yield of sugarcane crops in tropical and subtropical Queensland, Australia. Field Crop. Res. 92: 231–248.

Bhullar, M.S., U.S. Walia, S. Singh, M. Singh and A. Jhala. 2012. Control of morningglories (*Ipomoea* spp.) in sugarcane (*Saccharum* spp.). Weed Technol. 26: 77–82.

Braunack, M.V. and D. McGarry. 2006. Traffic control and tillage strategies for harvesting and planting sugarcane (*Saccharum officinarum*) in Australia. Soil Tillage Res. 89: 86–106.

Brecke, B.J. and W.B. Duke. 1980. Dormancy, germination, and emergence characteristics of fall panicum (*Panicum dichotomiflorum*) seed. Weed Sci. 28: 683–685.

Bullied, W.J., P.R. Bullock, G.N. Flerchinger and R.C. van Acker. 2014. Process-based modeling of temperature and water profiles in the seedling recruitment zone: Part II. Seedling emergence timing. Agr. Forest Meteorol. 188: 104–120.

Bullied, W.J., R.C. van Acker and P.R. Bullock. 2012. Hydrothermal modeling of seedling emergence timing across topography and soil depth. Agron. J. 104: 423–436.

Camp, C.R. 1998. Subsurface drip irrigation: A review. Trans. ASAE. 41: 1353–1367.

Chapman, L.S. 1966. Prolific nutgrass growth. Cane Grower's Quaterly Bulletin. 30: 1.

Chedzey, J. and J.B.R. Findlay. 1986. The use of acetochlor for weed control in sugarcane. Proc. S. Afr. Sug. Technol. Ass. 60: 183–190.

Cheeroo-Nayamuth, F.C., M.J. Robertson, M.K. Wegener and A.R.H. Nayamuth. 2000. Using a simulation model to assess potential and attainable sugar cane yield in Mauritius. Field Crop. Res. 66: 225–243.

Cherry, R., D.G. Hall, A. Wilson and L. Baucum. 2011. First report of damage by the sugarcane root weevil *Diaprepes abbreviatus* (Coleoptera: Curculionidae) to Florida sugarcane. Fla. Entomol. 94: 1063–1065.

Clement, A.A., J.S. Lammel, J.A. Filho and J.C. Barbosa. 1989. Weed control in sugar cane with hexazinone and its mixtures with diuron in pre-emergence. Planta Daninha 2: 85–88.

Clements, H.F. 1980. Sugarcane Crop Logging and Crop Control: Principles and Practices. The University Press of Hawaii, Honolulu. 520 p.

Dalley, C.D. and E.P. Richard Jr. 2008. Control of rhizome johnsongrass (*Sorghum halepense*) in sugarcane with trifloxysulfuron and asulam. Weed Technol. 22: 397–401.

De Figueiredo, D.B. and N. La Scala. 2011. Greenhouse gas balance due to the conversión of sugarcane áreas from burned to Green harvest in Brazil. Agr. Ecosyst. Environ. 141: 77–85.

Dias, M.O.S., M.P. Cunha, C.D.F. Jesus, G.J.M. Rocha, J.G.C. Pradella, C.E.V. Rossell, R.M. Filho and A. Bonomi. 2011. Second generation ethanol in Brazil: can it compete with electricity production? Bioresour. Technol. 102: 8964–8971.

Earle, F.S. 1928. Sugar Cane and Its Culture. John Wiley & Sons, New York. 355 p.

Etheredge, L.M., J.L. Griffin and J.M. Boudreaux. 2010. Nutsedge (*Cyperus* spp.) control programs in sugarcane. J. Am. Soc. Sugar Cane Technol. 30: 67–80.

Etheredge Jr, L.M., J.L. Griffin and M.E. Salassi. 2009. Efficacy and economics of summer fallow conventional and reduced-tillage programs for sugarcane. Weed Technol. 23: 274–279.

Fadayomi, O. 1988. Weed control in sugar cane with hexazinone alone or in combination with diuron. J. Agric. Sci. 111: 333–337.

(FAO) Food and Agriculture Organization of the United Nations. 2014. FAO Statistical Yearbook 2014. Europe and Central Asia Food and Agriculture. Food and Agriculture Organization of the United Nations Regional Office for Europe and Central Asia, Budapest. 113 p.

(FAO) Food and Agriculture Organization of the United Nations. 2015. FAO Statistical Pocketbook World Food and Agriculture. Food and Agriculture Organization of the United Nations, Rome. 231 p.

Fausey, J.C. and K.A. Renner. 1997. Germination, emergence, and growth of giant foxtail (_Setaria faberi_) and fall panicum (_Panicum dichotomiflorum_). Weed Sci. 45: 423–425.

Fernandez, J.V., D.C. Odero, G.E. MacDonald, J. Ferrell and L.A. Gettys. 2015a. Confirmation, characterization, and management of glyphosate-resistant ragweed parthenium (_Parthenium hysterophorus_ L.) in the Everglades agricultural area of South Florida. Weed Technol. 29: 233–242.

Fernandez, J.V., D.C. Odero and A.L. Wright. 2015b. Effects of _Parthenium hysterophorus_ L. residue on early sugarcane growth in organic and mineral soils. Crop Prot. 72: 31–35.

Firehun, Y. and T. Tamado. 2006. Weed flora in the Rift Valley sugarcane plantation of Ethiopia as influenced by soil types and agronomic practises. Weed Biol. Manag. 6: 139–150.

Firehun,Y., T. Tamado, T. Abera and Z. Yohannes. 2012. Competitive ability of sugarcane (_Saccharum officinarum_ L.) cultivars to weed interference in sugarcane plantations of Ethiopia. Crop Prot. 32: 138–143.

Forcella, F., R.L. Benech-Arnold, R. Sanchez and C.M. Ghersa. 2000. Modeling seedling emergence. Field Crop. Res. 67: 123–139.

Garside, A.L., M.J. Bell, J.E. Berthelsen and N.V. Halpin. 2002. Effect of fumigation, density and row spacing on the growth and yield of sugarcane in two diverse environments. Proc. Aust. Soc. Sugar Cane Technol. 24: 135–144.

Garside, A.L., M.J. Bell and B.G. Robotham. 2009. Row spacing and planting density effects on the growth and yield of sugarcane. 2. Strategies for the adoption of controlled traffic. Crop Pasture Sci. 60: 544–554.

Govinthasamy, K. and P.B. Cavers. 1995. The effects of smut (_Ustilago destruens_) on seed production, dormancy, and viability in fall panicum (_Panicum dichotomiflorum_). Can. J. Bot. 73: 1628–1634.

Grange, I. P. Prammanee and P. Prasertsak. 2005. Comparative analysis of different tillage systems used in sugarcane (Thailand). Austr. Farm Bus. Manag. J. 2: 46–50.

Grattan, S.T., L.J. Schwankl and W.T. Lanini. 1988. Weed control by subsurface drip irrigation. Calif. Agr. 42: 22–24.

Griffin, J.L. and W.A. Judice. 2009. Winter weed control in sugarcane. J. Am. Soc. Sugar Cane Technol. 29: 128–136.

Griffin, J.L., D.K. Miller and M.E. Salassi. 2006. Johnsongrass (_Sorghum halepense_) control and economics of using glyphosate-resistant soybean in fallowed sugarcane fields. Weed Technol. 20: 980–985.

Hanson, N.S. 1962. Weed control practices and research for sugar cane in Hawaii. Weeds 10: 192–200.

Harker, K.N. 2013. Slowing weed evolution with integrated weed management. Can. J. Plant Sci. 95: 759–764.

Hodnett, M.G., J.P. Bell, A.H. Koon, G.C. Soopramanien and C.H. Batchelor. 1990. The control of drip irrigation of sugarcane using 'index' tensiometers: some comparisons with control by the water budget method. Agric. Water Manag. 17: 189–207.

Holm, L.G., D.L. Blucknett., J.V. Pancho and J.P. Herberger. 1977. The world's worst weeds. Distribution and Biology. Univ. Press of Hawaii, Honolulu. 609 p.

Horowitz, M. 1973. Spatial growth of _Sorghum halepense_ (L.) Pers. Weed Res. 13: 200–208.

Inman-Bamber, N.G. 1994. Temperature and seasonal effects on canopy development and light interception of sugarcane. Field Crop. Res. 36: 41–51.

Jangaard, N.O., M.M. Sckerl and R.H. Schieferstein. 1971. The role of phenolics and abscisic acid in nutsedge tuber dormancy. Weed Sci. 19: 17–20.

Jannink, J.L., J.H. Orf, N.R. Jordan and R.G. Shaw. 2000. Index selection for weed suppressive ability in soybean. Crop Sci. 40: 1087–1094.

Jones, C.A. and J.L. Griffin. 2009. Red morningglory (_Ipomoea coccinea_) control and competition in sugarcane. J. Am. Soc. Sugar Cane Technol. 29: 25–35.

Judice, W.E., J.L. Griffin, C.A. Jones, L.M. Etheredge Jr. and M.E. Salassi. 2006. Weed control and economics using reduced tillage programs in sugarcane. Weed Technol. 20: 319–325.

Kanwar, R.S. and K.K. Sharma. 1974. Effect of interrow spacing on tiller mortality, stalk population and yield of sugarcane. Int. Soc. Sugar Cane Technol. Proc. 15: 751–755.

Keely, P.E. 1987. Interference and interaction of purple and yellow nutsedges (*Cyperus rotundus* and *C. esculentus*) with crops. Weed Technol. 1: 74–81.

Kohli, R.K., D.R. Batish, H.P. Singh and K.S. Dogra. 2006. Status, invasiveness and environmental threats of three tropical American invasive weeds (*Parthenium hysterophorus* L., *Ageratum conyzoides* L., *Lantana camara* L.) in India. Biol. Invasions. 8: 1501–1510.

Krutz, L.J., D.L. Shaner, W.A. Weaver, R.M.T. Webb, R.M. Zablotowicz, K.N. Reddy, Y. Huang and S.J. Thomson. 2010. Agronomic and environmental implications of enhanced s-triazine degradation. Pest Manag. Sci. 66: 461–481.

Kuva, M.A., R.A. Pitelli, P.L.C.A. Alves, T.P. Salgado and M.C.D.M. Pavani. 2008. Banco de sementes de plantas daninhas e sua corelaçã com a flora establecida no agroecossistema cana-crua. Planta Daninha. 26: 735–744.

Kuva, M.A., R.A. Pitelli, T.P. Salgado and P.L.C.A. Alves. 2007. Fitossociologia de comunidades de plantas daninhas em agroecossistema cana-crua. Planta Daninha. 25: 501–511.

Leal, M.R.L.V., M.V. Galdos, F.V. Scarpare, J.E.A. Seabra, A.Walter and C.O.F. Oliveira. 2013. Sugarcane straw availability, quality, recovery and energy use: A literature review. Biomass Bioenerg. 53: 11–19.

Lencse, R.J. and J.L. Griffin. 1991. Itchgrass (*Rottboellia cochinchinensis*) interference in sugarcane (*Saccharum* sp.). Weed Technol. 5: 396–399.

Leon, R. and R. Agüero. 2001a. Efecto de la profundidad del suelo en *Rottboellia cochinchinensis* (Lour) Clayton en caña de azúcar (*Saccharum officinarum* L.). Agronomia Mesoamericana. 12: 65–69.

Leon, R. and R. Agüero. 2001b. Efecto de tipos de labranza sobre la población de malezas en caña de azúcar. Agronomia Mesoamericana. 12: 71–77.

Leon, R.G., R. Agüero and D. Calderón. 2017a. Diversity of spatial heterogeneity of weed communities in a sugarcane cropping system in the dry tropics of Costa Rica. Weed Sci. 65: 128–140.

Leon, R.G., M.J. Mulvaney and B.L. Tillman. 2017b. Peanut cultivars differing in growth habit and canopy architecture respond similarly to weed interference. Peanut Sci. 43: 133–140.

Martins, D., E.D. Velini, C.C. Martins and L.S. De Souza. 1999. Broadleaf weed emergence in soil covered with sugar cane straw. Planta Daninha. 17: 151–161.

Matherne, R.J. 1974. Effects of inter-row spacing on sugarcane yields in Louisiana. Proc. Int. Soc. Sugar Cane Technol. 15: 746–750.

Mercado, B.L. 1978. Biology, problems and control of *Rottboellia exaltata* L.F. A monograph. Biotropical Bulletin No. 14. Seameo Regional Center for Tropical Biology, Bogor, Indonesia. 38 p.

Millhollon, R.W. 1965. Growth characteristics and control of *Rottboellia exaltata* L.F. a new weed in sugarcane. Sugar Bull. 44: 82–88.

Millhollon, R.W. 1980a. Johnsongrass competition and control in succession-planted sugarcane. Proc. Int. Soc. Sugar Cane Technol. 17: 85–92.

Millhollon, R.W. 1980b. Johnsongrass (*Sorghum halepense*) control and sugarcane tolerance from preemergence treatments with hexazinone. Proc. Int. Soc. Sugar Cane Technol. 17: 63–74.

Millhollon, R.W. 1990. Differential response of sugarcane cultivars to competition from johnsongrass (*Sorghum halepense*). Proc. Int. Soc. Sugar Cane Technol. 20: 577–584.

Millhollon, R.W. 1992. Effect of itchgrass (*Rottboellia cochinchinensis*) interference on growth and yield of sugarcane (*Saccharum* spp. hybrids). Weed Sci. 40: 48–53.

Millhollon, R.W. 1993. Preemergence control of itchgrass (*Rottboellia cochinchinensis*) and johnsongrass (*Sorghum halepense*) in sugarcane (*Saccharum* spp. hybrids) with pendimethalin and prodiamine. Weed Sci. 41: 621–626.

Millhollon, R.W. 1995. Growth and yield of sugarcane as affected by johnsongrass (*Sorghum halepense*) interference. J. Am. Soc. Sugar Cane Technol. 15: 32–40.

Monaghan, N. 1979. The biology of johnsongrass (*Sorghum halepense*) Weed Res. 19: 261–267.

Monquero, P.A., L.R. Amaral, D.P. Binha, P.V. Silva, A.C. Silva and F.R.A. Martins. 2008. Mapas de infestação de plantas daninhas em diferentes sistemas de colheita da cana-de-açúcar. Planta Daninha. 26: 47–55.

Moreira, J.R. and J. Goldemberg. 1999. The alcohol program. Energy Pol. 27: 229–245.

Mrini, M., F. Senhaji and D. Pimentel. 2001. Energy analysis of sugarcane production in Morocco. Environ. Dev. Sustain. 3: 109–126.

Odero, D.C. 2012. Response of ragweed parthenium (*Parthenium hysterophorus*) to saflufenacil and glyphosate. Weed Technol. 26: 443–448.

Odero, D.C., R.H. Cherry and D.G. Hall. 2013. Weedy host plants of the sugarcane root weevil (Coleoptera: Curculionidae) in Florida sugarcane. J. Entomol. Sci. 48: 81–89.

Odero, D.C., M. Duchrow and N. Havranek. 2016. Critical timing of fall panicum (*Panicum dichotomiflorum*) removal in sugarcane. Weed Technol. 30: 13–20.

Odero, D.C. and J.A. Dusky. 2014. Weed management in sugarcane. IFAS, Florida Coop Ext Serv, University of Florida, EDIS SS-AGR-09.

Odero, D.C. and D.L. Shaner. 2014a. Dissipation of pendimethalin in organic soils in Florida. Weed Technol. 28: 82–88.

Odero, D.C. and D.L. Shaner. 2014b. Field dissipation of atrazine and metribuzin in organic soils in Florida. Weed Technol. 28: 578–586.

Orgeron, A. 2016. Sugarcane weed management. pp. 67–93. *In:* Louisiana Suggested Weed Management Guide. LSU AgCenter.

Osgood, R.V., E. Floresca and H.W. Hilton. 1977. How important is nutsedge competition with sugarcane? pp. 42–43. *In:* Hawaiian Sugar Planters' Association Experimental Station Annual Report.

Overbeek, J.V. 1947. Use of synthetic hormones as weed killers in tropical agriculture. Eco. Bot. 1: 446–459.

Owen, M.D.K., H.J. Beckie, J.Y. Leeson, J.K. Norsworthy and L.E. Steckel. 2015. Integrated pest management and weed management in the United States and Canada. Pest Manag. Sci. 71: 357–376.

Perdomo, F., H. Vibrans, A. Romero, A. Domínguez and J.L. Medina. 2004. Análisis de SHE, una herramienta para estudiar la diversidad de maleza. Revista de Fitotecnia de México. 27: 57–61.

Quayyum, H.A., A.U. Mallik, D.M. Leach and C. Gottardo. 2000. Growth inhibitory effects of nutgrass (*Cyperus rotundus*) on rice (*Oryza sativa*) seedlings. J. Chem. Ecol. 26: 2211–2231.

Reinhardt, C., S. Karus, F. Walker, L. Foxcroft, P. Robbertse and K. Hurle. 2004. The allelochemical parthenin is sequestered at high level in capitate-sessile trichomes on the leaf surface of *Parthenium hysterophorus*. J. Plant Dis. Protect. 19: 253–261.

Richard Jr, E.P. 1990. Timing effects on johnsongrass (*Sorghum halepense*) control with asulam in sugarcane (*Saccharum* sp.). Weed Technol. 4: 81–86.

Richard Jr, E.P. 1992. Bermudagrass interference during a three-year sugarcane crop cycle. Proc. Int. Soc. Sugar Cane Technol. 21: 31–41.

Richard Jr, E.P. 1993. Preemergence herbicide effects on bermudagrass (*Cynodon dactylon*) interference in sugarcane (*Saccharum* spp. hybrids). Weed Technol. 7: 578–584.

Richard Jr, E.P. 1995. Johnsongrass (*Sorghum halepense*) control in fallow sugarcane (*Saccharum* spp. hybrids) fields. Weed Technol. 11: 410–416.

Richard Jr, E.P. and C.D. Dalley. 2007. Sugarcane response to bermudagrass interference. Weed Technol. 21: 941–946.

Richard Jr, E.P., J.W. Dunckleman and C.E. Carter. 1991. Productivity of sugarcane on narrow rows, as affected by mechanical harvesting. Field Crops Res. 26: 375–386.

Rodríguez Cuevas, C.N. and A.R. Romero Manzanares. 1994. Flora segetal cañera de la provincia de Holguín, Cuba. Biotam. 5(3): 39–50.

Sampietro, D.A., M.A. Sgariglia, J.O. Soberon, E.N. Quiroga and M.A. Vattuone. 2007. Role of sugarcane straw allelochemicals in the growth suppression of arrowleaf sida. Environ. Exp. Bot. 60: 495–503.

Sampietro, D.A. and M.A. Vattuone. 2006a. Nature of the interference mechanism of sugarcane (*Saccharum officinarum* L.) straw. Plant Soil. 280: 157–169.

Sampietro, D.A. and M.A. Vattuone. 2006b. Sugarcane straw and its phytochemicals as growth regulators of weed and crop plants. Plant Growth Regul. 48: 21–27.

Sanchez-Tames, R., M.D.V. Gesto and E. Vieitez. 1973. Growth substances isolated from tubers of *Cyperus esculentus* var. *aureus*. Physiol. Plant. 28: 195–200.

Segura, C. and R. Agüero. 1997. Combate de coyolillo (*Cyperus rotundus* L.) en caña de azúcar (*Saccharum officinarum* L.): Hacia un manejo integral. Agronomia Mesoamericana. 8: 101–106.

Shaner, D.L., L.J. Krutz, W. Henry, B.D. Hanson, M.D. Poteet, C.R. Rainbolt. 2010. Sugarcane soils exhibit enhanced atrazine degradation and cross adaptation to other s-triazines. J. Am. Soc. Sugar Cane Technol. 30: 1–10.

Sharom, M.S. and G.R. Stephenson. 1976. Behavior and fate of metribuzin in eight Ontario soils. Weed Sci. 24: 153–160.

Shresta, A., J.P. Mitchell and W.T. Lanini. 2007. Subsurface drip irrigation as a weed management tool for conventional and conservation tillage tomato (*Lycopersicon esculentum* Mill.) production in semi-arid agroecosystems. J. Sustain. Agr. 31: 91–112.

Shrivastava, P.K., M.M. Parikh, N.G. Sawani and S. Raman. 1994. Effect of drip irrigation and mulching on tomato yield. Agric. Water Manag. 25: 179–184.

Siebert, J.D., J.L. Griffin and C.A. Jones. 2004. Red morningglory (*Ipomoea coccinea*) control with 2,4-D and alternative herbicides. Weed Technol. 18: 38–44.

Silva Junior, A.C., C.C. Martins and D. Martins. 2016. Effects of sugarcane straw on grass weed emergence under field conditions. Bioscience J. 32: 863–872.

Silva-Olaya, A.M., C.E.P. Cerri, N. La Scala Jr., C.T.S. Dias and C.C. Cerry. 2013. Carbon dioxide emissions under different soil tillage systems in mechanically harvested sugarcane. Environ. Res. Lett. 8: 015014.

Simpson, S.E., H.N. Nigg, N.C. Coile and R.A. Adair. 1996. *Diaprepes abbreviatus* (Coleoptera: Curculionidae): Host plant associations. Environ. Entomol. 25: 333–349.

Singels, A. and M.A. Smit. 2009. Sugarcane response to row spacing-induced competition for light. Field Crop. Res. 113: 149–155.

Singels, A., M.A. Smit, K.A. Redshaw and R.A. Donaldson. 2005. The effect of crop start date, crop class and cultivar on sugarcane canopy development and radiation interception. Field Crop. Res. 92: 249–260.

Smith, D.T., E.P. Richard Jr. and L.T. Santo. 2002. Weed control in sugarcane and the role of triazine herbicides. pp. 185–198. *In:* LeBaron, H., McFarland, J. and Burnside, O. (eds.) The Triazine Herbicides 50 Years Revolutionizing Herbicides. Elsevier, Oxford.

Stoller, E.W. and R.D. Sweet. 1987. Biology and life cycle of purple and yellow nutsedges (*Cyperus rotundus* and *C. esculentus*). Weed Technol. 1: 66–73.

Surendran, U., M. Jayakumar and S. Marimuthu. 2016. Low cost drip irrigation: impact on sugarcane yield, water and energy saving in semiarid tropical agro ecosystem in India. Sci. Total Environ. 573: 1430–1440.

Tamado, T. and P. Milberg. 2004. Control of parthenium (*Parthenium hysterophorus*) in grain sorghum (*Sorghum bicolor*) in the smallholder farming system in Eastern Ethiopia. Weed Technol. 18: 100–105.

Thomas, P.E.L. and J.C.S. Allison. 1975. Seed dormancy and germination in *Rottboellia exaltata*. J. Agr. Sci. 85: 129–134.

Viator, B.J., J.L. Griffin and J.M. Ellis. 2002. Sugarcane (*Saccharum* spp.) response to azafeniden applied preemergence and postemergence. Weed Technol. 16: 444–451.

Viator, R.P., R.M. Johnson, C.C. Grimm and E.P. Richard Jr. 2006. Allelopathic, autotoxic, and hormetic effects of postharvest sugarcane residue. Agron. J. 98: 1526–1531.

Wallace, J.S., C.H. Batchelor, D.N. Dabeesing, M. Teeluck and G.C. Soopramanien. 1991. A comparison of the light interception and water use of plant and first ratoon sugar cane intercropped with maize. Agr. Forest. Meteorol. 57: 85–105.

Watson, P.R., D.A. Derksen and R.C. van Acker. 2006. The ability of 29 barley cultivars to compete and withstand competition. Weed Sci. 54: 783–792.

Weber, J.B. 1990. Behavior of dinitroaniline herbicides in soils. Weed Technol. 4: 394–406.

Wiedenfeld, B. 2004. Scheduling water application on drip irrigated sugarcane. Agric. Water Manag. 64: 169–181

Wood, A.W. 1991. Management of crop residues following green harvesting of sugarcane in north Queensland. Soil Tillage Res. 20: 69–85.

Yirefu, F., T. Tamado, A. Tafesse and Y. Zekarias. 2012. Competitive ability of sugarcane (*Saccharum officinarum* L.) cultivars to weed interference in sugarcane plantations in Ethiopia. Crop Prot. 32: 138–143.

Sustainable Weed Control in Pineapple

Victor Martins Maia[*][1], **Ignacio Aspiazú**[1] **and Rodinei Facco Pegoraro**[2]

[1] Department of Agricultural Sciences, State University of Montes Claros, 2630 Reinaldo Viana Avenue, BOX 91, Janaúba, MG, Brazil
[2] Institute of Agricultural Sciences, Federal University of Minas Gerais, Montes Claros, MG, Brazil

Introduction

Pineapple (*Ananas comosus* var. *comosus*) is among the five most important tropical fruits in the world with more than one million hectares of planted area and a production of 25.4 million tonnes in 2014. About 90 countries have pineapple cultivation areas; however, the 10 largest producers concentrate 70% of total world production. Main pineapple producers are Costa Rica, Brazil, Philippines, Thailand and Indonesia, which are responsible for 46% of the pineapple produced in the world (FAOSTAT 2017).

Costa Rica stands out as the world's leading exporter of fresh pineapple. The main importers are the North American countries, especially the United States, which is the world's leading importer of fresh pineapple, canned pineapple, concentrated pineapple juice and plain pineapple juice, followed by countries in Europe (mainly Netherlands, Belgium, Germany) and Japan. Countries, such as Brazil, India and China, despite the high production, have as main destination the domestic market (FAOSTAT 2017).

Although there are some mechanized or semi-mechanized farming practices, most of the activities of this crop require the intensive use of labor, which causes or requires the generation of many jobs in pineapple farming. Weed control is among the main practices that increase costs and hinder the cultivation of pineapple. In this case, both fully mechanized practices and those that rely exclusively on human labor can be used. The cost of weed control in pineapple fields has a deep impact in the total cost of production, therefore, the adoption of practices to reduce it with the least possible environmental impact are fundamental to the sustainability of the crops.

As it is a tropical fruit, the pineapple is cultivated commercially in the tropical and subtropical regions of the world, between latitudes 30° N and 30° S, which are characterized by the occurrence of high temperatures throughout most parts of the year. Commercial crops are dispersed in moist, semi-arid and arid areas and in different altitude conditions, which makes this agro-ecosystem very diversified.

Weed Impact

The pineapple is characterized as a slow growing plant and not very aggressive to compete with weeds. This slow growth results in long production cycles that vary, mostly, from 12 to 36

---

*Corresponding author: victor.maia@unimontes.br

months according to the location, the cultivar, the type of seedling and the cultural practices used. Weed competition can increase this production cycle even more. This slow growth is explained by the mechanisms that this species show to tolerate and adapt to water stress.

The plant shows some anatomical, physiological, and morphological adaptations which make its need for water inferior to other herbaceous species, and make it able to survive even in low rainfall conditions. However, in order to achieve high productivity, the use of irrigation in arid or semi-arid environments or with long periods of drought is fundamental. In crops grown in wetlands, the use of irrigation is dispensable.

One of the anatomical adaptations is the presence of a small number of stomata which are predominantly on the abaxial surface of the leaf. The stomata are located in depressions in the leaf epidermis, which also has a thick layer of epicuticular wax and many trichomes (Malézieux et al. 2003). The morphological adaptations are in the form, phyllotaxy and insertion angle of the leaves (Malézieux et al. 2003). However, the physiological adaptation that stands out is the carbon-fixing metabolism of pineapple plants. This species is one of the few commercially grown crops with the crassulacean acid metabolism (CAM). The carbon-fixing metabolism of pineapple plants has four phases, which are well described in the literature (Osmond 1978, Malézieux et al. 2003). The main characteristic of this mechanism is the inverted pattern of stomata opening. In CAM metabolism plants, the stomata are open at night, when atmospheric CO_2 is fixed by the phosphoenolpyruvate carboxylase (PEPcase) enzyme, forming malate, which is accumulated in the vacuoles. During the day, when temperatures are higher and there is a higher vapor pressure deficit, malate is decarboxylated, the stomata close, and carbon, the product of malate decarboxylation, is used by ribulose 1,5-bisphosphate carboxylase-oxygenase (Rubisco), thus, initiating the Calvin cycle (Osmond 1978, Cote et al. 1993, Malézieux et al. 2003, Taiz et al. 2015).

The result of these adaptations is the economy and greater efficiency in the use of water. A plant with the crassulacean acid metabolism transpires about 150 g of water for each g of fixed CO_2, while plants with C_4 and C_3 metabolism transpire 300 g and 600 g of water for each g of fixed CO_2, respectively, (Szarek and Ting 1975). However, as a consequence of this adaptation, the photosynthetic rate (g CO_2 cm^{-2} s^{-1}) is much lower than that of most plants with C_3 and C_4 metabolism. Compared to wheat (C_3 metabolism), pineapple plants fix only 25% of CO_2 per unit of soil per day (Cote et al. 1993). However, in some situations, daily carbon fixation can reach high values, which explains the large amount of dry matter produced by the pineapple (Nobel 1991, Pegoraro et al. 2014, Maia et al. 2016). All this causes long production cycles (12–36 months) and little competitiveness with weeds (Figure 25.1).

In regions with latitudes close to the equator, plants with C_4 metabolism have a higher daily carbon gain (g CO_2 m^{-2} day^{-1}) than plants with C_3 metabolism. This behavior is maintained until latitudes close to 40° (Cox and Moore 1985). Considering the previously described pineapple cultivation range, plants with C_4 metabolism and especially grasses (Model et al. 2008) will, in most cases, cause greater losses in pineapple farming. Some examples of C_4 weeds that occur in areas of pineapple cultivation are: *Digitaria abyssinica, Digitaria scalarum, Cyperus* spp. and *Cynodon* spp. As for the C_3 species we can mention *Bidens pilosa, Plantago lanceolata, Ageratum conyzoides, Galinsoga parviflora* and *Oxalis* spp. (Eshetu et al. 2007).

The occurrence of weed species in pineapple fields is very varied due to the distribution of such fields in the most diverse countries and in the most different agroecosystems. However, some major weed species can be cited, such as *Amaranthus spinosus, Bidens pilosa, Emilia sagitata, Mikania micrantha, Heliotropium indicum, Commelina benghalensis, Commelina diffusa, Murdania nudiflora, Convolvulus arvensis, Ipomoea cairica, Ipomoea indica, Ipomoea plebeia, Ipomoea purpurea, Ipomoea triloba, Cyperus difformis, Cyperus iria, Cyperus rotundus, Fimbristylis miliaceae, Chenopodium album, Chamaesyce hirta, Crotalaria mucronata, Mimosa invisa, Mimosa pudica, Sida acuta, Oxalis corniculata, Agrostis alba, Dactyloctenium aegyptium, Digitaria insularis, Digitaria sanguinalis, Echinochloa colonum, Eleusine indica, Imperata cylindrica, Melinis minutiflora, Panicum maximum, Panicum repens,Paspalum conjugatum, Paspalum dilatatum, Paspalum urville, Pennisetum purpureum, Rottboellia cochinchinensis, Saccharum spontaneum, Setaria verticillata, Sorghum*

Figure 25.1. Commercial pineapple fields with different levels of weed competition, Minas Gerais, Brazil. (Source: Victor Martins Maia)

halepense and *Solanum nigrum* (Holm et al. 1977, Py et al. 1987, Bartholomew et al. 2003, Brenes-Prendas and Agüero-Alvarado 2007, Model et al. 2008, Model and Favreto 2009).

The plants that produce seeds include monocotyledons and dicotyledons, with approximately 170,000 species. This group covers almost all plants considered to be weeds (about 30,000 species). Of these, about 1,800 are considered more harmful because of their characteristics and their behaviour, causing great losses every year in agriculture. As for the life cycle, these can be annual (germinate, develop, flourish, produce seeds and die within a year); biennials (in the first year, they germinate and grow, in the second, produce flowers, fruits, seeds and die, and must be controlled in the first year); and perennials (live more than two years, being characterized by the renewal of growth, year after year, from the same root system). Correct identification of the weed species is crucial because, in certain cases, herbicide selectivity is based on morphological and/or physiological differences between the weeds and the crop (Silva et al. 2007).

The *Poaceae* is the family with the most weedy species. About two-thirds of the worst weeds in the world are single-season or annual weeds. The rest are perennials in the temperate areas of the world, but in the tropics, they are accurately called several-season weeds. The categories annual and perennial do not have the same meaning in tropical climates, where growth is not limited by cold weather but may be limited by low rainfall. About two-thirds of the important weeds are broadleaved or dicotyledonous species. Most of the rest are grasses, sedges or ferns (Zimdahl 2007).

The losses caused by the presence of weeds in an area can be direct or indirect. The former are caused directly on the crop, such as competition for resources or decrease of product value, while the latter are caused by impediments or restrictions caused in cultivation and harvest operations (Concenço et al. 2014).

Examples of direct losses are lower productivity and product quality and additional water demand (Concenço et al. 2014). The latter is more important in tropical regions, where low rainfall is common, and where pineapples are usually grown. Plants, such as *Bidens pilosa*, for example, can tolerate stronger competition, since they have the capacity to extract water from the soil even in conditions of low water potential (Aspiazú et al. 2010).

Other factors may be considered indirect losses. The decrease in area use efficiency, for example, causes the available space for the crop to undergo limitations of use in function of the present weed species (Concenço et al. 2014). Some species may also act as alternative hosts to various pests and diseases. There are 47 weed species in 42 genera that infest the pineapple crop and are alternative hosts of *Heterodera marioni*, a root-knot nematode (Godfrey 1935, cited by Zimdahl 2007). Also, species, such as *Lantana camara* and *Paspalum conjugatum* may host symphylids (Rusydi et al. 2012). Those species should be controlled even after harvesting and in adjacent areas, which lead to increased production costs. Obstruction of irrigation is another problem that can occur, because plants, such as water hyacinth (*Eichornia crassipes*) and water lettuce (*Pistia stratiotes*) can clog water channels when in high densities. Furthermore, some venomous and poisonous animals may hide in certain weed species, increasing the chance of accidents (Concenço et al. 2014).

Although there is no direct relationship between C_3/C_4 plants being adapted to sun or shade conditions, due to the nature of the C_4 metabolism, it is expected that such plants would adapt themselves better to full sun conditions, while some C_3 species may have a better ability to adapt to shaded environments due to the lower energetic cost (and consequent light demand) of its photosynthetic metabolism. However, there are C_3 weed species that are completely adapted to full sun conditions. Plants can adapt themselves to different light regimes while growing and developing. In this sense, leaves of alexander grass plants may show characteristics of shaded leaves while growing between crop rows, or full sun leaves while growing in a pasture, being fully exposed to sunlight. Leaf thickness may, in some cases, be associated to the amount of wax on the surface, a barrier the herbicides must pass to penetrate (Concenço et al. 2014).

To be considered a true weed, the species must possess some characteristics, such as dormancy and uneven seed germination, high capacity to produce propagules, reproduction both sexual and by vegetative parts, efficient dispersal mechanisms, among others (Silva et al. 2007, Zimdahl, 2007).

Weed management has a strong economic impact on agricultural production systems. By stealing human's energy, they demand control, which ultimately increases production costs. Although it is difficult to determine the exact cost of weed control, it is estimated that US$ 8 billion are spent for this purpose in the United States (Zimdahl 2007), and US$ 2 billion in Australia (Llewellyn et al. 2016).

Weed competition with pineapple can cause losses of up to 80% in production (Eshetu et al. 2007, Sipes 2000 cited by Tachie-Menson et al. 2014). Productivity losses go from 30% to 80% when there is no weed control (Eshetu et al. 2007, Tachie-Menson et al. 2014) as well as a 50% reduction in the production of commercial fruits for export (Tachie-Menson et al. 2014). According to Wee and Ng (1970), weed competition can reduce yields by 41% and 21% in a plant crop and in a ratoon crop, respectively. Pineapple unweeded plots produced only 14 t ha^{-1}, while weeded plots achieved 79 t ha^{-1} and 83 t ha^{-1} in plots treated with herbicide and supplementary handweeding, respectively (Pinon 1976, Py et al. 1987).

The results of research and field observations indicate, due to the losses caused by competition with weeds, that the pineapple should be kept without competition throughout the vegetative phase, so that the plant reaches adequate size for floral induction (Brenes-Prendas and Agüero-Alvarado 2007), since fruit weight is proportional to plant mass (Zhang and Bartholomew 1997, Pegoraro et al. 2014, Vilela et al. 2015, Maia et al. 2016). Also, during much of the reproductive phase, competition is still detrimental to the plant and, consequently, to the weight of the produced fruit (Malézieux 1993) (Figure 25.2).

According to Reinhardt and Cunha (1984) the competition of pineapple plants with weeds causes reductions in fruit size when it happens between planting and floral induction of the pineapple, with a highlight to the first five months of the crop cycle. Weed control after flower

Figure 25.2. Pineapple commercial fields with weed control from planting to harvesting, Queensland, Australia. (Source: Victor Martins Maia)

induction does not result in a significant yield increase or an improvement in fruit quality. However, plants in unweeded areas did not produce commercial fruits.

Since pineapple harvesting is manual in most commercial fields and in some cases semi-mechanised, it is recommended that the area is left free of weeds at least until harvest, since their presence can make this operation even more time-consuming and expensive (Figure 25.2). Likewise, if the area is maintained for the propagation phase (seedling production), it should be left free of weeds due to the manual character of the seedlings' harvest. In addition to this, if the crop is extended to the ratoon, weed control in the first crop will aid in management, reducing infestation problems and control costs, and the same control practices can be applied to the ratoon.

Weed Control

Mechanical and Physical Weed Control

Handweeding or Mowing

Handweeding or mowing is a very common practice in low-tech crops and in developing countries with greater labor availability or in family farming. However, the high costs and difficulty of finding labor to work in the field, which is currently a global phenomenon, have reduced the use of this type of control. Up 150 to 500 work days per hectare per cycle could be necessary to maintain pineapple fields free of weed competition without the use of herbicides. It represents 75% of total labour needed to harvest one fruit from the plant (Py et al. 1987).

The use of a coastal mower increases the performance of this type of control; however, it is not suited to control weeds in the crop row. At least one weeding per month is required throughout the cycle, which may increase depending on the infestation of the area, the type of weed present (phytosociological survey) and the climatic conditions or irrigation system used (Reinhardt et al. 1981, Model et al. 2010). The adoption of irrigation systems that reduce wet area (drip irrigation) instead of systems that irrigate 100% of the area (sprinkles) decreases weed control frequency.

Although hand or manual weeding allows the production of bigger fruits, consequently higher productivity, when compared to areas without weeding, other practices provide better weed control and higher fruit weight and exportable fruit percentage (Tachie-Menson et al. 2014).

Soil Cultivation and Tillage

It was emphasized that land preparation should kill mature weeds, as the chemicals are most effective in controlling germinating weed seeds. However, this control practice alone is not sufficient and it is necessary to adopt other forms of control, such as weeding, chemical control or mulching.

Cover Crops and Mulching

The use of plastic or paper mulch, despite high costs, is a common practice in the more technically or organically grown pineapple crops and in the most diverse regions of the world. This technology significantly reduces the use of herbicides and labor cost, while promoting greater growth, reducing the production cycle and the need for irrigation of the pineapple plants (Botrel et. al 1990, Carr 2012, Tachie-Menson et al. 2014).

Tachie-Menson et al. (2014) studied five different weed control methods (T1-weedy check, T2-manual weed control (hoeing) only, T3-synthetic herbicide alone (bromacil + diuron), T4-manual weed control and plastic mulch and T5-herbicide and plastic mulch). There was no statistic difference of fruit weight between the two plastic mulched treatments, but both treatments were significantly higher than all other ones. Furthermore, the percentage of exportable fruits was similar among the synthetic and plastic mulch treatments and higher than weeding and control (weedy check). However, when the plant finds favorable climatic conditions, the positive effects of the use of plastic mulching, with a considerable increase in the cost of production, can be not observed (Reinhardt et. al 1981).

Alternatively, to the use of plastic mulch, there is the possibility of using cover crops for weed control (Eshetu et al. 2007, Matos et al. 2007). This practice can be used both in organic production as in certified pineapple production. The management of cover crops consists of mowing or spraying post-emergence herbicide at blooming stage and chopping the cover crops, allowing their residues to remain on the soil surface as mulch (Matos et al. 2007).

Several species can be used as cover crops in pineapple fields. It is necessary to choose the ones that are better adapted to the local conditions, and that are not hosts of pests and diseases of the pineapple and, if possible, biologically fix nitrogen, due to the high demand of the crop for this nutrient (Cardoso et al. 2013).

The use of *Cynodon dactylon* and *Pennisetum americanum* as cover crops (Figure 25.3) provides higher fruit weight and better fruit classification than conventional treatment (two herbicide sprays and six manual hoeings) with highlight and better results for *Pennisetum americanum* than to *Cynodon dactylon* (Matos et al. 2007). The increase in the production of pineapple fruits after using cover crops can also be linked to the increase in soil organic matter, as it contributes substantially to the improvement of the chemical and physical attributes of tropical soils, which are known to be poor in essential nutrients for the plants. Sunn hemp (*Crotalaria juncea*) can also be used as cover crop. The association between sunn hemp and solarization helps to suppress weeds and control nematodes (Wang et al. 2011).

Another possibility of weed control is the use of organic mulch. In this practice, plant residues or crop remains can be used, depending on their availability (Eshetu et al. 2007, Alwis and Herath 2012), or dead cover with allelopathic plants, such as *Dicranopteris linearis* (Ismail and Chong 2009).

Besides from its benefical effects on soil physical-chemical properties and moisture retention, reducing soil erosion, promoting vegetative growth and favorable environment to root development and overall increasing pineapple fruit yield, organic mulch allows weed control

Figure 25.3. Use of *Pennisetum americanum* as a cover crop in pineapple, Minas Gerais, Brazil.
(Source: Victor Martins Maia)

(Asoegwu 1998, EshETU et al. 2007, Alwis and Herath 2012). The application of a 5 cm thick layer of rice husk, wood chips and sawdust over the soil in pineapple cultivation, especially wood chips, reduces weed biomass and, as a consequence, weed competion (Asoegwu 1998). Other options of organic mulch are paddy husk, coconut husk (Alwis and Herath 2012), coffee husk (Eshetu et al. 2007) and ground cocoa husk. The application of organic mulch provides greater growth and productivity even compared to clean weeding all season (Eshetu et al. 2007).

The *Dicranopteris linearis* (Burm. f.) Underw. is a tropical fern which was reported to inhibit germination and growth of several common weeds in Malaysia (Ismail and Chong 2009). Use of air-dried *Dicranopteris* at 3.0 kg m^{-2} reduced 99% of weed emerged and did not reduce the growth of pineapple plants and fruit weight (Chong et al. 2011).

The use of cover crops and plastic mulch can also be simultaneous. Some pineapple growers place the plastic mulch in total area, which prevents the joint adoption of these two technologies. However, using plastic mulch only on the beds where the pineapple will be planted leaves a space between them, where weeds can germinate and grow. In this case, planting cover crops in this area will help to control weeds. The use of polyethylene and a cover crop in combination reduces weed germination and growth by 90% when compared to the absence of weed control (Mangara et al. 2009).

Cultural Weed Control

Planting Densities

The pineapple supports fairly high leaf area index (LAI) values that are uncommon to most cultivated species. These values can reach up to 12 at the time of floral induction, although the most common values are between 6 and 8 (Malézieux 1993). This indicates that the species supports high planting densities or populations. Older pineapple cultivations adopted populations close to 35,000 plants ha^{-1}. Currently, the recommended population for most cultivars is close to 50,000 plants ha^{-1}. However, super-dense fields can also be found in experimental and commercial areas with populations of up to 128,000 ha^{-1} plants (Zhang and Bartholomew 1997, Cardoso et al. 2013).

The use of high planting densities can help the pineapple in the competition with weeds (Brenes-Prendas and Agüero-Alvarado 2007), although making manual and/or mechanical control more difficult. However, it should be borne in mind that very high populations may result in fruit size reduction (Dass et al. 1978), requiring the intensive use of inputs to reduce or eliminate these effects (Zhang and Bartholomew 1997, Cardoso et al. 2013).

Rotations and Crop Residue Management

At the end of the production cycle, the amount of pineapple dry matter remaining in the area after the removal of the seedlings may exceed 50 t ha^{-1} (Pegoraro et al. 2014, Maia et al. 2016) and the residual of nutrients may be sufficient for the fertilization of an annual crop grown in sequence in the same area to be even suppressed. This residue left on the soil acts as a physical barrier to the sunlight, inhibiting weed germination.

The use of mulchers on the plants left in the field (Figure 25.4) after harvesting allows the possibility of sowing annual short cycle crops (90–100 days) with no tillage, while the pineapple seedlings are stored. This practice, associated with weed control on these crops, helps to reduce weed infestation in the area where there will be new pineapple cultivation in the future.

Fodder species, other fruit trees, such as passion fruit (Dias et al. 2017), or cover crops, especially nitrogen-fixing legume species, may also be included in the rotation system. The residue produced can also be incorporated into the soil while maintaining the crop rotation system. Care should be taken when choosing the species in the crop rotation system, avoiding those that are host to pests and diseases of the pineapple.

Planting Date

Planting dates must be respected in order to increase the competitive potential of the pineapple, creating the ideal conditions for it to establish itself as soon as possible, especially when in non-irrigated conditions. In this case, planting should coincide with the rainy season and higher temperatures. Under these conditions, rooting and new leaves emission will be stimulated and the crop can cover the soil soon.

Figure 25.4. Residue management after pineapple harvest in Paraíba, Brazil.
(Source: Victor Martins Maia)

Fertilizer Management

When it comes to analyzing the ability of a weed species to compete for nutrients, it should be considered, besides the amount extracted, the contents that it presents in its dry matter (Silva et al. 2007).

Besides the ability to extract nutrients from the soil, other species are also competitors in the use of this resource (Silva et al. 2007). *Bidens pilosa* and *Euphorbia heterophylla* showed higher efficiency in the use of N absorbed from the soil, compared to soybean and beans (Procópio et al. 2004). It can be affirmed that, in the field, the inadequate nutrient management, with the addition of subdoses, may favour plant species that use this resource more efficiently.

To reduce these effects, the localized application of fertilizers should be prioritized, which can be done with drip fertirrigation or application of solid fertilizers in the axils of older leaves.

Preventive Measures

The preventive control of weeds consists of the use of practices that aim to prevent the introduction, establishment and/or dissemination of certain problem species in areas that are not yet infested by them. These areas can be a country, a state, a municipality or a land area on the property. It is the responsibility of each farmer or cooperative to prevent the entry and dissemination of one or more weed species, which could become serious problems for a region. In short, the human element is the key to preventive control. The efficient occupation of the space of the agroecosystem by the crop reduces the availability of factors that are adequate to the growth and development of weeds, and can be considered an integration between the preventive and the cultural method (Silva et al. 2007). Some measures may prevent the introduction of the species in the area: carefully cleaning machines, harrows and harvesters; carefully inspecting seedlings purchased with clod and also all organic matter (manure and compost) provenient from other areas, and cleaning irrigation canals. The lack of such care has caused wide dissemination of the most diverse species. As examples can be cited the purple nutsegde (*Cyperus rotundus*), which has very small seeds and tubers that infest new areas with great ease, by means of manure, seedlings, etc., the hairy beggarticks (*Bidens pilosa*) and bur grass (*Cenchrus echinatus*), as well as other species, spread to new areas through clothes and shoes of the operators, animal hair, etc. (Silva et al. 2007).

Chemical Weed Control

Chemical weed control is probably the most commonly used practice for pineapple growers around the world. Among the reasons for the adoption of this form of control, it is worth noting the ease of execution, the efficiency and the costs that are much lower when compared to handweeding. Obviously, these costs vary within and between producing countries. However, the use of chemical weed control can be up to 78% cheaper than manual control (Model et al. 2010). However, most herbicides come with recommendations for weed control on the pineapple crop since the mid-20th century (Py et al. 1987), which may favour the occurrence of weed resistance. Furthermore, these are molecules with higher impact due to the higher risks of environmental contamination.

Active Ingredients

Although there are several works with different active principles and different modes of action (Mendoza Jr 1979, Reinhardt et al. 1981, Py et al. 1987, Sison and Mendoza Jr 1993, Jiménez 1999, Brenes-Prendas and Agüero-Alvarado 2007, Eshetu et al. 2007, Maia et al. 2012, Tachie-Menson et al. 2014), the legislative limitations of each producer country and of each pineapple consuming or importing country must always be taken into account. Because of this, the number of active principles allowed in producing countries is limited to a few products. The consequence of all this is the increased probability of the emergence of resistant weeds. For example, *Asystasia gangetica* cannot be effectively controlled by the common herbicides used for pineapple, such as atrazine, diuron or ametryn (Chong et al. 2011).

Among the most common and widely used herbicides in crops in the main producing countries are diuron and bromacil or the mixture of both (Reinhardt et al. 1981, Jiménez 1999, Brenes-Prendas and Agüero-Alvarado 2007, Maia et al. 2012, Tachie-Menson et al. 2014), being commonly applied in pre-emergence, with the diuron generally presenting the lowest control cost (Model et al. 2010). The application of bromacil, when necessary, should be limited to the moment of planting, since the use of this herbicide four months after planting causes reduction in fruit weight (Py et al. 1987).

The herbicide 3-(3,4-dichlorophenyl)-1,1-dimethylurea, known as diuron, acts as an electron transfer inhibitor in the photosystem II (FSII), which prevents the reduction of quinone A (QA), by competing with quinone B (QB) for the D1 protein binding site, causing the QB output and the interruption of the electron flow, not allowing the reduction of NADPH (Fuerst and Norman 1991). Diuron is normally applied in pre-emergence, although some producers apply during the whole phase of vegetative growth in blanket application. This type of application may cause some symptoms of phytotoxicity, although there are reports that these effects do not reduce the productivity (Maia et al. 2012). However, when applied during the reproductive phase, diuron causes reductions in CO_2 assimilation at all stages of CAM metabolism, transient reduction in stomatal conductance and transpiration, and transient increase in water use efficiency. Furthermore, the photochemical efficiency (Pv/Pm) is negatively affected by this herbicide, but the initial values are restored 36 days after the application (Carvalho et al. 2018).

Other active principles to control weeds can also be cited, while respecting the recommendations for use and the legislation. Sulfentrazone may also be mentioned in pre-emergence (Sison 2000). In post-planting, ametryne, quizalofop-P-ethyl (Jiménez 1999), fluazifop-p-butyl and atrazine (Sison and Mendoza Jr 1993, Maia et al. 2012) are also used in pre-emergence. In the interrows, in the carriers and roads or even in pre-planting to eradicate perennial weeds like *Saccharum spontaneum*, *Sorghum halepense*, *Imperata cylindrica*, with spot application, can also be mentioned glyphosate and dalapon (Eshetu et al. 2007), paraquat dichloride and ammonium-glufosinate. These herbicides should not be used directly on pineapple plants as they may cause their deaths (Sison and Mendoza Jr 1993, Suwanarak et al. 2000, Catunda et al. 2005, Brenes-Prendas and Agüero-Alvarado 2007).

The use of post-emergence herbicides should be very cautious in order to avoid phytotoxicity. Whenever possible, broadband spraying should be avoided, so that there is no contact of the herbicide with the leaves of the pineapple, being preferable the band-spraying of the lines between the crop. The exceptions are the herbicides fluazifop-p-butyl and ametryne, the latter being limited to the dose of 2 kg ha^{-1} of ai. These can be applied in broadband and in a repeated way, as well as mixed with some foliar fertilizers or pesticides to reduce operational costs, if there is compatibility among them. The exceptions are the herbicides fluazifop-p-butyl and ametryn, the latter being limited to a dose of 2 kg ha^{-1} of a.i. These can be applied in broadband and repeatedly, and can be mixed with some foliar fertilizers or pesticides to reduce operational costs, but only when the products are compatible among them (Py et al. 1987). Chemical control can be continued after harvesting the fruits, in order to facilitate harvesting the suckers, adopting the same care previously described.

The application of pre-emergent herbicides (e.g., diuron) should be done immediately after planting pineapple, with moist soil. Pre-emergent herbicides with post emergent effects (e.g., bromacil or ametryne) should be applied in the sequence, after planting, also with moist soil, but only if the weeds in the area show a maximum of two to four final leaves (Py et al. 1987). Most of the post-emergent herbicides must be applied only on vigorous plants, avoiding periods of drought and relative humidity of less than 70%.

Transgenics

Although it is not yet available for large-scale commercial cultivation, there is the possibility of using herbicide-resistant transgenic cultivars (Sripaoraya et al. 2001, Espinosa et al. 2002, Sripaoraya et al. 2006, Sripaoraya et al. 2011). Among the main studies, the development of a transgenic genotype resistant to the herbicide bialaphos stands out with stability results under field conditions (Sripaoraya et al. 2006) and in the production of F1 by hybridization, maintaining the characteristic of resistance to this active principle in the 1:1 proportion (Sripaoraya et al. 2010, Sripaoraya et al. 2011). The development of new genotypes with resistance to other herbicides may help to control weeds and reduce the costs of this practice.

Integrated Weed Management (IWM)

Strategies for integrated management in different weed species can be divided as short or long term. Measures, such as the use of weeding or direct use of herbicides (chemical control) can be considered as short duration, being responsible for only temporary control, requiring new applications at each growing season. In the case of measures considered as long term, the use of cultural practices and control by other biological agents, have a permanent character and take into account more pronounced changes in the different agronomic practices. This results in Integrated Weed Management, which should integrate prevention and other control methods that promote short-term (mechanical and chemical methods) and medium- and long-term (cultural and biological methods) control (Silva et al. 2007). The first step for the adoption of Integrated Weed Control is to identify the weeds in the area and their frequencies, so specific control practices and strategies can be employed in each area.

Because pineapples are slow growing and are not aggressive when competing with weeds, the use of practices that accelerate or promote their growth has contributed to integrated weed management by allowing the plant to achieve higher leaf area index values faster, thus shadowing the cultivated area and inhibiting the germination of some weed species. In addition to fertilization, irrigation and mulching practices that make daily carbon fixation comparable to C3 and C4 species, the use of some substances promotes plant growth. Among these substances, it is recommended to immerse the pineapple suckers, before planting, in solutions containing 2,000 mg L^{-1} of indol acetic acid (IAA) and 5,000 to 8,000 mg L^{-1} of purified monoammonium phosphate (MAP) to promote greater rooting and growth of such plants, consequently reducing the production cycle, making them more competitive with weeds (unpublished data).

Some pineapple organic farms use plastic or paper mulch to cover whole area and complete the weed control by handweeding. However, chemical control is most common around the world. The choice of the active ingredient, as well as the doses to be applied, should be based on the weed species existent in the area, which can be determined with a phytosociological survey, as well as in the other practices that will be adopted and in the type of soil where the plants are being cultivated. This caution should be emphasized, as there is the possibility of contamination of soil, groundwater and even water for human consumption. Residues of bromacil were found up to 3 m deep and about 18 months after application in a pineapple crop (Alavi et al. 2008).

Weed control is the main problem in pineapple cultivation on mineral soils. A combination of both herbicide applications and plastic mulching is recommended for its weed control. During the dry seasons, the pre-emergent herbicide Diuron was found to be the most effective providing control for four to six months while Gesapax for only one to one and a half months. The effectiveness of these pre-emergent herbicides was much reduced during the rainy season. Weed control using herbicides were compared with that using black plastic sheet mulch. Besides preventing weed growth, the usage of plastic mulch also had a positive effect on the growth and yield of pineapple (Abdul Rahman 1996).

According to DeFrank (1999), pre-emergence herbicides, such as bromacil and diuron should be applied after planting to the space between rows on a field-by-field basis. The application rates can be reduced by allowing weed growth after the plastic mulch is installed but before planting the pineapple. Some weeds, as the morning glory (*Ipomoea* spp.), are more difficult to control, and could be germinated by irrigation for a later control with contact herbicides. Rates may be significantly reduced if current use patterns are based on the control of weeds that could be easily removed by this pre-plant treatment. Remaining open areas should be treated with pre-emergence herbicides after the majority of the field has achieved a closed canopy. After harvesting the fruits, it is very important to prevent weeds to spread their seeds in order to reduce herbicide use in the next crops. Weeds should be killed with contact herbicides before the field is plowed or disked. Intact weeds will be more completely killed than weeds with disturbed roots caused by a disc harrow or similar tool. The plantation should consider chopping or mowing the standing pineapple to provide surface mulch for long-term weed suppression during fallow periods. Mulched fields also retain more moisture for subsequent crops, allow more rainfall infiltration, reduce runoff and help mitigating fugitive dust.

The correct calibration of the sprayers is also an important measure for IWM, as it allows obtaining the correct performance of the product, to be able to repeat successful applications and reduce costs and to identify failures in product application and/or injuries to the crop plants (DeFrank 1999).

The maintenance of the residue produced at the end of the pineapple harvest, associated to rotation with annual short-cycle crops under no-tillage system should also be part of an IWM system. These species can also be cover crops, chosen according to the weeds present in the area, to local climate adaptation or even to biologically fix N.

All of the above practices used in an integrated way will, over time, minimize weed problems, reducing the cost of control.

Evaluation of Weed Control Sustainability

The assumptions underlying IWM can be well synthesized in: quality assurance of the harvested product, including the exemption of pesticide residues in food; environmental sustainability, including non-degradation of soil and contamination of air and water; economic and social sustainability in production, while maintaining or increasing productivity; and guarantee of a better quality of life for the farmer in terms of economic return and greater safety in activities involving the use of pesticides (Silva et al. 2007) .

The idea of IWM is more understandable when weeds are treated not as a direct target that must be 'exterminated' but rather as an integral part of an ecosystem in which they are directly involved, among other things, in the cycling of nutrients in the soil. In addition, they form complex interactions with microorganisms, and by means of these associations guarantee the agronomic characteristics that give the environment a greater capacity to support a sustainable crop. With the exception of a few species that need to be eradicated from the area, a large part of the weed plant community controls soil nutrient dynamics, as well as being a key component in the formation and burning process of organic matter, mainly due to the role of rhizosphere in stimulating microbial activity (Silva et al. 2007).

Therefore, technical care is needed to achieve maximum efficiency with minimal negative impact on soil, water and non-target organisms. It should be emphasized that in IWM, the herbicide is considered only an additional tool in obtaining a control that is efficient and economical, preserving the quality of the harvested product, the environment and the human health. To do this, it is necessary to associate the different available control methods (preventive, mechanical, physical, cultural, biological and chemical), considering weed species, soil type, topography of the area, equipment available on the property, environmental conditions and the cultural level of the owner (Silva et al. 2007).

Concluding Remarks

Pineapple stands out as a fruit of great economic, social, and nutritional importance in several countries of the world. Weed management in pineapple fields is one of the most complex and costly practices due to the growth and competitiveness characteristics of the plant, the production cycle, the other cultural practices and harvest and the weeds found in this agroecosystem. Faced with this challenge, the adoption of practices in an integrated way and with lower cost and environmental impact is the great challenge for the pineapple grower. Adding to this is the fulfillment to the demands of the consumer market and the legal restrictions and commercial and phytosanitary barriers for the consumption and export of the fruit and its products. The current management techniques, associated to the knowledge about the weeds, allow the production of pineapple in a sustainable way and can be used by the most diverse producers around the world to serve different markets.

REFERENCES

Abdul Rahman, H. 1996. Weed control recommendations for pineapple (*Ananas comosus* [L.] Merr.) grown on mineral soils. pp. 261–266. *In:* Vijaysegaran, S., Pauziah, M., Mohamed, M.S. and Ahamad Tarmizi, S. (Eds.) International Conference on Tropical Fruits, Proceedings. Vol. I. Malaysian Agricultural Research and Development Institute, Kuala Lumpur.

Alwis, A., H.K.M.S.K. Herath. 2012. Impact of mulching on soil moisture, plant growth and yield of Mauritius Pineapple (*Ananas comosus*. L. Merr). J. Food Agr. 2: 15–21.

Asoegwu, S.N. 1998. Effect of vegetative cover, mulching and planting time on some soil physical properties and soil loss in pineaple plots. Agr Mech. Asia, Africa and Latin America 22: 39–43.

Aspiazú, I., T. Sediyama, J.I. Ribeiro Jr., A.A. Silva, G. Concenço, L. Galon, E.A. Ferreira, A.F. Silva, E.T. Borges and W.F. Araujo. 2010. Eficiéncia fotosintética y de uso del agua por malezas. Planta Daninha 28: 87–92.

Bartholomew, D.P., R.E. Paul and K.G. Roubach (Eds.). 2003. The pineapple, botany, production and uses. Honolulu: CAB, 301 p.

Botrel, N., D.L. Siqueira, F.A.D. Couto and V.H.V. Ramos. 1990. Plantio de Abacaxizeiro com Cobertura de Polietileno. Pesquisa Agropecuária Brasileira 25: 1483–1488.

Brenes-Prendas, S. and R. Agüero-Alvarado. 2007. Reconocimiento taxonómico de arvenses y descripción de su manejo, en cuatro fincas productoras de piña (*Ananas comosus* L.) en Costa Rica. Agronomía Mesoamericana 18: 239–246.

Cardoso, M.M., R.F. Pegoraro, V.M. Maia, M.K. Kondo and L.A. Fernandes. 2013. Crescimento do abacaxizeiro 'Vitória' irrigado sob diferentes densidades populacionais, fontes e doses de nitrogênio. Revista Brasileira de Fruticultura 35: 769–781.

Carr, M.K.V. 2012. The water relations and irrigation requirements of pineapple (*Ananas comosus* var. comosus): a review. Exp. Agr. 48: 488–501.

Carvalho, A.R.J., V.M. Maia, F.S. Oliveira, R.F. Pegoraro and I. Aspiazú. 2018. Physiological variables in pineapples submitted to the application of diuron. Planta Daninha. (In press)

Catunda, M.G., S.P. Freitas, J.G. Oliveira and C.M.M. Silva. 2005. Efeitos de herbicidas na atividade fotossintética e no crescimento de abacaxi (*Ananas comosus*). Planta Daninha 23: 115–121.

Chong, T.V., A. Nor Aris and A. Amarudin Amran. 2011. Effect of *Dicranopteris linearis* debris on weed emergence and pineapple growth in pineapple field. Acta Hort. 902: 377–380.

Concenço, G., A. Andres, A.F. Silva, L. Galon, E.A. Ferreira and I. Aspiazú. 2014. Ciência das Plantas Daninhas. *In:* Monquero, P.A. Aspectos da biologia e manejo das plantas daninhas. São Carlos: RiMa Editora, 430 p.

Cote, F.X., M. Folliot and M. Andre. 1993. Photosynthetic crassulacean acid metabolism in pineapple: diel rhythm of CO_2 fixation, water use, and effect of water stress. Acta Hort. 334: 113–130.

Cox, C.B. and P.D. Moore. 1985. Biogeography: An Ecological and Evolutionary Approach. Blackwell.

Dass, H.C., B.M.C. Reddy and G.S. Prakash. 1978. Plant-spacing studies with 'Kew' pineapple. Scientia Hort. 8: 273–277.

Dias, D.G., R.F. Pegoraro, V.M. Maia and A.C. Medeiros. 2017. Production and post-harvest quality of irrigated passion fruit after N-K fertilization. Revista Brasileira de Fruticultura 39, n. 3.

DeFrank, J. 1999. Weed management in pineapple. Pineapple News 6: 10.

Eshetu, T., W. Tefera and T. Kebede. 2007. Effect of weed management on pineapple growth and yield. Jimma Research Center. Ethiop. J. Weed Manag. 1: 29–40.

Espinosa, P., J.C. Lorenzo, A. Iglesias, L. Yabor, E. Menendez, Y. Borroto and L. Hernandez. 2002. Arencibia, A. Production of pineapple transgenic plants assisted by temporary immersion bioreactor. Plant Cell Rep. 21: 136–140.

FAO. 2014. Available at: http://www.fao.org/faostat/en/#data/QC (Accessed on February 2, 2017).

FAOSTAT. 2016. Production crops. Available at: https://www.fao.org.br (Accessed on February 2, 2017).

Fuerst, E.P. and M.A. Norman. 1991. Interactions of herbicides with photosynthetic electron transport. Weed Sci. 39: 458–464.

Holm, L., J. Pancho, J. Herberger and D. Plucknett. 1977. The world worst weeds: distribution and biology. The University Press of Hawaii. USA. 609 p.

Ismail, B.S. and T.V. Chong. 2009. Allelopathic effects of *Dicranopteris linearis* debris on common weeds of Malaysia. Allelopathy J. 23: 277–286.

Jiménez, J. 1999. Manual práctico para el cultivo de piña de exportación. Ed. Tecnológica. Cartago. Costa Rica. 224 p.

Kissan, K. 2004. Pineapple (*Ananas comosus* L.). Available at: www.kissankerala.net/kissan/kissancontents/pineapple.htm

Llewellyn, R.S., D. Ronning, J. Ouzman, S. Walker, A. Mayfield and M. Clarke. 2016. Impact of weeds on Australian grain production: the cost of weeds to Australian grain growers and the adoption of weed management and tillage practices. Kingston, ACT, Australia: Grains Research and Development Corporation (GRDC).

Maia, V.M., F.S. Oliveira, R.F. Pegoraro, I. Aspiazú and M.C.T. Pereira. 2016. 'Pérola' pineapple growth under semi-arid climate conditions. Acta Hort. 1111: 267–274.

Maia, L.C.B., V.M. Maia, M.H.M. Lima, I. Aspiazú and R.F. Pegoraro. 2012. Growth, production and quality of pineapple in response to herbicide use. Revista Brasileira de Fruticul. 34: 799–805.

Malézieux, E. 1993. Dry matter accumulation and yield elaboration of pineapple in Cote D'ivoire. Acta Hort. 334: 149–158.

Malézieux, E., F. Côte and D.P. Bartholomew. 2003. Crop environment, plant growth and physiology. pp. 69–107. *In*: Bartholomew, D.P., Paul, R.E., Roubach, K.G. (Eds.) The pineapple, botany, production and uses. Honolulu: CAB.

Mangara, A., A.A.N. Adopo, N.M.T. Kouame and M. Kehe. 2009. Effect of polythene and cover crop *Mucuna pruriens* (L.) DC. in the control of weeds in pineapple (*Ananas comosus* (L.) Merr.) in Côte d'Ivoire. J. App. Biosci. 22: 1326–1332.

Matos, A.P., N.F.S. Sanches, L.F. Souza, J. Elias Jr, F.A. Teixeira and S.C. Siebeneichler. 2007. Cover crops on weed management in integrated pineapple production plantings. Acta Hort. 822: 155–160.

Mendoza Jr, S.P. 1979. Weed management in pineapple. pp. 147–148. *In*: Moody, K. (Ed.) Weed Control in Tropical Crops. Manila: Weed Science Society of the Philippines.

Model, N.S., R. Favreto and A.E.C. Rodrigues. 2008. Weed species and biomass on pineapple culture under five weed control methods. Pesquisa Agropecuária Gaúcha 14: 95–104.

Model, N.S. and R. Favreto. 2009. Plantas espontâneas e daninhas identificadas em cinco épocas em área cultivada com abacaxizeiro em Maquiné, Rio Grande do sul, Brasil. Pesquisa Agropecuária Gaúcha 15: 57–64.

Model, N.S., R. Favreto and A.E.C. Rodrigues. 2010. Efeito de tratamentos de controle de plantas daninhas sobre produtividade, sanidade e qualidade de abacaxi. Pesquisa Agropecuária Gaúcha 16: 51–58.

Nobel, P. S. 1991. Achievable productivities of certain CAM plants: basis for high values compared with C3 and C4 plants. New Phytol. 119: 183–205.

Osmond, C.B. 1978. Crassulacean acid metabolism: a curiosity in context. An. Rev. Plant Physiol. 29: 379–414.

Pegoraro, R.F., B.A.M.D. Souza, V.M. Maia, U.D. Amaral and M.C.T. Pereira. 2014. Growth and production of irrigated 'Vitória' pineapple grown in semi-arid conditions. Revista Brasileira de Fruticultura 36: 693–703.

Procópio, S.O., J.B. Santos, F.R. Pires, A.A. Silva and E.S. Mendonça. 2004. Absorção e utilização do nitrogênio pelas culturas da soja e do feijão e por plantas daninhas. Planta Daninha 22: 365–374.

Py, C., J.J. Lacoeuilhe and C. Teisson. 1987. The pineapple. Cultivation and uses. G.-P. Maisonneuve et Larose, 568 p.

Reinhardt, D.H.R.C. and G.A.P Da Cunha. 1984. Determinação do período crítico de competição de ervas daninhas em cultura de abacaxi 'Pérola'. Pesquisa Agropecuária Brasileira 19: 461–467.

Reinhardt, D.H.R.C., N.F. Sanches and G.A.P. Cunha. 1981. Métodos de Controle de Ervas Daninhas na Cultura do Abacaxizeiro. Pesquisa Agropecuária Brasileira 16: 719–724.

Rusydi, N.E. and M. Basuki. 1999. Purwito. Symphilids control in pineapple fields in Indonesia. Pineapple News 6: 39–42.

Szarek, S.R. and I.P. Ting. 1975. Photosynthetic efficiency of CAM plants in relation to C3 and C4 plants. pp. 289–297. *In:* Environmental and Biological Control of Photosynthesis. Springer, Netherlands.

Silva, A.A., F.A. Ferreira, L.R. Ferreira and J.B. Santos. 2007. Biologia de Plantas Daninhas. *In:* Silva, A.A. and Silva, J.F. (Eds). Tópicos em manejo de plantas daninhas. Viçosa: Universidade Federal de Viçosa, 318 p.

Sison, C.H.E.S.E. 2000. Sulfentazone for preplant weed control in pineapple. Acta Hort. 529: 303–308.

Sison, C.M. and S.P. Mendoza Jr. 1993. Control of wild sugarcane in pineapple on the Del Monte Philippines, Inc. Plantation. Acta Hort. 334: 337–340.

Sripaoraya, S., R. Marchant, J.B. Power and M.R. Davey. 2001. Herbicide-tolerant transgenic pineapple (*Ananas comosus*) produced by microprojectile bombardment. Ann. Bot. 88: 597–603.

Sripaoraya, S., S. Keawsompong, P. Insupa, J.B. Power, M.R. Davey and P. Srinives. 2006. Transgene stability and expression of genetically modified pineapple (*Ananas comosus*) under experimental field conditions. Plant Breed. 125: 411–413.

Sripaoraya, S., M.R. Davey and P. Srinives. 2011. F_1 hybrid pineapple resistant to bialaphos herbicide. Acta Hort. 902: 201–207.

Sripaoraya, S., M.R. Davey and P. Srinives. 2010. Inheritance of the bialaphos resistance (Bar) gene from genetically modified pineapple (*Ananas comosus* L.) to commercial cultivars. Thai J. Agr. Sci. 43: 157–161.

Suwanarak, K., S. Kongsaengdao and S. Vasunun. 2000. Efficiency of pre-planting herbicides on weed control and growth of no tillage pineapple (*Ananas Comosus* L.). Acta Hort. 529: 293–302.

Tachie-Menson, J.W., J. Sarkodie-Addo and A.G. Carlson. 2014. Assessment of the impact of some common weed management methods on the growth and yield of pineapples in Ghana. J. Sci. Technol. 34: 1–10.

Taiz, L., E. Zeiger, I.M. Møller and A. Murphy. 2015. Plant physiology and development. Sinauer Associates, Inc. 761 p.

Vilela, G.B., R.F. Pegoraro and V.M. Maia. 2015. Predição de produção do abacaxizeiro 'Vitória' por meio de características fitotécnicas e nutricionais. Revista Ciência Agronômica 46: 724–732.

Wang, K.H., B.S. Sipes and C.R.R. Hooks. 2011. Sunn hemp cover cropping and solarization as alternatives to soil fumigants for pineapple production. Acta Hort. 902: 221–232.

Wee, Y.C. and J. Cl Ng. 1970. Weeds of pineapple areas. Research Bulletin, Pineapple Research Station. Malayan Pineapple Industry Board. Vol. 3, 1970.

Zhang, J.D. and P. Bartholomew. 1997. Effect of plant population density on growth and dry-matter partitioning of pineapple. Acta Hort 425: 363–376.

Zimdahl, R.L. 2007. Fundamentals of Weed Science (3rd edition). Academic Press, San Diego. 666 p.

Weed Management in Natural Rubber

Nilda Roma Burgos[*1] and Jesusa D. Ortuoste[2]

[1] Dept. of Crop, Soil, and Environmental Sciences University of Arkansas, Fayetteville, AR, USA 72704
[2] College of Agriculture, Sultan Kudarat State University, Tacurong City, Sultan Kudarat, Philippines

Introduction

The rubber industry is crucial to the current and future phase of human civilization. Uses of industrial rubber fall into five large categories in order of dominance: construction, manufacturing, aerospace, and automotive (primarily, tires) (https://www.futuremarketinsights.com/reports/industrial-rubber-market). Rubber is derived from plant latex or synthesized from petroleum by-products. Natural rubber is sourced from rubber trees in the genus *Hevea*, under the family Euphorbiaceae. There are ten species in this genus, but only the Para' rubber (*Hevea brasiliensis* Muell. Arg.) is grown commercially. Among the other nine species of rubber trees, only *H. benthamiana* produces a latex of decent quality, but this species has rarely been used in breeding programs. All are native to the Amazon region of South America encompassing Brazil, Venezuela, Ecuador, Colombia, Peru, and Bolivia (Schultes 1990; https://www.rainforest-alliance.org/species/rubber-tree). The utilization of *Hevea* rubber started with the ancient civilizations of the Olmec, Maya, and Aztec who utilized latex sap from the rubber tree to make rubber balls, among other things. The earliest records of rubber dated back to AD 600 in Mexico (Serier 1993) where the Aztecs and Mayans dominated at the pinnacle of their civilizations. Knowledge about the rubber plant was extended by Columbus to the Old World (Europe) toward the end of the 15th century, as he provided the first detailed description of the tree during his explorations of South America. Curiosity about the rubber plant prompted the first movement of seeds from the Americas (Peru) to Europe (France) in 1731 (Dijkman 1951). This started the long history of rubber research and expansion of rubber cultivation from South America to Europe, Asia, and Africa. There are other herbaceous and tree species that produce latex, including those in the genus *Castilla, Cryptostegia, Ficus, Funtumia, Holarrhena, Manihot, Parthenium,* and *Taraxacum* (Priyadarshan 2011). Of these, the exploration of Russian dandelion (*Taraxacum kok-saghyz* Rodin) as an alternative source of natural rubber is probably the most advanced (Hodgson-Kratky et al. 2017, McAssey et al. 2016), but is still a long way from commercialization.

The native habitat of Para' rubber lies between the equator and 15°S which is relatively flat, with a wet equatorial climate (Strahler 1969). The mean monthly temperature is 25°C to 28°C. Annual rainfall exceeds 2,000 mm. Therefore, conditions ideal for rubber cultivation are: 2,000 to 4,000 mm annual rainfall distributed over 100 to 150 d, mean annual temperature around 28

*Corresponding author: nburgos@uark.edu

± 2°C with a diurnal variation of about 7°C, and sunshine hours of about 2,000 h year[-1] (Barry and Chorley 2009, Ong et al. 1998). Thus, global rubber plantations occur along the equatorial zone (Figure 26.1), in wet tropical regions that meet these characteristics. In recent decades, however, rubber plantations have been extended beyond these ideal locations, to colder and more hilly regions (Fox and Castella 2013, Priyadarshan 2011), in an attempt to meet the global demand for natural rubber and alleviate local economic needs. Rubber plantations have been established as far away from the equatorial zone as 20°S in Brazil and 23°N in India, at altitudes of up to 671 m in China and in Vietnam where the mean temperature is 21°C (Priyadarshan 2011).

Natural Rubber Production

While there are other rubber tree species or other types of plants that produce latex (including the recent rubber crop of interest (Russian dandelion), only the Para' rubber tree is widely cultivated, or tapped from the wild, to produce natural rubber. In the rubber industry's infancy, natural rubber in the Amazon region was produced from tapped wild rubber trees. It was reported that in 1876, a British explorer Henry Wickham smuggled a large amount of seeds from the Santarem area in Brazil to the Royal Botanic Gardens (Kew Gardens) in London, England (Serier 1993). From there, the seeds were distributed to the British colonies such as Sri Lanka, Singapore, Malaysia, India, and parts of Africa. The rubber plantations established by the British were more efficient and productive than the predominantly wild rubber trees tapped for latex in the Amazon basin (Dean 1987).

Rubber production shifted from South America to Asia. Henry Wickham was credited for introducing Para' rubber to Asia in 1876 (Van Lam et al. 2009). Rubber eventually became an important perennial crop in this continent. Most plantation areas are located in Southeast Asia, especially in Thailand, Indonesia and Malaysia, with expanding production areas in Vietnam and China (Figure 26.1). Thus, the top five rubber-producing countries in 2015 were Thailand, Indonesia, Malaysia, India, and Vietnam (Figure 26.2). These five countries produced about 75% of the world's natural rubber, with Thailand producing 27%. In 2015, the global production

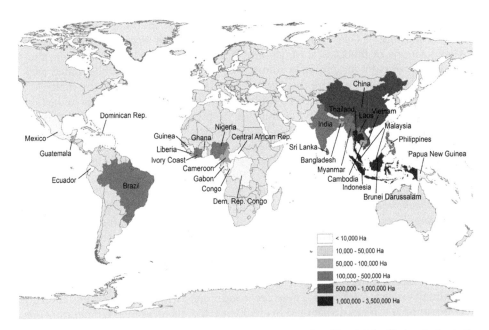

Figure 26.1. Geographical distribution of Para' rubber (*Hevea brasiliensis* Muell. Arg.) plantations. (Source: Warren-Thomas et al. (2015). Conservation Letters 8(4): 230-241)

of natural rubber was reported at 12,314,000 mt (Figure 26.3). In 2016, Thailand, Indonesia, and Vietnam exported $10.2 billion worth of natural rubber, comprising 80% of the global natural rubber trade (http://www.worldstopexports.com/natural-rubber-exports-country/). The current trend of rapid expansion of rubber plantations (Kou et al. 2015) indicate that China is going to become a consistent top player in the global rubber production and trade.

The global production of natural rubber has increased steadily, but slowly, in the last 16 years from 6,913 million mt in 2001 to 12,401 million mt in 2016 (Figure 26.3). This comprised 40% and 46% of total rubber production, respectively, at the beginning and end of this period. Considering that Para' rubber is native to the tropical rainforests of South America, and the use of rubber was discovered by the Aztec and Mayan cultures there, the tropical Americas did not remain to be the center of rubber production. Various factors contribute to the changing

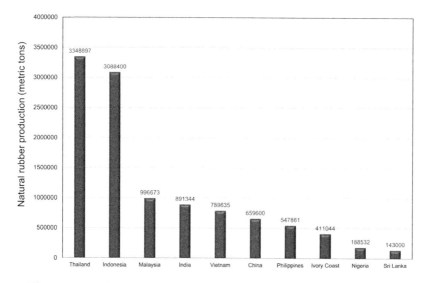

Figure 26.2. The top ten rubber-producing countries in the world, 2015. Available at: http://www. perfectinsider.com/top-ten-rubber-producing-countries-in-the-world/ (Accessed on February 20, 2018).

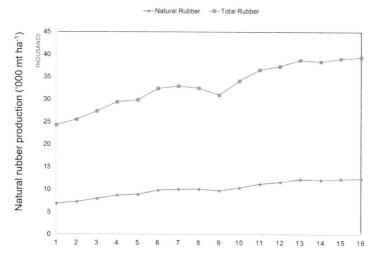

Figure 26.3. Global production of natural rubber, 2001–2016. (Source: IRSG 2017, Statista 2018)

dynamics of natural rubber production, one of these being the South American Leaf Blight (SALB) disease, which has reduced the productivity of rubber in the Americas (Lespinasse et al. 2000). The other was the predominantly extractive manner of harvesting of latex from wild trees in the Amazon basin. Brazil produces only about 1% of the world's natural rubber.

Natural Rubber Production Systems

Cultivated varieties are clones propagated vegetatively by grafting. The young rubber plants are raised in nurseries from clonal seeds for 6 to 8 months and budded with bud scion from a selected variety. The rootstocks are varieties with good root characteristics (resistant to soil pathogens, drought-tolerant). The ideal scion would be a variety with fast growth, high latex yield, resistant to foliar diseases, and cold-tolerant. The budded seedlings are then transplanted to the field at a population of 400 to 555 trees ha^{-1}. Tapping for latex starts when the rubber trees reach a girth of 45–50 cm. A rubber tree plantation can be productive for 40 years or more depending on the management practices, which determines tree health. Rubber can grow to a height of 18 to 39 m; thus, after some years the tree canopy covers the ground 100%. Trees beyond their productive years are cut for lumber and the scraps used for firewood or other purposes.

Intercropping Practices in Rubber Plantation

Rubber is usually planted at 500 plants ha^{-1}. The open spaces between young rubber trees is subject to erosion and weed invasion and sits unproductive if not planted with other crops. Intercropping young rubber trees with food crops can provide food and cash for the farmer, while waiting for the rubber to be tapped, and some ground cover to prevent soil erosion (Figure 26.4). Rubber trees may be intercropped with various other plant species including tea, cocoa, coffee, rattan, fruit trees, and cinnamon (Jessy et al. 2016, Pathiratna and Perera 2006, Penot and Ollivier 2009, Wu et al. 2016). Annual crops that could be intercropped with young rubber include upland rice, corn, mungbean, sorghum, field legumes, ramie, pineapple, and sweetpotato (Hondrade et al. 2017, Pamplona 1990). Starting in the third year, when the rubber canopy provides about 30% to 40% shade, crops which can be profitably grown include mungbean, sorghum, banana, and pineapple (Pamplona and Tinapay 1998).

Considering the annual revenue flow from latex production and the timber yield of rubber beyond the latex production years, the land expectation value of a rubber-tea intercropping system is higher than that of rubber or tea monoculture (Guo et al. 2006). Banana is another perennial intercrop, planted during the pre-tapping period, which can improve rubber production. Rubber trees intercropped with banana produced more latex and were ready for tapping about six months earlier than those without intercrop (Rodrigo et al. 2005). The benefit to rubber perhaps emanates from reduced competition from weeds when the rubber is young. Farmers in Nigeria intercrop rubber with soybean and melon, or with melon and maize to obtain food and income from the land while waiting for the rubber to produce latex (Esekhade et al. 2003). Pineapple, banana, and cassava are most commonly planted intercrops for rubber in India (Rajasekharan and Veeraputhran 2002).

In these systems, it is important to determine location-specific and crop-specific planting configurations to maximize the productivity of the intercrops and rubber. Under normal rubber planting population (i.e., 500 trees ha^{-1}) and planting configuration, the space between rubber rows is suitable for intercropping only in the first five years or so. Then the farmers will have to rely on other land for food production. To maximize the space suitable for longer-term intercropping, Rodrigo et al. (2004) in Sri Lanka recommended planting rubber in double rows spaced 18 m apart and with 3-m spacing of plants within the double rows. It should be noted that the wide open space between the double rows is also ideal for weeds to grow. Therefore, proper weed management for the intercrops is critical.

Figure 26.4. Intercropping rubber with banana (A), photo credit: www.iperca.org; cassava (B), photo credit: https://sites.google.com/site/hoangkimsite/Home/dayvahoc, Prof. Kazuo Kawano; coffee (C), photo credit: www.cirad.fr; and tea (D), photo credit: Tea Research Institute, Kandy, Sri Lanka, caringtea blogspot.com

If planned and managed well, intercropping generates income during the gestational period of the rubber plantation. It has been shown that all agroforestry systems, rubber intercropping being one, have stable internal microclimate environments (Wu et al. 2016). This means that this system, as a whole, is resilient to fluctuations in environmental conditions, such as soil moisture and temperature. Some intercrops, such as tea or coffee, improve the water use efficiency of rubber, curbing its tendency for wasteful water consumption (Wu et al. 2016). The limitation is that, intercropping cannot be done on steep slopes. In which case, only establishing a cover crop is practical and appropriate.

Weeds in Rubber Plantations

Weeds compete with young rubber for moisture and nutrients strongly enough to reduce the latex yield. Tall weeds infesting new rubber plantations, can also shade the young transplants, in the first year of growth. Shading by weeds can further retard the rubber growth and delay tapping. Logically, the effect of weed competition and the competitiveness of weeds decline with tree age as the tree canopy intercepts more light and limits weed growth in the understory. Weeds hamper farm operations, such as pruning, fertilizer application, spraying, tapping, and disease control (De Jorge 1962, Wycherly 1964). Weeds indirectly limit production by serving as hosts for organisms that are detrimental to rubber trees. Allelopathy (chemical production by living or decaying weedy plant tissues) may also adversely affect the growth of rubber plants.

One of the most destructive and difficult-to-control weeds in young rubber plantations (or any crop) is cogongrass (*Imperata cylindrica* [L.] P. Beauv.) (Table 26.1), a perennial, invasive

Table 26.1. Some common weeds in rubber plantations in the Philippines, India, and Malaysia[1]

Scientific name	English name	Invasiveness[2]	Category	Life cycle[3]
Ageratum conyzoides	Billygoat-weed	Yes	Broadleaf	Annual
Amaranthus spinosus	Spiny amaranth	No	Broadleaf	Annual
Asystasia coromandeliana/ gangetica	Chinese violet	Yes	Broadleaf	Perennial
Borreria laevis	Shaggy buttonweed	No	Broadleaf	Annual
Calopogonium mucunoides	Calopo, wild groundnut	Yes	Broadleaf	Annual
Chromolaena odorata	Siamweed	Yes	Broadleaf	Perennial
Commelina benghalensis/ Commelina diffusa	Dayflower	Yes	Broadleaf	A/P
Cynodon dactylon	Bermudagrass	Yes	Grass	Perennial
Cyperus kyllingia	Kyllingia, white Kyllinga	No	Sedge	Perennial
Cyperus rotundus	Purple nutsedge	Yes	Sedge	Perennial
Digitaria ciliaris	Southern crabgrass	Yes	Grass	Annual
Digitaria sanguinalis	Hairy/large crabgrass	No	Grass	Annual
Eleusine indica	Goosegrass	Yes	Grass	Annual
Heleotropium indicum	Indian heliotrope, Scorpionweed	No	Broadleaf	Annual
Imperata cylindrica	Cogongrass, speargrass	Yes	Grass	Perennial
Ipomoea triloba	Three-lobed Morningglory	Yes	Broadleaf	Annual
Ischaemum rugosum	Saramollagrass, Wrinklegrass	Yes	Grass	Annual
Mikania micrantha	Mile-a-minute	Yes	Broadleaf	Perennial
Mimosa invisa	Giant sensitive plant	Yes	Broadleaf	Perennial
Mimosa pudica	Sensitive plant, touch-me-not	No	Broadleaf	A/P
Murdannia nudiflora	Dove-weed	No	Sedge	Perennial
Oldenlandia verticillata	Whorled oldenlandia	Yes	Broadleaf	A/P
Panicum maximum	Guineagrass	Yes	Grass	Annual
Paspalum conjugatum	Sour paspalum	Yes	Grass	Perennial
Rottboellia cochinchinensis	Corngrass, Guineafowlgrass, Itchgrass	Yes	Sedge	Annual
Sorghum halepense	Johnsongrass	Yes	Grass	Perennial
Spermacoce alata	Broadleaf buttonweed	No	Broadleaf	Annual
Synedrella nodiflora	Nodeweed, synedrella	Yes	Broadleaf	Annual

[1] Source: Anonymous (2013), Asna and Ho (2005), Hondrade et al. (2017), Pancho (1978)

[2] Invasive category based on weed databases: Swearingen and Bargeron (2016); https://www.cabi.org/isc/; https://www.invasive.org; https://plants.usda.gov/

[3] Can be either annual or perennial depending on location.

grass that survives even in very poor soils because of its low nutrient requirement and high nutrient uptake efficiency (Brook 1989). Its supersharp, sturdy rhizomes can pierce the roots of coconuts and other trees, the corm of bananas, and storage roots of root crops. Inadequate control of cogongrass results in reduced growth of rubber and consequently, delayed tapping and reduced latex production (Riepma 1968).

In Thailand, 30% of the plantations established in the late 1960s failed due to severe cogongrass infestation while 60% of the plantations with 'lighter' infestation had unhealthy, unproductive trees (Harper 1973). Cogongrass does not only compete for resources but also secretes allelochemicals (Cerdeira et al. 2012, Mercado 1986), which inhibit the growth of rubber or other crops (Hussain et al. 1994). In addition, cogongrass shoots desiccate in the summer and its high amount of biomass creates a fire hazard for the plantation. Cogongrass is the epitome of a troublesome weed.

Another perennial grass weed that would rival cogongrass in notoriety is *Pennisetum polystachion* ssp. *setosum* (Swarz) Brunken (Noda et al. 1987), which is also an invasive species. It regrows quickly after mowing, sprouts from rhizomes if plowed, and germinates continuously from seeds all year round. At a full height of 2 m, it can shade and choke young rubber. Controlling this weed requires full integration of all weed control tools and strategies available to the farmer. Rubber plantations are ideal habitats for perennial weeds. Thus, several troublesome weeds in rubber are perennial and worse, the majority are invasive regardless of life cycle category (Table 26.1).

Mature rubber trees are not always free from competition with weeds. Weedy vines, such as *Mikania cordata* (Burn.) B.L. Robins and *Merremia tridentata* (L.) Hallier, if not controlled, will climb and cover the canopy of rubber trees. The productivity of rubber then declines as coverage by the vines increases. Eventually, the weed will 'choke' the tree and the rubber tree may die.

Weed Management in Rubber

Weeds are the most limiting factor in successful rubber establishment. Thus, proper weed management is an indispensable component of the best management practices for rubber production. Since the rubber canopy does not close 100% in the first five years, the farmer needs to implement long-term weed management strategies. Considering that rubber plantations are in tropical environments, weeds grow all year round without the periodic relief of winter kill in temperate climates. Therefore, weed management is the most expensive, iterative operation in the plantation for as long as the rubber canopy is open. The Rubber Research Institute of Malaysia (1975) estimates that 60% of expenses during the first 5–6 years of rubber establishment goes to weed control alone. The remaining 40% goes to fertilizer application and disease control. A recent review indicates weed management consumes up to 70% of input costs in the first two years of rubber establishment in Malaysia (Dilipkumar et al. 2017). Financial resources and availability of labor determine what weed control method farmers will use. Availability of inputs and new technologies, key weed problems, farm size, and availability of family labor are factors they take into account in making weed control decisions. Planning is important in making appropriate decisions on weed control. Unfortunately, weed control often is not planned. The decision to control is not made until the problem has become serious, when control may be uneconomical, ineffective or maybe impossible.

Critical Weed-free Period

In agronomic crops, the critical weed-free period is generally within 2–6 weeks after planting. This is modified by the weed species present, the crop being grown, and the planting time. Rubber takes a longer time to get established than annual crops and, therefore, may need a longer weed-free period. Spreading livergrass (*Urochloa decumbens* [Stapf.]) is one of the most problematic grass weeds in new rubber plantations in Brazil. In a location where the ground was covered >90% with this weed, the height, leaf area, and leaf biomass of rubber were reduced 98%, 96%, and 97%, respectively, during the first year of establishment (Guzzo et al. 2014). These are the most immediate, observable consequences of weed competition. The retardation of these plant variables was reversed very quickly when the weed-free period was extended, or when the rubber plants were kept weed-free during the second year of establishment, demonstrating high resilience of the rubber plant. The most noticeable improvement after weed removal was the 750% increase in rubber stem diameter, which bodes well for rubber productivity. This study showed that in Brazil, new rubber plantations have to be kept weed-free between 4 and 9½ months during the first year.

Weed Control Methods

Weed control methods can be grouped into mechanical, manual, thermal, cultural, chemical, and biological techniques. In rubber plantations, plowing, mowing, and herbicide application are the primary methods used to manage weeds (Webster and Baulkwill 1989). Each method has advantages and limitations. A single method is not adequate to attain sustainable weed management.

Mechanical/Physical Weed Control

Mechanical weed control refers to any technique that involves the use of farm equipment to control weeds. The two mechanical control strategies most often used are tillage and mowing. Tillage between the rubber rows is done only when preparing the soil for planting the intercrops and when cultivating the intercrops (Figure 26.5). Tillage is not advisable on sloping land because of the high rainfall in rubber-growing areas. Tillage aggravates soil erosion in formerly forested, sloping areas that had been opened for planting rubber. Intercrops that provide good soil cover should be used and supplemental leguminous cover crops should be planted to provide 100% ground cover quickly. In some cases, uncontrolled weeds in the interrows, especially in small farm holdings, are too dense and cannot be controlled by other means besides plowing or rotovation.

Tillage is effective on most small annual weeds, but only partially effective on weeds with rhizomes and roots that are capable of sprouting. The disadvantages of this method include: i) possible injury to the roots of rubber, ii) rapid loss of soil moisture, iii) loss of soil organic matter, iv) spread of weed propagules to other areas, and as mentioned above, v) increased soil erosion.

Mowing is not often practiced by smallholders. They cannot afford a mower. Mowing, however, is a quick way to suppress weeds in large plantations before cover crops are

Figure 26.5. Tillage of rubber interrows is done by animal or mechanical implement to prepare the soil for planting intercrops. Photo credits: (A) – Hondrade et al. (2017); B – Long-Sreng-International-Co-Ltd. Cambodia directory.com.

established, or if cover crops are not planted. This practice has the advantage of not disturbing the soil, leaving the cut vegetation on top of the soil as mulch, and being fast. However, cut weeds in tropical areas regrow very quickly during the rainy season so mowing has to be done perhaps weekly to keep weeds from getting too big and producing seeds. The mower could also spread weed propagules all over the plantation. Further, we know from pastures and managed turfgrass that mowing alters the morphology of several weed species, causing them to branch out at the base and grow close to the ground just below the mowing height. The primary obstacle in using a mower is the terrain. The majority of rubber in Asia, for example, are planted on sloping land or land with steep slopes. In such areas, only slashing and cover cropping are possible.

Manual or Handweeding

This type of weeding is employed particularly in the rubber nursery and in management of legume cover crops in the rubber understory. It includes hand-pulling, hoeing, and slashing. Repeated handweeding is necessary. When the rubber plants are still small, weeds within the planting strips are removed by hoeing. However, hoeing disturbs the soil, forms depressions in the soil, and may injure the rubber roots. Weeds growing too close to the rubber may be removed by hand. The base of newly planted rubber needs to be handweeded (or hoed) to avoid applying herbicides near the base of rubber plants, which may cause injury. Also, spot handweeding is employed in areas where problematic weeds grow in patches and for which herbicide use is uneconomical. Tall weeds in mature rubber plantations are slashed with a machete or scythe. One needs to be careful when hoeing or digging around the base of young rubber plants so as not to damage the roots. Damaged roots can expose the rubber to infection by soil-borne pathogens and weaken the tree.

Thermal

This is a method of killing weeds by heat. One way to do this is in conjunction with mowing. After mowing, the cut vegetation is allowed to dry, then the dry weed biomass is gathered in between the rubber rows, and burned. This is akin to narrow windrow burning in agronomic crops to kill weeds postharvest (Walsh and Newman 2007). In wheat fields in Australia, collecting cut wheat straw in a narrow band behind the combine, and burning it, generates enough heat to kill seeds of their most problematic weeds – rigid ryegrass (*Lolium rigidum* Gaud.) and wild radish (*Raphanus raphanistrum* L.). It takes 10 s of at least 400°C to kill ryegrass and 500°C to kill wild radish seeds (Walsh and Newman 2007). Although burning exposes the soil surface, thereby increasing the potential for erosion, strategic burning of narrow windrows significantly reduces the erosion risk where generally less than 10% of field area is exposed (Walsh and Newman 2007). The same principle can be applied to the inter-rows of rubber. The intense heat kills any live meristematic tissues from the cut weeds and prevents weed growth (at least along the burnt area) for some time. The success of this method depends on the amount of biomass that is burned because weed kill depends on the intensity and duration of fire.

Another means of killing weeds by heat treatment is with the use of flame weeders. The goal of flame weeding is not necessarily to burn the plant, but rather to apply enough heat to denature plant proteins and cause lethal wilting. Similarly, hot air weeders can heat up the seeds to the point of destroying them. Flame weeders can be combined with tillage techniques such as stale seedbed (preparing the seedbed early, then killing the flush of weeds that germinate before sowing the crop) and pre-emergence flaming (doing a flame pass over weed seedlings after the sowing the crop, but before crop emergence). Flamers are potentially useful in sensitive or riparian areas where traditional methods, such as chemical or mechanical control, are not practical nor advisable. Vitelli and Madigan (2004) evaluated a hand-held burner (Atarus Ranger) in North Queensland, Australia, targeting the woody species bellyache bush (*Jatropha gossypiifolia* L.), parkinsonia (*Parkinsonia aculeata* L.) and rubber vine (*Cryptostegia grandiflora* R. Br.). The best control for bellyache bush (92%) and parkinsonia (83%) was achieved with 10 s of flaming. Optimum control of rubber vine (76%) was achieved with 60 s of flaming. Flaming was least effective on rubber vine, which has the thickest bark, but was highly effective on bellyache bush, which has the highest bark moisture content. Just like herbicides, the efficacy of flaming varies across species. Logically, flaming is more effective on tender, herbaceous plants with high water content such as seedling or juvenile annual weeds. Even then, annual weeds still exhibit differential tolerance to flaming. It is more effective on broadleaf weeds than grasses. Data on annual crops are useful in planning the appropriate strategy to use this method for weed management in rubber interrows. For example, two-leaf seedlings of green foxtail (*Setaria viridis* L.) are controlled 100% by flaming at 6 km h^{-1}, but not at a faster speed (less heat exposure time) (Cisneros and Zandstra 2008). Larger seedlings are more tolerant to heat such that significant numbers escape even at the lowest speed of 2 km h^{-1}. Grass species also differ in sensitivity to heat. Barnyardgrass (*Echinochloa crus-galli* L. Beauv.) is more tolerant to flaming than green foxtail (Cisneros and Zandstra 2008). The same principle applies to broadleaf species. It is beneficial for rubber farmers to know the expected effectiveness of flaming on the spectrum of weed species in their plantation so that appropriate supplemental tactics are planned and implemented at the right time.

Weed burners heat up the soil quickly. Weed seeds are often heat-resistant; intense, quick, dry heat could break seed dormancy and encourage germination, instead of killing the seed. On the contrary, humid heat can destroy the plant cells and kill the plant more effectively. Since the 19th century, hot steam has been used to completely sterilize the soil from weed seeds. However, neither steam sterilization of soil nor soil solarization are practical for large-scale farms, let alone vast rubber plantations with highly variable terrains.

Cultural Weed Control

Cultural weed control refers to any technique that involves maintaining field conditions such that weeds are less likely to become established and/or increase in number. It includes non-chemical crop management practices, such as optimum planting time and plant population, competitive crops, crop rotation, cover cropping, and maintaining soil fertility and soil health. Cultural weed control is a crucial part of integrated weed management. As it applies to rubber, this entails choosing an appropriate location; planting disease-resistant, fast-growing, high-yielding, drought-tolerant variety; adopting the best planting configuration; planting intercrops and cover crops that would benefit (or harmless to) rubber; and improving soil health. Cultural weed control is cost-effective and easy to practice; acceptable and accessible to small- and large-scale farmers; environmentally friendly; and ecologically sound. Cover cropping and intercropping, which modify the crop environment to make it less favorable for emergence and growth of weeds, are examples of cultural approaches.

Cover Crops/Mulches

Cover cropping, if done properly, is very effective in suppressing weeds. For perennial plantation crops (such as rubber) grown in high-rainfall tropical areas with highly erodible soils, perennial, creeping, nitrogen-fixing cover crops are the best fit. This is arguably the most sustainable method of weed management in rubber plantations. Maintenance operations are still needed to keep the cover crop away from the base of the rubber trees to prevent competition. Thus, this method is best supplemented with slashing, mechanical vegetation trimming, mowing, or directed application of non-selective + residual herbicides around the base of trees (Figure 26.6).

The weed-smothering legume cover crops used in rubber plantations include *Mucuna bracteata* DC, *M. pruriens* (L.) DC, *Pueraria phaseoloides* Benth., *Centrosema pubescens* Benth., *Calopogonium muconoides* Desv., *Calopogonium caeruleum* (Benth.) Suav., and *Lablab purpureous* (L.) Sweet (Kobayashi et al. 2003, Kothandaraman et al. 1987, 1989). Successful establishment requires seed dormancy-breaking treatment because legume seeds have a hard seed coat.

Three techniques have been used to break seed dormancy: acid treatment, hot water treatment, and abrasion (Anonymous 2013). Treatment with concentrated sulfuric acid for 10 min breaks seed dormancy of *P. phaseoloides*, 30 min for *M. bracteata* and 20 to 30 min for *C. mucunoides*. Soaking in hot water (60–80°C) for 4 to 6 hours works for *P. phaseoloides* and *M. mucunoides*. Scarification with sand or sandpaper in a rotating drum, followed by an overnight soak in water, works for all these cover crop seeds. Planting of cover crops may commence before or after the establishment of rubber seedlings in the field. Generally, however, cover crops are planted after planting rubber. Cover crops grow quickly and form a thick cover on the soil surface. Rapid growth is an important trait because weeds germinate and grow fast in the tropics. For example, *M. pruriens* provides 80% ground cover in just 2 weeks after planting (Kobayashi et al. 2003). *M. bracteata* produces a large amount of biomass, i.e., 5.6 t ha^{-1} in three years and 12 t ha^{-1} in four years (Annie et al. 2005). This suppresses weed growth in addition to providing nitrogen, increasing organic matter, and preventing soil erosion. The growth of *Pennisetum* can be reduced 98% when growing with *M. pruriens*, which reduces the relative light intensity hitting the soil surface to as low as 10% (Kobayashi et al. 2003).

The advantages of using cover crops in rubber plantation aside from weed suppression are: i) prevention of soil erosion particularly when the plantation is hilly; ii) natural source of nitrogen; iii) enrichment of soil organic matter and humus from decomposing leaves; and iv) effective control of invasive perennial grasses, such as *I. cylindrica* and *P. polystachion*. In the Philippines, cover cropping plus intercropping provides more than enough income to cover the cost of weeding (Pamplona 1990). This reduces the cost of establishment and maintenance of young rubber plantations from P36, 000 to P12, 000 or from $720 to $240 ha^{-1}. On the other hand, if the cover crops are not maintained and are allowed to spread into the planting strips (Figure 26.6D), the cover crop will compete with rubber for nutrients and water especially during drought periods. Herbicides are used to control the growth of cover crops around rubber trees. In the 1960s, a urea herbicide (neburon) was found effective in controlling the growth of tropical legumes Pueraria phaseoloides and Centrosema pubescens (Riepma 1965). However, glyphosate and paraquat became the mainstays for maintaining clean strips chemically.

Figure 26.6. Cover crops are used to control weeds and minimize soil erosion in rubber plantations. Photo credits: (A) Chemically maintained clean strip, rubberplantation.net; (B) Manually maintained clean strip, Dr-plant.blogspot.co.id; (C) Mechanically mowed and mulched strip, You Tube; (D) unmaintained cover crop climbing onto rubber plants, www.mekarn.org.

Preventive Measures

Prevention is the most effective, and most difficult to implement, method of weed management. Major agents of weed dispersal (wind, water, animals) are beyond the farmers' control. Minimizing the factors that humans can control (spread by machinery, vehicles, clothing) is difficult, impractical, or uneconomical to practice. Once a weed has become established, eradication is almost always impossible. The only thing to do is mitigate weed spread and manage the farm to minimize weed population size.

Plant Residue Management

Residues of food crop, cover crop, and cut weeds, if distributed strategically in the plantation, will ultimately protect the soil from erosion and return organic matter and nutrients to the soil (Liu et al. 2017). Residue management includes mulching, composting (for smallholders), incorporation to soil by tillage, or burning. The method used to manage plant residues affect not only weed control, but also soil physical properties, nutrient, and water cycling in the plantation. Proper residue management reduces soil erosion, protects water quality, improves soil tilth, and sustains diversity of the soil biota. The farmer still has to monitor weed emergence and implement supplemental control methods, such as herbicide application, to prevent weed growth and seed production.

Biological Weed Control

This method utilizes living organisms to control weeds. Suitable domesticated farm animals such as poultry, goat, and sheep are allowed to graze in the rubber plantation to forage on some weedy species. This is economical and the animals can be consumed or sold for cash. However, these animals are selective in their feeding habit and leave unpalatable species untouched. This promotes shifts in weed populations; the ones left could be invasive or difficult to control with herbicides or with other non-chemical methods. Apart from this, care should be taken to prevent the animals from destroying the rubber trees. This aspect is difficult. Once the area is reserved for animal grazing, no herbicide may be used until the animals are moved to another area. Herbicides can then be applied for supplemental weed control.

Chemical Weed Control

Simply, this is control of weeds with herbicides. The use of herbicides is the most popular method of weed control in rubber cultivation across the globe because of its efficacy and efficiency. To illustrate, herbicides constitute 74% of total pesticides used in rubber plantations in Malaysia Cite Dilipkumar et al. (2017). Chemical weed control is cheaper and more effective than handweeding and mechanical cultivation. It also allows killing the weeds without disturbing the soil, which is an added advantage in sloping land. It minimizes soil erosion. Herbicides are necessary tools for perennial crop production; however, there is minimal published information on herbicide options for weed management in rubber starting from seedling phase to maintenance of mature trees. In India, for example, only four herbicides are listed in their online resource for rubber growers: diuron (with residual activity in soil); 2,4-D (for broadleaf weeds, foliar activity only); glyphosate and paraquat (both non-selective, foliar activity only) (Anonymous 2013). Accessible information on herbicides for rubber in Malaysia is old (Teoh et al. 1978). There is a dearth of publications on chemical weed management for rubber plantations. It should be clear to the farmer that herbicides are supplemental tools, rather than an alternative, to traditional methods of weed control.

Pre-emergence Herbicides

These are soil-applied chemicals used for killing weeds before they emerge. Also called residual herbicides, these are usually applied in newly prepared planting strips immediately after planting rubber to prevent weed growth for 6–7 weeks. Herbicide application is repeated, mixed with a non-selective foliar herbicide to desiccate emerged weeds, to maintain the planted strip weed-free. Preemergence herbicides for rubber have long, residual activity and are used at higher doses than in their respective labelled annual crops. Some examples are atrazine, diuron, EPTC, hexazinone, and oxyfluorfen.

Post-emergence Herbicides

This type of herbicides are applied to emerged weeds that are actively growing. Depending on the size of plantation and the farmer's resources, herbicides are applied using knapsack sprayers, spray booms attached to four-wheelers, or tractor sprayers. Various spray nozzle tips can be used depending on the spray volume desired and the type of application. Desiccating weeds/cover crops with thick foliage requires high spray volume and thorough coverage. This would need nozzles recommended for high-pressure spray application. Spraying on a band along newly planted rubber requires drift-reducing nozzle tips. Controlled droplet applications (CDA) are recommended for lower spray volumes of 15 to 30 L ha^{-1}, which farmers generally prefer, to reduce the cost of spraying. Foliar herbicides are either selective, non-selective, or mixed-spectrum. Some that are labeled for rubber are mentioned here. One example of a selective grass herbicide is fenoxaprop-P-ethyl. 2,4-D and MCPA are for broadleaf weed control only. Non-selective herbicides are glyphosate, glufosinate, and paraquat. These affect all plant species, although susceptibility to each herbicide varies across species. Glyphosate and glufosinate are systemic herbicides (translocate in the plant) while paraquat is a contact herbicide (does not translocate in the plant). Therefore, for paraquat to be highly effective, thorough coverage is necessary. Weeds will regrow. This also has implications on applying herbicides near the base of trees. Drift from glyphosate and glufosinate could injure young rubber severely. Paraquat is safer to use near young rubber plants. Some herbicides (i.e., metsulfuron-methyl, propanil) control some broadleaf and some grass species. Thus, mixing herbicides is recommended to achieve total weed control.

Weed species differ in their susceptibility to herbicides and herbicides differ in their spectrum of control and mechanism of action. A successful chemical weed control activity hinges on correct identification of weed species, proper selection of herbicides, and proper application procedure and timing. Mixing herbicides of different mechanisms of action will control a broader spectrum of weeds than each herbicide applied alone. Diversification of herbicide mixtures (different mechanisms of action) is necessary to avoid shifting the weed composition to more difficult ones, or selecting for herbicide-resistant genotypes as we have witnessed across all types of crops that rely heavily on herbicides for weed control. The international herbicide-resistant weed survey illustrates this problem (Heap 2018).

Various residual herbicides were tested in mixtures with paraquat on newly planted rubber in Bahia, Brazil (Lima and Pereira 1991). The most effective herbicides were: glyphosate (1.5 kg ae); diuron-hexazinone (2 kg ai) + paraquat (0.2 kg ai); and oxyfluorfen (1.5 kg ai) + paraquat. Of course, weed control lasts longer when non-selective herbicides (glyphosate, glufosinate, or paraquat) are used with residual herbicides such as diuron, hexazinone, or oxyfluorfen. Paraquat is safer to use between young trees because it is a contact herbicide. Therefore, if spray droplets are drifted onto the rubber leaves, it will not affect the whole plant as the herbicide will desiccate only tissues at the point of spray deposition. In contrast, glyphosate is a systemic herbicide. Glyphosate drift can injure the young rubber plants significantly as the herbicide translocates to the rest of the plant. When the trees are tall, glyphosate would have an advantage over paraquat because glyphosate can kill perennial weeds, which are dominant in plantation areas. In contrast, paraquat will desiccate only the green tissues and the weeds will regrow from the remaining live meristems.

Weed Management in Various Stages of Rubber Development

Weed Management in the Rubber Nursery

Prior to planting of rubber in the field, the seedlings are raised and budded in nursery beds

Figure 26.7. Typical arrangement of rubber seedlings raised in nurseries.

(Figure 26.7). Weeds infesting rubber nurseries are annuals because nursery areas are tilled and prepared thoroughly after raising every batch of seedlings. Weeds in the nursery are generally removed manually, as needed, with handheld implements. Mulch (wood chips, straw, others) could be spread on the beds and between the beds to reduce weed emergence and growth. Mulch materials should not carry weed propagules nor have disease-infested materials. The weeding frequency declines as the seedlings grow older. Increased shade from larger rubber canopy with time helps suppress weeds.

When the seedlings are at least two months old, pre-emergence herbicide may be sprayed, directed to the base of plants. For example, the use of diuron (3 kg ai ha^{-1}) can control weeds up to 3 months (Mathew et al. 1977). Directed spraying of herbicides can be done only if the rubber seedlings are arranged with gaps to allow passage of a sprayer. For post-emergence weed control, paraquat can be used on seedlings at least three months old, when the bark on the lower portion of the stems hardens. The herbicide then can be applied directed to the base of the seedling containers to avoid hitting green tissue. The herbicides must be sprayed between the rows of seedlings using a single, drift-reducing, flat fan spray nozzle. In place of paraquat, glufosinate ammonium may be used at the rate of 0.4 to 0.8 kg ha^{-1}. The same precautions on spray drift reduction must be followed. To control grass post-emergence, a selective grass herbicide (i.e., fenoxaprop-P-ethyl) can be applied broadcast over the top of rubber seedlings with the recommended adjuvant.

Weed Control in Immature Rubber

The spectrum of weed species infesting immature rubber plantations depends on whether the weeds are regularly slashed, intercropped with annual/perennial crops, or planted with leguminous cover crops. Regular slashing favors the perennial weeds, such as *I. cylindrica*, *Axonopus compressus* (Sw.) P. Beauv., *M. pudica*, and *Panicum maximum* Jacq. When areas between rows of rubber are cultivated and intercropped with annual crops, annual weeds become dominant. When the areas between the rows of rubber are planted with legumes as cover crop, weeds, such as *Ageratum conyzoides* L., *M. invisa*, *P. conjugatum, and A. compressus* proliferate before the legume crops close in. There are several options for farmers to control weeds in immature rubber. Weeds between the rows may be controlled by slashing, mowing, herbicide application, intercropping with annual or perennial crops or planting leguminous cover crops. A mixture of glyphosate (1.0 kg ai ha^{-1}) + metsulfuron-methyl (0.03 kg ai ha^{-1}) was reported to be safe to spray toward the base of rubber trees less than 1 yr old (Faiz 2006). Fluroxypyr, dicamba, or picloram + 2,4-D can also be mixed with glyphosate and applied along the planting strip to keep it weed-free (Faiz 2006). In any of these approaches, it is necessary to maintain a weed-free, legume-free strip of 1.25 m along the rows of rubber. Alternatively, a weed-free circular area around the base of each plant could be maintained. The goal is to prevent other plants from competing with rubber.

Weed Control in Mature Rubber

In a mature rubber plantation, thick tree canopies prevent sunlight from reaching the ground. This favors shade-tolerant species, such as *P. conjugatum, C. dactylon,* and *S. nudiflora.* Since many weed species are not tolerant to shade, weed control is easier in mature than immature rubber plantations. To control weeds in mature rubber plantations, either of the following would be effective: a) regular slashing of small shrubs and other remaining weeds; or b) herbicide application (i.e. glufosinate, glyphosate, paraquat) as needed, usually at 3 to 6 month intervals. The extremely problematic congongrass is sensitive to shade and will no longer be the dominant weed species in mature plantations.

Special Note About Cogongrass and Its Management

Cogongrass is the most widely distributed and the most difficult-to-control weed of rubber plantations in Asia. In fact, it would be the most difficult-to-control weed in any crop that it infests. It is among the top ten worst weeds in the world (Holm et al. 1977). It is widely and intensively distributed in Thailand, Malaysia, Indonesia, and the Philippines. It is invading even the southeastern USA, causing millions of dollars of crop losses and loss of biodiversity in infested areas. In the early 1980s, it was estimated that about 500 million ha were infested with cogongrass globally (Falvey 1981) and it has continued to expand its range, regrettably aided by human activities and international agricultural commerce.

The growth of cogongrass can be reduced by integrating methods including: i) monthly slashing during the dry season and at the onset of rainy season before new leaves attain full photosynthetic capacity; ii) tillage to a depth of 20–30 cm during the dry season, where tillage is appropriate; iii) planting fast-growing cover crops in young plantations to choke new growth; or iv) herbicide application. Glyphosate, glufosinate, and paraquat can desiccate the top growth of cogongrass. Of these, glyphosate is most effective because of its systemic action; but to achieve complete control, repeat application is necessary not only because one application cannot kill the plant, but also because none of these herbicides have residual activity. Unless all the rhizomes are killed, cogongrass will continue to sprout until all the vegetative propagule reserves are depleted. Dalapon, which has soil activity, is recommended in Asia, but this herbicide is no longer used in the USA because of its high toxicity and carcinogenic properties. Studies in the USA showed that only glyphosate and imazapyr are effective on cogongrass. Split application of each herbicide is needed (Ramsey et al. 2003) and at least two applications a year is necessary. With this process, it would take 2 to 3 years to deplete the rhizomes of older congongrass infestation (https://www.cogongrass.org/control/). Therefore, it is critical that cogongrass be controlled at the first phase of invasion, before the initial patch expands and the rhizomes build up. Tillage increases sprouting (Ramsey et al. 2003) as cutting the rhizomes breaks apical dominance and allows more rhizome buds to sprout. These young sprouts are easier to kill with glyphosate. Repeated disturbance of rhizomes, while continuing to control new sprouts with herbicides, hastens depletion of rhizomatous reserves and is the best technique to control cogongrass. Evidently, tillage is not advisable nor tenable in hilly plantations. In which case, prevention of infestation is key by planting weed-smothering legume cover crops with perennial intercrops. Rubber tappers and plantation managers need to be educated to identify and monitor any initial infestation of cogongrass or other perennial invasive weeds.

Challenges in Achieving Sustainable Natural Rubber Production

Impact on Biodiversity

The production of natural rubber is increasing, albeit gradually (Figure 26.3). There were about 10 million ha of rubber plantations globally in 2010 (Figure 26.8). Warren-Thomas et al. (2015) estimated that up to 8.5 million ha more rubber plantations are needed to meet demand by

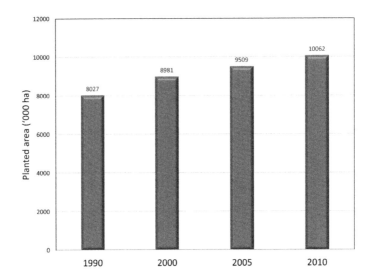

Figure 26.8. Rubber production area worldwide.
(Source: https://www.statista.com/statistics/238900/rubber-plantations-areas-worldwide/)

2024. Rubber plantations are expanding rapidly in Asia, specifically in Vietnam and China. This scenario may bode well for the economic development of the region (or the world) in the short-term, but it presents a significant level of risk in the long term that cannot be ignored. Data have been presented showing that conversion of natural forests into monoculture rubber plantations impacts biodiversity and species richness negatively (Warren-Thomas et al. 2015). Several endemic and endangered species live in the Asian region (Sodhi et al. 2004) that is now the rubber belt of the world (Figure 26.1). Clearing more forests for rubber plantations will destroy wildlife habitats. Conversely, research on rubber agroforestry (rubber intercropping) collectively show benefits from this system both economically and ecologically in certain areas. Well-planned and maintained rubber agroforests support some level of biodiversity, albeit different from that of natural forests.

Rubber agroforests are beneficial for areas that are already deforested. The diversity of soil biota and soil quality can be improved or maintained with cover cropping and the proper choice and management of intercrops. Agroforests support higher biodiversity of flora and fauna than rubber monoculture (Beukema et al. 2007). It does not, however, satisfy the specific habitat requirements of certain endemic wildlife species in natural forests. Areas planted with natural rubber are generally not deforested lands, but natural forests. As a point of reference, almost 80% of rubber plantations in the Central Highlands of Vietnam were established on what used to be natural forest land (Phuc and Nghi 2014). The same pattern is happening across the rubber belt of Asia (Warren-Thomas et al. 2015). By 2050, it is estimated that more than 4 million ha of natural forests will be cleared for rubber plantations in this region (Fox et al. 2012).

Mitigating Soil Erosion

As rubber plantations expand into steeper terrains, soil erosion becomes a serious problem. Not much can be done to mitigate this, except stopping the establishment of rubber in certain land classes. Doing this would require government intervention and strong collaboration among rubber growers nationally and internationally. This is a difficult proposition and the challenge becomes even more daunting when it comes to engaging smallholders, which comprise the majority of rubber growers. For now, the only recourse is to plant cover crops and intercrop rubber with tea, coffee, or cacao, which are adapted to hilly areas. The establishment of these perennial, shrubby crops in the rubber understory could help stabilize the soil. These crops are also important sources of cash for smallholders, which compose the majority of rubber producers. To make this work, having a strong technological and production input support system for rubber growers is very important. Locally adapted varieties of appropriate crops for the rubber agroforest need to be made available to growers. Educational, financial, and market support are crucial.

Weed Resistance to Herbicides

Resistance to herbicides is a global threat to agriculture. A resistant weed population is one which can no longer be controlled by the same dose of the herbicide that has been used to control it. In other words, what used to be susceptible to the herbicide is no longer affected by the same herbicide, after years of use. Weed resistance is evolving rapidly in agronomic crops (Burgos et al. 2018). Resistant weeds have also evolved in plantation crops where few herbicides are used multiple times every year at high doses. Weed population adaptation to herbicides occur where there is no diversity in crop management practices. This is exemplified by situations in rubber plantations where the crop is grown for about 40 years and weeds in the understory are managed primarily with slashing and herbicides. Besides these activities, the land is not disturbed generally beyond five or six years. To make matters worse, herbicides used for long-term weed management are only either glyphosate or paraquat and in some locations, glufosinate. The inevitable consequence of this persistent, strong selection pressure is resistance evolution to herbicides. A case in point is goosegrass (*Eleusine indica* [L]. Gaertn.) with multiple resistance to glufosinate and paraquat from what used to be a rubber plantation field in Malaysia (Seng et al. 2010). The resistance level was low (3.4- to 3.6-fold), but enough to cause problems in the field because the population is no longer controlled with the recommended field dose of each herbicide. While in rubber, the field had been sprayed with paraquat for more than 30 years and later, with glufosinate. Multiple resistance often results from successive selection with herbicides of different modes of action as observed with other global weeds, such as *Echinochloa* spp. (Rouse et al. 2018). Preventing and managing resistance to herbicides require adoption of best management practices for crop production, which is the integration of all sound agronomic and weed management practices and tools.

Concluding Remarks

The rubber industry has become indispensable to human civilization as we know it today. Although the majority of rubber products we use are from synthetic materials, it cannot substitute natural rubber 100%. Synthetic rubber production is also dependent on non-renewable petroleum products. The demand for natural rubber has long outpaced production. Since natural rubber can be grown well only in high-rainfall tropical zones, this has constrained the level of production and the sustainability of the rubber industry. Sustainable weed management is a great hurdle in rubber production. Preventive, cultural, mechanical, and chemical tools are available for farmers to build an effective, long-term weed management program. The choice of adopting one method instead of another depends on local social norms; farmer attitudes; availability of money to procure inputs; availability of labor; knowledge of, and access to, technology (e.g., improved varieties, herbicide application equipment); environmental factors; and factors that limit the range of feasible agronomic choices (e.g., length of the growing season, rainfall and temperature patterns, farm and market structure, and others). Diversification of the cropping system based on sound agro-ecological principles is key to sustainable weed management. Integration of preventive and cultural methods for weed management must always be pursued. Farmers must be educated to acquire a higher level of knowledge and technical skills. Comprehensive support for smallholders must be instituted and implemented. Global regulation of rubber plantation expansion is needed. The call for a robust sustainability initiative for natural rubber production is clear.

REFERENCES

Annie, P., S. Elsie and K.I. Punnoose. 2005. Comparative evaluation of dry matter producion and nutrient accumulation in the shoots of *Pueraria phaseoloides* Benth and *Mucuna bracteata* D.C. grown as cover crops in an immature rubber (*Hevea brasiliensis*). Plantation Natural Rubber Res., 18: 87–92.

Anonymous. 2013. Rubber (*Hevea brasiliensis*) crop management. KAU-Agri Infotech Portal, Centre for E-Learning, Kerala Agric.l Univ., India. Available at: http://www.celkau.in/Crops/ Plantation%20Crops/ Rubber/production.aspx (Accessed on February 28, 2018).

Anonymous. 2016. Natural Rubber Statistics 2016. Malaysian Rubber Board, Lembaga, Getah, Malaysia. Available at: http://www.lgm.gov.my/nrstat/nrstats.pdf (Accessed on February 24, 2018).

Asna, B.O. and H.L. Ho. 2005. Managing invasive species: the threat to oil palm and rubber – The Malaysian Plant Quarnatine Regulatory Perspective. Available at: http://www.fao.org/ docrep/008/ae944e/ ae944e05.htm#bm05/ (Accessed on February 20, 2018).

Barry, R.G. and R.J. Chorley. 2009. Atmosphere, Weather, and Climate, 1st ed'n, Routledge. 533 p.

BAR (Bureau of Agricultural Research Chronicle). 2012. BAR, UPLB conduct training on rubber production and processing. ISSN 1655-3942. 20 p.

Beukema, H., F. Danielsen, G. Vincent, S. Hardiwinoto and J. Andel. 2007. Plant and bird diversity in rubber agroforests in the lowlands of Sumatra, Indonesia. Agrofor. Syst. 70: 217–242.

Brook, R.M. 1989. Review of literature on *Imperata cylindrica* (L.) Raeuschel with particular reference to southeast Asia. Trop. Pest Manag. 35: 12–25.

Burgos, N.R., I. Heap, C.E. Rouse and A.L. Lawton-Rauh. 2018. Evolution of herbicide-resistant weeds. *In:* Korres, N., N.R. Burgos and S.O. Duke (Eds.) Weed Control: Sustainability, Hazards, and Risks in Cropping Systems Worldwide. Chapter 6.

Cerdeira, A.L., L.C.L. Cantrell, F.E. Dayan, J.D. Byrd and S.O. Duke. 2012. Tabanone, a new phytotoxic constituent of cogongrass (*Imperata cylindrica*). Weed Sci. 60: 212–218.

Cisneros, J.J. and B.H. Zandstra. 2008. Flame weeding effects on several weed species. Weed Technol. 22: 290–295. doi: 10.1614/WT-07-113.1

Dilipkumar, M., T.S. Chuah, S.S. Goh and I. Sahid. 2017. Weed management issues, challenges, and opportunities in Malaysia. doi.: 10.1016/j.cropro.2017.08.027

Guzzo, C.C., L.B. De Carvalho, P.R.F. Giancotti, P.L.C.A. Alves, E.C.P. Goncalves and J.V.F. Martins. 2014. Impact of the timing and duration of weed control on the establishment of a rubber tree plantation. Anais da Academia Brasileira de Ciências. 86: 495–504. doi:10.1590/0001-37652014119113

Dean, W. 1987. Brazil and the struggle for rubber. Cambridge University Press, Cambridge.

De Jorge, P. 1962. Report of botanical division. A Report by the Rubber Res. Inst. Malaya, 1961. pp. 71–73.

Dijkman, M.J. 1951. Hevea – Thirty years of research in the Far East. University of Miami Press, Coral Gables, Florida, 329 p.

Esekhade, T.U., J.R. Orimoloye, I.K. Ugwa and S.O. Idoko. 2003. Potentials of multiple cropping systems in young rubber plantations. J. Sust. Agric. 22: 79–94. doi:10.1300/J064v22n04_07

Faiz, M.A.A. 2006. Efficacy of glyphosate and its mixtures against weeds under young rubber forest plantation. J. Rubber Res. 9: 50–60.

Falvey, J.L. 1981. *Imperata cylindrica* and animal production in Southeast Asia: a review. Trop. Grasslands. 15: 52–56.

Fox, J. and J.-C. Castella. 2013. Expansion of rubber (*Hevea brasiliensis*) in mainland Southeast Asia: what are the prospects for smallholders? J. Peasant Stud. 40: 155–170.

Fox, J., J.B. Vogler, O.L. Sen, T.W. Giambelluca and A.D. Ziegler. 2012. Simulating land-cover change in montane mainland Southeast Asia. Environ. Manag. 49: 968–979.

Guo, Z.M., Y.Q. Zhang, P. Deegen and H. Uibrig. 2006. Economic analyses of rubber and tea plantations and rubber-tea intercropping in Hainan, China. Agroforestry Systems 66(2): 117–127 doi:10.1007/s10457-005-4676-2.

Harper, 1973. Efficiency of paraquat as weed control chemical for rubber nurseries. PANS 21: 401–405.

Heap, I. 2018. The international survey of herbicide-resistant weeds. Available at: www.weedscience. org. (Accessed on February 28, 2018).

Hodgson-Kratky, K.J.M., O.M. Stoffyn and D.J. Wolyn. 2017. Recurrent selection for rubber yield in Russian dandelion. J. Amer. Soc. Hort. Sci. 142: 470–475. doi: 10.21273/JASHS04252-17

Holm, L.G., D.L. Pucknett, J.B. Pancho and J.P. Herberger. 1977. The World's Worst Weeds. Distribution and Biology. University Press of Hawaii, Honolulu, HI, 609 p.

Hondrade, R.F., E. Hondrade, L. Zheng, F. Elazegui, J.L.J.E. Duque, C.C. Mundt, C.M. Vera Cruz and K.A. Garrett. 2017. Cropping system diversification for food production in Mindanao rubber plantations: a rice cultivar mixture and rice intercropped with mungbean. Peer J. http://dx.doi.org/10.7717/peerj.2975

Hussain, F., N. Abidi and Z.H. Malik. 1994. *Imperata cylindrica* (L.) P. Beauv. affects germination, early growth, and cell division and development in some crop species. Pakistan J. Sci. Ind. Res. 37: 100–103.

IRSG. 2017. World rubber production. International Rubber Study Group. Available at: http://www.rubberstudy.com/statistics.aspx. (Accessed on February 28, 2018).

Jessy, M.D., P. Joseph and S. George. 2016. Possibilities of diverse rubber based agroforestry systems for small holdings in India. Agroforestry Systems. doi: 10.1007/s10457-016-9953-8

Kothandaraman, R., D. Premakumari and P. Sivasankara. 1987. Studies on growth, nodulation and nitrogen fixation by *Mucuna bracteata*. pp. 283–288. *In:* Proc 6th Symposium Plantation Crops. Kottayam.

Kothandaraman, R., J. Mathew, A.K. Krishnakumar, K. Joseph, K. Jayarathnam and M.R. Sethuraj. 1989. Comparative efficiency of *Mucuna bracteata* D.C. and *Pueraria phaseoloides* Benth. on soil nutrient enrichment, microbial population, and growth of *Hevea*. Indian J. Natural Rubber Res. 2: 147–150.

Kou, W., X. Xiao, J. Dong, S. Gan, D. Zhai, G. Zhang, Y. Qin and L. Li. 2015. Mapping deciduous rubber plantation areas and stand ages with PALSAR and Landsat images. Remote Sens. 7: 1048–1073; doi:10.3390/rs70101048

Lespinasse, D., M. Rodier-Goud, L. Grivet, A. Leconte, H. Legnate and M. Seguin. 2000. A saturated genetic linkage map of rubber tree (*Hevea* spp.) based on RFLP, AFLP, microsatellite, and isozyme markers. Theor. Appl. Genet. 100: 127–138.

Lima, A.D. and R.J.D. Pereira. 1991. Weed control in rubber plantations. Pesquisa Agropecuaria Brasileira. 26: 163–167.

Liu, W., Q. Luo, H. Lu, J. Wu and W. Duan. 2017. The effect of litter layer on controlling surface runoff and erosion in rubber plantations on tropical mountain slopes, SW China. Catena 149: 167–175.

Mathew, M., K.I. Punnoose and S.N. Potty. 1977. Report on the results of chemical weed control experiments in the rubber plantations in South India. J. Rubber Res. Inst. Sri Lanka 54: 478–488.

McAssey, E.V., E.G. Gudger, M.P. Zuellig and J.M. Burke. 2016. Population genetics of the rubber-producing Russian dandelion (*Taraxacum kok-saghyz*). PLoS ONE 11(1): e0146417. doi:10.1371/journal.pone.0146417

Mercado, B.L. 1986. Control of *Imperata cylindrica*. pp. 268–278. *In:* K. Moody (Ed.) Weed Control in Tropical Crops. 2: 293. Published by the Weed Science Society of the Philippines (WSSP).

Noda, K., S. Kanjanajirawong, I. Chaiwiratnukul, T. Sangtong and M. Teeerawatsakul. 1987. Biological studies of Pennisetum species in Thailand as associated with its control. NWSRI Project by JICA, Ministry of Agriculture and Cooperative, Thailand. Proj. Res. Rep. 5: 1–29.

Ong, S.H., R. Othman and M. Benong. 1998. Breeding and selection of clonal genotypes for climatic stress condition. Proc. Int'l. Rubber Res. Dev. Board (IRRDB) Symposium on Natural Rubber (*Hevea brasiliensis*). Vol. I. General, Soils and Fertilization, and Breeding and Selection, October 14-15, 1997, Ho Chi Minh City, Vietnam. pp. 149–154.

Pamplona, P.P. 1990. Central Mindanao Agriculture Resources Research and Development Rubber Production and Management. Univ. Southern Mindanao, Kabacan, Cotabato, Philippines. Book 395 Series No. 1. 96 pp.

Pamplona, P.P. and S.S. Tinapay. 1998. Intercropping as an effective weed control tool in rubber plantation. Available at: http://agris.fao.org/agris-search/search.do?recordID=PH8910927 (Accessed on January 4, 2018).

Pancho, J. 1978. Systematic study of weeds in rubber plantations of the Philippines. PCARRD Res. Rep. No. 260. 104 p.

Pathiratna, L.S.S. and M.K.P. Perera. 2006. Effect of plant density on bark yield of cinnamon intercropped under mature rubber. Agroforestry Systems 68: 123–131. doi 10.1007/s10457-006-9003-z

Penot, E. and I. Ollivier. 2009. Rubber tree intercropping with food crops, perennial fruit, and tree crops: several examples in Asia, Africa, and America. Bois Et Forets Des Tropiques pp. 67–82.

Phuc, T.X. and T.H. Nghi. 2014. Rubber expansion and forest protection in Vietnam. Tropenbos International Vietnam and Forest Trends To, Hue City, Vietnam.

Priyadarshan, P.M. 2011. Introduction. Biology of Hevea Rubber. CABI. 221 p. SBN: ISBN number: 9781845936662, ProQuest Ebook Central. Available at: http://ebookcentral.proquest.com/ (Accessed on February 24, 2018).

Rajasekharan, P. and S. Veeraputhran. 2002. Adoption of intercropping in rubber small holdings in Kerala, India: a tobit analysis. Agroforestry Systems 56: 1–11. doi:10.1023/a:1021199928069

Ramsey, C.L., S. Josea, D.L. Miller, J. Cox, K.M. Portier, D.G. Shilling and S. Merritta. 2003. Cogongrass (*Imperata cylindrica* [L.] Beauv.) response to herbicides and disking on a cutover site and in a mid-rotation pine plantation in southern USA. Forest Ecol and Manag. 179: 195–207.

Riepma, P. 1965. A selective herbicide for use in tropical legumes. Weed Res. 5: 52–60.

Riepma, P. 1968. Weed control in rubber – a review. PAN (C) 14-43-61.

Rodrigo, V.H.L., C.M. Stirling, T.U.K. Silva and P.D. Pathirana. 2005. The growth and yield of rubber at maturity is improved by intercropping with banana during the early stage of rubber cultivation. Field Crops Res. 91: 23–33. doi:10.1016/j.jfcr.2004.05.005

Rodrigo, V.H.L., T.U.K Silva and E.S. Munasinghe. 2004. Improving the spatial arrangement of planting rubber (*Hevea brasiliensis* Muell. Arg.) for long-term intercropping. Field Crops Res. 89: 327–335. doi:10.1016/j.fcr.2004.02.013

Rouse, C.E., N.R. Burgos, J.K. Norsworthy, T.M. Tseng, C.E. Starkey and R.C. Scott. 2018. *Echinochloa* resistance to herbicides continues to increase in Arkansas rice fields. Weed Technol. https://doi.org/10.1017/wet.2017.82

RRIM. 1975. Course on Crop Protection in Rubber Plantations. Rubber Res. Inst. of Malaysia. May 12-17, 1975. Kuala Lumpur, Malaysia.

Schultes, R.E. 1990. A brief taxonomic view of the genus *Hevea*. In: MRRDB (Ed.) Monograph, Vol. 14, 57 pp. Kuala Lumpur, Malaysia.

Seng, C.H., L.V. Lun, C.T. San and I.B. Sahid. 2010. Initial report of glufosinate and paraquat multiple resistance that evolved in a biotype of goosegrass (*Eleusine indica*) in Malaysia. Weed Biol. Manag. 10: 229–233.

Serier, J.B. 1993. History of Rubber. Desjonqueres Editions, Paris, 273 pp.

Sodhi, N.S., L.P. Koh, B.W. Brook and P.K.L. Ng. 2004. Southeast Asian biodiversity: an impending disaster. Trends Ecol. Evol. 19: 654–660.

Statista. 2018. Global production of natural rubber. Available at: www.statista.com/statistics/275387/global-natural-rubber-production/ ([Accessed on February 28, 2018).

Strahler, A.N. 1969. Physical Geography. 3rd ed. John Wiley, New York.

Swearingen, J. and C. Bargeron. 2016. Invasive Plant Atlas of the United States. University of Georgia Center for Invasive Species and Ecosystem Health. Available at: http://www.invasiveplantatlas.org/ (Accessed on February 24, 2018).

Teoh, C.H., P.V. Toh, C.F. Chong and R.C. Evans. 1978. Recent developments in the use of herbicides in rubber and oil palm. PANS 24: 503–513.

Van Lam, L., T. Thanh, V.T. Quynh Chi and L.M. Tuy. 2009. Genetic diversity of Hevea IRRDB'81 collection assessed by RAPD Markers. Mol. Biotechnol. 42: 292–298. doi: 10.1007/s12033-009-9159-7

Vitelli, J.S. and B.A. Madigan. 2004. Evaluation of a hand-held burner for the control of woody weeds by flaming. Australian J Exp. Agric. 44: 75–81. doi: 10.1071/EA02096

Walsh, M. and P. Newman. 2007. Burning narrow windrows for weed seed destruction. Field Crops Res. 104: 24–30.

Warren-Thomas, E., P.M. Dolman and D.P. Edwards. 2015. Increasing demand for natural rubber necessitates a robust sustainability initiative to mitigate impacts on tropical biodiversity. Conservation Letters 8: 230–241. doi: 10.1111/conl.121

Webster, C.C. and W.J. Baulkwill. 1989. Rubber. Longman Scientific and Technical, New York.

Whycherly, P.R. 1964. Report of botanical division. A Report by the Rubber Inst. Malaya 1963. pp. 51–21.

Wu, J., W. Liu and C. Chen. 2016. Can intercropping with the world's three major beverage plants help improve the water use of rubber trees? J Appl. Ecol. 53: 1787–1799. doi:10.1111/1365-2664.12730

CHAPTER
27

Sustainable Weed Control in Orchards

Rakesh S. Chandran

West Virginia University, P.O. Box 6108, 3417 Agricultural Sciences Building,
Morgantown, WV 26506-6108
E-mail: RSChandran@mail.wvu.edu

Introduction

Sustainable Fruit Production and Weed Management

Commercial orchards are intensively managed for arthropod pests, plant diseases, and weeds. In a competitive market where consumer expectations are high for fruit quality and shelf life, commercial growers depend on pesticides in conventional systems to manage pests in a cost-effective manner. Although weeds seldom affect fruit quality directly, they compete successfully with trees to bring about yield losses. In terms of fruit quality, sustainable methods to manage weeds compared to those other pests are more feasible, although they incur higher costs to the grower (Himmelsbach 1992). Although profitability is an essential attribute of a sustainable orchard, it can sometimes be challenging to accomplish the ecological and economic goals of sustainability, simultaneously. Compared to profitability, sustainability is a more complex phenomenon determined by multiple factors that interact with one another. While the objectives of profitability are market-driven, those of sustainability are more long-term by nature. Therefore, it can be a formidable process to facilitate these opposing forces meeting at a desirable medium, yet not an impossible one. In studies to compare organic, conventional, and integrated systems on apple production Reganold (2001) concluded that the organic system ranked first in environmental and economic sustainability followed by integrated and conventional systems, respectively. An ideal sustainable approach requires decision making based on a knowledge-intensive process involving multiple disciplines.

The Orchard Floor System

Weeds compete with trees for water, nutrients, light, space, and pollinators, limiting tree development and yield as a result. Certain perennial creeping vines, such as poison ivy (*Toxicodendron radicans* [L.] Kuntze, and Virginia creeper (*Parthenocissus quinquefolia* [L.] Planch.) can also interfere with cultural practices in the orchard. Weeds, such as common lambsquarters (*Chenopodium album* L.) and pigweed (*Amaranthus retroflexus* L.) serve as alternate hosts for green peach aphids (*Myzus persicae* [Sulzer]), (Tamaki 1975), whereas common dandelion (*Taraxacum officinale* [L.] Weber ex F.H.Wigg) can host tomato ringspot virus that triggers orchard diseases, such as Prunus stem pitting and apple union necrosis (Mountain et al. 1983).

Orchard floor management typically employ varying practices based on soil, topography, planting density, and cultural practices. An ideal orchard floor should be conducive to tree-root development at a fast rate, possess a healthy and diverse flora that provide services to the ecosystem, provide habitat for beneficial organisms, and cause minimal adverse effects to fruit production especially in newly established orchards. In conventional commercial orchards, an herbicide strip is used to manage weeds in the tree rows, while the row middles are often under a permanent sod which is usually planted. By implementing this method of orchard floor management, approximately one-third of the orchard receives herbicide treatment. Atkinson (1980) found that herbicide strips can impede the rate of tree root development and soil microbial activity. It has also been documented that root system colonization with beneficial fungi can affect the rate of root growth; fast-growing roots associated with colonization of mycorrhizal fungi, and slow-growing roots to that without colonization of mycorrhizal fungi (Resendes et al. 2008). It is therefore critical to determine the composition of a sustainable orchard floor prior to establishment. A few options to manage the orchard floor are discussed in later sections of this chapter.

The orchard floor between tree rows is typically composed of a permanent ground cover that serves as a binding medium to conserve the soil, and to maintain optimal biotic and abiotic conditions in an orchard. It typically consists of a mixture of grasses and forbs suited for local climatic conditions given their ability to establish a fibrous root system to hold the soil while creating minimal bare spots for weed growth. Choosing an appropriate grass species as a sod for row middles is important to prevent troublesome weeds from subsequently encroaching this space, although most orchards have populations of resident weeds. Water-use efficiency and soil quality are other factors to be taken into consideration. Proper attention to the flora of immediate vicinities or the periphery of the orchard is also required to prevent the buildup and encroachment of undesirable vegetation. The hardiness zone where the orchard is located and other local environmental conditions should to be taken into consideration to determine the species used as a ground cover.

From the weed management point of view, the rate of establishment and the density of established stands are to be taken into consideration when choosing an appropriate mixture. Other factors to be considered include ability of the orchard floor to modulate temperature for frost control, fix nitrogen, suppress parasitic nematodes, mowing frequency (less mowing), relatively low water use in arid climates, and tolerance to traffic. Total vegetation management (both in and between rows) is not usually practiced due to soil erosion and water infiltration concerns.

Temporal Aspects of Weed Competition in an Orchard

The importance of managing weeds in newly planted orchards is well documented (Atkinson and White 1981, Robinson and O'Kennedy 1978). Weeds compete with young trees for nutrients resulting in retarded tree growth and development, thus trees on dwarfing rootstocks are considered to be highly susceptible to weed competition (Figure 27.1A, B, C). Based on long-term field experimentation using various ground covers, Merwin and Stiles (1994), concluded that ground vegetation and edaphic conditions can have complex interactions in a young orchard, and that various ground cover management systems could have important short- or long-term advantages or disadvantages which can impact the sustainability of orchards.

A good understanding of tree root growth patterns is essential to delineate appropriate weed management strategies in orchards. Some of the earlier research related to seasonal patterns of root growth in deciduous fruit trees revealed that root growth is closely related to soil temperature (Rogers 1939). Rogers noted that during winter months, with soil temperatures below 7°C, there was limited root growth and that root growth commenced when soil temperature rose above 7°C. It was also determined that root growth commenced before leaves unfolded and continued after shoot growth stopped. Subsequent research revealed that the initial flush of new root growth in apple occurred after bloom and a second period of new root

Figure 27.1. A. Perennial weeds emerging soon after planting an apple tree predisposing it to severe competition; such weeds should be controlled prior to planting since options are limited and cumbersome after planting; B. Newly planted apple trees facing severe weed competition; C. Common ragweed (*Ambrosia artemisiifolia*) leaf (on right) growing in close proximity to a newly planted apple tree showing dark green from nutrients applied to the tree. Leaf on left was from the middle of two trees (unfertilized area).

growth occurred in late summer and early autumn (Rogers and Head 1969). They considered that maximum root growth occurred in April and May in temperate climates followed by a rest during fruit growth and expansion followed by a second period of root growth during fall. Furthermore, in a review of carbohydrate partitioning patterns during different seasons of the year, Loescher et al. (1990) concluded that accumulation of reserves later on in the growing season was critical for the performance of the tree during the following year.

While most of the research indicates that periods of vigorous shoot and root growth rarely occur at the same time, Psarras et al. (2000) determined that peak periods of root growth on young apple trees on M.9 rootstocks coincided with that of leaf growth. The researchers observed minimal root growth prior to the onset of leaf growth around mid-May in New York with root growth continuing into mid-July. They concluded that for apple trees grown in the temperate region, competition for resources should be minimized during that period and from mid-May to mid-July. The researchers concluded that peak demand for water and nutrients for above-ground growth occurs during spring and early summer. In more recent review of root phenology, Eissenstat et al. (2006) concluded that root growth patterns in apple can vary from year to year and that historical data on bimodal growth patterns were questionable. They also indicated that growth patterns could vary based on location and varieties and rootstock.

Following a three-year field experiment to evaluate vegetation management in pecan trees, Smith et al. (2005) found that expansion of tree trunk was greatest when weeds were managed year-round compared to that when weeds were not controlled till 1 August (47% suppression), and not controlled after 1 June (37% suppression). In this experiment trunk diameter of trees kept weed-free from 1 June to fall frost or 1 April to 1 August were not different from that of trees kept weed-free year-round. Atkinson (1983) documented that new root growth apple trees were highest during the initial years after planting and that it slowed down once the trees were established and began fruiting. Based on the above findings, it could be inferred that weed management in orchards, especially in newly planted orchards, is necessary during the entire growing season and is critical during the period of peak demand for resources in order to provide optimal tree health.

Spatial Aspects of Weed Competition in an Orchard

In a review of spatial aspects related to tree root growth development Gilman (1990) noted that lateral roots grow parallel to the soil surface and are generally located in the top 30 cm with fine roots growing close to the surface. It was emphasized that lateral roots spread well beyond the end of the branches (dripline), and that root form and growth is governed by a number of factors, such as soil characteristics, moisture retention capacity, fertility, and competition from other plants. Similar findings were documented by Sokalska et al. (2009), where the highest percentage of roots were distributed between 20 cm and 40 cm soil layer in both irrigated and non-irrigated trees at the end of a three-year study. Young apple trees (non-dwarf varieties) are characterized by a root distribution that is bowl-shaped, whereas older trees display a more layered root distribution with higher root density further away from the tree trunk (De Silva et al. 1999).

Established orchards typically maintain a weed-free strip of 1.5 to 1.7 m (5 ft) on either side of the trunk by employing herbicides or other methods. Atkinson and White (1976) determined that majority of roots of young apple trees were produced within a 2-m herbicide strip which also met most of its nutrient needs. They also found that majority of the surface roots (0–20 cm) were in the herbicide strip in both 4-years-old and 7-years-old apple trees grown in fine sandy loam. Root growth in grassed alley was sparse but deeper. In a study to compare weed-free strips maintained around apple trees by using straw mulch, herbicide, or cultivation it was determined that 1.5 m mulched strip yielded twice compared to those maintained by cultivation (Baxter 1970). In this study herbicide strip of 1.5 m resulted in intermediate apple yield compared to cultivation and straw mulch. In some early extensive research on root growth, it was found that in mature apple and pear trees root growth extended the dripline by two to three times (Rogers 1934, Rogers and Vyvyan 1934). Dwarf and semi-dwarf rootstocks produce shorter roots. Ma et al. (2013) determined that use of a dwarf rootstock decreased root growth compared to that by using a vigorous rootstock, and that using a combination of dwarfing interstem on a vigorous rootstock resulted in a wider and deeper root system compared to that from a dwarfing rootstock alone. The growth of trees can also be affected by the width of the herbicide strip. In an experiment to determine the effect of herbicide strip width on mite populations, it was determined that the leaf N-concentrations were higher in 2-m herbicide strips compared to 0.5-m herbicide strips, while P and K levels responded in an opposite manner (Hardman et al. 2011).

During the year of planting, root growth is limited to a narrow area close to the tree trunk followed by more rapid growth during the subsequent years. In 2016, at the Kearneysville Tree Fruit Research and Education Center of West Virginia University, apple trees were excavated to observe root distribution of apple trees (Figures 27.2 and 27.3). The dripline of a 5-years-old tree extended to approximately 1.3 m on each side (Fig. 27.2A). The lateral roots were mostly confined to the top 20 cm of soil in this tree (Fig. 27.2B). The longest sinker-root recorded was over 2 m in length (Figure 27.2B, 27.2C) while majority of the roots were between 1.5 and 1.7 m long. In the 20-years-old tree, the dripline was approximately 2 m and the longest lateral root measured 2.3 m (Figure 27.3). During exhumation, this tree had lost majority of its root system; it is possible that longer lateral roots existed. These observations concur with findings documented in the literature and also reinforce the practice of managing weed strips in tree-rows to an average of 1.6 m from trunk where weed competition can affect tree performance.

Figure 27.2. A. A 5-years-old healthy apple tree prior to exhumation to examine root architecture; a 2.5 m scale indicates the dripline; B. An exhumed 5-years-old apple tree revealing the growth habit of lateral roots; C. Longest lateral root of an exhumed 5-yr old apple tree denoting a length of ~ 2 m from the tree-trunk.

Figure 27.3. An established apple tree displaying intact roots well over 2 m from the tree-trunk.

It was also evident through examination of a dying tree (due to non-descript biotic causes) that shoot-growth is merely a reflection of a tree's root-growth (Figure 27.4A, B). Establishment of a sound root system is of paramount importance to sustain the productivity of an orchard and all contributing factors, including weed management, should be taken into consideration to promote root growth.

Figure 27.4. A. A 5-yr old dying apple tree a before exhumation to examine root architecture; the scale indicates a dripline of 0.5 m from the trunk; B. A 5-yr old dying apple tree revealing a poor root system with lateral roots no longer than 25 cm.

Factors Affecting Management Decisions

Environmental Factors – Floral Diversity, Soil and Water Quality

Vascular plants provide services to an ecosystem and reductions in floral diversity can affect the ecosystem adversely (Chapin et al. 2000). Several experiments have attempted to mimic the flora of an orchard floor to that of natural ecosystems to attain sustainability. However, long-term benefits and other benefits that are not readily tangible or quantifiable associated with such approaches are seldom taken into consideration in commercial orchards. As illustrated in the following examples these alternative approaches come with benefits as well as inherent trade-offs to orchardists.

In an experiment carried out in a subtropical orchard ecosystem, Chen et al. (2004) compared the effects of increasing the weed species numbers on various soil health attributes, such as soil carbon, fertility, and beneficial fungi (arbuscular mycorrhizae). They were able to quantify that as the richness and diversity of weed species was increased from 1 to 12, soil carbon and nitrogen increased significantly, especially during the early growing season compared to the late growing season. They also found an increase in beneficial fungal spores in the soil with an increase in weed species richness. The competitive effects of weeds on fruit trees should be considered while designing sustainable orchard floors. Hoagland et al. (2008) used tilling as a method to manage weeds in a newly established apple orchard and compared its effect on leaf N levels and biological activity of the soils to that by a living understory cover. The living understory did improve soil biological activity and fertility, however, the tree growth was negatively affected as a result of competition for water and nutrients when compared to tillage. Interestingly, wood chip mulch in this experiment resulted in good tree growth but lower levels of soil and leaf N. The authors attribute improved tree growth to increased water availability in such systems. In this study, clove oil as an organic option resulted in poor weed control and lower fertility levels of soils.

Following an eight-year Canadian study to evaluate different types of orchard floors, Neilsen and Hogue (2000) were able to document similar findings. They compared a leguminous (white clover) orchard floor to a sod-grass vegetation and evaluated its effect on tree vigor and yield. Although a legume-based cover over the entire floor increased leaf N concentration compared to grass sod, a vegetation-free strip along tree rows resulted in most tree vigor based on trunk growth, and total fruit yield over a four-year harvest period regardless of floor vegetation beyond the strip. In a different long-term study (6-year), Neilsen et al. (2014) compared different non-chemical floor management options on soil health and apple yield. Bark mulch (10 cm depth of conifer wood waste) resulted in highest organic matter content and tree vigor compared to annual compost application or tree-rows treated with alfalfa or grass hay grown between rows.

St. Laurent et al. (2008) evaluated soils from tree rows maintained under different treatments (bark mulch, herbicide strip maintained by a combination of residual and post-emergence treatments, and that maintained by post-emergence herbicides alone, and mowed sod grass) for 14 years for the susceptibility of apple seedlings to Apple Replant Disease which is considered to be a chronic disease affecting tree yields. Secondary findings of this research revealed that bark chips resulted in superior mineral and organic matter content of soil as well as soil microbial respiration compared to other treatments. Interestingly, seedlings grown from soil collected from row middles kept under a grass lane had the highest dry weight compared to other treatments. This could be because of spatial variability of the disease prevalence; highest in tree-rows with high root density and lowest in tree-rows where there are few apple roots. From the tree-row treatments, the herbicide strip maintained by a combination of residual and post-emergence treatments resulted in higher seedling growth compared to other ground cover management systems.

Nitrate leaching can be considered as an indicator of soil quality in orchards. In a study to compare conventional orchard floor management methods to alternate methods, Sanchez et al. (2003) determined that a solid cover crop maintained in tree rows reduced nitrate leaching over 90% compared to conventional herbicide strips. They also noted that leaching was reduced when compost was used as a nitrogen source compared to fertilizers. Physical properties of the soil can also be affected by orchard floor management practices. In a Brazilian citrus orchard, Homma et al. (2012) evaluated alternative management approaches, such as mowing the orchard floor for weeds instead of using herbicides and applying the clippings as mulch along tree rows, and replacing fertilizers and pesticides by natural products. Soil compaction was lower in tree rows managed using the alternate approaches when compared to the same managed conventionally. They also noted improved root growth, better root colonization by beneficial fungi and reduced infestation of mealy bug, and even higher fruit yield during the third year, in favor of the alternative approach. Foshee et al. (1997) determined that pecan tree rows maintained with minimal soil disturbance by applying herbicides to manage grasses resulted in lower levels of soil compaction compared to that kept mowed. Regular traffic from mowing was attributed to higher levels of soil compaction in mowed plots.

Economic Factors

Economic factors involved in orchard floor management that directly or indirectly affect weed management are numerous and complex. As discussed earlier, it can be difficult to quantify the indirect costs to the environment associated with a certain management practice, whether it be beneficial or detrimental. The duration of weed control derived from a certain management method, fluctuations involved in the market prices during that time frame, and unpredictable events, such as atypical weather conditions, could make it more challenging to make decisions. Typically, a commercial grower considers material and labor costs associated with a management practice averaged over a certain time-period, value of the crop and specific varieties within a crop, the market value of the crop, potential injury (mechanical, vole damage) of a given practice to trees, availability of resources on the farm or locally to reduce costs, any cost-share programs that would subsidize out-of-pocket expenses, and replacement costs to make decisions.

In field experiments at two orchards over a three-year period, Merwin et al. (1995) carried out an economic analysis to compare conventional and alternate methods of orchard floor management. They concluded that for certain fruit varieties the increased returns justified higher cost involved with non-chemical management options, such as mulches, while on the same token, reduced fruit quality and lower market value of a crop managed by herbicides nullified any cost-savings to the grower. Long-term economic benefits could not be ascertained in this study. Based on the results of this experiment, mulching was considered to be more difficult and time-consuming to manage compared to conventional methods. The authors recommended that for low-cost, high-volume fruit growers who sold most of their produce to processing or utility grade markets, herbicide strips was considered to be the most practical ground management system. However, more recent field experiments have favored mulching as a viable method to manage weeds, especially in organic systems. Granatstein et al. (2014) compared tilling (5/yr) to mulching (wood chip over fabric mulch) along with supplemental flaming (0/yr, 5/yr, and 5/yr during the 3-yr period), and organically certified herbicide applications (4/yr, 1/yr, 0/yr during the 3-yr period) along with supplemental flaming (0/yr, 5/yr, 5/yr during the 3-yr period), in apple and pear orchards. They determined that mulching in combination with flaming performed the best in terms of gross revenue in all three years and in terms of net profits. Based on the economic analysis presented by the researchers, it could be extrapolated that profits would be higher from the mulch/flame treatment even compared to conventional herbicide strips used for weed management. The initial costs associated with materials could be a limiting factor in widespread adoption. The researchers concluded that a combination of techniques should be considered to maximize the benefits.

Sustainable Approaches

A sustainable weed management approach in an orchard employs cost-effective, yet environmentally and ecologically benign, methods while taking all aspects of the production system. It entails a multi-pronged or integrated approach which encompasses a sound understanding of weed biology, critical period/s of management and potential interactions with other cultural practices. Weed control methods could very well vary from block to block depending on tree-age, problem weeds, and marketability. The ultimate goal should take long-term viability of the orchard and minimization of costs beyond the farm gate into consideration. Weed management practices, or lack thereof, undertaken by a certain orchard could indirectly affect those of orchards in the vicinity (e.g., propagule dispersal resulting due to poor sanitation, development and spread of herbicide-resistant weed biotypes). The following sections include research findings related to different weed management options available to an orchardist.

Cover Crops

Use of sown-cover crops has generated a renewed interest in various cropping systems towards sustainable weed management (Teasdale 1996). In orchards, cover crops can enrich soil nutrients, reduce soil erosion, provide habitat for beneficial insects/arthropods, and can moderate the microclimate. However, they compete for water, could possess weedy attributes, attract vertebrate pests, and serve as alternate host for other pests occasionally. Cover crops are also utilized to establish an orchard where a grass cover crop is raised during the year prior to planting and trees are planted after killing the grass cover along rows using a systemic herbicide. The killed sod reduces erosion and provides a better medium for planting new trees.

The sandwich system was developed in Switzerland where an annual or perennial living mulch is maintained in strips in the tree rows and tilled on both sides to minimize competition while improving soil health. In an organically-maintained orchard in Sweden, among different approaches experimented for weed control for optimal tree growth, it was determined that year-round sandwich system resulted in improved leaf and fruit Ca content, produced fruits with good quality and high storability, increased soil respiration, and improved weed control compared to mechanical cultivation, living mulch, use of vinegar, and partial-year sandwich system (Tahir et al. 2015). The authors also determined that weed management during a 6-wk period from late May to mid-July improved apple yield compared to that in late summer and early fall. Personal communication, March 1, 2018. Granatstein[1], however, indicated that full tree-row tillage resulted in lower competition for resources and better tree performance compared to the sandwich system and that full cover crop in tree-rows could be too competitive, particularly for dwarf trees. Granatstein et al. (2009) also determined that wood chip mulch resulted in better tree performance compared to tillage and untreated control following a 3-yr study in Washington, U.S.A. The researchers also indicated that a leguminous living mulch improved soil quality and contributed to the N-pool but elevated vole populations. An integrated weed management approach was speculated as a viable option. In a global review of organic practices for floor management in orchards Granatstein and Sanchez (2009) concluded that a plant-based system would be ideal for sustainable orchard floor management but a widely-accepted approach was lacking compared to options available for conventional growers. Effective approaches in organic systems were considerably costlier compared to those in conventional orchards.

Asian ponysfoot (*Dichondra michrantha*), a prostrate herbaceous perennial native to New Zealand and Australia, was tested as a potential ground cover in an established apple orchard in Australia (Harrington et al. 1999). The researchers determined that there were no reductions in fruit yield compared to trees grown in bare ground. Ponysfoot established dense swards when established from seeds compared to stolon fragments. Nimblewill (*Muhlenbergia schreberi*) a shallow-rooted stoloniferous perennial grass typically considered to be a weed in several cropping systems, was compared to herbicide strip and other grasses for its effect on peach trunk growth over a period of five years (Parker and Meyer 1996). They determined that peach trees grown on nimble will vegetative cover and herbicide strip exhibited maximum trunk-diameter after five years which was higher than that of peach trees grown on centipede grass (*Eremochloa ophiuroides*), bahiagrass (*Paspalum notatum*), brome (*Bromus mollis*) or weedy control plots. Similarly, ground ivy (*Glechoma hederacea*) with similar growth habits, native to Europe and Asia but introduced to North America where it is considered a weed was tested for its potential as a ground cover in 'Elstar' apple in a German orchard (Hornig and Bunemann 1995). It was compared with a grass-mixture and while clover as ground covers, along with herbicide strip and mechanical methods to manage tree rows over a period of five years. Interestingly, ground ivy caused only a 3.5% cumulative yield reduction per tree during that period while grass-mixture resulted in the lowest yield. However, the grass-mixture improved fruit color and acid content compared to other treatments.

Tworkoski and Glenn (2012) determined from a four-year study that certain cool-season grasses grown in tree-rows successfully deterred weed competition without affecting apple and peach yield. The authors concluded that growing an annually-mowed grass in tree rows could be a viable option to reduce herbicide use in orchards but fruit size could be compromised. Black et al. (2017) recommended the use of cover crop mowed at 8 to 10 cm height to reduce competition with tree for resources and pollinators, and to radiate more energy back to the orchard compared to tall vegetation while interfering minimally with other orchard operations. In a study to compare nutrient uptake by different cover crops, it was determined that grass cover crop removed the greatest amount of primary and secondary mineral nutrients followed by clover and ground ivy (Hornig and Bunemann 1996). In this experiment, tree rows maintained by an herbicide strip resulted in higher P and K levels in the top soil with no differences in Mg level. Allowing residues of rye (*Secale cereale*), wheat (*Triticum aestivum*), sorghum (*Sorghum bicolor*), or barley (*Hordeum vulgare*), desiccated at a height of 40–50 cm, to remain on the soil provided up to 95% control of major weeds in an agroecosystem for 30 to 60 days (Putnam et al. 1983). The above species, except for sorghum which is a warm-season grass, could be evaluated in orchard systems for weed suppression and any allelopathic effects on apple root growth. Black et al. (2017) recommended a grass/legume mixture to take advantage of the benefits of

---

[1]Granatstein, D. Personal communication, March 1, 2018.

the two types of plants. In a study to compare weed control options in newly planted peaches grown organically, Reeve et al. (2017) determined that birdsfoot trefoil (*Lotus corniculatus*) grown as a living mulch in the alleyway provided good weed control during the establishment period. The tree-rows covered by either straw mulch or a living mulch (low-growing shallow rooted allysum, *Lobularia maritima*) using such the trefoil alleyway resulted in maximum trunk diameter gain by three years compared to grass alleyways with similar mulches in tree-rows. They concluded that trees grown in trefoil alleyways were able to access more resources than trees grown in grass alleyways.

Overall, cover crops could play a role in orchard floor management in conjunction with other weed management methods. Careful selection of plant material based on their performance under certain environmental conditions, and their ability to suppress weeds while competing minimally with trees for water and nutrients is essential for a successful approach. Barriers limiting widespread adoption include any vulnerability of the cover crop to attract pests, potential competition for pollinators, need for herbicides manage the cover crop, and potential allelopathic effects on trees. Research opportunities exist to generate more information and to examine the suitability of various cover crops as an integrated approach to manage an orchard floor.

Mulches

Mulches manage weeds by excluding light and by serving as a physical barrier to reduce weed germination and establishment. Several types of materials, such as plastic (including landscape fabrics), straw, wood chips, bark mulch, and newspaper or shredded paper are used as mulch. Apart from their ability to manage weeds, mulches also possess other desirable attributes, such as reduction of evaporative loss of moisture from the soil, moderation of soil temperatures, and enhancing fruit quality. By keeping soil covered, mulches can manage certain insect pests of apple, such as apple maggot, leafrollers, spotted tentiform leafminer, and codling moth that complete part of their life-cycle beneath the soil surface or among fallen leaves (Hogmire 1995, Howitt 1993). Lacey et al. (2006) were able to increase the efficacy and extend the residual activity of entomopathogenic nematodes to control over-wintering codling moth by using wood chip mulch.

One of the concerns of using mulches in orchards is the ability to serve as a habitat for rodents or other mammals that could damage the trees. Root asphyxiation in poorly drained soils and costs associated with installation and removal are considered to be other factors that limit their use. Mulches are not very effective in controlling perennial weeds, which are the most common and troublesome weeds in orchards. Landscape fabric mulches are durable and more effective to manage perennial weeds than polyethylene mulches but incur higher initial investment (~$2,200/Ha).

Plastic Mulch

Among different types of plastic mulches, permeable woven polypropylene plastic (landscape fabric) mulches are more commonly used than polyethylene mulches in orchards. In non-irrigated plots of Granny Smith on M.9 rootstocks in France, the yield under plastic mulch exceeded that in bare soil by 30% (Guiheneuf 1988). Results of this long-term study suggested that plastic mulching is beneficial on shallow soils with low available water reserves and no irrigation, but in established orchards on deep soils with good water supply and irrigation from the start, its use was not justified. In a different study, mulching was carried out using plastic foil (Stojanowska 1987). Results of this 7-yr trial showed improved weed control, increased soil water content, and yield increases of 18–24% compared with trees under the herbicide fallow. A 2-yr experiment comparing various mulches and an orchard floor vegetative cover, indicated that black polypropylene mulch increased apple yield in the second year compared to other treatments, while providing weed control (Marks 1993). While plastic mulch provided better weed control than bark mulch (Bootsma 1988), wood clippings, polypropylene-based mulch, and jute-coconut fiber sheet benefitted the soil structure and moisture content (Mantinger and

Gasser 1993). Yin et al. (2007) found that it took five years to recover costs associated with sweet cherry maintained under a polypropylene-based mulch compared to uncovered floor maintained using annual applications of glyphosate. Enhanced yield from mulched plots during subsequent years was expected to generate high net revenues compared to that from uncovered plots.

Applying plastic mulch over the fallen leaves could reduce the emergence of certain insect pests from the soil or leaves, during the growing season. Also, certain plastic mulches raise the temperature of the lower canopy from light reflection which speeds up drying of the dew of lower canopy making them less susceptible to disease pathogens. Potential drawbacks of such mulches include keeping the soil wet to induce root and collar rot in poorly drained soils, or serving as a habitat for rodents. Some physiological benefits have also been reported from the use of plastic mulches. Kasperbauer and Hunt (1998) tested colored plastic mulch that reflect more red light. They found that red plastic mulch increased tomato yield and concluded that it was caused by reflection of far-red light to the growing plants and its subsequent phytochrome-mediated regulation of photosynthate allocation to developing fruit. Similar effects on fruit quality have been documented through the use of reflective or colored plastic in orchards (Meinhold et al. 2011, M. M. Blanke 2008).

Organically Derived Mulches

Organically derived mulches, such as wood chip or bark mulch and straw mulch are also used in tree rows. In order to be effective, they have to be applied at a depth of 8 to 10 cm. Neilsen et al. (2003) compared different types of organic mulches to herbicide strip on yield and other growth parameters of 'Spartan' apple on M.9 rootstock. Shredded paper mulch treatment recorded highest average yield over a 5-yr period. Shredded paper along with biosolids mulch, and black plastic mulch resulted lower but similar yields, followed by biosolids mulch (alone) and alfalfa mulch. In this experiment, the herbicide strip treatment recorded lowest yield. In a long-term study, Niggli et al. (1988) found that uncomposted conifer, oak-bark and rape-straw controlled weeds in tree rows and resulted in apple yield and quality similar to herbicide-treated rows. They also determined that the nitrate concentration of soil water in herbicide-treated blocks was several times higher than that in mulched blocks. In newly planted 'Honey Crisp' apple, tree growth and fruit yield were compared following treatments comprised of bent grass cover, bare ground, green manure, reflective mulch, compost, and reflective mulch applied over compost (Reeke et al. 2012). In this research, reflective mulch applied over compost resulted in the highest growth rate and fruit yield while providing effective weed control.

In newly planted peaches, application of residual herbicides to maintain an herbicide strip was compared to organic mulches or a killed sod residue of tall or hard fescue plus ryegrass (Belding et al. 2004). It was determined that fruit yield was higher in herbicide-treated plots after four years and that weed interference in plots that received mulches had lower trunk diameters along with injury from voles compared to herbicide-treated plots. The use of composted poultry litter as a mulch in an orchard system was documented not only to reduce weed competition in apples but was also determined to be beneficial in an orchard ecosystem to manage certain tree fruit insect pests (Brown and Tworkoski 2006). The authors attributed this to increased predation of insect larvae in plots that received composted poultry litter and indicated that such animal wastes could be beneficial for sustainable biocontrol approaches as long as too much phosphorus was not added to the soil. Mulches can also affect the water-holding capacity of soils. In a 5-yr study, pine bark mulch and plastic mulch both reduced fluctuations in soil water content compared to herbicide strips. In this study, the bark mulch also resulted in highest cumulative yield over a period of five years with lowest yield recorded in the plots kept weed free by herbicides (Darbellay and Fournier 1996).

If resources are available locally to make certain types of mulches more cost-effective, such materials could be explored as potential options. In a study carried out in New Zealand, organic mulches, such as pine sawdust, barley straw, compost and wool dust applied to 2- and 3-years-old apple trees were compared to tree rows where residual and foliar herbicides were used to manage weeds (Hartley and Rahman 1997). It was determined that mulched plots maintained the soil temperature with the least fluctuations with lower overall mean temperatures during the growing season compared to non-mulched plots. Soil physical quality as determined by respiration rates and earthworm counts revealed that compost treatment performed the best. Interestingly, earthworm counts were lower in soil covered by wool dust mulch and managed for weeds using a residual herbicide. The researchers concluded that that sawdust and straw were the more effective alternatives to herbicides if such materials were available at a low cost locally.

Cultivation

Cultivation has been carried out to varying degrees of success by orchardists, using a range of equipment. While cultivation can be cost-effective and reduce rodent habitats, it can affect tree growth and cause injury to tree roots, trunks, and damage the irrigation system (Granatstein and Sánchez 2009). Research findings indicate cultivation to be effective in some instances and not so effective in others. Cultivation can also make orchards prone to soil erosion especially in plantings along undulated topography. Disturbance of soil to cultivate tree rows affected other attributes of the soil, such as loss of total nitrogen and carbon (Hornig and Bunemann 1996). Specialised in-row cultivation tools for orchards and vineyards, such as Wonder-Weeder™, Weed-Badger™, Bezzerides Orchard Berm Rake™, Weed Brush are available and have not been tested or compared side-by-side for efficiency. Such tools could potentially reduce or eliminate excessive soil disturbance and related injury to tree roots.

Mechanical tillage was compared to mulches and chemical weed control in two apple orchards over a period of three years in New York (Merwin et al. 1995). In this study, cultivation resulted in yields 29% higher than that of herbicide-treated plots but 12% lower yield compared to plots that received white plastic mulch, which performed the best overall. However, in a different experiment, regular tillage of the tree row strip achieved partial weed control, lowered the yield and reduced soil mineral nitrogen (Marks 1993). Tillage followed by compost addition has been documented to be beneficial for tree growth (Neilsen et al. 2014). In a South African study, Wooldridge and Harris (1989) recorded reduced trunk diameter and pruning mass, along with a decline in soil quality after two years of tillage. Apart from its deteriorating effects on soil quality and potential injury to tree trunks, cultivation by itself is not be a sustainable practice to control weeds in orchards but could be considered as one of the tools in the IPM toolbox.

Herbicides

Conventionally managed commercial orchards depend primarily on herbicides to control weeds for cost-effectiveness and simplicity. An ideal orchard herbicide should provide effective control of troublesome weeds, exhibit no phytotoxic effects on trees, be effective at low doses, and should provide a long duration and broad spectrum of weed control, apart from good applicator and fruit safety, and environmental attributes. It is a good practice to test a small area for potential tree injury before carrying out widespread applications especially if an herbicide is used for the first time.

A chemical control program depends on fruit type, age of the tree, location, problem weeds, and environmental attributes. It usually consists of applications of systemic herbicides prior to orchard establishment, and applications of residual (PRE) herbicide/s along with foliar herbicide/s (POST) during spring months. Depending on weed regrowth, additional application/s would be required during the growing season which is typically confined to use of a POST herbicide. Planting trees in an area free of perennial weeds is a prudent practice which will minimize the burden of managing them during the establishment period (Figure 27.1 A). Cost-effective control options to control perennial weeds are limited in young orchards. Perennial weeds can be prevented in a newly-planted orchard by applying a systemic herbicide, sequentially if needed, one or two years ahead of planting. To control susceptible annual weeds, an approved PRE herbicide can be applied to newly-planted trees once the soil settles. The choice of PRE and POST herbicides is limited in younger trees compared to established trees (Derr and Chandran 2016). Modes of action of herbicides labeled for use in pome and stone fruit crops (USA) are listed on Table 27.1. It can be rewarding to develop a spray program that rotates herbicide chemistries from year to year and utilizes tank-mixtures by employing different modes of action (along with other IPM methods) to minimize the buildup of herbicide-resistant weed biotypes.

Herbicide applications occasionally receive lower priority compared to other cultural practices, such as pruning, insecticide and fungicide applications during early season, especially when there is shortage of labor. Such a schedule often misses the window to avoid weed competition during critical times of the year. In commercial orchards, herbicide programs that fit within an overall sustainable model could be developed. A more systematic and aggressive approach should be directed towards younger and more productive blocks, especially in high-density plantings, where competition for resources is high. Older blocks could be managed less intensively for weeds due to differences in spatial distribution of roots and shading obtained from tree canopy which could deter weed seed germination compared to younger trees. Scouting and record-keeping data should be used to delineate spray programs as opposed to a pre-determined spray program. Adjustments have to be made to address emergent weed problems or other unforeseen events related to weather or market fluctuations. Ability to identify weeds and a general understanding of weed biology are useful skillsets to possess in this process. Proper sprayer calibration, labeling of herbicide storage containers to avoid tank contamination, using a dedicated sprayer for herbicide applications, choosing appropriate nozzles and adjusting sprayer pressure to minimize potential spray drift, and proper cleanup are other practices of a sound chemical weed control program.

Newer Technology

Novel approaches for weed control are being evaluated and implemented to limited extents in orchards. In a series of Canadian experiments that lasted two years, Rifai et al. (2002) compared flaming and the use of steam to manage weeds to more conventional methods, such as mulches and herbicides for weed control. At low speeds (2 km/h) flaming was effective to control annual weeds with <6 leaves. One or two more flaming events were required to kill weeds with more leaves. Steam was effective to control young weeds when applied sequentially twice at weekly interval. Perennial weeds could not be effectively controlled by either of these methods. The authors concluded that further research and development were necessary before commercial utility of such technologies. A potential risk includes injury to the trunk, roots, or branches. Operator safety and non-target burns, especially under dry conditions, are other concerns related to this technology. Electrical weed control systems have been developed and implemented in European and South American orchards. Some of the earlier research determined that spark discharges or continuous contact could be used to kill plants (Diprose and Benson 1984). Spark discharges utilized high-voltage short duration pulses (25 to 60 kV for 1-3 µs) to kill weeds and was considered to be effective for annuals. Continuous contact was considered to be more effective to control perennial weeds where an electrode conducting a high voltage (15 kV, 54 kW) of current was used to kill the plant as a result of rapid heating of plant parts. The use of microwave technology to manage weeds in orchard was developed by researchers in Australia (Hansen, 2013). The effect of such techniques on soil microbes is unclear. Melander et al. (2005) described the use of thermal and various mechanical devices to manage weeds in row crops. Other devices, such as harrows, torsion weeders, and finger weeders were also reviewed. The authors indicated that such implements are effective as an integrated approach to manage weeds that includes other approaches.

Technology related to directed application of herbicides has improved considerably in the recent years. Dammer and Wartenberg (2007) designed a sprayer capable of applying variable rates of herbicides by detecting weeds using a sensor. In field trials involving field crops, an average of 25% herbicide reduction was determined without causing any crop yield reduction. Such methods are yet to be tested extensively in orchard systems. Variability in weed spectrum,

Table 27.1. Modes of action of common herbicides labeled for use in pome and stone fruit crops (please refer with product label for more details)

Mode of action	Herbicide/s	Crops used (USA)	Weeds controlled	Remarks
Cellulose biosynthesis inhibitor	Dichlobenil	Apple, cherry	Annual grasses, broadleaves, and few perennials	Residual; to be applied between fall and early spring to reduce volatile loss and optimize efficacy; activation/incorporation recommended
	Indaziflam	Apple, all stone fruits	Annual grasses and broadleaves	Residual; provides extended duration of weed control; low water solubility; broad-spectrum weed control
Photosystem II inhibitor	Diuron	Apple, peach	Mostly annual grasses, few broad leaves	Residual; apply in early spring before fruit set; tank-mixing with terbacil broadens weed control spectrum
	Simazine	Apple, cherry, peach, plum	Mostly broadleaves, few annual grasses	Residual; not to be used in sandy or gravelly soils; pre-harvest interval for apple 150 days
	Terbacil	Apple, peach	Annual grasses, broadleaves	Residual; check for tree tolerance before widespread use; do not use in soils <1% OM and adjust application rate based on OM content
Microtubule/spindle apparatus (Root growth) inhibitor	Oryzalin	Apple, all stone fruits	Annual grasses, few broadleaves	Residual; can be used for new plantings after soil settling; high rate applied only in fall
	Pendimethalin	Apple, all stone fruits	Annual grasses, few broadleaves	Residual; can be used for new plantings after soil settling; pre-harvest interval of 60 days
	Pronamide	Apple, all stone fruits	Annual and few perennial grasses, few broadleaves	Residual; controls quackgrass and knotweed
Carotenoid synthesis inhibitor	Norflurazon	Apple, all stone fruits	Annual grasses, broadleaves	Residual; pre-harvest interval of 60 days
Protox inhibitor	Carfentrazone	Apple, all stone fruits	Annual broadleaves	Foliar; apply when weeds are < 15 cm tall; contact with green tissues or fruit causes injury; add crop oil or a non-ionic surfactant
	Flumioxazin	Apple, all stone fruits	Annual broadleaves and few annual grasses	Residual, foliar; can be used for new plantings after soil settling; contact with green tissues causes injury (apply when trees are dormant)
	Oxyfluorfen	Apple, all stone fruits	Annual broadleaves and few annual grasses	Residual, foliar; used for new plantings; contact with green tissues causes injury (apply when trees are dormant)

(Contd.)

Mode of action	Herbicide	Crops	Weeds controlled	Remarks
Auxin-type growth regulator	Clopyralid	Apple, all stone fruits	Annual and few perennial broadleaves	Foliar; suppresses thistles; pre-harvest interval of 30 days
	2,4-D (amine)	Apple, peach	Annual and few perennial broadleaves	Foliar; avoid spray contact with trees; use coarse sprays to reduce drift; pre-harvest interval of 40 days for cherry, peach, plum
	Fluroxypyr	Apple, pear	Annual and few perennial broadleaves	Foliar; suppresses hemp dogbane and certain brambles; do not apply during tree bloom
ACCase (lipid synthesis) inhibitor	Clethodim	Apple, all stone fruits	Annual and several perennial grasses	Foliar; to be used in non-bearing trees only (pre-harvest interval in peach is 14 days); add a non-ionic surfactant @ 0.25% vol/vol
	Fluazifop	Apple, all stone fruits	Annual and several perennial grasses	Foliar; to be used in non-bearing trees only; add crop-oil @ 1% vol/vol or non-ionic surfactant @ 0.25% vol/vol
	Sethoxydim	Apple, all stone fruits	Annual and several perennial grasses	Foliar; can be used in bearing trees; add crop-oil @ 1% vol/vol
ALS inhibitor (amino acid synthesis)	Halosulfuron	Apple	Annual broadleaves and nutsedge	Residual, foliar; apply to nutsedge during 3-5 lf stage; add a non-ionic surfactant
	Rimsulfuron	Apple, all stone fruits	Annual broadleaves and few annual grasses; nutsedge suppression	Residual, foliar; tank-mix with other residual or foliar herbicides to improve spectrum of weed control
EPSP synthase (amino acid synthesis) inhibitor	Glyphosate	Apple, all stone fruits	Non-selective; annual and perennial weeds	Foliar; systemic; only wiper application on stone fruits except cherry; avoid spray contact with green tissues or thin bark
Glutamate synthase inhibitor	Glufosinate	Apple, all stone fruits	Non-selective; perennial weeds regrow	Foliar; mostly contact (limited systemic activity); avoid spray contact with green tissues or thin bark
Photosystem I inhibitor (Cell membrane disrupter)	Paraquat	Apple, all stone fruits	Non-selective; perennial weeds regrow	Foliar; contact; restricted use pesticide; handle carefully; rinse tank thoroughly after use; contact with green tissues causes injury

age, and spatial distribution of emerged weeds are barriers to be overcome before widespread use of this technology. The use of robots for selective weed control is gaining interest in agriculture. In a review of robotic systems, Slaughter et al. (2007) indicated that detection and identification of weeds under a wide range of conditions was the greatest challenge in agricultural situations, however, they highlighted a high potential for the use this technology. They predicted that robots with onboard electronics to discriminate weeds from crops will be used for selective control of weeds. Discrimination of crop from weed is relatively easier in orchards compared to other crops, hence this technology would be more suited in orchards.

An IPM Approach

Sustainable systems seldom depend on single tactic methods to control pests. To manage weeds in orchards such methods include over-dependence on herbicides and mechanical methods, such as tilling or mowing. Although the use of cover crops and certain mulches are more sustainable they could be cost-prohibitive, compete with trees, or provide ineffective weed control. Management practices could also be dictated by factors such as dominant weed species in an orchard (grasses or broadleaves, annuals or perennials), type of fruit and its market value, age of the orchard, availability of local resources, climatic conditions, and whether the produce is sold as conventional or organic. Regardless, prevention is the key to successful management of perennial weeds in an orchard. Successful management of perennial weeds requires proper planning and application of systemic herbicides, sequentially if needed, one or two years prior to planting.

In fruit species or varieties that have higher market value, more inputs could go into the production system. Although it can be time consuming, materials that increase the organic matter content of soil to encourage root health, along with effective methods, such as use of appropriate herbicides or plastic mulches should be considered in younger orchards. Combinations of chemical and mechanical methods can be considered where herbicides are used to keep the tree rows weed-free during the first half of the growing season and mowing could be considered during late summer to carry the orchard relatively weed-free into the fall season. These are critical periods in terms of weed competition and root growth. Mowing weeds before they come to bloom will help to reduce the buildup of the weed seed bank as well. Moreover, mowing during early times of the year can encourage migration of pests, such as lygus bugs from row-middles to trees if the crop acts as an alternate host. Biological control methods have not been explored well in an IPM approach to manage weeds in orchards. Poor host-specificity, low shelf-life, lack of appropriate formulations, and long inoculation periods are barriers that limit the practical utility of mycoherbicides or other biocontrol agents in orchards. In a research trial in West Indies, poultry (goose and chicken) were integrated into an orchard to control weeds. Although this method was found to be effective to reduce total herbaceous weed cover in grazed plots, there was a shift in species towards less palatable weed species belonging to the sedge family (*Cyperaceae*). The researchers recommended the use of herbicides or other methods in conjunction with such biocontrol agents to provide effective weed control. Potential risks from *E. coli* and other produce contaminants should be considered before widespread adoption of this method.

Concluding Remarks

Current efforts to address sustainable weed management in various cropping systems including orchards, is expected to bring about positive outcomes in the long term. Given the long history of agriculture and the extremely brief history of modern crop production, such a conscientious effort could help rectify any brief disturbances in the equilibrium of sustainable orchard ecosystems. Weeds interfere with performance of fruit trees and other cultural operations in an orchard, therefore effective methods to manage them are critical for both the grower and the consumer. Cost-effective methods to manage weeds by virtue of scientific and technological advances over the past few decades have been accepted widely by growers. While herbicide-use increased productivity of orchards exponentially, it is questionable whether chemical weed control by itself is a sustainable approach in fruit production.

An examination of past research indicates that conventional weed management methods can have negative effects on the health of an orchard ecosystem, and that several options exist for consideration and evaluation. In order to spread risks while catering towards the goals of sustainability, a prototype orchard system is worthy of examination. In such an orchard, there would be blocks of newly-planted trees, established blocks under commercial production, blocks designed to harvest produce for niche-markets, and blocks with high-value fruit crops and/or varieties. Blocks with young trees would encourage practices that promote root-growth of the trees as outlined earlier in the chapter. Mulches that enrich the soil while providing weed control are highly desirable in a sustainable system. Such orchard floor management practices have to be carried out in conjunction with similar practices to manage other pests, such as insects and diseases. Established trees can be divided into blocks that depend on sustainable approaches, such as suitable ground covers, where applicable, to manage weeds within tree-rows and those that use conventional methods to manage weeds along with other pests. Such blocks would help mitigate the risks where a certain level of productivity can be expected with known inputs.

To be in harmony with various biotic and abiotic processes in an orchard ecosystem while maintaining its productivity and profitability can be a challenging target to attain. Before growers are able to invest in such orchards, substantial research should be carried out to generate supporting data. Obstacles faced by growers to address the above challenges are not confined by disciplinary boundaries. However, scholarly research currently tends to examine grower problems in a compartmentalized manner where each discipline, such as Entomology, Plant Pathology, Weed Science, Soil Science, Horticulture, Soil Microbiology, Ecology, Agricultural Economics, within their respective scopes. A deep understanding of component specific phenomena is crucial to generate solutions pertinent to problems within a specific discipline and to better understand systems and their relationships. However, research involving multiple disciplines should be nurtured to shed more light on complex interactions. A good understanding of such interactions or relationships is necessary to address the research questions encountered to foster sustainability. For instance, the roles of weeds as competitors, alternate hosts for other pests, potential cover crops or living mulch, service providers to the ecosystem, the role of fungicides and insecticides on weed seed predation by fauna present in the orchard floor, the impact of herbicides on beneficial insects or micro-climate, warrant further studies involving multiple disciplines in order to design sustainable orchards. Certain weeds, such as dandelion serve as alternate hosts of disease vectors such as tomato ringspot virus whereas other blooming weeds, such as clovers occasionally compete with the crop for pollinators.

A single sustainable model is not applicable for a wide range of orchard systems in operation. Practices have to be tailored based on economics, prevailing climate, soils, topography, pesticide regulation, availability of resources to carry out alternate practices, fruit species/variety, age of tree in a specific block, and availability and cost of labor. Weed management decisions could be based more on scouting data to prescribe block-specific methods. If incentives from the government are available to the grower, costs beyond the farm gate should be taken into consideration. Adjustments should be made during the growing season if outbreaks of other pests could have yield impacts. Climate zones experiencing high amounts of moisture and humidity experience more pest problems compared to drier zones where less aggressive approaches are feasible to manage pests. However, certain sustainable practices such as a sown-cover crop are not practical in dry climates due to competition for water with trees. Orchards in hilly or undulated terrains have to take soil erosion into consideration affecting the orchard floor management compared to those that are less vulnerable to erosion. Regulations related to pesticides can affect the options utilized to manage pests in an orchard. The outbreak of an invasive non-native pest may lead to sudden shifts in pest management methods offsetting IPM approaches practiced for years. Growers are willing to invest more inputs to manage pests when their products enjoy a premium market, such as organic outlets or farmers markets compared to processors or other marginal markets. Proximity to timbering areas could make it more cost-effective for a grower to utilize byproducts, such as mulches or composts to manage pests.

Acknowledgments

The author wishes to thank Mr. Garry Shanholtz and Mr. Kane Shanholtz of Shanholtz Orchards, Romney, West Virginia for their assistance with field research. He also wishes to acknowledge the assistance from Mr. Tim Winfield to study apple root architecture, and that from Dr. Mira Bulatovic-Danilovich and two anonymous reviewers for their valuable insight and inputs which improved the manuscript considerably.

REFERENCES

Atkinson, D. 1983. The growth, activity and distribution of the fruit tree root system. Plant and Soil. 71: 23–35.

Atkinson, D. 1980. The distribution and effectiveness of the roots of tree crops. Hort. Rev. (Amer. Soc. Hort. Sci.) 2: 424–490.

Atkinson, D. and G.C. White. 1981. The effects of weeds and weed control on temperate fruit orchards and their environment. pp. 415–428. *In:* J.M. Thresh (Ed.) Pests, Pathogens and Vegetation: The Role of Weeds and Wild Plants in the Ecology of Crop Pests and Diseases. Pittman, London.

Atkinson, D. and G.C. White. 1976. Effect of the herbicide strip system of management on root growth of young apple trees and the soil zones from which they take up mineral nutrients. Rep. East Malling Res Stn. 1975. E909.

Belding, R.D., B.A. Majek, G.R.W. Lokaj, J. Hammerstedt and A.O. Ayeni. 2004. Orchard floor management influence on summer annual weeds and young peach tree performance. Weed Technol. 18: 215–222.

Black, B., T. Roper, T. McCammon and M. Murray. 2017. *In:* T. McCammon and M. Murray (Eds.) Intermountain Tree Fruit Production Guide. Utah State University. Available at: http://www.intermountainfruit.org/contact-us (Accessed on August 31, 2017).

Blanke, M.M. 2008. Alternatives to reflective mulch cloth (Extenday™) for apple under hail net? Sci. Hort. 116: 223–226.

Bootsma, J. 1988. Are alternatives for weed control possible? Fruitteelt. 78: 23–24.

Brown, M.W. and T. Tworkoski. 2006. Pest management benefits of compost mulch in apple orchards. Agriculture, Ecosyst. & Environ. 103: 465–472.

Chapin III, F.S., E.S. Zavaleta, V.T. Eviner, R.L. Naylor, P.M. Vitousek, H.L. Reynolds, D.U. Hooper, S. Lavorel, O.E. Sala, S.E. Hobbie, M.C. Mack and S. Díaz. 2000. Consequences of changing biodiversity. Nature. 405: 234–242.

Chen, X., K. Shimizu, Z. Fang and J. Tang. 2004. Effects of weed communities with various species numbers on soil features in a subtropical orchard ecosystem. Agriculture, Ecosyst. & Environ. 102: 377–388.

Dammer. K.H. and G. Wartenberg. 2007. Sensor-based weed detection and application of variable herbicide rates in real time. Crop Prot. 26: 270–277.

Darbellay, C. and F. Fournier. 1996. Techniques of soil management in fruit growing. Revue-Suisse-de-Viticulture, -d'Arboriculture-et-d'Horticulture 28(2): 93–97.

De Silva, H.N., A.J. Hall, D.S. Tustin and P.W. Gandar. 1999. Analysis of distribution of root length density of apple trees on different dwarfing rootstocks. Ann. Bot. 83: 335–345.

Derr, J.F. and R.S. Chandran. 2016. Orchard Weed Control. pp. 126–142. *In:* Pfeiffer, D.G. (Ed.) Spray Bulletin for Commercial Tree Fruit Growers. Va. Coop. Ext. Serv. Publ. 456–419.

Diprose, M.F. and F.A. Benson. 1984. Electrical methods of killing plants. J. of Agrl. Engg. Res. 30: 197–209.

Eissenstat, D.M., T.L. Bauerle, L.H. Comas, A.N. Lakso, D. Neilsen, G.H. Neilsen and D.R. Smart. 2006. Seasonal patterns of root growth in relation to shoot phenology in grape and apple. Acta Hort. 721: 21–26.

Foshee, W.G., Patterson, M.G., Goff, W.D. and Raper, R.L. 1997. Orchard floor practices affect soil compaction around young pecan trees. HortScience 32: 871–873.

Gilman, E.F. 1990. Tree root growth and development. I. Form, spread, depth and periodicity. J. Envt. Hort. 8: 215–220.

Granatstein, D., P. Andrews and A. Groff. 2014. Productivity, economics, and fruit and soil quality of weed management systems in commercial organic orchards in Washington State, USA. Organic Ag. 4: 197–207.

Granatstein, D. and E. Sánchez. 2009. Research knowledge and needs for orchard floor management in organic tree fruit systems. Int. J. Fruit Sci. 9: 257–281.

Granatstein, D., M. Wiman, E. Kirby and K. Mullinix. 2009. Sustainability trade-offs in organic orchard floor management. Acta Horticulturae. 873: 115–122.

Guiheneuf, Y. 1988. Plastic mulching in fruit-tree growing. Fruit-Belge. 56: 66–72.

Hansen, M. 2013. Weed ZAPPER–New technology could offer growers a way to control herbicide-resistant weeds. Good Fruit Grower. April 15, 2013 Issue.

Hardman, J.M., J.L Franklin, N.J. Bostanian and H.M.A. Thistlewood. 2011. Effect of the width of the herbicides trip on mite dynamics in apple orchards. Experimental and Applied Acarology. 53(3): 215–234.

Harrington, K., T. Zhang, M. Osborne and A. Rahman. 1999. Orchard weed control with *Dichondra micrantha* ground covers. pp. 250–254. *In:* Proc. 12th Australian Weeds Conference.

Hartley, M.J. and A. Rahman. 1997. Organic mulches for weed control in apple orchards. Orchardist 70(10): 28–30.

Himmelsbach, J. 1992. Effect and economics of alternative soil management methods in intensive apple production. Erwerbsobstbau. 34: 47–52.

Hoagland, L., J. Smith, F. Peryea, J.P. Reganold, L. Carpenter-Boggs, D. Granatstein and M. Mazzola. 2008. Orchard floor management effects on nitrogen fertility and soil biological activity in a newly established organic apple orchard. Biol. & Fert. Soils 45(1): 11–18.

Hogmire, H.W. (Ed.). 1995. Mid-Atlantic Orchard Monitoring Guide. NRAES-75. Northeast Regional Agricultural Engineering Service. Ithaca, NY. 361 pp.

Homma, S.K., H. Tokeshi, L.W. Mendes and S. Tsai. 2012. Long-term application of biomass and reduced use of chemicals alleviate soil compaction and improve soil quality. Soil & Tillage Res. 120: 147–153.

Hornig, R. and G. Bunemann. 1996. Ground covers and fertigation in integrated apple production. III. Biomass yield and nutrient uptake of the ground covers and influence on soil moisture and nutrients in the soil. Gartenbauwissenschaft. 61(4): 164–173.

Hornig, R. and G. Bunemann. 1995. Alternative soil management and fertigation in apple orchards. Erwerbsobstbau. 37(6): 167–170.

Howitt, A.J. 1993. Common Tree Fruit Pests. NCR 63. Michigan State University. East Lansing, MI. 252 p.

Kasperbauer, M.J. and P.G. Hunt. 1998. Far-red light affects photosynthate allocation and yield of tomato over red mulch. Crop Sci. 38(4): 970–974; Erratum: 38(5): 1414.

Lacey, L.A., D. Granatstein, S.P. Arthurs, H. Headrick and R. Fritts Jr. 2006. Use of entomopathogenic nematodes (Steinernematidae) in conjunction with mulches for control of overwintering codling moth (Lepidoptera: Tortricidae). J. Ent. Sci. 41: 107–119.

Lavigne, A. 2012. Poultry for biological control of weeds in orchards. Fruits 67(5): 341–351.

Loescher, W.H., T. McCamant and J.D. Keller. 1990. Carbohydrate reserves, translocation, and storage in woody plant roots. Hort. Sci. 25: 274–281.

Ma, L., C.W. Hou, X.Z. Zhang, H.L. Li, Y. Wang and Z.H. Han. 2013. Seasonal growth and spatial distribution of apple tree roots on different rootstocks or interstems. J. Amer. Soc. Hort. Sci. 38: 79–87.

Mantinger, H. and H. Gasser. 1993. Further experiences with different treatments of tree strips in young apple. Erwerbsobstbau. 35: 188–193.

Marks, M.J. 1993. Preliminary results of an evaluation of alternatives to the use of herbicides in orchards. *In:* Proc. Brighton Crop Prot. Conf. Weeds 1: 461–466.

Meinhold, T., L. Damerow and M. Blanke. 2011. Reflective materials under hailnet improve orchard light utilisation, fruit quality and particularly fruit colouration. Hort. Sci. 127: 447–451.

Melander, B., I.A. Rasmussen and P. Bàrberi. 2005. Integrating physical and cultural methods of weed control: examples from European research. Weed Sci. 53: 369–381.

Merwin, I.A., D.A. Rosenberger, C.A. Engle, D.L. Rist and M. Fargione. 1995. Comparing mulches, herbicides, and cultivation as orchard groundcover management systems. Hort. Technol. 5(2): 151–158.

Merwin, I.A. and W.C. Stiles. 1994. Orchard groundcover management impacts on apple tree growth and yield, and nutrient availability and uptake. J. Amer. Soc. Hort. Sci. 119(2): 209–215.

Mountain, W.L., C.A. Powell, L.B. Forer and R.F. Stouffer. 1983. Transmission of tomato ringspot virus from dandelion via seed and dagger nematodes. Plant Dis. 67: 867–868.

Neilsen, G.H., E.J. Hogue, T. Forge and D. Neilsen. 2003. Mulches and biosolids affect vigor, yield and leaf nutrition of fertigated high density apple. Hort. Sci. 38: 41–45.

Neilsen, G.H., T. Forge, D. Angers, D. Neilsen and E. Hogue. 2014. Suitable orchard floor management strategies in organic apple orchards that augment soil organic matter and maintain tree performance. Plant Soil 378: 325–335.

Neilsen, G.H. and E.J. Hogue. 2000. Comparison of white clover and mixed sodgrass as orchard floor vegetation. Can. J. Plant Sci. 80: 617–622.

Niggli, U., F.P. Weibel and C.A. Potter. 1988. Weed control in a perennial crop using an organic mulch. Zeitschrift-fur-Pflanzenkrankheiten-und-Pflanzenschutz. Sonderheft 11: 357–365.

Parker, M.L. and J.R. Meyer. 1996. Peach tree vegetative and root growth respond to orchard floor management. Hort. Sci. 31: 330–333.

Psarras, G., I.A. Merwin, A.N. Lakso and J.A. RayRoot. 2000. Growth henology, root longevity, and rhizosphere respiration of field grown 'Mutsu' apple trees on 'Malling 9' rootstock. J. Amer. Soc. Hort. Sci. 125: 596–602.

Putnam, A.R., J. Defrank and J.P. Barnes. 1983. Exploitation of allelopathy for weed control in annual and perennial cropping systems. J. Chem. Ecol. 9: 1001–1010.

Reekie, J., E.G. Specht and B. Braun. 2012. Orchard floor management affecting the performance of young organic 'Honeycrisp' apple trees. pp. 127. *In:* Proc. Can. Organic Sci. Conf. and Sci. Cluster Strategic Meetings.

Reeve, J.R., C.M. Culumber, B.L. Black, A. Tebeau, C.V. Ransom, D. Alston, M. Rowley and T. Lindstrom. 2017. Establishing peach trees for organic production in Utah and the Intermountain West. Sci. Hort. 214: 242–251.

Reganold, J.P., J.D. Glover, P.K. Andrews and H.R. Hinman. 2001. Sustainability of three apple production systems. Nature 410(6831): 926.

Resendes, M.L., D.R. Bryla and D.M. Eissenstat. 2008. Early events in the life of appleroots: variation in root growth rate is linked to mycorrhizal and nonmycorrhizal fungal colonization. Plant and Soil 313: 175–186.

Rifai, M.N., T. Astatkie, M. Lacko-Bartosova and J. Gadus. 2002. Effect of two different thermal units and three types of mulch on weeds in apple orchards. J. Environ. Eng. Sci. 1: 331–338.

Robinson, D.W. and N.D. O'Kennedy. 1978. The effect of overall herbicide systems of soil management on the growth and yield of apple trees 'Golden Delicious'. Hort. Sci. 9: 127–136.

Rogers, W.S. 1934. Root studies III. Pears, gooseberry and black currant root systems under different soil fertility conditions with some observation on root stock and scion effect in pears. 1. Pomol. Hort. Sci. 11: 1–18.

Rogers, W.S. 1939. Root Studies XIII. Apple root growth in relation to rootstock, soil, seasonal and climatic factors. J. Pomol. Hort. Sci. 17: 99–130.

Rogers, W.S. and G.C. Head. 1969. Factors affecting the distribution and growth of roots of perennial woody species. Proceedings Easter School Agr. Sci., Univ. Nottingham 15: 280–295.

Rogers, W.S. and Vyvyan. 1934. Root studies V. Root stock and soil effect on apple root systems. 1. Pomol. Hort. Sci. 12: 110–150.

Sanchez, J.E., J.E. Nugent, K. Kizilkaya, W. Klein, T.L. Loudon, A. Middleton, G.W. Bird, C.E. Edson, M.E. Whalon, R.R. Harwood and T.C. Wilson. 2003. Orchard floor and nitrogen management influences soil and water quality and tart cherry yields. J. Am. Soc. Hort. Sci. 128: 277–284.

Slaughter, D.C., D.K. Giles and D. Downey. 2008. Autonomous robotic weed control systems: a review. Computers and Electronics in Agriculture 61: 63–78.

Smith, M.W., B.S. Cheary and B.L. Carroll. 2005. Temporal weed interference with young pecan trees. Hort. Sci. 40: 1723–1725.

Sokalska, D.I., D.Z. Haman, A. Szewczuk, J. Sobota and D. Dereń. 2009. Spatial root distribution of mature apple trees under drip irrigation system. Agri. Water Mgmt. 96: 917–924.

St. Laurent, A., I.A. Merwin and J.E. Thies. 2008. Long-term orchard groundcover management systems affect soil microbial communities and apple replant disease severity. Plant and Soil 304: 209–225.

Stojanowska, J. 1987. The influence of mulching with perforated black foil on growth and bearing of apple trees. Fruit-Science-Reports 14(2): 79–84.

Tahir, I.I., S.E. Svensson and D. Hansson. 2015. Floor management systems in an organic apple orchard affect fruit quality and storage life. Hort Science 50: 434–441.

Tamaki, G. 1975. Weeds in Orchards as Important Alternate Sources of Green Peach Aphids 1 in Late Spring 2. Env. Ento. 4: 958–960.

Teasdale, J.R. 1996. Contribution of cover crops to weed management in sustainable agricultural systems. J. Prod. Agri. 9: 475–479.

Tworkoski, T.J. and D.M. Glenn. 2012. Weed suppression by grasses for orchard floor management. Weed Technol. 26: 559–565.

Wooldridge, J. and R.E. Harris. 1989. Effect of ground covers on the performance of young apple trees and on certain topsoil characteristics. Deciduous Fruit Grower 39: 427–430.

Yin, X., C. Seavert, J. Turner, R. Núñez-Elisea and H. Cahn. 2007. Effects of polypropylene ground cover on soil nutrient availability, sweet cherry nutrition, and cash costs and returns. Hort. Sci. 42: 147–151.

CHAPTER

28

Sustainable Weed Control in Vineyards

Ilias S. Travlos[*1], Dimitrios J. Bilalis[1], Nikolaos Katsenios[1] and Rafael De Prado[2]

[1] Faculty of Crop Science, Agricultural University of Athens, 75, Iera Odos Str., GR11855, Athens, Greece
[2] Department of Agricultural Chemistry and Soil Science, University of Córdoba, Córdoba, Spain

Introduction

Among the cultivated lands, vineyards merit a particular attention, since they consist one of the most important crops in terms of income and employment (Anderson and Nelgen 2011). Viticulture is an important economic activity and a cultural legacy in many Mediterranean regions and other areas around the world (Jones et al. 2005). High economic and cultural value is attributed to the wine production and historical vineyards worldwide; thus, it is crucial to ensure that viticulture remains both economically and environmentally sustainable (Christ and Burritt 2013).

The productivity, longevity, and profitability of the vineyards is the result of the interaction of several factors like the variety, the agronomic practices, the grape and wine quality, and the production costs. Sustainable viticulture requires the balance between vegetative vigor and fruiting load and this is the first step for high yields and profitability (Gladstones 1992, Howell 2001).

Vineyards of several regions are often prone to high soil losses due to erosion (Kosmas et al. 1997, Cerdà and Doerr 2007, Raclot et al. 2009) since most of them are located in areas with high slopes (Arnáez et al. 2007) which are characterized by low organic matter contents (Ibáñez et al. 1996, Novara et al. 2011). Moreover, in most of the vineyards there are large areas without any plant coverage and therefore soil loss through wind or rainfall is significantly favored (Novara et al. 2011). From the ecological point of view, vineyards are considered to be permanent and heterogeneous systems with one of the highest biodiversities among all crops (Simon et al. 2010). However, this contribution to overall diversity is within a wide range, depending on the landscape and especially on the adopted agricultural system (i.e., conventional, integrated, organic) and the cultural practices (i.e., conventional, reduced-tillage, no-till) as stated by Nascimbene et al. (2012). Pest and weed control is of major importance for agriculture and the profitability of either annual or perennial crops. For perennials and particularly grapevines, weed competition (depending on density, weed species, available moisture and nutrients, etc.) is intense during the early crop growth stages and early years of establishment. In these periods, the effects on vines' growth, yield and quality could be significant and extremely negative.

*Corresponding author: travlos@aua.gr

Common Weeds and their Impact on Vineyards

Controlling weeds is a great task and often accounts for the majority of expenses in crop production (Fischer et al. 2002). Weeds compete with grapevines for plant resources including water, light and nutrients, thereby potentially reducing plant vigor and yield (Ingels et al. 2005, Wisler and Norris 2005, Hembree and Lanini 2006). While usually underestimated, weeds are often the most limiting factor of crop production (Elmore 1996). Weeds threaten grapevine performance and vineyard productivity-particularly in organic vineyards where herbicide use is not an option (Delate and Friedrich 2004; Sanguankeo et al. 2009). As a result, the direct yield losses resulted from weeds in vineyards have been accounted by 10.1% (Cramer 1967). Byrne and Howell (1978) have shown that weed competition could cause yield reductions by up to 37%, number of clusters per vine by 28%, and berry weight by 3%. Holm et al. (1997) reported that grapevines in weedy vineyards produced 28% fewer nodes, and in another study weeds reduced total vine dry matter by 80% (Bordelon and Weller 1997). According to Shrestha et al. (2010), mature 'Thompson Seedless' grapevines can tolerate relatively high weed densities, while the situation is significantly worst in newly planted vines due to their smaller shoot and root system and the intense weed competition (Bordelon and Weller 1997, Alcorta et al. 2011). Allelopathic effects of weeds like *Sonchus arvensis* on vines' growth have been also demonstrated (Racz and Siaba 1971). In the meantime the weed results in both difficulties in maintenance, harvest and become host to different diseases. Moreover, weeds can create difficulties in several agricultural works necessary for the vines (such as pruning, harvest, etc.), cause conditions of high relative humidity and thus favor several diseases, reduce the temperature in the vineyard and thus increase the risk for damages because of spring frosts.

Mediterranean vineyards with their high biodiversity are usually characterized by the presence of several annual and perennial weeds. Winter annual species, such as *Avena sterilis*, *Hordeum murinum*, *Bromus* spp., *Lolium* spp., *Calendula arvensis*, *Erodium* spp., *Senecio vulgaris*, *Conyza* spp., *Capsella bursa-pastoris* and *Stellaria media* are common in vineyards. Summer annual species like *Echnichloa crus-galli*, *Digitaria sanguinalis*, *Setaria* spp., *Amaranthus* spp., *Chenopodium album*, *Datura stramonium*, *Portulaca oleracea*, *Polygonum aviculare* and *Solanum nigrum* might be also found in high densities and strongly compete grapevines for water and nutrients. Concerning perennial weeds, most of them can be really troublesome (*Sorghum halepense*, *Cynodon dactylon*, *Cyperus rotundus*, *Cirsium arvense*, *Convolvulus arvensis* and *Malva sylvestris*), while others could have beneficial effects if properly treated and kept as natural vegetation (for example, *Lolium perenne*, *Trifolium* spp., *Taraxacum officinale*). It has to be noted that many of the above-mentioned species are widespread and common in several winegrowing areas as shown by Dujmović Purgar and Hulina (2004), Gago et al. (2007), Trivellone et al. (2014) and Mania et al. (2015). Asteraceae and Poaceae are considered to be among the most predominant families of weeds in vineyards (Dujmović Purgar and Hulina 2004, Mania et al. 2015). However, the weed flora of each vineyard is the combined result of soil-climatic conditions, neighbouring fields, previous weed management methods and other agronomic practices and consequently significant differences may arise. Primary weed species collected in some experiments in grapevines of Washington (USA) were common lambsquarters (*Chenopodium album*), shepherd's-purse (*C. bursa-pastoris*), ladysthumb (*Polygonum persicaria*), pale smartweed (*Persicaria lapathifolium*), and henbit (*Lamium amplexicaule*), while some of the previously mentioned species were also found (Olmstead et al. 2012). According to Donaldson et al. (1988) "vineyard weeds vary from location to location but often include a broad complex of 15 to 25 species".

Bermudagrass (*C. dactylon*) is a vigorous perennial weed that grows in the spring and summer and propagated both by seed and rhizomes-stolons. Like other perennials, it aggressively competes with grapevines for moisture and nutrients and therefore its control requires the combination of several methods (such as deep tillage and herbicide applications). Furthermore, the management of weeds like *Conyza* spp. (horseweed, tall and hairy fleabane) is even more challenging (Travlos and Chachalis 2010, 2012). These weeds produce more than 30-40,000 wind-disseminated small seeds per plant. Although it could be easily controlled by cultivation, the adoption of minimum and no-tillage systems along with growth plasticity (can be either annual or biennial) favor their dispersal. The development of herbicide resistance (for example, glyphosate) results to their further spread and dominance (Figure 28.1).

Figure 28.1. Dominance of *Conyza* spp. in Greek vineyards.

Weed Management Methods

Some key aspects of several weed management methods are given in the sections below. Weed management in conventionally grown vineyards mainly relies on the use of herbicides. Several herbicides are registered for use in grapevines. Cultural weed control in a perennial crop like vine is mainly related to the water and fertilizer management (for example, by means of the application in the vine rows), while preventive measures focus mostly at the prevention of entrance of vegetative organs and seeds of perennial and noxious, invasive weeds. However, in organic vineyards and integrated systems several other practices are widely used and suggested in order to ensure the maintenance of their efficacy and the sustainability of the vineyards.

Mechanical Control

Weed management is one of the most expensive and challenging practices for grape production (Guthman 2000, Dufour 2006), and especially many organic farmers traditionally rely on mechanical and hand cultivation for weed control. Although these methods are highly effective, they are also labor intensive, expensive, and their sustainability is questionable from a labor and environmental perspective (White 1996, Guthman 2000). Tillage can certainly provide effective weed control (Chauhan et al. 2006, Jackson 2000), especially for annual small-seeded weeds. On the contrary, when it is very frequent and in the same depth, it might have negative effects on soil structure and cause soil compaction (Baker et al. 2006). Mechanical weeding in vineyards can also decrease the soil organic carbon content (Six et al. 1999, Mazzoncini et al. 2011), induce soil degradation (Coulouma et al. 2006), or modify soil biological communities (Schreck et al. 2012).

The use of cultivators or ploughs for the management of vineyard inter-row area is very common. However, in case of a shallow cultivation and presence of perennial weeds vegetative organs like tubers, stolons and rhizomes may cut and spreads all over the vineyard. Concerning the area under the vines, this is cultivated with specialized equipment in order to minimize any damage to the vines; however in some cases shallow roots may be damaged. Several in-row cultivators specifically designed for orchard and vineyard use are the Wonder Weeder (Harris Manufacturing, Burbank, WA), Weed Badger (Weed Badger Division, Marion, ND), and the French plow. Although these tools can cultivate more closely to the vine row than the 'traditional' tools, a small area around them might require manual weeding (Zabadal 1999) and consequently increased economic cost. Although tillage may reduce potential competition between floor vegetation and the crop for limited resources like nutrients and water, it causes mineralization of inorganic nitrogen from recently incorporated and protected soil organic matter within aggregates (Calderón and Jackson 2002). Subsequently, soil's capacity to store organic nitrogen for later mineralization and plant uptake are reduced (Steenwerth and Belina 2008a, b). Changes in these soil parameters combined with physical disturbance like tillage may be the explanation for the observed shifts in seedbank and weed communities in vineyards and other cropping systems (Chauhan et al. 2006, Baumgartner et al. 2008, Steenwerth et al. 2010).

Mowing is another mechanical method often used by grape growers to keep weeds (and in some cases cover crops) to a manageable height and to avoid or delay seed production. It is a relatively quick and cheap agronomic practice that causes minimal soil disturbance. Depending on the mowing height, soil usually maintains its coverage, while vine-weeds competition is reduced and seedbank is not enriched. Therefore, mowing is widely adopted by grape growers, both in organic and conventional viticulture and is usually performed before seedset (end of spring-beginning of summer). The different response of broadleaved vs. grass species in mowing is quite noteworthy, with the latter generally showing a high regeneration ability even when mowing is performed very close to the ground. Mowing can also encourage regrowth of some broadleaved species, such as *Conyza* spp. (Travlos 2010).

Cover Crops and Mulching

Alternative practices that promote sustainability include the use of cover crops and mulches (Altieri 1995). Such techniques have the ability to suppress weeds between vine rows (Guerra and Steenwerth 2012, Salome et al. 2014) and result in wines with quality and prices comparable or superior to conventionally managed ones (Stolz and Schmid 2007, Delmas and Grant 2010). In particular, cover crops and their residues can suppress weeds by shading, altering soil temperature and soil moisture (Creamer et al. 1996). Moreover, they improve water infiltration, carbon sequestration, nutrient supply and retention, increase earthworm population, reduce water runoff and soil erosion (Tan and Crabtree 1990, Paoletti et al. 1998, Smith et al. 2008, Peregrina et al. 2010, Ruiz-Colmenero et al. 2013) and regulate vine vigor (Tesic et al. 2007, Hatch et al. 2011, Guerra and Steenwerth 2012). It is generally accepted that cover crops reduce the vigor of the grapevines without any significant effects on yield (Mercenaro et al. 2014). This reduction could be beneficial, especially in cases of vines with high vegetative growth and low quality. In another study, Giese and Wolf (2009) also showed that root pruning and complete vineyard floor cover crops favorably reduced grapevine vegetative growth and in most cases without any significant adverse effects on vine yields. A reduced number of shoots per vine as a result of cover crops use can significantly reduce labor cost of thining (Mercenaro et al. 2014). Although there are many benefits (and challenges) to using cover crops, their primary use has been and still is an alternative weed suppression technique in organic and conventional vineyards worldwide and in many cases this is adequately achieved (Dastgheib and Frampton 2000, Liebman and Davis 2000, Pardini et al. 2002).

In general, vineyard floor and cover crop management includes several cultural practices which improve soil organic matter content and aggregation, reduce weeds, manage nutrients and water, and enhance biodiversity (Guerra and Steenwerth 2012). The selection of the most appropriate practices for each vineyard should take into account factors like vine age, soil type, management, and climatic conditions like moisture in order to avoid any undesirable competition (Celette et al. 2008, Ripoche et al. 2010, Guerra and Steenwerth 2012). Moreover, the selection of the appropriate species of cover crop is crucial for the competition between plants (Smith et al. 2008). The use of an inappropriate cover crop mixture may negatively affect the productivity of the grapevine (Krohn and Ferree 2005), since cover crops have a variable impact on soil water content and grapevine nutrient uptake partly due to the climatic and soil conditions (Monteiro and Lopes 2007, Costello 2010). However, the high heterogeneity of vineyards (regarding spacing, row width, varieties, climate, soil, etc.) makes the choice difficult (Guerra and Steenwerth 2012). There are a number of factors affecting the choice of cover crops and their performance. It was found that cover crops could perform a better weed control in vineyards with clay soils and moderate acidic pH values, while many of them have a difficulty to establish in soils with high slopes (Miglécz et al. 2015). Moreover, soils rich in nitrogen can be desirable for many weed species during the first year and weed control may be ineffective (Miglécz et al. 2015). Grass species (Poaceae) are generally preferred in fertile soils in order to take some of the nitrogen, reduce vegetative growth and promote earliness of grapevines. On the contrary, in poor soils legume species are preferred due to N-fixation (Travlos 2013). In a study conducted by Sanguankeo et al. (2009), cover crop reduced yield of Zinfandel grapevine, it suppressed some important weed species such as horseweed (*Conyza canadensis* L. Cronquist), panicle willowherb (*Epilobium brachycarpum* K. Presl.), scarlet pimpernel (*Anagallis arvensis* L.), and sowthistle (Sanguankeo et al. 2009).

It has also to be noted that a cover crop (one or more species) can be seeded or consist of many different resident species that already exist in the vineyard as part of the natural vegetation (Guerra and Steenwerth 2012). Resident vegetation is cheaper and generally easier to manage (Smith et al. 2008) and therefore its maintenance is common, especially in the Mediterranean areas (Travlos and Chachalis 2010, 2012, Bilalis et al. 2014). Each species has specific advantages and disadvantages that have to be evaluated before the final decision. A mixture of short-lived species can suppress weed species but only for a limited period of time and this can lead to increased weeds' presence during the next period (Miglécz et al. 2015). On the other hand, legumes have the ability of nitrogen fixation, while grass species provide high biomass production and vigorous roots (Smith et al. 2008, Travlos 2010). Legumes often lack persistence but may also supply more nitrogen than desired in an already high vigour vineyard (Patrick-King and Berry 2005). Annual species establish well and suppress weeds from the first year, while perennial plants provide improved weed suppression during the later years (Miglécz et al. 2015); however, as previously indicated, the key requirement for the selection of perennial species is the high soil moisture and the irrigated conditions. The main competition between vineyard and cover crops or resident vegetation is for soil moisture and nutrients (Lopes et al. 2008, Hatch et al. 2011), with perennial species being more competitive for nutrients as shown by Stork and Jerie (2003) and Celette et al. (2009). Under hot and dry conditions, competition for water and nutrients, particularly if it occurs at early growth stages and flowering can lead to a substantial decrease in yield and vine capacity (Tesic et al. 2007). However, in many cases, excessive levels of soil moisture or soil nitrates can be a problem (Novara et al. 2011). The use of a complete vineyard floor of perennial grasses as cover crops effectively reduced excessive vegetative growth of Cabernet Sauvignon grapevines in a high rainfall environment (Giese et al. 2014). However, in most cases the presence of cover crops should not be vigorous during summer in order to avoid drought stress in grapevines. Therefore, perennial cover crops are suggested only for regions with high soil moisture (Travlos 2010), while in some cases several knockdown herbicides are used in order to kill the cover crop and reduce the competition to the grapevines.

In many regions, concerns of severe competition for water between cover crops and grapevines make wine growers rather cautious about cover crops (Tesic et al. 2007, Celette et al. 2009). Indeed, the use of cover crops in rainfed vineyards is clearly a big challenge. The vegetation cover although it reduces erosion regardless the soil type, increases the organic matter and prevents overall nutrient loss, it consumes an important amount of water valuable for the grapes

(Ruiz-Colmenero et al. 2013). Therefore, competition for resources has to be further studied and reduced (Sweet and Schreiner 2010, Ruiz-Colmenero et al. 2013) through adapted and case-specific cover crop management. Recent developments in soil biology and ecology contribute to the evaluation of soil quality, as they offer new indicators of soil functioning (Steenwerth and Belina 2008a, Coll et al. 2011, Salome et al. 2014). Similarly to what was mentioned above for weeds, cover crop selection is also particularly important in newly established vineyards as young vine growth can be more susceptible to cover crop competition (Bordelon and Weller 1997). Young vines have limited nutrient and carbohydrate reserves and therefore cover crops for young vineyards must be less competitive than those for established vineyards (Balerdi 1972, Eastham et al. 1996, Olmstead et al. 2012). Allelopathic effects of cover crops like rye or clovers or grapevines is also a challenge and therefore their establishment should be conducted after a preliminary testing and not in young vineyards. Seed cost, increased risk of spring frost and increased pest populations which may cause significant damage to the vines are also some of the concerns and risks that should be taken into account for cover crop's establishment.

Mulching is an alternative method commonly used in the landscape, which is progressively more and more adopted by many farmers. The use of mulch in vineyards has a dual role of weed control and soil moisture preservation. Mulching can be an effective weed suppression tool, as found when mulched winter cereal rye (*Secale cereale*) and crimson clover (*Trifolium incarnatum*) residues reduced in-row weed cover to 5% or less in a young 'Chardonnay' vineyard in western Oregon (Fredrikson et al. 2011). Mulch results in weed suppression in vineyards (Steinmaus et al. 2008), increase of organic matter and release of nitrogen and potassium into the soil (Ripoche et al. 2011, Snapp and Borden 2005) and a significant change in the soil temperature (Nachtergaele et al. 1998). In particular, Steinmaus et al. (2008) showed that mulched cover crops provided equal or greater weed control than tillage or synthetic herbicides in California vineyards. Barley straw cover of 59% resulted in a reduction of water runoff and thus significant water saving in a vineyard in Spain (Prosdocimi et al. 2016). Estimates of annual weed seed losses due to predation from several granivores and animals typically range from 50% to 90 % (Westerman et al. 2003, 2011) and this is a reason that makes grazing in some cases an alternative weed control method for vine growers.

In Mediterranean countries and USA, mulching is performed at the end of cover crop, usually at the beginning of summer. There is an increasing interest in using mulches, such as biomass grown between the vine rows in the vineyard to reduce weeds (Elmore et al. 1997). Mulches composed of legumes can break down more quickly, while those comprising grasses break down more slowly, providing sustained weed suppression (Teasdale and Mohler 1993). Ideal stand establishment of the cover crop and proper timing of mulching into vine rows can suppress weed populations similar to or even better than conventional tillage systems in vineyards (Steinmaus et al. 2008), which is predominantly achieved through the obstruction of light required for seed germination (Buhler 1997). However, limitations of mulching, such as high risk for plant diseases and introduction of new weed seeds and reduction of temperature do exist and should be taken into account.

Thermal Control

Thermal control of weeds in vineyards is another environmentally friendly technique, although it is not widely used practice for growers. The main thermal control methods are heating methods, such as flaming, hot water, steaming, etc., and indirect heating methods, such as microwaves, UV-light, etc. (Rask and Kristoffersen 2007). The results of the several studies on thermal weeding are rather incomparable due to the wide range of equipment and treatments (Rask and Kristoffersen 2007). Weed control also varies between species and therefore it is important that the calibration of heat applications for individual plant species (Vitelli and Madigan 2004). Overall, the effectiveness of thermal weed control is dependent on weed species, growth stage, and amount of heat and time of exposure (Guerra and Steenwerth 2012) and consequently preliminary tests are required with a special focus on the most dominant and troublesome weed species.

Some of the important benefits of thermal methods are that they are compatible with organic agriculture and that they do not disturb the soil or harm soil structure and soil organisms. On the contrary, most of them have a high energy cost (for example, diesel, propane) and require some repeats for an adequate weed control and cautions under dry conditions for fire avoidance. In all cases, further research is required in order to optimize such methods and make them more economical, feasible and easily adopted by grape growers. It has to be noted that there are significant differences between the several weeds regarding their susceptibility to thermal control like flaming. For instance, weeds like *Chenopodium album* and *Stellaria media* are very susceptible, while species like *Capsella bursa-pastoris* which have more protected growing points are less sensitive and may require repeated flaming applications. Moreover, in some cases several steam weeders have been successfully used against weeds in vineyards, while they pose the advantage of reduced fire risk without damaging the root stalks of vines. According to Sherstha et al. (2013), steam suppressed weeds for 2–3 weeks in a wine grape vineyard, though it had no effects on vine growth, water potential, petiole nitrate concentration, grape yield, or quality.

Cultural and Preventive Methods

Since grapevines have all the characteristics of perennial crops, many of the cultural methods used for the weed control in arable crops are not available in vines. This is certainly a difficulty, since in annual cropping systems solutions at many serious weed problems can often come from practices such as crop rotation, high seed rate, false or stale seedbed and use of competitive cultivars. The only methods which could be applied in grapevines are related with fertilizer and water management (in the row and not between the vine rows), in order to favor mainly vines and not the weeds (Travlos 2010, 2013).

Regarding prevention, clean machinery and manure are very important in order to avoid weed seed and vegetative organs' (rhizoms, stolons, tubers) introduction in the vineyard. In general, managing weeds before vines are planted helps to reduce the weeds' competitive pressure during the establishment (and most sensitive) phase of the new vines and avoid risks of vine damage resulting from cultivation and phytotoxicity from herbicides. Pre-planting weed control is even more important in the case of perennial weeds like bermudagrass (*Cynodon spp*), johnsongrass (*Sorghum halepense*) and purple nutsedge (*Cyperus rotundus*), since such weeds are very difficult to manage inside the crop (Elmore and Donaldson 1999). Vines are most susceptible to competition from weeds during their first three to four years of growth and inadequate weed management during this period can significantly delay the early growth and profitability of new vineyards (Elmore and Donaldson 1999). After this time, the vines' larger root systems permit them to compete better with weeds, while shade from the vine canopy also helps reduce seed germination and suppress weed growth under established vines.

Chemical Control

Chemical control along with soil tillage (mechanical weeding) clearly remains the most widely used and reliable option for weed control in grapevines (Travlos et al. 2015, Prosdocimi et al. 2016). Among several weed control methods, the use of pre-emergence herbicides and then a post-emergence herbicide as needed was the most effective and least expensive treatment (Elmore et al. 1997). Main advantages of herbicides are related with their quick and on-time action and the wide range of weeds that they suppress. Among these weeds are several troublesome, noxious, alien or perennial weeds (Elmore et al. 1997). Furthermore, their use especially in vineyards with high slopes reduces the risks of soil erosion and soil structure damages which could be the result of other weed management methods like tillage. Grape growers in most countries have many herbicides available, even pre- or post-emergent for weeds. Regarding the European countries and despite the withdrawal of many active ingredients, there are still several options, among them active ingredients, such as pendimethalin, diquat, glufosinate, flumioxazine, propyzamide, propaquizofop and flazasulfuron with knock-down or residual activity. Another interesting issue is related with the manipulation of vineyard weed flora (species sift) with herbicides. In a study conducted by Donaldson et al. (1988), the use of several selective postemergence herbicides shifted native weed populations quickly and economically to species that offered erosion control but would not compete with the crop (like *Poa annua* and *Cerastium vulgatum*). According to the authors "this technique would be cost effective where desired species exist and where cover crop seeding is impractical because of soil type or an excessive slope".

On the contrary, extended or wrong herbicide use is often accompanied by some problems. Common concerns are related to the reduced efficacy of some herbicides under specific conditions, phytotoxic symptoms and injuries especially for young vines, potential water pollution, emergence and dominance of alien, noxious or resistant weed species. However, the major issue for vineyards is herbicide resistance and particularly resistance to glyphosate. It is noticeable that the majority of European glyphosate resistance problems is focused on Southern European countries (France, Greece, Italy, Spain and Portugal) in vineyards and other perennial crops (olives, citrus, etc.) (Heap 2016). According to Travlos and Chachalis (2010, 2012) and De Prado (2012), the long history of extended glyphosate use (2–3 times per year and usually with improper spraying equipment) is typical in vineyards. Conservation tillage systems along with the absence of integrated management methods and alternative practices like cover cropping are considered to be some of the most important factors contributing to the development of glyphosate resistance in grapevines and orchards. In addition, weed control methods like tillage cannot be performed in grapevines in hilly areas with slopes, and in a number of those regions the glyphosate-resistant weed problems are more pronounced (Chachalis and Travlos 2012).

In particular, during the last few years and especially after the withdrawal of several herbicides in EU, there have been many reports from vine growers that *Conyza* spp. has become increasingly difficult to control, especially in no-tillage or minimum-tillage systems (Travlos et al. 2009, De Prado 2012) (Figure 28.2). Indeed, as tillage is declined in perennial crops, from moldboard to chisel and no-till, shifts on weed species have been documented favoring particular species, such as members of the *Conyza* spp. family (Davis et al. 2008). Rigid ryegrass (*Lolium rigidum* Gaud.), Italian ryegrass (*L.multiflorum* Lam.) and perennial ryegrass (*L. perenne* L.) are also three common species in vineyards. Resistance of *Lolium* species to herbicides is also a common problem (De Prado et al. 2005, Cirujeda and Taberner 2010, Owen et al. 2014, Travlos et al. 2015), especially in vineyards and other perennial crops (Figure 28.3). In countries like Italy, Spain, Greece and Portugal grapevines are considered as high risk cases (Chachalis and Travlos 2012, Collavo and Sattin 2012, Cottet and Favier 2012, De Prado 2012). Alcorta et al. (2011) showed that competition from horseweed can substantially reduce the growth of young grapevines and that the glyphosate-resistant biotype may be more competitive than the susceptible biotype.

For the management of this serious problem, several pro-active and re-active methods can be applied (Travlos and Chachalis 2010, Beckie 2011). Regarding the chemical options, herbicides with different modes of action can be applied in mixtures or sequentially in order to ensure a quick (knock-down) effect but also a residual activity, which are very crucial for the effective control of weeds, such as *Conyza* spp. and *Lolium* spp. The time of application should be also taken into account, while practices, such as incorporation or irrigation may be also required for maximizing efficacy of specific herbicides (Chachalis and Travlos 2012). Among 'best management practices', effective and frequent scouting is a vital component of any successful weed management program, including those focused on management of herbicide resistance cases. Periodical inspection of vineyards, careful selection of the right product and rate (according to the label's recommendations) and integration of chemical control with other methods are essential for the avoidance, elimination or delay of such problems. As stated by Norsworthy et al. (2012): *"mitigating the evolution of herbicide resistance depends on reducing selection through diversification of weed control techniques, minimizing the spread of resistance genes and genotypes via pollen or propagule dispersal, and eliminating additions of weed seed to the soil bank"*.

Figure 28.2. Glyphosate resistant *Conyza* spp. population in grapevines of Greece.

Figure 28.3. Glyphosate resistant *Lolium* spp. population in grapevines of Greece.

Integrated Weed Management

One of the major threats to the productivity of EU agricultural industry is the loss of key plant protection products due to the withdrawal of existing active ingredients and restrictions of the development of new plant protection products. Actually, the number of active substances lost under 91/414/EEC was dramatic since about 75% of them were lost due to safety or economical reasons (not supported for application for Annex I approval). Reliance on fewer herbicides will probably favor the development of herbicide-resistance and increase the production cost. Predicted yield decreases can be up to 50% in several crops after the loss of plant protection products. Compared to pests and diseases, where biological control and diseases can be an effective and reliable alternative; no such option exists for the vast majority of the most troublesome weeds. If it is to maintain or increase yields in a more environmentally friendly way, several alternative approaches ought to be included in management strategies. In particular, an integrated weed management program is essential for grapevines since usually there is plenty of space between vine rows and a careful weed management is therefore required.

Integrated weed management (IWM) is the strategy describing the integration of several methods to manage weeds. All available methods have their limitations and impacts. For instance, Elmore et al. (1997) found that there was no significant difference in sugar content (brix values) of the harvested grapes between any of the weed control treatments (herbicide application, cultivation, cover crop and mulching). Chemical control has environmental and other risks, such as herbicide resistance but also methods, such as mechanical weeding, mulching and hand-hoeing are often too expensive and sometimes difficult to perform without the suitable machinery. Furthermore, on organic farms where weed control is undertaken mainly mechanically, the problems of soil compaction, the disruption of soil structure and soil erosion can be exacerbated whilst at the same time increasing fuel consumption and risk of global warming. Consequently, these methods should be used in an integrated manner, since reliance on one or two particular methods may control some weeds or favor others. The case of a Greek vineyard shown in Figure 28.4 is noticeable but also very common for viticulture in several regions. In particular, conservation tillage systems may favor the dispersal of small-seeded weeds like *Conyza* spp. which are prone to resistance development. In order to control these weeds some farmers use light cultivators. However, their repeated use cut the rhizomes of perennial weeds like *Sorgum halepense* which can remain viable in the soil and result in their wide dispersal and strong competition for the grapevines. In such cases, monitoring, knowledge of weed ecology and biology, and choice of the most suitable combination of methods are vital in order to achieve the most effective weed management outcomes both in the short- and long-term and preventing noxious and competitive species from being dominant.

Figure 28.4. Time course of weed flora in grapevines depending on the cropping system and the weed management methods: (a) dense population of herbicide-resistant *Conyza* spp. population in no-till situations, (b) immediately after treatment with light cultivator and (c) progressive emergence of johnsongrass (*Sorghum halepense*) plants.

The early years and especially the first 3–4 months after vineyard establishment are very crucial for the longevity and productivity of the vineyard. Beginning with a weed-free situation is very important for the new vines. Therefore, in the case of dense populations of perennial or woody species it is suggested that several non-selective herbicides could be applied (along with other methods) some months or years before vine planting, since new vines are very sensitive to herbicide injuries. Moreover, any practices that enhance crop growth (for example, irrigation, fertilization, pruning, etc.) may be supplemented with herbicide use within the crop row during the dormant period (winter) and other methods between the rows (cover crops, tillage, mulching, etc.). These systems can be characterized as 'mixed systems' and lately are adopted by many vine growers (Steinmaus et al. 2008).

Nowadays, there is an urgent need for new technologies and tools to be developed and implemented with precision agriculture representing the most promising approach. Precision Agriculture (PA) is a whole-farm management approach using information technology and techniques like remote sensing. In simple terms, precision agriculture is a way to "apply the right treatment in the right place at the right time" (Gebbers and Adamchuk 2010). In particular, vineyards are characterized by a high heterogeneity due to soil parameters, climate conditions and cultural practices (Bramley 2003). The adoption of simple and cheap Precision Viticulture (PV) tools allowed a better understanding and exploitation of this high variability. Particularly, as stated by Proffit et al. (2006) "vineyards require a specific agronomic management to satisfy the real needs of the crop, in relation to the spatial variability within the vineyard". New technologies for supporting vineyard management reduce the environmental impact and enhance yield and quality (Matesse and Di Gennaro 2015). The high value of the crop and the importance of quality already resulted in the design and implementation of several research projects in wine producing areas (Ferreiro-Arman et al. 2006, Mazetto et al. 2010), while a further development of PA approaches, tools and applications is rather expected.

Evaluation of the Sustainability of Weed Management Methods

Today, the application of herbicides in an irrational way and with full reliance on specific active ingredients is impacting ecosystems (for example, runoff, drift, and ground water contamination) and causing entire cropping systems to fail (for example, herbicide-resistant weeds), signaling the need for the development and implementation of integrated weed management strategies in accordance with the concept of the EU's Sustainable Use Directive (EU Directive 2009/128/EC). In general, sustainability is a term used for production systems friendly to the environment and capable to meet the needs of the present without compromising the ability of future generations to meet their own needs (WCED 1987). Therefore, resources like soil, water, energy, biodiversity have to be used for the present production but maintained for the next generations with minimum adverse effects to the environment.

Under that point of view, the several weed management methods have either positive impacts or hazards for the sustainability of viticulture. Conservation-tillage practices that have been established to prevent soil erosion and promote water conservation, as part of anti-desertification national plans, restrict the wider application of tillage for weed management. In addition, the guidelines derived from the European Common Agricultural Policy (CAP) not only discourage tillage but in several cases prohibit its operations on vulnerable lands (Chachalis and Travlos 2014). Moreover, farmers are supported to adopt more environmentally sustainable agronomic practices (EU Regulation No. 1305/2013). On the other hand, herbicide use in vineyards should be reasoned, on time and kept to minimum. In grapevines, most of the herbicide applications are performed in winter time when the vines are in the dormant state prior to the initiation of the new growth and many months before harvest (Chachalis and Travlos 2014). This has clearly a tangible positive impact since it can probably minimize adverse effects of chemical control, such as high residues in grapes. In all cases, it has to be taken into account that the reduction of use of non-renewable resources like herbicides and the production of safe food and agricultural products are among the main objectives of sustainable weed management.

Sustainability of weed control in grapevines is ensured especially when proper and justified herbicide use is combined with other alternative methods for the weed management between rows (for example, cover crops, mulching, mowing, etc). Biodiversity needs to be protected, and possibly enriched, for its enormous ecological, agronomic and economic value. According to Mania et al. (2015), in order "to enhance biodiversity and viticultural sustainability at the landscape and field level", several strategies like cover crop between the vine rows and grass strips around the vineyards could be applied. Furthermore, sustainability should also characterize the management of troublesome and negative situations like herbicide-resistance and therefore it requires a longer-term perspective. For instance, reducing seed numbers in the soil seedbank reduces the number of future plants and consequently resistance risk (Neve et al. 2011). Additionally, by means of Precision Agriculture and technologically equipped machinery that can distinguish and target only individual weeds in real time (and not bare soil), environmental impacts (like drift, off-target movement, and herbicide resistance) and the high cost of inputs and labor are significantly reduced.

Furthermore, economic viability is also one of the major components of sustainability and consequently all our weed management strategies should guarantee the economic viability and profitability of vineyards. According to Matesse and Di Gennaro (2015), if it is to be truly sustainable, viticulture should aim to reduce inputs such as energy, fertilizers and chemicals (herbicides included) while preserving the environment. The major goal to be achieved is to compromise sufficient yield and quality and consequently several weed and crop management methods ought to be implemented.

Concluding Remarks

Weed management is one of the major challenges to viticulture, especially (but not solely) in the case of organic agriculture, as long as weeds can cause severe competition to vines and their significant yield losses. Overall, chemical weed control remains one of the most effective and cheap methods; however, several adverse effects and risks may arise. Alternative methods like cover crops, mulching, tillage, thermal method and cultural practices should be included in the overall management strategy, depending of course on the specific soil, crop and climatic conditions. In many cases, mechanical methods are still the most cost-effective weed management methods in organic vineyards. Priority should be always given to the crucial, early years of vineyard establishment. The use of integrated weed management strategies will ensure that these valuable tools remain available to and efficient for farmers in the long-term. Furthermore, the high heterogeneity of vineyards implies the need of site-specific agronomic practices in a context of precision viticulture; new technologies give all the necessary tools for such an approach. Sustainable weed management in grapevines is essential for many environmental, social and economic reasons and should be systematically promoted in order to exploit the high potential of viticulture for further development and to allow the grape and wine sector to substantially provide the basis for economic growth.

REFERENCES

Alcorta, M., M.W. Fidelibus, K.L. Steenwerth and A. Shrestha. 2011. Competitive effects of glyphosate-resistant and glyphosate-susceptible *Conyza canadensis* on young grapevines (*Vitis vinifera* L.) Weed Sci. 59: 489–494.

Altieri, M.A. 1995. Agroecology: the science of sustainable agriculture. Westview Press. Boulder, CO.

Anderson, K. and S. Nelgen. 2011. Global Wine Markets, 1961 to 2009: A Statistical Compendium. University of Adelaide Press, Adelaide.

Arnáez, J., T. Lasanta, P. Ruiz-Flaño and L. Ortigosa. 2007. Factors affecting runoff and erosion under simulated rainfall in Mediterranean vineyards. Soil Till. Res. 93: 324–334.

Baker, J.B., R.J. Southard and J.P. Mitchell. 2006. Agricultural dust production in standard and conservation tillage systems in San Joaquin Valley. J. Environ. Qual. 34: 1260–1269.

Balerdi, C.F. 1972. Weed control in young vineyards. Amer. J. Enol. Viticult. 23: 58–60.

Baumgartner, K., K.L. Steenwerth and L. Veilleux. 2008. Cover-crop systems affect weed communities in a California vineyard. Weed Sci. 56: 596–605.

Beckie, H.J. 2011. Herbicide-resistant weed management: focus on glyphosate. Pest Manag. Sci. 67: 1037–1048.

Bilalis, D.J., I.S. Travlos and P. Papastylianou. 2014. Natural vegetation as a key to sustainability of agroecosystems, Chapter 9. pp. 123–134. *In:* A.A. Zorpas (Ed.) Sustainability behind Sustainability. Nova Science Publishers, Inc. Hauppauge, New York.

Bordelon, B.P. and S.C. Weller. 1997. Preplant cover crops affect weed and vine growth in first-year vineyards. Hort.Sci. 32: 1040–1043.

Bramley, R. 2003. Smarter thinking on soil survey. Austral. New Zeal. Wine Industr. J. 18: 88–94.

Buhler, D.D. 1997. Effects of tillage and light environment on emergence of 13 annual weeds. Weed Technol. 11: 496–501.

Byrne, M.E. and Howell, G.S. 1978. Initial response of Baco noir grapevine to pruning severity, sucker removal, and weed control. Am. J. Enol. Viti. 29: 192–198.

Calderón, F.J. and L.E. Jackson. 2002. Roto-tillage, disking and subsequent irrigation: effects on soil nitrogen dynamics, microbial biomass and carbon dioxide efflux. J. Environ. Qual. 31: 752–758.

Celette, F., R. Gaudin and C. Gary. 2008. Spatial and temporal changes to the water regime of a Mediterranean vineyard due to the adoption of cover cropping. Eur. J. Agron. 29: 153–162.

Celette, F., A. Findeling and C. Gary. 2009. Competition for nitrogen in an unfertilized intercropping system: the case of an association of grapevine and grass cover in a Mediterranean climate. Eur. J. Agron. 30: 41–51.

Cerdà, A. and S.H. Doerr. 2007. Soil wettability, runoff and erodibility of major dry Mediterranean land use types on calcareous soils. Hydrol. Process. 21: 2325–2336.

Chachalis, D. and I.S. Travlos. 2012. Glyphosate resistance status and potential solutions in Greece. pp. 25–26. *In:* Proceedings of International Workshop on "European status and solutions for glyphosate resistance". Universidad de Cordoba, 3–4 May 2012, Cordoba, Spain.

Chachalis, D. and I.S. Travlos. 2014. Glyphosate resistant weeds in Southern Europe: current status, control strategies and future challenges. pp. 175–190. *In:* D. Kobayashi and E. Watanabe (Eds.) Handbook of Herbicides: Biological Activity, Classification, and Health and Environmental Iimplications. Nova Science Publishers, Inc. Hauppauge, New York.

Chauhan, B.S., G.S. Gill and C. Preston. 2006. Tillage system effects on weed ecology, herbicide activity and persistence: a review. Aust. J. Exp. Agric. 46: 1557–1570.

Christ, K.L. and R.L. Burritt. 2013. Critical environmental concerns in wine production: an integrative review. J. Cleaner Prod. 53: 232–242.

Cirujeda, A. and A. Taberner. 2010. Chemical control of herbicide-resistant *Lolium rigidum* Gaud. in north-eastern Spain. Pest Manag. Sci. 66: 1380–1388.

Coll, P., E. Le Cadre, E. Blanchart, P. Hinsinger and C. Villenave. 2011. Organic viticulture and soil quality: a long-term study in Southern France. Appl. Soil Ecol. 50: 37–44.

Collavo, A. and M. Sattin. 2012. Glyphosate resistance in Italy: status and potential solutions. pp. 29–30. *In:* Proceedings of International Workshop on "European status and solutions for glyphosate resistance". Universidad de Cordoba, 3–4 May 2012, Cordoba, Spain.

Costello, M.J. 2010. Grapevine and soil water relations with nodding needlegrass (*Nassella cernua*), a California native grass, as a cover crop. Hort. Sci. 45: 621–627.

Cottet, C. and T. Favier. 2012. Glyphosate resistance in French vineyard: current situation and perspectives. pp. 31–32. *In:* Proceedings of International Workshop on "European status and solutions for glyphosate resistance". Universidad de Cordoba, 3–4 May 2012, Cordoba, Spain.

Coulouma, G., H. Boizard, G. Trotoux, P. Lagacherie and G. Richard. 2006. Effect of deep tillage for vineyard establishment on soil structure: a case study in Southern France. Soil Till. Res. 88: 132–143.

Cramer, H.H. 1967. Plant protection and world crop production. Pflanzenschutz Nacrichten Bayer 1967. 1. Farben Fabriken Bayer A.G. Leverkusen, 524 p.

Creamer, N.G., M.A. Bennett, B.R. Stinner and J. Cardina. 1996. A comparison of four processing tomato production systems differing in cover crop and chemical inputs. J. Amer. Soc. Hort. Sci. 121: 559–568.

Dastgheib, F. and C. Frampton. 2000. Weed management practices in apple orchards and vineyards in the South Island of New Zealand. N. Z. J. Crop Hort. Sci. 28: 53–58.

Davis, V.M., K.D. Gibson and W.G. Johnson. 2008. A field survey to determine distribution and frequency of glyphosate-resistant horseweed (*Conyza canadensis*) in Indiana. Weed Technol. 22: 331–338.

Delate, K. and H. Friedrich. 2004. Organic apple and grape performance in the Midwestern U.S. Acta Hort. 638: 309–320.

Delmas, M.A. and L.E. Grant. 2010. Ecolabeling strategies and price-premium: the wine industry puzzle. Bus. Soc. 20(10): 1–39.

De Prado, R. 2012. Glyphosate impact on Mediterranean agriculture: expectations and solutions. pp. 11–14. *In:* Proceedings of International Workshop on "European status and solutions for glyphosate resistance". Universidad de Cordoba, 3–4 May 2012, Cordoba, Spain.

De Prado, J.L., M.D. Osuna, A. Heredia and R. De Prado. 2005. *Lolium rigidum*, a pool of resistance mechanisms to ACCase inhibitor herbicides. J. Agric. Food Chem. 53: 2185–2191.

Donaldson, D.R., C.L. Elmore, S.E. Gallagher and J.A. Roncoroni. 1988. Manipulating vineyard weeds with herbicides. California Agric. 42: 15–16.

Dufour, R. 2006. Grapes: organic production. Natl. Sustainable Agr. Info. Serv. Bul. IP031. October 7, 2011.

Dujmović Purgar, D. and N. Hulina. 2004. Vineyard weed flora in the Jastrebasko area (NW Croatia). Acta Bot. Croat. 63: 113–123.

Eastham, J., A. Cass, S. Gray and D. Hansen. 1996. Influence of raised beds, ground cover and irrigation on growth and survival of young grapevines. Acta Hort. 427: 37–44.

Elmore, C. 1996. A reintroduction to integrated weed management. Weed Sci. 44: 409–412.

Elmore, C., J.A. Roncoroni, L. Wade, and P.S. Verdegaal. 1997. Four weed management systems compared: mulch plus herbicides effectively control vineyard weeds. California Agric. 51: 14–18.

Elmore, C. and D. Donaldson. 1999. UC Pest Management Guidelines: Grape Integrated Weed Management. University of California. Available at: http://www.ipm.ucdavis.edu/PMG.

EU Directive 2009/128/EC. EU Directive of the European Parliament and of the Council of 21 October 2009 establishing a framework for Community action to achieve the sustainable use of pesticides (http://eur-lex.europa.eu/LexUriServ/LexUriServ.do?uri=OJ:L:2009:309:0071:0086:EN:PDF).

Ferreiro-Arman, M., J. Da Costa, S. Homayouni and J. Martin-Herrero. 2006. Hyperspectral image analysis for precision viticulture. pp. 730–741. *In:* Campilho A., Kamel M. (Eds.) Third International Conference, ICIAR 2006, Povoa de Varzim, Portugal, September 2006, Proceedings, Part II. Image Analysis and Recognition. Springer-Verlag, Berlin Heidelberg.

Fischer, B.B., E.A. Yeary and J.E. Marcroft. 2002. Vegetation management systems. *In:* Principle of Weed Control (3rd edition). Thomson Publication, Fresno, CA 93791.

Fredrikson, L., P.A. Skinkis and E. Peachey. 2011. Cover crop and floor management affect weed coverage and density in an establishing Oregon vineyard. Hort.Technol. 21: 208–216.

Gago, P., C. Cabaleiro and J. Garcia. 2007. Preliminary study of the effect of soil management systems on the adventitious flora of a vineyard in northwestern Spain. Crop Prot. 26: 584–591.

Gebbers, R. and V.I. Adamchuk. 2010. Precision agriculture and food security. Hort. Sci. 327(5967): 828–831.

Giese, G. and T.K. Wolf. 2009. Root pruning and groundcover to optimize vine vigor and berry composition in Cabernet Sauvignon grapevines. Abstr. Am. J. Enol. Vitic. 60: 551A.

Giese, G., C. Velasco-Cruz, L. Roberts, J. Heitman and T.K. Wolf. 2014. Complete vineyard floor cover crops favorably limit grapevine vegetative growth. Sci. Horti. 170: 256–266.

Gladstones, J.S. 1992. Viticulture and Environment. Wine Titles, Hyde Park Press, Adelaide, Australia.

Guerra, B. and K. Steenwerth. 2012. Influence of floor management technique on grapevines growth, disease pressure, and juice and wine composition: a review. Am. J. Enol. Vitic. 64: 149–164.

Guthman, J. 2000. Raising organic: an agro-ecological assessment of grower practices in California. Agr. Human Values 17: 257–266.

Hatch, T.A., C.C. Hickey and T.K. Wolf. 2011. Cover crop, rootstock and root restriction regulate vegetative growth of Cabernet Sauvignon in a humid environment. Am. J. Enol. Vitic. 62: 298–311.

Heap, I. 2016. The International Survey of Herbicide Resistant Weeds. Available at: http://www.weedscience.org (Accessed on August 11, 2016).

Hembree, K.J. and W.T. Lanini. 2006. Weeds. UC IPM Pest Management Guidelines: Grape. University of California Agriculture and Natural Resources. 3448: 90–108.

Holm, L.G., J. Doll, J.V. Pancho and J.P. Herberger. 1997. World Weeds: Natural Histories and Distribution. New York: John Wiley & Sons. pp. 226–235.

Howell, S. 2001. Sustainable grape productivity and the growth-yield relationship: a review. Am. J. Enol. Vit. 52: 165–175.

Ibáñez, J.J., G. Benito, A. García-Álvarez and A. Saldaña. 1996. Mediterranean soils and landscapes. An overview. pp. 7–36. *In:* J.L. Rubio and A. Calvo (Eds.) Soil Degradation and Desertification in Mediterranean Environments. Geoforma, Logroño.

Ingels, C.A., K.M. Scow, D.A. Whisson and R.E. Drenovsky. 2005. Effects of cover crops on grapevines, yield, juice composition, soil microbial ecology, and gopher activity. Am. J. Enol. Vitic. 56: 19–29.

Jackson, L.E. 2000. Fates and losses of nitrogen from a nitrogen-15-labeled cover crop in an intensively managed vegetable system. Soil Sci. Soc. Am. J. 64: 1404–1412.

Jones, G.V., M.A. White, R.C.C. Owen and K. Storchmann. 2005. Climate change and global wine quality. Clim. Change 73: 319–343.

Kosmas, C., N. Danalatos, L.H. Cammeraat, M. Chabart, J. Diamantopoulos, R. Farand, L. Gutierrez, A. Jacob, H. Marques, J. Martínez-Fernandez, A. Mizara, N. Moustakas, J.M. Nicolau, C. Oliveros, G. Pinna, R. Puddu, J. Puigdefabregas, M. Roxo, A. Simao, G. Stamou, N. Tomasi, D. Usai and A. Vacca. 1997. The effect of land use on runoff and soil erosion rates under Mediterranean conditions. Catena 29: 45–59.

Krohn, N.G. and D.C. Ferree. 2005. Effects of low-growing perennial ornamental groundcovers on the growth and fruiting of 'Seyval blanc' grapevines. Hort. Sci. 40: 561–568.

Liebman, M. and A.S. Davis. 2000. Integration of soil, crop and weed management in low-external-input farming systems. Weed Res. 40: 27–47.

Lopes, C.M., A. Monteiro, J.P. Machado, N. Fernandes and A. Araujo. 2008. Cover cropping in a sloping non-irrigated vineyard: II – effects on vegetative growth, yield, berry and wine quality of 'Cabernet Sauvignon' grapevines. Ciencia Tec. Vitiv. 23: 37–43.

Mazzetto, F., A. Calcante, A. Mena and A. Vercesi. 2010. Integration of optical and analogue sensors for monitoring canopy health and vigour in precision viticulture. Precision Agric. 11: 636–649.

Mazzoncini, M., T.B. Sapkota, P. Barberi, D. Antichi and R. Risaliti. 2011. Long-term effect of tillage, nitrogen fertilization and cover crops on soil organic carbon and total nitrogen content. Soil Till. Res. 114: 165–174.

Mania, E., D. Isocrono, M.L. Pedulla and S. Guidoni. 2015. Plant diversity in an intensively cultivated vineyard agroecosystem (Langhe, Nort-West Italy). S. Afr. J. Enol Vitic. 36: 378–388.

Matese, A. and S.F. Di Gennaro. 2015. Technology in precision viticulture: a state of the art review. Int. J. Wine Res. 7: 69–81.

Mercenaro, L., G. Nieddu, P. Pulina and C. Porqueddu. 2014. Sustainable management of an intercropped Mediterranean vineyard. Agric. Ecosyst. Environ. 192: 95–104.

Miglécz T., O. Valko, P. Torok, B. Deak, A. Kelemen, A. Donko, D. Drexler and B. Tothmeresz. 2015. Establishment of three cover crop mixtures in vineyards. Scientia Hortic. 197: 117–123.

Monteiro, A. and C.M. Lopes. 2007. Influence of cover crop on water use and performance of vineyard in Mediterranean Portugal. Agric. Ecosyst. Environ. 121: 336–342.

Nachtergaele, J., L. Poesen and B. Wesemael. 1998. Gravel mulching in vineyards of southern Switzerland. Soil Till. Res. 46: 51–59.

Nascimbene, J., L. Marini and M.G. Paoletti. 2012. Organic farming benefits local plant diversity in vineyard farms located in intensive agricultural landscapes. Environ. Manag. 49: 1054–1060.

Neve, P., J.K. Norsworthy, K.L. Smith and I.A. Zelaya. 2011. Modelling evolution and management of glyphosate resistance in *Amaranthus palmeri*. Weed Res. 51: 99–112.

Norsworthy, J.K., S.M. Ward, D.R. Shaw, R.S. Llewellyn, R.L. Nichols, T.M. Webster, K.W. Bradley, G. Frisvold, S.B. Powles, N.R. Burgos, W.W. Witt and M. Barett. 2012. Reducing the risks of herbicide resistance: best management practices and recommendations. Weed Sci. 60: 31–62.

Novara, A., L. Gristina, S.S. Saladino, A. Santoro and A. Cerdà. 2011. Soil erosion assessment on tillage and alternative soil managements in a Sicilian vineyard. Soil Till. Res. 117: 140–147.

Olmstead, M., T.W. Miller, C.S. Bolton and C.A. Miles. 2012. Weed control in a newly established organic vineyard. HortTechnol 22: 757–765.

Owen, M.J., N.J. Martinez and S.B. Powles. 2014. Multiple herbicide-resistant *Lolium rigidum* (annual ryegrass) now dominates across the Western Australian grain belt. Weed Res. 54: 314–324.

Paoletti, M.G., D. Sommaggio, M.R. Favretto, G. Petruzzelli, B. Pezzarossa and M. Barbafieri. 1998. Earthworms as useful bioindicators of agroecosystem sustainability in orchards and vineyards with different inputs. Appl. Soil Ecol. 10: 137–150.

Pardini, A., C. Faiello, F. Longhi, S. Mancuso and R. Snowball. 2002. Cover crop species and their management in vineyards and olive groves. Adv. Hort. Sci. 16: 225–234.

Patrick-King, A.P. and A.M. Berry. 2005. Vineyard δ^{15}N, nitrogen and water status in perennial clover and bunch grass cover crop systems of California's central valley. Agric. Ecosyst. Environ. 109: 262–272.

Peregrina, F., C. Larrieta, S. Ibanez and E. García-Escudero. 2010. Labile organic matter, aggregates, and stratification ratios in a semiarid vineyard with cover crops. Soil Sci. Soc. Am. J. 74: 2120–2130.

Proffit, T., R. Bramley, D. Lamb and E. Winter. 2006. Precision Viticulture – A New Era in Vineyard Management and Wine Production. Winetitles Pty Ltd., Ashford, South Australia.

Prosdocimi, M., A. Jordán, P. Tarolli, S. Keesstra, A. Novara and A. Cerdà. 2016. The immediate effectiveness of barley straw mulch in reducing soil erodibility and surface runoff generation in Mediterranean vineyards. Sci. Total Environ. 547: 323–330.

Raclot, D., Y. Le Bissonnais, X. Louchart, P. Andrieux, R. Moussa and M. Voltz. 2009. Soil tillage and scale effects on erosion from fields to catchment in a Mediterranean vineyard area. Agric. Ecosyst. Environ. 134: 201–210.

Racz, J. and K. Siaba. 1971. The allelopathic effect of weeds in the vineyards. Obstbau und Fruchtever Vertung 21: 264–268.

Rask, A.M. and P. Kristoffersen. 2007. A review of non-chemical weed control on hard surfaces. Weed Res. 47: 370–380.

Ripoche, A., F. Celette, J.P. Cinna and C. Gary. 2010. Design of intercrop management plans to fulfill production and environmental objectives in vineyards. Europ. J. Agron. 32: 30–39.

Ripoche, A., A. Metay, F. Celette and C. Gary. 2011. Changing the soil surface management in vineyards: immediate and delayed effects on the growth and yield of grapevine. Plant Soil 339: 259–271.

Ruiz-Colmenero, M., R. Bienes, D.J. Eldridge and M.J. Marques. 2013. Vegetation cover reduces erosion and enhances soil organic carbon in a vineyard in the central Spain. Catena 104: 153–160.

Salome, C., P. Coll, E. Lardo, C. Villenave, E. Blanchart, P. Hinsinger, C. Marsden and E. LeCadre. 2014. Relevance of use-invariant soil properties to assess soil quality of vulnerable ecosystems: the case of Mediterranean vineyards. Ecol. Indic. 43: 83–93.

Sanguankeo, P.P., R.G. Leon and J. Malone. 2009. Impact of weed management practices on grapevine growth and yield components. Weed Sci. 57: 103–107.

Schreck, E., L. Gontier, C. Dumat and F. Geret. 2012. Ecological and physiological effects of soil management practices on earthworm communities in French vineyards. Eur. J. Soil Biol. 52: 8–15.

Shrestha, A., M.W. Fidelibus, M.F. Alcorta and K. Cathline. 2010. Threshold of horseweed (*Conyza canadensis*) in an established Thompson Seedless vineyard in the San Joaquin Valley. Intl. J. Fruit Sci. 10: 301–308.

Shrestha, A., S.K. Kurtural, M.W. Fidelibus, G. Dervishian and S. Konduru. 2013. Efficacy and cost of cultivators, steam, or an organic herbicide for weed control in organic vineyards in the San Joaquin valley of California. Hort. Technology 23: 99–108.

Simon, S., J.C. Bouvier, J.F. Debras and B. Sauphanor. 2010. Biodiversity and pest management in orchard systems. A review. Agron. Sustainable Dev. 30: 139–152.

Six, J., E.T. Elliott and K. Paustian. 1999. Aggregate and soil organic matter dynamics under conventional and no-tillage systems. Soil Sci. Soc. Am. J. 63: 1350–1358.

Smith, R., L. Bettiga, M. Cahn, K. Baumgartner, L. Jackson and T. Bensen. 2008. Vineyard floor management affects soil, plant nutrition, and grape yield and quality. Calif. Agric. 62: 187–190.

Snapp, S.S. and H. Borden. 2005. Enhanced nitrogen mineralization in mowed or glyphosate treated cover crops compared to direct incorporation. Plant Soil 270: 101–112.

Steenwerth, K. and K.M. Belina. 2008a. Cover crops enhance soil organic matter, carbon dynamics and microbiological function in a vineyard agroecosystem. Appl. Soil Ecol. 40: 359–369.

Steenwerth, K. and K.M. Belina. 2008b. Cover crops and cultivation: impacts on soil N dynamics and microbiological function in a Mediterranean vineyard agroecosystem. Appl. Soil Ecol. 40: 370–380.

Steenwerth, K.L., K. Baumgartner, K. Belina and L. Veilleux. 2010. Vineyard weed seedbank composition responds to glyphosate and cultivation after three years. Weed Sci. 58: 310–316.

Steinmaus, S., C.L. Elmore, R.J. Smith, D. Donaldson, E.A. Weber, J.A. Roncoroni and P.R.M. Miller. 2008. Mulched cover crops as an alternative to conventional weed management systems in vineyards. Weed Res. 48: 273–281.

Stolz, H. and O. Schmid. 2007. Organic viticulture and winemaking: development of environment and consumer friendly technologies for organic wine quality improvement and scientifically based legislative framework. D 2.7 Public report about first round qualitative consumer research and market needs. Available at: http://orgprints.org/10608/

Stork, P.R. and P.H. Jerie. 2003. Initial studies of the growth, nitrogen sequestering, and de-watering potential of perennial grass selections for use as nitrogen catchcrops in orchards. Aust J. Agric. Res. 54: 27–37.

Sweet, R.M. and R.P. Schreiner. 2010. Alleyway cover crops have little influence on Pinot noir grapevines (*Vitis vinifera* L.) in two western Oregon vineyards. Amer. J. Enol. Viticult. 61: 240–252.

Tan, S. and G.D. Crabtree. 1990. Competition between perennial ryegrass sod and Chardonnay wine grapes for mineral nutrients. Hort. Sci. 25: 533–535.

Teasdale, J.R. and C.L. Mohler. 1993. Light transmittance, soil temperature, and soil moisture under residue of hairy vetch and rye. Agron. J. 85: 673–680.

Tesic, D., M. Keller and R.J. Hutton. 2007. Influence of vineyard floor management practices on grapevine vegetative growth, yield, and fruit composition. Am. J. Enol. Vitic. 58: 1–11.

Travlos, I.S. 2010. Legumes as cover crops or components of intercropping systems and their effects on weed populations and crop productivity. pp. 151–164. *In:* A.J. Greco (Ed.) Progress in Food Science and Technology. Nova Science Publishers, Inc. Hauppauge, New York.

Travlos, I.S. 2013. Weeds in perennial crops as an unexpected tool of integrated crop management. pp. 97–113. *In:* A. Taab (Ed.) Weeds and their Ecological Functions. Nova Science Publishers, Inc. Hauppauge, New York.

Travlos I.S. and D. Chachalis. 2010. Glyphosate-resistant hairy fleabane (*Conyza bonariensis*) is reported in Greece. Weed Technol. 24: 569–573.

Travlos I.S. and D. Chachalis. 2012. Relative competitiveness of glyphosate-resistant and glyphosate-susceptible populations of hairy fleabane (*Conyza bonariensis* L.). J Pest Sci. 86: 345–351.

Travlos, I.S., D. Chachalis and G. Economou. 2009. Characters for the *in situ* recognition of some *Conyza* species and glyphosate resistant populations from Greece. p. 63. *In:* Proceedings of the 2nd International Conference on Novel Sustainable Weed Management in Arid & Semi-arid Agro-ecosystems, EWRS, Santorini, Greece.

Travlos, I., I. Tabaxi, D. Papadimitriou, D. Bilalis and D. Chachalis. 2015. *Lolium rigidum* Gaud. biotypes from Greece with resistance to glyphosate and other herbicides. p. 542. *In:* Proceedings of the 14th International Symposium of University of Agricultural Sciences & Veterinary Medicine Cluj-Napoca "Prospects for the 3rd Millennium Agriculture", 24–26 September 2015, Cluj-Napoca.

Trivellone, V., N. Schoenenberger, B. Bellosi, M. Jermini, F. De Bello, E.A.D. Mitchell and M. Moretti. 2014. Indicators for taxonomic and functional aspects of biodiversity in the vineyard agroecosystem of Southern Switzerland. Biol. Conserv. 170: 103–109.

Vitelli, J.S. and B.A. Madagan. 2004. Evaluation of hand-held burner for the control of woody weeds by flaming. Austral. J. Exp. Agric. 44: 75–81.

WCED. 1987. Report of the World Commission on Environment and Development: Our 941 Common Future. United Nations, USA.

Westerman, P.R., J.S. Wes, M.J. Kropff and W. Van der Werf. 2003. Annual losses of weed seeds due to predation in organic cereal fields. J. Appl. Ecol. 40: 824–836.

Westerman, P.R., C.D. Luijendijk, J. Wevers and W. Van der Werf. 2011. Weed seed predation in a phonologically late crop. Weed Res. 51: 157–164.

White, G.B. 1996. The economics of converting conventionally managed vineyards to organic management practices. Acta Hort. 429: 377–384.

Wisler, G.C. and R.F. Norris. 2005. Interactions between weeds and cultivated plants as related to management of plant pathogens. Sym. Weed Sci. 53: 914–917.

Zabadal, T.J. 1999. Pest control in Small Vineyards. Michigan State Univ. Ext. Bul. 2698.

CHAPTER

29

Sustainable Potato Weed Management

Heidi Johnson[*1] and Jed Colquhoun[2]

[1] University of Wisconsin-Extension, 5201 Fen Oak Dr. Ste. 138, Madison, WI 53718
[2] University of Wisconsin-Madison, 1575 Linden Dr., Madison, WI 53706

Introduction (Importance of the Crop/Cropping System)

Three-hundred-eighty-five million metric tons of potatoes are produced worldwide. The top five potato producing countries are China, India, Russia, Ukraine and the United States. China is by far the top producing country, growing nearly 100 million metric tons annually, more than twice as much as India, the second biggest potato producer (National Potato Council 2016).

Potatoes are grown on about 420,000 hectares annually in the United States, producing over 20 million metric tons of potatoes (National Potato Council 2016). Over half of the potato acreage is in three states: Idaho, Washington and North Dakota. Potato production is a $3.85 billion industry. Over half of the potatoes produced are frozen, canned or processed into chips or fries, while most of the remainder is sold fresh.

Weed Impact (Major Weeds, Weed Life Cycle, Weed Strengths and Weaknesses, Impact on Yield Loss)

Weeds impact potato yield by competing with the crop for resources, such as sunlight, water and nutrients. Weeds can also be problematic for harvesting potatoes by becoming entangled in the harvest equipment, slowing down harvest operations and causing weed-entangled potato tubers to bypass harvest as they pass over the harvester machine belts and are dropped back on the ground behind harvest equipment.

Common problematic annual weeds in potato production include redroot pigweed (*Amaranthus retroflexus* L.), common lambsquarters (*Chenopodium album* L.), hairy galinsoga (*Galinsoga ciliata* [Raf.] Blake), large crabgrass (*Digitaria sanguinalis*), foxtail species (*Setaria* spp.), hairy nightshade (*Solanum sarrachoides* Sendtner), kochia (*Kochia scoparia*), wild buckwheat (*Polygonum convolvulus*), Pennsylvania smartweed (Polygonum pensylvanicum), wild oat (*Avena fatua*), common purslane (*Portulaca oleracea* L.) and barnyardgrass (*Echinochloa crus-galli*). Annual weeds need to be controlled during the growing season to not only reduce competition with potatoes but also to prevent seed production that builds the weed seedbank and impacts future potato crops or other crops in the rotation.

Perennial weeds that are problematic in potato production, such as nutsedge species (like *Cyperus* spp.) and quackgrass (*Agropyron repens*), are typically managed in other crops in the

---

*Corresponding author: heidi.johnson@ces.uwex.edu

potato rotation that have more weed management options, such as cereal grains, soy and corn. Quackgrass can be particularly problematic in potatoes, not only from a resource competition standpoint but also because the weed's rhizomes can grow within or become surrounded by bulking potato tubers, reducing crop quality.

Competition from weeds can cause significant yield loss in potatoes. Plots left unweeded in a study conducted by Thakral et al. (1989) resulted in a yield reduction of 40% to 43% compared to the weed free plots. Dvorak (2015) determined that tuber yield can be reduced by 20% to 30% at moderate weed densities but high weed densities can reduce tuber yield by up to 85% (depending on the weed species diversity). Baziramakenga and Leroux (1998) reported that potato tuber yield loss increased with quackgrass weed density and ranged from 19% to 73% over the two study years. Wall (1990) reported that potato yield was decreased with green foxtail populations of less than 75 plants per square meter. As few as one hairy nightshade plant per square meter reduced 'Russet Norkotah' potato tuber yield compared to weed-free potatoes (Hutchinson et al. 2011). These studies highlight the large potential reductions in potato yield when weed populations are not managed appropriately during the growing season. The specific diversity of the weed population in any given field may dictate the extent of potential yield reduction.

Timing of weed management in potatoes is also important. It takes six to eight weeks after planting for the potato vines to grow together between the rows and create a solid canopy that begin to shading out emerging weeds (British Potato Council 2007). The critical weed free period in potatoes varies based on region and the competitive ability of the predominant weed species. For example, a study from Iran found that the critical weed free period started 19 to 24 days after planting and continued until 43 to 51 days after planting to maintain yield losses less than 10% (Ahmadvand et al. 2009) while in Lithuania, researchers found that the critical weed free was from planting until 25 days after flowering (Ciuberkis et al. 2007). Weed species diversity also significantly influences the critical weed free period. In a study by Nelson and Thoreson (1981), potato yields were reduced by 54% by a mix of annual weeds that emerged one week after planting in comparison to a yield reduction of 19% when weeds emerged three weeks after planting. As few as one redroot pigweed or barnyardgrass per meter of row reduced potato yield by 19 to 33% when they were seeded in the row at the time of potato planting, while up to four of either weed species per meter row had no impact on potato yield when they were planted after the final hilling (VanGessel and Renner 1990). When foxtail management was delayed for two weeks after crop emergence, marketable yields were reduced by 29% (Wall and Friesen 1990). Ivany (1986) reported that quackgrass left unmanaged for the first two weeks after potatoes were planted reduced total potato yield by 21% and total marketable yield by 27%. These studies show a wide range a yield reductions caused by early season weed competition, depending on conditions and weed species diversity, but all demonstrate the importance of early-season weed management, particularly prior to canopy closure.

Weeds may also increase potato insect pests or diseases by providing additional insect habitat or hosts. Several weeds in the nightshade family (*Solanum* spp.) are closely related to potatoes and thus serve as an alternative host for potato diseases and insects. The most common is hairy nightshade, which is a host of three nematodes, nine pathogens and two insects that also can infect or feed on potatoes. Cultural practices, such as crop rotation can typically be used to manage levels of these diseases and pests but hairy nightshade can serve as an alternative host during non-potato years and nullify the rotational benefits (Nitzan et al. 2009, Boydston et al. 2007). Weeds, such as hairy nightshade should be carefully managed in potatoes and other crops in the rotation.

Cultural Weed Control

Seeding Rates and Utilizing Competitive Crops/Cultivars

Potatoes are inherently fairly competitive with weeds given their dense canopy for much of the growing season. It has been observed that potato cultivars differ in their growth rate and thus the amount of time between planting and canopy closure. Ten cultivars of potatoes were compared to evaluate differences in the amount of weed biomass accumulated through the growing season due to varietal differences in growth rate and, interestingly, no differences were observed (Colquhoun et al. 2009). However, some of the potato cultivars did maintain yield even under weedy conditions better than other cultivars. The researchers attributed this to earlier and denser canopy development. Therefore, selecting cultivars that develop an earlier canopy may compete more effectively with weeds for resources is likely beneficial in an integrated program.

Solan et al. (2011) evaluated several different weed management programs, including combinations of tillage and herbicide types and timings, on two potato cultivars. The selected cultivars had shown differential growth rates and differed in their ability to maintain yield under weedy conditions in Colquhoun et al. (2009). Neither cultivar maintained yield in the weedier cultivation-only treatment, differing from the results of Colquhoun et al. 2009. However, the cultivars did perform differently to other treatments, suggesting that differential treatment by cultivar may be warranted. Although, practically, growers are likely limited by equipment availability in how much they can vary their management by cultivar.

Experiences in other crops would suggest that manipulating the potato row spacing could decrease the time to canopy closure and shade out emerging weeds earlier. However when two row spacings were compared, no difference was observed in either weed and vine biomass or tuber yield (Love et al. 1995). The researchers also compared three within-row seed piece spacings and reported a minor decrease in weed biomass with decreasing in-row plant spacing. The researchers attributed this to rapid vine elongation and a greater impact on canopy closure (Love et al. 1995). Another study in Wisconsin found that decreasing row spacing actually led to an increase in weed biomass. The researchers attributed this to a specific weed, common lambsquarters, responding to the shady environment of the narrow row planting by increasing its growth rate (Conley et al. 2001). Thus, weed diversity differences between fields may dictate the effectiveness of using row spacing as a cultural weed management practice.

Rotations

A diversified crop rotation can disrupt weed life cycles by varying management practices associated with the different crops each year. To optimize this effect, growers can choose crops that vary in planting date, season length, competitiveness, and weed management strategies. A perennial crop in the rotation, such as hay or alfalfa can also help to reduce the weed seedbank if managed well with timely mowing (Boydston 2010).

Crop Residue Management

Leaving potato residue on the soil surface is not effective in reducing weed densities, likely because the residue is fragile and too decayed in the following year to provide any measurable in-season weed control. A study conducted in Prince Edward Island, Canada found no difference in weed densities when a zero tillage system (residue cover measured just after planting was 33%) was compared to full tillage which utilized a moldboard plow (with 2% residue cover) (Holmstrom et al. 2006).

False or Stale Seedbed

A false or a stale seedbed is a weed management technique where the area to be planted is prepared as if it will be planted, irrigated if possible to encourage weed germination and then new weed seedlings are terminated by either flaming or a shallow cultivation prior to planting the crop. There are no published studies establishing the effectiveness of this technique with potatoes but for growing regions that have a long enough season to devote time to conduct a false seedbed prior to planting, it may be an effective technique.

Mechanical/Physical Weed Control

Handweeding and Mowing

Handweeding, although effective in managing weeds in potato production, it is not practical or economical for large-scale production. Mowing is also not very effective, given the importance of a vegetative canopy for weed competition and as an energy source to support tuber bulking. In fact, mowing is used to desiccate potato vines in some regions and in organic production.

Soil Cultivation/Tillage (Chisel Plow, Disking, Field Cultivator, etc.)

Cultivation can be an effective way to manage weeds early in the season before the potato canopy has closed and when weed control is vital to maintain yield. Weeds emerging before the potatoes have been reported to be the most competitive with the crop (Love et al. 1995), thus pre-plant cultivation is important in an integrated management system.

Hilling is a mechanical field operation that is unique to potato production where soil is thrown onto the crop row to protect the tubers from sunlight and provide more area for below-ground tuber development. The process of hilling, although not primarily conducted as a weed control measure, can provide significant weed control, particularly when rolling cultivators are used behind the hilling blades to uproot weeds. Potatoes are generally hilled once or twice during a production season.

There are several types of in-season cultivators that have been evaluated in potato production, both alone and in combination. Rolling cultivators consist of sets, or gangs, of three to five metal wheels with strong curved teeth radiated from a center hub that roll over the soil surface between the crop rows, throwing soil in the direction they are set. This rolling movement and soil disruption is ideal for a crop like potatoes where hills are already being formed as a production practice. Shovel cultivators consist of larger shanks with shovel-shaped ends that are pulled through the soil between crop rows, turning the soil and ripping out young and established weeds. A flex-tine cultivator is a tool with a series of gangs with bent, wire tines that are laid out in a staggered fashion that scratch up the entire soil surface. A flex-tine cultivator works well to dislodge young weeds from the soil, prior to or just at crop emergence. Because shovel cultivators are more aggressive than the rolling or flex-tine cultivators, they have potential to damage crop roots and reduce yield when they are used as early as three weeks after planting (Nelson and Giles 1989). Given that a flex-tine cultivator works over the entire soil surface, it has the potential to damage the top of the crop if it is used later in the season.

Surprisingly, although the shovel and rolling cultivators are more aggressive, Bellinder et al. (2000) found that neither was more effective at reducing between-row weeds than the flex-tine cultivator. They also found that the flex-tine cultivator was not more effective than the between-row cultivators at reducing in-row weeds. They suggested that the flex-tine may have actually scratched the soil surface, uncovering more weed seeds that then germinated. Another surprising result of the Bellinder et al. (2000) study was that while none of the cultivation treatments, alone or in combination, reduced weed levels to the level of standard herbicide programs, there were not significant yield losses in the weedier cultivation treatments. This study included a late hilling operation (seven to eight weeks after planting), which the authors surmised contributed to the lack of difference in yield among the treatments. However, another study (VanGessel and Renner 1990a) found that cultivation alone did not result in equivalent yield to where herbicides were used, but this was when the final hilling operation occurred at four to seven weeks after planting. Thus, delaying the final hilling operation later into the season may be useful in systems where growers solely rely on tillage as a weed management tool.

Weed density and diversity also may impact the effectiveness of this weed control method. Late-emerging weeds, after the period when cultivation is a viable control measure, and larger and taller statured weeds could still compete with the potato crop later into the growing season.

Cultivation is also very weather dependent. Wet conditions can delay tillage field operations and allow weeds to become too large for cultivation to be effective. Arid production regions may be better for tillage-based potato production. Sandier soils may also lend themselves to tillage over heavier soils as they dry out more quickly and would be fit for tillage operations more days during the production season.

Thermal

Flame weeding is often used as a weed control method in organic potato production. Flame weeding, as the name suggests, consists of pointing a flame directly at emerging weeds to break cell structure. The danger with flame weeding is the flame can also desiccate the crop plant if timing and accuracy is not optimized. Mechanized flaming equipment often includes shields to direct the flame and protect the crop. Flame weeding in potatoes is possible if it is conducted immediately prior to or just after potato emergence (Boydston 2010). Flaming is only effective against young weed seedlings so this would be a management tool that would only be effective very early in the production season.

Cover Crops/Mulching

Cover crops can also be grown prior to potatoes as a weed control method. Plants from the Brassica family produce glucosinolates that can inhibit plant growth and seed germination. Fall-planted rapeseed (*Brassica napus*) incorporated in spring just before a potato crop reduced both weed density and biomass compared to potato after fallow (Boydston and Hang 1995). This could allow growers to reduce pre-emergence herbicides and rely more on post-emergence herbicides or cultivation for in-season weed control. Another study found that a rye cover, terminated with herbicides prior to planting, reduced weed density by 50% compared to where no cover crop was grown (Mehring et al. 2016).

Bellinder et al. (1996) reported significantly greater weed populations in plots that had a rye cover crop that was moldboard ploughed prior to planting compared to plots where the rye cover crop had not been soil-incorporated. This is likely due to the physical suppression of the weeds by the residue on the soil surface and possibly due to some allelopathic suppression from the rye. In general, for growers considering cover crops as a weed control strategy, leaving the cover crop residue on the soil surface is likely an important component of the effectiveness.

Mehring et al. (2016) observed that while cover crops reduced weed density and maintained potato tuber yield compared to conventionally managed potatoes, lower marketable yields were reported where cover crops were grown. The authors noted challenges in hilling potatoes where cover crop residue was present and subsequently exposed tuber sun-scalding. Any practices like cover crops that significantly increase surface residue will change the dynamics of mechanical soil movement and can result in insufficient hill formation.

The use of living cover crops between crop rows (living mulches) has been demonstrated to be effective in reducing weeds numbers in potato production but at the expense of potato yield. In one study, living mulches reduced weed populations as much as using a combination of mechanical and herbicides for weed control, however due to competition with the cover crop for resources, potato yield was significantly lower in the living mulch (white mustard, common vetch, Persian clover, and tansy phacelia) treatments (Kołodziejczyk 2015). Rajalahti et al. (1999) utilized interseeded cover crops (hairy or lana vetch, oats, barley, red clover or oats/hairy vetch), planted after hilling, but killed the cover crops with herbicides when they reached 25 cm to 30 cm tall and found that the mulched treatment controlled weeds as well as the herbicide standard but yields were still significantly reduced in the year of the study that had less precipitation. Boyd (2001) also planted living mulches (hairy vetch, Marino red clover and Kentucky bluegrass) after hilling and found no yield reduction in the potatoes; however they had very little growth on the mulches so they did not compete with the potato crop for water and other resources. These studies show that using living mulches can be an effective tool for reducing weed populations in potatoes but the living mulches can in turn become a weed in the system and compete with the crop for resources. More research is needed to determine a living mulch system that has no deleterious effects on the potato crop.

Non-living mulches, such as plastic or dried grass or hay can be used to suppress weeds. Given the labor intensity of applying mulch, they are not practical in large scale production but may have application at a small- to medium-scale. Mulches physically suppress weed emergence or even germination, but they can also serve other functions, such as warming spring soils and providing insect habitat. Dvorak et al. (2015) compared plastic and grass mulches for their effect on weed populations and potato yields. While the plastic mulch reduced weed densities more than the grass mulch or the cultivation-only control, yield under the plastic mulch was lower. The authors attributed this to higher densities of Colorado potato beetle on the plastic-mulched plants. They attributed higher insect densities to the plastic mulch being warmer than the other mulches.

Chemical Weed Control

Active Ingredients

The number of effective herbicide active ingredients is limited in potato production relative to agronomic crops, such as corn and soy, but a combination of pre- and post-emergent active ingredients alongside cultural practices can result in season-long weed management. It should be noted that pesticide labels change often and vary regionally, and potato producers should always read and follow the product labels prior to use.

Pre-emergent herbicides are applied after potato planting but prior to potato and weed emergence. In many production systems, the pre-emergent herbicides are applied after the first potato hilling such that soil disturbance is minimized after application. In this case, the hilling operation itself is used to eliminate any emerged weeds. In some production systems, potato fields are hilled and herbicides applied just after planting, with minimal soil disturbance until a single post-emergent cultivation event. Regardless of timing, pre-emergent herbicides often benefit from irrigation or precipitation soon after application to incorporate the herbicide into the soil zone where weed seeds germinate. Post-emergent herbicides are applied after weed and potato emergence, often to control those that have escaped hilling and any cultivation and pre-emergent herbicide use (Colquhoun et al. 2017, Peachey 2016, Shaner 2014).

Herbicide registrations and use vary among and within countries and should be considered prior to use of any pesticide. Common pre-emergent herbicide active ingredients include:

- **S-metolachlor and metolachlor** control primarily annual grass weeds and some annual broadleaves. The broadleaf weed control spectrum is often enhanced by tank-mixing with other active ingredients, such as metribuzin. Potato injury is sometimes observed when cool, wet weather occurs around the time of or soon after herbicide application, particularly in early-maturing potato varieties. These herbicides are members of the chloroacetamide family that inhibits very long-chain fatty acid synthesis. Minimal weed resistance development has been observed.
- **Pendimethalin** also controls many annual grasses and some annual broadleaves, and is often tank-mixed with other herbicides, such as metribuzin to broaden the weed control spectrum. Pendimethalin is applied after potato hilling and can also be applied after potato emergence but will not control emerged weeds. Despite over four decades of use, weed resistance development has been minimal.
- **Trifluralin** controls many annual grasses and certain broadleaf weeds. The herbicide is applied after potato planting but before emergence with incorporation within 24 hours to reduce volatilization risk and compromised weed control. It is often used in conjunction with a herbicide that controls broadleaf weeds. Trifluralin is a dinitroaniline herbicide that inhibits microtubule assembly. Resistance to this herbicide has been sporadically reported, particularly among grasses, such as the foxtail species.
- **EPTC** controls some annual grass and broadleaf weed species. The herbicide must be incorporated immediately after application to avoid volatilization, or conversion to a gaseous form. EPTC can also be applied as a directed spray, avoiding potato plant contact, after potato emergence but will not control emerged weeds. EPTC is a lipid synthesis

inhibitor from the thiocarbamate chemical family. No known cases of weed resistance have been documented.

- **Linuron** is applied after potato planting but prior to emergence, as emerged potatoes can be injured fairly severely. The herbicide controls several annual grass and broadleaf weeds common in potatoes, including some species when newly emerged. Linuron use is restricted on coarse-textured, low organic matter soils to mitigate crop injury and groundwater contamination risk. Linuron is a urea herbicide that inhibits photosystem II in susceptible plants. Limited populations of several weed species common in potatoes, such as redroot pigweed and common lambsquarters, have been identified with linuron resistance. However, confirmed linuron resistance in the U.S. has been limited to common purslane (Heap 2017).
- **Fomesafen** controls annual broadleaf weeds in potatoes, but is significantly restricted in use in areas where groundwater contamination is a risk and thus has been limited in commercial adoption on coarse-textured soils where potatoes are often grown. The herbicide is applied after potato planting but before emergence and activated by irrigation or with rainfall soon after application. Weed resistance to fomesafen has been documented, particularly in *Amaranthus* spp. and common ragweed. No cases of weed resistance to this herbicide have been documented.
- **Flumioxazin** controls several broadleaf weed species when applied to potatoes after planting but before emergence. Emerged potatoes and those yet to emerge but with less than 5 cm of soil cover can be severely injured. Potato varieties vary in their tolerance of flumioxazin. Flumioxazin the herbicide inhibits the same enzyme as fomesafen. Common ragweed resistance to fomesafen has been confirmed in one population (Heap 2017).

Herbicides used in potato that have both pre- and post-emergent activity include:

- **Metribuzin** is a commonly used herbicide for broadleaf weed control in potato production. It is often tank-mixed in a pre-emergent application with herbicides that primarily control grass weeds for broad spectrum weed control. Metribuzin is also used to control young annual broadleaf weeds in a post-emergent application. Several early-maturing, white-, smooth- or red-skinned potato varieties are susceptible to injury from metribuzin. The herbicide can be readily leached in sandy soils with low organic matter content. Metribuzin is a triazinone family herbicide that inhibits photosystem II in susceptible plants. Resistance to metribuzin is somewhat common among weed species, such as common lambsquarters and redroot pigweed, presumably related to the selection pressure from widespread use of related herbicides, such as atrazine in corn production.
- **Rimsulfuron** controls several broadleaf and grass species common in potato production and has exhibited good crop safety, and thus has been widely adopted as both a pre- and post-emergent tool. Additionally, rimsulfuron can suppress or control select perennial weeds such as quackgrass. Rimsulfuron is a sulfonylurea herbicide that inhibits the acetolactate synthase enzyme. This chemical family is at rather high risk for resistant weed selection, and thus there are several weed species common to potato production where resistance has developed, such as kochia and giant foxtail.

Common post-emergent herbicides include:

- **Sethoxydim and clethodim** control only emerged grass weeds that are actively growing, particularly when applied prior to tillering. A broad spectrum of annual and perennial grasses can be managed with these herbicides. The herbicides belong to the cyclohexanedione herbicide family that inhibits the acetyl CoA carboxylase enzyme. Unfortunately, resistance to such herbicides is rather common across several grass species, such as wild oat and the foxtail species.

Integrating Weed Control Strategies to Improve the Sustainability of Potato Production

The future of all agriculture depends on the evolution of practices that minimize negative impact on our natural resources while maintaining economically beneficial margins for farmers so they can remain profitable. Herbicides and tillage are the primary means of controlling weeds in potatoes and neither is without potential negative impacts on the environment.

Public pressure is mounting to reduce pesticide use in agriculture due to concerns about off-site movement which can cause surface and groundwater contamination and impact other organisms. Potatoes are typically grown on coarse-textured or sandy soils that allow for uniform root crop growth and generally need to be irrigated to maintain adequate moisture. This combination of a coarse soil and irrigation could make an easier route for soluble pesticides to move through the soil profile and leach into groundwater if not managed properly. A field study was conducted in Minnesota in 1994 to determine if metribuzin, a common potato herbicide, was detectable at significant depths in the soil profile for the growing season after it was applied (Burgard et al. 1994). In general, their results suggested that the herbicide did not readily move through the soil profile. It was detected below 30 cm in only 17 out of 432 samples and concentrations in those samples were within the EPA limits for drinking water. Similar results were reported when the herbicide metolachlor was studied with the same methodology by the same lead researcher (Burgard et al. 1993). They found metolachlor in 18 out of 409 samples below 30 cm in one year of the study and none in the other year. And those levels were again within EPA limits. While it may be possible for potato pesticides to leach in soil, these studies indicate that with prudent use and careful management, movement can be minimized.

Excessive tillage can degrade soil structure and leave soil without residue cover, both of which can increase soil erosion. Soil erosion can lead to the water quality degradation when nutrients, such as phosphorus that are adhered to the soil particles end up in surface water, causing algal blooms and offsetting ecosystem balance.

Many studies have evaluated various combinations of herbicides, herbicide rates, hilling timing, tillage implements and number of tillage passes and their impact on weed control and potato tuber yield. Integrated strategies need to be evaluated as a weed control program as they influence each other's effectiveness. For example, when cultivation and hilling are used in combination with herbicides, soil residual herbicide effectiveness may be reduced due to dilution of the chemical via soil mixing. Also, weed seeds can be moved into the germination zone during the hilling process resulting in additional weed flushes which may necessitate additional control strategies.

In an effort to reduce both herbicide use and tillage, many studies have evaluated whether herbicides can be banded over just the crop row to control in-row weeds while between-row weeds are managed with tillage. Banding herbicides allows the total amount of applied herbicides to be reduced by up to 66%, depending on the width of the band and the distance between crop rows. Bellinder et al. (2000) found that the banded herbicide controlled weeds well within the crop row, with in-row weed densities similar to the treatment where herbicides were broadcast-applied. This combination of a banded herbicide and tillage also resulted in similar yield to the broadcast herbicide treatment. Similarly, Ivany (2002) used a combination of a banded herbicide with tillage and achieved equivalent yield to the broadcast herbicide treatment. These studies show that herbicide use can be significantly reduced, while maintaining potato yield, by utilizing some tillage for between-row weed management. However, this tactic of herbicide banding will only be useful in reducing herbicide use in regions where herbicides are ground applied, rather than through chemigation (Solan et al. 2011).

Researchers have also evaluated whether herbicide rates could simply be reduced when used in combination with other weed tactics, such as varying hilling timing and cover crops. Bellinder et al. (1996) evaluated reduced and split herbicide applications in combination with full and reduced tillage, varying hilling timing and a cover crop. The authors found that under both tillage systems, weed control was adequate when they used a half rate of herbicide, as long as it was applied after hilling and hilling was conducted no later than five weeks after planting.

Bellinder et al. (1996) also reported that using a cover crop helped to manage weed densities. They compared a terminated rye cover crop, unincorporated, to a treatment where the cover crop had been moldboard ploughed prior to planting. However, although they had equivalent yield in both treatments, the unploughed cover crop treatment resulted in a greater proportion of green or unmarketable potatoes. The authors determined that the lack of tillage didn't allow them to get the potatoes planted deeply enough, resulting in more tubers exposed to sunlight. They felt that this could be solved by choosing a potato variety that was less sensitive to greening. Other studies have indicated mixed results with using a cover crop and reduced tillage. Wallace and Bellinder (1990) also compared an unincorporated cover crop to a treatment that utilized a moldboard plough and did not see a decrease in marketable potato yield while Wallace and Bellinder (1989) found a 22% yield reduction in the reduced tillage/ cover crop system. Overall, using an unincorporated cover crop as a part of an integrated weed management system has promised to aid in weed reduction while maintaining yield and minimizing potential environmental issues. However, growers would need to evaluate how cover crops work on their particular farm.

Manipulating row spacing in potatoes has also been evaluated in combination with other tactics in integrated weed management programs for potatoes. Conley et al. (2001) evaluated the effects of different combinations of potato cultivars, wide and narrow row spacing and weed control tactics (herbicide mixes, one-pass cultivation or a single hilling) on weed populations, yield and overall potato profitability. Surprisingly, wider row spacing in combination with the various weed control tactics yielded more in one year of the study and was more profitable. Although, chemical weed control treatments yielded more, there was no difference in net profitability in either year of the study among the weed control treatments. This suggests that when profitability is factored in, cultivation or hilling alone may be as profitable as using an herbicide treatment.

Concluding Remarks

These types of integrated research projects indicate that a combination of strategies can work together to reduce weed populations and maintain yield and profit, while reducing negative environmental impacts. Weed resistance, absent new herbicide active ingredients and modes of action, will continue to be significant threat to future weed management programs. Thus, future research should concentrate on integrated and alternative management strategies that are not only effective but also practical enough to be adopted in production and maintain grower economic solvency.

REFERENCES

Ahmadvand, G., F. Mondani and F. Golzardi. 2009. Effect of crop plant density on critical weed free period of potato. Sci. Hort. 121: 249–254.

Baziramakenga, R. and G.C. Leroux. 1998. Economic and interference threshold densities of quackgrass (*Elytrigia repens*) in potato (*Solanum tuberosum*). Weed Sci. 46(2): 176–180.

Bellinder, R.R., J.J. Kirkwyland, R.J. Wallace and J.B. Colquhoun. 2000. Weed control and potato (*Solanum tuberosum*) yield with banded herbicides and cultivation. Weed Technol. 14: 30–35.

Bellinder, R.R., R.J. Wallace and E.D. Wilkins. 1996. Reduced rates of herbicides following hilling controlled weeds in conventional and reduced tillage potato (*Solanum tuberosum*) production. Weed Technol. 10(2): 311–316.

Boyd, N.S. 2001. The effects of living mulches on tuber yield of potatoes (*Solarum tuberosum* L.). Biol. Agr. Hort. 18: 203–220.

Boydston, R.A. 2010. Managing weeds in potato rotations without herbicides. Amer. J. Potato Res. 87(5): 420–427.

Boydston, R.A. and A. Hang. 1995. Rapeseed (*Brassica napus*) green manure crop suppresses weeds in potato (*Solanum tuberosum*). Weed Technol. 9(4): 669–675.

Boydston, R.A., H. Mojtahedi, J.M. Crosslin, C.R. Brown and T. Anderson. 2007. Effect of hairy nightshade (*Solanum sarrachoides*) presence on potato nematodes, diseases, and insect pests. Weed Sci. 56(1): 151–154.

British Potato Council. 2007. Weed Control in Potatoes. Available at: https://potatoes.ahdb.org.uk/sites/default/files/publication_upload/Weed%20Control%20in%20Potatoes.pdf (Accessed on January 28, 2017).

Burgard, D.J., R.H. Dowdy, W.C. Koskinen and H.H. Cheng. 1994. Movement of metribuzin in a loamy sand soil under irrigated potato production. Weed Sci. 42(3): 446–452.

Burgard, D.J., R.H. Dowdy, W.C. Koskinen and H.H. Cheng. 1993. Metolachlor distribution in a sandy soil under irrigated potato production. Weed Sci. 41(4): 648–655.

Colquhoun, J.B., A.J. Gevens, R.L. Groves, D.J. Heider, B.M. Jensen, G.R.W. Nice and M.D. Ruark. 2017. A3422 Commercial Vegetable Production in Wisconsin. Madison, WI. University of Wisconsin Extension.

Colquhoun, J.B., C.M. Konieczka and R.A. Rittmeyer. 2009. Ability of potato cultivars to tolerate and suppress weeds. Weed Technol. 23(2): 287–291.

Conley, S.P., L.I. Binning and T.R. Connell. 2001. Effect of cultivar, row spacing, and weed management on weed biomass, potato yield, and net crop value. Amer. J. Potato Res. 78: 31–37.

Dvo ák, P., J. Tomášek, K. Hamouz and P. Kuchtová. 2015. Reply of mulch systems on weeds and yield components in potatoes. Plant Soil Env. 61: 322–327.

Heap, I. 2017. The International Survey of Herbicide Resistant Weeds. Available at: www.weedscience.org (Accessed on January 24, 2017).

Holmstrom, D., W. Arsenault, J. Ivany, J.B. Sanderson and A.J. Campbell. 2006. Effect of pre-plant tillage systems for potatoes in Prince Edward Island, Canada, on soil properties, weed control and potato yield. J. Soil Water Cons. 61: 370–380.

Hutchinson, P.J.S., B.R. Beutler and J. Farr. 2011. Hairy nightshade (*Solanum sarrachoides*) competition with two potato varieties. Weed Sci. 59(1): 37–42.

Ivany, J. 1986. Quackgrass competition effect on potato yield. Can. J. Plant Sci. 66: 185–187.

Ivany, J. 2002. Banded herbicides and cultivation for weed control in potatoes (*Solanum tuberosum* L.) Can. J. Plant Sci. 82(3): 617–620.

Kołodziejczyk. 2015. The effect of living mulches and conventional methods of weed control on weed infestation and potato yield. Sci. Hort. 191: 127–133.

Love, S.L., C.V. Eberlein, J.C. Stark and W.H. Bohl. 1995. Cultivar and seed piece spacing effects on potato competitiveness with weeds. Amer. Potato J. 72: 197–213.

Mehring, G.H., J.E. Stenger and H.M. Hatterman-Valenti. 2016. Weed control with cover crops in irrigated potatoes. Agronomy J. 6(1): 3.

National Potato Council. 2016. Potato Statistical Yearbook 2016. Available at: http://www.nationalpotatocouncil.org/files/7014/6919/7938/NPCyearbook2016_-_FINAL.pdf

Nelson, D.C. and J.F. Giles. 1989. Weed management in two potato (*Solanum tuberosum*) cultivars using tillage and pendimethalin. Weed Sci. 37: 228–232.

Nelson, D.C. and M.C. Thoreson. 1981. Competition between potatoes (*Solanum tuberosum*) and weeds. Weed Sci. 29(6): 672–677.

Nitzan, N., R. Boydston and D. Batchelor. 2009. Hairy nightshade is an alternative host of *Spongospora subterranea*, the potato powdery scab pathogen. Amer. J. Potato Res. 86(4): 297–303.

Peachey, E. 2016. Pacific Northwest Weed Management Handbook. Oregon State University, Corvallis, OR.

Rajalahti, R.M., R.P. Bellinder and M.P. Hoffmann. 1999. Time of hilling and interseeding affects weed control and potato yield. Weed Sci. 47(2): 215–225.

Shaner, D. 2014. Herbicide Handbook, Tenth Edition. Weed Science Society of America, Lawrence, KS. 513 pp.

Solan, R.L., J.B. Colquhoun and R.A. Rittmeyer. 2011. Cultivar-specific weed management programs for 'Russet Burbank' and 'Bannock Russet' potato production in Wisconsin. Hort. Technol. 21(4): 451–460.

Thakral, K.K., M.L. Pandita, S.C. Khutana and G. Kalloo. 1989. Effect of time of weed removal on growth and yield of potato. Weed Res. 29: 33–38.

VanGessel, M.J. and K.A. Renner. 1990a. Effect of soil type, hilling time, and weed interference on potato (*Solanum tuberosum*) development and yield. Weed Technol. 4(2): 299–305.

VanGessel, M.J. and K.A. Renner. 1990b. Redroot pigweed (*Amaranthus retroflexus*) and barnyardgrass (*Echinochloa crus-galli*) interference in potatoes (*Solanum tuberosum*). Weed Sci. 38: 338–343.

Wall, D. 1990. Green foxtail (*Setaria viridis*) competition in potato (*Solarum tuberosum*). Weed Sci. 38(4–5): 396–400.

Wall, D.A. and G.H. Friesen. 1990. Effect of duration of green foxtail (*Setaria viridis*) competition on potato (*Solanum tuberosum*) yield. Weed Technol. 4(3): 539–542.

Wallace, R.W. and R.R. Bellinder. 1989. Potato (*Solarum tuberosum*) yields and weed populations in conventional and reduced tillage systems. Weed Technol. 3(4): 590–595.

Wallace, R.W. and R.R. Bellinder. 1990. Low-rate applications of herbicides in conventional and reduced tillage potatoes (*Solanum tuberosum*). Weed Technol. 4: 509–513.

Sustainable Weed Control in Sugar Beet

Giovanni Campagna[1] and Daniele Rosini[*2]

[1] Agricultural Management, Experimentation Dep. Chief, COPROB, Minerbio, 40061 BO, Italy
[2] Agricultural Management, IT & Communication, COPROB, Minerbio, 40061 BO, Italy

Introduction

Reducing the negative impact of anthropogenic activities on agriculture by reduction of herbicide use is gaining more public attention. The rationalization of weed control practices through judicious use of pre-emergence residual herbicides remains a major issue in integrated weed management. The use of herbicides with residual activity tank-mixed with a broad-spectrum systemic herbicide as late pre-emergence application, usually delays the complementary interventions post-emergence. Sometimes only a single treatment of contact herbicides, mixed with triflusulfuron-methyl and low doses of lenacil, for example, completes the weed control program.

When higher doses of pre-emergence residual herbicides are used, it is possible to limit the post-emergence treatment with only one application of contact herbicides mixed with triflusulfuron-methyl and/or metamituron and/or lenacil. The synergistic action of a residual pre-emergence herbicide like clomazone, an inhibitor of photosynthetic pigments biosynthesis, improves the efficacy and broadens the spectrum of activity. In addition, mechanical weed control in fields with moderate weed infestations further reduces the use of residual herbicides. Currently, it is possible to integrate GIS technology for precision weed control to facilitate spraying of post-emergence herbicides and reduce the volume of herbicides used. Air-assisted spray systems optimize the distribution of post-emergence herbicides, thereby increasing efficacy and allowing a reduced number of herbicide applications.

Weed control in sugar beet through various herbicides programs is well established (Appendix 1) although the level of containment of some weeds is still poor. This leads to excessive herbicide use with subsequent phytotoxic effects, particularly at the early vegetative stages, that could lead to significant yield losses.

Crop-weed Interference

The weeds reduce biomass production of sugar beet and greatly reduce beetroot yield especially the marketable roots, since the small-sized roots cannot be picked up mechanically. The most competitive weed species are generally the species with allelopathic abilities and these weeds grow taller than the crop. It is estimated that the loss of sugar beet production caused by one volunteer sunflower plant m^{-2} is almost 50%, one *Chenopodium album* plant m^{-2} is

*Corresponding author: daniele.rosini@coprob.com

30%, *Amaranthus retroflexus* and *Abutilon theophrasti* is 20%, and *Echinochloa crus-galli* 5-10%. In addition to yield losses weeds also reduce the sugar content of beetroot. The reduction of the latter is proportional to the density (4–6 weeds per square meter result in a 5-10% decrease), whereas the reduction with parasitic species *Cuscuta campestris*, is higher (up to more than 25%).

The period of weed competition has a significant impact on the crop yield. Weeds infestation that occurs before or along with the crop cause more harm to the crop than the later weed infestations. For example, five to six *Amaranthus retroflexus* plants m^{-2} that emerge together with the sugar beet plants may cause up to 50% yield loss, but if these emerge later in the growing season the damage can be reduced to 1%. Usually, the critical period of weed control for a beet crop is about one month, from 2–3 weeks after emergence until the crop covers the ground with its foliage. Table 30.1 depicts potential yield losses in sugar beet due to weed competition.

Table 30.1. Yield loss due to weeds in sugarbeet

Species	Potential beet root yield loss %
Abutilon theophrasti	60
Amaranthus retroflexus	90
Chenopodium album	95
Echinochloa crus-galli	90
Setaria spp.	80
Sinapis arvensis	50
Solanum nigrum	40
Sorghum halepense, from rhizome	100

The Main Limiting Resources

The impact of weed competition on sugar beet crop depends on the characteristics of different weed species, weed density, duration, and the dynamics of competition and environmental interactions. Weed species that grow more in height and have bushy and expanded leaves, compete more for light. The development and expansion capacity of the root system can cause problems with water and nutrient uptake in the crop.

The competitive ability of weeds for light mainly depends on the height, leaf area, width of canopy, and rate of growth. Based on these biological characteristics *Amaranthus*, *Chenopodium*, *Xanthium*, *Datura*, *Abutilon* and *Sorghum* spp. are very competitive. Smaller weeds, such as *Stellaria media*, *Veronica persica*, *Capsella bursa-pastoris*, and *Poa annua* can be harmful, especially on beets, at the early stages of crop development (Table 30.2).

The more competitive species are typically those with C4 photosynthetic pathway, such as *Sorghum halepense*, *Cynodon dactylon*, *Cyperus* spp., and *Amaranthus* spp. These species, however, do not tolerate low light intensities, unlike the C3 species (e.g., *Avena* spp., *Lolium* spp., *Poa* spp., *Abutilon theophrasti*, *Chenopodium album*, *Solanum nigrum*, *Xanthium strumarium*). Shade-sensitive species can be managed effectively by maintaining optimum crop population density and row spacing, planting at optimum periods, and enhancing crop growth (by proper fertiliser use and irrigation if possible) to achieve quick and early canopy closure. A healthy crop is often able to smother the weeds below the crop canopy.

Competition for water is another aspect of weed-crop interference. Typically, C4 species have higher water use efficiency as compared to C3 species. The species with leaves covered with wax (e.g., *Chenopodium album*, *Polygonum aviculare*, *Euphorbia* spp. and *Fumaria officinalis*) are able to adjust leaf stomatal opening according to the water status (e.g., *Sorghum halepense*) are more competitive under drought conditions (Table 30.3).

Table 30.2. Effects of weeds on light interception in expense of the sugar beet crop

Weeds	Average height (cm)	Light reduction %
Echinochloa crus-galli	80	20
Avena sterilis	90	22
Sinapis arvensis	70	46
Chenopodium album	130	65

Table 30.3. Water consumption per unit of dry matter (d.m.) produced by several species of weeds in comparison with some crops

Weeds	Coefficient of evapotranspiration (L water consumption/kg dry matter)
Setaria spp.	250–300
Sorghum halepense	200–250
Amaranthus retroflexus	275–325
Chenopodium album	325–375
Sugar beet	350–450

The species that respire more, thus waste more water, typically cause greater damage to non-irrigated crops. Species with more expanded (*Xanthium* and *Gramineae*) and deep *(Amaranthus, Abutilon, Umbelliferae, Cruciferae* and *Leguminosae*) root system are generally more harmful than those exhibiting shallow and smaller root systems.

Regarding nutrient requirements and competition, weeds are able to accumulate considerable amounts and generally at absorption rates higher compared to crops, thus becoming more competitive. *Avena* spp. and most of the *Gramineae* are high consumers of nitrogen whereas *Sinapis* spp., *Amaranthus* spp., *Chenopodium album* and *Echinochloa crus-galli* are high consumers of phosphorus. Some weed species including *Amaranthus* spp., *Chenopodium album,* and *Portulaca oleracea* take up high amounts of potassium.

Weed families with high nitrogen uptake and utilization, in descending order, include *Gramineae, Chenopodiaceae, Amaranthaceae, Urticaceae, Solanaceae* and *Polygonaceae.* A greater amount of nitrogen is absorbed during the early vegetative stages, whereas most of the phosphorus and potassium are taken up just before flowering. Increasing the amount of fertilizer application cannot compensate the negative effects of competition for nutrients, since the weeds are generally more efficient in taking up nutrients and are more efficient in converting these into biomass than the crop. The localized placement of fertilizers on the row next to the crop optimizes the absorption of nutrients by sugar beet, giving it an advantage over the weeds (i.e., increased shading in the row and faster canopy closure), hence reducing the emergence and growth of weeds in the rows. This in turn reduces the man-hours needed to remove the weeds by mechanical hoeing.

The increase in weed density often determines production losses, even if the influence is not linear. The rate of weed development and the amount of biomass produced change depending on the space available. For example, a few weedy plants can cause high yield losses because of increased size per plant; conversely, high weed density results in less yield loss on a per plant basis because each plant is smaller (Figure 30.1). The largest total losses of production are observed with an intermediate number of weeds, the magnitude of which vary from species to species.

The intensity of weed competition is greater during the early stages of crop development and declines when the crop is actively growing (Figure 30.2). Weeds that grow more slowly, or those that emerge after the cultivation, are generally not able to cause significant damage.

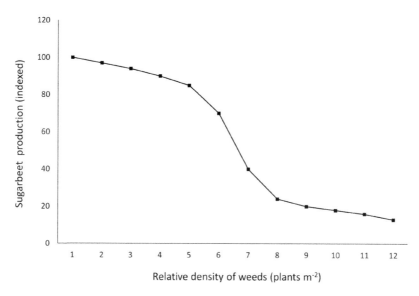

Figure 30.1. Example of production loss as a function of the relative density of weeds.

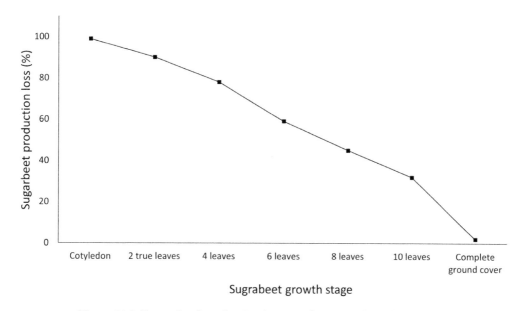

Figure 30.2. Example of production loss as a function of weed emergence and sugar beet development.

The duration of competition is in direct relation with the potential yield losses (Figure 30.3). For this purpose, it is necessary to clarify certain concepts, including the duration of tolerated competition (DTC), about 15 days from emergence, wherein the damage that can be caused is negligible and with no irreversible effects on production. This applies to weeds that have emerged at the same time with the crop.

The critical period (CP) is the phase in which the crop suffers the most damage from weed interference, within 30 days from DTC. When the crop is already developed, weeds no longer affect yield, although it may hinder crop harvest, facilitate dissemination of seed propagules, and significantly increase the soil seed bank. The required weed-free period may increase with

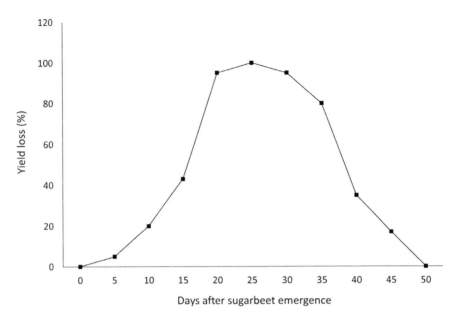

Figure 30.3. Example of sensitivity to the duration of the competition in function of the period of emergence of weeds.

early sowing or with cold and rainy weather during the early stages of crop development. During the critical period, production losses could exceed up to 50%.

In practice it is more appropriate to define the impact of weed competition timing on the basis of growing degree days (GDD), that determine the emergence and stage of development of the crop and weeds starting at the first pair of true leaves (Table 30.4, Figure 30.2). The importance of the GDD tool can be seen in the case of weed control strategies by post-emergence herbicide applications at very low doses (VLDs), the efficacy of which can be affected by the intervals of application (5–6 days up to 12–15 days), depending on the growth rate of weeds.

Based on this data it is possible to identify the most appropriate time to control the weeds, or even whether or not to intervene. Numerous studies have been carried out through the development of mathematical models, considering the competitive ability of individual species and their density, as well as the duration and the period in which competition is exercised.

Threshold, as a concept for appropriate weed control, includes: i) economic intervention threshold and ii) optimal economic threshold. Economic intervention threshold corresponds to a density of weeds for which the cost of treatment equals the benefits (Table 30.5). This threshold density varies according to the cost of the product, the efficacy, the value of the crop, and others. Optimal economic threshold takes into consideration the capacity of dissemination and the viability of seeds in the soil and, thus, of future damage. Therefore, the optimal economic weed density threshold is set lower than the economic intervention threshold. Surely the second is more accurate, but more complex and difficult to determine, and is used especially for hard-to-control weeds, which produce a large amount of persistent seeds (e.g., *Abutilon theophrasti*, *Ammi majus*, *Chenopodium* spp., etc.).

It is important to consider that the thresholds are determined by species, but many weeds infest a field. Furthermore, variable environmental conditions (light intesnity and duration, soil moisture, and temperature) affect the growth rate of species differently and also the crop-weed interaction. Therefore, using thresholds as a tool for decision-making on when to implement management practices is complex. In practice, farmers may use what they consider as the primary weeds as a trigger for intervention.

Table 30.4. Thermal requirements of some weeds and influences on development

Weeds	GDD to begin emergence (°C)	Min temperature for germination (°C)	Max temperature for germination (°C)	Average no. of days for emergence in the first half of March (soil temp. about 10°C)	Average no. of days for emergence in the first half of April (soil temp. about 15–20°C)	Optimal temperature for vegetative growth (°C)
Echinochloa crus-galli	135	10-13	45	-	>13	21–33
Avena sterilis	100	2-4	26	>10	>5	15–24
Amaranthus retroflexus	85	12	40	>28	>6	21–35
Chenopodium album	195	2-5	35	>20	>10	16–25
Solanum nigrum	170	10	33	-	>17	18–27
Sugar beet	90	3	27	>12	>5	15–24

Table 30.5. Production losses and economic intervention threshold for different weeds

Weed species	Average yield loss with 10 plants 10 m^{-2} (%)	Average no. of plants 10 m^{-2} to reduce the yield by 50%	Economic intervention threshold: No. of plants 10 m^{-2}*
Echinochloa crus-galli	5–10	50–100	7
Abutilon theophrasti	20	60–70	6
Amaranthus retroflexus	20	50–60	3
Chenopodium album	30	30–40	2–3
Sinapis arvensis	20	100–200	3
Sunflower	45	10	<1

*10 plants of *Solanum nigrum* and *Avena sterilis*

Chemical Weed Control in Sugar Beet

The early preparation of seedbed and preventive weed control before sowing, especially with heavy infestations, is the first step towards optimization of integrated weed management strategies. Where weed infestation is low, the use of pre-emergence herbicides tank-mixed with glyphosate for pre-plant vegetation desiccation might reduce the number of foliar herbicide applications. Residual herbicides, even when used with pre-conditioning products (metamitron + lenacilor metamitron + cloridazon), improve the efficacy of foliar herbicides, especially under unfavorable environmental conditions and with heavy infestations of *Abutilon theophrasti* Medik., *Polygonaceae* (*Fallopia convolvulus* (L.) A. Love, *Polygonum aviculare* L., *P. lapathifolium* L., and *P. persicaria* L.), *Amaranthaceae* (*Amaranthus retroflexus* L., *A. blitoides* S. Watson, *A. rudis* J. D. Sauer), large *Cruciferae*(e.g., *Sinapis arvensis, Rapistrum rugosum*), new emergence of rapeseed (*Brassica napus* L.) and coriander (*Coriandrum sativum* L.), and *Apiaceae* (*Ammi majus* and *Daucus carota* L.).

Weed control programs for summer annual grasses, involve the precautionary applications of *S*-metolachlor in small doses (0.3–0.5 litres ha^{-1}) in a mixture with the most selective metamitron, together with early post-emergence application of triflusulfuron-methyl Post-emergence. This strategy also prevents the evolution of resistant weed populations.

The post-emergence applications of herbicides at the weed cotyledon stage using VLD, and mixing it with triflusulfuron-methyl, increases the spectrum of activity and its efficacy against larger weeds (2 to 4 leaves for the most sensitive species).

Complex infestations of *Compositae*, including *Cirsium arvense, Sonchus arvensis, Sylibum marianum* and *Xanthium strumarium*, cannot be controlled with mixtures of phenmedipham + desmedipham + ethofumesate, or triflusulfuron-methyl + residual + oil. This weed spectrum requires the use of clopyralid to control early infestations.

Clopyralid can be used to control annual species of the *Compositae* family including *Matricaria* and *Picris* and volunteer sunflower (*Helianthus annuus* L.); *Umbelliferae* family, including *Ammi majus* and *Daucus carota* or new emergence of coriander, and legumes (e.g., *Vicia*) and regrowth of alfalfa. The formulations of clopyralid in split doses can be mixed with VLD of propyzamide and graminicides. Mixtures of clopyralid and triflusulfuron-methyl are antagonistic. The application of these herbicides must be separated by at least seven days.

The control of *Cuscuta* spp. is enhanced with split applications of propyzamide with VLD of the usual herbicides used in sugar beet production. Three applications (starting from 0.1–2.5 lha^{-1}) of propizamide (Kerb 80EDF) is necessary, with increasing doses until a total dose of at least 1 lha^{-1} or 0.5 kgha^{-1} is achieved.

Weed control programs consisting of a mixture of post-emergence herbicides, the amount of adjuvants, for example, crop oil should be reduced, or no adjuvant should be used at all, in order to maintain crop selectivity. This is in contrast with the application of triflusulfuron-methyl alone, which specifically requires crop oil.

Using multiple mechanisms of action of herbicides is expected to deter the evolution of herbicide-resistant weeds (Table 30.6). In addition, an integrated weed management approach, particularly for the species with continuous or late emergence (e.g., *Amaranthus* spp., *Ammi majus*, *Abutilon theophrasti* and *Setaria* spp.), will enhance the efficacy of the weed management program.

Weed Control in Autumn-planted Sugar Beet

Pre-sowing herbicide application is of minor importance in autumn-planted because the fields are freshly prepared. Usually, sugar beet in autumn follow some summer crop. Therefore, pre-emergence are primary weed control treatment.

Under irrigation, the use of residual herbicides is particularly important as unfavorable weather conditions (e.g., rain, frost or other adversities) could delay the post-emergence herbicde applications.

A combination of lenacil, chloridazon and *S*-metolachlor, and possibly ethofumesate + lenacil, is recommended for pre-emergence control of the most common cruciferous (*Brassicaceae*) broadleaf weeds (e.g., *Veronica*, *Papaver*, *Lamium* and *Fumaria)*. Ethofumesate is particularly effecive on *Galium* spp.

Post-emergence herbicide applications usually include split-applications of phenmedipham + desmedipham + ethofumesate followed by (fb)chloridazon + lenacil. Triflusulfuron-methyl can substitute cloridazon to control heavy infestations of *Asteraceae* species (e.g., *Chrysantemum* spp. or *Anthemis* spp.) and *Umbelliferae* species (e.g., *Bifora* spp., *Daucus* spp. and *Ammi* spp.).

Split applications are essential for preventive control of grasses and other broadleaf weeds during the critical period of weed control. This period typically lasts approximately 60 days between November and January, during which 2–3 herbicide applications are recommended. For the control of adverse infestations of *Avena* spp., *Phalaris* spp., and *Lolium* spp. at the early stages of crop development, it is necessary to include a low dose of a graminicide (with reduced adjuvant concentration to avoid phytotoxic effects on crop seedlings). In case of difficult weeds, such as *Umbelliferae* and *Compositeae*, including *Sylibum marianum*, the inclusion of clopyralid at half of the recommended dose is recommended in lieu of triflusulfuron-methyl.

Chemical Weed Control: Some Considerations

The speed of weed emergence depends on precipitation and temperature on or close to the soil surface. Weed growth and development on the other hand depends mainly on air, temperature and soil moisture. Weed control in sugar beet using post-emergence herbicide mixtures at (VLD) owes its success, in terms of effectiveness and crop selectivity, to multiple interventions starting at the very early stages of development (cotyledons to two leaves growth stage).

The interval between herbicide treatments is based on product label instructions. Nevertheless, scouting of weeds to detect weed emergence and their stage of development could alter the application interval and enhance the effectiveness of herbicide programs. Knowledge of GDD as it relates to the development and growth of weeds can be used to determine the optimum time for herbicide application.

For early weeding and under cold temperatures an interval of more than 20 days would be sufficient. Warmer periods require repeated applications of low doses of herbicide at approximately 10-days interval or increased herbicides split-doses are recommended as shown in Table 30.7.

Table 30.6. Herbicides used in sugar beet production

Chemical family	Active ingredient	Time for use recommended	Absorption mode	Mode of trans-location and mechanism of action (HRAC group)	Behavior in soil: absorption, mobility, persistence and degradation	Mechanism of selectivity and degradation in plants	Crop substitution and sensitive succession	Secondary aspects	Other targets
Organo phosphoric acid compounds	Glyphosate	Pre-sowing	Foliar: absorption in 6-8 hours.	Systemic; rapid translocation. Inhibits the synthesis of certain amino acids (G)	Quickly adsorbed to soil colloids and inactivated. It remains fixed, except in sandy soils. Fast degradation by microbes (10-15 days)	Selectivity by placement or shielding to avoid contact with the crop	-	Rain within 2 hours reduce the activity; water stress and heat stress also reduce activity	All
Residual herbicides									
Diazinepyridazinone	Cloridazon	Pre and Post-emergence	Root (foliar)	Systemic via xylem Inhibits photosynthesis (C1)	Adsorbed by colloids; inactivated by high organic matter. Mobile in soil. Microbial degradation between 6 and 30 °C; persists for 3-5 months.	Rapid inactivation by enzymes.	*Cruciferae*, tomato	Requires moisture for activation. Do not apply foliar at temperatures >25 °C. Do not apply to stressed plants	
Benzofuran	Ethofumesate		Seedling; coleoptile for grasses and roots for dicots (foliar secondary action, increased by the addition of solvents)	Systemic; limited translocation. Inhibits mitosis. Reduces cuticle thickness and increases sensitivity to other foliar herbicides (N)	Strongly adsorbed to clay, immobile in soil. Microbial degradation in 4-6 months.	Rapid degradation by hydrolysis.	Soybean	Requires soil moisture for activation. Do not use at temperatures >25 °C in Post-emergence	Tobacco
Diazine uracil	Lenacil		Roots of seedlings (foliar secondarily)	Systemic via xylem. Inhibits photosynthesis (C1)	Poor solubility. Strongly adsorbed, but not inactivated. Microbial degradation in 2-6 months.	Selectivity by stratigraphic and biological inactivation. Degradation by hydroxylation in about 1 month.	*Cucurbitaceae*	Requires moisture for activation. Do not apply foliar at temperatures >25 °C. Do not apply to stressed plants.	...

(Contd.)

Family	Active ingredient	Timing	Absorption	Mode of action / translocation	Soil behaviour / persistence	Selectivity	Sensitive crops	Conditions	Rotation crops
Asymmetrical triazines	Metamitron		Roots (foliar secondary)	Systemic via xylem and accumulation in the chloroplast (leaves). Contact action through leaf absorption. Inhibits photosynthesis (C1)	Strongly adsorbed, poor movable despite the high solubility. 3-4 months of persistence. microbial degradation, but also for road photolytic	Selectivity for biological inactivation. Fast degradation by enzymes.	*Cucurbitaceae*	It requires moist soil for radical activation or rain. Use with maximum temperatures of 25 °C in post-em. Avoid the vegetative stress in post-emergence.	Salads, artichoke, alfalfa, clover, sainfoin, the, lives, apple, pear
Amide benzamide	Propizamide	Post-emergence	Roots through the seedlings (foliar secondary)	Systemic via xylem. Very mobile for foliar absorption. It inhibits mitosis at the level of root tips (K1)	Adsorbed on average and not very movable. 2-6 months of persistence. Chemical decomposition and evaporation at high temperatures (greater persistence with low temperatures)	Selectivity for biological inactivation and by stratigraphy. Slow metabolism within the plant.	Corn, sorghum, grasses in general, strawberries, cruciferous vegetables, onion, garlic, Solanaceae, spinach, flax	It requires moist soil for radical activation or rain. .	
Amide Chlorine acetanilide	S-metolachlor	Pre-emergence	Seedlings (cotyledonous for dicotyledonous and coleoptile for grasses). but also by the young seedling roots	Poorly translocated. It inhibits the biosynthesis of lipid (K3)	Strongly adsorbed and inactivated by organic colloids; very mobile. 2-3 months of persistence. Main degradation microbial; and secondarily photolytic.	Selectivity by stratigraphic and biological inactivation. Hydrolytic degradation.	*Graminaceae*	It requires moist soil for radical activation or rain.	Corn, soybean, sunflower, tobacco
Foliar herbicides									
Ormone-like Carboxylic acid	Clopyralid	Post-emergence	Foliar: fast absorption in 1-2 hours. (radical)	Systemic accumulation in young meristematic tissues. (O)	Shortly adsorbed, subject to leaching. 3-5 months of persistence. Microbial degradation in less than 3-5 months.	Selectivity by biological inactivation. Is not degraded in cereals (risk of accumulation in the straw)	Soybeans and other legumes	Rain after one hour does not reduce its effectiveness. Treat preferably during the morning.	Corn, winter cereals, rapeseed and other cruciferous vegetables, onion, leek, beet vegetable garden

(Contd.)

Table 30.6. (*Contd.*)

Chemical family	Active ingredient	Time for use recommended	Absorption mode	Mode of translocation and mechanism of action (HRAC group)	Behavior in soil: absorption, mobility, persistence and degradation	Mechanism of selectivity and degradation in plants	Crop substitution and sensitive succession	Secondary aspects	Other targets
Carbamate	Desmedipham		Foliar	Contact. It interferes at the level of photosynthesis (C1)	Very adsorbed and very movable. Microbial degradation. 25-45 days maximum persistence, but no biological residual activity	Selectivity for biological inactivation. Hydrolysis in inactive compounds in beets.	-	Influence of temperature and brightness. Treat at evening with high temperatures.	
Carbamate	Phenmedipham		Foliar absorption in about 6 hours	Contact or poor movable. It interferes at the level of photosynthesis (C1)			-	Treat at evening with temperatures above 25 °C. Rains before 4-6 hours can reduce the effectiveness.	Spinach, vegetable and fodderbeet
	Trifulsulfuron-methyl		Foliar (Radical, restricted to limited period of persistence)	Systemic, is rapidly moved into acropetal and basipetal sense. It inhibits amino acid synthesis via inhibition of ALS (B)	Not strongly adsorbed and mobile, but quickly degraded, with limited possibility of leaching. Chemical and microbial degradation (slowed by acid pH). Persistence usually less than 1 month	Selectivity for biological inactivation. It is rapidly degraded into inactive compounds.	Horticultural and ornamental	Avoid temperatures higher than 23 °C or close to zero. Treat preferably in the morning.	-
Specific graminicides									
Aryloxy phenoxy propionic acid	Fluazifop-p-butyl Quizalofop-p-ethyl and Q. isomer D ... Propaquizafop	Post-emergence	Foliar: absorption in 1-3 hours (Secondary root absorption and at coleoptile level)	Systemic translocation to the meristems of the culms of tillering, roots and rhizomes which accumulate p.a. Inhibit the synthesis of fatty acids (the enzyme acetyl coenzimaA) essential for the formation of cell membranes at the meristematic level (A)	Little movable and moderately adsorbed colloids. Microbial degradation and secondary photolytic action. Persistence longer than few weeks, even if they are biologically active for much more limited periods.	Selectivity for biological inactivation. Hydrolysis, metabolism and conjugation to inactive compounds.	*Graminaceae* for max. period of 1 month after treatment	Absorption in 1-3 hours water stress, high temperatures and low relative humidity may reduce the effectiveness. Treat preferably in the morning. Rains after one hour does not affect the activity under optimal conditions.	Broadleaf crops
Cyclohexenone	Ciclossidim Clethodim	Post-emergence							

Table 30.7. Variation of herbicide rates in mixtures for post-emergence weed control

Conditions that warrant very low dose (VLD)	Conditions that warrant high doses (split applications)
When following Pre-emergence weed control, especially with high doses of broad-spectrum, and less selective herbicides (metamitron is the most selective)	No preventive weed control
Crop and weeds at the cotyledon growth stage	Beets and weeds are large
Risk of frost	Weeds in drought stress
After rainfall	Prolonged conditions of low relative humidity
Very sunny days (particularly after prolonged cloudy periods)	High density of weeds
High temperatures (> 25 °C)	Highly prolific weeds (e.g., POLAV, Cheal, AMASS, etc.)
Sudden temperature changes	
Short intervals between applications (175 GDD)	Long intervals between applications (225 GDD), or when conditions had prevented correct application timing

Herbicide Application Rates and Phytotoxicity

The selection of herbicide application rates for complex herbicidal programs (Figure 30.4; Table 30.8) requires certain precautions towards crop selectivity (Table 30.9). Spraying at cold periods or when temperatures are below 0 °C should be avoided. In the case of residual herbicides, doses should be reduced particularly for less selective products. Metamitron is the only fully selective herbicide for sugar beet.

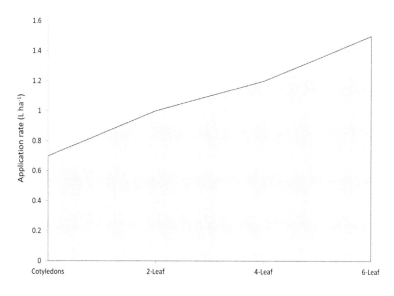

Figure 30.4. Example of dose adjustment (L ha^{-1}) as a function of the stage of crop and weed development (fenmedifam + desmedifam + ethofumesate).

Table 30.8. Herbicide application schemes depending on the weed species, weather and ground conditions (VLD: very low dose; sequential applications: increasing doses)

Target species	Base mixture; Very low dose	Sequential applications		
Larger weeds (e.g., Polygonaceae, Chenopodiaceae, etc.)	Phenmedipham+ Desmedipham+ Ethofumesate (e.g., Betanal Expert) 0.7-1 L/ha	+ metamitron (70%) 0.5-0.7 kg/ha Or Goltix Star 0.6-1 kg/ha	+ lenacil (e.g., Venzar) 80-100 g/ha	+ vegetable or mineral oil 0.3 L/ha
Cruciferous species at early stages of development (e.g., *Fallopia convolvulus*, etc.)	Phenmedipham + Desmedipham + Ethofumesate (e.g. Betanal Expert) 0.7-1 L/ha	+ metamitron (70%) 0.5-0.7 kg/ha Or Volcan Combi SC 1-1.3 L/ha	+ cloridazon (e.g., Better 400) 0.6-0.8 kg/ha	+ vegetable or mineral oil 0.3 L/ha
For greater flexibility of use and weeds difficult to control (*Abutilon theophrasti*, Cruciferae, Umbelliferae, Amaranthaceae, etc.)	Phenmedipham+ Desmedipham+ Ethofumesate (e.g.Betanal Expert) 0.7-1 L/ha	+ metamitron (70%) 0.5 kg/ha	+ triflusulfuron-metile (Safari) 30-40 g/ha	+ vegetable or mineral oil 0.5 L/ha
For *Cuscuta* spp., and also *Solanum nigrum*, *Polygonum aviculare*, etc.	Phenmedipham+ Desmedipham+ Ethofumesate (e.g.Betanal Expert) 0.7-1 L/ha	+ metamitron (70%) 0.5-0.7 kg/ha	+ propizamide (e.g Kerb 80EDF) 0.20-0.50 kg/ha*	+ vegetable or mineral oil 0-0.3 L/ha

Note:

[1] Graminicides are miscible in all combinations, although slight reduction of efficacy may occur.

[2] A base mixture (e.g., Betanal Expert) can be supplemented with phenmedifam (e.g., SE Betanal 0.5-1 L ha^{-1}) for large weeds (more than 4 leaves) and during dry periods, using greater than 200 L ha^{-1} spray volume.

[3] Possible other final operations after applying VLDor sequential applications:

—Safari + olio (30-40 g + 0.5 Lha^{-1}): in presence of sensitive weeds triflusulfuron-metile (e.g., *Abutilon theophrasti*, *Ammi majus*, *Daucus carota*, *Amaranthus* spp., developed Cruciferae, etc.)

—*Kerb 80 EDF for Cuscuta (total amount 1–1.2 kgha^{-1}) +/- eventual graminicide +/- Lontrel 72SG (100–120 gha^{-1})

[4] Lontrel 72SG (100–120 gha^{-1}) in presence of Umbelliferae (*Ammi majus*, *Daucus carrot*), legumes (*Vicia* spp., *Galega officinalis*), Compositae (*Xanthium strumarium*, *Cirsium arvense*, *Sonchus* spp.), alfalfa regrowth, volunteer sunflower, and coriander.

One of the business decisions that contribute to maximization of farm income is the choice of crop rotation, considering the suitability of the land, and the historcial knowledge of potential disease and insect pest infestation.

One must also consider that low yields are often due to excessive traffic (compaction), waterlogging, or high infestation of nematodes. In these cases, the beet plants are stunted and weeds take over, requiring a greater number of herbicide applications, which increases the degree of phytotoxicity and further delays canopy closure.

Integrated Pest Management (IPM)

The execution of farming practices at the right time allows the sugar beet crop to reach optimal development. The application of herbicide, for example, during the early stages of development, contributes to crop vigour and health, which are necessary for growth maximisation. Localised phosphate fertiliser application, accompanied by split application of reduced nitrogen fertiliser rates, provide a better 'starter effect' for the crop, making it more competitive with weeds.

Application of herbicide mixtures, using medium or low rates of residual herbicides, effectively removes present infestation, thereby reducing the number of follow-up herbicide applications or hoeing; thus, reducing production cost. Controlling weeds at the cotyledon-to 2-leaf growth stage is crucial for the success of weed management scheme. Subsequent operations could be performed knowing that the growth rate of the weeds is dictated primarily by air and soil temperatures. Hence, the timing and duration of weed control operations are different between spring/summer and fall/winter periods.

In recent years there has been a gradual and increasing establishment of ruderal weed species which are more difficult to control. These are characterised by prolonged emergence and increased aggressiveness even against developed crops. These include *Ammi majus, Daucus carota, Torilis arvensis, Xanthium strumarium, Bidenstripartita, B. frondosa, Abutilon theophrasti, Amaranthus retroflexus, A. blitoides, A. rudis, A. graecizans, A. deflexus, Chenopodium album* and other minor species like *C. ficifolium, C. vulvaria, C. opulifolium,* and some grass species *Echinochloa crus-galli, Setaria* spp. and *Sorghum halepense.*

In rotation systems with sugar beet, the most frequently used herbicides belong to a small number of chemical families under two mode-of-action groups: i) ALS inhibitors (i.e., sulfonylureas); and ii) ACCase inhibitors or selective grass herbicides (i.e., fluazifop, clethodim). Populations of *Echinochloa crus-galli* are resistant to one or both groups of herbicides and *Amaranthus* spp. are resistant to several ALS inhibitors. The use of residual herbicides, such as *S*-metolachlor, reduces the weed population size that is being exposed to selection pressure with post-emergence herbicides and can curtail resistance evolution. *S*-metolachlor is effective on *Amaranthus* species. and grasses and is also effective on other small-seeded weeds, including the troublesome *Ammi majus* (Tables 30.10 and 30.11).

Appropriate crop rotation and good cultivation technique with tillage done at appropriate times, optimum fertilizer application, early seeding and acceptable crop density (i.e., 9–12 plants m^{-2}) could reduce weed pressure significantly. Weed mapping, accurate identification, and characterization of weed infestation can be used to maximize the effectiveness of herbicides and minimize the degree of phytotoxicity to the crop. The ability to control weeds that get established ahead of the crop and are the most widespread (i.e., *Polygonum aviculare* and *Fallopia convolvulus*), as well as the ability to control the weeds that emerge during the winter, reduce the intervention costs.

Specific procedures performed before seeding, such as use of glyphosate in mixture with low doses of residual herbicides could reduce the number of chemical applications and lower the cost of production. When wet conditions prevail, it is possible to intervene effectively with early treatment of post-emergence herbicides, based on complex mixtures, such as phenmedipham + desmedipham + ethofumesate or phenmedipham + desmedipham + ethofumesate + lenacil.

Once the sugar beet plants have emerged, timing of chemical weed control (2- to 4-cotyledons or 2-leaf stage) must be carefully selected based on weeds present and the

Table 30.9. Miscibility of herbicides depending on the degree of effectiveness and selectivity

Herbicides	VLD	B	S	M	C	V	O	G	K	L
VLD: very low doses of contact products (e.g., Betanal Expert)										
B: Fenmedifam (e.g., Betanal SE)	=							*		
S: Safari (Triflusulfuron methyl, 50%)	++	+								
M: Metamitron	=	=	=							
C: Cloridazon (e.g., Better 400)	+	+	+	+						
V: Venzar (Lenacil, 80%)	+	+	+	+	+					
O: Mineral oils	+	+	+	=	+	+		**		
G: Specific graminicide	+	+	++	=	=	=	+			
K: Kerb 80EDF	++	+	+++	+	++	++	+++	+		
L: Lontrel 72SG (Clopiralid, 75 %)	++	+	X	+	++	++	++	+	++	

EFFECTIVENESS GRADE

- essential
- the same or improved by additives
- low action

* The graminicide manifesting the best degree of compatibility is ciclossidim

**with Stratos

Legend Phytotoxicity
= the same or improved by additives
da + a +++ from tolerated to elevate
X not tolerated or not recommended

Table 30.10. Integrated management of established or invasive weed species, and prevention of herbicide resistance development

Period	Agricultural and mechanical methods or tools	Chemical methods or tools
Pre-cultivation	Rotation with winter cereals	Use of herbicides with different modes of action
	Deep tillage for deep seed burial and reduction of the soil seed bank	Control of weeds herbicide mixtures or sequential applications before sowing or before blooming
	Stale seedbed	
	Mechanical removal of weeds before flowering	
During cultivation	Choice of varieties, growing techniques and optimum planting density to promote quick canopy closure	Using the best combination of pre- and post-emergence herbicides
	Mechanical weeding with minimal compaction of soil	Use of mixtures with different mechanism of action
	Localised weed under the canopy in combination with the use of GPS technology	Post-emergence applications during the early stages of development of weeds
	Destruction of remaining weed infestation by cutting bars or ropes	Using doses proportionate to the stage of development
	Monitoring and scouting before and after herbicide applications to detect escapes or new emergence	Avoid post-emergence applications in unfavorable climatic periods (vegetative stress)
		Using properly calibrated sprayers
Harvest	Cleaning of farm equipment to prevent the spread of seeds and vegetative propagules	
Post-cultivation	Shallow tillage to promote germination of weed seeds	Use of non-selective herbicides before blooming

synergistic relationships between pre- and post-emergence herbicides must be considered. The first treatment is typically consisting of a mixture of ethofumesate + phenmedipham, and sometimes desmedipham, with the addition of vegetable or mineral oil and a residual herbicide (i.e., metamitron, or metamitron + lenacil). Another herbicide that can be added to this mixture to further broaden the spectrum of control is triflusulfuron-methyl (a sulfonylurea, PS II inhibitor).

In the case of late-emerging species, such as *Amaranthaceae* (*Amaranthus retroflexus, A. albus,* and the increasingly widespread *A. rudis*) and *Abutilon theophrasti,* which emerge and grow simultaneously with sugar beet crop, the use of sulfonylurea herbicides should be considered. This is due to the fact that the crop must remain weed-free for about 5–7 weeks (critical period). The decision-making for intervention is becoming more complicated by the presence of *Cuscuta* spp., which requires the implementation of split applications of propyzamide, timed on specific growth stage of sugar beet, and especially in proper mixtures with triflusulfuron-methyl, to avoid phytotoxicity to the crop. When aggressive weeds have escaped control (i.e., *Abutilon theophrasti, Chenopodium album* and *Amaranthus retroflexus*), the use of glyphosate under the so-called 'humectants bars' is an option. This strategy allows for the control of floral scapes of wild sugar beet also.

Other problems may result from the presence of annual grasses not sufficiently controlled, unfavorable environmental conditions or due to uncontrolled reproductive organs, such as rhizomes of *Sorghum halepense.* In this case the most active graminicide should be applied. Selective grass herbicides may be applied with fungicides or foliar fertilizers.

Table 30.11. Recommended practices (agricultural, mechanical, and chemical) in relation to the biological characteristics of weed

Biological features	Species	Recommended practices
High seed production	*Abutilon theophrasti, Amaranthus* spp., *Ammi majus, Chenopodium* spp., *Papaver rhoeas, Polygonum aviculare,* Cruciferae	Rotation with different crop types Destruction of weeds before flowering Stale seedbed technique
Scalar germination	*Abutilon theophrasti, Ammi majus, Echinochloa crus-galli, Daucus carota,*	Hoeing Rotation with different crop types
Germination at soil surface or from shallow depths	*Cirsium vulgare, Epilobiumhirsutum, Erigeron canadensis, Daucus carota, Lactucaserriola, Matricariacamomilla, Picrisechioides, Plantagolanceolata, Senecio vulgaris, Sonchus* spp., *Torilis arvensis,*	Plowing Stale seedbed technique
Autumn-spring emergence	*Avena* spp., *Galiumaparine, Papaver rhoeas, Veronica* spp.	Earlier preparation of seedbeds Avoid the use of mixtures of glyphosate + residual pre-emergence herbicides
Spring-summer emergence	*Abutilon theophrasti, Amaranthus* spp., *Chenopodium* spp., *Cuscuta* spp., *Echinochloa crus-galli, Polygonum lapathifolium, Setaria* spp., *Solanum nigrum*	Rotation with winter cycle crops
High seed longevity in the soil	*Abutilon theophrasti, Amaranthus* spp.,*Capsella bursa-pastoris, Chenopodium* spp., *Fallopia convolvulus, Papaver rhoeas, Polygonum* spp., *Raphanus raphanistrum, Rapistrum rugosum, Sinapis arvensis, Thlaspi arvense, Veronica* spp.	Hoeing Cleaning equipment Shallow tillage after harvest
Short seed lifespan in the soil	*Bromus* spp., *Lolium* spp., *Poa* spp.	Plowing Stale seedbed technique Frequent shallow tillage
Perennial species with vegetative reproduction (tubers, rhizomes, bulbs)	*Cirsium arvense, Convolvulus arvensis, Calystegia sepium, Cyperus* spp., *Oxalis* spp., *Sorghum halepense*	Plowing Frequent tillage and/or shredding
Parasitic	*Cuscuta* spp.	Rotation with non-host species Eliminating host weeds Application of propyzamide Cleaning of the machines

The water volume for herbicide application is dictated by technical requirements of certain preparations. For example, phenmedipham cannot be applied below a certain level of concentration, because this will reduce its effectiveness. The quality of water is a significant factor not only in the use of glyphosate, but also for other herbicides. The presence of sodium bicarbonate in concentrations greater than 300 ppm seems to reduce the effectiveness of many herbicides; the same could be said about the presence of calcium, potassium, iron, and magnesium in water. The sulfate ion tends to antagonize the negative effect of calcium in particular, and this is why adding ammonium sulfate to the water before adding glyphosate is beneficial. The activity of sulfonylureas can be affected also by cations present in the water.

The use of pH-regulators could stabilize the level of herbicide action, although pH is not the only factor that alters the behavior of the herbicide in water and in the plant. Often, however, it is the clay particles or the organic matter suspended in the water that can compromise herbicide activity. This occurs if water is pumped from channels, ponds, or lakes.

The absorption period and the use of additives significantly affect the effectiveness of herbicides in sugar beet. The absorption period depends on soil and weather conditions, the biological characteristics of the weed, and the active ingredient. On an average, these times are about 5-6 hours for phenmedipham, desmedipham, clopyralid, and glyphosate and only 1–2 hours for graminicides. The use of additives or adjuvants (e.g., oils, wetting agents, fertilizers) can enhance retention and absorption, but can increase the risk of phytotoxicity. Mineral oils have been successfully, but in certain cases vegetable oils have provided better results. The oil concentration may also vary as a function of the herbicide rate used and the condition of the weeds. Wetting agents and sprayable fertilizers (ammonium nitrate, urea or ammonium sulfate) are seldom used because of phytotoxicity problems.

New Active Ingredients for Weed Chemical Control in Sugar Beet

With the advent of transgenic crops, research has not invested much on the discovery of new active ingredients. For many years, new herbicide molecules have not been tested on sugar beet. Clomazone and pyraflufen-ethyl, for example, are few of the active ingredients tested for preventive purposes to control difficult species, such as *Abutilon theophrasti* and *Polygonum aviculare*. Aspects to understand when new active ingredients are tested should include possible phytotoxic effects on the crop; drift to sensitive, non-target vegetation; persistent herbicide residue in sprayers causing injury to other crops; and persistence in soil (for residual herbicides).

Possible damage from residues of herbicides applied in previous crops (carry-over) is a major consideration in planning crop rotation systems. Given the sensitivity of the beet particularly during the very early growth stages, attention should be paid especially when using new herbicides, in particular the sulfonylureas. Carry-over also occurs when the same active ingredient is used repeatedly on crops in monoculture and there is not enough time for the compound to decay completely before the next application. Thus, progressively increasing amount of the compound remains in the soil to a level that could be injurious to the next crop. The residual amount after any given crop cycle is also expected to increase under drought conditions, which minimize degradation.

Proper cleanup of the sprayer and spray tank is important. Cleaning the sprayer system with just water is not enough because some compounds can dissolve the remnant residue in the tank, resulting in significant injury to a sensitive crop that happens to be sprayed with the contaminated sprayer. This could happen with certain sulfonylureas contaminating a spray mixture for sugar beet and could result in significant yield loss. Modern washing or cleaning materials are more expensive but are necessary to avoid this problem. Immediate rinsing of the spray tank is generally recommended. Recent sprayer models are equipped with an auxiliary tank 'circuit washing' to avoid these problems.

Several negative situations (drought, pests, etc.) can force farmers to substitute the beets with another crop. Typically, sorghum or maize crops seeded directly without disturbing the soil, are not affected by the residual activity of herbicides used in sugar beet (i.e., metamitron, lenacil, etc.). Other important crops are sensitive to various herbicides for example, soybean (ethofumesate), sunflower (chloridazon and lenacil), and tomato (to most pre-emergence herbicides, especially with rain around application time). Wheat seeded into minimum-tilled soil following sugar beet that has been treated with medium-high doses of propyzamide or ethofumesate is sometimes injured when there is excessive rain or when planted in clay soils where the herbicides could persist more.

Repeated exposure of a weed population to the same herbicide could quickly generate serious resistance problems. However, the probability of occurrence of resistant populations in sugar beet production is low because farmers apply complex mixtures of post-emergence herbicides with different mechanisms of action. The herbicides are also used in VLD in various split applications or full doses according to the crop and weed status, supplemented with other techniques to ensure that the crop is weed-free. The higher risk of resistance evolution lies with sulfonylureas because farmers use them on the main crops in rotation. Weeds that most likely could evolve resistance include *Amaranthus* spp., *Abutilon theophrasti*, *Polygonum aviculare*, *P. lapathifolium*, *Fallopia convolvulus*, and *Chenopodium* spp. This is based on the resistance histories of these species in the global herbicide-resistant weed survey.

Innovative Methods for Weed Control

Innovative strategies for containment of weeds attract considerable interest as an alternative to chemical weed control as well as part of integrated weed management. Biological tools are increasingly subjected to international study and experimentation, as alternatives to chemical control and more sustainable for the environment. These management strategies can be linked to advanced mechanical control methods.

The use of robots able to carry out a selective control of weeds in the row is another approach that has lately attracted considerable interest. This would restrict the chemical containment of the full field by integrating hoeing and other control methods, e.g., chemical, flaming only in the areas with the greatest presence of weeds. The study and the development of natural substances with highly specific herbicide action, such as mycoherbicides, is another interesting challenge. At present products of this type are mainly used in forestry systems.

Non-chemical Weed Control

The correct choice of seeding time, restricting the sources of seed contamination and taking advantage of the opportunity to promote germination of the weeds in certain periods outside of cropping cycles, sets the beet into a condition of competitive advantage.

Smoothing harrow can achieve good results especially with broadleaf weeds (up to 70% control). Hoeing is less effective on grasses as it is on broadleaf weeds. Grasses need to be hoed when small; otherwise, the plants are firmly rooted and tillered, and cannot be uprooted by hoeing. On the other hand, the small, uprooted grasses are easily reburied as the hoeing progresses, and are practically just being inadvertently 'transplanted'. Many hoed grasses regrow. For certain weeds, such as *Chenopodium*, *Amaranthus*, *Datura* and *Xanthium* which quickly develop in height, mechanical shearing with the aid of cutter bars or with micro-jets of water at high pressure have proven effective.

For the control of most problematic species, including perennials, such as *Cirsium arvense*, cultivation of field edges could reduce their presence and possible infestations. The early removal of weeds within a period of 3–5 weeks from emergence reduces the negative effects of weed competition.

Interesting perspectives have been provided by the use of machines equipped with rotary brushes, able to remove weeds in the crop rows, which are difficult to remove. New systems have also been tested through the use of microwave or heat sources, such as steam and foam to damage the weeds, or prevent weed seed production, or reduce the seed viability. Steam, for example, had been used successfully to kill the seeds of *Capsella bursa-pastoris*.

Stale seedbed is an approach that may contain the emergence of weeds and weed competition.

Biological Control of Some Important Weed Species in Sugar Beet

Convolvulus arvensis

The chemical and mechanical methods are not 100% effective on *Convolvulus arvensis*, which has an extraordinary ability to regenerate itself through rhizomatous roots. The search for a means of specific control has resulted in the identification of a fungus that produces metabolites, which can kill this weed. The inoculation of *Phomopsis convolvulus* conidia on seedlings at different stages of development (up to 7-leaf stage) under controlled environment, had reduced weed growth over 90%. The inoculum kills 3-leaf seedlings 100%. On large plants the reduction of in shoot biomass is higher than 70%, while the reduction in root biomass is less than 50%.

Senecio vulgaris

In Australia and Asia, as well as in Europe, *Puccinia lagenophorae*, is found effective on *Senecio vulgaris*. This is a potential agent of biological control, which can reduce seed production by 90% if the infection starts early.

Cuscuta spp.

Cuscuta spp. can be controlled with the employment of insects, in particular *Smicronyx* spp., a *Curculionidae coleoptera* enable to cause galls, limiting both weed development and reproduction. Other species of Diptera, such as *Agromyzidae* can destroy the seeds of *Cuscuta europaea*, as some species of *Smicronyx* are able to live at the expense of *Cuscuta epithymum* flowers and *C. europaea*.

Rumex obtusifolius

Interesting experiments were conducted on the control of *Rumex obtusifolius* with competition exerted by *Lolium perenne*, which strongly reduces the development of the former. The joint action by specific natural agents, such as fungi (e.g., *Uromycesrumicis*) and insects (e.g., *Gastrophysa viridula*), may also constitute a valid strategy.

Taraxacum officinale

Among the many insects that feed on this plant, there are two beetles, *Glocianus punctiger* and *Olibrus bicolor*, that feed preferentially on the inflorescences. Although period between flowering and seed production is short, the rapid cycle of development of these two insects would allow sufficient feeding on the flowers, which would prevent seed production.

Alliaria petiolata

This weed can be controlled by a weevil, *Ceutorhynchus scrobicollis*. During the autumn-spring period the weevil lays eggs at the base of the leaf collar of this weed, and the larvae that hatch dig tunnels into the basal roots. Depending on the severity of tunneling, the seedlings could die. Oftentimes, larval feeding reduces weed growth.

Polygonum aviculare

Gastroidea polygons is able to control the growth primarily of *Polygonum aviculare* and secondarily of *Fallopia convolvulus*. Both larvae and adults of these voracious insects feed selectively on these *Polygonaceae* species. The slow but progressive action of *Gastroidea polygons* hinders the emergence of other weeds, by virtue of the initial ground cover offered by the prostrate habit of *Polygonum aviculare* and *Fallopia convolvulus* in the initial period when the crop is still not sufficiently competitive.

REFERENCES

Amista, F. 2002. Experimental assessment of the elements for the design of a microwave prototype for weed control. 5th EWRS Workshop on Physical and Cultural Weed Control. Pisa.

Angoujard, G., N. Le Godec, P. Blanchet and L. Lefevre. 2001. Techniques alternatives au désherbagechimiqueen zone urbaine. Dix-huitièmeconférence du COLUMA. Toulouse.

Baerveldt, S. and J. Ascard. 2002. Effect of cutting height on weed regrowth. 5th EWRS Workshop on Physical and Cultural Weed Control. Pisa.

Balsari, P., G. Airoldi and A. Ferrero. 2002. Mechanical and physical weed control in maize. 5th EWRS Workshop on Physical and Cultural Weed Control. Pisa.

Bàrberi, P., A.C. Moonen, M. Raffaelli, A. Peruzzi, P. Belloni and M. Mainardi. 2002. Soil steaming with an innovative machine – effects on actual weed flora. 5th EWRS Workshop on Physical and Cultural Weed Control. Pisa.

Benvenuti, S., C. Falorni, G. Simonelli and M. Macchia. 2001. Weed seed bank evaluation and dynamics in *Matricaria chamomilla* grown with "organic" agricultural systems. Dix-huitièmeconférence du COLUMA. Toulouse.

Boari, A., A. Abouzeid Mohamed, Z.M. Chiara, V. Maurizio and E. Antonio. 2002. Pathogens and phytotoxins in biocontrol of the parasitic weed *Orobanche ramosa*. EWRS Workshop – Biological Control of Weeds. Reading, England.

Campagna, G. and G. Rapparini. 2003. Glifosate and glufosinate ammonium seed bed. *In:* Proceedings 7th EWRS Symp. Weed control in sustainable agriculture in the Mediterranean area – Adana. (Attisu CD).

Campagna, G. and G. Rapparini. 2008. Erbeinfestantidellecoltureagrarie – Riconoscimento, biologia e lotta. EdizioniL' Informatore Agrario.

Catizone, P. and G. Zanin. 2001. Malerbologia. Patron Editore.

Cloutier, D.C. and L. Leblanc Maryse. 2002. Effect of the combination of the stale seedbed technique with cultivation on weed control in maize. 5th EWRS Workshop on Physical and Cultural Weed Control. Pisa.

Cohen, B.A., Z. Amsellem and J. Gressel. 2002. Engineered *Fusarium* against *Orobanche*. EWRS Workshop – Biological Control of Weeds. Reading, England.

Dal, Re L. and A. Innocenti. 2002. Experiences related to the use of the weeding harrow and of the roll-star cultivator in Emilia-Romagna for weed control on hard and common wheat, sunflower and soyabean in organic agriculture. 5th EWRS Workshop on Physical and Cultural Weed Control. Pisa.

Dale, T.M., K.A. Renner and A.N. Kravchenko. 2006. Effect of herbicides on weed control and sugar beet (*Beta vulgaris*) yield and quality. Weed Technology 20: 150–156.

Davis, A.S. and M. Liebman. 2002. Response of giant foxtail (*Setaria faberi* Herrm.) demographic parameters in maize (*Zea mays* L.) to varied tillage and soil amendment practices: empirical and modeling studies. 5th EWRS Workshop on Physical and Cultural Weed Control. Pisa.

Delabays, N. and G. Mermillod. 2001. Mise enévidence, aux champs, des propriétésallélopathiques de l'*Artemisiaannua*L. Dix-huitièmeconférence du COLUMA. Toulouse.

Dor, E., Y. Kashman and J. Hershenhorn. 2002. Allelopathic compound from *Inula viscosa*. EWRS Workshop – Biological Control of Weeds. Reading, England.

El-Sayed, W., F. Walker and K. Hurle. 2001. The potential of *Phomopsis convolvulus* Ormeno for the control of field bindweed (*Convolvulus arvensis*). BCPC Conference Proceedings. Brighton.

Eveno, M.E. and A. Chabanne. 2001. Les effets allélopathiques de l'avoine (*Avena sativa*) sur différentes mauvaises herbes et plantes cultivées.Dix-huitièmeconférence du COLUMA. Toulouse.

Fenni, M., J. Maillet and A.H. Shakir. 2001. La viabilité des semences de *Bromus rigidus* Roth. et *Bromusrubens* L.. Dix-huitièmeconférence du COLUMA. Toulouse.

Fogelberg, F. and A. Blom. 2002. Water-jet cutting for weed control. 5th EWRS Workshop on Physical and Cultural Weed Control. Pisa.

Gange, A.C. 2002. Biological control of *Poa annua* in sports turf. EWRS Workshop – Biological Control of Weeds. Reading, England.

Gerber, E. and L. Hinz Hariet. 2002. Effect of herbivore density, timing of attack and plant size on the invasive weed *Alliaria petiolata* (Cruciferae). EWRS Workshop – Biological Control of Weeds. Reading, England.

Grace, B. and H. Muller-Scharer. 2002. Biological control of *Senecio vulgaris* in carrots – applying the theory. EWRS Workshop – Biological Control of Weeds. Reading, England.

Gressel, J. 2002. A proposed system for 'Bio-barcoding' mycoherbicides. EWRS Workshop – Biological Control of Weeds. Reading, England.

Hansson, D. 2002. Hot water for weed control on urban hard surface areas. 5th EWRS Workshop on Physical and Cultural Weed Control. Pisa.

Hartmann, K.M. and A. Mollwo. 2002. The action spectrum for maximal photosensitivity of germination and significance for lightless tillage. 5th EWRS Workshop on Physical and Cultural Weed Control. Pisa.

Heisel, T., C. Andreasen and S. Christensen. 2002. Yield effect of distance between plants and cutting of weeds. 5th EWRS Workshop on Physical and Cultural Weed Control. Pisa.

Honek, A. and Z. Martinkova. 2002. Development of two pre-dispersal predators of *Taraxacum officinale* seed. EWRS Workshop – Biological Control of Weeds. Reading, England.

Jensen, R.K., D. Archer and F. Forcella. 2002. A degree-day model of *Cirsium arvense* predicting shoot emergence from root buds. 5th EWRS Workshop on Physical and Cultural Weed Control. Pisa.

Juroszek, P., M. Berg, P. Lukashyk and U. Kopke. 2002. Thermal control of *Viciahirsuta* and *Viciatetrasperma* in winter cereals. 5th EWRS Workshop on Physical and Cultural Weed Control. Pisa.

Keary, I. and H. Paul. 2002. Prospects for the biological control of *Rumex obtusifolius* in competition with *Lolium perenne*: evidence from pot trials. EWRS Workshop – Biological Control of Weeds. Reading, England.

Leblanc, M.L. and C. Cloutier Daniel. 2002. Optimization of cultivation timing by using a weed emergence model. 5th EWRS Workshop on Physical and Cultural Weed Control. Pisa.

Mesbah, A. 1993. Interference of broadleaf and grassy weeds in sugar beets. Ph.D. Dissertation. University of Wyoming.

Meyer, J. 2002. Semi-automatic machine guidance system. 5th EWRS Workshop on Physical and Cultural Weed Control. Pisa.

Miller, S.D., K.J. Fornstrom, T. Neider, P. Renner, A. Mesbah, P. Koetz, W. York, J. Lauer and J. Krall. 1993. Progress report weed control, agronomic crops — sugar beet weed interaction. University Wyoming, Ag. Expt. 148 p.

Moonen, A.C. and P. Bàrberi. 2002. A system-oriented approach to the study of weed suppression by cover crops and their residues. 5th EWRS Workshop on Physical and Cultural Weed Control. Pisa.

Moonen, A.C., P. Bàrberi, M. Raffaelli, M. Mainardi, A. Peruzzi and M. Mazzoncini. 2002. Soil steaming with an innovative machine – effects on the weed seedbank. 5th EWRS Workshop on Physical and Cultural Weed Control. Pisa.

Morishita, D.W., M.J. Wille and S.L. Young. 2000. Weed thresholds and weed emergence patterns in sugar beet. Snake River Sugar Beet Conference.

Mouret, J.C., P. Marnotte, R. Hammand, G. Lannes and S. Roux. 2001. Effets du sarclagemécanique sur le peuplement végétal en riziculture biologique marguaise (France). Dix-huitième conférence du COLUMA. Toulouse.

Muller-Scharer, H. 2002. The genetic population structure of *Senecio vulgarisand Puccinia lagenophorae*: implications for biocontrol. EWRS Workshop – Biological control of weeds. Reading, England.

Muller-Stover, D. and J. Sauerborn. 2002. Biological control of *Orobanche cumana* in sunflower with a granular formulation of *Fusarium oxysporum* f. sp. *orthoceras*. EWRS Workshop – Biological Control of Weeds. Reading, England.

Naseema, A., R. Praveen, S. Balakrishnan and C.K. Peethambaran. 2001. Management of water hyacinth (*Eichornia crassipes*) with fungal pathogens. BCPC Conference Proceedings. Brighton.

Radics, L., I. Gal and P. Pusztai. 2002. Different combinations of weed management methods in organic carrot. 5th EWRS Workshop on Physical and Cultural Weed Control. Pisa.

Radics, L. and E. Szekelyne Bobnar. 2002. Comparison of different mulching methods for weed control in organic green and tomato. 5th EWRS Workshop on Physical and Cultural Weed Control. Pisa.

Radics, L. and E. Székelyné Bognar. 2001. Comparison of different mulching methods for weed control in organic green bean and tomato. Dix-huitième conférence du COLUMA. Toulouse.

Raffaelli, M., P. Bàrberi, A. Peruzzi and M. Ginanni. 2002. Options for mechanical weed control in grain maize – effects on weeds. 5th EWRS Workshop on Physical and Cultural Weed Control. Pisa.

Rasmussen, I.A., N. Holst, L. Petersen and K. Rasmussen. 2002. Computer model for simulating the long-term dynamics of annual weeds under different cultivation practices. 5th EWRS Workshop on Physical and Cultural Weed Control. Pisa.

Robson, M.C., D. Robinson, A.M. Litterick, C. Watson and M.X. Leitch. 2001. Investigations into allelopathic interactions of white lupin (*Lupinus albus*). BCPC Conference Proceedings. Brighton.

Rodriguez, A. 2001. Controle des adventices annueles en grandes cultures biologiques en region Midi-Pyrénées.Dix-huitième conférence du COLUMA. Toulouse.

Seier, M.K. and C.H. Evans. 2002. Indigenous fungal pathogens – a potential additional tool for the management of *Rhododendron ponticum* L. in the UK. EWRS Workshop – Biological Control of Weeds. Reading, England.

Shaw, D. 2002. Classical biological control of weeds in the UK: the challenges. EWRS Workshop – Biological Control of Weeds. Reading, England.

Singh, M. and S.D. Sharma. 2001. Bioecological factors affecting germination of weed seeds. BCPC Conference Proceedings. Brighton.

Toth, P. and C. Ludovit. 2002. Are there important natural enemies of parasitic weeds in Slovakia? EWRS Workshop – Biological Control of Weeds. Reading, England.

Turner, R.J., M. Lennartsson, M. Hesketh, A.C. Grundy and D. Whitehouse. 2001. Weed control in organically grown carrots. BCPC Conference Proceedings. Brighton.

Van der Weide, R.Y., P.O. Bleeker and L.A.P. Lotz. 2002. Simple innovations to improve the effect of the false seedbed technique. 5th EWRS Workshop on Physical and Cultural Weed Control. Pisa.

Vanhala, P., T. Lotjonen and J. Salonen. 2002. Effect of crop competition and cultural practices on the growth of *Sonchus arvensis*. 5th EWRS Workshop on Physical and Cultural Weed Control. Pisa.

Vidotto, F., A. Ferrero, R. Busi and A. Saglia. 2002. Weed growth and control as influenced by soyabean row spacing and soil tillage for seed bed preparation. 5th EWRS Workshop on Physical and Cultural Weed Control. Pisa.

Wilson, R.G., S.D. Miller and S.J. Nissen. 2001. Weed control. Sugar Beet Production Guide. pp. 117–119.

Zarina, L. 2002. Weediness in 40-year period without herbicide. 5th EWRS Workshop on Physical and Cultural Weed Control. Pisa.

Zwanenburg, B. 2002. Control of *Orobanche* and *Striga* using synthetic germination stimulants. EWRS Workshop – Biological Control of Weeds. Reading, England.

Recommended strategies for integrated weed control in sugar beet

Pre-sowing		Pre-emergence: precautionary treatments (residual) + further devitalizing		
Autumn-winter	Late winter	Active ingredients	Dose (L or kg/ha)	Note
Soil preparation	Eventual devitalisation	1) Metamitron (50%, Goltix 50 WG) (70% altri)	4-5 3-4	Maximum selectivity target: broadleaf weeds
(stale seedbed technique)	(glyphosate)	2) Metamitron (70%) +	3-4 +	Attention to the selectivity in loose soil (increased action)
Eventual devitalisation with products based on glyphosate on overdeveloped weeds		Lenacil (80%) or Metamitron+Lenacil (60%+5%)	0.2 3-4	
		3) Metamitron (70%) + S-metolachlor (960 g/L)	3-4 + 0.2-0.5	Attention to the selectivity in loose soil Increased action against summer grasses and other broadleaf weeds
		4) Metamitron (70%) + Cloridazon (413 g/L) or Metamitron+Cloridazon (40%+25%)	2-3 + 2-3 4-5	Good selectivity (increased action against Cruciferae and Fallopia convolvulus)

(*Contd.*)

Post-emergence: damping treatments (leaf + residual)					
Calendar and main treatments					Main targets
Sugar beet vegetative stage					
cot - 2 leaf	2 leaf	4 leaf	6 leaf	8 leaf	
VLD+M+/-PY 0.7+0.5+/-0.5	8-10 gg → VLD+M+V	10-12 gg → 1+0.7+0.1	VLD+M+/- B 1.2+1+/-0.5-1		Polygonaceae, Chenopodiaceae, SOLNI
	VLD+M+V 0.8+0.5+0.1	10-12 gg → VLD+M+V 1+0.7+0.1	VLD+M+V+/- B 1+0.7+0.1+/-0.5-1		Polygonaceae, FALCO, Cruciferae, SOLNI
o	VLD+M+V+S 0.8+0.5+0.1+30 g	12-15 → VLD+M+V+S	VLD+M+/-B+S 1+0.5+/-1+40 g		Polygonaceae, Cruciferae, AMIMA, Amaranthaceae
o	VLD+M+PY 0.8+0.5+0.5	10-12 gg → VLD+M+/-B+S 1+0.5+/-1+40 g	VLD+M+/-B+S 1+0.5+/-1+40 g		Polygonaceae, FALCO, Cruciferae, Chenopodiaceae
o	VLD+M+V 0.8+0.5+0.1	10-12 gg → VLD+M+/-B+S 1+0.5+/-1+40 g	VLD+M+/-B+S 1+0.5+/-1+40 g		FALCO, Cruciferae, Polygonaceae

(Contd.)

Additional control measures. Miscible with caution						Controlled weeds
Kerb (0.7-0.35)	10-12 gg	Kerb (1.2-0.6)				Cuscuta, SOLNI
or	Kerb (1-0.5)	15-20 gg	Kerb (1.5-0.75)			and others
L (0.06-0.08)	15-20 gg	L (0.06-0.12)				CIRAR and others
						AMIMA, DAUCA, ecc.
		or	L (0.16-0.2)			Regrowth of alfalfa
						Rebirths of sunflower, rapeseed, coriander
Specific graminicides						Perennial and annual Graminacea

31

Sweetpotato: Important Weeds and Sustainable Weed Management

David W. Monks[*1], **Katie M. Jennings**[1], **Stephen L. Meyers**[2], **Tara P. Smith**[3] **and Nicholas E. Korres**[4]

[1] North Carolina State University, 2721 Founders Dr., Raleigh, 27607, North Carolina, USA
[2] Mississippi State University, 8320 Hwy 15 S, Pontotoc, 38863, Mississippi, USA
[3] Louisiana State University Agricultural Center, 130 Sweet Potato Road, Chase, 71324, Louisiana, USA
[4] University of Arkansas, 1366 W. Altheimer Dr., Fayetteville, 72704, Arkansas, USA

Introduction

Sweetpotato (*Ipomoea batatas* [L.] Lam.) ranks sixth behind rice, wheat, potato, maize and cassava as most important food crop globally (CIP International Potato Center 2017) and its production is growing in many regions of the world including the U.S. and parts of Africa. Storage roots of sweetpotato are an important source of beta-carotene, vitamins B, C and E, and contain moderate amounts of iron and zinc. Sweetpotato shoot tips and leaves are consumed throughout the world both raw and cooked (Bouwkamp 1985). In addition to human consumption, sweetpotato is utilized for animal feed and processed products, such as starch, flour, syrup and dye used to add pigment to food and fiber. China is the world's largest producer and consumer of sweetpotato, where it is used for animal feed, human food, and processing products including ethanol (USDA-FAS 2017). In the U.S., sweetpotato is grown almost solely as human food, and storage roots are marketed nationally and internationally into fresh and processing venues. In addition, sweetpotato is increasingly used as an ingredient in high-end domesticated dog feed and treats. Sweetpotato in the U.S. is grown in rotation with other agronomic crops (cotton, corn, soybean, peanut, tobacco) and vegetable crops. Sweetpotato farmers depend on this crop for sustainability of their farming operation.

Sweetpotato production begins with shoot tip cuttings (transplants) 25.4 to 30.5 cm tall, which in propagation beds are cut above the soil surface to prevent disease transfer to the production field. Transplants, often containing no roots, are transplanted into fields previously plowed to form 30.5 to 40.6 cm tall ridged rows that are 0.9 to 1.1 m wide (Figure 31.1).

In-row plant spacing is commonly 20 cm to 38 cm (Schultheis et al. 1999, Stoddard et al. 2013). Sweetpotato yield per hectare is determined by the number of sweetpotato plants per hectare, the number of storage roots per plant and the size of each storage root at harvest (Meyers et al. 2014). These yield parameters are directly related to sweetpotato storage root initiation (14 to 30 days after transplanting), and the sizing up stage of storage roots that occurs during the last third of the growing season. Unfavorable environmental conditions (extreme soil moisture or extreme air/soil temperature or weed competition) during storage root initiation and the

*Corresponding author: David_monks@ncsu.edu

Figure 31.1. Sweetpotato trasplanting on ridged rows in Louisiana using mechanical transplanter. (Source: T.P. Smith)

sizing up stage of storage roots directly impact the number of storage roots produced per plant, the size of each storage root at harvest, and the resulting yield and quality (Gajanayake et al. 2013, 2014, 2015, Meyers et al. 2014, Pardales and Yamauchi 2003, Villordon et al. 2012). Likewise, unfavorable environmental conditions result from weeds being present during these two critical stages of sweetpotato growth (Harrison and Jackson 2011, Jose et al. 1994, Meyers et al. 2010, Meyers and Shankle 2015a, Nedunzhiyan et al. 1998, Seem et al. 2003, Workayehu et al. 2011).

Weed interference in sweetpotato can reduce yield and quality, usually exhibited as a reduction in number one (premium) grade yield (Meyers et al. 2010, Meyers and Shankle 2015a, Seem et al. 2003, Smith et al. 2017). To prevent yield and quality reductions, weeds must be controlled at critical times during the growing season. Fields are plowed to form ridged planting rows prior to sweetpotato transplanting. The period between plowing and transplanting can vary depending on weather related conditions and specific grower practices. As time between plowing and transplanting increases, weed emergence increases. If weeds are near emergence or emerged, they have a competitive advantage over the crop. It is critical to plant sweetpotato transplants in fields that are weed-free. If weeds are near emergence or present, fields should be re-plowed to ensure all emerged weeds are controlled. Transplants should be planted at least 12.7 cm deep (Figure 31.1) when soils have warmed to at least 18.3°C (65°F) to aid crop establishment, competitiveness with weeds and cultivation (Meyers et al. 2014). To prevent sweetpotato storage root yield and quality reductions, the critical weed-free period for a mixed weed population is 2 to 6 weeks after transplanting (Harrison and Jackson 2011, Jose et al. 1994, Nedunzhiyan et al. 1998, Seem et al. 2003, Smith et al. 2017). Weeds emerging six weeks after transplanting do not usually affect sweetpotato yield and quality (Seem et al. 2003). By six weeks after transplanting, the canopy of the decumbent sweetpotato vines is closed, preventing light from reaching weed seeds and seedlings between rows. Sweetpotato gains a competitive advantage over weeds when planted late (higher temperature) in the recommended period for sweetpotato planting (KM Jennings, NC State University, unpublished data, Seem et al. 2003). Thus, planting sweetpotato in fields with the highest weed density may be desirable when temperature is optimum for sweetpotato growth.

Weeds in Sweetpotato (Major Weeds, Weed Life Cycle, Weed Strengths and Weaknesses, Impact on Yield Loss)

Annual and perennial weeds can affect sweetpotato fields worldwide. Because sweetpotato vines grow along the soil and canopy height is often less than 0.5 m tall, weeds that grow up through and above the sweetpotato canopy are considered the most competitive. Amaranthaceae (pigweeds) species like Palmer amaranth (*Amaranthus palmeri* L.) (Figure 31.2) and Cyperaceae (sedges) like yellow nutsedge (*Cyperus esculentus* L.) (Figure 31.3) can compete and reduce sweetpotato yield and quality drastically without any other weed species present (Meyers et al. 2010, Meyers and Shankle 2015a).

Figure 31.2. Palmer amaranth in North Carolina surpassed sweetpotato canopy by 2 to 3 weeks after transplanting sweetpotato. (Source: K.M. Jennings)

Other weeds, even low growing annual weeds, can compete effectively with sweetpotato when they are present in mixed populations (Seem et al. 2003). Vining weeds like morningglory species (*Ipomoea* spp.) are low growing like sweetpotato, but have the ability to climb to the top of the sweetpotato canopy where they compete with sweetpotato for light resources (J.R. Schultheis, Sweetpotato Specialist, NC State University, personal communication).

Within the U.S. some weed species are common to fields in all states, while other weed species in sweetpotato are specific to fields in an individual state (Table 31.1). The following information is focused on the most troublesome weeds in U.S. sweetpotato.

Amaranthaceae – The Pigweeds

A number of annual species of the Amaranthaceae are troublesome weeds in sweetpotato including Palmer amaranth, redroot pigweed (*Amaranthus retroflexus* L.), spiny amaranth (*A. spinosus* L.), and smooth pigweed (*A. hybridus* L.) (Table 31.1). *Amaranthus* spp. grow rapidly, are capable of growth under varying environmental conditions including hot dry conditions, easily exceed the sweetpotato canopy height within two to three weeks after transplanting (Smith et al. 2017), and are prolific seed producers (Figures 31.2 and 31.4).

Figure 31.3. Yellow nutsedge growing in sweetpotato in Mississippi. (Source: S.L. Meyers)

Figure 31.4. Palmer amaranth (1.5 to 1.8 m tall) exceeding height of sweetpotato at harvest. (Source: K.M. Jennings)

The height of *Amaranthus* spp. from tallest to shortest is Palmer amaranth (>2 m) >redroot pigweed = smooth pigweed > spiny amaranth (1.5 m or less) (Sellers et al. 2003). Seed production for these species ranges from 100,000 to 1 million seeds per plant with seed production from most to least as Palmer amaranth > redroot pigweed = smooth pigweed > spiny amaranth (Sellers et al. 2003, Sosnoskie et al. 2014, A.C. York, NC State University unpublished data). Members of this family have demonstrated resistance to many herbicide sites of action including EPSP synthase inhibitors, acetolactate synthase inhibitors (imidazolinones and sulfonylureas),

HPPD inhibitors, microtubule inhibitors, protoporphyrinogen oxidase inhibitors, and triazines (Heap 2017, Ward et al. 2013). Uncontrolled smooth pigweed in mixed populations of weed species resulted in a 40% to 50% reduction in marketable sweetpotato yield (Seem et al. 2003). Uncontrolled Palmer amaranth at 0.5 to 6.5 plants per m of row resulted in 36 to 81% marketable sweetpotato yield reduction (Meyers et al. 2010). Palmer amaranth can grow more than 5 cm per day and produce viable seeds 30 days after germinating (K.M. Jennings unpublished data, Legleiter and Johnson 2013). In North Carolina when Palmer amaranth is allowed to compete with the sweetpotato crop from transplanting until three weeks after transplanting total yield loss was 10%. If Palmer amaranth is not controlled until six weeks after sweetpotato transplanting, total yield loss was 70% and loss of the premium number one grade was 90% (Smith et al. 2017). Prior to transplanting sweetpotato, established *Amaranthus* spp. are controlled by plowing or non-selective herbicides. Pre-emergence herbicides are applied pre-plant or after planting if available followed by shallow cultivation during the season (usually 2 to 3 shallow cultivations), hand-removal and/or hoeing when small. In late season, large broadleaf weeds are sometimes controlled by wicking with herbicide (glyphosate). Hand-removal of spiny amaranth is especially cumbersome due to sharp 5 mm to 10 mm long spines that appear at nodes along the stem. Escaped *Amaranthus* spp. should be rouged and removed from fields to prevent seed dispersal. Some *Amaranthus* spp. have the ability to produce roots and re-establish if left in the field. These plants should be cut at the soil surface to limit seed production (L.M. Sosnoskie, University of Georgia, unpublished data). In contrast when *Amaranthus* spp. are cut above the soil surface they are capable of regrowing rapidly and producing seeds.

Cyperaceae – The Sedges

Yellow nutsedge, purple nutsedge (*Cyperus rotundus* L.), annual sedge (*Cyperus compressus* L.) and rice flatsedge (*Cyperus iria* L.) are among the most troublesome weeds in sweetpotato (Table 31.1) (Meyers and Shankle 2015b, Webster 2014). Yellow and purple nutsedge are perennial weeds that spread by underground rhizomes and reproduce vegetatively by tubers (Meyers and Shankle 2015b). A single yellow nutsedge plant from a sprouted tuber can form a compact, densely populated patch (210 shoots/0.18 m²) after six months of growth (Webster 2005), and 3,000 shoots and 19,000 tubers in one year (Ransom et al. 2009). Meyers and Shankle (2015a) reported that yellow nutsedge shoot density in sweetpotato increased by 2.3 to 7.6 times in a single four month growing season, further documenting the ability of yellow nutsedge to expand rapidly by vegetative reproduction. Meyers and Shankle (2015a) reported marketable sweetpotato yield losses of 18% to 80% as yellow nutsedge density increased from 5 to 90 shoots per m². In studies by Webster (2005), purple nutsedge formed larger patches with less shoot density but had the ability to spread and distribute further than yellow nutsedge. Because of its high potential for vegetative growth and reproduction, management strategies for nutsedge should be focused on prevention (control prior to transplanting), early detection and containment, early treatment (cultivation) and integration of control strategies (optimum transplanting date, optimum sweetpotato growth, herbicides, cultivation, and crop rotation) to reduce weed competitive ability and spread (Meyers and Shankle 2015a, 2015b, Ransom et al. 2009). Equipment sanitation to prevent spread of tubers from field to field is a critical preventive method (Meyers and Shankle 2015b). Annual sedge and rice flatsedge are annual weeds and are not generally as competitive as the perennial nutsedges. However, in some sweetpotato producing areas, annual sedge and rice flatsedge occur in high density and/or with other weeds resulting in a highly competitive weed population with sweetpotato (Table 31.1).

Poaceae – The Grasses

Grasses infest sweetpotato worldwide. Annual grasses are of greater concern than perennial grasses as the latter do not frequently persist as a result of tillage operations utilized for sweetpotato production. In the U.S., some of the common annual grasses that infest sweetpotato

Table 31.1. Most troublesome weeds in sweetpotato in U.S. states

Alabama[1]	Arkansas[2]	California[3]	Georgia[4]	Louisiana[5]	Mississippi[6]	North Carolina[7]
Nutsedge, yellow	Amaranth, Palmer	Pigweed, redroot and other *Amaranthus* spp.	Nutsedge, yellow	Nutsedge, yellow	Nutsedge, yellow	Amaranth, Palmer
Morningglory, annual (Ipomoea spp.)	Morningglory, pitted	Lambsquarters, common	Nutsedge, purple	Nutsedge, purple	Amaranth, spiny	Nutsedge, yellow
Amaranth, Palmer	Morningglory, entireleaf	Nightshade (hairy; black)	Morningglory (*Ipomoea* spp.; *Jacquemontia* sp.)	Pigweed, smooth	Pigweed, redroot	Purslane (common; pink)
Pigweed, redroot	Barnyardgrass	Purslane, common	Sicklepod	Pigweed, spiny	Cocklebur, common	Pusley, Florida
Pigweed, smooth	Sedge, annual	Puncturevine	Amaranth, Palmer	Amaranth, Palmer	Sida, (prickly; arrowleaf)	Morningglory, annual (entireleaf; ivyleaf)
Pigweed, prostrate	Goosegrass	Knotweed, prostrate	Pusley, Florida	Groundcherry	Groundcherry, smooth	Pigweed, smooth
Amaranth, spiny	Crabgrass, large	Thistle, Russian	Purslane, pink	Smellmelon	Copperleaf, hophornbeam	Lambsquarters, common
Cocklebur, common	Sida, prickly	Barnyardgrass	Goosegrass	Alligatorweed	Morningglory, annual (*Ipomoea* spp.)	Smartweed, Pennsylvania
Sicklepod	Smartweed, Pennsylvania	Nutsedge, yellow	Beggarweed, Florida	Grass, annual (large crabgrass; barnyardgrass; Morningglory, annual (*Ipomoea* spp.) Smartweed, Pennsylvania bermudagrass, sicklepod	Flatsedge, rice	Radish, wild
Grass, annual	Nutsedge, yellow					

[1]Information provided by A. Caylor, Auburn University.
[2]Information provided by J. Norsworthy, University of Arkansas.
[3]Information provided by S. Stoddard, University of California.
[4]Information provided by S. Culpepper, University of Georgia.
[5]Information provided by D. Miller, Louisiana State University.
[6]Information provided by S. Meyers, Mississippi State University.
[7]Information provided by K. Jennings, North Carolina State University.

are large crabgrass (*Digitaria sanguinalis* [L.] Scop.), goosegrass (*Eleusine indica* [L.] Gaertn.), barnyardgrass (*Echinochloa crus-galli* [L.] Beauv.), broadleaf signalgrass (*Urochloa platyphylla* [Nash] R.D. Webster), and fall panicum (*Panicum dichotomiflorum* Michx.). The specific annual grass species that emerge in sweetpotato are dependent on weed field history, environment (temperature, rainfall) for that specific year, and transplanting date. These weeds typically emerge in the period of time between field preparation and the early growing season during crop establishment. Although the annual grasses have the ability to grow taller than the sweetpotato canopy, to do so they must establish early in the season before extensive sweetpotato vining occurs. Grasses not controlled early in the season must be controlled prior to the last third of the crop growing season when sweetpotato storage roots are sizing up, otherwise competition of annual grasses with sweetpotato will result in reduced crop vigor and yield. Members of *Poaceae* have demonstrated resistance to ALS inhibitors, ACCase inhibitors, cellulose inhibitors, EPSP synthase inhibitors, glutamine synthase inhibitors, lipid inhibitors, long chain fatty acid inhibitors, microtubule inhibitors, PSI Electron diverter, synthetic auxins and Photosystem II inhibitors (Heap 2017).

Convolvulaceae – The Annual Morningglories

Members of the morninglory family that are troublesome in sweetpotato include the annuals entireleaf morningglory (*Ipomoea hederacea* [L.] Jacq.), ivyleaf morningglory (*I. hederacea* var. *integriuscula* Gray), pitted morningglory (*I. lacunosa* L.), tall morningglory (*I. purpurea* [L.] Roth), and smallflower morningglory (*Jacquemontia tamnifolia* [L.] Griseb). These weeds are low growing vines with a taproot on which many fibrous roots are developed up to 3 m long (Bryson and DeFelice 2010, DeFelice 2001). They intertwine with sweetpotato and can compete with the crop for light, nutrients, and water. Morningglory species often not only grow throughout the sweetpotato canopy, they preferentially grow towards and on upright weeds where they can capture more sunlight resulting in enhanced growth and seed production (Price and Wilcut 2007). Seed production of morningglories on per plant basis ranges between 2,000 to 5,800 for ivyleaf morningglory (Gomes et al. 1978, Price and Wilcut 2007), 14,600 for entireleaf morningglory (Gomes et al. 1978), 15,400 for pitted morningglory (Gomes et al. 1978), and 26,000 for tall morningglory (DeFelice 2001). Morningglory seeds have a hard seed coat, which contributes to their ability to remain viable in the soil for many years (DeFelice 2001, Elmore et al. 1990). Removal or control is difficult as seeds of these weeds are large allowing them to emerge from soil quickly and making them difficult to control with pre-emergence herbicides. As they become established and intertwined with sweetpotato, they become increasingly difficult to cultivate or remove by hand.

Portulacaceae – The Purslanes

Common purslane (*Portulaca oleracea* L.) and pink purslane (*P. pilosa* L.) are predominately prostrate-growing annual plants with a thick taproot with many fibrous roots, stems up to 50 cm long and whole plant canopy diameter of 60 cm (Bryson and DeFelice 2001). However, height is dependent on the amount of light the plant receives. Common purslane grown under competitive conditions where light is limited will grow taller than in a non-competitive environment. The purslanes have succulent stems and leaves which contribute to their drought resistance. They can reproduce from seeds (greater than 100,000 per plant) or from fragmented stems with a node (Holm et al. 1977, Proctor 2013, Proctor et al. 2011). Purslane seeds have reportedly remained viable for as long as 40 years (Darlington 1941). Under favorable conditions (i.e., moist, 30–40°C), seeds begin to germinate in 12 hours and emergence is complete in 24 hours. Rapid growth occurs at about two weeks and seed production increases rapidly at 4 to 6 weeks after emergence (Haar and Fennimore 2003, Holm et al. 1977). To prevent seed production, cultivation or hand hoeing of common purslane should occur before three weeks post emergence or 125 growing degree days (Haar and Fennimore 2003, University of California 1990).

Pennsylvania Smartweed (*Polygonum pensylvanicum* L.)

Pennsylvania smartweed is an annual weed and can grow 1.2 to 1.8 m tall (Bryson and DeFelice 2010, Lorenzi and Jeffery 1987). It can exceed the height of sweetpotato canopy (0.5 m) by 40 to 60 days after emergence (Askew and Wilcut 2002). It prefers wet, poorly drained soils and those high in nitrogen or phosphorus but does not tolerate dry weather. Once established, this weed is adapted to a wide range of environments and its extensive root system in the upper and lower soil horizons allows for maximum nutrient uptake (Parrish and Bazzaz 1976). This weed reproduces by seed with as many as 20,000 seeds per plant (Anonymous 2017a). Its seeds are contained in buoyant achenes which move towards wetter regions of fields (Pickett and Bazzaz 1978). Seeds are viable for as many as 26 years in the soil (Anonymous 2017a). Members of this family have demonstrated resistance to ALS inhibitors and photosystem II inhibitors (Heap 2017). Pennsylvania smartweed should be controlled before exceeding 6.35 mm in height (Anonymous 2017a). Tillage at night can reduce smartweed emergence by 30% to 50% (Anonymous 2017a).

Florida Pusley (*Richardia scabra* L.)

Florida pusley is a very persistent summer annual weed with either a prostrate or upright growth habit and stems 15 to 50 cm long. This weed reproduces by seed with as many as 2,297 seeds per plant (Brewer and Oliver 2007). The germination and growing season for this weed is consistent with that of sweetpotato (Biswas et al. 1975).

Common Lambsquarters (*Chenopodium album* L.)

A member of the *Chenopodiaceae*, this summer annual begins to emerge prior to sweetpotato transplanting and emergence continues through the early transplanting season when temperatures tend to be cooler. This emergence period may be extended when above average rainfall or irrigation occurs during the season. Common lambsquarters has a short-branched taproot and is capable of growing 2 m tall (Bryson and DeFelice 2010). It is a prolific seed producer (as many as 70,000 seeds per plant) and can reach reproductive maturity six weeks after emergence (Curran et al. 2007). Members of this family have demonstrated resistance to ALS inhibitors, photosystem II inhibitors, and synthetic auxins (Heap 2017). Transplanting sweetpotato fields with a known history of this weed in the last half of the recommended transplanting season will aid in its management.

Malvaceae – The Mallows

Prickly sida (*Sida spinosa* L.) is an erect, branched annual weed that can grow 1 m tall (Bryson and DeFelice 2010). A related species, arrowleaf sida (*S. rhombifolia* L.), has a branching tap root with fibrous roots. Prickly sida seeds can germinate under limited soil moisture (Hoveland and Buchanan 1973) and germination is encouraged by high temperature and cycles of wet-dry soil moisture, and shifting of colder temperatures to a higher temperature regime (Baskin and Baskin 1984). Resistance of prickly sida to the ALS inhibitor herbicides has been reported (Heap 2017). Prickly sida can be suppressed by highly competitive crops (Green 2016).

Solanaceae – The Black Nightshade Complex and Groundcherries

Several *Solanum* species make up the black nightshade complex. They include black nightshade (*Solanum nigrum* L.), American black nightshade (*S. americanum* Mill.), and Eastern black nightshade (*S. ptycanthum* Dun.). Black nightshade and American black nightshade are low growing spreading, sometimes upright annual to short-lived perennials growing up to 1–1.5 m in height. The Eastern black nightshade is an annual weed (Thomson and Witt 1987). Nightshade species have a fibrous root system with a shallow taproot. Plants in the black nightshade complex can produce 30,000 (American black nightshade) to 100,000 (Eastern black nightshade) to 178,000 (black nightshade) seeds per plant (Holm 1977, Keeley and Thullen 1983, Werner et al. 1998). Tillage at night in the dark reduces Eastern black nightshade emergence by 50% to 75% (Anonymous 2017b). Members of this family have demonstrated resistance to ALS inhibitors, photosystem II inhibitors, and PSI electron diverter.

Clammy (*Physalis heterophylla*) and smooth (*P. subglabra*ta) groundcherries grow 0.3 to 0.9 m tall and have an upright, branching growth habit. Groundcherry plants have deep penetrating and creeping roots. The plants can produce as many as 30,000 seeds and can reproduce by root fragments. Seeds are contained in berries, each covered by a bladder-like husk that looks similar to a paper lantern. Root fragments that are moved to the soil surface do not usually survive (Anonymous 2017c). A related annual species, cutleaf groundcherry (*P. angulata* L.), can produce up to 4,200 seeds per plant and exceed 0.5 m by 30 to 40 days after emergence (Thomson and Witt 1987, Travlos 2012).

Hophornbeam Copperleaf (*Acalypha ostryifolia* Riddell)

A member of the *Euphorbiaceae* or spurge family, this summer annual can emerge over a wide range of environmental conditions and throughout the sweetpotato growing season. Hophornbeam copperleaf is an erect plant that can reach 1 m or more, reproduces by seed, and can produce as many as 12,500 seeds per plant (Harak et al. 1998, Steckel 2006). It can exceed the height of the sweetpotato canopy by 6 to 8 weeks after emergence (Harak et al. 1998). Management programs must continue all season because of this weed's prolonged germination period.

Fabaceae – The Bean Family

Sicklepod (*Senna obtusifolia* [L.] Irwin and Barnaby) is an upright, summer annual that grows to over 1.5 m and begins flowering at 50 to 84 days after germination followed by seed production (Retzinger 1984). It is a troublesome weed in crops like sweetpotato because of its high seed production (over 16,000 seeds per plant), ability to germinate under varying environmental conditions, and hard seed coat that contributes to seed dormancy (Nice et al. 2001). By 30 to 40 days after emergence, its height can exceed the canopy of sweetpotato (Smith 1992). Seeds of sicklepod can remain viable in the soil for a long period of time (Bararpour and Oliver 1998).

Florida Beggarweed (Desmodium tortuosum [Sweet] DC.)

Florida Beggarweed (*Desmodium tortuosum* [Sweet] DC.) germinates throughout the sweetpotato growing season (late May through September). It is an erect plant that can exceed the height of the sweetpotato canopy by 45 days after emergence and grow to 3.5 m tall (Cardina and Brecke 1991, Webster and Cardina 2004). Prior to exceeding the height of the crop, this weed remains unbranched and as it exceeds the crop canopy it begins rapid growth and produces branches with leaves that effectively shade the lower-growing sweetpotato by intercepting 30% of ambient sunlight. Flowering begins as early as 67 days after emergence and viable seeds can be produced within 10 days after flowering. Seeds can germinate and emerge throughout the growing season when soil is disturbed and sufficient soil moisture is present. Seeds have a hard seed coat that contributes to its persistence in the soil where it can remain viable for five years or more (Cardina and Brecke 1991, Webster and Cardina 2004). Late emerging Florida beggarweed is not very competitive with low-growing crops like sweetpotato (Cardina and Brecke 1991). Rotation with tall growing crops like corn, repeated shallow tillage/cultivation to deplete the soil seed bank, and hand removal are effective methods for controlling this weed.

Common Cocklebur (*Xanthium strumarium* L.)

A member of the *Asteraceae*, common cocklebur has a tall growth habit (up to 1.5 m), large canopy, and an extensive root system that provides an advantage in nutrient and water uptake (Anonymous 2017d, Crooks et al. 2005). Common cocklebur has potential to exceed the sweetpotato canopy within 30 to 40 days of emergence (Crooks et al. 2005). It is a summer annual weed that reproduces by flowering in late summer to early fall, and produces burs (fruit with external prickles) that contain seeds. Each bur contains two seeds that can survive up to three years in soil. One seed has the capacity to germinate the following year, whereas germination of the second seed is delayed for at least two years. Members of this family have demonstrated resistance to ALS inhibitors, and nucleic acid inhibitors (Heap 2017). Common cocklebur can thrive in varying soils and moisture conditions but will not tolerate shade.

Wild Radish (*Raphanus raphanistrum* L.)

A member of the *Brassicaceae*, this winter annual or summer annual can emerge any time during the year when moisture is sufficient for germination. Wild radish forms a basal rosette with upright, leafy inflorescences up to 1.5 m tall. It has an extensive fibrous root system capable of spreading 80 cm in all directions and a taproot capable of growing 1 m deep. As a result, it grows quickly and survives under varying environmental conditions. It reproduces by seeds (over 700 seeds per plant) that can remain viable in soil for more than 20 years (Anonymous 2015, Eslami et al. 2006, Peltzer and Douglas 2017). Wild radish seeds germinate in soils between 5°C and 35°C. Members of this family have demonstrated resistance to ALS inhibitors, carotenoid biosynthesis inhibitors, ESP synthase inhibitors, and synthetic auxins (Heap 2017).

Methods for Weed Control

Weed management in sweetpotato relies on the integration of multiple control methods including mechanical, chemical, and cultural approaches. Implementation of each method will vary by location, as access to labor, equipment, and registered herbicides vary from country-to-country, state-to-state and farm-to-farm. Below is an overview of control methods utilized by sweetpotato producers in the Southeast United States.

Mechanical Control

In the spring, land intended for sweetpotato production is cultivated with a disc or similar implement to remove winter annual weeds and loosen soil prior to ridged row formation. After transplanting, between-row cultivation is utilized. In the U.S., a tractor-mounted implement consisting of rolling cultivators followed by soil sweeps is used to remove emerged weeds between rows and deposit soil from between-row spaces on top of the small weeds growing within the row. Between-row cultivation can be used until vine closure, typically three to four weeks after transplanting. Timely cultivation should target small emerging weeds. Between-row cultivation often fails to completely remove weeds in the planted row.

Weeds that escape cultivation are often removed by hand. In the U.S. most commercial sweetpotato fields receive two hand-weeding operations per growing season. Hand-removal of weeds is labor intensive and, depending upon location, can be expensive. This practice is most appropriate for upright, annual weeds. Hand-removal of grasses and perennial sedges is difficult.

Mowing talls weeds in sweetpotato is practiced by some growers. However, this method provides limited control. Meyers et al. (2017) reported that Palmer amaranth mowed above the sweetpotato canopy produces extensive branching below the cut and results in a dense canopy of weedy vegetation just above the sweetpotato canopy. Mowing may be used as a salvage effort to reduce weed seed set, but is not a stand-alone control method.

Cultural Weed Control

Multiple cultural weed control methods have been investigated by researchers, but few are implemented on a commercial scale. Allelopathic sweetpotato cultivars, those that exude chemical substances capable of hindering weed growth and development, have been documented but often lack commercially desirable traits to be grown on a large scale (Harrison and Peterson 1986, 1991). La Bonte et al. (1999) reported that 11 sweetpotato clones with architecturally different canopies demonstrated a 2-fold and 3-fold difference in percentage canopy closure 42 days after transplanting and at harvest, respectively. The authors further identified five weed-tolerant clones, three of which were bunch or medium-internode type and concluded that while their research was conducted in Louisiana, more research was required to understand the interactions of canopy architecture and other production practices (row spacing, crop fertility) as well as investigations into competitive ability and tolerance to individual weeds or classes of weeds.

Crop rotation is used to manage pests including weeds. Producers can rotate to crops that are more competitive with weeds and those that have registered and efficacious herbicides for the target weed species. In Mississippi, producers rotate sweetpotato fields infested with yellow nutsedge to corn (*Zea mays* L.) or soybean (*Glycine max* [L.] Merr.). Both have more upright growth and compete for light resources more than the decumbent growing sweetpotato. Additionally, both corn and soybean have registered and efficacious herbicides that will control yellow nutsedge. A minimum of two growing seasons is often required to significantly reduce weed pressure. Of special consideration when utilizing crop rotation is plant-back interval, the required amount of time between an herbicide application and when sweetpotato may be safely transplanted. Plant-back restrictions for sweetpotato often err on the side of caution due to a lack of research to support reduced intervals.

Weed propagules (rhizomes and seeds) move with sweetpotato farming implements. Researchers recommend that implements, such as discs, mechanical transplanters, and harvesters (disc plows, chain diggers) have soil removed when being moved between fields. It is unclear what percentage of producers currently follow this recommendation.

Chemical Weed Control

Limited herbicides are registered for use in sweetpotato in the United States. Flumioxazin, a protoporphyrinogen oxidase inhibitor, is applied pre-transplanting to prepared production fields and provides pre-emergence control of numerous broadleaf weeds and suppression of grasses (Kelly et al. 2006). Clomazone, a carotenoid biosynthesis inhibitor, is broadcast-applied before or after transplanting for the control of grasses and select broadleaf weeds. For herbicides registered for application prior to transplanting, extreme care should be taken after herbicide application to limit soil movement by the mechanical transplanter, otherwise a narrow non-treated herbicide strip will result. *S*-metolachlor is a soil-applied chloroacetamide herbicide that inhibits the biosynthesis of fatty acids, lipids, proteins, isoprenoids, and flavonoids in susceptible plant species. It is broadcast-applied after transplanting for pre-emergence control of small-seeded broadleaf weeds, grasses, and yellow nutsedge (Meyers and Shankle 2017). Fluazifop, sethoxydim, and clethodim are applied post-emergence to selectively control grass weed species. These graminicides are applied with either a crop-based oil or non-ionic surfactant to improve efficacy. Glyphosate and carfentrazone-ethyl are used pre-plant to 'burn-down' existing weedy vegetation. Both can be applied with a shielded (hooded) application between rows for post-emergence weed control. Napropamide and DCPA are both registered for use in sweetpotato in the U.S., but are rarely utilized as they are highly dependent upon rainfall- or soil-incorporation and efficacy is often variable.

Herbicide wick and wiper applicators are used by a limited number of sweetpotato producers. Many models are available, but all have a similar function. A concentrated herbicide solution is placed into a reservoir and is soaked into canvas sleeves or absorbent ropes. The absorbent material is placed in contact with weeds that grow above the sweetpotato canopy. The systemic herbicide is translocated throughout the contacted weed. This application method has some limitations. As it is selective based on a weed-crop height differential, the target weeds must exceed the sweetpotato canopy, and in doing so compete with the crop before and between applications (Meyers et al. 2017).

Table 31.2. Suggestions for species specific weed control methods in sweetpotato

Weed species	Vigorous sweetpotato varieties[1]	After field bedding but prior to crop transplanting[2] — Tillage	After field bedding but prior to crop transplanting[2] — Post-emergence nonselective herbicide	At planting[3] / Pre-plant herbicide — Flumioxazin	At planting[3] / Pre-plant herbicide — Clomazone	At planting[3] / Post-plant herbicide — S-metolachlor	At planting[3] / Post-plant herbicide — Clomazone	Cultivation — Sweep	Cultivation — Rolling	Hand removal — Early season	Hand removal — Late season
Beggarweed, Florida	X	X	X	X	X		X	X	X	X	X
Cocklebur, common	X	X	X		X[5]		X[5]	X	X	X	X
Copperleaf, hophornbeam	X	X	X	X[5]	X[5]		X[5]	X	X		X
Grass, annual	X	X[4]	X[4]	X[5]	X	X	X	X	X	X	
Groundcherry	X	X	X					X	X	X	X
Lambsquarters, common	X	X	X	X	X		X	X	X	X	X
Morningglory, annual	X	X	X	X	X			X	X	X	X
Nightshade	X	X	X	X	X[5]	X	X[5]	X	X	X	X
Pigweed	X	X	X	X	X[5]	X	X[5]	X	X	X	X
Purslane, common/pink	X	X[4]	X[4]	X	X	X[4]	X	X	X	X	
Pusley, Florida	X	X	X	X		X	X	X	X	X	
Nutsedge, annual	X	X[4]	X[4]					X	X	X	
Nutsedge, purple	X	X[4]	X[4]					X			

(Contd.)

Table 31.2. (*Contd.*)

| Weed species | Vigorous sweetpotato varieties[1] | After field bedding but prior to crop transplanting[2] | | At planting[3] | | | | Cultivation | | Hand removal | |
| | | Tillage | Post-emergence nonselective herbicide | Pre-plant herbicide | | Post-plant herbicide | | Sweep | Rolling | Early season | Late season |
				Flumioxazin	Clomazone	S-metolachlor	Clomazone				
Nutsedge, yellow	X	X[4]	X[4]			X	X	X			
Sicklepod	X	X	X[4]					X	X	X	X
Sida, prickly	X	X	X[4]	X	X		X	X	X	X	
Smartweed, Pennsylvania	X	X[4]	X[4]	X[5]	X[5]		X[5]	X	X	X	X

[1] Research suggests that vigorous sweetpotato varieties are most beneficial in suppressing weeds compared to sweetpotato varieties that are slow to establish and grow/develop.

[2] Weeds emerge after field bedding but prior to transplanting sweetpotato. Control of small emerged weeds with tillage or Post- (non-selective) herbicide.

[3] Pre-emergence herbicide applied pre-plant (flumioxazin marketed under the trade name Valor herbicide by Valent or clomazone marketed under the trade name Command herbicide by FMC Corporation) to sweetpotato, or post-plant pre-emergence herbicide (S-metolachlor marketed under the trade name Dual Magnum herbicide by Syngenta or clomazone marketed under the trade name Command herbicide by FMC Corporation) applied after sweetpotato planting. Note: Valor is registered to only apply prior to planting sweetpotato. Dual Magnum is registered to apply only after sweetpotato transplanting.

[4] May require multiple tillage events or multiple applications of nonselective post-emergence herbicide(s).

[5] Suppression only.

Concluding Remarks

Weed control in sweetpotato, one of the most important crops worldwide, requires an integrated approach, particularly as the development of herbicide resistant weeds is accelerating. Utilization of hand removal, cultivation, and rotation of herbicides with different modes of action is critical to prevent development of herbicide resistant weed populations in sweetpotato. Amaranths, sedges, various grasses and morningglories can cause remarkable yield reductions of up to 80%. Sweetpotato fields should start weed-free and remain weed-free through canopy closure, approximately six weeks after transplanting. Weeds that establish after canopy closure are not likely to reduce yield but, depending on the weed species, may need to be removed to prevent weed seed production.

REFERENCES

Anonymous. 2015. *Raphanus raphanistrum* (wild radish) datasheet. CABI. Available at: www.cabi.org/isc/datasheet/46795.

Anonymous. 2017a. Michigan's Worst Weeds: Pennsylvania smartweeds and ladysthumb. Michigan State University. Available at: http://www.msuweeds.com/worst-weeds/smartweeds/.

Anonymous. 2017b. Michigan's Worst Weeds: Eastern black nightshade (*Solanum ptycanthum* Dun.) Michigan State University. Available at: http://www.msuweeds.com/worst-weeds/eastern-black-nightshade/.

Anonymous. 2017c. Ohio perennial and biennial weed guide – groundcherries (*Physalis* spp.). The Ohio State University. Available at: http://www.oardc.ohio-state.edu/weedguide/single_weed.php?id=50.

Anonymous. 2017d. *Xanthium strumarium*, common cocklebur. Available at: http://articles.extension.org/pages/65827/xanthium-strumarium-common-cocklebur.

Askew, S.D. and J.W. Wilcut. 2002. Pennsylvania smartweed interference and achene production in cotton. Weed Sci. 50: 350–356.

Bararpour, M.T. and L.R. Oliver. 1998. Effect of tillage and interference on common cocklebur (*Xanthium strumari*um) and sicklepod (*Senna obtusifolia*) population, seed production and seedbank. Weed Sci. 46: 424–431.

Baskin, J.M. and C.C. Baskin. 1984. Environmental conditions required for germination of prickly sida (*Sida spinosa*). Weed Sci. 32: 786–791.

Biswas, P.K., P.D. Bell, J.L. Crayton and K.B. Paul. 1975. Germination behavior of Florida pusley seeds. I. Effects of storage, light, temperature and planting depth on germination. Weed Sci. 23: 400–403.

Bouwkamp, J.C. (ed.). 1985. Sweet Potato Products: A Natural Resource for the Tropics. CRC Press, Inc.: Boca Raton, FL, pp. 175–183.

Brewer, C.E. and L.R. Oliver. 2007. Reducing weed seed rain with late-season glyphosate applications. Weed Technol. 21: 753–758.

Bryson, C.T. and M.S. DeFelice. 2010. Weeds of the Midwestern United States & Central Canada. University of Georgia Press, Athens, GA, 427 pp.

Cardina, J. and B.J. Brecke. 1991. Florida beggarweed (*Desmodium tortuosum*) growth and development in peanuts (*Arachis hypogaea*). Weed Technol. 5: 147–153.

CIP International Potato Center. 2017. Sweetpotato facts and figures. Available at: cipotato.org/sweetpotato/facts-2/

Crooks, H.L., M.G. Burton, A.C. York and C. Brownie. 2005. Vegetative growth and competitiveness of common cocklebur resistant and susceptible to acetolactate synthase-inhibiting herbicides. Journal of Cotton Sci. 9: 229–237.

Curran, B., C. Sprague, J. Stachler and M. Loux. 2007. Biology and management of common lambsquarters. Purdue Extension Publication GWC-112.

Darlington, H.T. 1941 The sixty-year period for Dr. Beal's seed viability experiment. Amer. J. Bot. 28: 271–273.

DeFelice, M.S. 2001. Tall morningglory, *Ipomoea purpurea* (L.) Roth - flower or foe? Weed Technol. 15: 601–606.

Elmore, C.D., H.R. Hurst and D.F. Austin. 1990. Biology and control of morningglories (*Ipomoea* spp.). Rev. Weed Sci. 5: 83–224.

Eslami, S.V., G.S. Gill, B. Belloti and G. McDonald. 2006. Wild radish (*Raphanus raphanistrum* L.) interference in wheat. Weed Sci. 54: 749–756.

Gajanayake, B., K.R. Reddy and Shankle. 2015. Quantifying growth and development responses of sweetpotato to mid- and late-season temperature. Agron. J. 107: 1854–1862.

Gajanayake, B., K.R. Reddy, M.W. Shankle, R.A. Arancibia and A.O. Villordon. 2014. Quantifying storage root initiation, growth, and development responses of sweetpotato to early season temperature. Agron. J. 106: 1795–1804.

Gajanayake, B., K.R. Reddy, M.W. Shankle and R.A. Arancibia. 2013. Early-season soil moisture deficit reduces sweetpotato storage root initiation and development. Hort. Sci. 48: 1457–1462.

Gomes, L.F., J.M. Chandler and C.E. Vaughan. 1978. Aspects of germination, emergence, and seed production of three *Ipomoea* taxa. Weed Sci. 26: 245–248.

Green, J.D. 2016. Prickly Sida (Teaweed) Management in Soybean, USB Take Action: Herbicide-resistance management series. Available at: http://takeactiononweeds.com/.

Haar, M.J. and S.A. Fennimore. 2003. Evaluation of integrated practices for common purslane (*Portulaca oleracea*) management in lettuce (*Latuca sativa*). Weed Technol. 17: 229–233.

Harak, M.T., Z. Gao, D.E. Peterson and L.D. Maddux. 1998. Hophornbeam copperleaf (*Acalypha ostryifolia*) biology and control. Weed Technol. 12: 515–521.

Harrison, H.F. and D.M. Jackson. 2011. Response of two sweet potato cultivars to weed interference. Crop Prot. 30: 1291–1296.

Harrison, H.F. Jr. and J.K. Peterson. 1986. Allelopathic effects of sweet potatoes (*Ipomoea batatas*) on yellow nutsedge (*Cyperus esculentus*) and alfalfa (*Medicago sativa*). Weed Sci. 34: 623–627.

Harrison, H.F. Jr and J.K. Peterson. 1991. Evidence that sweet potato (*Ipomoea batatas*) is allelopathic to yellow nutsedge (*Cyperus esculentus*). Weed Sci. 39: 308–312.

Heap, I. 2017. The international survey of herbicide resistant weeds. Available at: http://www.weedscience.org/default.aspx.

Holm, L.G., D.L. Pluchnett, J.V. Pancho and J.P. Herberger. 1977. The world's worst weeds: distribution and biology. The University Press of Hawaii, Honolulu. 609 p.

Hoveland, C.S. and G.A. Buchanan. 1973. Weed seed germination under simulated drought. Weed Sci. 21: 322–324.

Jose, J., A. Marcano, D. Omar, G. Colmenarez and G. Florencio Paredes. 1994. Periodo critico de competencia por maleza en el cultivo de batata (*Ipomoea batatas* [L.] Lam) cultivar UCV-7. Bioagro 6: 86–94.

Keeley, P.E. and R.J. Thullen. 1983. Influence of planting date on the growth of black nightshade (*Solanum nigrum* L.). Weed Sci. 31: 180–184.

Kelly, S.T., M.W. Shankle and D.K. Miller. 2006. Efficacy and tolerance of flumioxazin on sweetpotato (*Ipomoea batatas*). Weed Technol. 20: 334–339.

LaBonte, D.R., H.F. Harrison and C.E. Motsenbocker. 1999. Sweetpotato clone tolerance to weed interference. Hort. Sci. 34: 229–232.

Legleiter, T. and B. Johnson. 2013. Palmer amaranth biology, identification, and management. Purdue University Extension Publication WS-51.

Lorenzi, H.J. and L.S. Jeffery. 1987. Weeds of the United States and their control. Van Nostrand Reinhold Co., NY. 355 p.

Meyers, S.L., R.A. Arancibia, M.W. Shankle, J. Main and R.K. Reddy. 2014. Sweet potato storage root initiation. Mississippi State University Extension Publication 2809: 1–4.

Meyers, S.L., K.M. Jennings, J.R. Schultheis and D.W. Monks. 2010. Interference of Palmer amaranth (*Amaranthus palmeri*) in sweetpotato. Weed Sci. 58: 199–203.

Meyers, S.L., K.M. Jennings, J.R. Schultheis and D.W. Monks. 2017. Evaluation of wick-applied glyphosate for Palmer amaranth (*Amaranthus palmeri*) control in sweetpotato. Weed Technol. 30: 765–772.

Meyers, S.L. and M.W. Shankle. 2015a. Interference of yellow nutsedge (*Cyperus esculentus*) in 'Beauregard' sweet potato (*Ipomoea batatas*). Weed Technol. 29: 854–860.

Meyers, S.L. and M.W. Shankle. 2015b. Nutsedge management in Mississippi sweetpotatoes. Mississippi State University Extension Publication 2909.

Meyers, S.L. and M.W. Shankle. 2017. An evaluation of pre-emergence metam-potassium and *S*-metolachlor for yellow nutsedge (*Cyperus esculentus*) management in sweetpotato. Weed Technol. 31: 436–440.

Nedunzhiyan, M., S.P. Varma and R.C. Ray. 1998. Estimation of critical period of crop-weed competition in sweet potato (*Ipomoea batatas* L.). Adv. Hort. Sci. 12: 101–104.

Nice, G.R.W., N.W. Buehring and D.R. Shaw. 2001. Sicklepod (*Senna obtusifolia*) response to shading soybean (*Glycine max*) row spacing, and population in three management systems. Weed Technol. 15: 155–162.

Pardales, J.R. and A. Yamauchi. 2003. Regulation of root development in sweetpotato and cassava by soil moisture during their establishment period. Plant and Soil 255: 201–208.

Parrish, J.A.D. and F.A. Bazzaz. 1976. Underground niche separation in successional plants. Ecol. 57: 1281–1288.

Peltzer, S. and A. Douglas. 2017. Wild radish. Government of Western Australia Department of Agriculture and Food. Available at: https://www.agric.wa.gov.au/grains-research-development/wild-radish?page=0%2C0.

Pickett, S.T.A. and F.A. Bazzaz. 1978. Organization of an assemblage of early successional species of a soil moisture gradient. Ecol. 59: 1248–1255.

Price, A.J. and J.W. Wilcut. 2007. Response of ivyleaf morningglory (*Ipomoea hederacea*) to neighboring plants and objects. Weed Technol. 21: 922–927.

Proctor, C. 2013. Biology and control of common purslane (*Portulaca oleracea* L.). University of Nebraska – Lincoln, PhD thesis, 67 pp.

Proctor, C.A., R.E. Gaussoin and Z.J. Reicher. 2011. Vegetative reproduction potential of common purslane (*Portulaca oleracea*). Weed Technol. 25: 694–697.

Ransom, C.V., C.A. Rice and C.C. Shock. 2009. Yellow nutsedge (*Cyperus esculentus*) growth and reproduction in response to nitrogen and irrigation. Weed Sci. 57: 21–25.

Retzinger, E.J. 1984. Growth and development of sicklepod (*Cassia obtusifolia*) selections. Weed Sci. 32: 608–611.

Schultheis, J.R., S.A. Walters, D.E. Adams and E.A. Estes. 1999. In-row plant spacing and date of harvest of 'Beauregard' sweetpotato affect yield and return on investment. Hort. Sci. 34: 1229–1233.

Seem, J.E., N.G. Creamer and D.W. Monks. 2003. Critical weed-free period for 'Beauregard' sweetpotato (*Ipomoea batatas*). Weed Sci. 17: 686–695.

Sellers, B.A., R.J. Smeda, W.G. Johnson, J.A. Kendig and M.R. Ellersieck. 2003. Comparative growth of six *Amaranthus* species in Missouri. Weed Sci. 51: 329–333.

Smith, S., K.M. Jennings and D.W. Monks. 2017. Timing of Palmer amaranth control on sweetpotato yield and quality. Southern Region American Society for Horticultural Science meeting vegetable section abstracts, 14 p.

Smith, J.E. 1992. Shoot growth and form of *Senna obtusifolia* in response to soybean and intraspecific competition. LSU PhD dissertation, 101 p.

Sosnoskie, L.M., T.M. Webster, A.S. Culpepper and J. Kichler. 2014. The biology and ecology of Palmer amaranth: implications for control. UGA Extension Circular 1000. Available at: http://extension.uga.edu/publications/files/pdf/C%201000_2.PDF.

Steckel, L. 2006. Hophornbeam copperleaf. University of Tennessee Extension publication. Available at: https://extension.tennessee.edu/publications/Documents/W120.pdf.

Stoddard, C.S., R.M. Davis and M. Cantwell. 2013. Sweetpotato production in California. University of California Agriculture and Natural Resources Publication 7237.

Thomson, C.E. and W.W. Witt. 1987. Germination of cutleaf groundcherry (*Physalis angulata*), smooth groundcherry (*Physalis virginiana*) and eastern black nightshade (*Solanum ptycanthum*). Weed Sci. 35: 58–62.

Travlos, I.S. 2012. Invasiveness of cut-leaf ground-cherry (*Physalis angulata* L.) populations and impact of soil water and nutrient availability. Chilean J. Agr. Res. 72: 358–363.

US Department of Agriculture-Foreign Agricultural Service (USDA-FAS). 2017. Global Agriculture Information Network Report CH16067: Peoples Republic of China annual biofuels - biofuels demand expands, supply uncertain. Available at: https://gain.fas.usda.gov/Recent%20

GAIN%20Publications/Biofuels%20Annual_Beijing_China%20-%20Peoples%20Republic%20 of_1-18-2017.pdf.

University of California. 1990. Degree-Day Utility User's Guide Version 2.0. University of California Integrated Pest Management 9. Available at: http://www.ipm.ucdavis.edu/IPMPROJECT/ software.html.

Villordon, A., D. LaBonte, J. Solis and N. Firon. 2012. Characterization of lateral root development at the onset of storage root initiation in 'Beauregard' sweetpotato adventitious roots. Hort Science 47: 961–968.

Ward, S.M., T.M. Webster and L.E. Steckel. 2013. Palmer amaranth (*Amaranthus palmeri*): a review. Weed Technol. 27: 12–27.

Webster, T.M. 2014. Weed survey – southern states. 292 pp. *In:* Proceedings of the 67th Southern Weed Science Society. Southern Weed Science Society, Birmingham, AL.

Webster, T.M. 2005. Patch expansion of purple nutsedge (*Cyperus rotundus*) and yellow nutsedge (*Cyperus esculentus*) with and without polyethylene mulch. Weed Sci. 53: 839–845.

Webster, T.M. and J. Cardina. 2004. A review of the biology and ecology of Florida beggarweed (*Desmodium tortuosum*). Weed Sci. 52: 185–200.

Werner, E.L., W.S. Curran and D.D. Lingenfelter. 1998. Management of Eastern black nightshade in agronomic crops: an integrated approach. Penn State Extension Agronomy Facts 58: 1–6.

Workayehu, T., W. Mazengia and L. Hidoto. 2011. Growth habit, plant density and weed control on weed and root yield of sweet potato (*Ipomoea batatas* L.) Areka, Southern Ethiopia. J. Hort. For. 3: 251–258.

CHAPTER
32

Sustainable Weed Control with Aromatic Plants and Essential Oils

Thomas K. Gitsopoulos[*1], Kalliopi Kadoglidou[1] and Christos A. Damalas[2]

[1] Institute of Plant Breeding and Genetic Resources, Hellenic Agricultural Organization-Demeter, GR-57001 Thermi, Thessaloniki, Greece
[2] Department of Agricultural Development, Democritus University of Thrace, GR-68200 Orestiada, Greece

Introduction

Weed management in crop production systems has been a major concern possibly since the inception of agriculture. In fact, unmanaged weeds can produce the highest potential loss (34%), compared to animal pests and pathogens (18% and 16%, respectively), as stated by Oerke (2006) for wheat, rice, maize, potatoes, soybeans, and cotton. Although the most common method of direct weed control is the use of herbicides, there are also non-chemical options of weed control available to farmers, such as physical and cultural control practices, which have the potential to allow conventional farmers reduce herbicide use and risk. Recently, hot topics, such as environmental protection, ecological stability, herbicide resistance management, and organic agriculture lead to seeking for non-chemical weed control options (Dayan et al. 2009, Bajwa et al. 2015). Natural products, such as allelochemicals, attract great attention as a potential tool for weed control in production systems (Duke et al. 2000, Weston 2005, Narwal 2010), although they occupy only a small share of the agricultural market (Singh et al. 2002).

Research is now focussed on the isolation, the identification, and the bioactivity control of natural products of plants and on finding an appropriate application protocol (Duke et al. 2000). Plants are an ideal source of allelochemicals production, i.e., biologically active substances with a great diversity of chemical structure, normally identified as secondary allelopathic metabolites. Plants like alfalfa (*Medicago sativa* L.), sorghum (*Sorghum* spp.), rice (*Oryza sativa* L.) and various species of the genus *Eucalyptus* are well known for their allelopathic properties (Anaya 1999, Gealy et al. 2013). Also, weeds, such as *Sicyos deppei* of the Curcubitaceae family exhibit allelopathy, since it has been reported that this weed can inhibit the germination and growth of maize (*Zea mays* L.), bean (*Phaseolus vulgaris* L.), squash (*Cucurbita peppo* L.), lettuce (*Lactuca sativa* L.), tomato (*Lycopersicon esculentum* L.), prince's feather amaranth (*Amaranthus hypochondriacus* L.), and barnyardgrass (*Echinochloa crus-galli* L.) (Hernández-Bautista et al. 1996, Cruz-Ortega et al. 1998). Aromatic plants could play an important role in the promotion of sustainable agriculture due to their ability to produce essential oils with allelopathic properties

*Corresponding author: gitsopoulos@yahoo.gr

that could be used for sustainable weed control, exploiting their herbicidal activity. The last decades, essential oils of aromatic plants are increasingly studied for their use as pesticides in agriculture (Isman 2000, Daferera et al. 2003, Dayan et al. 2009).

The objective of this chapter is to summarize basic concepts of the role of the aromatic plants and their essential oils on weed management and the possibility of using them as potential eco-friendly methods of weed control in modern agriculture. The chapter focuses on practices for weed control with use of aromatic plants in integrated weed management and on the herbicidal activity of their essential oils for possible future use as natural herbicides. Limitations of these perspectives are also discussed.

Aromatic Plants and Essential Oils (EOs)

Aromatic herbs, mainly shrubs of the Lamiaceae, Asteraceae, Apiaceae (Bernath et al. 2009) and other botanical families, are rich in volatile, odorous, hydrophobic, and highly concentrated compounds widely known as essential oils (Christaki et al. 2012). These compounds are by-products of the secondary metabolism, liquids at room temperature, and easily transformed to gaseous stage without undergoing decomposition and are present as droplets of fluid found in different plant parts, such as roots, stems, bark, leaves, flowers, and fruits; they provide distinctive odour or flavor to a plant and they function as attractors or repellents of insects, protectors from cold and heat, and generally as plant defenders (Koul et al. 2008). There have been identified up to date approximately 200,000 secondary metabolites (Tulp and Bohlin 2005). These complex mixtures are naturally organic compounds and mainly contain terpenes (monoterpenes and sesquiterpenes in the form of alcohols, aldehydes, phenols, ketones, acids, esters, ethers, peroxides), aromatic and aliphatic constituents all at low molecular weight with different physical and chemical properties (Bakkali et al. 2008, Koul et al. 2008, Blazquez 2014). The essential oils may also contain nitrogenous and sulphured constituents, methyl anthranilate, coumarins, etc., products of secondary metabolism (Bakkali et al. 2008, Koul et al. 2008). Terpenes are derived from combination of several isoprenes (5-carbon-base) (Bakkali et al. 2008). Some of the most common constituents of the essential oils are: pinene, p-cymene, geraniol, linalool, citrol, citronellol, neral, geranial, citronellal, menthone, carvone, pulegone, camphor, linalyl acetate, menthyl, 1,8-cineole, thymol, carvacrol, chavicol, eugenol, anethole, estragole, etc. (Bakkali et al. 2008). Each essential oil may contain about 20–60 components at different concentrations; however, there are two to three major components found in high concentration (20%–70%) and the others are found in trace amounts (Bakkali et al. 2008). The amount of essential oil in aromatic plants can be in the range of 0.01 to 10%, with the majority between 1% to 2% (Koul et al. 2008).

Basil, citronella, coriander, eucalyptus, lavandin, levander, mint, peppermint, lemongrass, rosemary, sage, thyme, fennel, anise, parlsey, cinnamon, oregano, summer savory, common ragweed, common balm, dill, etc., are some aromatic plants containing essential oils (Tworkoski 2002, Bakkali et al. 2008, Koul et al. 2008, Dhima et al. 2010). The essential oils produced by aromatic plants have shown to exhibit microbial (Bassolé and Juliani 2012), insecticidal (Isman 2000) and herbicidal properties (Tworkoski 2002, Dayan et al. 2009). The wide range of herbicidal activity expressed by the essential oils is related to the qualitative and quantitative composition of the natural volatile mixtures. In the literature, there are numerous reports highlighting the allelopathic-herbicidal efficacy of the essential oils; however, the majority refers to petri dish bioassays as the most commonly used method (Scognamiglio et al. 2013). There are plenty reports of the inhibitory effect of essential oils on seed germination and seedling growth, as well as, of their allelopathic properties (Vaughn and Spencer 1993, Kohli et al. 1998, Dudai et al. 1999, Onen et al. 2002, Singh et al. 2002, Azirak and Karaman 2008, Ramezani et al. 2008, Singh et al. 2005, Dhima et al. 2009, De Martino et al. 2010, Zhang et al. 2012, Vasilakoglou et al. 2013). The effect of some essential oil components on weed germination is presented in Table 32.1.

Table 32.1. Effect of oil components on germination of weed species

Oil components	Germination rate (%)						
	Alcea pallida	*Amaranthus retoflexus*	*Centaurea salsotitialis*	*Sinapis arvensis*	*Sonchus oleraceus*	*Raphanus raphanistrum*	*Rumex nepalensis*
Thymol	23.0	1.2	10	0	0.2	1.5	26.6
Carvacrol	33.7	0.2	0	0	0.2	4.0	0.1
Limonene	91.7	92.2	23.3	68.7	94.2	31.8	99.6
Carvone	78.3	23.8	0	0	4.2	1.3	0.1
Control[a]	100	100	100	100	100	88	100

[a] Water+n-hexane (Azirak and Karaman 2008)

Terpenes are the largest group of secondary products of plant metabolism and the main components of essential oils expressing inhibitory activity (Muller et al. 1964, Kohli et al. 1998, Batish et al. 2004). Monoterpenes are the simplest group of terpenes (Ahmad and Misra 1994, Dudai et al. 1999, Singh et al. 2005, De Martino et al. 2010). Concerning the herbicidal effect of each essential oil, it is very important to know not only the botanical classification of the aromatic plant, but also its geographical origin (Zaouali et al. 2005), the plant growth stage, the date and time of collection, together with the drying and extraction processes.

Mode of Action of EOs

According to Macías et al. (2007), the approximately 270 herbicides currently available in the market have only 17 modes of action, with almost half of them acting on three sites: photosystem II (PSII), acetolactate synthase (ALS), and protoporphyrinogen oxidase (PPO). A major advantage of the EOs and their compounds is that they may offer opportunities for the discovery of new modes of action for herbicide design and this has been acknowledged and highlighted several times in the last decades (Duke et al. 2000). In this section, an attempt will be made to give an overview of the state-of-the-art in herbicide target site studies on EOs and compounds mentioned in previous sections.

Inhibition of mitosis was reported with the use of eugenol, 1,8-cineole, camphor, citral, and cinmethalin (El-Deek and Hess 1986, Grayson et al. 1987, Kriegs et al. 2010, Vaid et al. 2010, Chaimovitsh et al. 2012). Microtubules disruption and cell membrane leakage was revealed with the use of limonene and citral (Chaimovitsh et al. 2010, Chaimovitsh et al. 2017). Inhibitory effect on photosynthesis and decreased chlorophyll content was revealed with the compounds of 1,4-cineole, 1,8-cineole, citronellol, eugenol, cintronellal, limonene, β-pinene, p-cymene, and linalool (Singh et al. 2002, Kordali et al. 2008, Chowhan et al. 2011, Vaid et al. 2011, Graña et al. 2012). Mitochondrial respiration was affected by β-pinene, limonene, and pulegone (Mucciarelli et al. 2001, Abrahim et al. 2003, Vaid et al. 2010), while cell respiration was reported after the use of β-pinene, p-cymene, 1,8-cineole, citronellal, citronellol, eugenol, linalool, and limonene (Sing et al. 2002, Kordali et al. 2008, Chowhan et al. 2011, Vaid et al. 2011, Graña et al. 2012). Oxidative stress via the increase of malondialdehyde levels was revealed with the use of α-pinene, menthol, and thymol (Scrivanti et al. 2003, Zunino and Zygadlo 2004). Prevention of stomata closure, enhanced respiration, and swelling of protoplasts were detected after the use of menthol (Kriegs et al. 2010). DNA synthesis was inhibited by 1,8-cineole (Koitabashi et al. 1997), while inhibition of cell proliferation and DNA synthesis of the root apical meristem of *Brassica campestris* seedlings has been reported (Nishida et al. 2005). Also, the β-triketone leptospermone inhibits the enzyme p-hydroxyphenylpyruvate dioxygenase (HPPD) (Dayan et al. 2007).

It has been reported that there is a connection between the inhibitory effect and the water solubility of monoterpenes, with ketones being more soluble and active than alcohols and alcohols more soluble and active than hydrocarbons (Weidenhamer et al. 1993). However, in a recent study with 17 monoterpenes for microtubule disrupting activity in *Arabidopsis* plants, the hydrocarbon limonene was revealed to be the most active monoterpene; in particular, the aldehydes citral and citronellal exhibited lower inhibitory activity compared to limonene (Chaimovitch et al. 2017). Concerning efficacy, it has been reported different potency of EO compounds belonging to different chemical classes (De Martino et al. 2010). These compounds may interact with each other and the result of this interaction is unknown, particularly taking into account that the variability of the constituents is seasonal, intraspecific, originating from different population of the same species, or moreover, between individuals of the same population (Vokou et al. 2003). Additionally, it has been revealed that some monoterpenes have isomeric-specific activity, with the (+) enantiomers more potent than the (–) counterparts (Chaimovitch et al. 2017). Moreover, some essential oil constituents can express different target site according to their concentration. For example, cell division is the target site of lower concentration of the monoterpene citral (mixture of neral and geranial), however, at higher concentration the cell elongation is targeted (Chaimovitsh et al. 2012). Additionally, reports have shown that oxygenated monoterpenes were more active and phytotoxic compared to monoterpene hydrocarbons (Vaughn and Spencer 1993, Dudai et al. 2004). As mentioned by An et al. (1993), an allelochemical may express stimulating and inhibiting attributes, however, not necessarily having the same mode of action (Vokou et al. 2003).

From the above it is evident that there are oil constituents with different and more than one mode of action. This is very important for tackling with herbicide resistance and control resistant weeds. The mechanism of action of EOs for their herbicidal action is not yet fully known (Tworkoski 2002). EOs and their components are often characterized by multi-site action in plants without high specificity achieved in the case of synthetic herbicides.

Competitive Ability of Aromatic Plants

Competition is a means of weed suppression and it occurs directly between crops and weeds for nutrients, water, light and space, or indirectly through allelopathic chemicals production and exudation by the allelopathic crop with detrimental effects on weed species sharing the same habitat. Crops with high competitive ability and/or allelopahic properties can compete with certain weeds. Although the allelopathic effect of the aromatic plants has been investigated in laboratory experiments, there is limited information under field conditions and particularly in competition between aromatic plants and weeds. Certain aromatic plants with significant competitive ability could be cultivated in sustainable crop production systems with less reliance on synthetic herbicides. This was shown in competition field experiments where aromatic plants, such as lacy phacelia (*Phacelia tanacetifolia*), anise (*Pimpinella anisum*), coriander (*Coriandrum sativum*), and sweet fennel (*Foeniculum vulgare*) competed weeds, such as common purslane (*Portulaca oleracea*), black nightshade (*Solanum nigrum*), common lambsquarters (*Chenopodium album*) and barnyardgrass (*Echinochloa crus-galli*) (Dhima et al. 2010). More specifically, lacy phacelia caused significant reductions in plant number (53–75%) and fresh weight (63–82%) of the four abovementioned weed species at five weeks after planting (WAP); reductions in plant number of common purslane and black nightshade were caused by anise (39%) and coriander (62%), respectively, and by sweet fennel (42%) to common lambsquarters, compared to the corresponding number in crop-free treatment (Table 32.2). Lacy phacelia caused the greatest reduction (97–100%) in plant or stems number and fresh weight of common purslane, black nightshade, common lambsquarters, and barnyardgrass at 8 WAP (Table 32.2).

At the end of the experiment, lacy phacelia and sweet fennel produced the greatest biomass, with the former presenting the highest ability to withstand competition (AWC) and ability to compete (AC) (99.0% and 99.6%, respectively), followed by anise, sweet fennel, and coriander (Table 32.3).

However, this competitiveness may not be correlated with high essential oil phytotoxicity already detected in allelopathic bioassays (Dhima et al. 2010). In particular, sweet basil (*Ocimum basilicum*) although presented essential oil phytotoxicity in bioassays, it showed reduced competitiveness in the field; lacy phacelia and anise on the contrary exhibited high competitive ability in the field experiments, but reduced oil phytotoxicity in the laboratory, whereas sweet fennel exhibited both increased competitiveness and essential oil phytotoxicity, observed in the field and in the bioassays, respectively (Dhima et al. 2010).

Table 32.2. Plant or stem density and fresh weight of four weed species grown with or without competition of aromatic plants

Weed species	Aromatic plants	Plants[a]	Fresh weight[b]	Plants[a]	Stems[c]	Fresh weight[b]
		5 weeks after planting		8 weeks after planting		
	Control[d]	38	53.9	101	-	295.5
	Pacelia tanacetifolia	18	19.7	1	-	1.8
Portulaca oleracea	*Pimpinella anisum*	23	25.5	56	-	131.7
	Coriandrum sativum	23	39.4	52	-	117.9
	Foeniculum vulgare	38	53.8	88	-	229.1
	Control[d]	52	98.8	69	-	296.2
	Pacelia tanacetifolia	13	17.5	0	-	0.3
Solanum nigrum	*Pimpinella anisum*	20	36.8	27	-	118.1
	Coriandrum sativum	20	37.6	21	-	95.6
	Foeniculum vulgare	28	48.8	36	-	149.7
	Control[d]	120	259.0	192	-	1,075.5
	Pacelia tanacetifolia	55	65.8	3	-	6.0
Chenopodium album	*Pimpinella anisum*	105	178.1	140	-	528.7
	Coriandrum sativum	109	215.3	120	-	578.4
	Foeniculum vulgare	69	124.9	85	-	414.0
	Control[d]	64	88.5	-	176	368.0
	Pacelia tanacetifolia	29	21.8	-	6	7.9
Echinochloa crus-galli	*Pimpinella anisum*	58	78.8	-	65	142.9
	Coriandrum sativum	71	89.4	-	59	127.2
	Foeniculum vulgare	69	86.7	-	62	123.7

[a]Number m^{-2}; [b]g m^{-2}; [c]Number m^{-2} (only for *E. crus-galli*); [d]No aromatic plant planted (based on Dhima et al. 2010).

Table 32.3. Competition indices of annual aromatic plants

Aromatic plants	AWC[a]	AC[b]
	----------------------------------%----------------------------------	
Pimpinella anisum	91.7	72.9
Foeniculum vulgare	88.7	82.1
Coriandrum sativum	84.4	69.1
Phacelia tanacetifolia	99.0	99.6

[a]AWC, ability to withstand competition; [b]AC, ability to compete (based on Dhima et al. 2010)

This great competitiveness and the high biomass production of certain aromatic plants are not always correlated with high essential oil phytotoxicity, but could be attributed to early germination, vigorous and faster growth, earlier canopy closure, greater weed growth suppression as well as seed germination and root length inhibition properties (Serrato-Valenti et al. 1998, Zimdahl 2007, Dhima et al. 2009, Dhima et al. 2010). Due to limited research on competition between weeds and aromatic plants, it is not clear if the competitive ability of some aromatic plants is attributed to competition traits, such as higher growth rate, extensive soot system, etc., or to allelochemicals exuded through the root system as well. This is a question raised particularly by the results of the competitive ability of lacy phacelia that is devoid of essential oil constituents (Dhima et al. 2010).

Concerning allelopathic root extracts, little research has been conducted to make clear if root allelochemicals directly affect the weed germination/growth or they are primarily activated by soil microbes or the specific environmental conditions, or finally if these root extracts stimulate the soil microorganisms to produce allelochemicals (Soltys et al. 2013). Future increase of aromatic plant cultivation will probably force scientists to study in detail their competitive ability against weeds and provide more information. Selection of certain aromatic plants with high competitive ability or competitive biotypes or cultivars could be a matter of plant breeding. Additionally, any improvements, for example mycorrhizal inoculation, would improve the quality and bioactive phytoconstituents in aromatic plants, promoting their growth, nutrient uptake, productivity, and enhancing the chemical profile of their EOs (Kala 2015, Tarraf et al. 2017). Competitive aromatic plants could be used in crop rotation either as main crops or as cover crops in an integrated weed management plan.

Aromatic Plant Used as Green Manure Crops to Suppress Weeds

Another exploitation of the allelopathic properties of the aromatic plants for weed suppression in sustainable agriculture could be their use as green manure incorporated into the soil with the capacity to produce phytotoxic EOs (De Mastro et al. 2006, Dhima et al. 2009, Vasilakoglou et al. 2011). Particularly, the effect of different aromatic plants used as incorporated green manure on barnyardgrass, common purslane, puncturevine (*Tribulus terrestris*), common lambsquarters, and maize (*Zea mays*) was studied in field experiments (Dhima et al. 2009). Aromatic plants were mechanically incorporated into the soil and maize planting followed. The results revealed reductions in the emergence of barnyardgrass, common purslane, puncturevine, and common lambsquarters up to 50%, 59%, 79% and 83%, respectively, in the green manure treated plots compared to green manure-free plots used as control at 3 WAP (Table 32.4).

Table 32.4. Effect of aromatic plants incorporated as green manure (3 WAP) on plant number of four weed species grown in maize

Treatments	Plants m^{-2}			
	Echinochloa crus-galli	*Portulaca oleracea*	*Tribulus terrestris*	*Chenopodium album*
% of control[a]				
Pimpinella anisum	53	67	53	42
Foeniculum vulgare	62	41	21	17
Ocimum basilicum	60	72	63	25
Anethum graveolens	89	51	26	33
Coriandrum sativum	72	88	53	17
Petroselinum crispum	75	65	68	42
Phacelia tanacetifolia	51	77	47	25
Mentha × *verticillata*	68	72	68	25
Origanum vulgare	56	86	63	17
Melissa officinalis	50	49	74	25

[a] Green manure-free plots (based on Dhima et al. 2009)

Plant number and fresh weight of puncturevine and common lambsquarters (at 6 WAP) were lower in most cases and particularly in sweet fennel- or lacy phacelia-green manure plots by 83% and 83% or 83% and 75%, respectively, in comparison with those in the green manure-free plots; the corresponding reduction in fresh weight was 54% and 86% or 63% and 79%, respectively (Table 32.5). Stem number and fresh weight of barnyardgrass and common purslane (at 9 WAP) were lower in most cases; particularly in lacy phacelia-green manure plots, the stem number and fresh weight of barnyardgrass were reduced by 69% and 63%, respectively, compared to those in the green manure-free plots; moreover, in dill (*Anethum graveolens*)-green manure plots, there was 73% and 78% reduction in stem number and fresh weight of common purslane (Table 32.5).

In this study, it was also pointed that the aromatic plant green manure mainly affected weed germination and not the growth of the survived weeds, due to lack of significant differences between fresh weight over plant or stem number ratios of weeds emerged in the aromatic plant green manure. This could be attributed to rapid decomposition of the allelochemicals in the soil with consequent no further adverse effect on weed growth, or to increased ability of the survived weed to withstand and tolerate these allelochemicals (Dhima et al. 2009). Moreover, the same study indicated that the reduction in weed emergence was attributed more to allelochemicals observed in bioassays, and less to their physical impact, although as already stated lacy phacelia is devoid of essential oil constituents (Dhima et al. 2010). However, the phytotoxic effect of the allelochemicals was not always confirmed in the field (Dhima et al. 2009). Concerning maize, its emergence was not affected by any of the aromatic plants-green manure. Maize grain yield was 27–43% greater in anise, dill, oregano, and lacy phacelia green manure-herbicide untreated plots, compared to green manure-free herbicide untreated plots (Dhima et al. 2009).

In another study, Vasilakoglou et al. (2011) indicated also that soil incorporated green manure of oregano biotypes with high phenolic content (39 to 67 mg g^{-1} fresh weight) affected the emergence and suppressed the growth of barnyardgrass, common purslane, and bristly foxtail (*Setaria verticillata*) in subsequent cotton (*Gossypium hirsutum*) and corn (*Zea mays*) crops; particularly in cotton, in the oregano-green manure treatments the emerged plants of common purslane, barnyardgrass, and bristly foxtail were lower by 30–55%, 48–52%, and 43–86%, respectively, compared to oregano-green manure-free, weedy control. In corn, the corresponding reductions in weed emergence were 0–45%, 38–46%, and 60–80% (Table 32.6).

Table 32.5. Effect of 10 aromatic plants incorporated as green manure on plant or stem number and fresh weight of four weed species grown in maize

Weed species	Aromatic plants	Plants m⁻²	Fresh weight (g m⁻²)	Stems m⁻²	Fresh weight (g m⁻²)
		6 weeks after planting		9 weeks after planting	
		% of control			
	Pimpinella anisum	37	59	-	-
	Foeniculum vulgare	17	46	-	-
	Ocimum basilicum	40	61	-	-
Tribulus terrestris	*Anethum graveolens*	43	49	-	-
	Coriandrum sativum	50	107	-	-
	Petroselinum crispum	103	127	-	-
	Phacelia tanacetifolia	17	37	-	-
	Mentha × verticillata	47	59	-	-
	Origanum vulgare	90	117	-	-
	Melissa officinalis	70	83	-	-
		% of control			
	Pimpinella anisum	33	50	-	-
	Foeniculum vulgare	17	14	-	-
	Ocimum basilicum	38	42	-	-
Chenopodium album	*Anethum graveolens*	42	32	-	-
	Coriandrum sativum	21	12	-	-
	Petroselinum crispum	54	47	-	-
	Phacelia tanacetifolia	25	21	-	-
	Mentha × verticillata	29	29	-	-
	Origanum vulgare	17	16	-	-
	Melissa officinalis	46	60	-	-
				% of control	
	Pimpinella anisum	-	-	50	58
	Foeniculum vulgare	-	-	43	52
	Ocimum basilicum	-	-	38	46
Echinochloa crus-galli	*Anethum graveolens*	-	-	42	46
	Coriandrum sativum	-	-	49	59
	Petroselinum crispum	-	-	41	47
	Phacelia tanacetifolia	-	-	31	37
	Mentha × verticillata	-	-	52	62
	Origanum vulgare	-	-	50	43
	Melissa officinalis	-	-	43	44
				% of control	
	Pimpinella anisum	-	-	37	33
	Foeniculum vulgare	-	-	37	23
	Ocimum basilicum	-	-	49	39
Portulaca oleracea	*Anethum graveolens*	-	-	27	22
	Coriandrum sativum	-	-	33	35
	Petroselinum crispum	-	-	37	25
	Phacelia tanacetifolia	-	-	57	30
	Mentha × verticillata	-	-	43	26
	Origanum vulgare	-	-	49	59
	Melissa officinalis	-	-	41	37

Based on Dhima et al. 2009

Concerning the weed growth suppression, the fresh weight of common purslane was reduced by 9–39% in cotton and by 40–63% in corn (Table 32.6). In cotton, the reductions in number of stems and fresh weight of barnyardgrass in the oregano-green manure treatments were 53–73% and 69–76%, respectively. The corresponding reductions in corn were 21–82% and 36–81%. In bristly foxtail, the reductions in the number of stems, and fresh weight were 17–48%, and 57–76%, respectively, in cotton, whereas, in corn, the corresponding reductions were 0–80%, and 17–93% (Table 32.7).

Table 32.6. Effect of the four oregano biotypes (OBs) that were incorporated as green manure on the plant number (three weeks after planting) of common purslane, barnyardgrass and bristly foxtail that were grown in cotton or in corn

Treatment	*Portulaca oleracea*	*Echinochloa crus-galli*	*Setaria verticillata*
	No. of plants (m^{-2})		
Cotton			
Weedy control	121	48	7
OB-1	55	23	4
OB-2	75	23	2
OB-3	76	25	2
OB-4	85	24	1
Corn			
Weedy control	129	13	5
OB-1	71	8	2
OB-2	109	7	2
OB-3	80	8	2
OB-4	128	8	1

Based on Vasilakoglou et al. 2011

Concerning the effect on cotton and corn, the level of emergence was not affected. The cotton lint and corn grain yield in the oregano green manure treatments were greater by 24–88% and 5–16% than those in the oregano manure-free, weedy treatments. Such oregano biotypes should be established in the field in fall as cover crop incorporated into the soil before planting of cotton or corn during seed-bed preparation; however, such treatment should be accompanied by additional weed control method in order to enhance weed suppression, particularly in crops with low competitive ability such as cotton (Vasilakoglou et al. 2011).

In another field trial in processing tomato, two rates of oregano fresh biomass (1.7 and 3.5 kg m^{-2}) were incorporated into the soil top layer (10 cm) to evaluate their efficacy on weed germination and growth (De Mastro et al. 2006). More specifically, the lower biomass rate totally controlled weeds for at least 30 days and the weeds that finally emerged were suppressed by the tomato plants (Table 32.8); the higher rate of biomass, however, caused phytotoxic symptoms (delayed ripening and more green fruit yield) to the transplanted tomato plants (Table 32.8).

Finally, the effect of incorporated fresh and dry residues (10 and 20 g kg^{-1} of soil) of fennel, common rue (*Ruta graveolens*) and sage (*Salvia officinalis*) on the emergence, root length, shoot length, and fresh weight of hoary cress (*Lepidium draba*) was studied in a pot experiment (Ravlič et al. 2016).

Table 32.7. Effect of the four oregano biotypes (OBs) that were incorporated as green manure on the plant number (nine weeks after planting) of common purslane, barnyardgrass and bristly foxtail that were grown in cotton or in corn

Treatment	*Portulaca oleracea*	*Echinochloa crus-galli*		*Setaria verticillata*	
	Fresh weight (g m^{-2})	No. of stems (m^{-2})	Fresh weight (g m^{-2})	No. of stems (m^{-2})	Fresh weight (g m^{-2})
Cotton					
Weedy control	2710.2	260	2488.9	29	72.9
OB-1	1741.8	93	591.6	24	31
OB-2	2098.9	70	628.1	15	17.7
OB-3	2477.4	88	729.5	18	22.8
OB-4	1640.6	123	760.3	23	27.2
Corn					
Weedy control	1875.0	33.0	294.4	15	47.2
OB-1	1037.5	26.0	187.5	9	10.0
OB-2	889.2	6.0	55.6	11	13.9
OB-3	698.2	21.0	189.2	3	3.5
OB-4	1122.0	15.0	180.3	25	39.4

Based on Vasilakoglou et al. 2011

Table 32.8. Effects of oregano fresh biomass on weeds in transplanted tomato

Oregano biomass (kg m^{-2})	Weeds (n m^{-2})		
	Portulaca oleracea	*Amaranthus graecizans*	*Triticum durum*
1.7	0.0	0.0	0.0
3.5	0.0	0.0	0.0
Control	4.2	3.2	8.8

Based on De Mastro et al. 2006

The findings of this study revealed both inhibitory and stimulatory effect of the incorporated plant residues on weed emergence and growth, with fresh residues showing higher inhibitory effect on emergence, shoot length and fresh weight of hoary cress. Hoary cress emergence was significantly inhibited only by the highest rate of fresh residues of common rue (20 g kg^{-1}), while fennel residues caused reductions in shoot length, root length, and fresh weight (Table 32.10).

Table 32.9. Effect of oregano biomass on tomato yield and plant growth

Oregano biomass (kg m^{-2})	Plant height (cm)	Phytotoxicity (rank)	Fruit yield (t ha^{-1})
1.7	29.2	4.0	122.0
3.5	17.0	7.0	113.4
Control	27.1	0.0	110.8

Based on De Mastro et al. 2006

Table 32.10. Effect of aromatic and medicinal plant residues on germination
and seedling length of hoary cress

Treatments		Emergence (%)	Root length (cm)	Shoot length (cm)	Fresh weight (g)
Control		63.3	2.9	4.2	0.0221
Fresh residues					
Foeniculum vulgare	10 g kg⁻¹	47.9	1.9	3.5	0.0177
	20 g kg⁻¹	73.7	1.5	3.8	0.0182
Ruta graveolens	10 g kg⁻¹	46.9	3.0	3.8	0.0163
	20 g kg⁻¹	40.6	3.0	3.9	0.0189
Salvia officinalis	10 g kg⁻¹	51.2	2.9	4.1	0.0215
	20 g kg⁻¹	77.5	3.0	4.5	0.0249
Dry residues					
Foeniculum vulgare	10 g kg⁻¹	75.6	1.7	4.1	0.0217
	20 g kg⁻¹	93.2	2.0	4.2	0.0266
Ruta graveolens	10 g kg⁻¹	58.4	2.3	4.5	0.0268
	20 g kg⁻¹	61.8	1.7	4.2	0.0322
Salvia officinalis	10 g kg⁻¹	62.2	2.9	4.4	0.0279
	20 g kg⁻¹	69.3	2.7	4.2	0.0280

Based on Ravlič et al. 2016

The efficacy of common rue residues could be attributed to non-terpene compounds, such as furanocoumarins 5-methoxypsoralen (5-MOP), 8-methoxypsoralen (8-MOP) and the quinolone alkaloid graveoline isolated from common rue extracts. Graveoline and 8-MOP inhibited growth of lettuce seedlings and reduced chlorophyll content (Hale et al. 2004).

The results of the above studies suggest the use of certain aromatic plants as an additional measure for weed management, as part of an integrated weed management system aiming at minimising herbicide use. As already stated, although the herbicidal effect of the EOs on weed seed germination has been revealed in several bioassays, few field experiments have been conducted to confirm this effect on various weed species and under natural environmental conditions. The selectivity effect on the subsequent crop is another issue and has not been broadly documented, since there is a lack of information on this topic as well. Additionally, it is revealed that the aromatic green manure effect is mainly on weed germination and not on the weed growth or late weed germination, possibly due to the fast decomposition of the incorporated plant material or the ability of the survived weeds to tolerate these chemicals, as mentioned in many reports (Kobayashi 2004, Khanh et al. 2005, Dhima et al. 2009).

For these reasons, increased length of green manure decomposition would be of interest for more effective weed suppression. This is a matter of soil microbial population, environmental conditions, and timing of soil incorporation. Overall, the success of green manuring system to suppress weeds may be influenced by many factors including green manure crop species as reviewed by Mohammadi (2013). For this reason more field experiments with certain aromatic plants should be conducted to provide information about the allelopathic ability of the soil incorporated green manure to suppress weeds. Soil physicochemical characteristics and microbiological properties, plant species used as green manure, amount and conditions of residues can differentiate the allelopathic potential of the green manure, and consequently the weed suppression (Ravlič et al. 2016). Many times, bioassay results as already stated are different from results observed in the field (Blum et al. 1999, Dhima et al. 2009). This could be attributed to the environmental conditions effect on the amount of the allelochemicals released, the soil pH, the organic matter, and the available nitrogen (Blum et al. 1999, Inderjit and Keating 1999).

Limitations in the Commercial Production of Natural Herbicides from EOs of Aromatic Plants

Although there are many reports about the herbicidal activity of the EOs, the commercial production of such herbicides is limited. Indeed, there are many limitations in formulating a natural herbicide derived from essential oil of aromatic plants.

These types of herbicides are mainly contact, non-systemic herbicides, applied post-emergence with little or no soil activity, and they often require multiple applications in high amounts to be effective. Weed spectrum, little selectivity, lack of systemic activity, and the non-specific mechanism of action are some important issues (Dayan et al. 2009, Dayan and Duke 2010, Cai and Gu 2016).

Concerning weed spectrum, the efficacy of a certain compound of essential oil on a certain plant species cannot be surely maintained on another weed species. The non-systemic activity of the essential oil compounds reduces the ability for a long-term control of perennial weed species. Moreover, the efficacy of these materials falls short due to their high volatilization when compared to synthetic pesticides, although there are specific cases where weed control equivalent to that with conventional products has been observed (Dayan et al. 2011). Synergism between EOs components (Vasilakoglou et al. 2013) may increase efficacy and may broaden weed spectrum, probably due to greater transportation of the components with the greatest activity into the cells (Ultee et al. 2000).

Concerning selectivity, although in bioassay experiments EOs showed selectivity between weed and crop seeds, possibly due to different seed size (Williams and Hoagland 1982, Dhima et al. 2010, Chowhan et al. 2012, Gitsopoulos et al. 2013), the selectivity of natural herbicides applied post-emergence is a serious issue for the production of selective herbicides. The exact compound or compounds that cause the phytotoxic effect of the EOs are still under consideration, since the herbicidal activity can be attributed to the main component(s) of the essential oil; however, any synergistic or antagonistic effect with other compound(s) found in minor percentages cannot be ignored (Daferera et al. 2003). Unfortunately, most studies focus on the producer plants and not on the uptake, biotransformation, detoxification of the allelopathic compounds, and if these compounds finally reach their target or are subjected to any chemical and biochemical transformations (Macías et al. 2007).

The formulation of these products is another issue. Appropriate adjuvants or alternative formulations, such as microencapsulation, are being developed to provide the appropriate formulation of these oils, reduce the amounts applied, increase the duration of their effectiveness by reducing their volatilization, and slow down the rate of degradation in the environment. New formulation techniques, such as nanotechnology engaged with advanced lyophilisation encapsulation techniques as described in the patent of a new natural herbicide based on EOs for weed control (Symeonidou et al. 2014) can help the commercialization of such herbicides.

Additional challenges to the commercial application of plant essential-oil-based herbicides include availability of sufficient quantities of plant material when plant collection is unprofitable (Soltys et al. 2013), standardization and refinement of pesticide products, and protection of technology-patent. Although many EOs may be abundant and available all year round due to their use in perfume, food and beverage industries, application of essential-oil-based pesticides in a large commercial scale could require greater production of certain oils. In addition, as the chemical profile of plant species can vary naturally, depending on geographic, genetic, climatic, or seasonal factors, pesticide manufacturers must take additional steps to ensure that their products will perform consistently. This considerably increases the cost of production and thus companies may not be willing to invest, unless there is a high probability of recovering costs through some form of market exclusivity (for example, patent protection).

Finally, once all these issues are addressed, regulatory approval is required. This continues to be a barrier to commercialization and will likely continue to be a barrier until regulatory systems are adjusted to better accommodate these products (Isman and Machial 2006).

The use of EOs for weed control in organic agriculture seems promising, but these natural herbicides all act very rapidly and their efficacy is limited by the fact that they most likely volatilize relatively quickly as mentioned above, although low persistence in the field could be an advantage when compared with non-volatile herbicides (Auld et al. 2003). Alterations of the chemical structures of the constituents of the essential oils should be performed to overcome constrains of high volatility (Vaughn and Spencer 1993). According to Dayan et al. (1999), the potential difference in the monoterpenes mechanism of action than those of chemically synthesized pesticides, could be used as a template for the synthesis of new chemically synthesized herbicides in future.

Concluding Remarks

Considering all the above, aromatic plants could be used for sustainable weed control in integrated weed management exploiting their allelopathic activity. This can be achieved by using either the aromatic plants directly, i.e., as cover crop or green manure, or indirectly by producing natural herbicides based on their EOs. Concerning the former, this could be more applicable in small-scale farming and high-value crops rather in large-scale farming systems (Dayan et al. 2009) in a crop rotation strategy. Concerning the commercial production of these natural herbicides, the limitations described slow down this progress.

More field experiments should be carried out considering the competitive ability of different aromatic plants against various weed species, and their allopathic effects as green manure crop species. Additionally, more research is required on herbicidal effects of their EOs regarding commercialization purposes.

There is a necessity to transfer laboratory data into field experimental conditions together with new tools of molecular genetics, chemistry, and biochemistry aiming at the creation of selective and eco-friendly, more efficient, and cost-effective herbicides that could be used as components in an integrated weed management strategy (Soltys et al. 2013, Cai and Gu 2016). Future research should focus on the most drastic compounds for the various weed species, so that these compounds can be used for the production of natural herbicides (Shokouhian et al. 2016).

Finally, growers should consider aromatic plants as an alternative for integrated weed management and the chemical companies should shift towards new modes of action and 'greener' options to reduce environmental impacts and meet market and government demands (Dayan and Duke 2010). Natural herbicides from aromatic plants with low residues in soil, non-leaching effect in ground water and low toxicity in mammals (Isman 2000) meet these requirements and could play a significant role in herbicide industry the following years.

REFERENCES

Abrahim, D., A.C. Francischini, E.M. Pergo, A.M. Kelmer-Bracht and E.L. Ishii-Iwamoto. 2003. Effects of alpha-pinene on the mitochondrial respiration of maize seedlings. Plant Physiol. Biochem. 41: 985–991.

Ahmad, A. and L.N. Misra. 1994. Terpenoids from Artemisia annua and constituents of its essential oil. Phytochemistry 37: 183–186.

An, M., I.R. Jonhson and J.V. Lovett. 1993. Mathematical modelling of allelopathy: biological response to allelochemicals and its interpretation. J. Chem. Ecol. 19: 2379–2389.

Anaya, A.L. 1999. Allelopathy as a tool in the management of biotic resources in agroecosystems. Crit. Rev. Plant Sci. 18: 697–739.

Auld, B.A., S.D. Hethering and H.E. Smith. 2003 Advances in bioherbicide formulation. Weed Biol. Man. 3: 61–67.

Azirak, S. and S. Karaman. 2008. Allelopathic effect of some essential oils and components on germination of weed species. Acta Agric. Scand. Sect B. 58: 88–92.

Bajwa, A., G. Mahajan and B. Chauhan. 2015. Nonconventional weed management strategies for modern agriculture. Weed Sci. 63: 723–747.

Bakkali, F., S. Averbeck, D. Averbeck and M. Idaomar. 2008. Biological effects of essential oils – a review. Food Chem. Toxicol. 46: 446–475.

Bassolé, I.H.N. and H.R. Juliani. 2012. Essential oils in combination and their antimicrobial properties. Molecules 17: 3989–4006.

Batish, D.R., N. Setia, H.P. Singh and R.K. Kohli. 2004. Phytotoxicity of lemon-scented eucalypt oil and its potential use as a bioherbicide. Crop Prot. 23: 1209–1214.

Bernath, J. 2009. Aromatic plants. *In:* Fuleky, G. (Ed.) Cultivated Plants, Primarily as Food Sources. EOLSS, Paris, France.

Blazquez, M.A. 2014. Role of natural essential oils in sustainable agriculture and food preservation. J. Sci. Res. Rep. 3: 1843–1860.

Blum, U., S.R. Shafer and M.E. Lehman. 1999. Evidence for inhibitory allelopathic interactions involving phenolic acids in field soils: concepts vs. an experimental model. Crit. Rev. Plant Sci. 18: 673–693.

Cai, X. and M. Gu. 2016. Bioherbicides in organic horticulture. Horticulturae 2: 10. Available at: http://dx.doi.org/10.3390/horticulturae2020003.

Chaimovitsh, D., A. Shachter, M. Abu-Abied, B. Rubin, E. Sadot and N. Dudai. 2017. Herbicidal activity of monoterpenes is associated with disruption of microtubule functionality and membrane integrity. Weed Sci. 65: 19–30.

Chaimovitsh, D., M. Abu-Abied, E. Behausov, B. Rubin, N. Dudai and E. Sadot. 2010. Microtubules are an intracellular target of the plant terpene citral. Plant J. 61: 399–408.

Chaimovitsh, D., O. Rogovoy Stelmakh, O. Altshuler, E. Belausov, M. Abu-Abied, B. Rubin, E. Sadot and N. Dudai. 2012. The relative effect of citral on mitotic microtubules in wheat roots and BY2 cells. Plant Biol. 14: 354–364.

Chowhan N., H.P. Singh, D.R. Batish, S. Kaur, N. Ahuja and R.K. Kohli. 2012. β-pinene inhibited germination and early growth involves membrane peroxidation. Protoplasma 250: 691–700.

Chowhan, N., H.P. Singh, D.R. Batish and R.K. Kohli. 2011. Phytotoxic effects of β-pinene on early growth and associated biochemical changes in rice. Acta Physiol. Plant 33: 2369–2376.

Christaki, E., E. Bonos, I. Giannenas and P. Florou-Paneri. 2012. Aromatic plants as a source of bioactive compounds. Agriculture 2: 228–243.

Cruz-Ortega, R., A.L. Anaya, B.E. Hernández-Bautista and G. Laguna-Hernández. 1998. Effects of allelochemical stress produced by *Sicyos deppei* on seedling root ultrastructure of *Phaseolus vulgaris* and *Cucurbita ficifolia*. J. Chem. Ecol. 24: 2039.

Daferera, D.J., B.N. Ziogas and M.G. Polissiou. 2003. The effectiveness of plant essential oils on the growth of *Botrytis cinerea*, *Fusarium* sp. and *Clevibacter michiganensis* subsp. *michiganensis*. Crop Prot. 22: 39–44.

Dayan, F., J. Romagni, M. Tellez, A. Rimando and S. Duke. 1999. Managing weeds with natural products. Pesticide Outlook 10: 185–188.

Dayan, F.E., S.O. Duke, A. Sauldubois, N. Singh, C. McCurdy and C.L. Cantrell. 2007. p-Hydroxyphenylpyruvate dioxygenase is a herbicidal target site for b-triketones from *Leptospermum scoparium*. Phytochemistry 68: 2004–2014.

Dayan, F.E. and S.O. Duke. 2010. Natural products for weed management in organic farming in the USA. Outlooks Pest Manag. 21: 156–160.

Dayan, F.E., C.L. Cantrell and S.O. Duke. 2009. Natural products in crop protection. Bioorg. Med. Chem. 17: 4022–4034.

Dayan, F.E., J.L. Howell, J.P. Marais, D. Ferreira and M. Koivunen. 2011. Manuka essential oil, a natural herbicide with preemergence activity. Weed Sci. 59: 464–469.

De Martino, L., E. Mancini, L.F. De Almeida and V. De Feo. 2010. The antigerminative activity of twenty-seven monoterpenes. Molecules 15: 6630–6637.

De Mastro, G., M.F. Verdini and P. Montemurro. 2006. Oregano and its potential use as bioherbicide. Acta Hortic. 723: 335–345.

Dhima, K., I. Vasilakoglou, V. Garane, C. Ritzoulis, V. Lianopoulou and E. Panou-Philotheou. 2010. Competitiveness and essential oil phytotoxicity of seven annual aromatic plants. Weed Sci. 58: 457–465.

Dhima, K.V., I.B. Vasilakoglou, T.D. Gatsis, E. Panou-Philotheou and I.G. Eleftherohorinos. 2009. Effects of aromatic plants incorporated as green manure on weed and maize development. Field Crops Res. 110: 235–241.

Dudai, N., M. Ben-Ami, R. Chaimovich and D. Chaimovitsh. 2004. Essential oils as allelopathic agents: bioconversion of monoterpenes by germinating wheat seeds. Acta Hortic. 629: 505–508.

Dudai, N., A. Poljakoff-Mayber, A.M. Mayer, E. Putievsky and H.R. Lerner. 1999. Essential oils as allelochemicals and their potential use as bioherbicides. J. Chem. Ecol. 25: 1079–1089.

Duke, S.O., F.E. Dayan and J.G. Romagni. 2000. Natural products as sources for new mechanisms of herbicidal action. Crop Prot. 19: 583–589.

El-Deek, M.H. and F.D. Hess. 1986. Inhibited mitotic entry is the cause of growth inhibition by cinmethylin. Weed Sci. 34: 684–688.

Gealy, D.R., K.A. Moldenhauer and S. Duke. 2013. Root distribution and potential interactions between allelopathic rice, sprangletop (Leptochloa spp.), and barnyardgrass (Echinochloa crus-galli) based on 13C isotope discrimination analysis. J. Chem. Ecol. 39: 186–203.

Gitsopoulos, T., P. Chatzopoulou and I. Georgoulas. 2013. Effects of essential oils of *Lavandula* × *hybrida* Rev, *Foeniculum vulgare* Mill and *Thymus capitatus* L. on the germination and radical length of *Triticum aestivum* L., *Hordeum vulgare* L., *Lolium rigidum* L. and *Phalaris brachystachys* L. J. Essent. Oil Bearing Plants 16: 817–825.

Graña, E., C. Díaz-Tielas, A.M. Sánchez-Moreiras and M.J. Reigosa. 2012. Mode of action of monoterpenes in plant-plant interactions. Curr. Bioact. Comp. 8: 80–89.

Grayson, B.T., K.S. Williams, P.A. Freehauf, R.R. Pease, W.T. Ziesel, R.L. Sereno and R.E. Reinsfelder. 1987. The physical and chemical properties of the herbicide cinmethylin (SD 95481). Pest. Sci. 21: 143–153.

Hale, A.L., K.M. Meepagala, A. Oliva, G. Aliotta and S.O. Duke. 2004. Phytotoxins from the leaves of *Ruta graveolens*. J. Agric. Food Chem. 52: 3345–3349.

Hernández-Bautista, B.E., A. Torres-Barragán and A.L. Anaya. 1996. Evidence of allelopathy in *Sicyos deppei* (Cucurbitaceae). *In:* Proceedings of the First World Congress on Allelopathy. University of Cádiz, Puerto Real, Cádiz, Spain, 91 p.

Inderjit and K.I. Keating. 1999. Allelopathy: principles, procedures, progresses, and promises for biological control. Adv. Agron. 67: 141–231.

Isman, B.M. 2000. Plant essential oils for pest and disease management. Crop Prot. 19: 603–608.

Isman, M.B. and C.M. Machial. 2006. Pesticides based on plant essential oils: from traditional practice to commercialization. Adv. Phytomed. 3: 29–44.

Kala, C.P. 2015. Medicinal and aromatic plants: boon for enterprise development. J. Appl. Res. Med. Arom. Plants 2: 134–139.

Khanh, T.D., M.I. Chung, T.D. Xuan and S. Tawata. 2005. The exploitation of crop allelopathy in sustainable agricultural production. J. Agron. Crop Sci. 191: 172–184.

Kobayashi, K. 2004. Factors affecting phytotoxic activity of allelochemicals in soil. Weed Biol. Manag. 4: 1–7.

Kohli, R.K., D.R. Batish and H.P. Singh. 1998. Eucalypt oils for the control of parthenium (*Parthenium hysterophorus* L.). Crop Prot. 17: 119–122.

Koitabashi, R., T. Suzuki, T. Kawazu, A. Sakai, H. Kuroiwa and T. Kuroiwa. 1997. 1,8-Cineole inhibits root growth and DNA synthesis in the root apical meristem of *Brassica campestris* L. J. Plant Res. 110: 1–6.

Kordali, S., A. Cakir, H. Ozer, R. Cakmakci, M. Kesdek and E. Mete. 2008. Antifungal, phytotoxic and insecticidal properties of essential oil isolated from Turkish *Origanum acutidens* and its three components, carvacrol, thymol and p-cymene. Bioresour. Technol. 99: 8788–8795.

Koul, O., S. Walia and G.S. Dhaliwal. 2008. Essential oils as green pesticides: potential and constraints. Biopestic. Int. 4: 63–84.

Kriegs, B., M. Jansen, K. Hahn, H. Peisker, O. Samajová, M. Beck, S. Braun, A. Ulbrich, F. Baluska and M. Schulz. 2010. Cyclic monoterpene mediated modulations of *Arabidopsis thaliana* phenotype. Plant Signal Behav. 5: 832–838.

Macías, F.A., J.M.G. Molinillo, R.M. Varela and J.C.G. Galindo. 2007. Allelopathy – a natural alternative for weed control. Pest Manag. Sci. 63: 327–348.

Mohammadi, G.R. 2013. Alternative weed control methods: a review. Chapter 6. *In:* Soloneski, S. and M. Larramendy (Eds.) "Agricultural and Biological Sciences" Weed and Pest Control – Conventional and New Challenges. InTech Publications, Rijeka, Croatia.

Mucciarelli, M., W. Camusso, C.M. Bertea, S. Bossi and M. Maffei. 2001. Effect of (+)-pulegone other oil components of *Mentha × piperita* on cucumber respiration. Phytochemistry 57: 91–98.

Muller, C.H., W.H. Muller and B.L. Haines. 1964. Volatile growth inhibitors produced by aromatic shrubs. Science 143: 471–473.

Narwal, S. 2010. Allelopathy in ecological sustainable organic agriculture. Allelopathy J. 25: 51–72.

Nishida, N., S. Tamotsu, N. Nagata, C. Saito and A. Sakai. 2005. Allelopathic effects of volatile monoterpenoids produced by *Salvia leucophylla*: inhibition of cell proliferation and DNA synthesis in the root apical meristem of *Brassica campestris* seedlings. J. Chem. Ecol. 31: 1187–1203.

Oerke, E. 2006. Crop losses to pests. J. Agric. Sci. 144: 31–43.

Onen, H., Z. Ozer and I. Telci. 2002. Bioherbicidal effects of some plant essential oils on different weed species. J. Plant Dis. Prot. 18: 597–605.

Ramezani, S., M.J. Saharkhiz, F. Ramezani and M.H. Fotokian. 2008. Use of EOs as bioherbicides. J. Essent. Oil Bearing Plants 1: 319–327.

Ravlić, M., R. Baličević, M. Nikolić and A. Sarajlić. 2016. Assessment of allelopathic potential of fennel, rue and sage on weed species hoary cress (*Lepidium draba*). Not Bot Horti Agrobo. 44: 48–52.

Scognamiglio, M., B. D'Abrosca, A. Esposito, S. Pacifico, P. Monaco and A. Fiorentino. 2013. Plant growth inhibitors: allelopathic role or phytotoxic effects? Focus on Mediterranean biomes. Phytochem. Rev. 12: 803–830.

Scrivanti, L.R., M. Zunino and J.A. Zygadlo. 2003. *Targetes minuta* and *Shinus areira* essential oils as allelopathic agents. Biochem. Syst. Ecol. 31: 563–572.

Serrato-Valenti, G., M.G. Mariotti, L. Cornara and A. Corallo. 1998. A histological and structural study of *Phacelia tanacetifolia* endosperm in developing, mature and germinating seed. Int. J. Plant Sci. 159: 753–761.

Shokouhian, A., H. Hassan and A. Kayvan. 2016. Allelopatic effect of some medicinal plant essential oils on plant seeds germination. J. BioSci. Biotechnol. 5: 13–17.

Singh, H.P., D.R. Batish and R.K. Kohli. 2002. Allelopathic effect of two volatile monoterpenes against bill goat weed (*Ageratum conyzoides* L.). Crop Prot. 21: 347–350.

Singh, H.P., D.R. Batish, N. Setia and R.K. Kohli. 2005. Herbicidal activity of volatile oils from *Eucalyptus citriodora* against *Parthenium hysterophorus*. Ann. Appl. Biol. 146: 89–94.

Soltys, D., U. Krasuska, R. Bogatek and A. Gniazdowska. 2013. Allelochemicals as bioherbicides— present and perspectives (Chapter 20). *In:* Price, A.J. and J.A. Kelton (Eds.) Herbicides—Current Research and Case Studies in Use. InTech Publications, Rijeka, Croatia.

Symeonidou, A., K. Petrotos, I. Vasilakoglou, P. Gkoutsidis, F. Karkanta and A. Lazaridou. 2014. Natural herbicide based on essential oils and formulated as wettable powder. European patent application, EP 2 684 457 A1. Available at: http://www.google.ch/patents/EP2684457A1?cl=en.

Tarraf, W., C. Ruta, A. Taragelli, F. De Cillis and G. De Mastro. 2017. Influence of arbuscular mycorrhizae on plant growth, essential oil production and phosphorus uptake of *Salvia officinalis* L. Ind. Crops Prod. 10: 144–153.

Tulp, M. and L. Bohlin. 2005. Rediscovery of known natural compounds: nuisance or goldmine? Bioorg. Med. Chem. 13: 5274–5282.

Tworkoski, T. 2002. Herbicide effects of essential oils. Weed Sci. 50: 425–431.

Ultee, A., E.P.W. Kets, M. Alberda, F.A. Hoekstra and E.J. Smid. 2000. Adaptation of the food-borne pathogen *Bacillus cereus* to carvacrol. Arch. Microbiol. 174: 233–238.

Vaid, S., D.R. Batish, H.P. Singh and R.K. Kohli. 2011. Phytotoxicity of limonene against *Amaranthus viridis* L. The Bioscan 6: 163–165.

Vaid, S., D.R. Batish, H.P. Singh and R.K. Kohli. 2010. Phytotoxic effect of eugenol towards two weedy species. The Bioscan 5: 339–341.

Vasilakoglou, I., K. Dhima, E. Anastassopoulos, A. Lithourgidis, N. Gougoulias and N. Chouliaras. 2011. Oregano green manure for weed suppression in sustainable cotton and corn fields. Weed Biol. Manag. 11: 38–48.

Vasilakoglou, I., K. Dhima, K. Paschalidis and C. Ritzoulis. 2013. Herbicidal potential on *Lolium rigidum* of nineteen major essential oil components and their synergy. J. Essent. Oil Res. 25: 1–10.

Vaughn, S.F. and G.F. Spencer. 1993. Volatile monoterpenes as potential parent structures for new herbicides. Weed Sci. 41: 114–119.

Vokou, D., P. Douvli, G.J. Blionis and J.M. Halley. 2003. Effects of monoterpenoids, acting alone or in pairs, on seed germination and subsequent seedling growth. J. Chem. Ecol. 29: 2281–2301.

Weidenhamer, J.D., F.A. Macias, N.H. Fischer and G.B. Williamson. 1993. Just how insoluble are monoterpenes? J. Chem. Ecol. 19: 1799–1803.

Weston, L.A. 2005. History and current trends in the use of allelopathy for weed management. HortTechnol. 15: 529–534.

Williams, R.D. and R.E. Hoagland. 1982. The effects of naturally occurring phenolic compounds on seed germination. Weed Sci. 30: 206–212.

Zaouali, Y., C. Messaoud, A. Ben Salah and M. Boussaïd. 2005. Oil composition variability among populations in relationship with their ecological areas in Tunisian *Rosmarinus officinalis* L. Flav. Fragr. J. 20: 512–520.

Zimdahl, R.L. 2007. Fundamentals of Weed Science. 3rd ed. New York: Elsevier. 666 p.

Zhang, J., M. An and H. Wu. 2012. Chemical composition of essential oils of four *Eucalyptus* species and their phytotoxicity on silver leaf nightshade. Plant Growth Regul. 68: 231–237.

Zunino, M.P. and J.A. Zygadlo. 2004. Effect of monoterpenes on lipid oxidation in maize. Planta 219: 303–309.

Issues and Sustainability in Grassland Weed Control

Stevan Z. Knezevic[*1] and Nevin Lawrence[2]

[1] Professor of Integrated Weed Management, Northeast Research and Extension Center, Department of Agronomy and Horticulture, University of Nebraska-Lincoln, Concord, Nebraska, 68728, USA

[2] Integrated Weed Management Specialist, Panhandle Research and Extension Center, University of Nebraska-Lincoln, 4502 Avenue I, Scottsbluff, NE 69361-4939, USA

Introduction

Grasslands are large land areas covered mostly with grassy species, thus the name: grassland. Grasslands are usually covered by the types of grasses utilized for livestock grazing, which occur around the world in almost every eco-zone. Grass height can vary, from very short (for example, England and Ireland) to tall grasses, such as tallgrass prairies in North and South America and savannas in Africa (Gibson et al. 1991). Woody plants, shrubs or trees, may also grow on some grasslands. Grasslands are usually established in climates where annual rainfall ranges between 300 and 900 mm (10 and 35 in) and in colder (−20 °C) and hotter (30 °C) climatic conditions (Gibson 1995).

Much of the grasslands worldwide are extensively utilized for grazing by livestock. This activity led to creation of a discipline commonly known as the range management. Range management is dealing with the use of rangelands and grasslands for grazing by livestock and other purposes, such as wildlife habitat, recreation, and aesthetics. Rangeland management was also developed in response to rangeland deterioration due to overgrazing and other threats (Holechek et al. 2011).

One of the greatest threats to the natural grassland communities is the introduction of invasive weeds. Invasive weed species are non-native plants that can spread and invade new habitats. Thus, the term 'invasive' is most often utilized for any introduced species ('non-indigenous' or 'non-native') that can degrade the habitats they invade. Other commonly used synonyms are also: exotic pest plants, alien species, or invasive exotics (Knezevic 2017). These alien species usually have no natural enemies to control their spread, have substantial root systems and produce large quantities of seeds. They compete with native species for moisture and soil nutrients but won't be eaten by wildlife or livestock (Knezevic 2017). Non-native species reduce the biodiversity of the grasslands and, once established, are usually very difficult to control. Since there are many kinds of invasive weeds with different life cycles, a single control

*Corresponding author: sknezevic2@unl.edu

method is not effective. In addition, controlling weeds with one or two methods provides the weeds a chance to adapt to those practices. Therefore, integrated weed management on grassland is needed and should involve the use of various control techniques in a well-planned, coordinated, and organized program. Knezevic et al. (2017) described integrated weed management (IWM) as a combination of mutually supportive technologies based on the application of numerous alternative control measures. In practical terms, it means developing a weed management program using a combination of preventive, cultural, mechanical, and chemical practices. It does not mean abandoning chemical weed control, but relying on it less (Knezevic et al. 2017).

Therefore, instead of using a particular weed control method, IWM suggests the use of a mixture of control methods that can provide optimum economic returns and should be based on a few general principles:

- use land management practices that limit the introduction and spread of weeds (preventing weed problems before they start)
- help the grass compete with weeds (help 'choke out' weeds)
- use practices that keep weeds 'off balance' (do not allow weeds to adapt)
- documentation and record keeping.

Combining grassland management practices based on the above principles can help design weed control for any management operation. Also it is important to understand that an IWM program is not a 'recipe', it needs to be changed and adjusted to the particular operation, and from year to year. The goal is to manage, not eradicate weeds, as complete weed eradication is not possible for environmental and economic reasons.

Prevention from Invasion

Prevention should be the first line of defense against invasive weed species. Once an introduced species has become a widespread invasive, the economic and often environmental costs of eradication can be cost prohibitive, especially in grasslands that do not generate high economic returns.

As a result, many countries have established means and programs for preventing the import of invasive species. Also, a legal framework is essential to support efforts to prevent introduction of invasive species and especially to manage an introduced invasive species. There are general guidelines for designing legal and institutional frameworks on invasive alien species (Tanentzap et al. 2009). If an alien species is invasive, it will not stay within the boundaries of the ecosystem, municipality or region to which it was introduced. Thus, regional collaboration between states (or countries) in regard to invasive species management is essential. Also, numerous legal principles, approaches, and tools have been developed for dealing with problems of invasive alien species (Lass et al. 2005). Much efforts can be also done at the local level, including public education and early warning. Early warning is the ability to predict potential new invasion sites of a current invasive species, and/or predict potential new invasive species for a region (Lambert 2004).

Early Detection and Monitoring

The best management of any invasive species is to recognize potential weed problems early, control weeds before they reproduce and spread, and monitor sites regularly to maintain adequate follow-up control. Effective early detection efforts depend upon proper training of property owners, land managers, and pest management professionals. Understanding the potential threats that may exist on surrounding sites can provide an early warning system for weed invasion. One tool that can aid in preventing the invasion of weeds is to regularly conduct inventory of the area by field surveys, or aerial photography, and remove individual weed plants before they become well established (Lass et al. 2005).

Developing early detection methods for invasive species could result in substantial economic savings and circumvent negative ecological impacts. Components of early detection protocols should include some of the following aspects: i) knowledge on the current presence of a given plant species in management units, ii) knowing vectors and pathways of plant dispersal, and iii) understanding the likelihood of the plant establishing or spreading inside management units. This knowledge can then be used to plan a rapid response to remove or control the invasive plant. There are also computer models developed to predict weed spread based on remotely sensed and Geographic Information System (GIS) data. These models can help land managers in the early detection of invasive plants, especially over large landscapes. Once areas of invasive species occurrence are predicted, ground reconnaissance can be more effectively used for verification and control. For example, over the last decade in USA, there has been a substantial increase in the use of remotely sensed and GIS data to model invasive species distributions or potential habitats. This increase coincides with improved remote sensors and development of more powerful computer technology (Jarnevich et al. 2010).

To summarize, inventory and mapping should be the first steps in any integrated weed management program. Some land managers divide land area into management units and assign the level of infestation to each unit. A unit could be a section of land, which can be of a certain size or simply delineated by similar level of weed infestation. The second step would be to prioritize the control type in each management unit by choosing control techniques for a particular weed management unit. The third step is adopting proper range management practices but whatever is planned, it must fit into an overall range management plan.

Management

The goal of management is to reduce an invasive plant population below the threshold level. The timeframe of a management project may vary depending on the invasive plant and desired conservation outcome. For example, an invasive plant may be suppressed in a restoration effort for a few years in order for planted desired species to establish and become competitive.

In general, the management of invasive plants can be achieved by a variety of chemical and non-chemical tools, which might include mechanical, prescribed fire, and biological control.

Mechanical Weed Control

Mechanical weed control is one of the most commonly used non-chemical methods of weed control in grassland. Mowing, and cutting, as well as pulling and handweeding are the most common mechanical weed control techniques. Properly timed mowing or cutting will suppress weeds but with few exceptions will not kill them. Mowing tends to be more effective on annual broadleaf weeds than grasses since most grasses rapidly regrow from the crown. Mowing must be carefully timed to maximise damage to the weed and minimize damage to the grassland

Prescribed Fire

Prescribed burn of grassland is one of the oldest and cheapest methods for vegetation control or promoting desired vegetation. Many prescribed burns are designed to reduce the abundance of woody species that spread in savannas, prairies, and other grasslands. Repeated burns are sometimes necessary to effectively control invasive species (Schwartz and Helm 1996).

Efficacy of prescribed burns depends on the weather, topography, and available fuel to carry the fire, which is usually the dry plant biomass from the previous year(s). Spot-burning invasive weeds with a propane torch can be also done when the infestation is small. Spot-burning can be used to burn individual plants or groups of plants in a small area.

Before conducting prescribed burn, a fire management plan must be developed, which should contain information on how to start a burn program, and guidelines for conducting burns, with proper training and certification to conduct burns safely. Generally speaking, burning grassland safely is of the upmost importance, thus prescribed burns should be conducted by well-trained crew or certified fire agencies (Schwarzmeier 1984).

Chemical Weed Control

Chemicals used for managing weeds are commonly known as herbicides. Herbicides are the most powerful tools in the toolbox of integrated management. Depending on the application method, herbicides are used typically post-emergence in grassland. The cost of herbicides can be expensive, especially when utilized over large tracts of land. Effectiveness of herbicides can vary depending on the weed type and label directions and/or restrictions (Knezevic et al. 2017, Weed Guide).

Biological Weed Control

Biological weed control is the control of weeds by parasites, predators, or pathogens. Biological control reduces weed density but does not eliminate the target weed, as the biocontrol agent often requires the weed as a host or food source. In some instances biological control can be permanent as the biocontrol agent may be self-perpetuating and not require additional management. The target organism may then be controlled indefinitely without further human effort, a particular advantage in certain geographically or environmentally limiting settings. Biocontrol agents are rigorously evaluated prior to approval and release to avoid deleterious effects. The effect of biocontrol agents is limited to the target weed and perhaps a few of its close relatives. The economics of successful biocontrol can be favorable since following release the organism may perpetuate itself indefinitely and often disperses on its own. The response to a biocontrol program is often slow because the population of the organism must increase from the level of the initial release. As a result, most biocontrol agents are best suited to a stable long-term environment, i.e., grazing grassland or natural areas rather than an annual cropping system. Since most biocontrol agents by themselves do not reduce weed populations to an acceptable level, they must be used in conjunction with other approaches in an integrated weed management program (Knezevic et al. 2017, Weed Guide).

Several literature reviews have been published on the management of weeds in grasslands (DiTomaso 2000, Master and Sheley 2001). Rather than repeat the work of previous researchers, the authors of this book chapter choose two weed species, eastern redceder (*Juniperus virginiana* L.) and downy brome (*Bromus tectorum* L.), to use as case studies of IWM in grasslands. The two subjects were chosen because of their contrasting life histories (Stubbendieck et al. 2003, Thill et al. 1984). Eastern redceder is a once rare, but now common tree in grasslands of the US central plains, and downy brome is an invasive winter annual grass species native to Eurasia and the Mediterranean, and now found in six of seven continents. While both weed species differ considerably in their biology, each can reduce biodiversity, grazing and economic productivity of grasslands.

Case Example 1: Integrated Management of Eastern Redcedar

Introduction

Eastern redcedar is a problem on grasslands primarily because it reduces forage production (Stubbendieck et al. 2003). Developing trees alter the microclimate, which encourages a shift from desirable warm-season native grasses to introduced cool-season grasses, such as Kentucky bluegrass (*Poa pratensis* L.) (Gehring and Bragg 1992, Smith and Stubbendieck 1990). Heavy infestations make livestock handling more difficult. All these adverse effects can be reflected in lower rental rates or sale prices for infested grassland. Established infestations usually get worse over time due to over production of seeds and established trees get bigger, thus shading grass even more (Stritzke and Rollins 1984).

On many sites, complete coverage by eastern redcedar can be expected, resulting in total loss of grass production unless controlled. Control measures should be initiated as soon as possible, both to improve effectiveness and reduce total control costs (Knezevic et al. 2005). Management of infestations is best viewed as a long-term or ongoing effort, both to reduce the initial infestations and prevent them from redeveloping to economically damaging levels. Emphasis should be on management of the infestation, rather than eradication. Eradication is not economical and probably not physically possible in most cases. Instead, it should be recognized that some remaining larger trees, which are the most difficult and expensive to kill, do little damage. In fact, at low levels, eastern redcedars can be viewed as a potential resource, providing livestock shelter, wildlife habitat, timber products, and aesthetic values (Wilson and Schidt 1990). Most important, long-term selective management is considerably less expensive than a more intensive, short-term approach. If the goal is to reduce overall number of trees and stop further spreading (e.g., management of wildlife habitat), it is recommended to cut female trees only. Female trees are the ones that produce berry-like fruits with seeds. This would allow 'male trees' to grow and provide much needed cover for wildlife or land beautification, while reducing further spreading.

Manual and Mechanical Control
Manual and mechanical control involves methods, such as digging, cutting and mowing trees. It is very effective for small areas, and it is most efficient on trees up to 2 feet tall.

Cutting is an effective method of control because eastern redcedar is a non-sprouter. Trees cut below the lowest branches will not regrow (Crawthorne et al. 1982).A variety of handheld or motor-powered cutting tools can be utilized. For example, handheld tools (shears, saws, spade, shovel, heavyhoe) are effective on small trees (less than 3 feet tall),while larger trees require a chainsaw or vehicle-mounted shears. However, cutting is a method that can be time consuming and labor intensive. Also, all cut trees should be gathered and burned,or permanently removed from the grassland.

Fire
Ortman et al. (1998) suggested that prescribed fire is important both to initially reduce infestations and to maintain trees at economically tolerable levels. Burning as a method is inexpensive and very effective against smaller trees (Bragg et al. 1976). Its effectiveness declines as the tree size increases. Adequate fine fuel (usually last year's deadgrass) is necessary for the initiation of the fire.

Safety also is a concern since many land managers lack experience with fire and the equipment required to conduct fires (Eangle et al. 1992). If the pastures are isolated by roads, cultivated lands, and other fire breaks that will confine the fire and minimize risk. However, in central Nebraska, pastures often are located within large blocks of rangeland, making fire escape more likely and serious. This suggests the need for more planning and care on how to conduct the fire safely (Eangle et al. 1992). In some cases, the difficulty and risks of burning in areas of extensive grasslands can be greatly reduced by conducting 'landscape-scale' fires, rather than burning pastures individually. Under the landscape scale concept, the fire boundary is extended until adequate existing firebreaks are encountered. These maybe roads, watercourses, canals, cultivated lands, stands of broadleaf trees, relatively non-flammable canyon bottoms, or areas of short or green vegetation. Such large areas frequently contain the holdings of multiple land owners. Obviously, all land owners and managers within the area must be in agreement about the proposed burn (Buehring et al. 1971).

Chemical Control
Herbicides is also an important part of the integrated management program. Depending on the application method and chemical type, the use of herbicides can be time consuming and expensive, especially when utilized_on denser tree infestations or large tracts of land. Effectiveness also is variable depending on the tree size and label directions and/or restrictions. In general, herbicides for eastern redcedar control can be used for broadcast application or individual tree spraying (Knezevic et al. 2017, Weed Guide).

Broadcast Treatments
Broadcast application is the most common method of applying herbicides in agricultural settings. Knezevic et al. (2005) reported that the tree height was the most important factor influencing the level of chemical control (tree injury) with broadcast treatments. Recommended herbicides for trees that are up to 2 feet tall include: Surmount (Plenum), Grazon P&D, and

Tordon 22K. However, the same herbicides will not provide satisfactory broadcast control of trees taller than 2 feet, indicating the importance of tree height (Knezevic et al. 2017).

Individual Tree Treatments
Individual tree treatments can be applied directly to the tree foliage or to the soil around the tree base. Soil treatments can minimise the amount of herbicide used and the exposure on target species. However, soil treatments may not be effective unless applied before rainfall, preferably in spring or fall. Rainwater is needed to move the herbicide into the root zone, allowing uptake by a tree. Recommended herbicides for soil application around a tree base include Tordon 22K at the rate of 1 cc (ml) per every foot of tree height, and Velpar-Lat4 (cc) ml and Spike20 Pat 1 cc (ml) per every inch of tree diameter (Knezevic et al. 2017). Individual tree foliage also can be treated with various herbicides. Knezevic et al. (2005) reported that the best control of 2–10 feet tall trees was with Surmount (Plenum), Grazon P+D and Tordon 22K.

Therefore, the use of selective post-emergence herbicide treatments should be based on tree height. Broadcast treatments are effective only on short trees (upto 60 cm or 2 feet tall), while medium height trees (60 cm to 3 m or 2–10 feet) can be controlled with individual tree treatments (Knezevic et al. 2017).

Biological Control
Biological control is the use of natural enemies to reduce weed populations to economically acceptable levels (Ortman et al. 1998). In the case of eastern redcedar control, goats can be used as a helpful biocontrol agent for trees that are up to 3–4 feet tall as part of an integrated control approach. Most eastern redcedar trees less than 24 inches tall can be killed by goats in a paddock grazing system within the first year. The control level was reduced by 50 per cent on 4–8 feet tall trees; however the goats managed to defoliate bottom branches and strip bark from branches and trunks up to three inches in diameter. That size tree may take three to five years of browsing to kill.

Generally, goats are browsers with diets consisting of about 70% of non-grassy species, which indicates that they should not compete much with cattle for grass. Goats prefer non-grassy species, but they would eat grass if no other species were available. This also suggests that goats in general can help in controlling many plant species that cattle do not eat, including various noxious weeds (e.g., leafyspurge, thistles).

Important factors in managing goats include the use of appropriate stocking rates, quality fencing and protection from predators. In essence, the number of goats needs to be adjusted to the amount of plant material needing control. Younger animals will not eat eastern redcedar as well as older ones. Precise stocking rates for cedar control have not been established by research in Nebraska nor elsewhere. The bottom line is that goats must be fenced in the area where unwanted plants are to be controlled. Thus, per acre stocking rate should be at least 10 goats/acre of infested land. This stocking rate with moderate eastern redcedar infestation should result in significant damage to the trees within 30 days. Higher stocking rates would be better, but will require moving the fence more often. Trees and other perennial plants have high energy reserves in their root systems and repeated defoliation over several years is required to control them. Eastern redcedar trees, however, will not resprout and if the goats remove most of the needles and/or bark, the tree will eventually die.

Therefore, there are various integrated weed management tools for managing eastern redcedar. Small trees (<1 m tall) can be easily controlled by mechanical means, chemical and biocontrol. Trees that are 1–3 m tall can be easily controlled by cutting or using herbicides for individual tree treatments. As the trees get taller than 3 m (9 feet), the most viable tool is cutting and then burning the cut tress as a brush-pile. The take homes message is: control the eastern redcedar trees while they are small.

Case Example 2: Integrated Management of Downy Brome (*Bromus tectorum*) in Pasture

Introduction: Downy brome (*Bromus tectorum* L.) is an invasive winter annual grass. It is native to parts of Eurasia and North Africa, and naturalized across all continents except Antarctica.

Downy brome is an erect grass 5–60 cm tall, often growing in large tufts, with a few main stems and a finely divided fibrous root system (Upadhyaya et al. 1986). The leaf sheaths are light green and pubescent. Identifying characteristics of the *Bromus* genus present in downy brome include closed leaf sheaths and a drooping inflorescence (Morrow and Stahlman 1984). Downy brome panicles are dense and soft, and change from green to purple color in cold temperatures, water stress, or when the plant matures (Morrow and Stahlman 1984). Purple coloration can be a useful heuristic to assess seed maturity, as panicles are generally mature at the time leaf sheathes transition to purple (Hulbert 1955). Glumes are covered with short barbs that allow spikelets to attach to wool, hair, clothing, and other materials to aid its dispersion of panicles (Stewart and Hull 1949).

The spread of downy brome across the United States and Canada has been well described thanks to historic records and genetic evidence. The first record of downy brome in the United States occurred in Pennsylvania in 1790 (Muhlenberg 1793). Although the initial reports of downy brome described the species as rare, by 1861 it was a commonly reported species in the eastern United States (Stewart and Hull 1949). Expansion of downy brome into the western United States is largely thought to have occurred through movement of contaminated grain and hay. The distribution of downy brome was greatly increased by the depression of 1920 when abandoned wheat farms led to 'spectacular occupations' (Stewart and Hull 1949). Downy brome was first reported in Washington in 1893, Utah in 1894, Colorado in 1895, and Wyoming in 1900. While most of the expansion of downy brome in the West was accidental, it was deliberately sown on an experimental farm in Pullman, WA in 1898, and sold across the West as a '100 day forage grass' in 1915 (Upadhyaya et al. 1986). Genetic evidence suggests downy brome was introduced to North America through multiple immigrations rather than as a single event (Novak et al. 1993). By comparing genetic markers, Novak et al. (1993) were able to identify six unique founder events in the Pacific Northwest alone between 1889 and 1902. Downy brome currently exists in the western United States as one of the region's most abundant vascular plant species (Novak et al. 1993).

Downy brome typically behaves like a winter annual, but can germinate as a summer annual if conditions, such as drought favor delayed germination (Thill et al. 1984). Downy brome does not have very exacting habitat requirements and can be found in locations ranging from 15 to 56 cm in annual precipitations, and at elevations as high as 2,700 m (Klemmedson and Smith 1964, Morrow and Stahlman 1984). Stewart and Hull (1949) characterized downy brome as an opportunist, easily acclimating to local conditions thanks to rapid germination and establishment following variable fall or spring precipitation events. Mack and Pyke (1983) made similar observations during a two-year study of demography of isolated populations. A single population can respond simultaneously at the same site with different monocarpic life history strategies (Mack and Pyke 1983). While a prolonged dry period or an extreme frost after germination can threaten downy brome establishment in the fall, most populations experience a second flush of germination in the spring (Mack and Pyke 1983). Plants which emerge in the fall have far greater fecundity than spring-emerged downy brome (Mack and Pyke 1983, Stewart and Hull 1949), but the spring emergence ensures survival of a population even during fall drought or harsh winters. This second flush also complicates management as inputs used to control downy brome in the fall or winter will likely not impact late emerging individuals.

Vigorous initial growth and early seed productions compared to other grasses allows downy brome to establish in disturbed pasture and rangeland (Thill et al. 1984). Downy brome is capable of outcompeting native perennial grasses in rangelands and pastures early in the season, reducing soil moisture to the permanent wilting point to a depth of 0.7 m in natural stands (Hulbert 1955). Downy brome can establish in a wide range of rangeland habitats, but is particularly dominant in sagebrush steppe where it has established complete monocultures across thousands of acres within the Great Basin of the United States (Young and Evans 1973). Sagebrush steppes are particularly vulnerable to downy brome invasion when the land is grazed by domestic livestock, as the ecosystems did not evolve to support heavy grazing pressure and, when grazed, new niches open for downy brome to readily establish (Billings 1992).

In rangeland the two most associated disturbances with downy brome invasion are overgrazing and wildfire (Billings 1992, Knapp 1996). In areas of Western United States, where grazing has been limited or non-existent, downy brome exists as a rare species (Passey et al. 1982, Tisdale et al. 1965). When overgrazing occurs, native plants not adapted to heavy grazing decrease and downy brome increases in abundance. When downy brome increases to a high enough density in an invaded land the frequency of fire can also increase. After senescence, downy brome forms dry tufts of flammable dead plant material, an abundant fuel source for wildfires on rangeland. Downy brome invasion was recognised as a factor for increasing fire frequencies as early as the 1930s (Pickford 1932, Stewart and Hull 1949, Brooks et al. 2004). Increased fire frequencies can lead to further decreases in native vegetation and create more niches for downy brome invasion. This positive feedback cycle leads to drastic changes in diversity of plant communities, soil biota, and fauna at higher trophic levels (Knapp 1996, Levine et al. 2003).

Because downy brome can form monocultures, it is often the only available forage for grazing livestock and wildlife. Downy brome can be a valuable forage in the spring before entering the reproductive phase (Hull and Pechanec 1947). However, the quantity of forage it produces can be quite variable from year to year. Another problem is that downy brome is palatable for only a short time window compared to many other desirable native forages (Klemmedson and Smith 1964). As downy brome matures, the forage quality decreases significantly. Cook and Harris (1952) reported a decrease in crude protein content from 10%–12% to less than 5% as the plant transitioned from a vegetative to reproductive growth habit and finally senesced.

Fire

While wildfire often facilitates downy brome establishment in new areas or increase in abundance, controlled burns can also be an effective management tool for downy brome (Brooks 2002, DiTomaso et al. 2006). Most wildfires occur when conditions are hot and dry, a time when downy brome has already desiccated and set seed (Knapp 1996, Klemmedson and Smith 1964). Consequently, desirable vegetation cover is reduced by wildfires while downy brome is less impacted. In the following fall, downy brome can exploit the recent disturbance. A prescribed burn in the fall or winter can have the contrary effect, reducing downy brome abundance and litter at a time when most desirable species are dormant. A winter prescribed burn removes accumulated litter and reduces soil moisture needed for downy brome survival, creating niches for desirable vegetation in the spring (Adair et al. 2008).

However, under conditions of high downy brome abundance, limited desirable plants may be present in the soil seed bank, or conditions may not be appropriate for their germination following burning. In such situations, downy brome will simply reinvade and prescribed burning will have negligible long-term impacts (Jessop and Anderson 2007). Burning may bring the additional risk of invasion by noxious or invasive weeds other than downy brome (DiTomaso 2000). As is the case with herbicides, managing downy brome with fire often needs to be done concurrently with additional strategies, such as applying herbicides or seeding desirable vegetation following burning to prevent invasive weeds from establishing.

Chemical Control

Most herbicides recommended for control of downy in rangeland and pasture selectively target annual grass species while causing negligible injury to perennial grasses or forbs. Commonly recommended herbicides include imazapic, rimsulfuron, and others that belong to the acetolactate synthase (ALS) inhibiting class of herbicides. Herbicides are most effective at controlling downy brome when applied as early as possible (Kyser et al. 2013), often in the fall just prior to downy brome emergence or in the late fall to early winter. Herbicide efficacy varies depending on the size of the plant at the time of herbicide application or the environmental conditions before or after application (Lawrence 2014). When conditions and growth stage are optimal at the time of application, herbicides can consistently reduce downy brome abundance by over 90% (Kyser et al. 2013).

Herbicides do not require significant modifications in grazing or land use management practices and may be considered easier to integrate as a management option for invasive plants. Herbicides can be more selective than prescribed grazing or controlled burns, and can control targeted weeds while promoting beneficial plants. Such advantages might explain why herbicides remain the most used management strategy for invasive weed control in rangeland and pasture (Bussan & Dyer 1999).

As any other approach, the use of herbicides is not a panacea to control downy brome. Besides cost and the potential for environmental injury from improper use, Whitson and Koch (1998) found that even when herbicide treatments provided greater than 90% control, additional flushes of downy brome would emerge following applications.

Biological Control

Grazing Management: As many areas infested with downy brome are being managed to support cattle and other livestock, targeted grazing may limit cost and labour compared to other management strategies. Kelley et al. (2013) surveyed numerous land managers about downy brome management strategies and reported that ranchers prefer early spring grazing above other management strategies. Ranchers seemed to prefer management tools already available to them with limited additional cost or labor (Kelley et al. 2013).

Targeted grazing of downy brome can be an effective management strategy if timed correctly. As downy brome transitions from a vegetative to a reproductive growth stage, grazing can greatly reduce the number of seeds produced by destroying maturing flowers. To simulate the effectiveness of grazing for reducing downy brome and determine optimal timing of grazing, Hempty-Mayer and Pyke (2008) clipped downy brome plants either at the boot stage or as the plant began turning purple just prior to flowering. Unclipped plants produced 12,000 to 20,000 seed m^{-2}, while clipped treatments conducted when the plant was in the boot stage reduced seed production by 42% to 99%. Later timings also reduced seed production, but were not as effective at reducing seed production (Hempty-Mayer and Pyke 2008).

While grazing may be easier to integrate into existing practices, it would likely prove difficult to optimally time grazing to manage large areas infested with downy brome. Indeed, the effective window of time to reduce seed set through grazing is extremely narrow (Hempty-Mayer and Pyke 2008, Stewart and Hull 1949). Too early, and downy bromes produces additional culms, too late, and grazing becomes less effective and plants may become less palatable (Stewart and Hull 1949). An additional challenge is the preference of grazing animals for other species growing alongside downy brome. Stocking pastures at a high enough rate to significantly affect downy brome seed production is likely to affect desirable grasses and forbs. Timing, stocking rate, and the frequency of grazing are all critical factors to consider for managing downy brome or other invasive weeds (DiTomaso JM 2000).

Seeding Competitive Plants: In areas where a downy brome monoculture is present, seeding of desirable plants may be required to ensure downy brome does not reinvade after use of another control measure, such as fire or herbicides (Whitson and Koch 1998). Certain perennial grasses are the preferred choice for revegetation efforts as they can be competitive with downy brome and can tolerate heavy grazing pressure (DiTomaso 2000). If perennial grasses are able to establish, suppression of downy brome can occur for years following the initial planting.

In an evaluation of downy brome suppression, following seeding of competitive perennial grasses, Whitson and Koch (1998) compared two native and three introduced perennial species. Downy brome abundance was reduced in all treatments compared to an unseeded control after three years (Whitson and Koch 1998). While the results reported by Whitson and Koch (1998) are encouraging, other reports in the literature have had mixed results (Francis and Pyke 1996, Mazzola et al. 2008). Downy brome often outcompetes and suppresses the seeded perennials resulting in the failure of costly reseeding treatments.

Another challenge when sowing competitive perennials is ensuring the correct environmental conditions are met to ensure establishment. Jessop and Anderson (2007) had mixed success establishing competitive perennials to suppress downy brome following wildfire. Regardless of whether reseeding had occurred or not, downy brome increased in abundance in all burned plots compared to non-burned plots. The poor competitiveness of seeded plants was attributed to poor establishment because of high soil salinity and inadequate precipitation for germination (Jessop and Anderson 2007).

In a well-controlled experiment to establish the influence of plant density on competition, Francis and Pyke (1996) evaluated two perennial grasses species which are often cited in the literature for their ability to outcompete downy brome. Downy brome was the most competitive plant in the experiment, consistently suppressing the perennial competitors regardless of plant density or species mixture. At certain densities, a few cultivars were able to suppress downy brome and only under ideal conditions were perennial grasses effective competitors (Francis and Pyke 1996). The species identified in the literature as most competitive against downy brome are generally also non-native species (Whitson and Koch 1998, Francis and Pyke 1996).

Biological Control with Plant Pathogens: A number of efforts have been conducted to control downy brome using naturally occurring plant pathogens. In a review of literature Meyer et al. (2016) described five different fungal pathogens, *Fusarium* sp. n., *Pyrenophora seminperda*, *Rutstroemiaceae* sp. n., *Tilletia bromi,* and *Ustilago bullata*, which have been previously studied for their control of downy brome. Additionally, the soil bacteria *Pseudomonas fluorescens* has also been identified as another naturally occurring pathogen of downy brome (Kennedy et al. 2001). While the large number of known pathogens of downy brome is encouraging, there are great challenges in adapting these pathogens as management tools. To do so pathogens would need to be distributed as artificial inoculum and then applied a mycoherbicide. Such a strategy has been attempted for several pathogens, *Pyrenophora seminperda* and *Ustilago bullata*, however the results have not been efficacious (Meyer et al. 2010, 2016).

Roadblocks to using these pathogens to control downy brome include balancing control with persistence over multiple years, already present host-plant resistance, and narrow infection windows due to host-pathogen biology and weather conditions (Meyer et al. 2016). More research into the use of plant pathogens is needed, but the current literature does support the successful use of pathogens when integrated into other management approaches (Baughman et al. 2016, Ehlert et al. 2013).

Integration of Approaches

A variety of options are available for the management of downy brome, but all come with sometimes significant disadvantages. To maximize the probability of success, integrated approaches need to be implemented (DiTomaso 2000). Only a few examples of integrated management of downy brome have been exposed in the literature, but the available information supports the use of IWM.

Kessler et al. (2015) compared the herbicides glyphosate, imazapic, tebuthiuron, and a non-treated control in both burned and unburned plots. Burning alone and all herbicide treatments alone reduced downy brome biomass compared to non-treated plots. Control was greatest when burning and herbicides were combined. Kessler et al. (2015) attributed the improved control of burning followed by an herbicide treatment as a consequence of increased herbicide absorption. Burning prior to herbicide application increased herbicide absorption by 7% for every 50 g m^{-2} of biomass removed.

Ehlert et al. (2013) evaluated the ability of *Pyrenophora semeniperda* to suppress downy brome, compared to and in combination with the herbicide imazapic. The initial hypothesis was that both approaches would synergize to provide additional long-term suppression of downy brome exceeding either approach alone. Ehlert et al. (2013) found that both strategies did reduce downy brome alone, but in combination did not control downy brome better than imazapic alone. However, Ehlert et al. (2013) argued that there is still value in considering an integrated approach. Imazapic was the superior treatment at reducing biomass while *Pyrenophora semeniperda* was superior at limiting the emergence of downy brome. Further, imazapic control was inconsistent year-to-year and integrating both approaches may provide more reliable control over a wide range of conditions than relying on a single approach (Ehlert et al. 2013).

In the Great Basin of Utah, 'cheatgrass die-offs' have been reported where large stands of downy brome monocultures die leaving bare soil patches behind (Meyer et al. 2008). These die-offs are thought to be caused by several fungal pathogens, but the role of individual species has yet to be determined (Baughman and Meyer 2013, Meyer et al. 2014). While efforts to replicate naturally occurring die-offs with mycoherbicides have been unsuccessful, Baughman et al. (2016) leveraged a cheatgrass die-off to investigate the success of seeding native vegetation. Native grass species, from both local and non-local seed sources were seeded into plots where recent die-offs occurred, and nearby control plots. Native plants had lower emergence in die-off sites, but greater vigour and flowering activity in the first two years after establishment. Results indicate that die-off areas can be leveraged to improve restoration, however more data is needed on native seedling success after the first two years following seeding (Baughman et al. 2016).

Whitson and Koch (1998) compared the use of herbicides, grazing, and seeding of perennial grasses. Herbicides and grazing alone were successful in reducing downy brome abundance compared to a non-treated check. However, even after high levels of control from herbicides or grazing, reinvasion can occur. Seeding of competitive perennials was considered the best long-term strategy, and combining either grazing or herbicides with sowing of competitive species could result in the best of both approaches, providing initial high control coupled with long-term suppression (Whitson and Koch 1998).

'Cheatgrass management handbook: managing an invasive annual grass in the Rock Mountain region' is a recent and extensive resource on downy brome management. It is an extension guidebook published by the University of Wyoming, and available for free on the internet. Within the guidebook are several concise examples of downy brome management taken from actual land managers using integrated approaches to control downy brome. The information within the handbook may be useful to land managers interested in IWM approaches beyond managing downy brome (Mealor et al. 2013).

Concluding Remarks

Establishing realistic goals should be the first step in developing an integrated management plan. Complete eradication of downy brome (or any other invasive species), even in a single location, is likely not an attainable objective. However, preventing the further spread to new areas, reducing the abundance of invasive species, promoting greater diversity of beneficial plants, or improving forage quality over an area of land may be within the reach of our currently available management options. To be relevant, weed management plans in pasture and rangeland need to be applicable in the long-term, considering the economic costs, and focussing on prevention of further invasions (DiTomaso 2000).

REFERENCES

Adair, E.C., I.C. Burke and W.K. Lauenroth. 2008. Contrasting effects of resource availability and plant mortality on plant community invasion by *Bromus tectorum* L. Plant Soil 304: 103–115.

Billings, W.D. 1992. Ecological impacts of cheatgrass and the resultant fire on ecosystems in the western Great Basin. pp. 22–30. *In:* Symposium on Ecology, Management, and Restoration of Intermountain Annual Rangelands, Boise, ID.

Bragg, T.B. and L.C. Hulbert. 1976. Woody plant invasion of unburned Kansas bluestem prairie. J. Range Manage. 29: 19–24.

Brooks, M.L. 2002. Peak fire temperatures and effects on annual plants in the Mojave Desert. Ecol. Appl. 12: 1088–1102.

Brooks, M.L., C.M. D'Antonio, D.M. Richardson, J.B. Grace, J.E. Keeley, J.M. DiTomaso, R.J. Hobbs, M. Pellant and D. Pyke. 2004. Effects of invasive alien plants on fire regimes. BioScience 54: 677–688.

Baughman, O.W. and S.E. Meyer. 2013. Is *Pyrenophora semeniperda* the cause of downy brome brome (*Bromus tectorum*) die-offs? Invasive Plant Sci Manag. 6(1): 105–111.

Baughman, O.W., S.E. Meyer, Z.T. Aanderud and E.A. Leger. 2016. Cheatgrass die-offs as an opportunity for restoration in the Great Basin, USA: Will local or commercial native plants succeed where exotic invaders fail? J. Arid Environ. 124: 193–204.

Buehring, N.W., P.W. Santlemann and H.M. Elwell. 1971. Responses of eastern redcedar to various control procedures. J. Range Manage. 24: 378–382.

Bussan, A.J. and W.E. Dyer. 1999. Herbicides and rangeland. pp. 116–132. *In:* Sheley, R.L. and J.K. Petroff (Eds.) Biology and Management of Noxious Rangeland Weeds. Oregon State University Press, Corvallis, OR.

Cook, C.W. and L.E. Harris. 1952. Nutritive value of cheatgrass and crested wheatgrass on spring ranges of Utah. J Range Manage 5: 332–337.

Crawthorne, G.L., W.T. Scott and P.M. Ritty. 1982. Eastern redcedar control in Kansas. Down to Earth 38: 1–6.

DiTomaso, J.M. 2000. Invasive weeds in rangelands: species, impacts, and management. Weed Sci. 48(2): 255–265.

DiTomaso, J.M., M.L. Brooks, E.B. Allen, R. Minnich, P.M. Rice and G.B. Kyser. 2006. Control of invasive weeds with prescribed burning. Weed Technol. 20: 535–548.

Ehlert, K.A., J.M. Mangold and R.E. Engel. 2013. Integrating the herbicide imazapic and the fungal pathogen *Pyrenophora semeniperda* to control *Bromus tectorum*. Weed Res. 54: 418–424.

Engle, D.M. and J.F. Stritzke. 1992. Enhancing control of eastern redcedar through individual plant ignition following prescribed burning. J. Range Manage. 45: 493–495.

Francis, M.G. and D.A. Pyke. 1996. Crested wheatgrass-cheatgrass seedling competition in a mixed-density design. J Range Manage 49: 432–438.

Gehring, J.L. and T.B. Bragg. 1992. Changes in prairie vegetation under eastern redcedar (*Juniper virginiana* L.) in an eastern Nebraska bluestem prairie. Am. Midl. Nat. 128: 209–217.

Gibson, C.W.D. and V.K. Brown. 1991. The nature and rate of development of calcareous grassland in southern Britain. Biol. Con. 58: 297–316.

Gibson, C.W.D. 1995. Grasslands on former arable land: a review. Bioscan UK Ltd, Oxford.

Hempty-Meyer, K. and D.A. Pyke. 2008. Defoliation effects on *Bromus tectorum* seed production: implications for grazing. Rangeland Ecol Manage 61: 116–123.

Holechek, J.L., R.D. Pieper and C.H. Herbel. 2011. Range Management: Principles and Practices (6th Edition). Pearson, N.P.

Hulbert, L. 1955. Ecological studies of *Bromus tectorum* and other annual bromegrasses. Ecol Monogr 25: 181–213.

Hull, A.C. and J.F. Pehanec. 1947. Cheatgrass – a challenge to range research. J Forestry 45: 555–564.

Jarnevich, C.S., T.R. Holcombe, D.T. Barnett, T.J. Stohlgren and J.T. Kartesz. 2010. Forecasting weed distributions using climate data: a GIS early warning tool. Inv. Plant Sci. Manag. 3: 365–375.

Jessop, B.D. and V.J. Anderson. 2007. Cheatgrass invasion in salt desert shrublands: benefits of postfire reclamation. SRM 60(3): 235–243.

Kelley, W., M. Fernandez-Gimenez and C.S. Brown. 2013. Managing downy brome (*Bromus tectorum*) in the Central Rockies: land manager perspectives. Inv. Plant Sci. Manag. 6(4): 521–535.

Kennedy, A.C., B.N. Johnson and T.L. Stubbs. 2001. Host range of a deleterious rhizobacterium for biological control of downy brome. Weed Sci. 49(6): 792–797.

Klemmedson, J.O. and J.G. Smith. 1964. Cheatgrass (*Bromus tectorum* L.). Bot Rev. 30: 226–262.

Knapp, P.A. 1996. Cheatgrass (*Bromus tectorum* L.) dominance in the Great Basin Dessert: history, persistence, and influence on human activity. Global Environ Chang. 6(1): 37–52.

Knezevic, S.Z, S. Melvin, T. Gompert and S. Gramlich. 2005. Integrated Management of eastern redcedar. University Nebraska Lincoln-Extension Publication. EC-05-186.

Knezevic, S.Z. 2017. Invasive plant species. *In:* Brian Thomas, Brian G. Murray and Denis J. Murphy (Editors in Chief), Encyclopaedia of Applied Plant Sciences, 3: 300–303. Academic Press, Waltham, MA.

Knezevic, S.Z., A. Jhala and A. Datta. 2017. Integrated weed management. *In:* Brian Thomas, Brian G. Murray and Denis J. Murphy (Editors in Chief), Encyclopaedia of Applied Plant Sciences, 3: 459–462. Academic Press, Waltham, MA.

Knezevic et al. 2017. Guide for Weed, Disease and Insect Management in Nebraska. UNL-Extension Publication. 320 p. EC-130.

Kyser, G.B., R.G. Wilson, J. Zhang and J.M. DiTomaso. 2013. Herbicide-assisted restoration of great basin sagebrush steppe infested with medusahead and downy brome. Rangeland Ecol. Manag. 66: 588–596.

Lambert, J. (Deputy Under-Secretary). 2004. Prevention, early detection and rapid response to invasive plants. Weed Technol. 18: 1182–1184.

Lass, L.W., T.S. Prather, N.F. Glenn et al. 2005. A review of remote sensing of invasive weeds and example of the early detection of spotted knapweed (*Centaurea maculosa*) and babysbreath (*Gypsophila paniculata*) with a hyperspectral sensor. Weed Sci. 53: 242–251.

Lawrence, N.C., I.C. Burke and J.P. Yenish. 2014. 15 Years of Downy Brome Control in Eastern Washington. pp. 12–13. *In:* Proceedings of the 67th Annual Meeting of the WSWS. Western Society of Weed Science, Colorado Springs.

Levine, J.M., M. Vila, M. D'Antonio, J.S. Dukes, K. Grigulis and S. Lavorel. 2003. Mechanisms underlying the impacts of exotic plant invasions. Pro R Soc Lond B. 270: 775–781.

Mazzola, M.B., K.G. Allcock, C.J. Chambers, R.R. Blank, E.W. Schupp, P.S. Doescher and R.S. Nowak. 2008. Effects of nitrogen availability and cheatgrass competition on the establishment of vavilov Siberian wheatgrass. Rangeland Ecol. Manag. 61: 475–484.

Mack, R. and D. Pyke. 1983. The demography of *Bromus tectorum*: variation in time and space. J. Ecol. 71: 69–93.

Masters, R.A. and R.L. Sheley. 2001. Principles and practices for managing rangeland invasive plants. J. Range Manag. 54(5): 502–517.

Mealor, B.A., R.D. Mealor, W.K. Kelley, D.L. Bergman, S.A. Burnett, T.W. Decker, B. Fowers, M.E. Herget, C.E. Noseworthy, J.L. Richards, C.S. Brown, K.G. Beck and M. Fernandez-Gimenez. 2013. Cheatgrass management handbook: managing an invasive annual grass in the Rock Mountain region. University of Wyoming Extension, Laramie, WY.

Meyer, S.E., J. Beckstead and J. Pearce. 2016. Community ecology of fungal pathogens on *Bromus tectorum*. *In:* Germino, M., Chambers, J. and Brown, C. (Eds.) Exotic Brome-grasses in Arid and Semiarid Ecosystems of the Western US. Springer Series on Environmental Management. Springer, Cham.

Meyer, S.E., J.M. Franke, O.W. Baughman, J. Beckstead and B. Geary. 2014. Does *Fusarium*-caused seed mortality contribute to *Bromus tectorum* stand failure in the Great Basin? Weed Res. 54: 511–519.

Meyer, S.E., D.L. Nelson, S. Clement and J. Beckstead. 2008. Cheatgrass (*Bromus tectorum*) biocontrol using indigenous fungal pathogens. pp. 61–67. *In:* Proceedings of Shrublands under fire: disturbance and recovery in a changing world. Cedar City, UT: USDA Forest Service RMRS-P-52.

Meyer, S.E., T.E. Stewart and S. Clement. 2010. The quick and the deadly: growth vs. virulence in a seed bank pathogen. New Phytol. 187: 209–216.

Morrow, L.A. and P.W. Stahlman. 1984. The history and distribution of downy brome (*Bromus tectorum*) in North America. Weed Sci. 32: 2–6.

Muhlenberg, H. 1793. Index florae Lancastriensis. Trans Am Philos Soc Philadelphia 3: 157–184.

Novak, S., R. Mack and P. Soltis. 1993. Genetic variation in *Bromus tectorum* (Poaceae): introduction dynamics in North America. Can. J. Bot. 71: 1441–1448.

Ortman, J., J. Stubbendieck, R.A. Masters, G.H. Pfeiffer, T.B. Bragg. 1998. Efficacy and costs of controlling eastern redcedar. J. Range Manag. 51: 158–163.

Passey, H.B., V.K. Hugie, E.W. Williams and D.E. Ball. 1982. Relationships between soil, plant community, and climate on rangelands of the intermountain west. Washington, DC: USDA-NRCS Technical Bulletin No. 1669.

Pickford, G.D. 1932. The influence of continued heavy grazing and of promiscuous burning on spring-fall ranges in Utah. Ecol. 13: 159–171.

Schwartz, M.W. and J.R. Helm. 1996. Effects of a prescribed fire on degraded forest vegetation. Natural Areas J. 16(3): 184–190.

Schwarzmeier, J.A. 1984. Sweet clover control in planted prairies: refined mow/burn prescription tested (Wisconsin). Restor. Manag. Notes 2(1): 30–31.

Smith, S.D. and J. Stubbendieck. 1990. Production of tall-grass prairie herbs below eastern redcedar. Prairie Nature 22: 13–18.

Stewart, G., A. Hull. 1949. Cheatgrass (*Bromus tectorum* L.)—An ecologic intruder in southern Idaho. Ecol. 30: 58–74.

Stritzke, J.F. and D. Rollins. 1984. Eastern redcedar and its control. Weeds Today 15: 7–8.

Stubbendieck, J., M.J. Coffin and L.M. Landholt. 2003. Weeds of the Great Plains. Nebraska Dept. of Agric., PO Box 94756, Lincoln, NE, 605 p.

Tanentzap, A.J., D.R. Bazely, P.A. Williams and G. Hoogensen. 2009. A human security framework for the management of invasive nonindigenous plants. Inv. Plant Sci. Manag. 2: 99–109.

Thill, D.C., K.G. Beck and R.H. Callihan. 1984. The biology of downy brome (*Bromus tectorum*). Weed Sci. 32: 7–12.

Tisdale, E.W., M. Hironaka and M.A. Fosberg. 1965. An area of pristine vegetation in Craters of the Moon National Monument, Idaho. Ecol. 46: 349–352.

Upadhyaya, M., R. Turkington and D. McIlvride. 1986. The biology of Canadian weeds: 75. *Bromus tectorum* L. Can J Bot 66: 689–709.

Whitson, T.D. and D.W. Koch. 1998. Control of downy brome (*Bromus tectorum*) with herbicides and perennial grass competition. Weed Sci. 12: 391–396.

Wilson and Schmidt. 1990. Controlling eastern red cedar on range-lands and pastures. Rangelands. 12: 156–158.

Young, J.A. and R.A. Evans. 1973. Downy brome intruder in the plant succession of big sagebrush communities in the Great Basin. J. Range Manag. 26: 410–415.

34

Organic Farming and Sustainable Weed Control

Eric Gallandt[*1], **Sonja Birthisel**[2], **Bryan Brown**[2], **Margaret McCollough**[2] and **Margaret Pickoff**[2]

[1] Professor, School of Food and Agriculture, University of Maine, Orono, USA
[2] Graduate Students, School of Food and Agriculture, University of Maine, Orono, USA

Introduction

Organic farming remains a small sector of global agricultural production, occupying only 1% of cropland, but it continues to grow, and compared to conventional production, it offers greater profitability and environmental benefits (Crowder and Reganold 2015). Our aim in this chapter is to summarize recent and foundational research that may inform best practices for organic weed management in multiple farming systems. Organic farming systems present unique challenges related to sustainable weed control, but also notable opportunities, as organic farmers generally rely more on ecologically based management practices compared to their conventional peers. Organic farming is often defined not based on its foundational principles, but rather by prohibited inputs or practices, e.g., synthetic pesticides and fertilizers are not permitted. However, the organic farming community is diverse in motivation, philosophy and practices. Organic Certification requires that farmers follow precisely defined practices outlined by their certifying body. In the USA, regulations are developed by the United States Department of Agriculture's National Organic Program (USDA NOP 2017); in Canada, the Canadian Food Inspection Agency is the responsible entity; similar regulatory processes operate in the European Union, Australia, and in many other countries. While permitted inputs or practices may vary among certifying organizations, the regulations broadly address four core principles outlined by IFOAM-Organics International (Luttlkholt and Woodward 2005): (i) health of soil, plants, and humans; (ii) ecology, as the basis for system design and management; (iii) fairness, related to relationships with the environment and life opportunities; and (iv) care, to protect the health and wellbeing of current and future generations and the environment. Thus, while it is true that organic crops are managed without synthetic pesticides and fertilizers, there are many fundamental differences between organic and conventional farming systems, with many organic practices affecting weeds and weed management.

Organic farmers cite cultivation or physical weed control, crop rotation, and cover cropping as their foundational weed management practices (Baker and Mohler 2014). Crop competition is likewise a key strategy, readily manipulated by a wide array of species, genotype, density, arrangement, and resource variables. Despite a seemingly impressive toolkit to draw upon, and the high level of diversity in management practices reported by organic farmers (DeDecker et al. 2014), weeds remain their foremost production challenge. The transition from conventional

---

*Corresponding author: gallandt@maine.edu

to organic management commonly results in a large increase in weed pressure, evidenced in the weed seedbank (Albrecht 2005).

Organic farms are often more diverse than their conventional neighbors (Norton et al. 2009), using multiple crop species or enterprises for workload spreading, and to reduce economic risks associated with low prices or yields. This diversity provides opportunity for varied disturbance regimes that can contribute to multiple weed management benefits. Key to exploiting this diversity, however, is knowledge of weed biology, especially seedbank characteristics, emergence periodicity, and seed rain dynamics. While weeds remain an important production constraint in organic farming systems, there are many successful farmers who have either 'solved' or at least 'managed' their weed problems. Their success, at least in part, relies on knowledge of weed management principles, effective use of available physical weed control tools, and selected ecologically-based management tactics, many that are chosen to provide multiple benefits.

Principles

Weed management is guided by three principles: i) reduce weed density; ii) reduce damage caused by surviving weeds; and iii) avoid community shifts towards pernicious species. Conventional farmers often address this first principle using herbicides to reduce weed density and thus crop yield loss, a practice guided by the well-known weed-density and crop-yield loss relationships described by Cousens (1985). Liebman and Gallandt (1997) made a case for use of a diversity of tactics that individually may not provide acceptable levels of weed control, but could through their combined effects. Their 'many little hammers' analogy has been enthusiastically accepted by the organic farming community where physical weed control remains important, but is supported by diverse rotations, cover cropping, transplanting and high-density sowing, strategic handweeding, and other cultural practices that aim to reduce seedling densities, improve crop-weed competition, and reduce weed seedbanks. The increased complexity that accompanies a diverse, multi-tactic weed management approach is worth noting at the outset (Bastiaans et al. 2008); increased costs associated with implementation are borne by the producer, and ultimately the consumer. Notable as well, however, are environmental benefits from reduced pesticide use, improvements in soil quality and conservation, and increased biodiversity associated with organic farming systems (Reganold and Wachter 2016).

In the following review, we start with a discussion of Physical Weed Control, sometimes called *direct* control, an important strategy for most organic farmers. This is followed by discussion of Ecological Weed Management and so-called *indirect* or *cultural* weed management practices, broadly aiming to manipulate weed seedbank dynamics. Our focus is on the most important and effective practices that encourage or prevent germination and establishment, and preempt or reduce weed seed rain, aiming to maximize debiting while minimizing crediting of the weed seedbank (see Forcella 2003, Gallandt 2006, 2014). Next, we highlight diversity among organic cropping systems, with principles and practices both unique and broadly applied. We conclude with a presentation of case studies that consider the multiple effects, costs and benefits of contrasting organic weed management philosophies, specifically seedling *versus* seed-focussed management, as well as preventative, mulching practices.

Physical Weed Control

Physical weed control refers to direct, non-chemical methods intended to kill weeds. This may include flame or pneumatic weeding, use of a grit blaster (Forcella 2012), or weed seed milling (Walsh et al. 2012, Jacobs and Kingwell 2016), but usually refers to *cultivation*, i.e., shallow mechanical disturbances aiming to uproot, bury or sever seedlings (Mohler 2001). *Tillage* (discussed below under Ecological Weed Management) provides physical weed control, but its primary aim is related to seedbed preparation and planting.

Physical weed control is often criticized for its fossil fuel dependence and negative effects on soil quality. Certainly, tillage can negatively affect soil health (Browning et al. 1945, Rasmussen 1999, Busari et al. 2015) by depleting soil organic matter (Post and Kwon 2000), and contributing to erosion and compaction (Pagliai et al. 2004, Hamza and Anderson 2005). To our knowledge, the effect of cultivation on soil quality has not been explicitly studied. Furthermore, organic growers often implement soil-building practices that counteract potential negative effects of cultivation.

Inter- and Intra-Row Tools

Weed control tools typically target inter- (between-row) or intra-row (in-row) zones. Inter-row cultivation is commonly achieved with 'sweep' type hoes (see Bowman 2002). Inter-row tools may also be PTO-powered but such tools may not improve efficacy (Pullen and Cowell 1997) and likely require slower working rates. Although sweeps typically have high efficacy, newer inter-row prototypes have been designed to minimize variability resulting from changes in cultivation speed, soil moisture, and weed size (Evans et al. 2012). Additionally, sweeps designed for cultivation in high-residue conditions have been successful in conservation tillage corn (*Zea mays* L. ssp. *mays*) and soybean (*Glycine max* (L.) Merr.) systems (Keene and Curran 2016, Snyder et al. 2016). Other tools have been designed to work closer to the crop row, such as ground-driven rotary cultivators (e.g., Hillside Cultivator Co. LLC. or Spyder Weeders, BezzeridesBrothers Inc.). Flame weeding can also be used to control inter-row weeds, and may be advantageous if soils are too wet for cultivation (Ascard et al. 2007), but it is more typically used in other applications (see the *Stale Seedbed* section below).

Physical weed control in the intra-row zone is much more difficult. Tools need to selectively kill weeds without damaging the crop. Selectivity may be based on resistance to uprooting (Kurstjens et al. 2000), burial (Kurstjens and Perdok 2000, Jensen et al. 2004), or heat (Ascard 1994). Tools in this category include torsion, finger, pneumatic, and flame weeders, and each need to be carefully adjusted to achieve optimal selectivity (van der Schans et al. 2006). In high-value crops, intra-row weed control tools are often used to decrease handweeding labor. For example, torsion weeding allowed for a 73% reduction in handweeding labor in transplanted onions (*Allium cepa* L.) (Ascard and Fogelberg 2008). Intra-row flame weeding may also be successful, particularly against dicot weeds (Sivesind et al. 2009). Efficacy may be improved by following intra-row flaming with aggressive inter-row cultivation to bury small, intra-row weeds. In an experiment with organic maize, this technique provided over 90% weed control and highest yields compared to several cultivation- or flaming-only treatments (Stepanovic et al. 2016).

Harrowing

Harrowing refers to spring-tine implements that provide shallow disturbance across inter- and intra-row zones. Harrowing may be used to prepare a stale seedbed period prior to planting (see the *Stale Seedbed* discussed below), after seeding but prior to crop emergence, or post-emergence. Scheduling pre-emergence harrowing may be challenging, however, requiring small, white-thread-stage weeds, while the emerging crop should be below the zone of disturbance. Post-emergence harrowing is commonly used in small grains, row crops, and vegetables. Selectivity is based on the size differential between the larger, well-anchored crop, and the smaller, white-thread stage weeds (Rasmussen 1992, Kurstjens and Perdock 2000). Site conditions, including soil moisture and texture, also affect selectivity (Kurstjens and Kropff 2001), and may require adjustment of speed, tine angle, gauge, and arrangement to optimize performance (Cirujeda et al. 2003, Rasmussen et al. 2010). Harrow aggressiveness must be adjusted to maximize weed control while causing minimal crop injury (Rasmussen 2004).

Hand Tools

Weed management on small-scale organic vegetable farms is often achieved with handweeding

(Riemens et al. 2010, Jabbour et al. 2014b, Baker and Mohler 2014). Handweeding is labor intensive and thus costly (Ascard and Fogelberg 2008), but efficiency may be improved by use of appropriate hand tools (Figure 34.1). In field experiments comparing hand pulling, short-handled tools, long-handled tools, and wheeled tools (E.R.G., unpublished data), the smaller tools had the greatest efficacy but had the slowest working rates, and the reverse was true for the larger tools. In 2015, we focussed on the intra-row zone, by conducting similar experiments comparing the intra-row efficacy and working rates of several long-handled cultivation tools to that of several wheeled tools (B.B. and E.R.G., unpublished data). Overall, we found that efficacy of long-handled tools and wheeled tools was similar ($F_{1,38}$ = 1.300, P = 0.261), with a mean of 61%. However, the wheeled tools were four times faster (orthogonal contrast, $F_{1,36}$ = 108, P < 0.001) and killed three times as many weeds per second (orthogonal contrast, $F_{1,36}$ = 15.6, P < 0.001). Thus, with hand tools, it is important to maximise efficiency, wheel hoes and long-handled tools should be used first, followed by smaller tools if necessary, and hand pulling should only be used for intricate areas where high efficacy is desired (Figure 34.1).

Improving Efficacy

Efficacy of physical weed control is affected by four factors: (i) the implement or tool, including its design, adjustment, and speed of operation; (ii) the weed, species, and size; (iii) soil conditions, possibly texture, moisture, seedbed quality, and residues; and (iv) the crop species, size, and tolerance to physical disturbance (Gallandt et al. [in press] 2017). Given these many variables, it is not surprising that cultivation efficacy is often low and highly variable. In a review of 55 field studies of cultivation, efficacy ranged from 21 to 90%, with an overall mean of 66% (Gallandt et al. [in press] 2017). Farmers strive to optimise efficacy using some logical 'rules of thumb' to determine when and how to cultivate. For example, efficacy is generally greatest when weeds are small (Terpstra and Kouwenhoven 1981, Kurstjens et al. 2000, Cirujeda and Taberner 2004), but aggressive inter-row hoes may be unaffected by weed size (Pullen and Cowell 1997, Evans et al. 2012). Furthermore, efficacy is typically highest in dry conditions (Cirujeda and Taberner 2004, Evans et al. 2012) and weed survival increases with greater moisture levels (Terpstra and Kouwenhoven 1981, Kurstjens 2002, Mohler et al. 2016), but weed uprooting may be promoted by higher soil moisture content (Kurstjens et al. 2000). Soil texture is also an important factor, but results may vary with different tools. For example, harrowing was most effective on sandy soils (van der Weide and Kurstjens 1996), but when burying weeds, as with hilling, efficacy

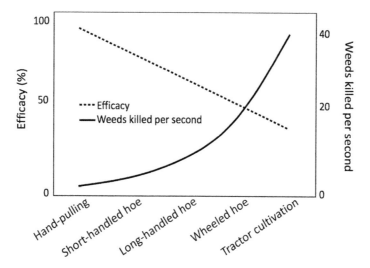

Figure 34.1. Schematic diagram portraying the contrasting efficacy and efficiency (weeds killed per second) of various intra-row tools.

was greater in heavier soils (Baerveldt and Ascard 1999). Increasing tool depth may increase efficacy (Terpstra and Kouwenhoven 1981, Kurstjens et al. 2000, Bleeker et al. 2002), but if cultivator blades are too deep, they may pass below the rooting zone of weeds, decreasing disturbance to weeds, while potentially causing greater injury to crop roots. Increased forward speed generally increases efficacy (Rydberg 1994, Rasmussen and Svenningsen 1995, Evans et al. 2012) possibly due to increased soil movement (Kouwenhoven and Terpstra 1979, Evans et al. 2012). However, speed effects may be inconsistent (Cirujeda et al. 2003, Rasmussen 1990), unaffected by forward speed if hoes cause minimal soil movement (Evans et al. 2012), or even negatively affected if hoes are powered (Pullen and Cowell 1997). Indeed, there are multiple factors affecting the efficacy of cultivation, and, unfortunately, a dearth of mechanistic studies related to this foundational practice.

Stacking Tools

Efficacy may be improved by combining or 'stacking' multiple tools, presumably exploiting their varied physical mechanisms of weed mortality. Early work showed that weed burial was the most important cause of mortality in spring tine harrowing (Habel 1954, Kees 1962, Koch 1964), a result mirrored by more recent studies (Kurstjens and Perdok 2000, Jensen et al. 2004). However, uprooting was found to be the dominant cause of mortality in the zone disturbed by hoeing, while burial was the sole cause of mortality on either side of the hoe (Terpestra and Kouwenhoven 1981, Fogelberg and Gustavsson 1999). In addition to uprooting and burial, severing of weed stems was also a factor in hoeing efficacy (Evans et al. 2012). Thus, it is likely that the relatively high efficacy achieved by inter-row hoeing reflects the implementation of multiple mechanisms of mortality. We recently evaluated the efficacy of several intra-row tools 'stacked', i.e., used in conjunction in the same pass (B.B. and E.R.G., unpublished data). We used torsion weeders, finger weeders, and a tine rake (like a tine harrow), each targeting the intra-row zone. Stacking consistently improved efficacy, and the magnitude of the effect in some cases was greater than the sum of individual tools, providing evidence of synergy. New tool designs should consider and evaluate multiple implements that utilize different mechanisms of action.

Guidance Systems

Efficacy can be increased, and crop damaged reduced with manual or autonomous guidance to accurately follow crop rows. There are low- and high-tech options for improved cultivator guidance. Rear-mounted cultivators require operators to steer while looking over their shoulder, while front- and belly-mounted equipment can both reduce operator strain and improve accuracy. Some rear-mounted cultivators allow an additional operator to more accurately steer by utilizing side-shifting hydraulics, however, this is a more labor intensive approach (Mohler 2001). Guidance of cultivation tools can also be achieved using cultivation equipment with sway in the three-point hitch and guide wheels that allow the cultivator to follow bed edges or furrows between rows (Bowman 2002).

GPS-, camera-, and sensor-guidance systems have improved working rates and accuracy of guidance, permitting closer working distance to the crop row (Van der Weide et al. 2008), and in some cases, improved efficacy (Lati et al. 2016). However, some intelligent intra-row weeding tools have not surpassed the performance of traditional spring tine harrow, finger, or torsion weeders in transplanted onion or cabbage (*Brassica oleracea* L.) (Melander et al. 2015), and failed to provide improved selectivity compared to a spring tine harrow in sugar beets (*Beta vulgaris* L.) (Rasmussen et al. 2012). Working rates are, however, impressive, with speeds of 5 to 12 km h^{-1}, respectively (www.garford.com). Robotic weed control machines may offer even greater labor savings by reducing the amount of handweeding required post-cultivation (Lati et al. 2016), however, detection and identification of weeds remains challenging (Slaughter et al. 2008). Although the degree of autonomy of contemporary robotic weeders may be somewhat limited (Merfield 2016), recent commercial launch of 'Tertill®' (Franklin Robotics), and 'Oz' (Naio Technologies) weeding robots suggests that this technology is advancing rapidly. Co-robots, or robots designed to work cooperatively with workers, may be a more attainable goal.

Indeed, such co-robots with intra-row hoes reduced handweeding labor requirements by 58% in tomato (*Solanum lycopersicum* L.) (Pérez-Ruízet al. 2014).

Site-Specific Management

The integration of geographic information systems, drone-mediated mapping, with guidance systems, and decision making software can optimize management practices based on within-field variability (Young and Pierce 2014). This may improve upon traditional farming practices, where weed management decisions and treatments are typically made at the field or farm scale (Grisso et al. 2009). To date, precision weed management has been utilized primarily in conventional systems for applying variable rates of herbicides, which has reduced excess application (Gomez-Casero et al. 2010, Joint Research Center of the European Commission et al. 2014). However, precision technologies might also benefit organic growers by optimizing both the physical and cultural management of weeds to favor the success of the crop. For example, a precision cultivator could account for soil conditions, crop characteristics, weather forecasts, as well as weed community composition and abundance to optimize tool choice, working speed, and working depth. Precision agriculture technologies have their limitations (Lopez-Granados 2010) and are expensive. In a 2016 report from the USDA, precision agriculture tools had only a small positive impact on profits for those corn producers utilizing them (Schimmelpfenning 2016). However, if precision and variable rate technologies continue to improve and become more affordable, they may be adapted for weed management applications in organic agriculture.

Ecological Weed Management

Ecological weed management seeks to benefit crop growth by manipulating the relationships between crops, weeds, soils, and other agroecosystem components such that weeds are disadvantaged. Weed management regimes that incorporate ecosystem thinking to include diverse practices—'many little hammers'—can result in effective non-chemical control (Liebman and Gallandt 1997, Figure 34.2). Further, ecological weed management may yield multiple benefits by incorporating practices that improve soil quality (Gallandt et al. 1999), preserve biodiversity (Benton et al. 2003), and support ecosystem services (Hooper et al. 2005).

Knowledge of crop and weed biology and ecology should underpin the development of ecological weed management approaches (Swanton et al. 2008). In a study surveying organic farmer knowledge and perceptions of weeds, those with a high level of understanding of ecological complexity had lower weed seedbank densities (Jabbour et al. 2014a). Diverse practices alone are insufficient to ensure effective weed control (Mirsky et al. 2010); considering the phenology of problem weeds is necessary to inform management that targets sensitive life stages (Swanton and Booth 2004). A mistimed application, such as harrowing that occurs later than the seedling stage, or cover crops that are terminated late, can dramatically increase weed seedbanks, contributing to high weed pressure in subsequent growing seasons (B.B. and E.R.G., unpublished data). Not only diversity, therefore, but timely utilization of diverse disturbance events is the key to implementation of successful organic weed management rotations (Smith 2006). Below we discuss several widely-used practices that can be leveraged to strategically disturb weeds at multiple life stages, with examples of successful application in vegetable, grain, and row crop systems.

Tillage

Tillage is the practice of using aggressive tools to invert or substantially agitate soil. Tillage is often used to break compaction, kill existing vegetation, and incorporate residues; these uses are typically referred to as primary tillage. Subsequently, secondary tillage utilizing less aggressive tools may be performed to kill weed seedlings and create a uniform seedbed prior to planting. Tillage tools including moldboard and chisel plows, field cultivators, sweeps, disks, and rotary tillers are aggressive and provide a high level of weed control (reviewed in Mohler 2001). A single tillage pass is often insufficient for control of perennial weeds, but the accumulated

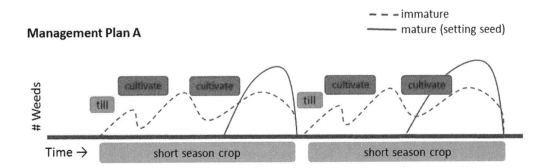

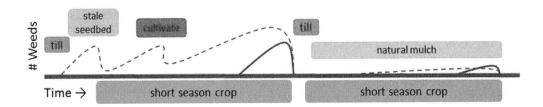

Figure 34.2. Timely utilisation of diverse disturbance events is the key to implementation of successful organic weed management rotations. Pictured here are hypothetical impacts of two weed management plans for back-to-back short season crops. Management Plan A is a cultivation-intensive approach that controls weeds during the critical period, but allows considerable seed rain prior to harvest. Management Plan B is an ecologically-based weed management plan that reduces seed rain through strategically employing diverse management tactics targeting the seedbank and seedling stage weeds.

stress of timely tillage events can deplete the carbohydrates in perenniating organs, allowing for control (Hakånsson 2003). The substantial soil disturbance resulting from tillage may bury weed seeds lying on the surface or bring seeds from deeper soil layers into the germination zone; tillage is therefore thought to 'reset' ecological succession, and its timing can act as a filter shaping weed community composition (Smith 2006).

Though tillage is relied upon to varying extents in most organic production systems, it can negatively impact soil health by depleting organic matter (Post and Kwon 2000) and contributing to nutrient leaching, erosion, and compaction (Pagliai et al. 2004, Hamza and Anderson 2005). No-till and conservation tillage systems offer numerous environmental benefits in comparison (Kurkalova et al. 2004), but are heavily dependent on herbicides, and organic agriculture analogues are difficult to implement. Strip tillage, in which tillage is spatially restricted to crop rows, is a promising practice for organic systems. Novel approaches aimed at reducing reliance on intensive tillage in organic systems are a subject of considerable interest among researchers (reviewed in Kurstjens 2007).

Stale Seedbed

A stale seedbed is one in which weeds in the top few centimeters of soil have been encouraged to germinate, perhaps establish, but are then killed, reducing the number of weeds that subsequently emerge to compete with the crop. Because germination is the most effective method to exhaust the weed seedbank (Gallandt 2006), shallow, uniform soil disturbance is used to encourage a 'flush' of germination, after which soil is left fallow to allow weeds to emerge so they can be killed by flaming (e.g., Rasmussen 2003) or shallow cultivation using a variety of tools (Johnson and Mullinix 2000). Flaming has the advantage of low soil disturbance, which reduces density of subsequent weed cohorts. Cultivation has the downside of potentially encouraging another 'flush' of weeds to emerge alongside the crop; this risk may be minimized by employing 'dark' cultivation (Riemans et al. 2007) to minimize light reaching buried seeds and providing cues necessary to break dormancy (Baskin and Baskin 1998). Punch planting, which minimizes soil disturbance to a small area during row crop planting, is another strategy that aims to minimize subsequent weed seedling emergence (Rasmussen 2003). Larger initial flushes of weeds, and therefore greater seedbank depletion, may be achieved by irrigating (Benvenuti and Macchia 2006) or using a roller or cultipacker to firm the soil and improve seed to soil-contact following tillage.

Mulching prepared beds for several weeks using clear plastic (solarization) or black plastic (occultation) can also be effective means of stale seedbed preparation (Fortier 2014, Birthisel and Gallandt, unpublished data, Marenco and Lustosa 2000). With either practice, crops may subsequently be planted in holes or slits cut into the mulch, or plastic can be removed prior to planting, allowing for re-use of materials. Materials used for implementing occultation on organic farms include durable silage tarps (Fortier 2014) and landscape fabrics. Previously used greenhouse plastic may be used for solarization (Avissar et al. 1985).

Stale seedbeds have improved weed control outcomes and crop yields in a wide variety of farming systems worldwide. In Northern India, a stale seedbed coupled with one handweeding pass reduced weed pressure by 77% and increased yield of organic garden pea (*Pisum sativum* L.) (Gopinath et al. 2009), while a stale seedbed created with three weeks of solarisation in Brazil decreased weed pressure and doubled carrot (*Daucus carota* L. var. *sativus* Hoffm.) yield (Marenco and Lustosa 2000) as compared to untreated controls. Further examples of the many systems in which non-chemical stale seedbeds have been found effective include direct seeded rice (*Oryza sativa* L.) (Rao et al. 2007), lettuce (*Lactuca sativa* L.) (Riemans et al. 2007), spinach (*Spinacia oleracea* L.) (Boyd et al. 2006), strawberry (*Fragaria* L.) (Stapleton et al. 2005), cucumber (*Cucumis sativus* L.) (Johnson and Mullinix 1998), and peanut (*Arachis hypogaea* L.) (Johnson and Mullinix 2000).

Efficacy of organic stale seedbed preparations varies (Riemans et al. 2007) but can rival that of herbicide preparations. Johnson and Mullinix (1998) found that weed control improved 50% and cucumber yields increased in a stale seedbed prepared by shallowly tilling twice (once two weeks prior to planting, and again directly prior to planting) as compared with a glyphosate stale seedbed preparation. However, Caldwell and Mohler (2001) found that mechanical weed control was less effective for stale seedbed preparation than flaming or glyphosate. Birthisel and Gallandt (unpublished data) compared stale seedbeds prepared by solarization plus flaming to a flaming-only treatment in northern New England, USA, and found that weed emergence was 78% less in solarized treatments.

Thermal Control

Thermal weed control tactics, including steaming, soil solarization, and flame weeding, can control weed seeds and seedlings. Response to thermal control, however, varies by weed species and soil conditions, especially at the seed stage (Baskin and Baskin 1998). For example, lower temperature thresholds are required in high moisture compared to dry conditions (Egley 1990), and lower temperature requirements may be needed for mortality when thermal control is integrated with incorporation of organic amendments (reviewed in Gamliel and Stapleton 2012) or crop residues (Mallek et al. 2007).

Steaming may be lethal not only to seedlings and weed seeds, but also to non-target organisms including beneficial microbiota (see reviews by Ascard et al. 2007, Kapulnik and Gamliel 2012). The history of this practice is reviewed in Runia (2012). The integration of soil steaming with rototilling may improve efficacy (Fennimore and Goodhue 2016), and mixing throughout the soil profile serves to minimize extreme temperatures at the soil surface, yet allowing lethal 65°C steam to penetrate to deeper soil layers (Miller et al. 2014). A recent prototype combination band steamer and rotovator (Raffaelli et al. 2016), showed promising results. Steaming is currently allowed under the National Organic Program in the USA (NOP 2016, Part 205, Subpart C, §205.206), but presumably due to concern about its impacts on soil ecology, its use is restricted or disallowed by organic standards in some European Union countries including the UK (http://organicrules.org/custom/differences.php?id=2aae).

Soil solarization is the practice of utilizing solar energy for pest control by covering irrigated soil with tightly secured clear plastic mulch (Mahrer and Shilo 2012, Katan et al. 1976). Its utility as a weed control technique in arid and Mediterranean climates is well documented (see reviews by Cohen and Rubin 2007, Rubin 2012), and successes in cooler climates have been reported as well (Peachey et al. 2001, Birthisel and Gallandt, unpublished data, but see Bond and Bursch 1989). Many of the world's most problematic weed species can be effectively controlled with solarization (Table 34.1), though efficacy depends on climate and treatment duration. Long-term impacts on overall weed seedbanks have not been reported, but based on high and accumulated temperatures typically achieved, and known weed seed thermo-tolerance thresholds (Egley 1990, Dahlquist et al. 2007), solarization likely causes considerable seed mortality (Cohen and Rubin 2007). It is also possible that solarization forces dormancy of some species (Marenco and Lustosa 2000), a less desirable outcome from a seedbank management standpoint. A consideration for organic farming is that soil microbial community and function may be altered because of solarization (Scopa and Dumontet 2007, Scopa et al. 2008). Effects on the microbial activity may, however, benefit crop growth as yields are often stimulated following solarization, perhaps due to disease suppression or changes in soil nutrient availability (Katan and Gamliel 2012).

Flaming to control weed seedlings is discussed elsewhere in this chapter (see *Inter- and Intra-Row Tools* and *Stale Seedbed* sections). Flaming to kill seeds laying on the soil surface (e.g., after seed rain in the fall) is effective, but requires relatively high heat exposure (E.R.G.,

Table 34.1. Responses of the world's 15 worst weeds (Ranked based on Holm et al. 1977) to solarization. Responses were assigned (+) = all or moderately controlled by solarization; or (0) = poorly controlled or no effect. This table is summarized from Cohen and Rubin's (2007) review of solarization susceptibility in over 150 weed species. Additional sources are identified with footnotes

Rank	Common name	Scientific name	Response
1	Purple nutsedge	*Cyperus rotundus*	+, 0
2	Bermudagrass	*Cynodon dactylon*	+
3	Barnyardgrass	*Echinochloa crus-galli*	+
4	Jungle rice	*Echinochloa colona*	+
5	Goosegrass	*Eleusine indica*	+
6	Johnsongrass	*Sorghum halepense*	+
7	Cogongrass	*Imperata cylindrica*	+ a
8	Water hyacinth	*Eichhornia crassipes*	0 b
9	Common purslane	*Portulaca oleracea*	+
10	Lambsquarters	*Chenopodium album*	+
11	Large crabgrass	*Digitaria sanguinalis*	+, 0
12	Field bindweed	*Convolvulus arvensis*	+, 0
13	Wild oat	*Avena fatua*	+
14	Smooth pigweed	*Amaranthus hybridus*	+
15	Spiny amaranth	*Amaranthus spinosus*	+ c

[a] Daelemans (1989)
[b] Black plastic, however, did result in weed mortality (Ogari and Van der Knaap 2002)
[c] Mushobozy et al. (1997)

unpublished data). While high (> 80%) mortality of wild mustard (*Sinapsis arvensis* L.), common crabgrass (*Digitaria sanguinalis* L.), and hairy galinsoga (*Galinsoga ciliata* (Raf.) Blake) seeds was achieved through seed flaming, slow tractor speeds of 1.3 km h⁻¹, and therefore high propane inputs, were required. Heat from flame weeding typically does not penetrate more than a few millimeters into the soil profile (review by Runia 2012), making this practice potentially less disruptive of soil ecology than either steaming or soil solarisation, though consequently less effective as a seedbank reduction technique. High CO_2 emissions resulting from flaming are a negative environmental impact resulting from its use (Ascard et al. 2007).

Mulching

Mulching can provide an alternative to season-long direct physical weed control operations, but it requires an early-season investment in materials and labor. Mulching is most often conducted in transplanted crops. In temperate climates, black polyethylene film is most common, likely due to its ease of use and its ability to warm the soil to promote early yield of heat-loving crops (Lamont 1993, Schonbeck and Evanylo 1998a). Polyethylene mulch can also reduce the amount of required irrigation (Abu-Awwad 1999) and conserve soil nitrate (Schonbeck and Evanylo 1998b). Different colored polyethylene mulches may be used for alternative management goals, such as improving arthropod pest management (Summers and Stapleton 1998). Since removal and disposal of polyethylene mulch can be expensive (DuPont et al. 2012), an active area of research is the development of biodegradable alternatives using starches (Halley et al. 2001), non-woven fabric (Miao et al. 2013), or food processing byproducts (Virtanen et al. 2017). Biodegradable mulches may perform as well as conventional management (Blick et al. 2010, Cirujeda et al. 2012, Cowan et al. 2014). Biodegradable spray mulches, that also form a physical barrier to weed establishment, may be used in a similar manner, perhaps with prolonged effectiveness (Immirzi et al. 2009, Sartore et al. 2013).

Natural mulches, such as hay, straw, or tree leaves may also control weeds if used in sufficient quantity (Teasdale and Mohler 2000, Abouziena et al. 2008). These mulches may improve water infiltration (Shock et al. 1999, Tindall et al. 1991), increase earthworm populations (B.B. and E.R.G., unpublished data), and replace seasonal carbon and nitrogen losses (Schonbeck and Evanylo 1998b). In temperate climates, natural mulches slow soil warming when applied in the spring. Mulches also contribute to management of selected arthropod pests (vanToor et al. 2004, Larentzaki et al. 2008, Quintanilla-Tornel et al. 2016) and pathogens (Hill et al. 1982). Net returns are generally similar with both natural and polyethylene mulches (Law et al. 2006, Schonbeck 1998, Schonbeck and Evanylo 1998a). In small-scale production in temperate areas, natural mulches are applied by hand after transplanting to allow more time for soil warming. One concern about natural mulches among farmers is the risk of bringing in weed seed (Zwickle 2011), which farmers address by knowing the mulch source or harvesting it themselves. Mulch hay may indeed contain viable weed seeds, but generally, weedy species in hay are not particularly troublesome in annual cropping systems. On a large scale, natural mulches can be applied prior to transplanting with a manure spreader or bale shredder (Cropp 2015). Over the past decade there has been an increasing interest in mulch-based reduced-tillage systems (see *Tillage* section above) in which a winter cereal is killed at peak biomass using a weighted roller with metal blades that crimp stems at several points (Ashford and Reeves 2003, Mirsky et al. 2009, Davis 2010). After 'roller-crimping' the residue acts as a mulch for subsequent large-seeded crops. Roller-crimped residue does not decompose as quickly as mowed residue (Creamer and Dabney 2002), allowing prolonged weed suppression.

Mowing

Repeated mowing may control summer annual weed species (Butler et al. 2013). Mowing is most effective for annual broadleaf weeds since it removes the apical meristem (Meiss et al. 2008). The timing of mowing should be based on weed phenology. For example, to prevent weed seed rain, mowing should occur before flowering, and subsequent mowings should aim to remove the flowers of subsequent lateral shoots (Milakovic et al. 2014). Mowing can also improve perennial weed control (Miller 2016) and reduce biomass of perennial weeds following autumn harvest (Ringselle et al. 2015). If only a single mowing is possible, perennial weeds should be mowed when the ratio of aboveground to belowground nutrient content is greatest (Jung et al. 2012). Inter-row mowing is being evaluated for weed control in no-till row crop systems (Donald 2007).

Alternatives to rotary or flail mowing include the new CombCut (www.justcommonsense. eu). A comb-like row of forward-pointing blades are spaced for flexible crop leaves or stems to pass through while more rigid weed stems, such as those of Canada thistle (*Cirsium arvense* [L.] Scop) or wild mustard (*Sinapsis arvensis* L.), are cut. The CombCut may also be lifted to sever the flowering heads of weeds above the height of the crop. Similarly, in field crops, harvest-time collection or destruction of chaff may prevent up to 85% of weed seeds from entering the weed seedbank (Walsh and Powles 2007, Walsh et al. 2012). As an additional alternative to mowing, livestock have been used for weed control in perennial and vegetable crop production (summarized in Hilimire 2011) and spring wheat (Lenssen et al. 2013), with the added benefit of manure and urine fertilizer (Maughan et al. 2009), though standards must be strictly followed to ensure food safety (USDA NOP 2017).

Crop-Weed Competition

Competition refers to mutually negative effects of organisms or species on one another (Bastiaans and Kropff 2017), that is, both intra- and inter-specific interactions for limited resources. In the case of plants, resources may be unidirectional, as with light, or generally available, as with water and nutrients (Liebman and Gallandt 1997). Crop-weed competition is considered by some to be the most important biological regulator of crop yield (Gallandt and Weiner 2007). Crop yield loss and weed density relationships are predictable and well described by a right, rectangular hyperbolic function; yield loss increasing with increasing weed density, reaching a maximal yield loss (Cousens 1985). The highly variable nature of this relationship (e.g., Lindquist et al., 1996), highlights the many opportunities which exist to manipulate this process.

Crop-weed competition may be affected by a great number of factors including: crop species, genotype, sowing density and arrangement, sowing large seed, transplanting, intercropping, selective resource placement, fertility source and timing of delivery, weed species, density and timing of emergence relative to the crop (Liebman and Gallandt 1997, Mohler 2001, Liebman and Davis 2009). Interestingly, the fundamental processes related to competitive outcomes may be different in organic as compared to conventional systems. Ryan et al. (2009) examined crop yield and weed data in a long-term organic cropping systems study in Pennsylvania, USA, providing evidence that organic maize maintained greater yield in the presence of weeds than did a conventional treatment. This was subsequently attributed to resource availability and crop growth rates in the organic system (Ryan et al. 2010a). However, Benaragama et al. (2016) observed that in cereal cropping systems, when fertility was low, the yield distinction between organic and conventional was not observed, resulting in severely limited yields in the organic system. Crop tolerance to weeds has, however, only short-term implications; more important in the long term is the weeds' response to competition, specifically seed rain.

Targeting nitrogen fertilizer placement to the area occupied by the crop's rooting zone is another effective method for offsetting resource availability in favor of the crop over weeds (Kirkland and Beckie 1998, Blackshaw et al. 2002). In organic cereals, Rasmussen (2002) found that when compared with a broadcast surface application of liquid manure, injection placement resulted in reduced weed density and biomass, and increased crop yield. Some farmers believe soil can be augmented by other mineral nutrients to achieve an 'ideal' makeup, known as basic cation saturation ratio, in which 60–85% of cation exchange cites are occupied by Ca, 6-12% by Mg and 2–5% by K (Kopittke and Menzies 2007). Proponents believe adjusting soil nutrients to achieve this specific ratio may contribute to increased yields and decreased weed growth (Albrecht 1975). However, field studies have not supported the base cation saturation hypothesis (Schonbeck 2001). Farmers' efforts may therefore be better spent focussing on the nutrient requirements of their crops, rather than working towards the basic cation saturation ratio (Kopittke and Menzies 2007).

While competition is important in all organic cropping systems, its value is perhaps most notable in extensive, organic cereal crops in which there are relatively few other options for weed management. In higher value vegetables or row crops, there are many physical weed control options. Organic cereals are routinely seeded at much higher rates than in conventional systems, a strategy that reduces tillering, but increases the competitive advantage of the crop (Kolb and Gallandt 2012). This effect can be extended by sowing the cereal in a more uniform, equidistant spacing, effectively optimizing both intra- and inter-specific effects (Olsen et al. 2012).

Diversity and Timing of Management Practices

Crop rotation and cover cropping are key cultural weed management practices, supporting multiple principals of organic agriculture. Both have the capacity to increase temporal and spatial on-farm diversity (Liebman and Dyck 1993), promote soil health and nutrient status (Gallandt 2003), and improve crop yields and quality by reducing pest-pressure from weeds, disease, and insects (Grubinger 1999). Additionally, these practices introduce variability in timing and frequency of tillage and cultivation events associated with planting, harvest, in-season management, and cover crop termination. Increased diversity in soil disturbance events help to prevent weed seedling establishment and reproduction (Liebman and Gallandt 1997).

Knowledge of physiological and cultural variables influencing crop-weed competition can be leveraged to optimize cropping strategies for weed suppression. For instance, weeds tend to thrive in crops with phenological traits like their own (Upadhyaya and Blackshaw 2007, Leibman and Dyck 1993). Therefore, a successful crop rotation or cover cropping sequence might place precedence on choosing crops with diverse planting dates, life cycles, growth habits and taxonomic affiliations to discourage weed establishment (Nordell and Nordell 2009, Mirsky et al. 2010). For example, rotating cool-season, warm-season, and overwintering crops with varying planting and harvest dates may help to stem repeated weed seed rain events from occurring (Colbach et al. 2014, Garrison et al. 2014, Anderson 2015).

Crop selection plays an important role in targeting weeds based on their emergence periodicity and life cycle traits. When paired with slow-growing crops, rapidly maturing weed species may flower and set seed prior to crop harvest, potentially damaging yield and contributing to future weed pressure. A farmer experiencing heavy pressure from early season weeds might therefore choose to shift production towards faster-maturing crops that produce ample leafy vegetation, like lettuces and spinach, to choke out weed seedlings and avoid yield loss (Smith 2012).

Soil disturbance events associated with field operations can provide effective weed management given the correct timing. The relative flexibility afforded to a farmer who is planting and terminating a cover crop, as opposed to a cash crop, means that the timing of soil disturbance can be fine-tuned to: i) encourage the germination of weed seeds at an appropriate time; and ii) terminate weeds before seed rain occurs, resulting in net losses to the weed seedbank (Sarrantonio and Gallandt 2003). Gallandt and Molloy (2008) found that cover cropping systems that included three or more soil disturbance events prior to weed seed rain experienced greatly reduced populations of lambsquarters (*Chenopodium album* L.), velvetleaf (*Abutilon theophrasti* Medik.), and yellow foxtail (*Setaria lutescens* [Weigel] Hubb.) relative to a system with only one disturbance. Cover crop termination method may also be selected to target problem weed species. Wortman et al. (2013) found that terminating spring cover crops with a sweep plow undercutter resulted in consistently lower densities of grassy weeds than termination with a field disk, which tended to stimulate grassy weed emergence.

Cover Cropping

Cover crops may decrease weed establishment by out-competing weeds for light, moisture, and nutrients. Weed suppression resulting from growing cover crops is largely driven by the development of a thick canopy, which decreases sunlight infiltration and soil temperature to a point that is unconducive to weed emergence (Liebman and Davis 2000). For this reason, cover crop species exhibiting rapid emergence and canopy closure are most suitable for minimising weed populations through competition (Sarrantonio and Gallandt 2003). Cereals like oats and rye *(Secalecereale* L.), as well as many brassica species, buckwheat *(Fagopyrum esculentum* Moench), and sorghum-sudangrass (*Sorghum bicolor* x *bicolor* var. sudanese) are some examples of highly competitive cover crops (Nelson et al. 1991, Gallandt 2003, Clark 2007). When sown into a fallow field, competitive cover crops can decrease the density and biomass of annual and perennial weeds reaching maturity and dispersing viable seeds (Blackshaw et al. 2001, Gallandt 2006, Teasdale et al. 2007). Doubling or tripling the seeding rate of cover crop mixtures can result in a significant boost in early season dry matter production and improved weed suppression (Brennan et al. 2009, Brennan and Boyd 2012).

Intercropping a cash crop with a cover crop can reduce the presence of in-season weeds (Hauggaard-Nielson et al. 2001, Mutch et al. 2003, Ringselle et al. 2015). For example, interseeding clover between rows of cereal grains can supply active, in-season weed control, while simultaneously establishing a green manure that will continue to grow after grain harvest (Mutch et al. 2003, Amossé et al. 2013). Mutch et al. (2003) found that frost seeding red clover (*Trifolium pratense* L.) into established winter wheat (*Triticum aestivum* L.) reduced common ragweed (*Ambrosia artemisiifolia* L.) density. The use of 'living mulches' may reduce the need for herbicides in vegetable (Kunz et al. 2016) and oilseed production (Lorin et al. 2015). Intercrops may better use available resources, leaving fewer resources available to support weed growth (Liebman and Dyck 1993, Hauggaard-Nielson et al. 2001, Smith et al. 2009). Mixtures of cover crop species can provide effective weed control by filling 'niche spaces' that may otherwise be occupied by weeds (Akemo et al. 2000, Gallandt 2006). Interseeded cover crops and living mulches may, however, cause crop yield losses (Uchino et al. 2009, Pfeiffer et al. 2015).

Cover crops and their residues can alter the soil environment in a way that delays or inhibits weed seed germination. Residues left on the soil surface block sunlight from reaching weed seeds and seedlings (Teasdale and Mohler 2000), and both surficial and incorporated residues can inhibit weed growth by altering soil nitrogen dynamics and releasing allelopathic compounds (Haramoto and Gallandt 2005). Cover crop residues on the soil surface can prevent light and heat from reaching the soil surface and physically impede emerging weeds (Teasdale and Mohler 2000, Liebman and Davis 2009). Mechanical termination techniques, such as sweep plow undercutting or roller crimping, can be effective at killing cover crops and leaving significant surface residue (Creamer et al. 1995, Mirsky et al. 2009, Davis 2010). The quality (e.g., C:N) and amount of surface residues are directly tied to their effectiveness at suppressing weeds (Teasdale and Mohler 1993, Teasdale and Mohler 2000). Consequently, cereal grains like rye, which decompose slowly and produce significant biomass, often create superior mulches for weed control (Blum et al. 1997, Smith et al. 2011). However, residues also increase soil moisture, and can encourage weed seed germination if they are not thick enough to prevent light and heat from reaching the soil surface (Teasdale and Mohler 1993).

The incorporation of leguminous green manure residue into the soil can affect the emergence of weeds by altering soil nitrogen dynamics to favor crops over weeds. Davis and Liebman (2001) found that corn following a wheat/clover intercrop experienced less weed pressure than corn following a sole wheat crop and treated with an early-season application of synthetic fertilizer. The authors hypothesized that the more gradual rate of nitrogen release in the cover cropping system was better suited to the growth and uptake of the crop plant than those of surrounding weeds (Liebman and Davis 2000, Davis and Liebman 2001). Hill et al. (2016) found that when high amounts of red clover biomass (>5 Mg ha $^{-1}$) were incorporated prior to organic dry bean planting, enough excess inorganic nitrogen was released by the clover tissue to significantly stimulate weed emergence. To avoid inadvertently stimulating weeds with excess nitrogen, green manure incorporation should be timed such that the residue's breakdown corresponds with the nutrient needs of the subsequent crop (Creamer and Baldwin 2000).

Conveniently, many of the strategies for reducing weeds using cover crops are also tied to an improvement in soil health. For example, the use of residues can provide erosion control (Sarrantonio and Gallandt 2003), increase soil organic matter and soil moisture (Unger and Vigil 1998), and improve soil nutrient cycling (Nordell and Nordell 2009).

Systems Comparison

To provide context for physical and cultural weed management techniques, we present an overview of possible organic weed management practices that may be used in five archetypal cropping systems (Table 34.2). The intensive nature of high-value vegetable production allows growers to use a wider pallet of weed control options than other systems. For example, in large-scale organic maize and soybean, hand-weeding is simply not economically feasible and direct weed control must be achieved with large, efficient equipment. Furthermore, these systems often have limited flexibility related to crop rotation, though diversified rotations may be as cost effective as conventional rotations and provide improved weed control (Davis et al. 2012). In small grains, high-density plantings (Kolb and Gallandt 2012) or harrowing (Rasmussen 1990) may further improve the competitive advantage of these crops. Similarly, the competitive ability of organic forage systems may be improved by well-timed mowing or selective grazing. However, crop rotation and the associated tillage may be needed to control perennial weed infestations. In orchards, weed management is imperative during crop establishment, and often involves use of cultivation, flaming, or mulching. Once orchards are established, weeds may be managed less intensively through living mulches, mowing, or grazing.

The extensive toolkit of organic weed management tactics presented above may appear overly complex, especially when compared to herbicide strategies in a conventional cropping system. Notable, however, is the fact that many organic weed management tactics may offer multiple benefits. Legume green manures reduce nitrogen fertilizer costs. Organic mulches add carbon and improve soil quality.

The relative costs and benefits of various weed management strategies is thus complex, with short- and long-term implications. To highlight some of the implications of weed management decisions so that growers may select the tactics most appropriate for their unique situation, we implemented a systems comparison of several distinct weed management strategies used by high-value vegetable growers. Using yellow onions as a test crop, we compared: i) 'Critical Period' weed control, which prioritizes cultivation during the crop's weed sensitive stage (Nieto et al. 1968, Knezevic et al. 2002); ii) a 'Zero Seed Rain' approach, which utilizes frequent cultivation to preempt seed rain so that over time weed seeds are depleted from the soil (Nordell and Nordell 2009, Gallandt 2014), and lastly, intensive mulching to suppress weeds with Polyethylene Mulch or Hay Mulch. Our hypothesis was that these contrasting strategies would have contrasting implications for farm economics and ecological health.

We quantified the economic (Figure 34.3a) and ecological (Figure 34.3b) tradeoffs involved with each weed management strategy. Overall, the Critical Period system required the least amount of labour while the Hay Mulch system required the most (B.B., E.R.G., and A.K.H., unpublished data). As expected, end-of-season weed biomass was greatest in the Critical Period system, as was weed seed production. The Hay Mulch system was generally most favorable for increasing soil organic matter, reducing compaction, and promoting earthworm populations, which are generally beneficial for soil health (Hopp and Hopkins 1946, Edwards and Lofty 1977). Soil nitrate (NO_3-N) was conserved in the Polyethylene Mulch system, reflecting less leaching during rainfall events. The two most labor-expensive strategies, Zero Seed Rain and Hay Mulch, ended up being the two most profitable strategies due to high yields. In addition, the weed seed rain of the Critical Period strategy caused a yield loss in the subsequent crop.

Farmer management is influenced by their perception of risk (Slovic 1987, Zwickle 2011, Jabbour et al. 2014a), thus the adoption of a particular weed management strategy likely relates to the ability of that strategy to reduce risk. This is reflected in our case studies of four farmers that have specialized in each strategy (B.B. and E.R.G., unpublished data). The case studies showed that the farmer who perceived high weeding labor costs as the dominant risk to production, used the Critical Period strategy, which required the least labor in our systems comparison. Similarly, the farmer whose largest perceived risk was establishing a large weed seedbank, specialized in a Zero Seed Rain approach.

Table 34.2. Weed management practices known to be applicable in five archetypal cropping systems. Practices applicable to each system are marked by √(applicable) or √√(highly applicable)

Organic weed management practice	High-value mixed vegetables	Maize/ Soybean	Small grains	Forage	Orchard
Cover cropping					
Summer cover crop	√√	√√	√		
Winter cover crop	√√	√	√		
Living mulch	√	√			
Weed seedbank reduction					
Summer fallow	√√	√	√√	√	
Stale seedbed periods	√√	√	√√	√	
Soil solarization	√√				
Soil steaming	√√				
Providing habitat for seed predators					
Rapid-fire cover crops			√		
Crop-weed competition					
Precise irrigation placement	√√				√√
Precise fertility placement	√√	√√			
Increasing planting density					
Adjusting planting dates to avoid peak weed emergence	√√	√	√√		
Mulching					
Plastic mulch	√√				√√
Natural mulch	√√				√√
Cover crop mulch	√√	√√			
Physical weed control					
Pre-emergence flaming	√√	√			
Pre-emergence harrowing	√√	√√	√√	√	
Post-emergence harrowing					
Inter-row cultivation	√√	√√	√		√√
Intra-row cultivation	√√	√			√
Intra-row flaming	√	√			
Hand-weeding	√√			√	
Mowing			√	√√	√√
Biological control					
Herbivorous insects				√√	
Grazing				√√	√√

The farmer utilizing Polyethylene Mulch for many crops placed nearly equal importance on multiple management criteria, while the farmer using Hay Mulch placed most importance on improving soil quality. These results concur with our systems comparison (B.B., E.R.G., and A.K.H., unpublished data) and indicate that while there may not be a single 'best' weed management strategy, small-scale organic vegetable growers need to understand the motivations and risks of each approach to select the most appropriate strategy for their situation.

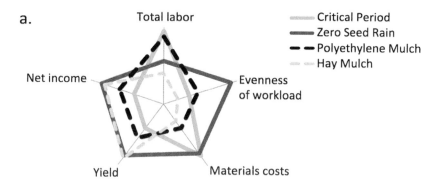

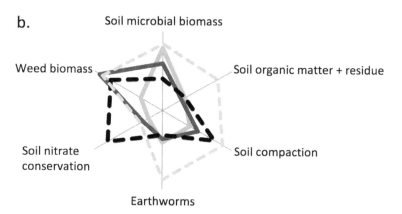

Figure 34.3. Radar plots of variables related to: (a) economics and (b) ecology of four weed management strategies used in organic onion. Each axis was oriented so that outermost values are most favourable. For example, the Critical Period strategy required the least 'Total labour,' which was most favourable. The axis of 'Evenness of workload' was oriented so that an even spread is most favourable, but the reverse may be true for farms with access to ample seasonal labour.

Concluding Remarks

Weeds remain a significant production problem in organic farming systems, though it is worth noting that the 'weed problem' is far from solved in contemporary GMO-herbicide-based cropping systems (Heap 2014). Viewed through the lens of sustainability, organic farming systems outperform conventional agriculture in many areas, especially regarding environmental impact and profitability, with the notable exception being lower yields (Reganold and Wachter 2016). In most developed countries, organic agriculture continues to experience impressive growth, but remains a relatively small sector of the overall food system, representing approximately 1% of global agricultural land (Crowder and Reganold 2015). When competing directly with conventional agriculture, lower yields and higher production costs put organic systems at a significant disadvantage, but one that is readily overcome with price premiums (Crowder and Reganold 2015). Break-even premiums to match conventional profits were surprisingly low, 5–7%, without consideration of ecosystem services associated with organic practices.

Sustainability of weed management in organic farming systems can be considered using the framework established by Hill and MacRae (1996): improving efficiency, input substitution and system redesign. Efficiency improvements are often motivated simply by economics, e.g., reduced rates of inputs, but have related environmental benefits. Substituting costly and environmentally harmful off-farm inputs for more benign, on-farm sources is often associated with greater change, more risks, but potentially greater benefits, e.g., legume green manures replacing purchased nitrogen. Many organic farmers start by substituting cultivation events for herbicide applications, a strategy appealing for its simplicity and lacking higher-level management. However, low and variable cultivation efficacy often result in increasing weed populations and more challenging management over time. Advances in physical weed control, especially camera guidance and robotics, promise increased efficiency, from both improved precision and greater working rates, but these advances need to be supported by basic research on physical weed control. We should not be satisfied that cultivation is as much 'art' as it is 'science'. Basic research should define efficacy over a wide range of tool, weed, crop and soil variables, defining operational parameters to make this practice more effective and reliable.

Redesigned systems may include an array of ecologically based weed management tactics employed to varying degrees. These approaches are most successful when diverse and complementary tactics (e.g., Liebman et al. 2016) are considered over an extended time domain (e.g., Bàrberi 2002), and with careful consideration of weed biology (Van Acker 2009). Indeed, there are exemplary farmers with such redesigned systems (Nordell and Nordell 2009). While conventional agricultural systems may be amenable to simplified prescription approaches to weed control, relying on highly reliable and effective herbicides, organic systems require a systems approach, based on knowledge of both management principles and the biology of problem weeds.

Acknowledgements

We would like to thank the two anonymous reviewers, and Richard Smith, for thoroughly reviewing an earlier version of this chapter. Publication Number 3560 from the Maine Agricultural and Forest Experiment Station. This project was supported by the USDA National Institute of Food and Agriculture, Hatch Project Number ME021606 through the Maine Agricultural & Forest Experiment Station.

REFERENCES

Abouziena, H.F., O.M. Hafez, S.D. Sharma, M. Singh, A. Sciences and L. Alfred. 2008. Comparison of weed suppression and mandarin fruit yield and quality obtained with organic mulches, synthetic mulches, cultivation, and glyphosate. HortScience 43: 795–799.

Abu-Awwad, A.M. 1999. Irrigation water management for efficient water use in mulched onion. J. of Agron. Crop. Sci. 183: 1–7.

Akemo, M., E.E. Regnier and M.A. Bennett. 2000. Weed suppression in spring-sown rye (*Secale cereale*): pea (*Pisum sativum*) cover crop mixes. Weed Sci. 14: 545–549.

Albrecht, W.A. 1975. The Albrecht papers. Vol. 1: Foundational concepts. Acres USA, Kansas City, Missouri.

Albrecht, H. 2005. Development of arable weed seedbanks during the 6 years after the change from conventional to organic farming. Weed Res. 45: 339–350.

Amossé, C., M.H. Jeuffroy, F. Celette and C. David. 2013. Relay-intercropped forage legumes help to control weeds in organic grain production. Eur. J. Agron. 49: 158–167.

Anderson, R.L. 2015. Integrating a complex rotation with no-till improves weed management in organic farming. A review. Agron. Sustain. Dev. 35: 967–974.

Ascard, J. 1994. Dose-response models for flame weeding in relation to plant size and density. Weed Res. 34: 377–385.

Ascard, J. and F. Fogelberg. 2008. Mechanical in-row weed control in transplanted and direct-sown bulb onions. Biol. Agric. Hortic. 25: 235–251.

Ascard, J., P.E. Hatcher, B. Melander and M.K. Upadhyaya. 2007. Thermal Weed Control. pp. 155–175. *In*: M.K. Upadhyaya and R.E. Blackshaw (Eds.) Non-Chemical Weed Management. CABI, Cambridge, MA.

Ashford, D.L. and D.W. Reeves. 2003. Use of a mechanical roller-crimper as an alternative kill method for cover crops. Am. J. Alternative Agr. 18: 37–45.

Avissar, R., O. Naot, Y. Mahrer and J. Katan. 1985. Field aging of transparent polyethylene mulches: II. Influence on the effectiveness of soil heating. Soil Sci. Soc. Am. J. 49: 205–209.

Baerveldt, S. and J. Ascard. 1999. Effect of soil cover on weeds. Biol. Agric. Hortic. 17: 101–111.

Baker, B.P. and C.L. Mohler. 2014. Weed management by upstate New York organic farmers: strategies, techniques and research priorities. Renew. Agr. Food Syst. 30: 1–10.

Bàrberi, P. 2002. Weed management in organic agriculture: are we addressing the right issues? Weed Res 42: 177–193.

Baskin, C.C. and J.M. Baskin. 1998. Seeds: Ecology, Biogeography, and Evolution of Dormancy and Germination. Academic Press, San Diego, CA.

Bastiaans, L. and M.J. Kropff. 2017. Weed competition. Second ed. Vol. 3, Encyclopedia of Applied Plant Sciences. Elsevier Ltd.

Bastiaans, L., R. Paolini and D.T. Baumann. 2008. Focus on ecological weed management: what is hindering adoption? Weed Res. 48: 481–491.

Benaragama, D., S.J. Shirtliffe, E.N. Johnson, H.S.N. Duddu, L.D. Syrovy and H. Albrecht. 2016. Does yield loss due to weed competition differ between organic and conventional cropping systems? Weed Res 56: 274–283.

Benton, T.G., J.A. Vickery and J.D. Wilson. 2003. Farmland biodiversity: is habitat heterogeneity the key? Trends Ecol. Evol. 18: 182–188.

Benvenuti, S. and M. Macchia. 2006. Seedbank reduction after different stale seedbed techniques in organic agricultural systems. Ital. J. Agron. 1: 11–21.

Bilck, A.P., M.V. Grossmann and F. Yamashita. 2010. Biodegradable mulch films for strawberry production. Polym. Test. 29: 471–476.

Blackshaw, R.E., J.R. Moyer, R.C. Doram, A.L. Boswell and R.E. Blackshaw. 2001. Yellow sweetclover, green manure, and its residues effectively suppress weeds during fallow. Weed Sci. 49: 406–413.

Blackshaw, R.E., G. Semach and H.H. Janzen. 2002. Fertilizer application method affects nitrogen uptake in weeds and wheat. Weed Sci. 50: 634–641.

Bleeker, P., R. van Der Weide and D.A.G. Kurstjens. 2002. Experiences and experiments with new intra-row weeders. pp. 97–100. *In:* The Proceeding of 5th EWRS Workshop on Physical Weed Control, Pisa, Italy.

Blum, U., L.D. King, T.M. Geric, M.E. Lehman and A.D. Worsham. 1997. Effects of clover and small grain cover crops and tillage techniques on seedling emergence of some dicotyledonous weed species. Am. J. Alternative Agr. 12: 146–161.

Bond, W. and P.J. Bursch. 1989. Weed control in carrots and salad onions under low-level polyethylene covers. Brighton Crop Prot. Conf. 1021–1026.

Bowman, G. 2002. Steel in The Field: A Farmer's Guide to Weed Management Tools. Handbook Series, Book 2. Sustainable Agriculture Network, Beltsville, Maryland.

Boyd, N.S., E.B. Brennan and S.A. Fennimore. 2006. Stale seedbed techniques for organic vegetable production. Weed Technol. 20: 1052–1057.

Brainard, D.C., E. Haramoto, M.M. Williams and S.B. Mirsky. 2013a. Towards a no-till no-spray future? Introduction to the symposium on non-chemical weed management in reduced tillage cropping systems. Weed Technol. 27: 190–192.

Brainard, D.C., E. Peachey, E. Haramoto, J. Luna and A. Rangarajan. 2013b. Weed ecology and management under strip-tillage: implications for Northern U.S. vegetable cropping systems. Weed Technol. 27: 218–230.

Brennan, E.B., N.S. Boyd, R.S. Smith and P. Foster. 2009. Seeding rate and planting effects on growth and weed suppression of a legume-oat cover crop for organic vegetable systems. Agron. J. 101: 979–988.

Brennan, E.B. and N.S. Boyd. 2012. Winter cover crop seeding rate and variety affects during eight years of organic vegetables: I. Cover crop biomass production. Agron. J. 104: 684–698.

Browning, G.M., R.A. Norton, E.V. Collins and H.A. Wilson. 1945. Tillage practices in relation to soil and water conservation and crop yields in Iowa. Soil Sci. Soc. Am. J. 9: 241–247.

Busari, M.A., S.S. Kukal, A. Kaur, R. Bhatt and A.A. Dulazi. 2015. Conservation tillage impacts on soil, crop and the environment. Int. Soil Water Conserv. Res. 3: 1–11.

Butler, R.A., S.M. Brouder, W.G. Johnson and K.D. Gibson. 2013. Response of four summer annual weed species to mowing frequency and height. Weed Technol. 27: 798–802.

Caldwell, B. and C.L. Mohler. 2001. Stale seedbed practices for vegetable production. HortScience 36: 703–705.

Cerdà, A., D.C. Flanagan, Y.L. Bissonnais and J. Bordman. 2009. Soil erosion and agriculture. Soil Till. Res. 106: 107–108.

Cirujeda, A. and A. Taberner. 2004. Defining optimal conditions for weed harrowing in winter cereals on *Papaver rhoeas* L. and other dicotyledoneous weeds. pp. 101–105. *In:* The Proceedings of 6th EWRS Workshop on Physical and Cultural Weed Control, Lilliehammer, Norway.

Cirujeda, A., B. Melander, K. Rasmussen and I.A. Rasmussen. 2003. Relationship between speed, soil movement into the cereal row and intra-row weed control efficacy by weed harrowing. Weed Res. 43: 285–296.

Cirujeda, A., J. Aibar, A. Anzalone, L. Martín-Closas, R. Meco, M.M. Moreno, A. Pardo, A.M. Pelacho, F. Rojo, A. Royo-Esnal, M.L. Suso and C. Zaragoza, C. 2012. Biodegradable mulch instead of polyethylene for weed control of processing tomato production. Agron. for Sustain. Dev. 32: 889–897.

Clark, A. 2007. Managing Cover Crops Profitably. Sustainable Agriculture Research and Education. DIANE Publishing, College Park, MD.

Cohen, O. and B. Rubin. 2007. Soil Solarization and Weed Management. pp. 177–200. *In:* M.K. Upadhyaya and R.E. Blackshaw (Eds.) Non-Chemical Weed Management. CABI, Cambridge, MA.

Colbach, N., L. Biju-Duval, A. Gardarin, S. Granger, S.H.M. Guyot, D. Meziere, N.M. Munier-Jolain and S. Petit. 2014. The role of models for multicriteria evaluation and multiobjective design of cropping systems for managing weeds. Weed Res. 54: 541–555.

Cousens, R. 1985. A simple model relating yield loss to weed density. Ann Appl Biol 107: 239–252.

Cowan, J.S., C.A. Miles, P.K. Andrews and D.A. Inglis. 2014. Biodegradable mulch performed comparably to polyethylene in high tunnel tomato (*Solanum lycopersicum* L.) production. J. Sci. Food Agr. 94: 1854–1864.

Creamer, N.G. and K.R. Baldwin. 2000. An evaluation of summer cover crops for use in vegetable production systems in North Carolina. HortScience 35: 600–603.

Creamer, N.G. and S.M. Dabney. 2002. Killing cover crops mechanically: review of recent literature and assessment of new research results. Am. J. Alternative Agr. 17: 32–40.

Creamer, N.G., B. Plassman, M.A. Bennett, R.K. Wood, B.R. Stinner and J. Cardina. 1995. A method for mechanically killing cover crops to optimize weed suppression. Am. J. Alternative Agr. 10: 157–162.

Cropp, J.H. 2015. Rotational no-till and mulching systems for organic vegetable farms. Under Cover Consultancy. Witzenhausen, Germany (Accessed on January 7, 2016)

Crowder, D.W. and J.P. Reganold. 2015. Financial competitiveness of organic agriculture on a global scale. Proc Natl Acad Sci USA 112: 7611–7616.

Daelemans, A. 1989. Soil solarization in West-Cameroon: effect on weed control, some chemical properties and pathogens of the soil. Acta Hort. 255: 169–176.

Dahlquist, R.M., T.S. Prather and J.J. Stapleton. 2007. Time and temperature requirements for weed seed thermal death. Weed Sci. 55: 619–625.

Davis, A.S. 2010. Cover-crop roller-crimper contributes to weed management in no-till soybean. Weed Sci. 58: 300–309.

Davis, A.S. and M. Liebman. 2001. Nitrogen source influences wild mustard growth and competitive effect on sweet corn. Weed Sci. 49: 558–566.

Davis, A.S., J.D. Hill, C.A. Chase, A.M. Johanns and M. Liebman. 2012. Increasing cropping system diversity balances productivity, profitability and environmental health. PloS One. 7: 1–8.

DeDecker, J.J., J.B. Masiunas, A.S. Davis and C.G. Flint. 2014. Weed management practice selection among Midwest U.S. organic growers. Weed Sci. 62: 520–531.

Donald, W.W. 2007. Between-row mowing systems control summer annual weeds in no-till grain sorghum. Weed Technol. 21: 511–517.

DuPont, S.T., A. Frankenfield, S. Guiser and J. Esslinger. 2012. Biodegradable mulches. Cooperative Extension, Pennsylvania State University (Accessed on January 5, 2016) http://extension.psu.edu/plants/vegetable-fruit/fact-sheets/Biodegradable%20Mulches.pdf

Edwards, C.A. and J.R. Lofty. 1977. Biology of earthworms (2nd ed.). Chapman & Hall, London.

Egley, G.H. 1990. High-temperature effects on germination and survival of weed seeds in soil. Weed Sci. 38: 429–435.

Evans, G.J., R.R. Bellinder and R.R. Hahn. 2012. An evaluation of two novel cultivation tools. Weed Technol. 26: 316–325.

Fennimore, S.A. and R.E. Goodhue. 2016. Soil disinfestation with steam: a review of economics, engineering, and soil pest control in California strawberry. Int. J. Fruit Sci. 16: 71–83.

Fogelberg, F. and A.M.D. Gustavsson. 1999. Mechanical damage to annual weeds and carrots by in-row brush weeding. Weed Res. 39: 469–479.

Forcella, F. 2003. Debiting the seedbank: priorities and predictions. Aspects of Applied Biology 69: 151–162.

Forcella, F. 2012. Air-propelled abrasive grit for postemergence in-row weed control in field corn. Weed Tech. 26: 161–164.

Fortier, J.-M. 2014. The Market Gardener. New Society Publishers, Vancouver.

Gainessi, L.P. and N.P. Reigner. 2003. The value of herbicides in U.S. crop production. Weed Technol. 21: 559–566.

Gallandt, E.R. 2003. Soil-improving practices for ecological weed management. pp. 267–284. *In:* Inderjit (Ed.) Weed Biology and Management. Springer Netherlands, Dordrecht, NL.

Gallandt, E.R. 2006. How can we target the weed seedbank? Weed Sci. 54: 588–596.

Gallandt, E.R. 2014. Weed management in organic farming. pp. 68–86. *In:* B.S. Chauhan and G. Mahajan (Eds.) Recent Advances in Weed Management. Springer, New York.

Gallandt, E.R. and T. Molloy. 2008. Exploiting weed management benefits of cover crops requires pre-emption of seed rain. *In:* The Proceedings of the 16th IFOAM Organic World Congress, Modena, Italy.

Gallandt, E.R., D.C. Brainard and B. Brown. In press. Developments in physical weed control. *In:* R. Zimdahl (Ed.) Integrated Weed Management for Sustainable Agriculture. Burleigh Dodds Science Publishing.

Gallandt, E.R., M. Liebman and D.R. Huggins. 1999. Improving soil quality: implications for weed management. J. Crop Prod. 2: 95–121.

Gallandt, E.R., T. Molloy, R.P. Lynch and F.A. Drummond. 2005. Effect of cover-cropping systems on invertebrate seed predation. Weed Sci. 53: 69–76.

Gallandt, E.R. and J. Weiner. 2007. Crop-weed competition. eLS. Wiley Online Library.

Gamliel, A. and J.J. Stapleton. 2012. Combining soil solarization with organic amendments for the control of soilborne pests. pp. 109–120. *In:* A. Gamliel and J. Katan (Eds.) Soil Solarization: Theory and Practice. American Phytopathological Society, St. Paul, MN.

Garrison, A.J., A.D. Miller, M.R. Ryan, S.H. Roxburgh and K. Shea. 2014. Stacked crop rotations exploit weed-weed competition for sustainable weed management. Weed Sci. 62: 166–176.

Gebbers, R. and V.I. Adamchuck. 2010. Precision agriculture and food security. Science 327: 828–831.

Gómez-Casero, M.T., I.L. Castillejo-González, A. García-Ferrer, J.M. Peña-Barragán, M. Jurado-Expósito, L. García-Torres and F. López-Granados. 2010. Spectral discrimination of wild oat and canary grass in wheat fields for less herbicide application. Agron. Sustain. Dev. 30: 689–699.

Gopinath, K.A., N. Kumar, B.L. Mina, A.V. Srivastva and H.S. Gupta. 2009. Evaluation of mulching, stale seedbed, hand weeding and hoeing for weed control in organic garden pea (*Pisum sativum* sub sp. Hortens L.). Arch. Agron. Soil Sci. 55: 115–123.

Griffin, R. Jayasekara and G. Lonergan. 2001. Developing biodegradable mulch films from starch-based polymers. Starch 53: 362–367.

Grisso, R., M. Alley, P. McClellan, D. Brann and S. Donnohue. 2009. Precision farming: a comprehensive approach. pp. 442-500. Virginia Polytechnic Institute and State University. Virginia Cooperative Extension Publication. http://pubs.ext.vt.edu/442/442-500/442-500_pdf.pdf [accessed in January 2017].

Grubinger, V.P. 1999. Crop rotation. pp. 69–77. *In:* V.P. Grubinger Sustainable Vegetable Production from Start-Up to Market. Natural Resource, Agriculture, and Engineering Service, Cooperative Extension (NRAES), Ithaca, New York.

Gunton, R.M, S. Petit and S. Gaba. 2011. Functional traits relating arable weed communities to crop characteristics. J. Veg. Sci. 22: 541–550.

Habel, W. 1954. Uber die wirkungsweise der eggen gegen samenunkrauter sowie dieempfndlichkeit der unkrautarten und ihrer altersstadien gegen den eggvorgang. Ph.D.

Hakånsson, S. 2003. Soil tillage effects on weeds. pp. 158–196. *In:* S. Hakånsson. Weeds and Management on Arable Land: An Ecological Approach. CAB International, Wallingford, UK.

Halley, P., R. Rutgers, S. Coombs, J. Kettels, J. Gralton, G. Christie, M. Jenkins, H. Beh, K. Griffin, R. Jayasekara and G. Lonergan. 2001. Developing biodegradable mulch films from starch-based polymers. Starch 53: 362–367.

Hamza, M.A. and W.K. Anderson. 2005. Soil compaction in cropping systems: a review of the nature, causes and possible solutions. Soil Till. Res. 82: 121–145.

Haramoto, E.R. and E.R. Gallandt. 2005. Brassica cover cropping I: effects on weed and crop establishment. Weed Sci. 53: 695–701.

Hauggaard-Nielsen, H., P. Ambus and E.S. Jensen. 2001. Interspecific competition, N use and interference with weeds in pea-barley intercropping. Field Crop. Res. 70: 101–109.

Heap, I. 2014. Global perspective of herbicide-resistant weeds. Pest Manage Sci 70: 1306–1315.

Hill, D.E., L. Hankin and G.R. Stephens. 1982. Mulches: their effect on fruit set, timing and yields of vegetables. The Connecticut Agricultural Experiment Station.

Hill, E.C., K.A. Renner, C.L. Sprague and A.S. Davis. 2016. Cover crop impact on weed dynamics in an organic dry bean system. Weed Sci. 64: 261–275.

Hill, S.B. and R.J. MacRae. 1996. Conceptual Framework for the transition from conventional to sustainable agriculture. J. Sustain. Agric. 7: 81–87.

Holm, L.G., D.L. Plucknett, J.V. Pancho and J.P. Herberger. 1977. The Worlds Worst Weeds. University of Hawaii Press, Honolulu, HI.

Hooper, D.U., F.S. Chapin, J.J. Ewel, A. Hector, P. Inchausti, S. Lavorel, J.H. Lawton, D.M. Lodge, M. Loreau, S. Naeem et al. 2005. Effects of biodiversity on ecosystem functioning: a consensus of current knowledge. Ecol. Monogr. 75: 3–35.

Hopp, H. and H.T. Hopkins. 1946. Earthworms as a factor in the formation of water-stable soil aggregates. J. Soil Water Conserv. 1: 11–13. http://articles.extension.org/pages/71822/rotational-no-till-and-mulching-systems-for-

ICROFS. 2017. Organic Rules and Certification (Accessed on January 30, 2017). http://organicrules.org/

Immirzi, B., G. Santagata, G. Vox and E. Schettini. 2009. Preparation, characterization and field-testing of a biodegradable sodium alginate-based spray mulch. Biosyst. Eng. 102: 461–472.

Jabbour, R., E.R. Gallandt, S. Zwickle, R.S. Wilson and D. Doohan. 2014a. Organic farmer knowledge and perceptions are associated with on-farm weed seedbank densities in Northern New England. Weed Sci. 62: 338–349.

Jabbour, R., S. Zwickle, E.R. Gallandt, K.E. Mcphee, R.S. Wilson and D. Doohan. 2014b. Mental models of organic weed management: comparison of New England US farmer and expert models. Renew. Agr. Food Syst. 29: 319–333.

Jacobs, A. and R. Kingwell 2016. The Harrington Seed Destructor: Its role and value in farming systems facing the challenge of herbicide-resistant weeds. Agr Syst. 142: 33–40.

Jensen, R.K., J. Rasmussen and B. Melander. 2004. Selectivity of weed harrowing in lupin. Weed Res. 44: 245–253.

Johnson, W.C. and B.G. Mullinix. 1998. Stale seedbed weed control in cucumber. Weed Sci. 46: 698–702.

Johnson, W.C. and B.G. Mullinix. 2000. Evaluation of tillage implements for stale seedbed tillage in peanut (*Arachis hypogaea*). Weed Technol. 14: 519–523.

Joint Research Centre of the European Commission, Monitoring Agriculture Resources Unit HO4: Zarco-Tejada, P.J., N. Hubbard and P. Loudjani. 2014. Precision agriculture: an opportunity for EU farmers – potential support with the CAP 2014-2020. Report prepared for the European Parliament's Committee on Agriculture and Rural Development.

Jung, L.S., R.L. Eckstein, A. Otte and T.W. Donath. 2012. Above- and below-ground nutrient and alkaloid dynamics in *Colchicum autumnale*: indications for optimal mowing dates for population control or low hay toxicity. Weed Res. 52: 348–357.

Kapulnik, Y. and A. Gamliel. 2012. Combining soil solarization with beneficial microbial agents. pp. 121–128. *In:* A. Gamliel and J. Katan (Eds.) Soil Solarization: Theory and Practice. American Phytopathological Society, St. Paul, MN.

Katan, J. and A. Gamliel. 2012. Mechanisms of pathogen and disease control and plant-growth improvement involved in soil solarization. pp. 135–145. *In:* A. Gamliel and J. Katan (Eds.) Soil Solarization: Theory and Practice. American Phytopathological Society, St. Paul, MN.

Katan, J., H. Greenberger, H. Alon and A. Grinstein. 1976. Solar heating by polyethylene mulching for the control of diseases caused by soil-borne pathogens. Phytopathology 66: 683–688.

Keene, C.L. and W.S. Curran. 2016. Optimizing high-residue cultivation timing and frequency in reduced-tillage soybean and corn. Agron. J. 108: 1897–1906.

Kees, H. 1962. Untersuchungen zur unkrautbekampfung durch netzegge undstoppelbearbeitungsmassnahmen unter besonderer berucksichtigung des leichtenbodens. Ph.D. Thesis, Universität Hohenheim, Stuttgart, Germany.

Kirkland, K.J. and H.J. Beckie. 1998. Contribution of nitrogen fertilizer placement to weed management in spring wheat (Triticum aestivum). Weed Tech. 12: 507–514.

Knezevic, S.Z., S.P. Evans, E.E. Blankenship, R.C. van Acker and J.L. Lindquist. 2002. Critical period for weed control: the concept and data analysis. Weed Sci. 50: 773–786.

Koch, W. 1964. Unkrautbekampfung durch eggen, hacken und meisseln in getreide I. Wirkungsweise und einsatzzeitpunkt von egge, hacke und bodenmeissel. J. Agron. Crop. Sci. 120: 369–382.

Kolb, L.N. and E.R. Gallandt. 2012. Weed management in organic cereals: advances and opportunities. Org. Agr. 2: 23–42.

Kopittke, P.M. and N.W. Menzies. 2007. A review of the use of the basic cation saturation ratio and the "ideal" soil. Soil Sci. Soc. Am. J. 71: 259–265.

Kouwenhoven, J.K. and R. Terpstra. 1979. Sorting action of tines and tine-like tools in the field. J. Agr. Eng. Res. 24: 95–113.

Kunz, C., D.J. Sturm, G.G. Peteinatos and R. Gerhards. 2016. Weed suppression of living mulch in sugar beets. Ges. Pflan. 68: 145–154.

Kurkalova, L., C.L. Kling and J. Zhao. 2004. Multiple benefits of carbon-friendly agricultural practices: empirical assessment of conservation tillage. Environ. Manage. 33: 519–527.

Kurstjens, D.A.G. 2002. Mechanisms of selective mechanical weed control by harrowing. Ph.D. Thesis, Wageningen University, Wageningen, Netherlands.

Kurstjens, D.A.G. 2007. Precise tillage systems for enhanced non-chemical weed management. Soil Tillage Res. 97: 293–305.

Kurstjens, D.A.G. and M.J. Kropff. 2001. The impact of uprooting and soil-covering on the effectiveness of weed harrowing. Weed Res. 41: 211–228.

Kurstjens, D.A.G. and U.D. Perdock. 2000. The selective soil covering mechanism of weed harrows on sandy soil. Soil Till. Res. 55: 193–206.

Kurstjens, D.A.G., U.D. Perdok and D. Goense. 2000. Selective uprooting by weed harrowing on sandy soils. Weed Res. 40: 431–447.

Lamont, W.J. 1993. Plastic mulch for production of vegetable crops. HortTechnology 3: 35–39.

Larentzaki, E., A.M. Shelton and J. Plate. 2008. Effect of kaolin particle film on *Thrips tabaci* (Thysanoptera: Thripidae), oviposition, feeding and development on onions: a lab and field case study. Crop Prot. 27: 727–734.

Lati, R.N., M M.C. Siemens, J.S. Rachuy and S.A. Fennimore. 2016. Intrarow weed removal in broccoli and transplanted lettuce with an intelligent cultivator. Weed Tech. 30: 655–663.

Law, D.M., A.B. Rowell, J.C. Snyder and M.A. Williams. 2006. Weed control efficacy of organic mulches in two organically managed bell pepper production systems. HortTechnology 16: 225–232.

Liebman, M. and A.S. Davis. 2000. Integration of soil, crop and weed management in low-external-input farming systems. Weed Res. 40: 27–48.

Liebman, M. and A.S. Davis. 2009. Managing weeds in organic farming systems: an ecological approach. pp. 173-195. *In:* C.A. Francis (Ed.) Organic Farming: The Ecological System. American Society of Agronomy, Madison, Wisconsin.

Liebman, M. and E. Dyck. 1993. Crop rotation and intercropping strategies for weed management. Ecol. Appl. 3: 92–122.

Liebman, M. and E.R. Gallandt. 1997. Many little hammers: ecological approaches for management of crop-weed interactions. pp. 291–330. *In:* L.E. Jackson (Ed.) Ecology in Agriculture. Academic Press, San Diego, CA.

Liebman, M., B. Baraibar, Y. Buckley et al. 2016. Ecologically sustainable weed management: how do we get from proof-of-concept to adoption? Ecol Appl 26: 1352–1369.

Lindquist, J.L., D.A. Mortensen, S.A. Clay et al. 1996. Stability of corn (zea mays)-velvetleaf (abutilon theophrasti) interference relationships. Weed Sci 44: 309–313.

López-Granados, F. 2010. Weed detection for site-specific weed management: mapping and real-time approaches. Weed Res. 51: 1–11.

Lorin, M., M.H. Jeuffroy, A. Butier and M. Valantin-Morison. 2015. Undersowing winter oilseed rape with frost-sensitive legume living mulches to improve weed control. Eur. J. Agron. 71: 96–105.

Luttikholt, L.W.M. and L. Woodward. 2006. Principles of organic agriculture worldwide participatory stakeholder process. orgprints.org.

Mahrer, Y. and E. Shilo. 2012. Physical principles of solar heating of soils. pp. 147–152. *In:* A. Gamliel and J. Katan (Eds.) Soil Solarization: Theory and Practice. American Phytopathological Society, St. Paul, MN.

Mallek, S.B., T.S. Prather and J.J. Stapleton. 2007. Interaction effects of *Allium* spp. residues, concentrations and soil temperature on seed germination of four weedy plant species. Appl. Soil Ecol. 37: 233–239.

Marenco, R.A. and D.C. Lustosa. 2000. Soil solarization for weed control in carrot. Pesqui. Agropecuária Bras. 35: 2025–2032.

Meiss, H., N. Munier-Jolain, F. Henriot and J. Caneill. 2008. Effects of biomass, age and functional traits on regrowth of arable weeds after cutting. J. Plant Dis. Plant Prot. 21: 493–500.

Melander, B., B. Lattanzi and E. Pannacci. 2015. Intelligent versus non-intelligent mechanical intra-row weed control in transplanted onion and cabbage. Crop Prot. 72: 1–8.

Melander, B., I.A. Rasmussen and J.E. Olesen. 2016. Incompatibility between fertility building measures and the management of perennial weeds in organic cropping systems. Agr. Ecosyst. Environ. 220: 184–192.

Melander, B., I.A. Rasmussen and P. Barberi. 2005. Integrating physical and cultural methods of weed control: examples from European research. Weed Sci. 53: 369–381.

Menalled, F., R. Peterson, R. Smith, W. Curran, D. Páez and B. Maxwell. 2016. The eco-evolutionary imperative: revisiting weed management in the midst of an herbicide resistance crisis. Sustainability 8: 1297.

Merfield, C.N. 2016. Robotic weeding's false dawn? Ten requirements for fully autonomous mechanical weed management. Weed Res. 56: 340–344.

Miao, M., A.P. Pierlot, K. Millington, S.G. Gordon, A. Best and M. Clarke. 2013. Biodegradable mulch fabric by surface fibrillation and entanglement of plant fibers. Text. Res. J. 83: 1906–1917.

Milakovic, I., K. Fiedler and G. Karrer. 2014. Management of roadside populations of invasive *Ambrosia artemisiifolia* by mowing. Weed Res. 54: 256–264.

Miller, T.C., J.B. Samtani and S.A. Fennimore. 2014. Mixing steam with soil increases heating rate compared to steam applied to still soil. Crop Prot. 64: 47–50.

Miller, T.W. 2016. Integrated strategies for management of perennial weeds. Invasive Plant Sci. Manag. 9: 148–158.

Mirsky, S.B., E.R. Gallandt, D.A. Mortensen, W.S. Curran and D.L. Shumway. 2010. Reducing the germinable weed seedbank with soil disturbance and cover crops. Weed Res. 50: 341–352.

Mirsky, S.B., W.S. Curran, D.A. Mortensen, M.R. Ryan and D.L. Shumway. 2009. Control of cereal rye with a roller/crimper as influenced by cover crop phenology. Agron. J. 101: 1589–1596.

Mohler, C.L. 1993. A model of the effects of tillage on emergence of weed seedlings. Ecol. Appl. 3: 53–73.

Mohler, C.L. 2001. Mechanical management of weeds. pp. 139–209. *In:* M. Liebman, C.L. Mohler and C.P. Staver (Eds.) Ecological Management of Weeds. Cambridge University Press, New York.

Mohler, C.L. 2001. Enhancing the competitive ability of crops. *In:* Ecological management of agricultural weeds. M. Liebman, C.L. Mohler and C.P. Staver (Eds.) Cambridge University Press, New York.

Mohler, C.L., J. Iqbal, J. Shen and A. DiTommaso. 2016. Effects of water on recovery of weed seedlings following burial. Weed Sci. 64: 285–293.

Mushobozy, D., V.A. Khan and C. Stevens. 1998. The use of soil solarization to control weeds, plant diseases, and integration of chicken litter amendment for tomato production in Tanzania. pp. 279-285. *In:* Proc. 27th Natl. Agric. Plastics Congr. Amer. Soc. for Plasticulture.

Mutch, D.R., T.E. Martin and K.R. Kosola. 2003. Red clover (*Trifolium pratense*) suppression of common ragweed (*Ambrosia artemisiifolia*) in winter wheat (*Triticum aestivum*). Weed Technol. 17: 181–185.

Nelson, W.A., B.A. Kahn and B.W. Roberts. 1991. Screening cover crops for use in conservation tillage systems for vegetables following spring plowing. HortScience 26: 860–862.

Nieto, H.J., M.A. Brondo and J.T. Gonzales. 1968. Critical periods of the crop growth cycle for competition from weeds. PANS(C) 14: 159–166.

Nordell, A. and E. Nordell. 2009. Weed the soil, not the crop. Acres USA, 40.

Norton, L., P. Johnson, A. Joys, R. Stuart, D. Chamberlain, R. Feber, L. Firbank, W. Manley, M. Wolfe, B. Hart, F. Mathews, D. Macdonald and R.J. Fuller. 2009. Consequences of organic and non-organic farming practices for field, farm and landscape complexity. Agr. Ecosyst. Environ. 129: 221–227.

Ogari, J. and M. van der Knaap. 2002. Solarization of water hyacinth, *Eichhornia crassipes*, on Lake Victoria. Fisheries Mgmt. Ecol. 9: 365–367.

Olsen, J., L. Kirstensen, J. Weiner and H.W. Griepentrog. 2005. Increased density and spatial uniformity increase weed suppression by spring wheat. Weed Res. 45: 316–321.

Olsen, J.M., H.W. Griepentrog, J. Nielsen and J. Weiner. 2012. How important are crop spatial pattern and density for weed suppression by spring wheat? Weed Sci. 60: 501–509.

Pagliai, M., N. Vignozzi and S. Pellegrini. 2004. Soil structure and the effect of management practices. Soil Till. Res. 79: 131–143.

Peachey, R.E., J.N. Pinkerton, K.L. Ivors, M.L. Miller and L.W. Moore. 2001. Effect of soil solarization, cover crops, and metham on field emergence and survival of buried annual bluegrass (*Poa annua*) seeds. Weed Technol. 15: 81–88.

Pelacho, F. Rojo, A. Royo-Esnal, M.L. Suso and C. Zaragoza. 2012. Biodegradablemulch instead of polyethylene for weed control of processing tomato production. Agron. Sustain. Dev. 32: 889–897.

Pérez-Ruíz, M., D.C. Slaughter, F.A. Fathallah, C.J. Gliever and B.J. Miller. 2014. Co-robotic intra-row weed control system. Biosystems Eng. 126: 45–55.

Pfeiffer, A., E. Silva and J. Colquhoun. 2015. Living mulch cover crops for weed control in small-scale applications. Renew. Agr. Food Syst. 31: 1–9.

Post, W.M. and K.C. Kwon. 2000. Soil carbon sequestration and land-use change: process and potential. Glob. Change Biology 6: 317–327.

Pullen, D.W.M. and P. A. Cowell. 1997. An evaluation of the performance of mechanical weeding mechanisms for use in high speed inter-row weeding of arable crops. J. Agr. Eng. Res. 67: 27–34.

Quintanilla-Tornel, M.A., K.H. Wang, J. Tavares and C.R.R. Hooks. 2016. Effects of mulching on above and below ground pests and beneficials in a green onion agroecosystem. Agr. Ecosyst. Environ. 224: 75–85.

Raffaelli, M., L. Martelloni, C. Frasconi, M. Fontanelli, S. Carlesi and A. Peruzzi. 2016. A prototype band-steaming machine: design and field application. Biosyst. Eng. 144: 61–71.

Rao, A.N., D.E. Johnson, B. Sivaprasad, J.K. Ladha and A.M. Mortimer. 2007. Weed management in direct-seeded rice. Adv. Agron. 93: 153–255.

Rasmussen, I.A. 2004. The effect of sowing date, stale seedbed, row width and mechanical weed control on weeds and yield of organic winter wheat. Weed Res. 44: 12–20.

Rasmussen, J. 1990. Selectivity: an important parameter on establishing the optimum harrowing technique for weed control in growing cereals. pp. 197–204. *In:* The Proceedings of The European Weed Research Society Symposium, Helsinki, Finland.

Rasmussen, J. 1992. Testing harrows for mechanical control of weed in agricultural crops. Weed Res. 32: 267–274.

Rasmussen, J. 2003. Punch planting, flame weeding and stale seedbed for weed control in row crops. Weed Res. 43: 393–403.

Rasmussen, J. and T. Svenningsen. 1995. Selective weed harrowing in cereals. Biol. Agric. Hortic. 12: 29–46.

Rasmussen, J., B.M. Bibby and A.P. Schou. 2008. Investigating the selectivity of weed harrowing with new methods. Weed Res. 48: 523–532.

Rasmussen, J., H. Mathiasen and B.M. Bibby. 2010. Timing of post-emergence weed harrowing. Weed Res. 50: 436–446.

Rasmussen, J., H.W. Griepentrog, J. Nielsen and C.B. Henriksen. 2012. Automated intelligent rototine cultivation and punch planting to improve the selectivity of mechanical intra-row weed control. Weed Res. 52: 327–337.

Rasmussen, K.J. 1999. Impact of ploughless soil tillage on yield and soil quality: a Scandinavian review. Soil Till. Res. 53: 3–14.

Rasmussen, K. 2002. Influence of liquid manure application method on weed control in spring cereals. Weed Res. 42: 12–20.

Reganold, J.P. and J.M. Wachter. 2016. Organic agriculture in the twenty-first century. Nature Plants 2: 15221.

Riemens, M., R. Groaneveld, M. Kropff, L. Lotz, R. Renes, W. Sukkel and R.Y. van Der Weide. 2010. Linking farmer weed management behavior with weed pressure: more than just technology. Weed Sci. 58: 490–496.

Riemens, M.M., R.Y. van der Weide, P.O. Bleeker and L.A.P. Lotz. 2007. Effect of stale seedbed preparations and subsequent weed control in lettuce (cv. Iceboll) on weed densities. Weed Res. 47: 149–156.

Ringselle, B., G. Bergkvist, H. Aronsson and L. Andersson. 2015. Under-sown cover crops and post-harvest mowing as measures to control *Elymus repens*. Weed Res. 55: 309–319.

Rubin, B. 2012. Soil Solarization as a Tool for Weed Management. pp. 71–76. *In:* A. Gamliel and J. Katan (Eds.) Soil Solarization: Theory and Practice. American Phytopathological Society, St. Paul, MN.

Runia, W.T. 2012. Soil Disinfestation by Soil Heating. pp. 23–31. *In:* A. Gamliel and J. Katan (Eds.) Soil Solarization: Theory and Practice. American Phytopathological Society, St. Paul, MN.

Ryan, M.R., D.A. Mortensen, L. Bastiaans et al. 2010a. Elucidating the apparent maize tolerance to weed competition in long-term organically managed systems. Weed Res 50: 25–36.

Ryan, M.R., R.G. Smith, S.B. Mirsky, D.A. Mortensen and R. Seidel. 2010b. Management filters and species traits: weed community assembly in long-term organic and conventional systems. Weed Sci. 58: 265–277.

Ryan, M.R., R.G. Smith, D.A. Mortensen et al. 2009. Weed-crop competition relationships differ between organic and conventional cropping systems. Weed Res 49: 572–580.

Rydberg, T. 1994. Weed harrowing: the influence of driving speed and driving direction on degree of soil covering and the growth of weed and crop plants. Biol. Agric. Hortic. 10: 197–205.

Sarrantonio, M. and E. Gallandt. 2003. The role of cover crops in North American cropping systems. J. Crop Prod. 8: 53–73.

Sartore, L., G. Vox and E. Schettini. 2013. Preparation and performance of novel biodegradable polymeric materials based on hydrolyzed proteins for agricultural application. J. Polym. Environ. 21: 718–725.

Schimmelpfenning, D. 2016. Farm Profits and Adoption of Precision Agriculture, Err-217. U.S. Department of Agriculture, Economic Research Service.

Schonbeck, M.W. 1998. Weed suppression and labor costs associated with organic, plastic, and paper mulches in small-scale vegetable production. J. Sustain. Agr. 13: 13–33.

Schonbeck, M.S. 2001. Balancing soil nutrients in organic vegetable production systems: testing Albrecht's base saturation theory in southeastern soils. Organic Farm. Res. Found. Inf. Bull. 10: 17–21.

Schonbeck, M.W. and G.K. Evanylo. 1998. Effects of mulches on soil properties and tomato production I. Soil temperature, soil moisture and marketable yield. J. Sustain. Agr. 13: 55–81.

Schonbeck, M.W. and G.K. Evanylo. 1998. Effects of mulches on soil properties and tomato production II. Plant-available nitrogen, organic matter input, and tilth-related properties. J. Sustain. Agr. 13: 83–100.

Scopa, A. and S. Dumontet. 2007. Soil solarization: effects on soil microbiological parameters. Effects on Soil. J. Plant Nutr. 30: 537–547.

Scopa, A., V. Candido, S. Dumontet and V. Miccolis. 2008. Greenhouse solarization: effects on soil microbiological parameters and agronomic aspects. Sci. Hortic. 116: 98–103.

Shock, C.C., M. Jensen, J.H. Hobson, M. Seddigh, B.M. Shock, L.D. Saunders and T.D. Stieber. 1999. Improving onion yield and market grade by mechanical straw application to irrigation furrows. HortTechnology 9: 251–253.

Sivesind, E.C., M.L. Leblanc, D.C. Cloutier, P. Seguin and K.A. Stewart. 2009. Weed response to flame weeding at different developmental stages. Weed Technol. 23: 438–443.

Slaughter, D.C., D.K. Giles and D. Downey. 2008. Autonomous robotic weed control systems: a review. Comput. Electron. Agr. 61: 63–78.

Slovic, P. 1987. Perception of risk. Science 236: 280–285.

Smith, A.N., S.C. Reberg-Horton, G.T. Place, A.D. Meijer, C. Arellano and J.P. Mueller. 2011. Rolled rye mulch for weed suppression in organic no-tillage soybeans. Weed Sci. 59: 224–231.

Smith, R.F. 2012. Weed control options for fresh market spinach. CAPCA Advisor 15: 31–33.

Smith, R.G. 2006. Timing of tillage is an important filter on the assembly of weed communities. Weed Sci. 54: 705–712.

Smith, R.G., D.A. Mortensen and M.R. Ryan. 2009. A new hypothesis for the functional role of diversity in mediating resource pools and weed-crop competition in agroecosystems. Weed Res. 50: 37–48.

Smith, R.G., L.W. Atwood, F.W. Pollnac and N.D. Warren. 2015. Cover-crop species as distinct biotic filters in weed community assembly. Weed Sci. 63: 282–295.

Snyder, E.M., W.S. Curran, H.D. Karsten, G.M. Malcolm, S.W. Duiker and J.A. Hyde. 2016. Assessment of an integrated weed management system in no-till soybean and corn. Weed Sci. 64: 712–726.

Stapleton, J.J., R.H. Molinar, K. Lynn-Patterson, S.K. Mcfeeters and A. Shrestha. 2005. Soil solarization provides weed control for limited-resource and organic growers in warmer climates. Calif. Agric. 59: 84–89.

Stepanovic, S., A. Datta, B. Neilson, C. Bruening, C.A. Shapiro, G. Gogos and S.Z. Knezevic. 2016. Effectiveness of flame weeding and cultivation for weed control in organic maize. Biol. Agric. Hortic. 32: 1–16.

Summers, C.G. and J.J. Stapleton. 1998. Management of vegetable insects using plastic mulch: 1997 season review. UC Plant Protection Quarterly 8: 9–11.

Swanton, C.J. and B.D. Booth. 2004. Management of weed seedbanks in the context of populations and communities. Weed Technol. 18: 1496–1502.

Swanton, C.J., K.J. Mahoney, K. Chandler and R.H. Gulden. 2008. Integrated weed management: knowledge-based weed management systems. Weed Sci. 56: 168–172.

Taufik Ahmad, M., L. Tang and B.L. Steward. 2014. Automated mechanical weeding. pp. 125–137. *In:* S.L. Young and F.J. Pierce. Automation: The Future of Weed Control in Cropping Systems. Springer Science and Business Media, Dordtecht.

Teasdale, J.R. and C.L. Mohler. 1993. Light transmittance, soil temperature, and soil moisture under residue of hairy vetch and rye. Agron. J. 85: 673–680.

Teasdale, J.R. and C.L. Mohler. 2000. The quantitative relationship between weed emergence and the physical properties of mulches, Weed Sci. 48: 385–392.

Teasdale, J.R., L. Brandsaeter, A. Calegari and F. Skora Neto. 2007. Cover crops and weed management. *In:* M.K. Upadhyaya and R.E. Blackshaw (Eds.) Non-Chemical Weed Management: Principles, Concepts and Technology. CAB International, Wallingford, Oxfordshire, UK.

Terpstra, R. and J.K. Kouwenhoven. 1981. Inter-row and intra-row weed control with a hoe-ridger. J. Agr. Eng. Res. 26: 127–134. Thesis, Universität Hohenheim, Stuttgart, Germany.

Tindall, J.A., R.B. Beverly and D.E. Radcliffe. 1991. Mulch effect on soil properties and tomato growth using micro-irrigation. Agron. J. 83: 1028–1034.

Uchino, H., K. Iwama, Y. Jitsuyama, T. Yudate and S. Nakamura. 2009. Yield losses of soybean and maize by competition with interseeded cover crops and weeds in organic-based cropping systems. Field Crop. Res. 113: 342–351.

Unger, P.W. and M.F. Vigil. 1998. Cover crop effects on soil water relationships. J. Soil and Water Cons. 53: 200–206.

Upadhyaya, M.K. and R.E. Blackshaw. 2007. Non-Chemical Weed Management: Principles, Concepts and Technology. CAB International, Wallingford, Great Britain.

USDA. 2017. National Organic Program Handbook (Accessed January 30, 2017). https://www.ams.usda.gov/rules-regulations/organic/handbook

Van Acker, R.C. 2009. Weed biology serves practical weed management. Weed Res 49: 1–5.

van der Schans, D., P.O. Bleeker, L.P.G. Moledijk, M.C. Plentinger, R.Y. van der Weide, L.A.P. Lotz, R. Bauermeiste, R. Total and D.T. Baumann. 2006. Practical weed control in arable farming and outdoor vegetable cultivation without chemicals. Lelystand, Netherlands: Applied Plant Research, Wageningen University.

van der Weide, R. and D. Kurstjens. 1996. Plant morphology and selective harrowing. *In:* The Proceedings of 2nd EWRS Workshop on Physical Weed Control, Copenhagen, Denmark.

van der Weide, R.Y, P.O. Bleeker, V.T.J.M. Achten, L.A.P. Lotz, F. Fogelberg and B. Melander. 2008. Innovation in mechanical weed control in crop rows. Weed Res. 48: 215–224.

van Toor, R.F., C.M. Till, D.E. James and D.A.J. Teulon. 2004. Evaluation of UV reflective mulches for protection against thrips (*Thrips tabaci*) in onions (*Allium cepa*) crops. New Zeal. Plant Prot. 57: 209–213.

Virtanen, S., R.R. Chowreddy, S. Irmak, K. Honkapaa and L. Isom. 2016. Food industry co-streams: potential raw materials for biodegradable mulch film applications. J. Polym. Environ. 25: 1110–1130.

Walsh, M.J. and S.B. Powles. 2007. Management strategies for herbicide resistant weed populations in Australian dryland crop production systems. Weed Technol. 21: 332–338.

Walsh, M.J., R.B. Harrington and S.B. Powles. 2012. Harrington seed destructor: a new nonchemical weed control tool for global grain crops. Crop Sci. 52: 1343–1347.

Westerman, P.R., J.S. Wes, M.J. Kropff and W. van der Werf. 2003. Annual losses of weed seeds due to predation in organic cereal fields. J. Appl. Ecol. 40: 824–836.

Wezel, A., M. Casagrande, F. Celette, J.-F.o. Vian, A.l. Ferrer and J.P. Peigné. 2014. Agroecological practices for sustainable agriculture. A review. Agronomy for Sustainable Development 34: 1–20.

Wortman, S.E., C.A. Francis, M.A. Bernards, E.E. Blankenship and J.L. Lindquist. 2013. Mechanical termination of diverse cover crop mixtures for improved weed suppression in organic cropping systems. Weed Sci. 61: 162–170.

Young, S.L. and F.J. Pierce. 2014. Automation: The Future of Weed Control in Cropping Systems. Springer, Dordrecht, the Netherlands.

Zang, C. and J.M. Kovacs. 2012. The application of small unmanned aerial systems for precision agriculture: a review. Precis. Agr. 13: 693–712.

Zwickle, S. 2011. Weeds and organic weed management: investigating farmer decisions with a mental models approach. M.S. Thesis, Ohio State University, Columbus, Ohio.

Index

Printed and bound by CPI Group (UK) Ltd, Croydon, CR0 4YY

24/10/2024

01778292-0013